D1281022

Engineering Mechanics

STATICS AND DYNAMICS

ARCHIE HIGDON

Engineering and Technology Education Consultant

WILLIAM B. STILES

Professor of Engineering Mechanics
University of Alabama

ARTHUR W. DAVIS

Professor of Engineering Mechanics
Iowa State University

CHARLES R. EVCES

Associate Professor of Mechanical Engineering
University of Alabama

PRENTICE-HALL, INC., ENGLEWOOD CLIFFS, NEW JERSEY

Library of Congress Cataloging in Publication Data

Main entry under title:

Engineering mechanics.

Includes index.
Earlier editions by A. Higdon and W. B. Stiles.
1.–Mechanics, Applied. I.–Higdon, Archie–(date) II.–Higdon, Archie–(date) Engineering mechanics.

TA350.H512–1976b 620.1 75-43663
ISBN 0-13-279380-6

Printed in the United States of America

10 9 8 7 6 5 4 3 2 1

PRENTICE-HALL INTERNATIONAL, INC., *London*
PRENTICE-HALL OF AUSTRALIA PTY. LIMITED, *Sydney*
PRENTICE-HALL OF CANADA, LTD., *Toronto*
PRENTICE-HALL OF INDIA PRIVATE LIMITED, *New Delhi*
PRENTICE-HALL OF JAPAN, INC., *Tokyo*
PRENTICE-HALL OF SOUTH-EAST ASIA PRIVATE LIMITED, *Singapore*

Contents

3 CENTROIDS AND CENTERS OF GRAVITY

4 EQUILIBRIUM

5 FRICTION

6 SECOND MOMENTS OR MOMENTS OF INERTIA

PART A—PLANE AREAS

Preface

We have prepared this second vector edition of *Engineering Mechanics* to reflect current and anticipated trends in introductory mechanics courses. Topics included in this revision are essentially the same as in the earlier editions, although many topics have been rearranged and expanded to facilitate adaptation to the numerous methods in which mechanics is integrated into modern engineering curricula. In particular, the material in dynamics is arranged to permit coverage of separate or combined courses in the dynamics of particles and of rigid bodies.

Use of the International System of Units (SI) is becoming more and more common in the United States, particularly among firms operating in several countries. Since it does not appear that there will be a complete and sudden changeover to SI units, between 20 and 25 per cent of the numerical examples and problems in this revision involve metric units. An asterisk by the problem number indicates a problem which uses SI units. When SI units are used they conform essentially to the recommendations of the American National Standards Institute.

The vector method is particularly useful in three-dimensional applications of mechanics, and in all cases it provides a systematic procedure for problem solutions. Physical concepts and visualization of forces and moments are first emphasized in the text without using vector algebra, however, to help develop a "feeling" for these basic concepts. Later it is shown that vector algebra can be used to obtain the same results, often with a considerable reduction in the mathematical manipulation required. Both scalar and vector solutions are presented for some examples to encourage the reader to compare procedures and to check his understanding of the subject matter. Vector methods frequently do not simplify the solution of two-dimensional problems, but such applications are essential to provide practice in using vector notation for the more involved three-dimensional problems.

The principles of mechanics are few and relatively simple; the applications, however, are infinite in their number and variety and frequently appear to be forbiddingly complicated. Since much of engineering is "applied mechanics" (just as mechanics is applied physics and applied mathematics), it is highly desirable that the engineering student grasp mechanics, not as a series of formulas and manipulative operations, but rather as a well-comprehended mechanism for stripping problems of their extraneous confusing aspects. This outlook will enable the student to break problems down into relatively simple, easily analyzed elements.

Emphasis is placed on an understanding of principles employed in the solution of problems rather than reliance on a rote process of substitution in numerous formulas. We believe that every teacher of engineering mechanics must be constantly aware of mechanics as a prelude to engineering design, in which innumerable complicating empirical considerations are superimposed on the relatively simple situations encountered in mechanics. Unless students are well grounded in the principles and well drilled in logical, orderly, step-by-step methods of analysis and procedure, they are likely to experience considerable difficulty as they try to include the design elements of judgment, approximation, and compromise in their solutions.

In an attempt to keep student attention focused on mechanics as a prelude to design, we see no justification for realism carried to the point of introducing time-consuming complexities merely because they are inherent in the problems of engineering. Thus we adhere to the common practice of freely using values for masses, weights, dimensions, forces, and angles that simplify the numerical work involved, in the belief that the time saved can be better employed in solving more problems and gaining a more complete mastery of the principles and their applications. Either a slide rule or a calculator can be used for the solution of problems, but it should be noted that, in general, the data available for engineering problems do not warrant results carried to more than three or four significant figures. In general, problem data are assumed to contain three significant figures regardless of the number of figures shown. Answers, therefore, are given to three significant figures unless the first significant digit is 1, in which case four figures are reported. We have endeavored to maintain a sense of engineering reasonableness by avoiding such absurdities as a wheel having a radius of gyration in excess of its radius, accelerations requiring abnormal coefficients of friction, and weights of bodies that are, for materials normally used, inconsistent with the volume or given dimensions.

Chapters 8 through 11 on kinematics and kinetics are arranged with A and B parts covering the dynamics of particles and of rigid bodies, respectively. Thus a course covering only particle dynamics would use the A part of each of the four chapters. An integrated course on dynamics can be taught by including both parts of each chapter as presented or by studying the A part of each chapter first followed by the B parts.

The selection of specific topics to be covered will vary from school to school. This revision includes those topics common to most courses and a number of additional topics which cannot be included in some instances due to time limitations. The inclusion of such topics in the book provides an opportunity to tailor the course to the needs of each school by the judicious selection of the topics to be covered. The number of problems is intentionally large, not with the idea that problem assignments should be burdensome, but rather to offer selection and afford variation in the problem assignments from term to term.

The closure article at the end of a chapter is intended as a brief review of the important topics covered and to emphasize the relationships of some of the topics after they have been covered individually. Review problems are included with some closures to provide an opportunity to develop judgment in selecting the appropriate methods or theories for solving selected problems.

For numerous suggestions, we are indebted to many users of the earlier editions of *Engineering Mechanics*, including colleagues at Iowa State University, the University of Alabama, the United States Air Force Academy, and the University of Denver. We want to express our particular appreciation to our wives for their patience, their encouragement, and their invaluable help in proofreading, and to Mrs. Afton Earnhart for her help in typing the manuscript.

A. HIGDON
W. B. STILES
A. W. DAVIS
C. R. EVCES

Symbols and Abbreviations*

$\mathbf{a}$, $\mathbf{a}_G$	linear acceleration, linear acceleration of mass center
A	area, amplitude
Art.	article
avg	average
b	breadth, width
$\mathbf{C}$	couple
C	compression, centroid of area
$C_1, C_2, \ldots$	constants of integration
cu	cubic
d	differential of (as dx), distance
D	diameter
e	coefficient of restitution
$\mathbf{e}_n$	unit vector in normal direction
$\mathbf{e}_t$	unit vector in tangential direction
$\mathbf{e}_r$	unit vector in radial direction
$\mathbf{e}_\theta$	unit vector in transverse direction

* The symbols and abbreviations used in this text conform essentially with those approved by the American National Standards Institute.

Eq.	equation
f	frequency, sag of a cable
$\mathbf{F}, \mathbf{F}_1, \mathbf{F}_A, \ldots$	forces or loads
$\mathbf{F}_x, \mathbf{F}_y, \mathbf{F}_n, \ldots$	components of force $\mathbf{F}$
Fig.	figure
fpm	feet per minute
fps	feet per second
fps^2	feet per second per second
ft	foot
$\mathbf{g}$	acceleration of gravity
$\mathbf{G}$	linear momentum
G	center of gravity, mass center
$\mathbf{H}$	angular momentum
h	height, length, depth
hp	horsepower
i	initial when used as subscript, as v_i
$\mathbf{i}, \mathbf{j}, \mathbf{k}$	unit vectors in x, y, z directions, respectively.
$\mathbf{i}', \mathbf{j}', \mathbf{k}'$	unit vectors in X, Y, Z directions, respectively.
$\mathbf{I}_L$	linear impulse
I	moment of inertia of area or mass
I_x, I_0	moment of inertia of area, or mass with respect to axis indicated by subscript
I_{xy}, I_{yz}	product of inertia of area or mass with respect to planes perpendicular to axes shown as subscripts
in.	inch
ips	inches per second
ips^2	inches per second per second
J_0, J_z	polar moment of inertia of area with respect to axis indicated by subscript
J	joule, newton · meter
k	spring constant (load per unit deflection), kip (1000 lb)
$k_x, k_y, \ldots$	radius of gyration with respect to axis indicated by subscript
K	constant
kg	kilogram
kip	kilopound (1000 lb)
l, L	length
lb	pound
ln	logarithm (natural)
log	logarithm (common)
m	mass, W/g, meter
$\mathbf{M}_0, \mathbf{M}_x$	moment of a force with respect to point or line shown as subscript
M_A, M_y	moment of an area or mass with respect to a point or line shown as subscript

m/s	meter per second
m/s^2	meter per second per second
max	maximum
min	minimum
mph	miles per hour
mphps	miles per hour per second
n	normal direction (used only as subscript)
$\mathbf{n}$	unit vector in general direction
$\mathbf{N}$	normal force
N	newton
O	origin of coordinates, instantaneous center of zero velocity
oz	ounce
p	pitch of threads, unit pressure, load per unit area
$\mathbf{P}$	force
P	power
psf	pounds per square foot
psi	pounds per square inch
$\mathbf{q}$	linear displacement
q	load per unit distance, distance from axis of rotation to center of percussion
$\mathbf{Q}$	force
Q	total distance traveled during a time interval
$\mathbf{r}$	position vector
r	radius, polar coordinate
$\mathbf{R}$	resultant force, reaction, position vector
R	radial direction, radius
rad	radian
rpm	revolutions per minute
rps	revolutions per second
s	function indicating position of a particle measured along a path from a fixed point on the path, arc length, length of cable, second
sec	second
sq	square
t	time, thickness, tangential direction
$\mathbf{T}$	torque (moment of a force or couple), tensile force
T	tension, period, transverse direction, kinetic energy
u, v	rectangular coordinates
U	work
$\mathbf{v}$, $\mathbf{v}_G$	linear velocity, linear velocity of mass center
V	volume, potential function
$\mathbf{W}$	total load, weight
W	watt, newton · meter per second
x, y, z	rectangular coordinates, moving or fixed
X, Y, Z	rectangular coordinates, fixed

x_c, y_c, z_c	rectangular coordinates of centroid
x_G, y_G, z_G	rectangular coordinates of mass center
$\dot{x}, \dot{y}, \dot{z}$	time derivatives of x, y, z
$\ddot{x}, \ddot{y}, \ddot{z}$	second derivatives of x, y, z with respect to time
y_{st}	static deflection
$\boldsymbol{\alpha}$	angular acceleration
α	angle
β	angle of contact for belt friction, angle
γ	specific weight (weight per unit volume)
δ	logarithmic decrement
$\delta x, \delta\theta, \ldots$	virtual displacement (linear or angular)
η	efficiency
θ	function indicating angular position of a line measured from a fixed reference line, a polar coordinate, angle
μ, μ_s, μ_k	coefficient of friction, static or kinetic coefficients
ρ	density (mass per unit volume)
ϕ	phase angle, angular displacement during a time interval
Φ	total angle turned through during a time interval
$\boldsymbol{\omega}$	angular velocity
ω	circular frequency of forced vibration
ω_d	circular frequency of damped vibration
ω_n	circular frequency of free vibration
$\longrightarrow\!\!\!\!\twoheadrightarrow$	vector representing the resultant force
$\Longrightarrow$	moment vector

SENSE (OF ROTATION)

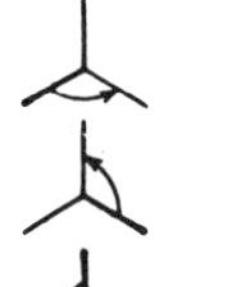

counterclockwise looking downward

counterclockwise looking inward

counterclockwise looking to the left

1 Basic Concepts

1-1 HISTORICAL BACKGROUND

Engineering mechanics is essentially a study of the effects of forces acting on bodies. That portion of mechanics known as *statics*, which deals with equilibrium of bodies—that is, force systems which produce no acceleration—is an old branch of science.[1] The builders of the pyramids of Egypt used some of the principles of mechanics. The writings of Archimedes of Syracuse (287–212 B.C.) show that he understood the necessary relationship between the forces acting on a lever to produce equilibrium. Stevinus (1548–1620) was the first to state the principle of the inclined plane and to employ the principle of the parallelogram of forces. Modern mechanics developed rapidly after the time of Stevinus. The experiments of Galileo Galilei (1564–1642) led to the development of the principles of dynamics by exploding

[1] For more historical information, see Ernst Mach, *Science of Mechanics* (Chicago: The Open Court Publishing Company, 1893); or Harvey F. Girvin, *A Historical Appraisal of Mechanics* (Scranton, Pa.: International Textbook Company, 1948).

some of the false theories of the Greek philosophers. He made investigations and obtained experimental proof of the laws of falling bodies even though he was handicapped by lack of adequate clocks for measuring small time intervals. Christian Huygens (1629–1695) continued the mechanical investigations started by Galileo. He invented the pendulum clock, determined the acceleration of gravity, and introduced the theorems respecting centrifugal force. Sir Isaac Newton (1642–1727) completed the formulation of the basic principles of engineering mechanics by his discovery of universal gravitation and by his statement of the laws of motion.

1-2 INTRODUCTION

Mechanics is that branch of physical science which considers the motion of bodies, with rest being considered a special case of motion. In engineering mechanics, attention is directed primarily to the external effects of a system of forces acting on a rigid body. The *external effect* of a force on a body is either acceleration of the body or development of resisting forces (reactions) on the body. *A force may be defined as the action of one body on another body which changes or tends to change the motion of the body acted on.*

When several forces act in a given situation, they are called a *system of forces* or a *force system.* Force systems can be classified according to the arrangement of the lines of action of the forces of the system. The forces may be coplanar or noncoplanar, parallel or nonparallel, concurrent or nonconcurrent, collinear or noncollinear. The most general system of forces is one in which the forces are noncoplanar, nonparallel, and nonconcurrent. This means that they do not all lie in a common plane; they are not all parallel to each other; and they do not all intersect at a common point.

The resultant of a force system is the simplest force system which can replace the original system without changing its external effect on a rigid body. The resultant of a force system can be a single force, a pair of parallel forces having the same magnitudes but opposite senses (called a *couple*), or a force and a couple. If the resultant is a force and a couple, the force will not be parallel to the plane containing the couple.

When the force system acting on a body is balanced, the system has no external effect on the body, the body is in equilibrium, and the problem is one of *statics.* When the force system has a resultant different from zero, the body will be accelerated, and the problem is one of *dynamics.* When the internal effects of a force system on a body are to be considered or when the changes in shape of the body are important, the problem becomes one of *mechanics of materials.* Only problems of statics and dynamics are considered in this text. The external effects of a force system on a physical body are not ordinarily altered appreciably by small distortions of the body. Many problems of statics would be unnecessarily complicated by taking such changes of shape into account. Thus most of the bodies in this text will be considered rigid bodies. *A body in which all particles remain at fixed distances from each other is called a rigid body.* No real body is absolutely rigid, but in many cases, the changes in shape of the body have a negligible effect upon the acceleration produced by a force system or upon the reactions required to maintain equilibrium. Whenever the changes in distance between the particles of a body can be neglected, the body is assumed to be rigid.

1-3 SCALAR AND VECTOR QUANTITIES—ELEMENTARY OPERATIONS

Physical quantities, such as force, mass, acceleration, volume, velocity, and time, used in engineering mechanics, can be classified as either scalar or vector quantities. A *scalar quantity* is one which has only magnitude, whereas a *vector quantity* has both magnitude and direction. More precisely, *any quantity that involves both magnitude and direction,* so that it can be represented by a directed line segment, *and that conforms to the parallelogram law of addition,* as explained later, *is defined as a vector quantity.* Vector quantities can be further divided into free and localized vectors. A *free vector* is one with a specified slope and sense but not acting through any particular point, whereas a *localized vector* has a definite or specific line of action. A localized vector may be further classified as a sliding vector or a fixed vector. A sliding vector may be applied at any point along its line of action, whereas a fixed vector must be applied at a specified point on its line of action.

Mass, volume, and time are examples of scalar quantities. It is evident that a time interval of 3 min has no direction and is therefore a scalar quantity. To say that the wind is blowing at a rate of 30 mph does not, however, tell a complete story. The direction of the wind is frequently just as important as the magnitude of its velocity. The velocity of the wind is an example of a vector quantity. Since the velocity of the wind does not act along one line, it is a free vector and not a localized vector. The velocity or acceleration of a particle can, however, be considered to be a localized vector through the particle. The moment of a couple, Art. 1-7, is another example of a free vector. Force, impulse, and momentum are examples of localized vectors. To illustrate the importance of the position of the action line of a force (localized vector), consider the beam shown in Fig. 1-1. It is simply supported at its ends and carries a 50-lb downward load. When the load is placed in position C at the center of the beam, the reactions of

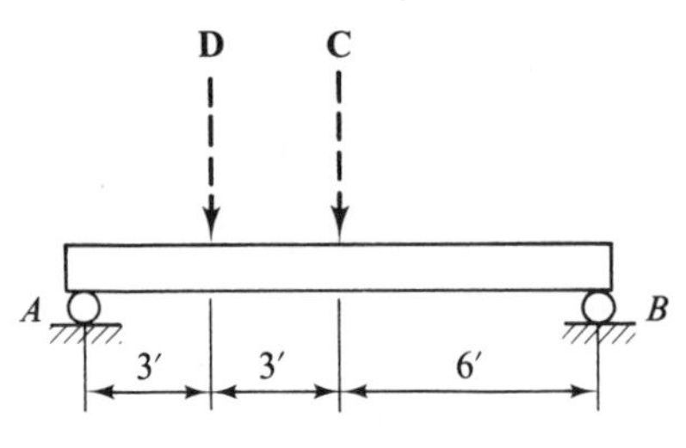

Figure 1-1

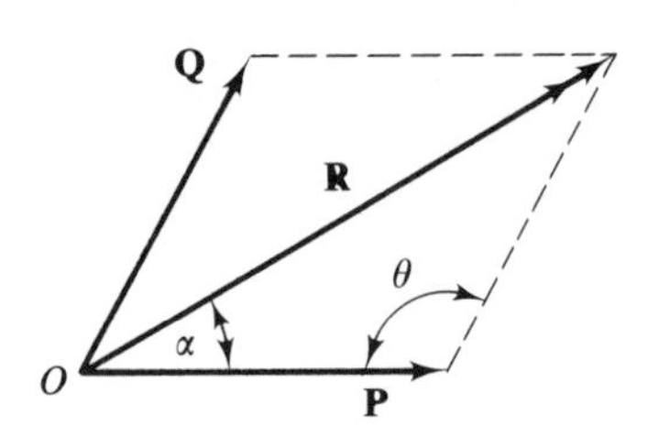

Figure 1-2

the supports at A and B on the beam are equal. If the load were moved to position D, the support at A would carry more of the load and the support at B would carry less. In other words, the effect of the supports on the beam (the external effect) depends on the position of the load it carries as well as on the magnitude and direction of the load.

Since both scalar and vector quantities are encountered repeatedly in mechanics, it is important to be able to distinguish between them. In these volumes, letters representing vector quantities are set in **boldface** type to indicate and emphasize the complete vector; letters representing magnitudes of vectors and other scalar quantities are set in regular lightface type.

The product of a positive scalar a and a vector $\mathbf{P}$ is defined as a vector of magnitude aP with the same direction as $\mathbf{P}$, where P is the magnitude or absolute value of the vector $\mathbf{P}$; hence $P=|\mathbf{P}|$. If the scalar multiplier is negative, the sense of the resulting vector is opposite to the sense of the original vector.

The sum of a pair of concurrent vectors can be determined by means of the *parallelogram law*, which states that the resultant sum is proportional to the diagonal of the parallelogram whose sides are proportional to the two vectors. *This law is not subject to an analytical proof but is rather the result of experimental observation.*

The parallelogram law is illustrated in Fig. 1-2. The sum of the two vectors $\mathbf{P}$ and $\mathbf{Q}$ is the single vector $\mathbf{R}$, which passes through O, the point of concurrence of $\mathbf{P}$ and $\mathbf{Q}$. The sum, called the *resultant*, can be determined graphically by drawing the parallelogram to scale. Its magnitude can also be determined algebraically by applying the cosine law for a general triangle, thus:

$$R=(P^2+Q^2-2PQ\cos\theta)^{1/2}$$

The angle that $\mathbf{R}$ makes with either $\mathbf{P}$ or $\mathbf{Q}$ can be determined by the law of sines, for example,

$$\frac{\sin\alpha}{Q}=\frac{\sin\theta}{R}.$$

Other methods of adding and subtracting vectors will be presented later.

As stated previously, any quantity that involves both magnitude and direction and conforms to the parallelogram law of addition is a vector. Note in Fig. 1-2 that vector addition is independent of the order of addition, so that

$$\mathbf{R} = \mathbf{P} + \mathbf{Q} = \mathbf{Q} + \mathbf{P},$$

and hence vector addition is *commutative.* Quantities which are not commutative, even though they can be represented by directed line segments, do not add according to the parallelogram law and therefore

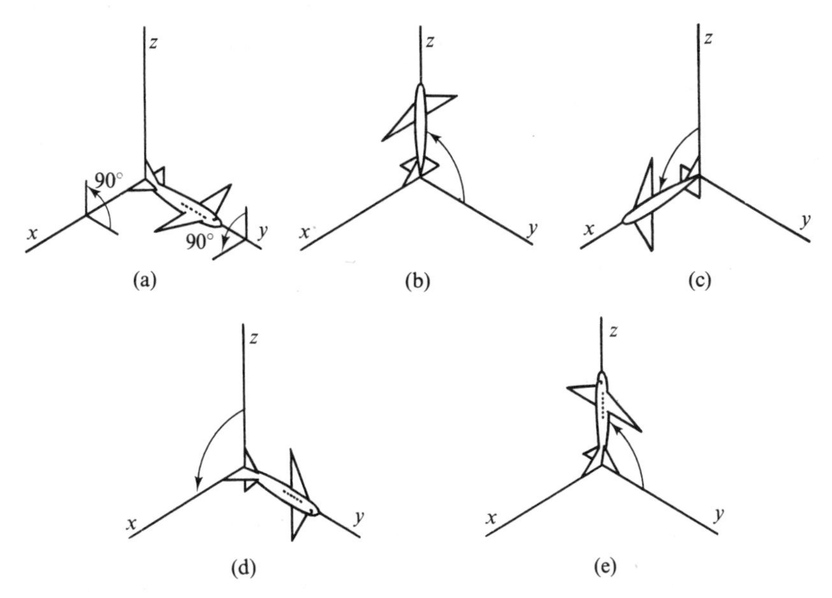

Figure 1-3

fail to qualify as vectors. Finite angular rotation has magnitude (rotation in radians) and direction (the axis around which the rotation takes place with a specification of clockwise or counterclockwise), but it is not a vector because the addition of finite angular rotations is not commutative. This is demonstrated in Fig. 1-3 in which the airplane is given two angular rotations, one 90° about the x axis and the other 90° about the y axis. If the first rotation is about the x axis and the second about the y axis, the final orientation is as shown in Fig. 1-3(c); the result of the reversed order is shown in Fig. 1-3(e). Since the results differ, the addition of finite angular rotations is not commutative, and therefore angular rotation is not a vector. It can be shown, however, that infinitesimal angular rotations are vectors.[2]

A *unit vector* is a vector which has a magnitude of unity. The

[2] For further information, see H. Goldstein, *Classical Mechanics* (Reading, Mass.: Addison-Wesley Publishing Company, Inc., 1950), pp. 124–132.

symbols **i**, **j**, and **k** are used to represent unit vectors along the positive x, y, and z axes, respectively, and the symbol **n** is used to designate a unit vector in the n direction. Thus the vector **P** may be written as $\mathbf{P} = P\mathbf{n}$, where **n** is the unit vector in the direction of **P**. Use of a consistent set of coordinate axes simplifies many of the laws of vector algebra. When three orthogonal (mutually perpendicular) axes x, y, and z are used, they should be arranged as a right-hand system so that if the system is caused to rotate about the z axis with the positive x axis turning toward the positive y axis, a right-hand screw would advance in the positive z direction. Figure 1-4(a) shows two arrangements of x, y, and z axes, both of which are right-hand systems. The unit vectors **i**, **j**, and **k** are also shown in Fig. 1-4(a).

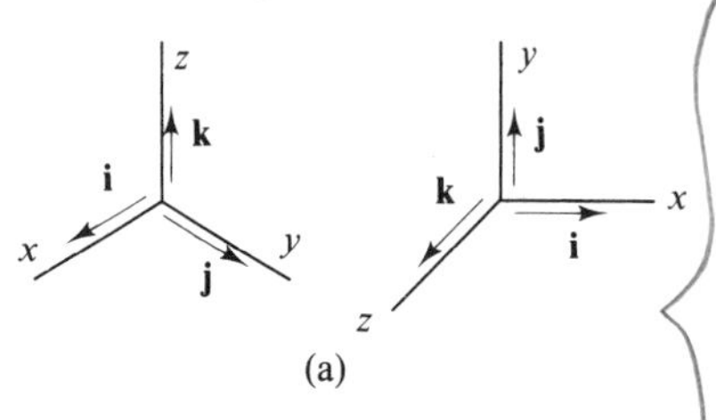

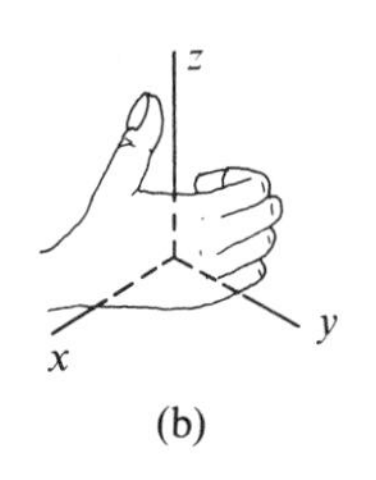

Figure 1-4

Another means of visualizing a right-hand system of axes is to imagine the fingers of the right hand wrapped around the z axis [see Fig. 1-4(b)], with the thumb pointing in the positive z direction. The fingers will curl in the direction the positive x axis would have to rotate to point in the direction of the y axis.

1-4 FORCES

Since mechanics is primarily a study of the effect of forces on bodies, it is important to have a clear understanding of the concept of a force. A force was defined in Art. 1-2. Because of the inertia possessed by all material bodies, they react or oppose any force which acts on them. When body A exerts a force on body B, body B exerts an equal opposite collinear force on body A, as stated in Newton's third law of motion. Thus forces never exist singly but always act in equal opposite collinear pairs. The effect of one body on another, however, is a single force (one of the pair) and will be studied as such.

Those properties which are necessary to distinguish one force from every other force are called its *characteristics*. The *characteristics of a force*, which describe its external effect on a rigid body, are (1) its magnitude, (2) its direction (sense and slope), and (3) the location of any point on its line of action. Magnitude and sense are mathematically inseparable in that the sense of a force can be indicated by the algebraic sign of its magnitude. Since a force is not completely determined or identified until all its characteristics are known, it is important to keep in mind all three characteristics as listed. Furthermore, to indicate completely the direction and a point on the action line of a force in a plane, it is necessary to locate two points on the action line of the force or to locate one point on the line and specify either the sense and the slope of the line or the sense and the angle the line makes with a reference axis in the plane. Figure 1-5 is a convenient way to show all the characteristics of a force in a plane.

If two forces have the same characteristics, they will produce the same external effect on the rigid body. This fact leads to the *principle of transmissibility*, which states that *the external effect of a force on a rigid body is independent of the point of application of the force along its line of action.* Until the early 1970s, engineers in the United States used the British gravitational system of units in which the unit of force is the pound (lb). Other units of force commonly used in this system are the kip or kilopound (1000 lb) and the ton (2000 lb). During the 1970s numerous American industries involving engineering began using a revised metric absolute system of units called the International System (SI). This system (SI) had already been adopted by most other industrial countries of the world. The unit of force is the newton (N) in SI units. Both British (gravitational) and SI (absolute) units are used in this text. The distinction between gravitational and absolute systems of units is discussed in Art. 1-8, and factors for conversion between British and SI units are given in Appendix A.

Since the sum of two vectors is a vector, as obtained by the parallelogram law, it is apparent that a single force or other vector can be resolved into two or more forces (components) by reversing the procedure. The process of replacing a force by its components is called *resolution.*

In the most general sense, a component of a force is any one of two or more forces which can be combined by the parallelogram law to produce the given force. These components need not be concurrent or even coplanar. Normally, however, the term "component" *is used to mean either one of two coplanar concurrent forces or any one of three noncoplanar concurrent forces which can be combined vectorially with the other components to produce the original force.* The point of concurrence must be on the action line of the original force. This more limited definition will be used in this text.

For most purposes, rectangular (mutually perpendicular) components of a force are more useful than general oblique components. In obtaining rectangular components, the parallelogram reduces to a rectangle. Consider, for example, the rectangle $OACB$ in Fig. 1-6 and

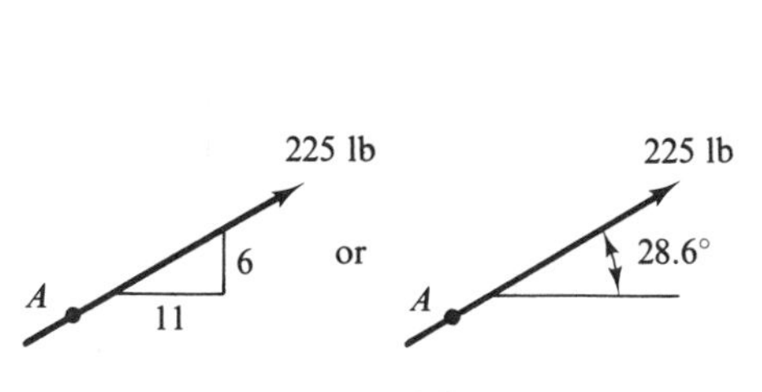

Figure 1-5

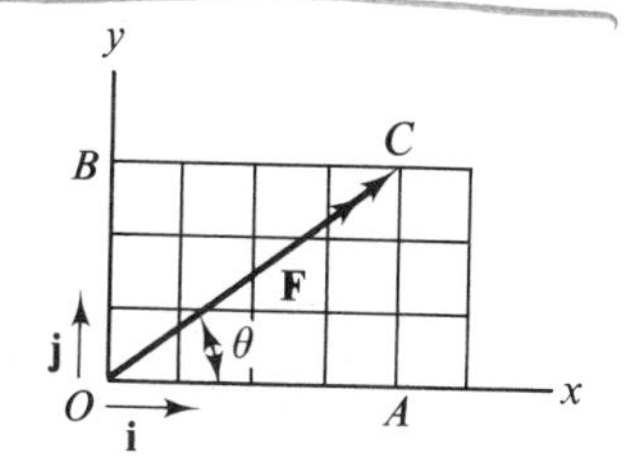

Figure 1-6

the force **F** along the diagonal OC. The force **F** may be written

$$\mathbf{F} = \mathbf{F}_x + \mathbf{F}_y = F_x\mathbf{i} + F_y\mathbf{j},$$

where

$$F_x = F\cos\theta = F\left(\frac{4}{5}\right),$$

$$F_y = F\sin\theta = F\left(\frac{3}{5}\right),$$

and one set of rectangular components is

$$\mathbf{F}_x = 0.8F\mathbf{i} \text{ (or } 0.8F \rightarrow\text{) through } O$$

and

$$\mathbf{F}_y = 0.6F\mathbf{j} \text{ (or } 0.6F\uparrow\text{) through } O.$$

Another correct set of rectangular components is

$$\mathbf{F}_x = 0.8F \rightarrow \text{ (or } 0.8F\mathbf{i}\text{) through } C,$$

and

$$\mathbf{F}_y = 0.6F\uparrow \text{ (or } 0.6F\mathbf{j}\text{) through } C.$$

Note that the two components of any force must intersect on the action line of the force. In other words, forces along **OA** and **AC** are not the components of **F**, but they are the components of an equal force through A parallel to **F**.

Frequently, it is convenient to resolve a force in space into three mutually perpendicular components parallel to three coordinate axes. The force can then readily be expressed as a function of the three unit vectors **i**, **j**, and **k**. As indicated in Fig. 1-7, the force **R** could first be

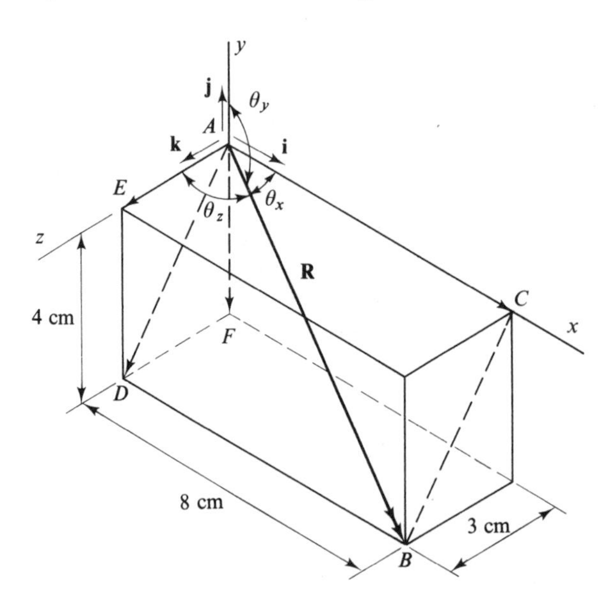

Figure 1-7

resolved into two components along AC and AD by means of the parallelogram law, and the component along AD could then be further resolved into components along AE and AF. From the figure, it is seen that the position vector from A to B is

$$\mathbf{AB} = \mathbf{AC} + \mathbf{AF} + \mathbf{AE}$$

$$= AC\mathbf{i} - AF\mathbf{j} + AE\mathbf{k}$$

and

$$\mathbf{n}_{AB} = \frac{AC\mathbf{i} - AF\mathbf{j} + AE\mathbf{k}}{[(AC)^2 + (AF)^2 + (AE)^2]^{1/2}}$$

$$= \cos\theta_x\mathbf{i} + \cos\theta_y\mathbf{j} + \cos\theta_z\mathbf{k}.$$

The angles θ_x, θ_y, and θ_z are the angles between the vector **AB** and the positive coordinate axes, and $\mathbf{n}_{AB}$ is a unit vector along **AB**. The cosines of the angles are called *direction cosines.* If the angle is more than 90°, as for θ_y, the cosine is negative. The force **R** can be expressed in terms of axial components by multiplying its magnitude by the unit vector, that is,

$$\mathbf{R} = R\mathbf{n}_{AB} = R\cos\theta_x\mathbf{i} + R\cos\theta_y\mathbf{j} + R\cos\theta_z\mathbf{k}.$$

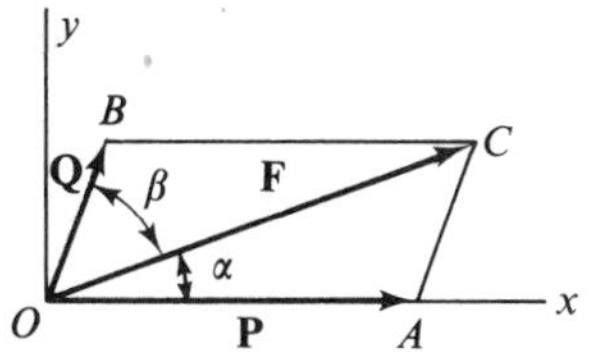

Figure 1-8

In either two or three dimensions, a rectangular component of a force is equal to the product of the force and the cosine of the angle between the force and the component. Notice again that the three components in Fig. 1-7 intersect at point A on the action line of the resultant force. The components of a force must always intersect at some point on the action line of the force. The vector sum of three forces along the lines AE, ED, and DB in Fig. 1-7 is not a single force but is a force and a couple, as indicated in Art. 2-4. Thus they are not the components of the single force **R**.

If nonrectangular components of a force are needed, several methods are available for determining them. The components of the force **F**, shown as **P** and **Q** along OA and OB, respectively, in Fig. 1-8, can be determined graphically by drawing the parallelogram to any convenient scale. The magnitude of the components can be determined algebraically from the law of sines; for example, in Fig. 1-8,

$$\frac{P}{\sin\beta} = \frac{F}{\sin(180° - \alpha - \beta)} = \frac{Q}{\sin\alpha}.$$

A third method is to resolve the force **F** into rectangular components and equate each of the rectangular components of **F** to the sum of the corresponding rectangular components of **P** and **Q** (Fig. 1-8). Example 1-1 illustrates this third method of procedure.

A convenient method for determining the components of a force, particularly in three dimensions, is to write an expression for a position vector parallel to the force from the geometry of the figure. A unit vector parallel to the force is obtained by dividing the position vector by its magnitude. When the magnitude of the force is multiplied by the unit vector the axial components are obtained directly.

EXAMPLE 1-1

In Fig. 1-8, the magnitude of the resultant force **F** is 300 lb and the angles α and β are 25° and 45°, respectively. Resolve the force **F** into a pair of components **P** along line *OA* and **Q** along line *OB*.

SOLUTION

The three forces **F**, **P**, and **Q** can be expressed in terms of **i** and **j** by multiplying each magnitude by its corresponding unit vector. Thus

$$\mathbf{F} = F\mathbf{n}_F = 300\mathbf{n}_F = 300(\mathbf{i}\cos 25° + \mathbf{j}\sin 25°),$$

$$\mathbf{P} = P\mathbf{n}_P = P\mathbf{i},$$

$$\mathbf{Q} = Q\mathbf{n}_Q = Q[\mathbf{i}\cos(45° + 25°) + \mathbf{j}\sin(45° + 25°)],$$

and

$$\mathbf{F} = \mathbf{P} + \mathbf{Q},$$

or

$$300(\mathbf{i}\cos 25° + \mathbf{j}\sin 25°) = P\mathbf{i} + Q(\mathbf{i}\cos 70° + \mathbf{j}\sin 70°)$$
$$= \mathbf{i}(P + Q\cos 70°) + \mathbf{j}Q\sin 70°.$$

These two vectors will be equal only if the corresponding components are equal. Therefore when the coefficients of **i** and **j** are equated

$$\mathbf{i}: \quad 300\cos 25° = P + Q\cos 70°$$

$$\mathbf{j}: \quad 300\sin 25° = Q\sin 70°$$

from which

$$Q = 300\frac{\sin 25°}{\sin 70°} = 134.9\text{ lb}$$

and

$$P = 300\cos 25° - 134.9\cos 70° = 226\text{ lb}.$$

The forces **P** and **Q** are

$$\mathbf{P} = \underline{226\mathbf{i}\text{ lb}} = \underline{226\text{ lb} \rightarrow \text{through } O} \qquad \text{Ans.}$$

and

$$\mathbf{Q} = 134.9(\mathbf{i}\cos 70° + \mathbf{j}\sin 70°)\text{ lb}$$

$$= \underline{134.9\text{ lb} \measuredangle 70° \text{ through } O.} \qquad \text{Ans.}$$

EXAMPLE 1-2

Determine the *x*, *y*, and *z* components of the 80-N force **R** of Fig. 1-7.

SOLUTION

The length of the diagonal *AB* is

$$\sqrt{3^2 + 4^2 + 8^2} = 9.43\text{ cm}$$

and the unit vector along AB is

$$\mathbf{n}_{AB} = \frac{\mathbf{AB}}{AB} = \frac{8\mathbf{i}-4\mathbf{j}+3\mathbf{k}}{9.43}$$

$$= 0.848\mathbf{i} - 0.424\mathbf{j} + 0.318\mathbf{k}.$$

Note that the coefficients in this equation are the direction cosines of **n** (and of **R**). The force **R** is

$$\mathbf{R} = R\mathbf{n}_{AB} = 80(0.848\mathbf{i} - 0.424\mathbf{j} + 0.318\mathbf{k})$$

$$= (67.8\mathbf{i} - 33.9\mathbf{j} + 25.4\mathbf{k})\text{ N},$$

and the components are

$$\mathbf{R}_x = 67.8\mathbf{i}\text{ N} = 67.8\text{ N} \searrow \text{ through } A \qquad \text{Ans.}$$

$$\mathbf{R}_y = -33.9\mathbf{j}\text{ N} = 33.9\text{ N} \downarrow \text{ through } A \qquad \text{Ans.}$$

and

$$\mathbf{R}_z = 25.4\mathbf{k}\text{ N} = 25.4\text{ N} \swarrow \text{ through } A \qquad \text{Ans.}$$

In both examples, to reduce possible confusion, the directions of the component forces are indicated by arrows or by means of unit vectors, rather than merely by algebraic signs. This practice is recommended and will be adhered to in this text.

PROBLEMS

1-1 Determine a pair of horizontal and vertical components of the 260-lb force of Fig. P1-1.

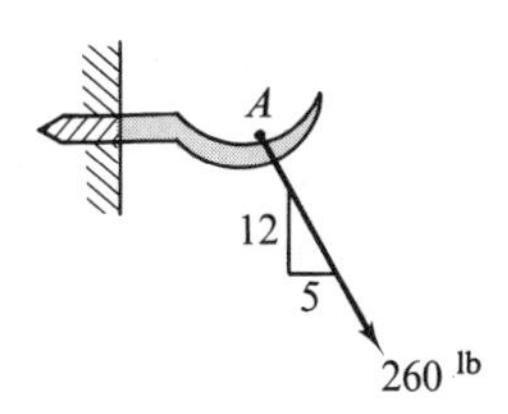

Figure P1-1

1-2 Determine a set of horizontal and vertical components of the 170-lb force in Fig. P1-2.

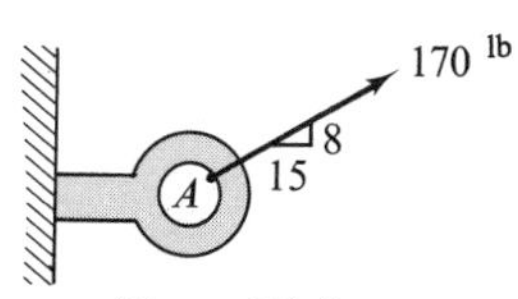

Figure P1-2

1-3 Determine the horizontal and vertical components of the 500-lb force of Fig. P1-3.

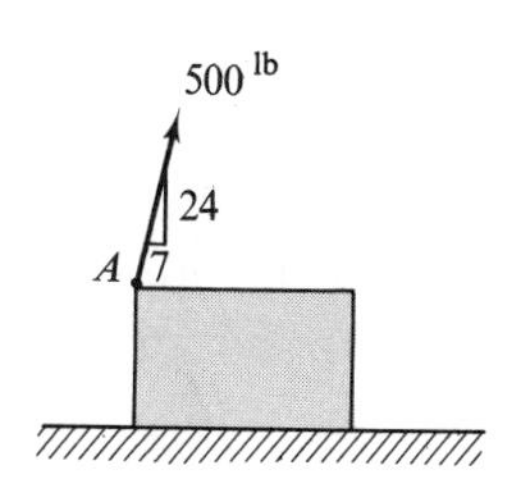

Figure P1-3

***1-4** Resolve the 400-kN force in Fig. P1-4 into two components, one acting along AB and the other perpendicular to AB.

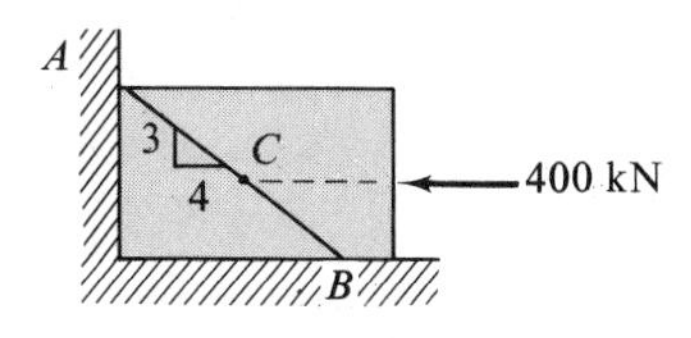

Figure P1-4

1-5 Resolve the 2600-lb force of Fig. P1-5 into two components, one acting along AB and the other perpendicular to AB. (See below.)

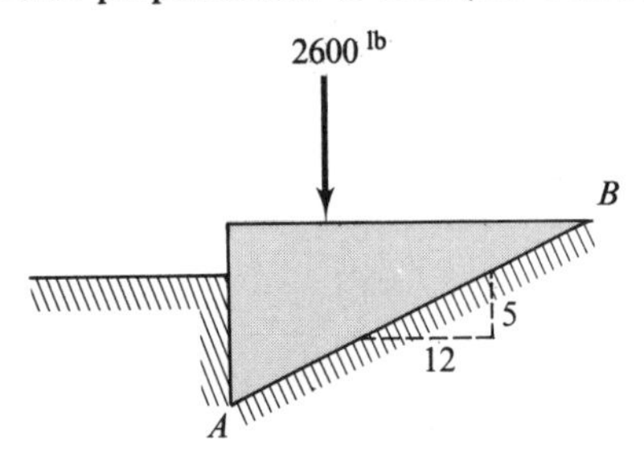

Figure P1-5

1-6 Resolve the 1000-lb force of Fig. P1-6 into x and y components when θ is (a) 20°, (b) 120°.

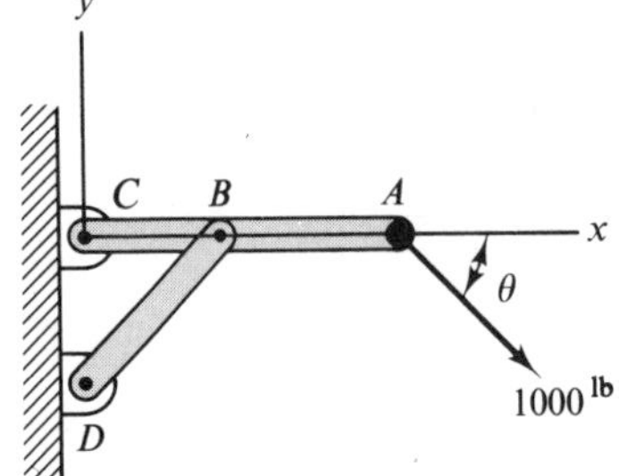

Figure P1-6

1-7 The vertical component of the force **F** of Fig. P1-7 is 150 lb upward through O. Determine the force **F** and its horizontal component.

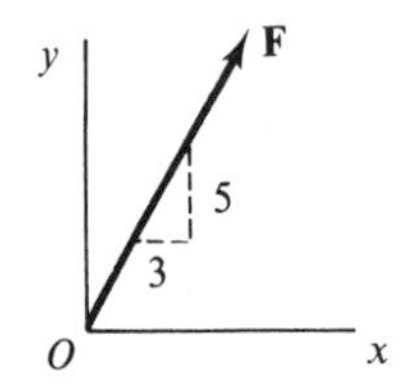

Figure P1-7

1-8 Determine a set of components of the 100-lb force in Fig. P1-8 with one acting along AB and the other parallel to BC.

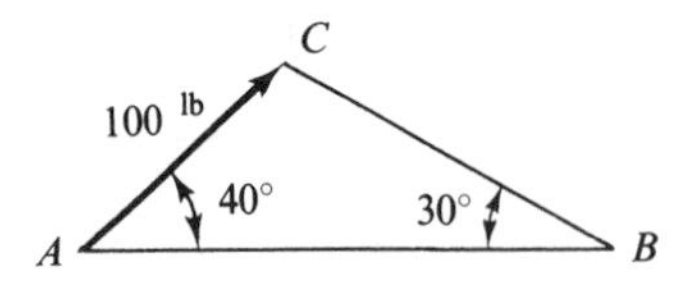

Figure P1-8

***1-9** (a) Can two forces which are the components of the 300-N force of Fig. P1-9 intersect at C?
(b) Resolve the 300-N force in Fig. P1-9 into a pair of horizontal and vertical components.
(c) Is each of two components of a force always less in magnitude than the force?

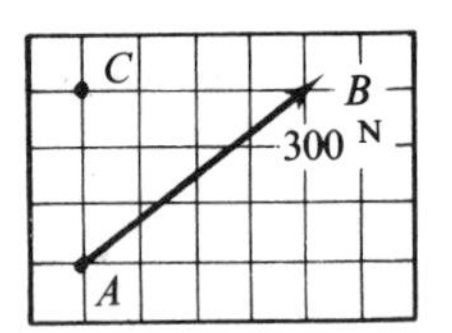

Figure P1-9

1-10 Resolve the 260-lb force of Fig. P1-10 into two components with one acting along AB and the other passing through the point C.

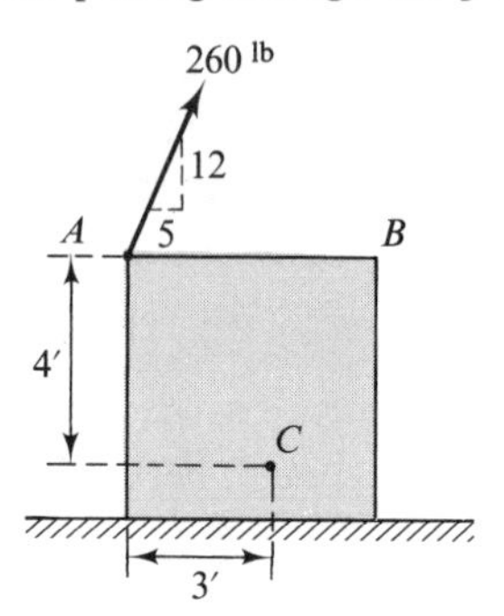

Figure P1-10

1-11 Resolve the 1000-lb force of Fig. P1-11 into two oblique components, one having a line of action along AB and the other acting parallel to BC.

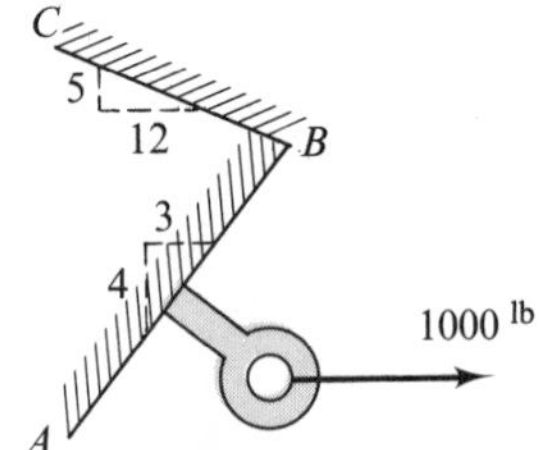

Figure P1-11

1-12 Resolve the 2000-lb force of Fig. P1-12 into two oblique components, one acting along AB and the other acting along BC.

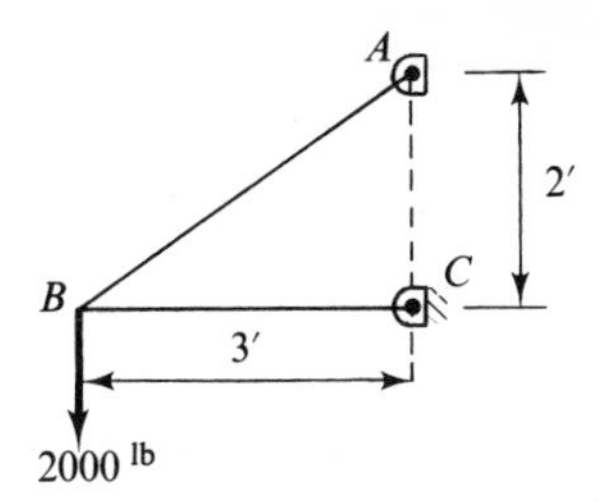

Figure P1-12

*1-13 Resolve the 200-N force of Fig. P1-13 into two oblique components, one having a line of action along AB and the second acting parallel to CD.

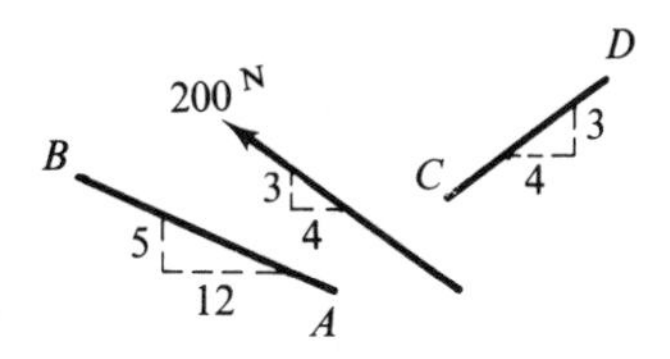

Figure P1-13

1-14 Determine two nonrectangular components of the 400-lb force of Fig. P1-14, one acting along AB and the second acting parallel to CD.

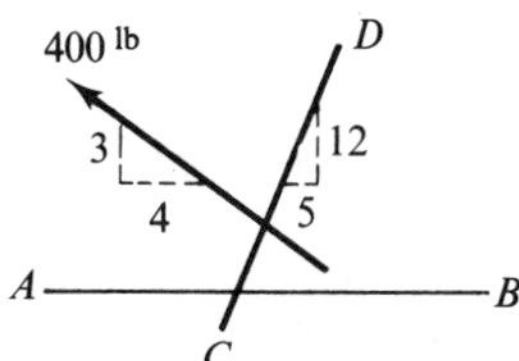

Figure P1-14

1-15 Resolve the 260-lb force of Fig. P1-15 into two components, one having a line of action along AB and the other parallel to CD.

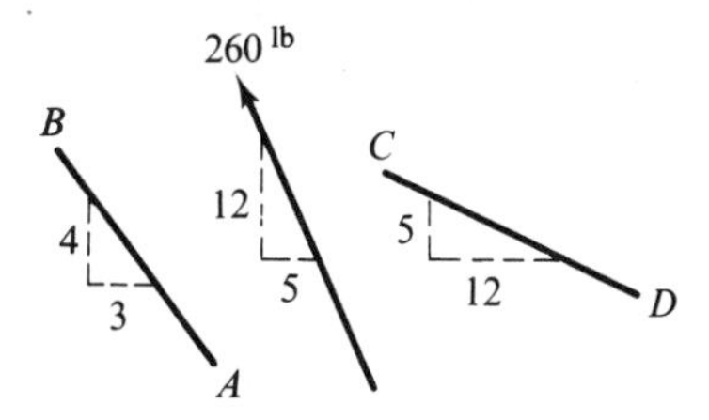

Figure P1-15

1-16 The force **F** in Fig. P1-16 is the resultant of a 100-lb force along BC and a force acting parallel to BD. Determine the component parallel to BD.

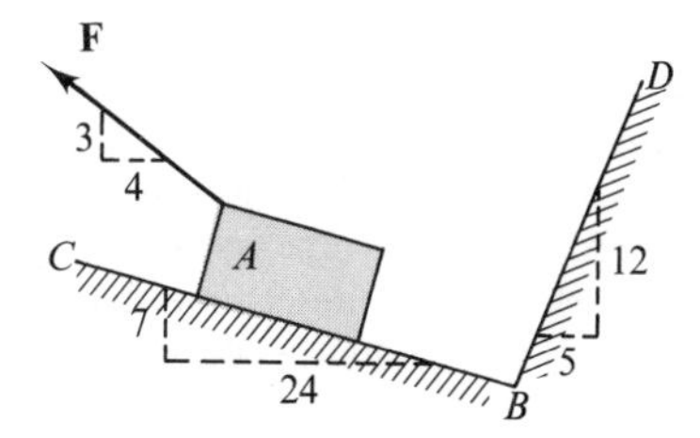

Figure P1-16

1-17 The wrecker in Fig. P1-17 is attempting to push the car up the incline AB. The force **P** exerted by the wrecker on the car is parallel to the plane BC. One component of **P** is 819 lb parallel to the incline AB, and the other component is perpendicular to AB. Determine the force **P**.

Figure P1-17

***1-18** The force **F** which acts upon the block resting on the inclined plane shown in Fig. P1-18 is known to have a horizontal rectangular component of 39.2 N to the right. Determine the magnitude and direction of the rectangular component of **F** perpendicular to the plane.

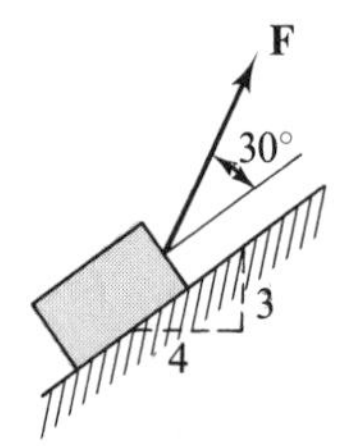

Figure P1-18

In Problems 1-19 and 1-20 the action line of the given force passes through the origin and the indicated point and is directed from the origin toward the point. The positive directions of the coordinate axes are as shown in Fig. P1-19. In each problem determine the direction cosines of the force and write the force in terms of the unit vectors **i**, **j**, and **k**.

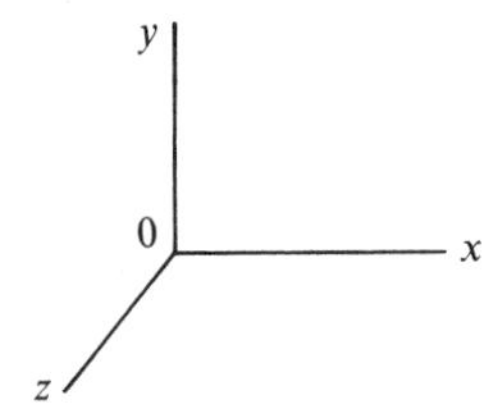

Figure P1-19

	Magnitude (lb)	Point (x, y, z)
1-19	300	2, 5, 14.
1-20	170	−8, 9, −12.

1-21 Determine the direction cosines of the 140-lb force in Fig. P1-21 and write the force in terms of **i**, **j**, and **k** components.

1-22 Determine the direction cosines of the 360-lb force in Fig. P1-21 and write the force in terms of **i**, **j**, and **k** components.

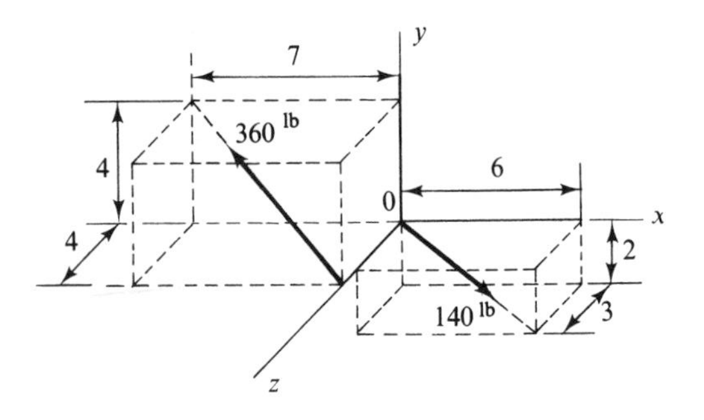

Figure P1-21

***1-23** The tension in the wire attached to the eyebolt in Fig. P1-23 is 522 N. Write the force in terms of the unit vectors **i**, **j**, and **k**.

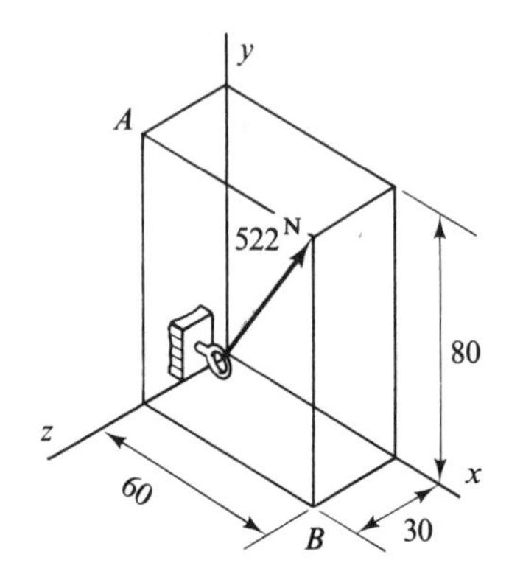

Figure P1-23

1-24 Determine a set of orthogonal components of the 200-lb force in Fig. P1-24 and express the force as a function of the unit vectors **i**, **j**, and **k**.

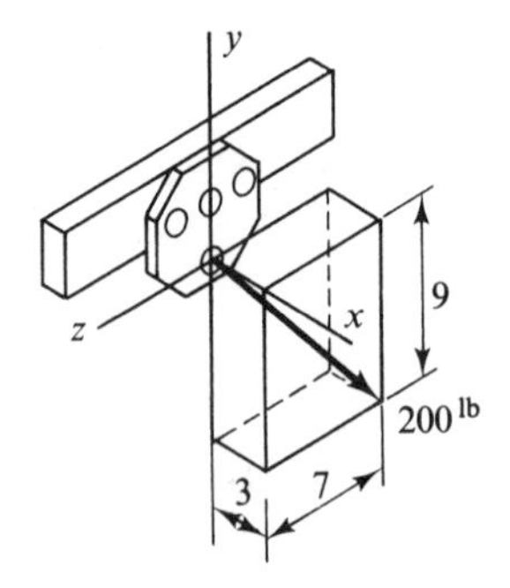

Figure P1-24

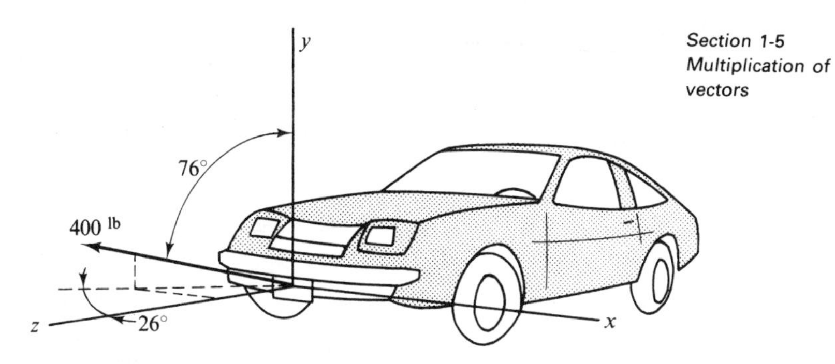

Figure P1-25

1-25 The automobile in Fig. P1-25 is towed by a rope with a tension of 400 lb. Determine a set of three rectangular components of the tension in the rope, and express the force as a function of the unit vectors **i**, **j**, and **k**.

1-26 Determine the direction cosines of the force in Fig. P1-25.

***1-27** The 1000-N force in Fig. P1-27 represents the force exerted on the rear wheel of an automobile in going around a curve. Determine a set of three rectangular components of the force. The xy plane is the horizontal plane.

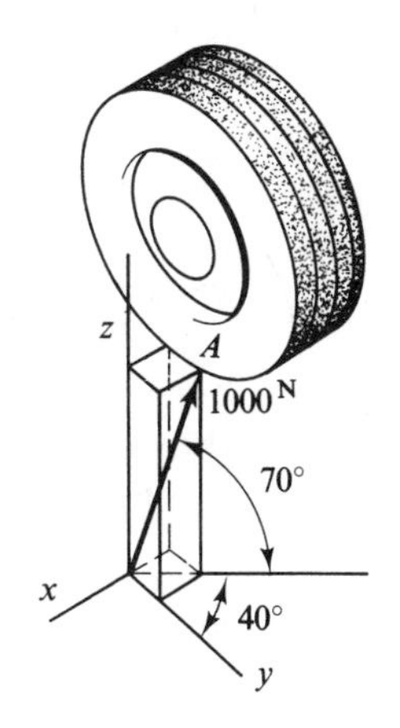

Figure P1-27

1-5 MULTIPLICATION OF VECTORS

Unlike scalar multiplication, there are two common types of vector products,[3] namely, the scalar or dot product (written as $\mathbf{A}\cdot\mathbf{B}$) and the vector or cross product (written as $\mathbf{A}\times\mathbf{B}$). These products have entirely different properties and are used for different purposes.

The dot product of two vectors **A** and **B** arranged with a common starting point as shown in Fig. 1-9 is defined as a scalar quantity whose magnitude is

$$\mathbf{A}\cdot\mathbf{B} = AB\cos\theta, \tag{1-1}$$

where θ is the smallest angle (0 to 180°) between

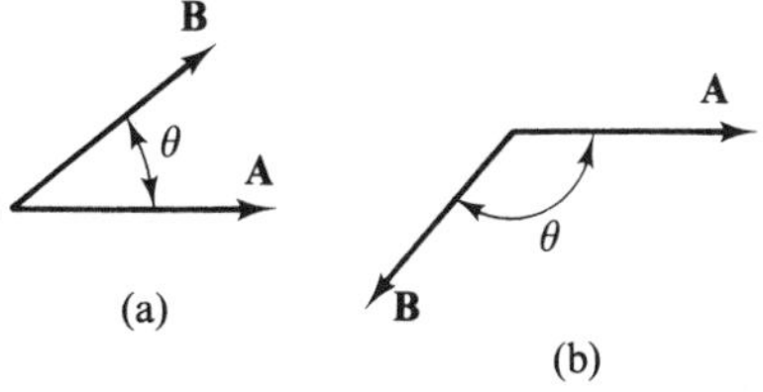

Figure 1-9

[3] A third type of product of vectors, the dyadic, **AB**, will not be discussed since it is seldom needed in solving elementary mechanics problems.

A and B as shown. Note that for $90° < \theta < 180°$, $\cos\theta$ is negative and the scalar product is negative. When $\theta = 90°$ (**A** perpendicular to **B**) $\mathbf{A} \cdot \mathbf{B} = 0$. Nonintersecting vectors are treated as free vectors which can be moved to parallel positions so that they intersect. From the definition, it is apparent that

$$\mathbf{A} \cdot \mathbf{B} = \mathbf{B} \cdot \mathbf{A};$$

that is, the dot product is commutative.

The scalar product also obeys the distributive law, that is,

$$\mathbf{A} \cdot (\mathbf{B} + \mathbf{C}) = \mathbf{A} \cdot \mathbf{B} + \mathbf{A} \cdot \mathbf{C}.$$

The expression $\mathbf{A} \cdot (\mathbf{B} \cdot \mathbf{C})$ is not defined since it implies the scalar product of a vector and a scalar. The expression $\mathbf{A}(\mathbf{B} \cdot \mathbf{C})$, however, is valid since it represents the multiplication of a vector by a scalar.

The scalar, or dot, product is useful for computing the rectangular component of a force or other vector in a particular direction. If **n** is a *unit vector*, the magnitude of the rectangular component of any vector **A** in the direction of **n** is $\mathbf{A} \cdot \mathbf{n}$ since $\mathbf{A} \cdot \mathbf{n} = A(1)\cos\theta$ and the component of **A** in the direction of **n** is $(\mathbf{A} \cdot \mathbf{n})\mathbf{n}$. The component of **A** perpendicular to **n** can be obtained by subtracting the component in the **n** direction from the original vector. Thus the two orthogonal components of **A** parallel and perpendicular to **n** are

$$\mathbf{A}_1 = (\mathbf{A} \cdot \mathbf{n})\mathbf{n}$$

and

$$\mathbf{A}_2 = \mathbf{A} - (\mathbf{A} \cdot \mathbf{n})\mathbf{n}.$$

The dot product can be readily computed once the vectors are expressed as functions of **i**, **j**, and **k**, since

$$\mathbf{i} \cdot \mathbf{i} = \mathbf{j} \cdot \mathbf{j} = \mathbf{k} \cdot \mathbf{k} = 1$$

and

$$\mathbf{i} \cdot \mathbf{j} = \mathbf{j} \cdot \mathbf{k} = \mathbf{k} \cdot \mathbf{i} = 0.$$

The first relations are true because θ is zero and $\cos\theta$ is unity; the others are true because θ is 90° and $\cos\theta$ is zero. With these relations, the dot product of two vectors

$$\mathbf{A} = A_x\mathbf{i} + A_y\mathbf{j} + A_z\mathbf{k}$$

and

$$\mathbf{B} = B_x\mathbf{i} + B_y\mathbf{j} + B_z\mathbf{k}$$

is found to be

$$\mathbf{A} \cdot \mathbf{B} = (A_x\mathbf{i} + A_y\mathbf{j} + A_z\mathbf{k}) \cdot (B_x\mathbf{i} + B_y\mathbf{j} + B_z\mathbf{k})$$
$$= A_xB_x + A_yB_y + A_zB_z.$$

EXAMPLE 1-3

Resolve the 390-lb force **F** in Fig. 1-10 into two components **P** and **Q** with **P** parallel to *OH* and **Q** perpendicular to *OH*.

SOLUTION

The vector from O to D is

$$\mathbf{OD} = (3\mathbf{i} + 12\mathbf{j} + 4\mathbf{k})\text{ ft}$$

and its magnitude is

$$|\mathbf{OD}| = [(3)^2 + (12)^2 + (4)^2]^{1/2} = 13\text{ ft}.$$

The unit vector along OD is

$$\mathbf{n}_{OD} = \frac{3\mathbf{i} + 12\mathbf{j} + 4\mathbf{k}}{13} = \frac{3}{13}\mathbf{i} + \frac{12}{13}\mathbf{j} + \frac{4}{13}\mathbf{k},$$

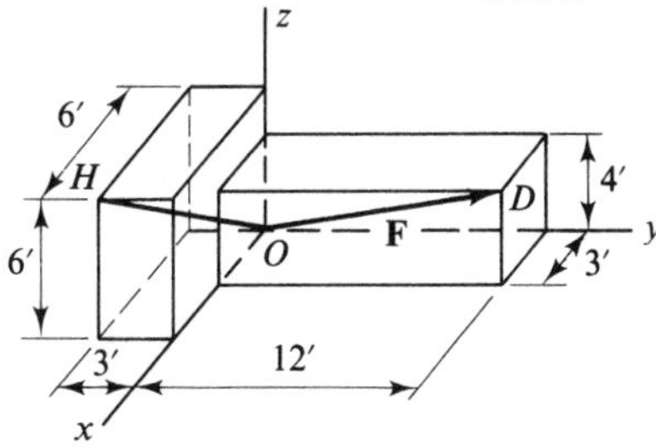

Figure 1-10

and the force $\mathbf{F}$ is

$$\mathbf{F} = 390\mathbf{n}_{OD} = 390\left[\frac{3}{13}\mathbf{i} + \frac{12}{13}\mathbf{j} + \frac{4}{13}\mathbf{k}\right]$$

$$= (90\mathbf{i} + 360\mathbf{j} + 120\mathbf{k})\text{ lb through } O.$$

Note that the coefficients of $\mathbf{i}$, $\mathbf{j}$, and $\mathbf{k}$ in the expression for $\mathbf{n}_{OD}$ are the direction cosines of $\mathbf{OD}$ and of $\mathbf{F}$.

The vector $\mathbf{OH}$ is

$$\mathbf{OH} = (6\mathbf{i} - 3\mathbf{j} + 6\mathbf{k})\text{ ft},$$

and the unit vector along OH is

$$\mathbf{n}_{OH} = \frac{6\mathbf{i} - 3\mathbf{j} + 6\mathbf{k}}{[6^2 + 3^2 + 6^2]^{1/2}} = \frac{6}{9}\mathbf{i} - \frac{3}{9}\mathbf{j} + \frac{6}{9}\mathbf{k}.$$

The orthogonal component, $\mathbf{P}$, of $\mathbf{F}$ along OH is

$$\mathbf{P} = (\mathbf{F} \cdot \mathbf{n}_{OH})\mathbf{n}_{OH}$$

$$= \left[(90\mathbf{i} + 360\mathbf{j} + 120\mathbf{k}) \cdot \left(\frac{2}{3}\mathbf{i} - \frac{1}{3}\mathbf{j} + \frac{2}{3}\mathbf{k}\right)\right]\mathbf{n}_{OH}$$

$$= (60 - 120 + 80)\mathbf{n}_{OH}$$

$$= \underline{20(0.667\mathbf{i} - 0.333\mathbf{j} + 0.667\mathbf{k})\text{ lb through } O.} \qquad \text{Ans.}$$

Since $\mathbf{F}$ is equal to the sum of $\mathbf{P}$ and $\mathbf{Q}$, the component $\mathbf{Q}$ perpendicular to OH is

$$\mathbf{Q} = \mathbf{F} - \mathbf{P}$$

$$= (90\mathbf{i} + 360\mathbf{j} + 120\mathbf{k}) - 20(0.667\mathbf{i} - 0.333\mathbf{j} + 0.667\mathbf{k})$$

$$= \underline{390(0.1969\mathbf{i} + 0.941\mathbf{j} + 0.274\mathbf{k})\text{ lb through } O.} \qquad \text{Ans.}$$

The cross or vector product of two intersecting vectors $\mathbf{A}$ and $\mathbf{B}$ arranged with a common starting point as shown in Fig. 1-9 is defined as the vector

$$\mathbf{C} = \mathbf{A} \times \mathbf{B} = \mathbf{n}AB \sin\theta, \qquad (1\text{-}2)$$

where $\mathbf{n}$ is a unit vector perpendicular to the plane containing $\mathbf{A}$ and $\mathbf{B}$ and θ is the smallest angle $(0 \le \theta \le 180°)$ measured from $\mathbf{A}$ to $\mathbf{B}$. The sense of $\mathbf{C}$, and of $\mathbf{n}$, is determined by the right-hand rule and is in the direction a right-hand screw would advance as **A** is rotated through the angle θ toward **B**; see Fig. 1-11. The right-hand rule indicates that the sense of $\mathbf{B}\times\mathbf{A}$ is opposite the sense of $\mathbf{A}\times\mathbf{B}$; that is,

$$\mathbf{B}\times\mathbf{A} = -\mathbf{A}\times\mathbf{B}.$$

Consequently, it is essential that the factors of a cross product be written in the correct order.

It has just been shown that the vector product is not commutative. The vector product does, however, obey the distributive law. That is,

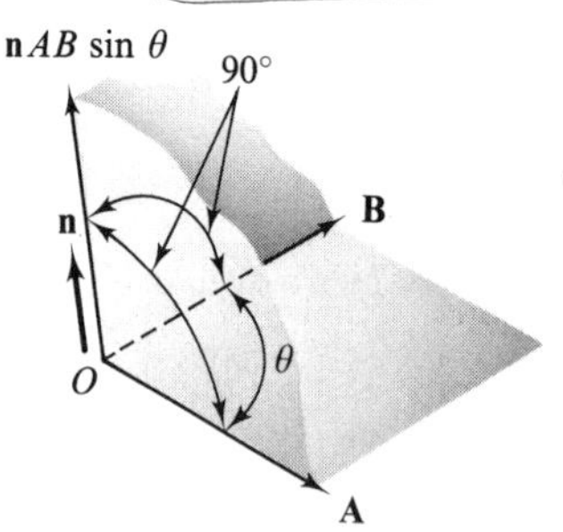

Figure 1-11

$$\mathbf{A}\times(\mathbf{B}+\mathbf{C}) = \mathbf{A}\times\mathbf{B}+\mathbf{A}\times\mathbf{C},$$

and

$$(\mathbf{D}+\mathbf{E})\times\mathbf{F} = \mathbf{D}\times\mathbf{F}+\mathbf{E}\times\mathbf{F}.$$

Note that the order of the factors must be preserved. The vector product does not in general satisfy the associative law. Thus

$$(\mathbf{A}\times\mathbf{B})\times\mathbf{C} \neq \mathbf{A}\times(\mathbf{B}\times\mathbf{C}).$$

As in the case of the dot product, it is advantageous to express vectors as functions of **i**, **j**, and **k** in order to determine their cross products. Since these unit vectors are mutually perpendicular and form a right-hand system as shown in Fig. 1-4, it follows that

$$\mathbf{i}\times\mathbf{j} = -\mathbf{j}\times\mathbf{i} = \mathbf{k},$$

$$\mathbf{j}\times\mathbf{k} = -\mathbf{k}\times\mathbf{j} = \mathbf{i},$$

$$\mathbf{k}\times\mathbf{i} = -\mathbf{i}\times\mathbf{k} = \mathbf{j},$$

and

$$\mathbf{i}\times\mathbf{i} = \mathbf{j}\times\mathbf{j} = \mathbf{k}\times\mathbf{k} = 0.$$

With these basic relations, the cross product of two vectors **A** and **B** can be obtained as follows:

$$\begin{aligned}\mathbf{A}\times\mathbf{B} &= (A_x\mathbf{i}+A_y\mathbf{j}+A_z\mathbf{k})\times(B_x\mathbf{i}+B_y\mathbf{j}+B_z\mathbf{k})\\
&= A_xB_x\mathbf{i}\times\mathbf{i}+A_xB_y\mathbf{i}\times\mathbf{j}+A_xB_z\mathbf{i}\times\mathbf{k}\\
&\quad+A_yB_x\mathbf{j}\times\mathbf{i}+A_yB_y\mathbf{j}\times\mathbf{j}+A_yB_z\mathbf{j}\times\mathbf{k}\\
&\quad+A_zB_x\mathbf{k}\times\mathbf{i}+A_zB_y\mathbf{k}\times\mathbf{j}+A_zB_z\mathbf{k}\times\mathbf{k}\\
&= (A_yB_z-A_zB_y)\mathbf{i}+(A_zB_x-A_xB_z)\mathbf{j}\\
&\quad+(A_xB_y-A_yB_x)\mathbf{k}. \qquad \text{(a)}\end{aligned}$$

Equation (a) can be expressed as the determinant

$$\mathbf{A}\times\mathbf{B} = \begin{vmatrix}\mathbf{i} & \mathbf{j} & \mathbf{k}\\ A_x & A_y & A_z\\ B_x & B_y & B_z\end{vmatrix},$$

as can be demonstrated by expanding the determinant. This avoids the necessity of multiplying the vectors term by term or of memorizing Eq. (a).

An important example of the cross product of two intersecting vectors is the moment of a force about a point as developed in the next section.

EXAMPLE 1-4

Determine the vector product of the two vectors **A** and **B** in Fig. 1-12, where $\mathbf{A}=3\mathbf{i}-4\mathbf{k}$ and $\mathbf{B}=-12\mathbf{i}+3\mathbf{j}+4\mathbf{k}$.

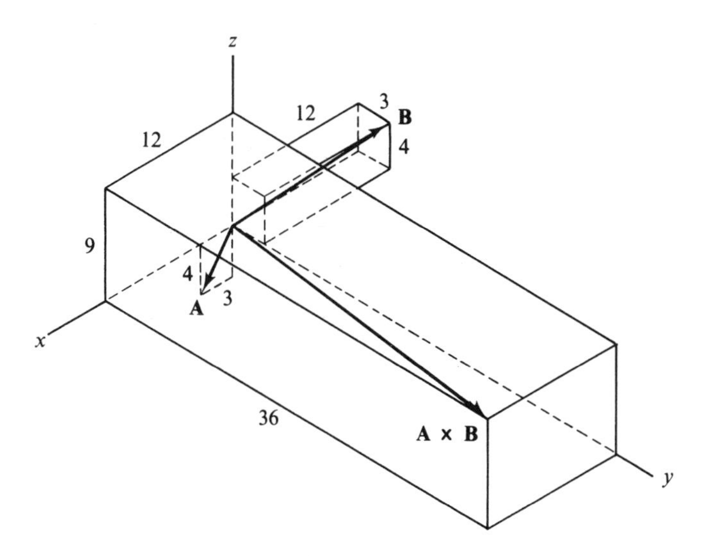

Figure 1-12

SOLUTION

The vector product is determined using the expansion of the determinant as indicated below:

$$\mathbf{A}\times\mathbf{B}=\begin{vmatrix} \mathbf{i} & \mathbf{j} & \mathbf{k} \\ 3 & 0 & -4 \\ -12 & 3 & 4 \end{vmatrix}$$

$$=\mathbf{i}[0(4)-3(-4)]-\mathbf{j}[3(4)-(-4)(-12)]$$

$$+\mathbf{k}[3(3)-(-12)0]=\underline{12\mathbf{i}+36\mathbf{j}+9\mathbf{k}}. \qquad \text{Ans.}$$

The result is shown in Fig. 1-12.

EXAMPLE 1-5

Use a vector triple product (two vector products in succession) to determine the component **Q** of the 390-lb force **F** in Example 1-3, Fig. 1-10, where **Q** is perpendicular to *OH*.

SOLUTION

The vector product of the unit vector $\mathbf{n}_{OH}$ and the force $\mathbf{F}$ is

$$\mathbf{A} = \mathbf{n}_{OH} \times \mathbf{F} = 1(F)\sin\theta\ \mathbf{n}_1, \tag{a}$$

where $\mathbf{n}_1$ is a unit vector perpendicular to the plane HOD in Fig. 1-10 and θ is the angle between $\mathbf{n}_{OH}$ and $\mathbf{F}$. The magnitude of this vector product is $F \sin\theta$, which is the magnitude of the desired component, $\mathbf{Q}$. Since the components $\mathbf{P}$ and $\mathbf{Q}$ are to be combined to give $\mathbf{F}$, all three vectors must lie in the plane HOD; therefore the vector $\mathbf{A}$ in Eq. (a) has the correct magnitude but is perpendicular to the correct direction. The vectors $\mathbf{A}$ and $\mathbf{n}_{OH}$ are perpendicular to each other, and their cross product will result in a new vector with the same magnitude which is perpendicular to both $\mathbf{A}$ and $\mathbf{n}_{OH}$. This new vector is the desired component $\mathbf{Q}$ since it has the correct magnitude, $F \sin\theta$, and the correct direction, perpendicular to OH, thus:

$$\mathbf{Q} = \mathbf{A} \times \mathbf{n}_{OH} = (\mathbf{n}_{OH} \times \mathbf{F}) \times \mathbf{n}_{OH}.$$

The values of $\mathbf{n}_{OH}$ and $\mathbf{F}$, as determined in Example 1-3, are

$$\mathbf{n}_{OH} = \frac{2}{3}\mathbf{i} - \frac{1}{3}\mathbf{j} + \frac{2}{3}\mathbf{k}$$

and

$$\mathbf{F} = (90\mathbf{i} + 360\mathbf{j} + 120\mathbf{k})\text{ lb}.$$

The first vector product is

$$\mathbf{n}_{OH} \times \mathbf{F} = \begin{vmatrix} \mathbf{i} & \mathbf{j} & \mathbf{k} \\ \frac{2}{3} & -\frac{1}{3} & \frac{2}{3} \\ 90 & 360 & 120 \end{vmatrix}$$

$$= \mathbf{i}(-40 - 240) - \mathbf{j}(80 - 60) + \mathbf{k}(240 + 30)$$

$$= -280\mathbf{i} - 20\mathbf{j} + 270\mathbf{k},$$

and the component $\mathbf{Q}$ is

$$\mathbf{Q} = (\mathbf{n}_{OH} \times \mathbf{F}) \times \mathbf{n}_{OH}$$

$$= \begin{vmatrix} \mathbf{i} & \mathbf{j} & \mathbf{k} \\ -280 & -20 & 270 \\ \frac{2}{3} & -\frac{1}{3} & \frac{2}{3} \end{vmatrix}$$

$$= \mathbf{i}(-13.33 + 90) - \mathbf{j}(-186.7 - 180) + \mathbf{k}(93.33 + 13.33)$$

$$= (76.7\mathbf{i} + 367\mathbf{j} + 106.7\mathbf{k})\text{ lb through } O. \qquad \text{Ans.}$$

This result agrees with Example 1-3, which provides a check of the calculations.

The vector triple product of three vectors $\mathbf{A}$, $\mathbf{B}$, and $\mathbf{C}$ can be expanded as follows (see any text on vector algebra):

$$(\mathbf{A} \times \mathbf{B}) \times \mathbf{C} = (\mathbf{A} \cdot \mathbf{C})\mathbf{B} - (\mathbf{B} \cdot \mathbf{C})\mathbf{A}.$$

When **A** and **C** are replaced by $\mathbf{n}_{OH}$ and **B** by **F** the expression becomes

$$(\mathbf{n}_{OH}\times\mathbf{F})\times\mathbf{n}_{OH}=(\mathbf{n}_{OH}\cdot\mathbf{n}_{OH})\mathbf{F}-(\mathbf{F}\cdot\mathbf{n}_{OH})\mathbf{n}_{OH}$$

$$=\mathbf{F}-\mathbf{F}_{OH},$$

which is the form used to obtain **Q** in Example 1-3.

PROBLEMS

1-28 Determine the dot and cross product of the vectors $\mathbf{A}=2\mathbf{i}-\mathbf{j}+3\mathbf{k}$ and $\mathbf{B}=-\mathbf{i}+5\mathbf{j}-2\mathbf{k}$.

1-29 For the vectors $\mathbf{A}=6\mathbf{i}-\mathbf{j}-2\mathbf{k}$ and $\mathbf{B}=3\mathbf{i}+2\mathbf{j}-5\mathbf{k}$, determine (a) the dot product, (b) the cross product.

***1-30** Determine the rectangular component of the 200-N force in Fig. P1-13 that is (a) along the line *AB*, (b) along the line *CD*.

1-31 Determine a rectangular component of the force **F** (magnitude = 100 lb) in Fig. P1-16 that is parallel to (a) the line *BC*, (b) the line *BD*.

1-32 Determine a rectangular component of the 300-lb force of Problem 1-19 in the direction $2\mathbf{i}+3\mathbf{j}+6\mathbf{k}$.

1-33 Determine the two rectangular components of the 110-lb force in Fig. P1-33 which are parallel and perpendicular to *AB*.

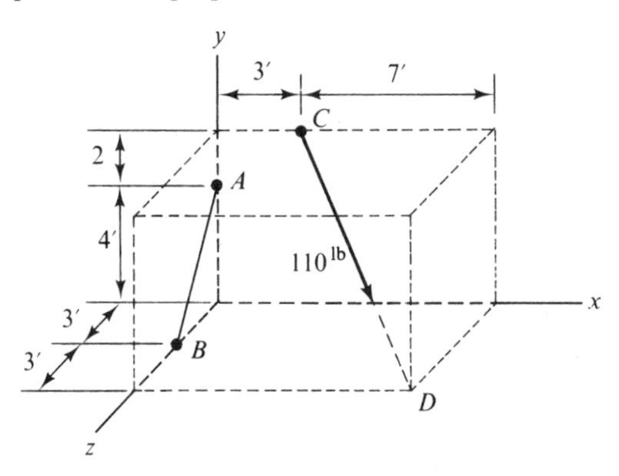

Figure P1-33

1-34 Determine the two rectangular components of the 90-lb force in Fig. P1-34 which are parallel and perpendicular to *AC*.

1-35 Determine a unit vector perpendicular to the two vectors $\mathbf{A}=2\mathbf{i}-\mathbf{j}+\mathbf{k}$ and $\mathbf{B}=3\mathbf{i}+4\mathbf{j}-5\mathbf{k}$.

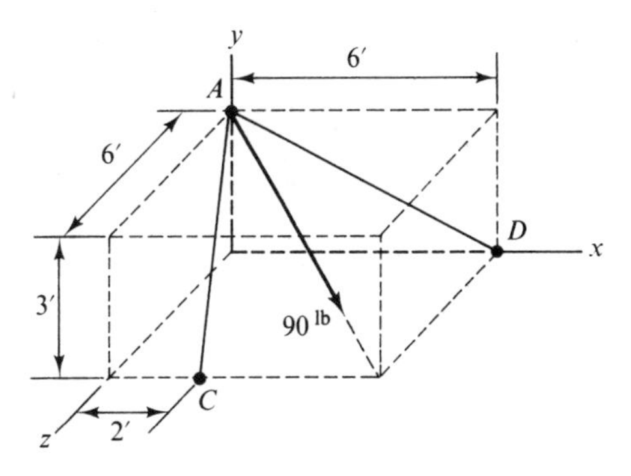

Figure P1-34

1-36 Derive a vector representation of a 35-lb force through *A* directed upward and perpendicular to the plane *ACD* of Fig. P1-34.

***1-37** Resolve the 140-N force of Fig. P1-37 into two orthogonal components, **P** and **Q**, where **P** is perpendicular to the plane *ABC*.

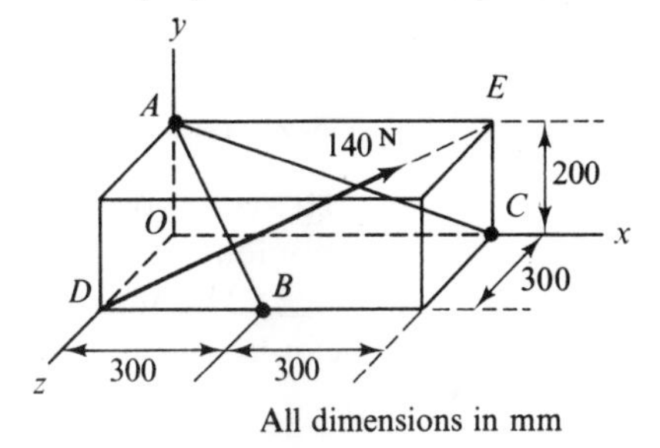

Figure P1-37

1-38 Resolve the force $\mathbf{F}=(300\mathbf{i}+400\mathbf{j}+500\mathbf{k})$ lb into three orthogonal components **P**, **Q**, and **R** defined as follows. Force **P** is to be parallel to the vector $\mathbf{A}=3\mathbf{i}-6\mathbf{j}+2\mathbf{k}$, force **Q** is to be parallel to the *xy* plane, and force **R** is (by definition) perpendicular to **P** and **Q**.

1-39 Three adjacent edges of the parallelopiped of Fig. P1-39 are represented by the vectors **a**, **b**, and **c**. Prove that

(a) The area of the base is equal to the magnitude of $\mathbf{a} \times \mathbf{b}$.

(b) The volume of the parallelopiped is equal to $\mathbf{c} \cdot (\mathbf{a} \times \mathbf{b})$.

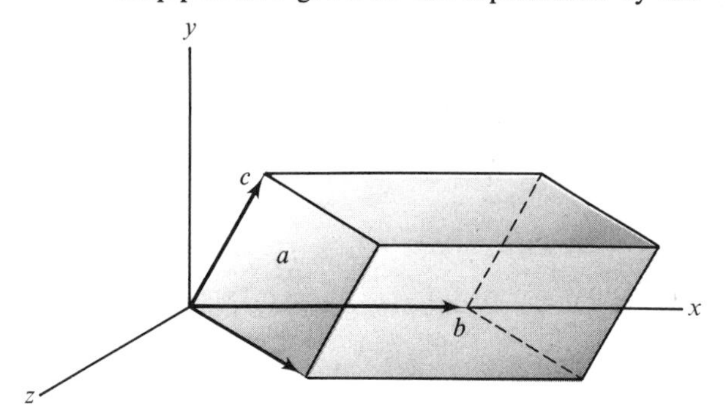

Figure P1-39

1-40 Given the vectors $\mathbf{A} = 2\mathbf{i} - \mathbf{j} - \mathbf{k}$, $\mathbf{B} = -\mathbf{i} + 2\mathbf{j} + \mathbf{k}$, and $\mathbf{C} = 3\mathbf{i} + 2\mathbf{j} + \mathbf{k}$, determine $\mathbf{A} \cdot \mathbf{B} \times \mathbf{C}$.

1-41 Determine $\mathbf{A} \times (\mathbf{B} \times \mathbf{C})$ for the vectors of Problem 1-40.

1-42 Determine $(\mathbf{A} \times \mathbf{B}) \times \mathbf{C}$ for the vectors of Problem 1-40.

1-6 MOMENT OF A FORCE

The moment of a force can be defined with respect to (about) a point and also with respect to a line or axis. *The moment of a force* **F** *with respect to a point A is defined as a vector with a magnitude equal to the product of the perpendicular distance from A to* **F** *and the magnitude of the force and with a direction perpendicular to the plane containing A and* **F**. The sense of the moment vector is given by the direction a right-hand screw would advance if turned about A in the direction indicated by **F** as shown in Fig. 1-13. The moment of a force is treated as a localized vector through the moment center (or axis) to emphasize the dependence of the moment of a force on the moment center (or moment axis). To avoid possible confusion, an open vector will be used to represent a moment of a force as shown in Fig. 1-13.

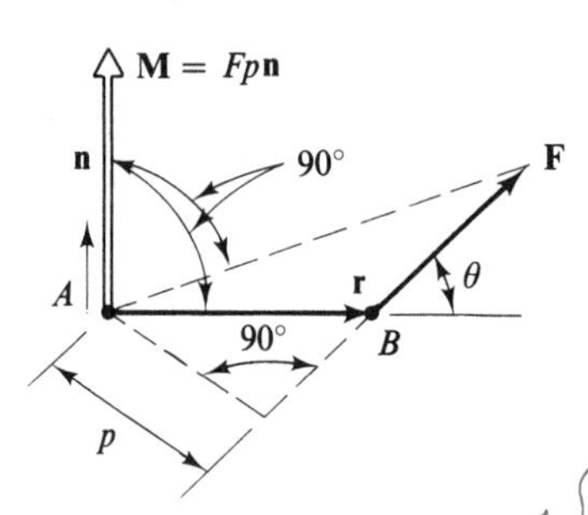

Figure 1-13

The moment of the force **F** in Fig. 1-13 with respect to point A can be written in terms of a vector product as

$$\mathbf{M} = \mathbf{r} \times \mathbf{F},$$

where **r** is a position vector from A to any point B on the line of action of **F**. The validity of the vector product representation can be demonstrated by expanding the cross product as follows:

$$\mathbf{M} = \mathbf{r} \times \mathbf{F} = Fr \sin \theta \mathbf{n} = Fp\mathbf{n}.$$

Notice that $p = r \sin \theta$ is the perpendicular distance from A to **F** for any point B on the line of action of **F**. The unit vector **n** is perpendicular to the plane of **r** and **F**, and its direction, as specified by the right-hand rule, has the same sense as the moment vector. The vector product method of calculating moments is, therefore, in agreement with the definition of a moment. This method is especially useful in three-dimensional problems in which the geometry becomes somewhat complicated and the perpendicular distances from moment centers to lines of action of forces are not apparent.

The moment of a force about a line perpendicular to a plane containing the force is defined as a vector with a magnitude equal to the product of the magnitude of the force and the perpendicular distance from the line to the force and with a direction along the line. Thus it is the same as the moment of the force about the point of intersection of the plane and the moment axis. The moment of a force about a point or axis is a measure of its tendency to turn or rotate a body about the point or axis.

The principle of moments as applied to a force system states that the moment of the resultant of the force system with respect to any axis or point is equal to the vector sum of the moments of the forces of the system with respect to the same axis or point. The application of this principle to a pair of concurrent forces is known as *Varignon's theorem.* Varignon's theorem may be demonstrated as follows: In Fig. 1-14, **R** is the resultant of the two forces **P** and **Q**; O is the moment center or the intersection of the moment axis, perpendicular to the plane of the forces, with the plane of the forces; and p, q, and r are the perpendicular distances from O to the corresponding forces. The angles θ_p, θ_q, and θ_r are measured as shown from the line connecting O with the point of concurrence of the forces, and a is the distance from O to the point of concurrence.

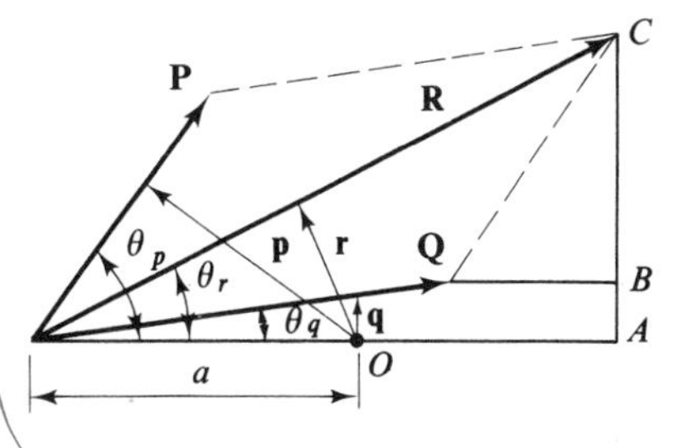

Figure 1-14

The magnitude of the moment of the resultant, **R**, with respect to O is

$$Rr = Ra \sin \theta_r.$$

From Fig. 1-14,

$$R \sin \theta_r = AC = AB + BC$$

$$= Q \sin \theta_q + P \sin \theta_p.$$

Therefore

$$Rr = a(Q \sin \theta_q + P \sin \theta_p)$$

$$= Qa \sin \theta_q + Pa \sin \theta_p$$

$$= Qq + Pp. \tag{a}$$

Equation (a) indicates that the moment of the resultant, **R**, with respect to O is equal to the sum of the moments of the component forces **P** and **Q** with respect to O.

The principle of moments is not restricted to two concurrent forces but may be extended to any force system.

The force **F** of Fig. 1-15 acts at point A and has components $\mathbf{F}_x = F_x\mathbf{i}$, $\mathbf{F}_y = F_y\mathbf{j}$, and $\mathbf{F}_z = F_z\mathbf{k}$. The moment of **F** with respect to point O may be expressed in component form by expanding the cross product,

$$\mathbf{M}_O = \mathbf{r}_{OA} \times \mathbf{F} = (x\mathbf{i}+y\mathbf{j}+z\mathbf{k}) \times (F_x\mathbf{i}+F_y\mathbf{j}+F_z\mathbf{k})$$

$$= \begin{vmatrix} \mathbf{i} & \mathbf{j} & \mathbf{k} \\ x & y & z \\ F_x & F_y & F_z \end{vmatrix}$$

$$= (yF_z - zF_y)\mathbf{i} + (zF_x - xF_z)\mathbf{j} + (xF_y - yF_x)\mathbf{k}$$

$$= M_x\mathbf{i} + M_y\mathbf{j} + M_z\mathbf{k},$$

where

$$M_x = yF_z - zF_y,$$

$$M_y = zF_x - xF_z,$$

$$M_z = xF_y - yF_x$$

are the magnitudes of the orthogonal components of $\mathbf{M}_O$ in the x, y, and z directions.

These values are also the magnitudes of the moment of **F** with respect to the x, y, and z axes. This can readily be seen from Fig. 1-15. The moment of **F** with respect to the x axis, from the principle of moments, is equal to the sum of the moments of the components of **F**. Since $\mathbf{F}_x$ is parallel to the x axis, its moment is zero. The x axis is perpendicular to the plane containing both $\mathbf{F}_y$ and $\mathbf{F}_z$; hence the magnitude of the moment about the x axis is

$$M_x = yF_z - zF_y,$$

where the algebraic signs are determined by the right-hand screw rule. The values of M_y and M_z can be determined by the same procedure.

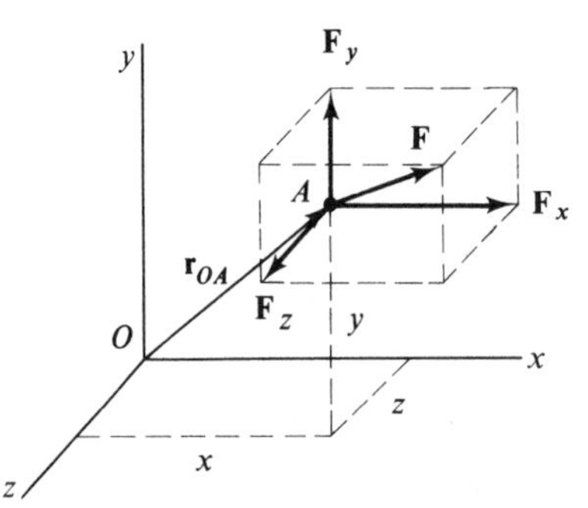

Figure 1-15

The moment of **F** with respect to O is thus seen to be the vector sum of the moments with respect to the three orthogonal axes x, y, and z through O. Conversely the orthogonal components of $\mathbf{M}_O$ are the moments with respect to the x, y, and z axes.

Since the x, y, and z axes may be chosen in any direction as long as they are perpendicular to each other, it follows that the moment of **F** with respect to any line through O is the orthogonal component of $\mathbf{M}_O$ parallel to this line. Therefore the following general statement may be made.

The moment of a force with respect to a line can be determined by calculating the moment of the force about some point on the line and then calculating the orthogonal component of the moment parallel to the reference line. These two operations, the vector product of the position vector and the force to obtain the moment about a point, followed by the scalar product of the moment about a point and a unit vector along the desired moment axis, can be performed in sequence or combined into one operation. If the moment of a force **F** with respect to a line AC is required, the moment about any point A on AC is $\mathbf{r} \times \mathbf{F}$, and

$$\mathbf{M}_{AC} = [\mathbf{n} \cdot (\mathbf{r} \times \mathbf{F})]\mathbf{n},$$

where **n** is a unit vector along the moment axis AC. The quantity inside the brackets is called a *scalar triple product* since it is the scalar product of two vectors **n** and $\mathbf{r} \times \mathbf{F}$. The scalar triple product can be computed by means of a determinant as

$$\mathbf{n} \cdot (\mathbf{r} \times \mathbf{F}) = \begin{vmatrix} n_x & n_y & n_z \\ r_x & r_y & r_z \\ F_x & F_y & F_z \end{vmatrix}.$$

The validity of this expression can be demonstrated by substituting the expanded forms of the three vectors and carrying out the indicated operations. The unit vector can be selected from A toward C or the reverse thereof. A positive coefficient of **n** in the expression for $\mathbf{M}_{AC}$ means that the moment vector has the same sense as that selected for **n**, whereas a negative sign indicates that $\mathbf{M}_{AC}$ is opposite to the sense of **n**.

EXAMPLE 1-6

(a) Determine the moment of the 1300-N force in Fig. 1-16 with respect to point O by using

1. The principle of moments.
2. The cross product.

(b) Determine the perpendicular distance from O to the line of action of **F** by using the results for part (a).

SOLUTION

(a) **1.** The 1300-N force is resolved into its horizontal and vertical components at A as shown in Fig. 1-16. The sum of the

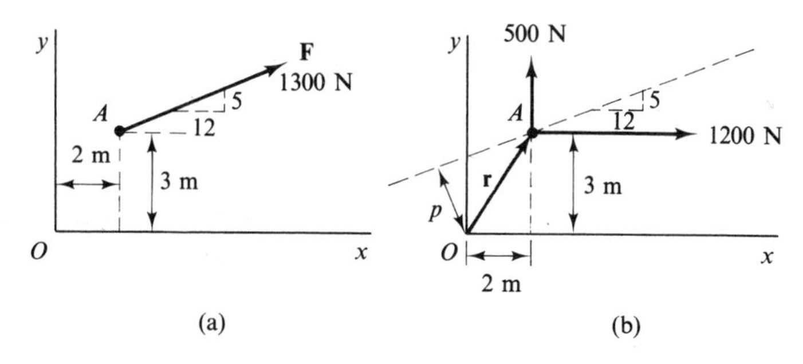

Figure 1-16

moments of these components is equal to the moment of **F**, and when counterclockwise is taken as positive according to the right-hand screw rule, the moment becomes

$$M_O = 500(2) - 1200(3) = 1000 - 3600 = -2600.$$

Therefore

$$\underline{\mathbf{M}_O = 2600\ \text{N} \cdot \text{m} \circlearrowright} \qquad \text{Ans.}$$

or the moment can be written

$$\mathbf{M}_O = \underline{-2600\mathbf{k}\ \text{N} \cdot \text{m}.} \qquad \text{Ans.}$$

2. The position vector from O to A is $\mathbf{r} = (2\mathbf{i} + 3\mathbf{j})$ m, and the force can be represented vectorially as $\mathbf{F} = (1200\mathbf{i} + 500\mathbf{j})$ N. The moment of **F** with respect to O is

$$\mathbf{M}_O = \mathbf{r} \times \mathbf{F} = \begin{vmatrix} \mathbf{i} & \mathbf{j} & \mathbf{k} \\ 2 & 3 & 0 \\ 1200 & 500 & 0 \end{vmatrix} = (1000 - 3600)\mathbf{k}$$

and

$$\mathbf{M}_O = \underline{-2600\mathbf{k}\ \text{N} \cdot \text{m}.} \qquad \text{Ans.}$$

(b) Let p be the perpendicular distance from O to the line of action of **F** as shown in Fig. 1-16(b). The magnitude of the moment of **F** about O is pF, by definition, and when this is equated to the result from part (a) it becomes

$$M_O = pF = p(1300) = 2600\ \text{N} \cdot \text{m}$$

from which

$$p = \underline{2.00\ \text{m}.} \qquad \text{Ans.}$$

EXAMPLE 1-7

The 130-lb force **F** in Fig. 1-17(a) is applied to point C as shown.

(a) Resolve **F** into three orthogonal components and determine the moment of **F** about the z axis using the principle of moments.
(b) Check part (a) by determining the z component of the moment of **F** about O.
(c) Resolve **F** into two orthogonal components, **P** and **Q**, with **P** parallel to the z axis and **Q** parallel to the xy plane.
(d) Determine the perpendicular distance from the z axis to the force **Q**.
(e) Determine the perpendicular distance from the origin to the force **F**.

SOLUTION

(a) The unit vector parallel to **F** is obtained by inspection as

$$\mathbf{n}_F = \frac{3\mathbf{i} + 4\mathbf{j} - 12\mathbf{k}}{(3^2 + 4^2 + 12^2)^{1/2}} = \frac{3}{13}\mathbf{i} + \frac{4}{13}\mathbf{j} - \frac{12}{13}\mathbf{k},$$

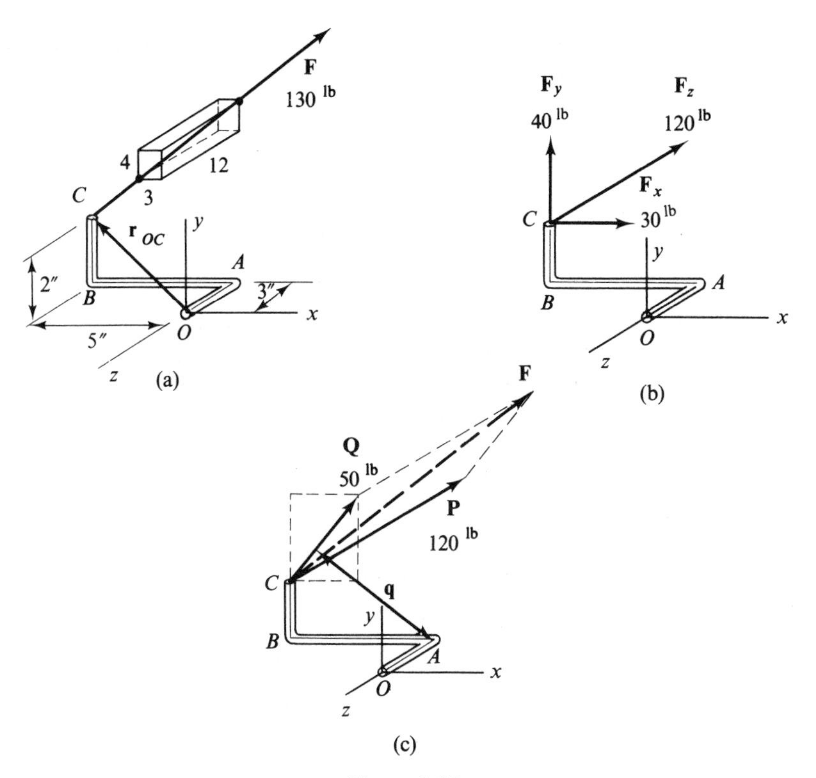

Figure 1-17

and the force **F** becomes

$$\mathbf{F} = F\mathbf{n}_F = 130\left(\frac{3}{13}\mathbf{i} + \frac{4}{13}\mathbf{j} - \frac{12}{13}\mathbf{k}\right)$$

$$= (30\mathbf{i} + 40\mathbf{j} - 120\mathbf{k}) \text{ lb through } C$$

as shown in Fig. 1-17(b). Since $\mathbf{F}_z$ is parallel to the z axis, its moment is zero, and the moment about the z axis is the sum of the moments of $\mathbf{F}_x$ and $\mathbf{F}_y$. Thus

$$M_z = (BC)F_x + (AB)F_y$$

$$= 2(30) + 5(40)$$

$$= 260 \text{ in-lb clockwise looking from } O \text{ toward } A.$$

The moment can be expressed as a vector as

$$\underline{\mathbf{M}_z = -260\mathbf{k} \text{ in-lb.}} \qquad \text{Ans.}$$

(b) The position vector from O to C is

$$\mathbf{r}_{OC} = (-5\mathbf{i} + 2\mathbf{j} - 3\mathbf{k}) \text{ in.,}$$

and the moment of **F** about O is

$$\mathbf{M}_O = \mathbf{r}_{OC} \times \mathbf{F}$$

$$= \begin{vmatrix} \mathbf{i} & \mathbf{j} & \mathbf{k} \\ -5 & 2 & -3 \\ 30 & 40 & -120 \end{vmatrix}$$

$$= \mathbf{i}(-240+120) - \mathbf{j}(600+90) + \mathbf{k}(-200-60)$$

$$= (-120\mathbf{i} - 690\mathbf{j} - 260\mathbf{k}) \text{ in-lb.}$$

The z component of $\mathbf{M}_O$ is $-260\mathbf{k}$ in-lb, which checks part (a).

(c) The two components **P** and **Q** are shown in Fig. 1-17(c). The force **P** is the same as $\mathbf{F}_z$, and **Q** is the resultant of $\mathbf{F}_x$ and $\mathbf{F}_y$. That is,

$$\mathbf{P} = \mathbf{F}_z = \underline{-120\mathbf{k} \text{ lb},} \qquad \text{Ans.}$$

$$\mathbf{Q} = \mathbf{F}_x + \mathbf{F}_y = 30\mathbf{i} + 40\mathbf{j} = \underline{50\mathbf{n}_Q \text{ lb},} \qquad \text{Ans.}$$

where $\mathbf{n}_Q = 0.6\mathbf{i} + 0.8\mathbf{j}$.

(d) Since the force **P** is parallel to the z axis, it has no moment about this axis, and the magnitude of the moment must be equal to the product of Q and the perpendicular distance, q, from the z axis to Q; see Fig. 1-17(c). Thus

$$M_z = qQ$$

$$260 = q(50)$$

and

$$q = \underline{5.20 \text{ in.}} \qquad \text{Ans.}$$

Note that q is the perpendicular distance from the z axis to the plane containing **P** and **Q**, and also **F**.

(e) The magnitude of the moment of **F** about O is equal to the product of the magnitude of **F** and the perpendicular distance from O to **F**. The magnitude of $\mathbf{M}_O$, from part (b), is

$$M_O = [(-120)^2 + (-690)^2 + (-260)^2]^{1/2}$$

$$= 747 \text{ in-lb,}$$

and the perpendicular distance, p, is

$$p = \frac{M_O}{F} = \frac{747}{130} = \underline{5.75 \text{ in.}} \qquad \text{Ans.}$$

Note that the distance p is measured from point O to the force **F** and that the distance q is measured from the z axis to the plane containing **P**, **Q**, and **F**. Neither of these distances is the perpendicular distance from the z axis to the force **F**. One method for obtaining this distance is described in Problem 1-57.

EXAMPLE 1-8

Determine the moment of the 150-lb force **F** in Fig. 1-18
(a) With respect to A.
(b) With respect to D.
(c) With respect to line AD.

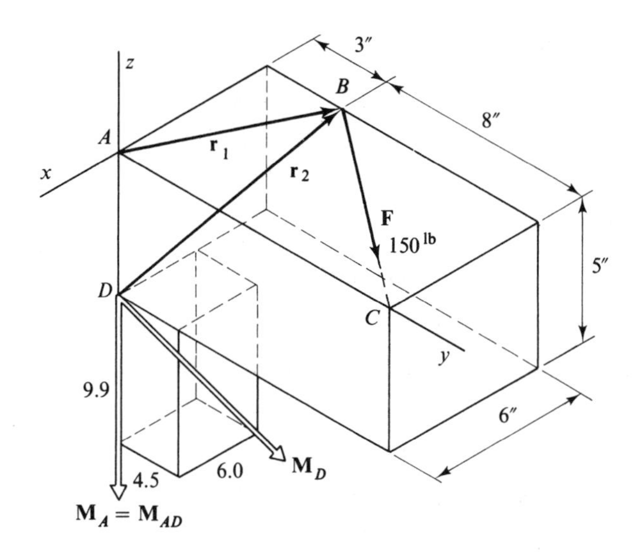

Figure 1-18

SOLUTION

(a) The force **F** can be expressed in terms of **i**, **j**, and **k** as

$$\mathbf{F} = 150\left[\left(\frac{6}{10}\right)\mathbf{i} + \left(\frac{8}{10}\right)\mathbf{j}\right]$$

$$= (90\mathbf{i} + 120\mathbf{j}) \text{ lb.}$$

The position vector from A to C is

$$\mathbf{r}_1 = 11\mathbf{j} \text{ in.}$$

The moment about A is

$$\mathbf{M}_A = \mathbf{r}_1 \times \mathbf{F} = 11\mathbf{j} \times (90\mathbf{i} + 120\mathbf{j})$$

$$= \underline{-990\mathbf{k} \text{ in-lb.}} \qquad \text{Ans.}$$

(b) The position vector from D to B is

$$\mathbf{r}_2 = (-6\mathbf{i} + 3\mathbf{j} + 5\mathbf{k}) \text{ in.,}$$

and the moment about D is

$$\mathbf{M}_D = \mathbf{r}_2 \times \mathbf{F} = (-6\mathbf{i} + 3\mathbf{j} + 5\mathbf{k}) \times (90\mathbf{i} + 120\mathbf{j})$$

$$= \begin{vmatrix} \mathbf{i} & \mathbf{j} & \mathbf{k} \\ -6 & 3 & 5 \\ 90 & 120 & 0 \end{vmatrix} = \mathbf{i}(0 - 600) - \mathbf{j}(0 - 450) + \mathbf{k}(-720 - 270)$$

$$= (-600\mathbf{i} + 450\mathbf{j} - 990\mathbf{k}) \text{ in-lb.}$$

If the position vector from D to C had been selected, the position vector would be

$$\mathbf{r} = 11\mathbf{j} + 5\mathbf{k},$$

and the moment becomes

$$\mathbf{M}_D = \mathbf{r} \times \mathbf{F} = \begin{vmatrix} \mathbf{i} & \mathbf{j} & \mathbf{k} \\ 0 & 11 & 5 \\ 90 & 120 & 0 \end{vmatrix}$$

$$= \underline{(-600\mathbf{i} + 450\mathbf{j} - 990\mathbf{k}) \text{ in-lb},} \qquad \text{Ans.}$$

which demonstrates that the position vector from D to any point on **F** can be used. The vector representing $\mathbf{M}_D$ is shown on Fig. 1-18. The magnitude of the resultant moment is

$$M_D = [600^2 + 450^2 + 990^2]^{1/2} = 1242 \text{ in-lb},$$

and the resultant moment can also be written as

$$\mathbf{M}_D = 1242(-0.483\mathbf{i} + 0.362\mathbf{j} - 0.797\mathbf{k}) \text{ in-lb}$$

$$= 1242\mathbf{n} \text{ in-lb},$$

where **n** is a unit vector in the direction of $\mathbf{M}_D$.

(c) Since the resultant moment with respect to point A lies along the line AD, the moment about line AD is the same as the moment about A. That is,

$$\mathbf{M}_{AD} = \mathbf{M}_A = \underline{-990\mathbf{k} \text{ in-lb}.} \qquad \text{Ans.}$$

This result can also be obtained by calculating the orthogonal component of $\mathbf{M}_D$ along line AD. The unit vector along AD is **k**, and the component of $\mathbf{M}_D$ along AD is

$$\mathbf{M}_{AD} = (\mathbf{n} \cdot \mathbf{M}_D)\mathbf{n} = [\mathbf{k} \cdot (-600\mathbf{i} + 450\mathbf{j} - 990\mathbf{k})]\mathbf{k}$$

$$= -990\mathbf{k} \text{ in-lb}.$$

EXAMPLE 1-9

Determine the moment of the 85-lb force **F** in Fig. 1-19
(a) With respect to point A.
(b) With respect to line AB.

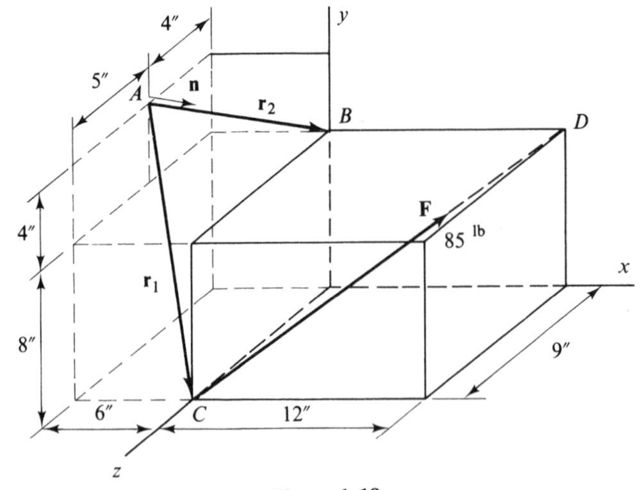

Figure 1-19

SOLUTION

(a) The position vector from C to D is

$$\mathbf{r}_{CD} = (12\mathbf{i} + 8\mathbf{j} - 9\mathbf{k}) \text{ in.},$$

and the unit vector along CD is

$$\mathbf{n}_{CD} = \frac{12\mathbf{i} + 8\mathbf{j} - 9\mathbf{k}}{(12^2 + 8^2 + 9^2)^{1/2}} = \frac{12}{17}\mathbf{i} + \frac{8}{17}\mathbf{j} - \frac{9}{17}\mathbf{k}.$$

The force $\mathbf{F}$, expressed in terms of $\mathbf{i}$, $\mathbf{j}$, and $\mathbf{k}$, is

$$\mathbf{F} = F\mathbf{n}_{CD} = 85\left(\frac{12}{17}\mathbf{i} + \frac{8}{17}\mathbf{j} - \frac{9}{17}\mathbf{k}\right) = (60\mathbf{i} + 40\mathbf{j} - 45\mathbf{k}) \text{ lb}.$$

The position vector $\mathbf{r}_1$ from A to C is

$$\mathbf{r}_1 = (6\mathbf{i} - 12\mathbf{j} + 5\mathbf{k}) \text{ in.}$$

The moment of $\mathbf{F}$ with respect to A can be obtained from the vector product of $\mathbf{r}_1$ and $\mathbf{F}$ and is

$$\mathbf{M}_A = \mathbf{r}_1 \times \mathbf{F} = \begin{vmatrix} \mathbf{i} & \mathbf{j} & \mathbf{k} \\ 6 & -12 & 5 \\ 60 & 40 & -45 \end{vmatrix}$$

$$= \mathbf{i}(540 - 200) - \mathbf{j}(-270 - 300) + \mathbf{k}(240 + 720)$$

$$= 340\mathbf{i} + 570\mathbf{j} + 960\mathbf{k}$$

$$= \underline{1167(0.291\mathbf{i} + 0.488\mathbf{j} + 0.823\mathbf{k}) \text{ in-lb.}} \qquad \text{Ans.}$$

(b) The position vector $\mathbf{r}_2$ from A to B is

$$\mathbf{r}_2 = 6\mathbf{i} - 4\mathbf{j} - 4\mathbf{k},$$

and the unit vector along AB is

$$\mathbf{n}_2 = \frac{\mathbf{r}_2}{r_2} = \frac{6\mathbf{i} - 4\mathbf{j} - 4\mathbf{k}}{(6^2 + 4^2 + 4^2)^{1/2}}$$

$$= \frac{1}{8.25}(6\mathbf{i} - 4\mathbf{j} - 4\mathbf{k}).$$

The moment of $\mathbf{F}$ with respect to AB is the component of $\mathbf{M}_A$ parallel to AB, and its magnitude can be determined by scalar multiplication of $\mathbf{n}_2$ and $\mathbf{M}_A$. Thus

$$\mathbf{M}_{AB} = (\mathbf{n}_2 \cdot \mathbf{M}_A)\mathbf{n}_2$$

$$= \left[\frac{1}{8.25}(6\mathbf{i} - 4\mathbf{j} - 4\mathbf{k}) \cdot (340\mathbf{i} + 570\mathbf{j} + 960\mathbf{k})\right]\mathbf{n}_2$$

$$= \frac{1}{8.25}[6(340) - 4(570) - 4(960)]\mathbf{n}_2$$

$$= \underline{-495(0.728\mathbf{i} - 0.485\mathbf{j} - 0.485\mathbf{k}) \text{ in-lb.}} \qquad \text{Ans.}$$

The negative sign indicates a moment vector from B toward A; that is, $\mathbf{F}$ tends to rotate the body clockwise about AB as viewed from B toward A.

The moment about AB can also be computed in a single step as indicated in the discussion. Thus

$$\mathbf{M}_{AB} = [\mathbf{n}_2 \cdot (\mathbf{r}_1 \times \mathbf{F})]\mathbf{n}_2,$$

where

$$\mathbf{n}_2 \cdot (\mathbf{r}_1 \times \mathbf{F}) = \frac{1}{8.25} \begin{vmatrix} 6 & -4 & -4 \\ 6 & -12 & 5 \\ 60 & 40 & -45 \end{vmatrix}$$

$$= \frac{1}{8.25} \{6[(-12)(-45) - 5(40)]$$

$$+ 4[6(-45) - 5(60)] - 4[6(40) - (-12)(60)]\}$$

$$= \frac{-4080}{8.25} = -495.$$

The moment is

$$\mathbf{M}_{AB} = \underline{-495\mathbf{n}_2 \text{ in-lb.}} \qquad \text{Ans.}$$

PROBLEMS

1-43 Determine the moment of each force in Fig. P1-43 with respect to point A.

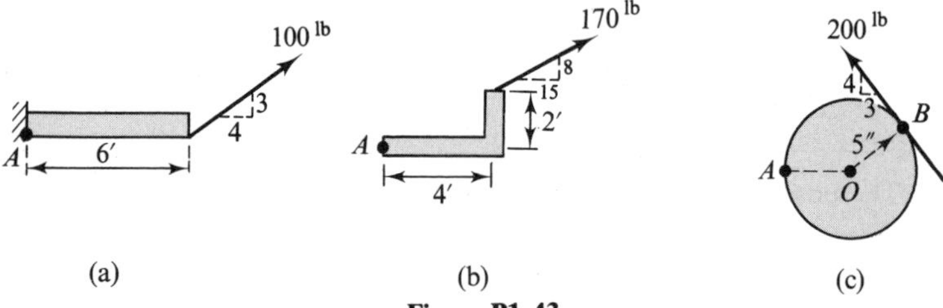

Figure P1-43

***1-44** The vertical 200-N force in Fig. P1-44 passes through the center of the rectangle. Determine
(a) The moment with respect to point A.
(b) The perpendicular distance from A to the line of action of the force.

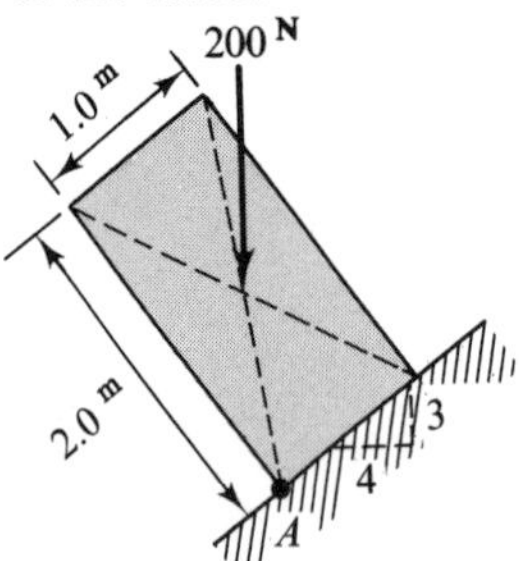

Figure P1-44

1-45 (a) Determine the moment of the 160-lb force of Fig. P1-45 with respect to point A.
(b) Apply the result of part (a) to determine where the 160-lb force intersects the y axis.

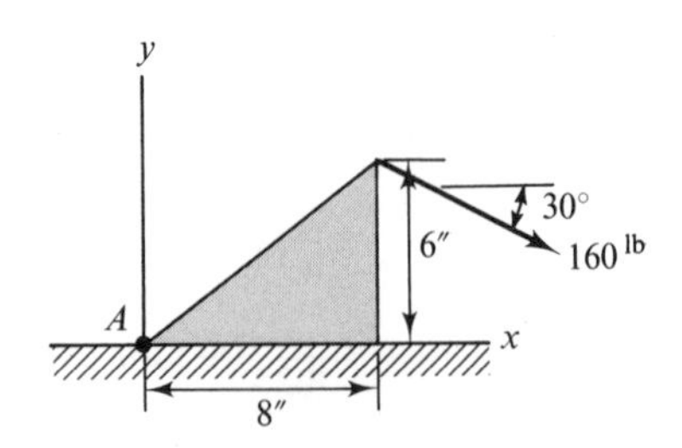

Figure P1-45

1-46 The magnitude of the horizontal component of **F** of Fig. P1-46 is 360 lb. Determine

(a) The force **F**.

(b) The moment of **F** with respect to point *A*.

(c) The perpendicular distance from *A* to the line of action of **F**.

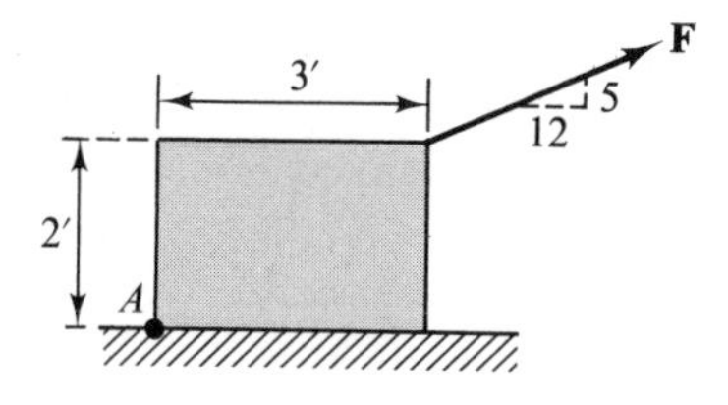

Figure P1-46

1-47 Determine the moment of the force system in Fig. P1-47 with respect to (a) point *O*, (b) point *A*.

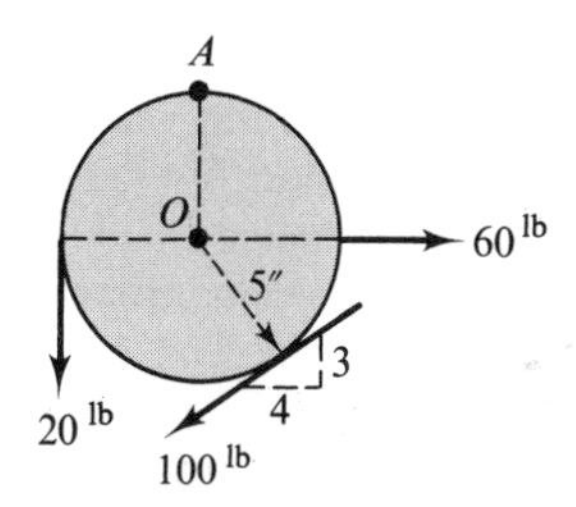

Figure P1-47

1-48 A 2-ft section of a concrete dam is shown in Fig. P1-48. The weight of the concrete is shown as the two downward forces. The resultant horizontal water pressure is the force **R**, and the uplift hydrostatic pressure is the force **Q**. Determine the resultant moment of all forces acting on the dam with respect to the toe A.

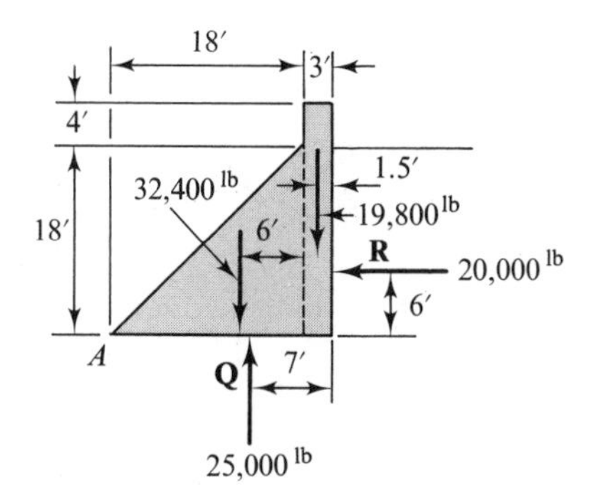

Figure P1-48

***1-49** By means of the principle of moments, determine the perpendicular distance from the action line of the 9.1-kN force in Fig. P1-49 to point *C*.

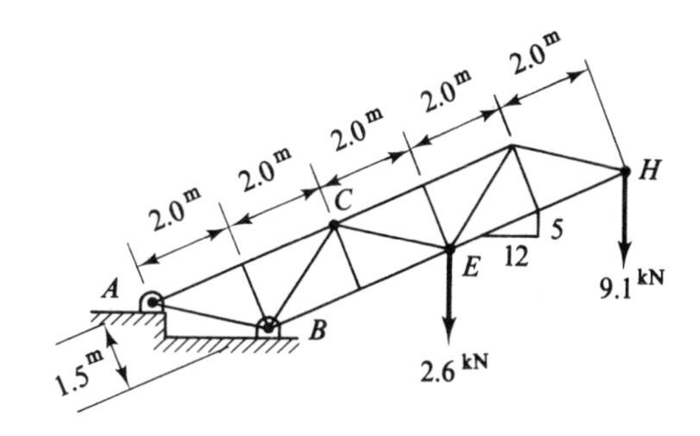

Figure P1-49

1-50 Determine the moment of the 90-lb force in Fig. P1-50 with respect to the *a* axis.

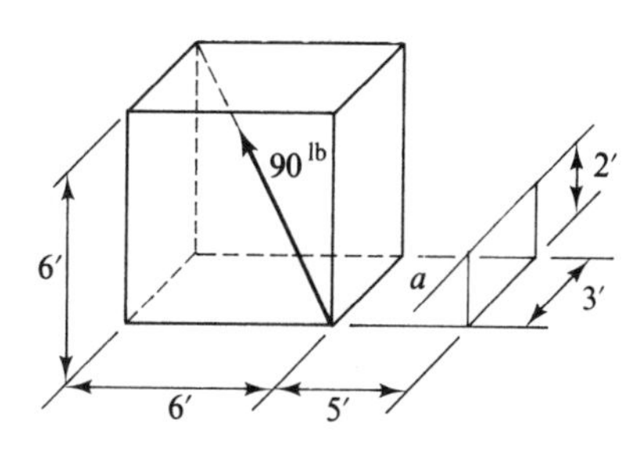

Figure P1-50

1-51 Determine the moment of the 280-lb force in Fig. P1-51 with respect to the line *ab*.

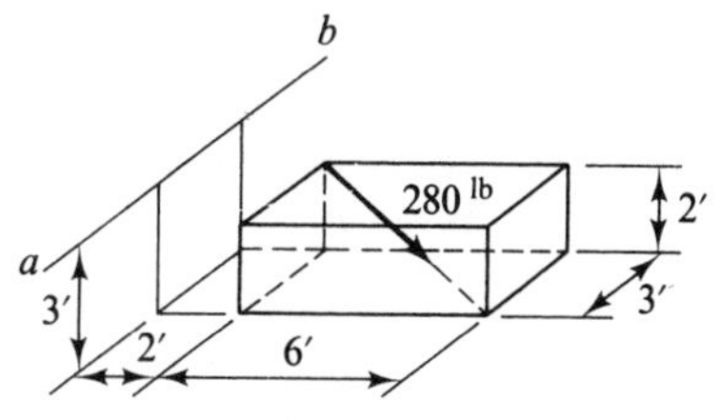

Figure P1-51

1-52 Determine the sum of the moments of the two forces in Fig. P1-52 with respect to the a axis.

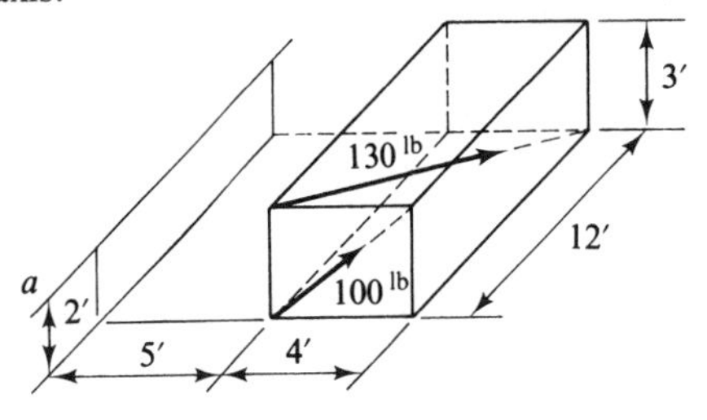

Figure P1-52

1-53 Determine the moment of the 110-lb force in Fig. P1-53 with respect to the line AB.

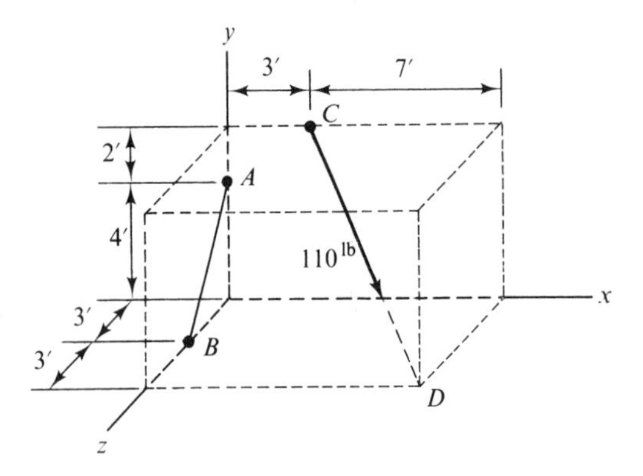

Figure P1-53

***1-54** Determine the moment of the 140-N force in Fig. P1-54 with respect to line AB and with respect to line AC.

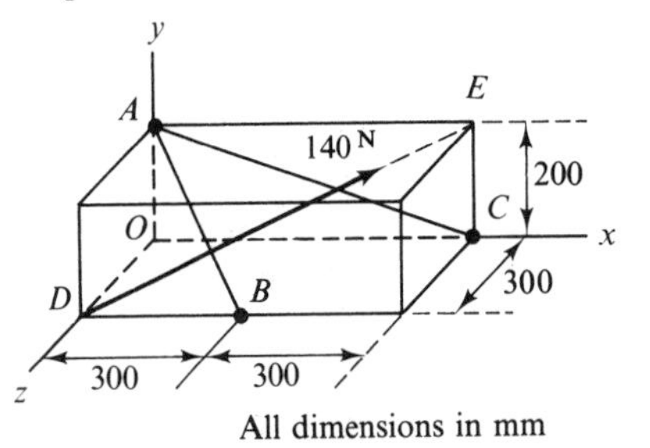

All dimensions in mm

Figure P1-54

1-55 The moment of the force **F** in Fig. P1-55 with respect to the line BC is 300 ft-lb clockwise looking from B to C. Determine the force **F**.

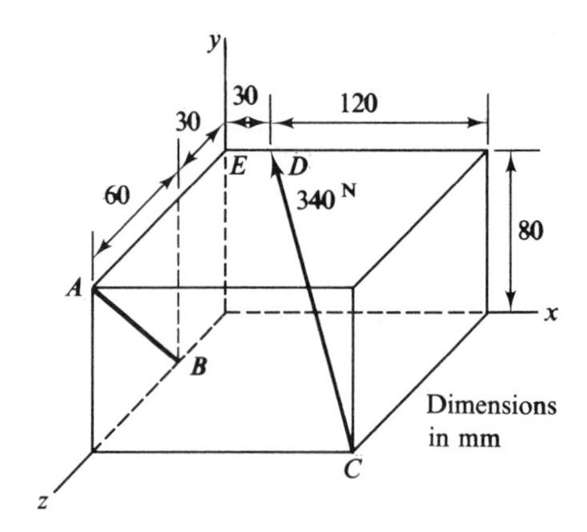

Figure P1-55

1-56 The force **F** in Fig. P1-55 has a magnitude of 140-lb.
(a) Determine the moment of **F** about point B.
(b) Determine the perpendicular distance from B to the force **F**.

1-57 (a) Determine a unit vector which is perpendicular to the lines OA and BC in Fig. P1-55.
(b) Determine the perpendicular distance from the line BC to the line OA in Fig. P1-55 by getting the projection, in the direction of the unit vector of part (a), of any line joining a point on BC to a point on OA.

***1-58** Determine the moment of the 340-N force in Fig. P1-58 with respect to
(a) Point B. (b) Line AE. (c) Line AB.

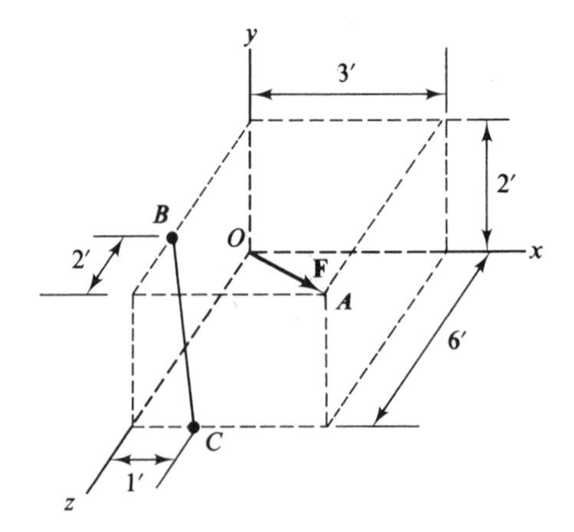

Figure P1-58

***1-59** Determine the perpendicular distance from the line AB to the line CD in Fig. P1-58.

1-60 Determine the moment of the force **F** in Fig. P1-60 about point D and the perpendicular distance from D to the force.

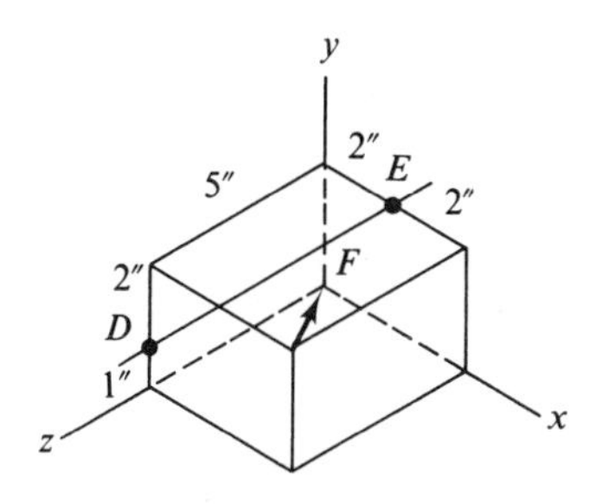

Figure P1-60

1-61 Determine the perpendicular distance between the line DE and the line of action of the force **F** in Fig. P1-60.

1-62 The rod in Fig. P1-62 is screwed into the rigid base at A and supports a force of 170 lb at D. The sections BC and CD are parallel to the y and z axes, respectively, and AB is perpendicular to the inclined face of the base. Determine
(a) The twisting moment of the force tending to screw the bracket into the base at A.
(b) The maximum bending moment at A tending to bend the bracket (about an axis in the inclined plane of the base).

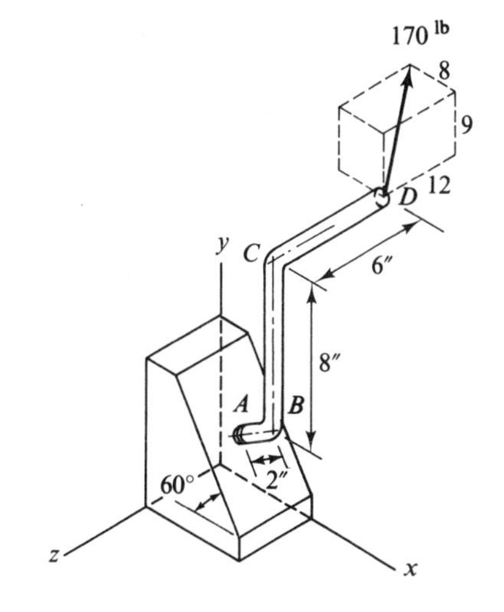

Figure P1-62

1-63 Determine the moment of the force **F** in Fig. P1-63 with respect to the line CD. Express the result in terms of the magnitude F.

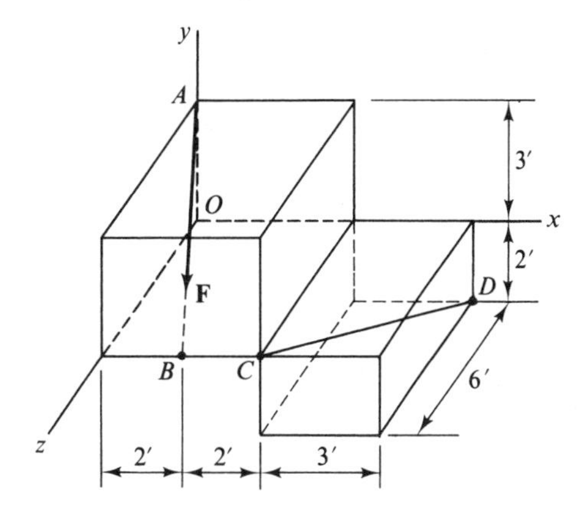

Figure P1-63

1-64 The x component of the moment of **F** in Fig. P1-63 about point D is 150**i** ft-lb. Determine

(a) The force **F**.
(b) The moment of **F** about point D.

1-7 COUPLES—DEFINITION AND APPLICATIONS

A couple consists of two forces which have equal magnitudes and parallel noncollinear lines of action but which are opposite in sense. Since the sum of the forces of a couple in any direction is zero, a couple has no tendency to translate a body in any direction but tends only to rotate the body on which it acts. As will be shown later, a couple cannot be reduced to a simpler force system.

A unique feature of a couple is that it has the same moment with respect to every point in space. In other words, the moment of a couple is independent of its moment center. This is demonstrated by Fig. 1-20 in which forces $\mathbf{F}_1$ and $\mathbf{F}_2$ lie in plane A and constitute a couple ($\mathbf{F}_1 = -\mathbf{F}_2$). The moment of the couple about any point, such as the origin at O, is the sum of the moments of the two forces. Thus

$$\mathbf{M}_O = \mathbf{r}_1 \times \mathbf{F}_1 + \mathbf{r}_2 \times \mathbf{F}_2 = \mathbf{r}_1 \times \mathbf{F}_1 + \mathbf{r}_2 \times (-\mathbf{F}_1) = (\mathbf{r}_1 - \mathbf{r}_2) \times \mathbf{F}_1,$$

where $\mathbf{r}_1$ and $\mathbf{r}_2$ are position vectors from the origin to arbitrary points on $\mathbf{F}_1$ and $\mathbf{F}_2$.

Since $\mathbf{r}_1 = \mathbf{r}_2 + \mathbf{r}_{21}$ in Fig. 1-20, from which $\mathbf{r}_1 - \mathbf{r}_2 = \mathbf{r}_{21}$, the moment becomes

$$\mathbf{M}_O = \mathbf{r}_{21} \times \mathbf{F}_1 = r_{21} F_1 \sin\theta\,\mathbf{n} = F_1 p\mathbf{n}$$

where $p = r_{21} \sin\theta$ is the perpendicular distance between the two forces and $\mathbf{n}$ is a unit vector perpendicular to the plane of the forces with its sense determined by the right-hand rule. The moment of the couple is independent of the moment center and depends only on the magnitude of the forces and the distance between them.

The moment of a couple and the moment of a force about a moment center (see Art. 1-6) are both vector quantities and can be resolved into components or combined into resultant vectors. It should be noted, however, that the moment of a force is dependent on the moment center and is thus a localized vector, whereas the moment of a couple is independent of the moment center and is a free vector.

The characteristics of a couple, which indicate its external effects on a rigid body, are (1) the magnitude of the moment of the couple and (2) the direction of the moment vector, which, by the right-hand rule, also specifies the sense of rotation of the couple. Thus a couple

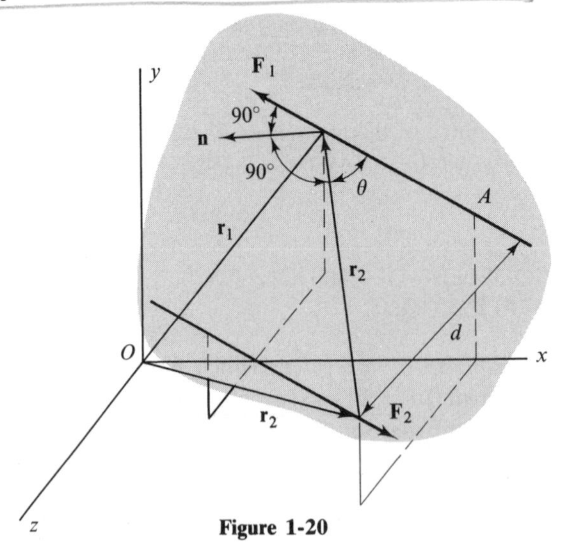

Figure 1-20

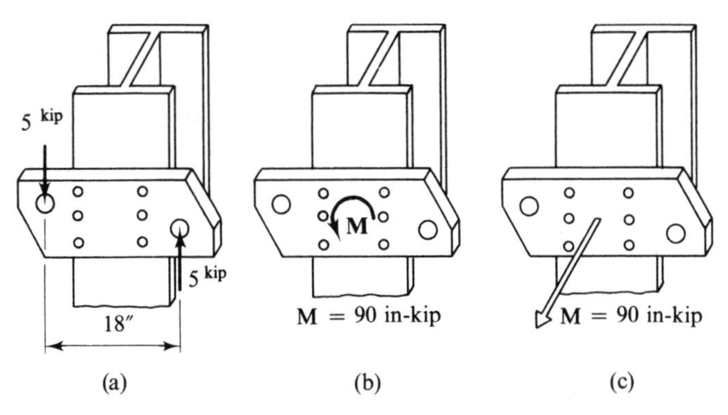

Figure 1-21

can be written as the magnitude of the couple times a unit vector in the direction of the moment vector. No point on the line of action of the moment vector can be specified since couples are free vectors and thus are independent of the moment center.

A couple is frequently indicated by a clockwise or counterclockwise arrow when coplanar force systems are involved instead of showing two separate forces. In this text, couple vectors are often shown as open-bodied arrows. Figure 1-21 illustrates three methods used to indicate a couple on a structural member. The moment of the couple is 5(18) = 90 in-kip = 90,000 in-lb.

EXAMPLE 1-10

Determine the moment of the couple in Fig. 1-22.

SOLUTION

The vector **AC** is

$$\mathbf{AC} = (-\mathbf{i} - 2\mathbf{j} + 2\mathbf{k})\ \text{m},$$

and the unit vector parallel to $\mathbf{F}_1$ is

$$\mathbf{n}_{AC} = \frac{1}{3}(-\mathbf{i} - 2\mathbf{j} + 2\mathbf{k}).$$

The force $\mathbf{F}_1$ becomes

$$\mathbf{F}_1 = F_1\mathbf{n}_{AC} = 60\left[\frac{1}{3}(-\mathbf{i} - 2\mathbf{j} + 2\mathbf{k})\right]$$
$$= (-20\mathbf{i} - 40\mathbf{j} + 40\mathbf{k})\ \text{N},$$

and the position vector from B to A is

$$\mathbf{r}_{BA} = (2\mathbf{i} - \mathbf{j})\ \text{m}.$$

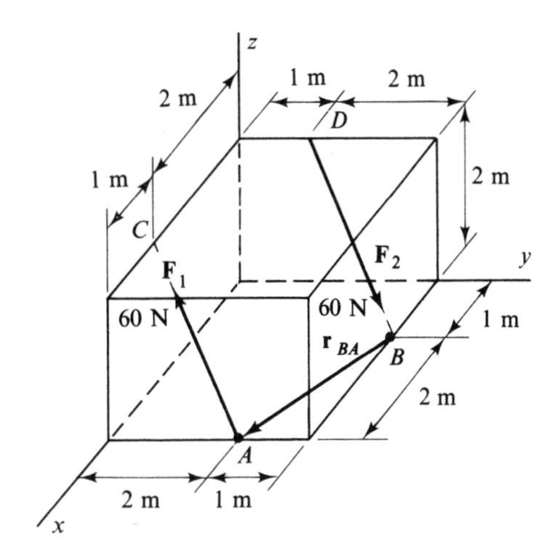

Figure 1-22

The moment of the couple is

$$\mathbf{M} = \mathbf{r}_{BA} \times \mathbf{F}_1 = \begin{vmatrix} \mathbf{i} & \mathbf{j} & \mathbf{k} \\ 2 & -1 & 0 \\ -20 & -40 & 40 \end{vmatrix}$$

$$= \mathbf{i}(-40) - \mathbf{j}(80) + \mathbf{k}(-80-20)$$

$$= (-40\mathbf{i} - 80\mathbf{j} - 100\mathbf{k})\ \text{N} \cdot \text{m}$$

$$= 134.2\mathbf{n}\ \text{N} \cdot \text{m}, \qquad \text{Ans.}$$

where $\mathbf{n}$ is a unit vector in the direction $(-2\mathbf{i} - 4\mathbf{j} - 5\mathbf{k})$.

Occasionally the effect of a force on a body may be more conveniently determined when the force is replaced by an equal parallel force through a different point and a couple. The addition of two equal collinear forces of opposite sense to a force system does not change the external effect of the system on a rigid body. By this means, a force may be replaced by an equal parallel force through any other point of the body and a couple. This procedure is demonstrated in the following example.

EXAMPLE 1-11

(a) Replace the 250-lb force in Fig. 1-23(a) by a force at A and a couple. The 250-lb force is parallel to the xy plane and acts at an angle of 60° with the horizontal plane.

(b) Resolve the couple in part (a) into a twisting couple parallel to AB and a bending couple perpendicular to AB.

SOLUTION

(a) Two forces $\mathbf{P}_1$ and $\mathbf{P}_2$, each equal to 250 lb, are placed at A as shown in Fig. 1-23(a). The original force and $\mathbf{P}_1$ constitute a couple. The position vector $\mathbf{r}_{AC}$ from A to C is

$$\mathbf{r}_{AC} = (6\mathbf{i} - 4\mathbf{k})\ \text{in.},$$

and the original force is

$$\mathbf{F} = 250(\mathbf{i} \cos 60° - \mathbf{j} \sin 60°)$$

$$= (125.0\mathbf{i} - 216.5\mathbf{j})\ \text{lb}.$$

The moment of the couple is

$$\mathbf{M} = \mathbf{r}_{AC} \times \mathbf{F} = \begin{vmatrix} \mathbf{i} & \mathbf{j} & \mathbf{k} \\ 6 & 0 & -4 \\ 125.0 & -216.5 & 0 \end{vmatrix}$$

$$= \underline{1640(-0.528\mathbf{i} - 0.305\mathbf{j} - 0.853\mathbf{k})\ \text{in-lb}}, \qquad \text{Ans.}$$

and the force at A is

$$\mathbf{P}_2 = \underline{250(0.500\mathbf{i} - 0.866\mathbf{j})\ \text{lb} \quad (\text{at } A)}. \qquad \text{Ans.}$$

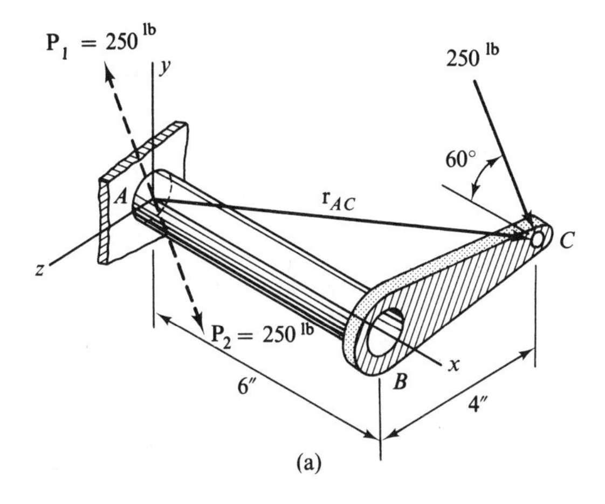

(a)

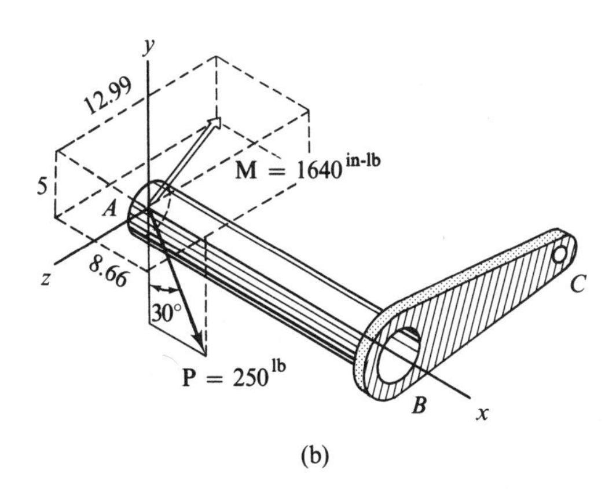

(b)

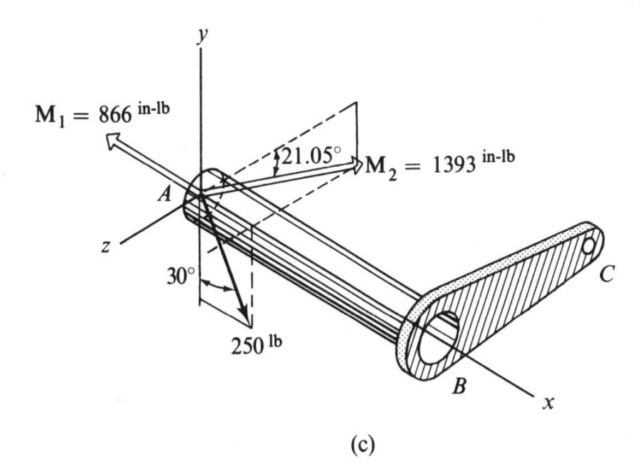

(c)

Figure 1-23

(b) The couple **M** can be resolved into two components by inspection in this problem. The twisting couple, $\mathbf{M}_T$, is

$$\mathbf{M}_T = \mathbf{M}_x = \underline{-866\mathbf{i} \text{ in-lb}}, \qquad \text{Ans.}$$

and the bending couple, $\mathbf{M}_B$, is

$$\mathbf{M}_B = \mathbf{M}_y + \mathbf{M}_z = -500\mathbf{j} - 1299\mathbf{k}$$
$$= \underline{1393\mathbf{n}_1 \text{ in-lb}}, \qquad \text{Ans.}$$

where $\mathbf{n}_1$ is a unit vector in the yz plane; see Fig. 1-23(c).

When a force is replaced by a force and a couple, the new force has the same magnitude and sense as the original force, and it has a parallel line of action. The couple vector will be perpendicular to the plane containing the initial and final positions of the force.

The foregoing procedure for replacing a force by a force and a couple can be reversed and used to replace a force and a couple by a single force, provided the force and the couple vector are perpendicular.

PROBLEMS

1-65 One force $\mathbf{F} = (-10\mathbf{i} + 20\mathbf{j})$ lb of a couple acts through the point $(-2, 1)$ and the other force acts through the point $(4, -2)$. The coordinates are in feet. Determine
(a) The other force.
(b) The moment of the couple.

1-66 The forces of a couple lie in the xy plane, and one force passes through the origin and has a magnitude of 250 lb and a slope of 3 upward and 4 to the right. The moment of the couple is 1000 ft-lb clockwise. Determine the other force and where it crosses the x axis.

***1-67** The 340-N force in Fig. P1-67 is one force of a couple; the other force acts through point A. Determine the other force and the moment of the couple.

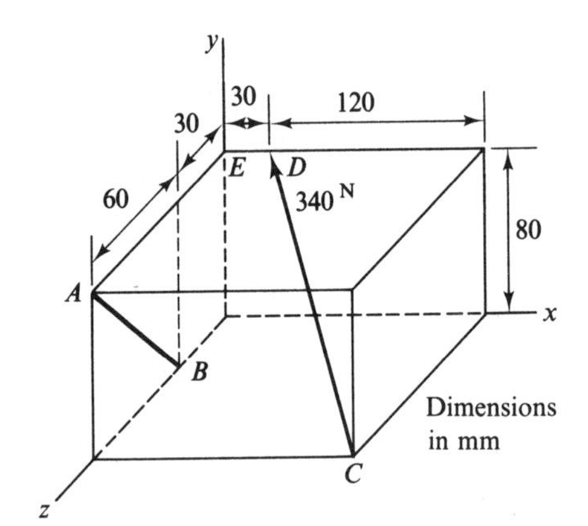

Figure P1-67

1-68 The force **F** in Fig. P1-68 is one force of a couple, and the second force acts through point C. The magnitude of the moment of the couple is 500 ft-lb. Determine the second force and the direction of the couple vector.

1-69 The force **F** in Fig. P1-68 has a magnitude of 140 lb and is one force of a couple whose moment is $(360\mathbf{j} - 120\mathbf{k})$ ft-lb. Determine the second force of the couple and locate its point of intersection with the xy plane.

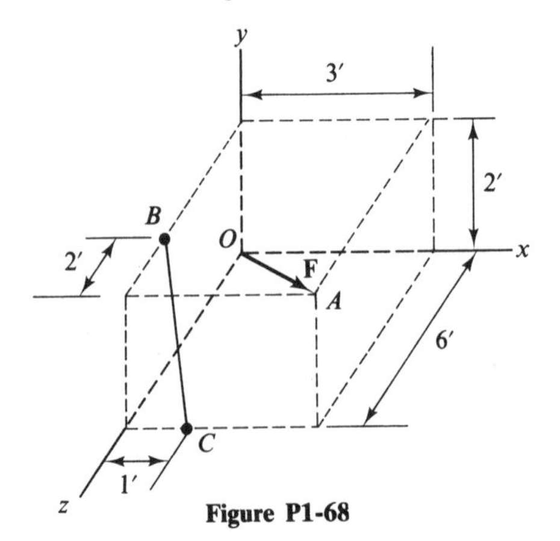

Figure P1-68

1-70 The force **F** in Problem 1-69 is one force of a couple that has a moment of $(200\mathbf{i}-60\mathbf{j}-80\mathbf{k})$ ft-lb. Determine the coordinates of the point where the second force of the couple intersects the xz plane.

***1-71** The 340-N force of Problem 1-67 is one force of a couple which has a moment of $(-16\mathbf{i}-15\mathbf{j}+8\mathbf{k})$ N · m. Determine the point of intersection of the other force with the xz plane.

***1-72** The 340-N force of Problem 1-67 is one force of a couple which has a moment of $(-9\mathbf{i}+9\mathbf{j}+20\mathbf{k})$ N · m. Determine the coordinates of the point of intersection of the other force with the xy plane.

1-73 Replace the 1000-lb force in Fig. P1-73 by a force through C and a couple.

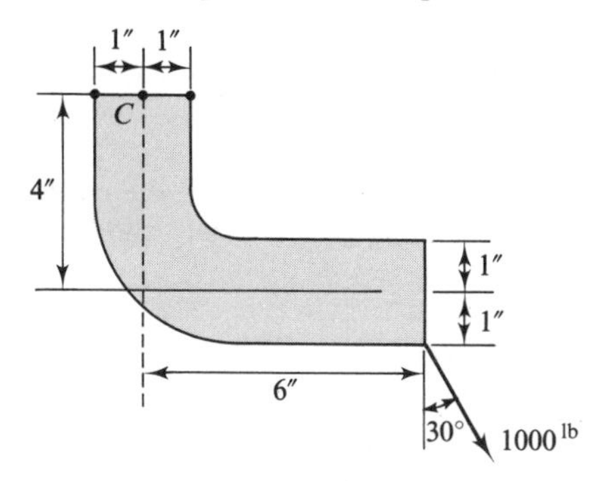

Figure P1-73

1-74 Replace the 8000-lb force in Fig. P1-74 by a force through A and a couple.

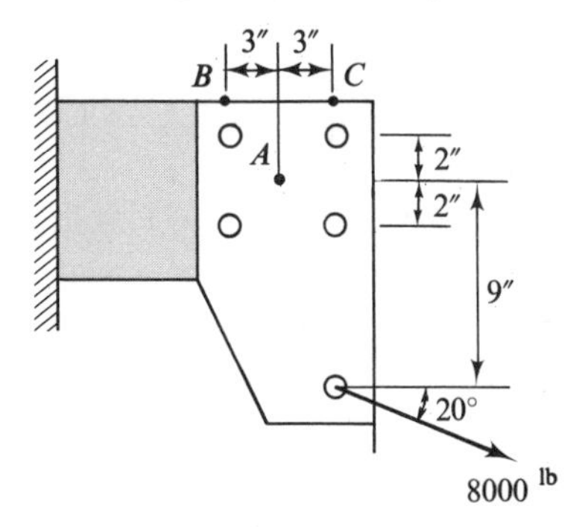

Figure P1-74

1-75 Replace the force and couple in Fig. P1-75 with a force at A and a couple.

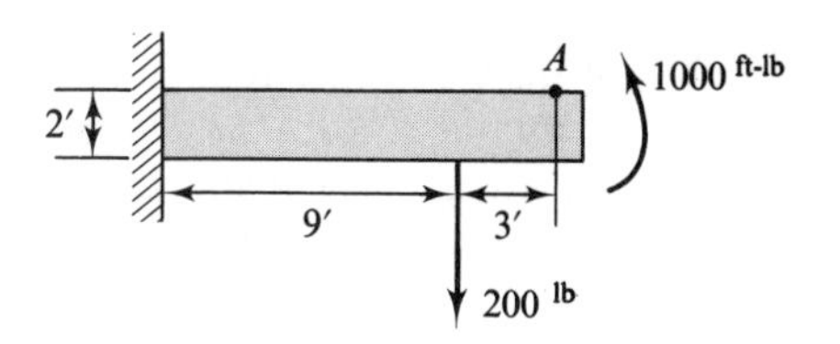

Figure P1-75

1-76 Replace the 1000-lb force in Fig. P1-76 by a force through O and a couple.

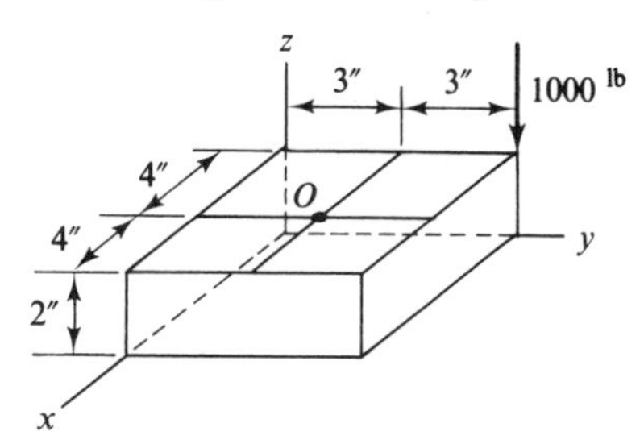

Figure P1-76

1-77 A 2000-lb jet engine is suspended from the wing of a jet aircraft on a 6-ft strut as shown in Fig. P1-77. If the engine is producing a 10,000-lb thrust, determine the resulting force and moment at A.

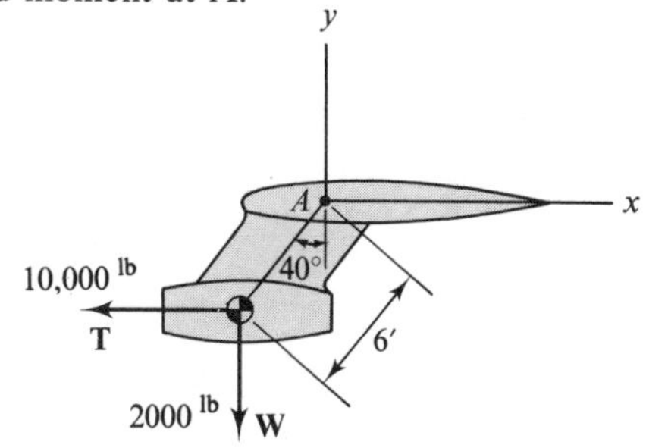

Figure P1-77

***1-78** Replace the 2.0-kN force in Fig. P1-78 by a force through A and a couple.

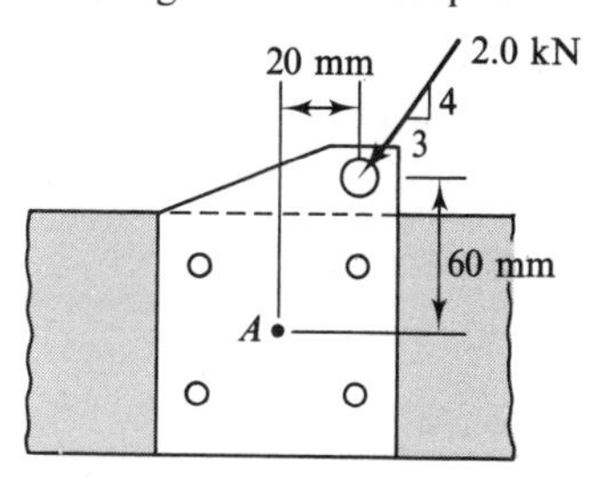

Figure P1-78

1-79 Replace the 75-lb force on the wrench at A in Fig. P1-79 by a force at O and two couples, one a twisting moment about the z axis and the second a bending moment about an axis perpendicular to the z axis.

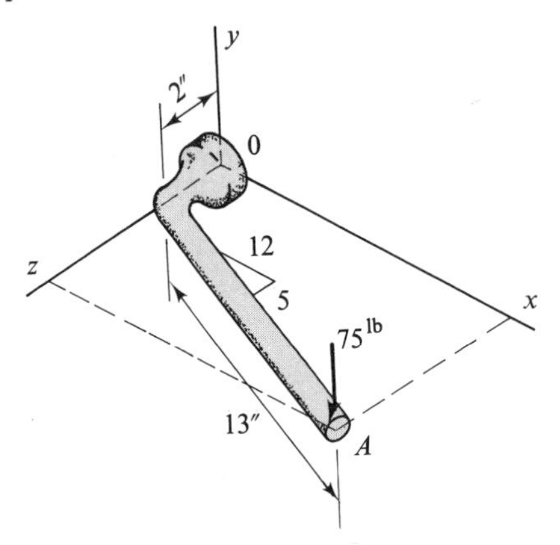

Figure P1-79

1-80 A 500-lb load is supported by three lengths of pipe arranged as shown in Fig. P1-80. The pipes are connected by 45° elbows, and all three sections lie in a common plane through the x axis. Determine the twisting moment (along the pipe axis), in each length of pipe.

1-81 The post AB in Fig. P1-81 is embedded at A and extends 10 ft from A in the yz plane. A cable extends from B to C and is tightened until it exerts a 1000-lb force on AB at point B. Replace the force at B by a force at A and a couple.

1-82 (a) Replace the force and couple at point C in Fig. P1-82 with a force at A and a couple.
(b) Resolve the couple from part (a) into a twisting couple along AB and a bending couple perpendicular to AB.

***1-83** Member OC in Fig. P1-83 is clamped in the slotted stud by the bolt AB. The stud is screwed into the fixed base at D. The 150-N force at C can be expressed as $\mathbf{P} = (100\mathbf{i} - 50\mathbf{j} + 100\mathbf{k})$ N. Member OC is 250 mm long.
(a) Replace the force at C by a force at O and a couple.
(b) Determine the magnitude of the moment of the couple tending to screw the stud into the base D.
(c) Determine the magnitude of the moment of the couple tending to rotate OC about the bolt AB.

1-84 Pipes AB and BC in Fig. P1-84 are screwed into a 45° elbow, and AB is screwed into the base plate at A. A pipe wrench with a 10-in. handle is placed on the pipe at C. The

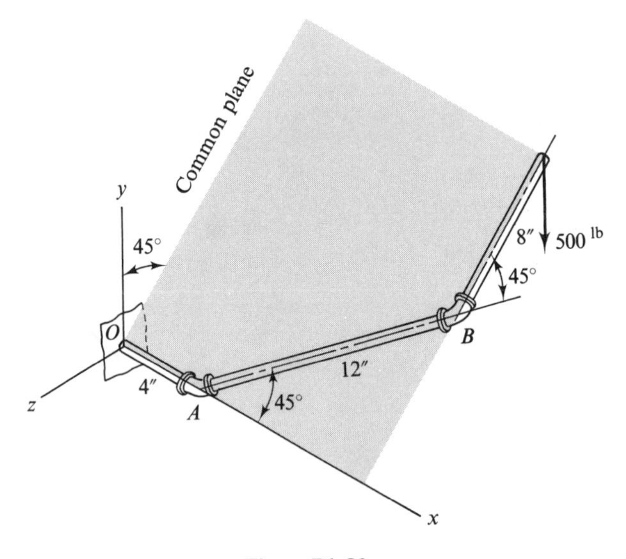

Figure P1-80

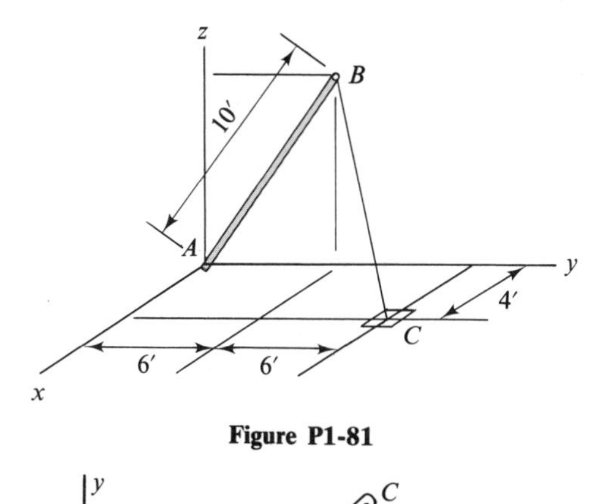

Figure P1-81

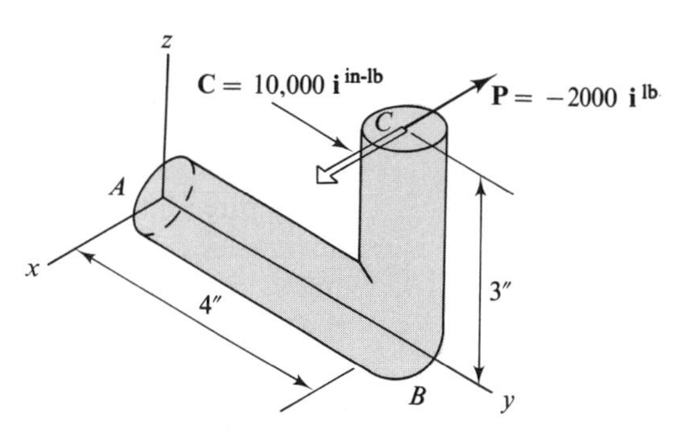

Figure P1-82

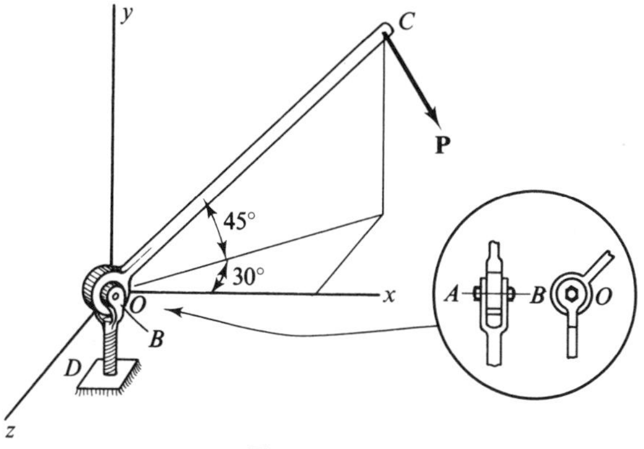

Figure P1-83

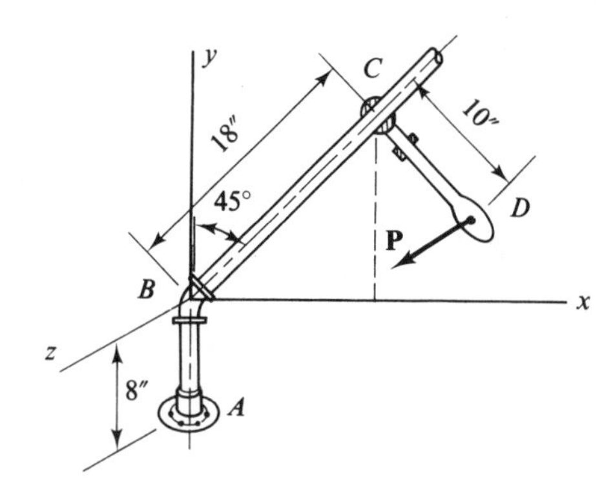

Figure P1-84

two lengths of pipe and the wrench handle all lie in the xy plane. A force **P** of 50 lb is applied at D perpendicular to the plane containing BCD. Replace the force **P** by a force at B and a couple and determine the twisting moment applied to each length of pipe.

1-85 The wrench in Problem 1-84 is rotated 30° about the pipe BC so that the z coordinate of D is +5 in. The force **P** is perpendicular to the new plane BCD. Replace **P** by a force at B and a couple and determine the twisting moment of the couple applied to each length of pipe.

1-86 The force $\mathbf{F} = (60\mathbf{i} + 40\mathbf{j} + 120\mathbf{k})$ lb and couple $\mathbf{M} = (300\mathbf{i} - 400\mathbf{j} + 600\mathbf{k})$ in-lb are shown at point O in Fig. P1-86. Replace **F** and **M** by a couple whose vector is parallel to the x axis and two forces, **P** through point O acting perpendicular to the x axis and **Q** through point A.

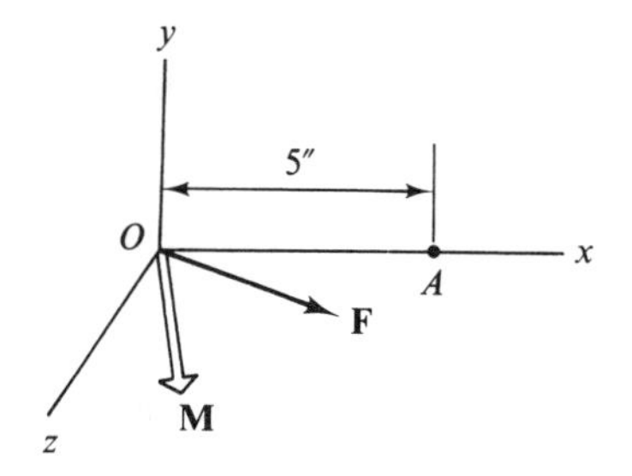

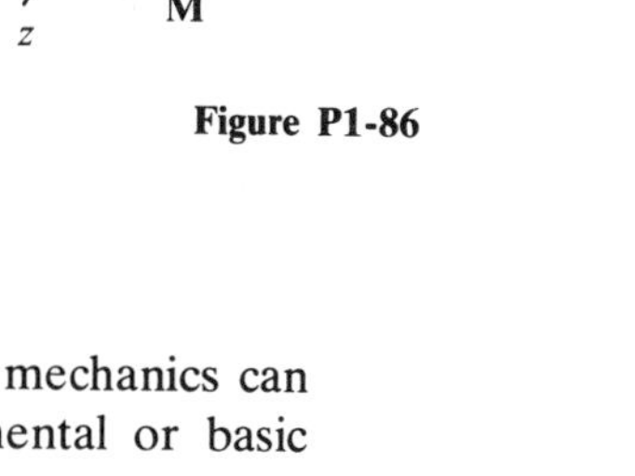

Figure P1-86

1-8 DIMENSIONAL EQUATIONS

Physical quantities usually studied in engineering mechanics can be expressed dimensionally in terms of three fundamental or basic dimensions. The three fundamental dimensions used in engineering mechanics are *force* (F), *length* (L), and *time* (T) in gravitational systems and *mass* (m), *length* (L), and *time* (T) in absolute systems. The dimensions of other physical quantities can be expressed in terms of the three basic dimensions selected. Temperature, electric charge or current, and luminous intensity are other fundamental quantities required is some cases, but they are not generally encountered in engineering mechanics.

Newton's second law of motion states that the acceleration of a particle acted on by an unbalanced force is proportional to the force and inversely proportional to the mass of the particle. This law can be expressed algebraically as

$$a = K\frac{F}{m}, \tag{1-3}$$

where K is a dimensionless constant of proportionality. Dimensionally the equation becomes

$$LT^{-2} = Fm^{-1}.$$

When force is considered to be the basic quantity, the dimensions of mass are FT^2L^{-1}, and when mass is considered to be the fundamental quantity the dimensions of force are mLT^{-2}.

When numerical values are substituted in a dimensionally homogeneous equation it is essential that consistent units be used. The British gravitational system uses feet (ft), pounds (lb), and seconds (sec) as units for length, force, and time, respectively. The metric absolute system (SI) uses the kilogram (kg), meter (m), and second (s) as units of mass, length, and time, respectively.

The constant K in Eq. (1-3) will be unity in either the gravitational or absolute system if the derived quantity is assigned units consistent with the units of the fundamental quantities. In the British system the unit of mass is the lb-sec^2 per ft, frequently called a *slug*. A body with a mass of 1 slug will be given an acceleration of 1 foot per second per second when acted on by a force of 1 pound. The unit force in SI is called a newton (N) and is the force which must be applied to a mass of 1 kilogram to produce an acceleration of 1 meter per second per second.

The mass of a body in slugs can be obtained by dividing the resultant force, in pounds, acting on the body by the acceleration of the body, in feet per second per second, produced by the force. When a body is falling freely its acceleration is the acceleration of gravity, g, and the only force acting on the body is its weight, W. Thus from Eq. (1-3)

$$g = (1)\frac{W}{m} \quad \text{or} \quad m = \frac{W}{g}.$$

The mass of a body in slugs is equal to its weight in pounds divided by the acceleration of gravity, approximately 32.2 fps^2 near the surface of the earth. Similarly, the weight of a body in newtons is equal to its mass in kilograms multiplied by the acceleration of gravity in meters per second per second, about 9.81 m/s^2.

Some examples of the dimensions of physical quantities commonly encountered in engineering mechanics are shown in the accompanying table together with units used by engineers in the British and metric (SI) systems.

When two or more quantities are to be added or subtracted, they must have the same dimensions if the result is to have any physical significance. Consider the expression

$$3\text{ ft}+4\text{ sec}+10\text{ lb},$$

written dimensionally as

$$L+T+F.$$

Physical Quantity	*Dimension*	*Common Engineering Units* British	*Common Engineering Units* SI
Force	F	lb (pound)	N (newton)
Length	L	ft (foot)	m (meter)
Time	T	sec (second)	s (second)
Mass	FT^2L^{-1}	lb-sec^2/ft, slug	kg, N · s^2/m
Area	L^2	ft^2	m^2
Volume	L^3	ft^3	m^3
Moment of a force	FL	ft-lb	N · m
Work	FL	ft-lb	N · m, J (joule)
Power	FLT^{-1}	ft-lb/sec, hp	W (watt), J/s, N · m/s
Angle	1 (dimensionless)	rad, degree	rad, degree
Sin θ	1 (dimensionless)	—	—
Linear velocity	LT^{-1}	ft/sec, fps	m/s
Linear acceleration	LT^{-2}	ft/sec^2, fps^2	m/s^2
Angular velocity	T^{-1}	rad/sec	rad/s
Angular acceleration	T^{-2}	rad/sec^2	rad/s^2
Linear impulse	FT	lb-sec	N · s

The terms to be added do not have the same dimensions, and the expression has no physical meaning. Every term of an equation must have the same dimensions if the equation is to be dimensionally homogeneous and have a real physical interpretation. This fact may be used to check the dimensional correctness of a derived equation when the dimensions of each quantity are known. It can also be used to determine the unknown dimensions of one or more quantities in an equation.

Dimensional equations are also useful in designing experiments involving model studies and in analyzing the results of such experiments. Such use is beyond the scope of this text.

Equations indicating the units used rather than the dimensions may also be used in checking the numerical values substituted in an equation, particularly when some quantity, such as length, is given in feet in one expression and as inches or miles in another.

EXAMPLE 1-12

The equation

$$C = I\alpha + mx_G a_y - my_G a_x$$

is dimensionally homogeneous when α is an angular acceleration, m is a mass, x_G and y_G are lengths, and a_x and a_y are linear accelerations. Determine the fundamental dimensions of C and I.

SOLUTION

Each of the quantities whose dimensions are given is expressed in fundamental dimensions, and the equation becomes

$$C = I\frac{1}{T^2} + \frac{FT^2}{L}(L)\frac{L}{T^2} - \frac{FT^2}{L}(L)\frac{L}{T^2}$$

or

$$C = \frac{I}{T^2} + FL - FL.$$

Since all terms must have the same dimensions, this equation is equivalent to the following separate equations:

$$C = \frac{I}{T^2} = FL = FL \quad \text{or} \quad C = \underline{FL} \qquad \text{Ans.}$$

and

$$I = \underline{FLT^2}. \qquad \text{Ans.}$$

EXAMPLE 1-13

Consider the equation

$$Ft \cos\theta = \frac{W}{g}v + I\omega,$$

where F is force, t is a time interval, θ is an angle, W is a weight (force), g is a linear acceleration, v is a linear velocity, I is (mass)$\times$(length)^2, and ω is an angular velocity. Is the equation dimensionally homogeneous?

SOLUTION

When the various quantities are expressed in terms of fundamental dimensions the equation becomes

$$FT = \left(\frac{FT^2}{L}\right)\left(\frac{L}{T}\right) + \left(\frac{FT^2}{L}\right)(L^2)\left(\frac{1}{T}\right)$$

or

$$FT = FT + FLT.$$

Since all terms do not have the same dimensions, the equation is not dimensionally homogeneous.

PROBLEMS

In the following problems, all numerical factors are dimensionless unless otherwise specified.

1-87 In the dimensionally homogeneous equation

$$F = \frac{KMm}{r^2}$$

F is a force, M and m are masses, and r is a distance. Determine the fundamental dimensions of K.

1-88 In the dimensionally homogeneous equation

$$R = cv + ag$$

R is a force, v is a velocity, and g is an acceleration. Determine the fundamental dimensions of c and a.

1-89 In the dimensionally correct equation

$$P = QwH$$

P is power, w is weight per unit of volume, and H is a length. Determine the dimensions of Q.

1-90 In the dimensionally homogeneous equation

$$Pd = \frac{1}{2}mv^2 + \frac{1}{2}I\omega^2$$

d is a length, m is a mass, v is a linear velocity, and ω is an angular velocity. Determine the fundamental dimensions of P and I.

1-91 In the dimensionally homogeneous equation

$$y = \frac{mg}{k} + \left(y_0 + \frac{mg}{k}\right)\cos\sqrt{\frac{k}{m}}\,t$$

y is a length, m is a mass, and g is an acceleration. Determine the fundamental dimensions of k, y_0, and t.

1-92 Determine the fundamental dimensions of y, b, a, and c in the dimensionally homogeneous equation

$$y = Ae^{-bt}\cos(\sqrt{1-a^2}\,bt + c)$$

in which A is a length and t is time.

1-93 In the dimensionally correct equation

$$Q = \sqrt{B}(W - ah)h^{3/2}$$

Q is a volume per unit of time and W and h are lengths. Determine the dimensions of B and a.

1-94 In the dimensionally homogeneous equation

$$v_f - v_i = Ig\ln\left(\frac{m_i}{m_f}\right)$$

m_i is a mass, v_i is a velocity, and g is an acceleration. Determine the fundamental dimensions of m_f and I.

1-95 In the equation

$$mv^2 = Rd + ma\cos\theta$$

m is a mass, v is a velocity, R is a force, d is a length, and a is a linear acceleration. Is the equation dimensionally correct?

1-96 In the equation

$$Pt = mv + I\alpha$$

P is a force, t is time, m is a mass, v is a linear velocity, I is (mass)(length)^2, and α is an angular acceleration. Is the equation dimensionally correct?

1-97 In the equation

$$y = y_0 + vt^2 + \frac{1}{2}at^2$$

y and y_0 are distances, v is a velocity, a is an acceleration, and t is time. Is the equation dimensionally homogeneous?

1-98 Determine the fundamental dimensions of c, ω, k, and P in the differential equation

$$m\frac{d^2x}{dt^2} + c\frac{dx}{dt} + kx = P\cos\omega t$$

in which m is mass, x is length, and t is time.

1-99 In the differential equation

$$\frac{d}{dt}[(qx + a)v] = -qgx + q\left(\frac{dx}{dt}\right)^2$$

q is a mass per unit length, a is a mass, and t is time. Determine the fundamental dimensions of x, g, and v.

1-9 NUMERICAL CALCULATIONS

In solving problems in mechanics, a slide rule will usually provide sufficient accuracy if used carefully. Most slide rules have scales for determining natural and common logarithms and trigonometric functions in addition to the scales used in multiplication and division. These scales make it unnecessary to refer to any tables of functions, and a thorough understanding of the slide rule will be extremely helpful in the following work. Slide rules will yield results accurate to three or four significant figures. It is suggested that all intermediate calculations be carried through with one more significant figure than is wanted or justified in the end result.

In general, the final result should contain no more significant figures than the least accurate figure used in obtaining it. For example, if the area of a rectangular figure is to be obtained by measuring the two sides, the result can be no more accurate than the least accurate measurement. Suppose that the long side were measured as 33.84 in. and the short side as 1.71 in.; the area would be 33.84(1.71) = 57.9 sq in. Including more figures gives a false sense of accuracy, since if the short side were anything from 1.705 to 1.715 in., it would still be given as 1.71 in. to the closest 0.01 in., which is all that is implied by the data. The two limiting widths would give areas from 57.7 to 58.0 sq in., so it would obviously be misleading to give more than three figures in the result. If two or more figures are to be added or subtracted, the final result should contain no more significant figures than the least accurate data. If a 65.4-in. piece of steel were added to a 0.887-in. piece, the total length would be

$$65.4 + 0.887 = 66.3 \text{ in.}$$

It would be meaningless to give the total length as 66.287 in., since one of the original lengths was measured only to the closest 0.1 in. If the original length had been measured more exactly, it could have been any length between 65.35 and 65.45 in. Thus without more precise measurements, the total length could be given only as 66.3 in.

The answers to problems in this text are usually given to four significant figures when the first digit is a 1 and to three figures for other numbers. *In problems with data given to only one or two significant figures, the answers are based on the assumption that all such data are exact.*

Neatness in setting up problems is not to be regarded as an end in itself but rather as a safeguard against careless mistakes. Unless a problem is to be solved graphically, it is unnecessary to draw figures to scale, and often a freehand sketch will be satisfactory. If the sketch and accompanying calculations are arranged in a neat and orderly manner, however, the problem can usually be solved faster, the calculations will be much easier to check, and the chances of error will be greatly reduced. Orderliness in setting up the problem also helps develop an orderly, systematic method of thinking, which is an additional benefit to be derived from a study of mechanics.[4]

[4] More specific suggestions on setting up problems are to be found in texts on engineering problems, such as R. G. Brown, *Introduction to Engineering Problems* (Englewood Cliffs, N.J.: Prentice-Hall, Inc., 1948), and F. C. Dana and L. R. Hillyard, *Engineering Problems Manual*, 5th ed. (New York: McGraw-Hill Book Company, 1958).

2

Resultants of Force Systems

2-1 INTRODUCTION

The *resultant* of a force system has been defined as the *simplest force system which can replace the original system without changing its external effect on a rigid body*. When the resultant of a force system is zero, the body on which the system acts is in equilibrium and the force system is said to be *balanced*. It is also common practice to say that a force system is in equilibrium when its resultant is zero. Consequently, the study of resultants is a desirable preliminary to determining the necessary equations of equilibrium for each type of force system. If an unbalanced force system acts on a body, the body will be accelerated, and the acceleration will depend, among other things, on the resultant of the force system. Therefore the study of resultants is a necessary preliminary to the study of dynamics. In this chapter, procedures will be developed for determining the resultants of various types of force systems.

The types of force systems, together with their possible resultants, are listed below for ready reference. To give complete information about a resultant, all characteristics of each resultant force and each resultant couple must be specified. A good method of giving this information is to show the resultant in a sketch. *In this text double arrowheads are used to indicate resultants.*

Type of Force System	*Possible Resultants*
Concurrent	Force
Coplanar	Force or a couple
Parallel	Force or a couple
General three-dimensional	Force, or a couple, or a force and a couple

2-2 RESULTANT OF A CONCURRENT FORCE SYSTEM

Consider the three noncoplanar forces $\mathbf{F}_1$, $\mathbf{F}_2$, and $\mathbf{F}_3$ in Fig. 2-1, all passing through point O. The resultant of the two forces $\mathbf{F}_1$ and $\mathbf{F}_2$ is the diagonal $\mathbf{R}_{12}$ of the parallelogram *oabc* as shown. The force $\mathbf{R}_{12}$ can next be combined with $\mathbf{F}_3$ by means of the parallelogram *obde*, giving the resultant of the three forces $\mathbf{F}_1$, $\mathbf{F}_2$, and $\mathbf{F}_3$ as $\mathbf{R}$. If there are more forces in the system, this process can be continued until all forces have been included. *Note that the resultant must pass through the point of concurrence.*

From the preceding discussion, it is apparent that the *resultant of a concurrent force system is a single force passing through the point of concurrence* and that the resultant force will be completely determined when its magnitude and direction are known. The resultant of a concurrent force system can thus be determined as the vector sum of the forces of the system.

The vector sum of the forces can be obtained most readily if each force is resolved into components parallel to a set of orthogonal axes as shown in Art. 1-4. Thus the vector sum of a system of concurrent

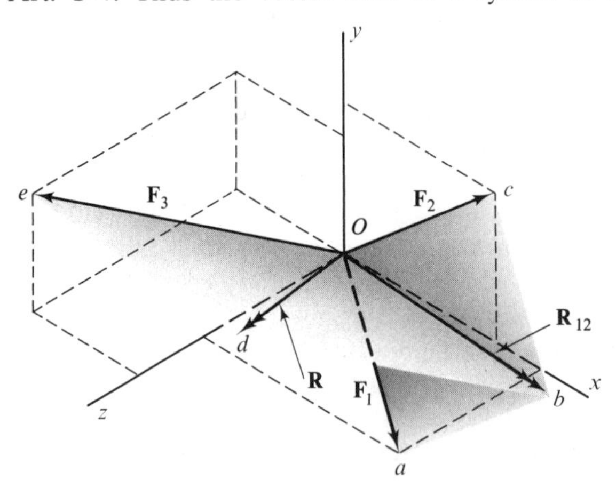

Figure 2-1

forces $\mathbf{F}_1$, $\mathbf{F}_2$, and $\mathbf{F}_3$ is

$$\mathbf{R} = \mathbf{F}_1 + \mathbf{F}_2 + \mathbf{F}_3 = \sum \mathbf{F}, \tag{2-1}$$

which can be written in component form as

$$R_x\mathbf{i} + R_y\mathbf{j} + R_z\mathbf{k} = \sum (F_x\mathbf{i} + F_y\mathbf{j} + F_z\mathbf{k}) = \sum F_x\mathbf{i} + \sum F_y\mathbf{j} + \sum F_z\mathbf{k}.$$

Therefore

$$R_x = \sum F_x, \quad R_y = \sum F_y, \quad \text{and} \quad R_z = \sum F_z. \tag{2-2}$$

The procedure for finding the resultant of a system of concurrent forces is illustrated in the following examples.

EXAMPLE 2-1

The magnitudes of the forces in Fig. 2-2 are $F_1 = 100$ lb, $F_2 = 50$ lb, and $F_3 = 60$ lb. Determine the resultant of the force system.

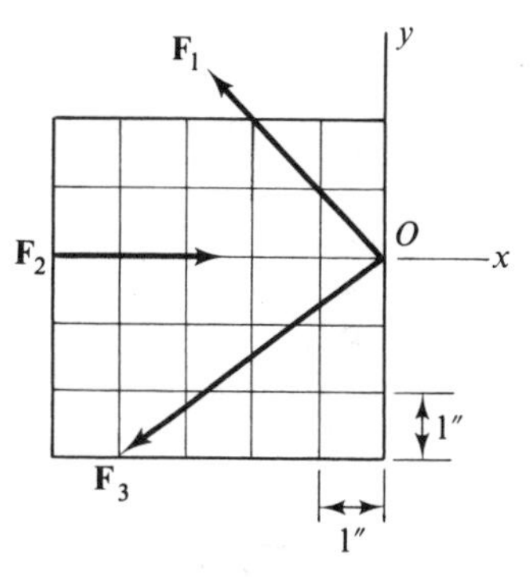

Figure 2-2

SOLUTION

Each force is expressed as a function of **i** and **j** (the **k** components are zero) and substituted in the vector equation as follows:

$$\mathbf{R} = \mathbf{F}_1 + \mathbf{F}_2 + \mathbf{F}_3$$
$$= 100(-0.707\mathbf{i} + 0.707\mathbf{j}) + (50\mathbf{i}) + 60(-0.8\mathbf{i} - 0.6\mathbf{j})$$
$$= -68.7\mathbf{i} + 34.7\mathbf{j}$$
$$= 77.0(-0.893\mathbf{i} + 0.451\mathbf{j}) \text{ lb through } O. \qquad \text{Ans.}$$

EXAMPLE 2-2

The following forces are concurrent at the origin and are directed away from it toward the points indicated. Determine the resultant of the system.

Force	*Magnitude (lb)*	*x, y, z, coordinates of point*
$\mathbf{F}_1$	90	1, −2, −2
$\mathbf{F}_2$	140	2, −6, 3
$\mathbf{F}_3$	60	−2, 3, −4

SOLUTION

The distance from the origin to the three points are 3, 7, and $\sqrt{29} = 5.39$, respectively; therefore the unit vectors and the orthogonal components of the forces are readily computed. The vector equation becomes

$$\mathbf{R} = \mathbf{F}_1 + \mathbf{F}_2 + \mathbf{F}_3$$
$$= 90\left(\frac{\mathbf{i} - 2\mathbf{j} - 2\mathbf{k}}{3}\right) + 140\left(\frac{2\mathbf{i} - 6\mathbf{j} + 3\mathbf{k}}{7}\right) + 60\left(\frac{-2\mathbf{i} + 3\mathbf{j} - 4\mathbf{k}}{5.39}\right)$$
$$= (47.7\mathbf{i} - 146.6\mathbf{j} - 44.6\mathbf{k})\text{ lb}$$
$$= \underline{160.5(0.297\mathbf{i} - 0.914\mathbf{j} - 0.278\mathbf{k})\text{ lb through the origin.}}$$

Ans.

PROBLEMS

2-1–2-4 Determine the resultant of each of the coplanar force systems in Figs. P2-1 through P2-4.

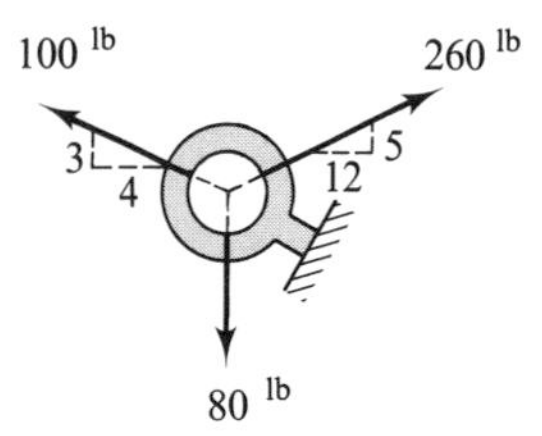

Figure P2-1

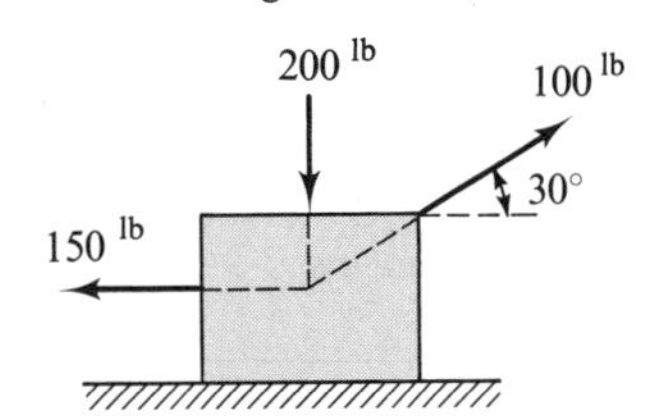

Figure P2-2

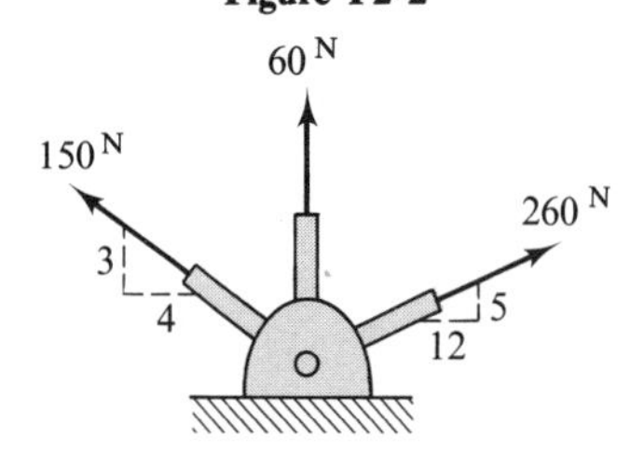

Figure P2-3

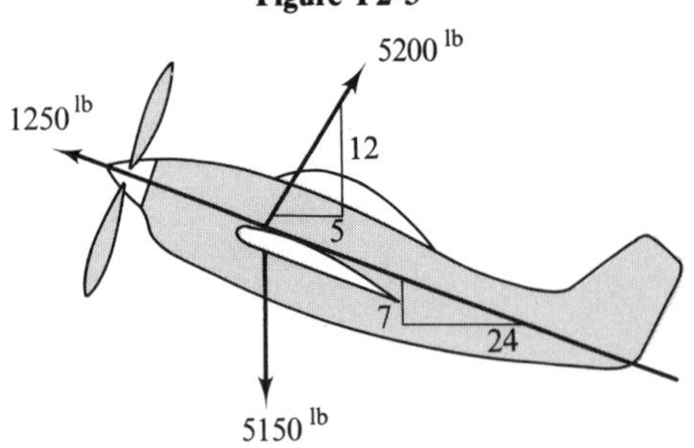

Figure P2-4

2-5 The resultant of the three forces of Fig. P2-5 is vertical. Determine the force **A** and the resultant.

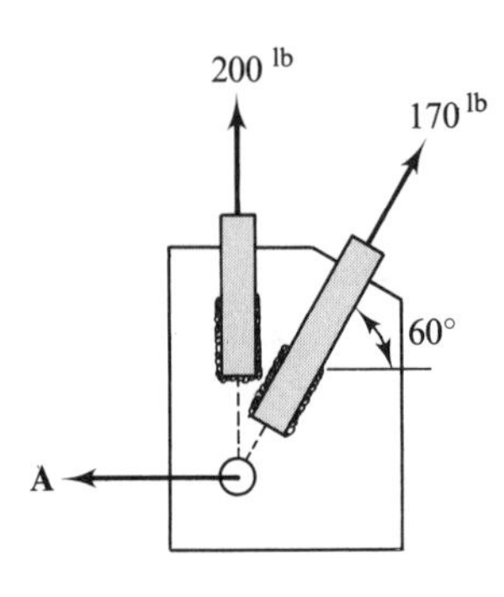

Figure P2-5

2-6 The 650-lb force in Fig. P2-6 is the resultant of the two forces **P** and **Q**. Determine the magnitudes of **P** and **Q**.

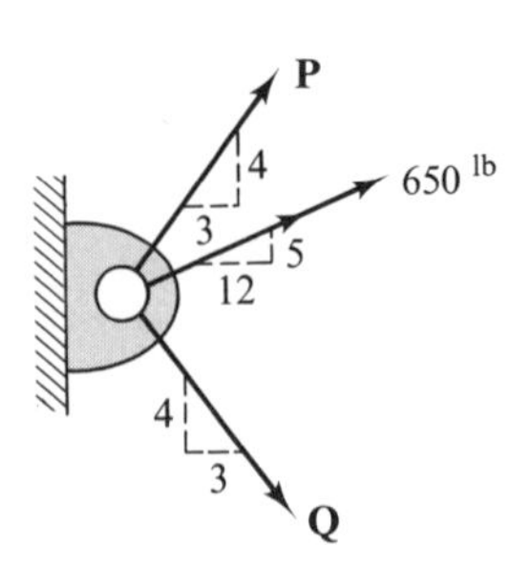

Figure P2-6

2-7 The 600-lb force in Fig. P2-7 is the resultant of four forces acting on the eyebolt, three of which are shown. Determine the fourth force.

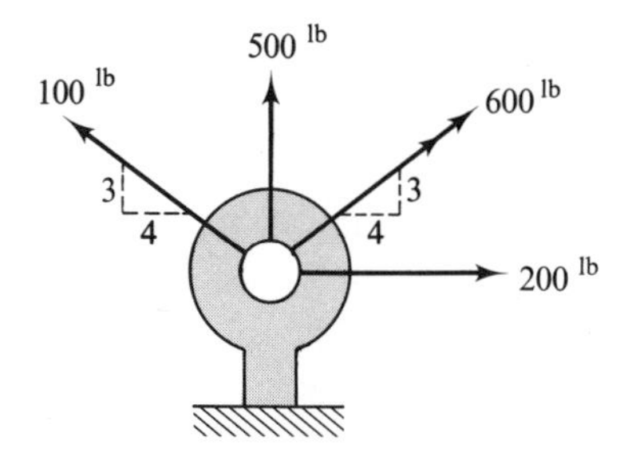

Figure P2-7

In Problems 2-8–2-10, the noncoplanar forces for each problem are concurrent at the origin and are directed from the origin toward the point indicated. Determine the resultant of each system of forces.

***2-8** 70 kN (6, 3, −2), 90 kN (−8, 4, −1), and 60 kN (1, −2, 2).

2-9 110 lb (−2, 6, 9), 260 lb (3, 4, −12), and 150 lb (−5, 2, 14).

2-10 170 lb (8, 9, 12), 110 lb (−6, 6, −7), and 100 lb (4, −6, 3).

2-11 (a) Determine the resultant of the three forces of Fig. P2-11.
(b) Determine the moment of the resultant force with respect to point A.

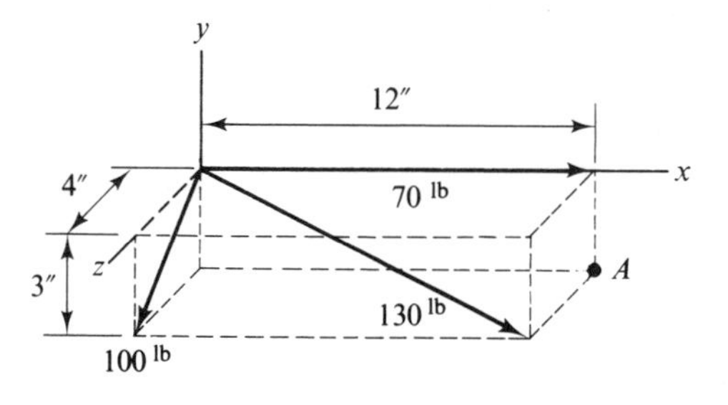

Figure P2-11

***2-12** (a) Determine the resultant of the force system shown in Fig. P2-12.
(b) Determine the moment of the resultant with respect to line AB.

2-13 Determine the resultant of the tensions in the guy wires of Fig. P2-13.

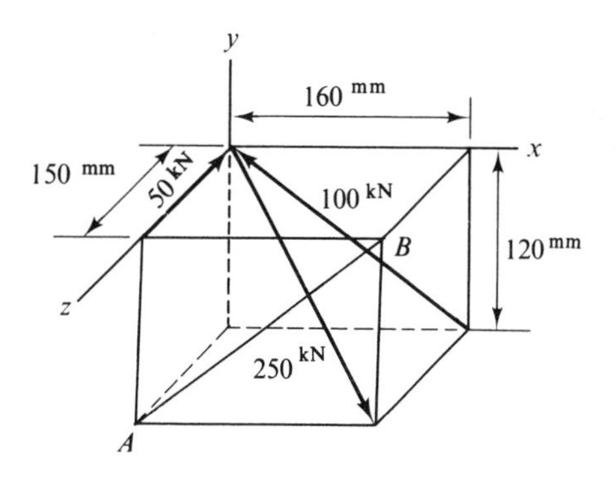

Figure P2-12

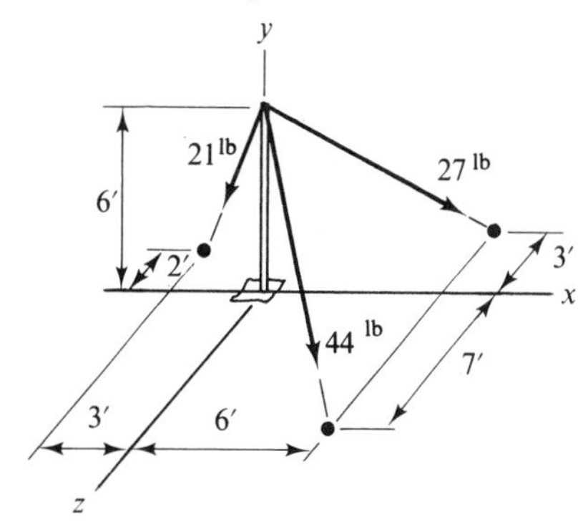

Figure P2-13

2-14 (a) Determine the resultant of the concurrent force system of Fig. P2-14.
(b) Determine the moment of the resultant with respect to line AB.

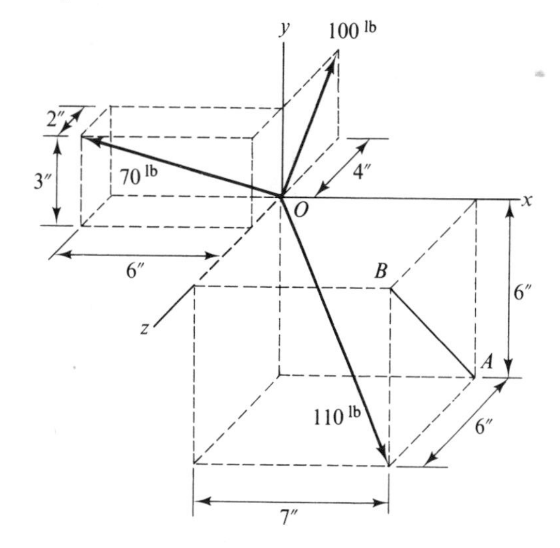

Figure P2-14

2-3 RESULTANT OF A COPLANAR FORCE SYSTEM

The resultant of the four coplanar forces $\mathbf{F}_1$, $\mathbf{F}_2$, $\mathbf{F}_3$, and $\mathbf{F}_4$ in Fig. 2-3 can be obtained by successive applications of the parallelogram law, and this method will also demonstrate the possible forms of the resultant. By making use of the principle of transmissibility the forces $\mathbf{F}_1$ and $\mathbf{F}_2$ are moved along their lines of action to their point of intersection. A parallelogram is then constructed to determine their resultant $\mathbf{R}_{12}$. The resultant $\mathbf{R}_{12}$ is then combined in the same manner with the force $\mathbf{F}_3$ to obtain the resultant $\mathbf{R}_{123}$, and finally the resultant of the force system, $\mathbf{R}_{1234}$, is determined by combining $\mathbf{R}_{123}$ with $\mathbf{F}_4$. The resultant is thus completely determined, since the three characteristics (magnitude, direction, and a point on the action line) are all included in the sketch. Note that the resultant force $\mathbf{R} = \mathbf{R}_{1234}$ is in the plane of the original system.

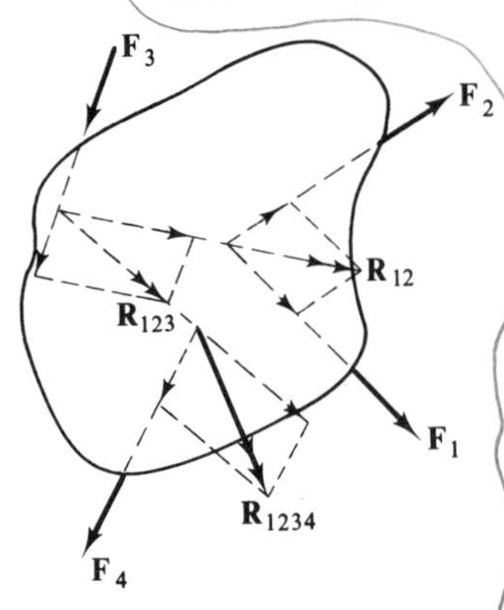

Figure 2-3

In the case just considered, the resultant was found to be a single force. The resultant will be a couple whenever the resultant of all but one of the forces of the system and the remaining force form a couple. Thus it is apparent that the *resultant of a nonconcurrent, coplanar force system is either a single force or a couple.*

The preceding method of obtaining the resultant of a nonconcurrent coplanar force system may become quite cumbersome when more forces are involved, and it is therefore not a practical method.

When the resultant is a force the procedure of summing the forces as described in Art. 2-2 can be used to determine the magnitude and direction of the resultant force. If $\sum \mathbf{F} = 0$, the resultant is either a couple or the system is in equilibrium. There remains only the problem of locating a point on the action line of the resultant force or of determining the magnitude of the moment and sense of rotation of a resultant couple.

The remaining problem can be solved by use of the principle of moments from Art. 1-6, which states that the moment of the resultant of a force system with respect to any point (moment center) or axis is equal to the vector sum of the moments of the forces of the system with respect to the same reference. With the magnitude and direction of the resultant known from the vector sum of the forces, the position vector from the moment center to a point on the resultant force will be the only unknown when the moment of the resultant is equated to the sum of the moments of the forces. The following examples illustrate the procedure.

EXAMPLE 2-3

The forces in Fig. 2-4(a) have the following magnitudes: $F_1 = 130$ lb, $F_2 = 200$ lb, and $F_3 = 100$ lb. Determine the resultant of the force system and locate it with respect to the origin O.

SOLUTION

Each force of the system is expressed in terms of the unit vectors **i** and **j** and substituted in the vector equation for **R** as follows:

$$\begin{aligned}\mathbf{R} &= \mathbf{F}_1 + \mathbf{F}_2 + \mathbf{F}_3 \\ &= (-50\mathbf{i} - 120\mathbf{j}) + (160\mathbf{i} + 120\mathbf{j}) + (-50\mathbf{i} + 86.6\mathbf{j}) \\ &= 60\mathbf{i} + 86.6\mathbf{j} = \underline{105.4(0.569\mathbf{i} + 0.821\mathbf{j})\text{ lb.}} \qquad \text{Ans.}\end{aligned}$$

To compute the sum of the moments of the forces of the system and of the resultant about the moment center (the origin), position vectors for each of the forces are needed. The vectors can be selected as

$$\mathbf{r}_1 = 0, \qquad \mathbf{r}_2 = \mathbf{OB} = 7\mathbf{i}, \qquad \mathbf{r}_3 = \mathbf{OC} = 3\mathbf{i} + 2\mathbf{j}$$

and

$$\mathbf{r}_R = x\mathbf{i} + y\mathbf{j},$$

where x and y are the coordinates of any point on **R**. When the moment of **R** is equated to the sum of the moments of the forces, the resulting expression in x and y is the equation of the line of action of **R**. If either x or y is specified, the other coordinate can be obtained from the equation. The principle of moments gives

$$\begin{aligned}\mathbf{M}_O &= \mathbf{r}_R \times \mathbf{R} = \mathbf{r}_1 \times \mathbf{F}_1 + \mathbf{r}_2 \times \mathbf{F}_2 + \mathbf{r}_3 \times \mathbf{F}_3 \\ &= (x\mathbf{i} + y\mathbf{j}) \times (60\mathbf{i} + 86.6\mathbf{j}) = 0 \times (-50\mathbf{i} - 120\mathbf{j}) \\ &\quad + (7\mathbf{i}) \times (160\mathbf{i} + 120\mathbf{j}) + (3\mathbf{i} + 2\mathbf{j}) \times (-50\mathbf{i} + 86.6\mathbf{j})\end{aligned}$$

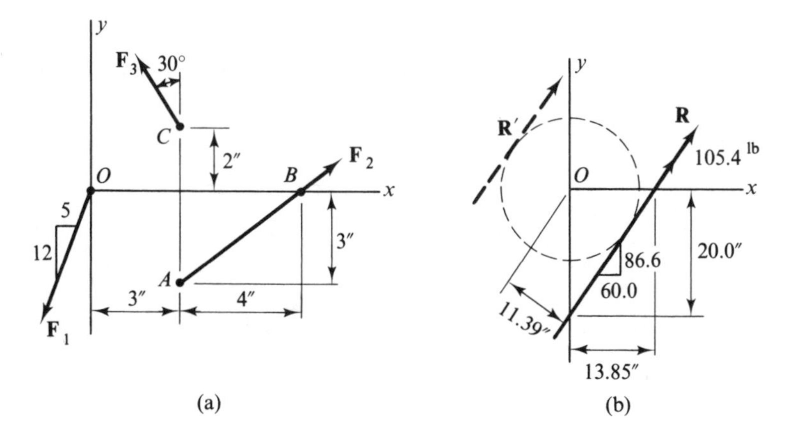

Figure 2-4

from which

$$(86.6x - 60y)\mathbf{k} = (840)\mathbf{k} + [3(86.6) - 2(-50)]\mathbf{k} = 1199.8\mathbf{k}$$

or

$$86.6x - 60y = 1200.$$

The intersection of **R** with the x axis can be obtained when y equals 0 is substituted in this equation. Thus,

$$x = \frac{1200}{86.6} = \underline{13.85 \text{ in.}} \qquad \text{Ans.}$$

The result is shown in Fig. 2-4(b).

If the intersection of **R** with the y axis is desired, x is set equal to zero in the equation, which gives

$$y = \frac{1200}{-60} = -20 \text{ in.}$$

This result is also shown in Fig. 2-4(b).

The perpendicular distance p from the moment center to the line of action of **R** can also be determined from the principle of moments.

The magnitude of $\mathbf{M}_O$ is defined as the product of the magnitude of the resultant and the perpendicular distance p from O to **R**. Thus

$$M_O = 1200 = pR = p(105.4)$$

and

$$p = \frac{1200}{105.4} = 11.39 \text{ in.}$$

The resultant must be located 11.39 in. from point O and is thus tangent to the circle in Fig. 2-4(b). There are two points where a force in the direction of **R** will be tangent to the circle. The correct point is the one which will result in a moment in the correct (counterclockwise) direction shown as **R**. The force **R′** would not give a moment in the correct direction.

EXAMPLE 2-4

The 100-N force of Fig. 2-5(a) is the resultant of the couple and three forces, two of which are shown in the diagram. Determine the third force **P** and locate it with respect to the point A.

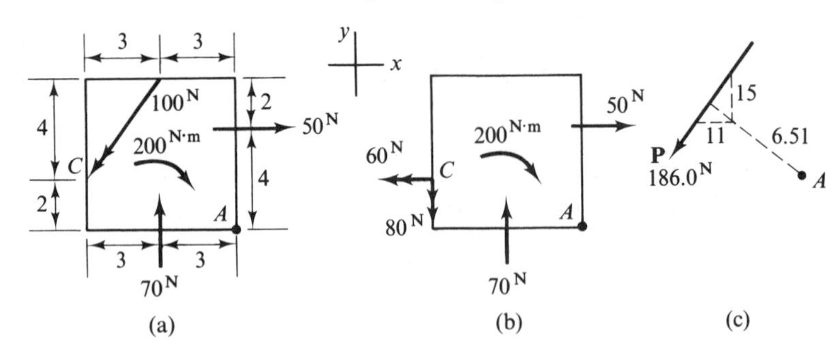

Figure 2-5 Dimensions in meters

SOLUTION

The 100-N resultant is resolved into two components at C as shown in Fig. 2-5(b). When each of the known forces is expressed in terms of the unit vectors and substituted into the equation $\mathbf{R}=\sum \mathbf{F}$, the equation becomes

$$-60\mathbf{i}-80\mathbf{j}=50\mathbf{i}+70\mathbf{j}+\mathbf{P}$$

from which

$$\mathbf{P}=-110\mathbf{i}-150\mathbf{j}$$
$$=186.0(-0.591\mathbf{i}-0.806\mathbf{j})\text{N}.$$

The principle of moments is used to locate the force $\mathbf{P}$ relative to point A; that is, the moment of the resultant with respect to point A is equated to the sum of the moments of the component forces. If p is the perpendicular distance from A to $\mathbf{P}$ and if $\mathbf{P}$ is assumed to be above and to the left of A, the moment of $\mathbf{P}$ about A will be $Pp\mathbf{k}$, and the moment equation becomes

$$60(2)\mathbf{k}+80(6)\mathbf{k}=-200\mathbf{k}-50(4)\mathbf{k}-70(3)\mathbf{k}+186.0p\mathbf{k}$$

from which

$$p=\frac{1210}{186.0}=6.51\text{ m}.$$

Since the value of p is positive, the assumption that $\mathbf{P}$ was above point A was correct. The location of $\mathbf{P}$ relative to A is shown in Fig. 2-5(c).

PROBLEMS

2-15 Determine the resultant of the force system shown in Fig. P2-15.

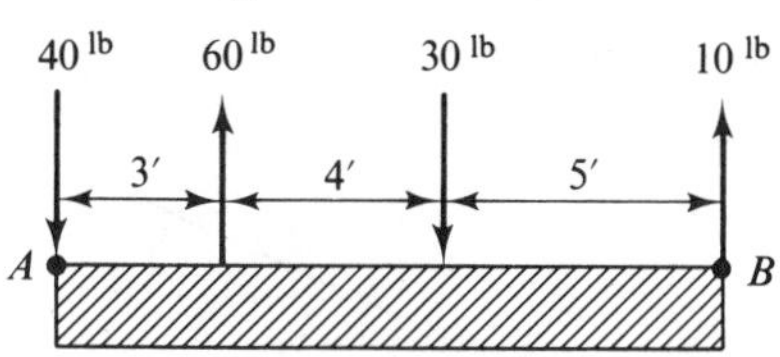

Figure P2-15

2-16 Determine the resultant of the forces and couple shown in Fig. P2-16 and locate it with respect to point A.

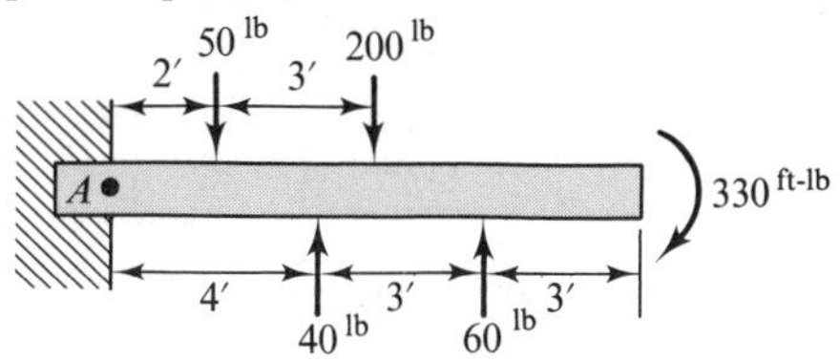

Figure P2-16

***2-17** Determine the resultant of the forces and couple shown in Fig. P2-17 and locate it with respect to point A.

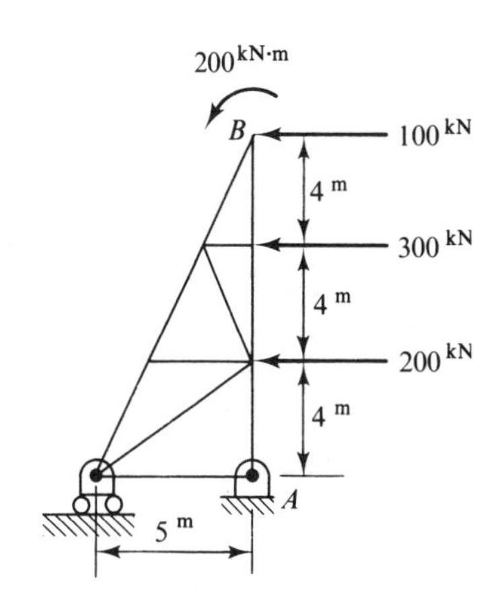

Figure P2-17

2-18 The 150-lb force in Fig. P2-18 is the resultant of the couple and four forces, three of which are shown. Determine the fourth force and locate it with respect to point A.

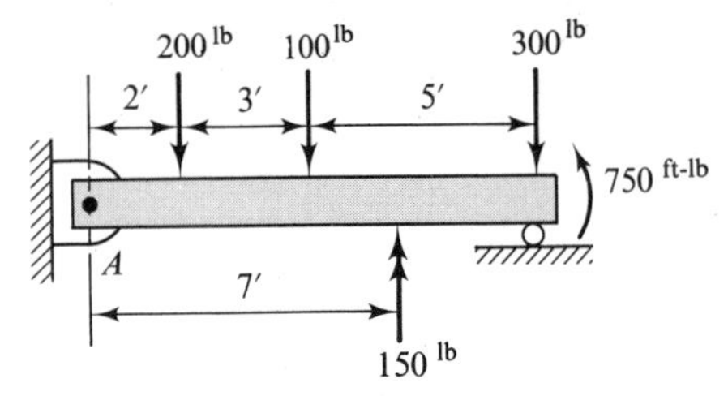

Figure P2-18

2-19 The position of the resultant **R** of the three forces and the couple **M** is shown in Fig. P2-19. Determine the magnitude of the resultant and the couple **M**.

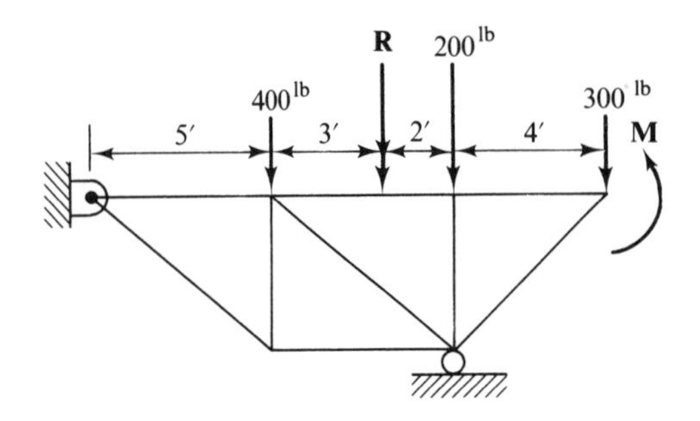

Figure P2-19

2-20 The cable AB in Fig. P2-20 supports four vertical loads. Determine the resultant and locate it with respect to point A.

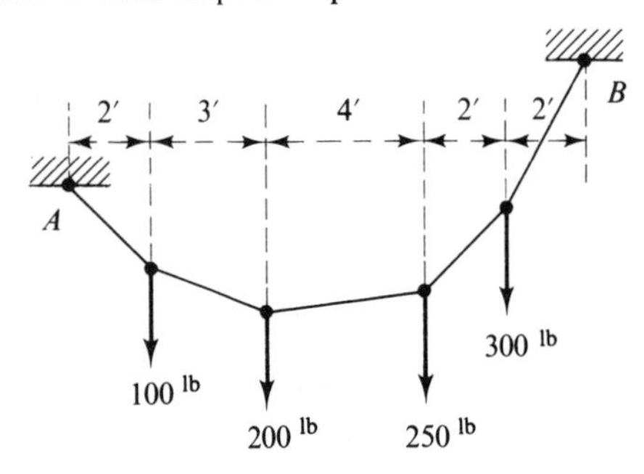

Figure P2-20

***2-21** Determine the resultant of the force system in Fig. P2-21 and locate it with respect to point A.

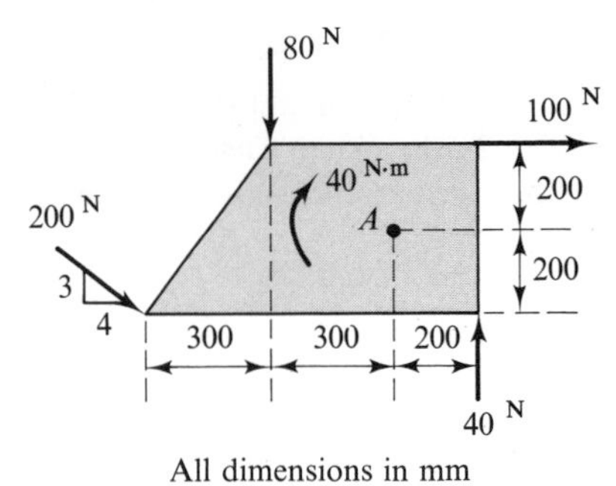

Figure P2-21

2-22 Determine the resultant of the force system shown in Fig. P2-22 and locate it with respect to point O.

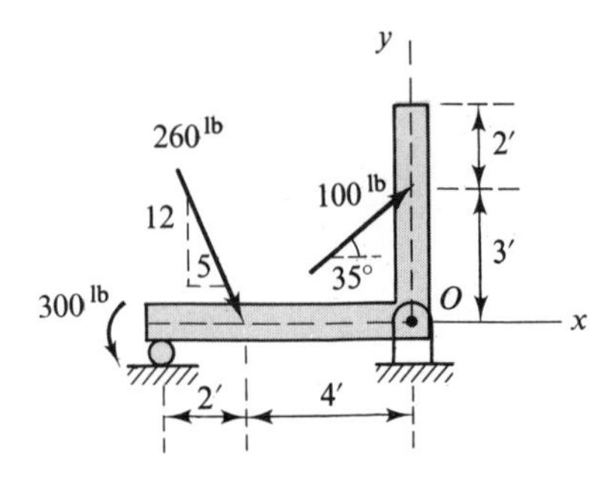

Figure P2-22

2-23 Determine the resultant of the force system shown in Fig. P2-23 and locate it with respect to point A.

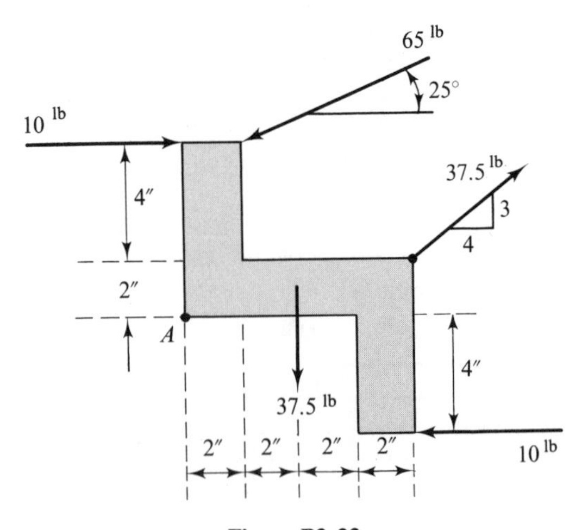

Figure P2-23

2-24 Determine the resultant of the force system in Fig. P2-24 and locate it with respect to the origin, O.

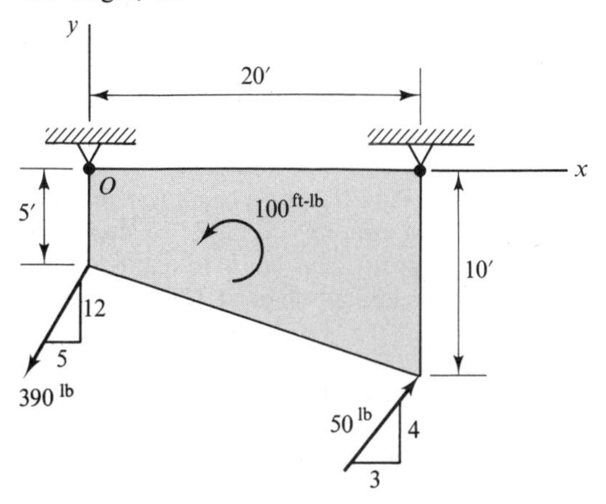

Figure P2-24

2-25 Four coplanar forces are shown in Fig. P2-25.

(a) Determine the angle θ of the 40-lb force which will cause the resultant force to have a zero component in the x direction.

(b) For an angle θ of 30°, determine the distance y such that the resultant of the forces passes through O.

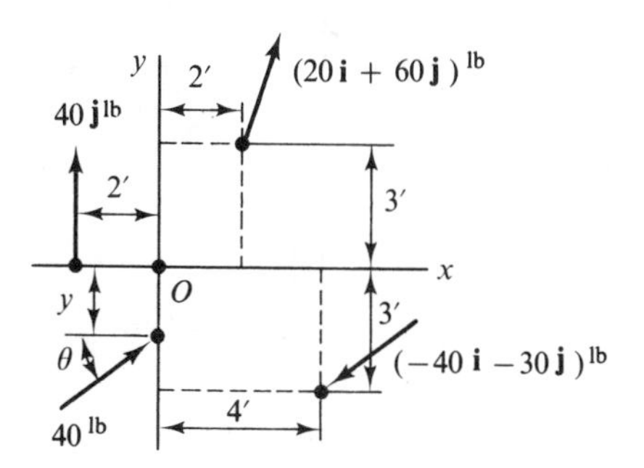

Figure P2-25

***2-26** The 200-N force in Fig. P2-26 is the resultant of the couple and three forces, two of which are shown in the diagram. Determine the third force and locate it with respect to point O.

2-27 The 200-lb force in Fig. P2-27 is the resultant of the 300-ft-lb couple and three forces, two of which are shown in the diagram.

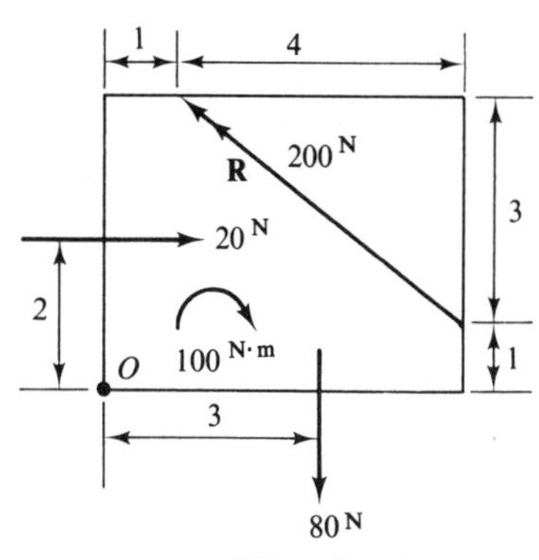

Figure P2-26

Determine the third force and locate it with respect to point A.

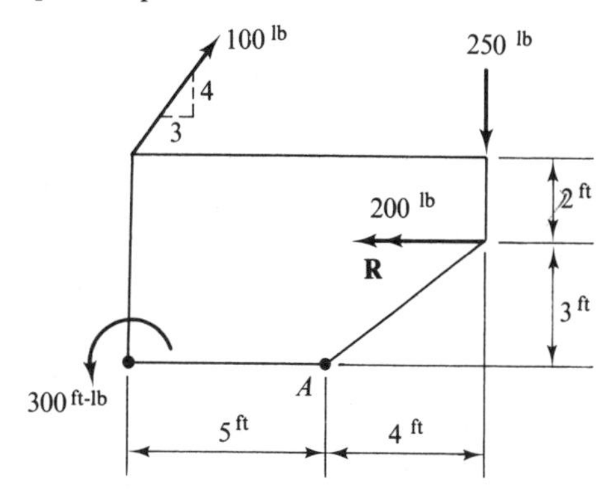

Figure P2-27

2-28 The 75-lb force in Fig. P2-28 is the resultant of the force system shown and two additional forces, a vertical force through B and a force through A. Determine the unknown forces.

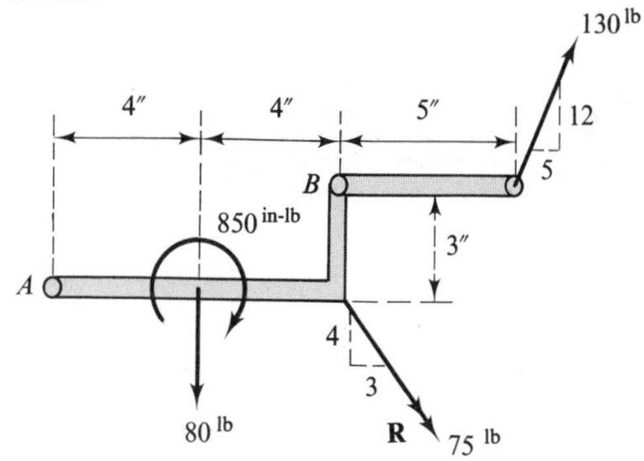

Figure P2-28

***2-29** The resultant of the three coplanar forces in Fig. P2-29 and an unknown force **P** through the point *B* is a vertical force through A. Determine the force **P** and the resultant. The three forces shown are tangent to the circle.

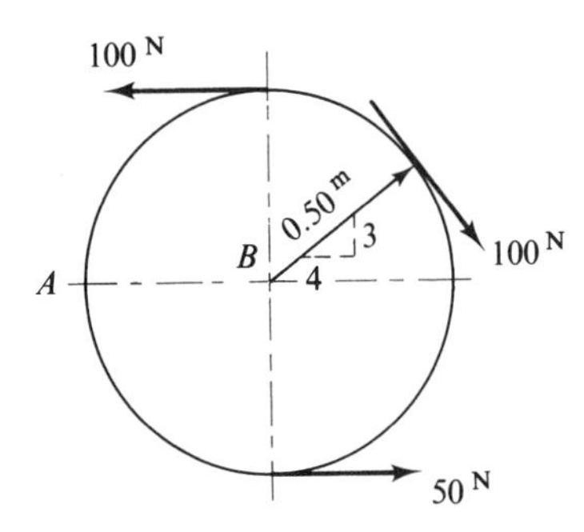

Figure P2-29

2-30 The car in Fig. P2-30 is climbing a 10% grade (slope of road). The 110-lb wind resistance force is parallel to the road, and the angle made by the rear wheel forces is measured from a normal to the road. Determine the resultant force acting on the car.

2-31 The resultant of the coplanar forces acting on the automobile shown in Fig. P2-31 is a horizontal force of 170 lb directed toward the left through the center of gravity. Determine all unknown forces shown acting on the car.

2-32 Figure P2-32(a) is a photograph of a Beechcraft Bonanza. Figure P2-32(b) shows the gross weight, thrust, and tail load acting on the plane in unaccelerated flight at 150 mph.
(a) Determine the resultant of these three forces.
(b) The resultant of these three forces must be balanced by the forces on the wing. It is convenient to show the wing force as a pair of horizontal and vertical components through the quarter-chord point (c/4) and a couple. Determine these components (called *drag* and *lift*) and the moment of the couple (called the *moment about the aerodynamic center*).

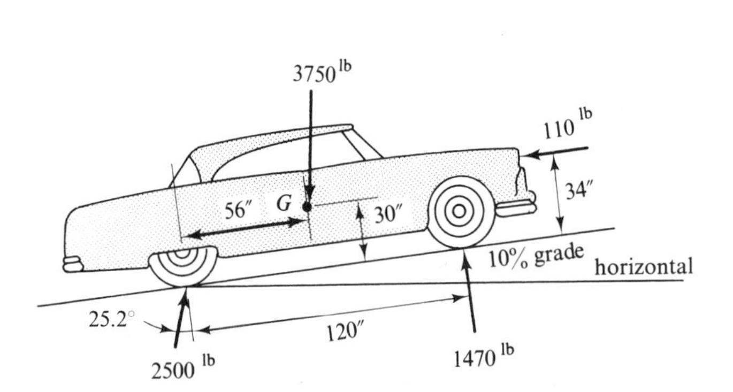

Figure P2-30

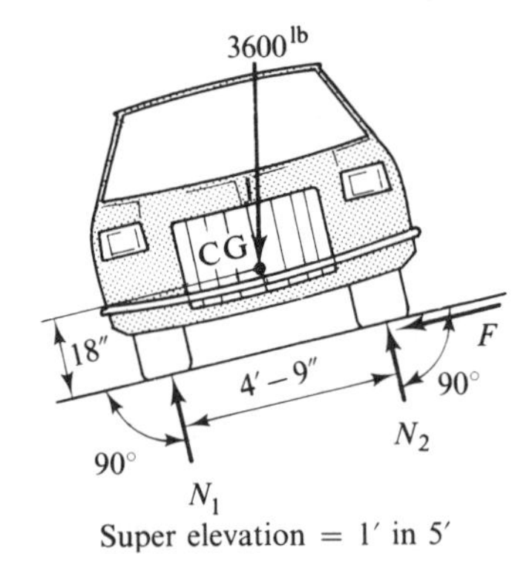

Figure P2-31

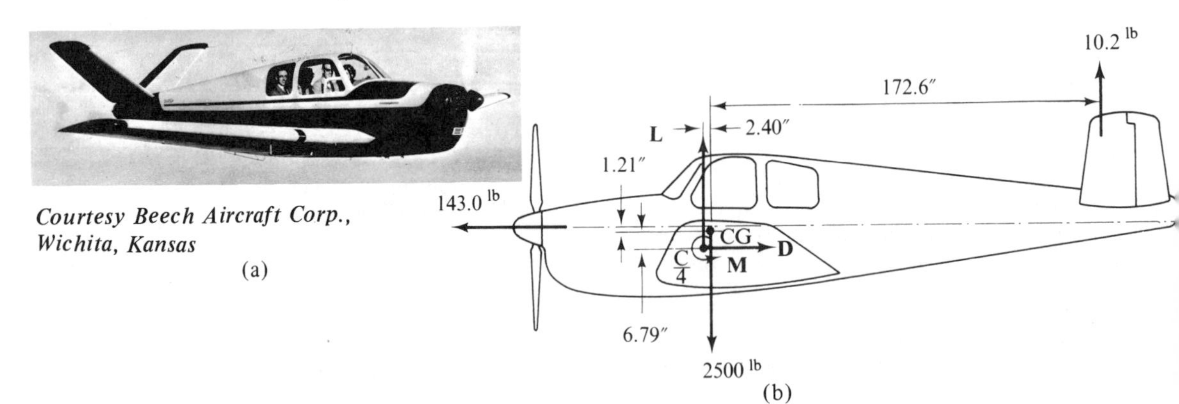

Courtesy Beech Aircraft Corp., Wichita, Kansas

(a)

(b)

Figure P2-32

2-4 RESULTANT OF A GENERAL FORCE SYSTEM IN SPACE

Section 2-4 Resultant of a general force system in space

The resultant of a general force system in space can be a single force or a couple, but in general it is a force and a couple.

When all forces of the system are parallel, the resultant will be a single force (parallel to the given forces) or a couple in a plane parallel to the given forces. The reasoning behind this statement will become apparent if the resultant of any two forces is first obtained and then combined with each added force in turn. The resultant is a force $\mathbf{R}=\sum \mathbf{F}$ when $\sum \mathbf{F}$ is different from zero. If $\sum \mathbf{F}$ is equal to zero, the resultant is a couple unless the sum of the moments of all the forces with respect to any point is zero, in which case the system is in equilibrium. When the resultant is a force its position can be determined by using the principle of moments as illustrated in Example 2-5.

The resultant of a system of couples must be a couple (or zero). From the definition of a couple, $\sum \mathbf{F}$ must be equal to zero. Therefore it follows that $\sum \mathbf{F}$ will be zero for any system of couples. Since the moment of a couple is independent of the moment center (the moment of a couple is a free vector), the vectors representing the moments of the couples can be taken to constitute a concurrent set of vectors in space through any convenient moment center. The resultant of a system of couples is thus the vector sum of the moments of the couples and must be a couple.

The resultant of any general force system can be obtained by resolving each force into a parallel force *through some common point* or origin and a couple. The general system is thus reduced to a set of concurrent forces and a set of couples. The resultant of the concurrent forces is obtained as the vector sum of the component forces, $\mathbf{R}=\sum \mathbf{F}$, as explained and demonstrated in Art. 2-2. The moment of each of the couples is $\mathbf{r}\times\mathbf{F}$, where $\mathbf{r}$ is the position vector from the common point to any point on $\mathbf{F}$, and the moment of the resultant of the set of couples is $\mathbf{M}=\sum \mathbf{r}\times\mathbf{F}$. If the forces of the resultant couple $\mathbf{M}$ are in a plane parallel to the resultant force $\mathbf{R}$ (the vectors $\mathbf{M}$ and $\mathbf{R}$ will be perpendicular), the couple and the force can be combined to give a single force as the resultant (see Art. 1-7). Otherwise, the resultant will consist of a force and a couple or of two noncoplanar forces. Such a system can be reduced to a force and a couple with forces in a plane perpendicular to the force; this is defined as a *screw* or *wrench*. Such a reduction is normally unnecessary. The vector representation of a couple is useful in identifying problems in which the resultant force is parallel to the plane containing the forces of the couple, in which case the force and couple can be combined to give a single force. Since the vector $\mathbf{M}$ representing the couple is perpendicular to the plane of the couple, it will also be perpendicular to the resultant force $\mathbf{R}$; therefore, $\mathbf{M}\cdot\mathbf{R}$ will be zero whenever $\mathbf{R}$ is parallel to the plane of the couple.

The following examples illustrate the procedure for determining the resultant of a general force system.

EXAMPLE 2-5

The magnitudes of the four parallel forces in Fig. 2-6(a) are $F_1 = 32$ lb, $F_2 = 60$ lb, $F_3 = 40$ lb, and $F_4 = 75$ lb. Determine the resultant of the forces and show it on a sketch.

SOLUTION

Each force is replaced by an equal force through the origin (along the y axis) and a couple. The choice of the origin is a matter of convenience since any arbitrary point could be used. The vector sum of the forces is

$$\mathbf{R} = 32\mathbf{j} + 60\mathbf{j} - 40\mathbf{j} - 75\mathbf{j} = -23\mathbf{j}$$

$$= 23 \text{ lb} \downarrow \text{ through the origin.}$$

The vector sum of the moments of the couples is

$$\mathbf{M}_O = \mathbf{r}_1 \times \mathbf{F}_1 + \mathbf{r}_2 \times \mathbf{F}_2 + \mathbf{r}_3 \times \mathbf{F}_3 + \mathbf{r}_4 \times \mathbf{F}_4$$

$$= (\mathbf{i} + \mathbf{k}) \times (32\mathbf{j}) + (3\mathbf{i} + 4\mathbf{k}) \times (60\mathbf{j})$$

$$+ (4\mathbf{i} + 2\mathbf{k}) \times (-40\mathbf{j}) + (3\mathbf{k}) \times (-75\mathbf{j})$$

$$= (33\mathbf{i} + 52\mathbf{k}) \text{ ft-lb.}$$

Since the vector representation of the resultant couple is in the xz plane, it is perpendicular to the resultant force $\mathbf{R}$, and therefore the system can be reduced to a single force parallel to $\mathbf{R}$. The vector $\mathbf{r} = x\mathbf{i} + z\mathbf{k}$, locating the single resultant force $\mathbf{R}$, can be determined by the principle of moments. Thus

$$\mathbf{M}_O = \mathbf{r} \times \mathbf{R}$$

or

$$33\mathbf{i} + 52\mathbf{k} = (x\mathbf{i} + z\mathbf{k}) \times (-23\mathbf{j})$$

$$= -23x\mathbf{k} + 23z\mathbf{i},$$

where x and z are the coordinates of the point of intersection of $\mathbf{R}$ and

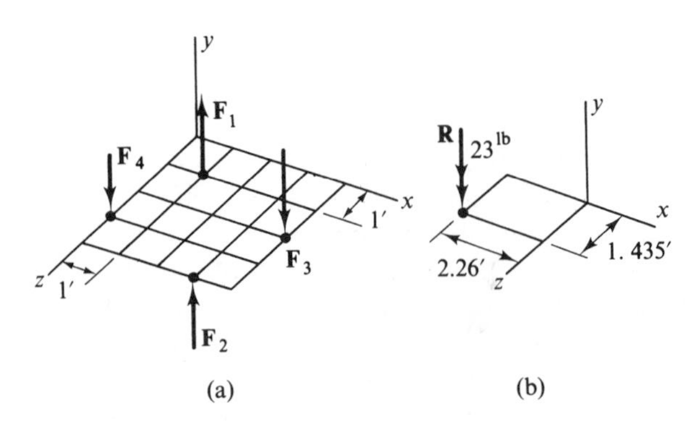

Figure 2-6

the xz plane. The solution of this vector equation gives

$$z=\frac{33}{23}=1.435 \text{ ft}$$

and

$$x=-\frac{52}{23}=-2.26 \text{ ft.}$$

The final resultant **R** is shown in Fig. 2-6(b).

EXAMPLE 2-6

Determine a resultant of the forces ($F_1=1200$ lb, $F_2=1100$ lb, $F_3=350$ lb, $F_4=878$ lb) applied to the bar of Fig. 2-7.

SOLUTION

Forces $\mathbf{F}_1$, $\mathbf{F}_2$, and $\mathbf{F}_3$ are each replaced by an equal force through the origin and a couple. The vector sum of the concurrent forces is

$$\mathbf{R}=\mathbf{F}_1+\mathbf{F}_2+\mathbf{F}_3+\mathbf{F}_4$$

$$=1200(-\mathbf{j})+1100\left(\frac{2\mathbf{i}+6\mathbf{j}-9\mathbf{k}}{\sqrt{4+36+81}}\right)$$

$$+350\left(\frac{-3\mathbf{i}+6\mathbf{j}-2\mathbf{k}}{\sqrt{9+36+4}}\right)+878\left(\frac{6\mathbf{i}+5\mathbf{j}+4\mathbf{k}}{\sqrt{36+25+16}}\right)$$

$$=-1200\mathbf{j}+(200\mathbf{i}+600\mathbf{j}-900\mathbf{k})$$

$$+(-150\mathbf{i}+300\mathbf{j}-100\mathbf{k})+(600\mathbf{i}+500\mathbf{j}+400\mathbf{k})$$

$$=650\mathbf{i}+200\mathbf{j}-600\mathbf{k}$$

$$=\underline{907(0.717\mathbf{i}+0.221\mathbf{j}-0.662\mathbf{k}) \text{ lb through the origin.}} \qquad \text{Ans.}$$

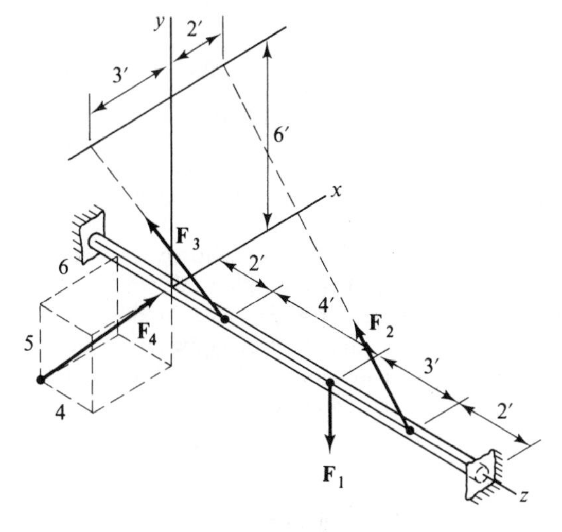

Figure 2-7

The moment of the resultant couple is

$$\begin{aligned}\mathbf{M}_O &= \mathbf{r}_1\times\mathbf{F}_1+\mathbf{r}_2\times\mathbf{F}_2+\mathbf{r}_3\times\mathbf{F}_3+\mathbf{r}_4\times\mathbf{F}_4\\ &= 6\mathbf{k}\times(-1200\mathbf{j})+(9\mathbf{k})\times(200\mathbf{i}+600\mathbf{j}-900\mathbf{k})\\ &\quad +(2\mathbf{k})\times(-150\mathbf{i}+300\mathbf{j}-100\mathbf{k})+0\\ &= 1200\mathbf{i}+1500\mathbf{j}+0\\ &= \underline{1921(0.625\mathbf{i}+0.781\mathbf{j})\text{ ft-lb.}}\end{aligned}\qquad \text{Ans.}$$

The scalar product of **R** and $\mathbf{M}_O$ is

$$\begin{aligned}\mathbf{R}\cdot\mathbf{M}_O &= (650\mathbf{i}+200\mathbf{j}-600\mathbf{k})\cdot(1200\mathbf{i}+1500\mathbf{j})\\ &= 780{,}000+300{,}000 = 1{,}080{,}000\end{aligned}$$

which is different from zero. Therefore, the resultant force and couple cannot be replaced by a single force.

EXAMPLE 2-7

A couple of $1800\mathbf{k}$ ft-lb is applied to the bar of Example 2-6 in addition to the forces shown in Fig. 2-7. Determine the resultant.

SOLUTION

The resultant force is the same as in the previous example since the couple does not enter into the summation of forces. Therefore

$$\mathbf{R} = (650\mathbf{i}+200\mathbf{j}-600\mathbf{k})\text{ lb.}$$

The sum of the moments with respect to O can be calculated by adding the moment of the couple to the sum of the moments of the forces calculated previously. Thus

$$\mathbf{M}_O = (1200\mathbf{i}+1500\mathbf{j}+1800\mathbf{k})\text{ ft-lb.}$$

The scalar product of **R** and $\mathbf{M}_O$ is

$$\begin{aligned}\mathbf{R}\cdot\mathbf{M}_O &= (650)(1200)+(200)(1500)-(600)(1800)\\ &= 780{,}000+300{,}000-1{,}080{,}000 = 0.\end{aligned}$$

Since this product is zero, the force and couple can be combined into a single force which has the same magnitude and direction as **R** but which does not pass through the origin.

The location of this single force can readily be determined by finding the point of intersection of the force **R** with one of the coordinate planes. Let $\mathbf{r} = x\mathbf{i}+y\mathbf{j}$ be the position vector of the point of intersection of **R** with the xy plane, where x and y are the coordinates of the point. The moment of **R** about O must equal the sum of the moments of the system of forces about O. Thus

$$\mathbf{r}\times\mathbf{R} = \mathbf{M}_O$$

$$(x\mathbf{i}+y\mathbf{i})\times(650\mathbf{i}+200\mathbf{j}-600\mathbf{k}) = 1200\mathbf{i}+1500\mathbf{j}+1800\mathbf{k},$$

$$-600y\mathbf{i}+600x\mathbf{j}+(200x-650y)\mathbf{k} = 1200\mathbf{i}+1500\mathbf{j}+1800\mathbf{k}.$$

The values of x and y can be determined by equating the coefficients of **i** and **j** which gives

$$-600y = 1200$$

$$y = -2 \text{ ft}$$

and

$$600x = 1500$$

$$x = 2.5 \text{ ft.}$$

A partial check can be made by equating the coefficients of **k**. This third relationship is satisfied by these values of x and y.

The resultant is

$$\underline{\mathbf{R} = 907(0.716\mathbf{i} + 0.220\mathbf{j} - 0.661\mathbf{k}) \text{ lb through point } (2.5, -2.0) \text{ ft.}}$$

Ans.

PROBLEMS

2-33–2-35 Determine the resultant of the parallel force systems in Figs. P2-33 through P2-35.

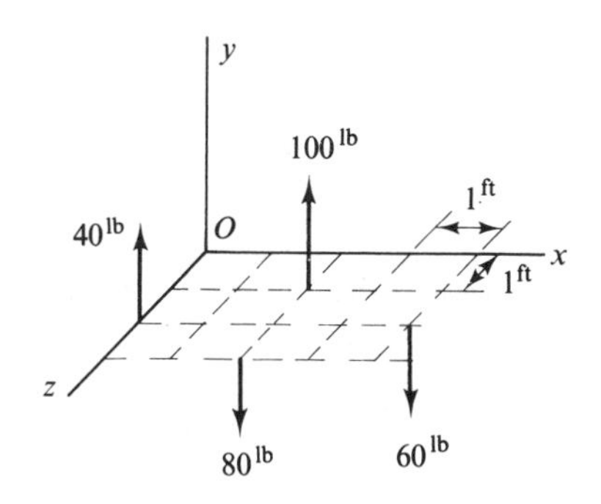

Figure P2-33

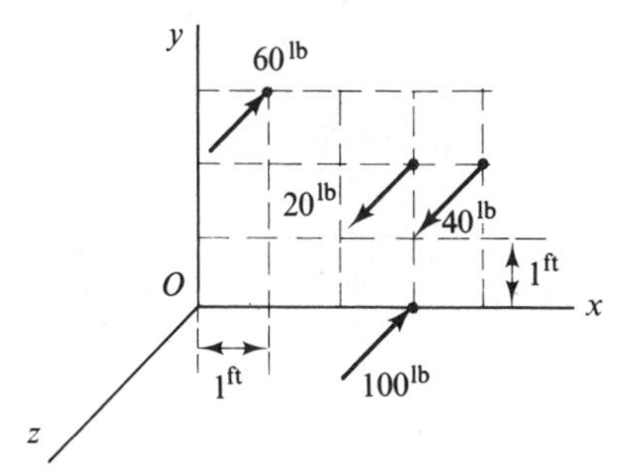

Figure P2-34

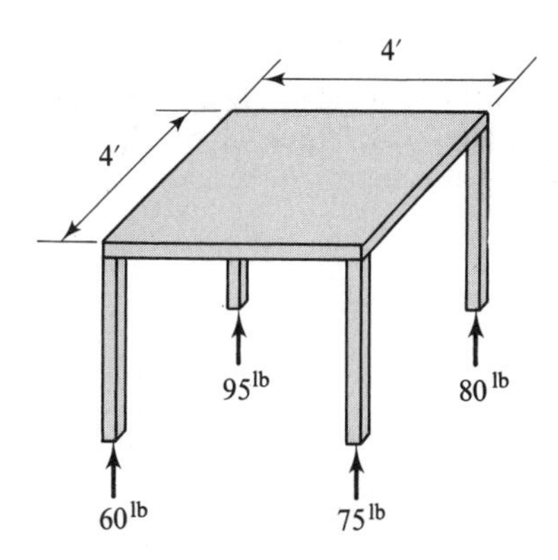

Figure P2-35

***2-36** The 30-N force in Fig. P2-36 is the resultant of three forces, two of which are shown. Determine the third force and locate it in a sketch.

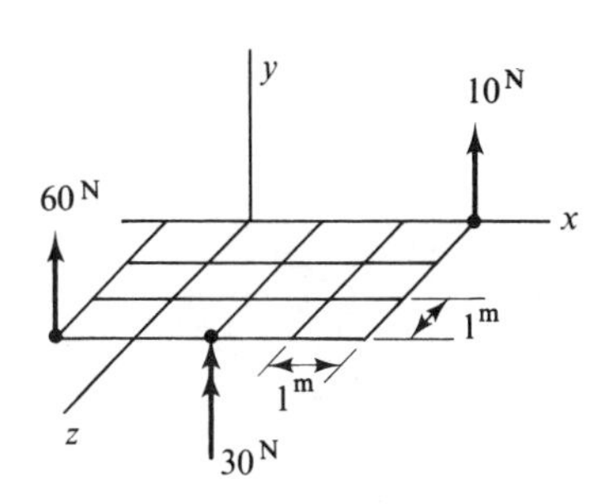

Figure P2-36

2-37 The 60-lb force in Fig. P2-37 is the resultant of three forces, two of which are shown. Determine the unknown force and locate it on the coordinate system.

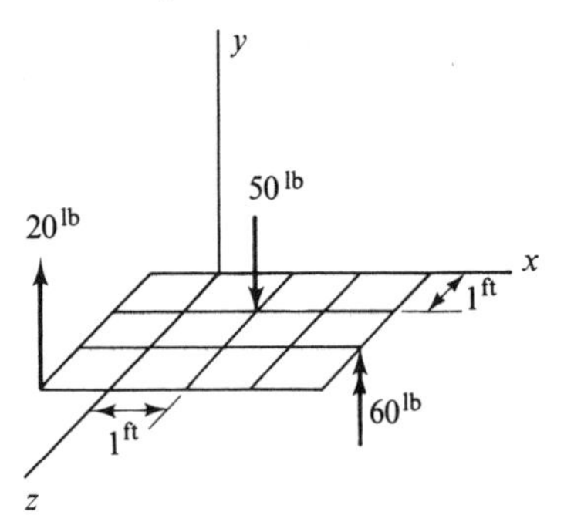

Figure P2-37

2-38 The 80-lb force of Fig. P2-38 is the resultant of four forces, three of which are shown. Determine the fourth force and show it on a sketch.

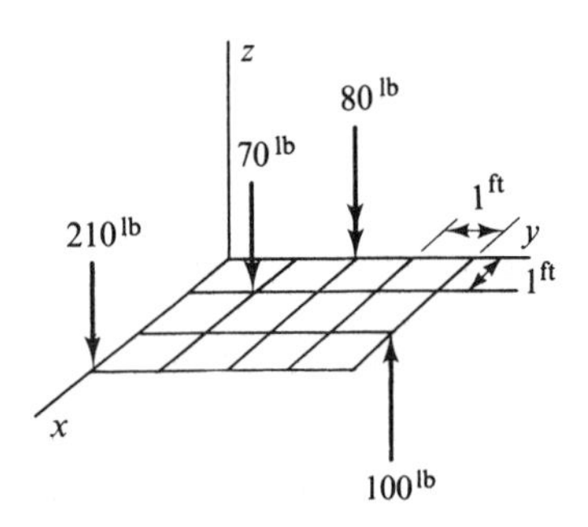

Figure P2-38

2-39 Determine the resultant of the force system shown in Fig. P2-39.

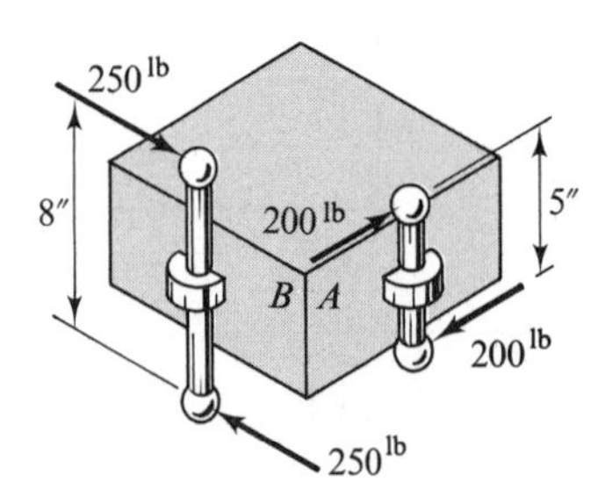

Figure P2-39

***2-40** Three control shafts are connected by the gear box in Fig. P2-40. The shafts are perpendicular to the faces of the box, and each is acted on by a couple as shown. Determine the resultant of the three couples.

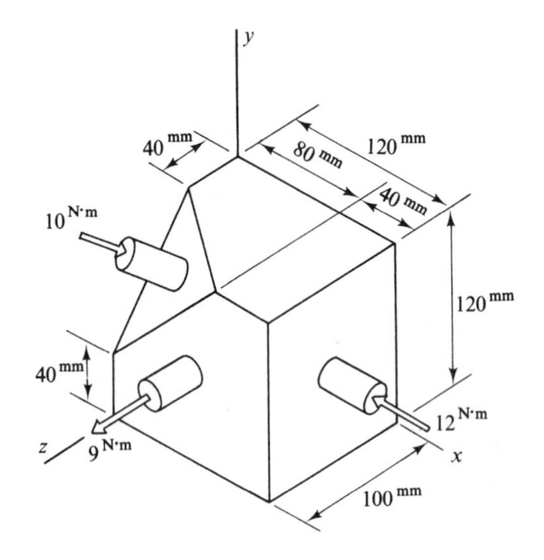

Figure P2-40

2-41 Determine the resultant of the three couples in Fig. P2-41.

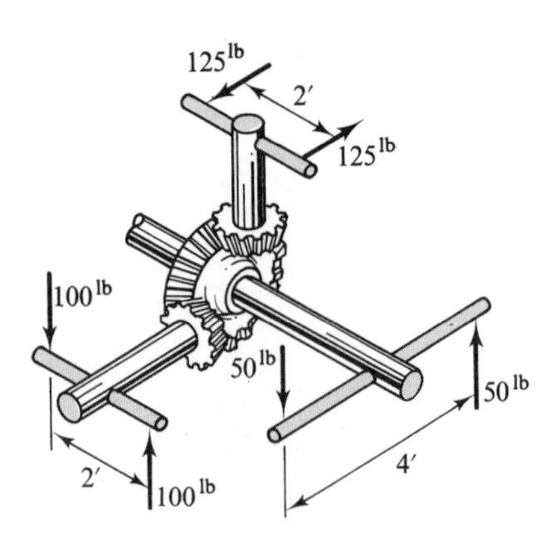

Figure P2-41

2-42 Determine the resultant of the force system in Fig. P2-42 and locate it with respect to point A.

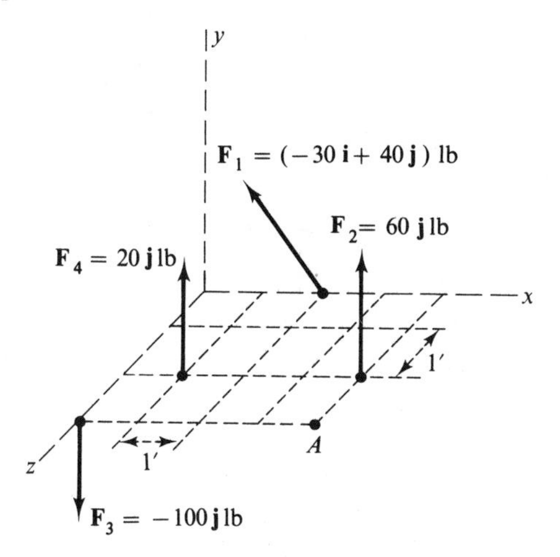

Figure P2-42

2-43 Determine the resultant of the two couples shown in Fig. P2-43.

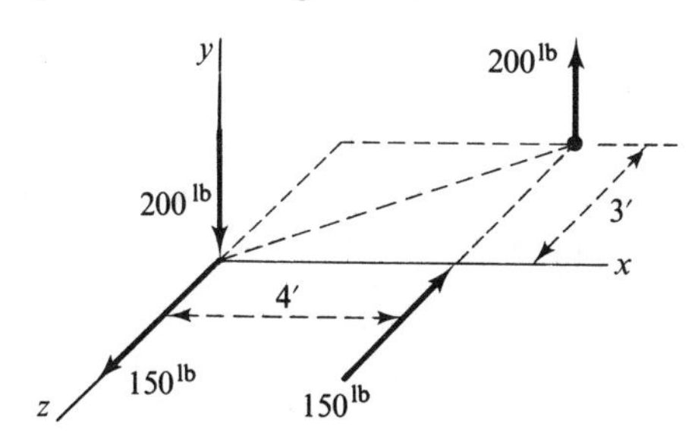

Figure P2-43

2-44 Determine the resultant of the force system shown in Fig. P2-44. The 50-lb force acts along edge EB.

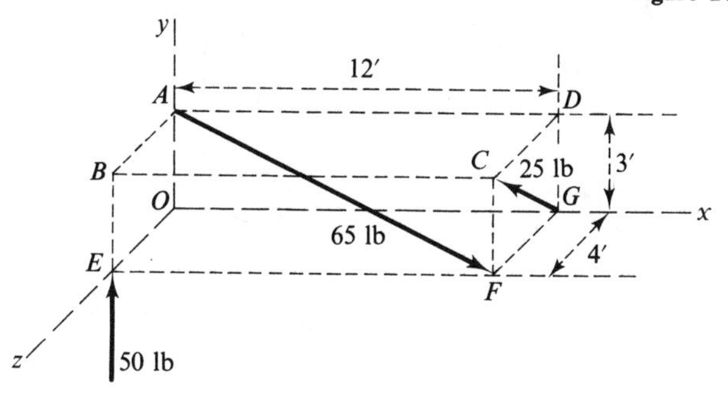

Figure P2-44

***2-45** The magnitudes of the forces in Fig. P2-45 are $F_1 = 20$ kN, $F_2 = 80$ kN, $F_3 = 57$ kN, and $F_4 = 36$ kN. Determine the resultant of the four forces.

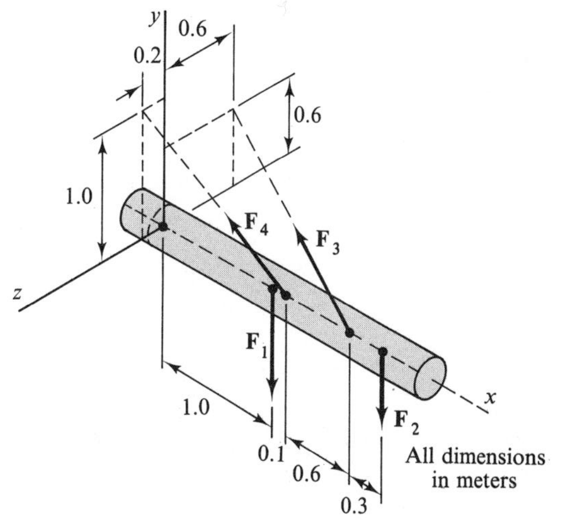

Figure P2-45

2-46 Determine the resultant of the force system shown in Fig. P2-46 and locate it with respect to the origin O. The 300-ft-lb couple is in the plane $ABDC$.

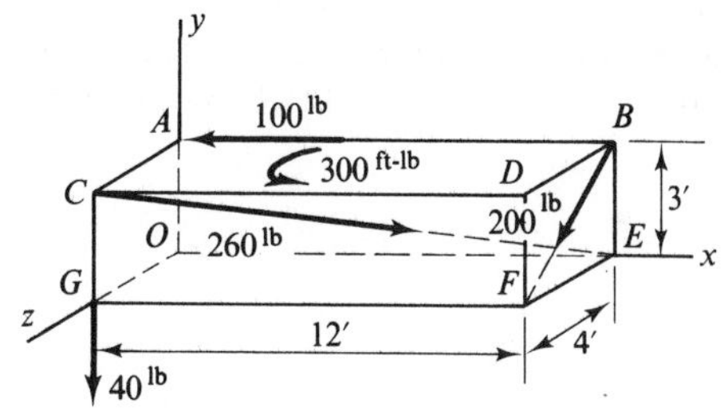

Figure P2-46

2-47 Determine the resultant of the force system in Fig. P2-47 and locate its point of intersection with the yz plane.

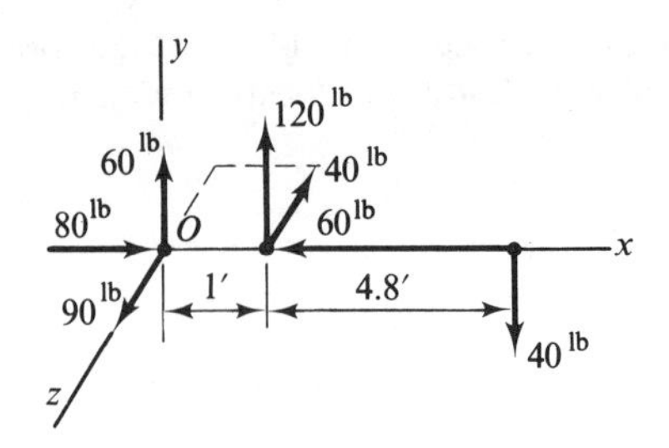

Figure P2-47

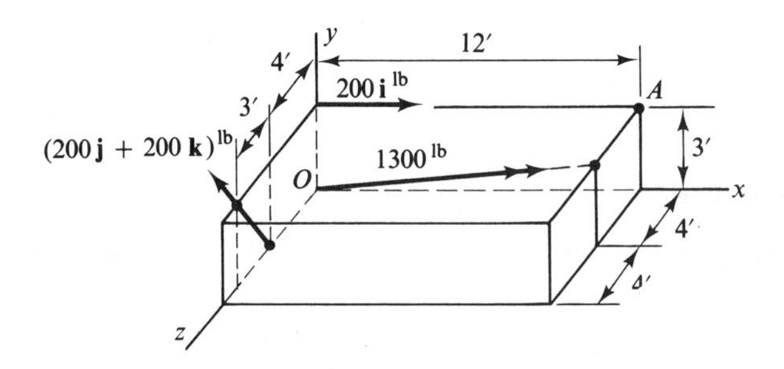

Figure P2-48

2-48 The 1300-lb force in Fig. P2-48 is the resultant of the two forces in the diagram and another force through point A and a couple neither of which is shown. Determine the unknown force and couple.

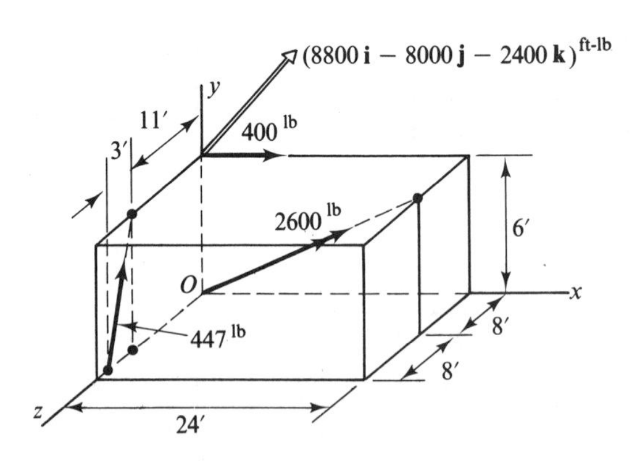

Figure P2-49

2-49 The 2600-lb force in Fig. P2-49 is the resultant of the couple and three forces, two of which are shown. Determine the missing force and locate its intersection with the xz plane.

2-5 CLOSURE

The possible resultants, other than zero, of force systems and the procedures for finding them can be summarized as follows.

The resultant of a concurrent force system is always a single force through the point of concurrence. Its component in any direction is equal to the sum of the components of the forces of the system in the given direction. No moment equations are needed to determine the resultant completely.

The resultant of a parallel force system, either coplanar or noncoplanar, and of a nonconcurrent, coplanar force system is either a force or a couple. It is a force when the vector sum of the forces is different from zero and either a couple or zero when the vector sum of the forces is equal to zero. A resultant force can be located by applying the principle of moments with respect to a point or to one or more axes. When the resultant is a couple, its moment can be determined by summing moments with respect to any point.

The resultant of any system of couples is a couple, and it can be determined as the vector sum of the moments of the couples.

The resultant of a general force system in space is a force and a couple. The resultant force through the origin is given by the vector sum of the forces, and the corresponding couple is given by the vector sum of the moments of the forces about the origin. Thus, two vector equations (which are equivalent to six scalar equations) are required to determine the resultant.

3 Centroids and Centers of Gravity

3-1 INTRODUCTION

The force of attraction of the earth for a particle is called the *weight* of the particle. A body consists of a number of particles, each of which has a weight, or force of attraction, directed toward the center of the earth. The resultant of this parallel[1] system of gravitational forces in space is the weight of the body. The position of the action line of the resultant weight when a body is in any position can be determined by means of the principle of moments for parallel forces in space as indicated in Art. 2-4. If the body is turned in space, the weights of the various particles do not retain the same relative positions, and therefore the resultant weight will be in a different position relative to the particles of the body.

The resultant weight does, however, pass through one point in the body, or the body extended, for all orientations of the body, and this point is defined as the center of gravity or center of mass[2] of the body.

[1] For ordinary bodies, the gravitational forces are essentially parallel.

[2] There is a negligible difference between the locations of the centers of gravity and mass of a body because the weight of a particle varies as its distance from the center of the earth changes.

Since the center of gravity of a body is a point, three coordinates are necessary to indicate its position completely. Two of these coordinates can be determined as the intersection of the line of action of the resultant of the parallel forces with a coordinate plane in the body which is not parallel to the forces. The third coordinate can be obtained from the position of the resultant weight for a different orientation of the body in space, as illustrated in Example 3-1.

Any symmetry possessed by a homogeneous body simplifies the problem of locating its center of gravity or center of mass. If the body has a plane of symmetry, the center of gravity lies in that plane. Similarly, the center of gravity of a homogeneous body is located on any line of symmetry or at a point of symmetry.

3-2 THE CENTER OF GRAVITY OF A SYSTEM OF PARTICLES

The procedure for determining the coordinates of the center of gravity of a system of particles is illustrated in the following example:

EXAMPLE 3-1

Locate the center of gravity of four small bodies (considered to be particles) arranged as shown in Fig. 3-1(a).

SOLUTION

The resultant force (weight) is the sum of the forces in Fig. 3-1(a), which is

$$\mathbf{R} = -2.00\mathbf{j} - 1.00\mathbf{j} - 0.50\mathbf{j} - 1.50\mathbf{j}$$
$$= -5.00\mathbf{j}\ \text{lb}.$$

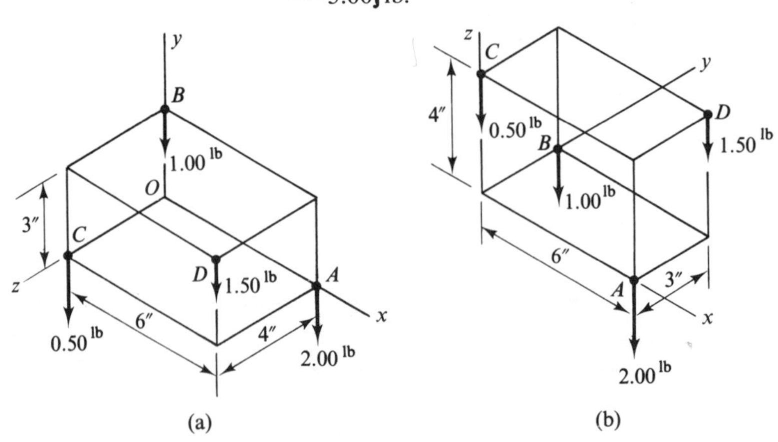

Figure 3-1

The moment of the resultant about the origin is

$$\begin{aligned}\mathbf{M}_O &= \sum \mathbf{r}\times\mathbf{F} = 6\mathbf{i}\times(-2.00\mathbf{j}) + 3\mathbf{j}\times(-1.00\mathbf{j}) \\ &\quad + 4\mathbf{k}\times(-0.50\mathbf{j}) + (6\mathbf{i}+3\mathbf{j}+4\mathbf{k})\times(-1.50\mathbf{j}) \\ &= -12.00\mathbf{k} + 2.00\mathbf{i} - 9.00\mathbf{k} + 6.00\mathbf{i} \\ &= (8.00\mathbf{i} - 21.00\mathbf{k}) \text{ in-lb.}\end{aligned}$$

The position $\mathbf{r}$ of the resultant $\mathbf{R}$ (in the xz plane) can be determined by applying the principle of moments. Thus

$$\begin{aligned}\mathbf{M}_O = \mathbf{r}\times\mathbf{R} &= (x\mathbf{i}+y\mathbf{j}+z\mathbf{k})\times(-5.00\mathbf{j}) \\ &= 5.00z\mathbf{i} - 5.00x\mathbf{k}.\end{aligned}$$

When these two values of $\mathbf{M}_O$ are equated they become

$$8.00\mathbf{i} - 21.00\mathbf{k} = 5.00z\mathbf{i} - 5.00x\mathbf{k},$$

and when the components of the two vectors are equated the results become

$$8.00 = 5.00z \quad \text{or} \quad z = 1.60 \text{ in.}$$

and

$$-21.00 = -5.00x \quad \text{or} \quad x = 4.20 \text{ in.}$$

The y coordinate of the center of gravity of the bodies can be obtained by rotating the system of bodies 90° as shown in Fig. 3-1(b) and again applying the principle of moments. Thus

$$\mathbf{R} = -5.00\mathbf{k} \text{ lb}$$

and

$$\begin{aligned}\mathbf{M}_O &= 6\mathbf{i}\times(-2.00\mathbf{k}) + 3\mathbf{j}\times(-1.00\mathbf{k}) + 4\mathbf{k}\times(-0.50\mathbf{k}) \\ &\quad + (6\mathbf{i}+3\mathbf{j}+4\mathbf{k})\times(-1.50\mathbf{k}) \\ &= -7.5\mathbf{i} + 21.0\mathbf{j}.\end{aligned}$$

Also

$$\begin{aligned}\mathbf{M}_O = \mathbf{r}\times\mathbf{R} &= (x\mathbf{i}+y\mathbf{j}+z\mathbf{k})\times(-5.00\mathbf{k}) \\ &= -5.00y\mathbf{i} + 5.00x\mathbf{j}.\end{aligned}$$

When these two expressions for $\mathbf{M}_O$ are equated they give $y = 1.50$ in. (and $x = 4.20$ in., which agrees with the previous calculation).

The center of gravity is located at

$$\underline{\mathbf{r} = (4.20\mathbf{i} + 1.50\mathbf{j} + 1.60\mathbf{k}) \text{ in.}} \qquad \text{Ans.}$$

The center of gravity of the system of bodies is shown at G in Fig. 3-2, located with respect to the coordinate axes. Such a sketch will frequently make it possible to detect gross errors.

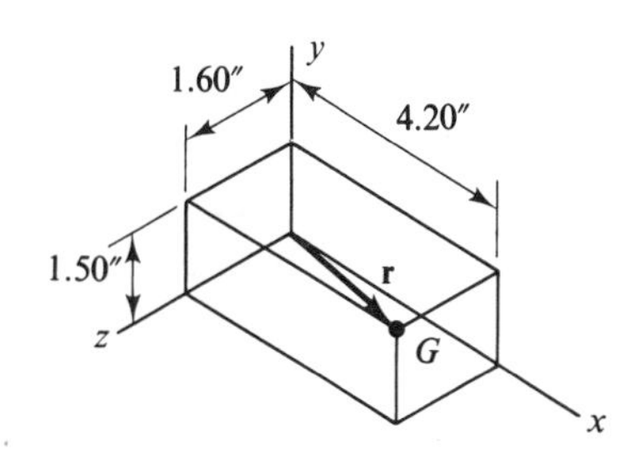

Figure 3-2

PROBLEMS

3-1 Determine the coordinates of the center of gravity of a coplanar system of particles which have weights of 4 lb, 7 lb, and 5 lb and are located at (3, 4), (4, −5), and (−3, 2), respectively. The coordinates are given in feet.

***3-2** Determine the coordinates of the mass center of the coplanar system of particles which have masses of 3 kg, 2 kg, 4 kg, and 6 kg and are located at (3, 5), (−2, 2), (5, 0), and (−5, −2), respectively. The coordinates are given in meters.

In the next two problems determine the coordinates of the center of gravity of the system of particles with weights and coordinates (positions) as listed.

	Particle	*Weight* (*lb*)	*x, y, z, Coordinates of Position of Particle* (*ft*)
3-3.	A	8	−2, −8, −3
	B	4	3, −5, −4
	C	5	6, 2, −1
3-4.	A	2	−1, 2, 7
	B	5	2, 5, −3
	C	3	6, −2, −5
	D	4	−3, −2, 6

3-5 Show that the center of gravity of three equal particles located at (x_1, y_1), (x_2, y_2), and (x_3, y_3) lies at the intersection of the medians of the triangle having the three points as vertices.

3-3 THE CENTER OF GRAVITY OF A BODY

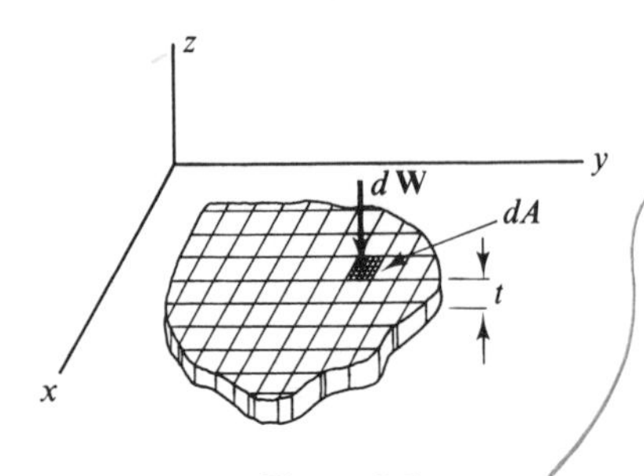

Figure 3-3

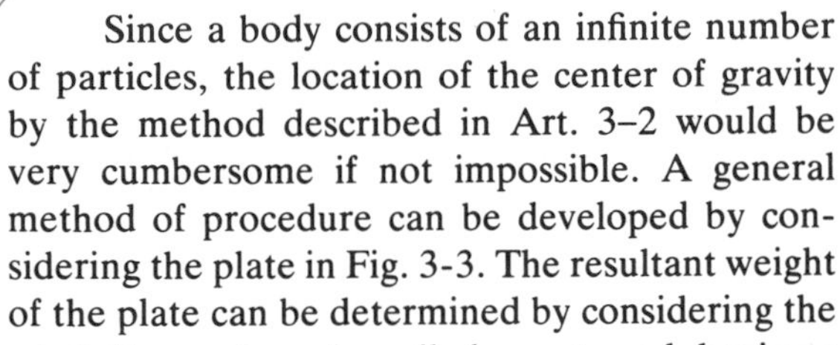

Since a body consists of an infinite number of particles, the location of the center of gravity by the method described in Art. 3–2 would be very cumbersome if not impossible. A general method of procedure can be developed by considering the plate in Fig. 3-3. The resultant weight of the plate can be determined by considering the plate to be made of an infinite number of small elements each having a weight given by the expression

$$d\mathbf{W} = -\mathbf{k}\gamma t\, dA,$$

where $d\mathbf{W}$ is the weight of an element, γ is the specific weight of the material (weight per unit volume), t is the thickness of the plate, and dA is the area of the element (surface area in the xy plane). The total weight of the plate is

$$\mathbf{W} = -\mathbf{k}W = -\mathbf{k}\int \gamma t\, dA,$$

where the integral is evaluated over the area of the plate.

The coordinates of a point on the line of action of the resultant weight can be determined by the principle of moments. The moment of the weight of the element about the origin is

$$\begin{aligned} d\mathbf{M}_O &= \mathbf{r} \times d\mathbf{W} \\ &= (x\mathbf{i} + y\mathbf{j}) \times (-\mathbf{k}\gamma t\, dA) \\ &= -y\gamma t\, dA\mathbf{i} + x\gamma t\, dA\mathbf{j}, \end{aligned}$$

and the resultant moment about the origin is

$$\mathbf{M}_O = -\mathbf{i}\int y\gamma t\,dA + \mathbf{j}\int x\gamma t\,dA. \qquad \text{(a)}$$

The intersection of the resultant weight with the xy plane is obtained by applying the principle of moments. That is,

$$\begin{aligned}\mathbf{M}_O &= \mathbf{r}_G \times \mathbf{W}\\ &= (x_G\mathbf{i} + y_G\mathbf{j}) \times (-W\mathbf{k})\\ &= -y_G W\mathbf{i} + x_G W\mathbf{j}. \qquad \text{(b)}\end{aligned}$$

Equations (a) and (b) will be equal only if the corresponding components are equal. Therefore

$$y_G W = \int y\gamma t\,dA \quad \text{or} \quad y_G = \frac{\int y\gamma t\,dA}{\int \gamma t\,dA}$$

$$x_G W = \int x\gamma t\,dA \quad \text{or} \quad x_G = \frac{\int x\gamma t\,dA}{\int \gamma t\,dA}. \qquad \text{(3-1a)}$$

The center of gravity of the body is on a line parallel to the z axis through the point (x_G, y_G). When t and γ are constant, z_G can be obtained by symmetry. If either t or γ is a variable, the plate can be rotated so that either the x axis or the y axis is vertical. In this case, the principle of moments gives

$$z_G = \frac{\int z\gamma\,dV}{\int \gamma\,dV} \qquad \text{(3-1b)}$$

where dV is equal to $dA\,dt$.

3-4 CENTROIDS

If the plate in Fig. 3-3 has a constant thickness and is homogeneous (has a constant specific weight), the product $t\gamma$ in Eq. (3-1a) can be taken outside the integral signs and eliminated from the equations, leaving

$$y_C = \frac{\int y\,dA}{\int dA}, \qquad x_C = \frac{\int x\,dA}{\int dA}. \qquad \text{(3-2)}$$

The terms x_G and y_G have been used to designate coordinates of the center of gravity (the line of action of the resultant weight passes through the center of gravity) of a body or series of bodies, and the terms x_C and y_C are used to indicate the centroidal coordinates of a geometrical shape, area, volume, etc. This distinction is pointed out in more detail later in this article. In each of these equations, the denominator is the surface area of the plate in the xy plane, and the numerator is called the first moment[3] of the area with respect to the x or the y axis.

The point located by the coordinates x_C and y_C is defined as the centroid of the area of the top surface of the plate. Since a plane area has no thickness, the centroid must lie in the plane of the area, and two coordinates are sufficient to locate it.

The first moment of an area about an axis is the product of the area and the coordinate of the centroid measured from that axis. Consequently, the moment of an area has the dimensions of L^2 times L, or L^3, and has units of in.^{3}, ft^3, m^3, and so on.

The moment of a force about a point is a vector quantity which is obtained as the vector product of the position vector and the force. The moment of an area about a point is also a vector which is equal to the product of the scalar area multiplied by the vector position of the centroid of the area. The moment (vector) can be resolved into components, and it is often convenient to determine the components directly. The numerators in Eq. (3-2) are the y and x components, respectively, of the moment of the area about (with respect to) the origin. Thus the moment about the x axis is the y component of the moment about the origin, and the moment about the y axis is the x component of the resultant moment.

If an area has an axis of symmetry, the centroid is on that axis, and if an area has two axes of symmetry, the centroid of the area is the point of intersection of the two axes.

The principle of moments can be used in a similar manner to locate the centroids of other geometrical figures, such as lines, volumes, and areas, which do not lie in a plane and also to locate the center of mass of a body.

The centroid of any geometrical figure (line, area, or volume) is a point in the figure, or the figure extended, located in such a manner that the first moment (vector) of the figure about any reference point is the product of the length, area, or volume multiplied by the position vector from the reference point to the centroid. The first moment of any figure with respect to its centroid is zero since the position vector is zero.

It was pointed out in Art. 3-1 that for all practical purposes the center of gravity and center of mass are identical. The equations for obtaining the center of mass of a system of n particles can be

[3] The term *first moment* is used to distinguish this quantity from the second moment, or moment of inertia, which is discussed in Chapter 6.

conveniently expressed in vector form by

$$\mathbf{r}_G \sum_{i=1}^{n} m_i = \sum_{i=1}^{n} \mathbf{r}_i m_i, \tag{3-3a}$$

where $\mathbf{r}_G$ is the position of the mass center and $\mathbf{r}_i$ is the position of the ith particle of mass m_i.

If the particles form a continuous body, the summations can be replaced by integrations to give

$$\mathbf{r}_G \int dm = \int \mathbf{r}\, dm = \int \mathbf{r}\rho\, dV, \tag{3-3b}$$

where ρ is the density (mass per unit volume) of the element and dV is its volume.

Equations (3-3a) and (3-3b) can be resolved into components by replacing $\mathbf{r}_G$ by $x_G\mathbf{i}+y_G\mathbf{j}+z_G\mathbf{k}$ and $\mathbf{r}$ by $x\mathbf{i}+y\mathbf{j}+z\mathbf{k}$. Thus each component of the centroid can be calculated independently. Since x_G and x represent the distance from the yz plane, the expression $\int x\, dm$ is the moment of the mass with respect to the yz plane, and similar statements apply to $\int y\, dm$ and $\int z\, dm$.

The center of gravity of a body and the centroid of the corresponding volume will coincide when the material of the body is homogeneous. If the specific weight is variable, the center of gravity of the body and the centroid of the volume will usually be different. For example, in a sphere composed of wooden and steel hemispheres as indicated in Fig. 3-4, the centroid of the volume is at C, whereas the center of gravity is at G, some distance to the right of C. The term *centroid* is used here in connection with geometrical figures, whereas *mass center* and *center of gravity* refer to physical bodies.

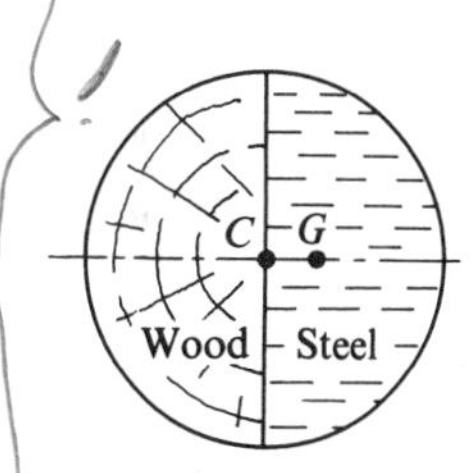

Figure 3-4

3-5 CENTROIDS AND CENTERS OF GRAVITY BY INTEGRATION

Identical principles are involved in the determination, by integration, of the coordinates of the centroid of a line, area, or volume of a geometrical figure or of the center of mass or center of gravity of a body. For this reason, the following discussion will be limited principally to areas, and the areas will be assumed to lie in the xy plane. Sometimes it is possible to determine one or both of the coordinates of the centroid from symmetry, as, for example, the coordinates of the centroid of a rectangular or a circular area.

It was pointed out in Art. 3-4 that the moment of an area with respect to any reference point is a vector. The determination of this vector, however, is usually accomplished by calculating its scalar components, and the following discussion will be concerned with these components. If an area lies in the xy plane, its moment about the origin is

$$\mathbf{M}_O = \int \mathbf{r}\, dA$$

$$= \int (x\mathbf{i} + y\mathbf{j})\, dA$$

$$= \mathbf{i}\int x\, dA + \mathbf{j}\int y\, dA.$$

The quantity $\int x\, dA$ is called the moment of the area with respect to the y axis since x is the distance from the y axis to the element. Similarly, $\int y\, dA$ is the moment of the area about the x axis.

The location of the centroid of an area by integration involves the selection and use of an element of that area. The element can usually be selected in such a manner that only one integration is necessary. Sometimes, however, it may be necessary to use double integration or perhaps triple integration for volumes. The element must be selected in such a manner that the distance from its centroid to the moment axis or plane is known. These two rules may help in selecting an appropriate element:

1. When the element is chosen in such a manner that all parts of it are the same distance from the reference axis or plane this common distance is obviously the centroidal distance.
2. When the parts of the element selected are at different distances from the reference axis or plane the location of the centroid of the element must either be known or be readily obtainable by symmetry.

In the analysis of any problem, a logical, orderly procedure helps to develop both a thorough understanding of the subject matter and efficiency in solving problems. The following steps are suggested as outlining such a procedure for determining the coordinates of the centroid of a line, area, or volume of a geometrical figure or of the center of mass or center of gravity of a body. The steps listed apply to all cases even though area is referred to in the following discussion.

1. On a sketch of the figure draw the selected element and dimension it completely as shown in the following illustrative examples. When using double integration, sketch the element and then extend it, with dashed lines, to indicate the result of the first integration as indicated in Fig. 3-7.
2. Write an expression for the area of the element and simplify the expression as much as possible.
3. Integrate the expression from step 2 to determine the area. If double integration is used, particular care in setting up the limits should be employed.

4. Write an expression for the moment of the element of area with respect to any desired reference point, axis, or plane and simplify the expression as much as possible.

5. Integrate the expression from step 4 to determine the moment of the area.

6. Divide the moment from step 5 by the area from step 3 to obtain the position of the centroid with respect to the reference point, axis, or plane.

If the centroid is shown on a sketch approximately to scale, gross errors will often be detected.

In some of the following problems equations for the boundaries of areas and solids are given by equations and contain statements such as "$3y = 4x^2$, where x and y are in feet." For this type of equation to be dimensionally homogeneous it should be understood that some of the coefficients of the variables x and y may have dimensions. Thus, the above equation is dimensionally homogeneous if the coefficient 3 is dimensionless and 4 has dimensions of feet^{-1}.

The following examples illustrate the two types of elements and the sequence of steps used to locate the centroid of a figure or the center of gravity or mass of a body.

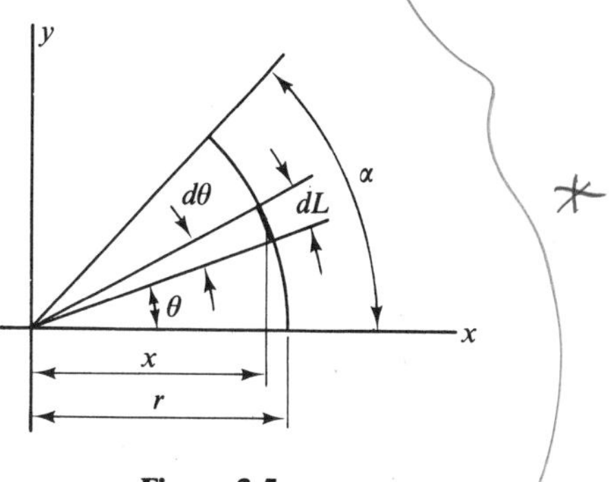

Figure 3-5

EXAMPLE 3-2

Locate the centroid of the circular arc (line) in Fig. 3-5.

SOLUTION

The element of length is shown as dL and is dimensioned on the sketch according to step 1. In this solution, the steps are indicated by numbers for emphasis.

The length of the element is

$$dL = r\,d\theta \qquad \text{(step 2)},$$

and the length of the arc is

$$L = \int_0^\alpha r\,d\theta = [r\theta]_0^\alpha = r\alpha \qquad \text{(step 3)}.$$

The angle α is expressed in radians, and the length L could have been determined directly from the definition of a radian. The moment of the element with respect to the origin is

$$d\mathbf{M}_O = (r\cos\theta\,\mathbf{i} + r\sin\theta\,\mathbf{j})r\,d\theta$$
$$= r^2(\mathbf{i}\cos\theta + \mathbf{j}\sin\theta)\,d\theta \qquad \text{(step 4)},$$

and the moment of the arc is

$$\mathbf{M}_O = r^2 \int_0^\alpha (\mathbf{i} \cos\theta + \mathbf{j} \sin\theta)\, d\theta$$

$$= r^2[\mathbf{i} \sin\theta - \mathbf{j} \cos\theta]_0^\alpha$$

$$= r^2[\mathbf{i} \sin\alpha - \mathbf{j}(\cos\alpha - 1)] \qquad \text{(step 5).}$$

The position of the centroid is determined by dividing the moment of the element by the length of the element, thus:

$$\mathbf{r}_c = \frac{\mathbf{M}_O}{L} = \frac{r^2}{r\alpha}[\mathbf{i} \sin\alpha + \mathbf{j}(1 - \cos\alpha)] \qquad \text{(step 6)}$$

$$= \frac{r}{\alpha}[\mathbf{i} \sin\alpha + \mathbf{j}(1 - \cos\alpha)]. \qquad \text{Ans.}$$

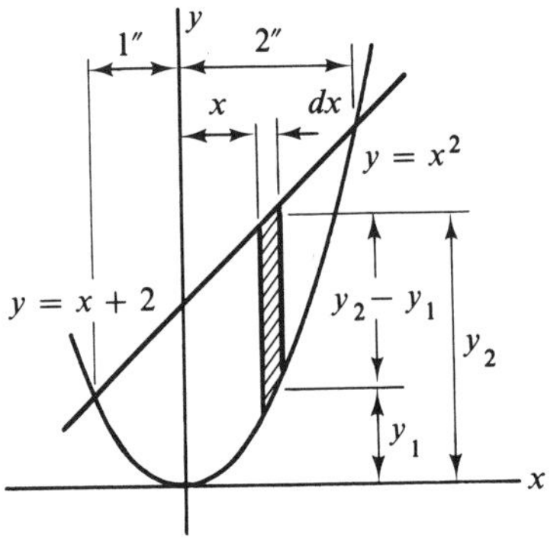

Figure 3-6

EXAMPLE 3-3

Locate the centroid of the area in Fig. 3-6 bounded by the curve $y = x^2$ and the line $y = x + 2$. In the equations, x and y are in inches.

SOLUTION

A vertical element of area is selected because the same element can be used for the entire area. If a horizontal element were chosen, it would be necessary to use two different elements. Note that the element is completely dimensioned. The area of the element is

$$dA = (y_2 - y_1)\, dx = (x + 2 - x^2)\, dx,$$

and the total area is

$$A = \int_{-1}^{2} (x + 2 - x^2)\, dx = \left[\frac{x^2}{2} + 2x - \frac{x^3}{3}\right]_{-1}^{2} = 4.50 \text{ in.}^2$$

The element is a rectangle, and its centroid is located at

$$\mathbf{r} = x\mathbf{i} + \left(y_1 + \frac{y_2 - y_1}{2}\right)\mathbf{j} = x\mathbf{i} + \left(\frac{y_1 + y_2}{2}\right)\mathbf{j}.$$

The moment of the element about the origin is

$$d\mathbf{M}_O = \mathbf{r}\, dA = \left(x\mathbf{i} + \frac{y_1 + y_2}{2}\mathbf{j}\right)(y_2 - y_1)\, dx,$$

which, when y_1 and y_2 are expressed in terms of x becomes

$$d\mathbf{M}_O = [\mathbf{i}(x^2 + 2x - x^3) + \mathbf{j}(\tfrac{1}{2})(x^2 + 4x + 4 - x^4)]\, dx.$$

The moment of the area about the origin is

$$\mathbf{M}_O = \int d\mathbf{M}_O = \int_{-1}^{2} \left[\mathbf{i}(x^2 + 2x - x^3) + \mathbf{j}\frac{x^2 + 4x + 4 - x^4}{2}\right] dx$$

$$= \mathbf{i}\left[\frac{x^3}{3} + \frac{2x^2}{2} - \frac{x^4}{4}\right]_{-1}^{2} + \frac{\mathbf{j}}{2}\left[\frac{x^3}{3} + \frac{4x^2}{2} + 4x - \frac{x^5}{5}\right]_{-1}^{2}$$

$$= (2.25\mathbf{i} + 7.20\mathbf{j}) \text{ in.}^3.$$

The position of the centroid is

$$\mathbf{r}_C = \frac{\mathbf{M}_O}{A} = \frac{2.25\mathbf{i} + 7.20\mathbf{j}}{4.50} = \underline{(0.500\mathbf{i} + 1.600\mathbf{j})\text{ in.}} \qquad \text{Ans.}$$

When only one coordinate of the centroid of an area is needed the moment can be obtained about a reference axis in the plane, and the solution may be shortened as shown in the next example.

EXAMPLE 3-4

Locate the centroid of the semicircular area in Fig. 3-7.

SOLUTION

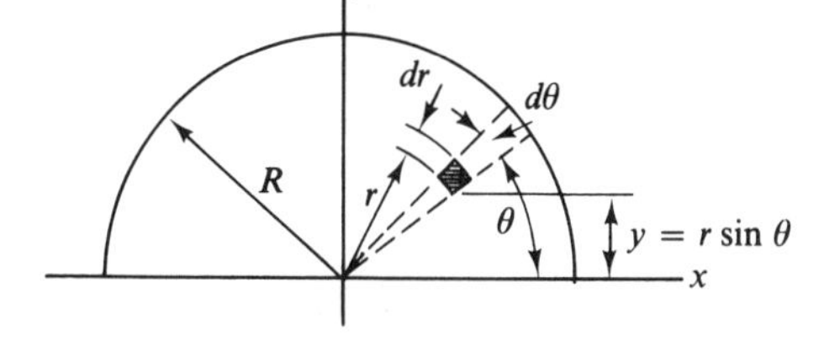

Figure 3-7

By symmetry, the centroid must be on the y axis; therefore, it is necessary only to determine y_C. The element of area is approximately a rectangle of length dr and width $r\,d\theta$, and the area of the element is

$$dA = r\,d\theta\,dr.$$

It is not necessary to integrate this expression for the area, since the area of a semicircle is

$$A = \frac{\pi R^2}{2}.$$

The centroidal distance from the x axis to the element is

$$y = r \sin\theta,$$

and the moment of the element with respect to the x axis is

$$dM_x = y\,dA = (r\sin\theta)(r\,d\theta\,dr).$$

The moment of the entire area about the x axis is

$$M_x = \int_0^{\pi}\int_0^{R} r^2 \sin\theta\,dr\,d\theta$$

$$= \int_0^{\pi}\left[\frac{r^3}{3}\right]_0^R \sin\theta\,d\theta$$

$$= \left[\frac{R^3}{3}(-\cos\theta)\right]_0^{\pi} = \frac{2R^3}{3}.$$

From the principle of moments, the y coordinate of the centroid is

$$y_C = \frac{M_x}{A} = \underline{\frac{4R}{3\pi}}. \qquad \text{Ans.}$$

This problem can be solved with a single integration by selecting the wedge-shaped element shown by the dashed lines. The area of the element is $R(R\,d\theta)/2$, and the centroid of the element is $\frac{2}{3}R\sin\theta$ from the x axis. The product of these quantities is the result obtained by the first integration above.

EXAMPLE 3-5

Determine the coordinates of the center of mass of the frustum of the homogeneous right circular cone shown in Fig. 3-8.

SOLUTION

By symmetry, x_G and y_G are both zero. To determine the z coordinate of the mass center, the moment of the mass of the frustum of the cone with respect to the xy plane is needed. A thin circular disk of radius x and thickness dz is selected as the element of mass because all parts of it are the same distance from the xy plane. The mass of this element is

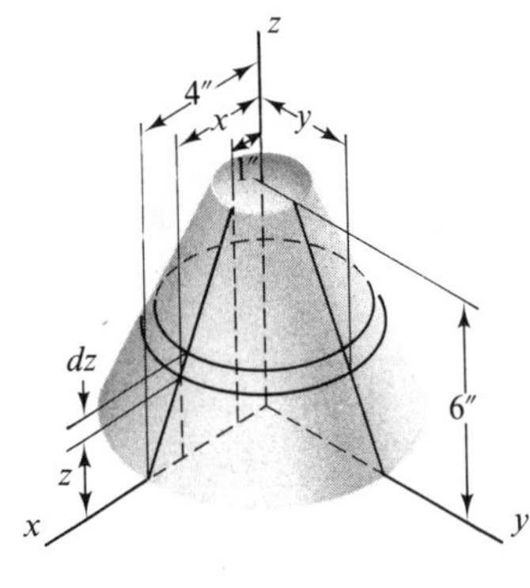

Figure 3-8

$$dm = \rho\, dV = \rho\pi x^2\, dz,$$

where ρ is the density (mass per unit volume) of the material of the body and dV is the volume of the element. The relationship between x and z may be obtained from similar triangles as

$$\frac{z}{4-x} = \frac{6}{4-1} \quad \text{from which} \quad x = 4 - \frac{z}{2}.$$

The mass of the element then becomes

$$dm = \rho\pi\left(16 - 4z + \frac{z^2}{4}\right) dz,$$

and the total mass of the body is

$$m = \rho\pi \int_0^6 \left(16 - 4z + \frac{z^2}{4}\right) dz = 42\rho\pi.$$

The moment of the mass of the element with respect to the xy plane is

$$dM_{xy} = z\, dm = \rho\pi\left(16z - 4z^2 + \frac{z^3}{4}\right) dz,$$

and the moment of the body is

$$M_{xy} = \rho\pi \int_0^6 \left(16z - 4z^2 + \frac{z^3}{4}\right) dz = 81\rho\pi.$$

From the principle of moments, the z coordinate of the mass center is

$$z_G = \frac{M_{xy}}{m} = \frac{81\rho\pi}{42\rho\pi} = \underline{1.929 \text{ in. above the } xy \text{ plane.}} \quad \text{Ans.}$$

If the material in the body had not been homogeneous, the density, ρ, would have been a variable and could not have been taken outside the integral sign. In this case, the problem could not be solved without first obtaining the law relating the density and the coordinates of any point in the body.

EXAMPLE 3-6

Locate the mass center of the body generated by revolving the shaded area of Fig. 3-9(a) about the y axis. The equation of the curve is $x^2 = y$, and the density of the material varies directly as the square of the distance from the y axis.

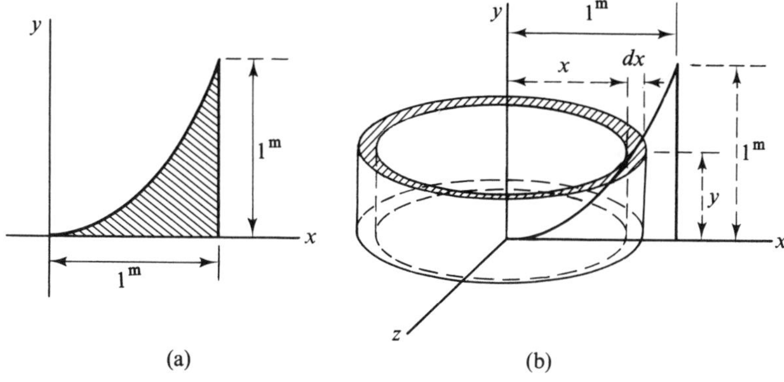

Figure 3-9

SOLUTION

By symmetry x_G and z_G are both zero. To determine y_G, the third coordinate of the mass center, the moment of the mass of the body with respect to the xz plane is needed. A differential element of mass in the shape of a cylindrical shell [Fig. 3-9(b)] is used for integration since the entire mass of the shell has the same density. The density of the material varies directly as the square of the distance from the axis of revolution; hence $\rho = kx^2$, where k is a constant and x is the radius of the element.

The area of the cross section of the shell [shaded in Fig. 3-9(b)] is

$$dA = 2\pi x\, dx,$$

and the volume of the element is

$$dV = y\, dA = 2\pi xy\, dx.$$

The mass of the element is

$$dm = \rho\, dV = kx^2(2\pi xy\, dx) = 2\pi kx^5\, dx$$

from which the mass of the body is

$$m = \int_0^1 2\pi kx^5\, dx = \left[2\pi k\frac{x^6}{6}\right]_0^1 = \frac{\pi k}{3}.$$

The moment of the element of mass about the xz plane is

$$dM_{xz} = \frac{y}{2}\, dm = \frac{x^2}{2}(2\pi kx^5\, dx) = \pi kx^7\, dx$$

and

$$M_{xz} = \int_0^1 \pi kx^7\, dx = \left[k\frac{x^8}{8}\right]_0^1 = \frac{\pi k}{8}.$$

The y coordinate of the mass center thus becomes

$$y_G = \frac{M_{xz}}{m} = \frac{\pi k}{8}\frac{3}{\pi k} = 0.375 \text{ m},$$

and the mass center is located at

$$\mathbf{r}_G = \underline{0.375\mathbf{j} \text{ m}}. \qquad \text{Ans.}$$

PROBLEMS

Note: The following problems are to be solved by integration.

3-6 Locate the center of gravity of a thin homogeneous triangular plate bounded by the y axis and the lines $y=10$ in. and $5x-3y=15$, where x and y are measured in inches.

3-7 Determine the y coordinate of the centroid of the area of the general triangle in Fig. P3-7.

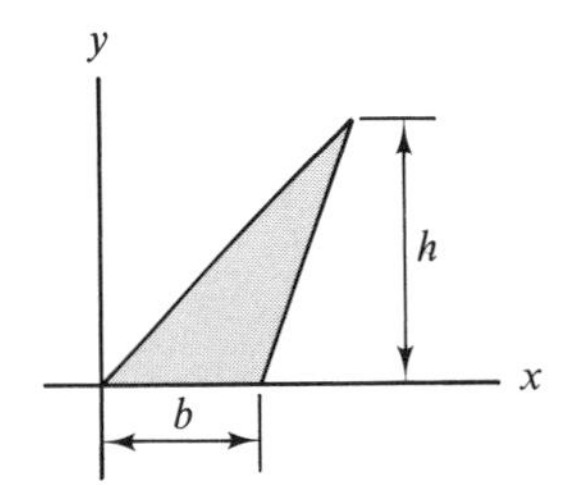

Figure P3-7

***3-8** Locate the centroid of the shaded area in Fig. P3-8. The equation of the curve is $y^3=2x$, where x and y are in centimeters.

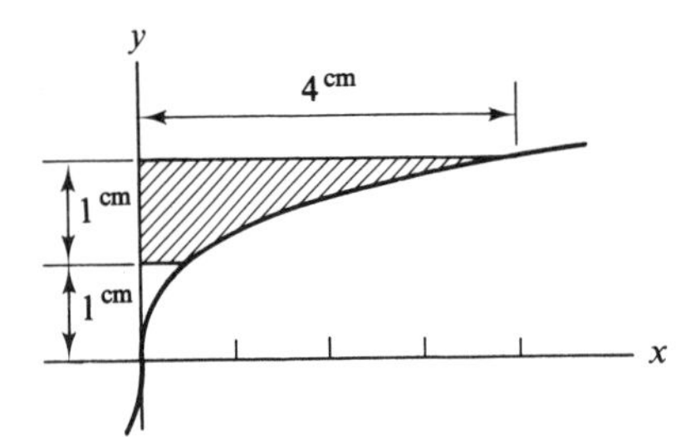

Figure P3-8

3-9 Determine the y coordinate of the centroid of the arc of the upper half of the hypocycloid in Fig. P3-9.

3-10 Determine the coordinates of the centroid of the arc of the parabola $y=x^2$ (x and y are measured in feet) from the origin to the point (1, 1). The length of the arc is $0.25[2\sqrt{5}+\ln(2+\sqrt{5})]$ ft.

3-11 Determine the first moment of the shaded area in Fig. P3-11 with respect to the a axis. The equation of the curve is $8y=x^3$.

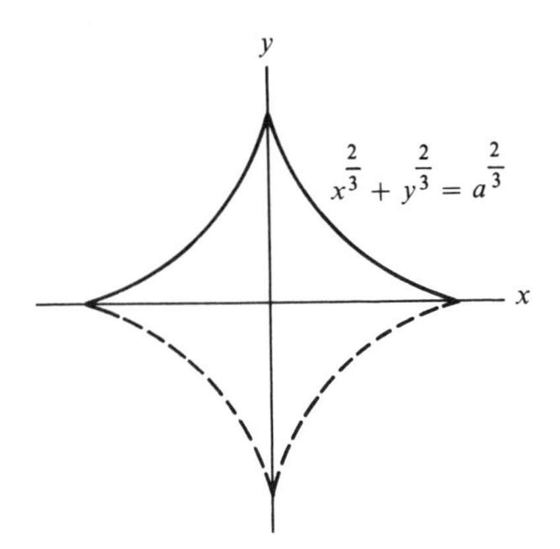

Figure P3-9

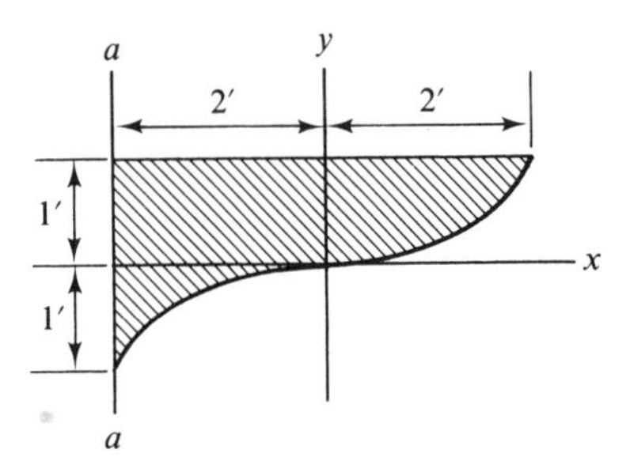

Figure P3-11

3-12 Determine the first moment of the shaded area in Fig. P3-12 with respect to the y axis. The equation of the curve is $y=x^2-2x$, with x and y in inches.

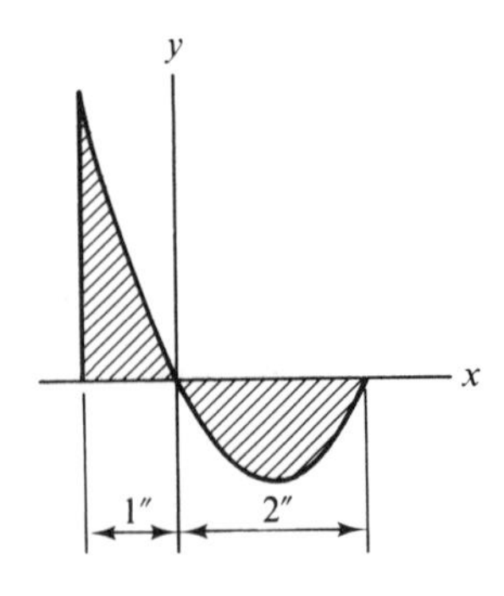

Figure P3-12

***3-13** Determine the first moment of the shaded area in Fig. P3-13 with respect to the line a. The equation of the curve is $y = x^2 - 2x$, where x and y are in centimeters.

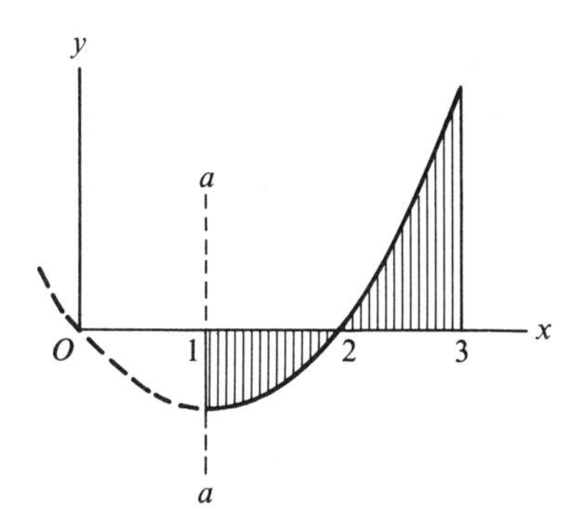

Figure P3-13

3-14 Locate the centroid of the area enclosed by the parabola $bx = 2by - y^2$ and the y axis.

3-15 The line $2y = 4a - x$ and the parabola $4ay = x^2$ are shown in Fig. P3-15. Locate the centroid of the shaded area I.

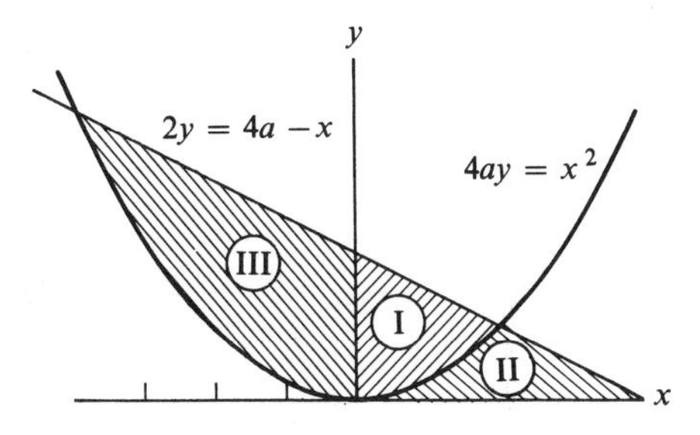

Figure P3-15

3-16 Locate the centroid of the shaded area II in Fig. P3-15.

3-17 Locate the centroid of the shaded area III in Fig. P3-15.

***3-18** Determine the x coordinate of the centroid of the area bounded by the curves $2y = x^2$ and $8y = x^3$, where x and y are measured in meters.

3-19 Determine the first moment of the shaded area in Fig. P3-19 with respect to the line $x = 3$ in. The equation of the curve is $xy = 6$.

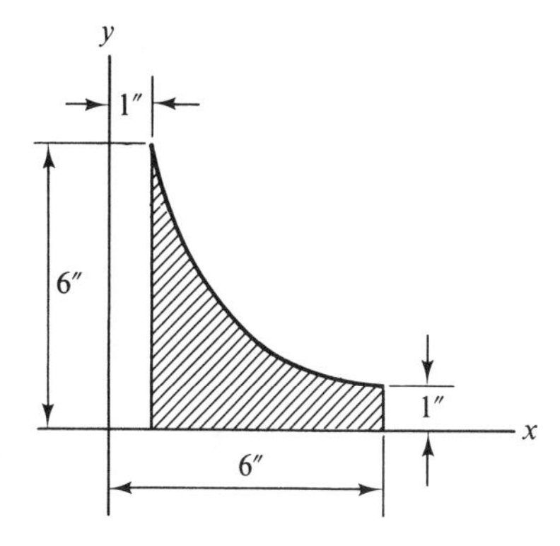

Figure P3-19

3-20 Locate the centroid of the shaded area in the first quadrant bounded by the curve $r = 10 \sin 2\theta$ (see Fig. P3-20).

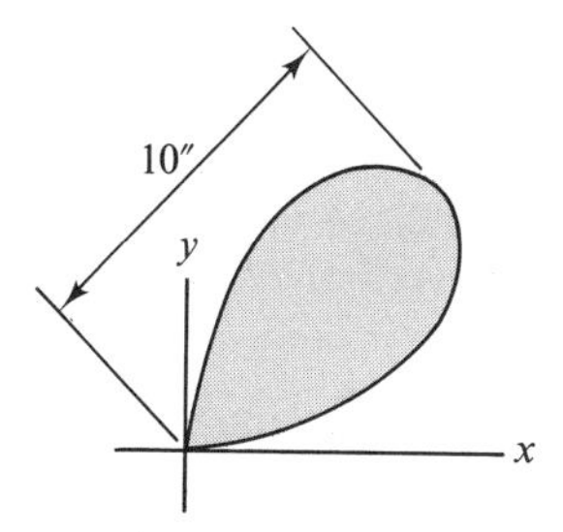

Figure P3-20

3-21 Determine the y coordinate of the centroid of the shaded area in Fig. P3-21. The equation of the curve is $r = \cos^2 \theta$ with r in inches.

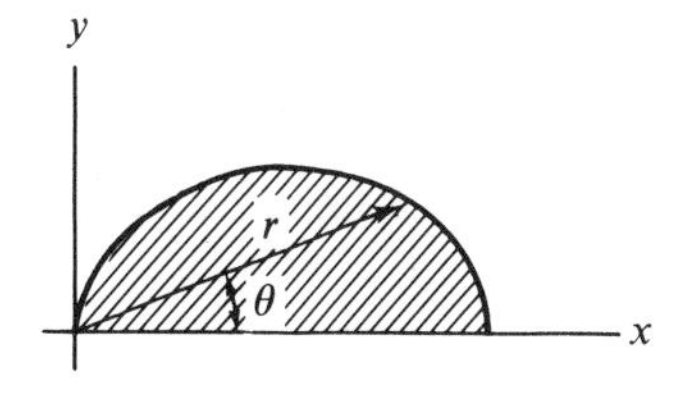

Figure P3-21

3-22 Locate the center of mass of a solid homogeneous hemispherical body of radius r.

3-23 The lower half of the area in Problem 3-14 is revolved about the line $y = b$. Locate the centroid of the resulting volume.

3-24 The homogeneous cone in Fig. P3-24 has a circular base in the xy plane and the element OA lies along the z axis. Locate the mass center of the cone.

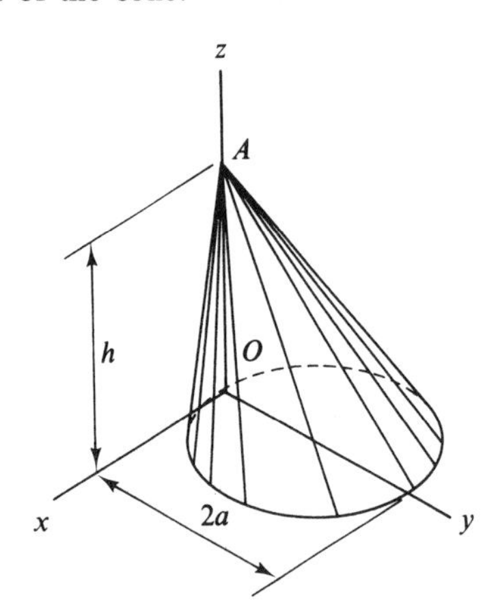

Figure P3-24

3-25 Solve Problem 3-24 if the density of the material in the oblique cone varies directly as the square of the vertical distance from the apex at A.

3-26 Locate the mass center of a homogeneous body generated by revolving the shaded area in Fig. P3-26 about the line $y = 1$ ft. The equation of the curve is $y^2 = x$.

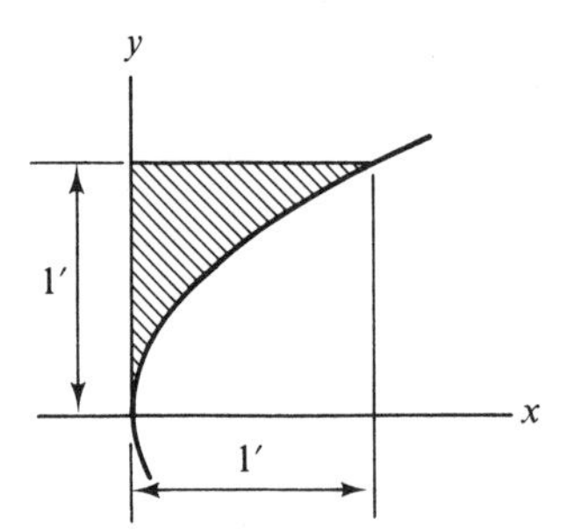

Figure P3-26

3-27 Solve Problem 3-26 if the area is revolved about the line $x = 1$ ft. All other data are unchanged.

3-28 Locate the mass center of a homogeneous body having the volume generated by revolving the area in Fig. P3-28 about the line $x = 5$ in.

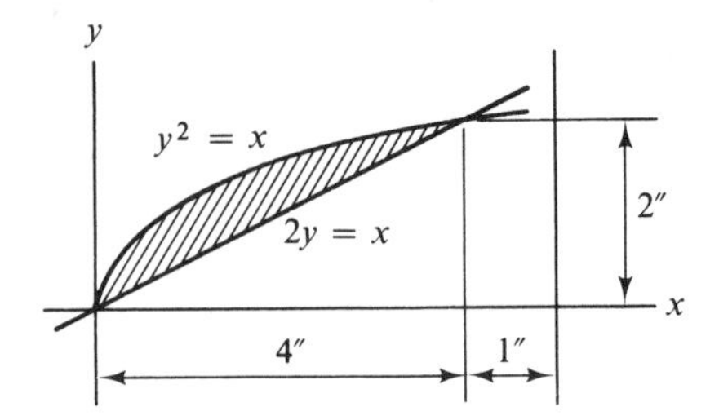

Figure P3-28

***3-29** The shaded area in Fig. P3-29 is rotated about the line $y = 10$ cm to form a homogeneous solid of revolution. Locate the mass center of the body.

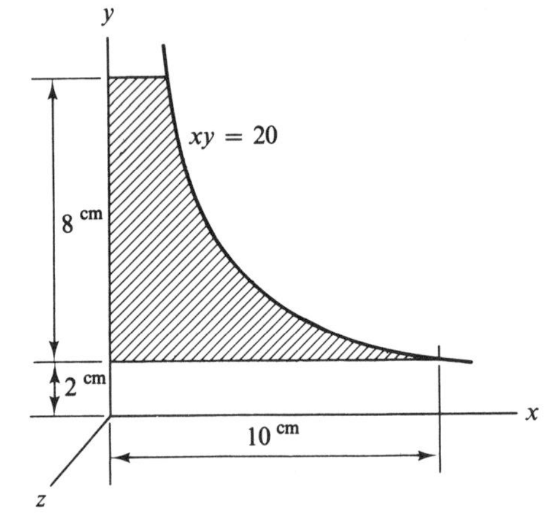

Figure P3-29

***3-30** Solve Problem 3-29 if the density of the material of the body is directly proportional to the distance from the axis of rotation.

***3-31** Solve Problem 3-29 if the density of the material in the body is directly proportional to the distance from the yz plane.

3-32 The shaded area in Fig. P3-32 is rotated around the x axis to generate a homogeneous solid of revolution. The equation of the curve is $y = a \sin \pi x/2b$. Locate the mass center of the body.

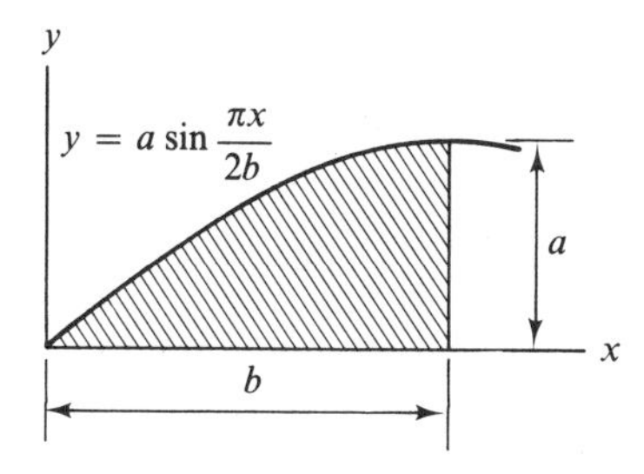

Figure P3-32

3-33 The area between the y axis, the line $y = a$, and the curve ($y = a \sin \pi x/2b$) in Fig. P3-32 is rotated around the y axis to generate a solid of revolution. Locate the centroid of the volume.

3-34 The specific weight of the material of a long slender rod of uniform cross section and of length L varies directly as the square of the distance from the left end. Locate the center of gravity of the rod.

3-6 CENTROIDS AND CENTERS OF GRAVITY OF COMPOSITE BODIES

The centroid or center of gravity of any figure or body can be obtained by means of the principle of moments if the area, weight, and so on and the moment of these quantities about any axis or plane can be determined. Frequently an area can be divided into two or more simple rectangles, triangles, circles, or other shapes whose areas and centroidal coordinates can be readily obtained. Structural shapes, such as I beams, channels, angles, and others, are frequently assembled to form more complicated members, such as the one in Fig. 3-10, which is made up of a plate and two channels. Handbooks list properties of such shapes, including the areas and location of the centroids. When an area can be divided into a number of simple areas, the resultant or total area is the sum of the separate areas (if some parts are removed, the corresponding areas must be subtracted), and the resultant moment about any point, axis, or plane is the algebraic sum of the moments of the component areas. The centroid of the area can be calculated from

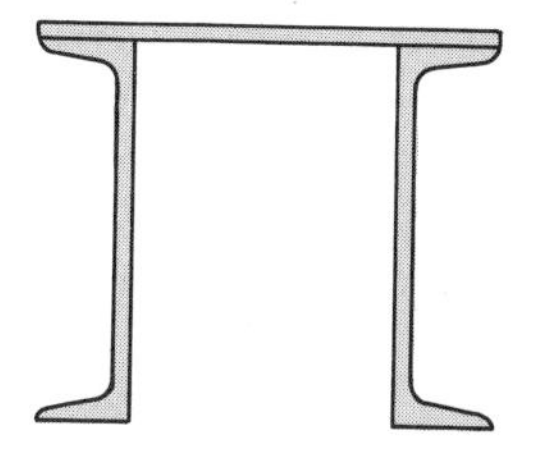

Figure 3-10

$$\mathbf{r}_C = \frac{\mathbf{M}_O}{A} = \frac{\sum \mathbf{r}A}{\sum A}$$

in which the summation is carried out for all component parts of the area. The equation can also be written in component form as

$$x_C = \frac{\sum xA}{\sum A}, \quad y_C = \frac{\sum yA}{\sum A}, \quad \text{and} \quad z_C = \frac{\sum zA}{\sum A}.$$

This method avoids the necessity for integration, provided the area and centroid of each part of the figure are known, and it also applies to lines, volumes, weights, and masses. In using this method, it is desirable to indicate clearly on a sketch the separate parts into which the figure is resolved. The following examples illustrate the method.

EXAMPLE 3-7

Locate the center of gravity of the homogeneous uniform wire *ABCD* shown in Fig. 3-11.

SOLUTION

The wire is made up of the three straight sections *AB*, *BC*, and *CD*. Since the wire is uniform and homogeneous, the center of gravity is the same as the centroid of the line. The length and coordinates of the centroid of each section are shown in Fig. 3-11. The total length is

$$L = 3 + 5 + 2 = 10 \text{ in.}$$

The first moment of the length of the wire with respect to the origin is

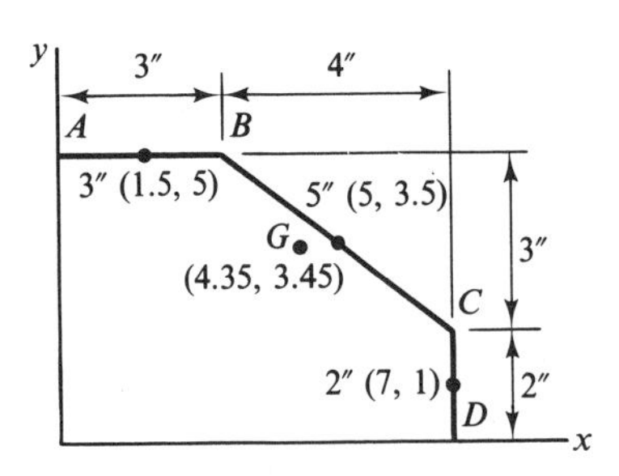

Figure 3-11

$$\begin{aligned}\mathbf{M}_O &= \sum L_i\mathbf{r}_i \\ &= 3(1.5\mathbf{i}+5\mathbf{j})+5(5\mathbf{i}+3.5\mathbf{j})+2(7\mathbf{i}+\mathbf{j}) \\ &= 4.5\mathbf{i}+15\mathbf{j}+25\mathbf{i}+17.5\mathbf{j}+14\mathbf{i}+2\mathbf{j} \\ &= 43.5\mathbf{i}+34.5\mathbf{j}.\end{aligned}$$

Also

$$\mathbf{M}_O = \mathbf{r}_C L = (x_C\mathbf{i} + y_C\mathbf{j})10,$$

from which

$$x_C = \frac{43.5}{10} = 4.35 \text{ in.} \quad \text{and} \quad y_C = \frac{34.5}{10} = 3.45 \text{ in.}$$

and

$$\underline{\mathbf{r}_C = (4.35\mathbf{i} + 3.45\mathbf{j}) \text{ in.}} \qquad \text{Ans.}$$

The center of gravity of the wire is shown as point *G* in Fig. 3-11.

EXAMPLE 3-8

Locate the centroid of the shaded area in Fig. 3-12.

SOLUTION

The composite area in Fig. 3-12 is divided into two right triangles *A* and *B*, a large rectangle *C*, and a small rectangle *D* whose area is to be removed. All necessary data for each area are arranged in compact and convenient form in the accompanying table:

Symbol Used in Sketch	*Area (in.²)*	*Position of Centroid (in.)*	*Moment about Origin (in.³)*
A	27	$3\mathbf{i}+2\mathbf{j}$	$81\mathbf{i}+54\mathbf{j}$
B	9	$7\mathbf{i}+2\mathbf{j}$	$63\mathbf{i}+18\mathbf{j}$
C	48	$3\mathbf{i}-2\mathbf{j}$	$144\mathbf{i}-96\mathbf{j}$
D	−12	$4.5\mathbf{i}-\mathbf{j}$	$-54\mathbf{i}+12\mathbf{j}$
Totals	72		$234\mathbf{i}-12\mathbf{j}$

The moment of the area about the origin is

$$\mathbf{M}_O = (234\mathbf{i} - 12\mathbf{j})\text{ in.}^3,$$

and it can be written as

$$\mathbf{M}_O = \mathbf{r}_C A = (x_C\mathbf{i} + y_C\mathbf{j})72.$$

When these two expressions are equated they give

$$x_C = 3.25\text{ in.} \quad \text{and} \quad y_C = -0.1667\text{ in.}$$

and

$$\mathbf{r}_C = (3.25\mathbf{i} - 0.1667\mathbf{j})\text{ in.} \qquad \text{Ans.}$$

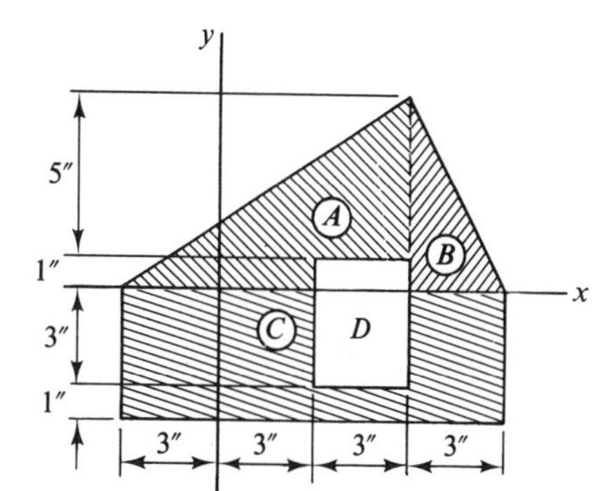

Figure 3-12

EXAMPLE 3-9

Figure 3-13 represents a 30-kg homogeneous cylinder attached to three rods A, B, and C which have masses of 10, 5, and 8 kg, respectively. Locate the mass center of the resulting body.

SOLUTION

The calculations for determining the mass center of the body are conveniently arranged in tabular form as follows:

Member	*Mass* (kg)	*Position of Mass Center* (mm)	*Moment of Mass About Origin* (g · m)
A	10	$100\mathbf{i}$	$1000\mathbf{i}$
B	5	$200\mathbf{i}+80\mathbf{j}-60\mathbf{k}$	$1000\mathbf{i}+400\mathbf{j}-300\mathbf{k}$
C	8	$200\mathbf{i}-60\mathbf{k}$	$1600\mathbf{i}-480\mathbf{k}$
D	30	$200\mathbf{i}+80\mathbf{j}-170\mathbf{k}$	$6000\mathbf{i}+2400\mathbf{j}-5100\mathbf{k}$
Totals	53		$9600\mathbf{i}+2800\mathbf{j}-5880\mathbf{k}$

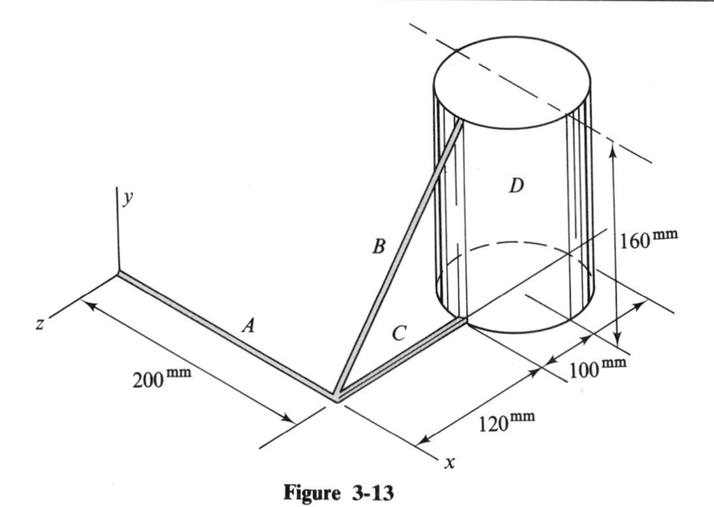

Figure 3-13

The position vector locating the mass center is obtained by means of the principle of moments:

$$\mathbf{M}_O = m\mathbf{r}_G = m(x_G\mathbf{i}+y_G\mathbf{j}+z_G\mathbf{k})$$

from which

$$x_G = \frac{9600}{53} = 181.1 \text{ mm},$$

$$y_G = \frac{2800}{53} = 52.8 \text{ mm},$$

and

$$z_G = -\frac{5880}{53} = -110.9 \text{ mm}.$$

The position of the mass center is

$$\mathbf{r}_G = \underline{(181.1\mathbf{i}+52.8\mathbf{j}-110.9\mathbf{k}) \text{ mm}.} \qquad \text{Ans.}$$

The accompanying tabulations give the area and centroidal coordinates of a number of simple figures and a few selected structural shapes for ready reference.

CENTROIDS OF SELECTED COMMON SHAPES

	Shape	*Length Area, Volume*	x_c	y_c
Quadrant of a circular arc	y, R, y_c, x, x_c	$\frac{\pi R}{2}$	$\frac{2R}{\pi}$	$\frac{2R}{\pi}$
Triangle	y, c, x_c, h, y_c, x, b	$\frac{bh}{2}$	$\frac{b+c}{3}$	$\frac{h}{3}$
Quadrant of a circle	y, R, y_c, x, x_c	$\frac{\pi R^2}{4}$	$\frac{4R}{3\pi}$	$\frac{4R}{3\pi}$

CENTROIDS OF SELECTED COMMON SHAPES (Cont.)

	Shape	Length, Area, Volume	x_c	y_c	z_c
Quadrant of an ellipse	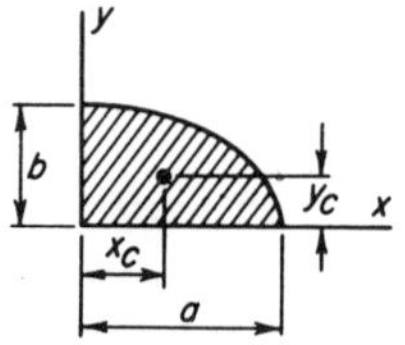	$\frac{\pi ab}{4}$	$\frac{4a}{3\pi}$	$\frac{4b}{3\pi}$	
Quadrant of a parabola	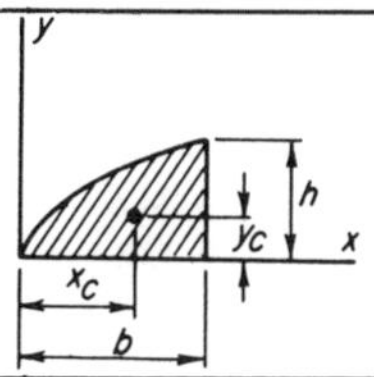	$\frac{2bh}{3}$	$\frac{3b}{5}$	$\frac{3h}{8}$	
Parabolic spandrel	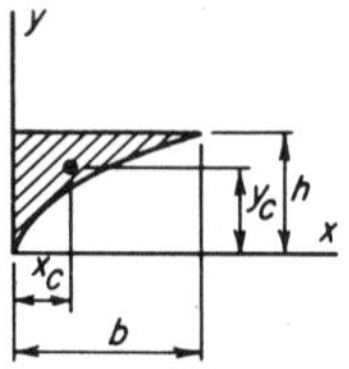	$\frac{bh}{3}$	$\frac{3b}{10}$	$\frac{3h}{4}$	
Right circular cone	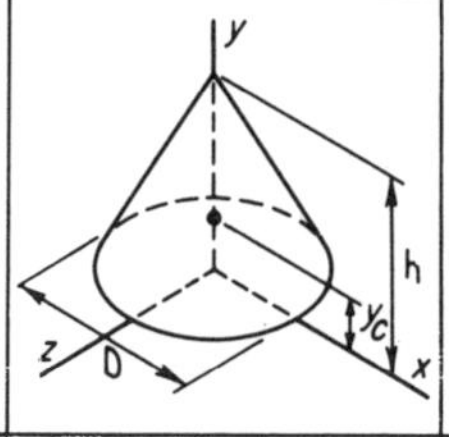	$\frac{\pi R^2 h}{3}$	0	$\frac{h}{4}$	0
Hemisphere	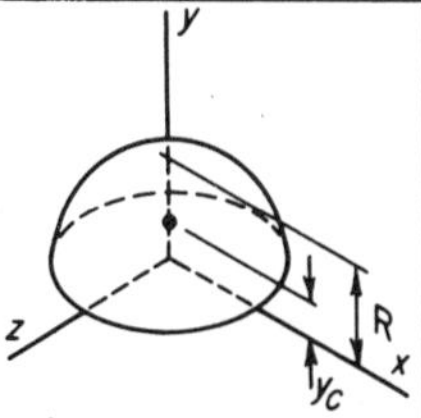	$\frac{2\pi R^3}{3}$	0	$\frac{3R}{8}$	0

PROPERTIES OF SELECTED STRUCTURAL SHAPES

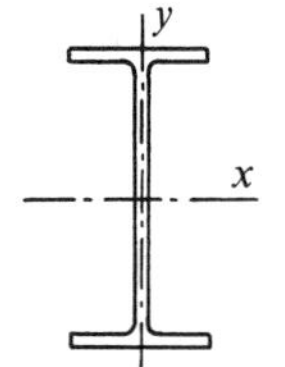

WIDE-FLANGE STEEL BEAM

Depth (in.)	*Weight Per Foot* (lb)	*Area* (in.^{2})	*Flange Width* (in.)	*Web Thickness* (in.)
24.00	100	29.43	12.00	0.468
14.00	34	10.00	6.750	0.287

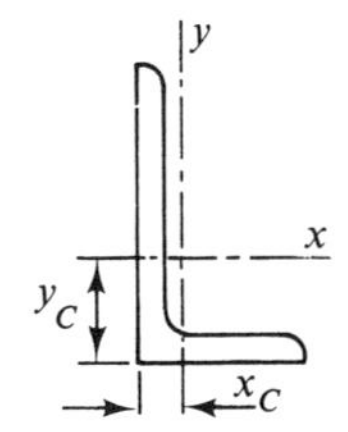

AMERICAN STANDARD STEEL CHANNEL

Depth (in.)	*Weight Per Foot* (lb)	*Area* (in.^{2})	*Flange Width* (in.)	*Web Thickness* (in.)	x_c (in.)
12.00	25.0	7.32	3.047	0.387	0.68
9.00	13.4	3.89	2.430	0.230	0.61

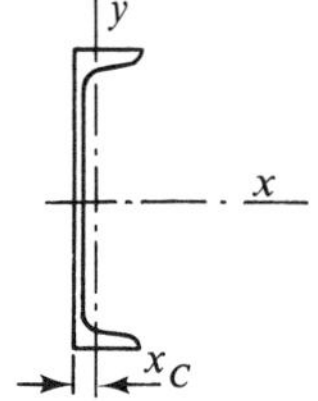

UNEQUAL LEG STEEL ANGLE

Size (in.)	*Thickness* (in.)	*Weight Per Foot* (lb)	*Area* (in.^{2})	x_c (in.)	y_c (in.)
7×4	$\frac{3}{4}$	26.2	7.69	1.01	2.51
5×3	$\frac{1}{2}$	12.8	3.75	0.75	1.75

PROBLEMS

3-35 Locate the centroid of the shaded area in Fig. P3-35.

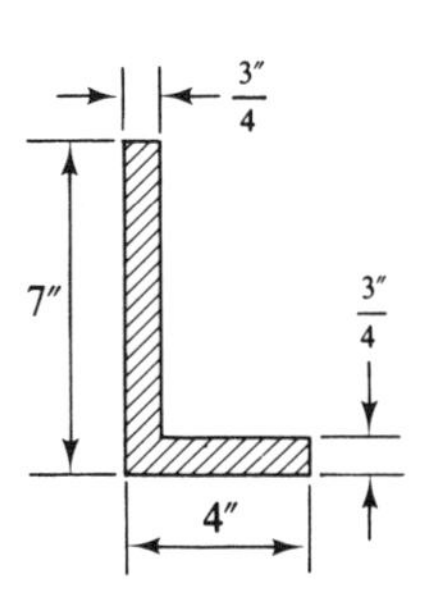

Figure P3-35

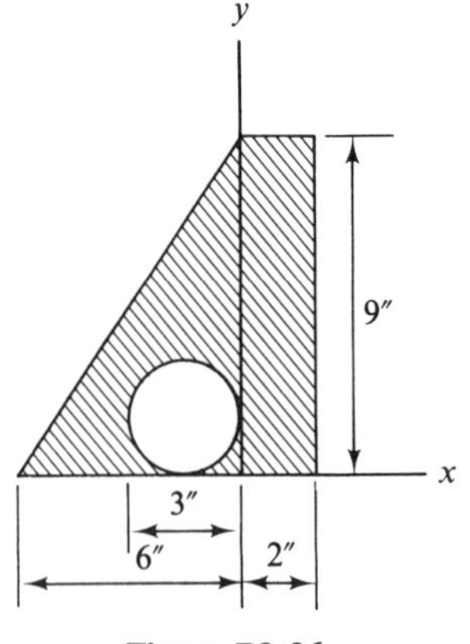

Figure P3-36

3-36 Locate the centroid of the shaded area in Fig. P3-36.

3-37 Locate the centroid of the shaded area in Fig. P3-37.

3-38 The uniform slender homogeneous wire in Fig. P3-38 is bent in the form of three sides of a square and a semicircle. Locate the mass center.

***3-39** Locate the centroid of the shaded area in Fig. P3-39.

***3-40** Locate the centroid of the perimeter of the shaded area in Fig. P3-39.

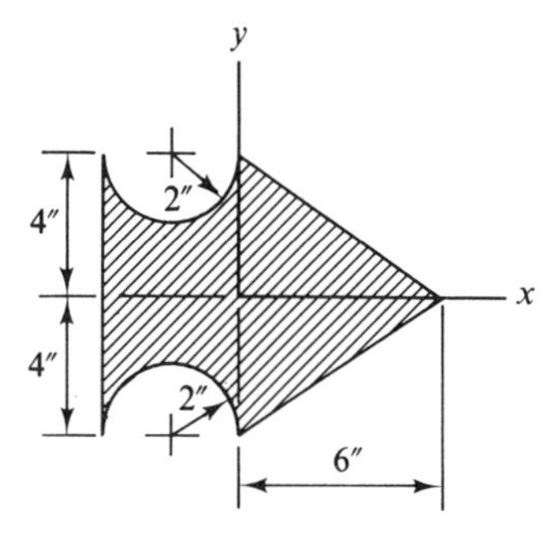

Figure P3-37

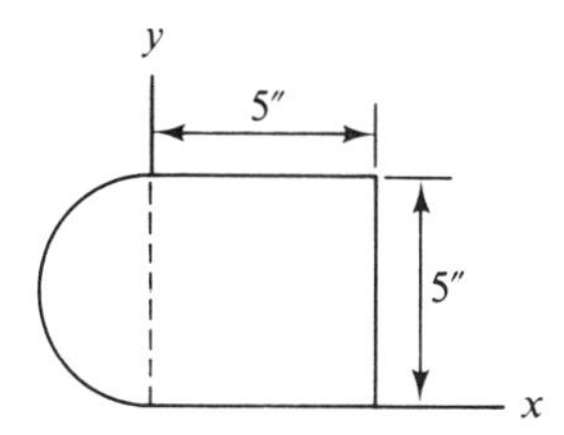

Figure P3-38

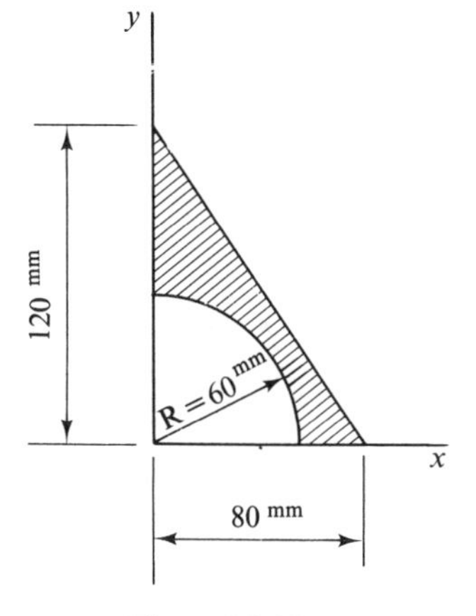

Figure P3-39

3-41 The cross section of a built-up flexural member is shown in Fig. P3-41. It is composed of three steel plates and four 5-in. by 3-in. by $\frac{1}{2}$-in. steel angles. Locate the centroid of the cross-sectional area.

3-42 A 24WF100 (24-in. wide-flange 100-lb steel) beam is reinforced by riveting a $\frac{3}{4}$-in. steel plate, having the same width as the flange, to the top flange. Locate the centroid of the area of the cross section of the composite beam.

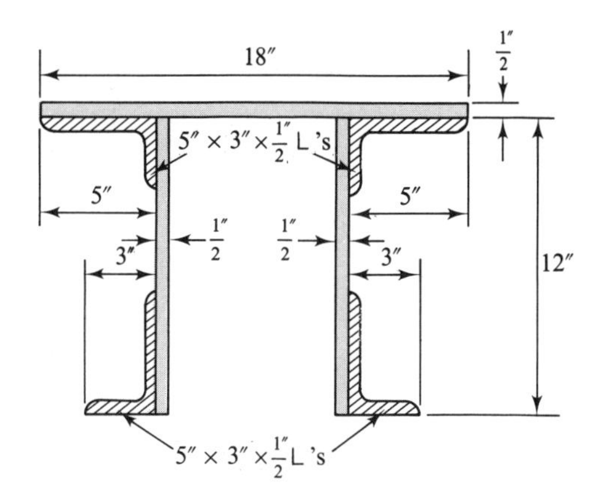

Figure P3-41

3-43 A beam is constructed of two 9-in., 13.4-lb steel channels and a steel plate. The cross section of the beam is shown in Fig. P3-43. Locate the centroid of the cross-sectional area.

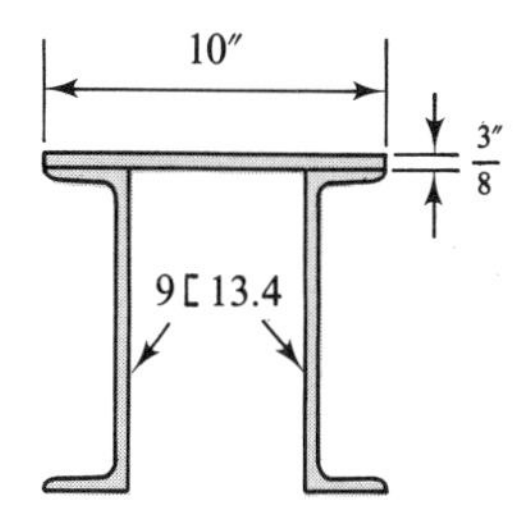

Figure P3-43

3-44 Determine the y coordinate of the centroid of the shaded area shown in Fig. P3-44.

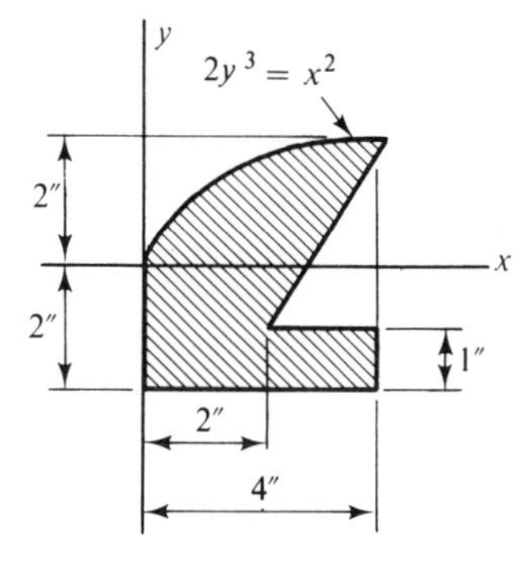

Figure P3-44

3-45 Determine the x coordinate of the centroid of the shaded area of Fig. P3-45. The equation of the curve is $x^2 = 40y$.

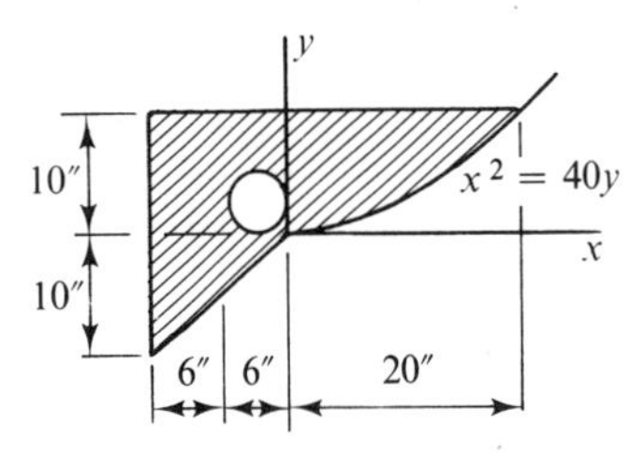

Figure P3-45

3-46 A triangle has the three vertices located at (x_1, y_1), (x_2, y_2), and (x_3, y_3). Determine the x coordinate of the centroid of the area.

3-47 A piece of sheet metal weighing 5 lb per ft^2 is cut into a rectangle, a triangle, and a parabolic shape. After cutting a 2-in.-diameter hole in the rectangular piece, all members are welded to form the arrangement in Fig. P3-47. The equation of the parabola is $2y = x^2$, with x and y in inches. Locate the mass center of the assembly.

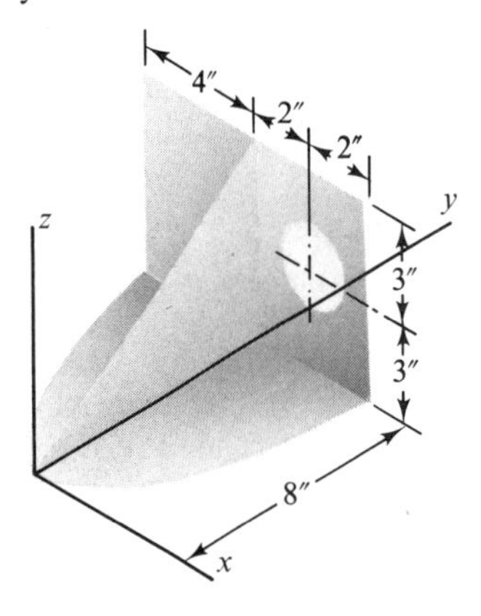

Figure P3-47

***3-48** The tetrahedron in Fig. P3-48 has all four sides covered with a thin sheet of metal with a uniform thickness. Locate the mass center of the metal covering.

3-49 A uniform slender homogeneous wire is bent to the shape shown in Fig. P3-49. The semicircular section ABC is in the yz plane, and

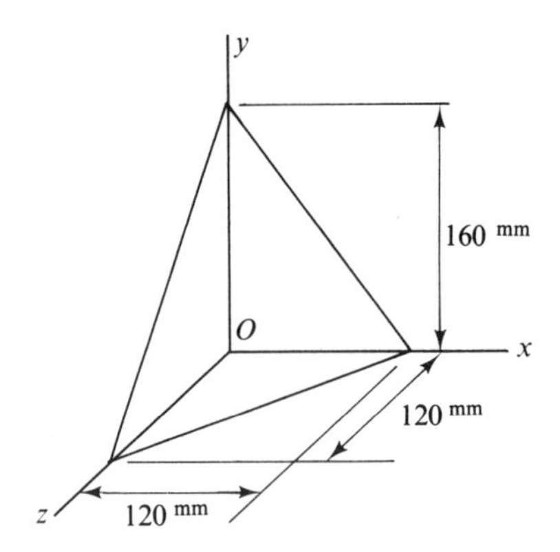

Figure P3-48

the straight section CD is in the xz plane. Locate the mass center of the wire.

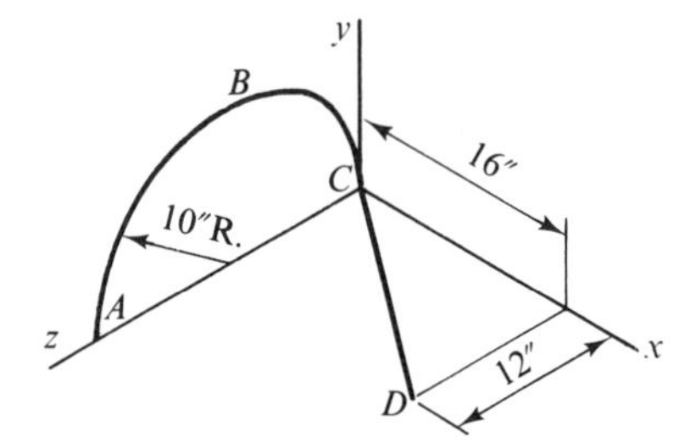

Figure P3-49

3-50 Solve Problem 3-10 by dividing the curved line into five straight-line segments with equal projections on the x axis. Select segments so that the ends are on the curve.

***3-51** Bodies A, B, and C in Fig. P3-51 are connected in such positions that the xy plane is a plane of symmetry for the composite body.

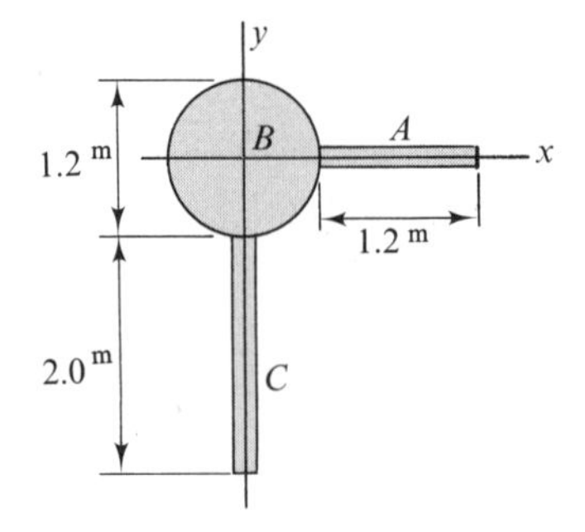

Figure P3-51

The masses of the three homogeneous bodies are 15 kg for bar A, 100 kg for cylinder B, and 60 kg for bar C. Locate the mass center of the composite body.

3-52 The homogeneous body in Fig. P3-52 consists of a half-cylinder and an eccentric pyramid. The edge AB is parallel to the x axis. Locate the centroid of the combined body.

3-53 If the body in Fig. P3-52 is hollow and is made by soldering together pieces of metal of uniform thickness, locate the center of mass of the body.

***3-54** Locate the centroid of the arc of the curve $y = a \sin \pi x/2b$ from (0, 0) to (b, a) when $a = b = 5$ cm. Obtain an approximate solution by dividing the curve into five straight-line segments with equal projection on the x axis.

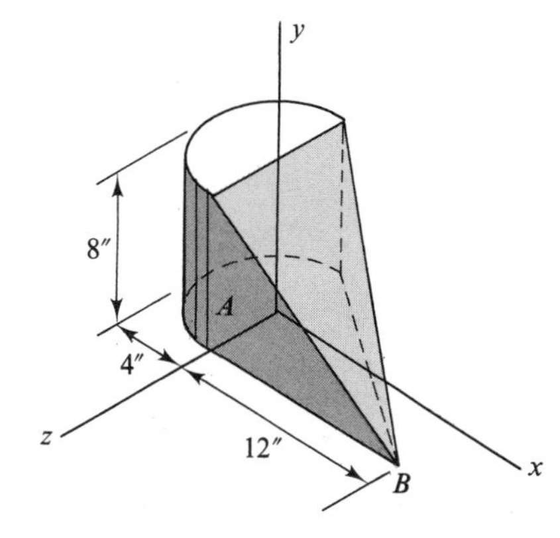

Figure P3-52

3-7 THEOREMS OR PROPOSITIONS OF PAPPUS[4]

Sometimes it is necessary to determine the surface area or volume of a solid of revolution, that is, a figure generated by rotating a plane curve or area about a line in the plane. Disk wheels and pulleys are familiar examples of solids of revolution. To compute the amount of paint required to cover the pulley or the weight of the pulley, it would be necessary to obtain the surface area or the volume of the pulley. The theorems or propositions of Pappus can be used to determine the surface area or volume of any solid of revolution.

The first proposition of Pappus states that *the surface area of any solid of revolution is the product of the length of the generating curve multiplied by the distance traveled by the centroid of the arc of the curve.*

The second proposition of Pappus states that *the volume of any solid of revolution is the product of the generating area and the distance traveled by the centroid of the area.*

In each case, the generating figure must not cross the axis of rotation.

The first proposition of Pappus can be proved as follows: The plane curve BC of length L in Fig. 3-14 is rotated about the x axis through any angle θ of 2π rad or less to generate a surface of revolution, and y_C is the distance of the centroid of the arc BC from the x axis. As the arc BC is rotated, the element of length dL

[4] Sometimes called the *theorems of Guldinus*, after Paul Guldin (1577–1643), a Swiss mathematician who republished them without reference to, and possibly without knowledge of, the original work of Pappus (about A.D. 380).

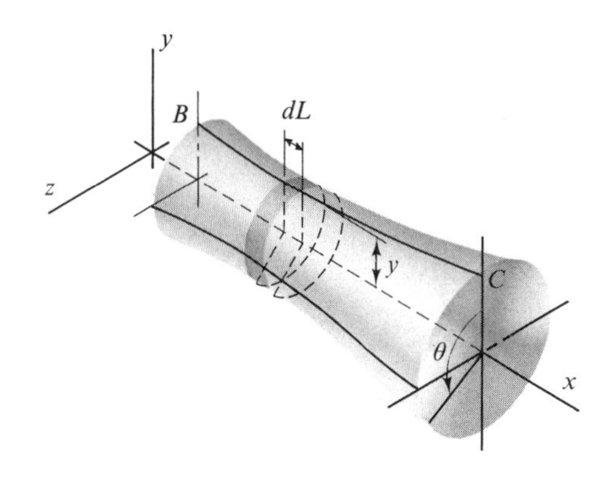

Figure 3-14

generates the shaded strip whose length is θy and whose area is

$$dA = \theta y\, dL.$$

The total area generated is

$$A = \theta \int y\, dL = \theta y_C L$$

since $\int y\, dL$ is equal to $y_C L$ from the principle of moments. The quantity θy_C is equal to the distance traveled by the centroid of the generating arc BC. Thus the first proposition is demonstrated; that is,

$$\begin{pmatrix}\text{surface}\\ \text{area}\end{pmatrix} = \begin{pmatrix}\text{length}\\ \text{of arc}\end{pmatrix}\begin{pmatrix}\text{distance traveled by}\\ \text{the centroid of the arc}\end{pmatrix}.$$

The second proposition of Pappus can be proved as follows: The plane area A in Fig. 3-15 is revolved about the x axis through any angle θ less than, or equal to, 2π rad. The element of area dA generates an element of volume whose length is θy and whose volume is

$$dV = \theta y\, dA$$

from which

$$V = \theta \int y\, dA = \theta y_C A.$$

The factor y_C in the preceding expression is the distance from the x axis to the centroid of the generating area, and the quantity θy_C is the distance traveled by the centroid, thus demonstrating the second proposition.

Although the theorems of Pappus generally are used to determine the surface area or volume of a solid of revolution when the properties of the generating figure are known, they are also useful for locating the centroid of an arc or area when the length or area of the generating figure and the surface area or volume of the generated figure are known. The following examples illustrate the use of the theorems of Pappus.

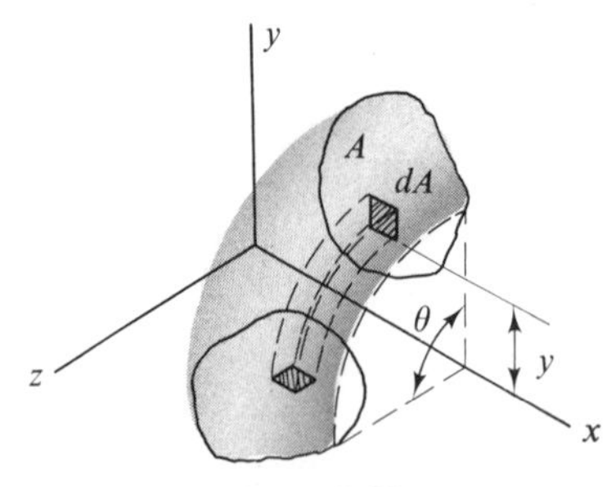

Figure 3-15

EXAMPLE 3-10

By means of the first proposition of Pappus, determine the lateral area of the frustum of the right circular cone in Fig. 3-16.

SOLUTION

The lateral area of the cone is generated by revolving the line AB (the generating line) around the y axis. By symmetry, the x coordinate of the centroid of the line AB is 2.5 in., and the y coordinate is 3.0 in. The length of the generating line is $\sqrt{3^2+6^2}$ in.

From the theorem, the lateral area is

$$A = L(2\pi x_C)$$

$$= \sqrt{3^2+6^2}(2\pi 2.5) = \underline{105.4 \text{ in.}^2}. \quad \text{Ans.}$$

This result gives only the curved surface area and excludes the area of the bases (the ends) of the frustum.

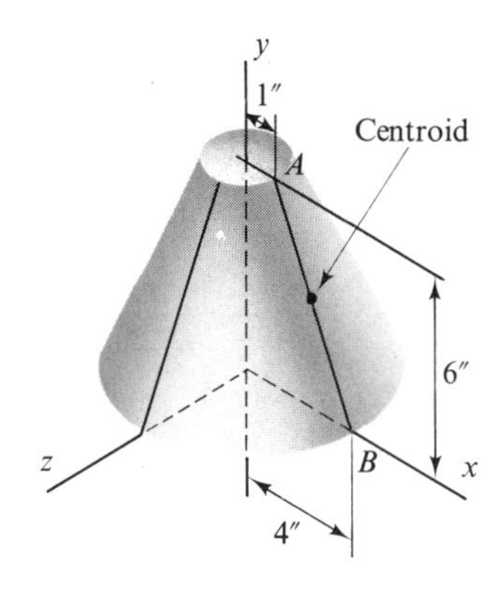

Figure 3-16

EXAMPLE 3-11

By means of the second proposition of Pappus, locate the centroid of a semicircular area.

SOLUTION

Let the semicircular area of Fig. 3-17 be revolved around the x axis to generate a sphere. The volume of the sphere is $(\frac{4}{3})\pi R^3$, and the area of the semicircle is $(\frac{1}{2})\pi R^2$. By the second theorem, the volume is

$$V = A 2\pi y_C$$

from which

$$(\tfrac{4}{3})\pi R^3 = (\tfrac{1}{2})\pi R^2 2\pi y_C$$

or

$$\underline{y_C = \frac{4R}{3\pi}}, \qquad \text{Ans.}$$

and by symmetry,

$$\underline{x_C = 0}. \qquad \text{Ans.}$$

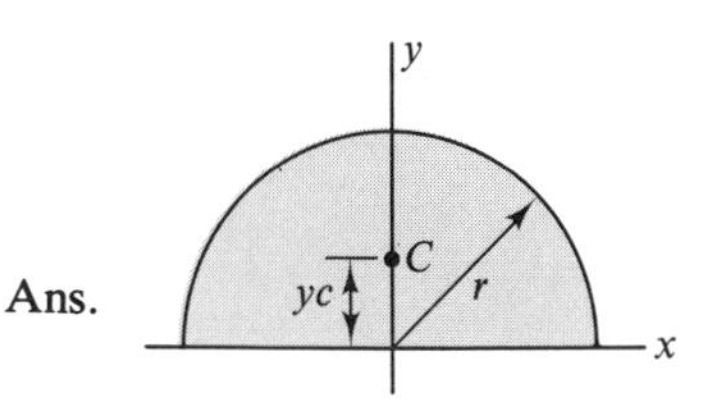

Figure 3-17

PROBLEMS

3-55 By means of one of the theorems of Pappus, locate the centroid of a semicircular arc.

3-56 The circular area in Fig. P3-56 is revolved about the x axis, generating a torus or anchor ring. By use of the first proposition of Pappus, determine the surface area of the torus.

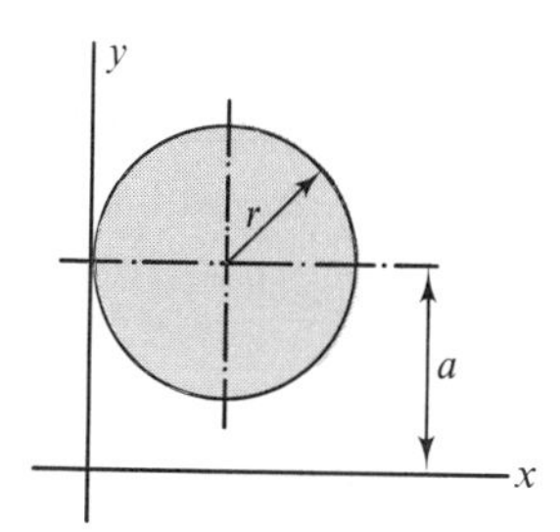

Figure P3-56

3-57 A ring is generated by revolving the right-hand half of the circular area in Fig. P3-56 around the y axis. Determine the volume of the ring.

3-58 Determine the surface area of the ring in Problem 3-57.

3-59 The shaded area in Fig. P3-59 represents the cross section of the rim of a steel flywheel. Steel has a specific weight of 0.283 lb per in.^{3}. Determine the weight of the rim of the flywheel.

3-60 The shaded area in Fig. P3-60 is revolved about the line a to form a volume of 51.2π in.^{3}. The equation of the curve is $9y = x^2$. Determine
(a) The y coordinate of the centroid of the area by means of a theorem of Pappus.
(b) The location of the centroid of the volume of revolution.

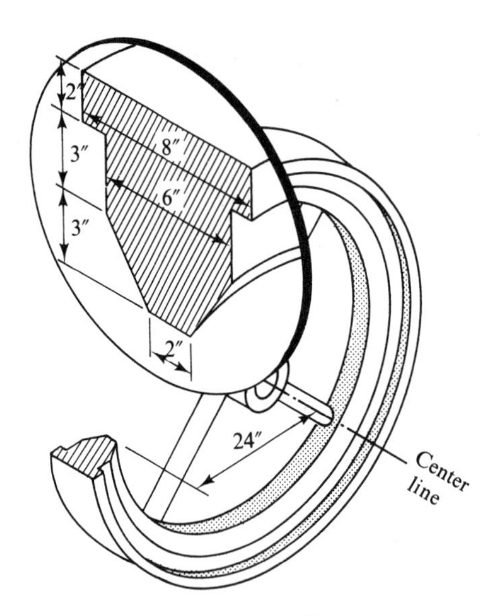

Figure P3-59

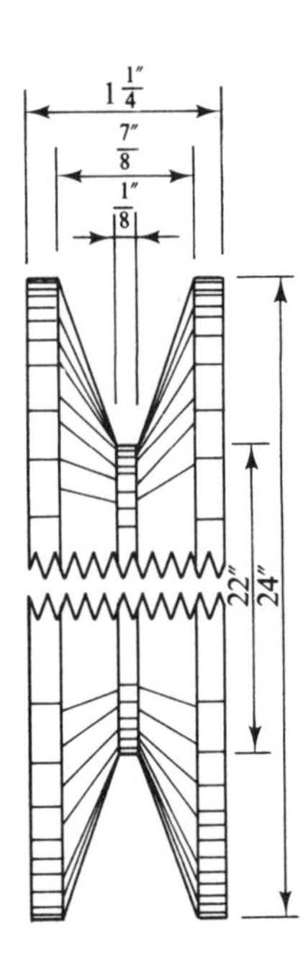

Figure P3-61

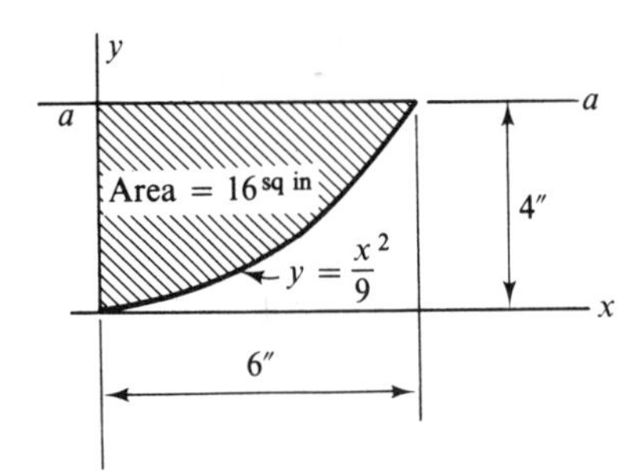

Figure P3-60

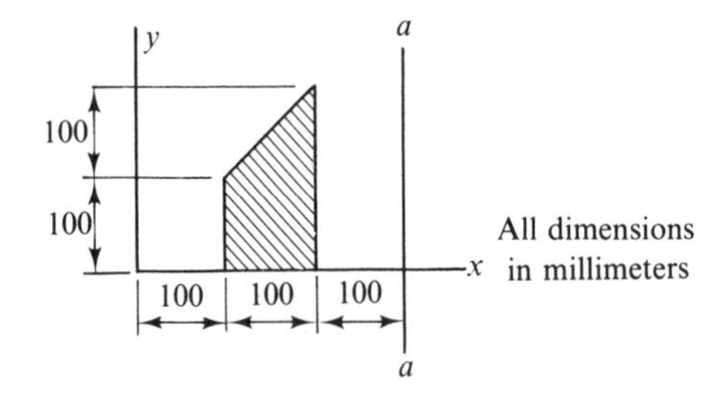

Figure P3-63

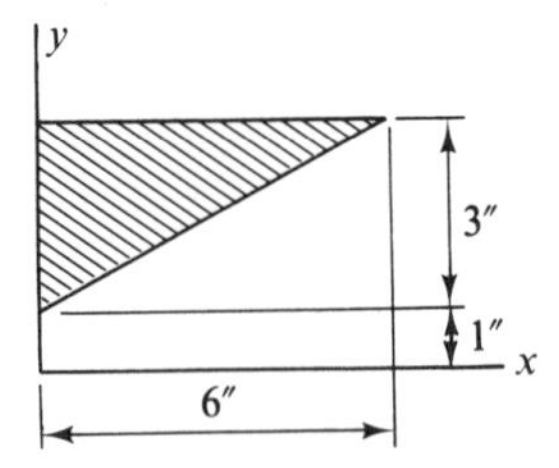

Figure P3-64

3-61 The dimensions of the belt groove for a V-belt pulley are given in Fig. P3-61. Determine the volume of metal which is removed in cutting the groove.

3-62 Determine the surface area of the rim of the flywheel in Problem 3-59.

***3-63** The shaded area in Fig. P3-63 is revolved about the a axis.
(a) By means of a theorem of Pappus, determine the volume generated.
(b) Locate the centroid of the generated volume.

3-64 The shaded area in Fig. P3-64 is revolved about the x axis to generate a volume.
(a) Determine the volume generated by using a theorem of Pappus.
(b) Locate the centroid of the generated volume.

3-8 CENTER OF PRESSURE

When a force is exerted on a body over a small area, the force is called a *concentrated force* and is assumed to act at a point. Frequently, a force is exerted on a body over a large area, and in this case it is called a *distributed force* or *pressure*. The effect of grain on the floor and walls of a grain bin, of water on a dam, and of wind on the side of a building are examples of distributed forces. The distributed force or pressure can be considered to be a number of concentrated forces each acting on a small part of the surface. Usually, these forces act normal to the surface, and for a plane surface they constitute a parallel force system. The resultant pressure can, therefore, be obtained by the method described in Art. 2-4. If the force is the same on all equal elements of area no matter how small the elements, the pressure is said to be *uniform*, and the resultant force acts at the centroid of the plane area. If the force is not the same on equal elements of area, the pressure is *variable*, and the line of action of the resultant force must be determined by the principle of moments. *The intersection of the line of action of the resultant of the distributed force system and the plane on which it acts is known as the center of pressure.*

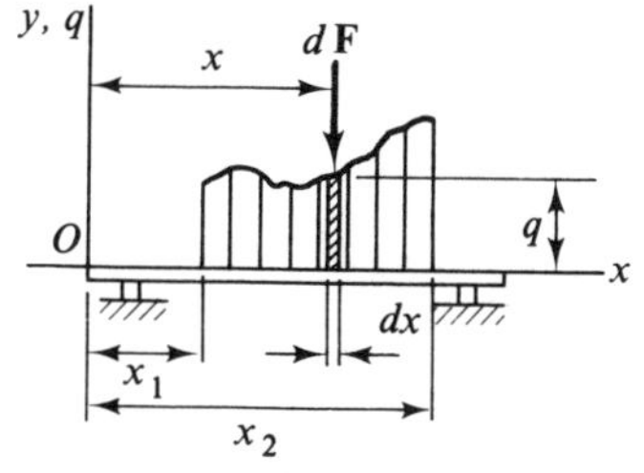

Figure 3-18

A load-distribution or pressure diagram showing the variation of the intensity of the distributed force is often useful in determining the resultant force. When the pressure varies in only one direction, as, for example, the load on a floor beam, the load-distribution diagram can be drawn as shown in Fig. 3-18. The ordinate q of the diagram indicates the intensity of the load in pounds per foot or similar units. The load $d\mathbf{F}$ on an element of the beam of length dx is

$$d\mathbf{F} = -q\,dx\,\mathbf{j},$$

and the total load is

$$\mathbf{F} = -\int_{x_1}^{x_2} q\,dx\,\mathbf{j} = -\mathbf{j}\int_{x_1}^{x_2} q\,dx$$

since $\mathbf{j}$ is constant in magnitude and direction.

The moment of the resultant force about an axis through O is

$$\mathbf{M}_O = \int \mathbf{x}\times d\mathbf{F} = \int_{x_1}^{x_2} x\mathbf{i}\times(-q\,dx)\mathbf{j} = -\mathbf{k}\int_{x_1}^{x_2} xq\,dx$$

since $\mathbf{i}\times\mathbf{j}=\mathbf{k}$ is a constant. The x coordinate, x_p, of the resultant can be determined by means of the principle of moments as follows:

$$\mathbf{M}_O = -\mathbf{k}\int_{x_1}^{x_2} xq\,dx = \mathbf{x}_p\times\mathbf{F} = x_p\mathbf{i}\times\left(-\mathbf{j}\int_{x_1}^{x_2} q\,dx\right) = -x_p\mathbf{k}\int_{x_1}^{x_2} q\,dx$$

from which

$$x_p = \frac{\int_{x_1}^{x_2} xq\,dx}{\int_{x_1}^{x_2} q\,dx} = \frac{M_O}{F}.$$

By analogy, it is apparent that the equation for the magnitude of the resultant force would also give the area under the load-distribution diagram and that the moment of the resultant force about an axis through O is the same as the moment of the area of the pressure diagram with respect to the y axis. Therefore, the resultant force is represented by the area under the load-distribution diagram and passes through the centroid of the area of the diagram.

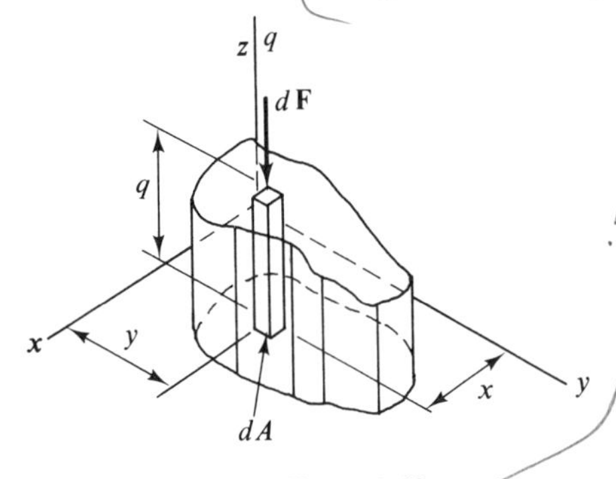

Figure 3-19

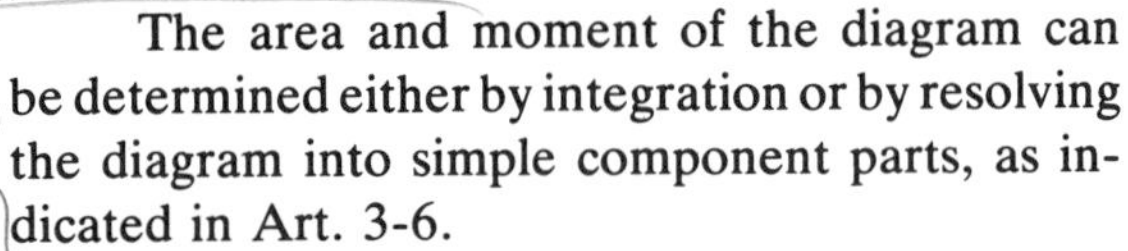

The area and moment of the diagram can be determined either by integration or by resolving the diagram into simple component parts, as indicated in Art. 3-6.

When the intensity of the distributed force varies over an area instead of in a single direction, for example, the pressure on the floor of the grain bin, the pressure diagram becomes a volume as in Fig. 3-19 instead of an area. The distributed pressure is applied to the area in the xy plane and the ordinate, q, along the z axis represents the intensity of the force, that is, the force per unit area, in pounds per square foot or similar units. The element of force dF on the element of area dA is equal to the product of the element of area and the intensity of the force, that is,

$$d\mathbf{F} = -q\,dA\mathbf{k},$$

and the total force is

$$\mathbf{F} = \int -q\,dA\mathbf{k} = -\mathbf{k}\int q\,dA$$

in which the quantity $\int q\,dA$ is equal to the volume of the pressure solid. The resultant force can be located by means of the principle of moments, and it passes through the centroid of the volume that represents the pressure diagram.

The hydrostatic force resulting from a liquid in contact with a surface is another example of a distributed force. The pressure, p (force per unit area), in a liquid varies with the distance from the surface of the liquid and is given by the equation $p = \gamma h$, where γ is the specific weight of the liquid and h is the vertical distance of the point below the liquid surface.

The following examples illustrate the procedure for determining the resultant of a distributed force system.

EXAMPLE 3-12

The beam in Fig. 3-20(a) weighs 100 lb per ft and is subjected to a pressure that varies as indicated in the figure. The pressure $q = x^2/0.48$ is in pounds per linear foot, and the distance x, in feet, is measured from the left support. Determine the resultant of the weight and pressure acting on the beam.

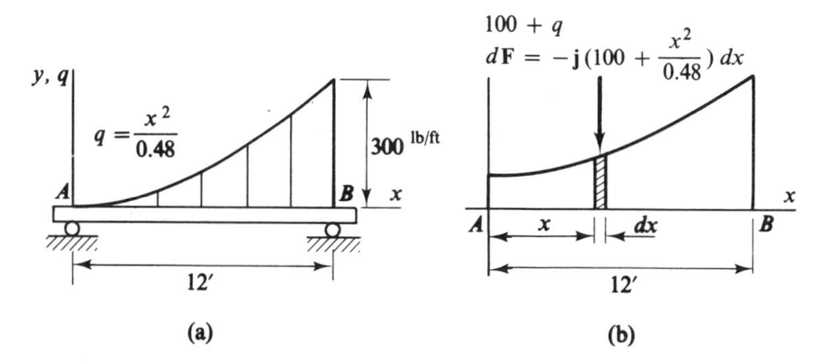

Figure 3-20

SOLUTION

The intensity of the load on the beam, including its weight, is shown in Fig. 3-20(b). The force $d\mathbf{F}$ at a distance x from the left support is equal to the product of the resultant load intensity (in pounds per foot) and the length dx (in feet):

$$d\mathbf{F} = -\mathbf{j}(100 + q)\,dx = -\mathbf{j}\left(100 + \frac{x^2}{0.48}\right)dx.$$

The resultant force on the beam is

$$\mathbf{F} = -\mathbf{j}\int_0^{12}\left(100 + \frac{x^2}{0.48}\right)dx$$

$$= -\mathbf{j}\left[100x + \frac{x^3}{1.44}\right]_0^{12} = \underline{2400 \text{ lb}\downarrow}. \qquad \text{Ans.}$$

The moment of the force $d\mathbf{F}$ with respect to point A is

$$d\mathbf{M}_A = \mathbf{x}\times d\mathbf{F} = x\mathbf{i}\times(-\mathbf{j})\left(100 + \frac{x^2}{0.48}\right)dx,$$

and the resultant moment is

$$\mathbf{M}_A = -\mathbf{k}\int_0^{12}\left(100x + \frac{x^3}{0.48}\right)dx$$

$$= -\mathbf{k}\left[50x^2 + \frac{x^4}{1.92}\right]_0^{12} = 18{,}000 \text{ ft-lb}\circlearrowright.$$

The moment of the resultant force about A can also be written as

$$\mathbf{M}_A = \mathbf{x}_p \times \mathbf{F} = x_p\mathbf{i}\times(-2400\mathbf{j}) = -x_p(2400)\mathbf{k}.$$

When these two expressions for $\mathbf{M}_x$ are equated they become

$$-18{,}000\mathbf{k} = -x_p(2400)\mathbf{k}$$

from which

$$x_p = \frac{18{,}000}{2400} = 7.50 \text{ ft}$$

and

$$\mathbf{x}_p = \underline{7.50\mathbf{i}\text{ ft.}} \qquad \text{Ans.}$$

EXAMPLE 3-13

The floor load in a warehouse is distributed in such a manner that the load on a beam varies as shown in Fig. 3-21(a). Determine the resultant load on the beam. Neglect the mass of the beam.

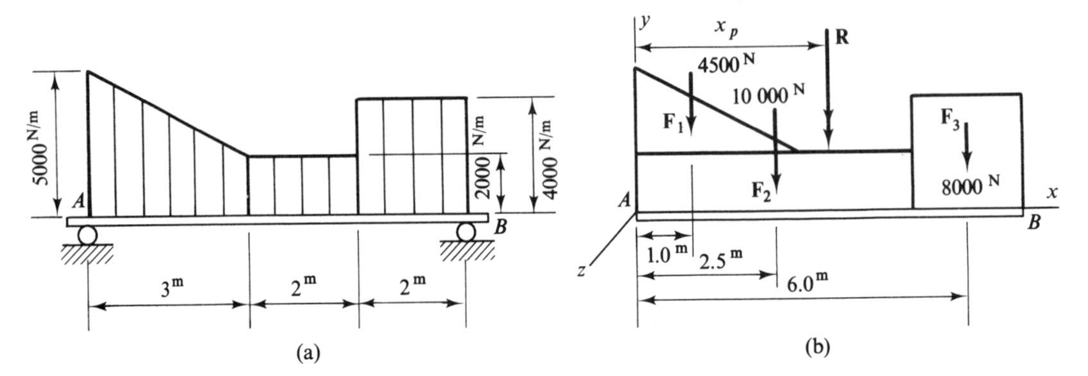

Figure 3-21

SOLUTION

The distributed load can be resolved into three component loads as indicated in Fig. 3-21(b). The load F_1 is given by the area of the triangle $(5000-2000)(3)/2 = 4500$ N, and it acts through the centroid of the triangle. Similarly, $F_2 = 10\,000$ N and $F_3 = 8000$ N. The resultant load is

$$\mathbf{R} = -4500\mathbf{j} - 10\,000\mathbf{j} - 8000\mathbf{j} = -22\,500\mathbf{j}\text{ N}$$

$$= \underline{22.5 \text{ kN}\downarrow.} \qquad \text{Ans.}$$

The moment of the resultant load about the z axis is

$$\mathbf{M}_z = -1(4500)\mathbf{k} - 2.5(10\,000)\mathbf{k} - 6(8000)\mathbf{k} = -77\,500\mathbf{k}\text{ N}\cdot\text{m}$$

$$= 77.5 \text{ kN}\cdot\text{m}\curvearrowright.$$

The distance from A to the resultant force is

$$x_p = \frac{M_x}{R} = \frac{77.5}{22.5} = 3.44 \text{ m}$$

and

$$\mathbf{x}_p = \underline{3.44 \text{ m to the right of A.}} \qquad \text{Ans.}$$

EXAMPLE 3-14

Plate AB of Fig. 3-22(a) closes a 2-ft by 5-ft opening in the sloping side of an oil tank. The specific weight of the oil in the tank is 50 lb per ft^3 (pcf). Determine the resultant force of the oil on the plate.

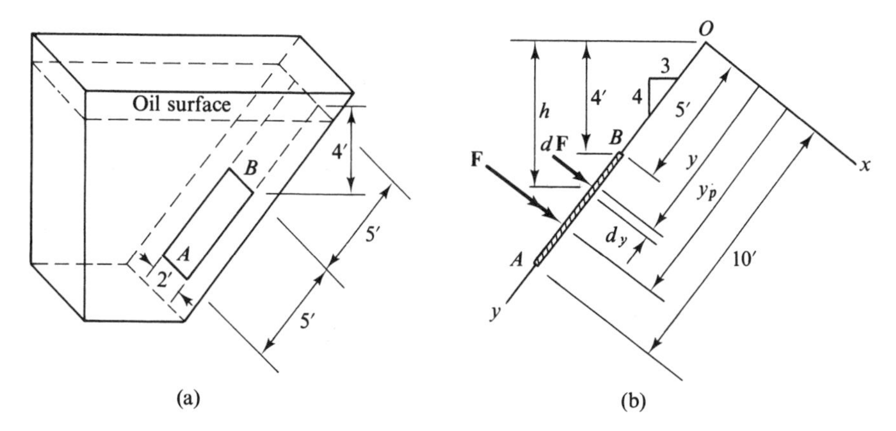

Figure 3-22(a) and (b)

SOLUTION

Axes are selected as shown in Fig. 3-22(b) with the intersection of the plane area and the oil surface as the z axis. The pressure (in pounds per square foot) at a point h feet below the fluid surface is $p = \gamma h$, where γ is the specific weight of the fluid. The force $d\mathbf{F}$ in Fig. 3-22(b) is the fluid force on an element of area 2 ft wide and dy ft high and is

$$d\mathbf{F} = \mathbf{i}p\,dA = \mathbf{i}\gamma h(2\,dy).$$

Since $h = \frac{4}{5}y$ and $\gamma = 50$ pcf, the differential force becomes

$$d\mathbf{F} = \mathbf{i}(50)(0.8y)(2\,dy) = \mathbf{i}80y\,dy.$$

The resultant force on the plate is

$$\mathbf{F} = \mathbf{i}80\int_5^{10} y\,dy$$

$$= \mathbf{i}80\left.\frac{y^2}{2}\right]_5^{10} = \underline{\mathbf{i}3000\text{ lb.}} \qquad \text{Ans.}$$

The principle of moments can be used to locate the resultant force. The magnitude of the moment of $d\mathbf{F}$ about point O is

$$dM_O = y(dF) = 80y^2\,dy,$$

and the moment of force is

$$M_O = 80\int_5^{10} y^2\,dy$$

$$= 80\left.\frac{y^3}{3}\right]_5^{10} = 23{,}330\text{ ft-lb.}$$

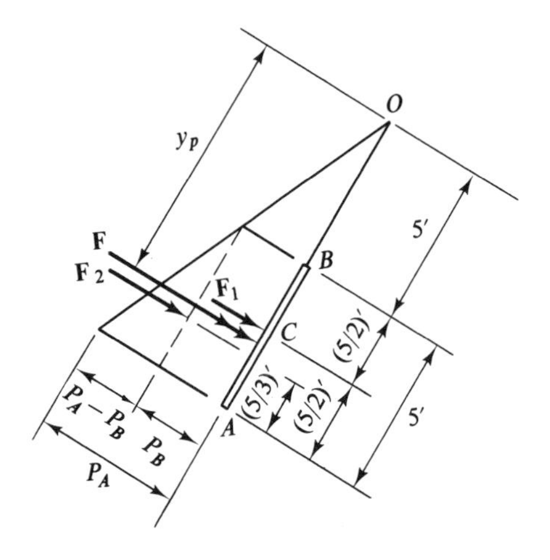

Figure 3-22(c)

The distance y_p from O to the resultant force is

$$y_p = \frac{M_O}{F} = \frac{23{,}330}{(3000)} = 7.78 \text{ ft},$$

and $\mathbf{y}_p = \underline{7.78\mathbf{j}\text{ ft.}}$ Ans.

An alternate solution using a *pressure solid* is illustrated in Fig. 3-22(c). The pressures at A and B are

$$p_A = \gamma h_A = 50[(0.8)10] = 400 \text{ psf}$$

and

$$p_B = \gamma h_B = 50[(0.8)5] = 200 \text{ psf}.$$

The pressure solid is divided into a rectangular prism (representing a uniform pressure of 200 psf) and a triangular prism (representing a pressure which varies from 0 to 200 psf). The magnitudes of the two component forces, equal to the volumes of the two parts of the pressure solid, are

$$F_1 = p_B A = 200(2)(5) = 2000 \text{ lb},$$

and

$$F_2 = \frac{p_A - p_B}{2} A = \frac{400 - 200}{2}(2)(5) = 1000 \text{ lb}.$$

The resultant force is

$$\mathbf{F} = F_1\mathbf{i} + F_2\mathbf{i} = \underline{3000\mathbf{i}\text{ lb.}}$$ Ans.

The resultant can be located by summing moments about any convenient moment center. When moments are summed about point C the moment of F_1 is zero and

$$F(y_p - 7.5) = F_2(\tfrac{5}{2} - \tfrac{5}{3})$$

or

$$3000(y_p - 7.5) = 1000(2.500 - 1.667)$$

from which

$$y_p = 7.5 + \frac{0.833}{3} = \underline{7.778 \text{ ft.}}$$ Ans.

PROBLEMS

3-65 Determine the resultant of the distributed load in Fig. P3-65.

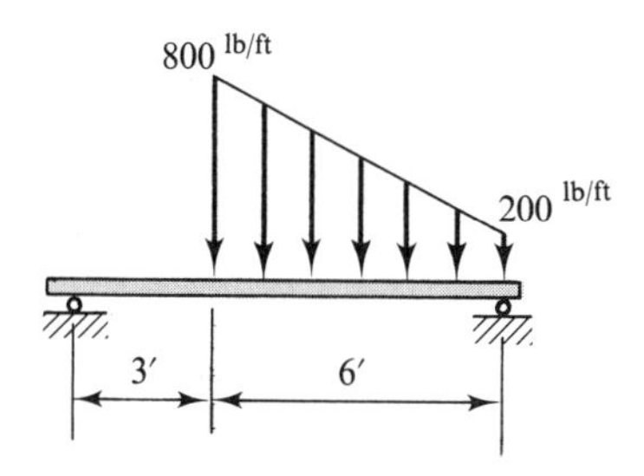

Figure P3-65

3-66 The beam in Fig. P3-66 is subjected to a distributed load which varies according to the equation $q = kx^{1/2}$, where q is in pounds per linear foot when x is in feet.
(a) Determine the value of k which will make $q = 150$ lb per ft when $x = 9$ ft.
(b) Determine the resultant of the distributed load.

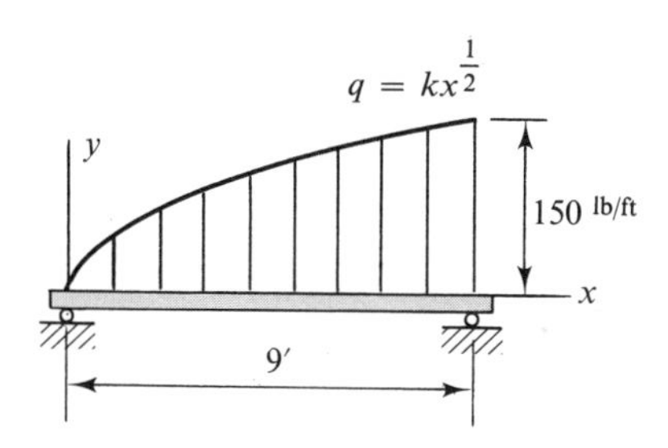

Figure P3-66

3-67 The load on the beam in Fig. P3-66 is replaced by one which varies according to the equation $q = Cx(18 - x)$, where again q is in pounds per foot and x is in feet.
(a) Determine the value of C which will make $q = 150$ lb per ft when $x = 9$ ft.
(b) Determine the resultant of the distributed load.

***3-68** Determine the resultant of the forces in Fig. P3-68.

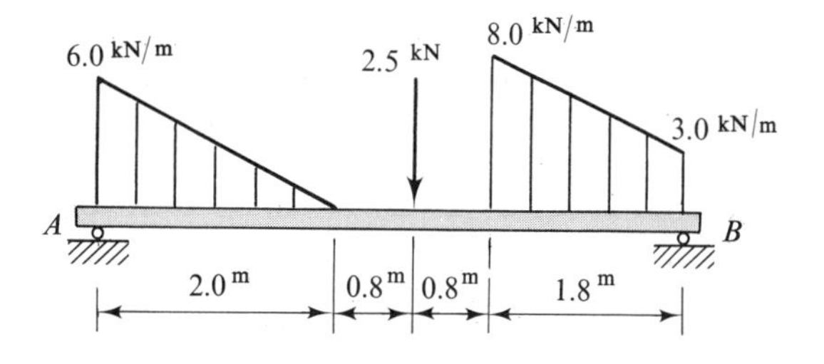

Figure P3-68

3-69 Determine the resultant of the coplanar force system of Fig. P3-69 and locate it with respect to point A.

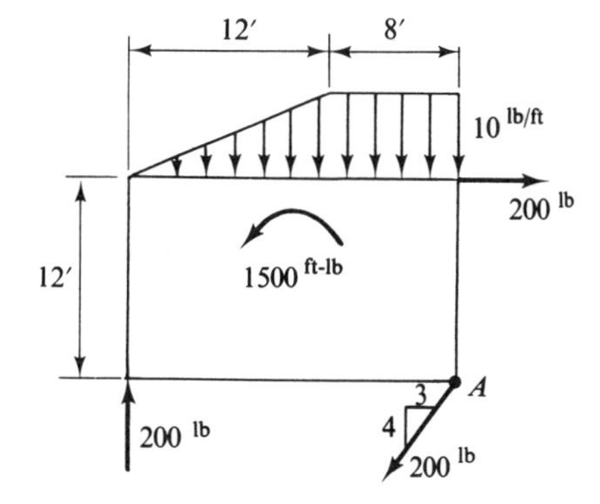

Figure P3-69

3-70 Determine the resultant of the force system of Fig. P3-70 and locate it with respect to point A.

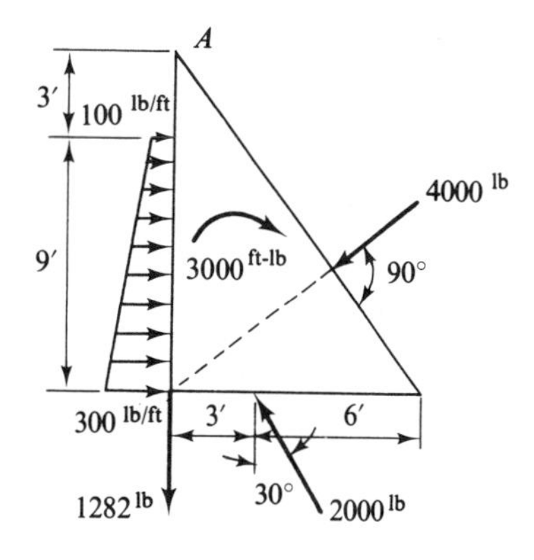

Figure P3-70

3-71 The structure in Fig. P3-71 is subjected to two distributed loads as shown. The intensity, q, of the vertical load is in pounds per foot when x is in feet. Determine the resultant of the two loads and locate it with respect to the origin at O.

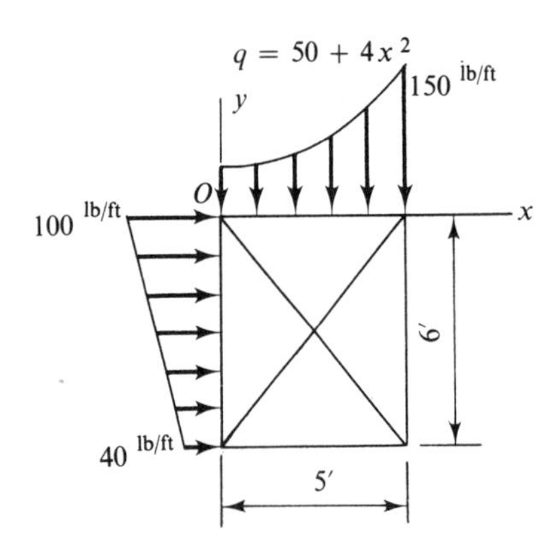

Figure P3-71

***3-72** The force system in Fig. P3-72 consists of the forces shown and one additional force **P**, which passes through point B. The resultant, **R**, passes through A as shown. Determine the force **P**.

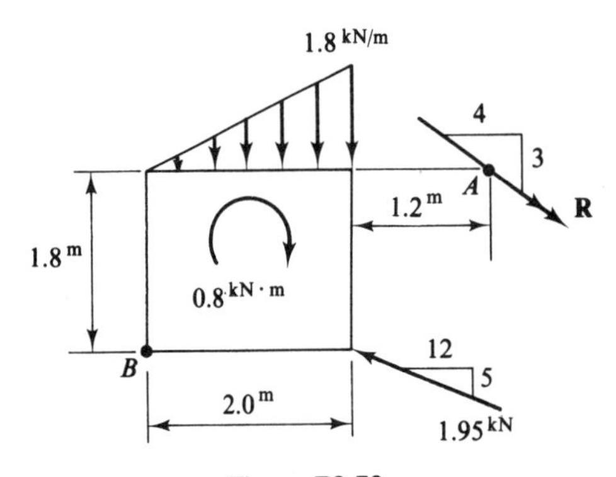

Figure P3-72

3-73 A distributed pressure acts on plane $ABCD$ in Fig. P3-73 and varies linearly from AB to CD. The pressure is constant along any horizontal line. The resultant force on the plane is 4500 lb, and it acts 9.00 in. above AB. Determine the pressures p_1 and p_2.

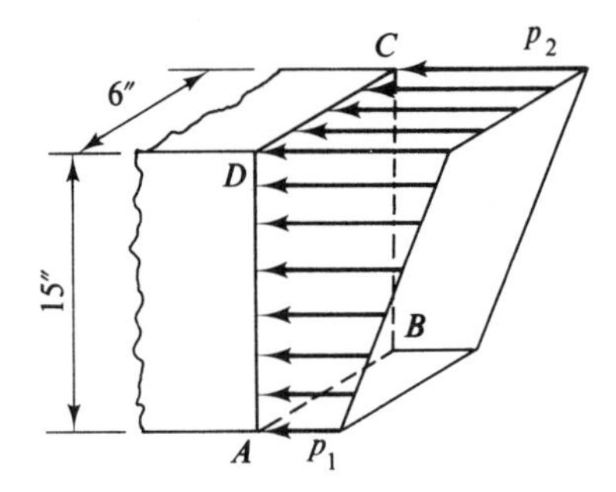

Figure P3-73

3-74 The flat triangular plate in Fig. P3-74 supports a distributed pressure which varies according to the equation $p = 6x^2$, where p is the pressure in pounds per square inch and the x coordinate is measured in inches. Determine the resultant of the distributed load.

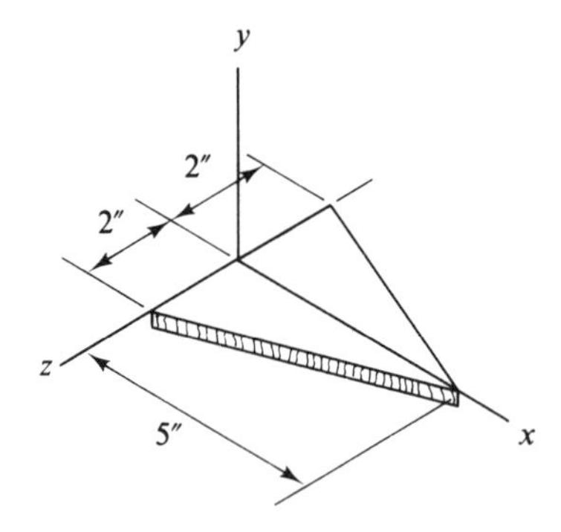

Figure P3-74

3-75 Determine the resultant fluid force acting on an isosceles-triangular plate in the vertical side of a tank of water. The triangle has a base (horizontal) of 4 ft and an altitude of 6 ft. The vertex is above the base and is 2 ft below the water surface.

3-76 Solve Problem 3-75 if the base of the triangle is above the vertex and is 2 ft below the water surface.

***3-77** The cross section of a beam [see Fig. P3-77(b)] is subjected to a variable pressure as shown in Fig. P3-77(a). Determine the resultant of the pressure on the end of the beam.

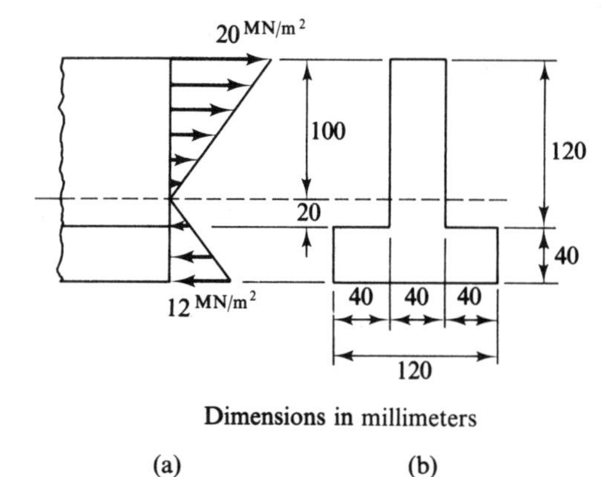

Figure P3-77

3-78 The air pressure on the vertical stabilizer of a large airplane is assumed to vary as shown in Fig. P3-78. The pressure along the leading edge *AB* is 4.00 psi, and the pressure along the trailing edge *CD* is 1.00 psi. The pressure varies linearly along any horizontal line. Determine the resultant air force on the surface.

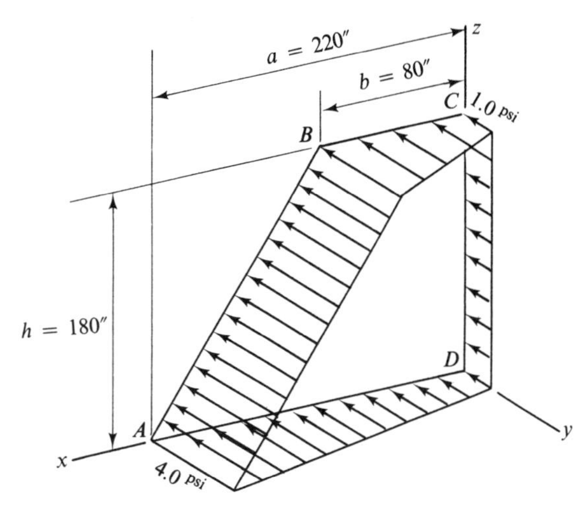

Figure P3-78

REVIEW PROBLEMS

3-79 Locate the centroid of the shaded area in Fig. P3-79.

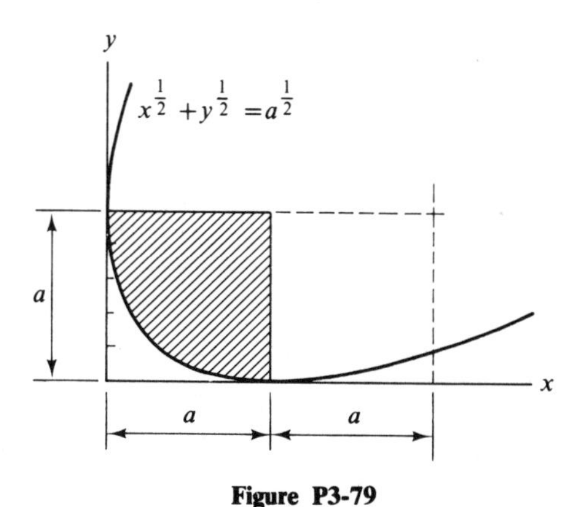

Figure P3-79

3-80 Determine the x coordinate of the centroid of the area bounded by the curve in Problem 3-79 and the lines $y = a$ and $x = 2a$.

3-81 The 500-lb force of Fig. P3-81 is the resultant of the force system shown and one unknown force, **F**, not shown in the diagram. Determine the force **F** and locate it with respect of point *A*.

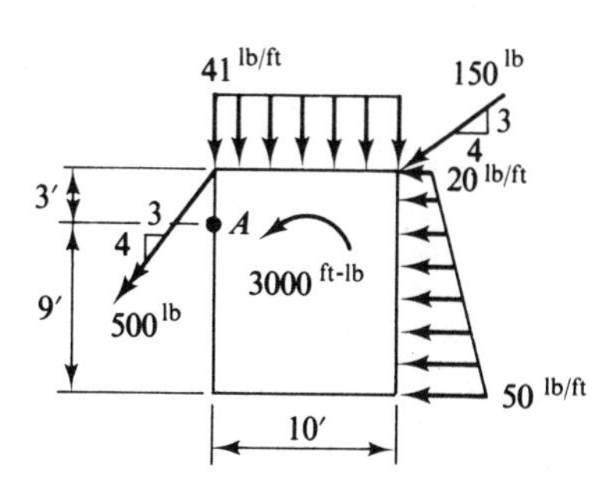

Figure P3-81

3-82 A structural member is constructed of a 12-in., 25-lb steel channel and two 5-in. by 3-in. by $\frac{1}{2}$-in. steel angles arranged as shown in Fig. P3-82. Locate the centroid of the built-up section.

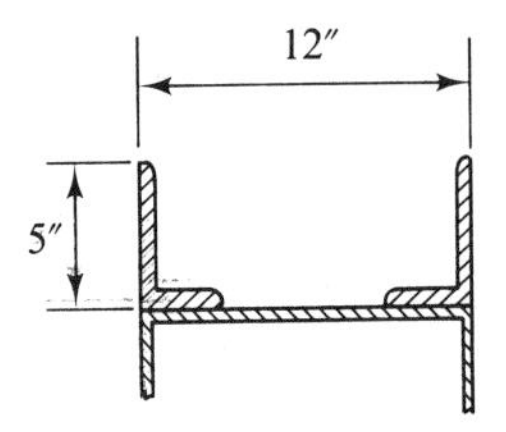

Figure P3-82

3-83 A 7-in. by 4-in. by $\frac{3}{4}$-in. angle is welded to the web of a 14WF34 section as indicated in Fig. P3-83. Locate the centroid of the cross-sectional area.

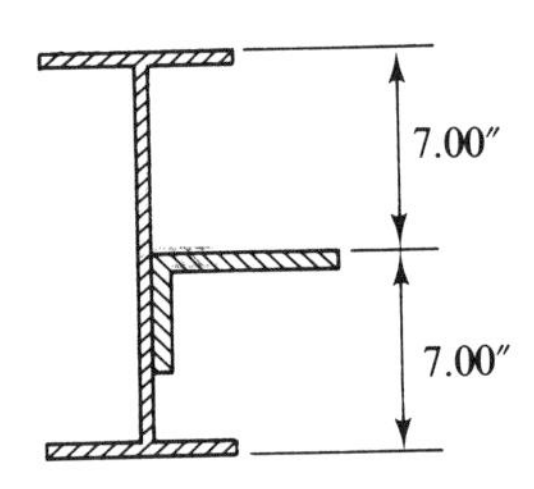

Figure P3-83

***3-84** Determine the first moment of the shaded area in Fig. P3-84 with respect to the a axis.

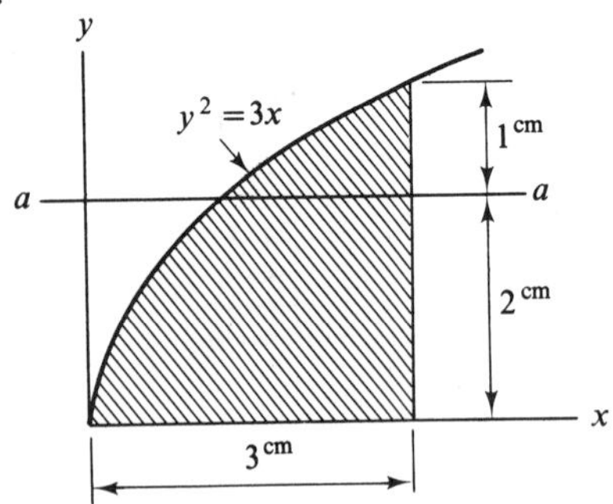

Figure P3-84

3-85 Determine the x coordinate of the centroid of the shaded area in Fig. P3-85.

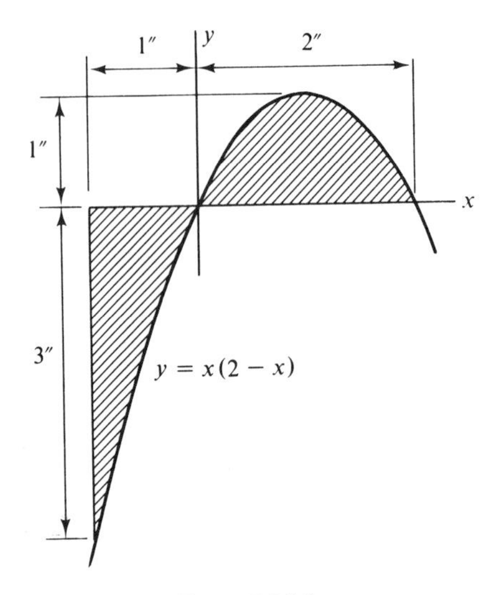

Figure P3-85

3-86 Determine the y coordinate of the centroid of the shaded area in Fig. P3-85.

3-87 Locate the centroid of the shaded area in Fig. P3-87.

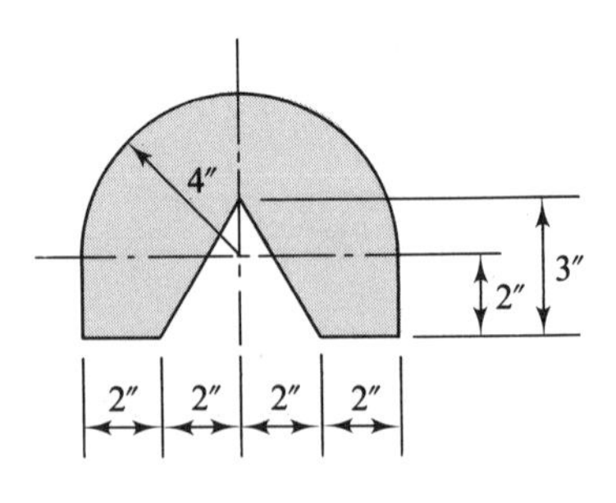

Figure P3-87

***3-88** Figure P3-88 is a sectional view of a solid of revolution. Determine the volume of the body.

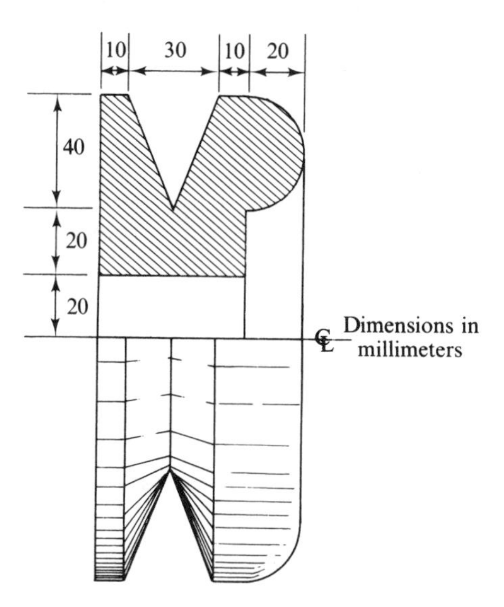

Figure P3-88

***3-89** Determine the total surface area of the body in Problem 3-88.

3-90 The triangle in Fig. P3-90 is rotated about the line $x=4$ in. to generate a solid of revolution. Locate the centroid of the solid.

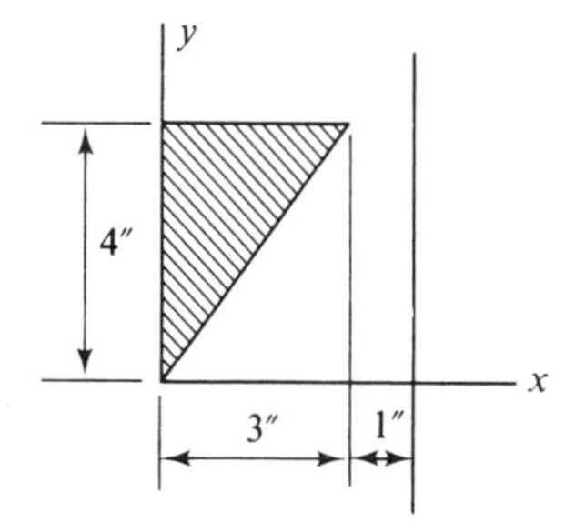

Figure P3-90

3-91 The body in Fig. P3-91 is composed of a cylinder and a hemisphere made of a homogeneous material. The body will be in stable equilibrium as shown if the mass center of the body is below the plane between the cylinder and the hemisphere. Determine the maximum value of h (in terms of r) for stability.

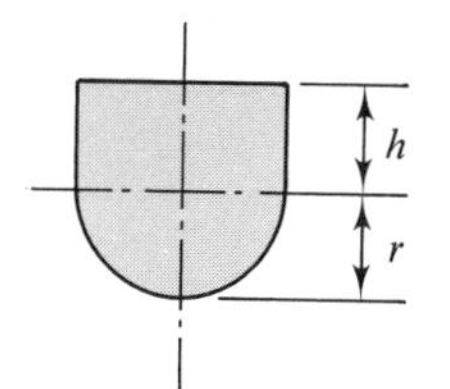

Figure P3-91

3-92 A plate in the vertical side of a water tank has the shape indicated in Fig. P3-92. Determine the center of pressure of the force of the water on the plate.

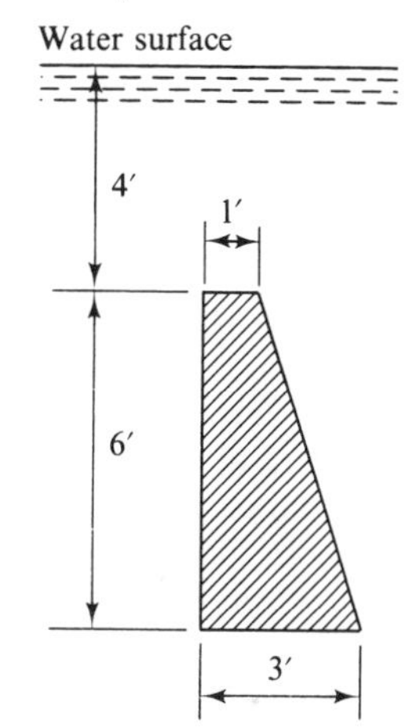

Figure P3-92

***3-93** The equation of the cardioid in Fig. P3-93 is $r=2(1+\cos\theta)$. Locate the centroid of the area bounded by the cardioid.

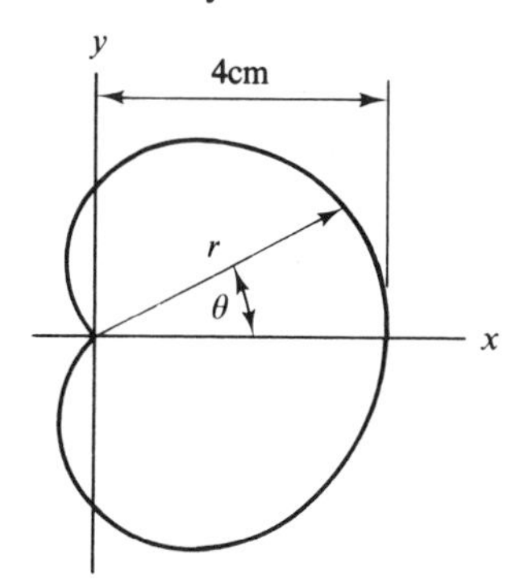

Figure P3-93

3-94 The homogeneous body in Fig. P3-94 consists of a rectangular prism and half of a cylinder. The axis of the hole is parallel to the x axis. Locate the mass center of the body.

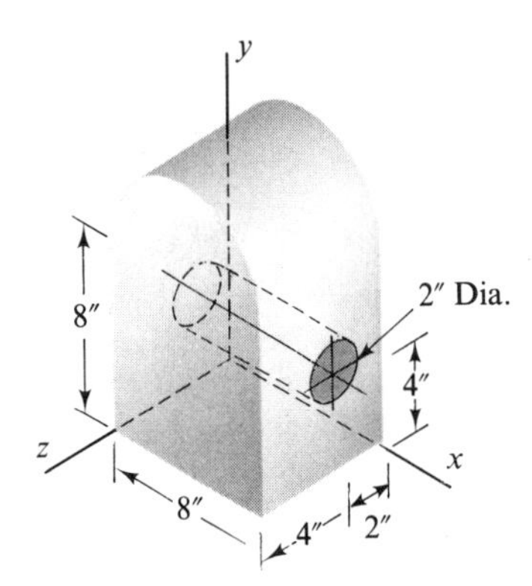

Figure P3-94

3-95 Sand weighing 100 lb per cu ft is placed in a rectangular bin as shown in Fig. P3-95. The surface of the sand forms a plane surface.
(a) Write an expression for the pressure at any point on the bottom of the bin.
(b) Locate the center of pressure of the weight of the sand on the bottom of the bin.

3-96 Locate the mass center of the body generated by revolving the shaded area of Fig.

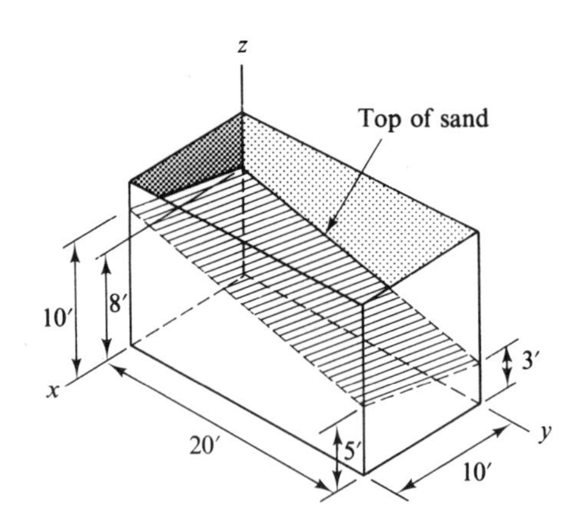

Figure P3-95

P3-96 about the line $x = 1$ ft. The equation of the curve is $2x^2 = y$, and the density of the material varies directly as its distance from the axis of revolution.

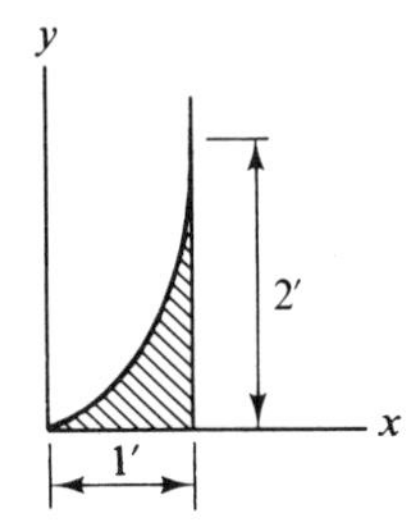

Figure P3-96

4 Equilibrium

4-1 INTRODUCTION

When a system of forces acting on a body has no resultant, the body is in equilibrium. Newton's first law of motion states that if the resultant force acting on a particle is zero, the particle will remain at rest or move with a constant velocity. This law provides the basis for the equations of equilibrium which are developed in this chapter.

In Chapter 2, the resultant of each type of force system studied was determined by obtaining the sum of the forces of the system and the sum of the moments of the forces with respect to a point (or to certain axes). When all these sums are zero for any force system, its resultant is zero, and the body on which the system acts is in equilibrium. The conditions assuring equilibrium of a body with a particular type of force system can therefore be expressed as a set of algebraic or vector equations which must be satisfied. By means of these conditions, it is possible to determine one or more unknown forces acting on a body which is in equilibrium. The number of forces which can be determined depends on the type of force system involved.

To study the force system acting upon any body or any portion of a body, it is first necessary to recognize completely what forces, both known and unknown, act on the body. After the forces acting on a given body have been identified, it is possible to determine the type of force system involved as well as the number of unknowns in the force system. The omission of existing forces or the inclusion of nonexisting forces in a force system is both a serious and a common source of difficulty in statics. The development of a method for the correct identification of every force acting on a body in a given situation is essential for success in the study of equilibrium. The next article is devoted to an explanation of a procedure for recognizing correctly and showing on a diagram all the forces acting on a body under any existing set of conditions.

4-2 FREE-BODY DIAGRAMS

A *free-body diagram* is a sketch of a body, a portion of a body, or two or more connected bodies completely isolated or free from all other bodies, showing the forces exerted by all other bodies on the one being considered. A free-body diagram has three essential characteristics: (1) it is a diagram or sketch of the body; (2) the body is shown completely separated (isolated, cut free) from all other bodies including foundations, supports, and so on; (3) the action on the body of each body removed in the isolating process is shown as a force or forces on the diagram.

Each force in a complete free-body diagram should be labeled either with its known magnitude or with a letter when it is unknown. *The sense of unknown forces, when not obvious at a glance, may be assumed and corrected later if found to be incorrect.* The slope or angle of inclination of all forces not obviously horizontal or vertical should be indicated. Too many dimensions on a free-body diagram may be confusing; consequently, it may be advantageous to show on the diagram only unknown distances to be determined in the solution.

A free-body diagram as defined and explained in the preceding paragraphs is an invaluable aid in identifying correctly all the forces that act on a given body in a given case. To draw a free-body diagram, the type of force exerted on a body by each type of connection or contact must be known. The accompanying tabulation indicates how the action of a few bodies or connections can be represented by a force or forces on a free-body diagram when the bodies or connections are removed.

A knife edge 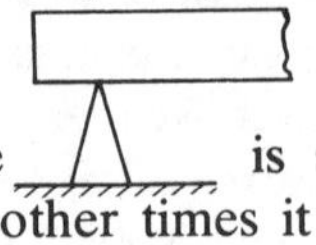 is sometimes used to represent a pin connection, and at other times it represents a smooth surface. Either convention is acceptable if the desired interpretation is understood.

TABLE 4-1

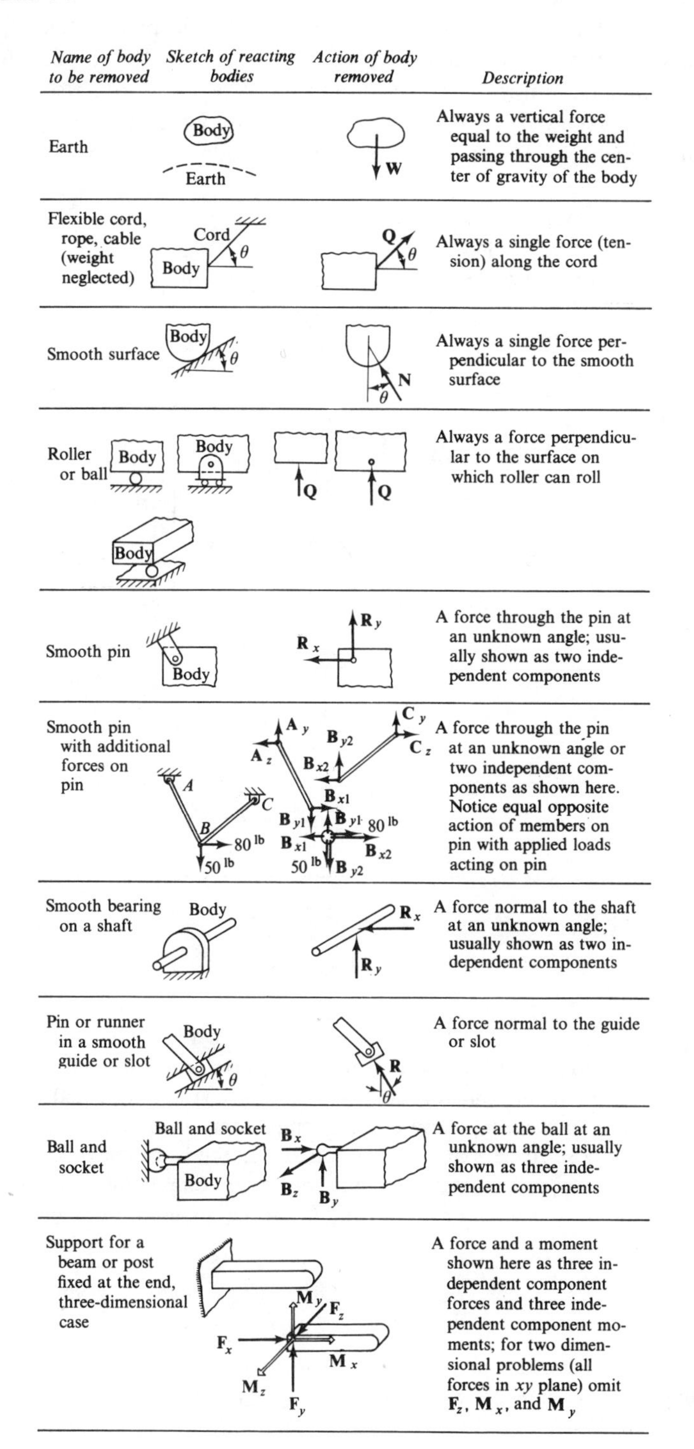

Name of body to be removed	*Sketch of reacting bodies*	*Action of body removed*	*Description*
Earth			Always a vertical force equal to the weight and passing through the center of gravity of the body
Flexible cord, rope, cable (weight neglected)			Always a single force (tension) along the cord
Smooth surface			Always a single force perpendicular to the smooth surface
Roller or ball			Always a force perpendicular to the surface on which roller can roll
Smooth pin			A force through the pin at an unknown angle; usually shown as two independent components
Smooth pin with additional forces on pin			A force through the pin at an unknown angle or two independent components as shown here. Notice equal opposite action of members on pin with applied loads acting on pin
Smooth bearing on a shaft			A force normal to the shaft at an unknown angle; usually shown as two independent components
Pin or runner in a smooth guide or slot			A force normal to the guide or slot
Ball and socket			A force at the ball at an unknown angle; usually shown as three independent components
Support for a beam or post fixed at the end, three-dimensional case			A force and a moment shown here as three independent component forces and three independent component moments; for two dimensional problems (all forces in xy plane) omit $\mathbf{F}_z$, $\mathbf{M}_x$, and $\mathbf{M}_y$

Because of possible confusion, the knife edge is not used in this text. Other types of supports such as thrust bearings, rockers, journal bearings which may resist moments perpendicular to the shaft, etc., may be encountered, and an analysis of possible reactions should be made in such cases to attempt to determine which forces are significant and which can be neglected without introducing serious errors.

EXAMPLE 4-1

Body A in Fig. 4-1(a) weighs 200 lb and the homogeneous bar B weighs 40 lb. Draw a free-body diagram of each of the two bodies.

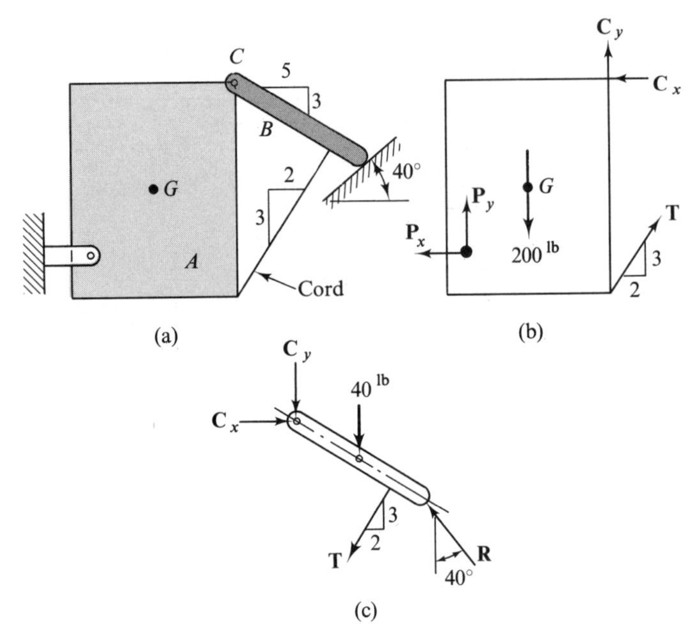

Figure 4-1

SOLUTION

Figures 4-1(b) and 4-1(c) are the two free-body diagrams. The components of the pin reaction at C on body A are assumed to be upward and to the left. These are the components of the reaction of body B on A. The force exerted by A on B has the same magnitude but acts in the opposite direction. Thus the same components are shown downward and to the right on the free-body diagram of B in Fig. 4-1(c).

PROBLEMS

In the following problems, all surfaces are smooth, and weights of members are negligible unless otherwise stated.

4-1 Draw a free-body diagram of the 200-lb rod AB in Fig. P4-1.

4-2 A force P is exerted on each handle of the tongs in Fig. P4-2. Draw a free-body diagram of member AB.

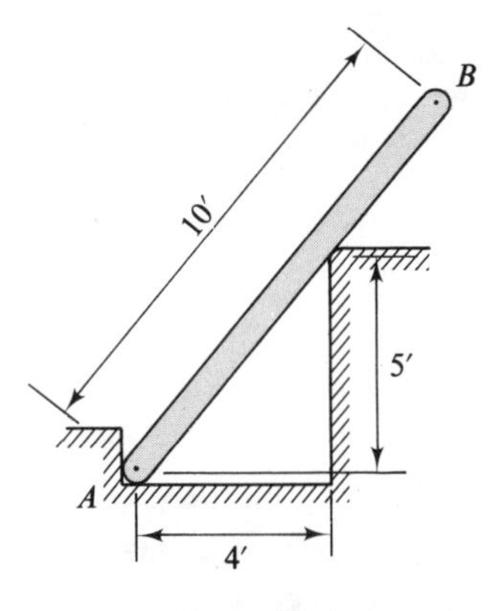

Figure P4-1

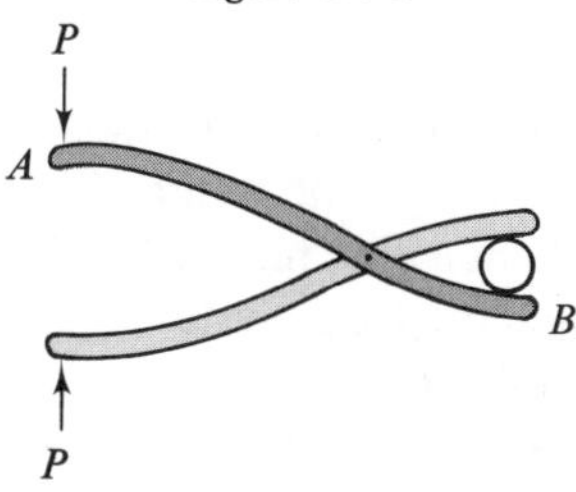

Figure P4-2

4-3 Pipes *A*, *B*, *C*, and *D* in Fig. P4-3 weigh 30, 80, 40, and 60 lb, respectively. The pipes are stacked in the rack as shown. Draw a free-body diagram of (a) member *EF*, (b) pipe *C*.

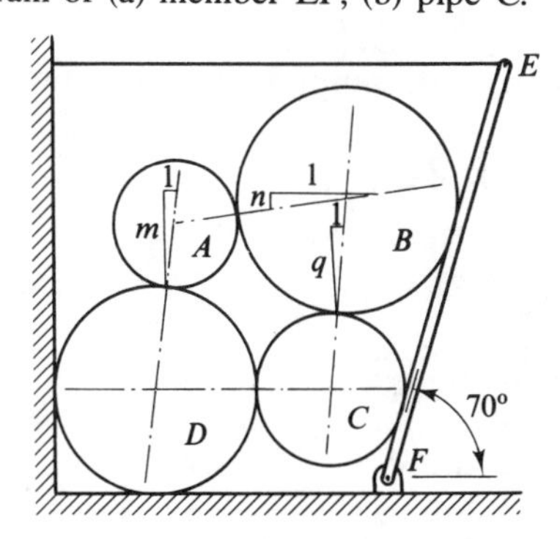

Figure P4-3

4-4 The 100-lb homogeneous rod *AB* in Fig. P4-4 is supported by the 75-lb post *CD* and a cord. Draw a free-body diagram of each of the two bodies.

4-5 Members *AB*, *AC*, and *AD* in Fig. P4-5 each weigh 200 lb, and each is 12 ft long. The three members are joined by a common pin at *A*. Draw free-body diagrams of each of the three members and of the pin at *A*.

***4-6** In Fig. P4-6 body *A* has a mass of 120 kg and *BC* has a mass of 20 kg. Draw free-body diagrams of body *A* and of member *BC*.

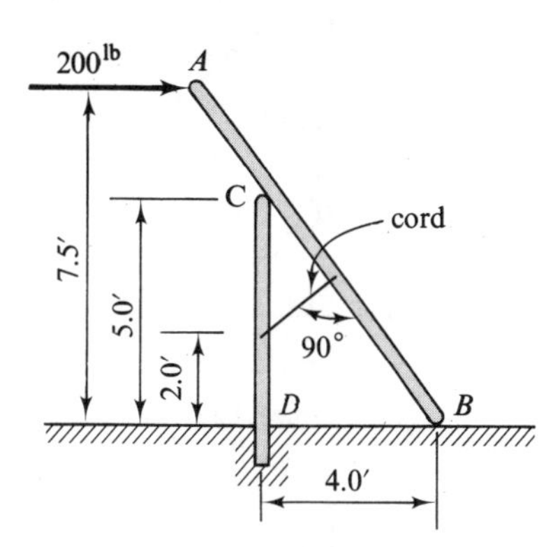

Figure P4-4

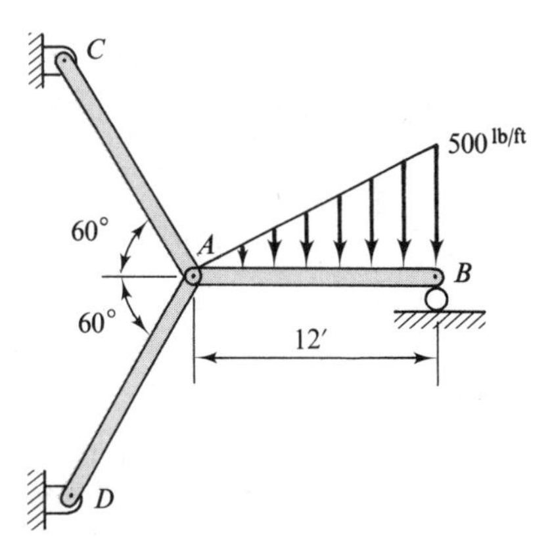

Figure P4-5

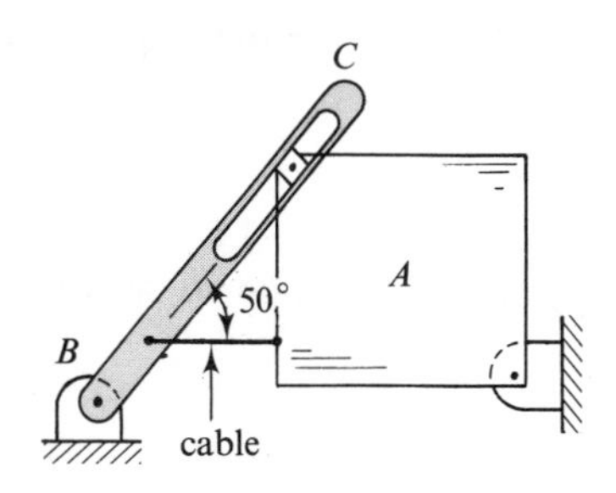

Figure P4-6

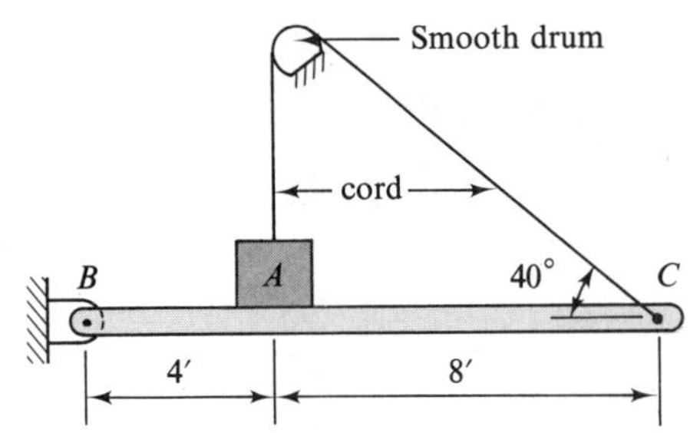

Figure P4-7

4-7 Body A in Fig. P4-7 weighs 500 lb, and the homogeneous member BC weighs 80 lb. Draw free-body diagrams of A and of BC.

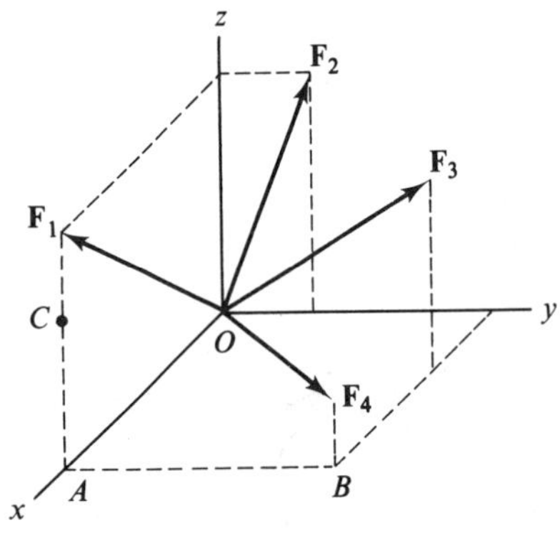

Figure 4-2

4-3 EQUATIONS OF EQUILIBRIUM FOR A CONCURRENT FORCE SYSTEM

The resultant of a concurrent force system is a single force through the point of concurrence (see Art. 2-2). When this resultant force is zero, the body on which the force system acts is in equilibrium. The equation or equations necessary to ensure a zero resultant are the equations of equilibrium.

Consider the concurrent force system shown in Fig. 4-2. The resultant is

$$\mathbf{R} = \mathbf{F}_1 + \mathbf{F}_2 + \mathbf{F}_3 + \mathbf{F}_4,$$

as shown in Art. 2-2. The system is in equilibrium when $\mathbf{R} = 0$; therefore the vector equation of equilibrium is

$$\mathbf{R} = \mathbf{F}_1 + \mathbf{F}_2 + \mathbf{F}_3 + \mathbf{F}_4 = \sum \mathbf{F} = 0. \tag{4-1}$$

Because two equal vectors have equal components, the component equations of equilibrium are

$$\begin{aligned} R_x\mathbf{i} &= (\textstyle\sum F_x)\mathbf{i} = 0, \\ R_y\mathbf{j} &= (\textstyle\sum F_y)\mathbf{j} = 0, \\ R_z\mathbf{k} &= (\textstyle\sum F_z)\mathbf{k} = 0, \end{aligned} \tag{4-2a}$$

and in scalar form, these equations become

$$\begin{aligned} R_x &= \textstyle\sum F_x = 0, \\ R_y &= \textstyle\sum F_y = 0, \\ R_z &= \textstyle\sum F_z = 0. \end{aligned} \tag{4-2b}$$

These three equations can be used to determine three unknown quantities (three magnitudes, three slopes, or any combination of three magnitudes and slopes) for a general (three-dimensional) concurrent force system. When the forces are coplanar, two of the axes (say, the x and y axes) can be selected in the plane of the forces and the equation $\sum F_z = 0$ will not provide any useful information. Consequently there are only two independent equations of equilibrium for a coplanar, concurrent force system, and only two unknowns can be determined.

Similarly, only one unknown can be determined for a collinear force system. When more unknowns are involved than the number of independent equations of equilibrium, the force system is said to be *statically indeterminate.*

It is sometimes convenient to replace one or more of the force equations of equilibrium [Eq. (4-2)] by moment equations, because with properly selected moment centers the necessity of solving simultaneous equations may be avoided. Since the resultant force is zero, the sum of the moments of the forces about any point or line must also be zero. All three component force equations can be replaced by three component moment equations (provided the three moment axes are not mutually concurrent or parallel and none of them passes through the point of concurrence of the force system), although such a procedure is rarely advantageous. To prove that force equations can be replaced by moment equations consider the concurrent force system of Fig. 4-2. If the sum of the moments of the forces of the system about a line, such as AB in the xy plane of Fig. 4-2, is zero, any possible resultant must lie in the plane OAB (the xy plane). Thus the equation $\sum M_{AB} = 0$ provides the same information as the equation $\sum F_z = 0$; that is, there can be no z component of the resultant force. In the same manner, the equation $\sum M_{AC} = 0$ is equivalent to the force equation $\sum F_y = 0$. If both of the preceding moment equations are satisfied, the only possible resultant is a force along the line OA. The axis for the third moment equation of equilibrium could be any line not parallel to OA and which does not intersect OA. For example, the moment axis might be a line through B parallel to the z axis or one through C parallel to the y axis.

Equilibrium can also be assured by means of two moment equations where moments are summed with respect to a point instead of a line. The two moment centers must not be collinear with the point of concurrence. Since the moment about a point can be represented by a vector, each moment equation would be equivalent to three component equations. The three component equations for one point are not independent, however, and they would not eliminate the possibility of a force through the reference point. Thus, although equating the moments of the force system to zero about each of two points is equivalent to six component equations, only three of the resulting equations are independent, and only three unknowns can be determined for any concurrent force system.

4-4 GENERAL PROCEDURE FOR THE SOLUTION OF PROBLEMS IN EQUILIBRIUM

One of the important abilities the successful engineer must develop is a logical, orderly method of attacking a problem. The following sequence of steps is designed to aid in organizing the analysis and solution of any problem in equilibrium. It is not considered desirable that these steps be memorized but rather that they be used as a guide to help develop an orderly procedure in analyzing equilibrium problems. The steps are listed in the order in which they should be performed in the solution.

STEP-BY-STEP PROCEDURE

1. Determine carefully what data are given and what results are required.
2. Draw a *free-body diagram* of the member or group of members on which some or all of the unknown forces are acting.
3. Observe the type of force system which acts on the free-body diagram.
4. Note the *number* of independent equations of equilibrium available for the type of force system involved.
5. Compare the number of unknowns on the free-body diagram with the number of *independent* equations of equilibrium available for the force system.
6. (a) If there are as many independent equations of equilibrium as unknowns, proceed with the solution by writing and solving the equations of equilibrium.

(b) If there are more unknowns[1] to be evaluated than *independent* equations of equilibrium available, *draw a free-body diagram* of another body and repeat steps 3, 4, and 5 for the second free-body diagram drawn.

7. (a) If there are as many independent equations of equilibrium as unknowns for the second free-body diagram, proceed with the solution by writing and solving the necessary equations of equilibrium.

(b) If there are more unknowns to be evaluated than independent equations of equilibrium for the second free-body diagram, compare the *total* number of unknowns on both free-body diagrams with the *total* number of independent equations of equilibrium available for both diagrams.

8. If there are as many independent equations of equilibrium as unknowns for both diagrams, proceed to solve the problem by writing and solving the equations of equilibrium. If there are more unknowns than independent equations, repeat steps 6(b) and 7. If there are still too many unknowns after as many free-body diagrams have been drawn as there are individual bodies in the problem, then the problem is statically indeterminate; that is, not all of the unknowns can be evaluated by statics alone.

The selection of the most desirable member or combination of members for step 2 is largely a matter of experience. Sometimes free-body diagrams of each separate member are required, and the initial choice is of no consequence. In other instances, a free-body

[1] Special cases may arise in which certain unknowns (but never all of them) may be evaluated even though there are more unknowns than independent equations of equilibrium.

diagram of two or more members as a unit may provide the best approach. Free-body diagrams can be quickly and easily drawn, and when the most direct approach is not readily apparent, it is probably better to draw several free-body diagrams and check steps 3, 4, and 5 than to spend too much time trying to decide where to start.

Many students become confused and fail to obtain the solution of a problem in equilibrium by writing equations that are not independent before making the analysis indicated in steps 3, 4, and 5 to see that additional free-body diagrams are necessary. This frequently means a waste of time and effort in attempting the impossible.

The procedure for determining the unknown forces in a balanced concurrent force system is illustrated in the following examples, in which the steps outlined above will be referred to by number. The sense of each unknown force is assumed where not readily apparent. The magnitude of a force is independent of the assumed sense. A positive result indicates that the assumed sense is correct; a negative result indicates that the sense of the force is opposite to that assumed.

EXAMPLE 4-2

The 300-lb shaft *M* and the 500-lb shaft *N* are supported as shown in Fig. 4-3(a). Neglecting friction at the contact surfaces *P*, *Q*, *R*, and *S*, determine the reactions at *R* and *S* on shaft *N*. Assume that the resultant weights of *M* and *N* and all reactions of the surfaces lie in one plane.

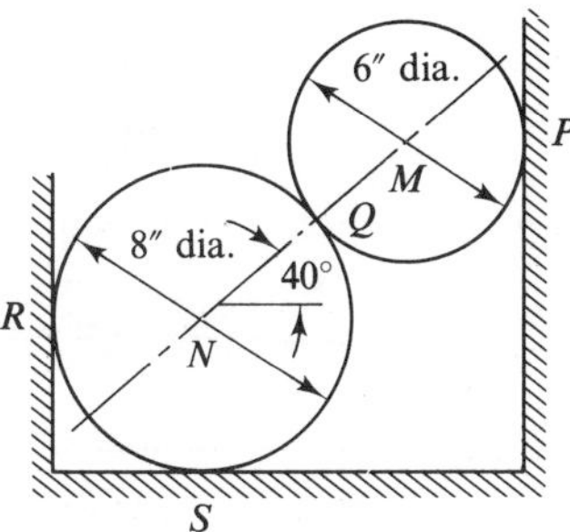

Figure 4-3(a)

SOLUTION

Step 1 is contained in the statement of the problem. Since the desired forces act on body *N*, a free-body diagram, Fig. 4-3(b), of body *N* is drawn for step 2. The force system acting on *N* is a concurrent, coplanar system (step 3), and only two of the equations of equilibrium [Eq. (4-2)] are independent for such a system (step 4). The system contains three unknowns (step 5), the magnitudes of the forces **Q**, **R**, and **S**. Since there are more unknowns than independent equations of equilibrium, the unknown magnitudes cannot all be determined from this free-body diagram.

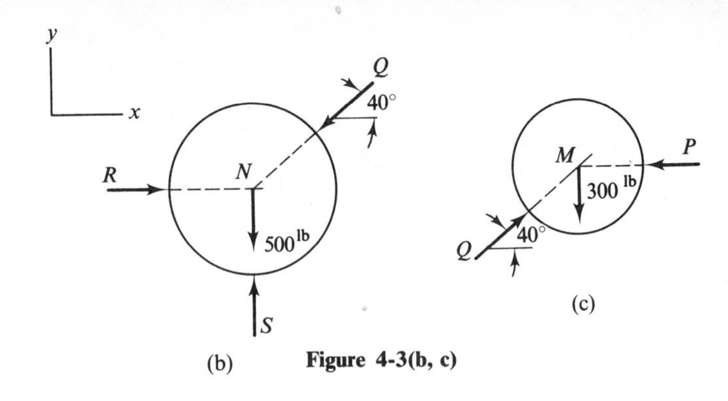

Figure 4-3(b, c)

The next step, 6(b), is drawing a free-body diagram of body M, as in Fig. 4-3(c). Notice that the force exerted by shaft M on N is shown as the force $\mathbf{Q}$ downward to the left in Fig. 4-3(b), whereas the force exerted by shaft N on M is shown as the force $\mathbf{Q}$ upward to the right in Fig. 4-3(c). These forces are equal in magnitude, collinear, and opposite in sense as indicated by Newton's third law of motion, which states that for every action there is an equal and opposite reaction.

The concurrent, coplanar force system of body M has two independent equations of equilibrium and contains only two unknowns, the magnitudes of forces $\mathbf{P}$ and $\mathbf{Q}$. Force $\mathbf{Q}$ can therefore be determined from the free-body diagram in Fig. 4-3(c), and there will be only two unknown forces remaining on the free-body diagram of body N with two equations of equilibrium available. Thus the analysis of the problem is complete.

Positive axes are selected as shown, and the equation of equilibrium for body M, Fig. 4-3(c), is

$$\sum \mathbf{F} = 0 = Q(\mathbf{i}\cos 40° + \mathbf{j}\sin 40°) - P\mathbf{i} - 300\mathbf{j}$$
$$= \mathbf{i}(Q\cos 40° - P) + \mathbf{j}(Q\sin 40° - 300).$$

This equation is satisfied only if both components are equal to zero. Therefore

$$Q\cos 40° - P = 0$$

and

$$Q\sin 40° - 300 = 0$$

from which $Q = 467$ lb and $P = 358$ lb. Note that since the force $\mathbf{P}$ is not needed for later calculations and was not asked for in the statement of the problem, the first of the two equations need not be solved.

The equation of equilibrium for body N, Fig. 4-3(b), is

$$\sum \mathbf{F} = 0 = R\mathbf{i} + S\mathbf{j} - 500\mathbf{j} + Q(-\mathbf{i}\cos 40° - \mathbf{j}\sin 40°)$$
$$= \mathbf{i}(R - Q\cos 40°) + \mathbf{j}(S - 500 - Q\sin 40°).$$

When 467 lb is substituted for Q and each of the component expressions is equated to zero the results are

$$R = 358 \text{ lb} \quad \text{and} \quad S = 800 \text{ lb}.$$

The reactions on the shaft N are

$$\mathbf{R} = \underline{358\mathbf{i}\text{ lb}} \quad \text{and} \quad \mathbf{S} = \underline{800\mathbf{j}\text{ lb}}. \qquad \text{Ans.}$$

EXAMPLE 4-3

The block W in Fig. 4-4(a) has a mass of 500 kg and is supported by three wires as shown. Determine the tension in wire AB.

SOLUTION

Step 1 is contained in the statement of the problem. A wire is a flexible body and therefore exerts a single force (tension) along the wire. A free-body diagram of the connection at A is shown in Fig. 4-4(b) (step 2). The force system acting on the connection at A is a concurrent system in space (step 3), and either the general vector equation of equilibrium, Eq. (4-1), or the three component equations of equilibrium, Eq. (4-2), can be used for the solution (step 4). The free-body diagram has three unknown magnitudes, T_B, T_C, and T_D (step 5). Since there are three unknowns and three independent equations of equilibrium (from the vector equation), the analysis is complete [step 6(a)]. To apply Eq. (4-1), the forces in each of the wires should be resolved into components by multiplying the magnitudes by the corresponding unit vectors along each of the wires.

The unit vectors are

$$\mathbf{n}_{AB} = \frac{-1.5\mathbf{i}+3.0\mathbf{j}+3.0\mathbf{k}}{[(-1.5)^2+(3.0)^2+(3.0)^2]^{1/2}} = \frac{-1.5\mathbf{i}+3.0\mathbf{j}+3.0\mathbf{k}}{4.5}$$

$$= \frac{-\mathbf{i}+2\mathbf{j}+2\mathbf{k}}{3}.$$

Similarly,

$$\mathbf{n}_{AC} = \frac{-1.5\mathbf{i}+6.0\mathbf{j}-2.0\mathbf{k}}{6.5} = \frac{-3\mathbf{i}+12\mathbf{j}-4\mathbf{k}}{13},$$

$$\mathbf{n}_{AD} = \frac{1.5\mathbf{i}+2.5\mathbf{j}-2.0\mathbf{k}}{3.535} = \frac{3\mathbf{i}+5\mathbf{j}-4\mathbf{k}}{7.07}.$$

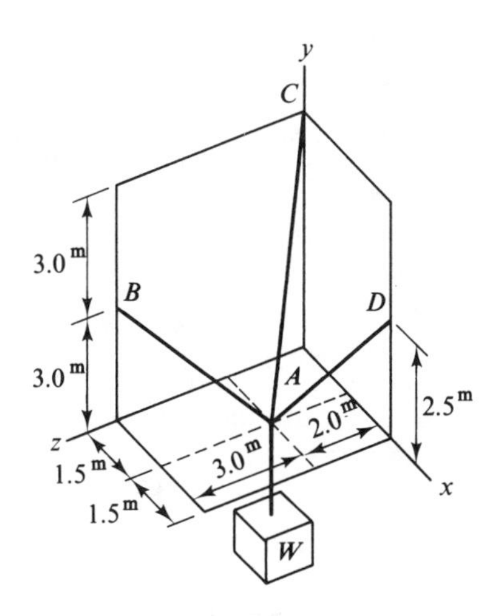

Figure 4-4(a)

The equation of equilibrium is

$$\mathbf{R}=\sum \mathbf{F}=\mathbf{T}_B+\mathbf{T}_C+\mathbf{T}_D+\mathbf{W}=0,$$

where $\mathbf{W}=-mg\mathbf{j}=-500(9.81)\mathbf{j}\,\text{N}=-4905\mathbf{j}\,\text{N}$, as shown in Art. 1–8. When the forces are expressed in terms of unit vectors the equilibrium equation becomes

$$\mathbf{R}=\frac{T_B(-\mathbf{i}+2\mathbf{j}+2\mathbf{k})}{3}+\frac{T_C(-3\mathbf{i}+12\mathbf{j}-4\mathbf{k})}{13}+\frac{T_D(3\mathbf{i}+5\mathbf{j}-4\mathbf{k})}{7.07}$$

$$-4905\mathbf{j}$$

$$=\mathbf{i}\left(-\frac{T_B}{3}-\frac{3T_C}{13}+\frac{3T_D}{7.07}\right)+\mathbf{j}\left(\frac{2T_B}{3}+\frac{12T_C}{13}+\frac{5T_D}{7.07}-4905\right)$$

$$+\mathbf{k}\left(\frac{2T_B}{3}-\frac{4T_C}{13}-\frac{4T_D}{7.07}\right)=0.$$

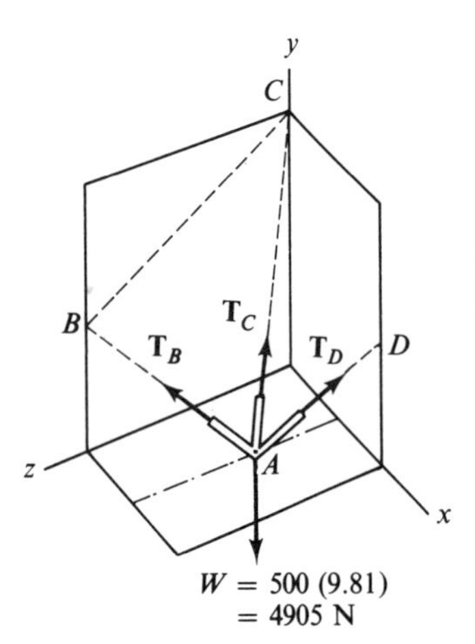

Figure 4-4(b)

Each of the components of **R** must be zero if the system is to be in equilibrium; thus the following equations must be satisfied:

$$-\frac{T_B}{3}-\frac{3T_C}{13}+\frac{3T_D}{7.07}=0,$$

$$\frac{2T_B}{3}+\frac{12T_C}{13}+\frac{5T_D}{7.07}-4905=0,$$

$$\frac{2T_B}{3}-\frac{4T_C}{13}-\frac{4T_D}{7.07}=0.$$

The simultaneous solution of these equations gives

$$T_B=2890\text{ N},\qquad T_C=1045\text{ N},\qquad T_D=2840\text{ N},$$

and the tension of wire AB is

$$T_{AB}=\underline{2890\text{ N.}}\qquad\qquad \text{Ans.}$$

ALTERNATE SOLUTION

In the preceding solution, it was necessary to solve three simultaneous equations in three unknowns. It is possible to avoid the solution of simultaneous equations by equating the sum of the components of the forces in a properly selected direction to zero. The force T_B is to be determined, and any equation which will eliminate the other two unknowns can be used to calculate $\mathbf{T}_B$. The equation

$$\sum \mathbf{F}_n=0,$$

where the **n** direction is perpendicular to the plane containing $\mathbf{T}_C$ and $\mathbf{T}_D$, will not involve either of these forces and thus can be used to calculate $\mathbf{T}_B$. The vector

$$\mathbf{N}=\mathbf{AC}\times\mathbf{AD}$$

is perpendicular to $\mathbf{T}_C$ and $\mathbf{T}_D$, and the equation

$$\mathbf{n}\cdot\sum \mathbf{F}=\frac{\mathbf{N}}{N}\cdot\sum \mathbf{F}=0$$

can be used to calculate T_B. If both sides of the preceding equation are multiplied by N, it becomes

$$\mathbf{N}\cdot\sum\mathbf{F}=(\mathbf{AC}\times\mathbf{AD})\cdot\sum\mathbf{F}=0. \qquad \text{(a)}$$

From the diagram

$$\mathbf{AC}=-1.5\mathbf{i}+6.0\mathbf{j}-2.0\mathbf{k},$$

$$\mathbf{AD}=1.5\mathbf{i}+2.5\mathbf{j}-2.0\mathbf{k},$$

$$\sum\mathbf{F}=\mathbf{T}_B+\mathbf{W}$$

$$=\frac{T_B(-\mathbf{i}+2\mathbf{j}+2\mathbf{k})}{3}-4905\mathbf{j},$$

and Eq. (a), in determinant form, becomes

$$\begin{vmatrix} -1.5 & 6.0 & -2.0 \\ 1.5 & 2.5 & -2.0 \\ -\frac{T_B}{3} & \frac{2T_B}{3}-4905 & \frac{2T_B}{3} \end{vmatrix}=0.$$

Note that T_C and T_D are omitted since they have no component in the $\mathbf{N}$ direction. The determinant can be expanded to give

$$-1.5\left(\frac{5T_B}{3}+\frac{4T_B}{3}-9810\right)-6.0\left(\frac{3T_B}{3}-\frac{2T_B}{3}\right)-2.0\left(\frac{3T_B}{3}-7358+\frac{2.5T_B}{3}\right)=0,$$

which reduces to

$$T_B=\underline{2890\text{ N.}} \qquad \text{Ans.}$$

PROBLEMS

4-8 Determine the forces in cables A and B in Fig. P4-8.

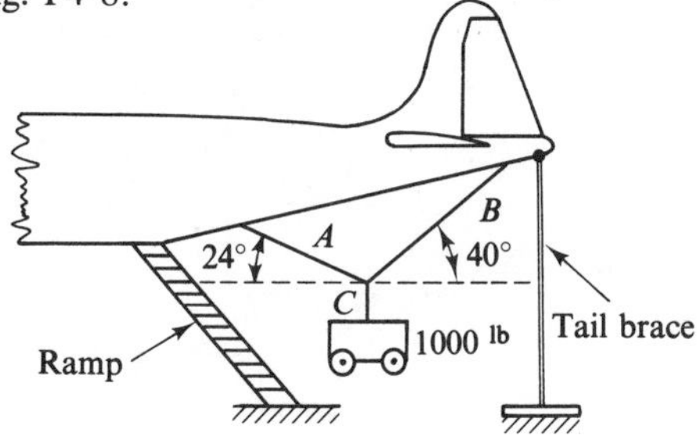

Figure P4-8

4-9 The smooth cylinders A and B in Fig. P4-9 weigh 100 and 30 lb, respectively.
(a) Determine all forces acting on body A when the magnitude of $\mathbf{P}$ is 200 lb.
(b) Determine the maximum magnitude $\mathbf{P}$ can have without lifting body A off the floor.

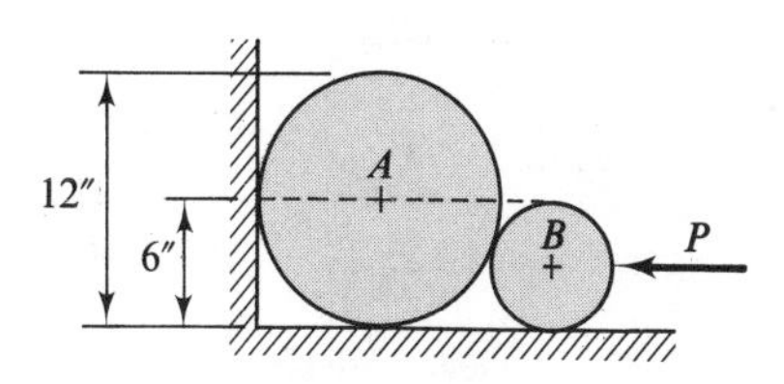

Figure P4-9

4-10 In Fig. P4-10, A and B are identical smooth cylinders weighing 100 lb each. Determine the maximum force $\mathbf{P}$ which can be applied as shown without causing A to leave the floor.

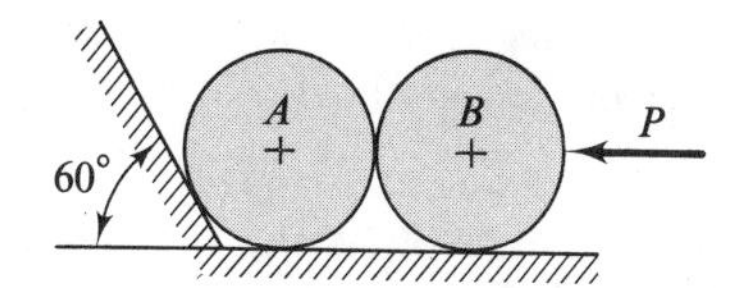

Figure P4-10

*4-11 Cables A and B in Fig. P4-11 support bodies W_1 and W_2 as shown. Body W_1 has a mass of 450 kg and W_2 has a mass of 90 kg. Determine the forces cables A and B exert on the support.

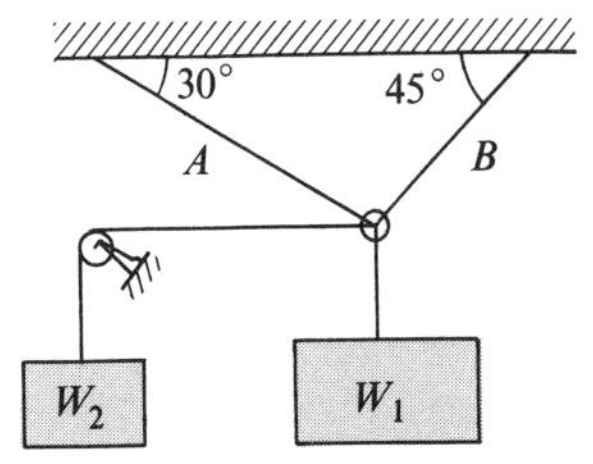

Figure P4-11

*4-12 The maximum safe forces in cables A and B in Fig. P4-11 are 3.6 and 4.4 kN, respectively. Block W_2 has a mass of 90 kg. Determine the maximum mass of W_1 if neither of the cable forces is to exceed its safe value.

4-13 The 40-lb weight in Fig. P4-13 is attached to a frictionless pulley which rides on the cable. The cable is attached to a wall support at A and runs over a second frictionless pulley at B at the same level. A 60-lb weight hangs from the end of the cable. Determine the angle α for equilibrium.

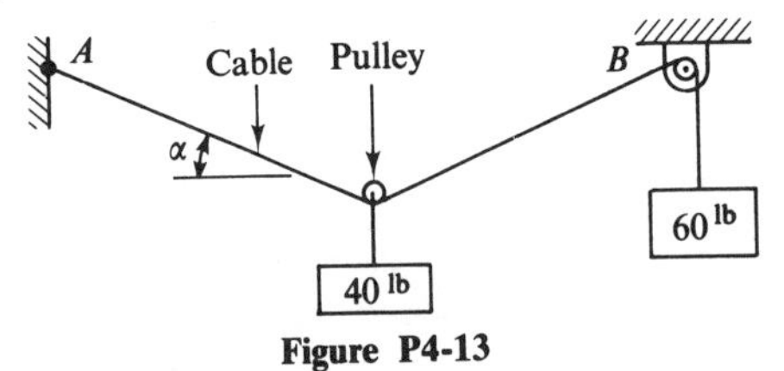

Figure P4-13

4-14 A workman attempting to pull a pile out of the ground tied a cable to it at A (see Fig. P4-14). He fixed the other end of the cable at B, attached another cable to the point C, and fixed this cable at D. Then he exerted a pull of 200 lb on the second cable at E. Before the pile began to move, AC was vertical, EC was horizontal, BC made an angle of 18° with the vertical, and DE made an angle of 11° with the horizontal. Determine the tension in AC.

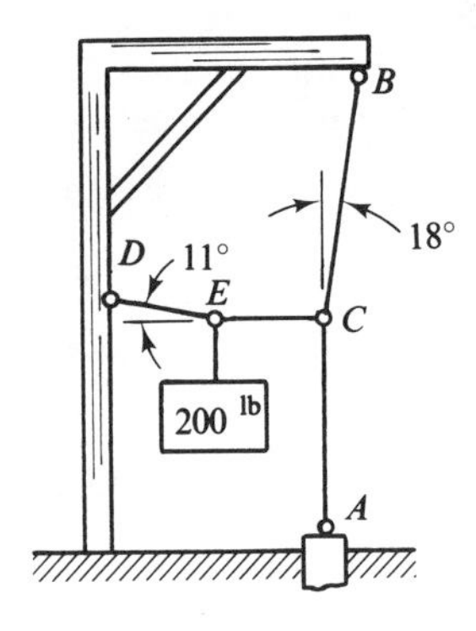

Figure P4-14

4-15 In Fig. P4-15, A and B are identical rollers with a weight of W, and C is a roller with a weight of $W/3$. If the force **P** is increased gradually, either A or B will be forced off the floor. Determine the force **P** when this occurs and the forces acting on the other two rollers. All surfaces are smooth.

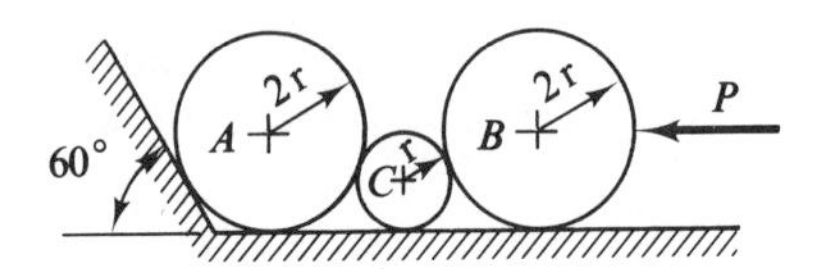

Figure P4-15

4-16 Figure P4-16 shows several identical smooth rollers weighing W lb each stacked on an inclined plane. Determine

(a) The maximum number of rollers which will lie in a single row as shown.

(b) All forces acting on roller A under condition (a).

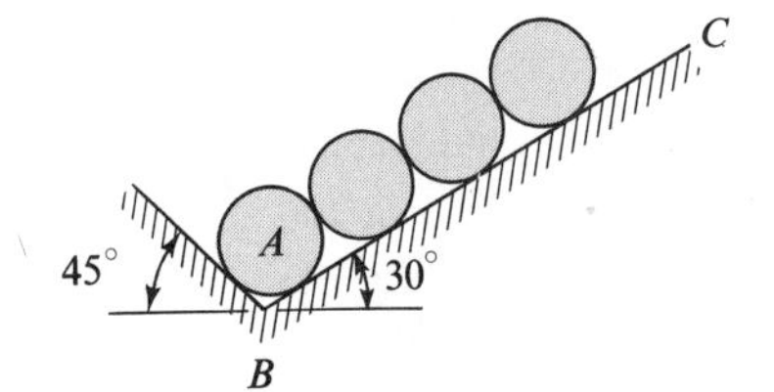

Figure P4-16

4-17 Solve Problem 4-16 if the plane BC is replaced by a V-groove with a 120° angle between the planes forming the groove and if the rollers are replaced with identical smooth spheres weighing W lb each; see Fig. P4-17. (It is suggested that one axis be selected along the line of centers of the balls.)

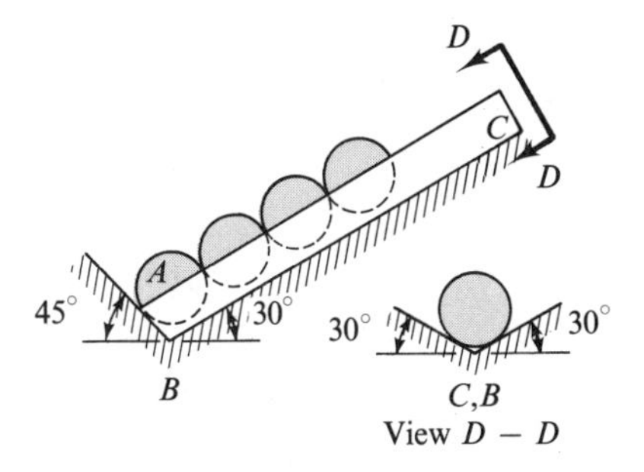

Figure P4-17

4-18 A blimp is held by three cables as shown in Fig. P4-18. The resultant upward force developed on the blimp is 8000 lb, and there is a 1000-lb drag force due to the wind which is horizontal and perpendicular to cable AB. The guide vanes keep the blimp headed into the wind. Determine the tension in cable AB.

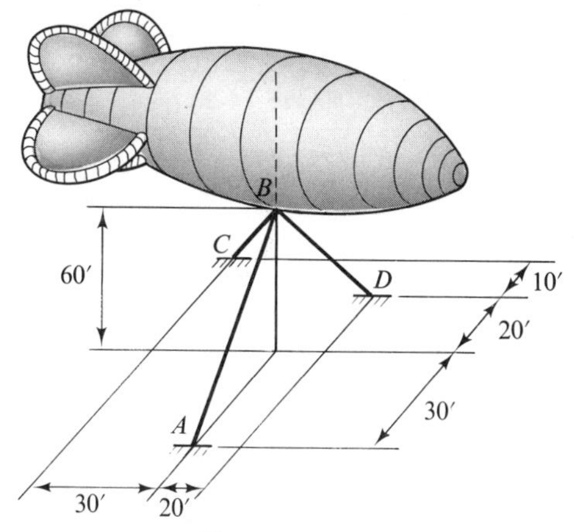

Figure P4-18

*4-19 A smooth homogeneous 150-kg sphere is held in an upper corner of a rectangular room by a force of 3.6 kN that has a line of action through the mass center of the sphere and the corner. The sphere touches the ceiling and the two intersecting walls. Determine the reaction of either wall and of the ceiling on the sphere.

4-20 The 500-lb roller C in Fig. P4-20 is connected to bodies A (260 lb) and B (300 lb) by ropes passing over smooth pulleys. Determine

(a) The force $\mathbf{P}$ needed to maintain equilibrium when $\theta = 0°$ and the corresponding reaction of the floor on C.

(b) The maximum angle θ for which equilibrium can be maintained and the corresponding magnitude of $\mathbf{P}$.

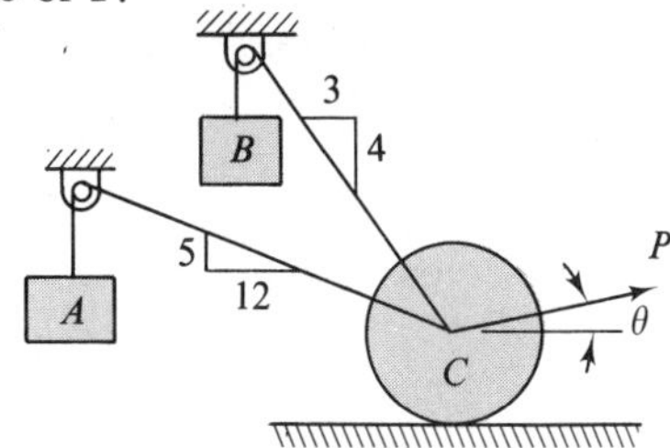

Figure P4-20

4-21 A 60-in. cord is connected to points A and B in Fig. P4-21. A load W is suspended from a small pulley which rides along the cord. Determine

(a) The force in the cord between A and the pulley.

(b) The horizontal distance x from the vertical wall to the pulley.

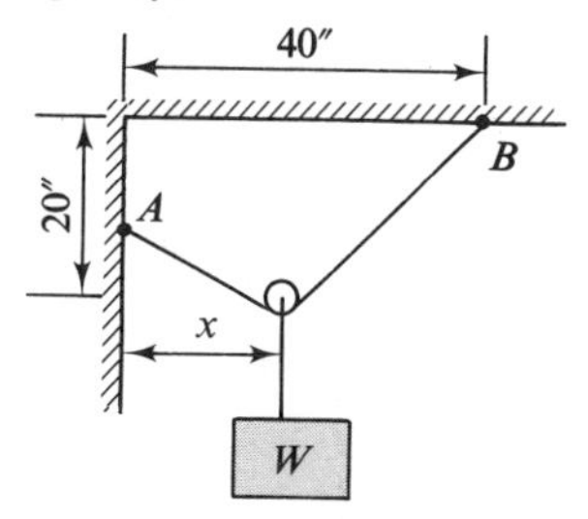

Figure P4-21

4-22 The 500-lb body W in Fig. P4-22 is supported by three wires A, B, and C. Determine the force in each of the three wires.

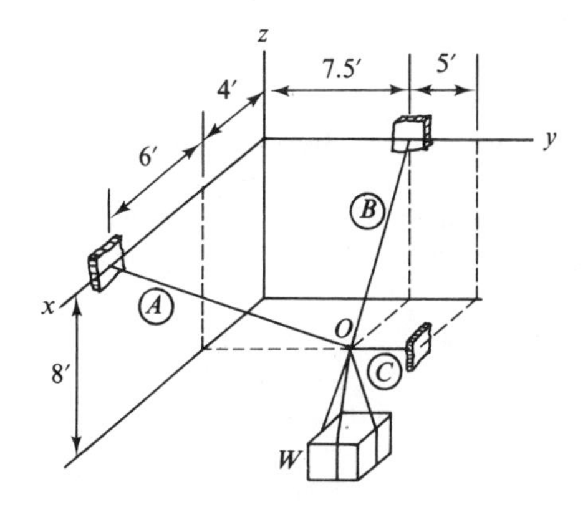

Figure P4-22

4-23 An additional force **P** is applied to point O in Fig. P4-22 parallel to the x axis. Determine the force **P** which will reduce the force in wire A to zero and the corresponding forces in wires B and C.

4-24 Solve Problem 4-23 if the force is to be zero in wire B (instead of in wire A).

4-25 An amateur woodsman plans to cut down tree A in Fig. P4-25 and wants it to fall toward tree B. He cuts a notch at C and makes a saw cut at D. As an added precaution he ties a rope between trees A and B as shown. His "helper" applies a force of 75 lb at E, which is 4 ft above the ground and 2 ft from the xy plane containing the trees. The 75-lb force is in the plane of the two sections of rope and is also perpendicular to the x axis. Determine the force in each section of the rope.

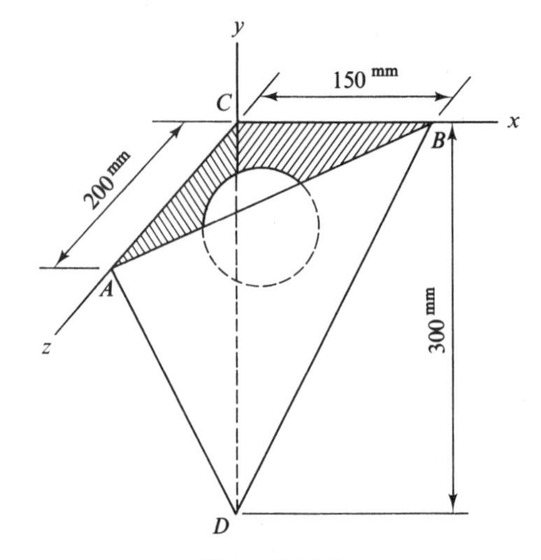

Figure P4-26

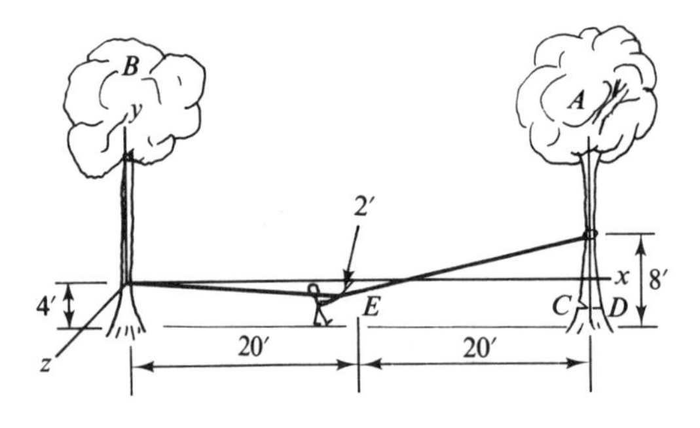

Figure P4-25

*__4-26__ Three triangular plates are fastened together to form the pyramidal-shaped container in Fig. P4-26. The edge CD is vertical, and points A, B, and C lie in the xz plane. A 50-kg sphere is placed in the container. Determine the reaction of each of the planes on the sphere.

*__4-27__ Solve Problem 4-26 if plate ABD is vertical [the coordinates of D are $(9, -30, 8)$] and plates ACD and BCD are both inclined with respect to the vertical as shown in Fig. P4-27.

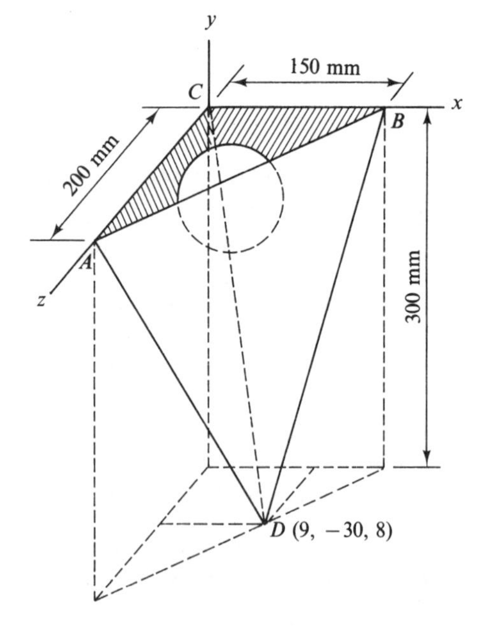

Figure P4-27

4-5 EQUILIBRIUM OF BODIES ACTED ON BY TWO OR THREE FORCES

When a body is acted on by only two forces, it is called a *two-force* body or member. A *three-force* member is one which is acted on by three, and only three, nonparallel forces.

If a body is held in equilibrium by two forces, the forces must be collinear, equal in magnitude, and opposite in sense. The body will be in equilibrium only if the resultant force and the resultant moment are both zero. For the vector sum of two forces to be zero, the forces must be equal in magnitude and opposite in sense. The forces must be collinear to have a zero moment; otherwise they would form a couple. The preceding fact or principle will frequently simplify the solution of problems in equilibrium, and it is particularly useful in the solution of truss problems, as will be shown later.

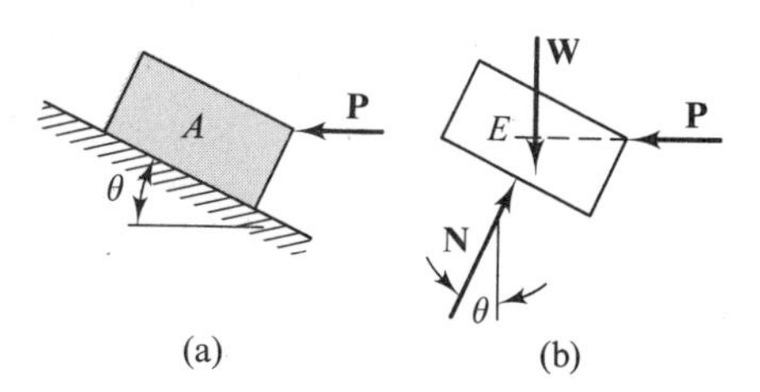

Figure 4-5

When a body is held in equilibrium by three nonparallel forces they must be concurrent and coplanar. To prove the statement, consider the block A of Fig. 4-5(a), which is held in equilibrium on the smooth plane by three forces as indicated in Fig. 4-5(b). If the resultant of the force system is to be zero, the resultant of two of the forces, such as **P** and **W**, must be collinear with the third force and also have the same magnitude but be opposite in sense. The resultant of the two nonparallel forces **P** and **W** will be a single force only if they are concurrent, which means that they must be coplanar (see Art. 2-4); consequently, the force **N** must lie in the same plane and pass through the point of concurrence. In general, the resultant reaction of a plane on a block can act anywhere along the block, but in this case it must act through point E. If the force **P** were moved up or down, the line of action of **N** would be changed in such a manner that **N** would pass through the new position of E. The fact that the three forces are concurrent can often be used to advantage in locating the point of intersection of three forces that hold a body in equilibrium and thus provide a simple solution to some equilibrium problems. More general methods of solving this type of problem are developed in Art. 4-6.

EXAMPLE 4-4

The uniform bar AB of Fig. 4-6(a) weighs W lb and has a length of L ft. The bar is placed with its ends on smooth inclined planes. Determine the angle θ at which the bar will be in equilibrium.

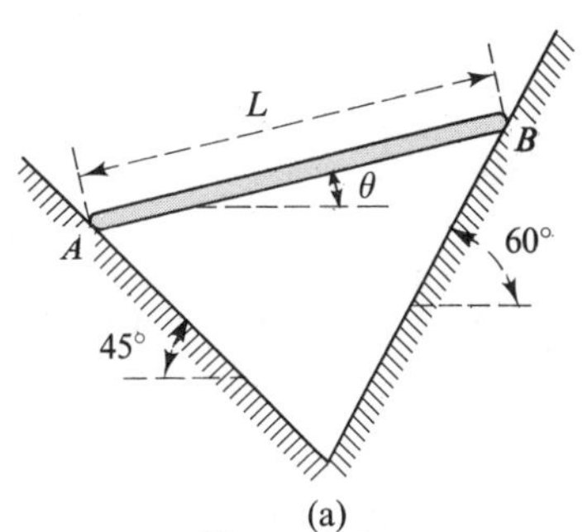

Figure 4-6(a)

SOLUTION

A free-body diagram of the bar is drawn in Fig. 4-6(b). Since there are only three forces acting on the bar, they must intersect at a common

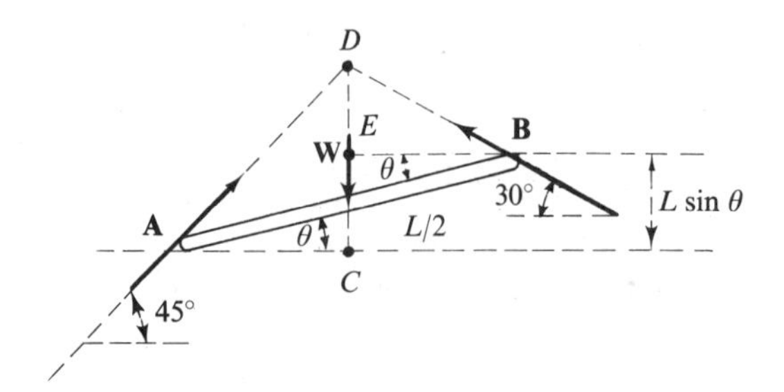

Figure 4-6(b)

point, D. There are two unknown forces $\mathbf{F}_A$ and $\mathbf{F}_B$. This is a concurrent force system, and there are two equations of equilibrium. The unknown forces can be determined in terms of the weight W from the equilibrium equations without knowing the value of the angle θ. The angle θ can be determined from the geometry of the diagram. The length $AC = (L/2)\cos\theta$, and $DC = AC$ since the triangle ACD is a 45° isosceles right triangle. Also $EB = (L/2)\cos\theta$, and

$$DE = EB \tan 30° = \frac{L}{2}\cos\theta\left(\frac{1}{\sqrt{3}}\right).$$

From the diagram

$$DC - DE = L \sin\theta$$

or

$$\frac{L}{2}\cos\theta - \frac{L}{2}\cos\theta\left(\frac{1}{\sqrt{3}}\right) = L\sin\theta$$

from which

$$(\sqrt{3}-1)\cos\theta = 2\sqrt{3}\sin\theta,$$

$$\tan\theta = \frac{\sqrt{3}-1}{2\sqrt{3}} = \frac{1.732-1}{2(1.732)} = 0.211,$$

and

$$\underline{\theta = 11.9°.} \qquad \text{Ans.}$$

PROBLEMS

All connections are smooth, and all members are of negligible weight unless otherwise specified.

4-28 The center of gravity of the 50-lb ladder in Fig. P4-28 is 6 ft from the left wall. Determine all unknown forces acting on the ladder.

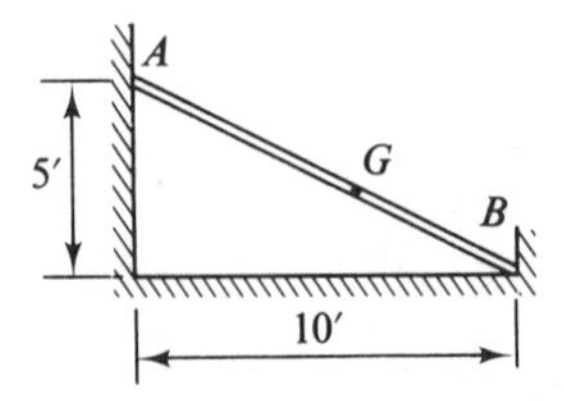

Figure P4-28

4-29 The homogeneous rod AB in Fig. P4-29 weighs W pounds and is supported as shown by the horizontal cable BC. Determine the tension in the cable and the reaction at A on AB.

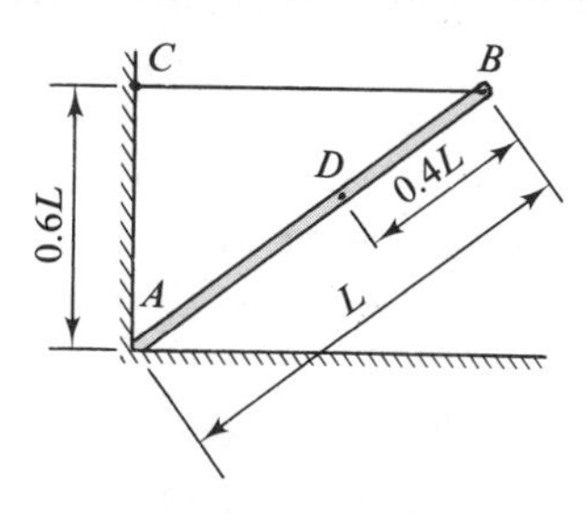

Figure P4-29

4-30 Solve Problem 4-29 if the cable from C to B is replaced by one from C to D.

4-31 Determine the point where the cable should be attached to AB in Problem 4-29 which will result in a horizontal reaction at A.

***4-32** A passenger vessel has an arrangement for suspending its lifeboats by supports at their ends similar to that shown in Fig. P4-32. There is a socket at O and a smooth bearing through the deck rail at M. A flexible cable passes over the locked pulley. The boat fully loaded has a mass of 2.1 Mg. Determine the resultant reaction at M and at O on one of the two supports for the boat. Consider the load concentrated at the midpoint of the boat.

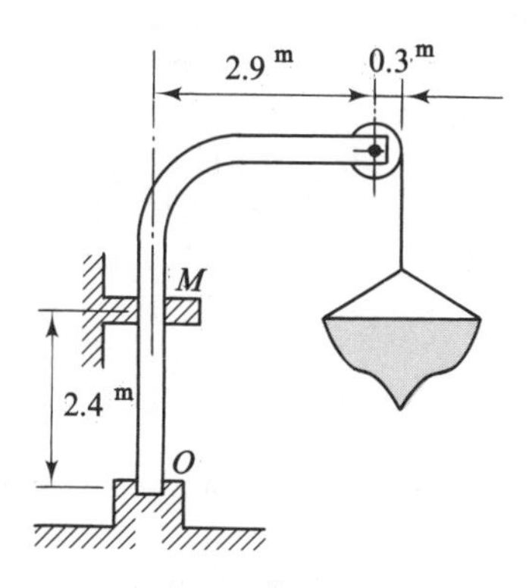

Figure P4-32

4-33 Determine all unknown forces acting on member ABC in Fig. P4-33.

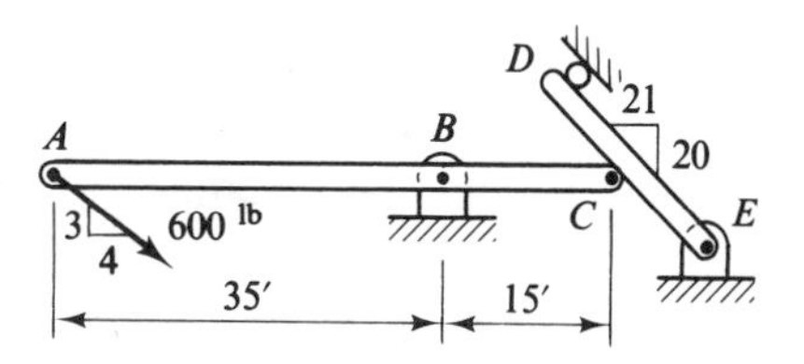

Figure P4-33

4-34 The homogeneous rod AB in Fig. P4-34 weighs W pounds and is supported by the half-cylinder C and the floor and wall at A. Determine the angles made by the reactions at A and at E with the horizontal when $L = 12$ ft, $D = 6$ ft, and $R = 3$ ft.

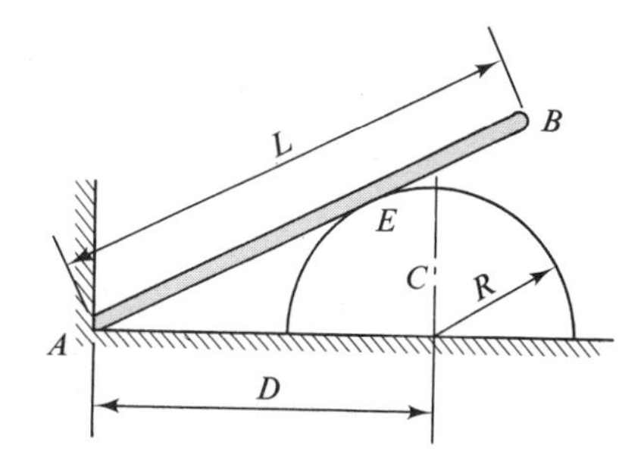

Figure P4-34

4-35 Solve Problem 4-34 if D is adjusted so that E is at the middle of AB. The distances L and R are unchanged.

4-36 Determine the distance D in Problem 4-34 which will cause the reaction at A to be horizontal when $L = 12$ ft and $R = 2$ ft and calculate the resulting angle of the reaction of C on AB.

4-6 EQUILIBRIUM OF GENERAL COPLANAR FORCE SYSTEMS

The resultant of a general coplanar force system is either a single force in the plane of the system or a couple (see Art. 2-3). The equations which are satisfied only if all possible resultants are zero constitute a complete set of equations of equilibrium. If the sum of the moments of the forces of the system with respect to any point A, not in the plane of the forces, is zero, the resultant must be zero for the following reasons. If the resultant were a couple, it would have a moment about any point, and if it were a force, it would have a moment about any point not on its line of action. Since the resultant would have to lie in the plane of the forces and since A is not in this plane, the resultant must be zero if it has no moment about A. Thus the equation

$$\sum \mathbf{M}_A = 0 \qquad (4\text{-}3)$$

is sufficient to assure equilibrium of a coplanar force system if A does not lie in the plane of the forces. This vector equation can be resolved into three component equations which can be solved for three unknown quantities (forces, distances, or directions).

If the sum of the moments of the forces of the system with respect to any point B, in the plane of the forces, is zero, the resultant cannot be a couple, but it can be a force through point B. If, in addition, the sum of the forces is also equal to zero, the resultant cannot be a force, and the system must be in equilibrium. Consequently, the equations

$$\sum \mathbf{M}_B = 0 \quad \text{and} \quad \sum \mathbf{F} = 0 \qquad (4\text{-}4)$$

are also sufficient to assure equilibrium of a general coplanar force system where B is a point in the plane of the forces.

Equations (4-4) are vector equations, and each can be resolved into three component equations; however, only three of these six equations will be independent for a coplanar force system. If the plane of the forces is selected as the xy plane with point B as the origin, the equations $\sum \mathbf{M}_x = 0$, $\sum \mathbf{M}_y = 0$, and $\sum \mathbf{F}_z = 0$ will be identically zero and will yield no useful information. Thus the two vector equations in Eq. (4-4) are equivalent to the following component equations:

$$\sum \mathbf{M}_z = 0, \qquad \sum \mathbf{F}_x = 0, \quad \text{and} \quad \sum \mathbf{F}_y = 0. \qquad (4\text{-}5)$$

The moment axis (or center) can be any line not parallel to the plane of the forces (or any point), but it is usually convenient to select an axis perpendicular to the plane of the forces (or a point in the plane).

Either or both of the preceding force equations can be replaced by moment equations provided the moment centers (in the plane of the forces) are properly selected. Two such sets of equilibrium equations are

$$\sum \mathbf{M}_A = 0, \qquad \sum \mathbf{M}_B = 0, \qquad \sum \mathbf{F}_x = 0, \qquad (4\text{-}6)$$

where A and B are in the plane of the forces and the line connecting

A and B is not perpendicular to the x axis, and

$$\sum \mathbf{M}_A = 0, \qquad \sum \mathbf{M}_B = 0, \qquad \sum \mathbf{M}_C = 0, \tag{4-7}$$

where points A, B, and C (in the plane of the forces) are not collinear. (The proof that these equations are sufficient to insure equilibrium is left as an exercise.)

Regardless of which set of equations is used, there are only three independent scalar equations of equilibrium for a general, coplanar force system because only three equations are required to insure equilibrium in each case. This information is needed for step 4 of the procedure outlined in Art. 4-4.

When all forces of the system are parallel, only two independent scalar equations of equilibrium are available. In the case where the y axis is parallel to the forces, the equation $\sum F_x = 0$ in Eq. (4-5) or (4-6) gives no information. In Eq. (4-7), one of the moment equations can be obtained by combining the other two and thus is not independent.

The following examples illustrate the use of the preceding equations, together with the procedure outlined in Art. 4-4.

EXAMPLE 4-5

The tension in the spring in the pin-connected structure in Fig. 4-7(a) is 540 lb. The weights of the members and friction at all contact surfaces can be neglected. Determine the horizontal and vertical components of the pin reaction at B on member EB.

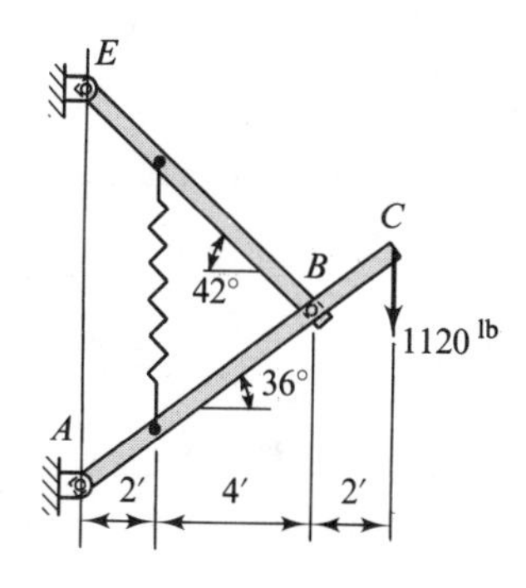

Figure 4-7(a)
(a)

SOLUTION

The steps outlined in the procedure explained in Art. 4-4 will be referred to by number for convenience.

All angles, dimensions, and applied loads are given or can be readily obtained from the given data, and the horizontal and vertical components of force **B** on member EB are to be determined (step 1).

The free-body diagram of body EB is shown in Fig. 4-7(b) (step 2). The nonconcurrent, coplanar force system (step 3) has three independent equations of equilibrium (step 4) and four unknowns (step 5). The free-body diagram of the entire structure is shown in Fig. 4-7(c)—step 6(b). The entire body is used for the free-body diagram rather than bar AC because it leads to a simpler solution. Each free-body diagram has four unknowns, but three of them can be eliminated by using the proper moment equation for the entire body, thus eliminating the necessity of solving simultaneous equations. The nonconcurrent, coplanar force system in Fig. 4-7(c) has three equations of equilibrium and four unknowns.

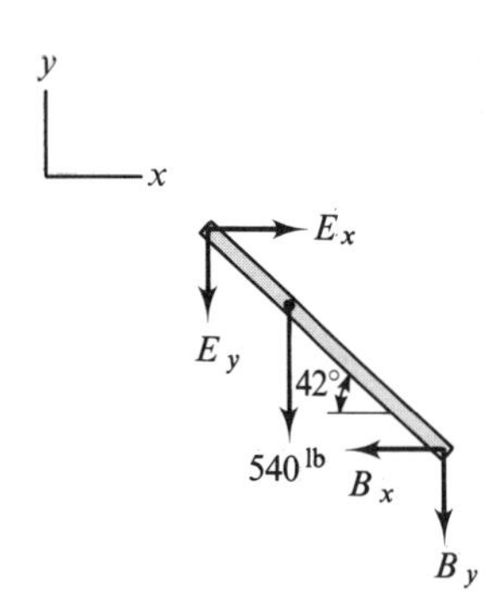

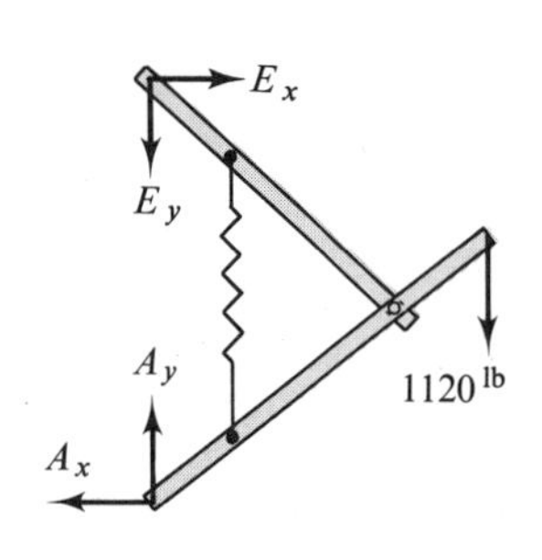

Figure 4-7(b) and (c)

The two free-body diagrams contain a total of six unknowns, and six equations of equilibrium are available; therefore the analysis is complete (step 7). The remainder of the solution follows. From the free-body diagram of the entire structure, the equation

$$\sum \mathbf{M}_A = \sum \mathbf{r} \times \mathbf{F} = 0$$

becomes

$$(6 \tan 36° + 6 \tan 42°)\mathbf{j} \times E_x \mathbf{i} + 8\mathbf{i} \times (-1120\mathbf{j}) = 0$$

or

$$-(4.36 + 5.40)E_x \mathbf{k} - 8960\mathbf{k} = 0$$

from which

$$E_x = -918 \text{ lb}$$

and

$$\mathbf{E}_x = -918\mathbf{i} \text{ lb} = 918 \text{ lb} \leftarrow \text{on } BE.$$

The force equation of equilibrium can be applied to the free-body diagram of BE to give

$$\sum \mathbf{F} = E_x\mathbf{i} - E_y\mathbf{j} - 540\mathbf{j} - B_x\mathbf{i} - B_y\mathbf{j}$$
$$= \mathbf{i}(E_x - B_x) + \mathbf{j}(-E_y - 540 - B_y) = 0. \qquad \text{(a)}$$

The coefficients of both **i** and **j** must each be zero, thus:

$$E_x - B_x = 0$$

or

$$B_x = E_x = -918 \text{ lb}$$

and

$$\mathbf{B}_x = -(-918)\mathbf{i} \text{ lb} = \underline{918\mathbf{i} \text{ lb}} = \underline{918 \text{ lb} \rightarrow \text{on } BE.} \qquad \text{Ans.}$$

Note that the magnitudes of both $\mathbf{B}_x$ and $\mathbf{E}_x$ were found to be negative. This indicates that they were assumed wrong on the free-body diagrams. A negative sign indicates that the force is in the negative direction only when the unknown force was assumed to be in the positive direction (see E_x). If the unknown force was assumed to be in the negative direction, as was B_x, a negative sign indicates that the force was assumed incorrectly, and thus it acts in the positive direction. When the value of E_x was used to obtain B_x, the negative value was used as calculated, and the free-body diagrams *were not* altered. In case either of these results is used in a subsequent equation, the negative sign must be used as calculated.

An alternate procedure when the sense of one or more unknown forces has been assumed wrong is to circle the incorrect arrowheads and show the correct sense on all free-body diagrams involved. In this procedure, the negative result must be used in any equations written before the changes were made, and positive values must be used in equations written after the changes were made. The second component of Eq. (a) contains two unknowns and thus cannot be used to determine B_y.

The equation $\sum \mathbf{M}_E = 0$ can be applied to the free-body diagram in Fig. 4-7(b) to obtain B_y. Although the moments can be calculated as vector products, it is frequently more direct to use the principle of moments and write the moment equation by inspection. Thus

$$\sum \mathbf{M}_E = \mathbf{k}[-2(540) - 6B_y - (6\tan 42°)B_x] = 0$$

from which

$$6B_y = -2(540) - (6\tan 42°)(-918)$$

or

$$B_y = 647 \text{ lb}$$

and

$$\mathbf{B}_y = \underline{-647\mathbf{j}\text{ lb}} = \underline{647 \text{ lb}\downarrow \text{on } BE.} \qquad \text{Ans.}$$

EXAMPLE 4-6

Body G, in Fig. 4-8(a), weighs 1500 lb, and the weights of all other members can be neglected. Determine the force exerted by the pin at A on member AB. Friction may be neglected at all contact surfaces. The roller E is pinned to the member AB.

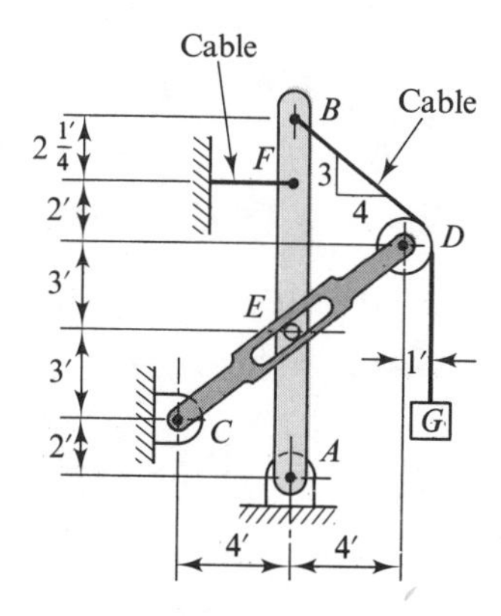

Figure 4-8(a)

SOLUTION

Again the steps suggested in Art. 4-4 will be indicated for ready reference.

All weights, slopes, and dimensions are given, and the reaction at A on AB is wanted (step 1). The free-body diagram of member AB is shown in Fig. 4-8(b) (step 2). The nonconcurrent, coplanar force system (step 3) has three independent equations of equilibrium (step 4)

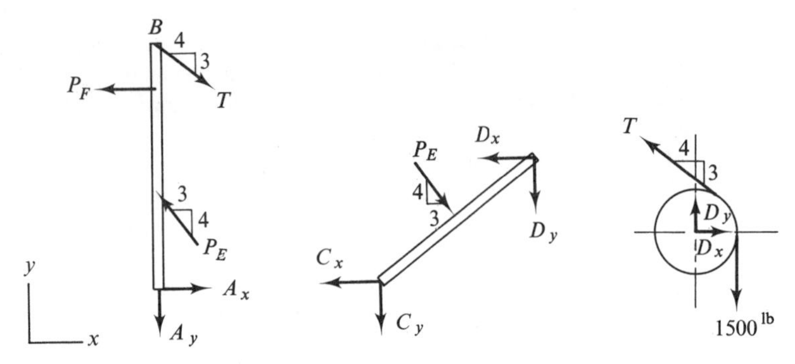

Figure 4-8(b), (c), and (d)

and contains five unknown forces (step 5). Therefore another free-body diagram is required.

The force system in the free-body diagram of member *CD* in Fig. 4-8(c) is a nonconcurrent, coplanar system having three equations of equilibrium and five unknowns (step 6). The two free-body diagrams contain nine unknowns with only six equations of equilibrium—step 7(b); therefore a free-body diagram of the pulley *D*, Fig. 4-8(d), is needed (step 8). The force system in Fig. 4-8(d) is a nonconcurrent, coplanar system having three equations of equilibrium and three unknowns. These three unknowns can be determined from the free-body diagram of the pulley, leaving only three unknowns, $\mathbf{C}_x$, $\mathbf{C}_y$, and $\mathbf{E}$, on the free-body diagram of *CD*. After these three unknowns are determined, only three unknowns, $\mathbf{A}_x$, $\mathbf{A}_y$, and $\mathbf{F}$, will remain on the free-body diagram of *AB*. Therefore the analysis is complete, and the remainder of the solution follows.

On the free-body diagram of Fig. 4-8(d),

$$\sum \mathbf{M}_D = \mathbf{k}[1(T) - 1(1500)] = 0$$

from which

$$T = 1500 \text{ lb} \quad \text{and} \quad \mathbf{T} = 1500(-0.8\mathbf{i} + 0.6\mathbf{j}) \text{ lb on the pulley}$$

and

$$\sum \mathbf{F} = \mathbf{i}D_x + \mathbf{j}D_y - \mathbf{j}1500 - \mathbf{i}1200 + \mathbf{j}900$$

$$= \mathbf{i}(D_x - 1200) + \mathbf{j}(D_y - 1500 + 900) = 0$$

from which

$$D_x = 1200 \text{ lb}, \qquad D_y = 600 \text{ lb},$$

and

$$\mathbf{D}_x = 1200\mathbf{i} \text{ lb} \quad \text{and} \quad \mathbf{D}_y = 600\mathbf{j} \text{ lb on the pulley.}$$

On the free-body diagram of Fig. 4-8(c),

$$\sum \mathbf{M}_C = \mathbf{k}(6D_x - 8D_y - 5P_E) = 0,$$

$$5P_E = 6(1200) - 8(600),$$

$$P_E = 480 \text{ lb} \quad \text{and} \quad \mathbf{P}_E = 480(0.6\mathbf{i} - 0.8\mathbf{j}) \text{ lb on } CD.$$

The equations of equilibrium are applied to the free-body diagram of Fig. 4-8(b) as follows:

$$\sum \mathbf{M}_F = \mathbf{k}[10A_X - 5(0.6P_E) - 2.25(0.8T)] = 0$$

from which

$$10A_x = 5(0.6)(480) + 2.25(0.8)(1500)$$

or

$$A_x = 414 \text{ lb} \quad \text{and} \quad \mathbf{A}_x = 414\mathbf{i} \text{ lb on } AB.$$

Also

$$\sum \mathbf{F} = A_x\mathbf{i} - A_y\mathbf{j} - P_F\mathbf{i} + 0.8T\mathbf{i} - 0.6T\mathbf{j} - 0.6P_E\mathbf{i} + 0.8P_E\mathbf{j}$$

$$= \mathbf{i}(A_x - P_F + 0.8T - 0.6P_E) + \mathbf{j}(-A_y - 0.6T + 0.8P_E) = 0.$$

The second of the preceding coefficient equations gives

$$A_y = -0.6(1500) + 0.8(480) = -516 \text{ lb}$$

and

$$\mathbf{A}_y = -(-516)\mathbf{j} = 516\mathbf{j} \text{ lb on } AB.$$

The reaction at A is

$$\mathbf{F}_A = \mathbf{A}_x + \mathbf{A}_y = \underline{662 \text{ lb} \ \angle 51.3^\circ} \qquad \text{Ans.}$$

PROBLEMS

All connections are smooth and all members are of negligible weight unless otherwise specified.

***4-37** Determine the reactions at A and B on the beam in Fig. P4-37.

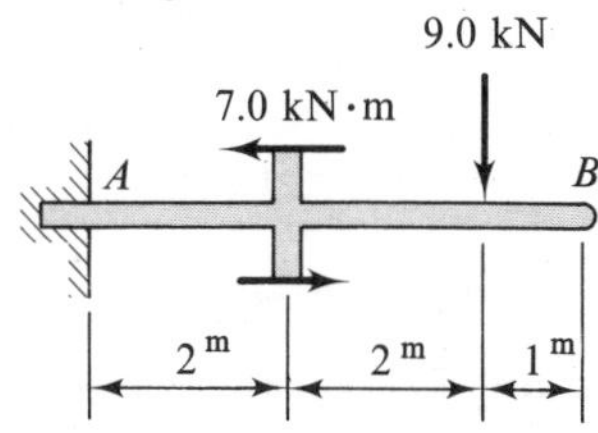

Figure P4-37

***4-38** Solve Problem 4-37 for the beam in Fig. P4-38.

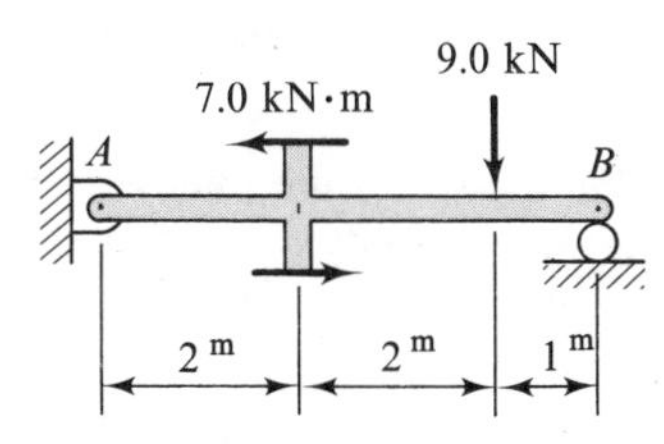

Figure P4-38

4-39 Determine the distance x in Fig. P4-39 which will cause the reactions at B and C to have equal magnitudes.

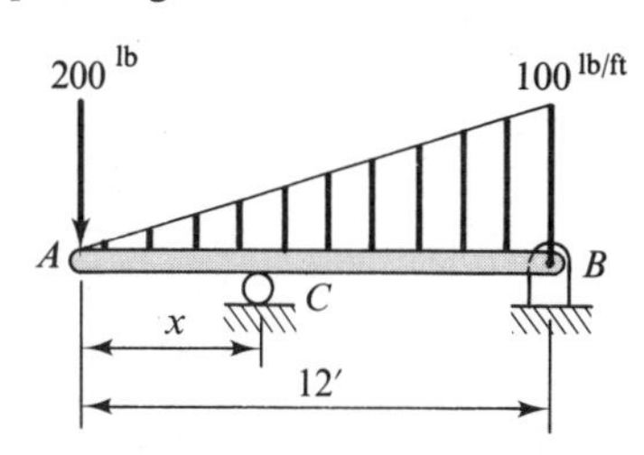

Figure P4-39

4-40 In Fig. P4-40 the load P is 1000 lb and Q is 2000 lb. Determine all reactions on the two beams.

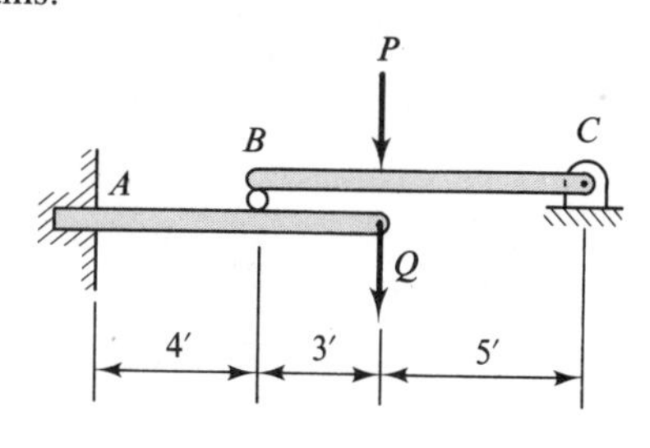

Figure P4-40

4-41 Solve Problem 4-40 if the loads P and Q are interchanged.

4-42 (a) Determine the force exerted by the pliers on the nut in Fig. P4-42 when each of the forces $\mathbf{P}_1$ and $\mathbf{P}_2$ has a magnitude of 30 lb.
(b) Determine the bearing reaction at E on member AB when the forces in part (a) are applied.

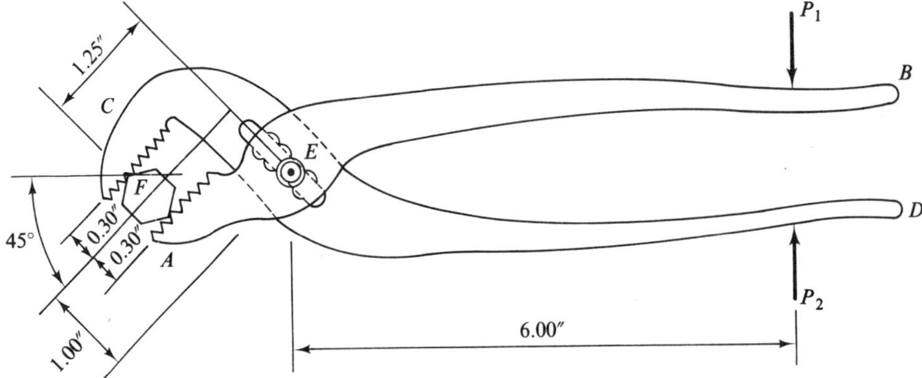

Figure P4-42

***4-43** The 140-kg body C in Fig. P4-43 is supported by two cords as shown. Determine the reactions at A and B on member AB.

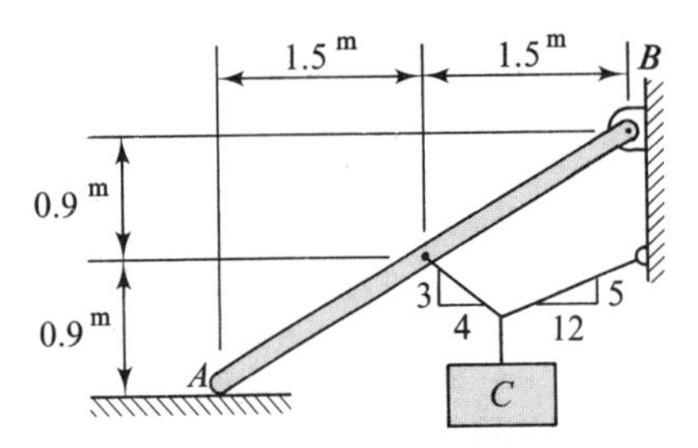

Figure P4-43

4-44 Body Q in Fig. P4-44 weighs 1000 lb. Determine the reactions at A and B on member AB.

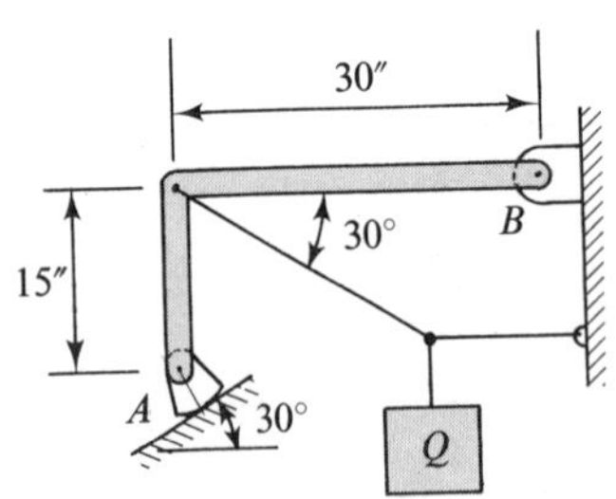

Figure P4-44

4-45 The required pressure on the nut in Fig. P4-45 is 50 lb. Determine the necessary force P on the handle of the pliers.

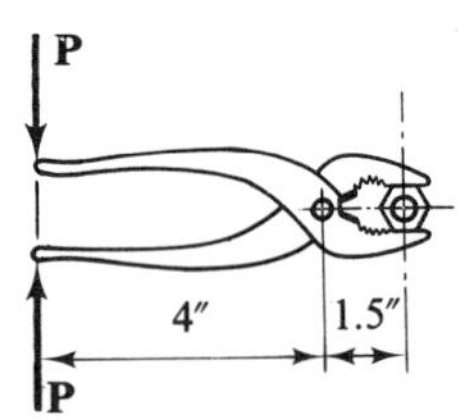

Figure P4-45

4-46 The 330-lb body in Fig. P4-46 is held in equilibrium by the weight W and the system of pulleys. The rope over pulleys B and C is continuous. The two pulleys at A are fastened together and act as a unit. The ropes from A to B and C are fastened to the rims of the pulleys at A. Determine the weight of W for equilibrium.

***4-47** The masses of the pulleys in Fig. P4-47 may be neglected. Body B has a mass of 120 kg. Determine the mass of A necessary to maintain equilibrium.

4-48 The fork lift truck in Fig. P4-48(a) is used for lifting and stacking materials. The truck

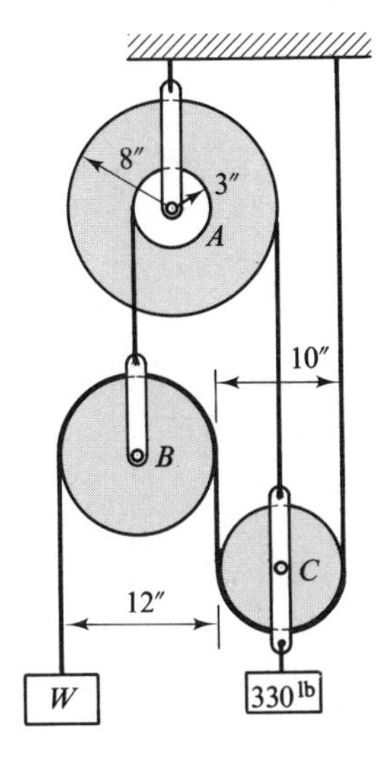

Figure P4-46

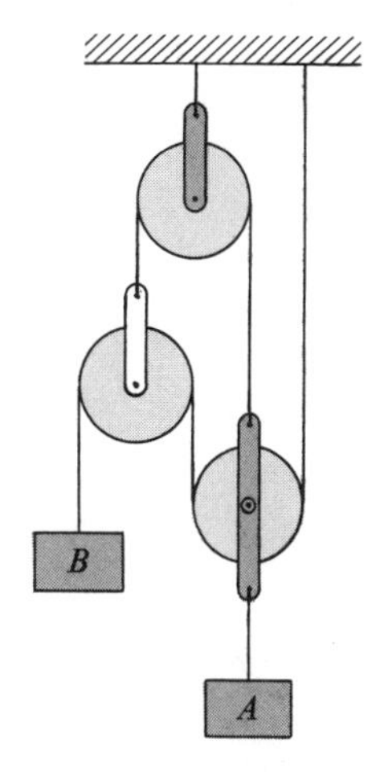

Figure P4-47

Figure P4-48(a)

Courtesy Hyster Co., Portland, Ore.

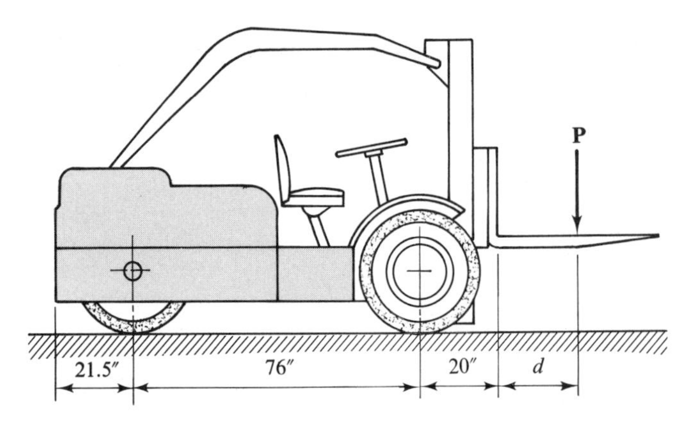

Figure P4-48(b)

weighs 9800 lb and has a rated capacity of 7500 lb when the distance, d, from the mass center of the load to the back of the fork is 21 in. [see Fig. P4-48(b)]. Assume that a load 30% greater than the rated load, applied as indicated, would cause the rear wheel to lift off the ground.

(a) Determine the horizontal coordinate of the center of gravity of the unloaded truck.

(b) Determine the reaction of the ground on each of the two front wheels when the truck is loaded to its rated capacity.

4-49 The wood plane in Fig. P4-49 has forces applied as shown when planing a piece of wood. Assume that the plane is traveling at a constant velocity. Determine the resultant normal force of the wood on the plane and state where it occurs.

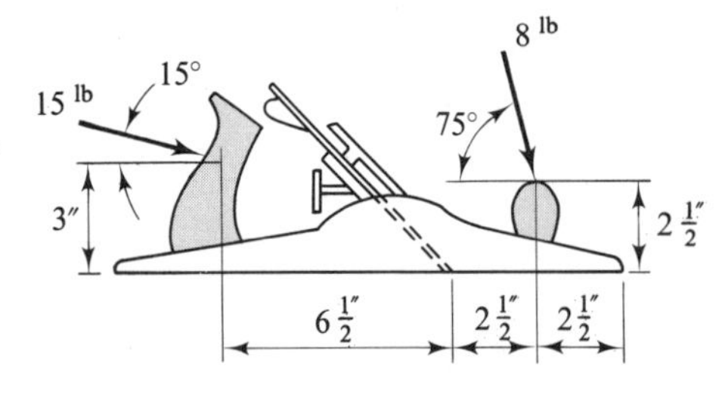

Figure P4-49

4-50 Figure P4-50(a) shows a 3900-lb Cessna Model 337 Super Skymaster. When the velocity of the plane is constant in level flight, it is held in equilibrium by its weight **W**, the propeller thrust **T**, the aerodynamic forces on the wing, and a balancing tail load **Q**. It is convenient to indicate the wing forces as lift and drag components, **L** and **D**, through the quarter-chord point A and a couple **M** [see Fig. P4-50(b)]. When the velocity is 160 mph, the lift is 4062 lb, the drag is 400 lb, and the couple M is 1590 ft-lb. Determine the propeller thrust, the balancing tail load, and the distance d.

4-51 The vice-grip pliers in Fig. P4-51 are used to produce pressures between the jaws much greater than can be achieved with ordinary pliers. They can also be used as a clamp or wrench, since they lock in place when the handles are forced together. Determine the clamping force **P** which will occur when the force **Q** on each handle has a magnitude of 50 lb.

4-52 The utility jack in Fig. P4-52 is used to support laboratory equipment. The diagram shows one of a pair of identical supports. The height of the jack is controlled by a screw between members B and E midway between the two frames. The slots at D and F permit adjustment of the jack while maintaining the top of the jack parallel with the bottom. The load **W** is centered between the two frames and between A and D. Show that the tension in the adjusting screw is equal to $2W/\tan\theta$ when the jack is

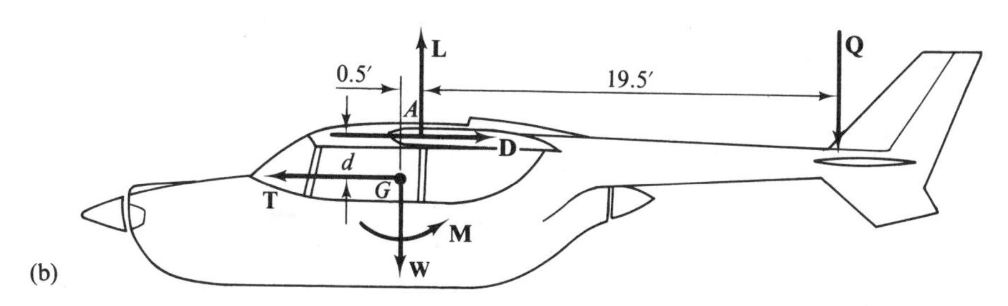

Figure P4-50

Courtesy Cessna Airplane Co., Wichita, Kan.

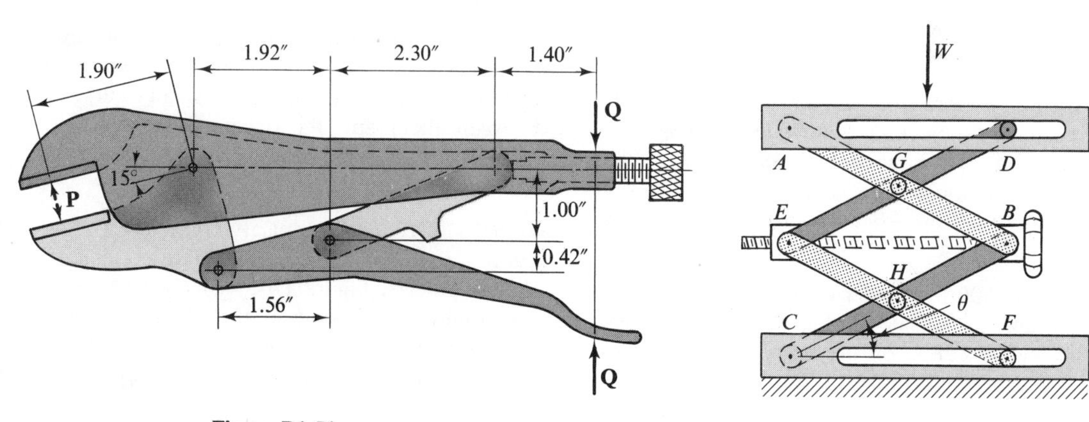

Figure P4-51

Figure P4-52

partially extended. All members of the supporting structure are the same length, and friction at the pins is neglected.

4-53 The instrument jack in Problem 4-52 relies on the sliding pins or rollers at D and F to keep the top level. An alternate design which eliminates the sliding friction is shown in Fig. P4-53. All connections are pinned, and AB is not pinned to DE (nor is BC pinned to EF). To keep the top of the jack level, two *floating* links GH and JK are pinned to AB and EF and keep these main members parallel. This in turn keeps the top platform horizontal. The *floating* links are assumed to carry no force. Derive an expression for the tension in the screw in terms of the load and the angle θ.

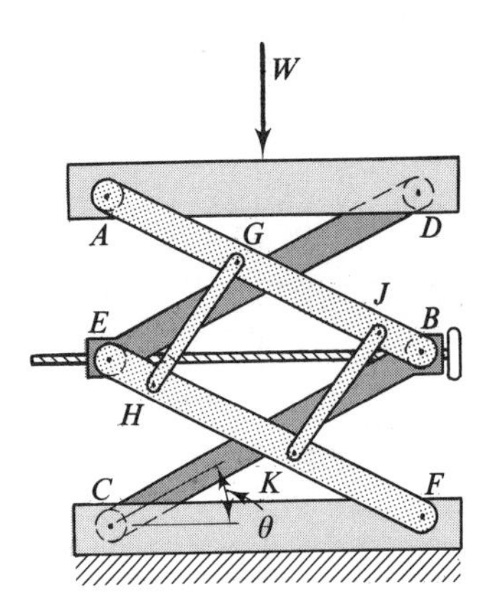

Figure P4-53

***4-54** The cord from B over the smooth peg D to the 50-kg block C is adjusted in length so that beam AB is horizontal, as shown in Fig. P4-54. Determine the components of the pin reaction at A on AB and the force exerted by AB on block C.

4-55 A 500-lb cylinder is supported by a beam and cable as indicated in Fig. P4-55. Determine the reaction at A on the beam and the forces exerted by the cylinder on the beam.

4-56 An F-110 jet fighter is acted upon by the force system in Fig. P4-56. The aircraft is climbing at a constant velocity. Determine the magnitudes of L and D and locate the line of action of D with respect to the G axis.

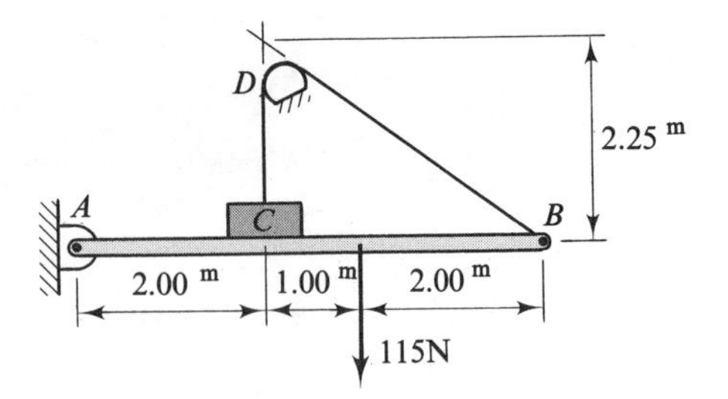

Figure P4-54

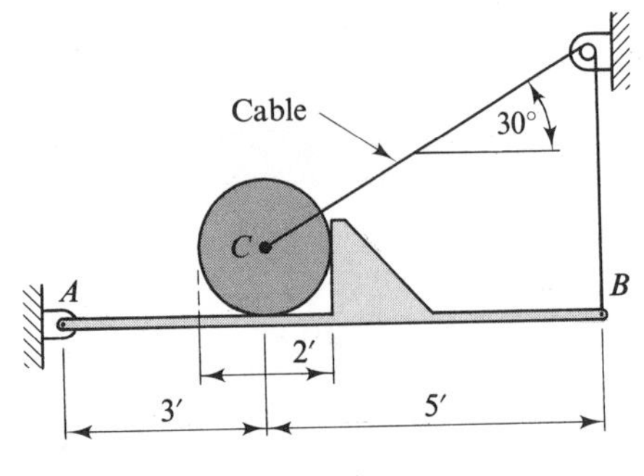

Figure P4-55

***4-57** The homogeneous triangular plate ABC of Fig. P4-57 has a uniform thickness, has a mass of 1.2 Mg, and is supported in a vertical plane as shown. The force **F** is applied to the pin at C. Determine the reaction of the pin at C on each member.

4-58 The 500-lb drum in Fig. P4-58 is supported as shown. The supports on the two sides of C are spaced so that there is a small clearance between them and the drum. Determine the forces exerted on the beam by the reaction at A and by the drum C.

4-59 The 10,000-lb-capacity crane in Fig. P4-59(a) is used for moving or loading crates and similar objects. Figure P4-59(b) gives some of the details of the boom of the crane as it supports a load of 5000 lb. Determine the reactions of the bearings at A and B on the boom.

4-60 Solve Problem 4-59 if the 5000-lb load is lifted by the hook on the cable as shown in Fig. P4-59(a). The cable does *not* loop over a pulley at the load as shown in Fig. P4-59(b) but is connected directly to the load.

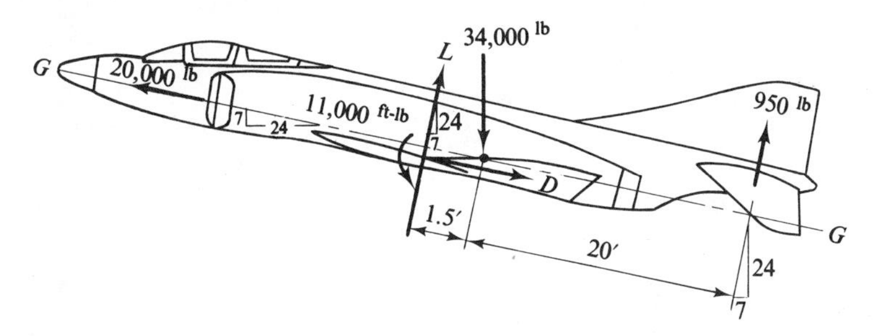

Figure P4-56

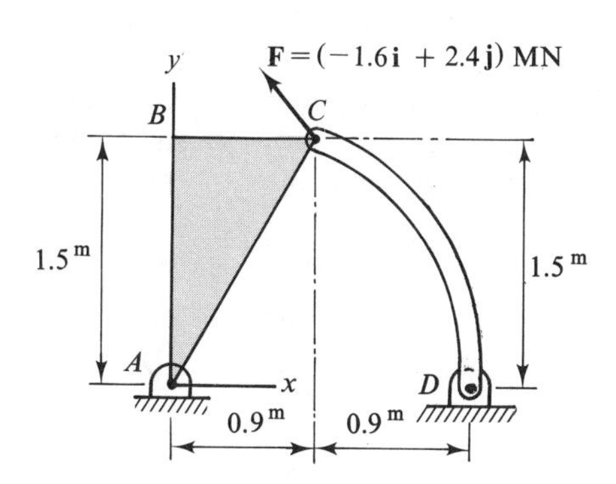

Figure P4-57

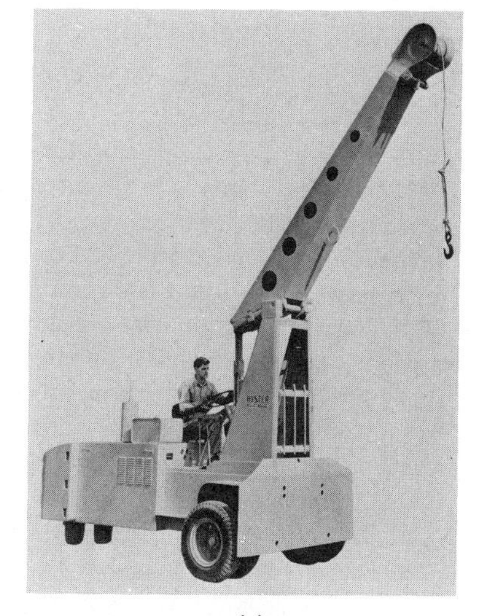

(a)

Courtesy Hyster Co., Portland, Ore.

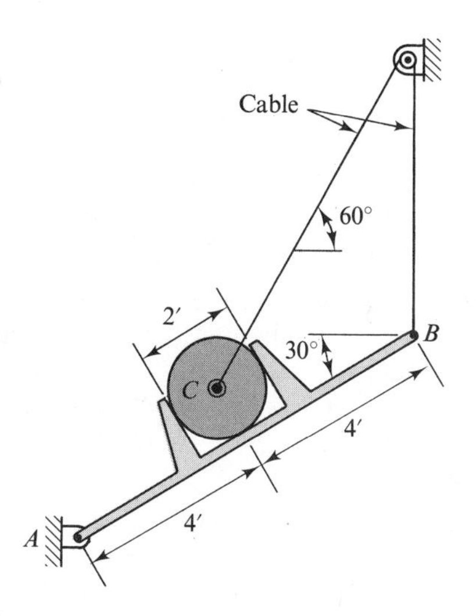

Figure P4-58

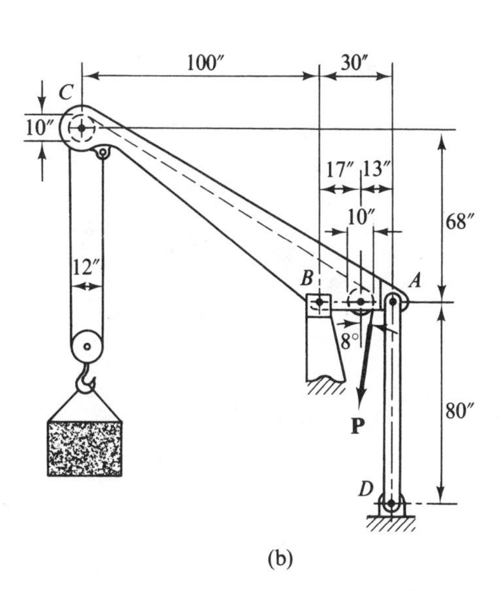

(b)

Figure P4-59

4-61 Is it possible to place the ladder of Fig. P4-61 in a position of equilibrium when planes AB and BC are both smooth? If so, for what value or values of θ?

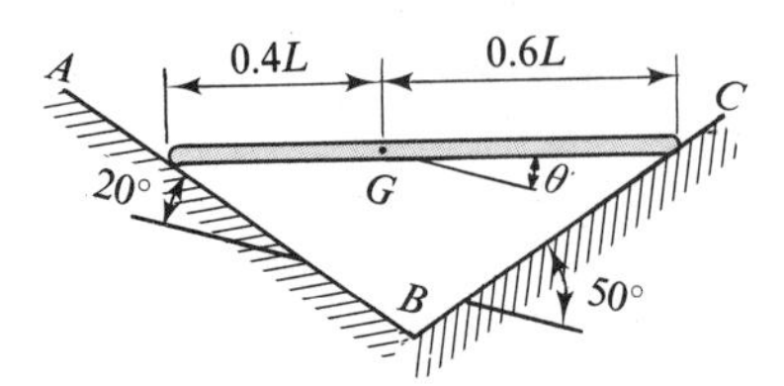

Figure P4-61

4-62 The homogeneous trap door AB in Fig. P4-62 weighs 30 lb and is mounted on a smooth hinge at A. The arrangement of the cord and weight W is intended to hold the door open or let it stay closed if that is desired.

(a) Determine the greatest weight W can have without causing the door to open by itself.

(b) If the weight is 14 lb, how much force must be applied upward at C to start the door to open?

(c) At what angle θ will the door continue to open due to the 14-lb weight?

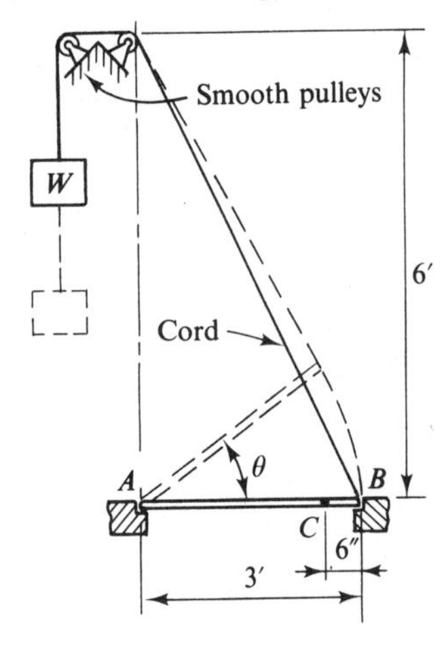

Figure P4-62

4-63 Determine the maximum weight the homogeneous triangular plate in Fig. P4-63 can have if the tension in the cable and the bearing reaction at A on the plate are each limited to 1000 lb.

***4-64** The 760-kg body W in Fig. P4-64 is

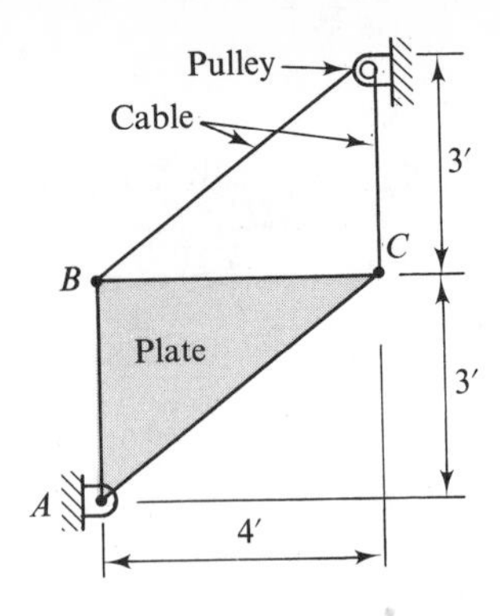

Figure P4-63

supported by the frame as shown. Member CD is free to slide in the block E, which is pinned to AB. Both ends of the cable over the pulley are attached to member AB. Determine the reaction at E on member CD and the tension in the cable over the pulley.

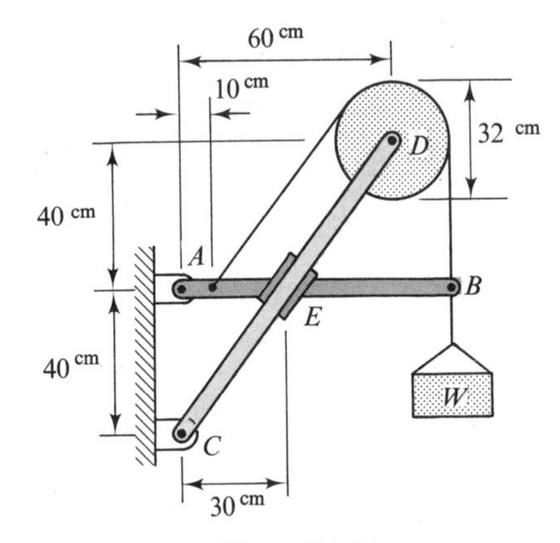

Figure P4-64

***4-65** Solve Problem 4-64 if the pinned connection at C is replaced by a roller which bears on the vertical surface and if members AB and CD are connected at E by a pin.

4-66 The homogeneous triangular plate A in Fig. P4-66 weighs 600 lb and is supported by members BC and DE. Member DE is free to slide in block F, which is pinned to member BC. Determine the forces acting on the plate at C and E.

4-67 In the pin-connected structure of Fig. P4-67, member AD fits inside the clevis on AB, which in turn fits inside the larger clevis on AC. Determine

(a) The pin reaction at A on member AC.

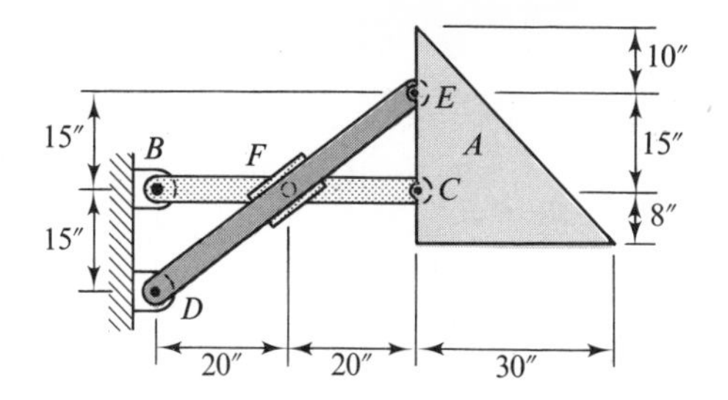

Figure P4-66

(b) The maximum shearing force (internal force perpendicular to the axis of the pin) in the pin at A.

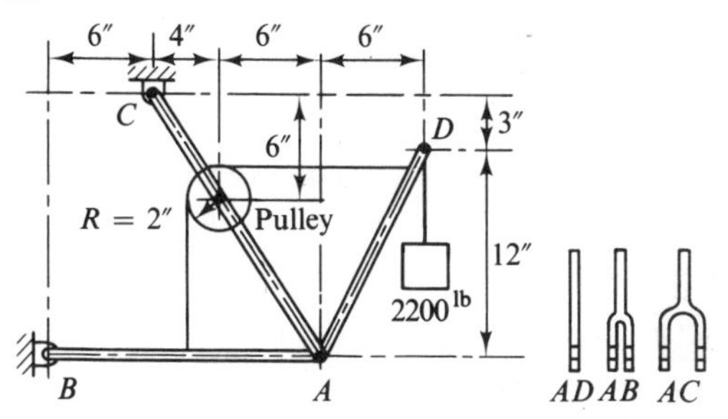

Figure P4-67

4-68 The two plates A and B in Fig. P4-68 each weigh 0.30 lb per in.^{2} of area and are supported by pin connections at C, D, and E. The 100-lb force at D is applied to the pin. Determine the reaction at D on each of the plates.

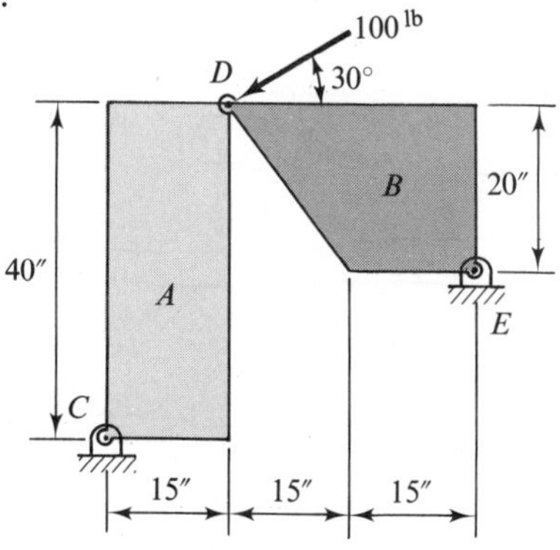

Figure P4-68

4-69 The frame in Fig. P4-69 supports the two forces and one couple as shown. Determine the pin reaction at B on AB and on BC.

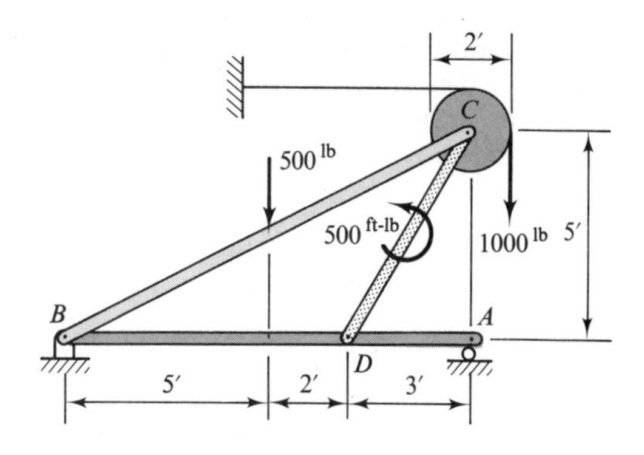

Figure P4-69

4-70 Body W in Fig. P4-70 weighs 500 lb. Determine the reaction of the pins at A and D on each of the three members AB, AE, and CD.

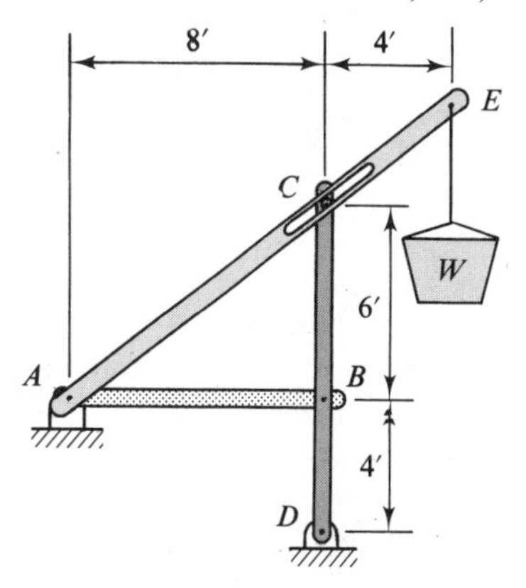

Figure P4-70

4-71 The 1000-lb crate W in Fig. P4-71 is suspended from a rope which passes over a pulley and is attached to the pin at B. Determine the force exerted by the pin at B on member BD.

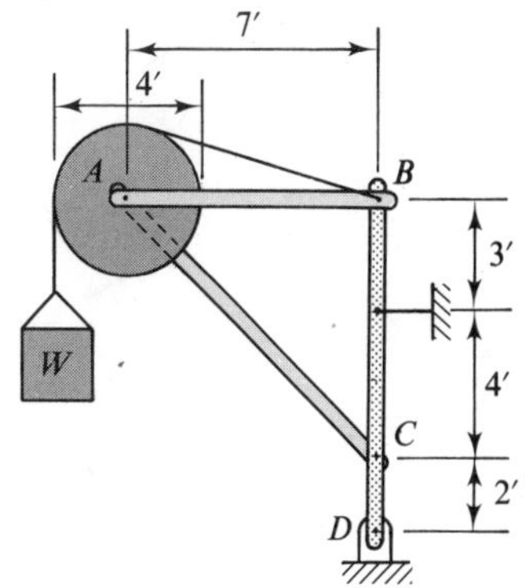

Figure P4-71

*4-72 A 500-kg crate is suspended from cables attached to pins at A and B as shown in Fig. P4-72. Determine the forces exerted by the pin at B on AB and on BC.

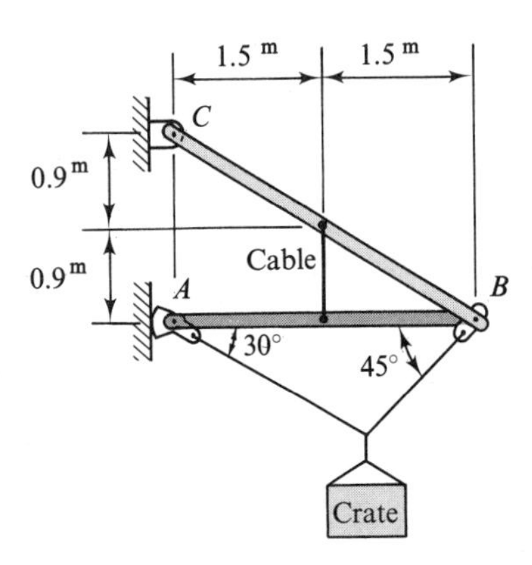

Figure P4-72

4-73 A motor-driven blower is shown in Fig. P4-73. Tension in the belt is maintained by mounting the 20-lb motor on a bearing at A.
(a) Determine the tension in the belt when the blower is stopped and the components of the reaction at A on the motor.
(b) When the blower is operated in a clockwise direction, a 10-in-lb moment is developed in the motor shaft, causing the tension in the upper part of the belt to increase and the tension in the lower part to decrease. Determine the two belt tensions and the components of the reaction at A on the motor.

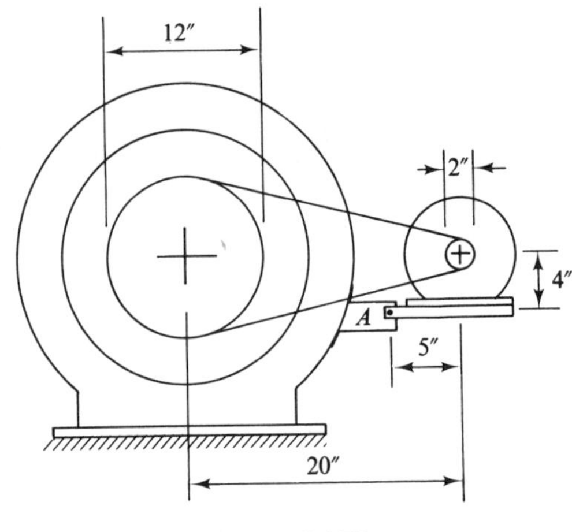

Figure P4-73

4-74 Solve Problem 4-73(b) when the blower is rotating counterclockwise. Assume that the same torque magnitude is developed for either direction of rotation.

4-75 The forces $\mathbf{P}_1$ and $\mathbf{P}_2$ in Fig. P4-75 have magnitudes of 50 and 20 lb, respectively.
(a) Replace forces $\mathbf{P}_1$ and $\mathbf{P}_2$ by a force at the center of bolt F and a couple.
(b) Determine the forces exerted by the pliers on the bolt at the points of contact.

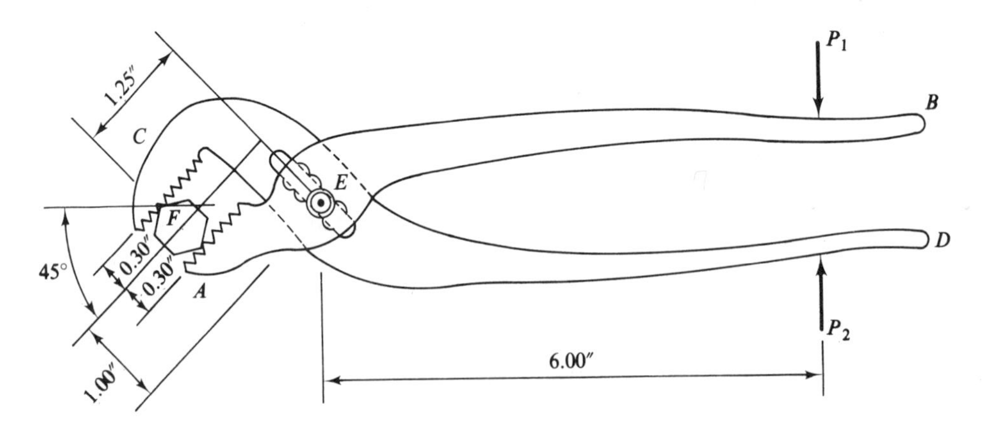

Figure P4-75

4-7 TRUSSES

A truss is a structure made up of a number of members fastened together at their ends in such a manner as to form a rigid body. A truss may be used to support a larger load or to span a greater distance than can be done effectively by a single beam or column. Trusses are frequently used in bridge and in roof construction. Trusses can be classified as coplanar (all members in a common plane) or as space trusses in which members are located in more than one plane. The discussion in this text is limited to trusses in which the members, loads, and reactions are all in the same plane.

When three members are connected by smooth pins at their ends as indicated in Fig. 4-9, they form a rigid structure and will support a load **P**. If four members are connected by smooth pins at their ends, as shown in Fig. 4-10(a), the resulting structure will not be rigid and will collapse when a load **P** is applied. If an additional member from B to D or from A to C is added to the structure, as indicated in Fig. 4-10(b), the five members form a rigid body and will support the load **P** without collapsing. The addition of the member BD in Fig. 4-10(b) reduces the structure to two triangles, and, in general, a truss can be constructed by starting with three members arranged in the form of a triangle and adding members in noncollinear pairs for each additional joint. The connections at the ends of the members are usually called *joints* or *panel points*.

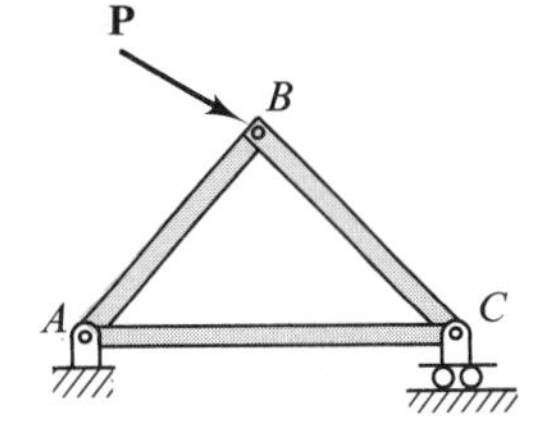

Figure 4-9

If a truss is rigid, as in Fig. 4-10(b), the addition of a single member between two pins (for example, from A to C) will cause the structure to be indeterminate, and one of the members is redundant. The loads in the members of an indeterminate truss cannot be computed by the method of statics alone. The solution of such problems is presented in texts on structural analysis.

When the internal forces (stresses) in the individual members of a truss are to be determined, the calculations are usually based on the following assumptions:

1. The members of the truss are joined together by means of smooth pins at their ends.

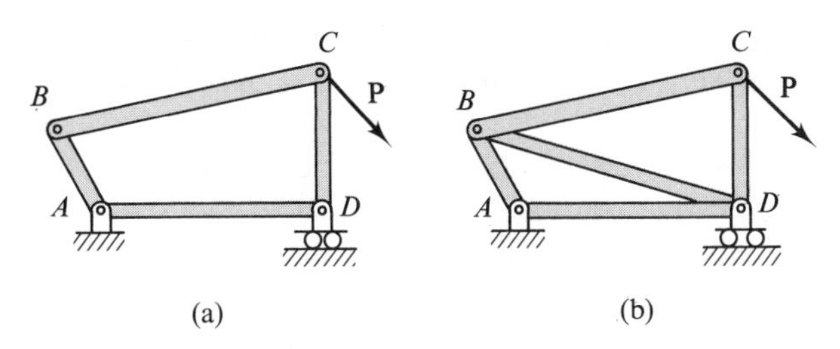

Figure 4-10

2. The loads and reactions act only at the joints.
3. The weights of the individual members can be neglected.

In justifying the truss assumptions consider the highway bridge in Fig. 4-11(a). The bridge crosses the Mississippi River from Helena, Arkansas, to Mississippi. To justify the first assumption, the joints must consist of smooth pins at the ends of the members only. In practice, the connections are usually made by riveting or welding the members to gusset plates, as shown by the joint in Fig. 4-11(b) (one of the upper joints), and the main top and bottom chords are frequently continuous members instead of a series of shorter members between joints. If in erecting the truss, however, reasonable care is observed to make sure that the center lines of all members intersect at a common point at each joint, the error in assuming pin connections and noncontinuous members is usually small enough to justify the first assumption.

The second assumption is valid for most bridges, since the floor or deck of the bridge is usually supported by cross beams between the two trusses, and the cross beams are connected to the truss at the joints. The loads on other trusses, such as roof trusses, are also applied at the joints in many cases.

The third assumption requires careful consideration, because it is rather obvious that the total weight of the members of a large railroad or highway bridge truss cannot be neglected in comparison to the combined weight of the floor system and the trains or the cars and trucks on the bridge at any one time. Even in small bridges, roof trusses, and similar structures, the weight of the truss cannot be neglected. Structural designers have found, however, that generally the weight of the truss can be apportioned as loads at the joints of the truss without serious error. Therefore, it is common practice to assume that the weight of the truss acts with all other loads on the truss at the joints. No attempt will be made here to explain the distribution of the weight of a truss to the joints, since the procedure of calculating the weight is beyond the scope of this text. Furthermore, the methods developed here for determining the stresses produced by loads at one or more joints apply equally well to a truss with loads at most or all of the joints.

When the preceding assumptions are justified, each member of the truss is held in equilibrium by only two forces, the actions of the pins at the two ends of the member, and it is thus a two-force member (see Art. 4-5).

When a two-force member is cut, a free-body diagram of either piece shows that the internal force is equal to the load applied at the end of the member and is collinear with the member's axis. When the applied load tends to stretch the member, as indicated in Fig. 4-12, the internal force is called a *tensile* force, but if it tends to shorten or compress the member, it is known as a *compressive* force. A complete

(a)

(b)

Figure 4-11 *Courtesy Arkansas Highway Department.*

specification for an internal force always includes an indication of whether the force is tensile or compressive, since the design problems are different for members subjected to these two types of forces.

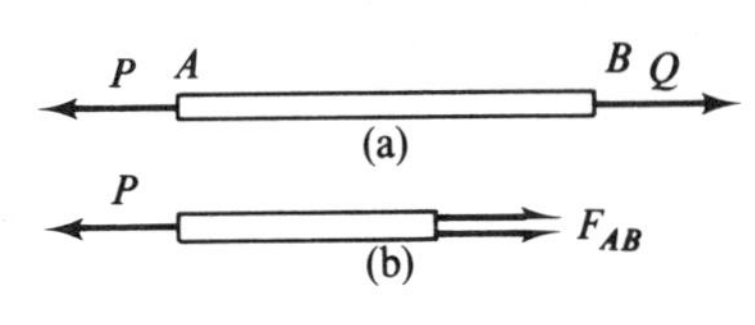

Figure 4-12

Internal forces are frequently called *stresses*, but the term *stress* may be used to indicate either the intensity of the force (the force per unit area) or the total force. To avoid this possible ambiguity, the term is not used in the discussion of truss analysis.

4-8 ANALYSIS OF TRUSSES

The forces in the members of a truss can be determined by drawing a free-body diagram of a portion of the truss involving one or more of the unknown forces and applying the equations of equilibrium. When a single joint in the truss is isolated as a free body, the forces are said to be determined by the *method of joints*. If two or more nonconcurrent members are cut to obtain the free body, two or more joints will be included in the free-body diagram, and the process of obtaining the internal forces is referred to as the *method of sections*. These two methods are illustrated in the following discussion.

An inclined-chord Pratt bridge truss is shown in Fig. 4-13. The reactions at A and L can be determined by use of a free-body diagram of the entire truss or by symmetry. In either case, the reactions are each 5000 lb acting vertically upward. The portion of the truss inside the dashed line a can be isolated as a free body as shown in Fig. 4-14(a). The action of the parts of the members AB and AC removed from the isolated joint are shown as the forces $\mathbf{F}_{AB}$ and $\mathbf{F}_{AC}$ along the axes of the members, since they are two-force members. The other force on the free-body diagram is the 5000-lb reaction. The force system is concurrent and coplanar. Since such a system has two independent equations of equilibrium and since two unknown forces are involved, the unknowns can be determined. The force $\mathbf{F}_{AB}$ is

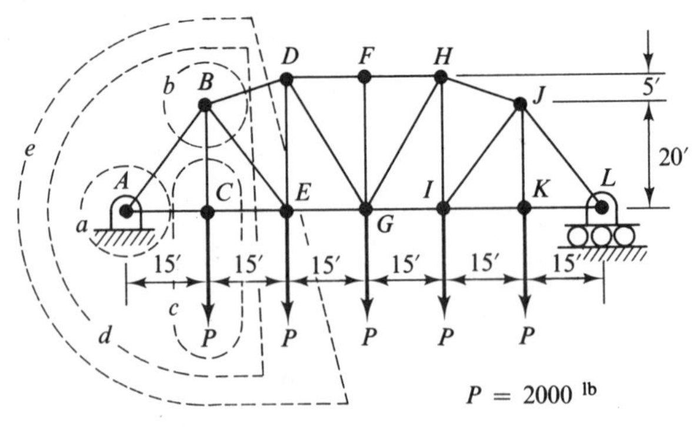

Figure 4-13

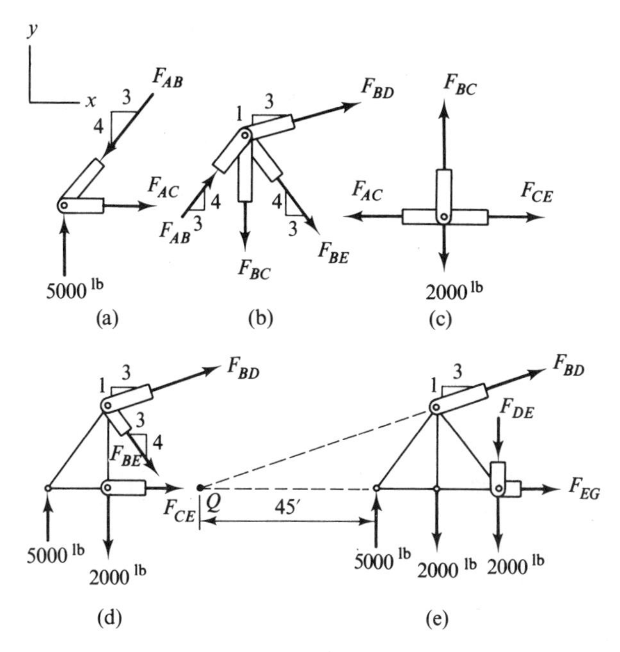

Figure 4-14

assumed to be a compressive force, and $\mathbf{F}_{AC}$ is assumed to be a tensile force. If a negative answer is obtained for any force, the result will be correct in magnitude and opposite in sense to that assumed. If all forces are assumed to be tensile, as is frequently done, a negative answer indicates a compressive force.

A free-body diagram of the joint at B is shown in Fig. 4-14(b). There are four forces, representing the four members that are cut away in isolating the body. Force $\mathbf{F}_{AB}$ can be determined from the free-body diagram in Fig. 4-14(a). Since the system is concurrent at B, however, there are only two independent equations of equilibrium, and the three remaining unknown forces cannot be determined from the free-body diagram of Fig. 4-14(b). A diagram of the joint at C in Fig. 4-14(c) has two unknown forces, $\mathbf{F}_{BC}$ and $\mathbf{F}_{CG}$, since $\mathbf{F}_{AC}$ can be determined from the diagram in Fig. 4-14(a). When $\mathbf{F}_{BC}$ is determined from Fig. 4-14(c) there will be only two unknown forces in the free-body diagram of joint B.

When the method of joints is used to determine the forces in the members of a truss, the first joint considered should have not more than two unknown forces, and each successive joint must introduce not more than two new unknowns. Thus, to determine the force in a member near the center of the truss, it is usually necessary to determine the forces in all the members from one end of the truss to the member in question. In the special case in which all but one of the members at a joint are collinear, the force in the noncollinear member can be determined by summing forces perpendicular to the unknown collinear members.

When the isolated portion of the truss involves two or more joints, as in the portion included in the dashed lines *d* or *e* in Fig. 4-13, the resulting free-body diagram depicts a nonconcurrent, coplanar force system, as indicated in Fig. 4-14(d) or 4-14(e). There are three independent equations of equilibrium for this type of force system, and therefore three unknown forces can be determined. The method of sections has the advantage that the force in a member near the center of the truss usually can be determined without first obtaining the forces in all the other members of the truss to one side of the member, and as a result the force is independent of any errors in internal forces previously calculated.

Whether the forces are determined by the method of joints or by the method of sections, the free-body diagrams represent bodies which are in equilibrium, and the principles involved in determining the unknown forces are the same. As with other problems of equilibrium, a systematic approach to the solution of problems involving the forces in members of a truss is important. It is desirable to draw a free-body diagram of an isolated portion of the truss and, in the isolating process, to cut one or more of the members in which the force is desired. The remaining procedure for the analysis and solution of the problem is outlined in Art. 4-4.

Experience in selecting the proper free bodies will frequently reduce the number of calculations required for determining some of the forces. In designing a truss, it is necessary to determine the forces in all members, but the fundamental principles can be demonstrated by obtaining the forces in only a few of the members. *The following example and problems are chosen to demonstrate the principles rather than to determine all the forces in each truss.*

EXAMPLE 4-7

Determine the forces in members *AB*, *AC*, *DE*, and *EG* of the pin-connected truss in Fig. 4-13.

SOLUTION

As an aid in deciding which free-body diagrams to draw, it is often helpful to identify the members in which the forces are to be determined by check marks on the original figure. In this example, the free-body diagram of the joint at A in Fig. 4-14(a) involves two of the desired forces and the 5000-lb reaction (determined from a free-body diagram of the entire truss). The equation of equilibrium

$$\sum \mathbf{F} = 0$$

gives

$$F_{AB}(-0.6\mathbf{i} - 0.8\mathbf{j}) + F_{AC}\mathbf{i} + 5000\mathbf{j} = 0$$

or

$$\mathbf{i}(-0.6F_{AB}+F_{AC})+\mathbf{j}(-0.8F_{AB}+5000)=0.$$

When each of the components of the equation are set equal to zero the following equations result:

$$-0.6F_{AB}+F_{AC}=0$$

$$-0.8F_{AB}+5000=0$$

from which

$$F_{AB}=\frac{5000}{0.8}=\underline{6250\text{ lb }C,} \qquad \text{Ans.}$$

and

$$F_{AC}=0.6(6250)=\underline{3750\text{ lb }T.} \qquad \text{Ans.}$$

The results are labeled C (compression) or T (tension) rather than indicating directions since the same force acts on each half of the member but in opposite directions. Both of the forces had positive signs, which indicates that they were assumed correctly. A negative sign would indicate that the sense, tension or compression, had been assumed wrong.

The other two forces to be determined appear in the free-body diagram in Fig. 4-14(e). There are three independent equations of equilibrium for the force system and three unknown forces. Since forces BD and DE act through D, they have no moment about D. Therefore, the force EG can be determined from the equation

$$\sum \mathbf{M}_D=\mathbf{k}[25F_{EG}+15(2000)-30(5000)]=0$$

from which

$$F_{EG}=\underline{4800\text{ lb }T.} \qquad \text{Ans.}$$

The force DE can be determined by summing moments with respect to any point on the action line of the unknown BD. Whenever possible, it is desirable to use equations which do not require the use of computed results, to prevent any errors in previous calculations from affecting subsequent results. The equation

$$\sum \mathbf{M}_Q=\mathbf{k}[-75F_{DE}+45(5000)-60(2000)-75(2000)]=0$$

does not involve the forces in members BD and EG and can be solved for F_{DE}, giving

$$F_{DE}=-600\text{ lb}=\underline{600\text{ lb }T.} \qquad \text{Ans.}$$

The negative sign indicates that the assumed sense of the force (compression) was incorrect and that the force is actually a tensile force.

PROBLEMS

***4-76** Determine the forces in all members of the king-post truss in Fig. P4-76.

4-77 The 2000-lb homogeneous crate W in Fig. P4-77 is supported as shown. Determine the forces in members AD, BD, and DE of the truss.

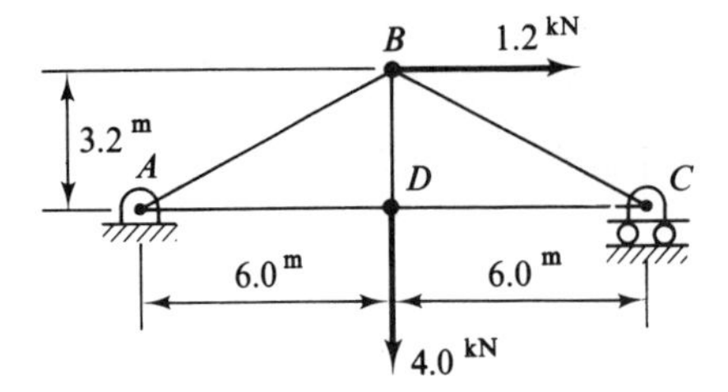

Figure P4-76

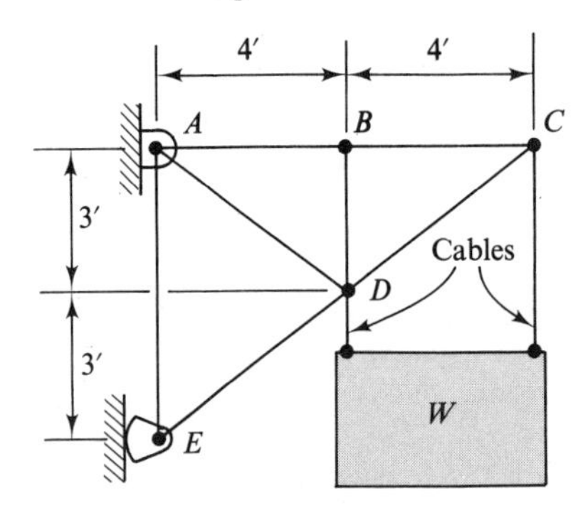

Figure P4-77

4-78 The cables supporting the crate in Fig. P4-77 are rearranged as shown in Fig. P4-78. Determine the forces in members *AB*, *AE*, and *CD* of the truss.

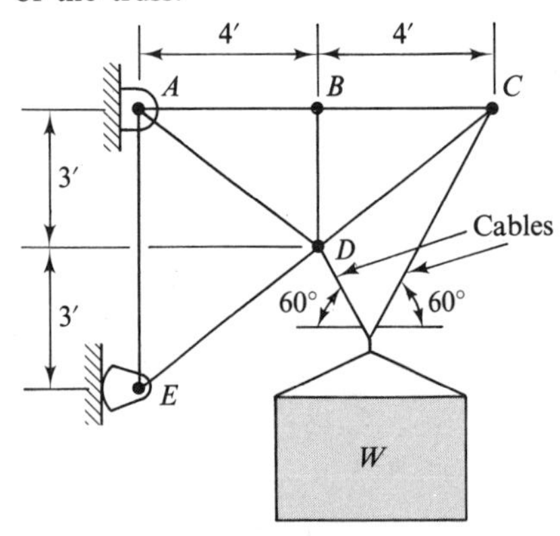

Figure P4-78

4-79 Determine the forces in all members of the truss in Fig. P4-79 in terms of the magnitude of *P*.

4-80 Solve Problem 4-79 if the diagonal members *BF* and *DF* are replaced by members from *A* to *C* and from *C* to *E*, respectively.

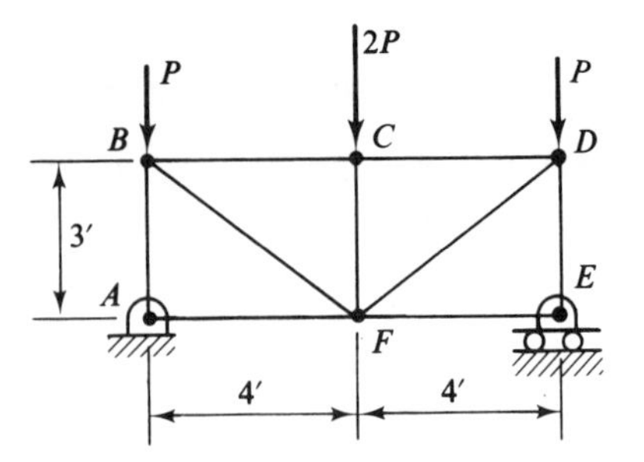

Figure P4-79

4-81 Determine the forces in all members of the truss in Fig. P4-81.

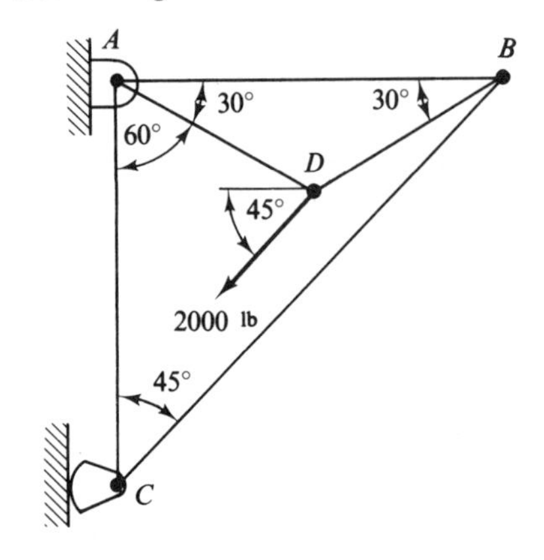

Figure P4-81

4-82 Determine the forces in members *BC*, *BF*, and *BG* of the truss in Fig. P4-82.

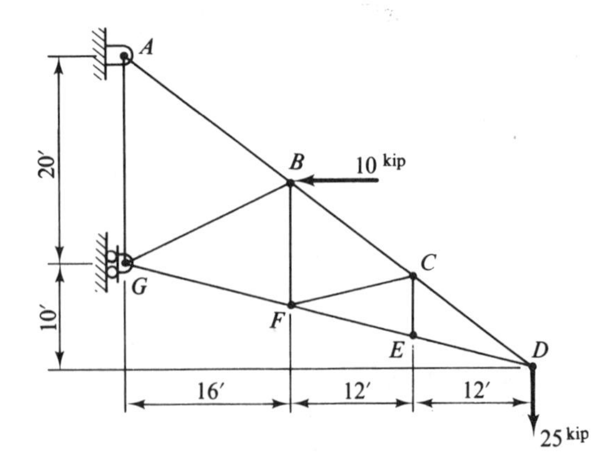

Figure P4-82

***4-83** Determine the forces in members AB, BG, and CG of the truss in Fig. P4-83.

4-84 The gravel chute of Fig. P4-84 is supported by a truss at points A and F. The chute and contents, when full, weigh 300 lb per running foot. Determine the forces in members CE and EF of the truss.

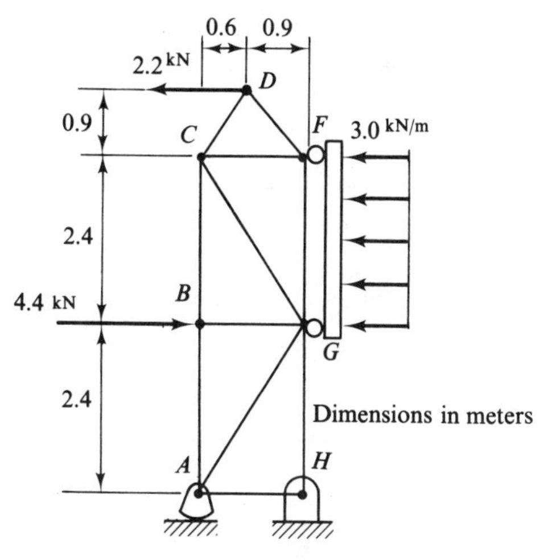

Figure P4-83

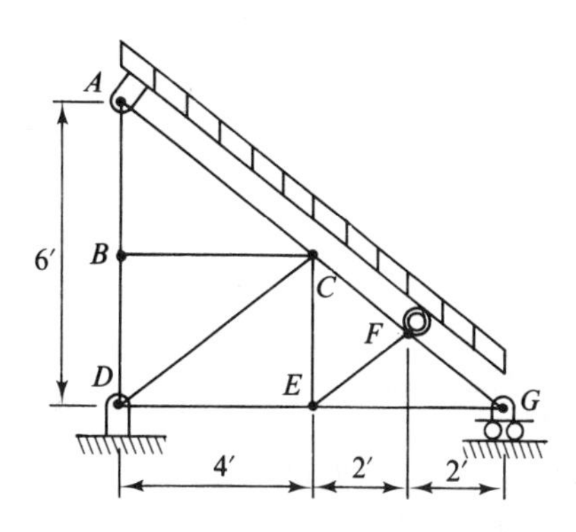

Figure P4-84

4-85 The signal truss in Fig. P4-85 is subjected to a distributed load and supports a signal which weighs 3000 lb. The signal is connected to the truss at joints D, E, and F, and the lengths of the connecting links are adjusted so that each one carries a third of the load. Determine the forces in members AB, BC, and CD.

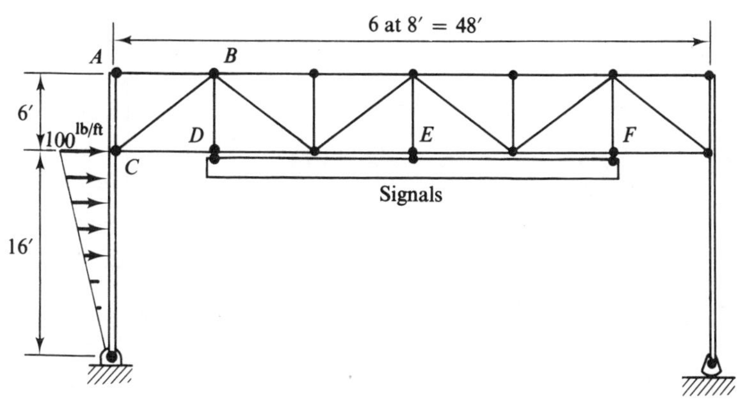

Figure P4-85

***4-86** Determine the forces in members AC, BC, DF, and EF of the truss in Fig. P4-86.

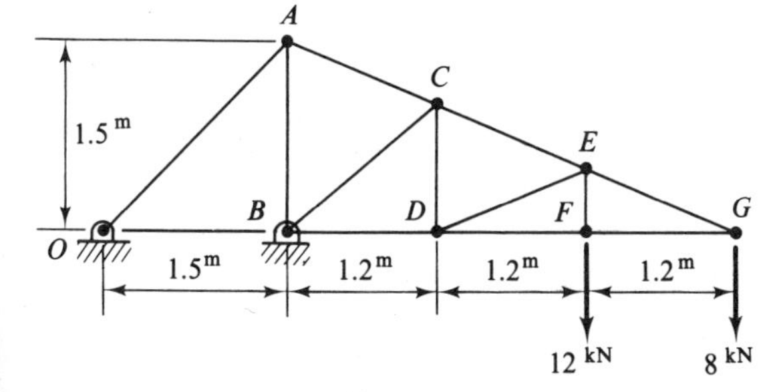

Figure P4-86

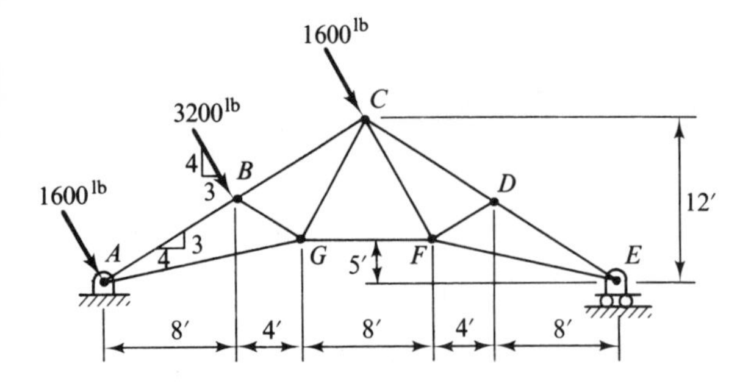

Figure P4-87

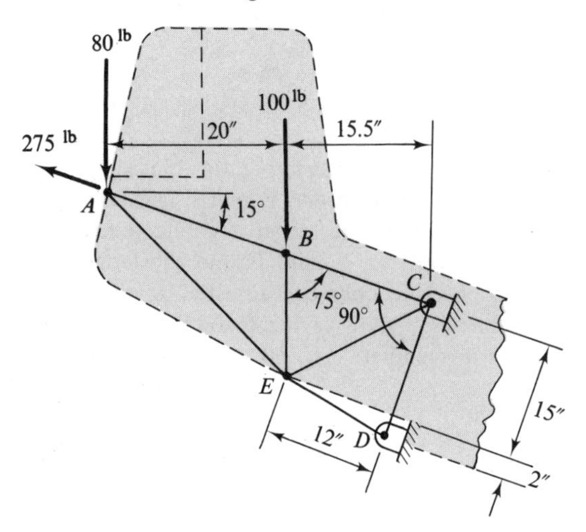

Figure P4-88

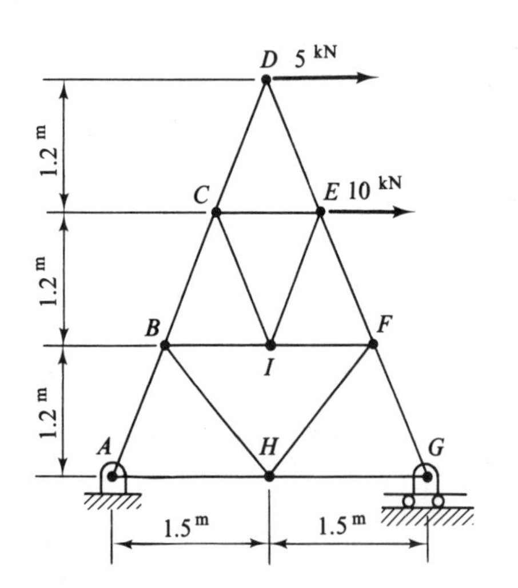

Figure P4-91

4-87 Determine the forces in members *AG*, *FG*, and *BC* of the truss in Fig. P4-87.

4-88 A scale model of an aircraft tail section is loaded as shown in Fig. P4-88. Determine the forces in members *BE* and *DE*.

4-89 Calculate the forces in members *AD*, *BD*, and *CD* of the truss in Fig. P4-89.

4-90 Calculate the forces in members *EG*, *EH*, and *FH* of the truss in Fig. P4-90.

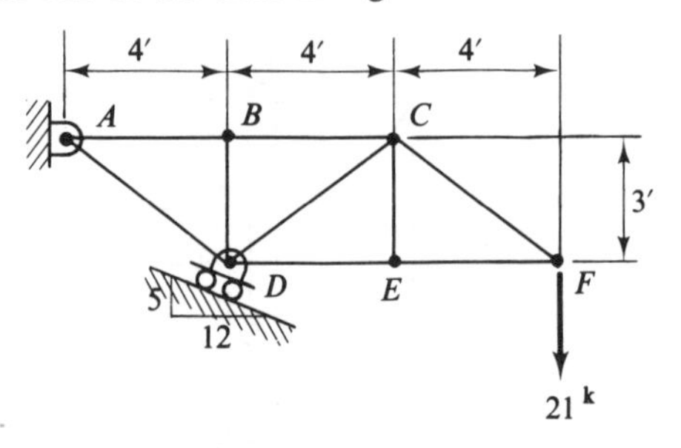

Figure P4-89

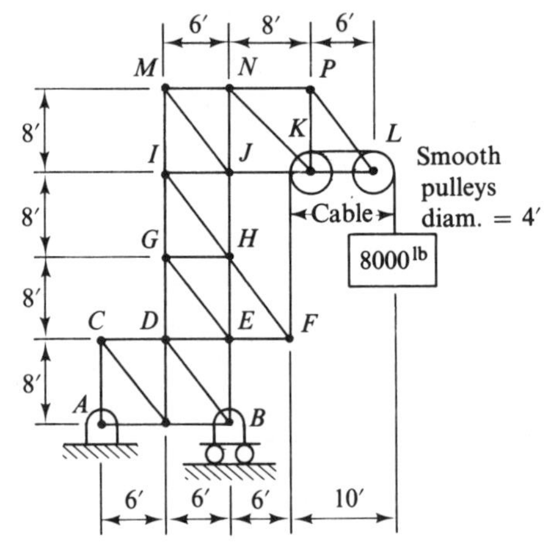

Figure P4-90

***4-91** Determine the forces in members *AB*, *AH*, and *DE* of the truss in Fig. P4-91.

***4-92** Compute the forces in members *BC* and *BI* of the truss in Fig. P4-91.

4-93 (a) Determine the maximum load *Q* which can be supported by the truss in Fig. P4-93 without producing a force in member *CE* of more than 10,000 lb *C*.
(b) Calculate the forces in members *AD* and *AE* corresponding to the load *Q*.

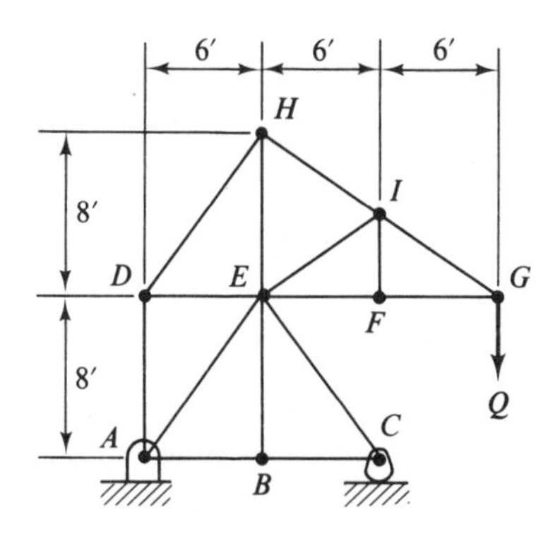

Figure P4-93

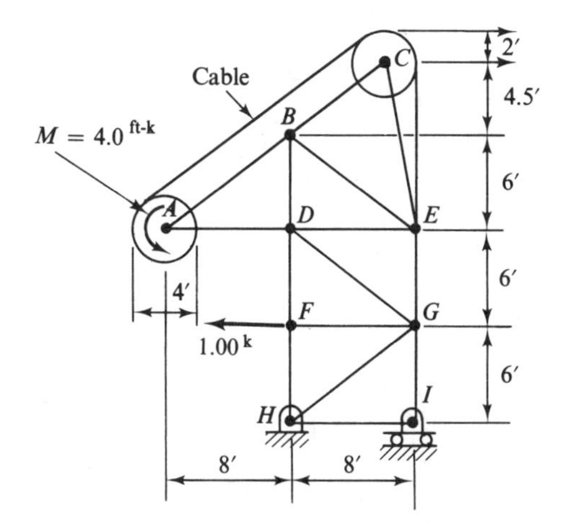

Figure P4-94

4-94 The cable in Fig. P4-94 is connected to the pin at E and passes over the pulley C to the drum at A on which it is wound. The 4.00-ft-kip couple is applied to the drum. Calculate the forces in members BD, BE, and DG of the truss.

***4-95** Compute the forces in members CH, CI, and HI of the truss in Fig. P4-95.

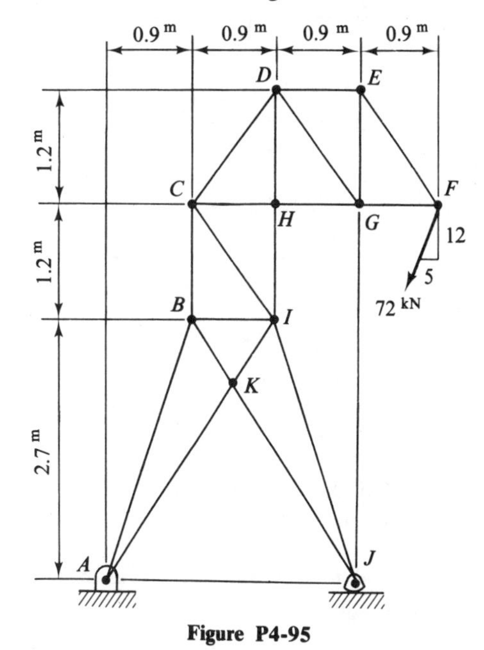

Figure P4-95

***4-96** Determine the forces in members BI and BK of the truss in Fig. P4-95.

4-97 Compute the force in member L_2m of the Fink roof truss of Fig. P4-97.

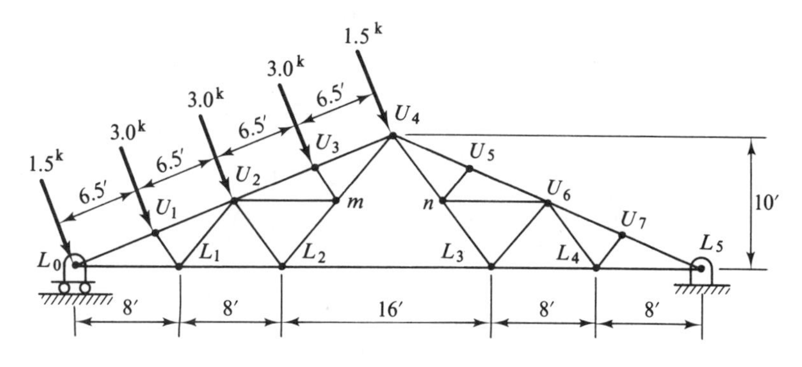

Figure P4-97

4-98 The 4000-lb load W in Fig. P4-98 is being raised at a constant velocity by the winch at D. The radii of the winch and the pulley at A are both 2 ft. Determine the required torque T and the forces in members BE, DG, and FG.

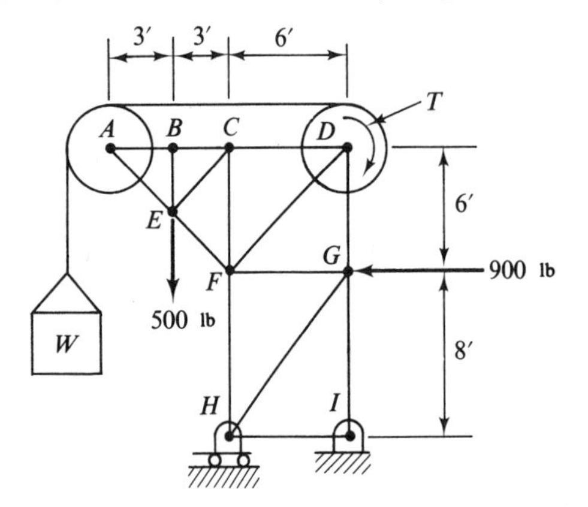

Figure P4-98

4-99 Determine the forces in members DJ, FI, and HI of the truss in Fig. P4-99.

4-100 The 5000-lb drum M in Fig. P4-100 is supported by the truss and a cable which is wrapped around the drum and fastened to the truss at B and D. Compute the forces in members BC, DE, and GH.

4-101 Determine the weight of the drum M in Problem 4-100 which will cause the force in member CG to be 2500 lb T.

***4-102** Two 56-kN loads are supported by the truss in Fig. P4-102. Determine the forces in members AC, BC, and DF.

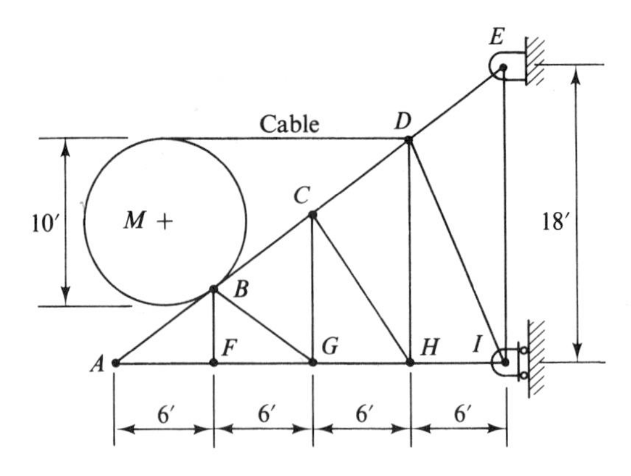

Figure P4-100

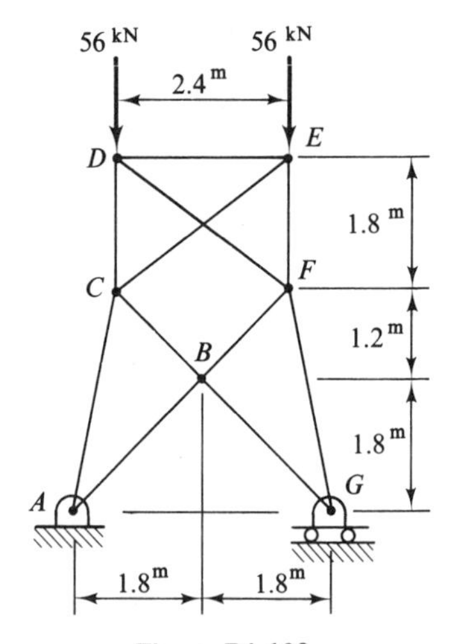

Figure P4-102

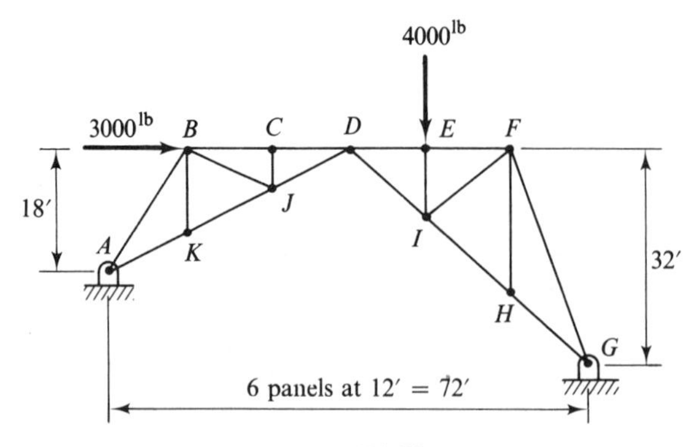

Figure P4-99

4-103 Compute the forces in members U_0L_1 and U_0L_2 of the truss in Fig. P4-103.

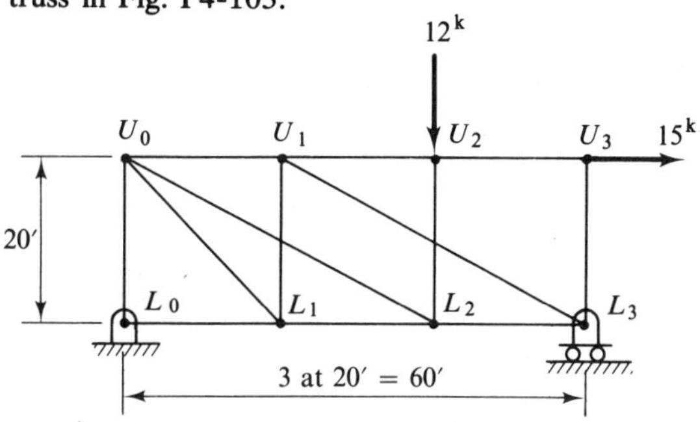

Figure P4-103

4-104 Determine the force in member DM of the K-truss in Fig. P4-104.

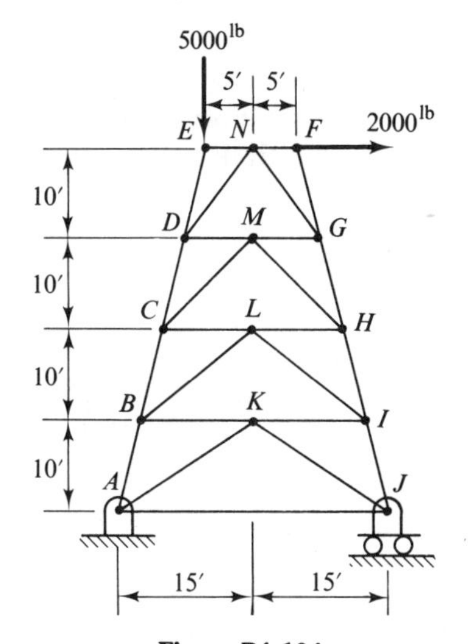

Figure P4-104

4-105 Members AB and CD of the truss in Fig. P4-105 are tie rods and can carry only tensile loads. (They will buckle if subjected to compressive loads.) Show that AB will be in tension when P_1 is the only load acting and that CD will be in tension when P_2 acts alone. Determine the ratio of P_1 to P_2 which will cause both AB and CD to carry no load.

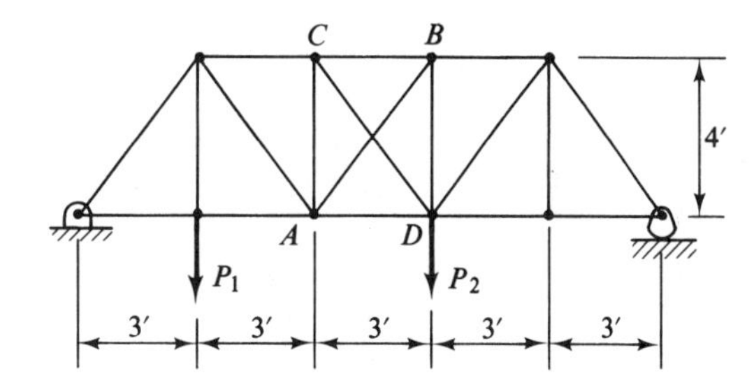

Figure P4-105

4-106 The 3200-lb load W in Fig. P4-106 is supported by a cable which is attached to the truss at B and passes around the pulley F, over the pulley C, back to pulley F, and finally to D. Determine the force in member BC and the reaction of the support at A on the truss.

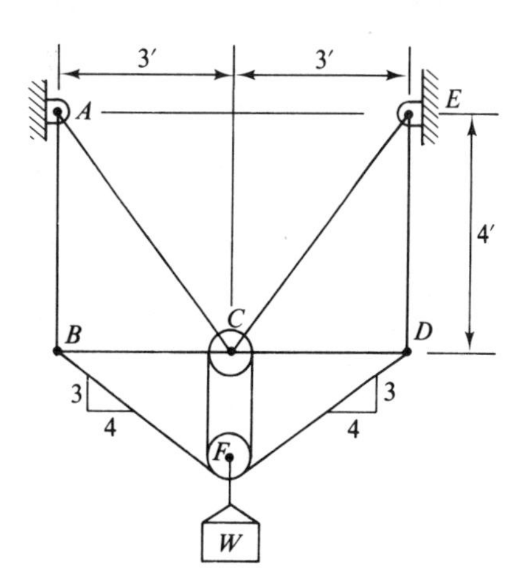

Figure P4-106

4-9 FLEXIBLE CABLES

In many engineering structures, such as suspension bridges and transmission lines, cables which are assumed to be flexible are suspended between supports and subjected to vertical loads. Since a flexible cable offers no resistance to bending, the resultant internal force on any cross section of the cable must act along the tangent to the cable at that section. The resistance to bending offered by actual cables is usually relatively small and can be neglected without serious error.

The two common types of assumed load distribution on cables are (1) load uniformly distributed along the horizontal span and (2) load uniformly distributed along the cable. The weight of a suspension-bridge roadway is an example of a load which is uniformly distributed along the horizontal span. Cables loaded uniformly along the span hang in the shape of parabolas, as will be shown in Art. 4-10. The weight of a homogeneous cable of constant cross section is an example of a load distributed uniformly along the cable. If the cable is stretched so tightly that its sag is small compared to its length, its weight can be assumed to be distributed uniformly along the horizontal span instead of along the length of the cable without introducing an appreciable error. A cable loaded uniformly along the cable hangs in the shape of a catenary, as will be shown in Art. 4-11. The term *sag*, as used here, *is the difference in elevation between the lowest point on the cable and a support.* When the supports are not at the same elevation, the sag measured from one support will be different from the sag measured from the other support. *The span is the horizontal distance between supports.*

4-10 PARABOLIC CABLES

Let ADB in Fig. 4-15(a) be a flexible cable supported and loaded as shown. The lowest point of the cable, D, is selected as the

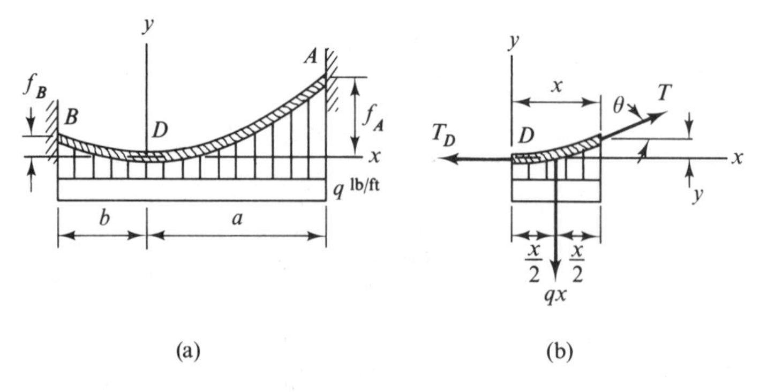

Figure 4-15

origin for the coordinate axes. A free-body diagram of a portion of the cable immediately to the right of point D is shown in Fig. 4-15(b). The horizontal tension in the cable at D is denoted by T_D, and the tension at any other point (coordinates x, y) by T.

The equations of equilibrium

$$\sum F_x = 0, \qquad \sum F_y = 0$$

give

$$T \cos\theta = T_D, \qquad T \sin\theta = qx, \tag{a}$$

where q is the load per unit length (measured horizontally). Elimination of T from these equations gives

$$\tan\theta = \frac{qx}{T_D}. \tag{b}$$

Since the slope of the curve is

$$\frac{dy}{dx} = \tan\theta,$$

Eq. (b) becomes

$$\frac{dy}{dx} = \frac{qx}{T_D}. \tag{c}$$

When Eq. (c) is integrated, the result is

$$y = \frac{qx^2}{2T_D} + C.$$

The constant C is determined by the choice of axes. In this case, when x is equal to zero, y is equal to zero; therefore, C is equal to zero, and the equation of the loaded cable is

$$y = \frac{qx^2}{2T_D}, \tag{4-8}$$

which is a parabola with its vertex at the lowest point of the cable. The elimination of θ from Eq. (a) gives

$$T = (T_D^2 + q^2x^2)^{1/2}. \tag{d}$$

Equation (d) shows that the tension in the cable varies from the minimum value, T_D, at the lowest point in the cable when x is equal to zero, to maximum values at the supports or towers. By substituting the values of x and y at the supports in Eq. (4-8) and using Eq. (d), the tensions at the supports are

$$T_A = qa\sqrt{1 + \frac{a^2}{4f_A^2}}, \qquad T_B = qb\sqrt{1 + \frac{b^2}{4f_B^2}}. \tag{4-9}$$

One method of construction of power lines is by pulling the wire tight enough to give the prescribed tension in the cable and then checking the sag in certain spans to be sure the line is ready to fasten in place.

Equations (4-9) can be used to determine the sag for a given tension and span.

Sometimes it is desirable to determine the length of cable, s, between the origin and one of the supports. The length of the curve from D to A, s_A, is given by the following expression, as in any calculus text:

$$s_A = \int_0^a \sqrt{1+\left(\frac{dy}{dx}\right)^2}\, dx = \int_0^a \sqrt{1+\frac{q^2x^2}{T_D^2}}\, dx.$$

From Eq. (4-8), $T_D = qa^2/(2f_A)$, and thus

$$s_A = \int_0^a \sqrt{1+\frac{q^2x^2}{[qa^2/(2f_A)]^2}}\, dx = \int_0^a \sqrt{1+\frac{4f_A^2x^2}{a^4}}\, dx. \tag{e}$$

When Eq. (e) is integrated directly, the result is a cumbersome hyperbolic or logarithmic function. A more useful expression for s_A can be obtained by expanding the integrand by the binomial expansion and integrating term by term as follows:

$$s_A = \int_0^a \left(1+\frac{2f_A^2x^2}{a^4}-\frac{2f_A^4x^4}{a^8}+\frac{4f_A^6x^6}{a^{12}}-\cdots\right) dx$$

$$= a\left(1+\frac{2f_A^2}{3a^2}-\frac{2f_A^4}{5a^4}+\frac{4f_A^6}{7a^6}-\cdots\right). \tag{4-10}$$

This series converges to the correct length of s_A for all values of f_A/a less than $\frac{1}{2}$. Similarly, the length of the cable s_B from the origin to the left support is

$$s_B = b\left(1+\frac{2f_B^2}{3b^2}-\frac{2f_B^4}{5b^4}+\frac{4f_B^6}{7b^6}-\cdots\right). \tag{4-11}$$

The series converges to the correct length of s_B for all values of f_B/b less than $\frac{1}{2}$. Usually, the ratio f_A/a or f_B/b is quite small, and the first two terms of the series give a sufficiently close approximation to the actual length.

EXAMPLE 4-8

A piece of chain weighing 0.500 lb per ft is suspended between two poles with the same elevation, spaced 20 ft apart. The pull of the chain on the poles is 8.00 lb. Determine the sag and the length of the chain, assuming the weight to be uniformly distributed along a horizontal axis.

SOLUTION

In Eq. (4-9),

$$q = 0.500 \text{ lb per ft}, \qquad a = \tfrac{20}{2} = 10 \text{ ft},$$

and

$$T_A = 8.00 \text{ lb}$$

Therefore,

$$8.00 = 0.50(10)\sqrt{1+\frac{(10)^2}{4f_A^2}}$$

and

$$f_A = \underline{4.00 \text{ ft.}} \qquad \text{Ans.}$$

The value of f_A/a is 0.400, which assures that the series in Eq. (4-10) is convergent. Thus

$$\begin{aligned} s_A &= 10[1+(\tfrac{2}{3})(0.400)^2-(\tfrac{2}{5})(0.400)^4 \\ &\quad +(\tfrac{4}{7})(0.400)^6-\cdots] \\ &= 10(1+0.1067-0.0102+0.0023-\cdots) \\ &= 10(1.099) = 10.99 \text{ ft} \end{aligned}$$

for half the chain, and the total length of the chain is

$$2s_A = \underline{22.0 \text{ ft.}} \qquad \text{Ans.}$$

PROBLEMS

4-107 The two cables of a suspension bridge support a load of $1.25(10^5)$ kips which is uniformly distributed along the span. The span is 3000 ft, and the sag is 350 ft. Compute the angle the cables make with the horizontal at the towers and the tension in each cable at the center of the span.

4-108 A cable carries a load of 40 lb per horizontal ft. The left support is 10 ft lower than the right support, and the sag, measured from the left support, is 10 ft. The supports are 150 ft apart. Determine the minimum and maximum tensions in the cable and the length of the cable.

4-109 The center span of the Mid-Hudson Suspension Bridge is 1500 ft. The design tension in the 16.75-in.-diameter cables is 6000 tons at the towers, and the sag is 150 ft. Determine the design load per foot on the cable and the length of the cable for the center span.

4-110 A 5-ton load is uniformly distributed over a span of 200 ft. It is to be supported by a single flexible cable hung from two towers at the same level. The design strength of the cable is 10 tons. Determine the minimum sag and length of cable which can be used.

***4-111** A 60-m cable is suspended between two points at the same elevation and 55.5 m apart. The cable supports a large load which is uniformly distributed along the span. Determine the sag, f, in the cable.

4-112 The cable shown in Fig. P4-112 carries a linearly varying, distributed load. Apply the same method used for deriving the shape of parabolic cables to show that the equation for the shape of loaded cable is

$$y = \frac{q_0}{2T_D}\left(x^2 - \frac{x^3}{3L}\right),$$

and derive an equation for the maximum tension in the cable in terms of q_0, f, and L.

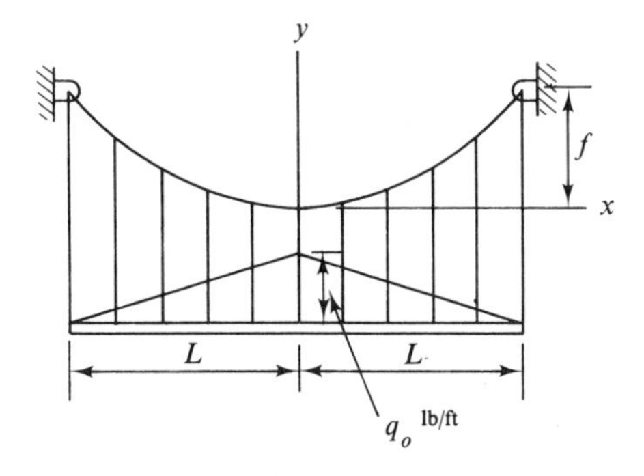

Figure P4-112

4-11 CATENARY CABLES

Let ADB in Fig. 4-16(a) represent a flexible cable having a weight of q lb per ft length of cable. The origin is selected at a distance C below point D to simplify the final results. The distance C will be determined later. A free-body diagram of the portion of the cable of length s from D to any point P is shown in Fig. 4-16(b). When the equations of equilibrium are applied to the free-body diagram, the results are

$$T \cos \theta = T_D, \qquad T \sin \theta = qs. \tag{a}$$

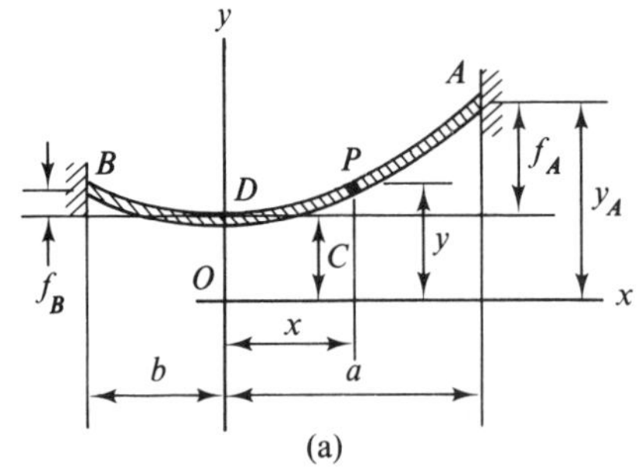

(a)

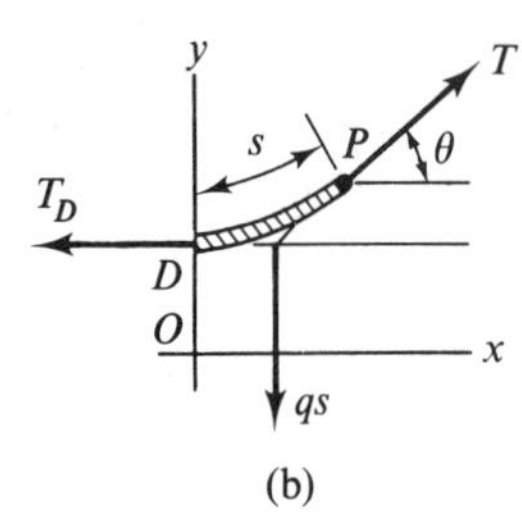

(b)

Figure 4-16

When T is eliminated from Eq. (a), the result is

$$\tan \theta = \frac{qs}{T_D} \quad \text{or} \quad s = \frac{T_D}{q} \tan \theta, \tag{b}$$

which is the equation of the curve in terms of s and θ and the constants q and T_D. The equation may be written in terms of cartesian coordinates by using the following relationship, which applies to any plane curve:

$$\frac{dy}{d\theta} = \frac{dy}{ds}\frac{ds}{d\theta} = \sin \theta \frac{ds}{d\theta},$$

since dy/ds is equal to $\sin \theta$ for the curve in Fig. 4-16(b). From Eq. (b)

$$\frac{ds}{d\theta} = \frac{T_D}{q} \sec^2 \theta,$$

and thus

$$\frac{dy}{d\theta} = \sin \theta \frac{T_D}{q} \sec^2 \theta$$

from which

$$y = \int \frac{T_D}{q} \sec \theta \tan \theta \, d\theta = \frac{T_D}{q} \sec \theta + C_1.$$

At the point $x = 0$, $y = C$, the angle θ is zero; therefore

$$C = \frac{T_D}{q}(1) + C_1.$$

By selecting $C = T_D/q$, the constant of integration C_1 becomes zero and

$$y = \frac{C}{\cos \theta} \quad \text{or} \quad \cos \theta = \frac{C}{y}. \tag{4-12}$$

The slope of the curve is

$$\frac{dy}{dx} = \tan \theta = \frac{\sqrt{y^2 - C^2}}{C},$$

and

$$dx = \frac{C\,dy}{\sqrt{y^2 - C^2}}$$

from which

$$x = C \cosh^{-1} \frac{y}{C} + C_2.$$

When x is equal to zero, y is equal to C and $\cosh^{-1} 1$ is equal to zero; therefore C_2 is equal to zero, and the preceding equation can be written

$$y = C \cosh \frac{x}{C}. \tag{4-13}$$

Equation (4-13), the equation of the curve in cartesian coordinates, is called a *catenary*.

When C is substituted for T_D/q in Eq. (b), the resulting equation can be combined with Eq. (4-12) to obtain the relationship between y and s as follows:

$$C = y \cos \theta,$$

$$s = C \tan \theta.$$

From the accompanying figure representing the equation

$$\cos \theta = \frac{C}{y},$$

it is evident that

$$\tan \theta = \frac{\sqrt{y^2 - C^2}}{C}.$$

Thus

$$s = C \frac{\sqrt{y^2 - C^2}}{C} = \sqrt{y^2 - C^2},$$

or

$$y = \sqrt{s^2 + C^2}. \tag{4-14}$$

An equation relating x and s can be obtained by combining Eq. (4-13) and Eq. (4-14) to eliminate y as follows:

$$y = C \cosh \frac{x}{C} = \sqrt{s^2 + C^2}$$

from which

$$s^2 + C^2 = C^2 \cosh^2 \frac{x}{C}$$

or

$$s^2 = C^2\left(\cosh^2\frac{x}{C} - 1\right) = C^2\sinh^2\frac{x}{C}$$

and

$$s = C\sinh\frac{x}{C}. \qquad (4\text{-}15)$$

The sag f in the cable is the difference between the value of y at the support (y_A) and C. For the right-hand section of the cable in Fig. 4-16(a), the sag is

$$f_A = y_A - C = C\left(\cosh\frac{a}{C} - 1\right). \qquad (4\text{-}16)$$

Also

$$f_A = y_A - C = \sqrt{s_A^2 + C^2} - C.$$

If θ is eliminated from Eq. (a), the result is

$$T = \sqrt{T_D^2 + q^2 s^2} = \sqrt{C^2 q^2 + q^2 s^2}$$
$$= q\sqrt{C^2 + s^2} = qy = qC\cosh\frac{x}{C}. \qquad (4\text{-}17)$$

Equation (4-17) indicates that the tension in the cable at any point is equal to the product of the load per foot of cable and the vertical distance of the point from the x axis.

The use of Eq. (4-13) to Eq. (4-17) requires the determination of the length C, which must usually be done by a trial-and-error process or a graphical solution. Various tabular and semigraphical procedures have been devised to facilitate the solution of these equations.[2] The following example illustrates one procedure.

EXAMPLE 4-9

Solve Example 4-8 using the exact solution.

SOLUTION

In Eq. (4-16),

$$y = y_A = f_A + C$$

is the distance from the x axis to the support, and in Eq. (4-17), $T = T_A$ is the tension in the chain at the support. Therefore

$$8.00 = 0.500 y_A$$

and

$$y_A = C + f_A = \frac{8.00}{0.500} = 16.00\text{ ft} = C\cosh\frac{x}{C}.$$

---

[2] Elihu Greer, "Simplified Numerical Solution of a Catenary," *Industrial Mathematics* (1956), Vol. 7.

In this equation, x is equal to 10 ft when y is equal to y_A. Therefore,

$$C \cosh \frac{10}{C} = 16$$

or

$$\frac{16}{C} = \cosh \frac{10}{C}.$$

The value of C can be determined graphically by plotting values of $16/C$ against C and also by plotting values of $\cosh 10/C$ against C on the same graph. The intersection of the curves will determine the value of C. The accompanying table of values is plotted in Fig. 4-17, and the value of C as determined from the point of intersection of the curves is approximately 11.27 ft.

C	$\frac{16}{C}$	$\cosh \frac{10}{C}$
10.0	1.600	1.543
10.5	1.524	1.489
11.0	1.454	1.442
11.5	1.391	1.403
12.0	1.333	1.368

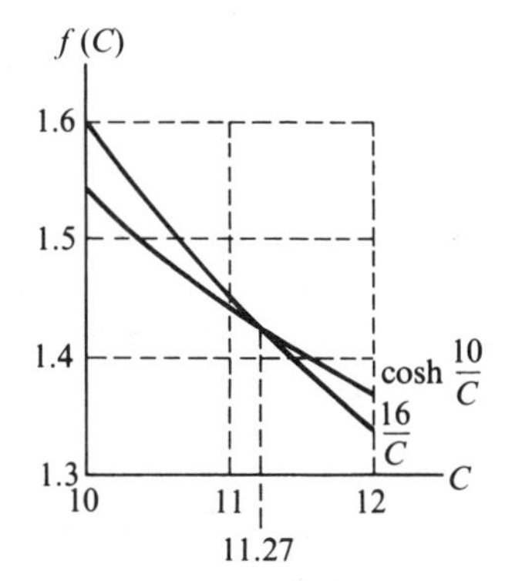

Figure 4-17

The value of C can also be determined by successive trials as shown in the following table:

C	$\frac{10}{C}$	$\cosh \frac{10}{C}$	$C \cosh \frac{10}{C}$
10.0	1.000	1.543	15.43
11.0	0.909	1.442	15.86
11.4	0.877	1.410	16.07
11.3	0.885	1.418	16.02
11.27	0.887	1.420	16.00

The value of C for which $C \cosh(10/C) = 16$ is the solution of the equation and is seen to be approximately 11.27 ft, which checks the graphical solution.

The sag is

$$f_A = y_A - C = 16.00 - 11.27 = 4.73 \text{ ft}$$

as compared with a value of 4.00 ft in Example 4-8, indicating an error of 0.73/4.73, or 15.43%. If the chain is drawn tighter so that the sag is decreased, the difference in the two results will decrease, and for small ratios of sag to span (less than 5%) the difference in the two results can usually be neglected.

In Eq. (4-15), s is the length of the chain from D to any point (x, y). When

$$x = a = 10 \text{ ft}$$

the corresponding value of s, which equals s_A, is one-half the length of the chain. The value of s_A is

$$s_A = C \sinh \frac{a}{C} = 11.27 \sinh \frac{10.00}{11.27} = 11.36 \text{ ft}.$$

Therefore, the length of the chain is 22.7 ft as compared with a length of 22.0 ft from the approximate solution.

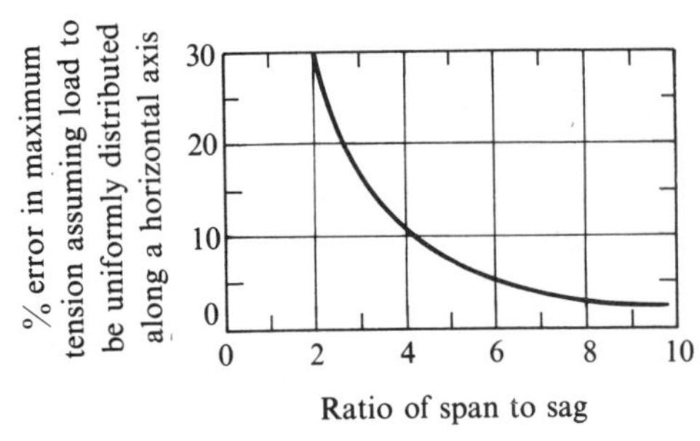

Figure 4-18

The diagram in Fig. 4-18 will aid in deciding when it is necessary to use the exact method for computing tensions, sags, and similar results for flexible cables. The error in the maximum tension is the difference between the approximate tension, computed by means of Eq. (4-9), and the exact value, computed by means of Eq. (4-17). The variations of the percentage error with the ratio of the span to sag is indicated in the figure.

PROBLEMS

4-113 An electrical cable weighing 0.40 lb per ft is to be suspended between two points 200 ft apart, and the maximum sag is to be 40 ft. Determine the maximum tension in the cable and the length of cable required.

4-114 A uniform chain 120 ft long and weighing 0.50 lb per ft has a span of 100 ft. The supports are at the same elevation. Determine the sag and the maximum and minimum tensions in the chain.

***4-115** The towers that support a steel cable with wrapped telephone lines are 750 m apart. The combination of cable and wires has a mass of 0.75 kg/m. The maximum allowable tension in the cable is 4.5 kN. Calculate the length of the wire and the sag necessary to produce this tension.

4-116 A 300-ft cable weighing 150 lb is hung between two supports at the same elevation. Determine the span which will result in a maximum cable tension of 125 lb.

4-12 EQUILIBRIUM OF FORCE SYSTEMS IN SPACE

The resultant of a nonconcurrent force system in space is a single force, a couple, or a force and a couple (see Art. 2-4). Any set of equations which insures the absence of all possible resultants is a set of equations of equilibrium for the system. From Art. 2-4, the resultant force through the origin and resultant moment about the origin are given by the equations

$$\mathbf{R} = \sum \mathbf{F}_i \quad \text{and} \quad \mathbf{M}_0 = \sum \mathbf{r}_i \times \mathbf{F}_i.$$

The equations of equilibrium can therefore be stated in vector form as

$$\sum \mathbf{F}_i = 0 \quad \text{and} \quad \sum \mathbf{r}_i \times \mathbf{F}_i = 0, \tag{4-18}$$

where $\mathbf{r}_i$ is the position vector of a point on the force $\mathbf{F}_i$ from any point selected as the origin. These two vector equations can be resolved into the following six component equations:

$$\begin{aligned} \sum F_x &= 0, & \sum F_y &= 0, & \sum F_z &= 0, \\ \sum M_x &= 0, & \sum M_y &= 0, & \sum M_z &= 0, \end{aligned} \qquad (4\text{-}19)$$

where F_x, F_y, and F_z represent magnitudes of the orthogonal components of the forces, and M_x, M_y, and M_z are the magnitudes of the moments of the forces about a set of x, y, and z axes. The moment axes are usually selected as being mutually orthogonal, although various other axes may be chosen subject to certain limitations. The quantities M_x, M_y, and M_z may also be considered to be the orthogonal components of the moments of the forces about the origin (see Art. 1-6).

It is evident that if the three component force equations are satisfied, the resultant cannot be a force. Likewise, if the three moment equations are satisfied, the resultant cannot be a couple. Consequently, if all six scalar equations (or two vector equations) are satisfied for any force system in space, the body on which the force system acts must be in equilibrium.

Any or all of the component force equations of Eq. (4-19) can be replaced by additional moment equations, provided the moment axes are so selected that six independent equations are obtained. There is no simple or convenient way to state the necessary restrictions for selecting these additional moment axes; therefore the best procedure is to use any convenient axes and discard any redundant equations. The procedure just described will be used in the following illustrative examples. Not more than six unknowns (magnitudes, distances, or angles) can be determined from one free-body diagram of a body in space, since there are only six independent equations of equilibrium for the general case.

In many problems, fewer than six independent equations involve the forces of the system, and therefore fewer than six unknowns can be determined. When all forces of a system are parallel to one axis, say the y axis, only three independent equations of equilibrium are useful, and only three unknowns can be determined with such a system. The equations

$$\sum F_x = 0, \qquad \sum F_z = 0$$

provide no useful information because there are no forces in these directions. Likewise, the equation

$$\sum M_y = 0$$

provides no information concerning the unknowns because none of the forces has a moment about the y axis. This leaves

$$\sum F_y = 0, \qquad \sum M_x = 0, \qquad \sum M_z = 0$$

as a complete set of equations of equilibrium for a parallel force system in space where the y axis is parallel to the forces.

Another special case is a system of couples in space. In this problem, the three component force equations give no useful information, and only three unknowns can be obtained from the three moment equations.

Still another special case resulting in less than six independent equations of equilibrium is a system where all forces intersect one line in space, say the x axis. In such a system, the equation

$$\sum M_x = 0$$

will yield no useful information, and only five unknowns can be evaluated by means of the equations of equilibrium.

In some cases (see Example 4-12) where a smooth bearing or socket is used, it is easier to show three mutually perpendicular components of a force or moment even though only two components are necessary to describe the vector provided they are properly selected. Such a procedure will normally give one more unknown than the number of available equations of equilibrium. An additional relation between the three components is needed to solve the problem. For situations of this type, one procedure is to write the three components as a resultant vector and dot it into a unit vector directed along the line of zero force or moment. Because this dot product gives the component of the force or moment along the selected axis and because the component force or moment in this direction must be zero, the product can be equated to zero for the necessary additional equation relating the selected components.

The following examples illustrate the procedures for evaluating unknown forces, distances, and slopes in a nonconcurrent force system in space where equilibrium is specified.

EXAMPLE 4-10

The body in Fig. 4-19(a) is supported by a ball and socket at A, two vertical cables B and C, and the horizontal cable D. The weight of the body can be neglected. Determine the tension in each of the cables and the reaction on the body at A.

SOLUTION

The free-body diagram in Fig. 4-19(b) shows three known forces ($\mathbf{F}_1$, $\mathbf{F}_2$, and $\mathbf{F}_3$) and four unknown forces ($\mathbf{F}_A$, $\mathbf{F}_B$, $\mathbf{F}_C$, and $\mathbf{F}_D$). Since both the magnitude and direction of $\mathbf{F}_A$ are unknown, this is equivalent to three unknown quantities (the magnitudes of the components of $\mathbf{F}_A$ or some combination of magnitude and angles). Thus there are six

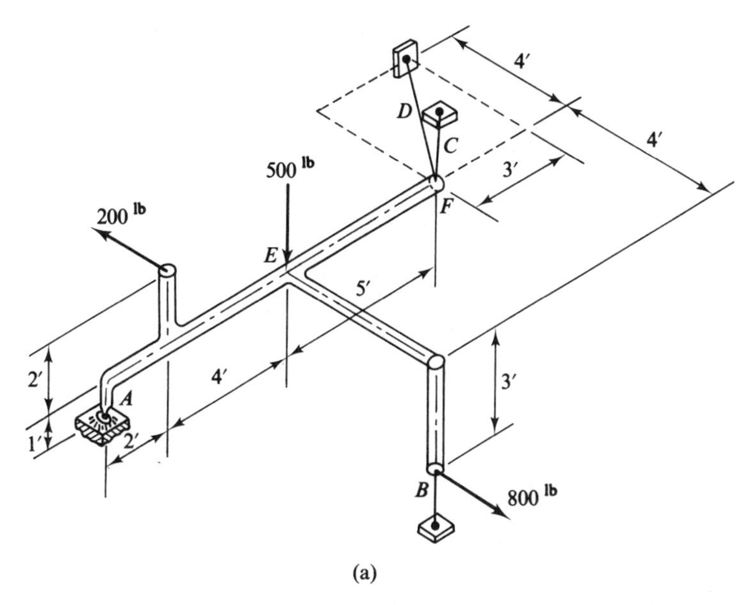

(a)

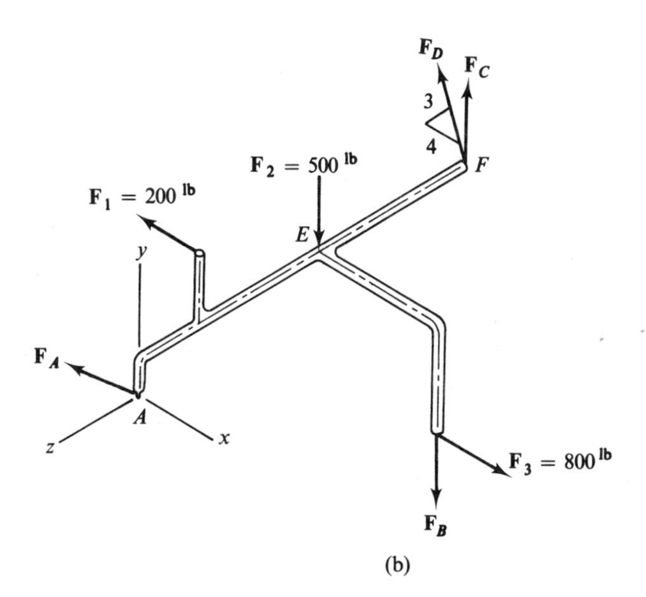

(b)

Figure 4-19

unknowns, and they can be evaluated from the six equations of equilibrium. In vector form, the forces are

$$\mathbf{F}_A = A_x\mathbf{i} + A_y\mathbf{j} + A_z\mathbf{k}; \qquad \mathbf{F}_B = -B\mathbf{j}; \qquad \mathbf{F}_C = C\mathbf{j};$$

$$\mathbf{F}_D = -0.8D\mathbf{i} - 0.6D\mathbf{k}; \qquad \mathbf{F}_1 = -200\mathbf{i};$$

$$\mathbf{F}_2 = -500\mathbf{j}; \qquad \mathbf{F}_3 = 800\mathbf{i}.$$

When using the equation $\sum \mathbf{r} \times \mathbf{F} = 0$, it is convenient to select a moment center which will eliminate one or more of the unknowns, such as point A in this problem. The position vectors to points on the forces (not necessarily the points of application) from A are

$$\mathbf{r}_A = 0; \quad \mathbf{r}_B = 4\mathbf{i} - 6\mathbf{k}; \quad \mathbf{r}_C = -11\mathbf{k}; \quad \mathbf{r}_D = 1\mathbf{j} - 11\mathbf{k};$$

$$\mathbf{r}_1 = 3\mathbf{j} - 2\mathbf{k}; \quad \mathbf{r}_2 = -6\mathbf{k}; \quad \mathbf{r}_3 = -2\mathbf{j} - 6\mathbf{k}.$$

The equation $\sum \mathbf{F} = 0$ becomes

$$\sum \mathbf{F} = \mathbf{i}(A_x - 0.8D - 200 + 800) + \mathbf{j}(A_y - B + C - 500) + \mathbf{k}(A_z - 0.6D) = 0,$$

which is equivalent to three scalar equations:

$$A_x - 0.8D + 600 = 0,$$

$$A_y - B + C - 500 = 0,$$

and

$$A_z - 0.6D = 0.$$

The moment equation is

$$\begin{aligned}
\mathbf{M}_A &= \sum \mathbf{r} \times \mathbf{F} \\
&= 0 + (4\mathbf{i} - 6\mathbf{k}) \times (-B\mathbf{j}) + (-11\mathbf{k}) \times (C\mathbf{j}) \\
&\quad + (\mathbf{j} - 11\mathbf{k}) \times (-0.8D\mathbf{i} - 0.6D\mathbf{k}) \\
&\quad + (3\mathbf{j} - 2\mathbf{k}) \times (-200\mathbf{i}) + (-6\mathbf{k}) \times (-500\mathbf{j}) \\
&\quad + (-2\mathbf{j} - 6\mathbf{k}) \times (800\mathbf{i}) \\
&= 0.
\end{aligned}$$

When these products are expanded the equation becomes

$$\begin{aligned}
\mathbf{M}_A &= -4B\mathbf{k} - 6B\mathbf{i} + 11C\mathbf{i} + 0.8D\mathbf{k} - 0.6D\mathbf{i} + 8.8D\mathbf{j} \\
&\quad + 600\mathbf{k} + 400\mathbf{j} - 3000\mathbf{i} + 1600\mathbf{k} - 4800\mathbf{j} \\
&= \mathbf{i}(-6B + 11C - 0.6D - 3000) \\
&\quad + \mathbf{j}(8.8D + 400 - 4800) \\
&\quad + \mathbf{k}(-4B + 0.8D + 600 + 1600) \\
&= 0
\end{aligned}$$

which is equivalent to three more scalar equations:

$$-6B + 11C - 0.6D - 3000 = 0,$$

$$8.8D - 4400 = 0,$$

and

$$-4B + 0.8D + 2200 = 0.$$

The last three equations contain only three unknowns (as a result of the selection of the moment center), and when they are solved simultaneously, they give

$$B = \underline{650 \text{ lb},} \qquad \text{Ans.}$$

$$C = \underline{655 \text{ lb},} \qquad \text{Ans.}$$

$$D = \underline{500 \text{ lb}.} \qquad \text{Ans.}$$

When the values for B, C, and D are substituted in the force equations, the magnitudes of the components of $\mathbf{F}_A$ are found to be

$$A_x = -200 \text{ lb}, \qquad A_y = 495 \text{ lb}, \qquad A_z = 300 \text{ lb}.$$

The reaction at A is

$$\mathbf{F}_A = -200\mathbf{i} + 495\mathbf{j} + 300\mathbf{k}$$

$$= \underline{612(-0.327\mathbf{i} + 0.808\mathbf{j} + 0.490\mathbf{k}) \text{ lb}}$$

$$\underline{\text{through } A.} \qquad \text{Ans.}$$

EXAMPLE 4-11

A shaft with two 90° bends is supported by three bearings as indicated in Fig. 4-20(a). Determine the reaction of the bearing at C on the shaft.

SOLUTION

The free-body diagram in Fig. 4-20(b) contains six unknown components of the bearing reactions, and the two vector equations of equilibrium are equivalent to six scalar equations. If the moment center is selected at points A or B, the moment equation will contain only four unknowns, two of which are the components of the reaction at C, and may thus yield a more direct solution. The reaction components can be

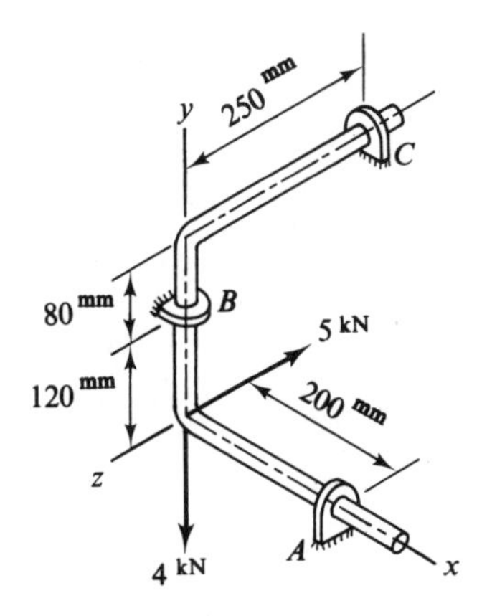

Figure 4-20(a)

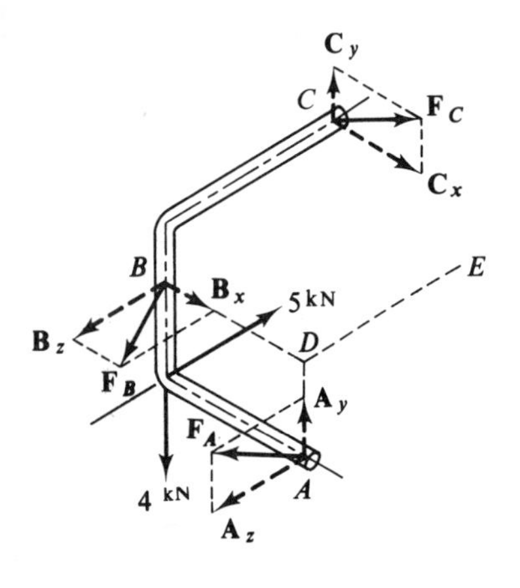

Figure 4-20(b)

determined from the following equations:

$$\sum \mathbf{F} = A_y\mathbf{j} + A_z\mathbf{k} + B_x\mathbf{i} + B_z\mathbf{k} + C_x\mathbf{i} + C_y\mathbf{j} - 4\mathbf{j} - 5\mathbf{k}$$
$$= \mathbf{i}(B_x + C_x) + \mathbf{j}(A_y + C_y - 4) + \mathbf{k}(A_z + B_z - 5)$$
$$= 0$$

and

$$\sum \mathbf{M}_A = \sum \mathbf{r}\times\mathbf{F} = (-200\mathbf{i} + 120\mathbf{j})\times(B_x\mathbf{i} + B_z\mathbf{k})$$
$$+ (-200\mathbf{i} + 200\mathbf{j} - 250\mathbf{k})\times(C_x\mathbf{i} + C_y\mathbf{j})$$
$$+ (-200\mathbf{i})\times(-4\mathbf{j} - 5\mathbf{k})$$
$$= \mathbf{i}(120B_z + 250C_y) + \mathbf{j}(200B_z - 250C_x - 1000)$$
$$+ \mathbf{k}(-120B_x - 200C_x - 200C_y + 800)$$
$$= 0.$$

These two vector equations are equivalent to the following six scalar equations:

$$B_x + C_x = 0, \tag{a}$$
$$A_y + C_y = 4, \tag{b}$$
$$A_z + B_z = 5, \tag{c}$$
$$120B_z + 250C_y = 0, \tag{d}$$
$$200B_z - 250C_x = 1000, \tag{e}$$
$$-120B_x - 200C_x - 200C_y = -800. \tag{f}$$

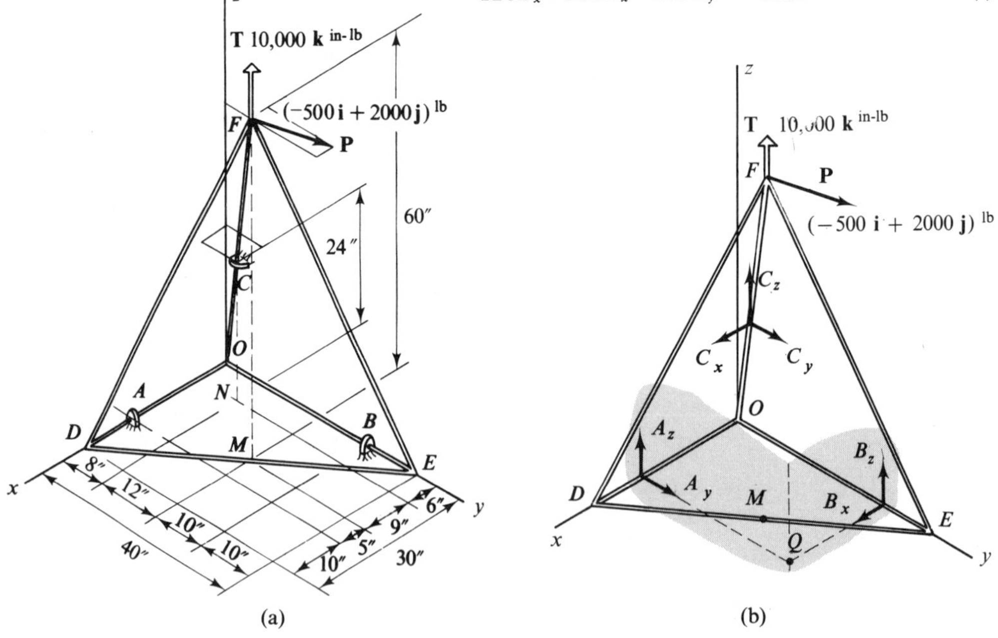

(a)

(b)

Figure 4-21

Equations (a), (d), (e), and (f) can be solved simultaneously for four unknowns, two of which are

$$C_x = -32.0 \text{ kN} \quad \text{and} \quad C_y = 16.8 \text{ kN}$$

from which

$$\mathbf{F}_C = \underline{36.1(-0.885\mathbf{i} + 0.465\mathbf{j}) \text{ kN through point } C.} \quad \text{Ans.}$$

EXAMPLE 4-12

The six rods in Fig. 4-21(a) are welded to form a rigid frame which is supported and loaded as shown. Rods *OD*, *OE*, and *DE* lie in the *xy* plane. Members *OF*, *DF*, and *EF* are each 65 in. long, and point *F* is directly above *M*, the midpoint of *DE*. The three journal bearings at *A*, *B*, and *C* support the frame, and *C* is directly above point *N*. Determine the force exerted by the bearing at *C* on member *OF* due to the force and twisting couple applied at *F*.

SOLUTION

Figure 4-21(b) is a free-body diagram of the frame. Although the force at *C* must be perpendicular to the rod and therefore can be shown as two components, it is frequently simpler to show it as three components. This procedure results in seven unknown components, and since there are only six independent equations of equilibrium, one additional relationship is needed. The component of the force at *C* along the rod must be zero, and since this component is given by the scalar (dot) product of the force **C** and the unit vector along *OF*, an equation relating the components of **C** can be obtained. The force is

$$\mathbf{C} = C_x\mathbf{i} + C_y\mathbf{j} + C_z\mathbf{k},$$

and the unit vector along *OF* is

$$\mathbf{e}_{OF} = \frac{6\mathbf{i} + 8\mathbf{j} + 24\mathbf{k}}{26} = \frac{3\mathbf{i} + 4\mathbf{j} + 12\mathbf{k}}{13}.$$

The component of **C** along *OF* is

$$\mathbf{C} \cdot \mathbf{e}_{OF} = (C_x\mathbf{i} + C_y\mathbf{j} + C_z\mathbf{k}) \cdot \left(\frac{3\mathbf{i} + 4\mathbf{j} + 12\mathbf{k}}{13}\right) = 0,$$

which reduces to

$$3C_x + 4C_y + 12C_z = 0. \tag{a}$$

Two additional equations in these three unknowns can be obtained from the equilibrium equations. The components at *A* will have no moment about any line in the plane of the forces (the plane through *A* perpendicular to *OD*). Similarly, the components at *B* will have no moment about any line in the plane through *B* perpendicular to *OE*.

Therefore all four forces at A and B will be eliminated by summing moments about the line of intersection of these planes (through Q). Equating the sum of the moments about line AB will also eliminate the forces at A and B.

The magnitude of the moment about line AB is

$$\begin{aligned} M_{AB} &= \sum \mathbf{M}_A \cdot \mathbf{e}_{AB} \\ &= (\sum \mathbf{r}\times\mathbf{F}+\mathbf{T})\cdot\mathbf{e}_{AB} \\ &= \mathbf{r}_C\times\mathbf{C}\cdot\mathbf{e}_{AB}+\mathbf{r}_P\times\mathbf{P}\cdot\mathbf{e}_{AB}+\mathbf{T}\cdot\mathbf{e}_{AB}, \end{aligned}$$

where $\mathbf{r}$ is the position vector from A (or any other point on AB) to the corresponding force. The unit vector along AB is

$$\mathbf{e}_{AB} = \frac{-2\mathbf{i}+3\mathbf{j}}{\sqrt{13}},$$

and the moment about AB is

$$M_{AB} = \frac{1}{\sqrt{13}}\begin{vmatrix} -14 & 8 & 24 \\ C_x & C_y & C_z \\ -2 & 3 & 0 \end{vmatrix}$$

$$+\frac{1}{\sqrt{13}}\begin{vmatrix} -5 & 20 & 60 \\ -500 & 2000 & 0 \\ -2 & 3 & 0 \end{vmatrix} + 10{,}000\mathbf{k}\cdot\left(\frac{-2\mathbf{i}+3\mathbf{j}}{\sqrt{13}}\right)$$

$$=0,$$

which reduces to

$$72C_x+48C_y+26C_z=-150{,}000. \qquad \text{(b)}$$

The z component of the moment about Q is

$$\mathbf{M}_Q\cdot\mathbf{k}=\sum \mathbf{r}\times\mathbf{F}\cdot\mathbf{k}+\mathbf{T}\cdot\mathbf{k}=0,$$

where each $\mathbf{r}$ is measured from Q to the corresponding force. This equation becomes

$$\mathbf{M}_Q\cdot\mathbf{k}=\begin{vmatrix} -14 & -22 & 24 \\ C_x & C_y & C_z \\ 0 & 0 & 1 \end{vmatrix}+\begin{vmatrix} -5 & -10 & 60 \\ -500 & 2000 & 0 \\ 0 & 0 & 1 \end{vmatrix}+10{,}000\mathbf{k}\cdot\mathbf{k}=0,$$

which reduces to

$$22C_x-14C_y=5000. \qquad \text{(c)}$$

When Eq. (a), (b), and (c) are solved simultaneously the force C is found to be

$$\begin{aligned} \mathbf{C} &= (-1068\mathbf{i}-2035\mathbf{j}+945\mathbf{k})\text{ lb} \\ &= \underline{2485(-0.430\mathbf{i}-0.818\mathbf{j}+0.380\mathbf{k})\text{ lb through } C.} \end{aligned}$$

Ans.

PROBLEMS

Unless otherwise stated, neglect the weights of the members and assume no friction.

4-117 The three-wheeled, warehouse pushcart in Fig. P4-117 has a homogeneous platform which weighs 250 lb. A crate (not shown) weighing 1500 lb is placed with its center of gravity above point G, which is on line DE. The reaction of the floor on wheel B is 550 lb. Locate point G and determine the force exerted by the floor on wheel C.

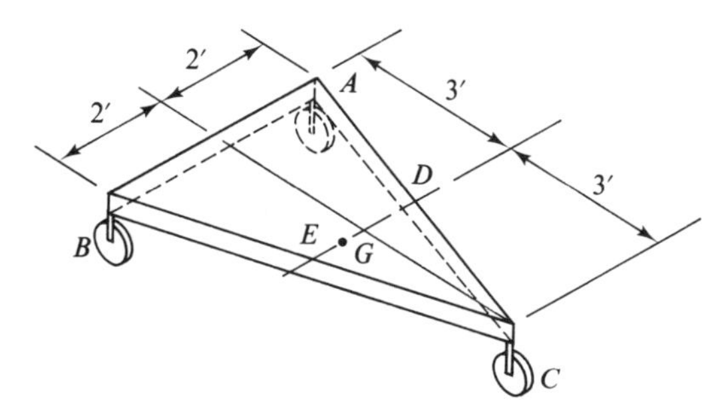

Figure P4-117

4-118 A tier table with three feet or legs as shown in Fig. P4-118 weighs 15 lb. To check its design for stability a 12-lb weight A is placed as shown. For the greatest danger of tipping, should the weight be placed directly above a foot or at the midpoint between two feet? Calculate the force on each foot for both cases to support your conclusion.

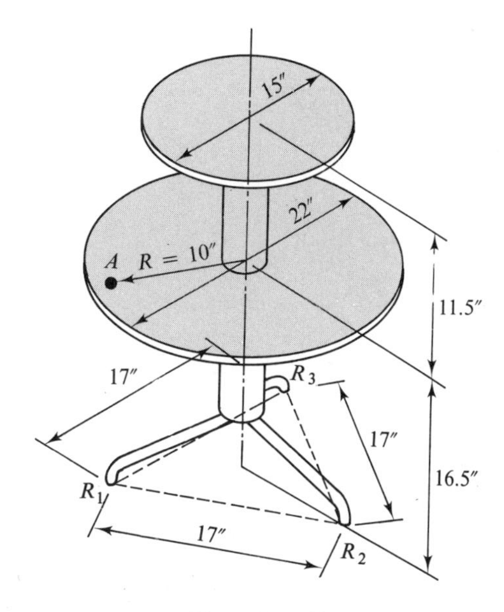

Figure P4-118

4-119 The wheelbarrow in Fig. P4-119 together with a load of sand and rocks weighs 250 lb. The mass center of the composite weight is at G.

(a) Determine the vertical force on each handle necessary to lift legs A and B off the ground.

(b) Determine the vertical force on handle D necessary to lift leg A off the ground.

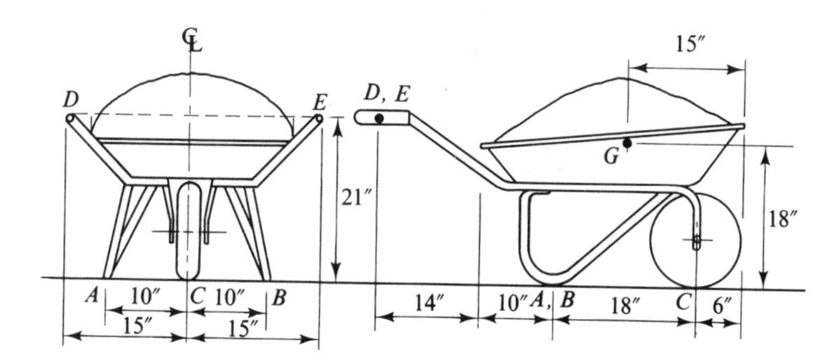

Figure P4-119

***4-120** The magnitudes of the crankpin forces $\mathbf{P}_1$ and $\mathbf{P}_2$ on the crankshaft in Fig. P4-120 are 26 and 15 kN, respectively, in the position shown. Determine the torque T required for equilibrium and the bearing reactions $\mathbf{R}_1$ and $\mathbf{R}_2$.

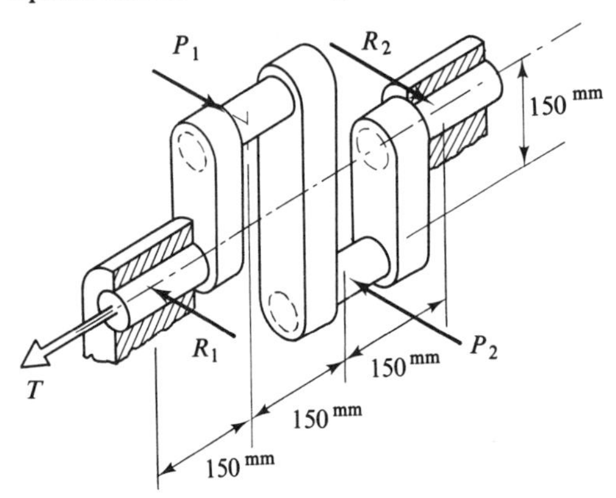

Figure P4-120

4-121 The refueling tanker boom in Fig. P4-121 is mounted on a test stand so that it is perpendicular to plane ABC and is supported by a ball and socket at C and by cables AD and BD. Determine all unknown forces acting on the boom.

4-122 The boom AB in Fig. P4-122 is supported by a ball and socket at B and by cables AC and AD. Determine the reaction of the support at B on the boom.

4-123 Members AB and AC in Fig. P4-123 are connected by ball and socket joints at A, B, and C, and member AB rests on the rigid bar EF at D. Bar EF is parallel to the x axis. Determine the reaction of EF on AB at D.

4-124 Block A of Fig. P4-124 slides on a smooth vertical cylindrical rod and is attached to the inclined rod by a clevis with a horizontal pin. Block B slides on a smooth horizontal rod and is connected to the inclined rod by a ball and socket joint. Determine the vertical force $\mathbf{F}$ on block A to maintain equilibrium and the reaction of the vertical rod on A.

4-125 The force $\mathbf{F}$ in Problem 4-124 is replaced by a couple $\mathbf{C}$ whose axis is vertical. Determine the couple $\mathbf{C}$ necessary to maintain equilibrium and the force exerted by the vertical rod on A.

4-126 Figure P4-126 shows two rods in a rectangular box. Rods AB (25 in. long) and CD (20 in. long) weigh 0.20 lb per ft. Rod AB is placed in the box first, and CD has one end in the corner at C and also rests on rod AB and the vertical wall.

(a) Show that the unit vector from C to D is $\mathbf{n} = -0.800\mathbf{i} + 0.375\mathbf{j} - 0.469\mathbf{k}$ and locate E, the point of intersection of the rods (neglect the thickness of the rods).

(b) Determine the force at E on rod AB.

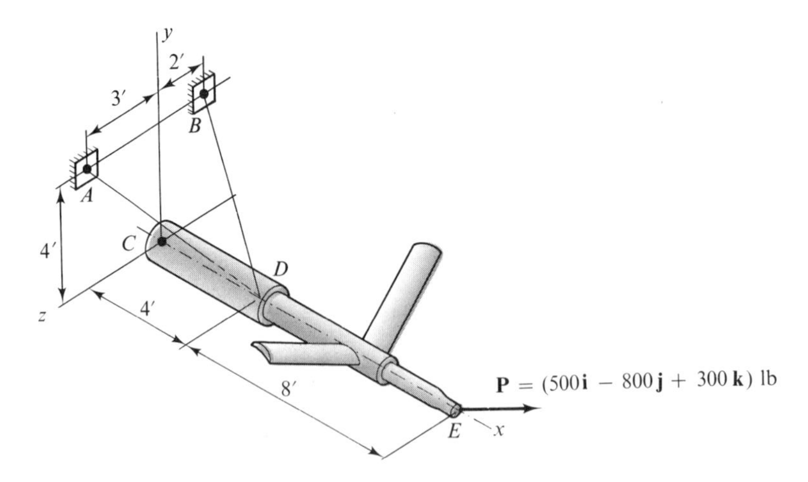

Figure P4-121

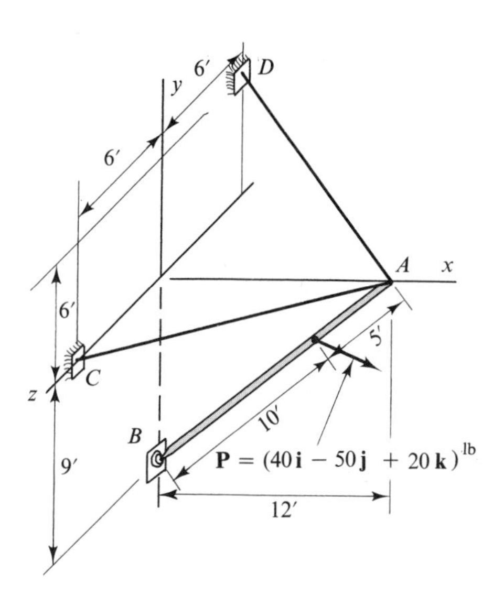

Figure P4-122

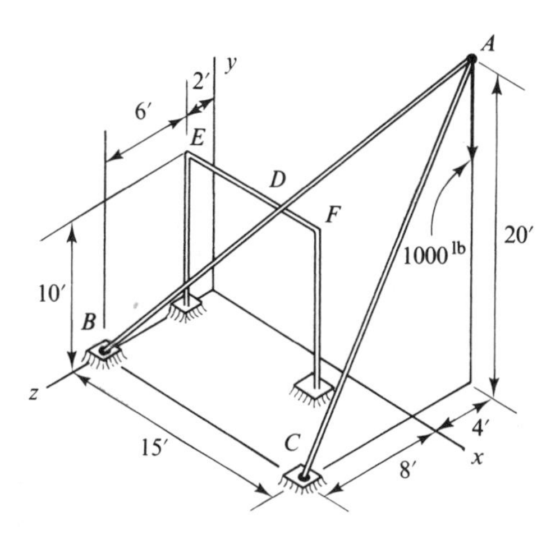

Figure P4-123

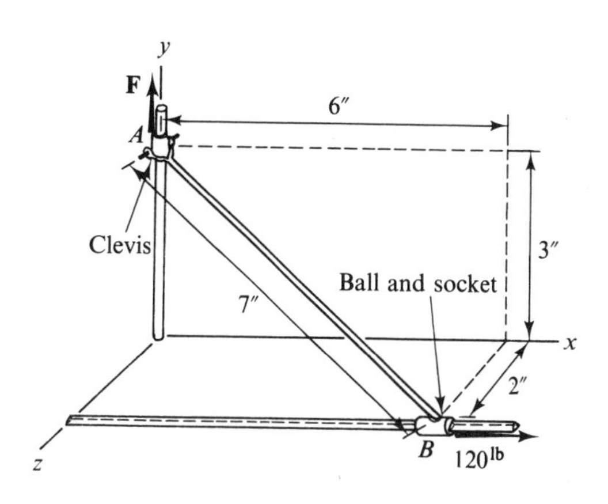

Figure P4-124

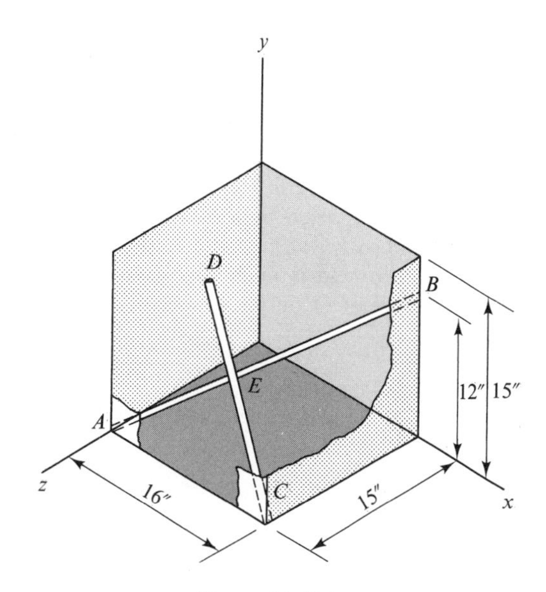

Figure P4-126

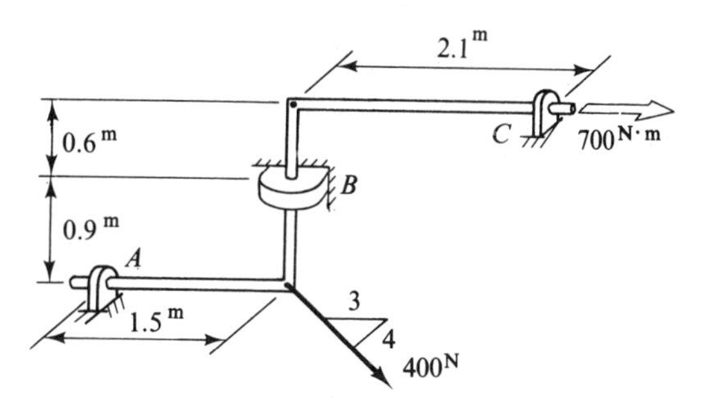

Figure P4-127

***4-127** The bent rod in Fig. P4-127 is supported by three journal bearings at A, B, and C and lies in a vertical plane. The 400-N force is in a horizontal plane, and the vector of the 700-N · m couple is horizontal and in the plane of the rod. Determine the bearing reaction at A on the rod.

4-128 Rod *ABCD* in Fig. P4-128 is supported by three bearings at *E, F,* and *G*. The 200-lb force at *C* lies in the plane of *BCD*. Determine the reaction at *F* on the rod.

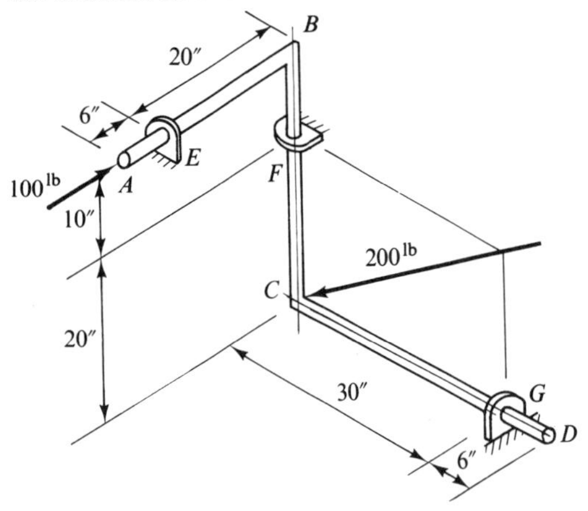

Figure P4-128

4-129 Frame *ABCD* in Fig. P4-129 is made of pipes and is supported by a journal bearing at *A*, a ball and socket at *D*, and a cord from *C* to *E*. The 500-lb load is in the yz plane. Determine the force at *D* on the frame.

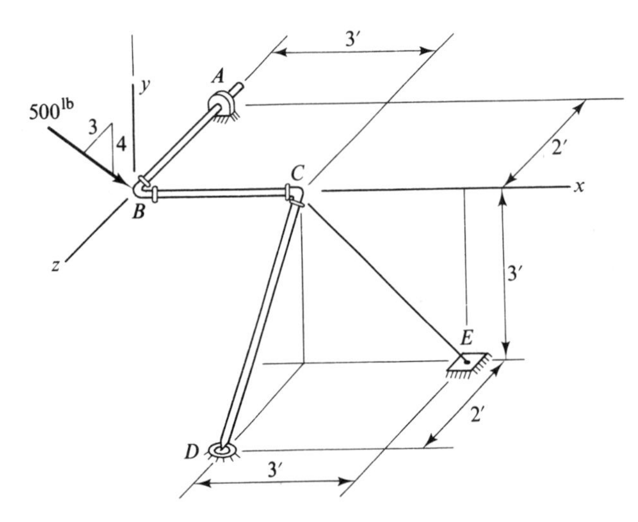

Figure P4-129

4-130 A rod is bent to the shape *ABCD* shown in Fig. P4-130, and end *A* is placed in a snug-fitting vertical hole which keeps *AB* vertical but permits the frame to rotate about *AB*. Rotation about *AB* is prevented by the horizontal rod *EF*. Determine the force and couple which must be developed at *A* to maintain equilibrium.

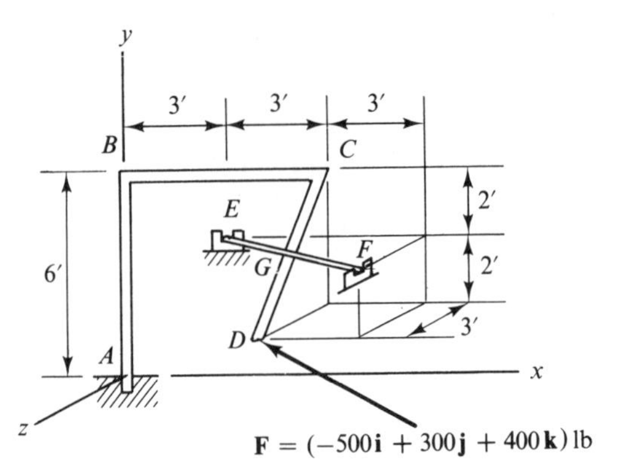

$\mathbf{F} = (-500\mathbf{i} + 300\mathbf{j} + 400\mathbf{k})$ lb

Figure P4-130

4-131 Solve Problem 4-130 if the rod *EF* is removed and the rod *ABCD* is welded in the hole at *A*.

***4-132** The homogeneous wedge-shaped block in Fig. P4-132 is supported as indicated in the sketch. The block has a mass of 500 kg, and the force **P** has a magnitude of 12 kN and acts along *EF*. Determine the reaction on the block at *A*.

4-133 The 100-lb homogeneous block in Fig. P4-133 is held in equilibrium by the ball and socket at *A*, the journal bearing at *B*, and the cord at *C* in the yz plane. Determine the forces on the block at *B* and *C*.

4-134 The homogeneous rectangular trapdoor in Fig. P4-134 weighs 500 lb and is propped open by a slender rod. The connections at *E* and *F* are smooth hinges, and only *F* can exert an axial force along the hinge pin. The ends of rod *BH* rest in small indentations. Determine the hinge reaction at *F* on the door.

4-135 The 100-lb rectangular plate *A* in Fig. P4-135 is suspended from the horizontal support by rings at *B* and *C*. A collar on the rod at *B* prevents sliding along *DE*. A cord runs from *F* over a pulley at the wall. Determine the magnitude of **P** necessary to cause *A* to be inclined 30° with the vertical.

4-136 Solve Problem 4-135 if the inclined supporting wires are shortened so that *DE* makes

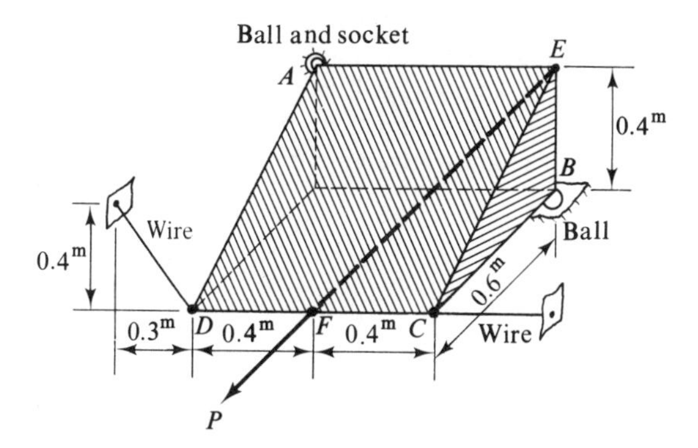

Figure P4-132

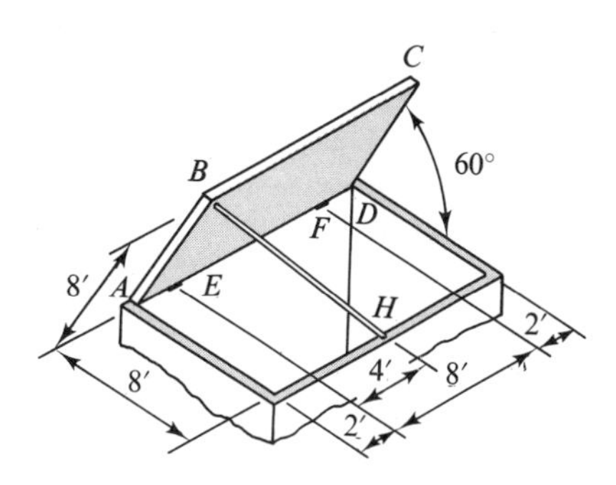

Figure P4-134

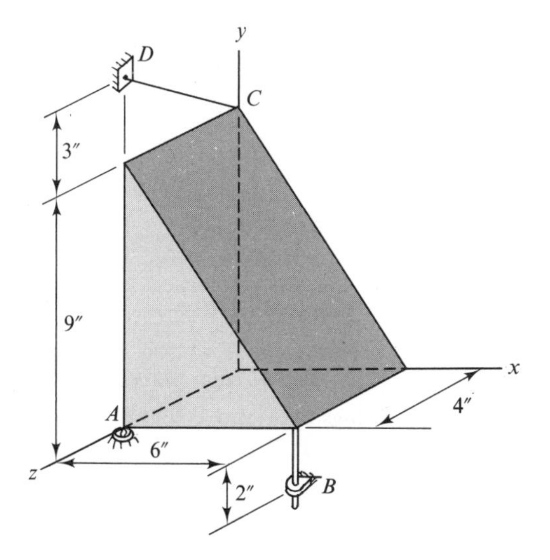

Figure P4-133

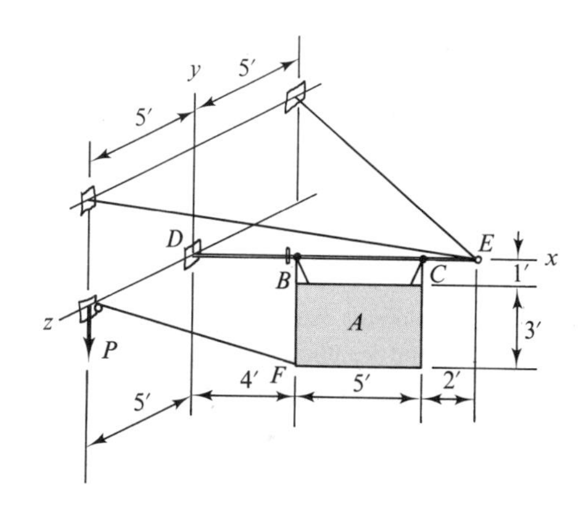

Figure P4-135

an angle of 30° with the x axis in the xy plane. Plate A is to make an angle of 30° with the xy plane.

***4-137** The hoist in Fig. P4-137 consists of four struts AC, AD, BC, and BD and two cables CD and CE. All joints are assumed to be ball and sockets. Determine the tension in cable CE when body m has a mass of 2.0 Mg.

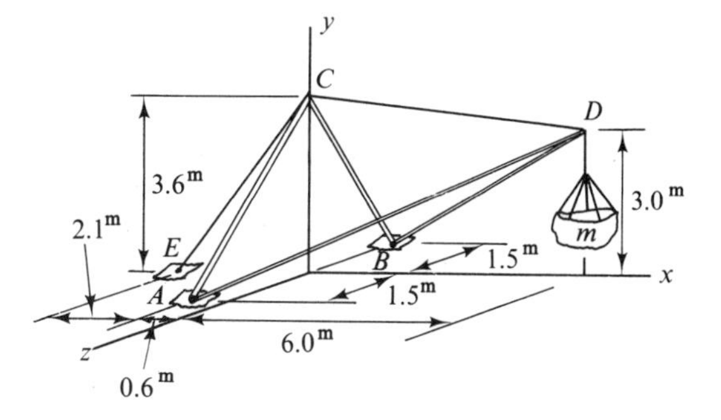

Figure P4-137

4-138 The boom AD of the simplified stiff-leg derrick in Fig. P4-138 is in the xy plane, and the force $\mathbf{E}$ is $(5000\mathbf{i}-9000\mathbf{j}+6000\mathbf{k})$ lb. The support at A is a ball and socket joint; the legs FB and GB are pinned to the foundation at F and G in the xz plane. Determine

(a) The minimum force at C (CD is 30 ft) for equilibrium of the boom.

(b) The force carried by each leg.

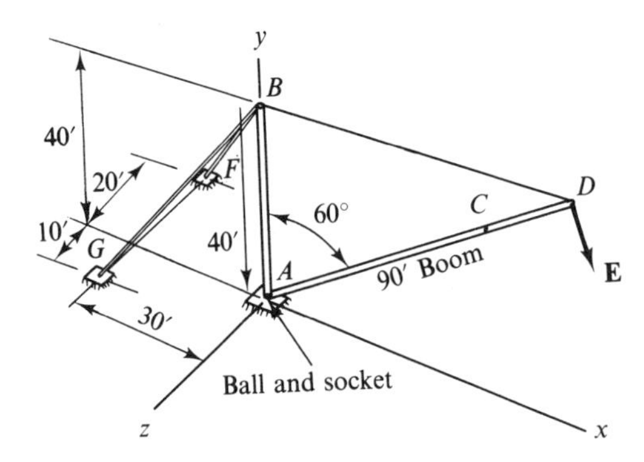

Figure P4-138

4-139 The slender bar in Fig. P4-139 weighs 30 lb per ft and is supported on a smooth floor at A, by smooth walls at C, and by horizontal cords DB and BE. Determine the tensions in the cords.

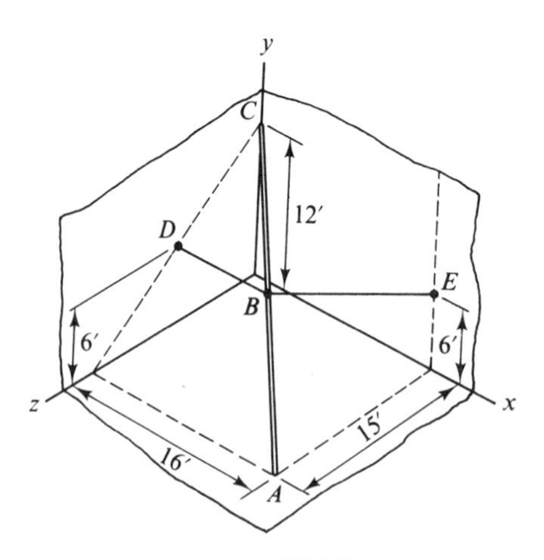

Figure P4-139

4-140 The slender, homogeneous, 400-lb bar in Fig. P4-140 is supported in a smooth hemispherical depression in the horizontal plane at A and by a smooth wall at B. The force $\mathbf{C}=C(-0.8\mathbf{j}+0.6\mathbf{k})$ also acts on the bar at B. Determine all unknown forces acting on the bar.

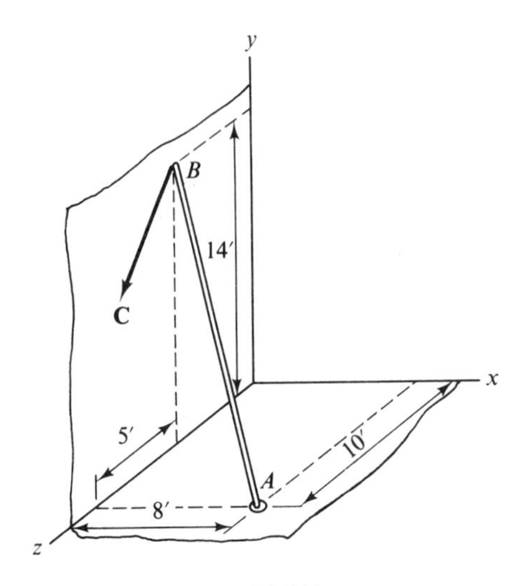

Figure P4-140

4-141 The bent rod in Fig. P4-141 is supported by bearings at A and B and by horizontal cords CE and DE. The bearings at A and B

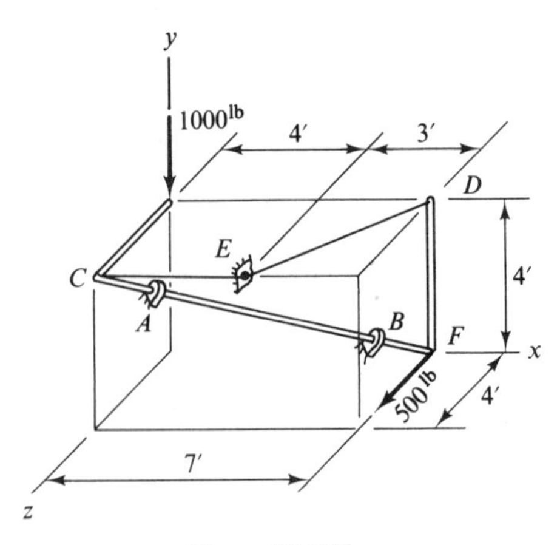

Figure P4-141

are 2 ft from C and F, respectively. Determine the tension in each of the cords.

***4-142** The homogeneous 0.80-kg block of Fig. P4-142 is supported by a cord at C in the yz plane and a pin at A which fits firmly into a hole in a fixed base. The pin is free to turn about its axis, AB, in the hole, but no other movement is possible. Determine the resultant force and moment of the base on the pin at A.

4-143 Example 4-12 was solved by assuming three dependent components of the force at C and deriving a constraint equation for the components. Show that the two forces $\mathbf{C}_1 = C_1(4\mathbf{i}-3\mathbf{j})/5$ and $\mathbf{C}_2 = C_2(-36\mathbf{i}-48\mathbf{j}+25\mathbf{k})/65$ are perpendicular to each other and to the rod OF and thus represent a possible set of orthogonal components of the reaction. Determine $\mathbf{C}_1$ and $\mathbf{C}_2$ and show that their resultant agrees with the solution of the example.

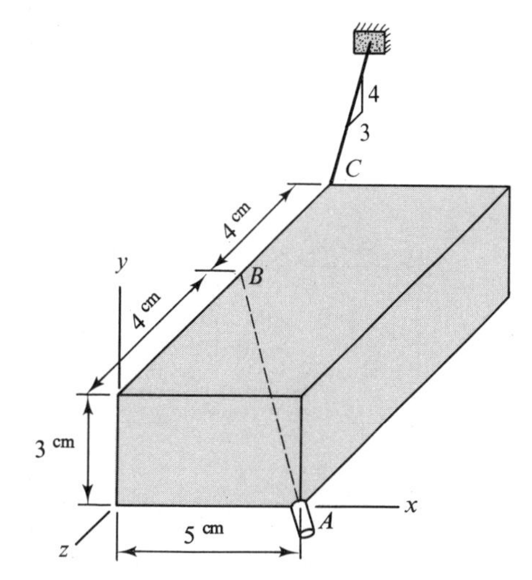

Figure P4-142

4-13 HYDROSTATICS AND BUOYANCY

Structures such as dams, retaining walls, canal gates, valves, and tanks may be acted on by constant or variable fluid pressures. The determination of the resultants of such pressures was discussed in Art. 3-8, and these resultants can be shown on free-body diagrams, together with other pertinent forces, to analyze equilibrium problems.

When a body is completely or partially submerged in a fluid (usually a liquid), it is subjected to such forces, but the analysis of the problem is usually more direct if the concept of *buoyancy* is used.

The tank of fluid in Fig. 4-22(a) is assumed to be in static equilibrium. An arbitrary volume, A, of the fluid is contained in the dashed line and is, of course, also in equilibrium. A free-body diagram

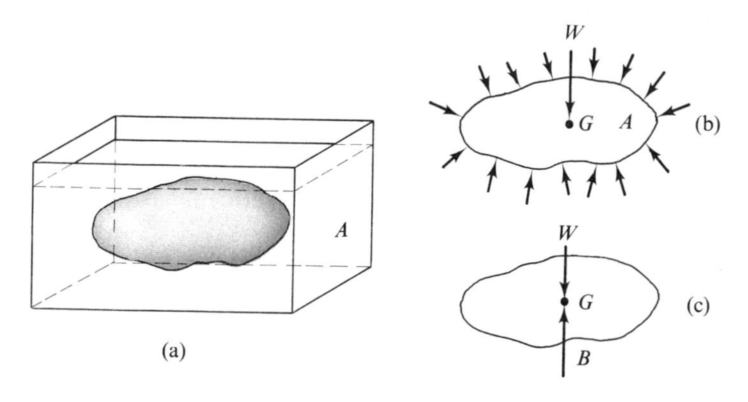

Figure 4-22

of A is shown in Fig. 4-22(b) in which the fluid forces act on the boundary of A and the weight acts through G. Since the fluid is in equilibrium, the resultant of the fluid forces must balance the weight **W** and thus must pass through G. The resultant of the fluid forces is called the buoyancy and is shown as **B** in Fig. 4-22(c). The buoyant force is an upward vertical force equal to the weight of the fluid contained in A, and it acts through the center of gravity of A. If the fluid in A is replaced by any other body having the same volume and shape, the fluid forces on its boundary, and thus the buoyant force, will be unchanged. Therefore, the buoyant force on any submerged body is equal to the weight of the displaced fluid and passes through the center of gravity of the displaced fluid.

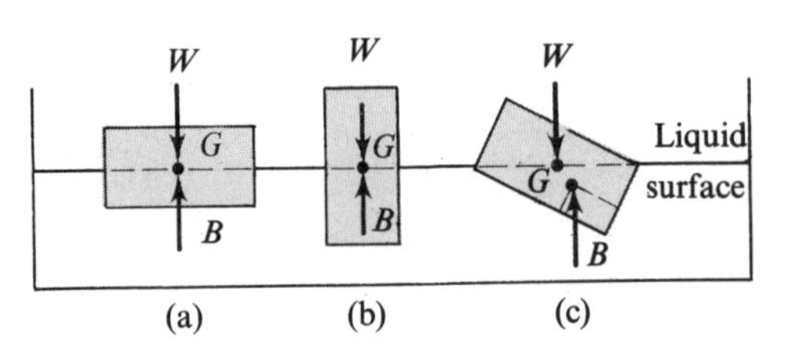

Figure 4-23

Liquids are fluids which are relatively incompressible. When this assumption is justified the specific weight of the liquid is constant for all depths, and the buoyant force on any submerged body is independent of the depth of the body below the surface. When the fluid is compressible, such as the earth's atmosphere, the specific weight of the fluid will vary with the elevation, and this variation may have to be considered.

When part of a body is submerged in a liquid and the rest in air, the buoyant effect of the air is usually negligible when compared with the buoyancy of the liquid. Sometimes a body displaces two different liquids, for example, when oil is floating on water and the floating body extends into both liquids. In this case the buoyant force will equal the sum of the weights of the fluids displaced.

A floating body (one which is partially submerged) may be in equilibrium but may not be stable. The specific weight of the rectangular block in Fig. 4-23 is assumed to be half that of the liquid in which it is floating. Thus the buoyant force **B** will equal the weight **W** when half of the block is submerged. The block is in equilibrium in either position *a* or *b*, but experience indicates that the block is not stable in position *b* even though both force and moment equations of equilibrium are satisfied. When the block is in position *c* the two forces **W** and **B** are seen to constitute a counterclockwise couple (**B** acts through the centroid of the lower triangle), which would tend to rotate the block to position *a*. The stability of a floating body can thus be determined by calculating the righting (or overturning) moment of the buoyant force when the body is displaced from an equilibrium position. The method of virtual work will be utilized in Chapter 7 to analyze the stability of any body in equilibrium. The stability of ships is of prime importance, and extensive analyses are made of proposed designs.

The following examples illustrate the use of buoyant and hydrostatic forces for equilibrium problems.

EXAMPLE 4-13

The cross section of a concrete gravity dam is shown in Fig. 4-24(a). The specific weight of the concrete is 150 lb per ft^3 (pcf). Due to seepage there is a hydrostatic uplift pressure under the dam which varies linearly from full hydrostatic pressure at A to zero at B. Consider a 1-ft-long section of the dam. Determine
(a) The reaction of the ground on the dam.
(b) The factor of safety against overturning which is defined as the ratio of the resisting (counterclockwise) moment about the toe B to the overturning moment of the hydrostatic forces about B.

SOLUTION

A free-body diagram of a 1-ft section of the dam is shown in Fig. 4-24(b). The triangular wedge of water is included as part of the body as a matter of convenience since the resultant of **H** and $\mathbf{W}_1$ is the same as the force of the water on the inclined surface of the dam. The force **U** is the uplift force due to seepage, which is assumed to vary linearly. The hydrostatic pressure at A is the same on either the inclined face or the bottom. The force $\mathbf{W}_2$ is the weight of the dam, and **R** is the reaction of the earth on the dam.

(a) The pressure at point A is (see Art. 3-8)

$$p_A = \gamma h = 62.4(18) = 1123 \text{ lb per ft}^2,$$

and the pressure varies uniformly from zero at the surface to p_A at point A. The pressure volume for the force on the surface AC is a triangular wedge which is 1 ft thick. The horizontal force H is equal to the volume of the wedge and acts through the centroid of the wedge.

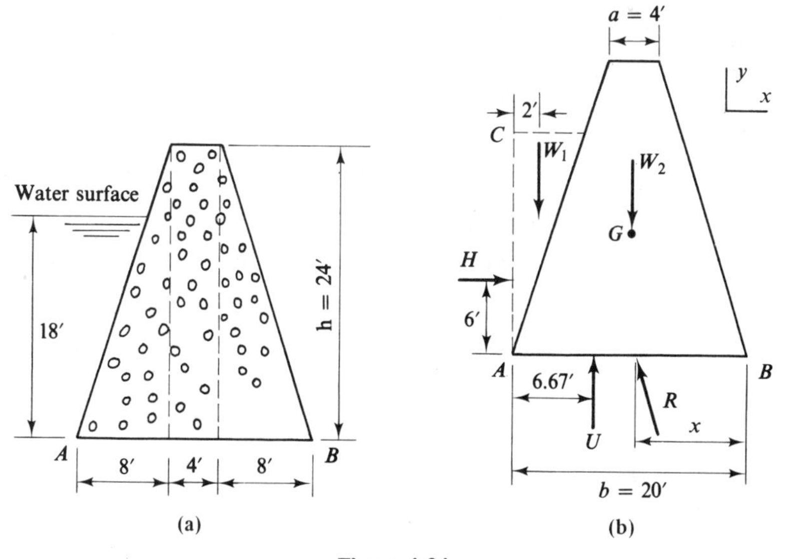

Figure 4-24

That is,

$$H = \frac{p_A(AC)(1)}{2} = \frac{1123(18)(1)}{2} = 10{,}110\text{ lb},$$

$$\mathbf{H} = 10{,}110\mathbf{i}\text{ lb as shown.}$$

In a similar manner, the force U due to the uplift pressure is given by the triangular pressure volume on the base AB. The volume of the pressure diagram is

$$U = \frac{p_A(AB)(1)}{2} = \frac{1123(20)(1)}{2} = 11{,}230\text{ lb},$$

$$\mathbf{U} = 11{,}230\mathbf{j}\text{ lb as shown.}$$

The force W_1, the weight of the water between the face of the dam and the vertical plane AC, is

$$W_1 = \gamma(\text{vol}) = 62.4\,\frac{18(6)(1)}{2} = 3370\text{ lb},$$

$$\mathbf{W}_1 = -3370\mathbf{j}\text{ lb}.$$

Similarly, the force W_2 is

$$W_2 = \gamma_{\text{dam}}(\text{vol}) = \gamma_{\text{dam}}\,\frac{a+b}{2}\,h(1)$$

$$= 150\left(\frac{4+20}{2}\right)(24)(1) = 43{,}200\text{ lb},$$

$$\mathbf{W}_2 = -43{,}200\mathbf{j}\text{ lb}.$$

The force equation $\sum \mathbf{F} = 0$ gives

$$\mathbf{R} + 10{,}110\mathbf{i} - 3370\mathbf{j} - 43{,}200\mathbf{j} + 11{,}230\mathbf{j} = 0$$

from which

$$\mathbf{R} = (-10{,}110\mathbf{i} + 35{,}340\mathbf{j})\text{ lb}.$$

The position of $\mathbf{R}$ is determined from the equation $\sum \mathbf{M}_B = \sum \mathbf{r}\times\mathbf{F} = 0$. Thus

$$6\mathbf{j}\times\mathbf{H} + (-18\mathbf{i})\times\mathbf{W}_1 + (-10\mathbf{i})\times\mathbf{W}_2 + (-13.33\mathbf{i})\times\mathbf{U} + (-x\mathbf{i})\times\mathbf{R} = 0,$$

which becomes

$$6\mathbf{j}\times(10{,}110\mathbf{i}) - 18\mathbf{i}\times(-3370\mathbf{j}) - 10\mathbf{i}\times(-43{,}200\mathbf{j}) - 13.33\mathbf{i}$$
$$\times 11{,}230\mathbf{j} - x\mathbf{i}\times(-10{,}110\mathbf{i} + 35{,}340\mathbf{j}) = 0$$

from which

$$-60{,}660\mathbf{k} + 60{,}660\mathbf{k} + 432{,}000\mathbf{k} - 149{,}800\mathbf{k} - 35{,}340x\mathbf{k} = 0$$

and

$$x = 7.99\text{ ft}.$$

The resultant force of the earth on the dam is

$$\underline{\mathbf{R} = 36{,}700\text{ lb} \quad 74^\circ \quad \text{through a point 7.99 ft left of } B.} \quad \text{Ans.}$$

(b) The moment about B of the forces $\mathbf{W}_1$ and $\mathbf{W}_2$ (which resist overturning) is

$$(M_B)_{resist} = 60{,}660 + 432{,}000 = 492{,}700 \text{ ft-lb},$$

and the moment of **H** and **U** (the overturning hydrostatic forces) is

$$(M_B)_{overturn} = 60{,}660 + 149{,}800 = 210{,}460 \text{ ft-lb}.$$

The factor of safety against overturning is

$$\text{f.s.} = \frac{(M_B)_{resist}}{(M_B)_{overturn}} = \frac{492{,}700}{210{,}500} = \underline{2.34.} \qquad \text{Ans.}$$

Note that **R** is not considered to be an overturning force. If the dam were to start to overturn, **R** would move to point B and thus have no moment about the toe. If **R** were considered to be an overturning force, the resisting and overturning moments would be the same, and the factor of safety would always be unity.

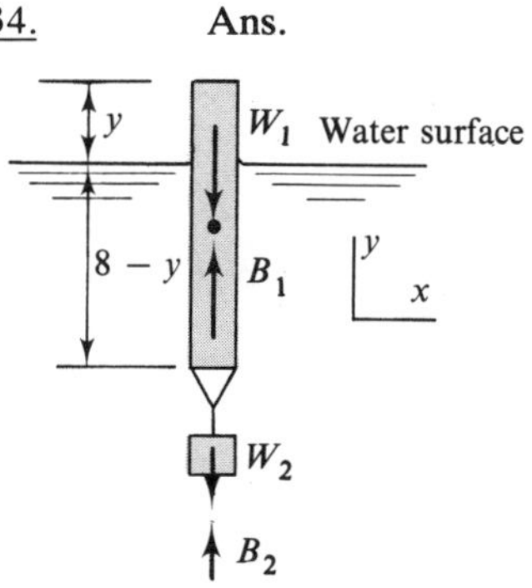

Figure 4-25

EXAMPLE 4-14

One cubic foot of concrete, weighing 150 lb, is attached by a cable to the end of a 15-in. by 15-in. by 8-ft timber which weighs 40 lb per ft^3. The timber and concrete are placed in a deep tank of water. Determine the length of timber above the water surface when equilibrium is established.

SOLUTION

Figure 4-25 is a free-body diagram of the two bodies. The weight of the timber is

$$W_1 = \gamma_{timber}(\text{vol}) = 40(1.25)(1.25)(8) = 500 \text{ lb}.$$

The buoyant force on the timber is

$$B_1 = \gamma_{water}(\text{vol}) = 62.4(1.25)(1.25)(8 - y),$$

and the buoyant force on the concrete is

$$B_2 = \gamma_{water}(\text{vol}) = 62.4(1) = 62.4 \text{ lb}.$$

The equation of equilibrium in the vertical direction gives

$$\sum \mathbf{F}_y = \mathbf{B}_1 + \mathbf{B}_2 + \mathbf{W}_1 + \mathbf{W}_2$$
$$= \mathbf{j}[62.4(1.25)(1.25)(8 - y) + 62.4 - 500 - 150] = 0$$

from which

$$y = \underline{1.973 \text{ ft.}} \qquad \text{Ans.}$$

EXAMPLE 4-15

The wooden pole in Fig. 4-26(a) has a length L of 10 ft and a specific weight which is 0.70 the specific weight of water. The thickness t is small as compared to the length L. As the distance h is increased the angle θ will also increase. Determine the minimum value of h which will cause the pole to become vertical.

SOLUTION

A free-body diagram of the pole is shown in Fig. 4-26(b). The weight of the pole is

$$W = \gamma_{\text{pole}}(\text{vol}) = \gamma_{\text{pole}}(LA),$$

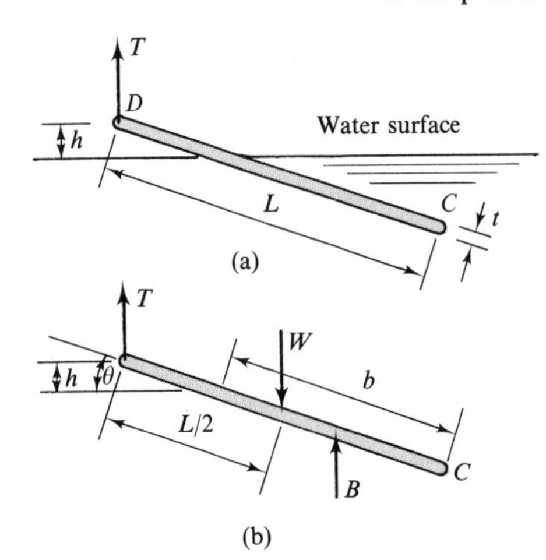

Figure 4-26

where A is the cross-sectional area of the pole, and the buoyant force is

$$B = \gamma_W(\text{vol submerged}) = \gamma_W bA.$$

When moments are summed about D (counterclockwise is positive) the results are

$$\sum M_D = \left(L - \frac{b}{2}\right)(\cos\theta)B - \frac{L}{2}(\cos\theta)W = 0,$$

which becomes

$$\left(L - \frac{b}{2}\right)(\cos\theta)\gamma_W bA - \frac{L}{2}(\cos\theta)\gamma_{\text{pole}}LA = 0,$$

which reduces to

$$b^2 - 2Lb + L^2\left(\frac{\gamma_{\text{pole}}}{\gamma_W}\right) = 0,$$

and finally

$$b = L(1 \pm \sqrt{1 - \gamma_{\text{pole}}/\gamma_W}) = 0.$$

Since b must be less than L, the negative sign must be used, and the value of b becomes

$$b = 10(1 - \sqrt{1 - 0.7}) = 4.52 \text{ ft}.$$

The value of b is constant for all values of h less than $L - b$ as long as the thickness t is small enough that the centroid of the displaced water can be assumed to be a distance $b/2$ from C.

The angle θ is, from Fig. 4-26(b),

$$\theta = \sin^{-1}\frac{h}{L - b} \tag{a}$$

and will be 90° when $h = L - b$; that is,

$$h = 10 - 4.52 = \underline{5.48 \text{ ft.}} \qquad \text{Ans.}$$

Note that if h is more than 5.48 ft, Eq. (a) is not valid. In this case the distance b will decrease, and the force T will increase.

PROBLEMS

Note: Water has a specific weight of 62.4 lb per ft^3 and a density of 1.00 Mg/m^3.

4-144 A concrete (specific weight = 150 pcf) wall having the cross section in Fig. P4-144 has been proposed for a certain site. Determine the factor of safety with respect to overturning if water were to stand at a depth of 21 ft as indicated. Assume no uplift due to seepage of water under the wall.

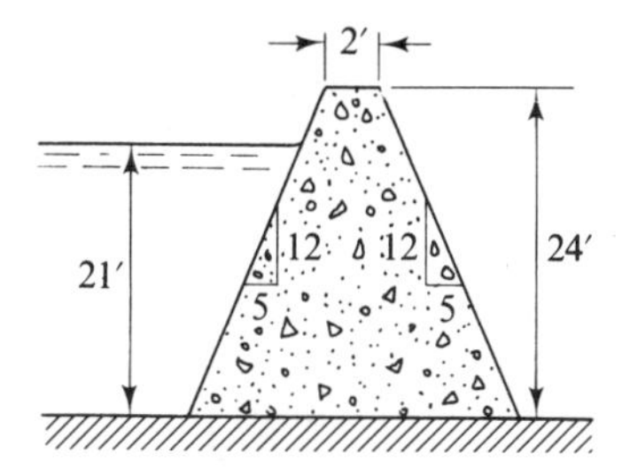

Figure P4-144

***4-145** The cross section of a concrete wall (density = 2.4 Mg/m^3) is shown in Fig. P4-145. Assume that the fluid uplift pressure varies uniformly from full hydrostatic pressure at point *A* to zero at *B*. Determine the factor of safety with respect to failure by overturning.

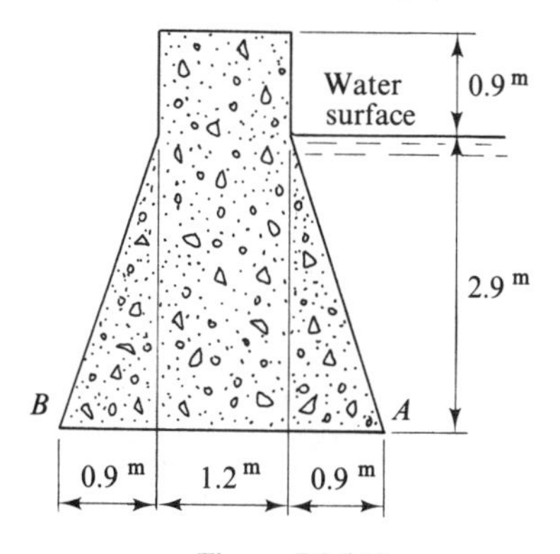

Figure P4-145

4-146 Figure P4-146 is a cross section of a concrete wall (specific weight = 150 pcf). The hydrostatic uplift pressure on the base varies from full pressure at *A* to zero at *B*. Determine the factor of safety against overturning when the base *b* is 15 ft long.

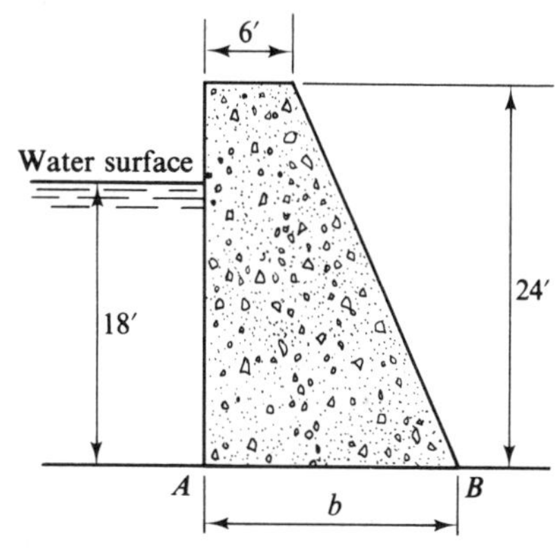

Figure P4-146

4-147 Determine the width *b* of the base of the wall in Problem 4-146 which will result in a factor of safety against overturning of 2.00.

4-148 Determine the width *b* of the base of the wall in Problem 4-146 for which the resultant force of the earth on the base will intersect the base one-third of the base width from *B*.

4-149 The rectangular gate in Fig. P4-149 is 4 ft wide and is held in the closed position by the spring. The compressive force in the spring is 3000 lb when the gate is closed. Determine the height *H* when the gate starts to open.

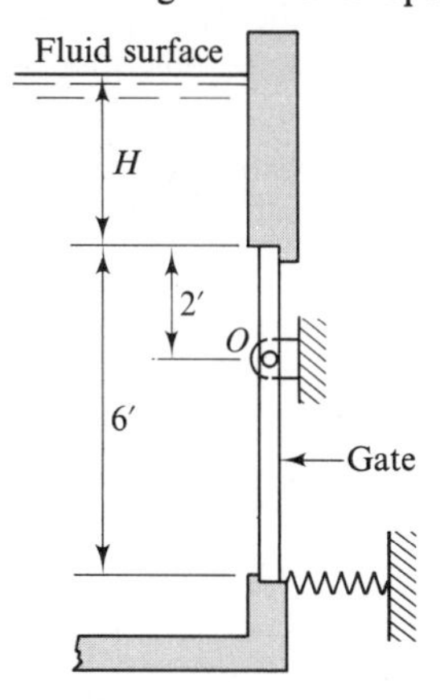

Figure P4-149

4-150 Solve Problem 4-149 if the tank contains both water and oil, which do not mix. The water extends from the bottom of the tank to the center of the gate, and the oil (specific weight = 50 pcf) extends from the water to the fluid surface.

4-151 A gate having the cross section shown in Fig. P4-151 closes an opening 2.0 ft wide by 4.0 ft high. The gate is homogeneous and weighs 1200 lb. Determine the force **P** necessary to open the gate.

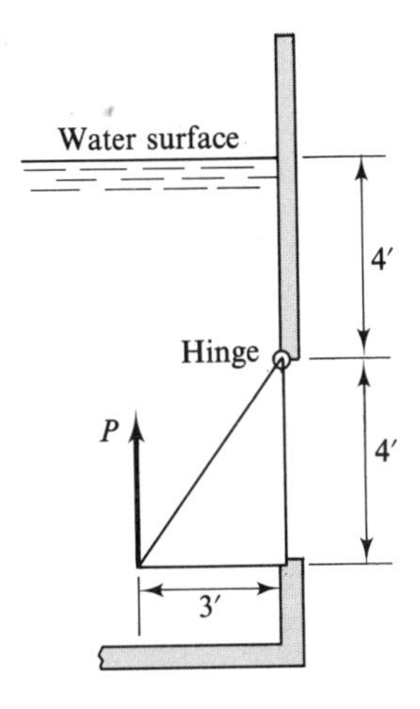

Figure P4-151

*4-152 A 1.2-m-high rectangular gate is to be mounted in the vertical side of a tank as shown in Fig. P4-152. If the bearing at O is below the center of the gate, it will open automatically

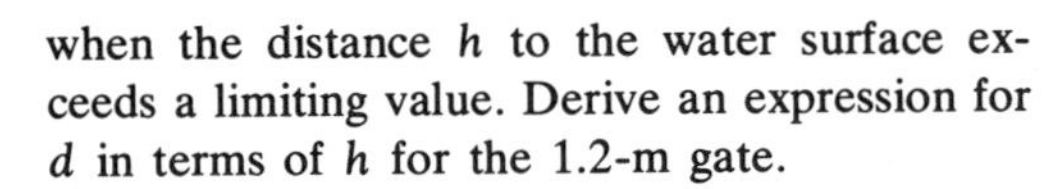

when the distance h to the water surface exceeds a limiting value. Derive an expression for d in terms of h for the 1.2-m gate.

4-153 An empty rectangular barge with vertical sides draws 1 ft of water. When fully loaded with 10,000 ft^3 of oil (specific weight = 53 pcf) the barge draws 3 ft of water. Determine the weight of the barge.

4-154 A 10-ft square raft consists of 3-in. planking fastened to a 10-ft length of 12-in. by 12-in. timber at one end and a 10-ft length of 12-in. by 24-in. timber at the other end; see Fig. P4-154. The timber weighs 28 pcf, and the raft floats in water. A 300-lb load W is placed on the raft. Determine

(a) The location of W which will cause the raft to float level.

(b) The distance from the top of the raft to the water surface.

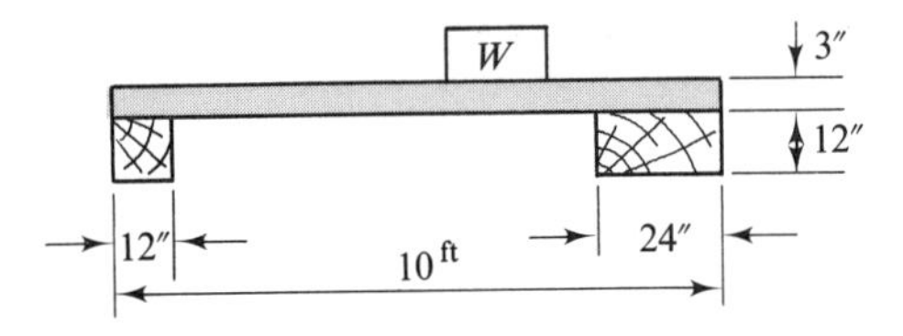

Figure P4-154

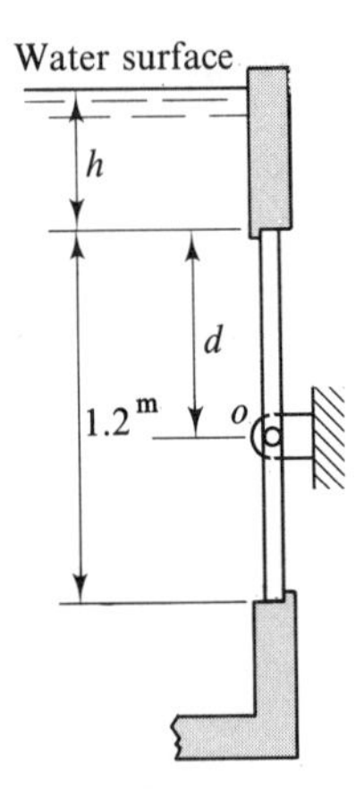

Figure P4-152

4-155 The homogeneous block A in Fig. P4-155 is 5 ft by 6 ft by 10 ft long and weighs 25 pcf. Block B is made of concrete weighing 150 pcf and is attached to A in such a position that

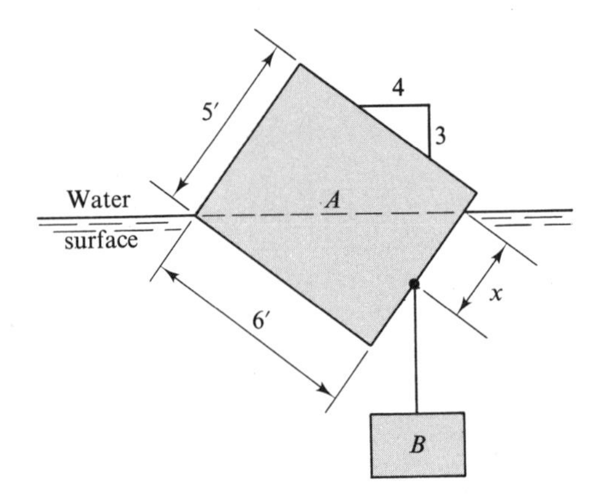

Figure P4-155

the system is in equilibrium in the position shown. Determine

(a) The volume of block B.
(b) The distance x.

4-156 The two prismatic blocks A and B in Fig. P4-156 are made of wood weighing 38 pcf. The cross-sectional areas are 0.50 ft² for A and 1.20 ft² for B. Bar CD is made of the same wood and has a cross-sectional area of 0.20 ft². Determine the distance B must rise or sink to bring the system to equilibrium.

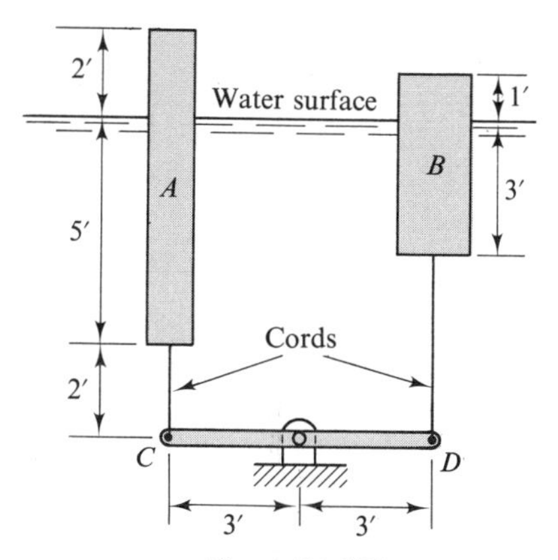

Figure P4-156

4-157 Bodies A and B in Fig. P4-157 are solid homogeneous cylinders with cross sections of 1.0 ft² each. Cylinder A weighs 110 pcf, and B weighs 160 pcf. A tension spring (one which can act only in tension) connects A to the bottom of the tank. The force in the spring is equal to the spring modulus multiplied by the extension of the spring. The spring is unstretched in the position shown. Determine the final position of the top of A relative to the top of B when

(a) The spring modulus is 20 lb per ft.
(b) The spring modulus is 60 lb per ft.

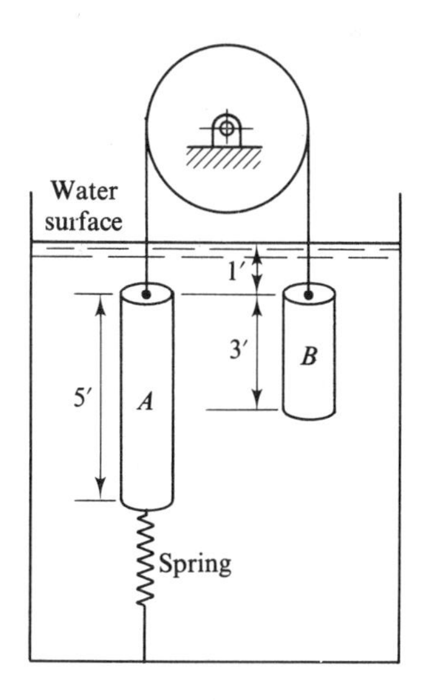

Figure P4-157

***4-158** A rectangular tank has inside dimensions of 1.20 m by 1.50 m by 0.75 m deep. The tank contains water to a depth of 0.45 m; see Fig. P4-158. A solid homogeneous rectangular parallelopiped 0.90 m by 1.20 m by 0.75 m high is placed in the tank. The block has a density of 800 kg/m³. Determine where the top of the block will come to rest with respect to the top of the tank.

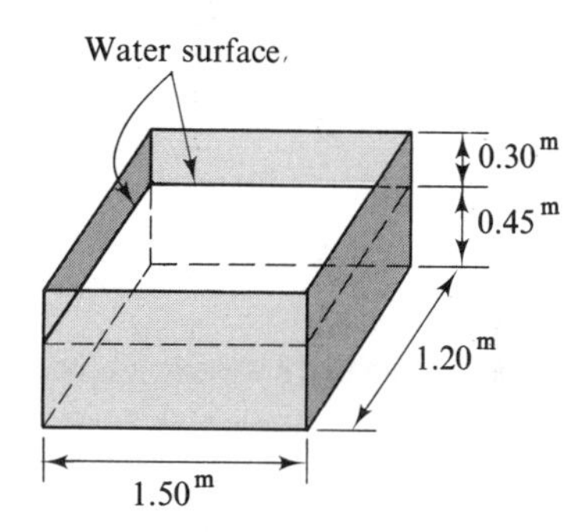

Figure P4-158

***4-159** Solve Problem 4-158 if the block has a density of 640 kg/m³.

4-160 The uniform bars AB and CD in Fig. P4-160 have the same cross-sectional areas and are connected by a cord which runs over the frictionless pulley. The specific weights are 40 pcf for AB and 50 pcf for CD. The length of the cord from B to C is adjusted so that when AB

is horizontal at the water surface end D is just touching the water as in Fig. P4-160(a). When the bars are released they will come to rest in the position shown in Fig. P4-160(b). Determine the angle θ.

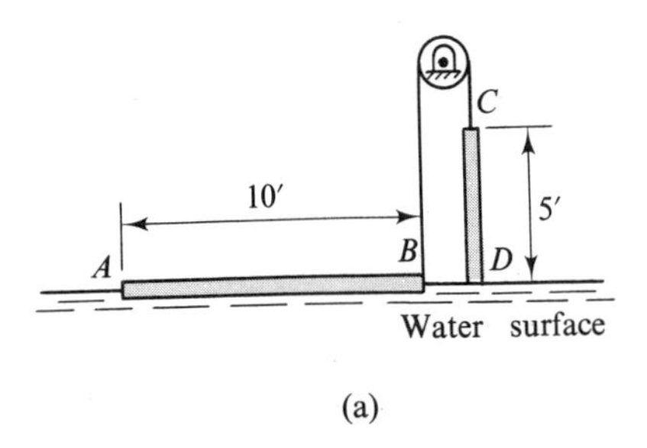

(a)

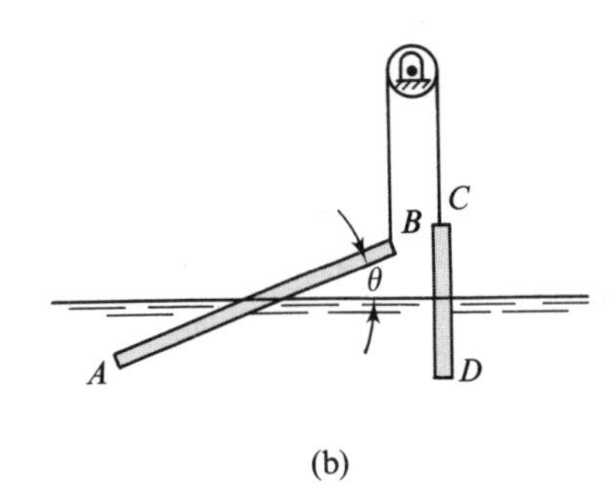

(b)

Figure P4-160

4-161 Body A in Fig. P4-161 is a homogeneous triangular prism which weighs 40 pcf. The 4.0 ft³ body B weighs 600 lb and is suspended from A by cords as shown. Determine the depth of submergence, h, of body A for equilibrium.

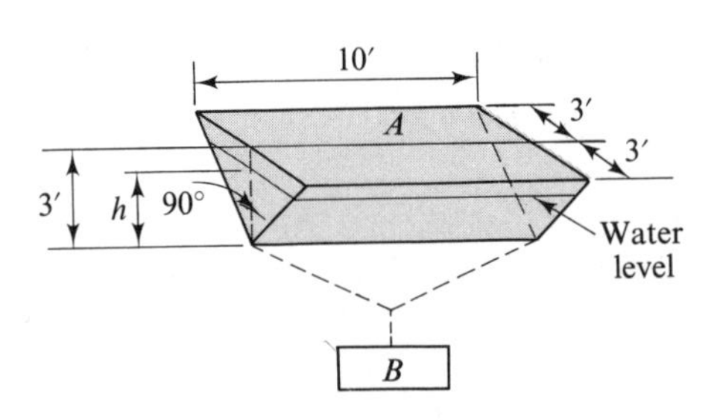

Figure P4-161

4-162 The homogeneous body A in Fig. P4-162 is a right circular cone weighing 40 pcf. Body B weighs 150 pcf and is attached to A by a wire. The bodies are in equilibrium in the position shown. Determine

(a) The volume of block B.

(b) The resultant fluid force on the curved surface of A.

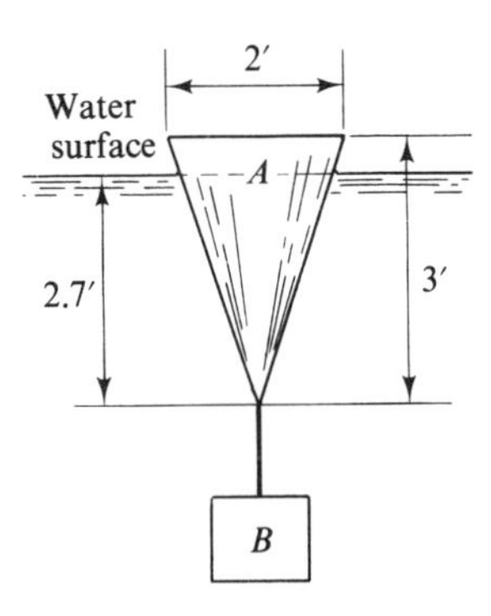

Figure P4-162

4-163 The cylindrical float in Fig. P4-163 is 4 ft long and weighs 1060 lb. The mass center of the float is at G, and the float is held completely submerged by the cable at O. Body W weighs 450 lb, has a volume of 1.00 ft³, and is suspended from a cable which is wrapped around the float. Determine the angle θ when the system is in equilibrium.

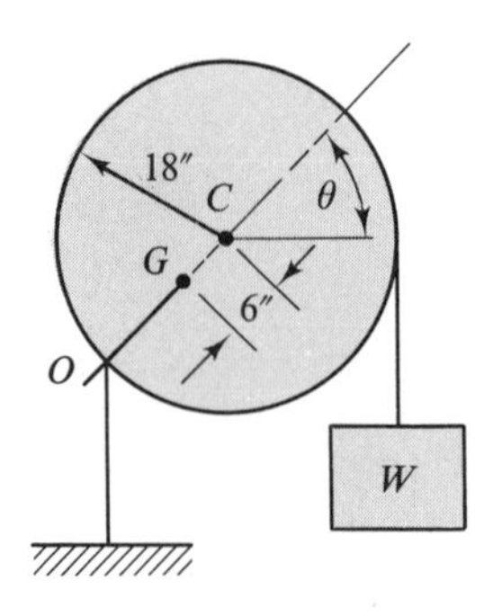

Figure P4-163

4-164 A piece of lumber 2 in. by 2 in. by 10 ft weighs 25 pcf and is held in the position shown in Fig. P4-164 by means of a cord at A. Determine

(a) The angle θ when $h = 3$ ft.

(b) The least value of h for θ to be 90°.

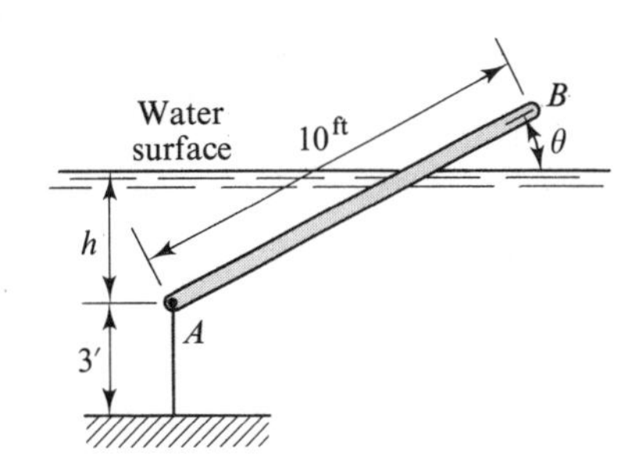

Figure P4-164

4-165 In Problem 4-164 the cord at A is attached to a 10-lb block of concrete ($\gamma = 150$ pcf) which rests on the bottom of the tank. Determine

(a) The force exerted by the tank on the concrete block when $h = 3$ ft.

(b) The least value of h which will lift the concrete block off the bottom.

***4-166** When the 360-kg crate A is placed on the 3.6-m long raft B in Fig. P4-166 the raft floats half submerged, as in position (a). Assume that the crate and raft are homogeneous. Determine the distance d the crate must be moved to cause the raft to float in position (b).

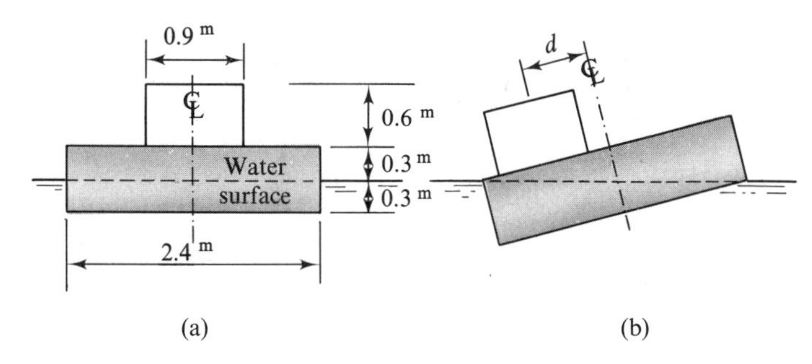

Figure P4-166

4-167 The 2-ft by 2-ft by 8-ft piece of timber in Fig. P4-167 floats half submerged with no tension in the cable in position (a). The water level rises until the timber is in position (b). Determine the distance d from corner B' to the water surface.

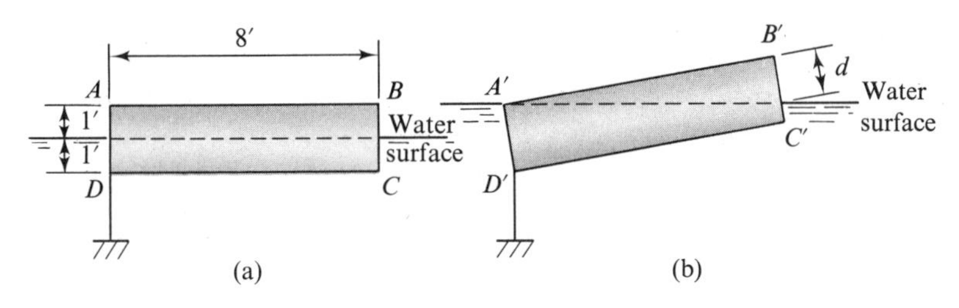

Figure P4-167

4-14 CLOSURE

The solution of problems in equilibrium of bodies requires

1. A means of identifying correctly and easily all the forces that act on a given body or portion of a body. A free-body diagram is an excellent tool for this purpose.
2. A knowledge of the available equations of equilibrium for each type of force system. These equations can be derived as the necessary conditions for the elimination of all possible resultants.
3. A logical, orderly method of procedure for analyzing and solving problems. One such method is outlined in Art. 4-4.

It is important to remember that the forces acting on a two-force member are collinear and that the forces on a three-force member are concurrent.

PROBLEMS

In the following problems all surfaces are assumed to be smooth and all members weightless unless otherwise noted.

4-168 Figure P4-168 represents a chain hoist used to lift heavy bodies. The two pulleys of radii r_1 and r_2 are rigidly fastened together and have teeth on their circumferences which prevent the chain from slipping on either of them. The 1000-lb load B can be lifted by a 50-lb pull on the chain at A. Determine the ratio of r_1 to r_2.

4-169 In Problem 4-168 friction was neglected, and if the tension in the chain at A were removed, the load B would move downward. This motion can be prevented by either frictional forces or a stop to prevent the upper pulley from rotating in its support. Discuss what would occur so that the system would be in equilibrium.

4-170 The cross section of a hydraulic gate is shown in Fig. P4-170. The gate weighs 600 lb, and its mass center is at G. The gate closes an opening which is 3 ft wide and 4 ft high, and the

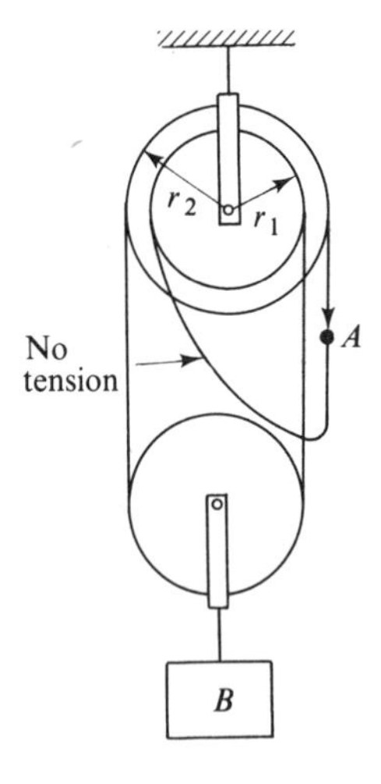

Figure P4-168

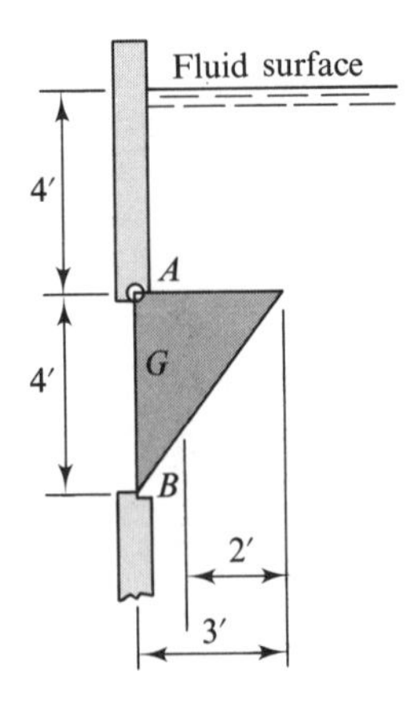

Figure P4-170

specific weight of the liquid is 55 lb per ft³. Determine the reactions on the gate at A and B.

***4-171** Determine the maximum mass A which can be supported as shown in Fig. P4-171 without causing the tension in cord B to exceed 720 N.

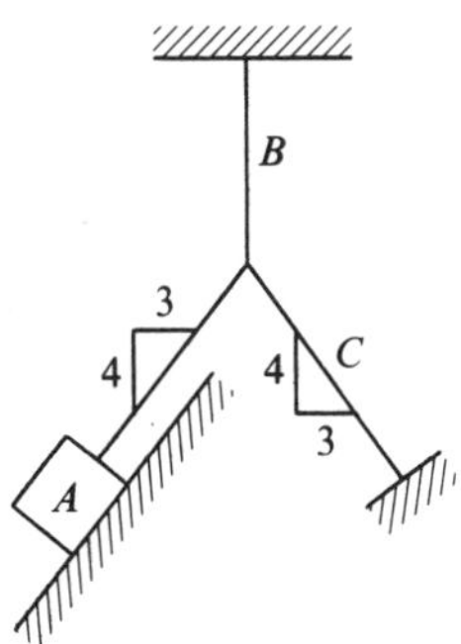

Figure P4-171

***4-172** Body A in Fig. P4-172 has a mass of 100 kg and is supported as shown. Determine the reactions of the bar BC on body A.

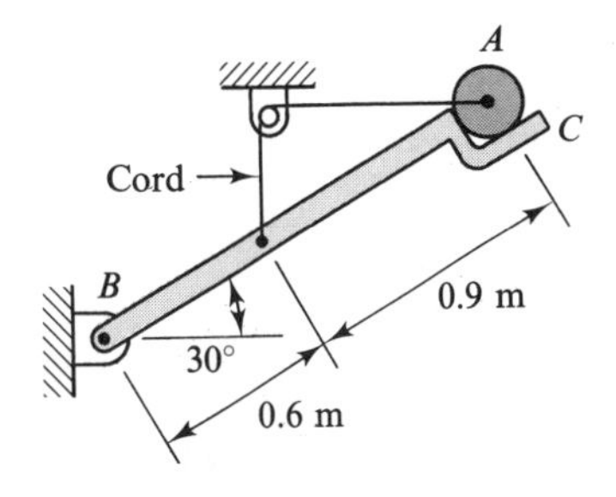

Figure P4-172

4-173 Determine the forces in numbers, DE, DF, and FG of the wing truss in Fig. P4-173.

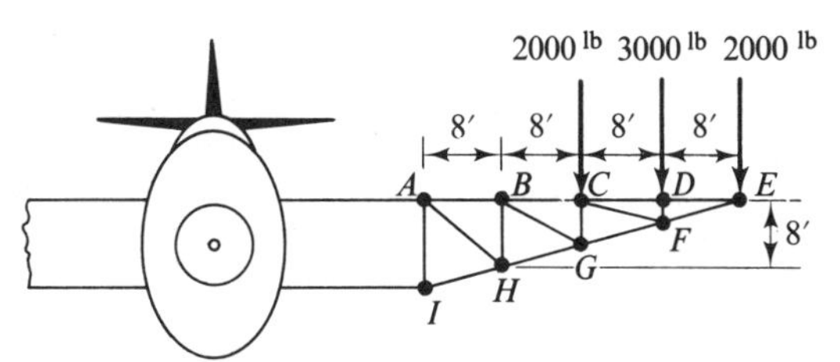

Figure P4-173

4-174 The cross section of a concrete gravity dam is shown in Fig. P4-174. Assume that a vertical uplift pressure, due to seepage, is developed with a full hydrostatic pressure at A and varies linearly to zero at B. The specific weight of the concrete is 150 pcf. Determine

(a) The reaction of the ground on the dam per linear foot.
(b) The factor of safety of the dam with respect to overturning.

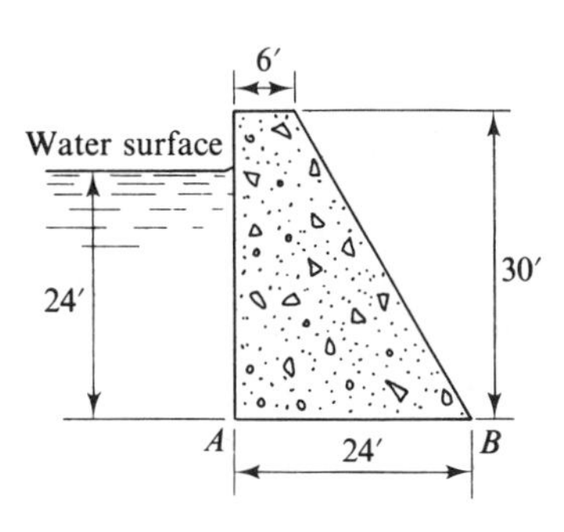

Figure P4-174

4-175 The locomotive crane illustrated in Fig. P4-175 weighs 88,000 lb and has a rated capacity of 7900 lb. The maximum permissible load

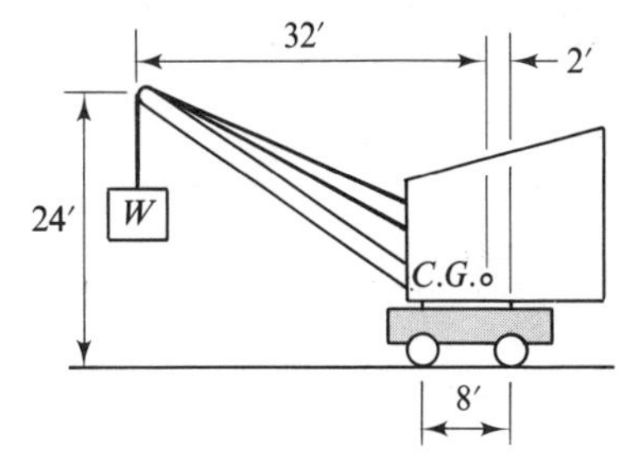

Figure P4-175

on any one of the four wheels (two wheels on each track) is 6100 lb.

(a) Determine the load on each wheel when W is 7000 lb.

(b) Determine the ratio of the moments of the weight of the crane and of the rated load with respect to the left axle. This ratio is called the *factor of safety* against overturning.

(c) Will the greatest load on any one wheel occur when the crane is unloaded or when it is carrying its rated load?

4-176 The homogeneous body A in Fig. P4-176 is supported as indicated. Determine the minimum distance d which will permit equilibrium.

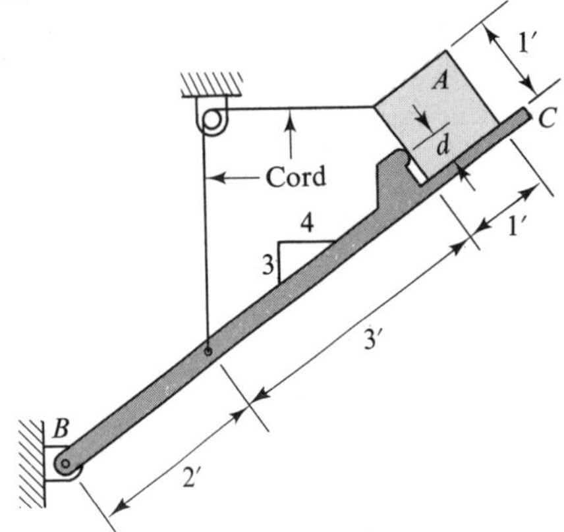

Figure P4-176

4-177 (a) Determine the ratio P/Q which will maintain equilibrium of the system in Fig. P4-177.

(b) Determine the components of the forces at D and F on member CD in terms of the load Q and the dimensions of the system.

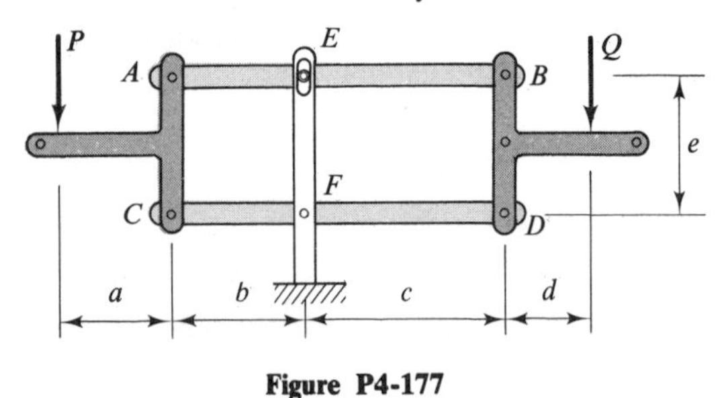

Figure P4-177

4-178 The homogeneous plate CEK of uniform thickness shown in Fig. P4-178 weighs 60 psf. It is pinned to AD at C, and it rests against the roller on BD at E. The weights of all other members can be neglected. Determine the horizontal and vertical components of the pin reaction at B on member BD.

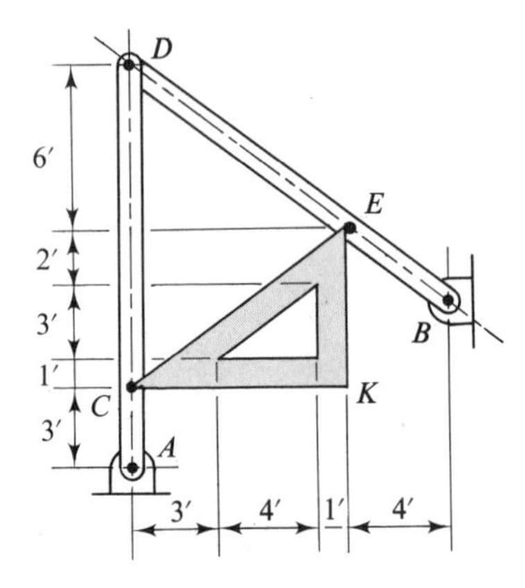

Figure P4-178

***4-179** The 150-kg homogeneous plate A in Fig. P4-179 is supported as shown. Determine the components of the forces at C and F on member BC.

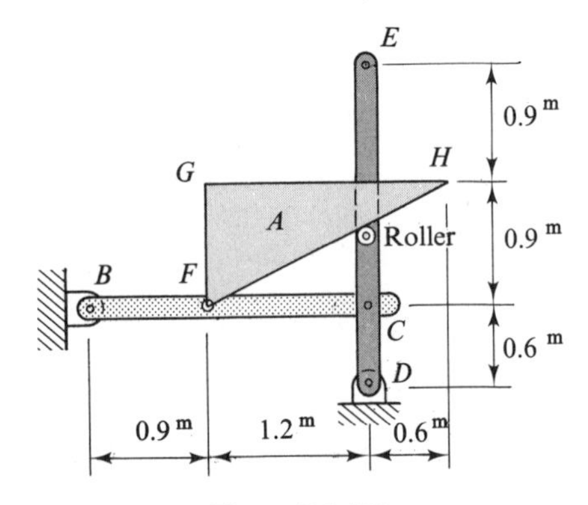

Figure P4-179

***4-180** If the pin at B in Problem 4-179 is replaced by a roller which bears against the vertical wall, the structure will collapse. It is proposed to make the frame stable by connecting a wire from E to either G or H. Determine which point the wire should be connected to and the resulting forces of the wire and the roller on body A.

4-181 Determine the reactions at C and D on the beam BD in Fig. P4-181.

4-182 The mechanism in Fig. P4-182 is part of a system for determining the weight W on the platform P. The lever system is supported on rollers at A, E, and F, and the force R is to be proportional to the weight. The platform is suspended from the levers by three rods, and the tensions in the rods will vary with the position of W on P. Determine the length e, in terms of lengths a, b, c, d, and f, which will make R independent of the position of W on P.

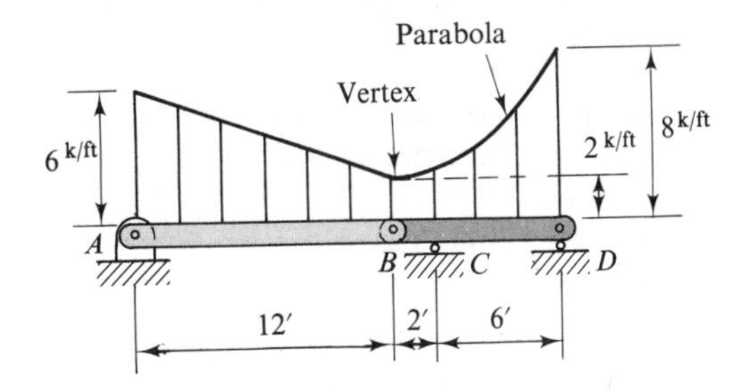

Figure P4-181

4-183 Figure P4-183 represents a hydraulic automobile hoist which consists of two identical

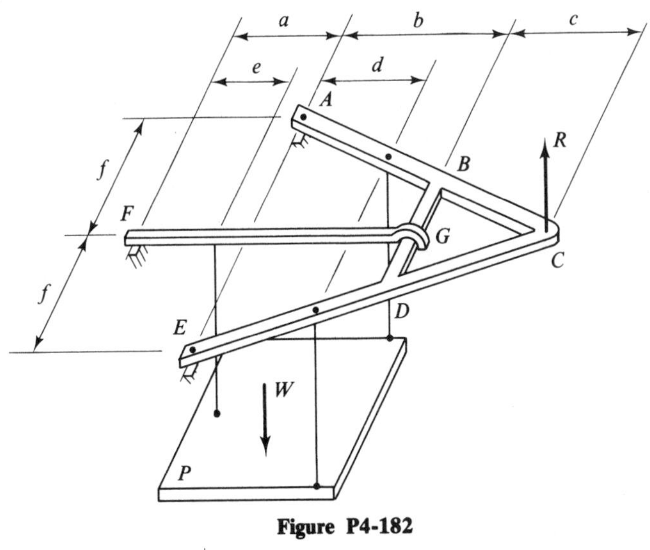

Figure P4-182

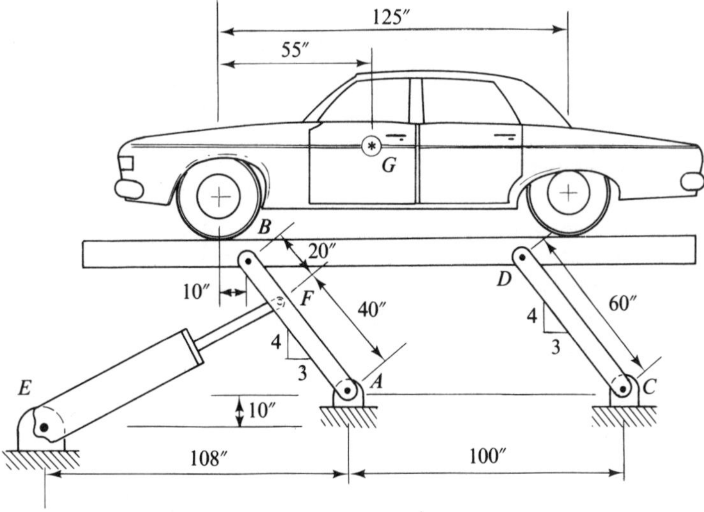

Figure P4-183

frames like the one shown. The hydraulic piston is centered between the two frames. A 4000-lb automobile is supported in the position shown (2000 lb on each frame). Determine
(a) The force on the piston rod EF on each of the two members AB.
(b) The bearing reaction at A on AB.

4-184 Determine the horizontal and vertical components of the reactions of the pins B and E on member DE of the pin-connected structure of Fig. P4-184.

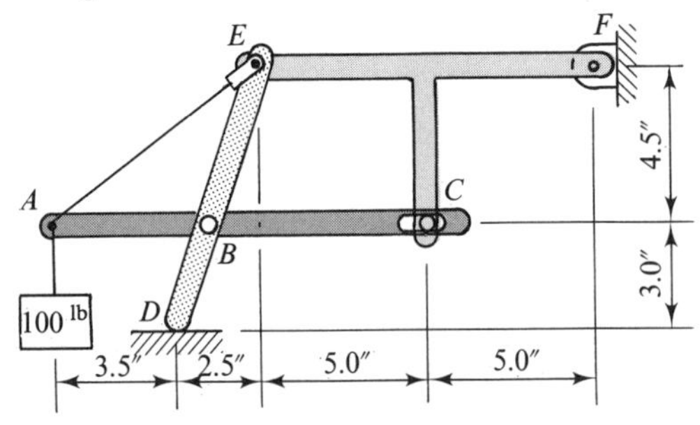

Figure P4-184

***4-185** Determine the forces in members U_2L_2, U_3L_2, U_3L_3, and L_3L_4 of the Warren truss with verticals shown in Fig. P4-185.

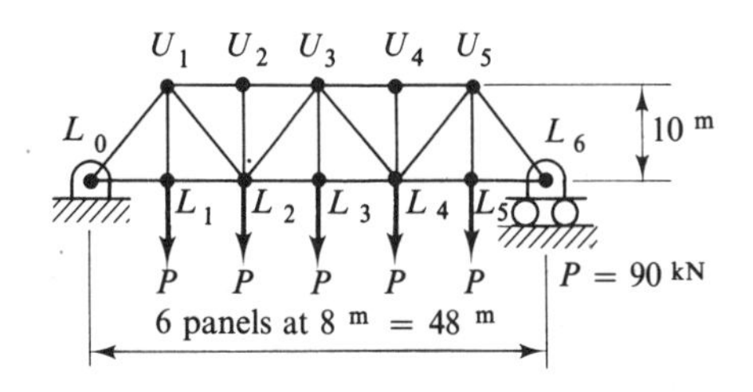

Figure P4-185

4-186 Figure P4-186 represents a coplanar pin-connected truss for an engine mount. The loads shown are due to the engine weight, thrust, and torque. Determine algebraically the forces in members BE and DE.

4-187 Determine the forces in members BK, CD, and DK of the pin-connected truss of Fig. P4-187.

4-188 Determine the forces in members AK, CJ, and FH of the truss shown in Fig. P4-188.

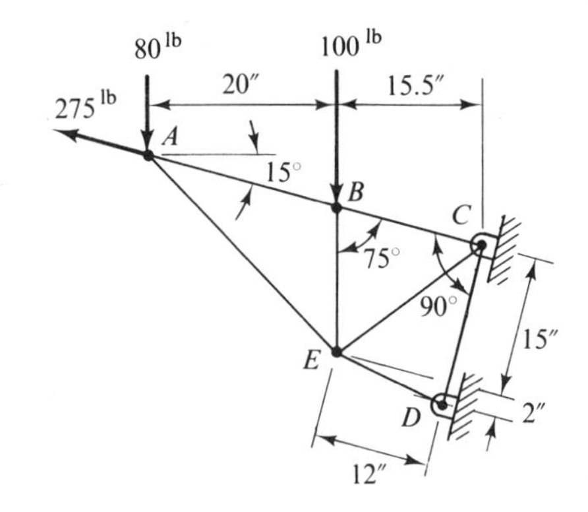

Figure P4-186

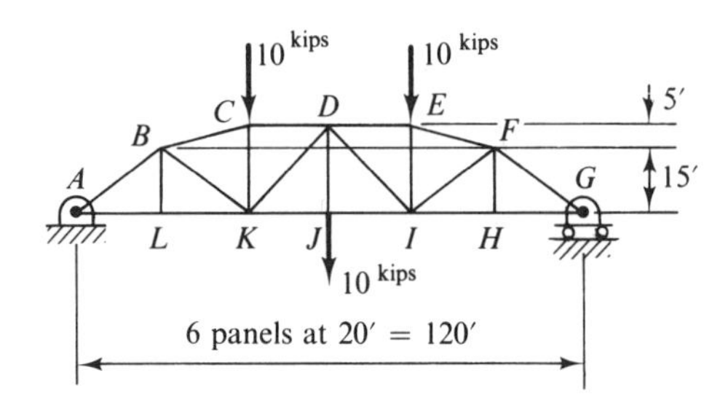

Figure P4-187

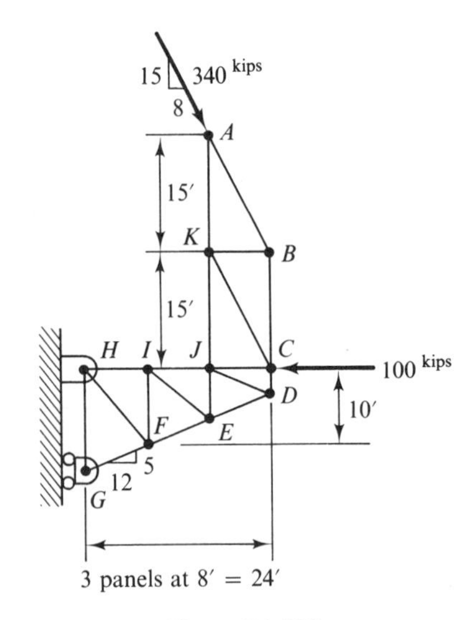

Figure P4-188

4-189 Determine the forces in members AB, AJ, and BJ of the truss shown in Fig. P4-189.

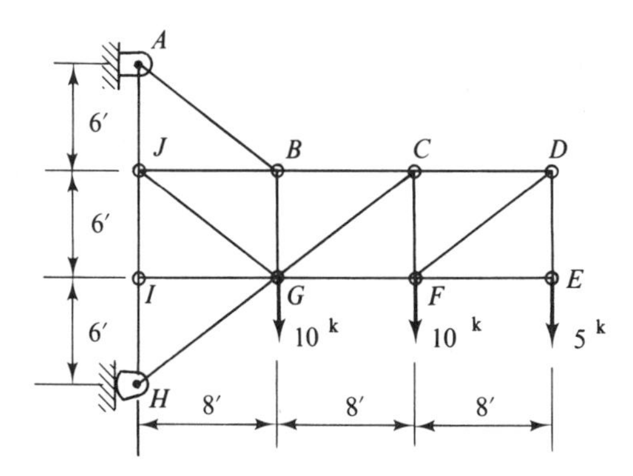

Figure P4-189

*4-190 The 7.5-m bar AB in Fig. P4-190 has a mass of 500 kg and is supported by the smooth vertical wall at B and the rope from B to C. End A rests in the corner of the room. Determine the tension in the rope BC and the reaction of the wall at B on AB.

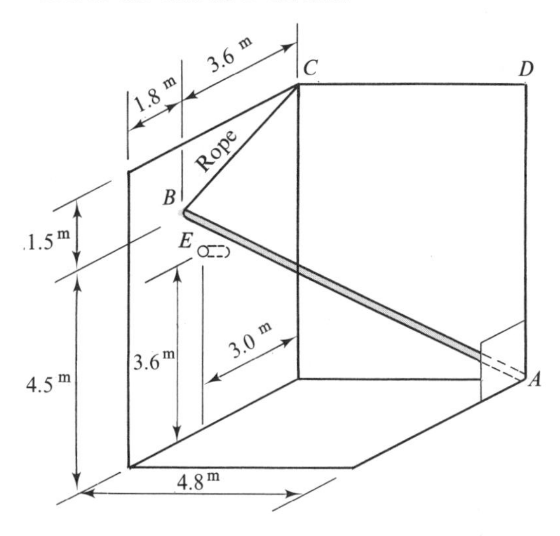

Figure P4-190

*4-191 Solve Problem 4-190 if the rope from B to C is replaced by a rope from B to D.

*4-192 In Problem 4-190 a horizontal rod, parallel to CD, is installed from E to the opposite side of the room. The rope at B is removed so that AB rests on the fixed rod from E. (End B will contact the vertical wall at a different point.) Determine

(a) The force exerted by the fixed rod on AB.
(b) The force exerted by the wall at B on AB.

4-193 A small blimp is to be anchored with three cables as shown in Fig. P4-193. The nose of the blimp is to be 100 ft above ground level, and the cable supports are to be placed symmetrically on a circle with a radius of R ft. The lifting force of the blimp is 2000 lb. It is estimated that the maximum horizontal wind force will be 1500 lb and that it can act in any direction. Determine

(a) The minimum value of R that will prevent any of the cables from becoming slack for any wind direction.
(b) The cable tensions resulting from the assumed wind direction and value of R from part (a).

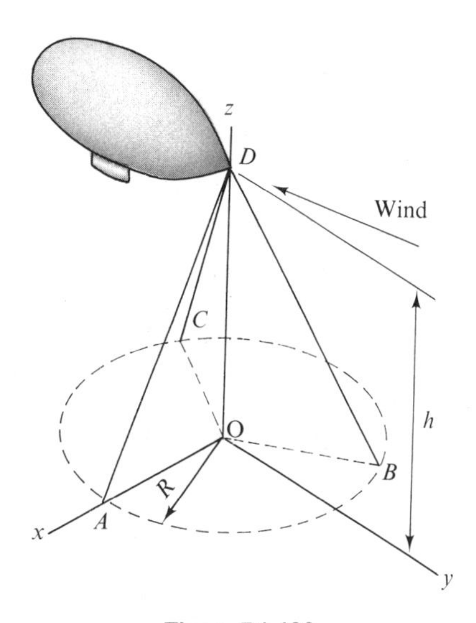

Figure P4-193

4-194 Solve Problem 4-193 if anchor points B and C are located on the circle in such a manner that angles AOB and AOC are each 135°.

4-195 If the wind direction varies 180° from the direction resulting in zero tension in one of the cables in Problem 4-193, that cable will carry the maximum load. Determine this cable tension.

4-196 Solve Problem 4-195 for the cable location in Problem 4-194.

4-197 The triangular bin in Fig. P4-197 is supported by vertical piling at C, D, and E and is partitioned to form two storage spaces. The grain in section A weighs 400 lb per ft² of floor space, and the grain in section B weighs 300 lb per ft² of floor space. Determine the load on pile C due to the weight of the grain.

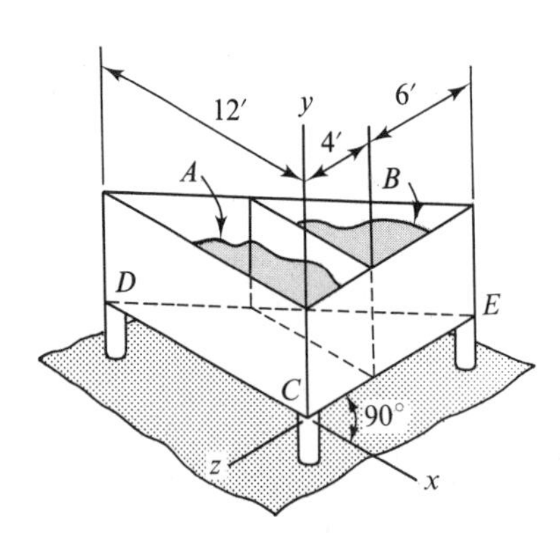

Figure P4-197

4-198 The floodgate of a dam is to be mounted so that it opens automatically when the water level behind the dam becomes 60 ft deep; see Fig. P4-198. The bottom of the spillway is 5 ft above the base of the dam. Determine the distance d at which the hinge pin should be located above the spillway floor.

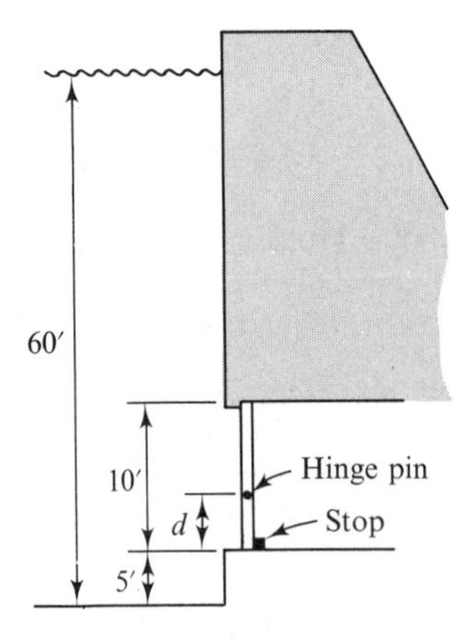

Figure P4-198

***4-199** A shaft with two bends is supported by three smooth bearings along the portions of the shaft as shown in Fig. P4-199. Determine the components of the reaction of the bearing at C on the shaft.

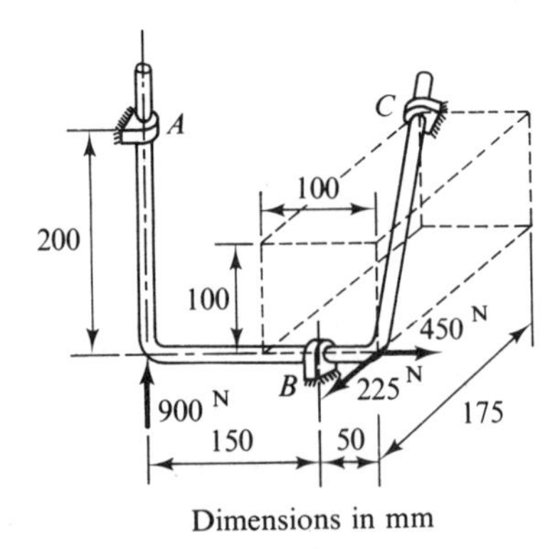

Figure P4-199

4-200 Determine the components of the reactions at B, C, and D on the space frame in Fig. P4-200. The pin D is parallel to DE.

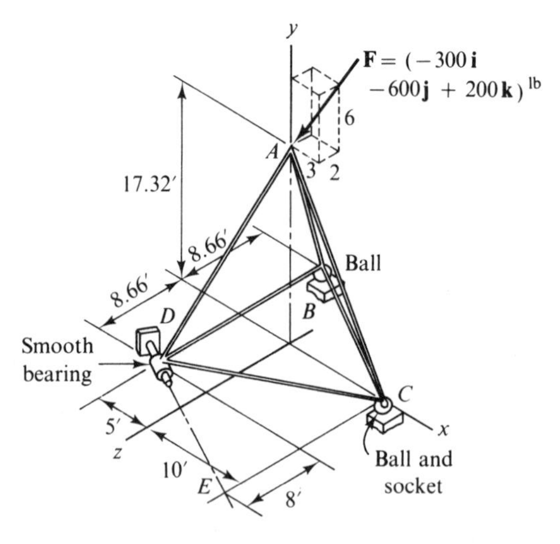

Figure P4-200

4-201 A bent steel shaft is supported by three bearings as shown in Fig. P4-201. The 1000-lb

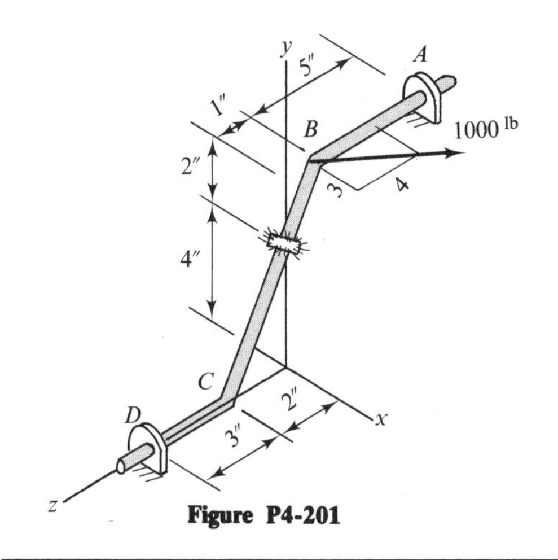

Figure P4-201

force at B is in a horizontal plane. Determine the reactions of each of the bearings on the shaft.

4-202 Rod $ABCD$ in Fig. P4-202 is used to lift the 100-lb weight by applying a force $\mathbf{P}$ parallel to the z axis. Sections BC and CD lie in the horizontal (xz) plane. The shaft is supported by journal bearings at E and F, and collars are fastened to the shaft on either side of the bearing at E to prevent axial motion. Determine the force $\mathbf{P}$ and the components of the reaction at E on the rod.

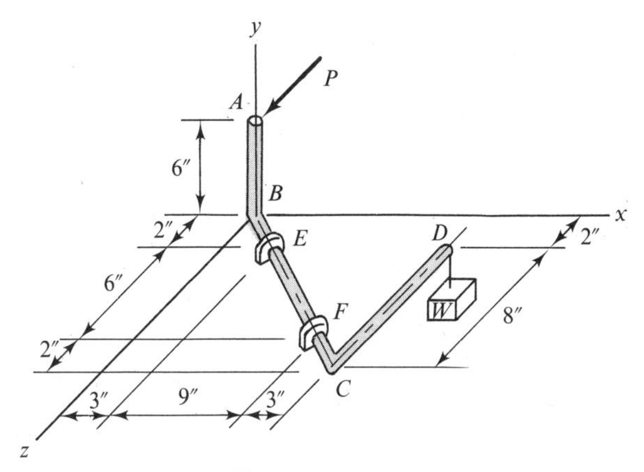

Figure P4-202

4-203 A cable weighing 0.90 lb per ft is stretched between the tops of towers at A, B, and C; see Fig. P4-203. Points A, B, and C are at the same elevation. The spans and sags are indicated on the figure. The anchor cable at B is to be placed at an angle ϕ, which will balance the cable tensions and prevent any horizontal force on the tower. Determine the angle ϕ and the tension in the anchor cable.

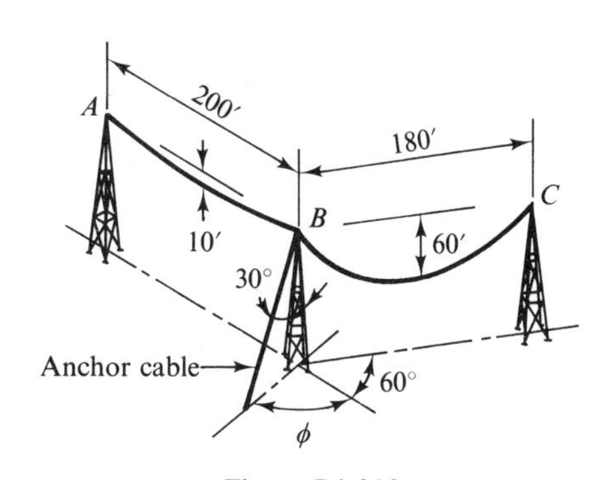

Figure P4-203

5 Friction

5-1 NATURE OF FRICTION

When a body slides or tends to slide on another body, the force tangent to the contact surface which resists the motion, or the tendency toward motion, of one body relative to the other is defined as *friction*.

People could not walk or skate or drive automobiles without the beneficial effects of friction to make tractive forces possible. Belt drives, friction clutches, and brakes all require frictional forces in order to function.

In contrast, friction can produce detrimental effects. There is always a change of mechanical energy into heat energy when one body slides on a second. The tremendous amount of money spent for lubricants to reduce the friction between rubbing surfaces indicates a disadvantage of friction. Wheels and rollers are also used to limit frictional resistance to desired movements.

In the preceding chapters, when two bodies were in contact, the surface of contact was assumed to be smooth, and the reaction of one body on the other was a force normal to the contact surface. In most instances in actual practice, the contact surface is not smooth, and the reaction of one body on the other is not normal to the contact surface. When the reaction is resolved into two components, one perpendicular and the other tangent to the contact surface, *the component tangent to the surface is called the frictional force or the friction.* Therefore, free-body diagrams for problems involving friction are the same as those for problems with smooth surfaces except that a frictional force tangent to the contact surface must be included.

When there is no relative motion between two bodies, that is, when neither body moves or both bodies move as one, the resistance to any tendency toward relative motion is called *static friction.* When one body moves relative to another body, the resisting force between the bodies tangent to the contact surface is called *kinetic friction.* The kinetic frictional force is less than the maximum static frictional force for any given pair of surfaces with the same normal force. It is rather difficult to obtain reliable data on either the kinetic frictional force or the maximum static friction between bodies of any two materials because any slight variation in the condition of the contact surfaces has an appreciable effect on the resulting frictional force. Different investigators seldom obtain identical results for the same materials, probably because of varying traces of moisture, dirt, or lubricant on the contact surfaces. The average of a series of tests, however, indicates that the maximum static friction is greater than the kinetic friction of any pair of surfaces if the normal force is constant.

In dealing with frictional forces, it is important to realize that *the frictional force between two bodies always opposes the relative motion of the bodies or the tendency to move.* Furthermore, *with static friction the frictional force between two bodies will increase as the force tending to cause sliding between the bodies increases.* In other words, the static frictional force is always the minimum force required to maintain equilibrium or prevent relative motion between the bodies. It is an adjustable force that changes with the situation. The kinetic friction

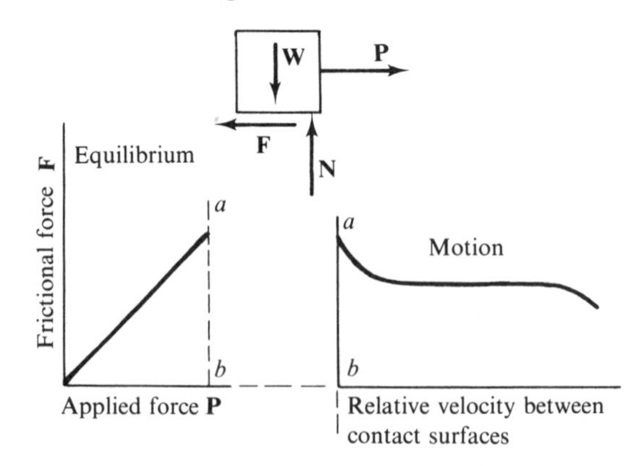

Figure 5-1

varies somewhat with the velocity. This variation is discussed more in detail in Art. 5-2. The variation of the frictional force with the applied force for static friction and with the relative velocity of the sliding surfaces for kinetic friction is shown in a general way in Fig. 5-1.

As the force **P** in Fig. 5-1 is increased from zero, the frictional force, **F**, increases just enough to prevent motion. This balance is maintained as **P** increases until the force **F** reaches a maximum or limiting value. *When the frictional force has reached its maximum value, motion impends, or a condition of unstable equilibrium exists.* Any slight increase in **P** will produce motion, and the force **F** becomes a function of the velocity. As indicated in the figure, the frictional force decreases rapidly for a small increase in velocity until it reaches a fairly constant value. Another decrease of **F** is shown for high velocities.

5-2 LAWS OF FRICTION

The results of a rather large number of experiments on friction of *dry surfaces* published by C. A. de Coulomb in 1781 provided some of the earliest information on the laws of friction. A. J. Morin's experiments published in 1831 confirmed Coulomb's results. Their work led to the following laws of friction for dry surfaces:

1. The maximum frictional force which can be developed is proportional to the normal force.
2. The maximum frictional force which can be developed is independent of the size of the contact area.
3. The limiting static frictional force is greater than the kinetic frictional force.
4. The kinetic frictional force is independent of the relative velocity of the bodies in contact.

As a result of more recent experiments, the following additions and modifications of these laws should be made:

1. For extremely low normal forces and for normal forces high enough to produce excessive deformation, the coefficient of static friction increases somewhat.
2. For extremely low relative velocities, the coefficient of kinetic friction increases and apparently becomes equal to the coefficient of static friction without any mathematical discontinuity.
3. For very high velocities, the coefficient of kinetic friction decreases appreciably.
4. Ordinary changes in temperature do not materially affect the coefficient of friction.

These laws of friction are based on experimental evidence and as such are empirical in nature. Many efforts have been made to arrive at a theoretical explanation of the variation of frictional forces when motion impends or when relative motion exists. None of these efforts has been successful, and at present the laws as stated above are assumed to be valid within the limits of accuracy of the measurements.

The laws of friction for lubricated surfaces are different from the laws for clean dry surfaces. For lubricated surfaces, the frictional force depends primarily on the lubricant instead of on the magnitude of the normal pressure and the type of material of the contacting bodies.

The only safe policy for selecting a coefficient of friction for any given situation is to make tests approximating as closely as possible the surface conditions, materials, pressures, and other factors which are to exist in the machine or structure in question.

5-3 COEFFICIENT OF FRICTION

The maximum static frictional force which exists when motion impends between any two surfaces is denoted by $\mathbf{F}'$. *The coefficient of static friction, μ, is defined as the ratio of the magnitude of the maximum static frictional force, F', to the magnitude of the normal force, N, between the two surfaces.* In mathematical form,

$$\mu = \frac{F'}{N}. \tag{5-1}$$

The coefficient of static friction is an experimentally determined constant which depends on the materials from which the contacting bodies are made and on the condition of the contact surfaces. The variation caused by the condition of the surfaces was reported by Campbell[1] to be from 0.78 for clean steel on steel to 0.11 for steel on steel with a solid film of oleic acid between the steel surfaces. He cites published values of the coefficient of static friction for steel from 0.146 to 0.82 and concludes that this variation is probably due to the presence of different amounts of grease or other film on the contact surfaces.

The values of the coefficients of static friction listed in the following table indicate the general range of published results obtained by experiments on dry surfaces.

In general, when the contact surfaces are moving with reference to each other, the coefficient of friction decreases. For this situation the ratio of the frictional force to the normal force is defined as the *coefficient of kinetic friction.* Tests on square-threaded screws at the Illinois Experiment Station[2] indicate that the coefficient of kinetic

[1] W. E. Campbell, "Variables Influencing the Coefficient of Static Friction Between Clean and Lubricated Metal Surfaces," *Trans. A.S.M.E.*, Vol. 61 (1939), 633–41.

[2] University of Illinois, *Engineering Experiment Station Bulletin* 247, 1932.

APPROXIMATE COEFFICIENTS OF STATIC FRICTION FOR DRY SURFACES

Surfaces	Coefficient
Steel on steel	0.4–0.8
Wood on wood	0.2–0.5
Wood on metal	0.2–0.6
Metal on stone	0.3–0.7
Metal on leather	0.3–0.6
Wood on leather	0.2–0.5
Earth on earth	0.2–1.0
Cast iron on cast iron	0.3–0.4
Rubber on concrete	0.6–0.8
Rubber on ice	0.05–0.2

friction is approximately three-fourths the coefficient of static friction for the materials tested. *Since the range of values for the coefficient of static friction is usually greater than the difference between the average coefficients of static and kinetic friction for given materials, the values given in problems in this text are to be used for either static or kinetic friction as needed, unless both values are given in one problem.*

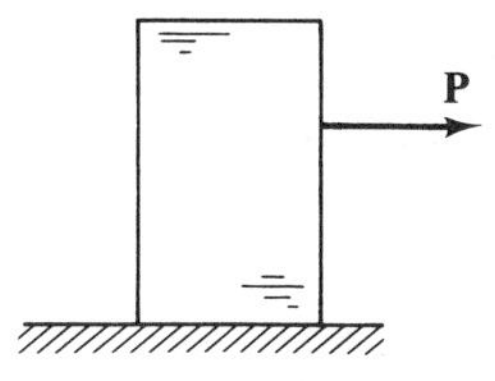

(a)

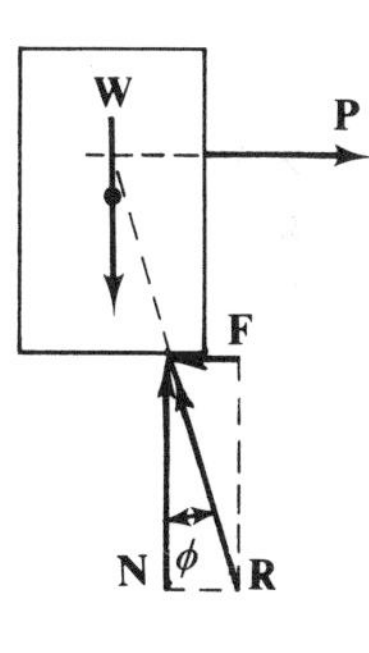

(b)

Figure 5-2

5-4 ANGLE OF FRICTION

The block in Fig. 5-2(a) weighs W lb and is supported by a rough plane. Figure 5-2(b) is a free-body diagram of the block, and the forces **N** and **F** of the plane on the block will maintain equilibrium if the force **P** is not too large. As the force **P** increases, the resisting force **F** must also increase if equilibrium is to be maintained, and the angle ϕ will increase. Note that the resultant normal force **N** will move to the right so that the resultant of **F** and **N** will be collinear with the resultant of **P** and **W**; hence **R** must pass through the point of intersection of **P** and **W**. If **P** is increased until motion impends, the magnitude of the friction force will become F' or μN. From the figure it is seen that

$$N = R \cos \phi \quad \text{and} \quad F = F' = R \sin \phi$$

from which

$$\tan \phi = \frac{F'}{N} = \mu. \tag{5-2}$$

If the line of action of **P** were high enough and the block narrow enough, the point of intersection of **F** and **N** might move off the base of the block before F became equal to F'. In this case the block would tip over before it would slide.

When motion is impending, the reaction of one body on the

other at the surface where motion impends can be shown on the free-body diagram as two separate forces (normal and tangent to the plane) or as a single resultant force inclined at the angle of friction with the normal. For algebraic solutions, the use of two component forces is usually simpler and will be emphasized in the following work. A single inclined reaction is particularly useful when the solution is to be obtained graphically.

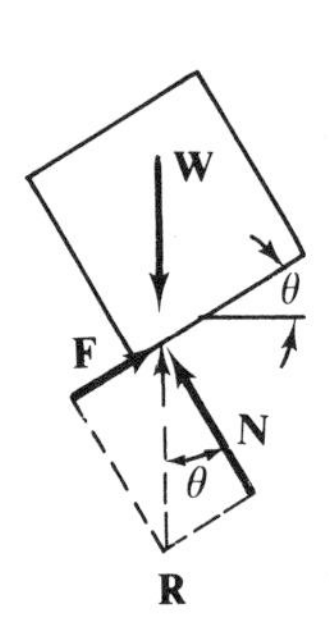

Figure 5-3

When a block of weight **W** is placed on a rough plane inclined upward to the right at an angle θ with the horizontal, it will be held in equilibrium by the reaction, **R**, of the plane as shown in Fig. 5-3, provided the angle θ is small enough.

As the angle between the plane and the horizontal is increased to some limiting value, motion of the block will impend down the plane. *The angle at which motion of the block impends is defined as the angle of repose.* From the equations of equilibrium,

$$N = W\cos\theta \quad \text{and} \quad F = F' = W\sin\theta$$

from which

$$\tan\theta = \frac{F'}{N} = \mu. \tag{5-3}$$

By comparing Eqs. (5-2) and (5-3), it is evident that the angle of friction is equal to the angle of repose. The coefficient of static friction can be obtained experimentally by measuring the angle of repose for any pair of surfaces.

5-5 TYPES OF PROBLEMS INVOLVING FRICTIONAL FORCES

When analyzing a problem involving frictional forces the question arises as to whether the magnitude of the friction is equal to or is less than the limiting or maximum friction μN. The answer depends on the motion condition that exists. It is usually desirable to classify problems involving frictional forces according to the available data and information to be obtained, and most problems can be identified by one of the following descriptions:

1. Impending motion is not assured from the statement of the problem.
2. Impending or relative motion is specified at all contact surfaces where there are frictional forces.
3. Impending motion is known to exist, but either the type of impending movement (slipping or tipping) is not specified or the surface or surfaces where motion impends are not specified.

The third type of problem includes all those in which there is more than one possible way or place for motion to start. Examples are a block that can either slide or tip, a block that moves either by sliding on a supporting block or by causing the supporting block to slide with it, and a wheel which can either roll or slip on a plane surface.

The following examples illustrate the general types of problems just described and a method of solution for each type.

EXAMPLE 5-1

A 26-ft homogeneous ladder weighs 50 lb and is placed against a smooth vertical wall with its lower end 10 ft from the wall [see Fig. 5-4(a)]. The coefficient of friction between the ladder and the floor is 0.30. Determine the frictional force at A acting on the ladder. The mass center of the ladder is assumed to be at its midpoint.

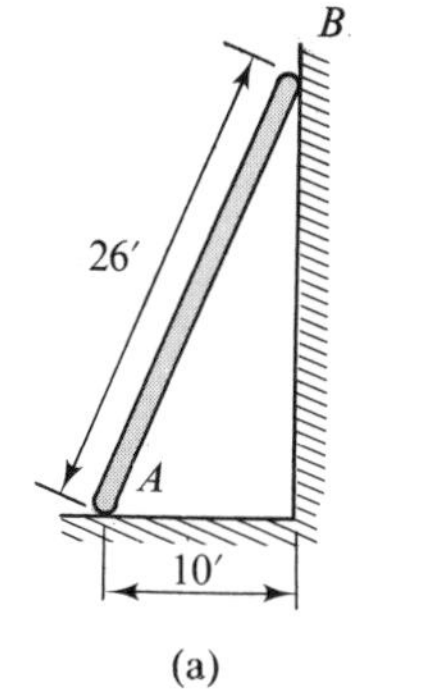

(a)

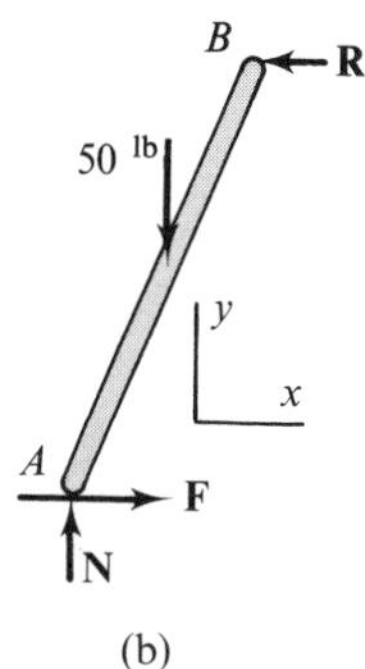

(b)

Figure 5-4

SOLUTION

This problem belongs to the first group since the statement of the problem does not specify that motion is impending or exists. It is assumed that motion does not impend and that the ladder is in equilibrium, but this assumption must be checked before the solution can be accepted. The free-body diagram is shown in Fig. 5-4(b), and the force equation of equilibrium gives

$$\sum \mathbf{F} = F\mathbf{i} + N\mathbf{j} - 50\mathbf{j} - R\mathbf{i}$$
$$= \mathbf{i}(F - R) + \mathbf{j}(N - 50) = 0$$

from which

$$N = 50 \text{ lb} \quad \text{and} \quad F = R.$$

The moment equation of equilibrium about B gives

$$\sum \mathbf{M}_B = \sum \mathbf{r} \times \mathbf{F}$$
$$= -10\mathbf{i} \times N\mathbf{j} + (-5\mathbf{i}) \times (-50\mathbf{j}) + (-24\mathbf{j}) \times F\mathbf{i}$$
$$= \mathbf{k}(-10N + 250 + 24F) = 0$$

or

$$24F = 10(50) - 250 = 250$$

and

$$F = 10.42 \text{ lb}.$$

The maximum value the frictional force can have with the 50-lb normal force is

$$F' = \mu N = 0.30(50) = 15.0 \text{ lb},$$

which is more than the 10.42-lb force required for equilibrium. Therefore equilibrium is assured, and the frictional force is

$$\mathbf{F} = \underline{10.42 \text{ lb to the right at } A.} \qquad \text{Ans.}$$

EXAMPLE 5-2

A 150-lb man starts to climb the ladder in Example 5-1. Determine the distance q from the man to the wall when the ladder starts to slip [see Fig. 5-5(a)].

SOLUTION

This problem belongs to the second group because the distance q is to be found for impending slipping of the ladder. When motion impends, the sense of the frictional force must be correctly shown on the free-body diagram to oppose the motion, to the right in this case.

There are four unknowns on the free-body diagram in Fig. 5-5(b); three forces and the unknown distance q. There are three equations of equilibrium and one equation of friction, $F' = \mu N$, since the problem states that motion impends. The force equation of equilibrium is

$$\sum \mathbf{F} = F\mathbf{i} + N\mathbf{j} - 50\mathbf{j} - 150\mathbf{j} - R\mathbf{i}$$
$$= \mathbf{i}(F - R) + \mathbf{j}(N - 50 - 150) = 0$$

from which

$$N = 200 \text{ lb} \quad \text{and} \quad F = R.$$

Since motion is specified, the frictional force is equal to the limiting value, that is,

$$F = F' = \mu N = 0.30(200) = 60 \text{ lb},$$
$$\mathbf{F} = 60\mathbf{i} \text{ lb}.$$

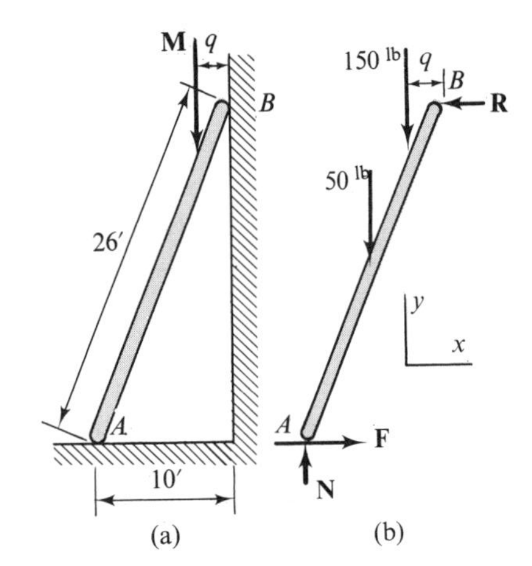

Figure 5-5

The moment equation of equilibrium about point B is

$$\sum \mathbf{M}_B = \sum \mathbf{r} \times \mathbf{F}$$
$$= -10\mathbf{i} \times N\mathbf{j} + (-24\mathbf{j}) \times F\mathbf{i} + (-5\mathbf{i}) \times (-50\mathbf{j}) + (-q\mathbf{i}) \times (-150\mathbf{j})$$
$$= \mathbf{k}[-10(200) - (-24)(60) + 250 + 150q] = 0$$

and

$$q = \underline{2.07 \text{ ft (distance from the wall).}} \qquad \text{Ans.}$$

EXAMPLE 5-3

To reach a little higher with the ladder in the preceding examples, the lower end is placed on a 100-lb box which is 2.0 ft high, as indicated in Fig. 5-6(a). The coefficients of friction are 0.20 between B and the vertical wall, 0.35 between A and the box C, and 0.25 between C and the floor. The weights of the man and the ladder are the same as in the preceding example. Determine the distance q when motion of the ladder impends.

SOLUTION

The problem states that motion of the ladder impends, and from the figure it is apparent that B will slip on the wall and also that either the

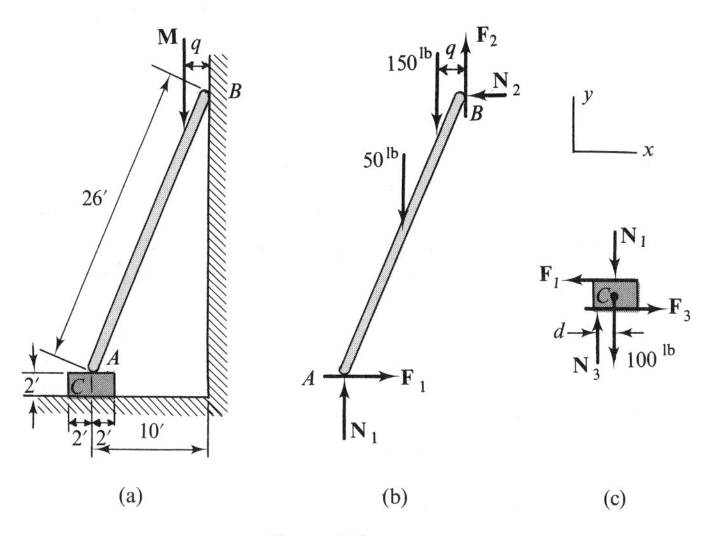

Figure 5-6

ladder will slide on C or that A and C will move together as C slides on the horizontal surface. Since the problem specifies that motion of the ladder impends but not where it impends, the problem comes under the third classification. The two free-body diagrams are drawn in Fig. 5-6(b) and (c). There are eight unknowns, six forces and two distances, on the two diagrams, and six equations of equilibrium are available. Motion impends at B; therefore $F_2' = 0.20N_2$. The eighth equation is obtained by assuming that motion impends at one surface of C and checking the second surface to make certain the frictional force does not exceed its limiting value. If the friction at the second surface does exceed the value of μN, the wrong assumption was made, and the problem should be solved on the basis of impending motion at the second surface.

The solution is a little simpler if motion impends at A (since enough equations will be available to solve for all unknowns on the ladder without using the free-body diagram of the block); therefore it is assumed that F_1 is equal to its limiting value and that F_3 is less than its limiting value. The equations of friction and equilibrium give the following results:

$$F_1 = F_1' = \mu_1 N_1 = 0.35N_1, \tag{a}$$

$$F_2 = F_2' = \mu_2 N_2 = 0.20N_2, \tag{b}$$

and

$$\sum \mathbf{F} = F_1\mathbf{i} + N_1\mathbf{j} - 50\mathbf{j} - 150\mathbf{j} + F_2\mathbf{j} - N_2\mathbf{i}$$

$$= \mathbf{i}(F_1 - N_2) + \mathbf{j}(N_1 - 50 - 150 + F_2) = 0,$$

which reduces to

$$F_1 - N_2 = 0 \tag{c}$$

and

$$N_1 + F_2 - 200 = 0. \tag{d}$$

Equations (a) and (b) can be substituted into Eqs. (c) and (d) to give

$$0.35N_1 - N_2 = 0$$

and

$$N_1 + 0.20N_2 - 200 = 0$$

from which

$$N_1 = 186.9 \text{ lb} \quad \text{and} \quad F_1 = 65.4 \text{ lb}.$$

Before determining the distance q it is desirable to check the assumption that motion does not impend between C and the horizontal plane. The force equation of equilibrium for Fig. 5-6(c) is

$$\sum \mathbf{F} = -F_1\mathbf{i} - N_1\mathbf{j} + F_3\mathbf{i} + N_3\mathbf{j} - 100\mathbf{j}$$

$$= \mathbf{i}(-F_1 + F_3) + \mathbf{j}(-N_1 + N_3 - 100) = 0,$$

which gives

$$F_3 = F_1 = 65.4 \text{ lb} \quad \text{and} \quad N_3 = N_1 + 100 = 286.9 \text{ lb}.$$

The limiting value of F_3 is

$$F_3' = \mu_3 N_3 = 0.25(286.9) = 71.7 \text{ lb}.$$

Since this value is more than the force required for equilibrium, the assumption that the ladder slides on the box is correct. The distance q can be determined from Fig. 5-6(b) by

$$\sum \mathbf{M}_B = \sum \mathbf{r} \times \mathbf{F}$$

$$= -q\mathbf{i} \times (-150\mathbf{j}) + (-5\mathbf{i}) \times (-50\mathbf{j}) + (-10\mathbf{i}) \times (186.9\mathbf{j})$$

$$+ (-24\mathbf{j}) \times (65.4\mathbf{i})$$

$$= \mathbf{k}(150q + 250 - 1869 + 1570) = 0,$$

which gives

$$q = \underline{0.327 \text{ ft.}} \qquad \text{Ans.}$$

If the block C had been narrow, it might have tipped over before the ladder would slide on it. This possibility can be checked by determining the distance d in Fig. 5-6(c). If d is small enough that the resultant normal force does not lie at, or beyond, the corner of the block, it will not tip. In this case d is equal to 0.456 ft, which indicates that there is no danger of C tipping.

EXAMPLE 5-4

The solid homogeneous 200-kg crate A in Fig. 5-7(a) rests on a rough floor. The coefficient of friction between the crate and the floor is 0.25. The workman attempts to move the crate by pulling on the rope over the smooth pulley as indicated. Determine the tension in the rope when motion impends.

SOLUTION

The problem states that motion impends, but it is not evident whether the crate slides on the floor or whether it tips about the lower right

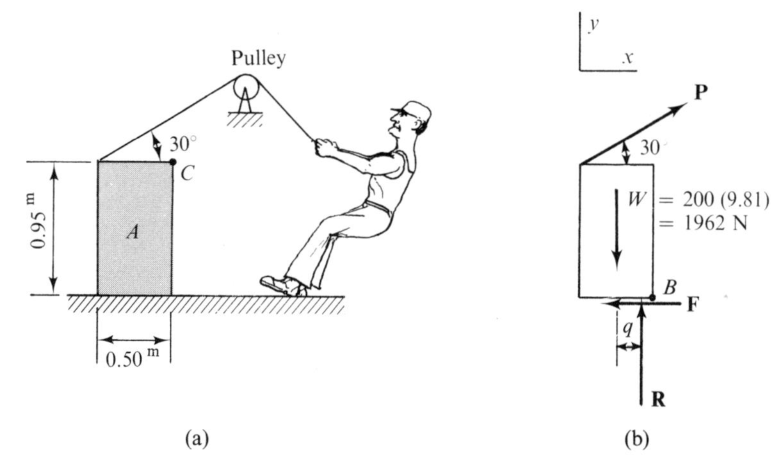

Figure 5-7

corner. Thus the problem comes under the third classification. The free-body diagram of the crate in Fig. 5-7(b) has three unknown forces, **F**, **R**, and **P**, and the unknown distance q. Since there are only three independent equations of equilibrium for this force system, a fourth equation must be obtained, and this equation will depend on whether the crate tips or slides.

If the body tips before it slides, the distance q in Fig. 5-7(b) will equal 0.25 m (**R** acts at the corner of the crate when it tips), and the magnitude of the force **F** will be less than the limiting value of μR. If the crate slides before it tips, the friction F will equal its limiting value, and the distance q will be less than 0.25 m.

In this example it is convenient to determine two values of **P**, first the force which will cause the crate to slide (assuming it does not tip) and second the force which will cause the crate to tip (assuming it does not slide). The smaller value of **P** is the required result.

If sliding is assumed, the force **P** can be determined from the following equations:

$$\sum \mathbf{F} = P(0.866\mathbf{i} + 0.500\mathbf{j}) - 1962\mathbf{j} - F\mathbf{i} + R\mathbf{j}$$

$$= \mathbf{i}(0.866P - F) + \mathbf{j}(0.500P - 1962 + R) = 0$$

from which

$$0.866P - F = 0 \qquad \text{(a)}$$

and

$$0.500P + R - 1962 = 0. \qquad \text{(b)}$$

Since the crate is assumed to slide, F and R are related by

$$F = \mu R = 0.25R. \qquad \text{(c)}$$

The simultaneous solution of Eqs. (a), (b), and (c) gives

$$P = 495\text{N}$$

and

$$\mathbf{P} = 495\text{N} \quad \angle 30°.$$

If tipping is assumed to occur, R will act at point B on Fig. 5-7(b), and the magnitude of $\mathbf{P}$ can be obtained from

$$\sum \mathbf{M}_B = \sum \mathbf{r} \times \mathbf{F} = 0.$$

Thus

$$\sum \mathbf{M}_B = (-0.50\mathbf{i} + 0.95\mathbf{j}) \times P(0.866\mathbf{i} + 0.500\mathbf{j}) + (-0.25\mathbf{i}) \times (-1962\mathbf{j})$$

$$= \mathbf{k}[(-0.50)(0.500P) - 0.95(0.866P) + (-0.25)(-1962)] = 0,$$

which gives

$$\mathrm{P} = 457\mathrm{N}$$

and

$$\mathbf{P} = 457\mathrm{N} \quad \angle 30^\circ.$$

Since the crate will tip when $P = 457\mathrm{N}$ and will not slip until $P = 495\mathrm{N}$, the force to cause impending motion is

$$\mathbf{P} = 457\mathrm{N} \quad \angle 30^\circ \qquad \text{Ans.}$$

The preceding discussion of friction problems can be summarized by outlining a suggested procedure for solving each type of problem. It is important to determine the type of problem from the statement of the problem before attempting the solution.

Procedure when impending motion is not assured:

1. Assume the system to be in equilibrium.
2. Determine the frictional and normal forces, using the equations of equilibrium.
3. Check the initial assumption by comparing F with μN, where μ is the coefficient of static friction for each contact surface. If F is less than or equal to μN, the assumption is correct, and the problem is solved. If F is greater than μN, equilibrium does not exist, and a complete solution will involve the principles of dynamics.

Procedure when impending or relative motion is specified at all contact surfaces with frictional forces:

1. Write the equation $F = F' = \mu N$ for all surfaces where motion impends. Make sure that the sense of the friction force is correct, that is, that the friction force opposes the motion or the tendency to move.
2. Determine the unknown quantities, using the equations of equilibrium together with the friction equations (accelerated motion is not considered here). No check is necessary for this type of problem because no assumptions are required.

Procedure when impending motion is specified but either the type of impending motion or the surface or surfaces where motion impends is not specified:

1. The number of unknown quantities always exceeds the number of equations of equilibrium available; therefore it is necessary to assume that one or more of the bodies will start to move in one of the two or more possible ways. Set $F = F' = \mu N$ at the pair or pairs of surfaces where sliding is assumed or known to impend, or place the

normal force at the corner about which the body is assumed to tip when tipping is assumed. At all other surfaces, motion is assumed *not* to impend.

2. Determine the frictional and normal forces at all surfaces where motion is assumed *not* to impend.

3. Check the initial assumption by comparing F and μN for all surfaces where slipping was assumed not to impend, or check the position of the normal where tipping was assumed not to impend. If F is less than, or equal to, μN for all such surfaces or if none of the normal forces acts outside the base of the body, the inital assumption was correct. If F is greater than μN for any of the surfaces or if any of the normal forces acts outside of the base of the body, the original assumption was incorrect, and a different assumption must be made as to where motion impends.

Alternate procedure for the third type of problem just discussed:

1. Determine the unknown quantity (usually a force) required to produce the specified impending motion in each one of the possible ways the motion can exist.

2. Select as the correct solution the value which satisfies the minimum or maximum condition as stated in the problem.

PROBLEMS

5-1 Block A in Fig. P5-1 weighs 8 lb, and B weighs 65 lb. The coefficient of friction between B and the plane is 0.25. Determine all unknown forces acting on B.

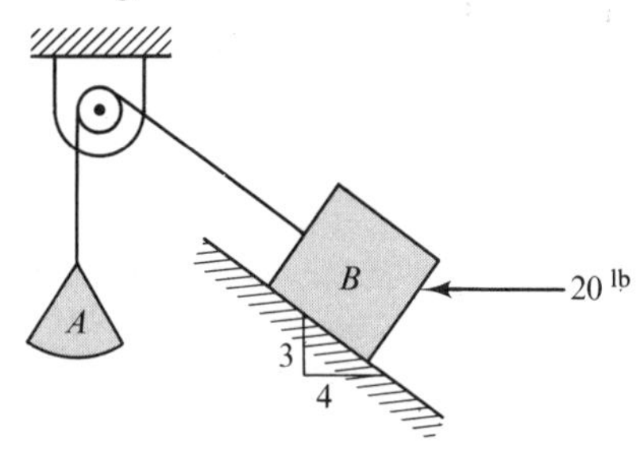

Figure P5-1

5-2 Block W in Fig. P5-2 weighs 26 lb, and the coefficient of friction between W and the plane is 1/3.

(a) Determine the minimum force **P** required to slide W up the plane.

(b) If **P** is removed, will W slide down the plane? Explain.

***5-3** The mass of the nonhomogeneous rod AB in Fig. P5-3 is 15 kg, and its center of gravity, G, is located one-third of the distance from A to B. The coefficient of friction between

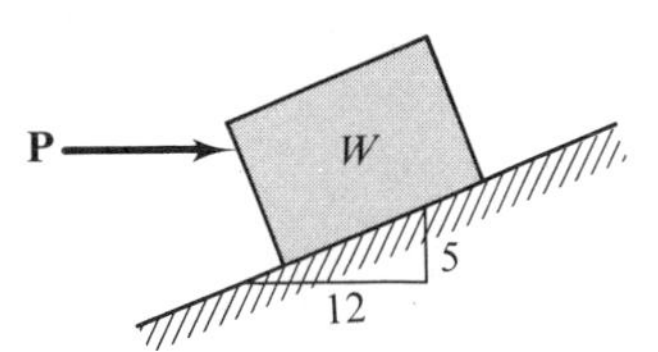

Figure P5-2

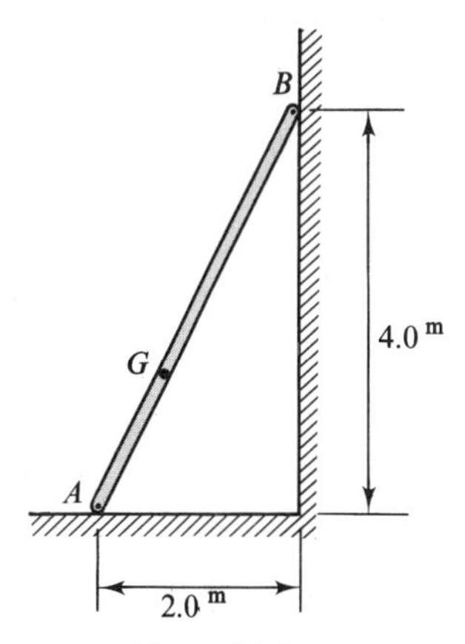

Figure P5-3

A and the floor is 0.20, and the vertical wall is smooth. Determine the frictional force at A on the rod.

5-4 The mass center of the nonhomogeneous half-cylinder in Fig. P5-4 is located at G. The coefficient of friction between the body and plane is 0.20. Determine, in terms of the weight W, the magnitude of **P** to maintain equilibrium in the position shown when
(a) The angle θ is zero.
(b) The angle θ is 90°.

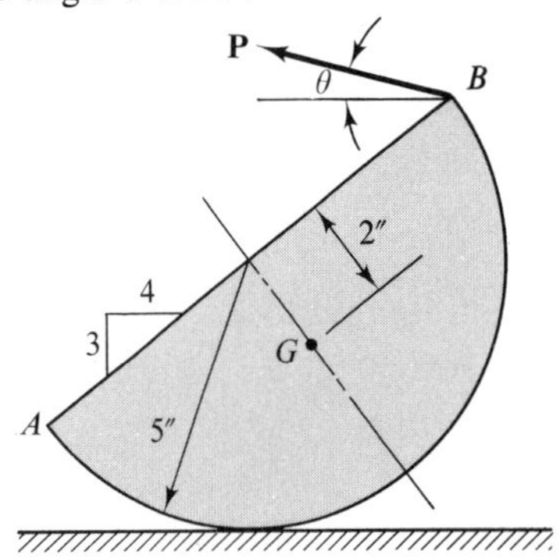

Figure P5-4

5-5 Solve Problem 5-4 when the force **P** is perpendicular to AB.

5-6 Determine the angle θ in Problem 5-4 which will result in a minimum magnitude of **P** to maintain equilibrium in the position shown and the minimum required value of the coefficient of friction.

***5-7** The mass of the wheel and drum in Fig. P5-7 is 12 kg, and the mass of body Q is 9 kg. The horizontal plane is smooth, the coefficient of friction at B is 0.50, and $r_2 = 2r_1$. Determine all unknown forces acting on the wheel and drum.

***5-8** Determine the minimum coefficient of friction between wheel A and the surface at B in Fig. P5-7 which will maintain equilibrium. The surface at C is smooth.

5-9 The wheel and drum in Fig. P5-9 weighs 12 lb and rests on a smooth horizontal plane. The coefficient of friction is 0.50 between A and the inclined plane at B. Body Q weighs 9 lb, and $r_2 = 2r_1$. Determine all unknown forces acting on body A.

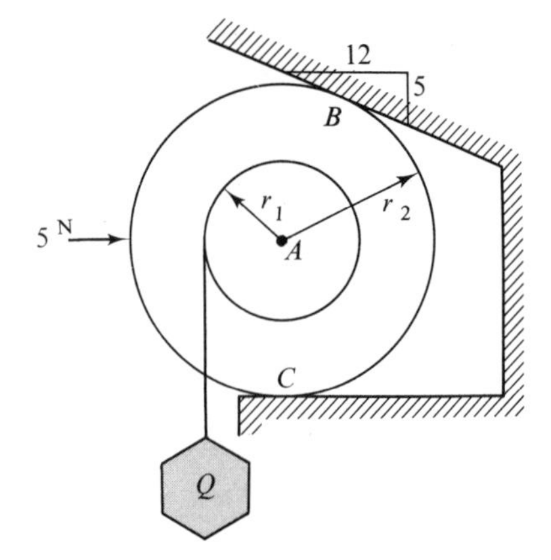

Figure P5-7

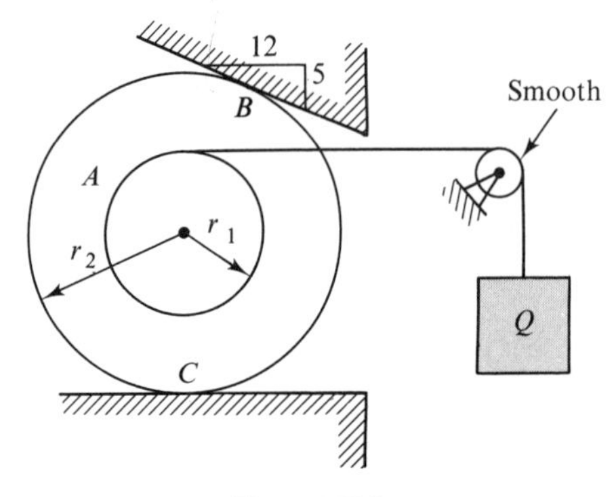

Figure P5-9

5-10 Determine the minimum coefficient of friction between wheel A and the surface at B in Fig. P5-9 which will maintain equilibrium. The surface at C is smooth.

5-11 Body A in Fig. P5-11 weighs 200 lb, and the coefficient of friction between A and each of the surfaces is 0.25. Determine the maximum weight B can have without disturbing the equilibrium of the system.

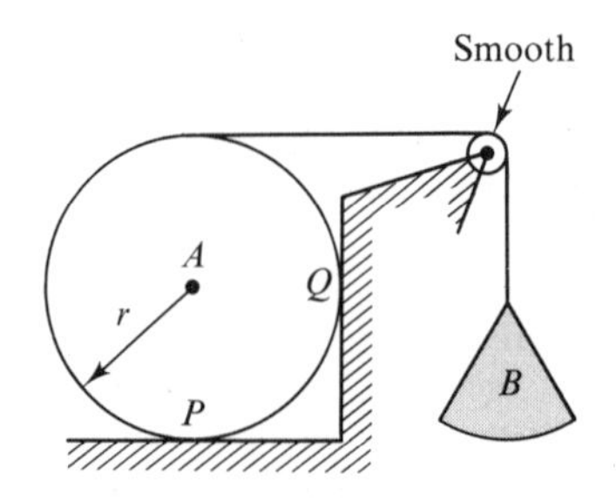

Figure P5-11

5-12 If the weight of body B in Problem 5-11 is less than the maximum for equilibrium, the problem is statically indeterminate. Discuss the possible ranges of the frictional forces at P and Q when the weight of B is 60 lb.

***5-13** The mass of body B in Fig. P5-13 is 100 kg. The coefficients of friction are 0.40 between the homogeneous bodies A and B and 0.30 between B and the plane. Determine the smallest mass that A may have if the system is to remain in equilibrium.

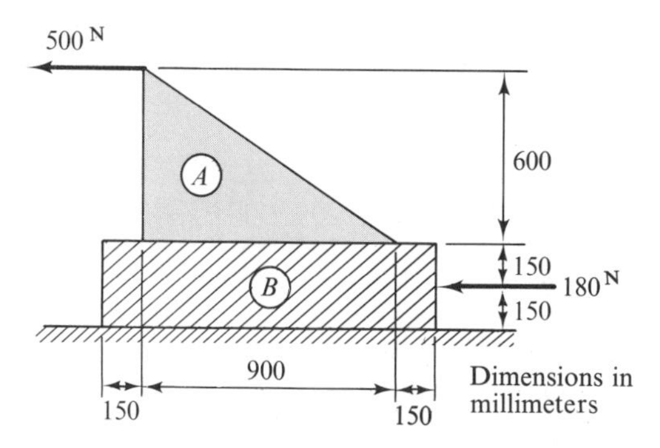

Figure P5-13

***5-14** Solve Example 5-4 if the rope is attached to the crate at point C and makes the same angle with the horizontal.

5-15 Wheel A in Fig. P5-15 weighs 40 lb, and B weighs 15 lb. The horizontal plane is smooth, and the coefficient of friction between A and the inclined plane is 0.30. Determine all unknown forces acting on A.

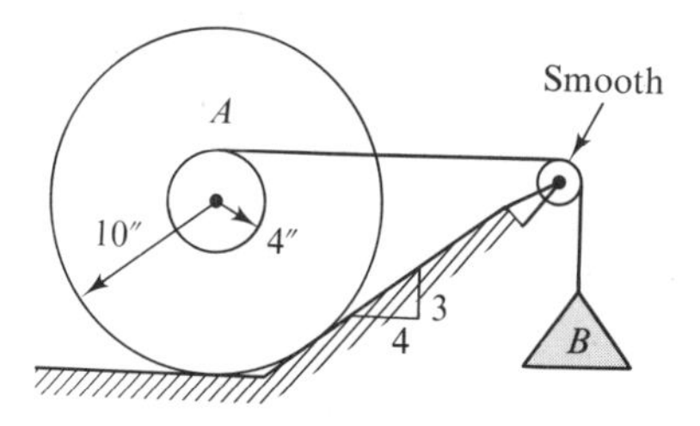

Figure P5-15

5-16 Determine the minimum weight of the wheel A in Problem 5-15 to maintain equilibrium. All other data are the same as in Problem 5-15.

5-17 The homogeneous block B in Fig. P5-17 weighs 50 lb. The coefficient of friction between B and the plane is 0.30. Determine the range of the weight of A which will maintain equilibrium of the system.

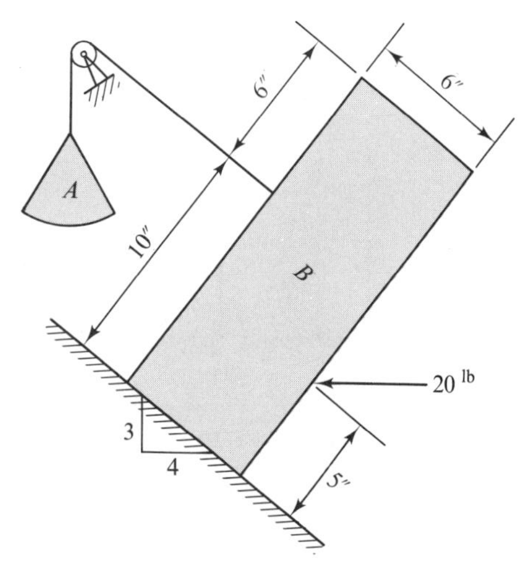

Figure P5-17

***5-18** Rod AB and block C in Fig. P5-18 have masses of 25 and 10 kg, respectively. The vertical wall is smooth, and the coefficients of friction for the other surfaces are 0.30 at the top of C and 0.20 at the bottom of C. Determine the magnitudes of all forces acting on C.

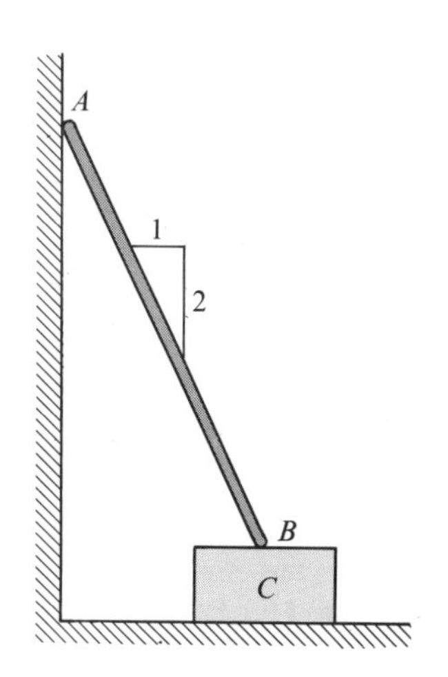

Figure P5-18

5-19 The homogeneous block in Fig. P5-19 weighs W lb, and the coefficient of friction between the block and the plane is 0.40. If the force **P** increases gradually until motion ensues, will the block slide or tip and for what value of **P**?

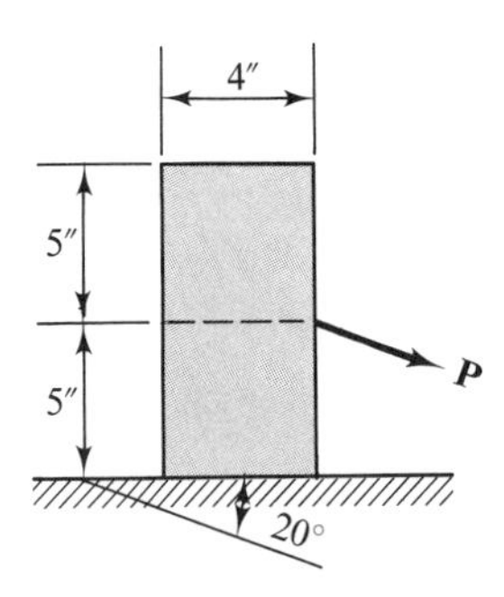

Figure P5-19

5-20 The homogeneous block in Fig. P5-20 weighs 20 lb and rests on a projection on the floor. Determine the minimum coefficient of friction between the block and the support which will maintain equilibrium.

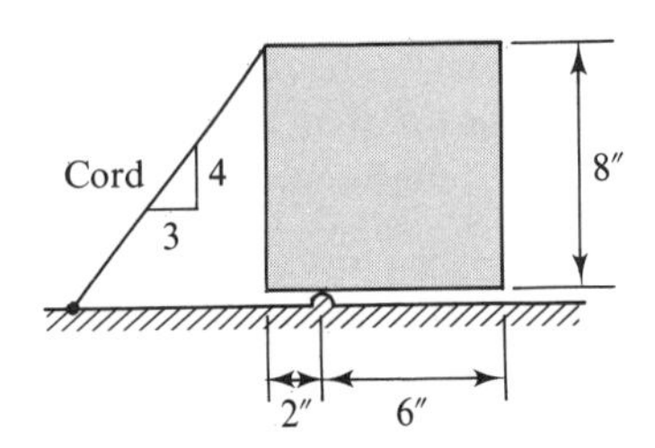

Figure P5-20

5-21 Determine the magnitude of the force **P** in Fig. P5-21 to produce impending motion of the 40-lb block W. Neglect the weight of ABC. The friction coefficient is 0.15 for both sliding surfaces.

5-22 Determine the maximum value of the coefficient of friction for Problem 5-21 for which motion is possible.

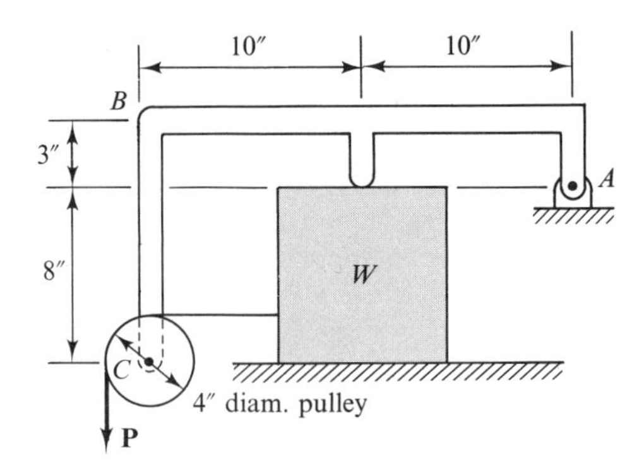

Figure P5-21

5-23 Investigation of the aircraft gear in Fig. P5-23 showed that the down locks were not fully in place. How much force **P** will the mechanic have to exert to force the down locks into place? Assume the coefficients of friction between the down locks and B to be 0.25 and between the down locks and A to be 0.20.

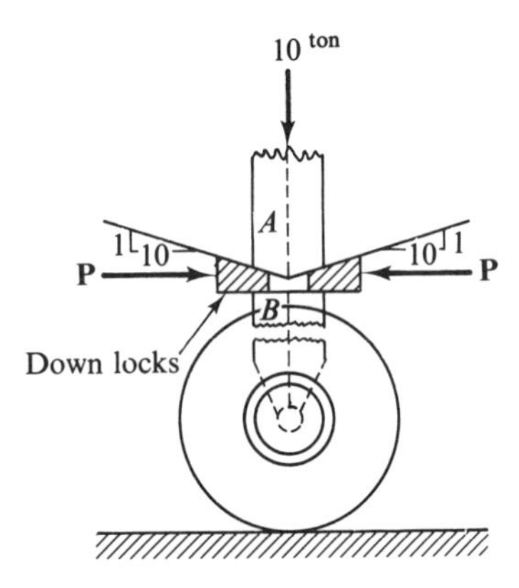

Figure P5-23

***5-24** Body A in Fig. P5-24 has a mass of 30 kg and is forced to slide along the plane by the force **P** applied to the structure. The coefficient of friction for all surfaces is 0.20. Neglect the weight of the lever. Determine the magnitude of **P** when slipping occurs.

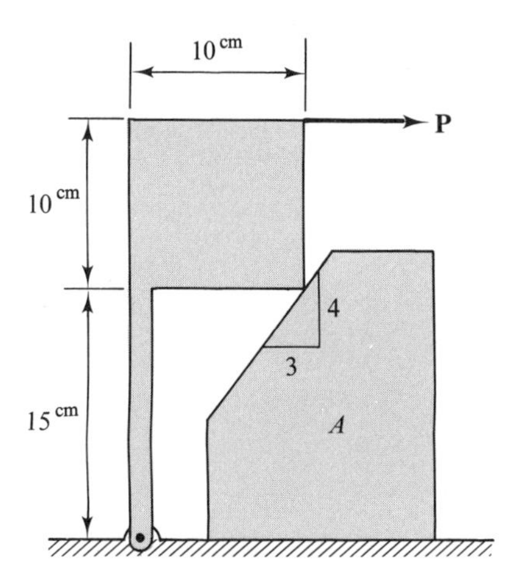

Figure P5-24

5-25 The wheel in Fig. P5-25 weighs W lb and has a radius of r in. The friction coefficient between the wheel and the inclined floor is 0.50, and the vertical wall is smooth. Determine the couple $\mathbf{C}$ acting on the wheel that will result in impending motion

(a) If the couple is clockwise.

(b) If the couple is counterclockwise.

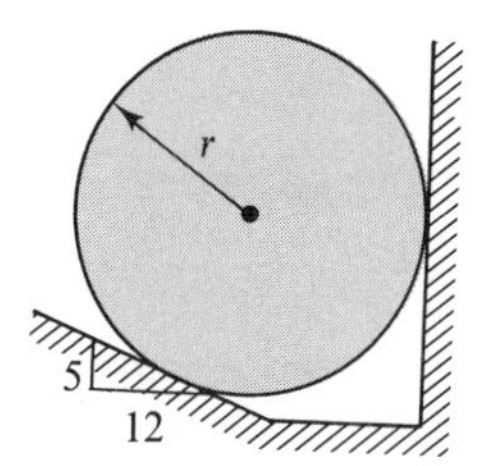

Figure P5-25

5-26 The drum in Fig. P5-26 weighs 10 lb, and the block B weighs 18 lb. The coefficient of friction between B and the drum is 0.45. Determine the components of the reaction at A on block B after the cord C is cut.

5-27 The rod AC in Fig. P5-27 weighs 120 lb. The contact surfaces at A are smooth, and at B

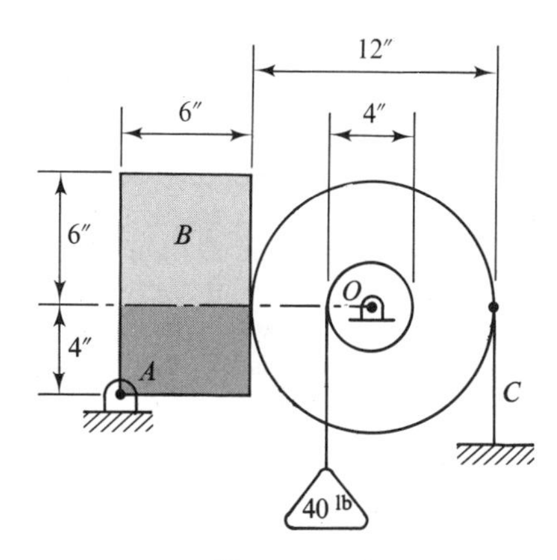

Figure P5-26

the coefficient of static friction is 0.30. Determine the components of the force on the rod at B.

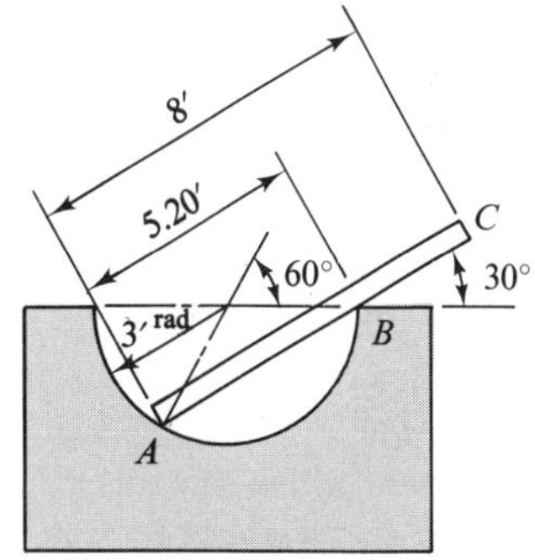

Figure P5-27

5-28 Body A in Fig. P5-28 weighs twice as much as body B. The coefficients of friction are 0.30 between A and B and 0.20 between B and the rotating frame. Determine the angle θ at which motion impends.

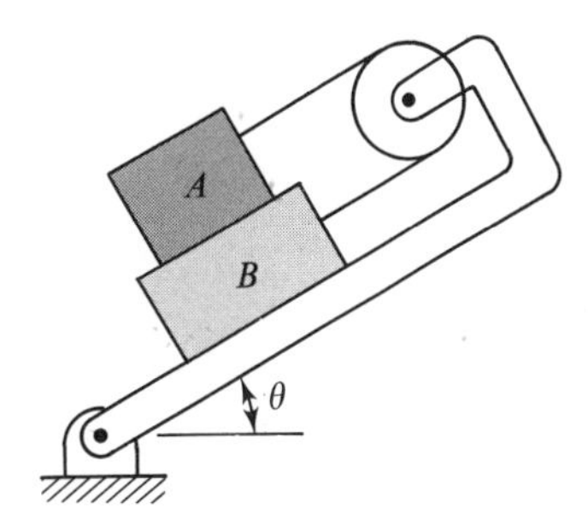

Figure P5-28

5-29 Solve Problem 5-28 if body A weighs W lb and B weighs $2W$ lb. The coefficients of friction are unchanged.

***5-30** Figure P5-30 is a proposed design for an articulated brake. Both braking surfaces have the same friction coefficient. Derive an expression relating the magnitude of the torque **T** to the magnitude of the force **P** when motion of the drum impends clockwise.

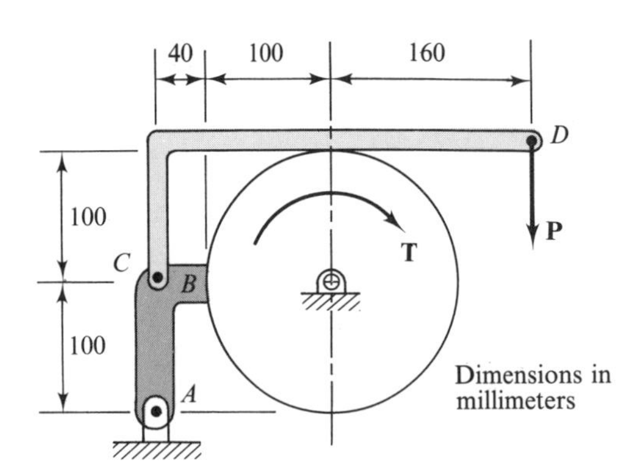

Figure P5-30

***5-31** Determine the force **P** in Fig. P5-30 which will produce a braking torque on the drum of 20 N·m when the drum is rotating clockwise. The coefficient of friction is 0.40.

5-32 The homogeneous bodies A and B in Fig. P5-32 weigh 50 and 100 lb, respectively. The coefficients of friction are 0.30 between B and the plane and 0.15 between A and the plane. The surface between A and B is smooth. Determine the maximum force **P** for which the system will be in equilibrium.

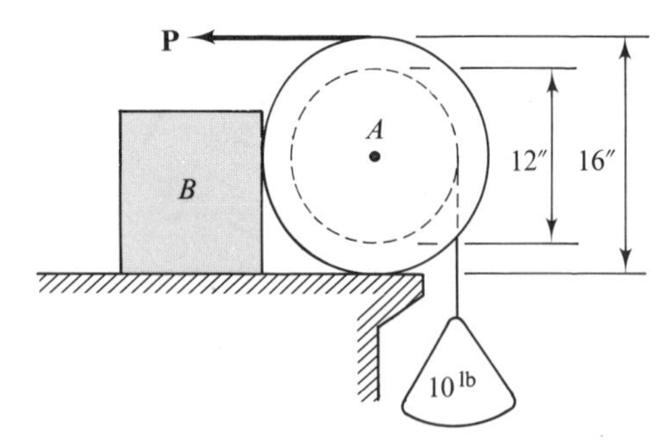

Figure P5-32

5-33 The 10-lb block A in Fig. P5-33 rests on the 20-lb block B. The coefficients of friction are 0.40 between A and B and 0.25 between B and the floor. Determine the force **P** necessary to cause impending motion of the system.

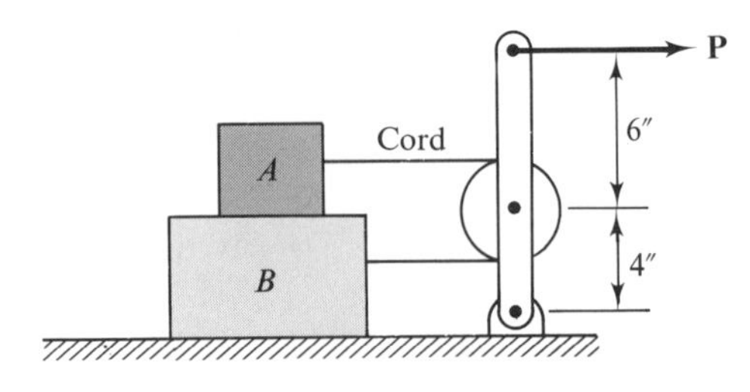

Figure P5-33

5-34 Solve Problem 5-33 if the cord connecting A and B is arranged as shown in Fig. P5-34. All other data are unchanged.

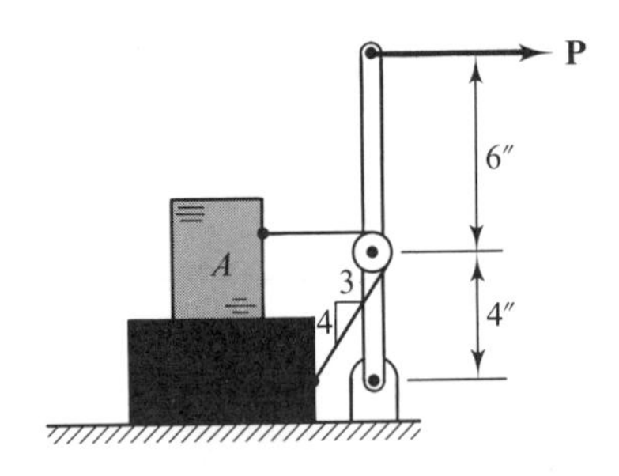

Figure P5-34

***5-35** The 10-kg homogeneous bar AB in Fig. P5-35 rests on the 15-kg block C. The coefficient of friction for both sliding surfaces is 0.20. Determine the force **P** to produce impending motion to the right.

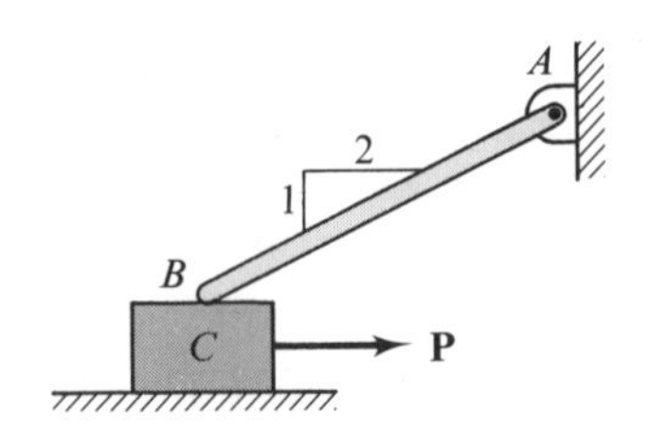

Figure P5-35

5-36 Block D in Fig. P5-36 weighs 5 lb, and the weights of members AB and BC can be neglected. The coefficient of friction between D and the plane is μ. Show that motion of D cannot impend to the right for any value of P if $\tan\theta > 1/\mu$.

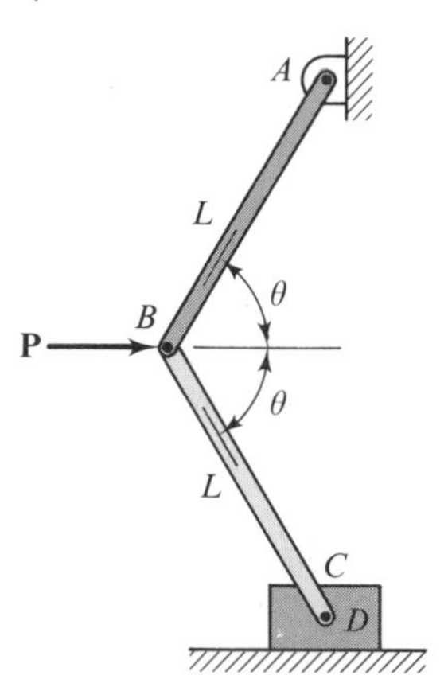

Figure P5-36

5-37 Bar AB and blocks C and D in Fig. P5-37 each weigh 10 lb. The coefficient of friction between C and AB and between C and the plane is 0.20. Block D slides in a smooth slot. Determine the magnitude of **P** to cause D to have impending motion up the plane.

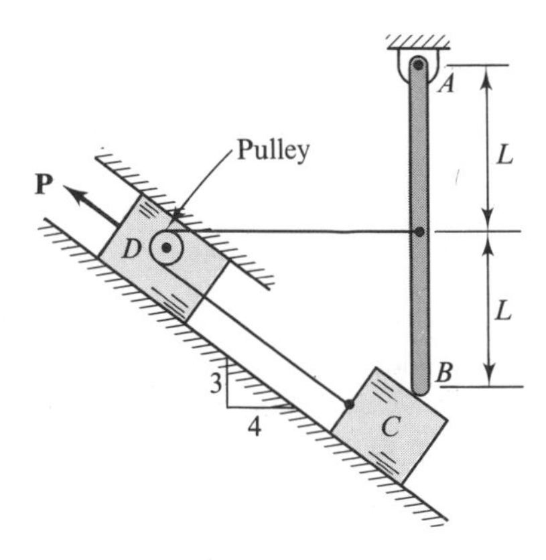

Figure P5-37

5-38 The homogeneous block in Fig. P5-38 weighs 10 lb and rests on a horizontal plane. The coefficient of friction between the block and the plane is 0.50. Determine the value of $\mathbf{P}_2$ (parallel to the y axis) to produce impending motion when $\mathbf{P}_1 = 4\mathbf{i}$ lb.

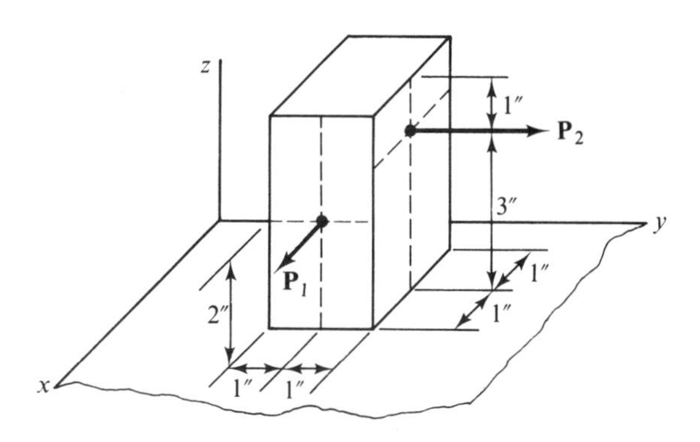

Figure P5-38

5-39 Solve Problem 5-38 when $\mathbf{P}_1 = 3\mathbf{i}$ lb.

***5-40** The block on the inclined plane in Fig. P5-40 has a mass of 10 kg. Lines ab and cd on the plane are horizontal (parallel to the z axis). The force $\mathbf{P}_2$, parallel to the x axis, has a magnitude of 40 N. The coefficient of friction between the block and the plane is 0.40. Determine the force $\mathbf{P}_1$, parallel to the z axis, which will cause motion of the block to impend.

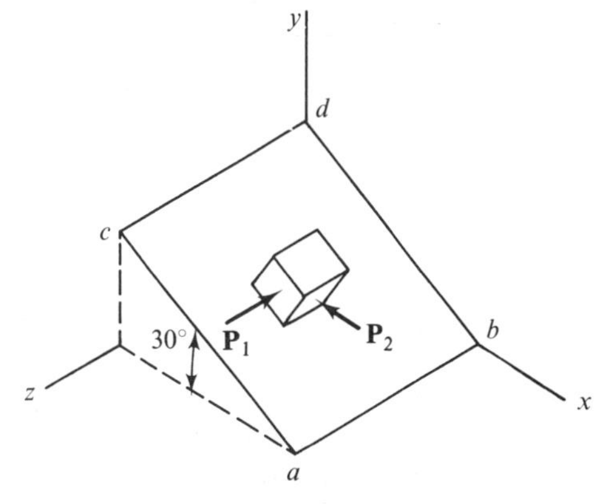

Figure P5-40

***5-41** If the force $\mathbf{P}_1$ in Problem 5-40 is 40 N as shown, what values of $\mathbf{P}_2$, if any, will maintain equilibrium? Other data are as given for Problem 5-40.

5-42 In Fig. P5-42 the 30-lb body A and the 40-lb body B are connected by a cord which passes over a smooth pulley on the 30-lb body C. The coefficients of friction are 0.20 between A and B, 0.30 between B and the plane, and 0.40 between C and the plane. Determine the maximum value **P** can have without disturbing the equilibrium of the system.

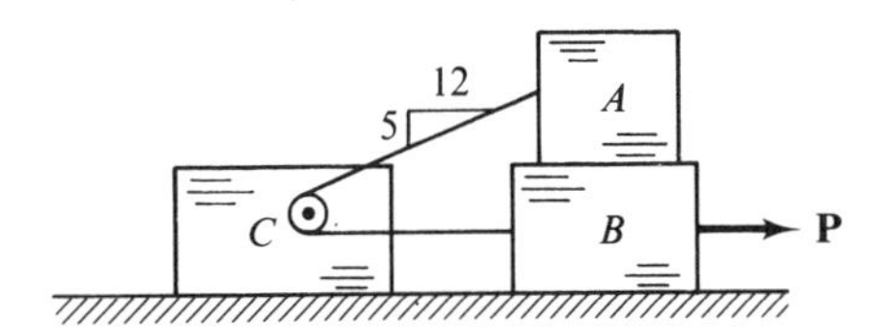

Figure P5-42

5-43 Bodies A and B in Fig. P5-43 weigh 900 and 340 lb, respectively. The coefficient of friction at all surfaces is 0.60. Couple **C** is applied to body A in an attempt to shove B up the inclined plane. Describe what happens when motion impends. The connecting bar is horizontal and of negligible weight.

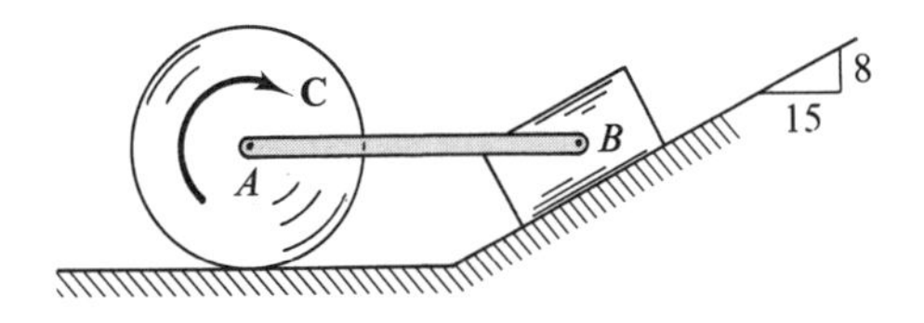

Figure P5-43

5-44 The 800-lb aircraft tractor in Fig. P5-44 can be used as a bulldozer in an emergency to move the 600-lb crate C to the right. The coefficients of friction are 0.20 for both surfaces of the crate and 0.50 for the tire and the ground. The mass center of the tractor is at G, and power is applied only to wheels A. Assume that both front wheels are in symmetrical contact with C. Determine the moment of the couple of the axle on the rear wheels when they start to move.

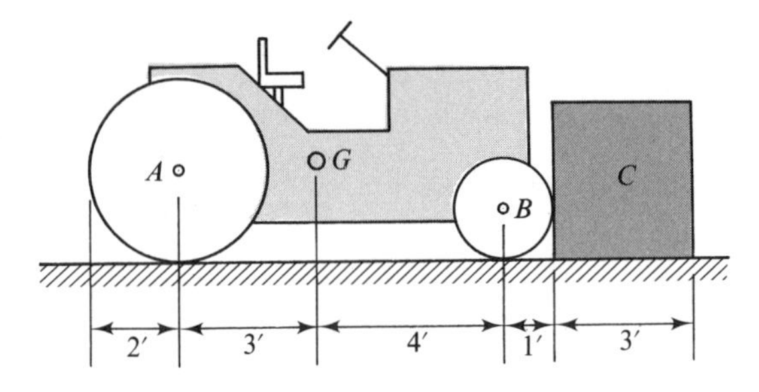

Figure P5-44

5-45 Solve Problem 5-44 if the tractor is turned around and backed against the crate so that wheels A are in contact with the crate.

5-46 The homogeneous bar in Fig. P5-46 has a ball and socket at A, and end B rests against a vertical wall. The coefficient of friction between B and the wall is μ. Determine the angle β, in terms of θ and μ, at which motion will impend.

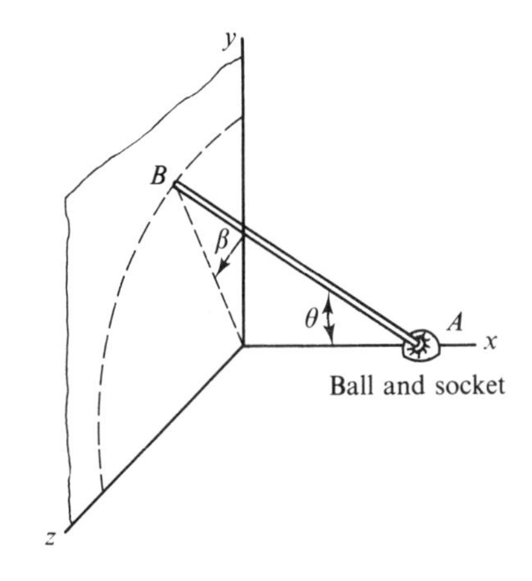

Figure P5-46

***5-47** In Fig. P5-46 bar AB has a mass of 5 kg and is 500 mm long. The angle θ is 60°, and the angle β is 15°. Determine the minimum coefficient of friction at B to maintain equilibrium.

5-48 The homogeneous rod AB in Fig. P5-48 has end B on a horizontal plane, and end A rests in a groove as shown. The coefficient of friction for all surfaces is 0.40. Determine the minimum angle θ for equilibrium when $\beta = 0$.

5-49 Solve Problem 5-48 when $\beta = 60°$.

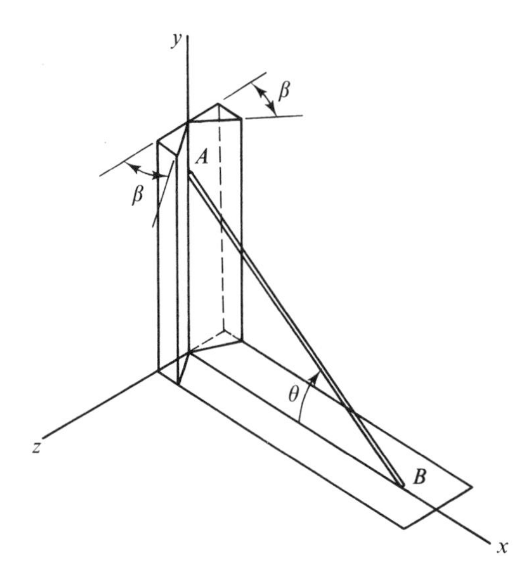

Figure P5-48

5-50 The 10-in. homogeneous rod *AB* in Fig. P5-50 rests against the *xz* and *yz* planes as shown. An 8-in. cord connects *B* to *C*. The coefficient of friction between A and the vertical wall is 0.50. Determine the maximum angle θ which will maintain equilibrium of the rod.

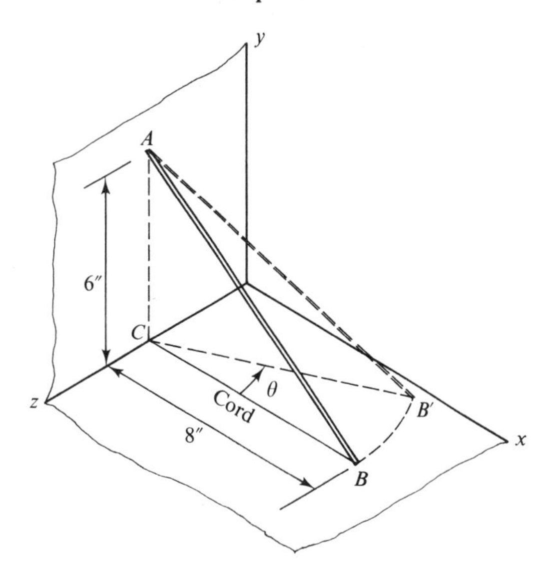

Figure P5-50

5-6 FRICTIONAL FORCES ON FLEXIBLE BANDS AND FLAT BELTS

The friction that is developed between a flexible band or belt and a flat pulley or drum can be utilized for the transmission of power and also for brakes. The relationship between the two belt tensions *when slipping of a belt on a pulley impends* can be developed by use of the conditions of equilibrium and the relationship between the normal and frictional forces. The weight *W* in Fig. 5-8(a) is suspended from the belt which passes over the fixed drum and is acted on by a force $\mathbf{T}_L$. If $\mathbf{T}_L$ is large enough, motion will impend between the belt and the fixed

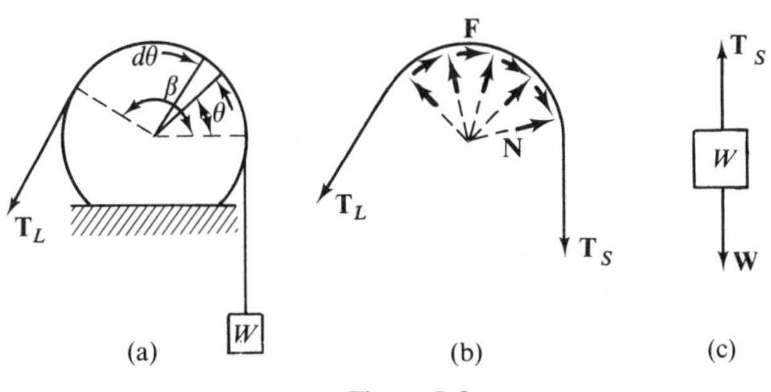

Figure 5-8

drum, and the body W will have impending motion upward. From the free-body diagram in Fig. 5-8(c), it is evident that the weight **W** is equal to the belt tension $\mathbf{T}_S$. The free-body diagram of the belt in Fig. 5-8(b) shows that the belt tension $\mathbf{T}_L$ is greater than **W** (or $\mathbf{T}_S$), since it must overcome the frictional force **F** developed between the belt and drum as well as lift the weight. A free-body diagram of the element of the belt subtended by the angle $d\theta$ in Fig. 5-8(a) is shown in Fig. 5-9.

The thickness of the belt is considered to be negligible in comparison with the radius, r, of the pulley. Likewise, the weight of the belt is neglected in comparison with the other forces acting upon it. The following results are obtained from Fig. 5-9 and the equations of equilibrium:

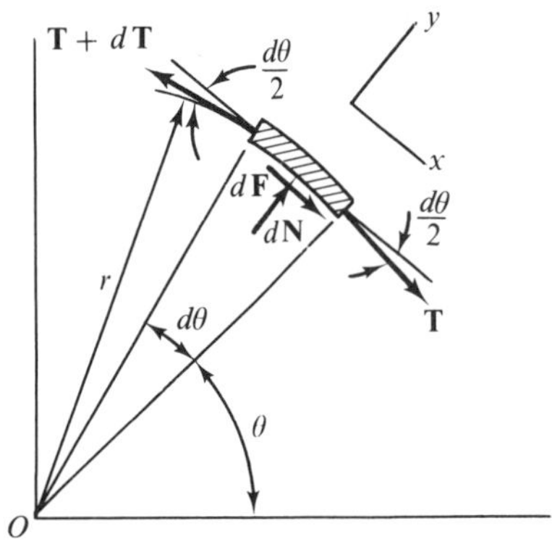

Figure 5-9

$$\sum \mathbf{M}_O = r(T+dT)\mathbf{k} - rT\mathbf{k} - r\,dF\mathbf{k} = 0,$$

which reduces to

$$dT = dF. \tag{a}$$

$$\sum \mathbf{F}_y = dN\mathbf{j} - (T+dT)\sin\frac{d\theta}{2}\mathbf{j} - T\sin\frac{d\theta}{2}\mathbf{j} = 0,$$

which reduces to

$$dN = 2T\sin\frac{d\theta}{2} + dT\sin\frac{d\theta}{2}.$$

The sine of an angle approaches the angle in radians as the angle becomes small. Consequently the expression for dN reduces to

$$dN = T\,d\theta + dT\frac{d\theta}{2} = T\,d\theta \tag{b}$$

when second-order infinitesimals are neglected in comparison with first-order infinitesimals.

When motion between the belt and drum is impending, $dF = \mu\,dN$ and the expression for dT becomes

$$dT = dF = \mu\,dN = \mu T\,d\theta \quad \text{or} \quad \frac{dT}{T} = \mu\,d\theta.$$

This differential equation can be integrated in the following manner:

$$\int_{T_S}^{T_L} \frac{dT}{T} = \mu \int_0^\beta d\theta$$

and

$$[\ln T]_{T_S}^{T_L} = \mu[\theta]_0^\beta,$$

or

$$\ln T_L - \ln T_S = \ln\frac{T_L}{T_S} = \mu\beta. \tag{5-4}$$

The symbol "ln" is the abbreviation for "natural logarithm." Equation

(5-4) can also be written

$$T_L = T_S e^{\mu\beta}, \tag{5-5}$$

where e, equal to 2.718. . . ., is the base of natural logarithms.

The relationship between the large tension, T_L, and the small tension, T_S, when motion impends between the belt and drum can thus be expressed as Eq. (5-4) or (5-5). The angle β is the angle of contact, in radians, between the belt and drum. When slipping occurs between the belt and drum the coefficient of kinetic friction must be used for μ.

The preceding discussion does not take into account the inertial effect of the mass of the belt when the belt and pulley are moving. Since each particle of the belt has mass, it tends to move in a straight line instead of traveling in a curved path around the pulley (see Newton's first law of motion in Chapter 9). This tendency of the belt particles reduces the normal and frictional forces between the belt and pulley and may seriously reduce the effectiveness of a belt drive if the pulleys are operated at large angular velocities. See Problem 9-49 (Chapter 9) for information on the effect of the inertia of the belt.

EXAMPLE 5-5

The mass center of the 25-lb body B of the braking mechanism in Fig. 5-10(a) is at G. The pulley A weighs 75 lb, and the coefficients of static friction are 0.40 between A and B and 0.30 between A and the brake band. Determine the moment of the couple **C** required to produce impending motion of A counterclockwise.

SOLUTION

In this problem, motion impends simultaneously between A and the band and between A and B. When motion impends, it is necessary to determine the correct sense of the frictional forces as they act on each body. After these directions have been determined, it is possible to distinguish between the large and small belt tensions.

The correct sense of the frictional forces can be determined by drawing a sketch of the body which has impending motion, indicating the direction of motion and showing the frictional forces opposing the motion. In this example, pulley A has impending motion counterclockwise, and therefore the frictional forces on A have a clockwise sense as indicated in Fig. 5-10(b). The friction of A on the belt has a counterclockwise sense as shown in Fig. 5-10(c) and thus assists the tension on the left side. Therefore, the left-hand tension is T_s in Eq. (5-5), and

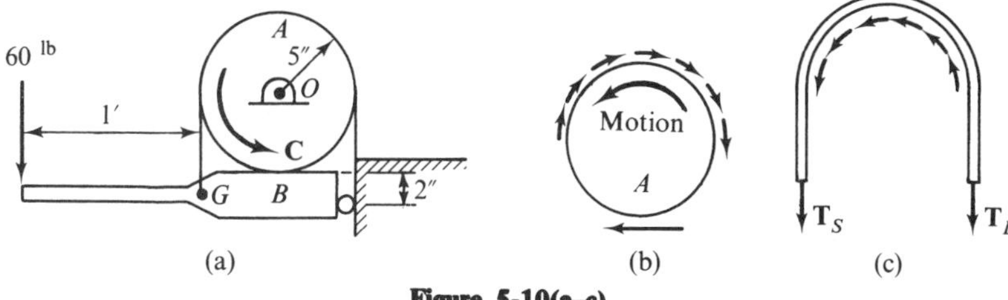

Figure 5-10(a–c)

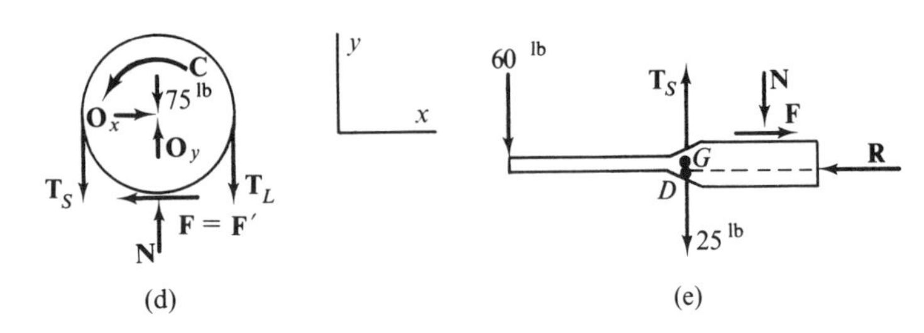

Figure 5-10(d, e)

the right-hand tension is the large tension. After it has been determined which tension is the larger, the free-body diagrams of the belt and body A and of body B in Fig. 5-10(d) and Fig. 5-10(e) can be drawn with the frictional forces in the proper directions. There are eight unknown quantities in the two diagrams: the two tensions, the components of the reaction at O, the frictional and normal forces between A and B, the moment of couple **C**, and the reaction **R**. The eight unknowns can be determined from the six equations of equilibrium and two friction equations, one between F and N and one between T_L and T_S.

The forces **F** and **N** can be determined from a moment and a friction equation if the moment center is chosen as the intersection of the lines of action of $\mathbf{T}_S$ and **R** (point D). Thus,

$$\sum \mathbf{M}_D = -5N\mathbf{k} - 2F\mathbf{k} + 12(60)\mathbf{k} = 0.$$

Since motion impends, $F = 0.40N$; therefore

$$5N + 2(0.40N) = 720$$

and $N = 124.2$ lb as shown.

$$\sum \mathbf{F}_y = T_S\mathbf{j} - 60\mathbf{j} - 25\mathbf{j} - 124.2\mathbf{j} = 0$$

from which $T_S = 209$ lb as shown. The large tension T_L is determined from Eq. (5-5) as

$$T_L = T_S e^{\mu\beta} = 209(2.718)^{0.30\pi} = 536 \text{ lb}.$$

From Fig. 5-10(d),

$$\sum \mathbf{M}_O = C\mathbf{k} + 5(209)\mathbf{k} - 5(536)\mathbf{k} - 5(0.40)(124.2)\mathbf{k} = 0$$

from which

$$C = \underline{1883 \text{ in-lb}}$$

and

$$\mathbf{C} = \underline{1883\mathbf{k} \text{ in-lb} = 1883 \text{ in-lb.}} \quad ↺ \quad \text{Ans.}$$

PROBLEMS

5-51 Determine the range of values of the weight of Q in Fig. P5-51 which will hold a weight P of 500 lb in equilibrium. The coefficient of friction between the cable and the fixed drum is 0.40.

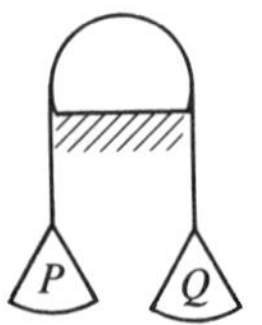

Figure P5-51

5-52 The 100-lb body A in Fig. P5-52 is supported by a rope wrapped around the fixed drum and attached to the 2-lb body B. The coefficient of friction between the rope and drum is 0.15. Determine the minimum number of turns of the rope on the drum if the system is to be in equilibrium.

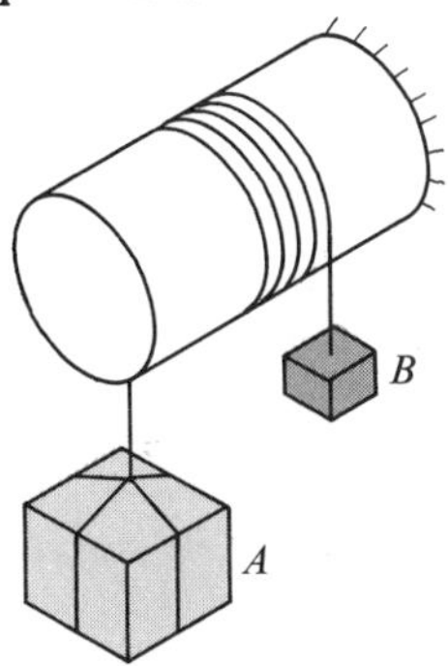

Figure P5-52

***5-53** The 25-kg motor in Fig. P5-53 drives a machine by means of a belt around two 200-mm pulleys. The motor is mounted so that its weight will produce a tension in the belt. Determine the maximum torque which can be transmitted under the conditions shown with the motor rotating clockwise. The coefficient of friction is 0.40.

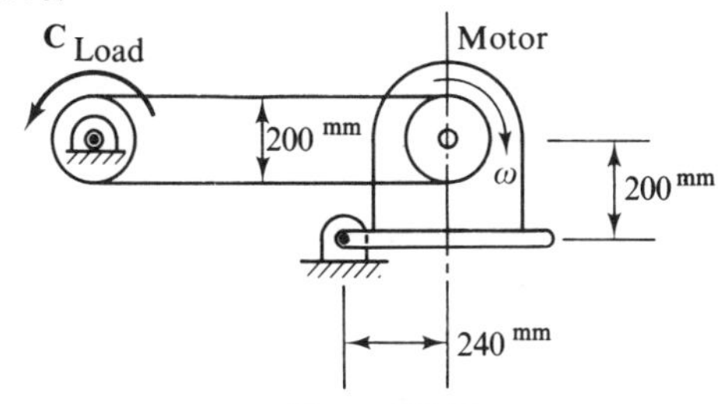

Figure P5-53

***5-54** Solve Problem 5-53 if the motor is rotating counterclockwise.

5-55 In Fig. P5-55, body A weighs 40 lb, and B weighs 70 lb. The length of CD is twice the radius of the drum. As the force $\mathbf{P}$ at the midpoint of CD is increased the angle θ decreases. When P becomes 30 lb the cord slips on the drum. Determine the coefficient of friction between the cord and the drum.

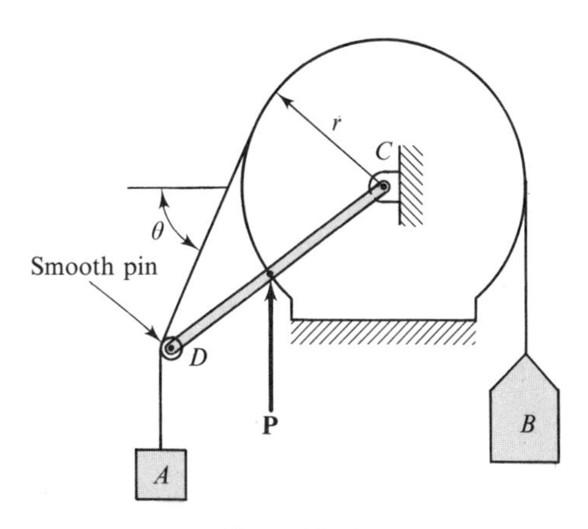

Figure P5-55

5-56 The homogeneous block A in Fig. P5-56 weighs 100 lb and rests on a rough ($\mu = 0.30$) plane. The coefficient of friction between the cord and the fixed drum is $2/\pi$. Determine the maximum weight B can have without moving.

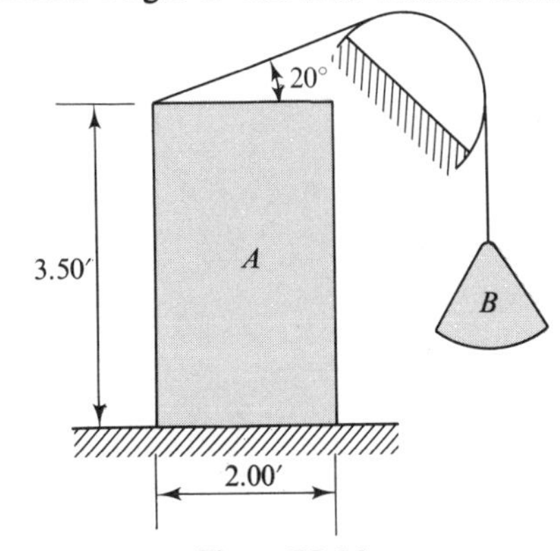

Figure P5-56

5-57 In Fig. P5-57 the weights of the bodies are A, 140 lb; B, 175 lb; C, 225 lb; and D,

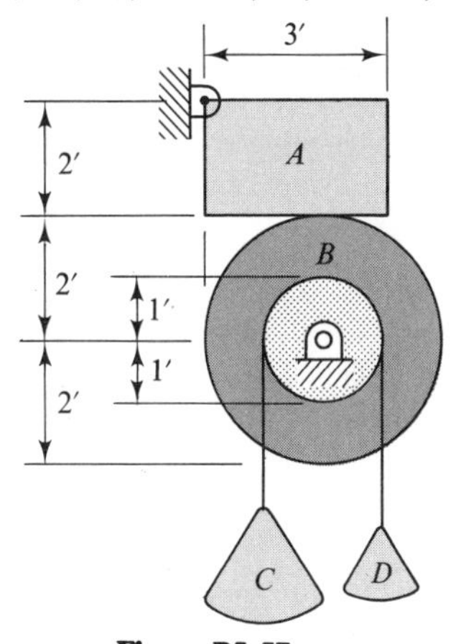

Figure P5-57

100 lb. The coefficients of friction are 0.40 between A and B and 0.30 between B and the cord. Is the system in equilibrium? If not, describe the motion.

5-58 The homogeneous block A in Fig. P5-58 weighs 85 lb. The coefficients of friction are 0.50 between A and the plane and 0.25 between the cord and the fixed drum. Determine the least weight of B which will cause it to move downward.

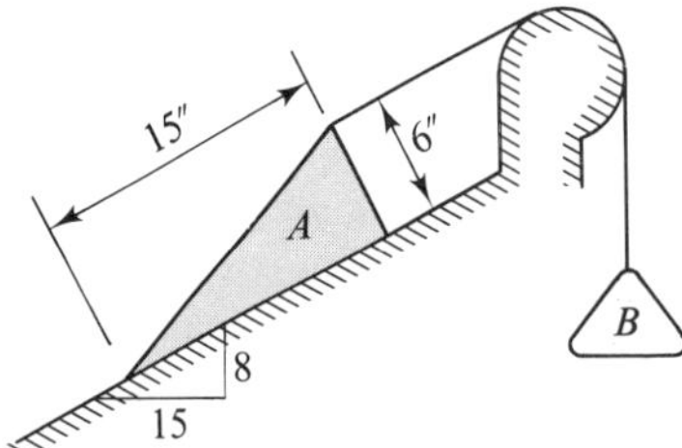

Figure P5-58

5-59 Body A in Fig. P5-59 weighs 20 lb, and the coefficient of friction for all surfaces is 0.20. For what values of W_B is the system in equilibrium?

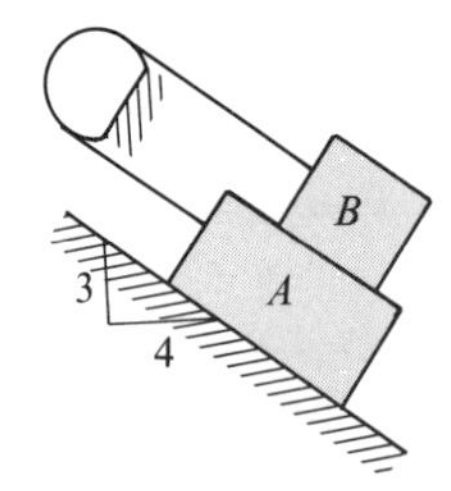

Figure P5-59

5-60 Determine the maximum coefficient of friction between the cord and drum in Problem 5-59 for which motion of B down the plane is possible. All other data are the same as in Problem 5-59.

***5-61** The 20-kg body A in Fig. P5-61 is suspended by a flexible belt which passes over the 30-kg pulley D and is anchored at B. The cord supporting the 15-kg body C is fastened to the pulley at E. The coefficient of friction between the belt and drum is 0.40. Determine the tension in the belt at B.

5-62 The drum in Fig. P5-62 is fixed to member AB and weighs W_1 lb. The coefficient of friction at C is 0.30. Determine the least coefficient of friction between B and the belt for which equilibrium is possible for a large enough weight W_2.

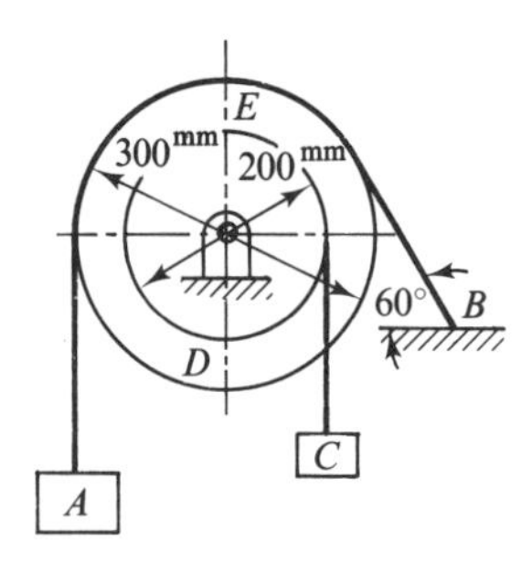

Figure P5-61

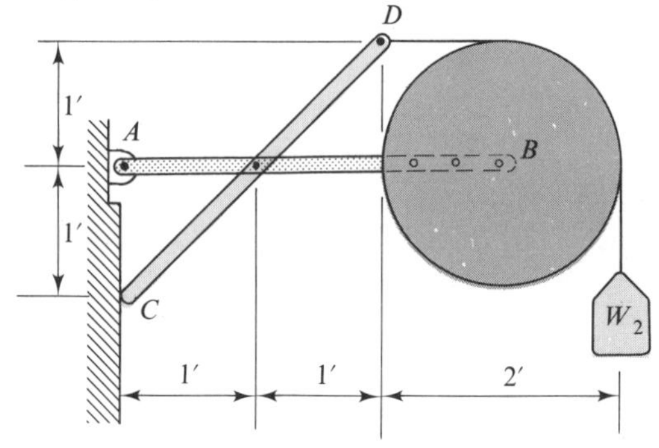

Figure P5-62

5-63 Body A in Fig. P5-63 weighs 50 lb, and B weighs 30 lb. Body B consists of a drum rigidly mounted between two blocks. The coefficient of friction for all sliding surfaces is 0.30. Determine the least weight of C which will cause it to move downward.

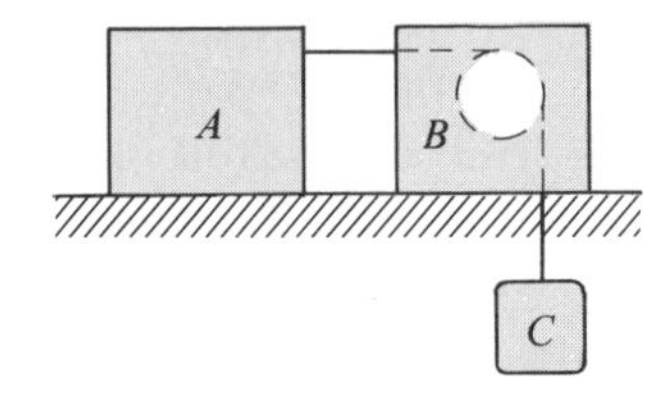

Figure P5-63

5-64 Body A in Fig. P5-64 weighs 10 lb, and the symmetrical body B weighs 30 lb. A cord from A passes over a fixed drum and is attached to the wheel B. The coefficients of friction are 0.25 between the cord and the drum and 0.40

between B and the floor. Determine the moment of the couple $\mathbf{C}$ which will cause B to have impending motion.

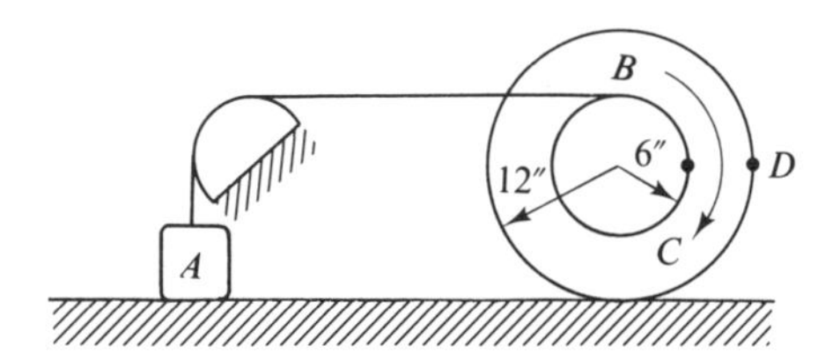

Figure P5-64

5-65 In Problem 5-64 the couple $\mathbf{C}$ is replaced by a downward force $\mathbf{P}$ at point D. Determine the magnitude of $\mathbf{P}$ which will cause B to have impending motion.

***5-66** A 5-m beam with a mass of 120 kg is supported by a cable passing over two fixed drums as shown in Fig. P5-66. The coefficient of friction between the drums and the cable is 0.40. A 90-kg man stands at the center of the beam. Determine the distance he can walk toward either end of the beam before the cable slips.

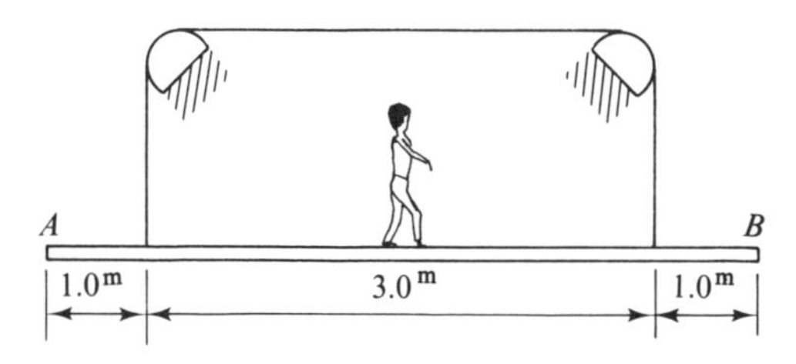

Figure P5-66

5-67 The 20-lb load B in Fig. P5-67 is lifted by winding the cable over the fixed drum on a winch at the top of body A. The coefficient of friction at both sliding surfaces is 0.20. Determine the minimum weight of body A to keep it from moving.

5-68 The 15-in. by 20-in. homogeneous block A in Fig. P5-68 weighs 100 lb, and body B weighs 50 lb. The coefficients of friction are $1/\pi$ between the cord and drum D and 0.25 between A and the plane. Determine what values of the weight of C, if any, will maintain the system in equilibrium.

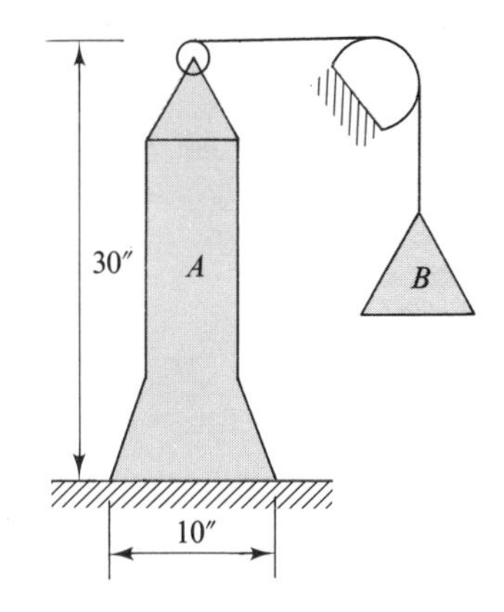

Figure P5-67

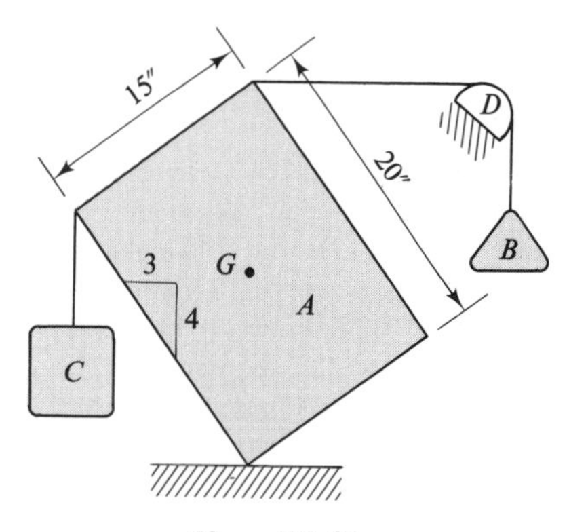

Figure P5-68

5-69 Bodies A and B in Fig. P5-69 weigh 60 and 12 lb, respectively. The coefficients of friction between A and the plane and between the cord and drum D are 0.20 and $1/2\pi$, respectively. Determine the magnitude of the force $\mathbf{P}$ which will cause impending motion.

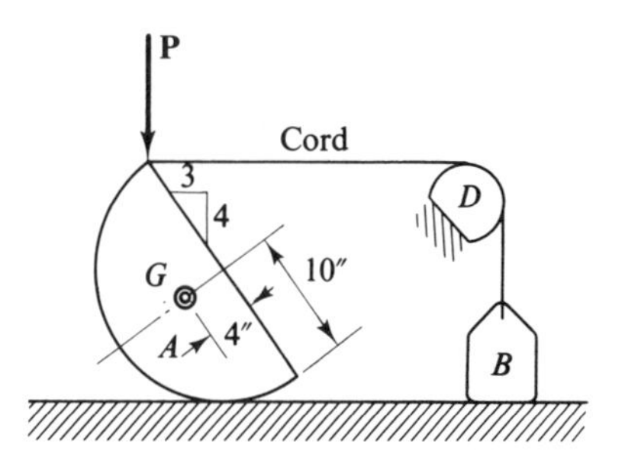

Figure P5-69

*__5-70__ Body B in Fig. P5-70 has a mass of 50 kg, and the masses of the pulley and brake may be neglected. The coefficients of friction are 0.60 between the belt and pulley and 0.40 between the brake and pulley. Determine the mass of body A that will cause it to have impending motion downward.

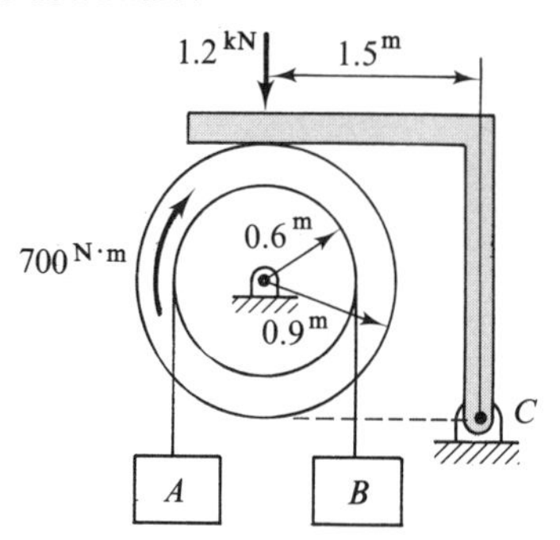

Figure P5-70

5-71 Body A in Fig. P5-71 consists of a drum mounted rigidly between two rectangular blocks and weighs 140 lb. Body B weighs 70 lb. The friction coefficients are 0.20 between the cord and drum and 0.30 between bodies A and B and the corresponding planes. Determine the weight of C to cause impending motion.

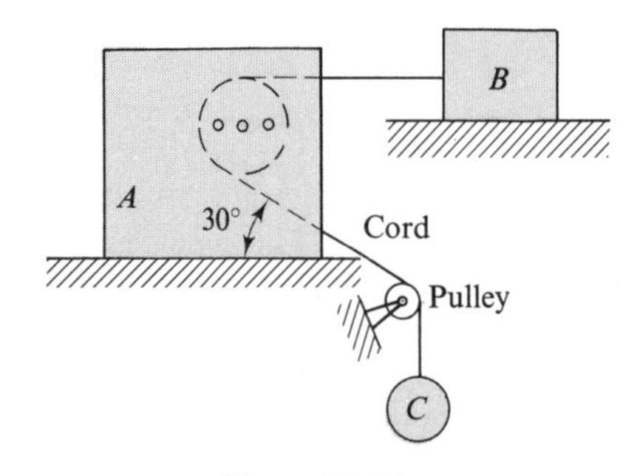

Figure P5-71

5-72 Blocks A and B in Fig. P5-72 weigh 10 lb each. The coefficient of friction is 0.25 for all sliding surfaces. Determine the minimum force **P** which will cause B to slide to the right.

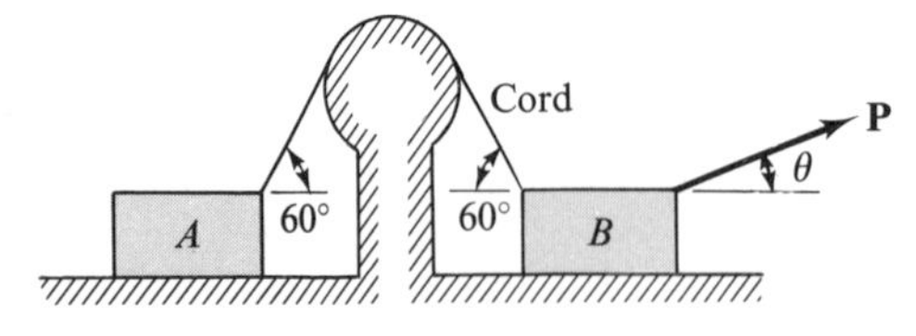

Figure P5-72

5-73 The weights of blocks A, B, and C in Fig. P5-73 are 20, 40, and 60 lb, respectively. The coefficients of friction are 0.40 between A and B and between B and the plane and 0.20 between the cord and drum fixed in C. The plane is smooth under block C. Determine the magnitude of **P** to cause impending motion of C to the right.

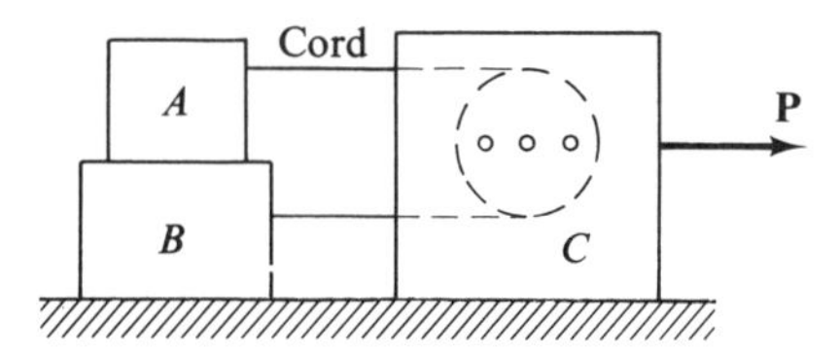

Figure P5-73

5-74 In Fig. P5-74, A weighs 100 lb, B weighs 200 lb, and C weighs 25 lb. The coefficients of friction are 0.30 between A and B and between B and the plane and 0.12 between the cord and the fixed drum D. Determine the minimum force **P** necessary to cause the block B to move to the left.

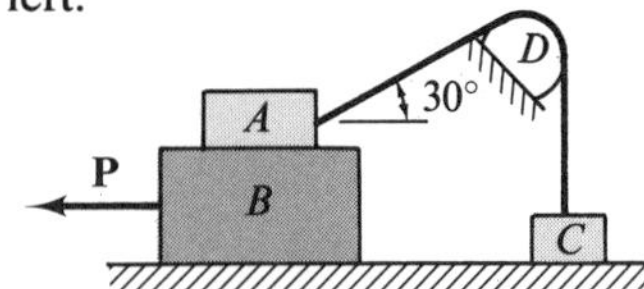

Figure P5-74

*__5-75__ The 15-kg body C in Fig. P5-75 is to be lifted by a cord wrapped three-fourths of a turn around the inner drum of the 60-kg body A.

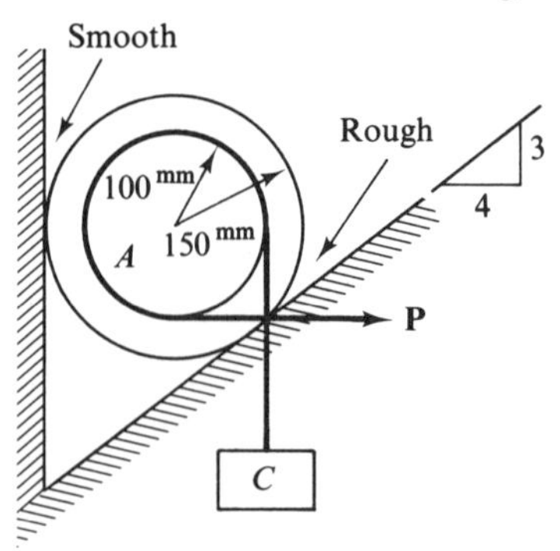

Figure P5-75

The coefficient of friction between the cord and drum is $1/\pi$. Determine the minimum coefficient of friction between A and the inclined plane if the wheel is not to turn. The vertical wall is smooth.

5-76 The mechanism in Fig. P5-76 is used as a braking device. The wheel D weighs 50 lb and is acted on by the 200-in-lb couple C. The coefficient of friction for all sliding surfaces is 0.30. Determine the least weight of E to prevent rotation of D.

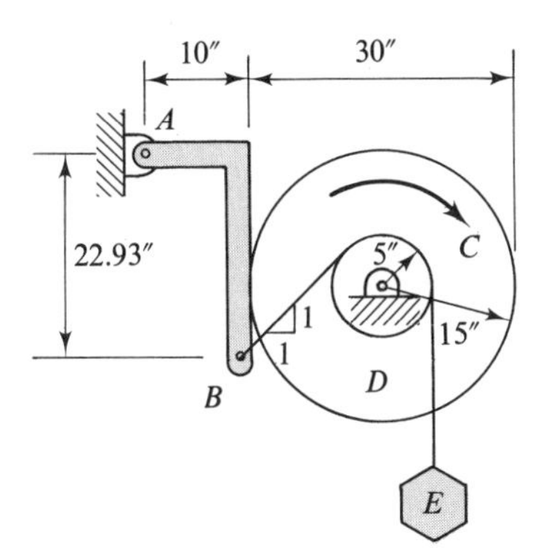

Figure P5-76

5-77 The homogeneous triangular plate ABC in Fig. P5-77 weighs 300 lb and is supported by the 100-lb member BE. The cord attached to the midpoint of BE passes over the drum D to body F. The coefficients of friction at D and E are 0.20. Determine the range of values of the weight of F which will maintain equilibrium.

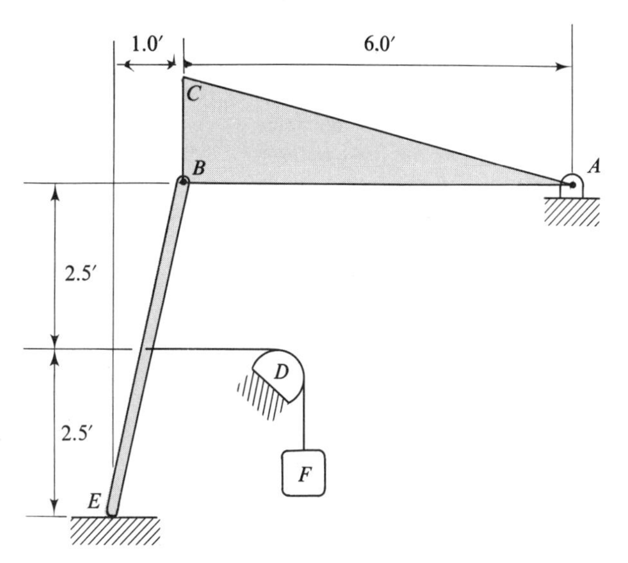

Figure P5-77

***5-78** The 10-kg body B in Fig. P5-78 is to be lifted by the force **P**. The mass of A is 5 kg, and the masses of pulley C and of member AD can be neglected. The coefficients of friction are 0.60 between the cord and A and 0.30 between A and the horizontal plane. Determine the magnitude of **P**.

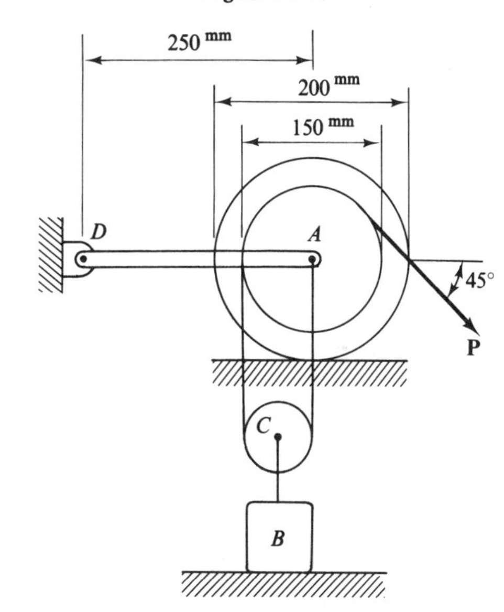

Figure P5-78

5-79 In Fig. P5-79 body A weighs 65 lb and B weighs 50 lb. The friction coefficients are 0.30 between the cord and drum fixed to B, 0.20 between B and the plane, and zero between A and the plane. Determine the maximum weight C can have without moving.

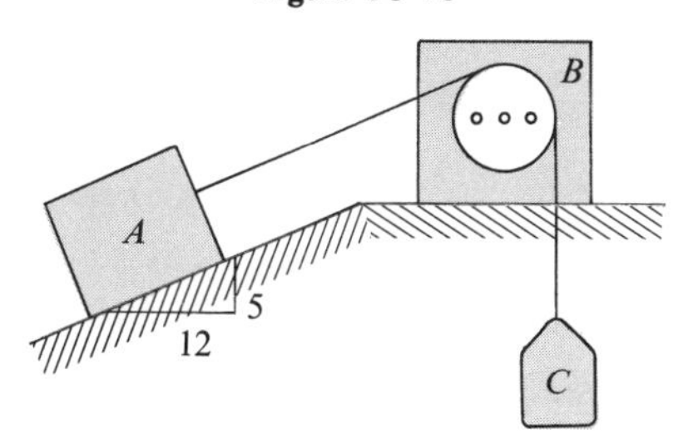

Figure P5-79

5-80 Body A in Fig. P5-80 weighs 18 lb and B weighs 14 lb. The friction coefficients are $1/\pi$ between the cord and drum D, 0.40 between C and B, and 0.30 between B and the plane. Determine the least weight of the homogeneous body C to maintain equilibrium.

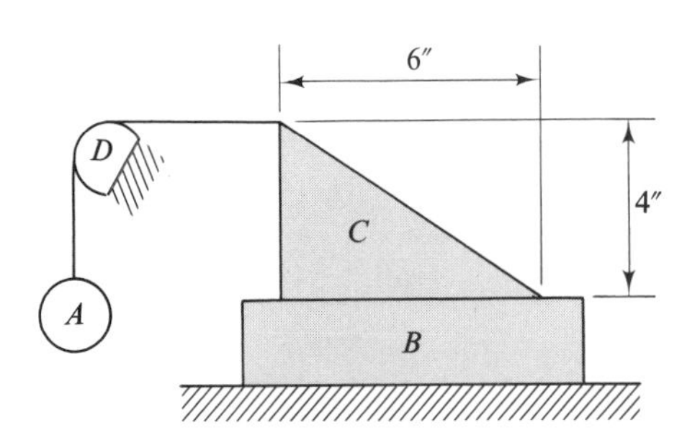

Figure P5-80

5-81 Neglect the weight of body A in Fig. P5-81. Body B weighs 30.0 lb and body C weighs 40.0 lb. The coefficients of friction are $1/\pi$ between the rope and body A, 0.40 between C and B, and 0.20 between B and the plane. The surface between A and B is smooth. Determine the force **P** that will cause body C to have impending motion.

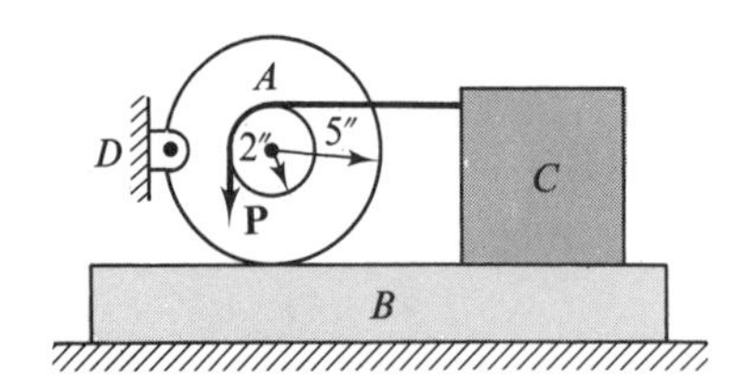

Figure P5-81

5-82 The weight W in Fig. P5-82 is to be lifted by a rope over a pipe which is supported by two sets of planks arranged as shown. The coefficient of friction between the pipe and the planks is 0.30. Determine the least coefficient of friction between the rope and pipe which will cause the pipe to rotate. Neglect the weight of the pipe.

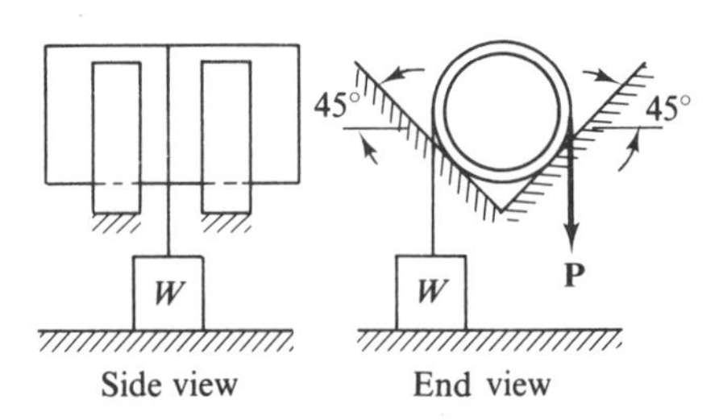

Figure P5-82

5-83 The unbalanced drum in Fig. P5-83 is to be lifted by a flat belt which is wound on a winch. Show that the drum will not slip on the belt as the drum is lifted provided the friction coefficient between the belt and drum is greater than $(1/\pi)\ln[(R+r)/(R-r)]$.

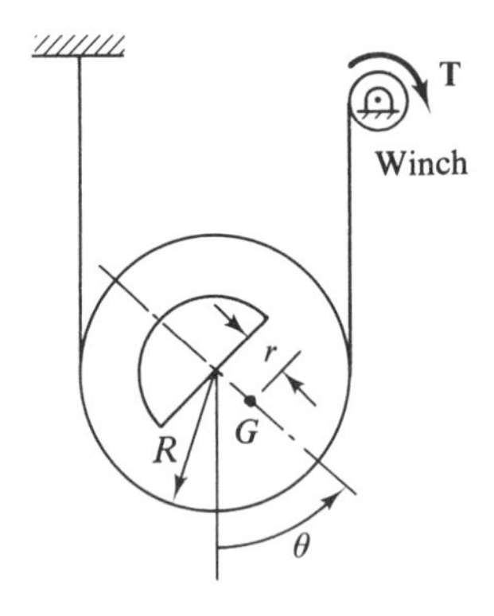

Figure P5-83

5-84 In Problem 5-83 $R=3r$ and $\mu=0.20$. Determine the angle θ, if any, at which the drum starts to slip on the belt.

5-7 FRICTIONAL FORCES ON V BELTS

The belts used for belt drives are frequently V-shaped. When V belts are used, they are in contact with the sides of the grooves of the pulley, and they do not touch the bottom of the grooves. Thus the normal and frictional forces act on the inclined sides of the belt.

The relationship between the two belt tensions *when motion impends* between the belt and pulley is derived in a manner similar to that used for flat belts. Figure 5-11 represents a V belt and pulley in which motion of the belt impends clockwise on the pulley; that is, $\mathbf{T}_L$ is greater than $\mathbf{T}_S$. A free-body diagram of an element of the belt is shown in Fig. 5-12. The weight of the belt and inertial effect due to its motion are neglected. The angle between the two sides of the belt is 2ϕ. From symmetry, the two normal forces $d\mathbf{N}_1$ and $d\mathbf{N}_2$ are equal. The following results are obtained from the equations of equilibrium:

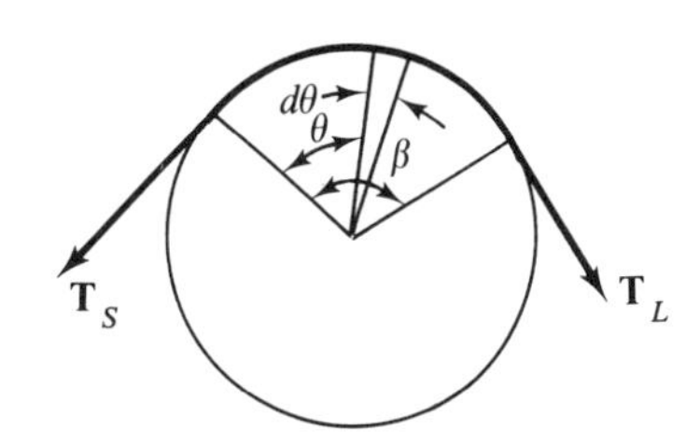

Figure 5-11

$$\sum F_y = dN_1 \sin\phi + dN_2 \sin\phi$$

$$- T\sin\left(\frac{d\theta}{2}\right) - (T + dT)\sin\left(\frac{d\theta}{2}\right) = 0.$$

Since $dN_1 = dN_2$, this equation becomes

$$dN_1 = \frac{T\,d\theta}{2\sin\phi} \qquad \text{(a)}$$

when second-order differentials are neglected in comparison with first-order differentials and $\sin(d\theta/2)$ is replaced by $d\theta/2$.

$$\sum F_x = (T + dT)\cos\frac{d\theta}{2} - T\cos\frac{d\theta}{2} - dF_1 - dF_2 = 0$$

from which

$$dT\cos\frac{d\theta}{2} = dT = dF_1 + dF_2 \qquad \text{(b)}$$

since the angle $(d\theta/2)$ is small and $\cos(d\theta/2)$ is approximately equal to 1. When motion is impending

$$dF_1 = \mu\,dN_1, \qquad dF_2 = \mu\,dN_2 = \mu\,dN_1,$$

and Eq. (b) becomes

$$dT = 2\mu\,dN_1 = \frac{\mu T\,d\theta}{\sin\phi}$$

from Eq. (a).

The last equation can be written

$$\frac{dT}{T} = \frac{\mu}{\sin\phi}\,d\theta,$$

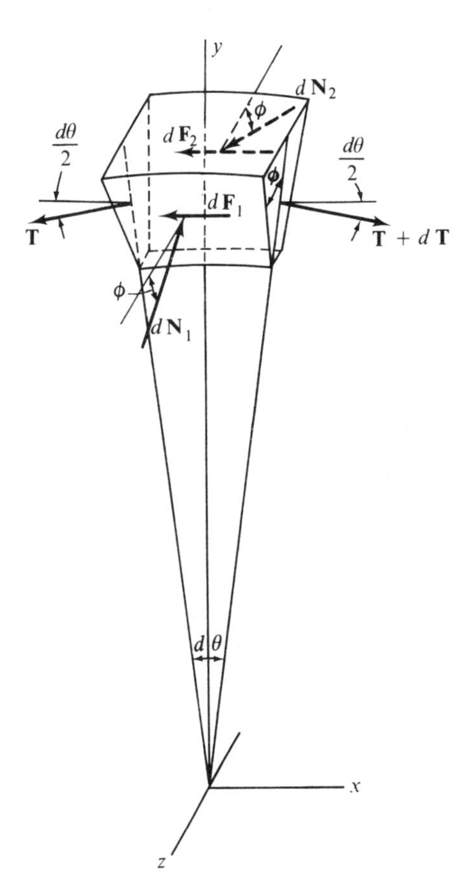

Figure 5-12

and it can be integrated as follows:

$$\int_{T_S}^{T_L} \frac{dT}{T} = \int_0^{\beta} \frac{\mu}{\sin \phi}\, d\theta \ln \frac{T_L}{T_S} = \frac{\mu\beta}{\sin \phi} \tag{5-6}$$

or

$$T_L = T_S e^{\mu\beta/\sin \phi}. \tag{5-7}$$

When the angle ϕ is 90°, Eq. (5-7) reduces to Eq. (5-5) for a flat belt. The angle ϕ is usually between 17 and 20°, although for small pulleys it may be as small as 13°.

PROBLEMS

5-85 A lathe is driven by a V belt passing over a pulley attached to a jack shaft as shown in Fig. P5-85. The resisting torque developed by the lathe while in operation is 30 ft-lb. Determine the minimum tension required in the portion of the belt from A to B if no slippage is to occur between the belt and pulleys. The coefficient of friction between the belt and pulleys is 0.30, and the angle between the side of the belt and the axis of symmetry of the belt section is 19°.

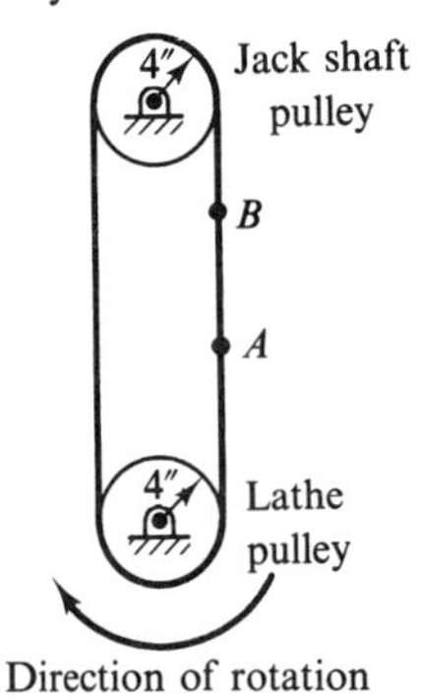

Figure P5-85

***5-86** The belt drive in Fig. P5-86 consists of a flat pulley A, a grooved pulley B, and four V belts. The angles between the sides of the grooves in B are 40°, and the size of the belt and groove is such that the radius of the inside of the belt is 50 mm where it wraps around B. The coefficients of friction are 0.40 between the belt and A and 0.30 between the belt and B.

(a) If pulley A is locked in place and a large couple is applied to B, will the belt slip on pulley A or on pulley B?

(b) The resisting couple acting on A when motion impends is 550 N·m clockwise. Determine the maximum belt tension.

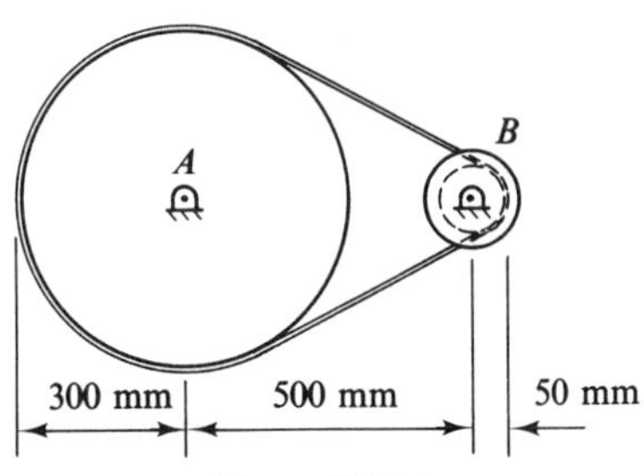

Figure P5-86

5-87 The 2-hp motor in Fig. P5-87 weighs 50 lb and delivers 100 in-lb of torque to pulley A by means of a V belt. The two pulleys have effective diameters (to the point of contact with the belt) of 6 in., and the half-angle between the sides of the belt is 18°. The friction coefficient between the belt and pulleys is 0.30. Determine the minimum distance a to prevent slipping when the rotation is clockwise.

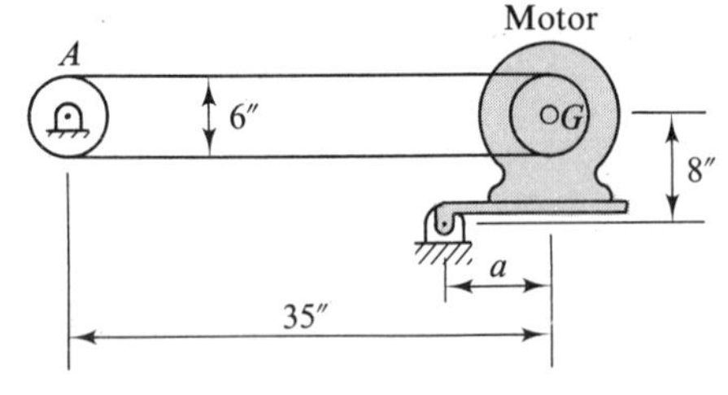

Figure P5-87

5-88 In Problem 5-87 assume that $a = 5$ in. and that the rotation is clockwise. Determine the maximum moment which can be transmitted to pulley A.

5-8 SQUARE-THREADED SCREWS

There are three common types of threads used for power transmission and jackscrews, namely, square, acme, and buttress, as shown in Fig. 5-13. V-Shaped threads are widely used for bolts and other fastenings. A square-threaded screw or a buttress screw can be regarded as an inclined plane wound around a cylinder on which a block

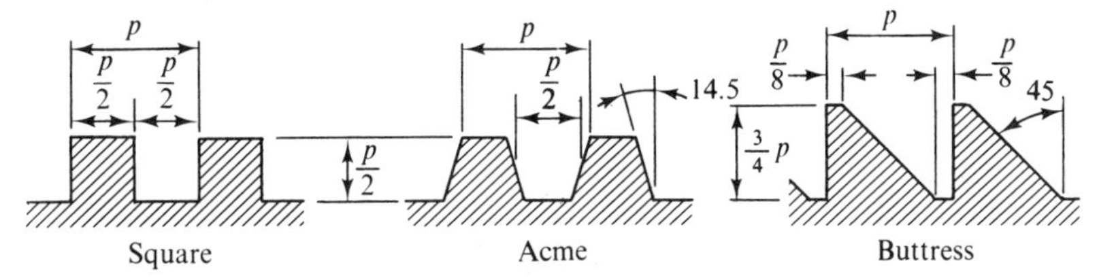

Figure 5-13

is moved by a horizontal force. Acme and V threads require a somewhat more involved analysis than square threads and will not be considered in this text.

The lead, L, *of a screw is the distance the nut will advance in the direction of the axis of the screw in one revolution. The pitch,* p, *is the distance between similar points on adjacent threads.* The jackscrews considered here are single-threaded, and L is equal to p. For multiple-threaded screws, L is equal to mp, where m is the multiplicity of the threads. Figure 5-14(a) shows a portion of a square-threaded jackscrew. The thread A of the screw is represented in the inset figure by a

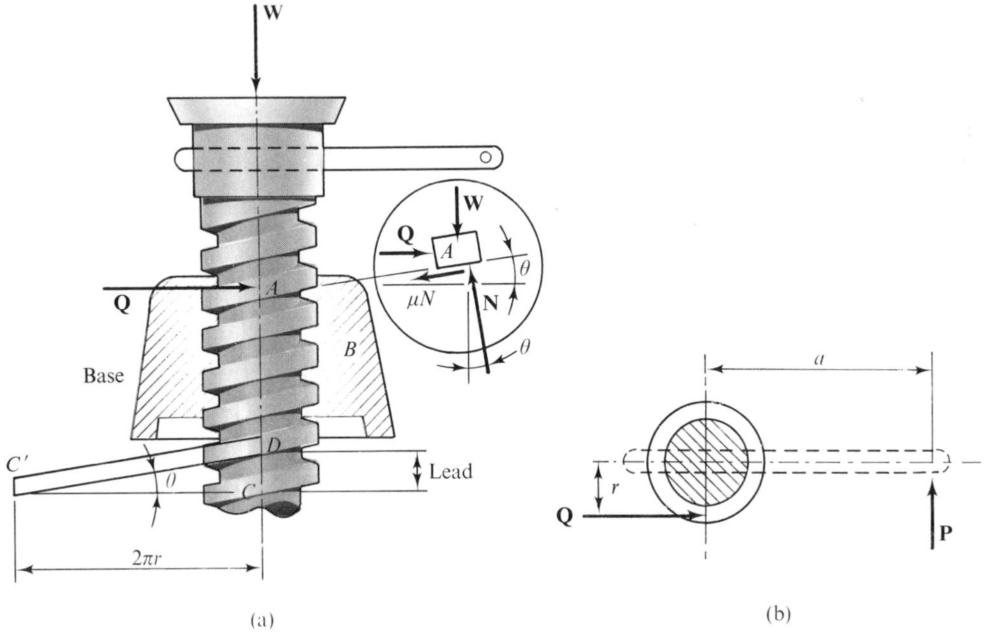

Figure 5-14

block that slides up or down the inclined plane representing the thread of the base B. The horizontal force **Q** is the force which, if applied at the mean radius r of the screw as in Fig. 5-14(b), would be necessary to slide the block up the plane, thus raising the weight **W**. If the weight is to be lowered, the friction shown in the free-body diagram inset in Fig. 5-14(a), and usually the force **Q**, are reversed. By unwinding one turn of the thread from C to D in Fig. 5-14(a), it is seen that the slope of the equivalent inclined plane is equal to the lead divided by the mean circumference of the screw; that is,

$$\tan \theta = \frac{L}{2\pi r} = \frac{p}{2\pi r}. \tag{a}$$

With the angle θ determined from Eq. (a), the force **Q** shown in the free-body diagram inset of Fig. 5-14(a) can be determined for any given load **W** from the equations of equilibrium and friction for block A. After the force **Q** has been determined, the actual force **P** required at the end of the jack handle as shown in Fig. 5-14(b) can be evaluated from the relation for equivalent moments; that is,

$$Pa = Qr. \tag{b}$$

The angle θ is usually small for jackscrews. When the angle is less than the angle of friction ϕ, the jackscrew is *self-locking* and will not turn down by itself under the action of the load; that is, a force must be applied to the handle to lower the load.

When the slope of the plane ($\tan \theta$) is less than 0.1, the approximations $\cos \theta = 1$ and $\sin \theta = \tan \theta$ give results to within sufficient accuracy for most calculations.

EXAMPLE 5-6

The mean diameter of the screw of a square-threaded jackscrew is 2 in., the pitch of the thread is 0.50 in., and the coefficient of static friction for the screw and nut is 0.30. Determine the force which is required at the end of a 15-in. lever to raise a 2000-lb load.

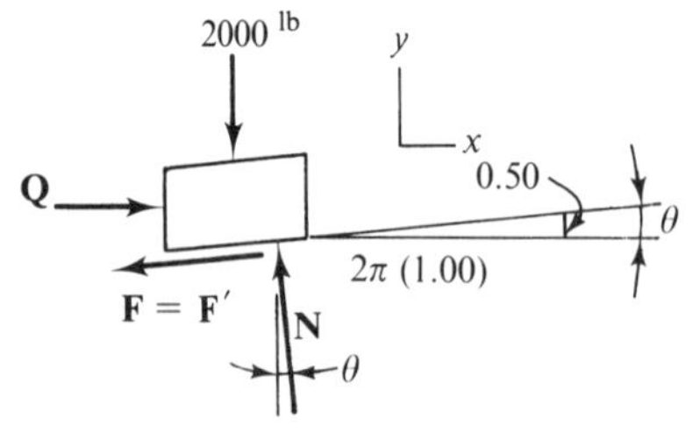

Figure 5-15

SOLUTION

Figure 5-15 is the free-body diagram showing a 2000-lb load on a block which is being pushed up an inclined plane by the horizontal force **Q**. From the figure,

$$\tan \theta = \frac{0.50}{2\pi} = 0.0796,$$

and the approximations

$$\sin \theta = 0.0796 \quad \text{and} \quad \cos \theta = 1.000$$

are justified for slide-rule computations. The equations of equilibrium and friction for the body give the following results:

$$\sum F_y = N \cos \theta - 2000 - F \sin \theta = 0,$$

and

$$F = F' = 0.30N.$$

Therefore

$$N[1.000 - 0.30(0.0796)] = 2000,$$

and

$$\mathbf{N} = 2050 \text{ lb as shown.}$$

$$\sum F_x = Q - N \sin \theta - F \cos \theta = 0$$

from which

$$Q = 2050[0.0796 + 0.30(1.000)]$$

and

$$\mathbf{Q} = 778 \text{ lb as shown.}$$

Equation (b) gives

$$P(15) = 778(1.0)$$

or

$$\mathbf{P} = \underline{51.9 \text{ lb as shown in Fig. 5-14(b).}} \qquad \text{Ans.}$$

PROBLEMS

5-89 The mean diameter of a square-threaded jackscrew is 3.0 in., and there are three threads per inch. The coefficient of friction is 0.10. Determine the magnitude of the force which must be applied to the end of a 20-in. lever to raise a 5-ton load.

5-90 The mean diameter of a jackscrew is 4.0 in., the pitch is 0.40 in., and the coefficient of friction is 0.15. A force of 15 lb at the end of a 16-in. handle is required to lower the load on the jack. Determine the load.

***5-91** The screw of the small press in Fig. P5-91 has a mean diameter of 40 mm and a square thread with a lead of 5.0 mm. The coefficient of friction is 0.30. A 5.0-kN force **F** is to be applied by a 250-N force **P** as shown. Determine the length d of the moment arm of **P**.

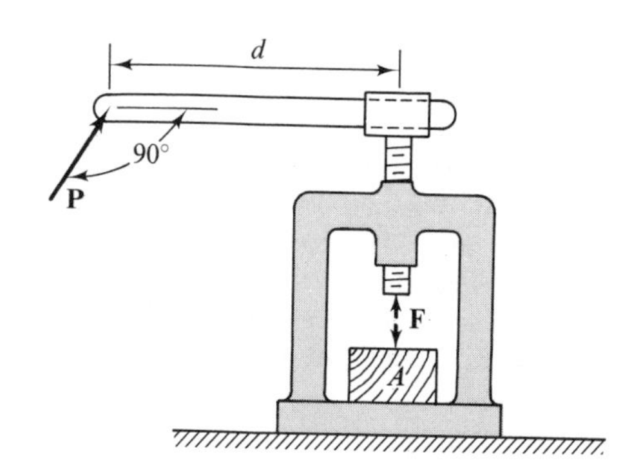

Figure P5-91

5-92 A square-threaded jackscrew has a mean diameter of 2.00 in. and there are five threads per inch. A 50-lb force is required at the end of an 18-in. handle to raise a 2500-lb load. Determine the coefficient of friction between the screw and nut.

5-93 The small bench vise in Fig. P5-93 is used to compress a spring with a force of 300 lb. The guide bar E slides in a slot in the base and is in contact with points A and B when under load. The screw has a mean diameter of $\frac{3}{4}$ in., and there are six threads per inch. The coefficient of friction between the base and both the screw and guide bar is 0.20. Determine the forces of the couple which must be applied to C and D to compress the spring as indicated.

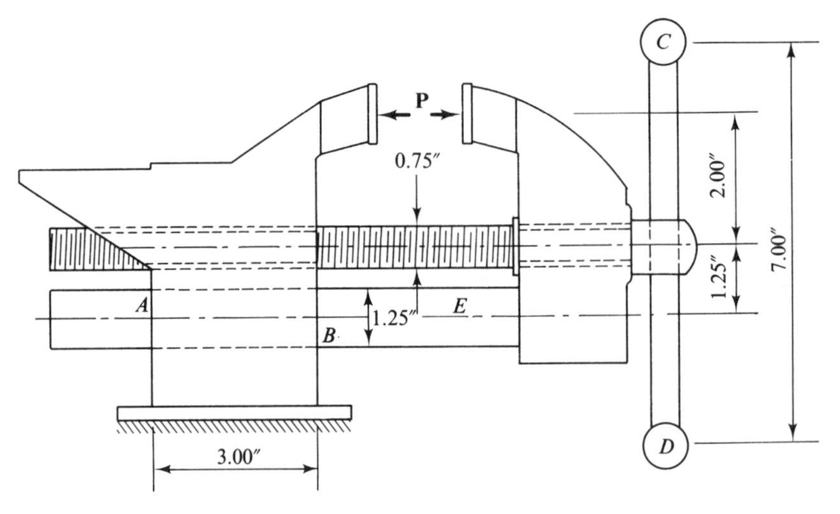

Figure P5-93

5-94 Derive an expression for the ratio of lead to diameter for a square-threaded screw in terms of the coefficient of friction and the ratio of the load to the tangential force (Q in Fig. 5-14) needed to raise the load.

5-95 The screw of the bench jack in Fig. P5-95 has a mean diameter of 0.50 in. with eight threads per inch. The coefficient of friction is 0.15. Determine the couple which must be applied to the screw to lift a 200-lb load when the angle θ is 30°. [*Hint:* See Problem 4-52.]

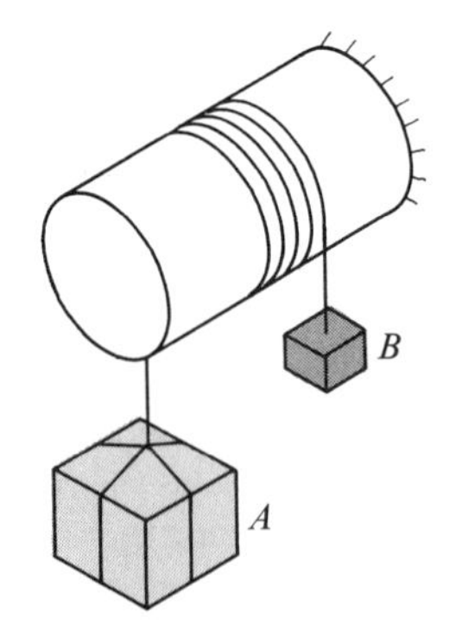

Figure P5-95

5-9 FRICTIONAL MOMENTS ON THRUST BEARINGS AND DISK CLUTCHES

Frictional moments are developed in thrust bearings, such as step or end bearings, collar bearings, and disk-type clutches, as a result of the normal pressure exerted by one plane circular area on another. The end of a shaft with a vertical load **P** is shown in a step or end bearing in Fig. 5-16(a). The center portion of the shaft is frequently cut away to avoid excessive pressure there; the outer portion wears faster because it travels farther each revolution. A collar bearing is shown in Fig. 5-16(b).

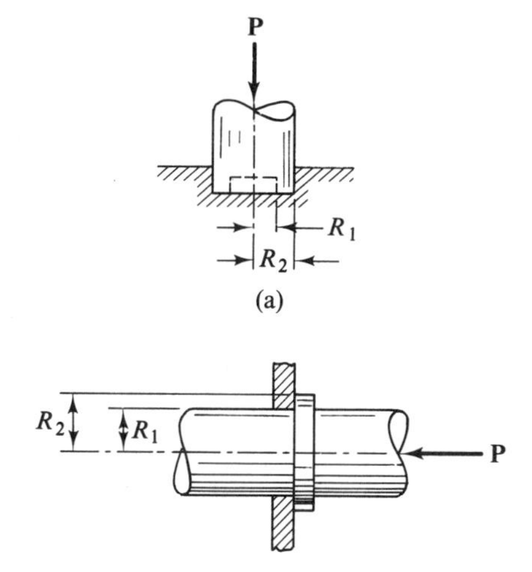

Figure 5-16

The moment of the couple **T** required to produce impending rotation of the hollow shaft in Fig. 5-17 on a step bearing depends on the thrust **P**, the coefficient of friction, and the distribution of the normal pressure over the contact area. The pressure will be assumed to be uniformly distributed over the contact area, although wear of bearings and flexibility of shaft collars usually alter this pressure distribution. The normal force per unit area is equal to the total thrust, **P**, divided by the area of the end of the shaft. The normal force on the element of area is equal to the product of the unit pressure and the element of area; that is,

$$dN = \frac{P}{\pi(R_2^2 - R_1^2)}dA = \frac{P}{\pi(R_2^2 - R_1^2)}r\,d\theta\,dr.$$

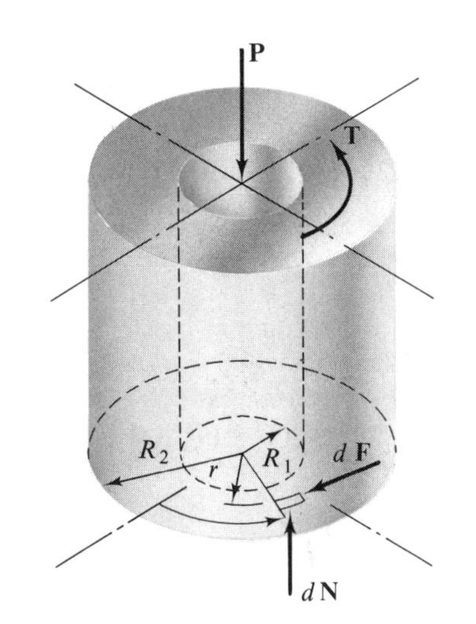

Figure 5-17

The frictional force on the element of area when motion impends is equal to the normal force multiplied by the coefficient of static friction. Thus

$$dF = \mu\,dN = \frac{\mu Pr\,d\theta\,dr}{\pi(R_2^2 - R_1^2)}.$$

The moment of the frictional force on the element of area about the axis of the shaft is

$$dM_0 = r\,dF = \frac{\mu Pr^2\,dr\,d\theta}{\pi(R_2^2 - R_1^2)}.$$

The magnitude of the total frictional moment on the shaft is

$$M_0 = \int_0^{2\pi}\int_{R_1}^{R_2} \frac{\mu Pr^2\,dr\,d\theta}{\pi(R_2^2 - R_1^2)} = \int_0^{2\pi} \frac{\mu P(R_2^3 - R_1^3)\,d\theta}{3\pi(R_2^2 - R_1^2)}$$

$$= \frac{2\mu P}{3}\left(\frac{R_2^3 - R_1^3}{R_2^2 - R_1^2}\right) \tag{5-8}$$

Equation (5-8) gives the frictional moment on the shaft when motion impends which is equal to the torque **T** necessary to produce impending motion. For a solid shaft, the inner radius, R_1, is zero, and

Eq. (5-8) reduces to

$$M_0 = \tfrac{2}{3}\mu P R_2. \qquad (5\text{-}9)$$

When μ represents the coefficient of kinetic friction, these equations give the resisting moment due to kinetic friction.

Multiple-disk clutches and brakes, such as those used for bicycle coaster brakes, make use of the frictional moment occurring between each pair of a series of disks. Half the disks turn with one shaft (the bicycle wheel hub, for example), whereas the other disks, which are alternated with the first set, turn with the second shaft (or are stationary, as with the bicycle brake). The normal forces applied to the outside disks cause a frictional moment to be developed between each pair of disks as given by Eq. (5-8). The total frictional moment is equal to the sum of the moments developed by each pair of disks, and thus a small normal force can cause a relatively large frictional moment to be developed by the clutch or brake.

PROBLEMS

5-96 The three disks in Fig. P5-96 act as a brake on the shaft. Disk A is fastened rigidly to the end of the shaft, disk B is keyed to the shaft but is free to slide on it, and disk C is restrained from turning by the two lugs. The brake is actuated by a spring which forces B and C against A. The outer diameter of the disks is 10.0 in., and the inner diameter of disk C is 2.00 in. The coefficient of friction for the disks is 0.25. Determine the compressive force in the spring necessary to produce a frictional moment of 200 in-lb.

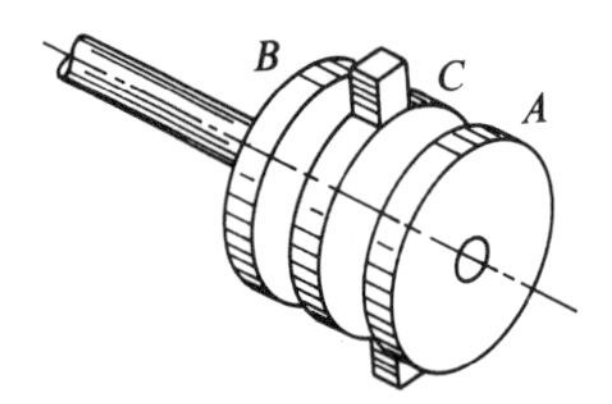

Figure P5-96

5-97 Using methods similar to those presented in Art. 5-9, derive an expression in terms of $\mathbf{P}$, R_1, R_2, ϕ, and μ for the maximum torque $\mathbf{T}$ that can be transmitted by the conical clutch in Fig. P5-97. Assume uniform contact pressure.

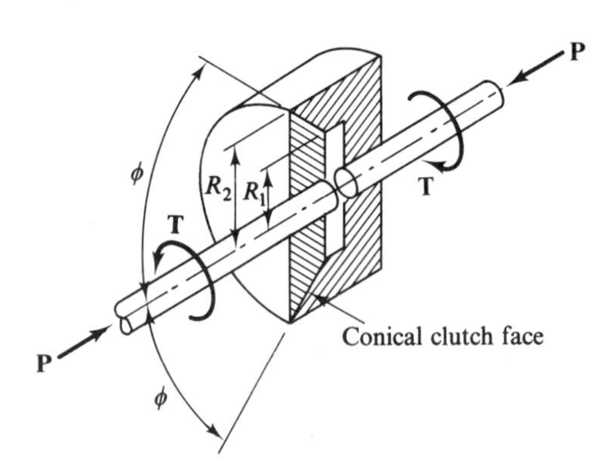

Figure P5-97

5-98 In a step bearing such as the one in Fig. 5-17 the outer section travels farther than the inner section and is thus subject to more wear. Assume that as a result of this wear the normal pressure at any point varies inversely as the distance of the point from the center of the shaft; that is, $p = k/r$, where p is the force per unit area and k is a constant. Derive an expression for the frictional moment similar to Eq. (5-8).

5-10 ROLLING RESISTANCE

When a rigid cylinder or wheel rolls along a rigid horizontal plane with no forces acting on it other than its weight and the reaction of the plane, it will continue to roll indefinitely, according to theory, since there will be no resisting force acting on the cylinder (see Art. 9-11). Experience indicates, however, that the cylinder will slow down and eventually come to rest unless a force is applied to the wheel to keep it rolling. The resistance to the motion occurs because no wheel or plane is *rigid*, so that there will always be some deformation of the wheel and plane as indicated in Fig. 5-18. The horizontal force **P** is the force required to keep the wheel moving with a constant velocity.

The free-body diagram of the wheel in Fig. 5-19 shows the reaction of the plane at an angle θ with the vertical. The equations of equilibrium give

$$P = R \sin \theta \quad \text{and} \quad W = R \cos \theta$$

from which

$$P = W \tan \theta = W\frac{b}{c}.$$

For most materials, the angle θ is found to be small, and the value of c is approximately equal to the radius r. Hence, the preceding equation can be written

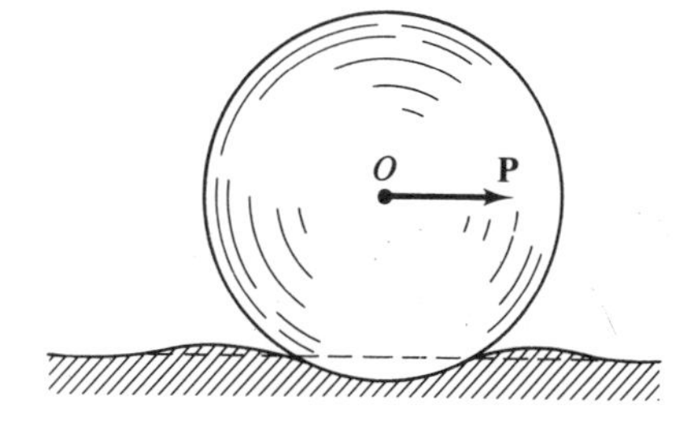

Figure 5-18

$$P = \frac{b}{r} W.$$

The distance b is called the *coefficient of rolling resistance.* Unlike the coefficient of friction, however, b is not a dimensionless constant but has the dimensions of length. Coefficients of rolling resistance for various materials have been obtained experimentally and published, although they are not generally used. For most materials which are fairly rigid, the rolling resistance is usually much smaller than other forces acting on the wheel, and it is therefore neglected.

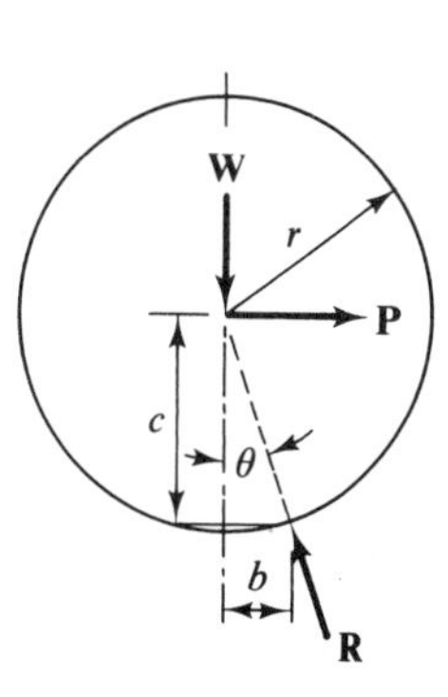

Figure 5-19

6 Second Moments or Moments of Inertia

6-1 INTRODUCTION

In the analysis of the angular motion of rigid bodies, an expression is developed which Euler[1] called the *moment of inertia of mass* or more simply the *moment of inertia of the body*. In determining stresses and deflections in beams and shafts and in locating the center of pressure on submerged areas, an expression is encountered which is similar to the moment of inertia of mass except that it it involves an area rather than a mass. Since the two expressions are similar in form, it is common practice to refer to both as moments of inertia. An area has no mass and hence no inertia; therefore, a more appropriate term would be the *second moment of an area*, as used in the following article. For either a mass or an area, the moment of inertia is a mathematical expression, and it is difficult to obtain a physical concept of the quantity. Because the moments of inertia of areas and masses appear so frequently in analytical equations, it is desirable that the engineer be familiar with and able to determine them.

[1] Leonhard Euler (1707–1783), a noted Swiss mathematician and physicist.

PART A—PLANE AREAS

6-2 DEFINITIONS

The second moment or moment of inertia of an element of area, such as dA in Fig. 6-1, with respect to any axis is defined as the product of the area of the element and the square of the distance from the axis to the element. For example,

$$dI_y = x^2\, dA,$$

where dI_y is the second moment of the element of area dA with respect to the y axis. The sum of the second moments of all the elements of an area is defined as the moment of inertia of the area; that is,

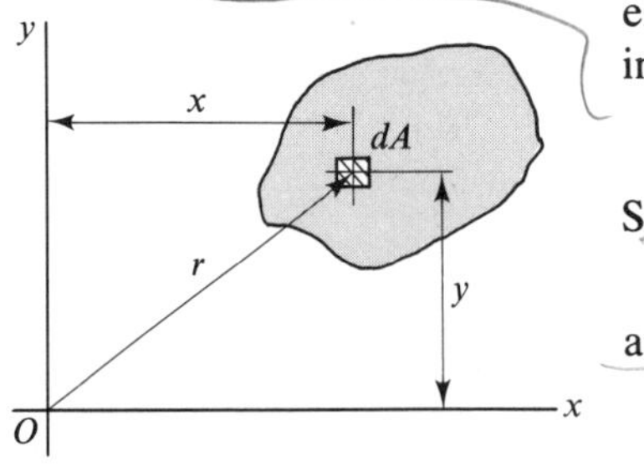

Figure 6-1

$$I_y = \int x^2\, dA.$$

Similarly,

$$dI_x = y^2\, dA$$

and

$$I_x = \int y^2\, dA.$$

When the moment axis is in the plane of the area, the second moment of the area is called the *rectangular moment of inertia*. If the moment axis is perpendicular to the plane of the area, the second moment of the area is called the *polar moment of inertia*. The second moment of the element of area in Fig. 6-1 with respect to an axis through O perpendicular to the plane of the area is

$$dJ_O = r^2\, dA = (x^2 + y^2)\, dA,$$

where dJ_O is the polar moment of inertia of the element with respect to an axis through O perpendicular to the plane of the area. The polar moment of inertia of the area is

$$J_O = \int (x^2 + y^2)\, dA = \int x^2\, dA + \int y^2\, dA = I_y + I_x. \tag{6-1}$$

Thus the polar moment of inertia of an area is equal to the sum of the rectangular moments of inertia with respect to any two perpendicular axes intersecting the polar axis.

The second moment of an area has dimensions of length raised to the fourth power, L^4, and common units of in.^{4}, ft^4, cm^4, and so on.

An element of area is inherently positive. Since the square of the length of its moment arm is also positive, the second moment of an element of area is always a positive quantity. The moment of inertia of an area is the sum of the second moments of the elements of the area; consequently, it is always positive.

6-3 THE PARALLEL-AXIS THEOREM FOR AREAS

After the moment of inertia of an area has been obtained with respect to one axis, it is frequently necessary to determine the moment of inertia of the area with respect to a parallel axis.

The *parallel-axis theorem* (sometimes called the *transfer formula*) provides a convenient relationship between the moments of inertia of an area with respect to two parallel axes, one of which passes through the centroid of the area.

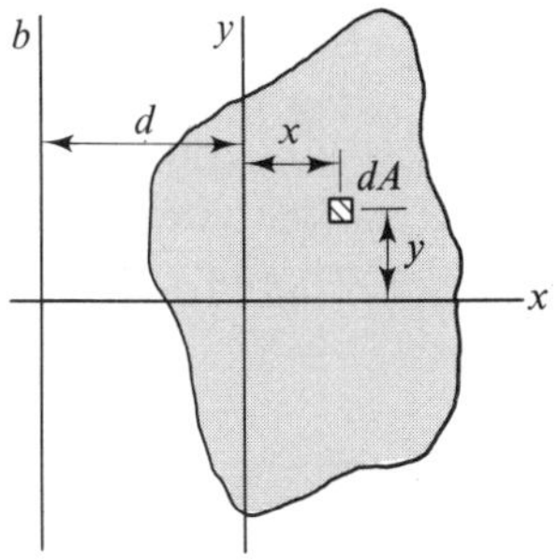

The moment of inertia of the element of area dA in Fig. 6-2 with respect to the b axis is

$$dI_b = (d + x)^2 \, dA,$$

Figure 6-2

where d is the distance between the y axis and the parallel b axis. When the preceding equation is expanded and integrated over the area, the result is

$$\begin{aligned} I_b &= \int (d+x)^2 \, dA \\ &= d^2 \int dA + 2d \int x \, dA + \int x^2 \, dA \\ &= Ad^2 + 2d \int x \, dA + I_y. \end{aligned} \quad \text{(a)}$$

The integral of $x \, dA$ evaluated over the area is the first moment of the area with respect to the y axis. If the y axis passes through the centroid of the area, the expression $\int x \, dA$ is zero. Consequently, Eq. (a) reduces to

$$I_b = Ad^2 + I_c, \quad \text{(6-2)}$$

where I_c is the second moment of the area with respect to an axis through the centroid parallel to the b axis. In a similar manner, it can be shown that

$$J_b = Ad^2 + J_c, \quad \text{(6-3)}$$

where J_b is the polar moment of inertia of the area with respect to an axis through point B, J_c is the polar moment of inertia of the area with respect to an axis through the centroid C, and d is the distance between the centroid and point B.

The parallel-axis theorem can be stated as follows: *The moment of inertia of an area with respect to any axis is equal to the moment of inertia with respect to a parallel axis through the centroid of the area plus the product of the area and the square of the distance between the two axes.*

From the preceding statement, it is apparent that the moment of inertia of any area with respect to an axis through its centroid is less than that for any parallel axis.

PROBLEMS

6-1 The a and b axes in Fig. P6-1 are parallel. The moment of inertia of the shaded area with respect to the a axis is 1005 in.^{4}. Determine the moment of inertia of the area with respect to the b axis.

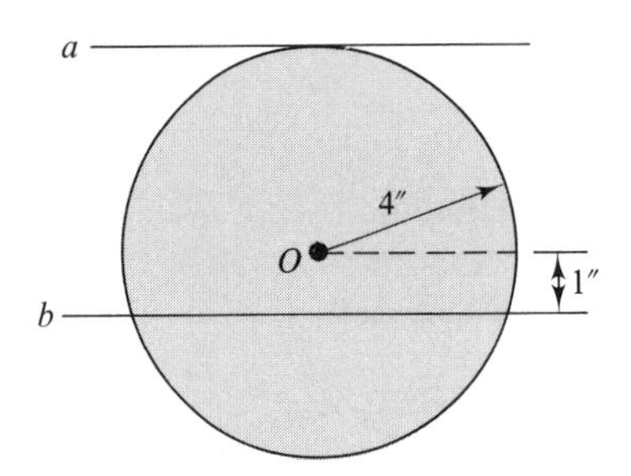

Figure P6-1

6-2 The 6-in.^{2} shaded area in Fig. P6-2 has a moment of inertia of 16 in.^{4} with respect to the a axis and a polar moment of inertia of 24 in.^{4} with respect to an axis through its centroid C. Determine the moment of inertia of the area with respect to the b axis.

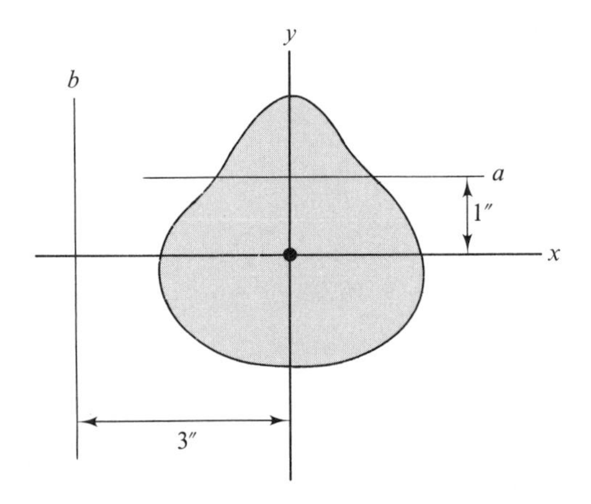

Figure P6-2

6-3 Determine the moment of inertia of the shaded area in Fig. P6-3 with respect to the b axis in terms of the moment of inertia with respect to the a axis, the area A, and the distances d_a and d_b. The centroid of the area is at C.

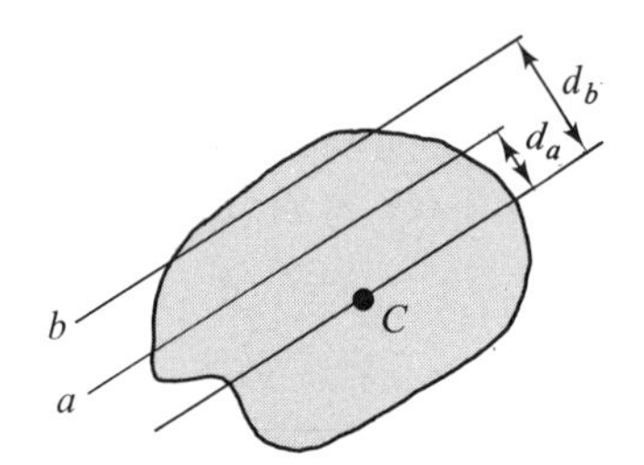

Figure P6-3

***6-4** The 35-cm^2 shaded area in Fig. P6-4 has a polar moment of inertia of 500 cm^4 with respect to an axis through point A. The centroid of the area is at the origin. Determine the polar moment of inertia of the area with respect to an axis through point B.

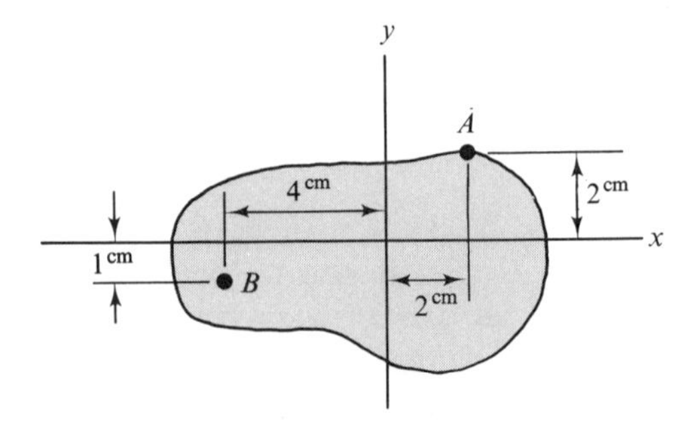

Figure P6-4

6-4 SECOND MOMENTS OF AREAS BY INTEGRATION

In determining the moment of inertia of an area by integration, it is necessary to select an element of area. When all parts of the element selected are the same distance from the moment axis, the moment of inertia of the element of area is obtained directly from the definition of a second moment as the product of the square of this distance and the area of the element. When any other element of area is selected, the moment of inertia of that element must either be known or obtainable from a known result by the parallel-axis theorem. In summary, the element of area chosen should satisfy one of the following requirements:

1. All parts of the element of area are the same distance from the moment axis.
2. The second moment of the element of area with respect to the moment axis is known.
3. The second moment of the element of area with respect to an axis through the centroid of the element parallel to the moment axis is known, and both the area of the element and the distance between the parallel axes are known.

When selecting an element of area, the choice between double and single integration should be made. If double integration is used, all parts of the element will always be the same distance from the moment axis. The expression for the moment of inertia of the element can usually be set up more readily by the use of double integration, whereas the limits can be selected with less possibility of error with single integration. In general, the choice between single and double integrals is a matter of either personal preference or previous training.

In some problems there will also be a choice of coordinate systems. For circular areas or portions thereof, polar coordinates are frequently advantageous.

The following examples illustrate the procedure for determining second moments of plane areas by integration.

EXAMPLE 6-1

Determine the second moment of a rectangular area with respect to an axis through the centroid parallel to the base of the rectangle.

SOLUTION

The rectangular area is shown in Fig. 6-3 with the x axis as the moment axis. The element of area is selected as indicated in Fig. 6-3, and all necessary dimensions are on the figure. The following sequences of steps, similar to those listed in Art. 3-5, is recommended:

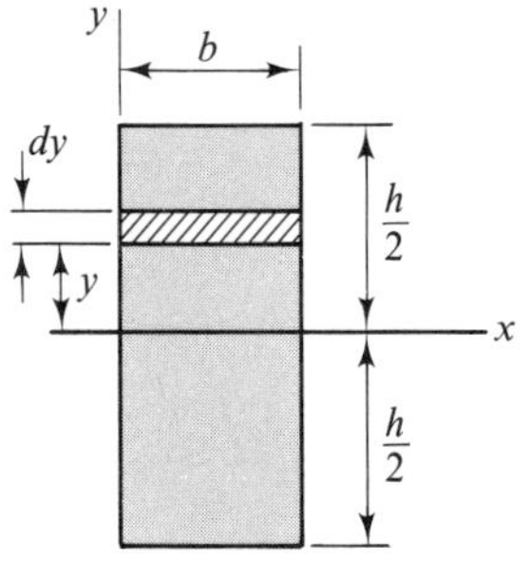

$$dA = b\,dy,$$

$$dI_x = y^2\,dA = y^2 b\,dy,$$

and

$$I_x = b\int_{-h/2}^{h/2} y^2\,dy = \left[\frac{by^3}{3}\right]_{-h/2}^{h/2}$$

$$= \frac{b}{3}\left[\frac{h^3}{8} - \left(\frac{-h^3}{8}\right)\right] = \underline{\frac{bh^3}{12}}. \qquad \text{Ans.}$$

Figure 6-3

EXAMPLE 6-2

Determine the moment of inertia of a circular area of radius R with respect to a diameter of the circle.

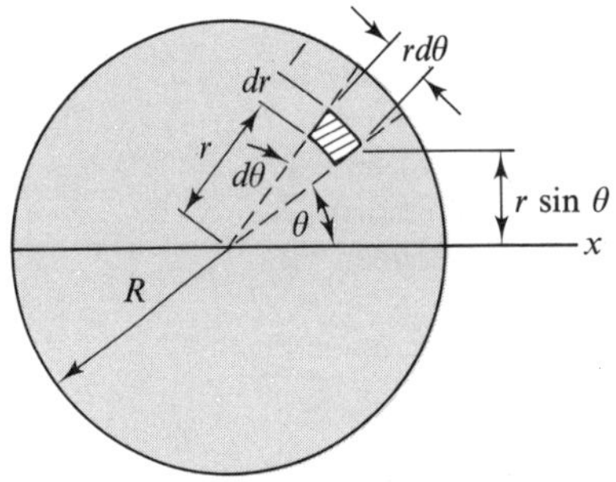

SOLUTION

A circular area of radius R is shown in Fig. 6-4. The x axis is the moment axis. Double integration and polar coordinates are convenient for this problem, although rectangular coordinates could also be used. The element is indicated with the necessary dimensions in the figure. The dashed lines indicate that the first integration is to be performed with respect to r to complete the sector of the circle.

Again it is recommended that the expressions for the element of area and the second moment of the element of area be written first as follows:

Figure 6-4

$$dA = dr(r\,d\theta),$$

$$dI_x = (r\sin\theta)^2\,dA = r^3\sin^2\theta\,dr\,d\theta,$$

and

$$I_x = \int_0^{2\pi}\int_0^R r^3\sin^2\theta\,dr\,d\theta = \int_0^{2\pi}\left[\frac{r^4}{4}\right]_0^R \sin^2\theta\,d\theta$$

$$= \frac{R^4}{4}\int_0^{2\pi}\sin^2\theta\,d\theta = \frac{R^4}{4}\int_0^{2\pi}\frac{1-\cos 2\theta}{2}\,d\theta$$

$$= \frac{R^4}{4}\left[\frac{\theta}{2} - \frac{\sin 2\theta}{4}\right]_0^{2\pi} = \underline{\frac{\pi R^4}{4}}. \qquad \text{Ans.}$$

EXAMPLE 6-3

Determine the moment of inertia of the area bounded by the line $x = 9a$ and the parabola $y^2 = 4ax$ with respect to the x axis.

SOLUTION

The element of area is selected as shown in Fig. 6-5, and all necessary dimensions are included. The element is a rectangle of height $2y$ and width dx, and the x axis is the centroidal axis of the rectangle. Thus, from Example 6-1, the moment of inertia of the element is

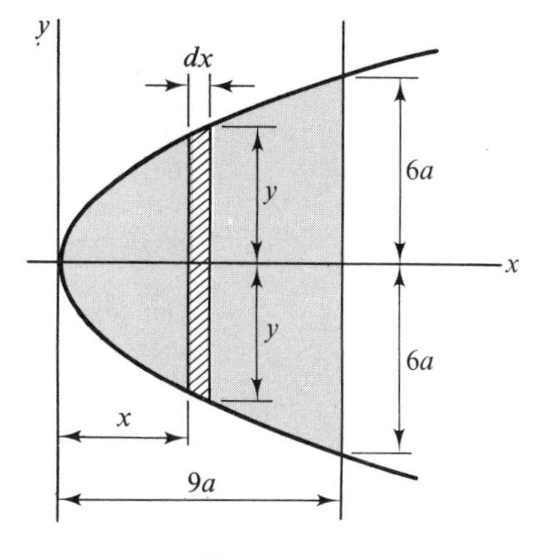

Figure 6-5

$$dI_x = \frac{bh^3}{12} = \frac{dx}{12}(2y)^3 = \frac{2}{3}y^3\,dx.$$

Also

$$y = 2a^{1/2}x^{1/2}.$$

Therefore

$$dI_x = \frac{2}{3}8a^{3/2}x^{3/2}\,dx$$

and

$$I_x = \frac{16}{3}a^{3/2}\int_0^{9a} x^{3/2}\,dx = \frac{16}{3}a^{3/2}\left[\frac{2}{5}x^{5/2}\right]_0^{9a}$$
$$= \underline{518a^4}. \qquad \text{Ans.}$$

EXAMPLE 6-4

Determine the second moment of the area in Fig. 6-6 with respect to the x axis. In the equation of the curve, $a = 1$ in. and $p = 1$ rad per in.

SOLUTION

In this problem, either a double integral or a single integral with a vertical element of area can be used advantageously. A single integral with a horizontal element is not used because $y = a \sin px$ has an infinite number of values of x for a given value of y. The shaded element of area selected is shown in Fig. 6-6, and all necessary dimensions are included. The equations for the element of area and the necessary moments of inertia follow:

$$dA = y\,dx = a \sin px\,dx,$$

$$dI_c = \frac{bh^3}{12} = \frac{dx(y)^3}{12} = \frac{1}{12}a^3 \sin^3 px\,dx,$$

where the c axis is parallel to the x axis and passes through the centroid of the element. The moment of inertia of the element about the x axis is

$$dI_x = dI_c + (dA)d^2$$
$$= \frac{1}{12}a^3\sin^3 px\,dx + a \sin px\,dx\left(\frac{y}{2}\right)^2 = \frac{1}{3}a^3\sin^3 px\,dx$$

and

$$I_x = \frac{a^3}{3}\int_0^{\pi/p} \sin^3 px\,dx$$
$$= \frac{a^3}{3}\int_0^{\pi/p}(1-\cos^2 px)\sin px\,dx$$
$$= \frac{a^3}{3p}\left[-\cos px + \frac{\cos^3 px}{3}\right]_0^{\pi/p} = \frac{4a^3}{9p}.$$

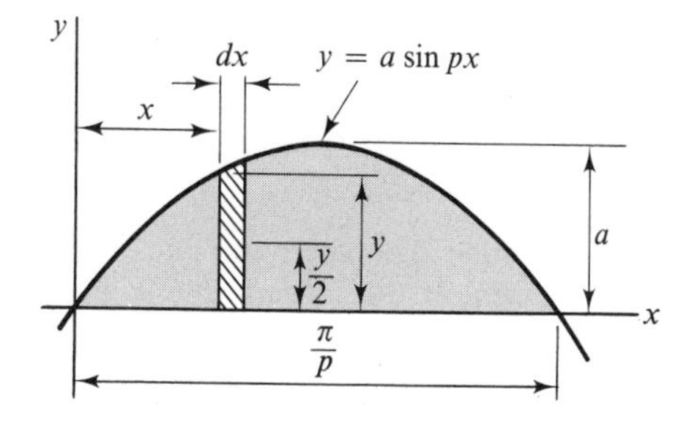

Figure 6-6

When $a = 1$ in. and $p = 1$ rad per in., the second moment becomes

$$I_x = \frac{4(1)^3}{9(1)} = \underline{0.444 \text{ in.}^4} \qquad \text{Ans.}$$

EXAMPLE 6-5

Determine the polar moment of inertia of a hollow circular area of inner radius R_1 and outer radius R_2 with respect to the axis through the center of the area.

SOLUTION

The shaded element of area selected is shown in Fig. 6-7 with all necessary dimensions. The polar moment of inertia can be obtained as follows:

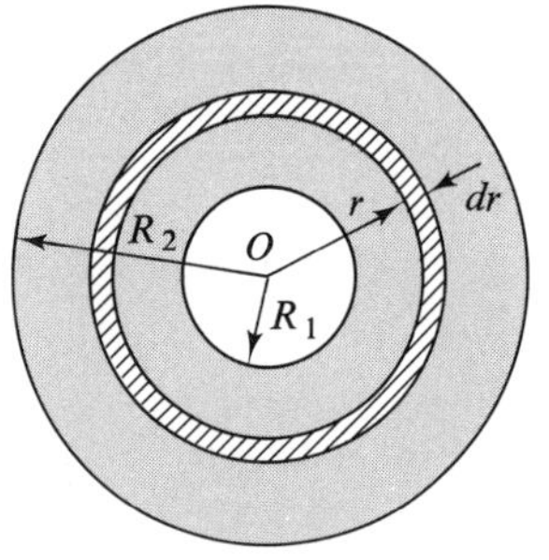

Figure 6-7

$$dA = 2\pi r\, dr$$

$$dJ_O = r^2\, dA = 2\pi r^3\, dr$$

$$J_O = 2\pi \int_{R_1}^{R_2} r^3\, dr = \frac{\pi}{2}\Big[r^4\Big]_{R_1}^{R_2} = \underline{\frac{\pi}{2}(R_2^4 - R_1^4)}. \quad \text{Ans.}$$

Sometimes it is desirable to express a second moment of an area in terms of the area. In this example,

$$A = 2\pi \int_{R_1}^{R_2} r\, dr = \pi\Big[r^2\Big]_{R_1}^{R_2} = \pi(R_2^2 - R_1^2).$$

A convenient way to express J_O as a function of A is to multiply J_O by

$$1 = \frac{A}{A} = \frac{A}{\pi(R_2^2 - R_1^2)}.$$

Therefore

$$J_O = \frac{\pi}{2}(R_2^4 - R_1^4)\frac{A}{\pi(R_2^2 - R_1^2)} = \underline{\frac{A}{2}(R_2^2 + R_1^2)}. \quad \text{Ans.}$$

PROBLEMS

6-5 Determine the moment of inertia of a general triangular area of altitude h and base b with respect to an axis through the centroid parallel to the base.

6-6 The equation of the curve in Fig. P6-6 is $y = 2x^2$. Determine the polar moment of inertia of the shaded area with respect to an axis through the origin.

6-7 Determine by integration the polar moment of inertia of the shaded area in Fig. P6-7 with respect to an axis through the origin.

***6-8** The equation of the curve in Fig. P6-8 is $xy = 4$. Determine the moment of inertia of the shaded area with respect to
(a) The y axis.
(b) The x axis.

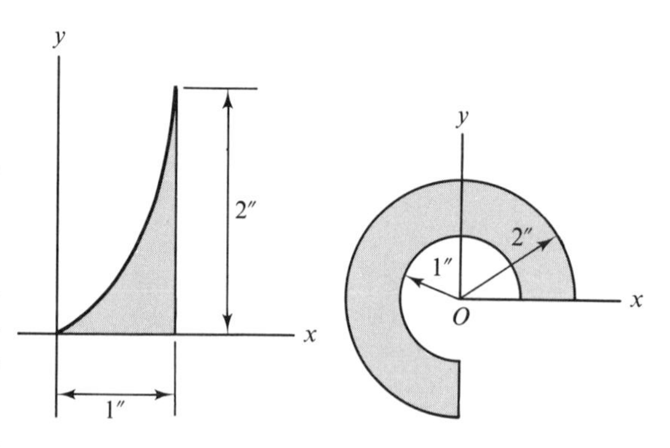

Figure P6-6

Figure P6-7

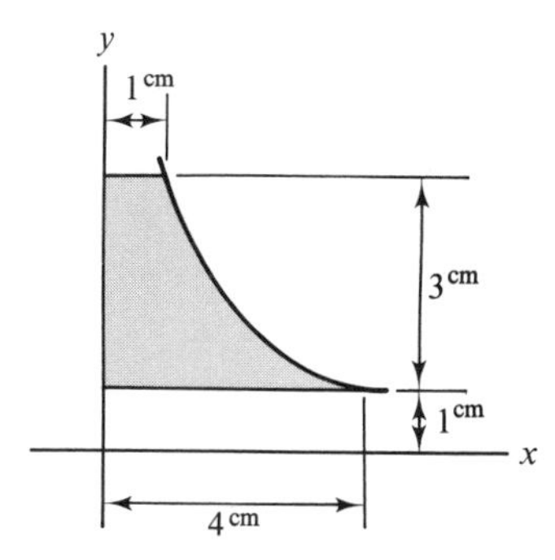

Figure P6-8

*6-9 Determine the moment of inertia of the shaded area of Problem 6-8 with respect to the line $y = 1$ cm.

6-10 Determine the moment of inertia of the shaded area in Fig. P6-10 with respect to the a axis. The equation of the curve is $y^2 = 4x$.

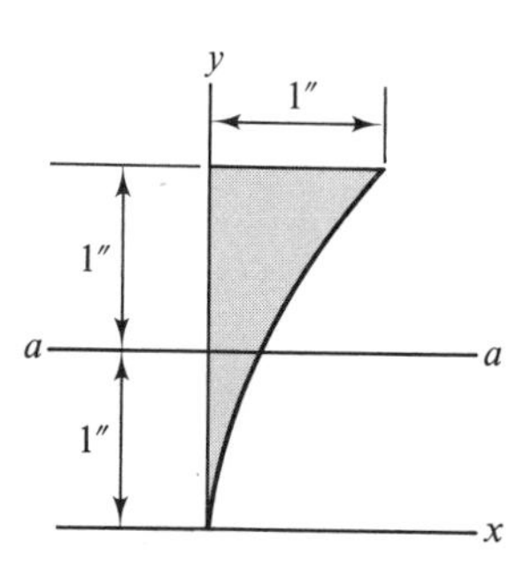

Figure P6-10

6-11 Determine the second moment of the area bounded by the curve $y = x^2$, the line $2x + y = 8$, and the x axis with respect to the x axis. All lengths are in inches.

6-12 Determine the polar moment of inertia of the area bounded by one loop of the curve $r^2 = \cos\theta$ (r is in inches) with respect to an axis through the origin.

6-13 Determine the second moment of the shaded area in Fig. P6-13 with respect to the x axis. The equation of the curve is $y = \cos x$ with lengths measured in inches.

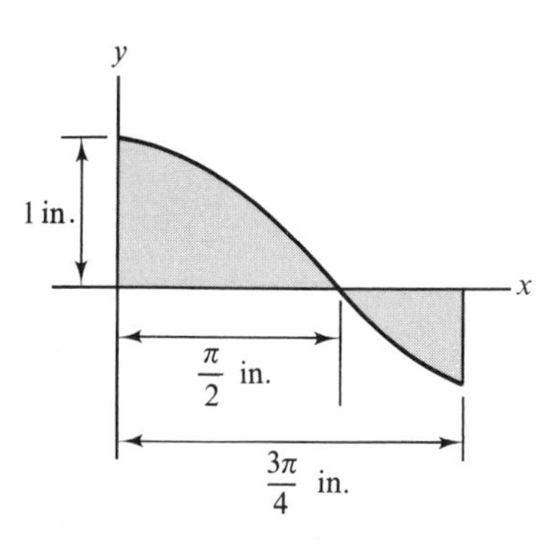

Figure P6-13

***6-14** Determine the moment of inertia of the shaded area in Fig. P6-14 with respect to the x axis. The equation of the curve is $y = x^2 - 2x$.

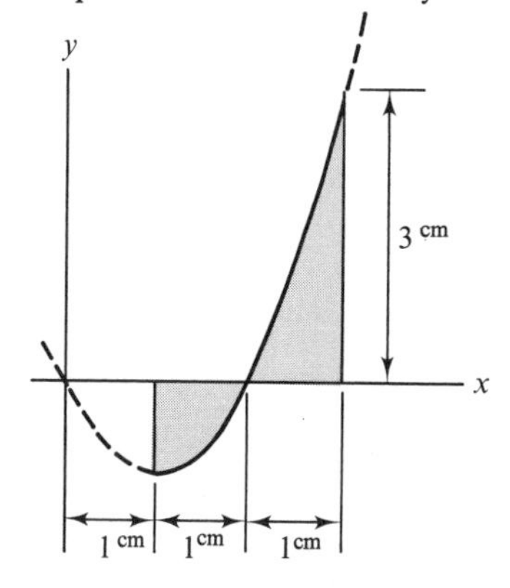

Figure P6-14

6-15 Determine the moment of inertia of the shaded area in Fig. P6-15 with respect to the y axis. The equation of the curve is $x^2 - 4x = 2y$, with x and y in inches.

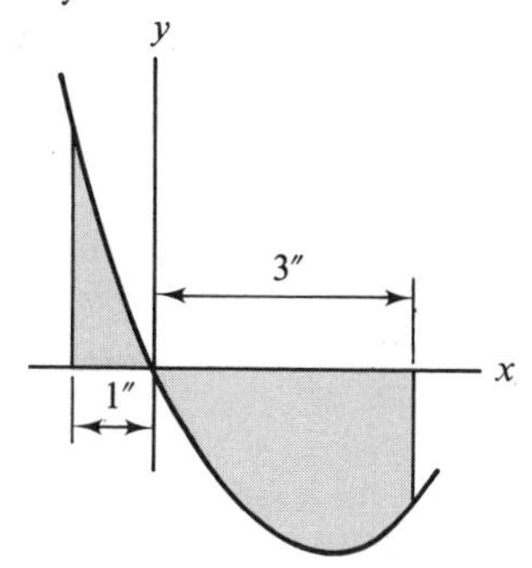

Figure P6-15

6-16 Determine the second moment of the shaded area in Fig. P6-16 with respect to the a axis. The equation of the curve is $x^2=12y$.

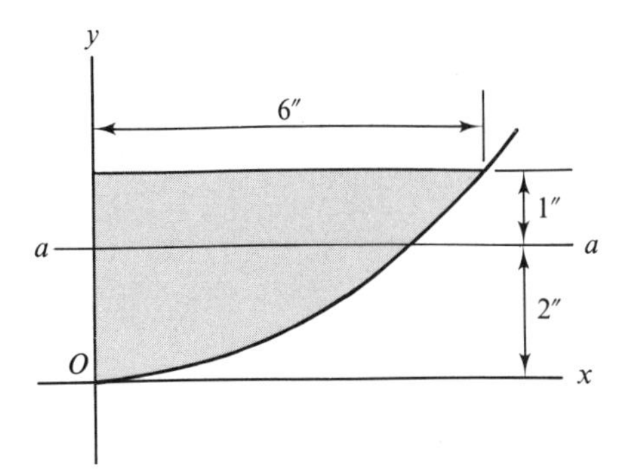

Figure P6-16

6-17 Determine the second moment of the shaded elliptical area in Fig. P6-17 with respect to the x axis. The equation of the curve is $b^2x^2+a^2y^2=a^2b^2$.

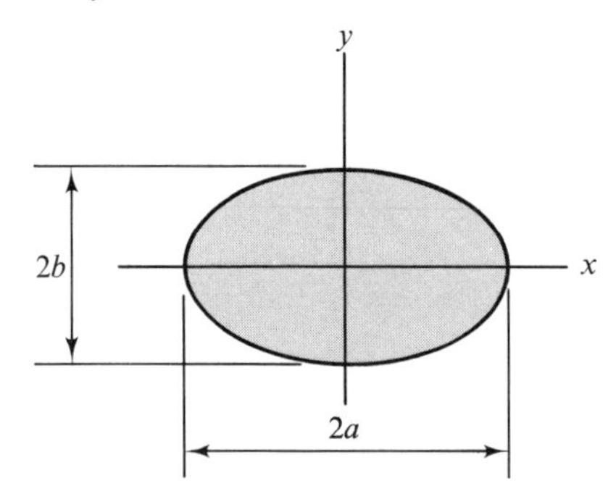

Figure 6-17

6-18 The equation of the curve in Fig. P6-18 is $b^3y^2=a^2x^3$. Determine the second moment of the shaded area with respect to the x axis.

6-19 Determine the second moment of the shaded area in Fig. P6-19 with respect to the y axis. The equation of the curve is $8y=x^3+2x^2+4x$.

6-20 Determine the moment of inertia of the shaded area in Fig. P6-20 with respect to the a axis. The equation of the curve is $x^2=y$.

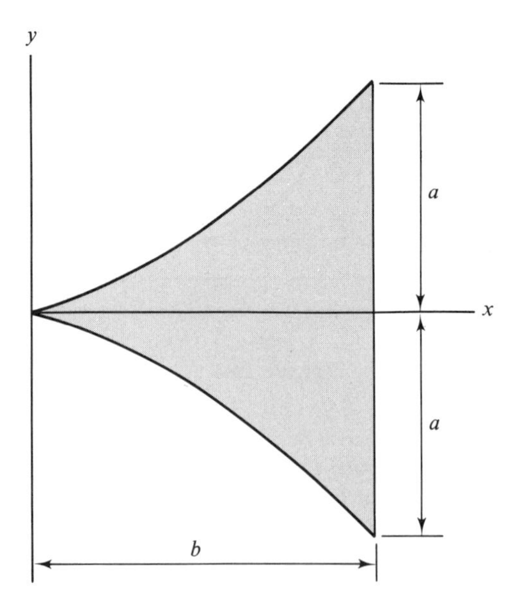

Figure P6-18

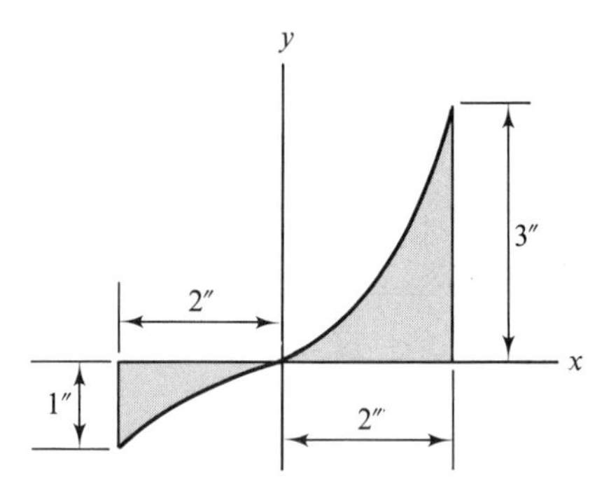

Figure P6-19

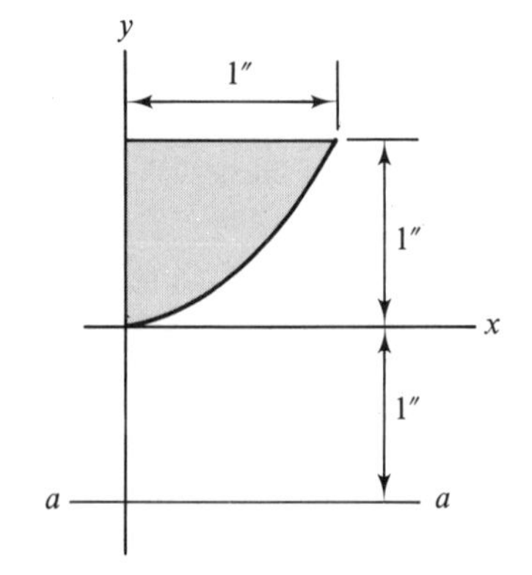

Figure P6-20

6-5 RADIUS OF GYRATION OF AREAS

It is frequently desirable to express the moment of inertia of an area as a function of the area and a length. Since the second moment of an area has dimensions of length to the fourth power, it can be written as the area multiplied by a length squared. *The radius of gyration of an area with respect to an axis is defined as the length which, when squared and multiplied by the area, will give the moment of inertia of the area with respect to the given axis.* This definition is expressed mathematically in the following equations:

$$I_b = Ak_b^2 \quad \text{and} \quad J_z = Ak_z^2.$$

The radius of gyration is not the distance from the reference axis to some specific fixed point in the area (such as the centroid), but it is a useful mathematical concept. *The radius of gyration for any axis is always greater than the distance from the axis to the centroid.* The proof of this statement is apparent from the parallel-axis theorem. In Fig. 6-8 the c axis is a centroidal axis, and the following equations are valid:

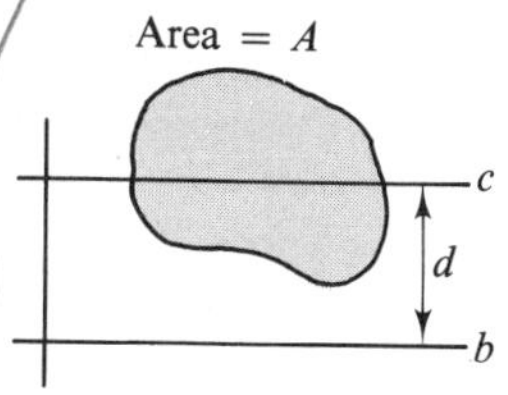

$$I_b = I_c + Ad^2,$$

$$k_b^2 A = k_c^2 A + Ad^2,$$

$$k_b^2 = k_c^2 + d^2,$$

Area = A

Figure 6-8

and

$$k_b = (d^2 + k_c^2)^{1/2}.$$

Therefore, k_b must always be greater than d. If the b axis is moved up 1 in. in Fig. 6-8, k_b will be decreased by an amount somewhat less than 1 in., indicating that the radius of gyration is not the distance to the same point in the area from different axes.

The radius of gyration of an area can be considered as the distance from a given axis at which the entire area can be conceived to be concentrated without changing the second moment of the area about the given axis. When the area is so concentrated, its first moment with respect to the given axis will not be the same as the first moment of the actual area. Usually, however, there is no advantage in associating a radius of gyration with any particular distance in an area.

The radius of gyration of an area is used for convenience in many problems in mechanics, as in the formulas for the determination of the strength of columns.

PROBLEMS

***6-21** Determine the radius of gyration of the shaded area in Fig. P6-21 with respect to the x axis. The equation of the curve is $x^2 = 4y$.

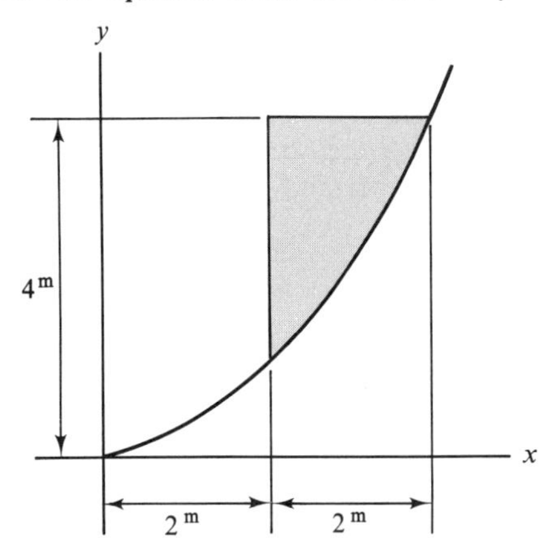

Figure P6-21

6-22 The equation of the curve in Fig. P6-22 is $xy^2 = 4$. Determine the radius of gyration of the shaded area with respect to

(a) The x axis.
(b) The y axis.

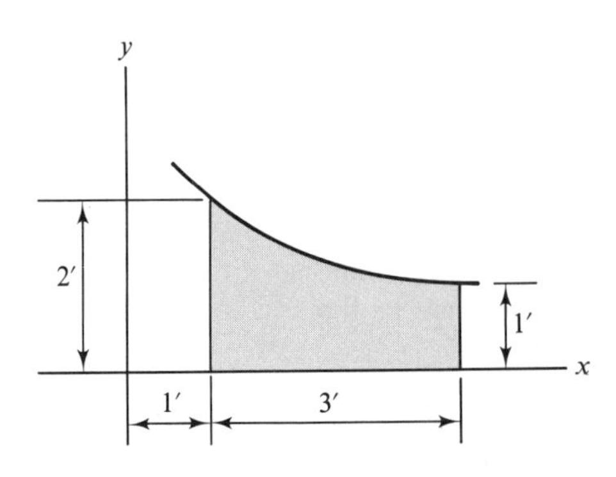

Figure P6-22

6-23 Determine the polar radius of gyration of the shaded area in Fig. P6-23 with respect to the axis through the origin. The equation of the curve is $y^2 = x$.

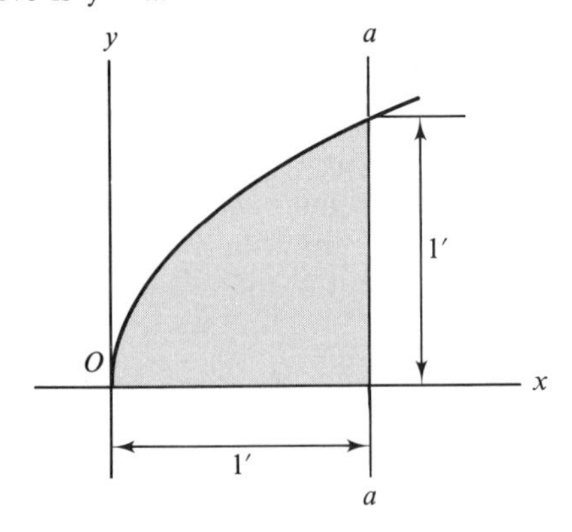

Figure P6-23

6-24 Determine the radius of gyration of the shaded area in Fig. P6-23 with respect to the a axis.

6-25 Determine the radius of gyration of the shaded area in Fig. P6-25 with respect to the a axis.

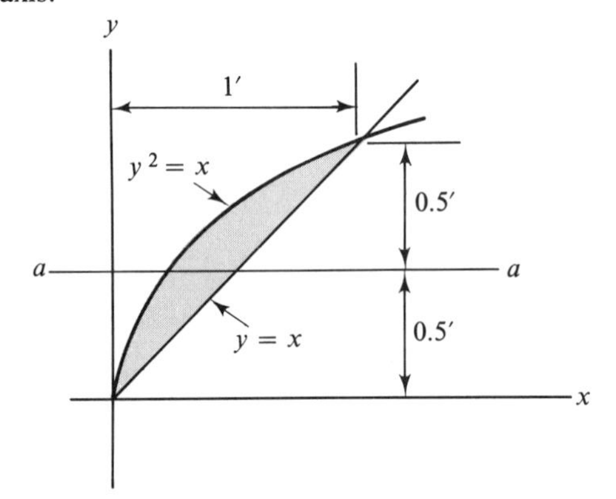

Figure P6-25

6-6 MOMENTS OF INERTIA OF COMPOSITE AREAS

A *composite area* consists of two or more simple areas, such as rectangles, triangles, and circles. The cross-sectional areas of standard structural elements, such as channels, I beams, and angles, are

Moments and Products of Inertia of Common Geometric Areas

Geometric area	Moment of inertia	Radius of gyration
Rectangle 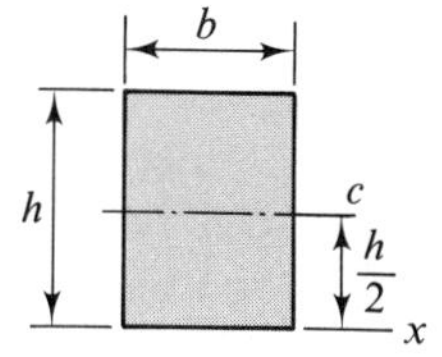	$I_c = \dfrac{bh^3}{12}$ $I_x = \dfrac{bh^3}{3}$	$k_c = \dfrac{h}{2\sqrt{3}} = 0.2887h$ $k_x = \dfrac{h}{\sqrt{3}} = 0.5774h$
Any triangle 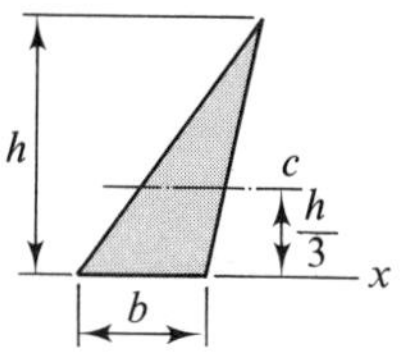	$I_c = \dfrac{bh^3}{36}$ $I_x = \dfrac{bh^3}{12}$	$k_c = \dfrac{h}{3.\sqrt{2}} = 0.2357h$ $k_x = \dfrac{h}{\sqrt{6}} = 0.4082h$
Circle 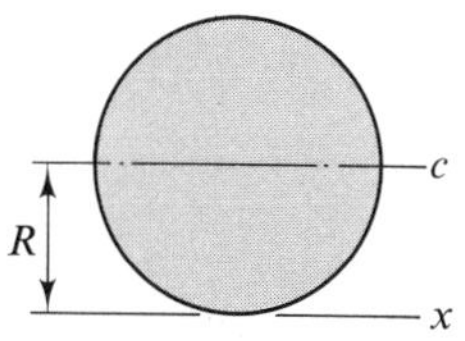	$I_c = \dfrac{\pi R^4}{4}$ $I_x = \dfrac{5\pi R^4}{4}$	$k_c = \dfrac{R}{2} = 0.5000R$ $k_x = \dfrac{R\sqrt{5}}{2} = 1.118R$
Quarter circle 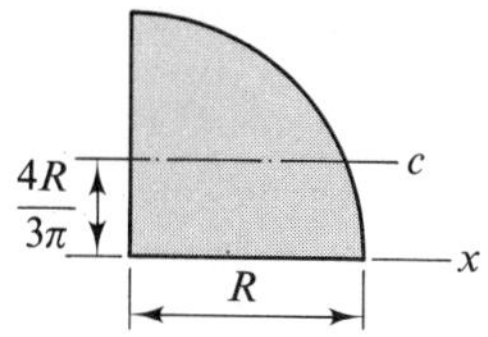	$I_c = 0.0549R^4$ $I_x = \dfrac{\pi R^4}{16} = 0.1963R^4$	$k_c = 0.2643R$ $k_x = \dfrac{R}{2} = 0.5000R$
Ellipse 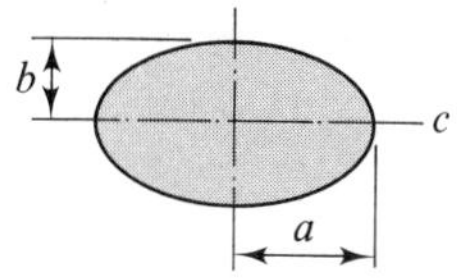	$I_c = \dfrac{\pi ab^3}{4}$	$k_c = \dfrac{b}{2}$

frequently included in composite areas. The moment of inertia of a composite area with respect to any axis is equal to the sum of the moments of inertia of its component areas with respect to the same axis. When an area, such as a hole, is removed from a larger area, its moment of inertia is subtracted from the moment of inertia of the larger area to obtain the net moment of inertia.

It is unnecessary to use integration for computing the moment of inertia of a composite area provided the moments of inertia of the component parts are readily available. It is advisable to have formulas available for common areas frequently encountered in routine work. A representative tabulation is included for ready reference. Extensive tables listing values of moments of inertia of the cross-sectional areas of various structural shapes will be found in engineering handbooks and in data books prepared by industrial organizations such as the American Institute of Steel Construction and the Aluminum Company of America. A few of these values are listed for information and for use in the problems.

The following examples illustrate the use of formulas and handbook data in determining moments of inertia of composite areas.

EXAMPLE 6-6

Determine the moment of inertia of the shaded area in Fig. 6-9 with respect to the x axis.

SOLUTION

The composite area can be divided into a 10- by 6-cm rectangle A and a semicircle B with the triangular area C removed. The locations of the centroidal axes of three areas are indicated by the dimensions to the right of the area. The position of the centroid of area B was determined from Example 3-4 and that of area C from Problem 3-7. These

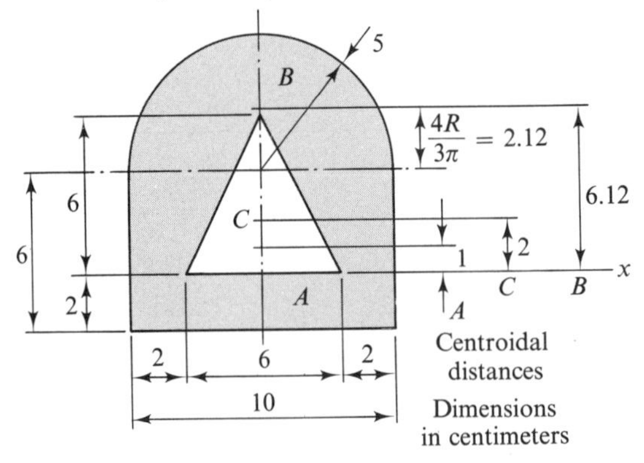

Figure 6-9

values are also given in the accompanying tabulation of moments of inertia of common geometrical areas. The moments of inertia of the three areas with respect to the x axis can be obtained as follows:

For area A:

$$I_x = I_c + Ad^2 = \frac{bh^3}{12} + Ad^2$$

$$= \frac{10(6)^3}{12} + 10(6)(1)^2 = 240\ \text{cm}^4.$$

For area B: I_c is not known directly and must be computed. By symmetry, the moment of inertia of the semicircle with respect to the diameter is one-half the moment of inertia of a full circle. Thus

$$I_{\text{diam}} = I_c + Ad^2$$

from which

$$I_c = \frac{\pi R^4}{8} - Ad^2$$

$$= \frac{\pi(5)^4}{8} - \frac{\pi(5)^2}{2}(2.12)^2 = 68\ \text{cm}^4$$

and

$$I_x = I_c + Ad^2$$

$$= 68 + \frac{\pi(5)^2}{2}(6.12)^2 = 1539\ \text{cm}^4.$$

For area C:

$$I_x = \frac{bh^3}{36} + Ad^2 = \frac{6(6)^3}{36} + \frac{6(6)}{2}(2)^2 = 108\ \text{cm}^4.$$

For the composite area:

$$I_x = 240 + 1539 - 108 = \underline{1671\ \text{cm}^4.} \qquad \text{Ans.}$$

EXAMPLE 6-7

The column in Fig. 6-10 is constructed of two 12-in., 25.0-lb American Standard channels and a $\frac{1}{2}$-in. by 14.00-in. steel plate. Determine the moments of inertia and radii of gyration of the cross-sectional area with respect to axes through the centroid parallel and perpendicular to the plate. Neglect the effect of the lattice.

SOLUTION

Figure 6-10(b) shows the section with significant dimensions. The centroidal dimensions are placed above and to the right of the sketch.

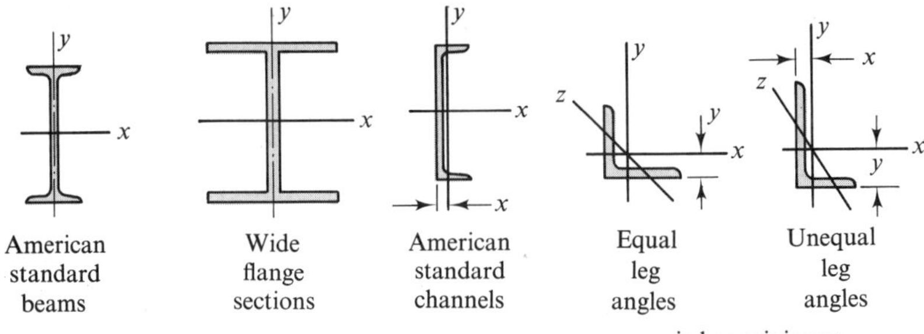

Nominal size in.	Weight per ft lb	Area of section sq in.	Flange width in.	x axis		y axis		
				I in.^{4}	k in.	I in.^{4}	k in.	x in.
American Standard Beams								
$10\times4\frac{5}{8}$	35.0	10.22	4.944	145.8	3.78	8.5	0.91	
Wide Flange Sections								
$18\times7\frac{1}{2}$	50.0	14.71	7.500	800.6	7.38	37.2	1.59	
14×12	84.0	24.71	12.023	928.4	6.13	225.5	3.02	
American Standard Channels								
$15\times3\frac{3}{8}$	33.9	9.90	3.400	312.6	5.62	8.2	0.91	0.79
12×3	25.0	7.32	3.047	143.5	4.43	4.5	0.79	0.68
$8\times2\frac{1}{4}$	11.5	3.36	2.260	32.3	3.10	1.3	0.63	0.58

Size in.	Thickness in.	Weight per ft lb	Area of section sq in.	x axis			y axis			z axis
				I in.^{4}	k in.	y in.	I in.^{4}	k in.	x in.	k in.
Equal Leg Angles										
8×8	1	51.0	15.00	89.0	2.44	2.37				1.56
3×3	$\frac{1}{4}$	4.9	1.44	1.2	0.93	0.84				0.59
Unequal Leg Angles										
7×4	$\frac{1}{2}$	17.9	5.25	26.7	2.25	2.42	6.5	1.11	0.92	0.87
4×3	$\frac{1}{2}$	11.1	3.25	5.1	1.25	1.33	2.4	0.86	0.83	0.64
$3\times2\frac{1}{2}$	$\frac{1}{4}$	4.5	1.31	1.2	0.95	0.91	0.74	0.75	0.66	0.53

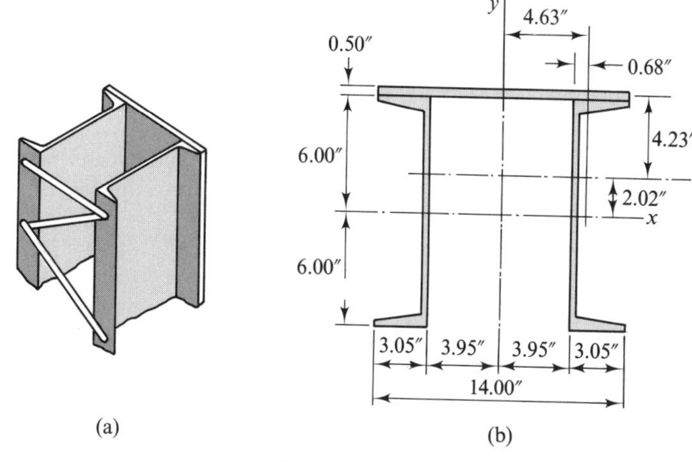

Figure 6-10

The required axes through the centroid are the y axis (from symmetry) and x_c axis, which must be located. The table on p. 258 gives the area of one channel as 7.32 in.^{2} and locates its centroid as indicated on the figure. The x_c axis is located by means of the principle of moments. The total area is

$$A = 14.00(0.500) + 2(7.32)$$
$$= 21.64 \text{ in.}^2.$$

The moment of the area with respect to the x axis is

$$M_x = \sum Ay_c = 14.00(0.50)(6.00 + 0.25) + 2(7.32)0$$
$$= 43.75 \text{ in.}^3.$$

The distance from the x axis to the centroid is

$$y_c = \frac{M_x}{A} = \frac{43.75}{21.64} = 2.02 \text{ in.}$$

The moment of inertia and radius of gyration of the area with respect to the x_c axis are obtained as follows:

For the plate:

$$I_{x_c} = I_c + Ad^2$$
$$= \frac{14.00(0.500)^3}{12} + 14.00(0.500)(4.23)^2$$
$$= 125.4 \text{ in.}^4$$

For the two channels:

$$I_{x_c} = 2(I_c + Ad^2)$$
$$= 2[143.5 + 7.32(2.02^2)] = 346.7 \text{ in.}^4$$

For the composite area:

$$I_{x_c} = 125.4 + 346.7 = 472.1 = \underline{472 \text{ in.}^4}. \qquad \text{Ans.}$$

and

$$k_{x_c} = \sqrt{\frac{I_{x_c}}{A}} = \sqrt{\frac{472.1}{21.64}} = \underline{4.67 \text{ in.}} \qquad \text{Ans.}$$

In a similar manner the moment of inertia and radius of gyration of the area with respect to the y axis are as follows:

For the plate:

$$I_y = I_c = \frac{0.500(14.00)^3}{12} = 114.3 \text{ in.}^4$$

For the two channels:

$$I_y = 2(I_c + Ad^2)$$

$$= 2[4.5 + 7.32(4.63)^2] = 322.8 \text{ in.}^4$$

For the composite area:

$$I_y = 114.3 + 322.8 = 437.1 = \underline{437 \text{ in.}^4}. \qquad \text{Ans.}$$

and

$$k_y = \sqrt{\frac{437.1}{21.64}} = \underline{4.49 \text{ in.}} \qquad \text{Ans.}$$

PROBLEMS

***6-26** Determine the moments of inertia of the shaded area in Fig. P6-26 with respect to horizontal and vertical axes through the centroid of the area.

6-27 Determine the radii of gyration with respect to the horizontal and vertical centroidal axes of the shaded area in Fig. P6-27.

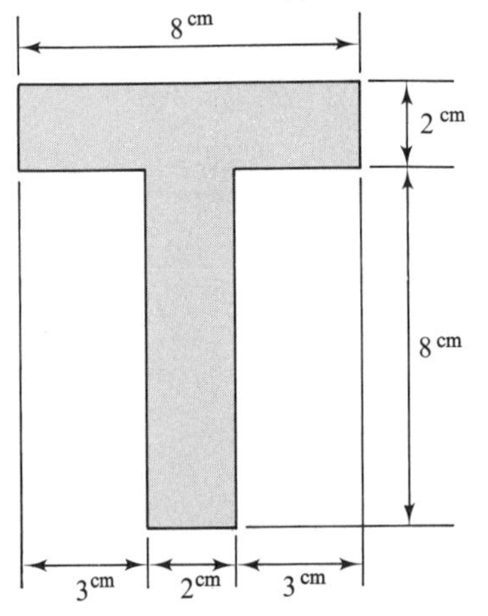

Figure P6-26

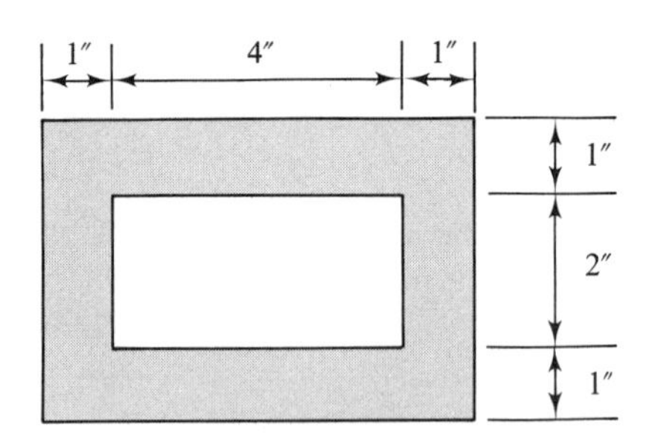

Figure P6-27

6-28 Determine the second moment of the shaded area in Fig. P6-28 with respect to the x axis.

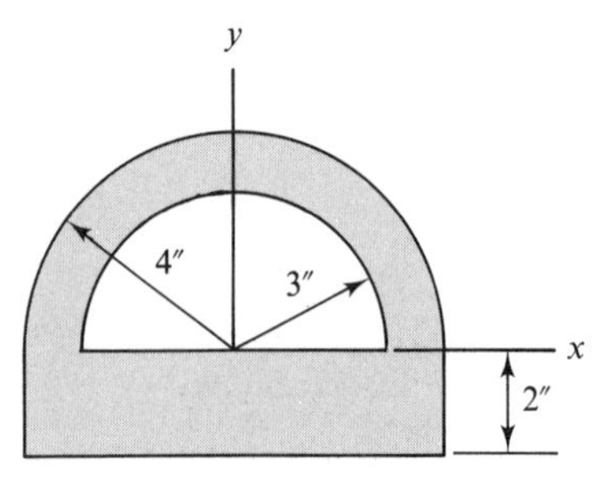

Figure P6-28

6-29 Determine the moment of inertia of the shaded area in Fig. P6-29 with respect to the x axis.

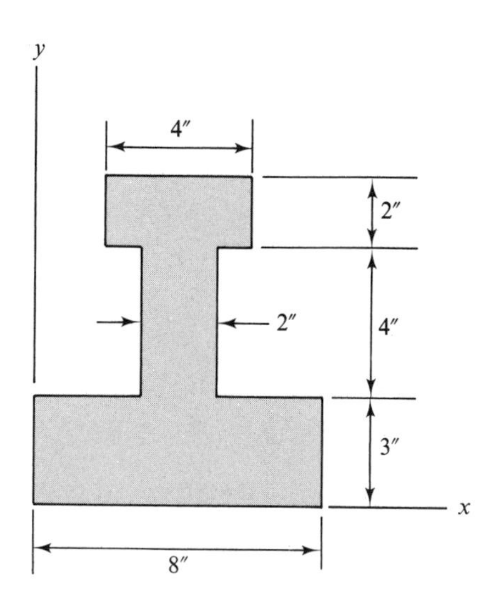

Figure P6-29

6-30 Determine the radius of gyration of the shaded area in Fig. P6-30 with respect to the x axis.

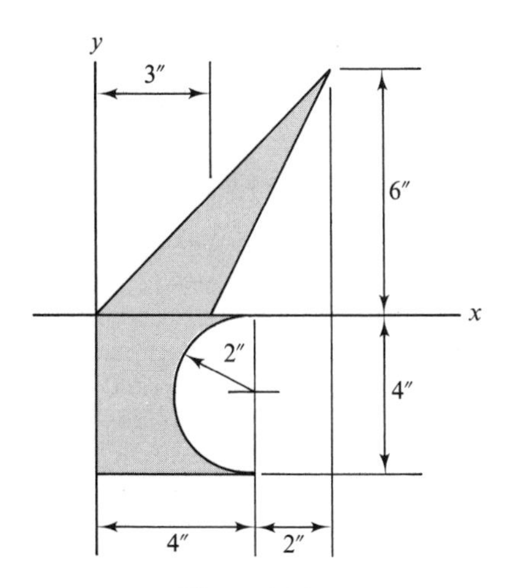

Figure P6-30

***6-31** Determine, with respect to the a axis, the radius of gyration of the shaded area in Fig. P6-31. The curve is a semicircle.

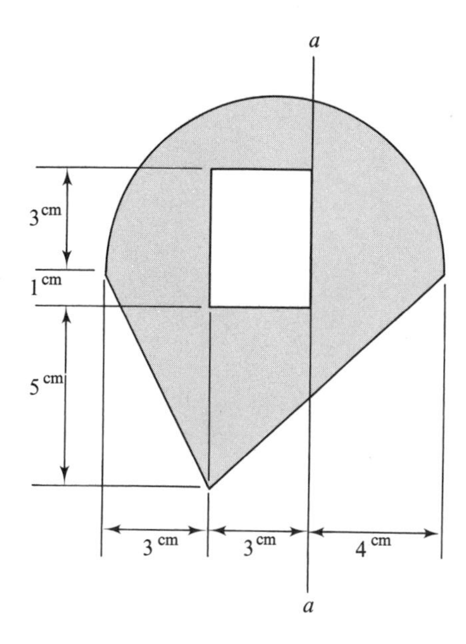

Figure P6-31

6-32 Determine the polar moment of inertia of the shaded area in Fig. P6-32 with respect to an axis through the origin.

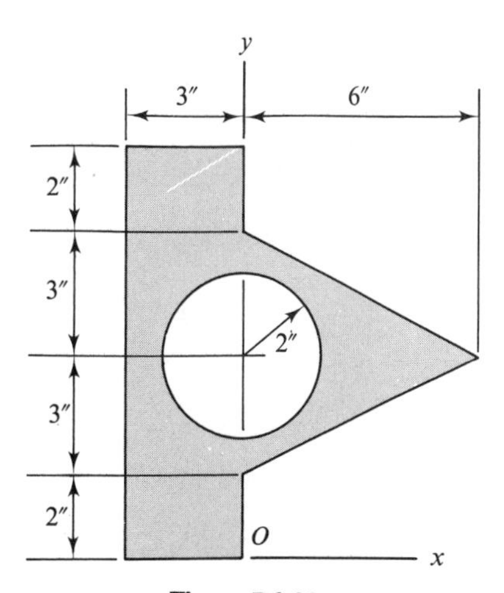

Figure P6-32

***6-33** Determine the radius of gyration of the shaded area in Fig. P6-33 with respect to the a axis.

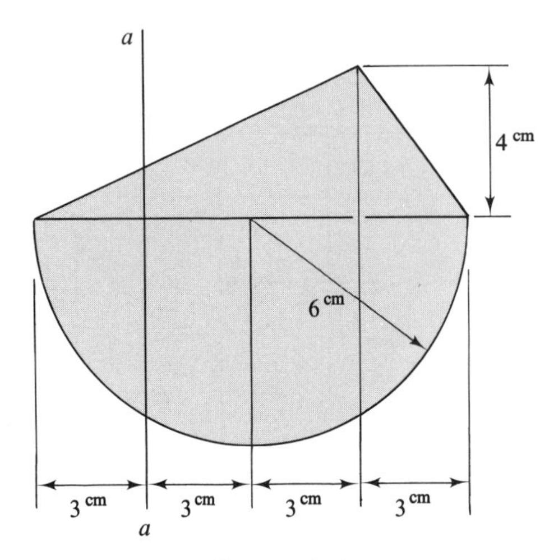

Figure P6-33

6-34 Determine the moment of inertia of the shaded area in Fig. P6-34 with respect to the y axis. The 15-ft^2 area C has a radius of gyration with respect to the a axis of 1.3 ft, and the centroid of area C is at point D.

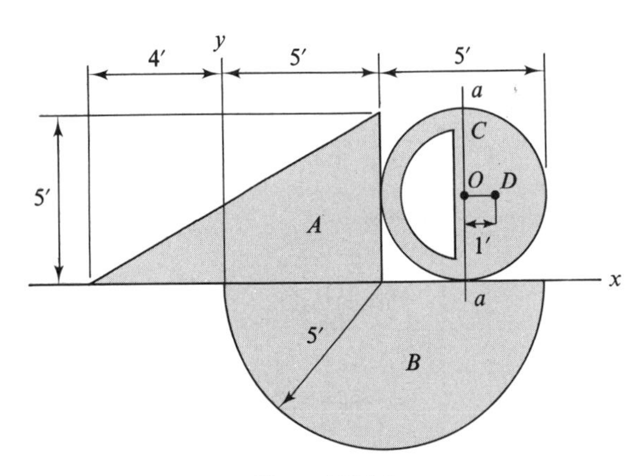

Figure P6-34

6-35 Determine the moment of inertia of the shaded area in Fig. P6-35 with respect to the x axis. The area to the right of the y axis is a semicircle, the 10-in.^{2} area A which is to the left of the y axis and above the x axis has a radius of gyration of 1.4 in. with respect to the a axis, and its centroid is at C.

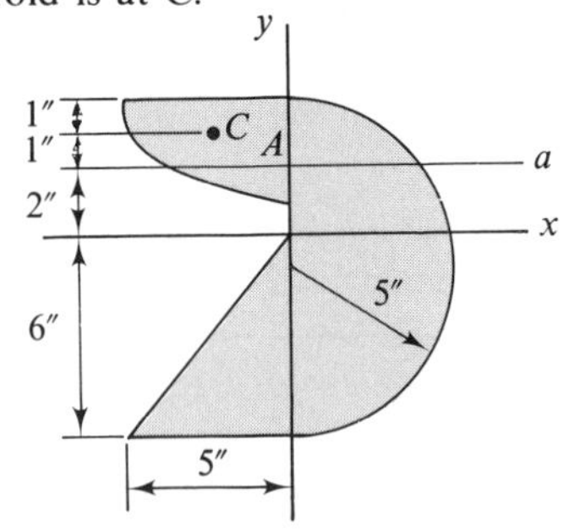

Figure P6-35

6-36 An 18WF50 beam is strengthened by adding a 15-in., 33.9-lb channel as shown in Fig. P6-36. Determine

(a) The distance y_1 to the horizontal centroidal axis of the composite section.

(b) The moments of inertia of the composite section with respect to the x and y axes.

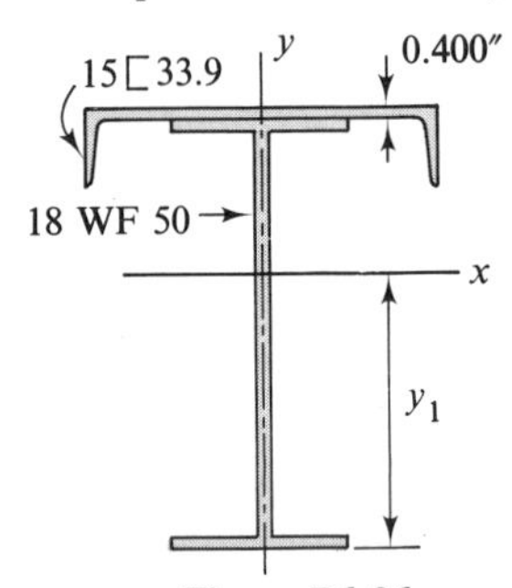

Figure P6-36

6-37 A plate-and-angle girder, selected from the tables prepared by the AISC (American Institute of Steel Construction) for an average span and relatively light load, consists of four 4- by 3- by $\frac{1}{2}$-in. angles, a web plate, and two cover plates arranged as shown in Fig. P6-37. Determine the moment of inertia of the composite section with respect to the horizontal axis of symmetry.

6-38 A plate-and-angle column consists of four 8- by 8- by 1-in. angles, a 16- by $1\frac{1}{8}$-in. web plate, and two cover plates arranged as shown in Fig. P6-38. Determine the moment of inertia of the composite cross section with respect to the x and y axes of symmetry.

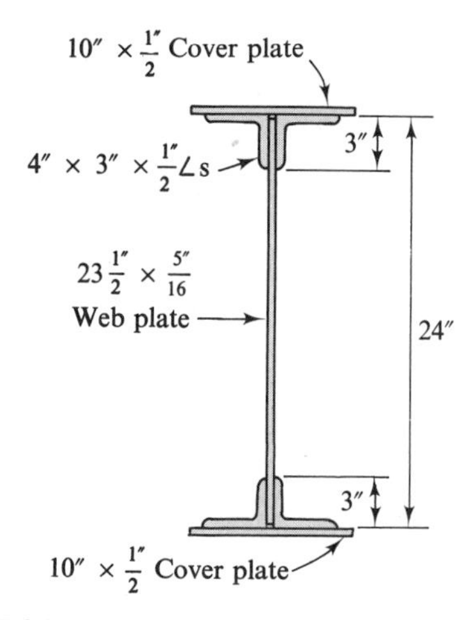

Figure P6-37

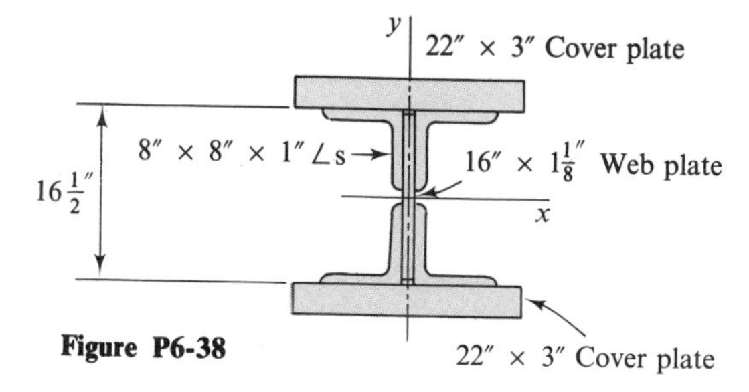

Figure P6-38

6-39 An eave strut is made of an 8-in., 11.5-lb channel, a 3- by 3- by $\frac{1}{4}$-in. angle, and a 3- by $2\frac{1}{2}$- by $\frac{1}{4}$-in. angle as shown in Fig. P6-39. Determine (a) The distances x_1 and y_1 to the x and y axes through the centroid of the section.
(b) The moment of inertia of the cross section with respect to the x and y axes.

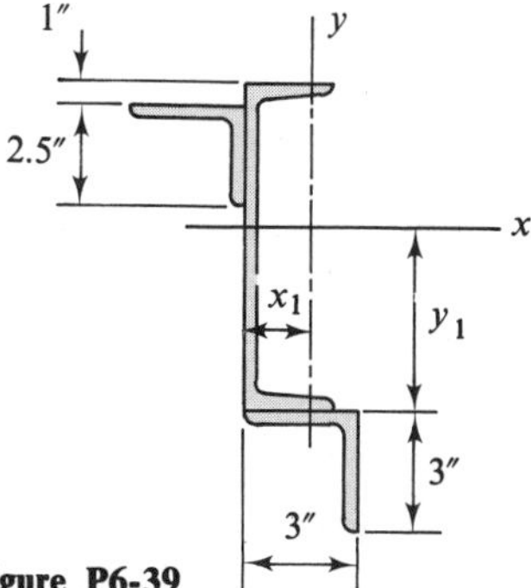

Figure P6-39

6-40 Determine the moment of inertia of the shaded area in Fig. P6-40 with respect to the y axis. The equation of the curve is $x^2 = 4y$.

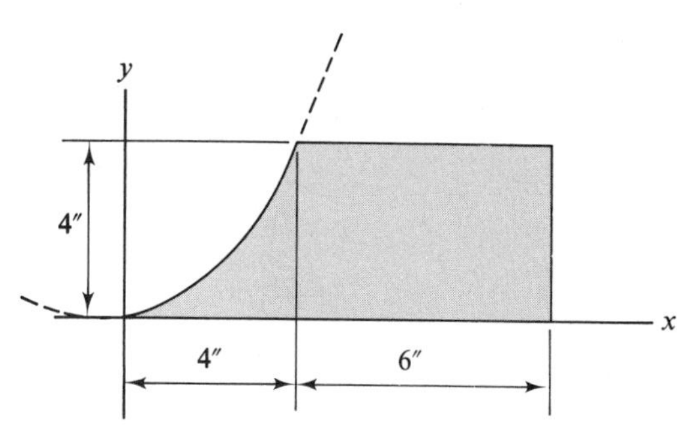

Figure P6-40

6-41 Determine the radius of gyration with respect to the x axis of the shaded area in Fig. P6-41. The equation of the curve is $x = y^2$.

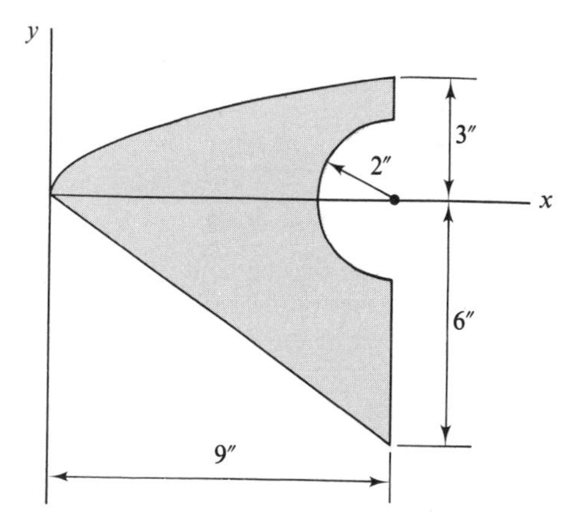

Figure P6-41

***6-42** Determine the moment of inertia of the shaded area in Fig. P6-42 with respect to the a axis. The equation of the curve is $y^2 = 3x$.

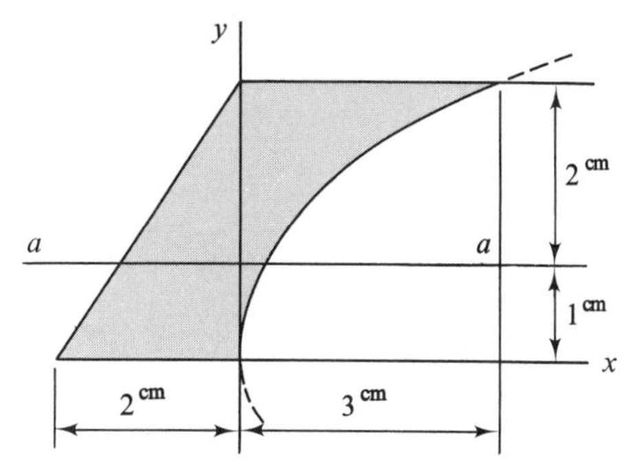

Figure P6-42

6-43 Determine the moment of inertia of the shaded area in Fig. P6-43 with respect to the x axis. The equation of the curve is $4y = x^2$.

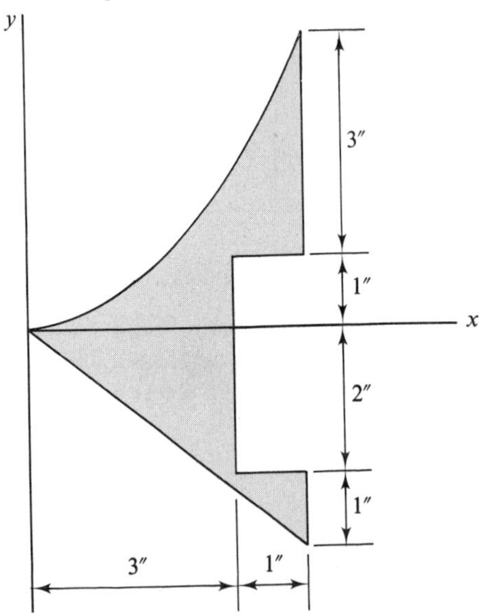

Figure P6-43

***6-44** Determine the moment of inertia of the shaded area in Fig. P6-44 with respect to the a axis. The quation of the curve is $x^2 = 3y$.

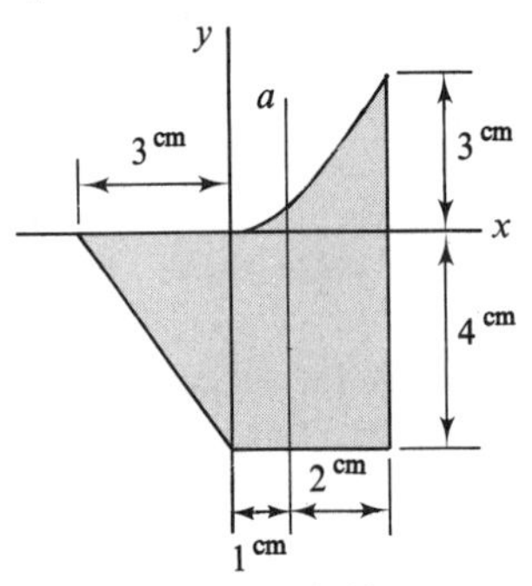

Figure P6-44

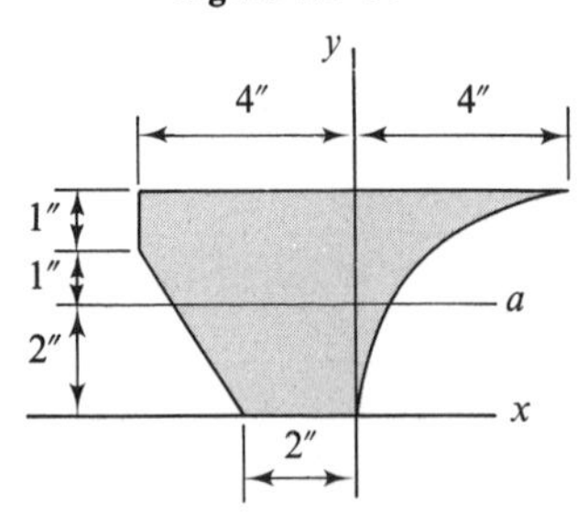

Figure P6-45

6-45 Determine the moment of inertia of the shaded area in Fig. P6-45 with respect to the a axis. The equation of the curve is $y^2 = 4x$.

6-46 Determine the second moment of the shaded area in Fig. P6-46 with respect to the y axis.

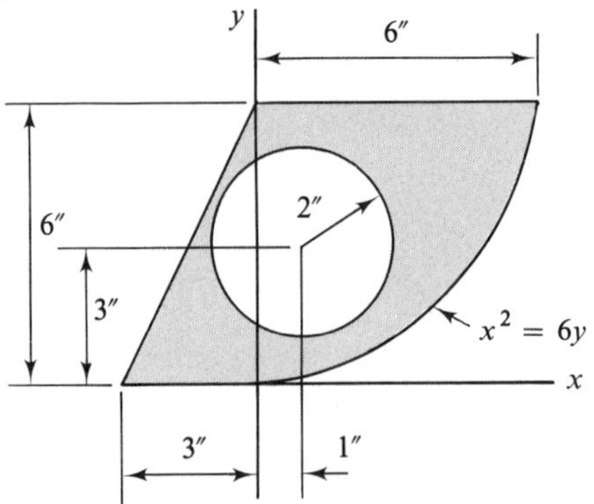

Figure P6-46

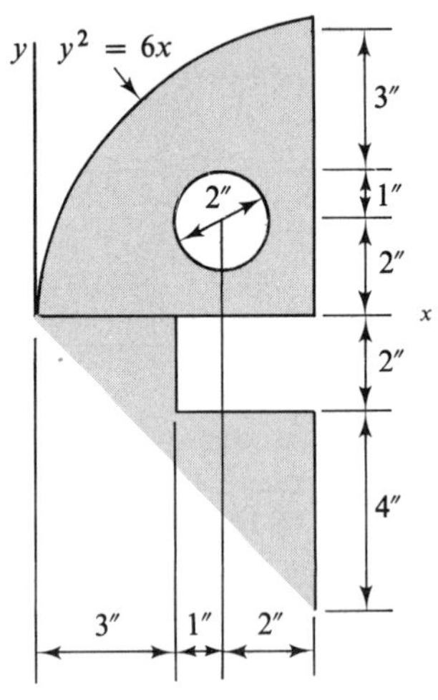

Figure P6-47

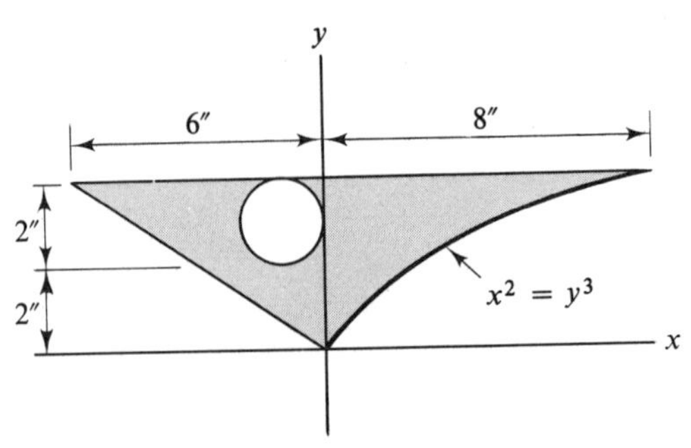

Figure P6-48

6-47 Determine the moment of inertia of the shaded area in Fig. P6-47 with respect to the x axis.

6-48 Determine the moment of inertia of the shaded area in Fig. P6-48 with respect to the y axis.

6-7 PRODUCTS OF INERTIA OF AREAS

The rectangular moment of inertia of an area with respect to an axis passing through a fixed point in the area usually varies with the orientation of the axis. For many applications it is necessary to determine (1) the direction of the axis through a point in the area for which the moment of inertia is a maximum or a minimum and (2) the corresponding second moment. The moment of inertia of an area with respect to any inclined axis can be determined by integration, but usually it is simpler to express it in terms of the moments of inertia with respect to two perpendicular axes (x and y), the product of inertia of the area with respect to the x and y axes, and the angle between the inclined axis and the x axis.

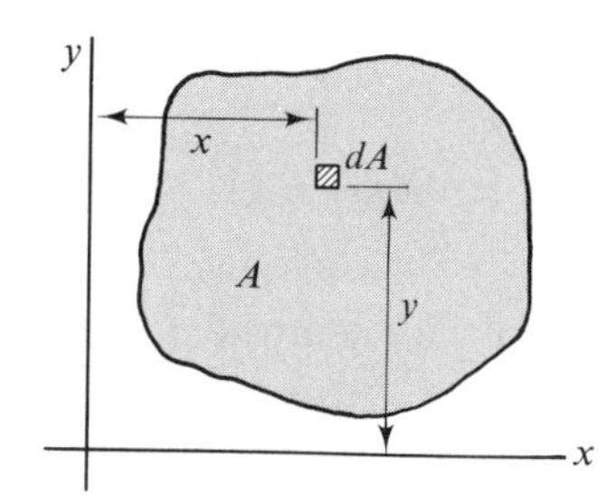

Figure 6-11

The product of inertia, dI_{xy}, of the element of area dA in Fig. 6-11 with respect to the x and y axes is defined as the product of the area and the two coordinates of the element; that is,

$$dI_{xy} = xy\, dA.$$

The product of inertia of the total area in Fig. 6-11 with respect to the x and y axes is the sum of the products of inertia of all the elements of the area. Hence

$$I_{xy} = \int xy\, dA.$$

The dimensions of the product of inertia of an area are length raised to the fourth power, $(L)^4$, and common units are in.^{4}, ft^4, cm^4, and so on. The element of area dA is inherently positive, but the product xy may be either positive or negative. Consequently, the product of inertia of an area may be positive, negative, or zero, as contrasted to a moment of inertia, which is always positive.

The product of inertia with respect to any two rectangular axes is zero when either or both of the axes is an axis of symmetry. To prove this statement, consider the area in Fig. 6-12, which is symmetrical with respect to the y axis. For every element dA there is an element dA' which has a product of inertia equal in magnitude but opposite in sign to that for dA because the x coordinates are opposite in sign. Therefore the sum of the products of inertia of all the elements is zero with respect to the x and y axes.

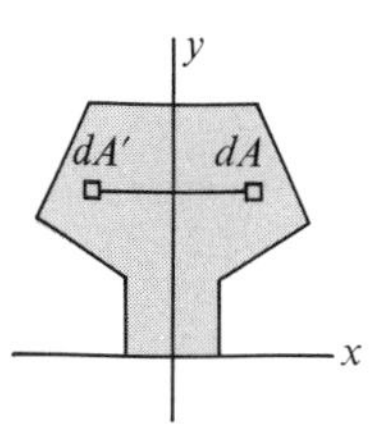

Figure 6-12

The parallel-axis theorem for products of inertia of areas can be derived as follows. Let x' and y' in Fig. 6-13 be a set of rectangular axes through the centroid of the area A, and let the x and y axes be

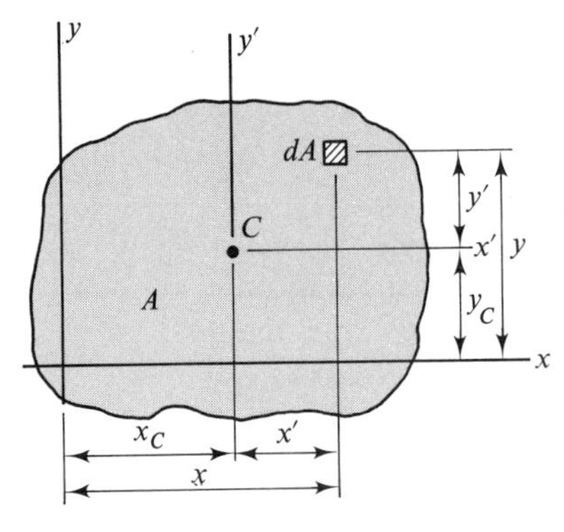

Figure 6-13

any set of rectangular axes parallel to x' and y', respectively. The product of inertia with respect to the x and y axes is

$$dI_{xy} = xy\,dA = (x_C + x')(y_C + y')\,dA$$
$$= x_C y_C\,dA + x'y'\,dA + x_C y'\,dA + x'y_C\,dA$$

and

$$I_{xy} = x_C y_C \int dA + \int x'y'\,dA + x_C \int y'\,dA + y_C \int x'\,dA$$
$$= x_C y_C A + I_{x'y'} + x_C \int y'\,dA + y_C \int x'\,dA,$$

where $I_{x'y'}$ is the product of inertia of the area with respect to the centroidal axes x' and y'. The integrals of $y'\,dA$ and $x'\,dA$ over the area are zero because they are the first moments of the area A with respect to the centroidal axes x' and y'. Consequently,

$$I_{xy} = I_{x'y'} + x_C y_C A. \tag{6-4}$$

The parallel-axis theorem for products of inertia can be stated as follows: *The product of inertia of an area with respect to any two perpendicular axes, x and y, is equal to the sum of the product of inertia of the area with respect to a set of axes through the centroid of the area (parallel to the x and y axes) and the product of the area and the two centroidal coordinates of the area from the x and y axes.*

Either single or double integration can be used for products of inertia. The product of inertia of the element is simpler to set up using double integration, whereas the limits are usually simpler when single integration is used. The following examples illustrate the procedure for determining products of inertia of areas.

EXAMPLE 6-8

Determine the products of inertia of the triangular area in Fig. 6-14 with respect to (a) the x and y axes and (b) a pair of axes through the centroid parallel to the x and y axes.

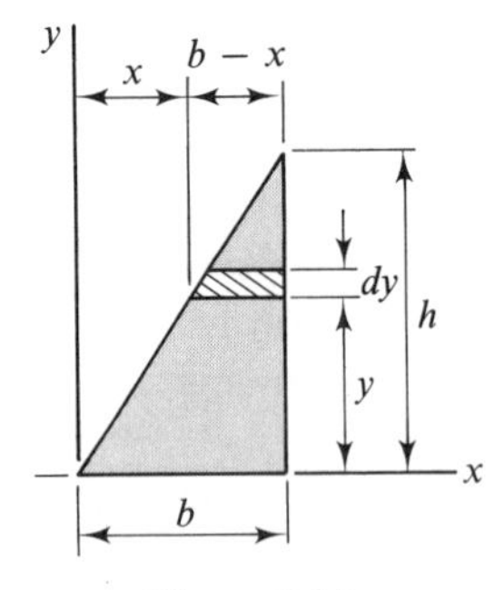

Figure 6-14

SOLUTION

(a) The element of area shown in Fig. 6-14 is a rectangle of length $(b-x)$ and height dy. From symmetry, the product of inertia of this element with respect to a pair of axes through the centroid of the element parallel to the x and y axes is zero. The product of inertia of the element with respect to the x and y axes is thus the product of the area and the two coordinates of its centroid, $(b+x)/2$ and y; that is,

$$dI_{xy} = (b-x)\,dy\left(\frac{b+x}{2}\right)y = \frac{(b^2-x^2)y}{2}\,dy.$$

From similar triangles, x is equal to $(b/h)y$, which can be substituted in the preceding expression to give

$$dI_{xy} = \frac{b^2}{2}\left(y - \frac{y^3}{h^2}\right) dy$$

$$I_{xy} = \frac{b^2}{2}\int_0^h \left(y - \frac{y^3}{h^2}\right) dy = \frac{b^2}{2}\left[\frac{y^2}{2} - \frac{y^4}{4h^2}\right]_0^h = \underline{\frac{b^2h^2}{8}}. \qquad \text{Ans.}$$

(b) The parallel-axis theorem can be used to determine the product of inertia with respect to the centroidal axes as follows:

$$I_{xy} = I_{x'y'} + Ax_Cy_C$$

or

$$\frac{b^2h^2}{8} = I_{x'y'} + \frac{bh}{2}\left(\frac{2}{3}b\right)\left(\frac{1}{3}h\right).$$

Therefore

$$I_{x'y'} = \frac{b^2h^2}{8} - \frac{b^2h^2}{9} = \underline{\frac{b^2h^2}{72}}. \qquad \text{Ans.}$$

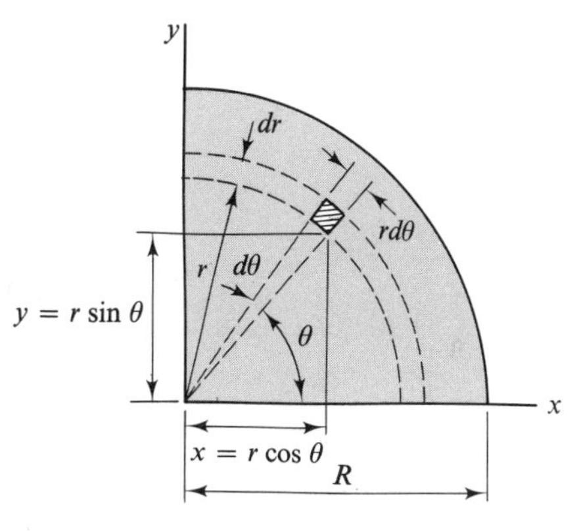

Figure 6-15

The product of inertia of a right triangle with respect to a pair of centroidal axes parallel and perpendicular to the base will have a magnitude of $b^2h^2/72$ no matter how the triangle is oriented, but the sign will be positive only when the right angle of the triangle is in the second or fourth quadrants of the centroidal axes, being negative when the right angle is in the first or third quadrants (see Problem 6-49 below).

EXAMPLE 6-9

Determine the product of inertia of the area of the quadrant of the circle in Fig. 6-15 with respect to the x and y axes.

SOLUTION

Double integration and polar coordinates will be used with the element of area selected as indicated in Fig. 6-15. Since the limits of integration with respect to r and θ are independent, the order of integration is immaterial. In the following solution, the first integration is made with respect to θ:

$$dA = r\,d\theta\,dr,$$

$$dI_{xy} = xy\,dA = r^3 \sin\theta\cos\theta\,d\theta\,dr,$$

and

$$I_{xy} = \int_0^R \int_0^{\pi/2} r^3 \sin\theta \cos\theta\,d\theta\,dr = \int_0^R r^3\left[\frac{\sin^2\theta}{2}\right]_0^{\pi/2} dr$$

$$= \frac{1}{2}\int_0^R r^3\,dr = \left[\frac{r^4}{8}\right]_0^R = \underline{\frac{R^4}{8}}. \qquad \text{Ans.}$$

PROBLEMS

6-49 (a) Determine by integration the product of inertia of the area of the right triangle shown in Fig. P6-49 with respect to the x and y axes. (b) Using the results of part (a) and the parallel-axis theorem, determine the product of inertia with respect to axes through the centroid of the triangle parallel to the x and y axes.

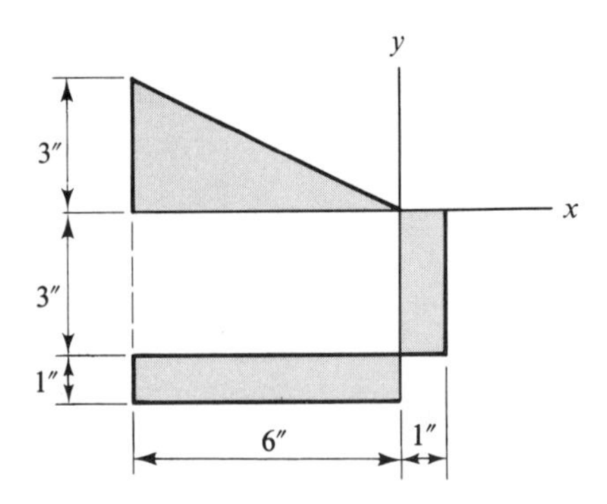

Figure P6-51

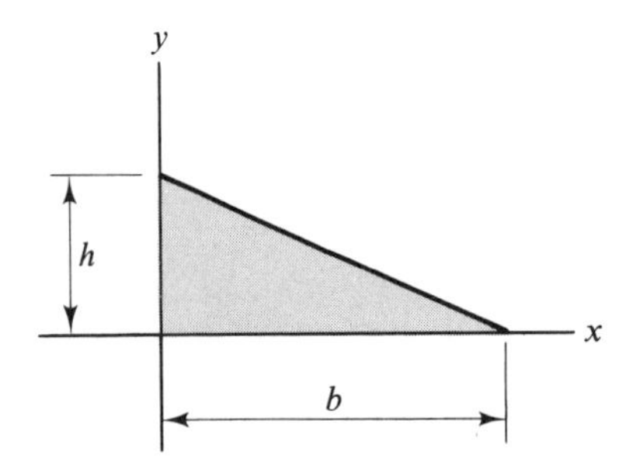

Figure P6-49

6-50 Determine the product of inertia of the shaded area in Fig. P6-50 with respect to the x and y axes.

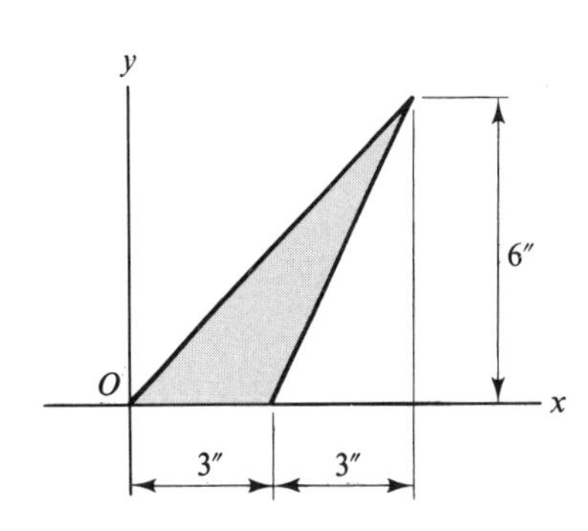

Figure P6-50

6-51 Determine the product of inertia of the shaded area shown in Fig. P6-51 with respect to the x and y axes.

***6-52** The equation of the curve in Fig. P6-52 is $y=2x^2$. Determine the product of inertia of the shaded area with respect to the x and y axes.

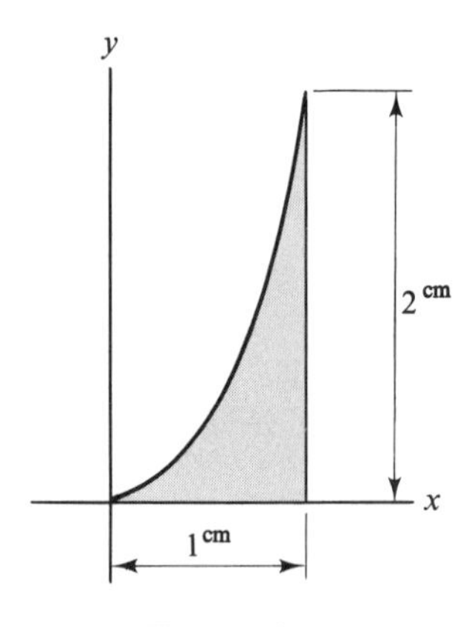

Figure P6-52

6-53 Determine the product of inertia of the shaded area in Fig. P6-53 with respect to the x and y axes. The equation of the curve is $y^2=4x$.

***6-54** For the shaded area in Fig. P6-54, determine the product of inertia with respect to the x and y axes.

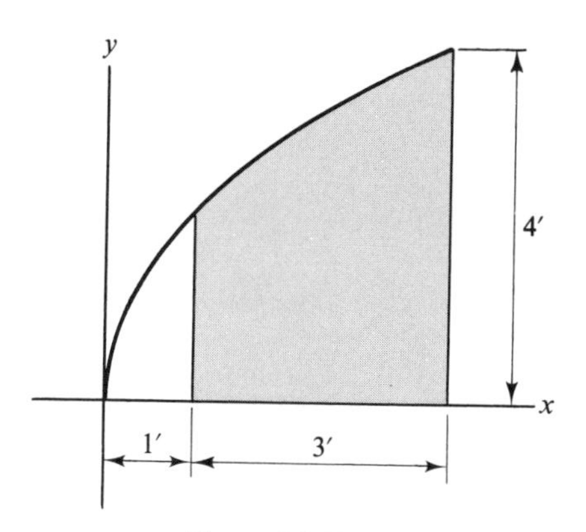

Figure P6-53

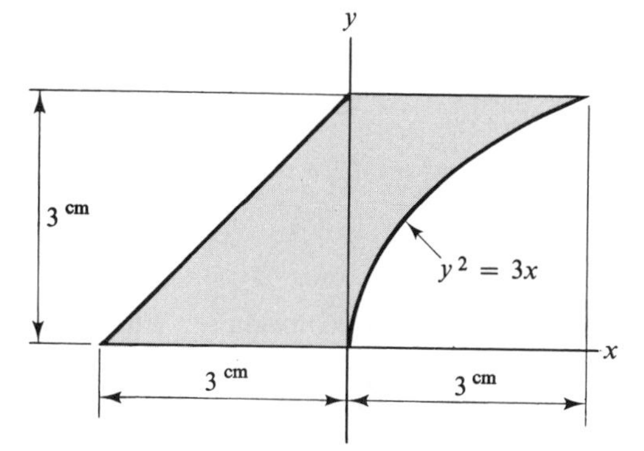

Figure P6-54

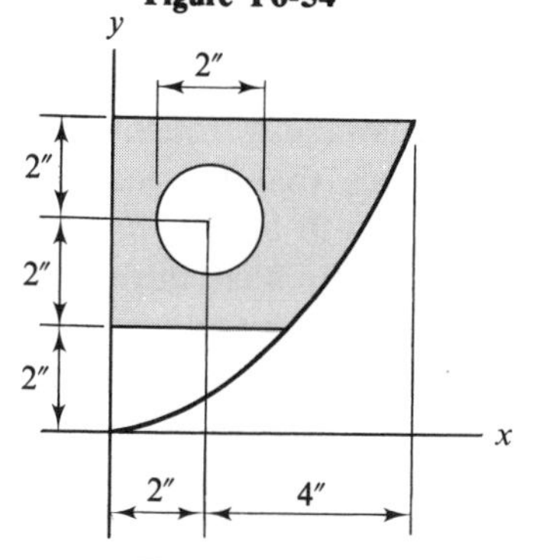

Figure P6-55

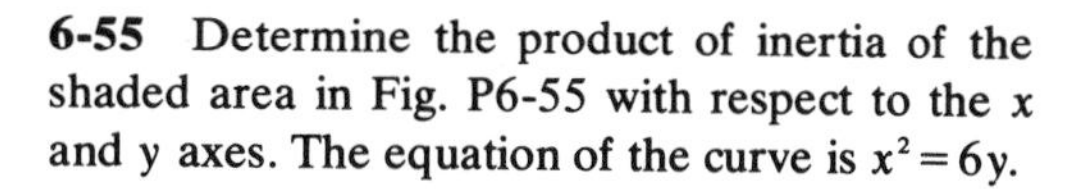

6-55 Determine the product of inertia of the shaded area in Fig. P6-55 with respect to the x and y axes. The equation of the curve is $x^2 = 6y$.

6-56 Determine the product of inertia of the area in Fig. P6-56 with respect to the x and y axes.

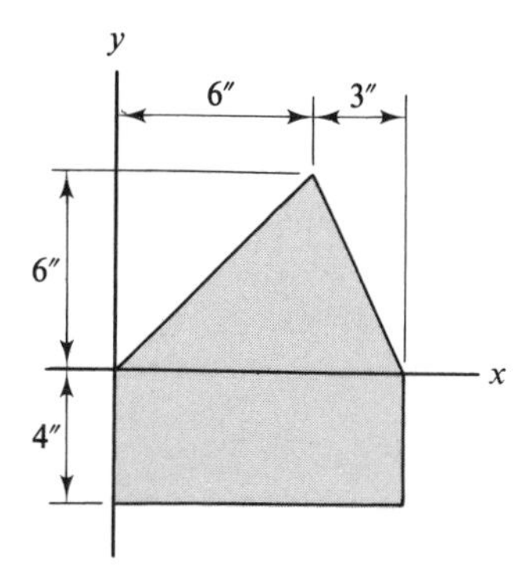

Figure P6-56

6-57 Determine the product of inertia of the area shown in Fig. P6-57 with respect to the x and y axes. The equation of the curve is $xy = 8$.

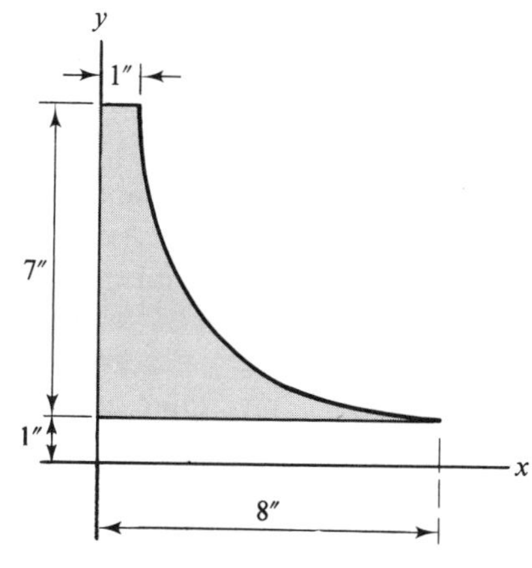

Figure P6-57

6-8 MAXIMUM AND MINIMUM SECOND MOMENTS OF AREAS

The moment of inertia of the area A in Fig. 6-16 with respect to the x' axis through the point O varies with the orientation of that axis; that is, $I_{x'}$ will, in general, have a different value for each value of θ. In stress analysis and in other situations, it is sometimes necessary to

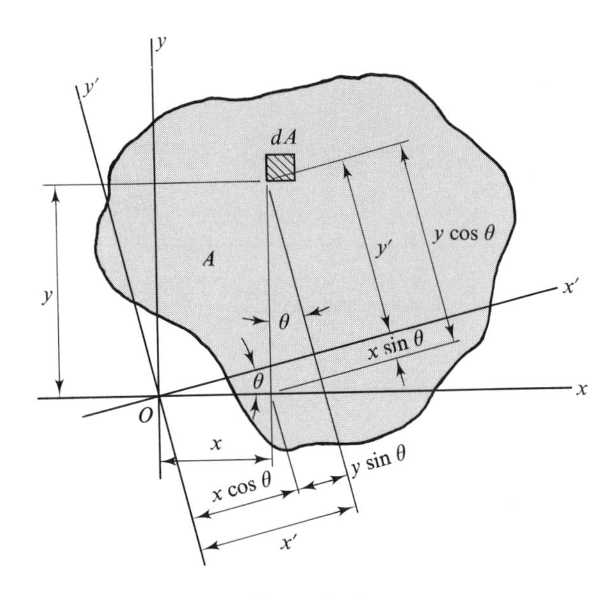

Figure 6-16

determine the maximum and minimum second moments with respect to axes through some point O. The x and y axes used in obtaining Eq. (6-1) were any pair of rectangular axes in the plane of the area which pass through point O; therefore the equation can be written

$$J_O = I_x + I_y = I_{x'} + I_{y'},$$

where x' and y' are any pair of rectangular axes through point O. Since the sum of $I_{x'}$ and $I_{y'}$ is equal to a constant, that is, equal to J_O, $I_{x'}$ will be the maximum second moment, and $I_{y'}$ will be the minimum second moment for one particular orientation of the axes through point O.

The particular rectangular axes for which the second moment is maximum and minimum are designated as the u and v axes, and they are called the *principal axes* of the area through the point O. The moments of inertia with respect to these axes are called the *principal moments of inertia* of the area and are designated as I_u and I_v. There is only one pair of principal axes for any point in an area unless all axes have the same second moment, such as the diameters of a circle. The moment of inertia of the area with respect to one of these axes is the maximum for all axes in the plane of the area passing through the point, and the moment of inertia with respect to the other axis is the minimum.

A convenient way to determine the maximum and minimum second moments for an area is to express $I_{x'}$ in terms of θ, I_x, I_y, and I_{xy} and then set the derivative of $I_{x'}$ with respect to θ equal to zero to obtain the values of θ which give the maximum and minimum values of the second moment. From Fig. 6-16,

$$dI_{x'} = y'^2\, dA = (y \cos\theta - x \sin\theta)^2\, dA$$

and

$$\begin{aligned} I_{x'} &= \cos^2\theta \int y^2\, dA - 2\sin\theta\cos\theta \int xy\, dA + \sin^2\theta \int x^2\, dA \\ &= I_x \cos^2\theta - I_{xy}\, 2\sin\theta\cos\theta + I_y \sin^2\theta \\ &= \frac{I_x + I_y}{2} + \frac{I_x - I_y}{2}\cos 2\theta - I_{xy}\sin 2\theta. \end{aligned} \tag{6-5}$$

The second form of the expression in Eq. (6-5) is obtained when the functions of the angle θ are replaced by the equivalent functions of the angle 2θ. The angle θ or 2θ for which $I_{x'}$ is a maximum can be determined by differentiating $I_{x'}$ with respect to θ and setting the derivative equal to zero; thus

$$\frac{dI_{x'}}{d\theta} = -2\frac{I_x - I_y}{2}\sin 2\theta - 2I_{xy}\cos 2\theta = 0$$

from which

$$\tan 2\beta = -\frac{I_{xy}}{(I_x - I_y)/2}, \tag{6-6}$$

where β represents the two values of θ that locate the principal axes u and v.

Equation (6-6) gives two values of 2β that are 180° apart and thus two values of β that are 90° apart. The maximum and minimum values of the moment of inertia can be obtained by substituting these values of β in Eq. (6-5). From Eq. (6-6),

$$\cos 2\beta = \mp\frac{(I_x - I_y)/2}{\sqrt{[(I_x - I_y)/2]^2 + I_{xy}^2}},$$

$$\sin 2\beta = \pm\frac{I_{xy}}{\sqrt{[(I_x - I_y)/2]^2 + I_{xy}^2}},$$

and when these values are substituted in Eq. (6-5), the result reduces to

$$I_{u_{\substack{\min\\\max}}} = \frac{I_x + I_y}{2} \mp \sqrt{\left(\frac{I_x - I_y}{2}\right)^2 + I_{xy}^2}. \tag{6-7}$$

If I_{xy} is positive and I_x is greater than I_y, the first value of 2β from Eq. (6-6) will be in the second quadrant and the other value in the fourth quadrant. The first value of β is between 45 and 90° and is represented by the top signs in the preceding equations. From Eq. (6-7) this value of β is seen to give the minimum moment of inertia of the area. The other value of β, of course, gives the maximum moment of inertia. If

I_{xy} is negative or if I_x is less than I_y, the signs are reversed, but if both these conditions exist, the above signs are valid.

The product of inertia of the element of the area in Fig. 6-16 with respect ot the x' and y' axes is

$$dI_{x'y'} = x'y'\,dA$$
$$= (x\cos\theta + y\sin\theta)(y\cos\theta - x\sin\theta)\,dA,$$

and

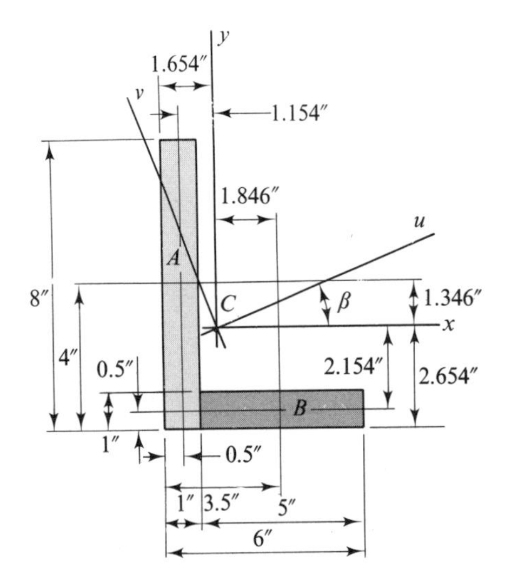

Figure 6-17

$$I_{x'y'} = (\cos^2\theta - \sin^2\theta)\int xy\,dA - \sin\theta\cos\theta\int x^2\,dA$$
$$+ \sin\theta\cos\theta\int y^2\,dA$$
$$= I_{xy}\cos 2\theta - \tfrac{1}{2}I_y\sin 2\theta + \tfrac{1}{2}I_x\sin 2\theta. \qquad (6\text{-}8)$$

The angle θ for which the product of inertia is zero can be obtained by setting $I_{x'y'}$ in Eq. (6-8) equal to zero and solving for θ. The result is

$$\tan 2\theta = -\frac{I_{xy}}{(I_x - I_y)/2},$$

which is the same as Eq. (6-6) for the angle locating the principal axes.

Therefore *the product of inertia is zero with respect to the principal axes.* Since the product of inertia is zero with respect to any axis of symmetry, it follows that any axis of symmetry of an area is a principal axis for any point on the line of symmetry.

The following example illustrates the procedure for determining the second moments with respect to the principal axes.

EXAMPLE 6-10

Determine the maximum and minimum second moments of the area of the unequal-leg angle in Fig. 6-17 with respect to axes through the centroid of the area.

SOLUTION

The area is divided into two rectangles A and B, and the location of the centroid of each area is indicated by dimensions. The coordinates of the centroid at C are determined from the principle of moments as presented in Chapter 3. The values of I_x and I_y for the two areas are obtained by use of Eq. (6-2) and the table on p. 255.

For part A:

$$I_x = I_c + Ad^2 = \frac{1(8)^3}{12} + 1(8)(1.346)^2 = 57.2 \text{ in.}^4.$$

For part B:

$$I_x = \frac{5(1)^3}{12} + 5(1)(2.154)^2 = 23.6 \text{ in.}^4$$

For the total area:

$$I_x = 57.2 + 23.6 = 80.8 \text{ in.}^4.$$

Similarly, *for part A:*

$$I_y = \frac{8(1)^3}{12} + 8(1)(1.154)^2 = 11.32 \text{ in.}^4.$$

For part B:

$$I_y = \frac{1(5)^3}{12} + 1(5)(1.846)^2 = 27.5 \text{ in.}^4.$$

For the total area:

$$I_y = 38.8 \text{ in.}^4.$$

The products of inertia of the areas are determined as indicated in Art. 6-7.

For part A:

$$I_{xy} = I_{x'y'} + x_C y_C A = 0 + (-1.154)(1.346)(1)(8) = -12.43 \text{ in.}^4.$$

For part B:

$$I_{xy} = 0 + (1.846)(-2.154)(5)(1) = -19.88 \text{ in.}^4.$$

For the composite area:

$$I_{xy} = -32.3 \text{ in.}^4.$$

From Eq. (6-6)

$$\tan 2\beta = -\frac{I_{xy}}{(I_x - I_y)/2}$$

$$= -\frac{-32.3}{(80.8 - 38.8)/2} = 1.538,$$

$$2\beta = 57.0° \quad \text{or} \quad 237° \quad \text{and} \quad \beta = 28.5° \quad \text{or} \quad 118.5°.$$

From Eq. (6-5) with $\beta = 28.5°$, the maximum second moment is

$$I_u = I_x \cos^2\beta - I_{xy}\, 2\sin\beta\cos\beta + I_y \sin^2\beta$$

$$= 80.8(\cos^2 28.5) - (-32.3)(2)(\sin 28.5)(\cos 28.5) + 38.8(\sin^2 28.5)$$

$$= 62.4 + 27.1 + 8.83 = \underline{98.3 \text{ in.}^4}. \qquad \text{Ans.}$$

With $\beta = 118.5°$, the minimum second moment is

$$I_v = 80.8(\cos^2 118.5) - (-32.3)(2)(\sin 118.5)(\cos 118.5) + 38.8(\sin^2 118.5)$$

$$= 18.38 - 27.1 + 30.0 = \underline{21.3 \text{ in.}^4}. \qquad \text{Ans.}$$

The principal moments of inertia can also be determined by means of Eq. (6-7).

$$I_{\substack{max\\min}} = \frac{I_x + I_y}{2} \pm \sqrt{\left(\frac{I_x - I_y}{2}\right)^2 + I_{xy}^2}$$

$$= \frac{80.8 + 38.8}{2} \pm \sqrt{\left(\frac{80.8 - 38.8}{2}\right)^2 + (-32.3)^2}$$

$$= \underline{98.3 \text{ in.}^4 \text{ and } 21.3 \text{ in.}^4} \qquad \text{Ans.}$$

PROBLEMS

6-58 The x and y axes in Fig. P6-58 are centroidal axes. Determine the moment of inertia of the area with respect to the x' axis.

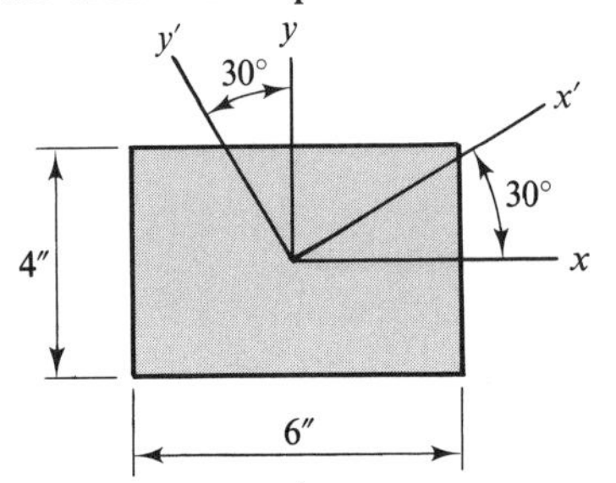

Figure P6-58

***6-59** Determine the minimum moment of inertia of the shaded area in Fig. P6-59 with respect to an axis through the origin.

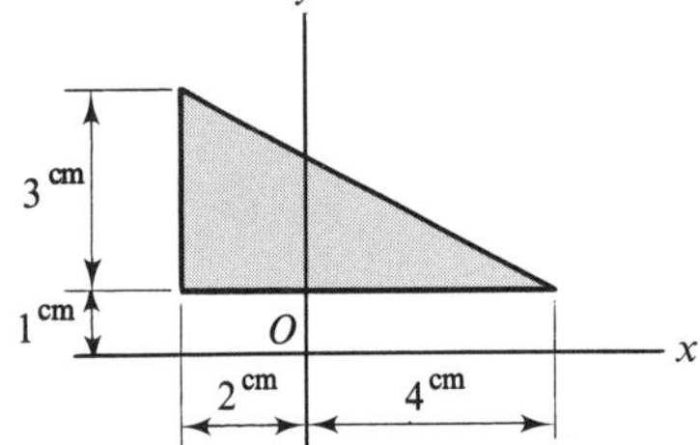

Figure P6-59

6-60 Determine the second moments of the cross section of a 10-in., 35-lb American Standard I beam with respect to the x' and y' axes shown in Fig. P6-60.

6-61 Determine the minimum rectangular moment of inertia of the shaded area in Fig. P6-61 with respect to an axis through the origin and show the location of this axis.

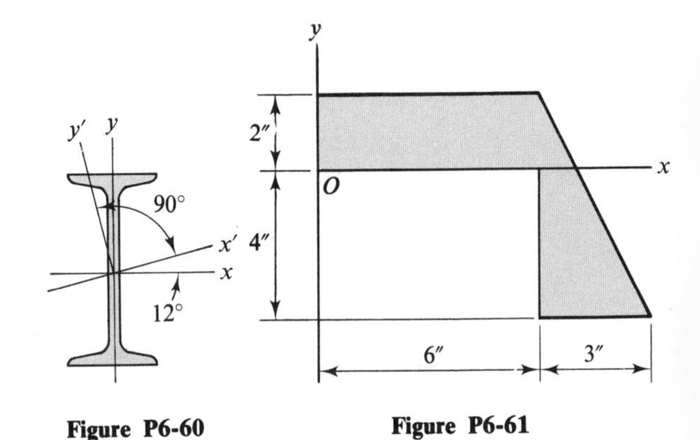

Figure P6-60 **Figure P6-61**

***6-62** Determine the maximum rectangular moment of inertia of the area shown in Fig. P6-62 with respect to an axis through the origin. The equation of the curve is $y^2 = x$.

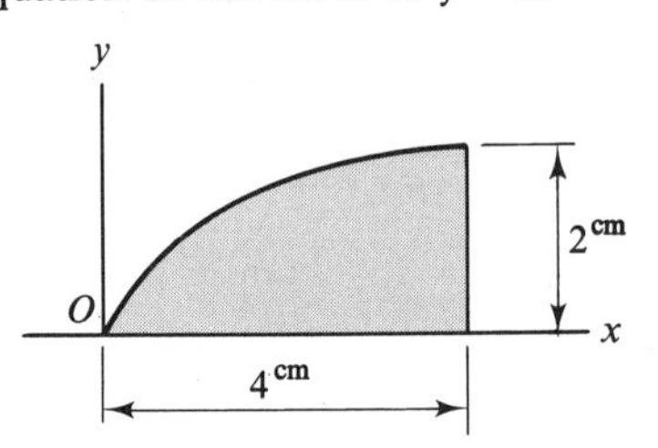

Figure P6-62

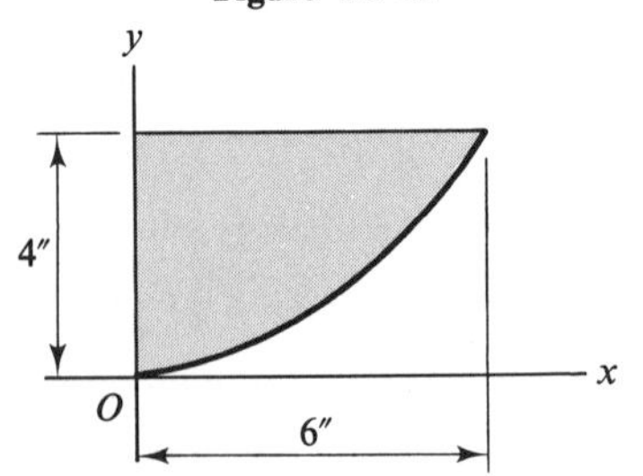

Figure P6-63

6-63 Determine the maximum and minimum second moments of the shaded area in Fig. P6-63 with respect to axes through the origin. The equation of the curve is $x^2 = 9y$.

6-9 MOHR'S CIRCLE FOR SECOND MOMENTS OF AREAS

The German engineer Otto Mohr (1835–1918) developed a useful pictorial or graphical interpretation of Eqs. (6-5)–(6-8). This method, frequently called *Mohr's circle*,[2] involves the construction of a circle in such a manner that coordinates of each point on the circle represent $I_{x'}$ and $I_{x'y'}$ for one orientation of the x' and y' axes.

I_x, I_y, and I_{xy} must be computed if not already known. Assume that I_x is greater than I_y and that I_{xy} is positive. Draw a set of rectangular axes and label the horizontal axis I_x and the vertical axis I_{xy} as indicated in Fig. 6-18. Second moments of areas are always positive and are plotted to the right of the origin. Products of inertia may be either positive or negative, and it is customary to plot positive values above the I_x axis and negative values below. Lay off the distance OA' along the I_x axis equal to I_x and $A'A$ parallel to the I_{xy} axis equal to I_{xy}. In a similar manner locate B by making OB' equal to I_y and $B'B$ equal to $-I_{xy}$, the value of I_{xy} with the algebraic sign reversed. Draw the line AB, intersecting the I_x axis at C, and draw a circle with AB as its diameter. This circle is Mohr's circle for moments of inertia, and each point on the circle represents $I_{x'}$ and $I_{x'y'}$ for one particular orientation of the x' and y' axes, the abscissa representing $I_{x'}$ and the ordinate representing $I_{x'y'}$. To demonstrate this statement draw the diameter DE at an angle 2θ counterclockwise from line AB and note that CD is equal to CA. From the figure, it is apparent that

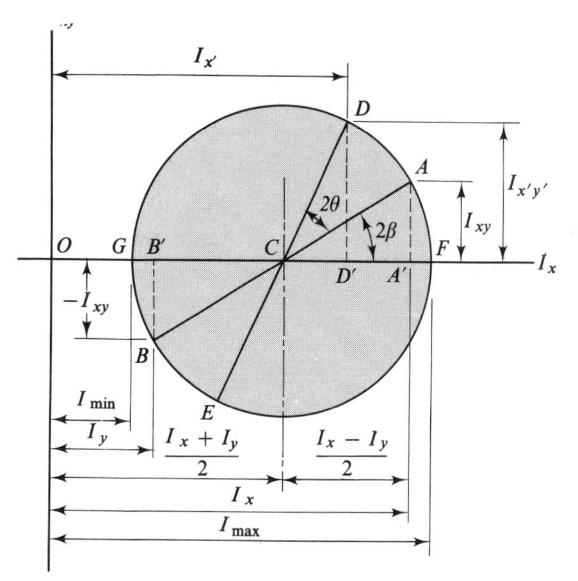

Figure 6-18

$$OD' = OC + CD \cos(2\beta + 2\theta),$$

which, from trigonometry, reduces to

$$OD' = OC + CA \cos 2\beta \cos 2\theta - CA \sin 2\beta \sin 2\theta.$$

---

[2] Mohr's circle is also, in fact more commonly, used as a pictorial interpretation of the formulas relating the stresses or strains on various planes through a point in a body subjected to "plane stress" and "plane strain." This application is developed in the study of mechanics of materials, theory of elasticity, and similar fields.

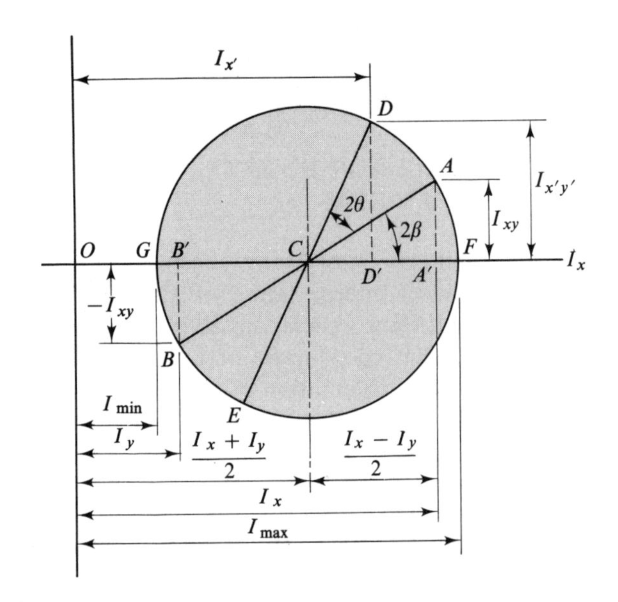

Figure 6-18

Again referring to the figure,

$$OD' = OC + CA' \cos 2\theta - A'A \sin 2\theta$$

$$= \frac{I_x + I_y}{2} + \frac{I_x - I_y}{2} \cos 2\theta - I_{xy} \sin 2\theta = I_{x'}$$

from Eq. (6-5). In a similar manner,

$$D'D = CD \sin(2\beta + 2\theta)$$

$$= CA \sin 2\beta \cos 2\theta + CA \cos 2\beta \sin 2\theta$$

$$= A'A \cos 2\theta + CA' \sin 2\theta$$

$$= I_{xy} \cos 2\theta + \frac{I_x - I_y}{2} \sin 2\theta = I_{x'y'}$$

from Eq. (6-8).

Since the horizontal coordinate of each point on the circle represents a particular value of $I_{x'}$, the maximum moment of inertia is represented by OF, and its value is

$$I_{max} = OF = OC + CF = OC + CA$$

$$= \frac{I_x + I_y}{2} + \sqrt{\left(\frac{I_x - I_y}{2}\right)^2 + I_{xy}^2},$$

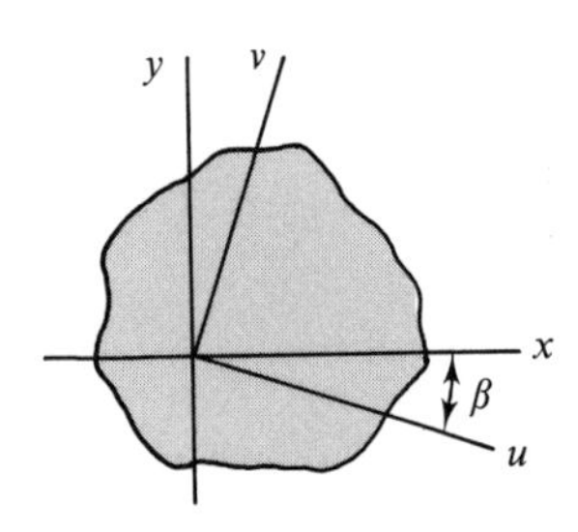

Figure 6-19

which agrees with Eq. (6-7). Figure 6-19 represents an area and a set of axes for which the data used in constructing Fig. 6-18 are valid. In deriving Eq. (6-5) the angle between the x and x' axes was θ. In obtaining the same equation from Mohr's circle, the angle between the radii to I_x, I_{xy} (line CA) and $I_{x'}$, $I_{x'y'}$ (line CD) is 2θ. In other words, all angles in Mohr's circle are twice the corresponding angles for the actual area. The angle 2β in Fig. 6-18 is twice the angle between the x axis represented by CA and the principal axis (u axis) represented by CF. Therefore, as shown in Fig. 6-19, the principal (u) axis is at an angle β clockwise from the x axis. From Fig. 6-18,

$$\tan 2\beta = \frac{I_{xy}}{(I_x - I_y)/2},$$

which is the same as Eq. (6-6) except for sign. The negative sign indicates that the principal axis is at an angle β clockwise (counterclockwise is positive) from the x axis.

EXAMPLE 6-11

Solve Example 6-10 by means of Mohr's circle.

SOLUTION

Values of I_x, I_y, and I_{xy} must be computed in the same manner as in Example 6-10. These values are

$$I_x = 80.8 \text{ in.}^4, \qquad I_y = 38.8 \text{ in.}^4, \qquad I_{xy} = -32.3 \text{ in.}^4.$$

The circle in Fig. 6-20 is constructed as indicated from the values of I_x, I_y, and I_{xy}. Point A is located at I_x and I_{xy} (which is negative), and B is

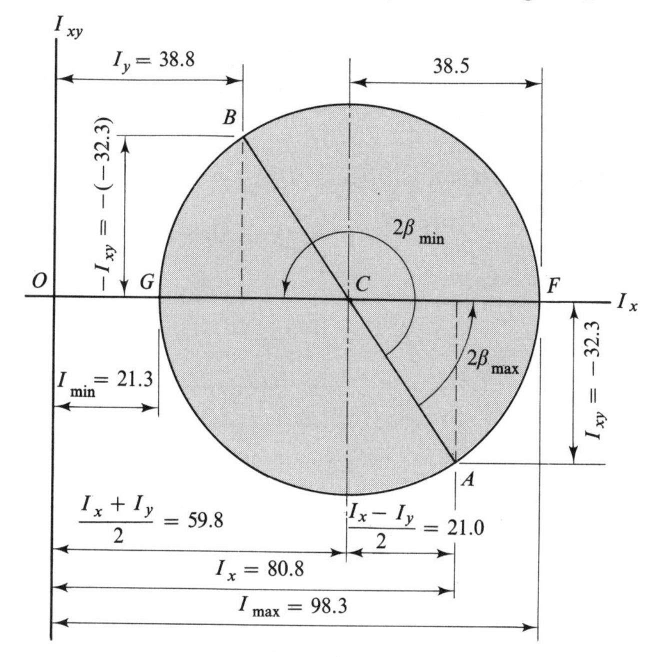

Figure 6-20

at I_y and $-I_{xy}$. Point C is the center of the circle and is 59.8 units from O. The radius of the circle is

$$CA = (21.0^2 + 32.3^2)^{1/2} = 38.5 \text{ in.}^4,$$

and the principal moments of inertia are

$$I_{max} = OC + CF = 59.8 + 38.5 = \underline{98.3 \text{ in.}^4,} \qquad \text{Ans.}$$

$$I_{min} = OC - CG = 59.8 - 38.5 = \underline{21.3 \text{ in.}^4.} \qquad \text{Ans.}$$

Twice the angle from the x axis to the principal axis with the maximum moment of inertia is shown on the sketch as $2\beta_{max}$, and

$$\beta_{max} = \frac{1}{2} \tan^{-1} \frac{32.3}{21.0}$$

$$= 28.5° \text{ counterclockwise from the } x \text{ axis.}$$

The angle from the x axis to the principal axis with the minimum moment of inertia is β_{min} and is

$$\beta_{min} = \beta_{max} + 90°$$

$$= 118.5° \text{ counterclockwise from the } x \text{ axis.}$$

These values were obtained analytically from the geometry of Mohr's circle. They could also have been measured directly from the figure. The accuracy of the results obtained by scaling the distances from the figure would, of course, depend on the scale used and the care employed in constructing the figure.

PROBLEMS

6-64 Determine the maximum and minimum moments of inertia of the shaded area in Fig. P6-64 with respect to axes through the origin. Locate the principal axes on a sketch.

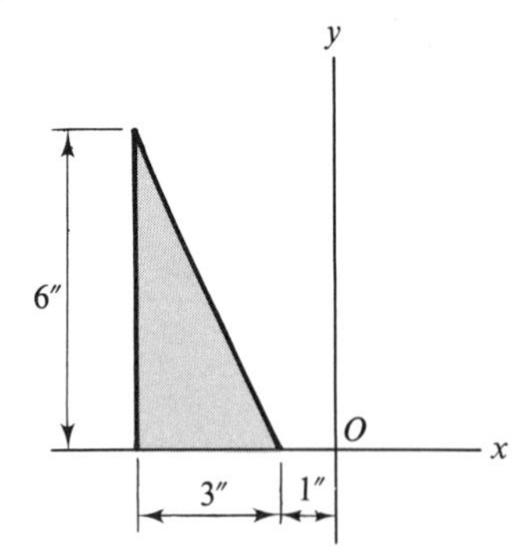

Figure P6-64

***6-65** Determine the maximum and minimum rectangular moments of inertia of the shaded area in Fig. P6-65 with respect to a set of axes through the origin. Show on a sketch the location of the principal axes.

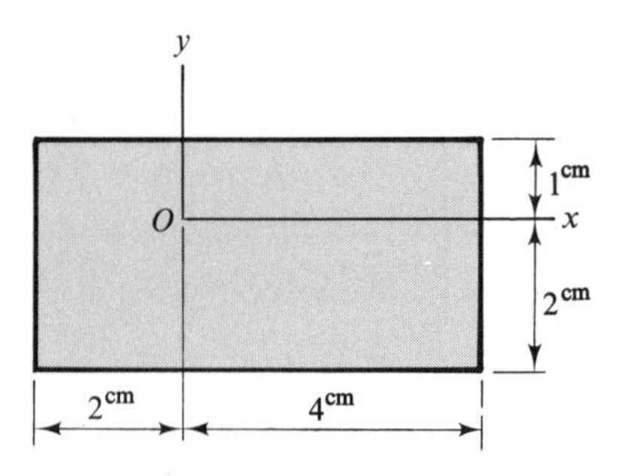

Figure P6-65

6-66 Determine the principal moments of inertia of the area shown in Fig. P6-66 with respect to axes through the origin. Show the location of the corresponding axes on a sketch. The equation of the curve is $y^2 = x$.

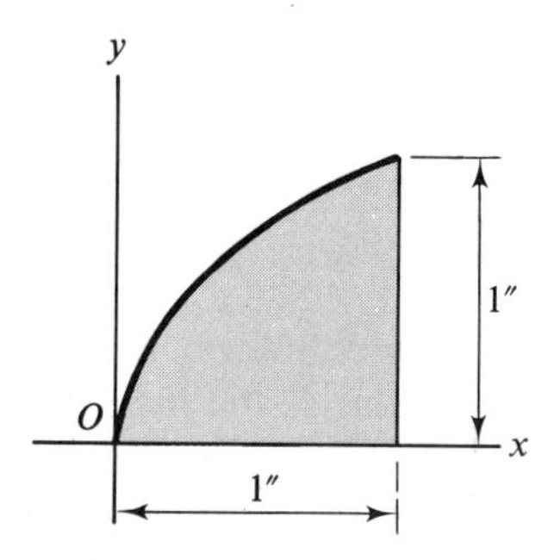

Figure P6-66

6-67 (a) Use Mohr's circle to determine the moment of inertia of the shaded area in Fig. P6-67 with respect to the x' axis.
(b) Determine the maximum product of inertia of this area with respect to axes through the origin.

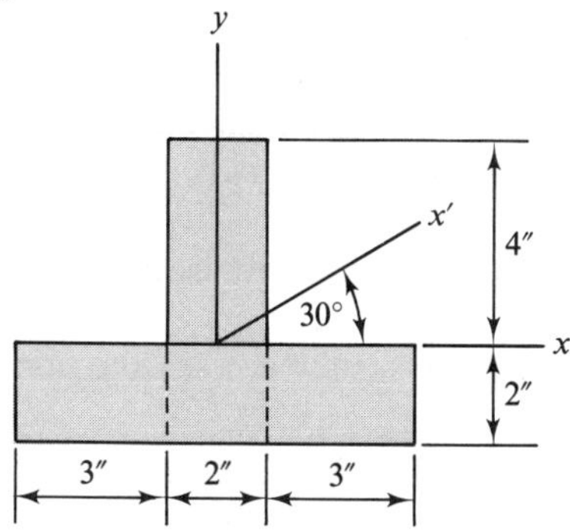

Figure P6-67

***6-68** Determine the moment of inertia of the area shown in Fig. P6-68 with respect to the principal axes through the origin. Show the location of the corresponding axes. The equation of the curve is $x^2 = 6y$.

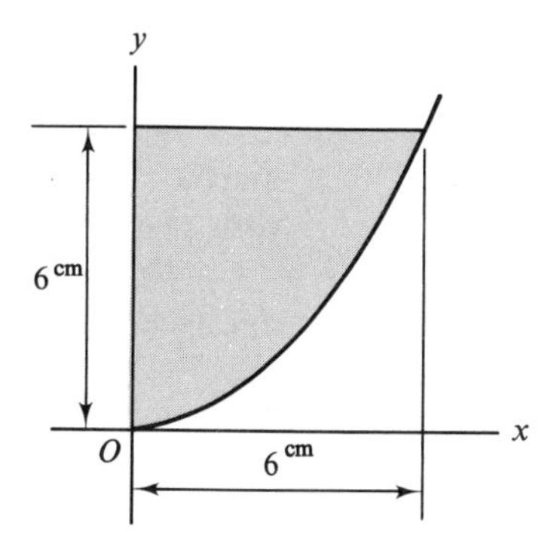

Figure P6-68

6-69 A lintel is made by welding a 7 by 4 by $\frac{1}{2}$-in. angle to a 12-in., 25-lb channel as shown in Fig. P6-69. Determine the minimum radius of gyration of the area with respect to an axis through the centroid.

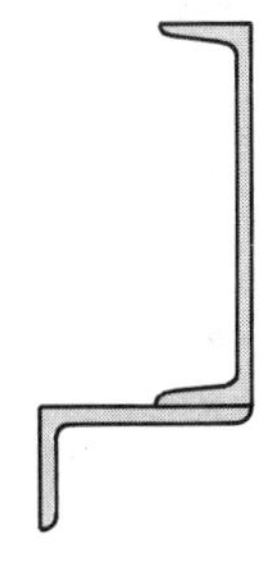

Figure P6-69

PART B–MASSES

6-10 DEFINITIONS

In the analysis of the motion of rigid bodies, the moment of inertia (and sometimes the product of inertia) of the mass of the body with respect to an axis (or a pair of planes) is frequently required.

The moment of inertia or second moment of an element of mass dm (see Fig. 6-21) *with respect to any axis or line is defined as the product of the mass of the element and the square of the distance from the axis to the element.* Thus the moment of inertia of the element of mass with respect to the x axis is

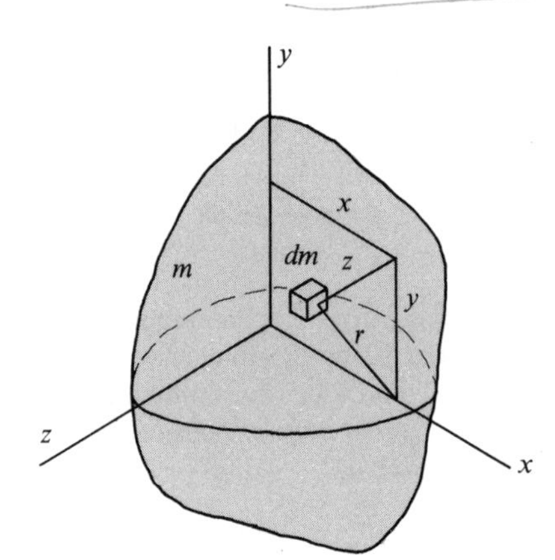

Figure 6-21

$$dI_x = r^2\,dm = (y^2 + z^2)\,dm.$$

Similar expressions can be written for the y and z axes. The quantity $y^2\,dm$ is sometimes referred to as the *second moment* of the mass of the element with respect to the xz plane. Similarly, $z^2\,dm$ is called the second moment of the mass of the element with respect to the xy plane. These quantities are useful primarily as part of the expression for dI_x. Since they do not arise individually in physical equations, they are seldom computed separately.

The product of inertia of an element of mass dm (as in Fig. 6-21) *with respect to a pair of orthogonal coordinate planes is defined as the product of the mass of the element and the coordinate distances from the planes to the element.* For example, the product of inertia of the element with respect to the xz and yz planes is

$$dI_{xy} = xy\,dm.$$

The sum of the moments (or products) of inertia of all elements of mass of a body with respect to any axis (or pair of planes) is defined as the moment (or product) of inertia of the body with respect to the indicated reference. The expressions for the axes of Fig. 6-21 are

$$I_x = \int (y^2 + z^2)\,dm, \qquad I_y = \int (z^2 + x^2)\,dm, \qquad I_z = \int (x^2 + y^2)\,dm \qquad (6\text{-}9)$$

and

$$I_{xy} = \int xy\,dm, \qquad I_{yz} = \int yz\,dm, \qquad I_{zx} = \int zx\,dm.$$

The equation for I_x may also be written as

$$I_x = \int y^2\,dm + \int z^2\,dm = I_{xz\ \text{plane}} + I_{xy\ \text{plane}}, \qquad \text{(a)}$$

and similar relationships exist for I_y and I_z. It is important to note the difference between the moment of inertia of mass with respect to the xy plane, $I_{xy\ \mathrm{plane}}$, and the product of inertia I_{xy}. Equation (a) can be stated as follows: *The moment of inertia of the mass of a body with respect to any axis is equal to the sum of its moments of inertia with respect to two perpendicular planes which intersect along the axis.* The moments of inertia with respect to planes are useful only in that they sometimes simplify the determination of the moments of inertia of bodies with respect to the line of intersection of two planes.

Moments and products of inertia of mass have the dimensions of mass multiplied by length squared, mL^2. When force, length, and time are selected as the fundamental quantities, mass is FT^2L^{-1} (see Art. 1-8), and the second moment of mass has dimensions of $(FT^2L^{-1})L^2$ or FT^2L. Common units for moment of inertia are lb-sec^2-ft, kg·m^2, or other combinations having the same dimensions. If the mass of the body, W/g, is expressed in slugs (lb-sec^2/ft), it is usually desirable to express the moment of inertia in slug-ft^2 instead of slug-in.^{2} to avoid a combination of feet and inches.

The moment of inertia of mass is always a positive quantity, since it is the sum of the products of elements of mass, which are inherently positive, and distances squared. The product of inertia of a body may be positive, negative, or zero, since the two coordinates have independent signs, and the product will be positive for coordinates with the same sign and negative when they have opposite signs.

The product of inertia of the mass of a body with respect to two perpendicular planes is zero if either of the planes is a plane of symmetry. This is true because the elements will always occur in pairs on opposite sides of the plane of symmetry one of which will have a positive and one a negative product of inertia. Any plane of symmetry will, of course, pass through the mass center of the body, and the perpendicular plane used in computing the product of inertia may or may not pass through the mass center. It should be noted, however, that planes through the mass center are not necessarily planes of symmetry and that the product of inertia of the body with respect to such planes will not necessarily be zero.

6-11 PARALLEL-AXIS AND PARALLEL-PLANE THEOREMS FOR MASSES

The moment of inertia of the mass of a body with respect to any axis can be expressed in terms of the moment of inertia of mass of the body with respect to a parallel axis through the mass center as indicated in the following development. In Fig. 6-22, the x', y', and z' axes are parallel to the x, y, and z axes and pass through the mass center, G, of the body. The moment of inertia of the element of mass,

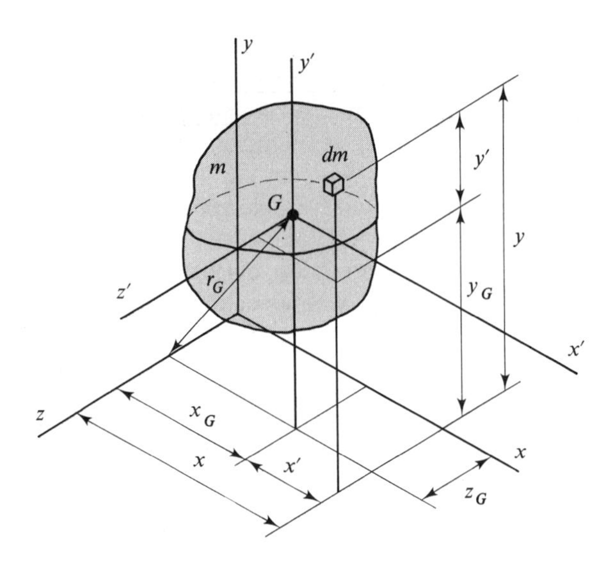

Figure 6-22

dm, with respect to the z axis is

$$dI_z = (x^2 + y^2)\,dm$$
$$= [(x_G + x')^2 + (y_G + y')^2]\,dm$$
$$= [(x_G^2 + y_G^2) + (x'^2 + y'^2) + 2x_G x' + 2y_G y']\,dm$$

and

$$I_z = (x_G^2 + y_G^2)\int dm + \int (x'^2 + y'^2)\,dm + 2x_G \int x'\,dm + 2y_G \int y'\,dm.$$

From Fig. 6-22, $(x_G^2 + y_G^2)$ is equal to r_G^2. Also, $\int (x'^2 + y'^2)\,dm$ is the moment of inertia of the mass of the body with respect to the z' (centroidal) axis, and the two expressions $\int x'\,dm$ and $\int y'\,dm$ are both equal to zero, since they represent the first moment of the mass with respect to the centroidal planes parallel to the yz and xz planes, respectively. Thus the equation can be written

$$I_z = I_{z'} + mr_G^2 = I_G + mr_G^2 \tag{6-10}$$

where I_G is the moment of inertia of the mass of the body with respect to an axis through the mass center parallel to the z axis. In general, *the moment of inertia of the mass of a body with respect to any axis is equal to the sum of the moment of inertia of the mass with respect to a parallel axis through the mass center and the product of the mass and the square of the distance between the two axes.*

A similar theorem can be proved for the moment of inertia of the mass of a body with respect to a *plane*. The theorem states that the moment of inertia of the mass of a body with respect to any plane is equal to the sum of the moment of inertia of the mass with respect to a parallel plane through the mass center of the body and the product of the mass of the body and the square of the distance between the two planes.

Section 6-11 Parallel-axis parallel-plane theorems for masses

A similar relationship can be developed for products of inertia. The product of inertia of the element of mass in Fig. 6-22 with respect to the xz and yz coordinate planes is

$$dI_{xy} = xy\,dm = (x_G + x')(y_G + y')\,dm$$
$$= (x_G y_G + x_G y' + x' y_G + x' y')\,dm.$$

The product of inertia of the body is

$$I_{xy} = x_G y_G \int dm + x_G \int y'\,dm + y_G \int x'\,dm + \int x'y'\,dm$$
$$= x_G y_G m + I_{xy_G} \tag{6-11}$$

since $\int y'\,dm = \int x'\,dm = 0$. Equation (6-11) indicates that *the product of inertia of the mass of a body with respect to any two perpendicular planes is equal to the sum of the product of inertia of the mass with respect to the two parallel planes through the mass center of the body and the product of the mass and the two coordinates of the mass center from the given planes.*

PROBLEMS

6-70 The moment of inertia of mass of the 322-lb homogeneous sphere in Fig. P6-70 with respect to the x axis is 134 slug-ft². Determine its moment of inertia with respect to the a axis.

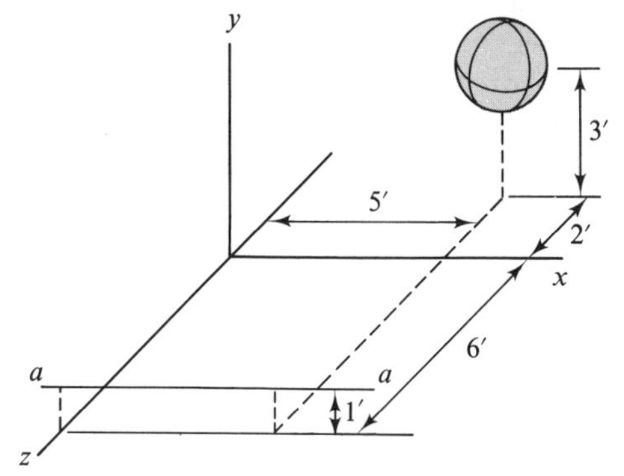

Figure P6-70

6-71 Determine the moment of inertia of mass of the sphere of Problem 6-70 with respect to
(a) The y axis.
(b) The z axis.

6-72 For the sphere of Problem 6-70, determine the following products of inertia of mass:
(a) I_{xy}, (b) I_{xz}, (c) I_{yz}.

6-73 Determine the second moment of mass of the sphere of Problem 6-70 with respect to
(a) The xy plane.
(b) The xz plane.

6-74 The homogeneous tetrahedron or pyramid in Fig. P6-74 weighs 5.80 lb and has a product of inertia of mass I_{xy} (with respect to the yz and xz planes) of 0.0180 lb-sec²-in. Determine the product of inertia $I_{x'y'}$ (with respect to the $y'z'$ and $x'z'$ planes).

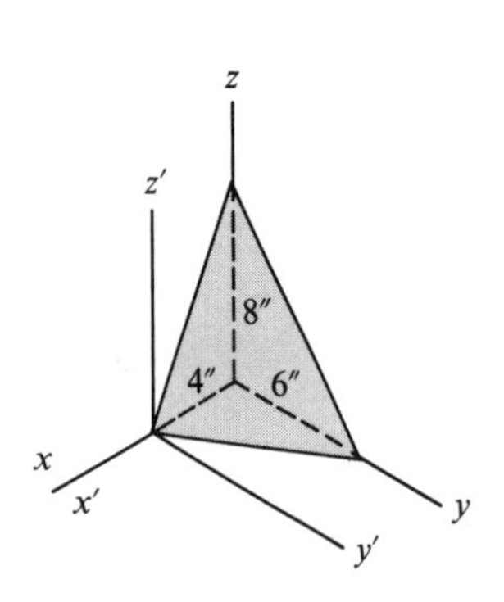

Figure P6-74

6-75 Figure P6-75 represents a homogeneous right circular cone. Which, if any, of the quantities I_{xy}, I_{yz}, and I_{zx} are zero? Determine in terms of the mass m and the dimensions a and h any of the quantities $I_{x'y'}$, $I_{y'z'}$, and $I_{z'x'}$ which can be calculated without integration.

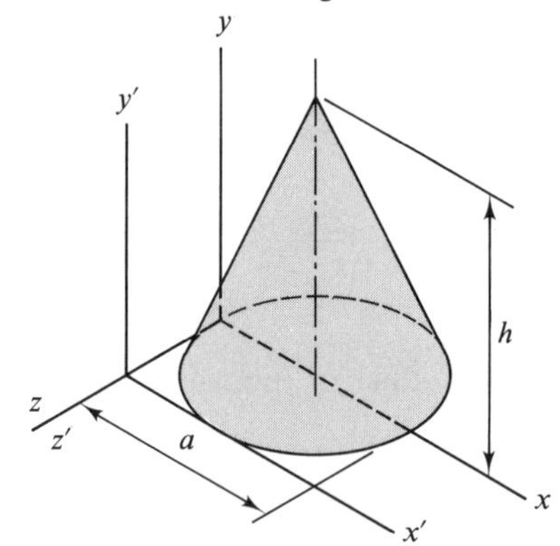

Figure P6-75

6-12 MOMENTS AND PRODUCTS OF INERTIA BY INTEGRATION

In determining moments and products of inertia of mass by integration, the element can be selected for single, double, or triple integration. The choice of element depends on whether the moment or product of inertia is to be obtained and on the orientation of the reference axis or planes. The element should be selected in such a manner that either (1) all parts of the element are the same distance from the reference axis (or from the reference planes for products of inertia), or (2) the moment or product of inertia of the element with respect to the reference axis or planes is known or can be determined. When triple integration is used, the element always satisfies the first requirement, but this condition is not necessarily satisfied by elements requiring single or double integration.

It will be helpful if the element selected is shown on a sketch of the body and dimensioned completely.

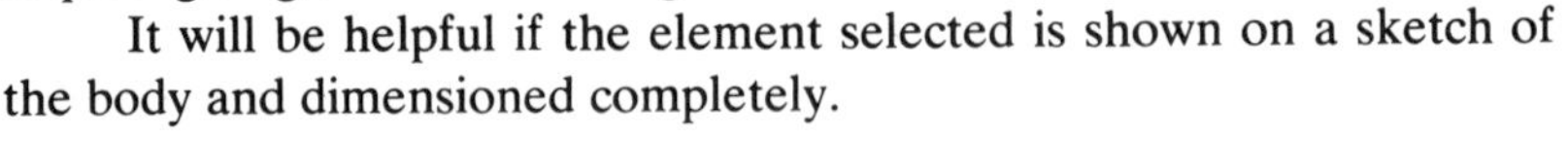

EXAMPLE 6-12

Determine the moment of inertia of the mass of a homogeneous right circular cylinder with respect to the geometrical axis. Express the result in terms of the mass of the cylinder.

SOLUTION

A cylinder of radius R and height h is shown in Fig. 6-23. A thin cylindrical element, all parts of which are the same distance from the geometrical axis, is selected, and its dimensions are shown in the figure. The mass of the element can be expressed as the product of the volume and the density (mass per unit volume):

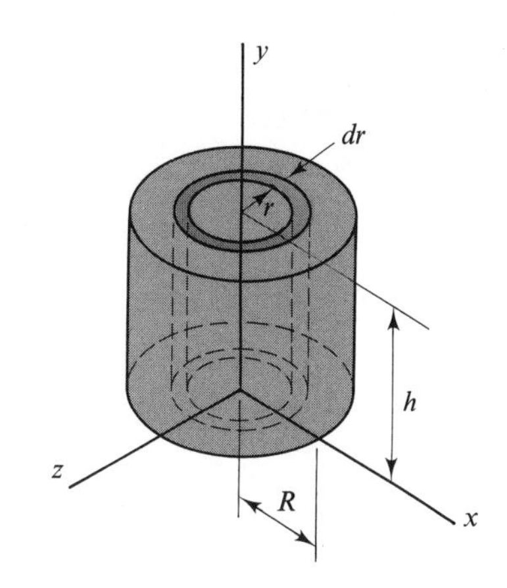

Figure 6-23

$$dm = \rho\, dV = \rho 2\pi r h\, dr,$$

where ρ is the density of the material.

The mass of the cylinder is

$$m = \int_0^R \rho 2\pi r h\, dr = \rho\pi R^2 h.$$

The moment of inertia of the mass of the element with respect to the y axis is

$$dI_y = r^2\, dm = \rho 2\pi r^3 h\, dr$$

and

$$I_y = \int_0^R \rho 2\pi r^3 h\, dr = \frac{\rho\pi R^4 h}{2}.$$

The moment of inertia of the mass is thus expressed in terms of the density and dimensions of the body. The moment of inertia can be expressed in terms of the mass by multiplying by m and dividing by its equivalent in terms of the density and dimensions of the body as follows:

$$I_y = \frac{\rho\pi R^4 h}{2}\frac{m}{\rho\pi R^2 h} = \frac{1}{2} mR^2. \qquad \text{Ans.}$$

EXAMPLE 6-13

Determine the moment of inertia of the mass of a homogeneous right circular cylinder with respect to

(a) A diameter in the base of the cylinder.

(b) A diameter through the mass center of the cylinder.

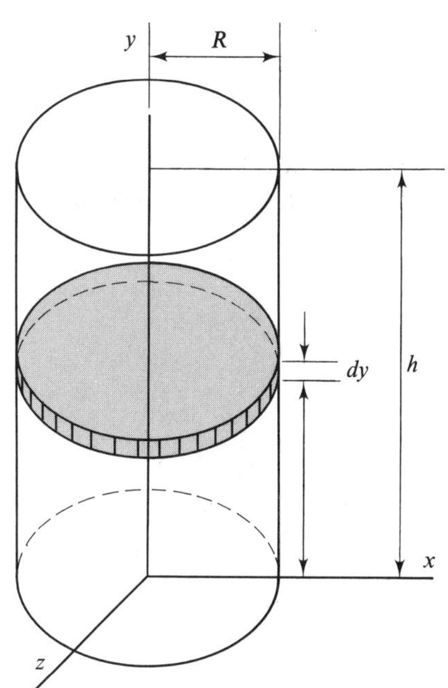

Figure 6-24

SOLUTION

(a) A cylinder of radius R and height h is shown in Fig. 6-24. The x axis is a diameter in the base. Unless double or triple integration is used, an element having all points the same distance from the x axis would be quite involved. However, the moment of inertia with respect to the x axis is equal to the sum of the moments of inertia with respect to the xy and xz planes;

that is

$$I_x = I_{xy\ \text{plane}} + I_{xz\ \text{plane}}$$

The value of $I_{xy\ \text{plane}}$ is equal to $I_{yz\ \text{plane}}$ from symmetry, and

$$I_y = I_{xy\ \text{plane}} + I_{yz\ \text{plane}} = 2I_{xy\ \text{plane}} = \frac{1}{2}mR^2 \quad \text{(from Example 6-12)}.$$

Therefore

$$I_{xy\ \text{plane}} = \frac{1}{4}mR^2.$$

All points of the element in the shape of a flat disk as shown are the same distance from the xz plane. The mass of the element is

$$dm = \rho\, dV = \rho\pi R^2\, dy,$$

and its moment of inertia with respect to the xz plane is

$$dI_{xz\ \text{plane}} = y^2\, dm = \rho\pi R^2 y^2\, dy.$$

The moment of inertia of the mass of the cylinder with respect to the xz plane is

$$I_{xz\ \text{plane}} = \int_0^h \rho\pi R^2 y^2\, dy = \rho\pi R^2 \frac{y^3}{3}\bigg]_0^h = \frac{\rho\pi R^2 h^3}{3},$$

which can be expressed in terms of the mass as

$$I_{xz\ \text{plane}} = \frac{\rho\pi R^2 h^3}{3}\frac{m}{\rho\pi R^2 h} = \frac{1}{3}mh^2.$$

The moment of inertia about the x axis is

$$I_x = I_{xy\ \text{plane}} + I_{xz\ \text{plane}} = \frac{mR^2}{4} + \frac{mh^2}{3} = \frac{m}{12}(3R^2 + 4h^2). \quad \text{Ans.}$$

(b) The moment of inertia of the cylinder with respect to an axis through the mass center parallel to the x axis is

$$I_G = I_x - m\left(\frac{h}{2}\right)^2 = \frac{m}{12}(3R^2 + 4h^2 - 3h^2) = \frac{m(3R^2 + h^2)}{12}. \quad \text{Ans.}$$

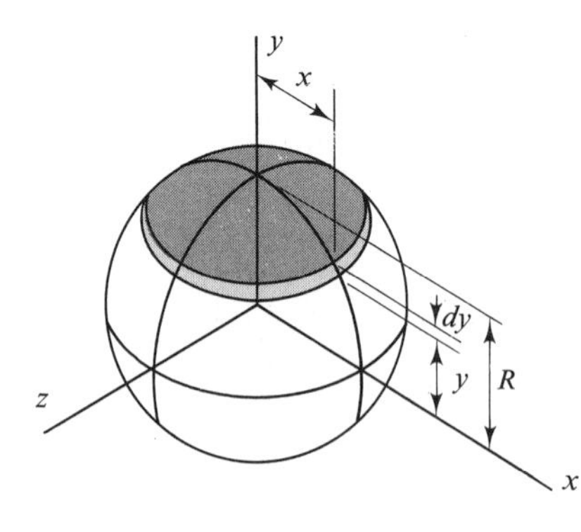

Figure 6-25

EXAMPLE 6-14

Determine the moment of inertia of the mass of a solid homogeneous sphere with respect to any centroidal axis.

SOLUTION

Figure 6-25 represents a sphere with a radius R. The y axis is selected as the centroidal axis for this example. The element shown is a thin cylindrical disk. The moment of inertia of the element of mass with respect to the y axis is (see Example 6-12) $dm\,r^2/2$. The mass of the element is

$$dm = \rho\pi x^2\, dy = \rho\pi(R^2 - y^2)\, dy$$

since the equation of the circle in the xy plane is $x^2+y^2=R^2$. The mass of the sphere can be obtained by integration or by multiplying the volume of the sphere by the density of the material. In either case,

$$m=\frac{4\rho\pi R^3}{3}.$$

The moment of inertia of the mass of the cylindrical element is

$$dI_y=\tfrac{1}{2}(dm)x^2=\tfrac{1}{2}(\rho\pi x^2\,dy)x^2=\tfrac{1}{2}\rho\pi(R^4-2R^2y^2+y^4)\,dy,$$

and the total moment of inertia is

$$I_y=\int_{-R}^{R}\frac{1}{2}\rho\pi(R^4-2R^2y^2+y^4)\,dy=\frac{8\rho\pi R^5}{15}.$$

The moment of inertia can be expressed in terms of the mass of the sphere as follows:

$$I_y=\frac{8\rho\pi R^5}{15}\,\frac{m}{(4\rho\pi R^3/3)}=\frac{2mR^2}{5}. \qquad \text{Ans.}$$

EXAMPLE 6-15

A solid of revolution is formed by revolving the area bounded by the curve $y^2=ax$, the line $x=a$ and the x axis around the x axis (see Fig. 6-26). The density of the material in the solid is proportional to the distance from the x axis.

(a) Determine the moment of inertia of the mass of the body with respect to the x axis. Express the result in terms of the mass of the body.
(b) The body weighs 16 lb and a is 4 in. Determine the moment of inertia of the mass of the body with respect to the x axis.

SOLUTION

(a) The hollow cylindrical element of mass in Fig. 6-26 is selected because the density of the element is constant and all parts of the element are the same distance from the moment axis. The mass of the element is

$$dm=\rho\,dV=\rho 2\pi y(a-x)\,dy,$$

and since

$$\rho=Ky \quad \text{and} \quad x=\frac{y^2}{a},$$

$$dm=K2\pi y^2\left(a-\frac{y^2}{a}\right)dy=K2\pi\left(ay^2-\frac{y^4}{a}\right)dy.$$

The mass of the body is

$$m=K2\pi\int_0^a\left(ay^2-\frac{y^4}{a}\right)dy=K2\pi\left[\frac{ay^3}{3}-\frac{y^5}{5a}\right]_0^a=\frac{4}{15}\pi Ka^4.$$

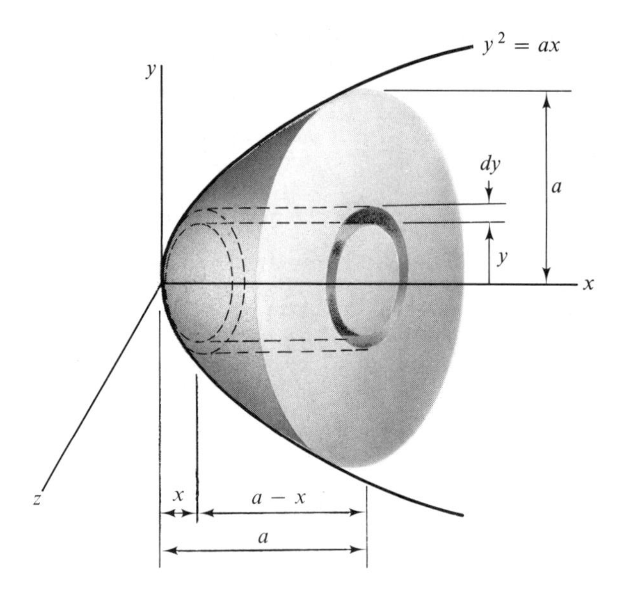

Figure 6-26

The moment of inertia of the mass of the element with respect to the x axis is

$$dI_x = y^2\,dm = K2\pi\left(ay^4 - \frac{y^6}{a}\right)dy,$$

and the moment of inertia of the body is

$$I_x = K2\pi\int_0^a\left(ay^4 - \frac{y^6}{a}\right)dy = K2\pi\left[\frac{ay^5}{5} - \frac{y^7}{7a}\right]_0^a$$

$$= K2\pi\left(\frac{2a^6}{35}\right) = \frac{K4\pi a^6}{35}\,\frac{m}{\frac{4}{15}\pi K a^4} = \underline{\frac{3}{7}ma^2}. \qquad \text{Ans.}$$

(b) The mass of the body is equal to its weight divided 'by the acceleration of gravity; that is,

$$m = \frac{W}{g} = \frac{16}{32.2} = 0.497\,\frac{\text{lb-sec}^2}{\text{ft}} = 0.497\text{ slug}.$$

The length a should have the same units of length as the acceleration g; thus $a = 4$ in. $= \frac{1}{3}$ ft, and the moment of inertia becomes

$$I_x = \frac{3}{7}(0.497)\left(\frac{1}{3}\right)^2 = \underline{0.0236\text{ lb-sec}^2\text{-ft}} = \underline{0.0236\text{ slug-ft}^2}. \qquad \text{Ans.}$$

If g and a were both expressed in inches, the result would be

$$I_x = \frac{3}{7}\,\frac{16}{12(32.2)}(4)^2 = \underline{0.284\text{ lb-sec}^2\text{-in.}} \qquad \text{Ans.}$$

EXAMPLE 6-16

Determine the product of inertia of the mass of the homogeneous tetrahedron or pyramid in Fig. 6-27 with respect to the xz and yz coordinate planes.

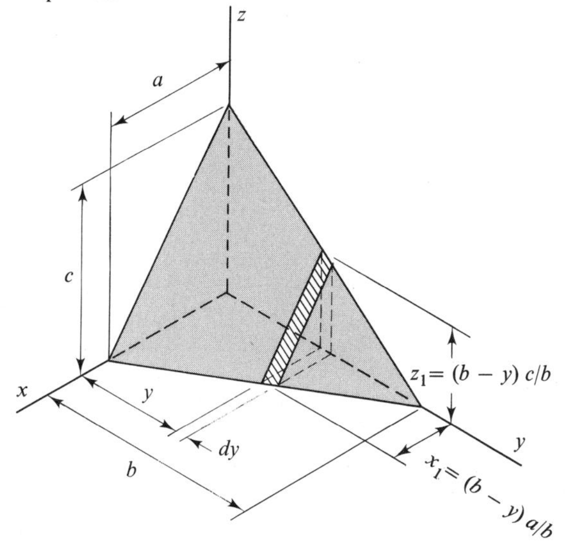

Figure 6-27

SOLUTION

An element parallel to one of the reference planes (the xz plane) is selected, since the plane of the element will be a plane of symmetry for the element. The product of inertia of the mass of the element with respect to a set of planes through the mass center of the element (parallel to the specified planes) is zero since one of the planes is a plane of symmetry. The mass of the element is

$$dm = \rho\, dV = \frac{\rho x_1 z_1}{2}\, dy.$$

The expressions $x_1 = (a/b)(b-y)$ and $z_1 = (c/b)(b-y)$ can be obtained from similar triangles and the mass of the element becomes

$$dm = \frac{\rho ac}{2b^2}(b-y)^2\, dy.$$

The mass of the body is

$$m = \frac{\rho ac}{2b^2}\int_0^b (b-y)^2\, dy$$

$$= -\frac{\rho ac}{2b^2}\frac{(b-y)^3}{3}\Big]_0^b = \frac{\rho abc}{6}.$$

The product of inertia of the mass of the element is

$$dI_{xy} = dI_{xyG} + x_G y_G \, dm$$

$$= 0 + y \frac{x_1}{3} \, dm$$

$$= y \frac{a}{3b}(b-y) \frac{\rho ac}{2b^2}(b-y)^2 \, dy$$

$$= \frac{\rho a^2 c}{6b^3}(b^3 y - 3b^2 y^2 + 3by^3 - y^4) \, dy$$

and the product of inertia of the mass of the body is

$$I_{xy} = \frac{\rho a^2 c}{6b^3} \int_0^b (b^3 y - 3b^2 y^2 + 3by^3 - y^4) \, dy$$

$$= \frac{\rho a^2 c}{6b^3}\left[\frac{b^3 y^2}{2} - b^2 y^3 + \frac{3by^4}{4} - \frac{y^5}{5}\right]_0^b$$

$$= \left(\frac{\rho a^2 b^2 c}{120}\right)\left(\frac{m}{\rho abc/6}\right) = \underline{\frac{mab}{20}}. \qquad \text{Ans.}$$

EXAMPLE 6-17

A homogeneous solid of revolution is formed by revolving the shaded area in Fig. 6-28(a) about the a axis. The equation of the curve is $y^2 = b(b-x)$, where $b = 0.300$ m, and the density of the material is 800 kg/m³. Determine the moment of inertia of the mass of the body with respect to the z axis.

SOLUTION

The differential element of mass shown in Fig. 6-28(b) is a thin cylindrical disk with the a axis as its geometrical axis. The c axis through the mass center of the differential disk is parallel to the z axis. The mass of the element is the product of the density, ρ, and the volume, dV; hence

$$dm = \rho \, dV = \rho\pi(b-x)^2 \, dy = \frac{\rho\pi y^4}{b^2} \, dy.$$

From Example 6-13, the moment of inertia of the mass of the disk with respect to the c axis is

$$dI_c = \frac{dm(3R^2 + h^2)}{12}$$

$$= \frac{dm(b-x)^2}{4} + \frac{dm(dy)^2}{12}.$$

Since the last term is a differential of third order, it can be neglected when it is added to the first-order differential. Therefore

$$dI_c = \frac{\rho\pi(b-x)^4 \, dy}{4} = \frac{\rho\pi y^8 \, dy}{4b^4}.$$

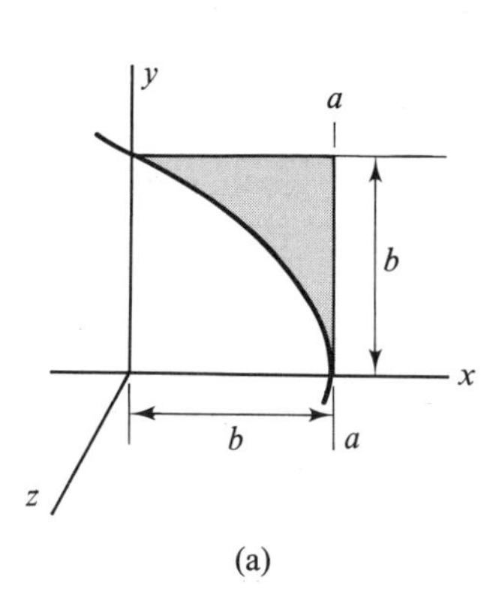

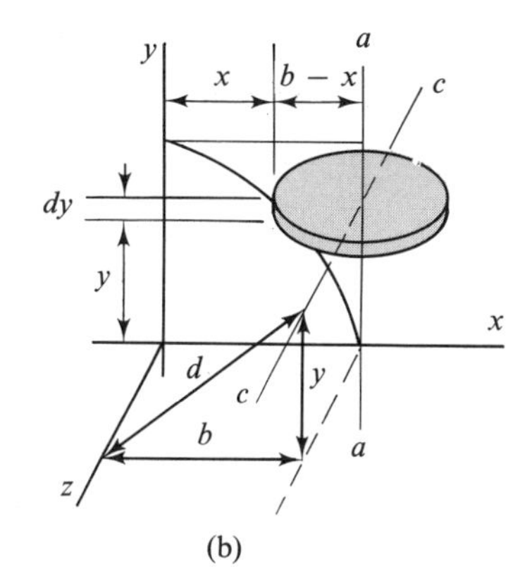

Figure 6-28

From Eq. (6-10), the moment of inertia of the differential element with respect to the z axis becomes

$$dI_z = dI_c + dm(b^2 + y^2)$$

$$= \frac{\rho\pi y^8}{4b^4}\,dy + \frac{\rho\pi y^4}{b^2}(b^2 + y^2)\,dy.$$

The moment of interia of the mass of the entire body can be determined by integration as follows:

$$I_z = \frac{\rho\pi}{4b^4}\int_0^b y^8\,dy + \frac{\rho\pi}{b^2}\int_0^b (b^2y^4 + y^6)\,dy$$

$$= \rho\pi b^5\left(\frac{1}{36} + \frac{1}{5} + \frac{1}{7}\right) = 800(\pi)(0.300)^5(0.3706)$$

$$= \underline{2.26\ \text{kg}\cdot\text{m}^2}. \qquad \text{Ans.}$$

PROBLEMS

6-76 In Example 6-13 it was shown that the moment of inertia of mass of a solid homogeneous right circular cylinder with respect to an axis through the mass center and perpendicular to the axis of the cylinder is $m(3R^2 + h^2)/12$. For a slender rod, an approximate value for this moment of inertia is $mh^2/12$. Determine the minimum ratio of the height, h, to the diameter, D, of the cylinder if the error when using the approximate formula is not to exceed 1%.

6-77 Determine the moment of inertia of the mass, m, of the homogeneous slender rod in Fig. P6-77 with respect to the x axis. Assume that the rod lies in the xy plane.

6-78 Determine the product of inertia of mass, I_{xy}, of the rod in Problem 6-77.

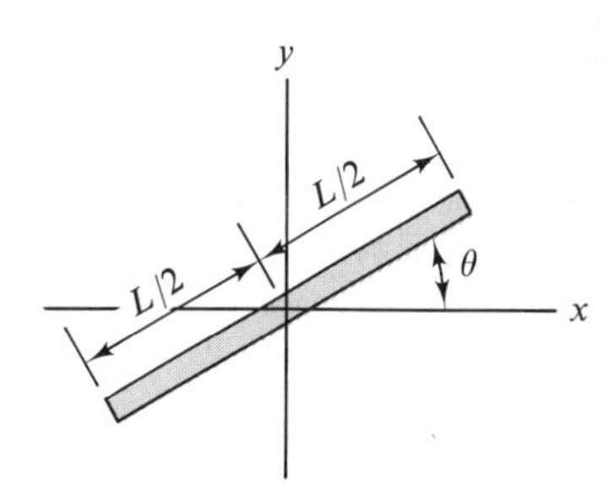

Figure P6-77

6-79 The density of a slender 6-ft rod of constant cross section varies directly as the square root of the distance from the left end. The rod weighs 128.8 lb. Determine the moment of inertia of mass with respect to an axis (or plane) perpendicular to the rod at the left end.

6-80 The cross section of a homogeneous hollow cylinder of mass m is shown in Fig. P6-80. Determine by integration the moment of inertia of mass with respect to the axis of the cylinder.

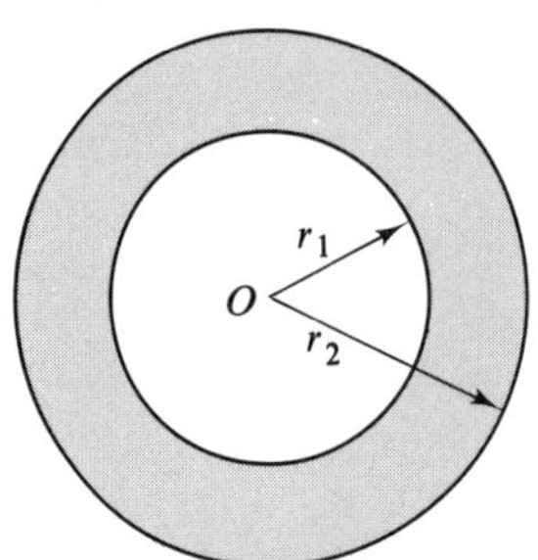

Figure P6-80

***6-81** The shaded area in Fig. P6-81 is revolved about the x axis to form a homogeneous cone with a density of 2.0 Mg/m³. Determine the moment of inertia of the mass with respect to
(a) The x axis. (b) The y axis.

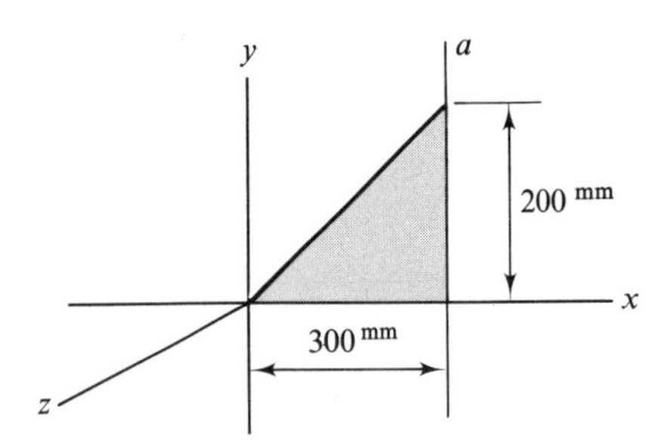

Figure P6-81

***6-82** The shaded area in Fig. P6-81 is revolved about the a axis to form a homogeneous solid with a density of 1.2 Mg/m³. Determine the moment of inertia of mass with respect to
(a) The a axis. (b) The x axis.

***6-83** The mass generated in Problem 6-81 has a variable density. Determine the moment of inertia of mass, in terms of the mass, with respect to
(a) The x axis when the density varies directly as the distance from the x axis.
(b) The y axis when the density varies directly as the distance from the yz plane.

***6-84** The mass generated in Problem 6-82 has a variable density. Determine the moment of inertia of mass, in terms of the mass, with respect to
(a) The a axis when the density varies directly as the distance from the a axis.
(b) The x axis when the density varies directly as the distance from the xz plane.

6-85 The solid homogeneous pyramid in Fig. P6-85 weighs 150 lb/ft³.
(a) Determine, by integration, the product of inertia I_{xy} (with respect to the yz and xz planes).
(b) Check the solution of part (a) by using the result of Example 6-16 and the parallel-axis theorem.

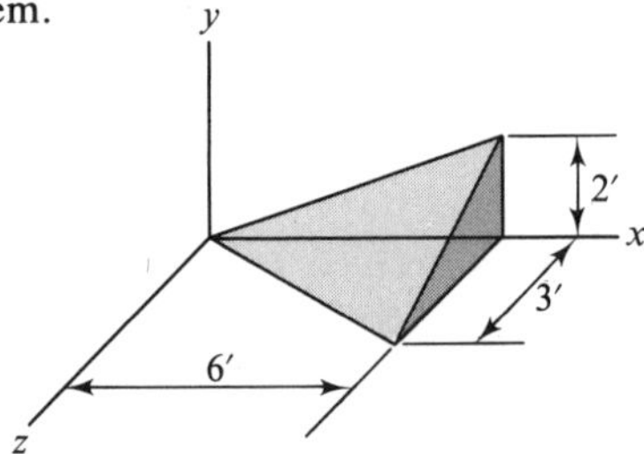

Figure P6-85

6-86 Determine I_y for the pyramid in Problem 6-85.

6-87 A homogeneous solid is formed by revolving the shaded area in Fig. P6-87 about the y axis. The material of the body weighs 48.3 lb/ft³. Determine the moment of inertia of mass of the solid with respect to the x axis. The equation of the curve is $x^2 = y$.

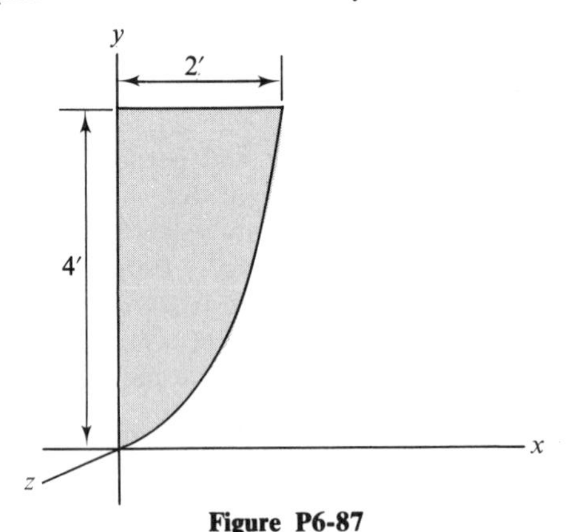

Figure P6-87

6-88 Solve Problem 6-87 if the density of the material varies as the square of its distance from the xz plane. Express the moment of inertia in terms of the mass of the body.

6-89 The shaded area in Fig. P6-89 is revolved about the a axis generating a homogeneous solid weighing 64.4 lb/ft³. Determine the moment of inertia of mass of the solid with respect to the a axis. The equation of the curve is $y^2 = 4x$.

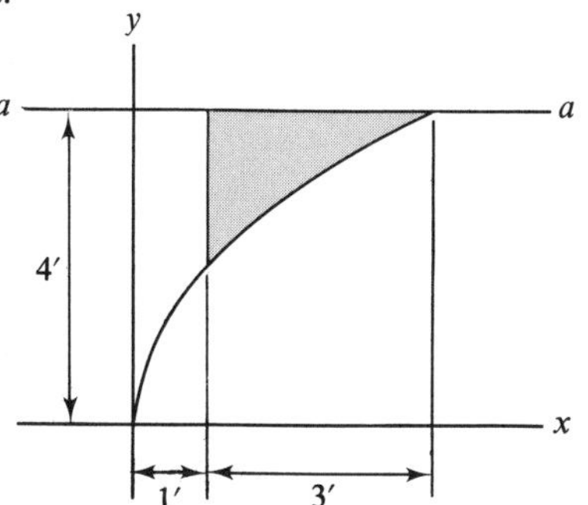

Figure P6-89

6-90 Determine the product of inertia, I_{xy}, for the mass of Problem 6-89.

6-91 Determine the moment of inertia of mass with respect to the y axis of the homogeneous body formed by revolving the shaded area in Fig. P6-91 about the y axis. The equation of the curve is $xy = 4b^2$.

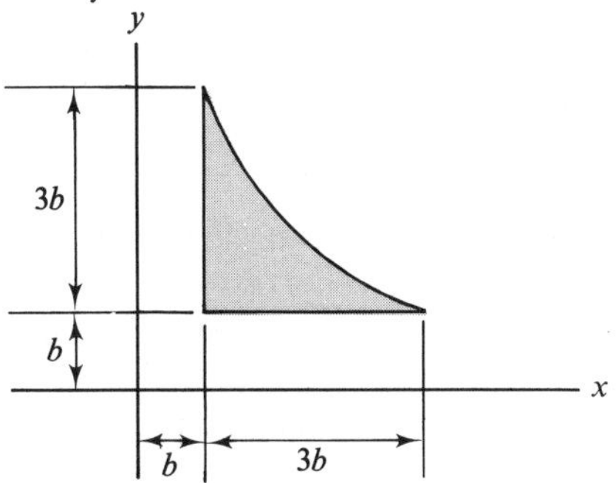

Figure P6-91

6-92 The density of the mass of Problem 6-91 varies directly as the distance from the axis of revolution. Determine the moment of inertia with respect to the y axis.

6-93 One element of an eccentric cone with a circular base lies along the z axis as shown in Fig. P6-93. The material in the cone is homogeneous. Determine the moment of inertia of the mass of the body with respect to the z axis.

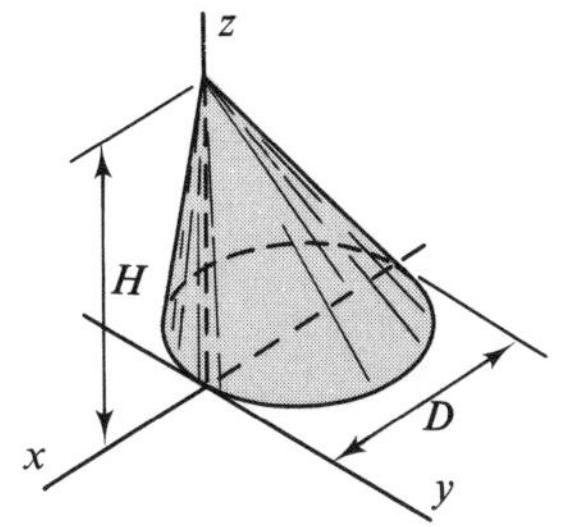

Figure P6-93

6-94 The triangular area in Fig. P6-94 is revolved about the center line to generate a homogeneous body. Determine the moment of inertia of the mass of the body with respect to the axis of rotation. The body is made of aluminum weighing 0.128 lb/in.³

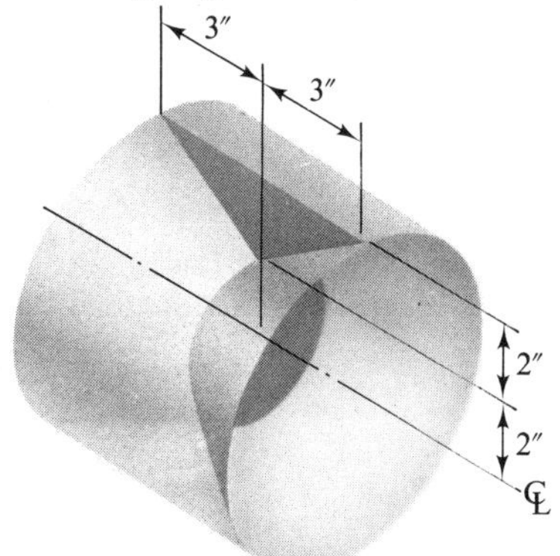

Figure P6-94

6-95 Figure P6-95 represents a homogeneous right triangular prism. Determine the products of inertia I_{xy} and I_{xz} of the mass of the body in terms of the mass and dimensions.

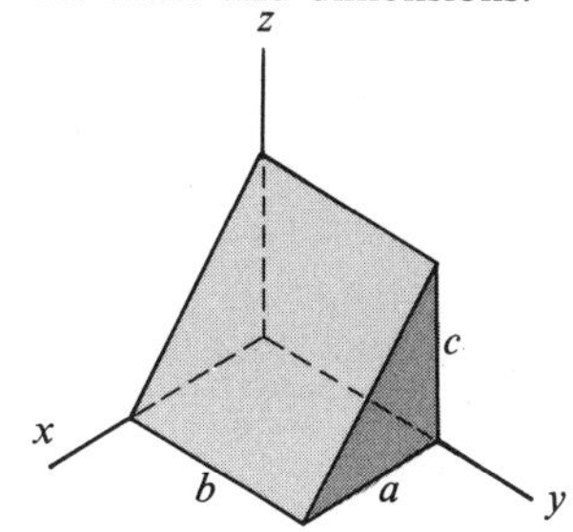

Figure P6-95

6-96 The homogeneous body in Fig. P6-96 is one-fourth of a circular cylinder. Determine the products of inertia I_{xy} and I_{yz} of the mass of the body.

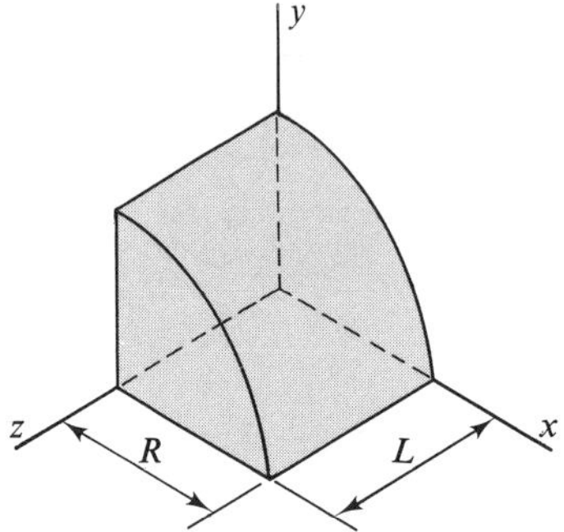

Figure P6-96

6-97 The shaded area in Fig. P6-97 is revolved about the y axis forming a solid whose density is proportional to the square of the distance from the xz plane. Determine the moment of inertia of the mass with respect to the a axis, which lies in the xz plane. The equation of the curve is $y^2 = x$.

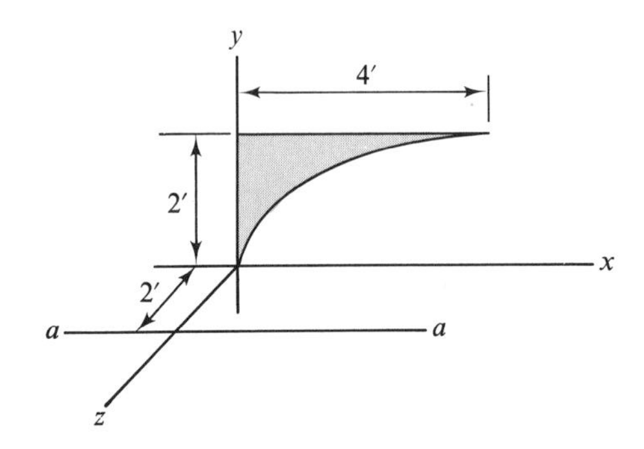

Figure P6-97

6-13 RADIUS OF GYRATION OF MASS

It was pointed out in Art. 6-10 that the dimensions of the moment of inertia of mass are mass multiplied by length squared. *The radius of gyration of the mass of a body with respect to an axis is defined as the length which must be squared and multiplied by the mass of the body to give the moment of inertia of the body with respect to the axis.* This definition is expressed mathematically in the following equation:

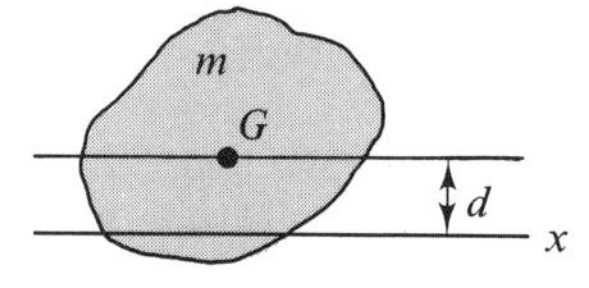

Figure 6-29

$$I_x = k_x^2 m.$$

The radius of gyration does not indicate the distance from the given axis to a fixed point in the body (such as the mass center). The parallel-axis theorem can be used to demonstrate that the radius of gyration of the mass of a body with respect to any axis is greater than the distance from the axis to the mass center of the body. From Fig. 6-29

$$I_x = I_G + md^2,$$

$$k_x^2 m = k_G^2 m + md^2,$$

$$k_x^2 = k_G^2 + d^2,$$

and

$$k_x = (k_G^2 + d^2)^{1/2},$$

which indicates that k_x is greater than d. The radius of gyration of the mass of a body with respect to an axis is the distance from the axis at which the entire mass of the body can be imagined to be concentrated and still have the same moment of inertia as the distributed mass.

The radii of gyration of the mass of a body with respect to two parallel lines 1 ft apart do not in general differ by 1 ft. In fact, there is no particularly useful physical interpretation or meaning of a radius of gyration. It is merely a convenient means of expressing the moment of inertia of the mass of a body in terms of its mass and a length.

PROBLEMS

6-98 Determine the radius of gyration of mass of a homogeneous slender rod of length L with respect to an axis
(a) Perpendicular to the axis of the rod through the mass center.
(b) Perpendicular to the axis of the rod at one end.
(c) Parallel to the rod at a distance d from the axis of the rod. Assume that d is large compared to the radius of the rod.

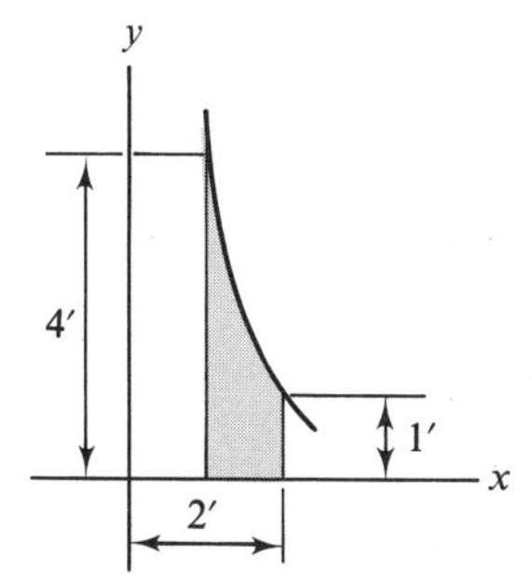

Figure P6-99

6-99 A homogeneous solid of revolution is generated by revolving the shaded area in Fig. P6-99 about the y axis. Determine the radius of gyration of the mass of the body with respect to the y axis. The equation of the curve is $x^2y = 4$.

6-100 The shaded area in Fig. P6-100 is revolved about the x axis forming a solid the density of which varies directly as the square of the distance from the yz plane. Determine the radius of gyration of mass with respect to the x axis.

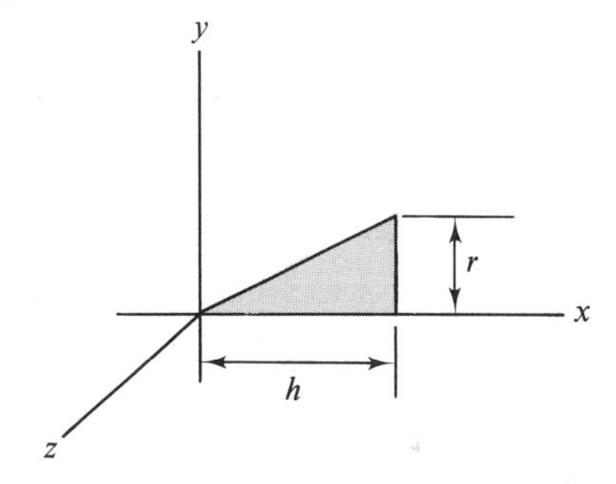

Figure P6-100

6-14 MOMENTS AND PRODUCTS OF INERTIA OF COMPOSITE MASSES

When the moment of inertia of a body is to be determined, it is frequently convenient to resolve the body into a number of simple shapes, such as cylinders, spheres, and rods. The parallel-axis theorem can be used to determine the moment of inertia of each part if the moment of inertia of each part with respect to a parallel axis through the mass center is known. The moment of inertia of the composite body is the algebraic sum of the values for each part, all calculated for the same axis. A similar statement applies to products of inertia. A table listing moments and products of inertia of several bodies is included on p. 296.

Moments and Products of Inertia of Homogeneous Bodies

Body	Moment of inertia	Product of inertia
Cylinder 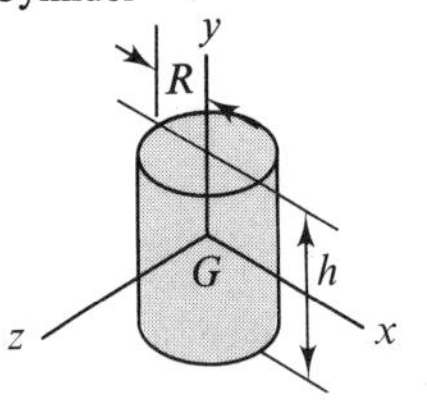	$I_x = I_z = \frac{m}{12}(3R^2 + h^2)$ $I_y = mR^2/2$	$I_{xy} = I_{yz} = I_{zx} = 0$
Sphere 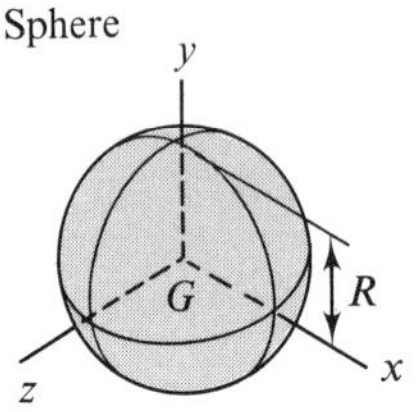	$I_x = I_y = I_z = 2mR^2/5$	$I_{xy} = I_{yz} = I_{zx} = 0$
Rectangular prism 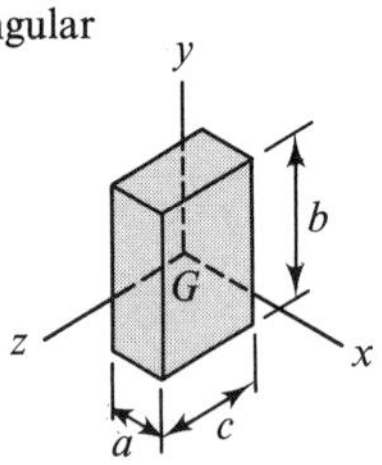	$I_x = m(b^2 + c^2)/12$ $I_y = m(c^2 + a^2)/12$ $I_z = m(a^2 + b^2)/12$	$I_{xy} = I_{yz} = I_{zx} = 0$
Triangular wedge 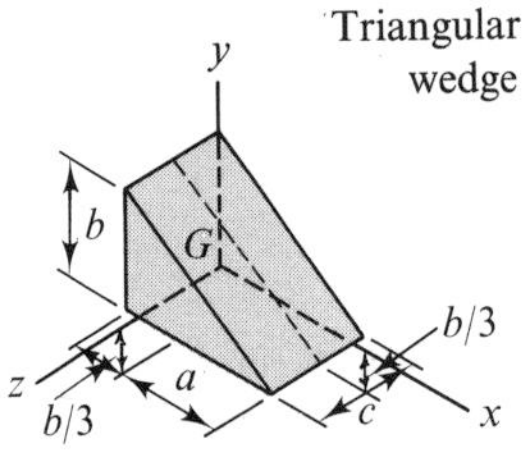	$I_x = m(2b^2 + 3c^2)/36$ $I_y = m(2a^2 + 3c^2)/36$ $I_z = m(a^2 + b^2)/18$	$I_{xy} = -mab/36$ $I_{yz} = 0$ $I_{zx} = 0$
Cone 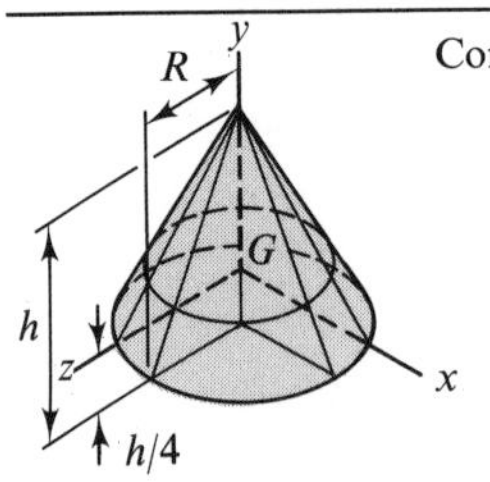	$I_x = I_z = 3m(h^2 + 4R^2)/80$ $I_y = 3mR^2/10$	$I_{zy} = I_{yz} = I_{zx} = 0$

EXAMPLE 6-18

The flywheel in Fig. 6-30 is made of steel weighing 490 lb/ft³. The spokes have elliptic cross sections. Determine the moment of inertia of the mass of the wheel with respect to the axis of the wheel.

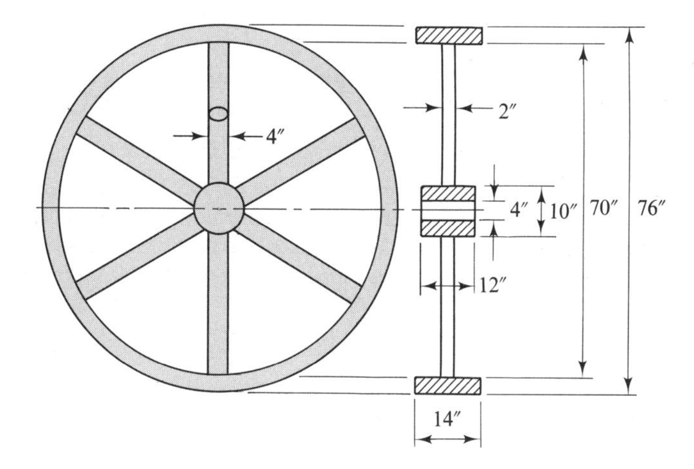

Figure 6-30

SOLUTION

The rim and hub can be considered solid cylinders with smaller cylinders removed, and the spokes can be assumed to be slender rods. The density of the material is

$$\rho = \frac{490}{32.2} = 15.22 \text{ slugs/ft}^3.$$

The moment of inertia of the mass of the rim with respect to the axis is

$$I_O = \tfrac{1}{2} m_{\text{solid}} (r_{\text{solid}})^2 - \tfrac{1}{2} m_{\text{hole}} (r_{\text{hole}})^2$$

$$= \frac{1}{2}\left[\pi\left(\frac{38}{12}\right)^2 \frac{14}{12}(15.22)\right]\left(\frac{38}{12}\right)^2 - \frac{1}{2}\left[\pi\left(\frac{35}{12}\right)^2 \frac{14}{12}(15.22)\right]\left(\frac{35}{12}\right)^2$$

$$= 2804 - 2018 = 786 \text{ slug-ft}^2.$$

The moment of inertia of the mass of the hub is

$$I_O = \frac{1}{2}\left[\pi\left(\frac{5}{12}\right)^2\left(\frac{12}{12}\right) 15.22\right]\left(\frac{5}{12}\right)^2 - \frac{1}{2}\left[\pi\left(\frac{2}{12}\right)^2\left(\frac{12}{12}\right) 15.22\right]\left(\frac{2}{12}\right)^2$$

$$= 0.721 - 0.0185 = 0.703 \text{ slug-ft}^2.$$

The area of the elliptical cross section of the spokes is $\pi ab/4$, where a and b are the major and minor axes. The moment of inertia of the spokes is

$$I_O = 6\left(\frac{ml^2}{12} + md^2\right)$$

$$= 6\left\{\frac{1}{12}\left[\frac{\pi}{4}\left(\frac{2}{12}\right)\left(\frac{4}{12}\right)\left(\frac{30}{12}\right)15.22\right]\left(\frac{30}{12}\right)^2\right.$$

$$\left. + \left[\frac{\pi}{4}\left(\frac{2}{12}\right)\left(\frac{4}{12}\right)\left(\frac{30}{12}\right)(15.22)\right]\left(\frac{20}{12}\right)^2\right\}$$

$$= 6[0.865 + 4.61] = 32.8 \text{ slug-ft}^2.$$

The total moment of inertia of the mass of the flywheel is

$$I_O = 786 + 0.7 + 32.8 = \underline{820 \text{ slug-ft}^2}. \qquad \text{Ans.}$$

Notice that the moment of inertia of the hub has a negligible effect on the result and that the spokes contribute only about 4 per cent of the total moment of inertia. It is thus apparent that the shape of the hub and spokes is relatively unimportant in calculating the moment of inertia of the mass of the flywheel.

PROBLEMS

6-101 The homogeneous semicircular cylinder of height h and radius R shown in Fig. P6-101 has a mass m. Determine its moment of inertia of mass with respect to the x axis.

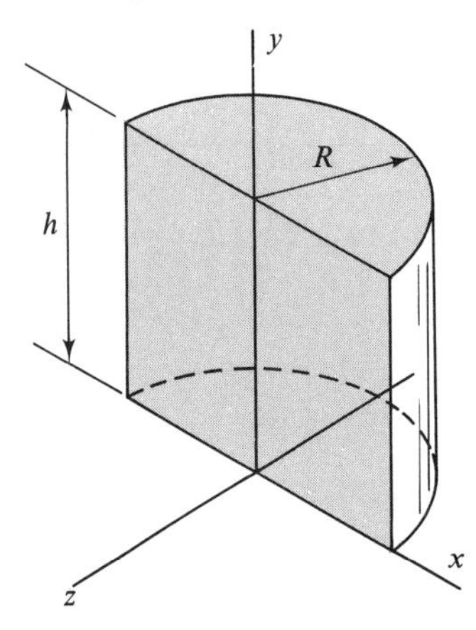

Figure P6-101

6-102 Determine the moment of inertia of the composite mass shown in Fig. P6-102 with respect to the line ab. The disks A and B weigh 48.3 lb each and are parallel to the yz plane, and the uniform slender rods C and D weigh 322 and 161 lb, respectively.

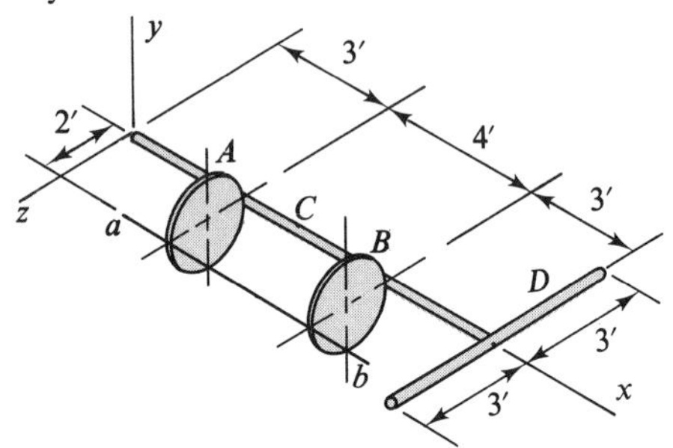

Figure P6-102

6-103 Body A of Fig. P6-103 weighs 322 lb, has its mass center at G, and has a radius of gyration of mass with respect to an axis through O perpendicular to the plane of the figure of 2.5 ft. The homogeneous bar BC weighs 96.6 lb, and the thin cylindrical shell D weighs 128.8 lb. Determine the radius of gyration of the mass of the composite body with respect to an axis through E perpendicular to the plane of the figure.

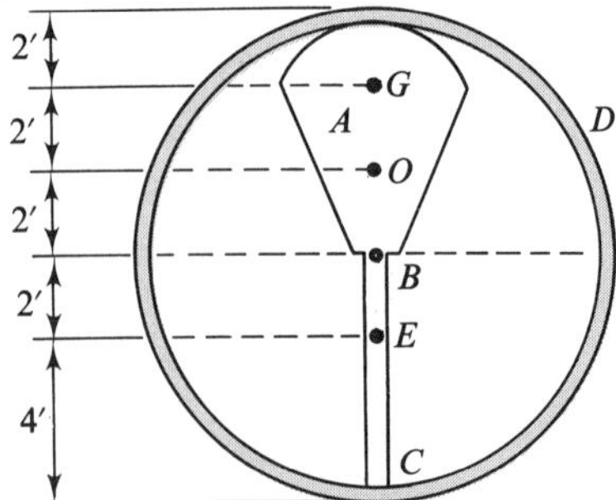

Figure P6-103

6-104 In Fig. P6-104, the slender homogeneous rods A and B and the thin homogeneous half-ring C (mass center at G) are welded together to form a rigid frame which rotates about the horizontal axis at O. The weights of A, B, and C are 64.4, 32.2, and 64.4 lb, respectively. Determine the moment of inertia of the mass of the assembly with respect to the axis of roatation.

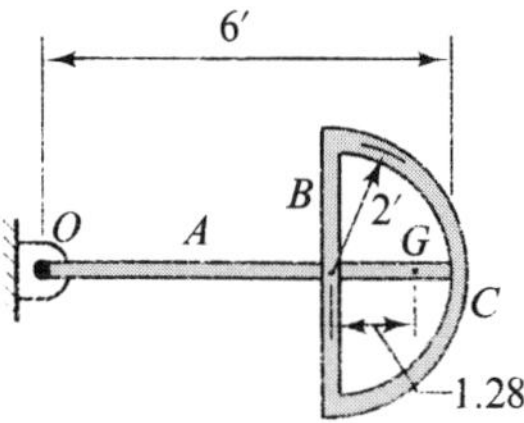

Figure P6-104

***6-105** A solid homogeneous cylinder with a density of 800 kg/m³ has an off-center 300-mm-diameter hole bored in it as shown in Fig. P6-105. Determine the moment of inertia of the body with respect to the a axis.

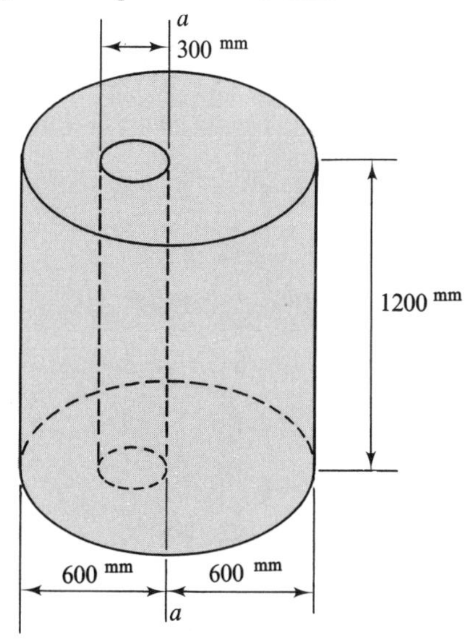

Figure P6-105

6-106 Determine the moment of inertia of the mass of the 322-lb homogeneous slender rod AB and the 966-lb nonhomogeneous cylinder C in Fig. P6-106 with respect to an axis through A perpendicular to the plane of the figure. The radius of gyration of the mass of body C with respect to an axis through O perpendicular to the plane of the figure is 3 ft, and the mass center of the cylinder is at G.

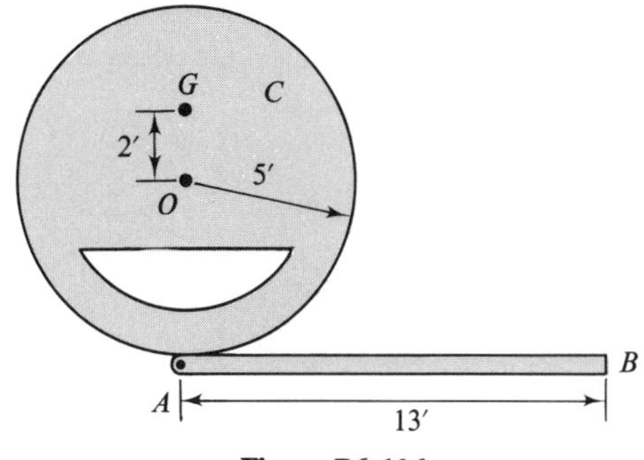

Figure P6-106

6-107 The rigid assembly in Fig. P6-107 is composed of a homogeneous 161-lb sphere A, homogeneous slender rods B and C weighing 16.1 lb per linear ft, and a homogeneous 128.8-lb cylinder D. The yz plane is a plane of symmetry. Determine

(a) The moment of inertia of the mass with respect to the z axis.

(b) The product of inertia I_{yz} for the assembly.

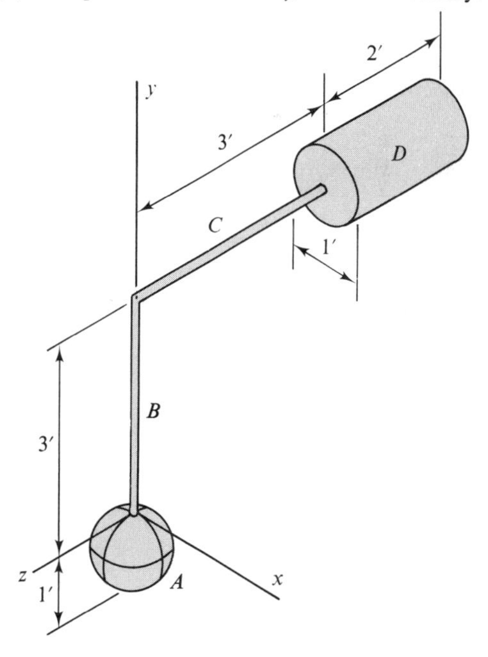

Figure P6-107

6-108 The homogeneous sphere C in Fig. P6-108 weighs 322 lb. The homogeneous bar OA lies in the xz plane and weighs 193.2 lb. The homogeneous bar AB is parallel to the y axis and weighs 161 lb. Determine

(a) The radius of gyration of mass of the composite body with respect to the y axis.

(b) The product of inertia I_{xy} of the mass of the body.

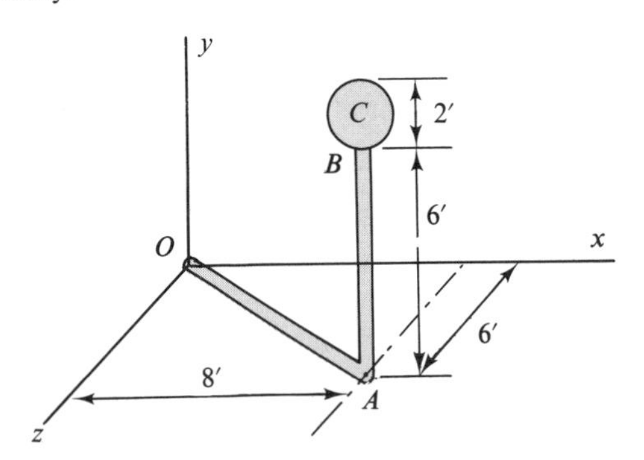

Figure P6-108

*6-109 The mass of the homogeneous hollow cylinder A in Fig. P6-109 is 4.5 Mg. Bodies B and C are homogeneous slender bars with masses of 2.0 and 1.5 Mg, respectively. Determine the radius of gyration of the body with respect to the a axis.

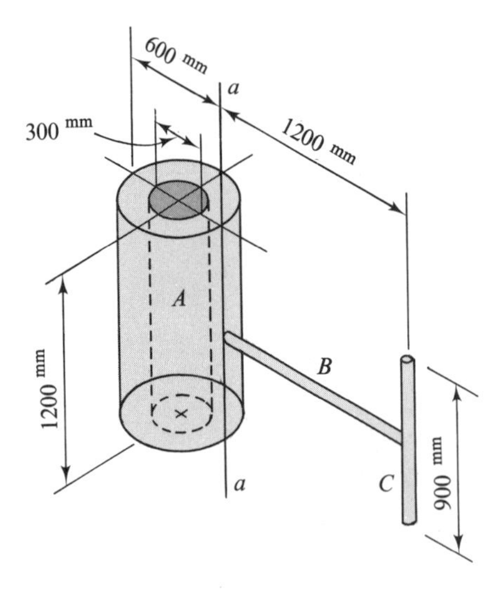

Figure P6-109

6-110 Body A in Fig. P6-110 is a sphere, B is a slender rod, and C is a hollow cylinder. The three bodies are homogeneous and weigh 161, 96.6, and 644 lb, respectively. Determine the moment of inertia of mass of the composite body with respect to the a axis.

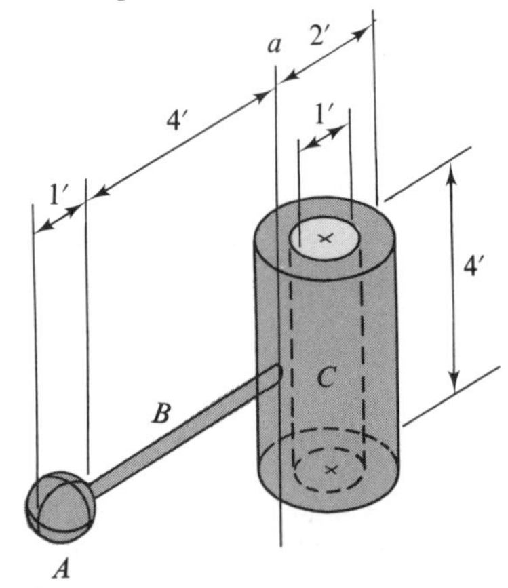

Figure P6-110

6-111 The 644-lb body A in Fig. P6-111 is a homogeneous sphere; bar BC and the cylinder D are homogeneous and weigh 128.8 and 966 lb, respectively. The xy plane is a plane of symmetry. Determine

(a) The radius of gyration of mass with respect to the y axis.

(b) The product of inertia of mass, I_{xy}.

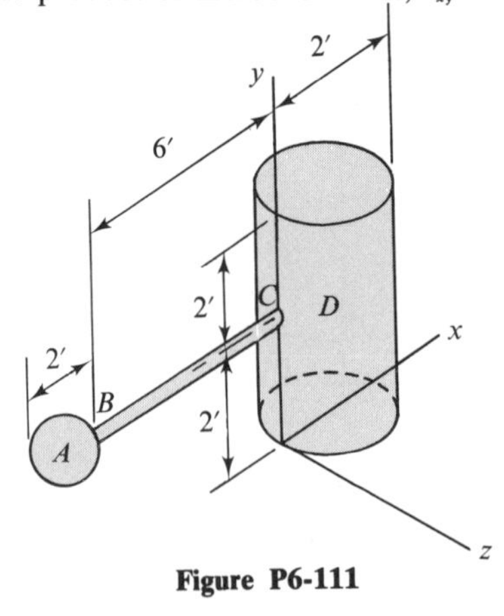

Figure P6-111

6-112 The frustum of the right circular cone in Fig. P6-112 weighs 64.4 lb per ft³. Determine the moment of inertia of mass with respect to the axis of symmetry.

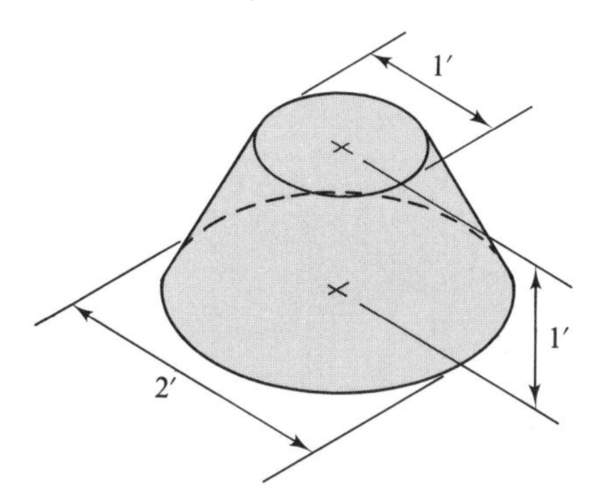

Figure P6-112

6-113 Determine the radius of gyration of mass, with respect to the x axis, of the homogeneous body in Fig. P6-113.

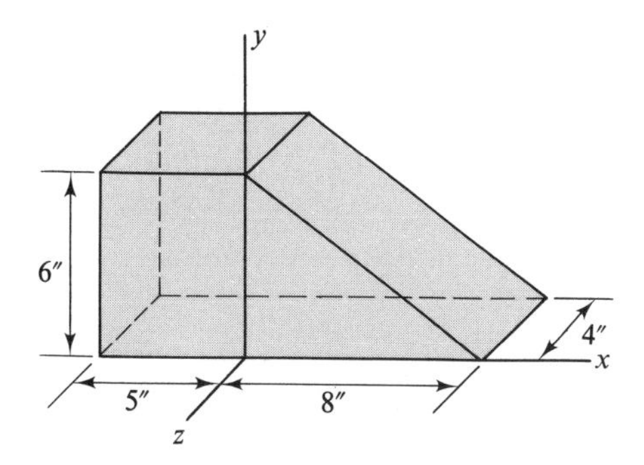

Figure P6-113

6-114 The homogeneous thin cylindrical shell C, weighing 32.2 lb, is connected by a slender homogeneous bar weighing 32.2 lb to the nonhomogeneous body A in Fig. P6-114. Body A weighs 96.6 lb (mass center at G) and has a

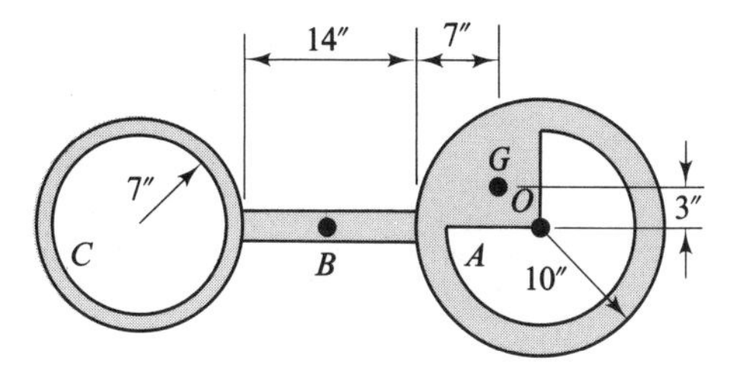

Figure P6-114

radius of gyration of mass with respect to an axis through O perpendicular to the plane of the figure of 9 in. Determine the radius of gyration of the mass of the system with respect to an axis perpendicular to the plane of the figure through the mass center, B, of the rod.

6-115 The assembly in Fig. P6-115 is composed of three homogeneous bodies, a 1932-lb cylinder, a 64.4-lb slender rod, and a 322-lb sphere. The xy plane is the plane of symmetry. Determine

(a) The radius of gyration of mass with respect to the x axis.

(b) The product of inertia of mass, I_{xy}.

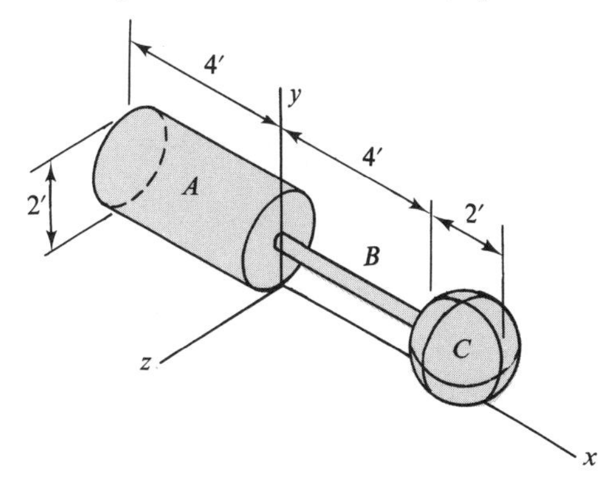

Figure P6-115

***6-116** The two balls and rods of a governor shown in Fig. P6-116 are made of brass with a

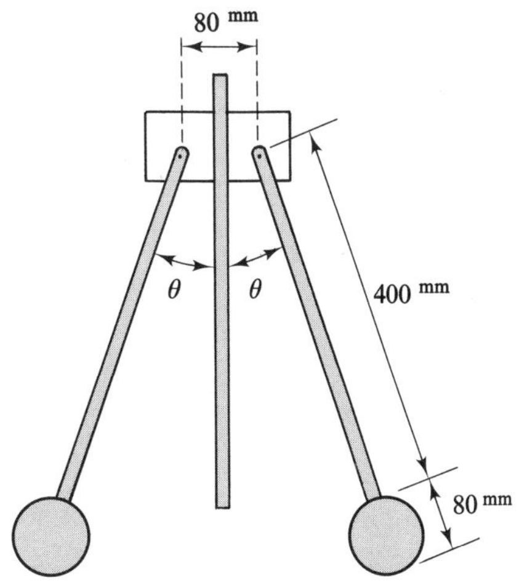

Figure P6-116

density of 8.6 Mg/m³. The rods have diameters 10 mm, and the balls have diameters of 80 mm. Determine the moment of inertia, with respect to the vertical axis, of the mass of the balls and rods when the angle θ is 20°. How great an error is introduced if the mass of the rods is neglected?

6-117 All bodies in Fig. P6-117 are homogeneous, and the system is symmetrical with respect to the xy plane. Body A is a sphere with a mass of 5 slugs, C is a disk having a mass of 14 slugs, and bars DE and BO have masses of 3 slugs and 2 slugs, respectively. Determine
(a) The radius of gyration of mass with respect to the z axis.
(b) The product of inertia of mass I_{xy}.

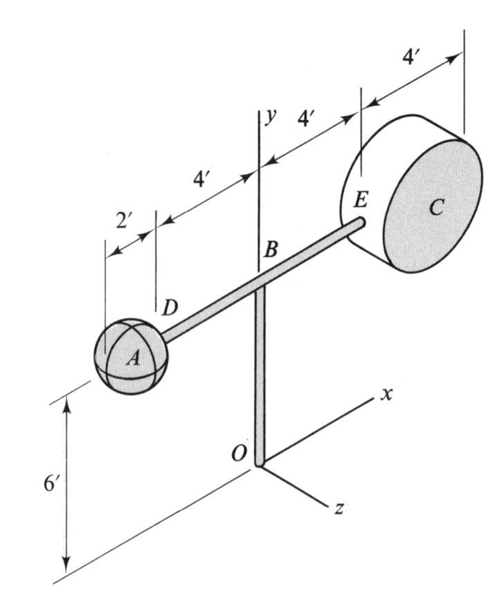

Figure P6-117

7 Virtual Work

7-1 METHODS OF SOLVING PROBLEMS IN EQUILIBRIUM

In the preceding chapters, the concepts of the parallelogram of forces and the moment of a force were used as a basis for the solution of problems in equilibrium. The idea of the parallelogram of forces was first employed, but never expressed formally by Stevinus (1548–1620).[1] The principle of the parallelogram of forces was first clearly stated by Newton in his *Principles of Natural Philosophy* in 1687. In this same year, Varignon also expressed the concept of the force parallelogram.

Although the parallelogram of forces provides an important approach to the solution of problems in equilibrium, other methods have been proposed and used at various times. Archimedes of Syracuse (287–212 B.C.) discussed the principle of the lever some two

[1] See Art. 1-1 and Ernst Mach, *Science of Mechanics* (Chicago: Open Court Publishing Company, 1893).

hundred years before Christ, and Galileo and others made a more complete analysis of the lever some 1800 years later. The actions of bodies on inclined planes were first investigated by Stevinus, who also studied the equilibrium of pulleys. While working with pulleys, Stevinus suggested the *method of virtual work*, and Galileo (1564–1642) also made use of this method in working with bodies on inclined planes. John Bernoulli, in a letter to Varignon in 1717, was the first to give a formal statement of the method which he called the *principle of virtual velocities*. The name *virtual velocities* has been replaced by the names *virtual work*, *virtual displacement*, and *method of work*, all of which are more indicative of the method used. The method of virtual work, as applied to the solution of problems in equilibrium, is developed in the following articles.

The closely related methods of virtual work and minimum potential energy are commonly used to investigate the stability of equilibrium positions of a body or system of bodies. The unique feature of these methods is that they frequently can be applied without considering all the forces acting on the system. These methods are also often used to prove theorems and to derive equations used in advanced mechanics.

Although it is possible to make allowances for the presence of frictional forces while applying either the method of virtual work or the method of minimum potential energy, the procedure frequently becomes so involved that such problems yield more readily to a solution using equilibrium equations. Therefore in this text all developments regarding virtual work or minimum potential energy will assume *no frictional forces*.

7-2 DISPLACEMENT

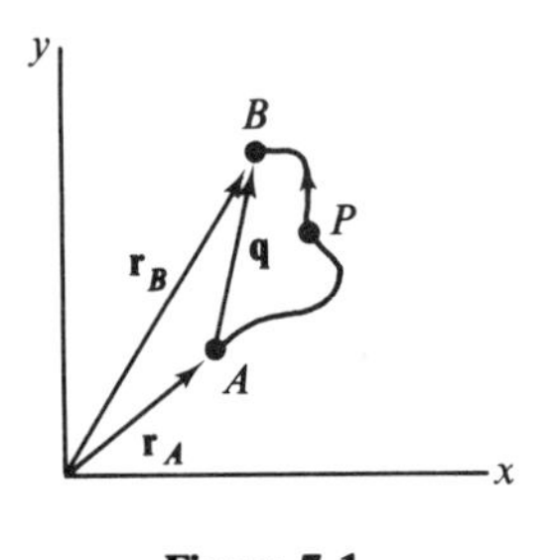

Figure 7-1

The *displacement of a point is defined as the change of position of that point and is the vector from the initial to the final position.* Thus the displacement of point P in Fig. 7-1 as it moves along the path from position A to B is

$$\mathbf{q} = \mathbf{r}_B - \mathbf{r}_A.$$

The displacement does not depend on the path; it depends only on the initial and final positions.

When particles are connected in a rigid body, their displacements must be consistent with the fact that the body is rigid. Any two points on a rigid body must remain a fixed distance apart. The bar AB in Fig. 7-2(a) and 7-2(b) is an example of a rigid body. If the end A is assumed to have no displacement, the only possible displacement of B will be from B to B' due to a rotation of the bar, as shown in Fig. 7-2(a). Figure 7-2(b) shows another possible displacement of the bar in

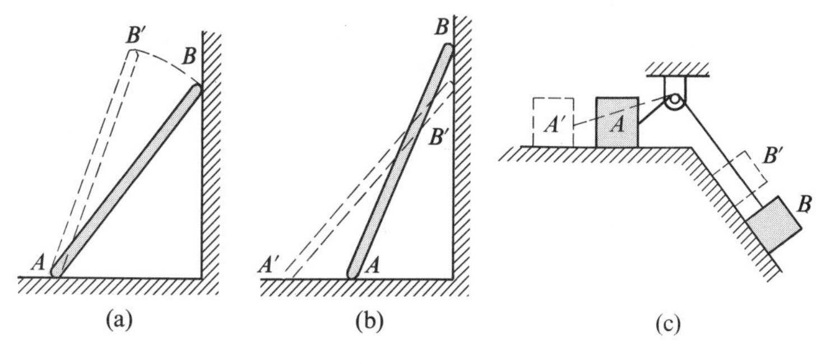

Figure 7-2

which the two ends remain in contact with the two surfaces, and A has a displacement from A to A' to the left, whereas the displacement of B is from B to B' downward. The distance from A' to B' must be the same as the distance from A to B, and the relationship between the two displacements is governed by this fact. When two particles are connected by an inextensible cord, their displacements are governed by the fact that the cord is inextensible even though it is not a rigid body. In Fig. 7-2(c), a displacement of A to the left must be accompanied by a corresponding displacement of B, upward along the plane if the cord is inextensible.

A line has angular motion when the angle between it and a fixed reference line changes. The angle between a line moving in a plane and a fixed reference line in the plane defines the angular position of the moving line. The *angular displacement of a rigid body is defined as the change of the angular position of a line in the body.*

7-3 WORK

Before discussing the method of virtual work, it is necessary to define the work done by forces and couples.

The work done by a constant force **F** *during a displacement* **q** *of the force* is given by the expression

$$U = \mathbf{F} \cdot \mathbf{q} = Fq \cos \theta, \tag{7-1}$$

where θ is the angle between **F** and **q**. Figure 7-3 shows that the work done by the force **F** during the displacement **q** as the block moves from

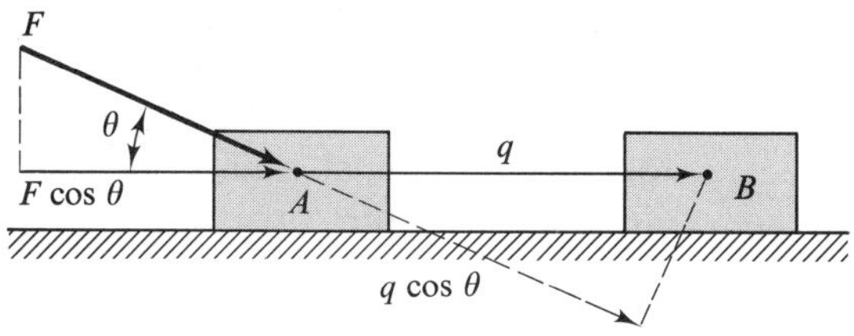

Figure 7-3

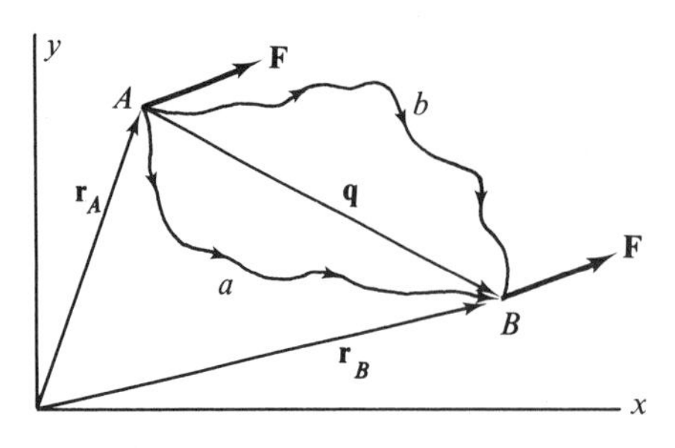

Figure 7-4

position A to position B is the product of the magnitude of the rectangular component of $\mathbf{F}$ in the direction of the displacement ($F \cos \theta$) and the magnitude of the displacement. Thus, U is equal to $(F \cos \theta)q$. The work done may also be thought of as the product of the magnitude of $\mathbf{F}$ and the magnitude of the rectangular component of the displacement in the direction of $\mathbf{F}$; that is, $U = (F)(q \cos \theta)$.

Work is defined as the *scalar product of two vectors;* therefore it is a scalar and has only magnitude and algebraic sign. The quantities $\mathbf{F}$ and $\mathbf{q}$ can be written with vector notation as

$$\mathbf{F} = (F_x\mathbf{i} + F_y\mathbf{j} + F_z\mathbf{k}) \text{ lb}$$

and

$$\mathbf{q} = (q_x\mathbf{i} + q_y\mathbf{j} + q_z\mathbf{k}) \text{ ft.}$$

The work done by $\mathbf{F}$ can be obtained by means of the dot product which determines both the magnitude and algebraic sign; thus

$$U = \mathbf{F} \cdot \mathbf{q} = F_x q_x + F_y q_y + F_z q_z.$$

From the definition of work—see Eq. (7-1)—it may be noted that the work done by a force is positive when the displacement and the component of the force along the displacement have the same sense ($0 \leq \theta < 90°$). When the displacement and the component of the force in the direction of the displacement are in opposite directions ($90° < \theta \leq 180°$) the force does negative work. When $\theta = 90°$ the force has no component along the displacement, and the work done by the force is zero.

Note that by definition the work done by a *constant* force does not depend on the path traveled by the point of application of the force during the displacement. Thus the work done by the *constant* force $\mathbf{F}$ in Fig. 7-4 is the same if it is moved from position A to position B along path a, along path b, or along the straight line from A to B. The preceding statement is particularly useful when computing the work done by the weight of a body, since the work is Wh, where W is the weight and h is the vertical distance the center of gravity moves. If the center of gravity moves downward, the weight will do positive work, and if it moves upward, the work done will be negative.

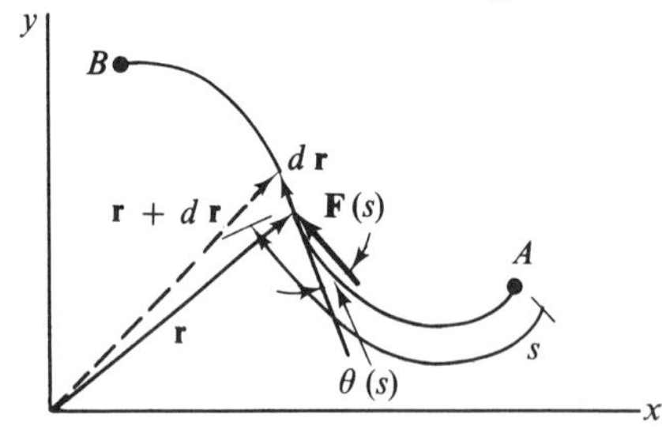

Figure 7-5

When the force varies either in magnitude or direction, as in Fig. 7-5, Eq. (7-1) is valid only for an infinitesimal change in position, and the work done by the force as it moves from A to B is

$$U = \int \mathbf{F} \cdot d\mathbf{r} = \int_{s_A}^{s_B} F \cos \theta \, ds. \qquad (7\text{-}2)$$

Note that ds is the magnitude of the infinitesimal displacement $d\mathbf{r}$ and that θ is the angle between $\mathbf{F}$ and $d\mathbf{r}$. Equation (7-2) is valid for any general force and displacement, and it can be used to compute the work done if relationships between F, θ, and s can be established. Although $\mathbf{F}$ and $d\mathbf{r}$ are vector quantities, the dot product is a scalar quantity, and ordinary scalar integration can be used to determine the work done.

Sometimes a graphical or semigraphical method for determining the work done by a force is useful, especially when the mathematical relation between F, θ, and s in Eq. (7-2) is either unknown or cumbersome. When the component of the force in the direction of the displacement, $F \cos \theta$, is plotted against the position function s as indicated in Fig. 7-6, the area under the curve represents the work done—see Eq. (7-2). The indicator card used to determine the work done by the steam on the piston of a steam engine is an example of an F-s diagram.

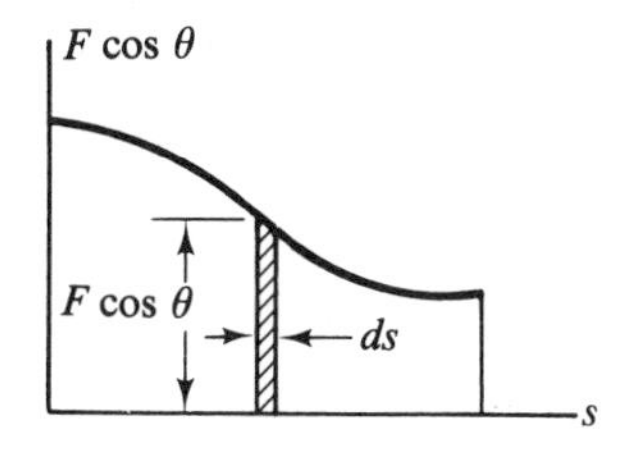

Figure 7-6

A spring is a body which deforms when acted on by a force. The relationship of the force and deformation depends on the material used and on the dimensions and shape of the spring (helical, spiral, conical, cantilever, or other). Only springs which exhibit a linear relationship between force and displacement or deformation are considered in this section. In this case, the magnitude, F, of the force is given by the expression

$$F = ks,$$

where s is the deformation of the spring from its unloaded position and k is a constant known as the *modulus* of the spring. The force-displacement diagram for a spring is shown in Fig. 7-7. As the spring is

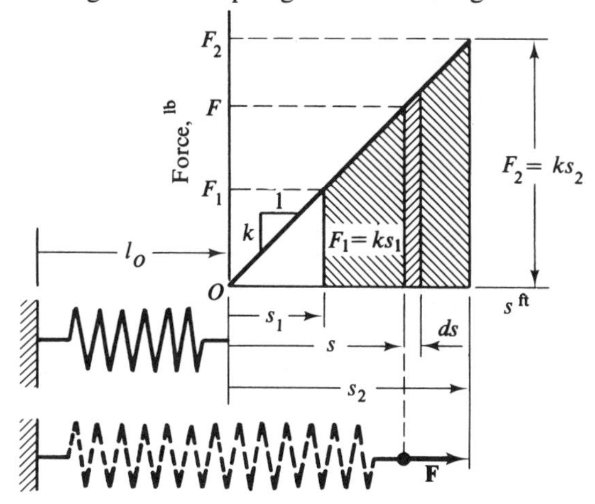

Figure 7-7

stretched an amount s from its unstretched length l_0, the force increases from zero to $F = ks$. The spring modulus, k, is the slope of the F-s diagram and has dimensions of force divided by length, often expressed as pounds per foot, pounds per inch, or newtons per meter.

The work done in stretching the spring in Fig. 7-7 from an initial length $l_0 + s_1$ to a final length $l_0 + s_2$ can be determined from Eq. (7-2). Since the force and displacement are in the same direction, the angle θ is zero, and the equation becomes

$$U = \int_{s_1}^{s_2} F \cos \theta \, ds = \int_{s_1}^{s_2} ks \, ds = \frac{k(s_2^2 - s_1^2)}{2}.$$

The preceding expression can also be obtained as the area under the F-s diagram between s_1 and s_2, the shaded trapezoidal area. This area is

$$U = \text{area} = \frac{F_1 + F_2}{2}(s_2 - s_1).$$

When $F_1 = ks_1$ and $F_2 = ks_2$ are substituted in the expression, it becomes

$$U = \frac{ks_1 + ks_2}{2}(s_2 - s_1) = \frac{k(s_2^2 - s_1^2)}{2}$$

as before. When the spring is initially unstretched, the trapezoid becomes a triangle, and the work done on the spring is $ks_2^2/2$.

It should be noted that, as a spring is being deformed (either stretched or compressed), the force on the spring and the displacement are in the same direction, and the work done *on the spring* is positive. If a spring is initially deformed and released gradually, the force and displacement are in opposite directions, and the work done *on the spring* is negative. The force of the spring on a body is opposite the force of the body on the spring, but both the body and the end of the spring have the same displacement; therefore, the work done *by a body on a spring* and the work done *by the spring on the body* will have the same magnitudes but opposite signs.

When a spring is constructed so that it can act in either tension or compression, such as an automobile suspension spring, it is usually assumed that the spring modulus is the same in tension and compression. The force displacement diagram will be similar to Fig. 7-8, and it is apparent that during a change of length from tension to compression (or the reverse) the force will change direction while the displacement will continue in the same direction. Consequently, the area on one side of the s axis will represent positive work and that on the other side will give negative work.

A couple does work when it turns through an angle in the plane of the couple (about the axis of the couple). *The work done by a constant couple as it rotates through an angle ϕ in the plane of the couple* *is equal to the product of the magnitude of the moment of the couple and the angular displacement* (in radians); that is,

$$U = T\phi. \qquad (7\text{-}3)$$

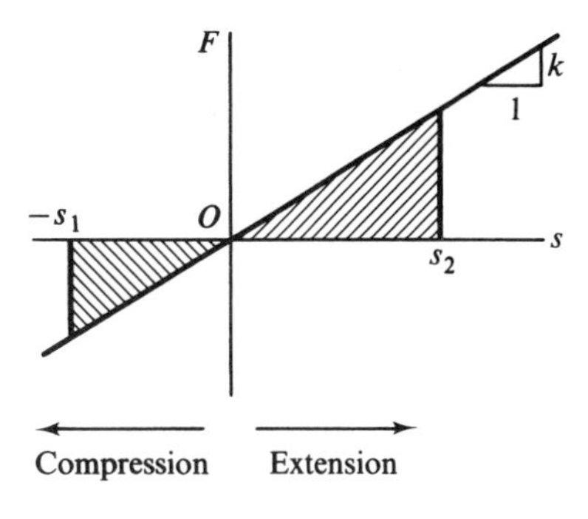

Figure 7-8

The work is positive if the angular displacement is in the same direction as the sense of rotation of the couple and negative if the displacement is in the opposite direction. No work is done if the couple is translated or rotated about an axis parallel to the plane of the couple.

The *work done by a force was defined as the dot product of two vectors: force and linear displacement.* The work done by a couple can be obtained by applying this definition to each force of the couple. Finite angular displacements, however, are not vectors since they do not obey the parallelogram law of addition (see Art. 1-3); therefore, they cannot be used to form scalar products to express the work done by couples. Infinitesimal angular displacements do satisfy the definition of vectors and can be used to determine the work done by variable couples.

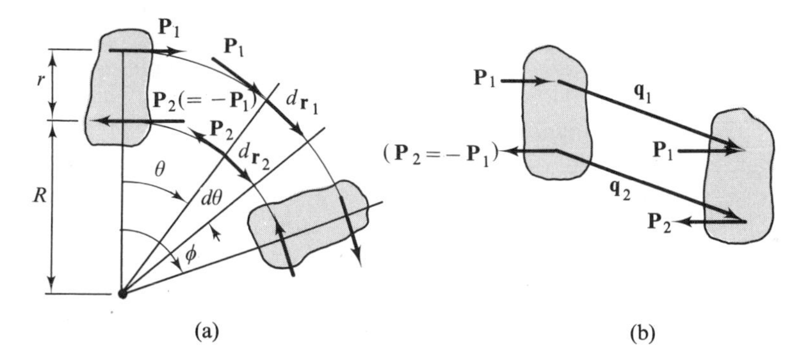

Figure 7-9

Consider the work done by the couple in Fig. 7-9(a) as it rotates through the infinitesimal angle $d\theta$. The work done by the two forces is

$$U = \mathbf{P}_1 \cdot d\mathbf{r}_1 + \mathbf{P}_2 \cdot d\mathbf{r}_2,$$

where $\mathbf{r}_1$ and $\mathbf{r}_2$ are the position vectors to forces $\mathbf{P}_1$ and $\mathbf{P}_2$, respectively, and $d\mathbf{r}_1$ and $d\mathbf{r}_2$ are the displacements of the two forces. Since $\mathbf{P}_1$ and $d\mathbf{r}_1$ are in the same direction, $\mathbf{P}_1$ does positive work, while $\mathbf{P}_2$ does negative work by the same reasoning. Thus the work can be written as

$$dU = P_1\, dr_1 - P_2\, dr_2.$$

From the definition of a couple, P_1 is equal to P_2, and from the geometry of the figure

$$dr_1 = (R + r)\, d\theta \quad \text{and} \quad dr_2 = R\, d\theta.$$

Therefore, the work of the couple is

$$dU = P_1(R + r)\, d\theta - P_1 R\, d\theta = P_1 r\, d\theta = T\, d\theta.$$

The work done during a finite angular displacement can be obtained by integration of the expression

$$U = \int_{\theta_i}^{\theta_f} T\, d\theta, \tag{7-4}$$

and if T is constant, the work is

$$U = T(\theta_f - \theta_i) = T\phi,$$

where ϕ is the change of θ, that is, the angular displacement of the couple; see Eq. (7-3).

If the body is translated as shown in Fig. 7-9(b), the work done by the couple is

$$U = \mathbf{P}_1 \cdot \mathbf{q}_1 + \mathbf{P}_2 \cdot \mathbf{q}_2 = \mathbf{P}_1 \cdot \mathbf{q}_1 - \mathbf{P}_1 \cdot \mathbf{q}_1 = 0$$

since $\mathbf{P}_2 = -\mathbf{P}_1$ and $\mathbf{q}_2 = \mathbf{q}_1$. Thus a couple does no work due to a translation of the body on which it acts. When a couple acts on a body which is simultaneously translated and rotated, the couple does work only as a result of the rotation.

When the magnitude of T is a variable, Eq. (7-4) must be integrated to obtain the work done. Moment-angular position diagrams (similar to F-s diagrams) are frequently useful for problems involving variable couples, particularly when the T-θ diagram is a straight line. Linear variations occur for some torsion springs (such as a watch spring), in which case the moment and angular position are related by

$$T = k\theta,$$

where k is the torsional spring constant and is often expressed in such units as foot-pounds per radian, inch-ounces per radian, or newton · meters/radian.

The work done on a *rigid* body by a system of external forces and couples is the algebraic sum of works done by the individual forces and couples. Since the internal forces occur in equal and opposite collinear pairs, each of which has the same displacement if the body is rigid, the resultant work done by the internal forces will be zero. *If the body is not rigid, this statement is not valid since work will be done by the internal forces.* When two or more rigid bodies are connected by smooth pins or by flexible, inextensible cables, the resultant work done by the connecting members of the system is also zero. The two forces at the ends of a cable have the same magnitude (the mass of the cable is assumed to be negligible), and if the cable is inextensible, the components of the displacements of the two ends in the direction of the forces must have the same magnitudes; therefore the resultant work done by the cable must be zero. The following examples demonstrate the calculation of the work done by various forces.

EXAMPLE 7-1

The uniform 6-ft, 40-lb bar in Fig. 7-10 rotates about a fixed horizontal axis at O. Compute the work done by the weight of the bar as the bar rotates from the horizontal position OA to position OB.

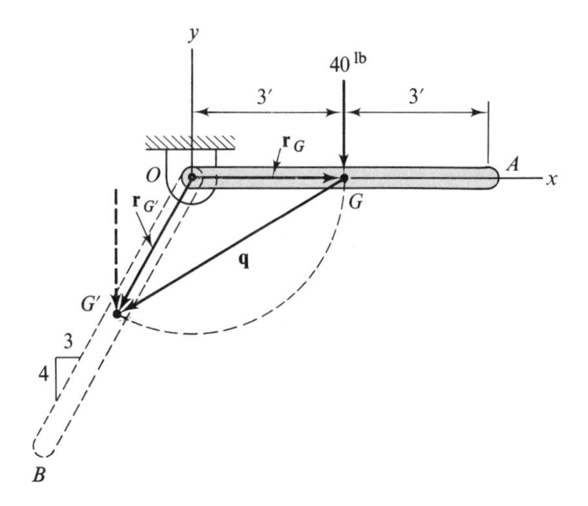

Figure 7-10

SOLUTION

The force is a constant and can be written in vector notation as

$$\mathbf{F} = -40\mathbf{j}.$$

The displacement $\mathbf{q}$ of the center of gravity can be determined by subtracting the initial position vector $\mathbf{r}_G$ from the final position vector $\mathbf{r}_{G'}$. The position vectors are

$$\mathbf{r}_G = 3\mathbf{i} \text{ ft}$$

and

$$\mathbf{r}_{G'} = 3(-0.6\mathbf{i} - 0.8\mathbf{j}) = (-1.8\mathbf{i} - 2.4\mathbf{j}) \text{ ft},$$

and the displacement of the center of gravity G is

$$\mathbf{q} = \mathbf{r}_{G'} - \mathbf{r}_G = -1.8\mathbf{i} - 2.4\mathbf{j} - 3\mathbf{i} = (-4.8\mathbf{i} - 2.4\mathbf{j}) \text{ ft}.$$

The work done by the force $\mathbf{F}$ is

$$U = \mathbf{F} \cdot \mathbf{q} = (-40\mathbf{j}) \cdot (-4.8\mathbf{i} - 2.4\mathbf{j}) = \underline{96.0 \text{ ft-lb.}} \qquad \text{Ans.}$$

As noted earlier, the work done by a constant force, such as the weight of the bar, can be obtained as the product of the force and the component of the displacement in the direction of the force. This analysis gives a more direct solution, since it is not necessary to consider the horizontal movement of G. The resultant vertical displacement of G is 0.8(3), or 2.4 ft, which can be multiplied by the weight to give the work done as

$$U = 2.4(40) = 96.0 \text{ ft-lb.}$$

Since both the force and displacement are in the same direction (downward), the force does positive work on the bar.

EXAMPLE 7-2

The bar AB in Fig. 7-11(a) weighs 20 lb and is connected to blocks at the ends which slide in smooth horizontal and vertical slots as indicated. The spring OA is arranged so that it can act in either tension or compression. It has a modulus of 5.0 lb per in. and an unstressed length of 7 in. Determine the work done by all forces acting on the bar as it moves to the horizontal position $A'B'$ as indicated by dashed lines. The weights of the blocks of A and B may be neglected.

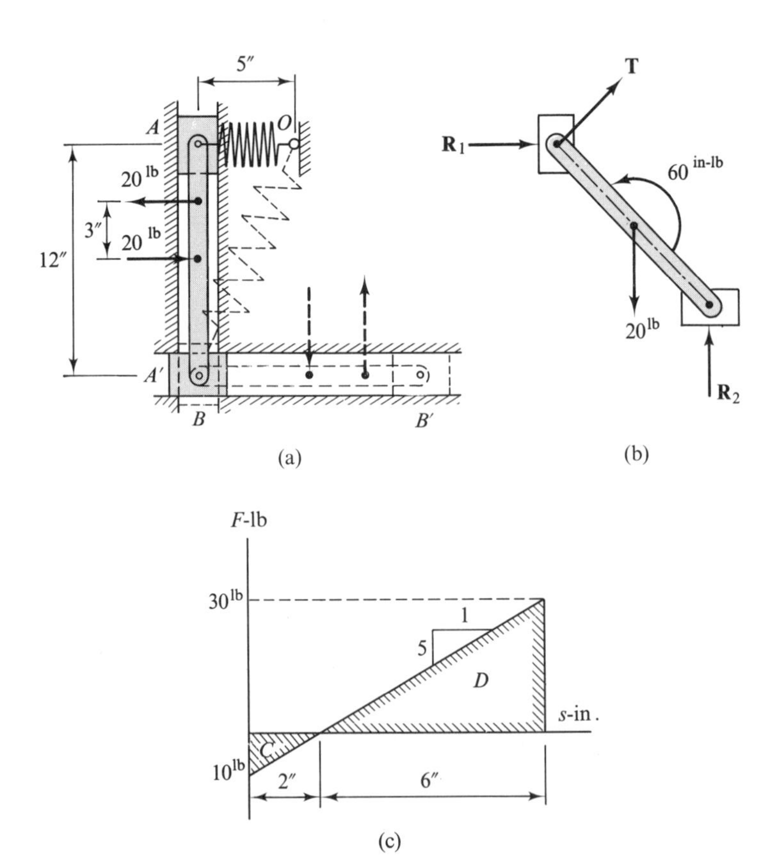

Figure 7-11

SOLUTION

A free-body diagram of the bar in an intermediate position is shown in Fig. 7-11(b). There are four forces and a couple acting on the body. Two of the forces, $\mathbf{R}_1$ and $\mathbf{R}_2$, are normal to their corresponding displacements, and as a result they do no work. The resultant work on the bar is thus

$$U = U_{\text{spring}} + U_{\text{weight}} + U_{\text{couple}}.$$

An F-s diagram for the spring is shown in Fig. 7-11(c). The length of the spring changes 8 in., from 5 to 13 in. The spring is initially compressed 2 in. and exerts a 10-lb force on the block A. In the final position, the spring is stretched 6 in. and has a tensile force of 30 lb. The area C represents negative work *on the spring* since the spring is in compression but is increasing in length. The area C represents positive work on block A since the vertical component of the force on A and the displacement are in the same (downward) direction. In a similar manner, the area D represents negative work of the spring on block A. The work of the spring on AB is

$$U_{\text{spring}} = \frac{10(2)}{2} - \frac{30(6)}{2} = -80 \text{ in-lb on } AB.$$

The work done by the weight is

$$U_{\text{weight}} = 20(6) = 120 \text{ in-lb on } AB,$$

and the work done by the couple is

$$U_{\text{couple}} = 60\frac{\pi}{2} = 94.2 \text{ in-lb on } AB.$$

The total work done on AB is

$$U = -80 + 120 + 94.2 = \underline{134.2 \text{ in-lb.}} \qquad \text{Ans.}$$

PROBLEMS

7-1 A spring with a modulus of 100 lb per ft is initially compressed by a load of 50 lb. The load is increased to 125 lb causing the spring to be further compressed. Determine the work done on the spring as the load is increased.

7-2 Block A in Fig. P7-2 weighs 50 lb and the plane is smooth. Determine the work done on the body as it is moved 10 ft up the plane.

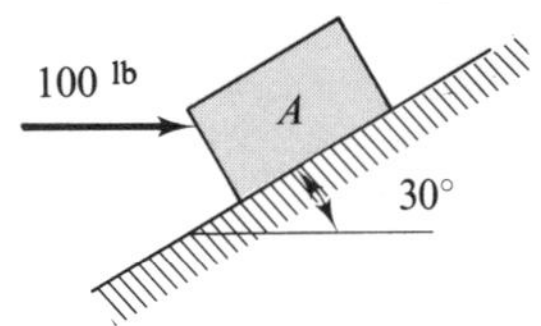

Figure P7-2

7-3 The force $\mathbf{P}$ which acts on the 40-lb block in Fig. P7-3 has a constant direction as shown, but its magnitude varies according to the equation $P = (20s - 6s^2)$ lb, where s is the position of the block in feet. Determine the work done on the body as it moves from $s = 1$ ft to $s = 3$ ft. The plane is smooth.

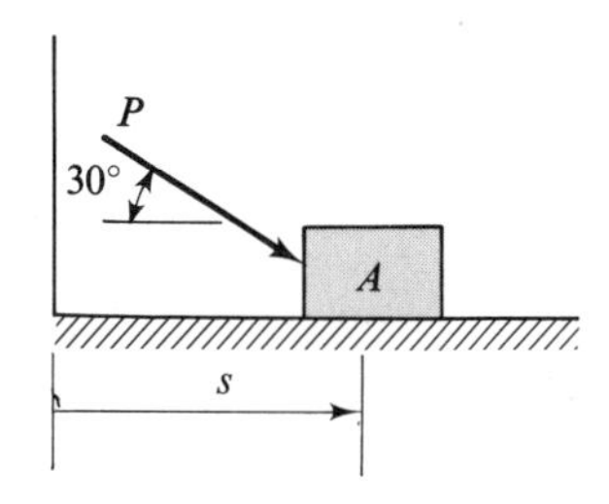

Figure P7-3

7-4 The cable in Fig. P7-4 passes through smooth pulleys B and C and is attached to block D. A 30-lb body E is supported by pulley B.

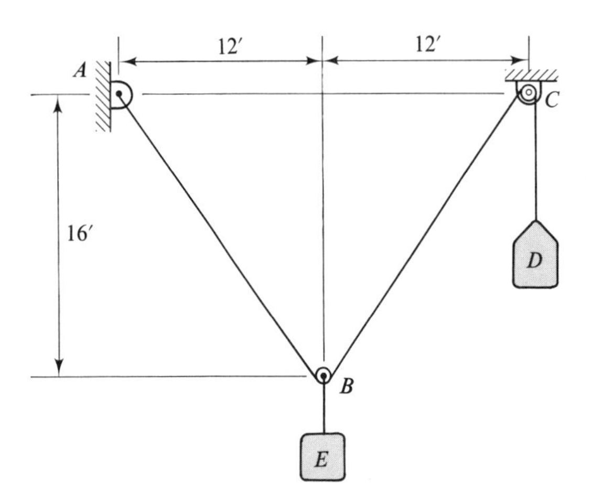

Figure P7-4

The work done by the weights of bodies D and E is 300 ft-lb as pulley B is raised 11 ft from the position shown. Determine the weight of body D.

7-5 The 8-ft uniform rod AB in Fig. P7-5 weighs 50 lb. A 30-lb weight D is attached to AB by a cable which passes over a smooth pulley at C. Determine the work done by the two weights as the rod rotates 90° counterclockwise from the indicated position.

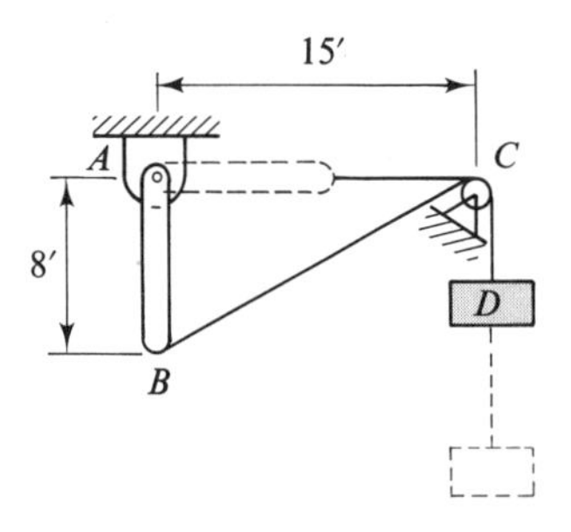

Figure P7-5

7-6 Bar AB in Fig. P7-6 weighs 100 lb and rotates in a vertical plane about a horizontal axis through A. When the bar is in the vertical position the tension in the spring is 120 lb. As the bar rotates 90° counterclockwise the work done on the bar is 200 ft-lb. Determine the modulus of the spring.

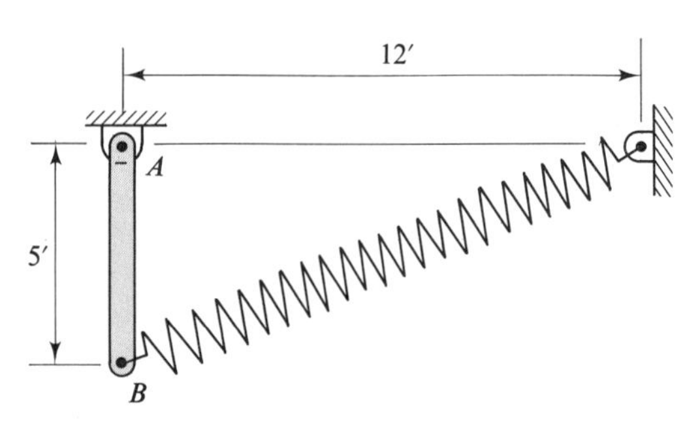

Figure P7-6

7-7 The 21-in. uniform rod AB in Fig. P7-7 weighs 20 lb. The 30-lb body D is attached to AB by a cable which passes over a smooth pulley at C. Determine the work done on the two bodies as the rod rotates from a downward vertical position to the upward vertical position shown by the dashed lines.

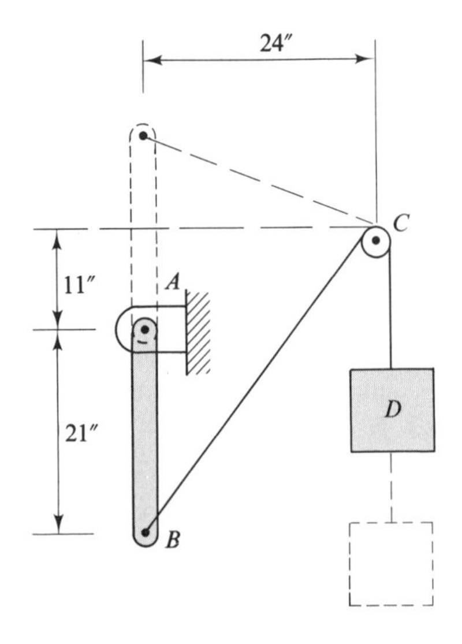

Figure P7-7

7-8 The 10-lb block in Fig. P7-8 slides in a smooth vertical slot. The spring is connected to A and to the block. The modulus of the spring is 20 lb per ft, and its unstretched length is 2 ft. Determine the work done on the block as it moves from position B to position C.

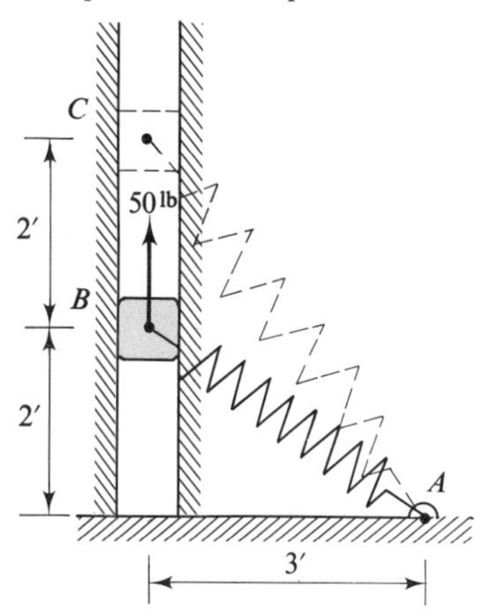

Figure P7-8

7-9 Bars AB and BC in Fig. P7-9 weigh 12 lb each. The spring S has a modulus of 10 lb per in. and is connected to the midpoint of AB and to C. The spring tension is 50 lb in the position shown. Determine the work done on the two bars as block C moves 10 in. downward to point D.

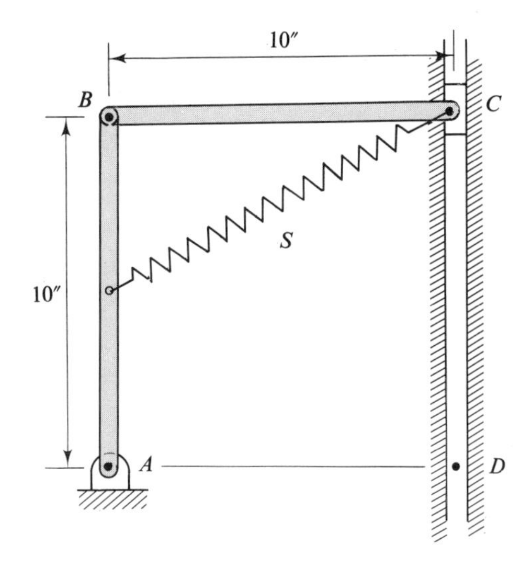

Figure P7-9

7-10 The 30-lb body A in Fig. P7-10 is pulled up the smooth incline and along the smooth horizontal plane to the dashed position by means of the constant 20-lb force. Determine the work done on the body during the displacement.

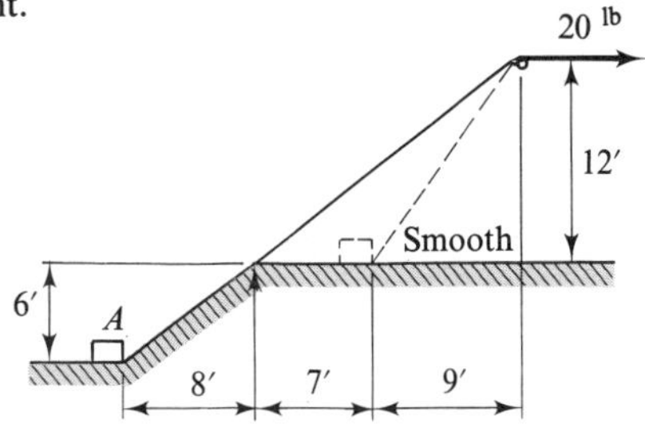

Figure P7-10

7-4 THE METHOD OF VIRTUAL WORK

The method of virtual work might be compared to a man who shakes a ladder to see whether it is steady before he starts to climb. The method determines equilibrium positions by giving the system an infinitesimal displacement (corresponding to the shake) and examines the work done by the external forces. The infinitesimal displacement is called a *virtual displacement* and may be any *arbitrary infinitesimal displacement consistent with the constraints of the system.* An example of a virtual displacement is given in Fig. 7-12 which shows a uniform ladder AB of weight W placed against a vertical wall. The virtual displacement consists of letting A, the bottom end of the ladder, move an infinitesimal distance, $\delta x_A \mathbf{i}$, to the right causing the upper end B to

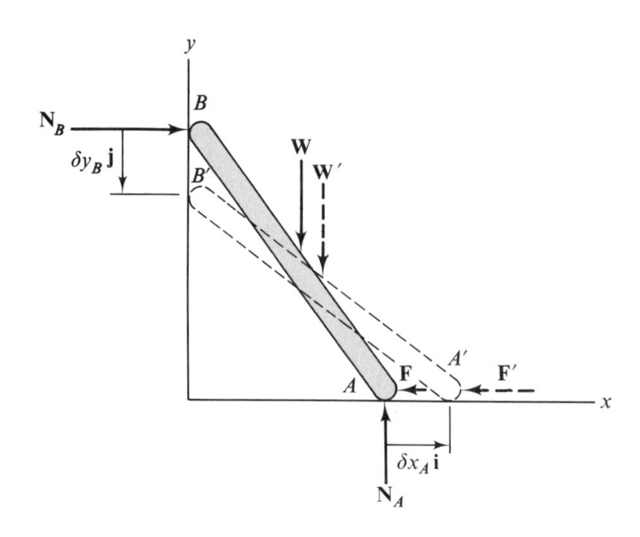

Figure 7-12

move an infinitesimal distance, $\delta y_B \mathbf{j}$, down (δy will be a negative quantity). The constraints of the system are satisfied by requiring the ends A and B to remain in contact with the horizontal and vertical surfaces, respectively, while they also remain a fixed distance apart.

The principle of virtual work may be stated as follows: *If the virtual work done by all external forces and couples on a frictionless (ideal) system of particles or rigid bodies is zero for all virtual displacements, the system is in equilibrium.* The virtual work is defined as the work done by a force (or system of forces) on a body during a virtual displacement of the body. The principle of virtual work may be expressed in equation form as

$$\delta U = \sum_{i=1}^{m} \mathbf{F}_i \cdot \delta \mathbf{r}_i + \sum_{j=1}^{n} \mathbf{T}_j \cdot \delta \boldsymbol{\theta}_j = 0,$$

where δU is virtual work, $\delta \mathbf{r}_i$ is the virtual displacement of the ith force, $\mathbf{F}_i$, and $\delta \boldsymbol{\theta}_j$ is the virtual angular displacement of the jth couple, $\mathbf{T}_j$. The use of the dot product in writing the virtual work done by a couple is justified since *infinitesimal* angular displacements add according to the parallelogram law and hence are vectors, although *finite* angular displacements do not satisfy the definition of vectors.

For a particle, the principle of virtual work can be proved simply. Consider the particle P in Fig. 7-13 under the action of three forces. If this particle is given a virtual displacement $\delta \mathbf{r}$ in the $\mathbf{n}$ direction from P to P', the work done is

$$\delta U = \sum_{i=1}^{3} \mathbf{F}_i \cdot \delta \mathbf{r}_i = \mathbf{F}_1 \cdot \delta \mathbf{r}_1 + \mathbf{F}_2 \cdot \delta \mathbf{r}_2 + \mathbf{F}_3 \cdot \delta \mathbf{r}_3.$$

In this case, however,

$$\delta \mathbf{r}_1 = \delta \mathbf{r}_2 = \delta \mathbf{r}_3 = \delta r \mathbf{n},$$

and the virtual work may be written as

$$\delta U = (\mathbf{F}_1 + \mathbf{F}_2 + \mathbf{F}_3) \cdot \delta r\mathbf{n}.$$

If the virtual work is to be zero, either

$$\mathbf{F}_1 + \mathbf{F}_2 + \mathbf{F}_3 = 0$$

or $(\mathbf{F}_1 + \mathbf{F}_2 + \mathbf{F}_3)$ must be perpendicular to the direction of the virtual displacement. The virtual work must be zero, however, for *any* virtual displacement; therefore it follows that $(\mathbf{F}_1 + \mathbf{F}_2 + \mathbf{F}_3)$ must be zero, since this sum could not be perpendicular to all possible directions of $\mathbf{n}$. If the vector sum of all forces acting on the particle is zero, the particle must be in equilibrium.

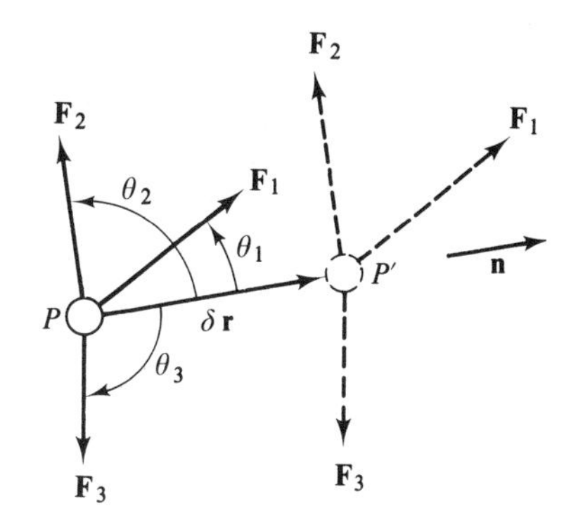

Figure 7-13

The virtual work done on a rigid body in equilibrium during any virtual displacement is also zero, since it is the sum of the virtual works done on each of the particles of the body. The internal forces between particles occur in equal, opposite, collinear pairs and do no work on a *rigid* body during any displacement, as explained in the preceding article. Consequently, only the work done by the external forces needs to be considered for *rigid* bodies.

The virtual work done on a system of rigid bodies in equilibrium during any virtual displacement is zero, provided that the forces exerted by the connections do positive virtual work on one body and the same amount of negative virtual work on the other body of each connected pair. Such a condition exists when the connections are smooth pins, smooth rollers, and inextensible cables.

The method of virtual work for a particle or a single rigid body usually provides no advantage over using the equations of equilibrium. One advantage of the method of work comes in the analysis of a system of bodies where this method frequently makes it unnecessary to analyze the force system acting on each body separately. The method of work is also effective as a means of determining the equilibrium position for a body or system of bodies.

When the force exerted by a pin or other support is to be determined by the method of work, the restraint is replaced by a force, the body is given a virtual displacement, and the virtual work done by the reaction and all other forces acting on the body is computed. If several forces are to be determined, the body (or system of bodies) can be given a series of separate virtual displacements in which only one of the unknown forces does virtual work during each displacement. This discussion is limited to cases in which the system has a single degree of freedom, that is, to systems for which the virtual displacements of all points can be expressed in terms of a single variable (displacement).

The methods of using the principle of virtual work can best be explained by the following examples.

EXAMPLE 7-3

A 12-ft bar under the action of two forces of 40 and 80 lb is to be supported by a smooth pin as shown in Fig. 7-14(a). Determine the distance b necessary for equilibrium.

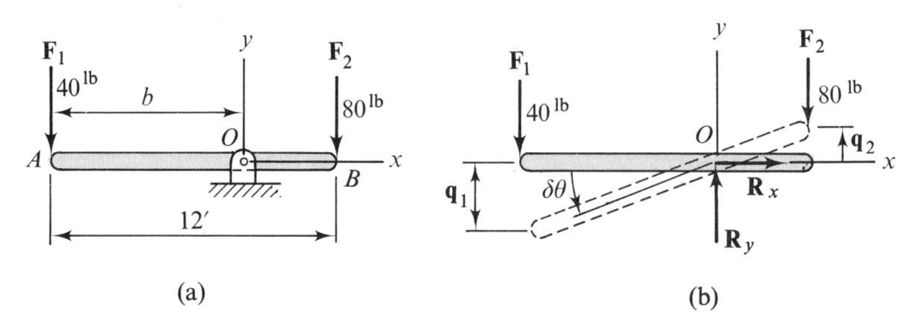

Figure 7-14

SOLUTION

A logical virtual displacement in this case is to rotate the beam an angle $\delta\theta$ from the horizontal as shown in Fig. 7-14(b) since the reaction forces, $\mathbf{R}_x$ and $\mathbf{R}_y$, do no work during the displacement. The displacements of the applied loads are

$$\mathbf{q}_1 = -b\,\delta\theta\mathbf{j}; \qquad \mathbf{q}_2 = (12-b)\,\delta\theta\mathbf{j}.$$

The principle of virtual work may be applied as follows:

$$\delta U = (-40\mathbf{j})\cdot(-b\,\delta\theta\mathbf{j}) + (-80\mathbf{j})\cdot(12-b)\,\delta\theta\mathbf{j}$$

$$= 40b\,\delta\theta - 80(12-b)\,\delta\theta = [40b - 80(12-b)]\,\delta\theta = 0.$$

Since

$$\delta\theta \neq 0,$$

$$40b - 80(12-b) = 0$$

from which

$$b = \underline{8\text{ ft.}} \qquad \text{Ans.}$$

EXAMPLE 7-4

An 80-lb ladder 25 ft long, with its center of gravity at its midpoint, rests against a vertical wall and is acted on by a horizontal 30-lb force at the bottom as shown in Fig. 7-15(a). All contact surfaces are smooth. Determine (a) the distance x for which the ladder will be in equilibrium and (b) the reactions at the ends of the ladder on the ladder.

SOLUTION

(a) A free-body diagram is shown in Fig. 7-15(b). If the bottom end of the ladder is given a virtual displacement $\delta x\mathbf{i}$ (to the left so $\delta x < 0$), the top of the ladder will move up $\delta y\mathbf{j}$, and the center of gravity of the

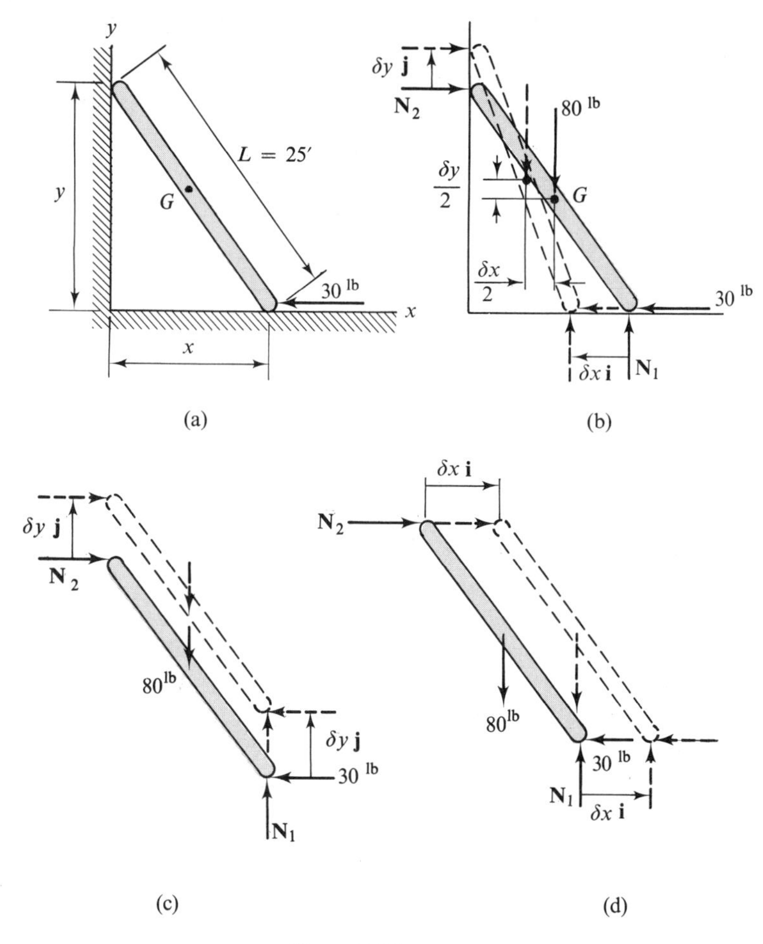

Figure 7-15

ladder will move $(\delta x\mathbf{i}+\delta y\mathbf{j})/2$. The virtual work, which must be zero for equilibrium, is

$$\delta U=(-30\mathbf{i})\cdot(\delta x\mathbf{i})+(N_1\mathbf{j})\cdot(\delta x\mathbf{i})+(N_2\mathbf{i})\cdot(\delta y\mathbf{j})$$

$$+(-80\mathbf{j})\cdot\frac{\delta x\mathbf{i}+\delta y\mathbf{j}}{2}=0$$

or

$$\delta U=-30\delta x-40\delta y=0. \tag{a}$$

This last equation contains two variables, and a solution cannot be obtained until a second equation is written in terms of the same two variables. The length of the ladder must be the same before and after the virtual displacement; therefore

$$x^2+y^2=(x+\delta x)^2+(y+\delta y)^2=x^2+2x\,\delta x+(\delta x)^2+y^2+2y\,\delta y+(\delta y)^2.$$

When this equation is simplified and the squares of virtual displacements neglected when added to first-order terms, the result is

$$+2x\,\delta x + 2y\,\delta y = 0$$

or

$$\delta y = -\frac{x}{y}\,\delta x.$$

This last equation can be combined with Eq. (a) to give

$$-30\delta x + 40\frac{x}{y}\,\delta x = -10\delta x\left(3 - 4\frac{x}{y}\right) = 0$$

from which

$$\frac{x}{y} = \frac{3}{4} \quad \text{or} \quad y = \frac{4x}{3}.$$

This expression can be substituted into the equation relating x and y and the length of the ladder to give

$$x^2 + y^2 = (25)^2 = x^2 + \left(\frac{4x}{3}\right)^2$$

from which

$$x = \underline{15\text{ ft.}} \qquad \text{Ans.}$$

(b) The reaction $\mathbf{N}_1$ can be determined by giving the ladder a virtual displacement in which $\mathbf{N}_1$ will do work. Such a displacement is shown in Fig. 7-15(c) in which all parts of the ladder are displaced $\delta y\mathbf{j}$. The virtual work is

$$\delta U = (N_1\mathbf{j})\cdot(\delta y\mathbf{j}) + (-30\mathbf{i})\cdot(\delta y\mathbf{j}) + (-80\mathbf{j})\cdot(\delta y\mathbf{j}) + (N_2\mathbf{i})\cdot(\delta y\mathbf{j}) = 0$$

or

$$\delta U = N_1\,\delta y - 80\,\delta y = 0$$

and

$$\mathbf{N}_1 = \underline{80\text{ lb upward.}} \qquad \text{Ans.}$$

The reaction $\mathbf{N}_2$ can be determined by giving the ladder a horizontal virtual displacement $+\delta x\mathbf{i}$, as shown in Fig. 7-15(d). The virtual work is

$$\delta U = (N_1\mathbf{j})\cdot(+\delta x\mathbf{i}) + (-30\mathbf{i})\cdot(+\delta x\mathbf{i}) + (-80\mathbf{j})\cdot(+\delta x\mathbf{i}) + (N_2\mathbf{i})\cdot(+\delta x\mathbf{i}) = (-30 + N_2)\,\delta x = 0$$

from which

$$\mathbf{N}_2 = \underline{30\text{ lb to the right.}} \qquad \text{Ans.}$$

Note that part (b) of this problem could have been solved by inspection, using the equations of equilibrium. The method of virtual work was used to illustrate the possibility of using various virtual displacements. The Pythagorean theorem was used to relate virtual displacements in part (a) of the solution. Trigonometric relations can also be used advantageously for such problems.

EXAMPLE 7-5

A slider crank mechanism with a spring attached to the slider is shown in Fig. 7-16(a). The crank OA is 6 in. long, the connecting rod AB is

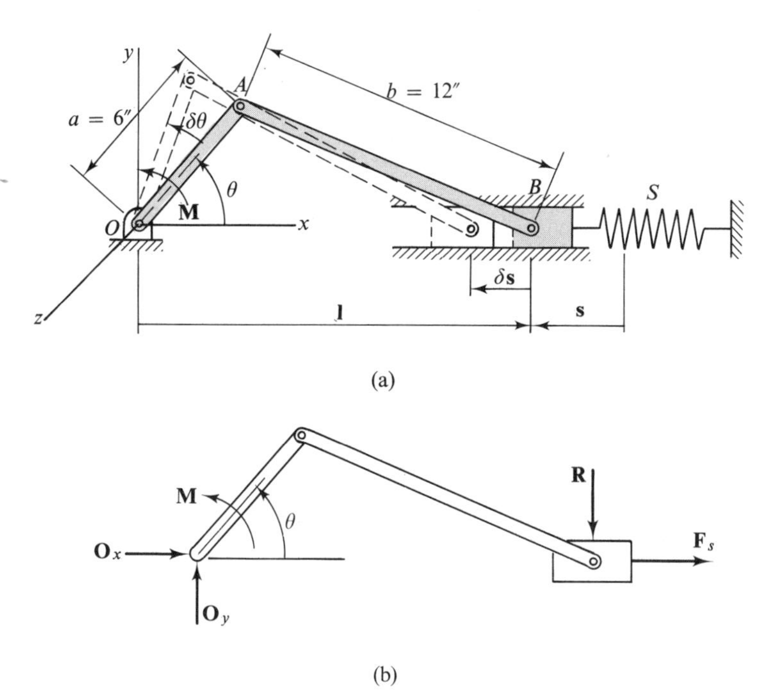

Figure 7-16

12 in. long, and the spring S has a modulus of 100 lb per in. and is unstretched when $\theta = 0$. Determine the moment **M** which must be applied to the crank OA in order to produce equilibrium when $\theta = 60°$. All surfaces are smooth, and weights of the members can be neglected.

SOLUTION

Although the problem asks for **M** for a particular value of θ, the best procedure is to write an equation for the work in terms of θ and solve the equation using the particular value of θ. Symbols for other forces and lengths, rather than actual numbers, will also be used in setting up the problem since this helps detect errors in dimensions and units.

A free-body diagram of the system is shown in Fig. 7-16(b). The virtual displacement is produced by giving the crank OA an infinitesimal angular displacement $\delta\theta\mathbf{k}$ which causes the slider B to have a horizontal displacement $-\delta s\mathbf{i}$; see Fig. 7-16(a) (**s** is the position vector of B measured from its extreme position to the right). The forces **R**, $\mathbf{O}_x$, and $\mathbf{O}_y$ do no work during the virtual displacement since the displacement of each force is either zero or perpendicular to the force. The virtual work equation is

$$\delta U = (M\mathbf{k}) \cdot (\delta\theta\mathbf{k}) + (F_S\mathbf{i}) \cdot (-\delta s\mathbf{i}) = 0. \qquad \text{(a)}$$

Since the force $\mathbf{F}_S$ is zero when θ (and s) are zero, it can be expressed in terms of θ as

$$\mathbf{F}_S = ks\mathbf{i} = k(a + b - l)\mathbf{i} = k[a + b - (a \cos\theta + \sqrt{b^2 - a^2 \sin^2\theta})]\mathbf{i}.$$

The magnitude of the virtual displacement, δs, can be obtained from the expression for s (in terms of θ) as follows;

$$\delta s = \frac{ds}{d\theta}\,\delta\theta = \frac{d}{d\theta}[a + b - (a \cos\theta + \sqrt{b^2 - a^2 \sin^2\theta})]\,\delta\theta$$

$$= \left[a \sin\theta + \frac{a^2 \sin\theta \cos\theta}{\sqrt{b^2 - a^2 \sin^2\theta}}\right]\delta\theta.$$

When Eq. (a) is expanded using the expressions for F_s and δs, it becomes

$$\delta U = M\,\delta\theta - k[a + b - (a \cos\theta + \sqrt{b^2 - a^2 \sin^2\theta})]$$

$$\left(a \sin\theta + \frac{a^2 \sin\theta \cos\theta}{\sqrt{b^2 - a^2 \sin^2\theta}}\right)\delta\theta = 0.$$

Since δU must be zero for any value of $\delta\theta$, the moment $\mathbf{M}$ can be obtained by substituting numerical values for k, a, b, and θ into the preceding expression; thus the magnitude of M is

$$M = 100[6 + 12 - (6 \cos 60° + \sqrt{12^2 - 6^2 \sin^2 60°})]$$

$$\left(6 \sin 60° + \frac{6^2 \sin 60° \cos 60°}{\sqrt{12^2 - 6^2 \sin^2 60°}}\right),$$

and

$$\mathbf{M} = \underline{2770\mathbf{k}\text{ in-lb.}} \qquad \text{Ans.}$$

PROBLEMS

The following problems should be solved by the method of virtual work. Friction is assumed to be negligible in all problems. The method of solution discussed in Chapter 4 can be used to check the solutions and in some problems may provide a shorter solution.

7-11 The mechanism in Fig. P7-11 is connected by smooth pins, and the weights of the members are negligible. Determine the force **P** for equilibrium when the distance b is
(a) 2 in.
(b) 4 in.

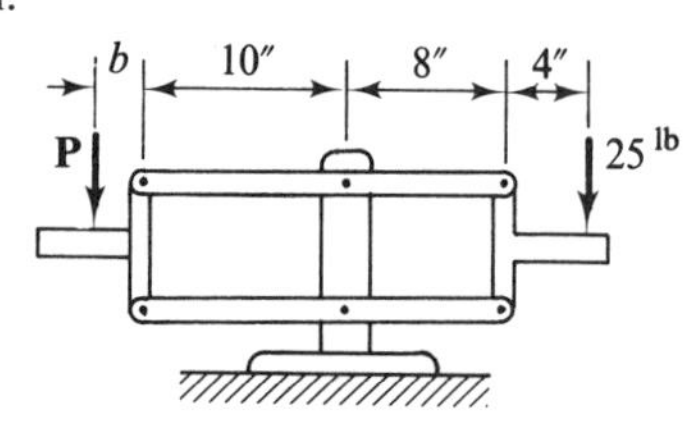

Figure P7-11

7-12 A safety gate used to close porches, doors, and stairways is shown in Fig. P7-12. A force **Q** with a magnitude of 20 lb is applied as indicated. Determine the horizontal force acting at point A necessary to hold the gate in equilibrium.

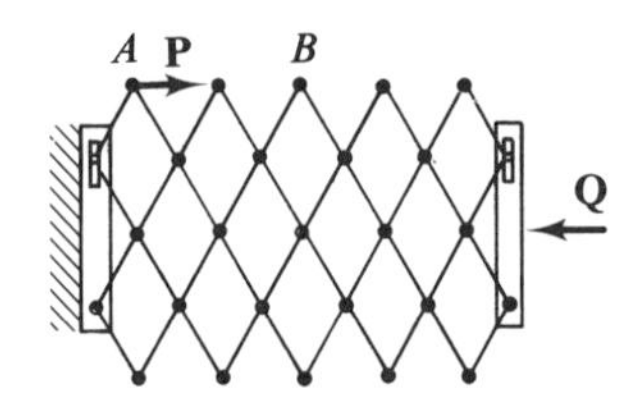

Figure P7-12

7-13 Determine the vertical force at B necessary to maintain equilibrium of the gate in Problem 7-12 when the members of the gate make angles of 50° with the horizontal.

***7-14** In the pin-connected structure of Fig. P7-14 the spring modulus is 2.0 kN/m, and the length b is 20 cm. The spring is unstressed when

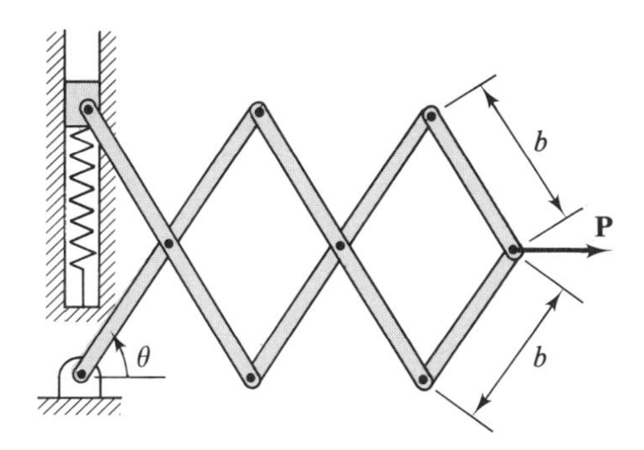

Figure P7-14

the angle θ is 90°. Determine the force **P** required to maintain equilibrium when $\theta = 60°$.

*7-15 Determine the angle θ in Problem 7-14 for which the structure will be in equilibrium when $P = 40$ N.

7-16 The mechanism in Fig. P7-16 is so constrained that member *A* remains horizontal as it moves downward when the force **P** is applied. The modulus of the spring is 25 lb per in., and its free length is 14 in. Determine the value (or values) of the angle θ for equilibrium when $P = 120$ lb.

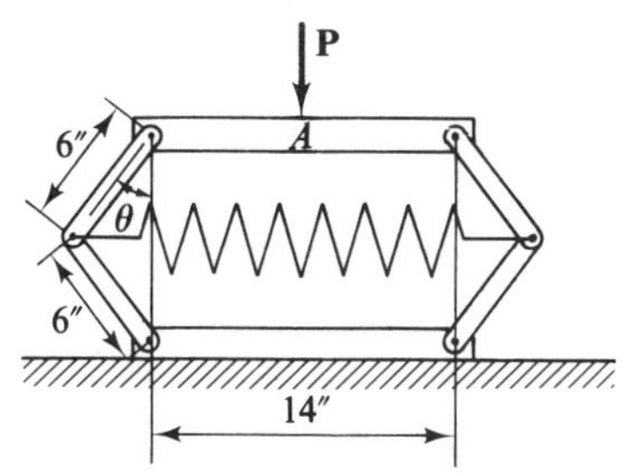

Figure P7-16

7-17 Bar AB in Fig. P7-17 is homogeneous and weighs 20 lb. Determine the distance x for which the bar will be in equilibrium.

7-18 Block *D* of the mechanism in Fig. P7-18 slides in the vertical slot and comes in contact with the spring when *D* is 5 in. below *B*. The system is in equilibrium when $P = 20$ lb and $\theta = 30°$. Determine the spring modulus.

7-19 The force **P** in Problem 7-18 has a magnitude of 50 lb, and the spring modulus is 5.0 lb per in. The spring is unstressed but makes contact with block *D* when $\theta = 0$. Determine the value(s) of θ for which the system will be in equilibrium.

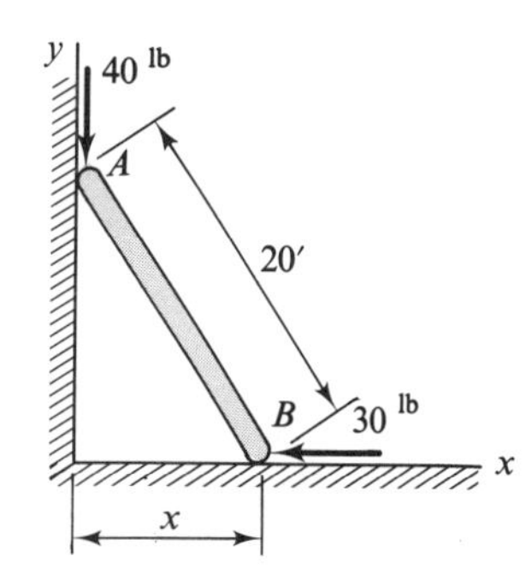

Figure P7-17

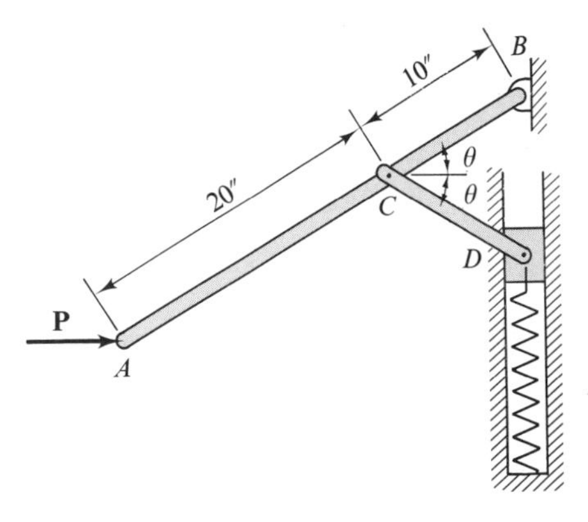

Figure P7-18

*7-20 The two bodies in Fig. P7-20 are supported by an inextensible cord which passes over pulleys which are assumed to have negligible diameters and weights. Determine the distance h for equilibrium.

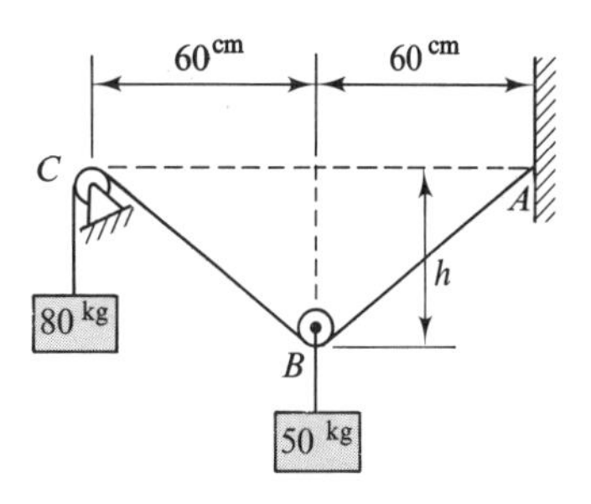

Figure P7-20

7-21 Determine the weight W required to keep the system in Fig. P7-21 in equilibrium for the position shown. Neglect the weights of bars AB and BC.

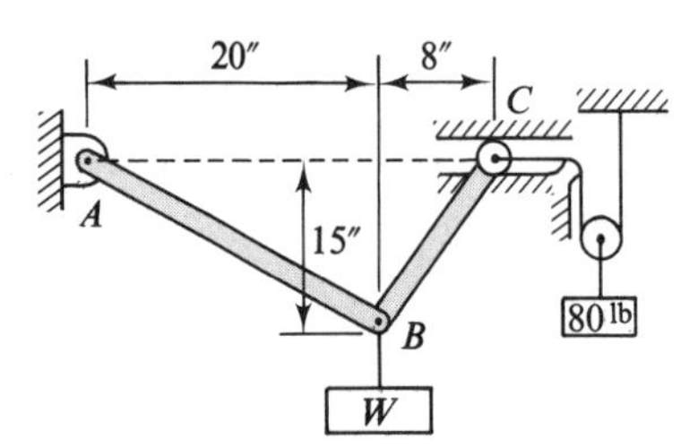

Figure P7-21

7-22 The system of weights and pulleys in Fig. P7-22 is in equilibrium. The rope over pulleys B and C is continuous. Ropes from A to B and C are fastened to pulley A. The weights of the pulleys are 25 lb for A, 15 lb for B, and 10 lb for C. Determine the weight of W.

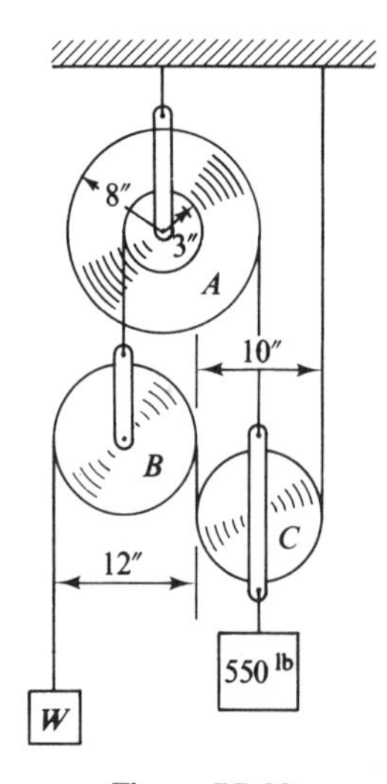

Figure P7-22

7-23 The two uniform bars AB and BC in Fig. P7-23 have the same cross section and are welded together at B. The small blocks at A and B slide in smooth slots. Determine any values of θ for which the bars will be in equilibrium and show the results on a sketch.

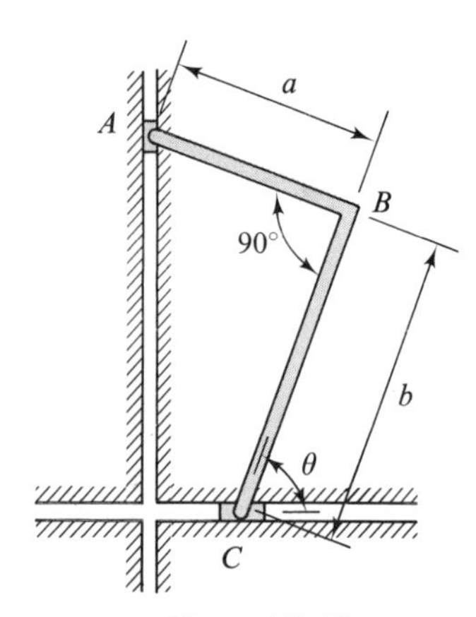

Figure P7-23

7-24 The uniform square block A in Fig. P7-24 weighs 30 lb, and body B weighs 20 lb. The two rollers at C and D move in smooth slots. Body B is connected to D by means of a cord which passes over a smooth peg at E. Determine two values of θ for which the system will be in equilibrium.

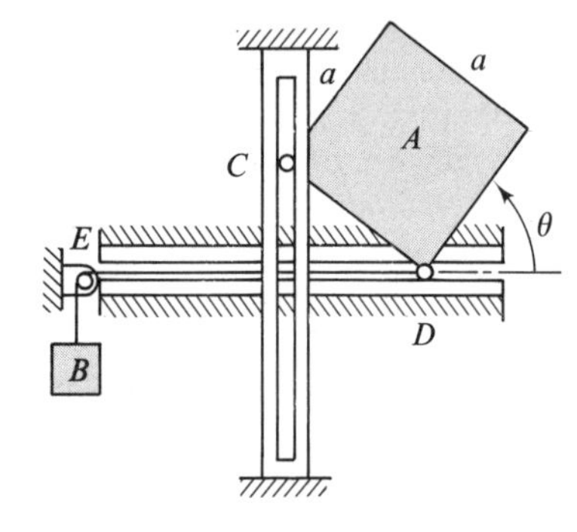

Figure P7-24

7-25 Determine the weight of W which will hold the system in Fig. P7-25 in equilibrium when the 300-lb uniform bar AB is horizontal.

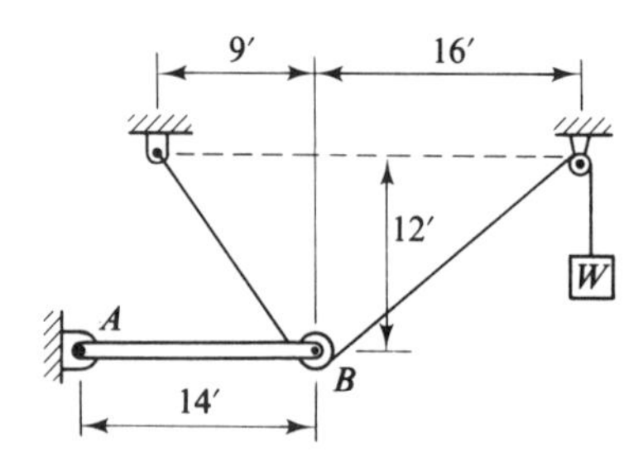

Figure P7-25

7-26 Determine the magnitude of the force **P** which will maintain equilibrium of the mechanism in Fig. P7-26 for the indicated position. Bars AB and BC weigh 10 lb each.

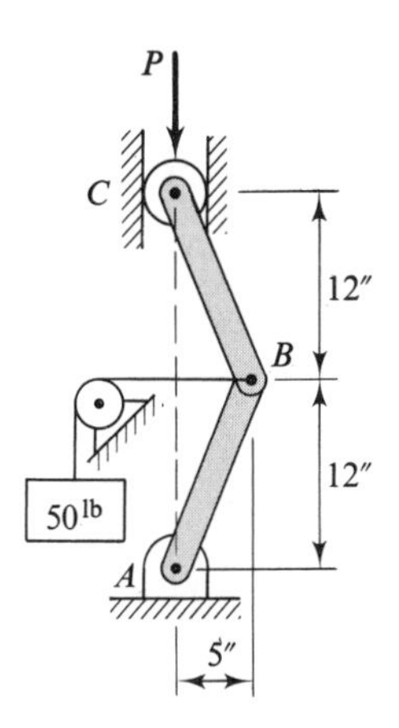

Figure P7-26

*7-27 Determine the magnitude of the force **P** which will maintain equilibrium of the system in Fig. P7-27 for the indicated position. The mass of AB is 5.0 kg, and the mass of BC is 3.0 kg.

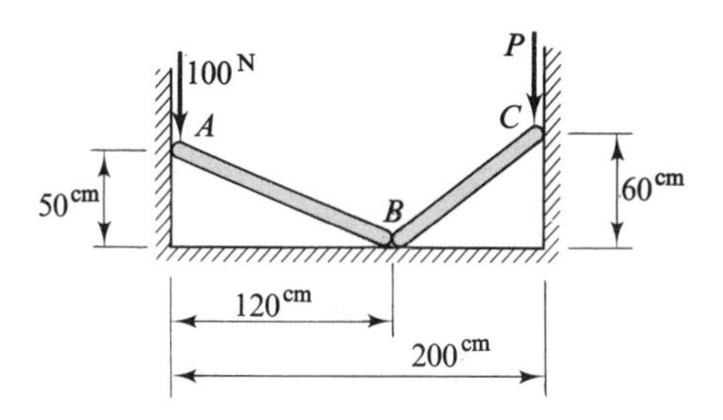

Figure P7-27

*7-28 The 1.0-m bar AB in Fig. P7-28 has a mass of 5.0 kg, and the 1.7-m bar BC has a mass of 3.4 kg. The locations of the centers of gravity of the bars are given on the sketch. Determine the distance y_1 for which the bars will be in equilibrium.

7-29 The three identical blocks in Fig. P7-29 weigh W lb each and are connected by pins at C

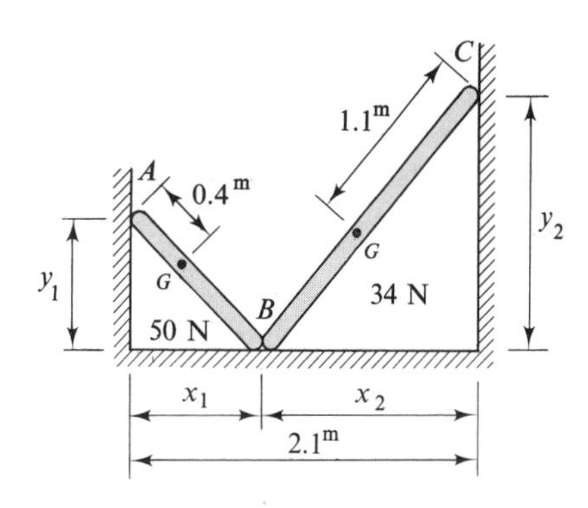

Figure P7-28

and E. Corners A, D, and F slide along the horizontal plane, and G slides in the smooth vertical slot. The system is held in equilibrium by a horizontal force **P** at point B. The length c is equal to $2b$. Determine the force **P** when
(a) $\theta = 20°$.
(b) $\theta = 35°$.

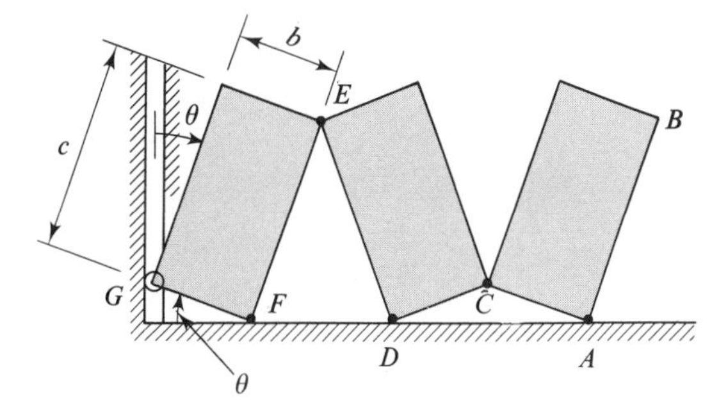

Figure P7-29

7-30 In Problem 7-29 length c is $1.5b$, and the horizontal force **P** is applied to point A. Determine the angle θ for equilibrium for $P = 5W$ when **P** is
(a) To the left.
(b) To the right.

7-31 The mechanism in Fig. P7-31 consists of four members of length $b = 5$ in. connected to blocks A and B and to the horizontal member EF. The modulus of each spring is 20 lb per in., and when blocks A and B are in contact the angle θ is 30° and the force in each spring is 100 lb. Determine
(a) The maximum force **P** for which the angle θ will be 30°.
(b) The value of θ when $\mathbf{P} = 200$ lb downward.

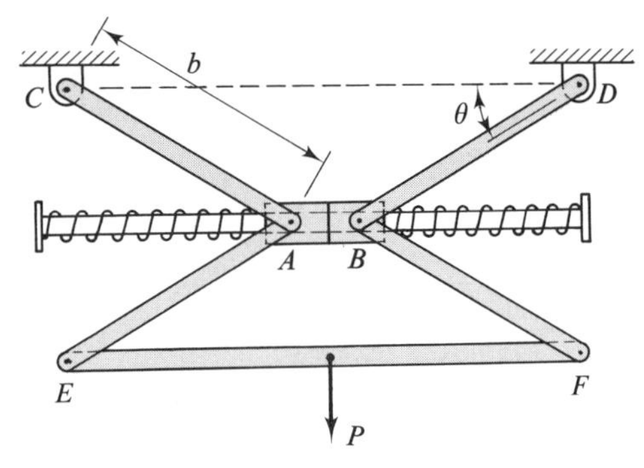

Figure P7-31

7-5 STABILITY OF EQUILIBRIUM

In Art. 7-4, the method of virtual work was used to determine the position of a body or system of bodies for which the system would be in equilibrium. The method can be extended to determine whether the resulting equilibrium is stable or unstable. Examples of stable and unstable equilibrium and also of neutral or indifferent equilibrium are shown in Fig. 7-17. A body is in stable equilibrium if, when it is

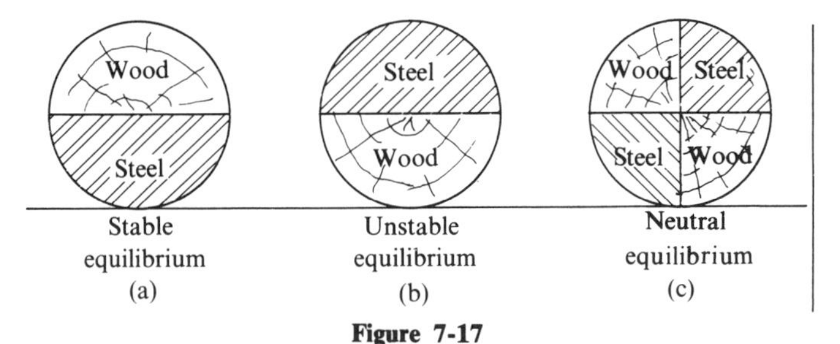

Figure 7-17

displaced slightly from its equilibrium position and released, it returns to the original position. If the body tends to continue to move farther from its equilibrium position when displaced slightly and released, the body is in unstable equilibrium. When a body remains in any position in which it is placed, it is in neutral equilibrium. In Fig. 7-17 the three cylinders are made of wood and steel. The cylinder in Fig. 7-17(a) is in stable equilibrium, since it will return to the position shown if it is rotated slightly. If the cylinder in Fig. 7-17(b) is rotated slightly and released, it will continue to move until the steel half is on the bottom. The cylinder in Fig. 7-17(c) is in neutral equilibrium since it will remain in any position to which it may be rolled.

When a force is applied to a body in stable equilibrium causing a small displacement, the force system holding the body in equilibrium does negative work on the body during this small displacement. When the disturbing force is removed, the force system will do positive work, and the body will return to the position of stable equilibrium. When a force is applied to a body in unstable equilibrium, thereby causing a small displacement, the force system holding the body in equilibrium does positive work on the body, and even though the disturbing force is removed, the body, once displaced, will continue to move away from the position of unstable equilibrium. When a disturbing force is applied to a body in neutral equilibrium and a small displacement results, the force system holding the body in neutral equilibrium does no work during this displacement. In this case, when the disturbing force is removed, the body does not return to its original position, nor does it tend to move away from the original equilibrium position; it remains in any position in which it is placed.

When both the equilibrium position and the type of equilibrium are to be determined, it is advantageous to modify the procedure of Art. 7-4 by introducing a work or potential energy function. For ideal systems with one degree of freedom, the work done on the system during a finite displacement can be expressed in terms of a single variable representing the position of the system measured from some initial position. The work done on a body during a finite displacement from any convenient initial position to a new position, defined by the variable x, can be expressed as

$$U = f(x),$$

where U is called the *work function.*

The virtual work done during a virtual (infinitesimal) displacement of the body from the position defined by x is the change in the work function during an infinitesimal change in x, or, in other words, it is the differential of U, which can be written

$$\delta U = \delta f(x) = \frac{df(x)}{dx}\,\delta x.$$

Thus the virtual work can be obtained by differentiation of the work function as well as by the method illustrated in Art. 7-4. The criterion for equilibrium, as shown in Art. 7-4, is

$$\delta U = \frac{df(x)}{dx}\,\delta x = 0,$$

which reduces to

$$\frac{df(x)}{dx} = \frac{dU}{dx} = 0.$$

This equation indicates that the work function must be a maximum or a minimum if the system is in equilibrium.

Since the forces acting on a body in stable equilibrium do negative work on the body during a virtual displacement, the amount of work done on the body will be less when the body is displaced from a position of stable equilibrium than when it is not displaced. Therefore, the work function will be maximum when the body is in stable equilibrium. Similarly, the forces holding a body in unstable equilibrium do positive work during a virtual displacement from the unstable position, which indicates that the work function is a minimum when the body is in unstable equilibrium. The mathematical requirements for the function to be maximum, that is, for stable equilibrium, are that the first derivative be zero and the second derivative be negative. If the second derivative is positive, the work function is minimum and the equilibrium is unstable. A zero second derivative indicates neutral equilibrium.

The preceding discussion is based on the use of the work function. The entire discussion can also be based on the potential function. *The potential energy of a body or system of bodies is defined as the ability of the body or system of bodies to do work as a result of their position or configuration with respect to some original position.*

When a body of weight W is raised a distance h above some reference plane, work is done to overcome the gravitational pull of the earth, and the potential energy of the body is increased by an amount

$$V = Wh,$$

where V represents the potential function. The work done by the weight is

$$U = -Wh,$$

which is the same as the potential function with the sign reversed.

When a spring of modulus k is compressed an amount x from the unstressed position, its potential energy is increased by an amount

$$V = \int F\,ds = \int_0^x ks\,ds = \frac{kx^2}{2}.$$

The work done by the spring on the body which compresses it is the product of the average force and the distance it moves; that is,

$$U = -\frac{kx}{2}x = \frac{-kx^2}{2}.$$

For an ideal system of bodies (bodies with no dissipation of energy) in equilibrium, the work done by all the forces acting on the system during any displacement is equal to the decrease in the potential energy of the system. That is, the work done plus the potential energy change must be a constant, or

$$U + V = \text{constant}.$$

In a more general sense, it is customary to consider the potential at infinity to be zero and to define the potential at any point in a field of force as the work done in bringing a particle of unit mass from infinity to the point. In a gravitational field of force, the potential function is thus the potential energy of a particle of unit mass at the point. In electrostatics, the potential function is the potential energy of a unit charge at the point, and so on.[2] When potential energy is used instead of the work function, it is apparent that V will be maximum when U is minimum. The condition for equilibrium is that the first derivative of the potential function with respect to the position variable be zero. The second derivative for the potential function will be positive if the equilibrium is stable and negative if it is unstable. It is a matter of personal preference whether the work function or the potential energy function is used for problems in equilibrium.

The procedure for determining the stability of equilibrium positions may be summarized as follows:

1. Draw a free-body diagram of the system.
2. Write an expression for the work function U (or potential energy V) and express it in terms of one variable, say, x.
3. Use the equation $dU/dx = 0$ (or $dV/dx = 0$) to determine the value or values of x which correspond to equilibrium positions.
4. Substitute the values of x found in step 3 into the second derivative of U, or of V, with respect to x and examine the algebraic sign of the result to determine the state of stability of the equilibrium position. The tests are listed in the following table:

	Method	
State of Stability	*Work Function*	*Potential Energy*
Stable	$\frac{d^2U}{dx^2}<0$	$\frac{d^2V}{dx^2}>0$
Unstable	$\frac{d^2U}{dx^2}>0$	$\frac{d^2V}{dx^2}<0$
Neutral	$\frac{d^2U}{dx^2}=0$	$\frac{d^2V}{dx^2}=0$

There are many problems in which the physical constraints, instead of the minimum potential energy, determine one or more (perhaps all) of the possible equilibrium positions. Consider the bar AB and the weight C in Fig. 7-18. If the slots are constructed so that A cannot move below D and B cannot move to the left of D, the system will have two rather obvious equilibrium positions: one when AB is horizontal, and one when it is vertical. There will also be a third equilibrium position when the bar is inclined as shown, the angle depending on the weights of the two bodies. This third equilibrium position is rather obviously unstable. Other equilibrium

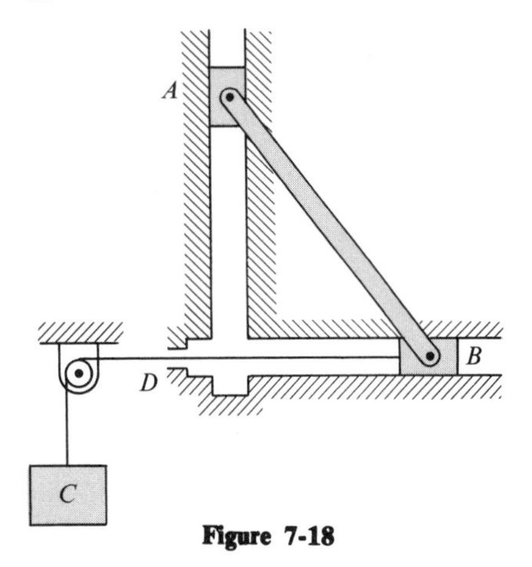

Figure 7-18

[2] For more information on the potential function, see Leigh Page, *Introduction to Theoretical Physics*, 3rd ed. (New York: Van Nostrand Reinhold Company, 1952).

positions could be established by placing a rigid stop in one of the slots, by replacing the slot at A with a pin, or by introducing some other physical restraint on the system. Such a stop introduces a discontinuity in the work or potential function (a *rigid* stop gives an *infinite* discontinuity), and the function must be modified to include the effect of the stop. The rigid stop can be replaced with a spring and a solution obtained in terms of the spring modulus k. The solution for a rigid stop can be obtained from the limit as k approaches infinity. Such a procedure is unnecessary for determining the equilibrium position, since the stop or other restraint gives the position. The procedure may be useful, however, in determining the stability of the position. This concept is illustrated in Example 7-7.

EXAMPLE 7-6

The 30-lb uniform rod AB in Fig. 7-19(a) and the 6.0-lb body D are connected by an inextensible cord which passes over a smooth peg at C. Determine the equilibrium position or positions of the system and the type of equilibrium.

SOLUTION

The free-body diagram in Fig. 7-19(b) indicates that there are five forces acting on the system, but three of these forces do not move and hence do no work. The work done by the other two forces moving from their positions when $\theta = 0$ is

$$U = 30y_1 - 6y_2.$$

Also

$$y_1 = 2 \sin \theta$$

and

$$y_2 = BC = 2\left[6 \sin \frac{\theta}{2}\right].$$

Therefore

$$U = 60 \sin \theta - 72 \sin \frac{\theta}{2}.$$

The equilibrium position is determined as the value of θ for which $dU/d\theta = 0$; that is,

$$\frac{dU}{d\theta} = 60 \cos \theta - 36 \cos \frac{\theta}{2} = 0$$

or

$$\cos \theta = 0.6 \cos \frac{\theta}{2}.$$

From the trigonometric identity $\cos^2(\theta/2) = 0.5(1 + \cos \theta)$, this equation, when squared, becomes

$$\cos^2 \theta = 0.36(0.5)(1 + \cos \theta)$$

or

$$\cos^2 \theta - 0.18 \cos \theta - 0.18 = 0$$

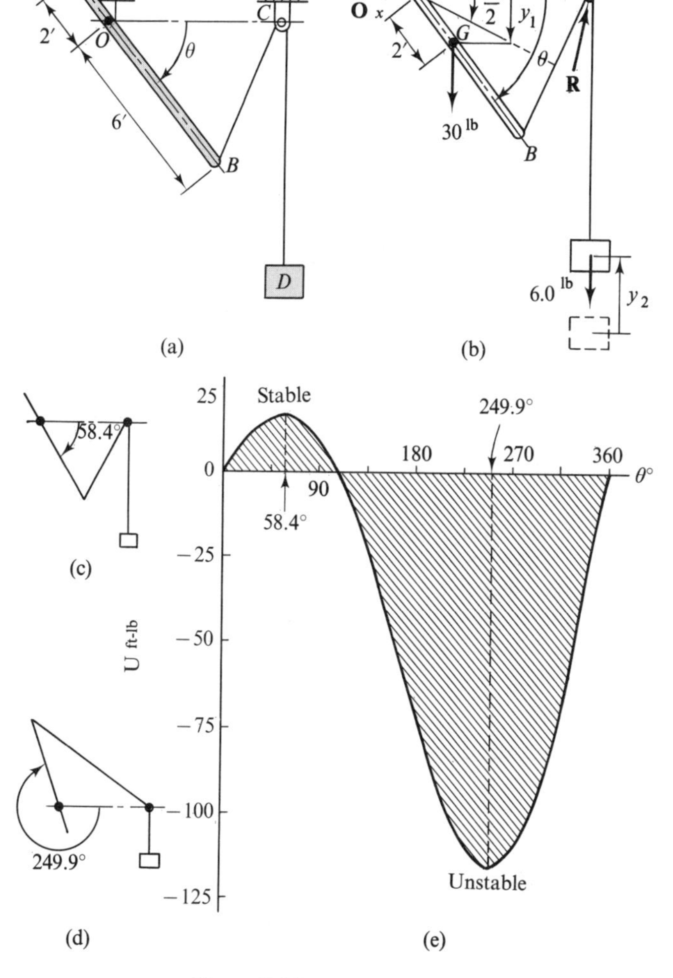

Figure 7-19

from which

$$\cos\theta = 0.524 \quad \text{or} \quad -0.344.$$

These two values of $\cos\theta$ are satisfied with four different values of θ between 0 and 360°. These values are 58.4 and 301.6° for $\cos\theta = 0.524$, and 110.1 and 249.9° for $\cos\theta = -0.344$. Two of these values are extraneous roots introduced by squaring the original equation and will not satisfy the equation

$$\cos\theta = 0.6\cos\frac{\theta}{2}.$$

These values are $\theta = 110.1$ and 301.6°, and in this instance observation of the system drawn in these two positions shows that the system is not in equilibrium for these values of θ. The system is in equilibrium for the other two values of θ which are given in Fig. 7-19(c) and (d). It appears obvious from the figures that the equilibrium will be stable for $\theta = 58.4°$ and unstable for $\theta = 249.9°$. This conclusion can be checked by determining the value of the second derivative of the work function for these values of θ.

$$\frac{d^2U}{d\theta^2} = -60 \sin \theta + 18 \sin \frac{\theta}{2}.$$

For $\theta = 58.4°$, this becomes

$$\frac{d^2U}{d\theta^2} = -60(0.852) + 18(0.488) = -42.3.$$

The negative sign indicates that the work done by the forces is a maximum and that the equilibrium is stable. When $\theta = 249.9°$,

$$\frac{d^2U}{d\theta^2} = -60(-0.939) + 18(0.820) = +71.1,$$

and the equilibrium is unstable since the second derivative is positive.

The results of the preceding analysis can also be verified by constructing a graph of the work function versus θ as shown in Fig. 7-19(e). Note that at the equilibrium positions of $\theta = 58.4$ and 249.9° the values of U are maximum and minimum. The position at 58.4° is the stable one since it corresponds to the maximum value of U. The extraneous roots of $\theta = 110.1$ and 301.6° have no significance because $dU/d\theta$ is not zero for these values.

EXAMPLE 7-7

The uniform 100-lb, 4-ft bar AB of Fig. 7-20(a) is constrained so that ends A and B can move only in vertical and horizontal slots, respectively. A spring with a modulus of 25 lb per ft acts on end B of the rod and has an unstrained length of 8 ft. Neglecting the weights of the blocks at A and B, determine the value or values of θ corresponding to equilibrium positions and investigate the stability of each position.

SOLUTION

The use of potential energy is illustrated in this solution. The datum plane for the potential energy of the weight is selected through the lower end of the bar, and the potential energy due to gravity is

$$V_{wt} = W\frac{L}{2} \cos \theta.$$

The potential energy stored in a spring is $k\Delta^2/2$, where k is the modulus and Δ is the distance the spring is compressed, $L \sin \theta$ for this problem, and thus

$$V_S = \frac{kL^2 \sin^2 \theta}{2}.$$

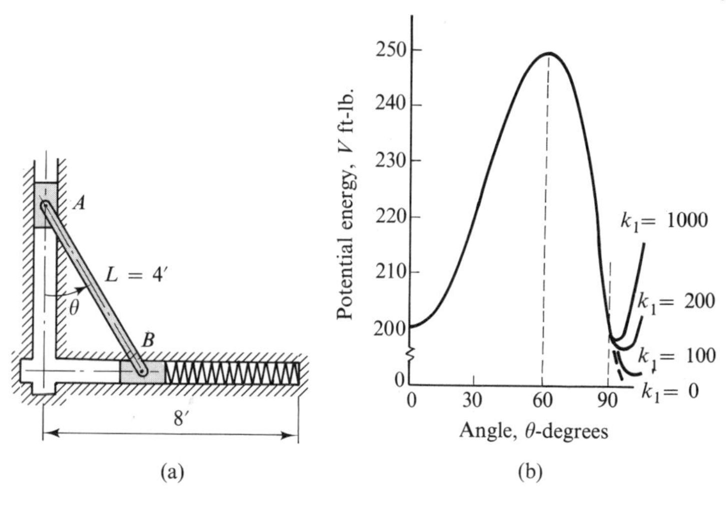

Figure 7-20

These are the only two forces contributing to the potential energy of the system, as long as the blocks do not contact the ends of the slots; therefore, the potential function is

$$V = \frac{WL\cos\theta}{2} + \frac{kL^2\sin^2\theta}{2}. \qquad \text{(a)}$$

The derivative of the potential function with respect to θ is

$$\frac{dV}{d\theta} = -\frac{WL}{2}\sin\theta + kL^2\sin\theta\cos\theta$$

$$= L\sin\theta\left(kL\cos\theta - \frac{W}{2}\right).$$

The equilibrium positions are obtained by equating the derivative to zero. That is,

$$\sin\theta = 0 \quad \text{and} \quad \cos\theta = \frac{W}{2kL}.$$

When numerical values are substituted for W, k, and L, the angles are found to be

$$\theta = \underline{0} \quad \text{and} \quad \theta = \cos^{-1}\frac{100}{2(25)4} = \cos^{-1} 0.5 = \underline{60°}. \qquad \text{Ans.}$$

The first derivative is also zero for $\theta = 180°$ and for $\theta = -60°$, but the physical constraints of the system limit θ to values from 0 to 90°.

The stability of the equilibrium positions can be examined by evaluating the second derivative of the potential function at the two values of θ. The second derivative is

$$\frac{d^2V}{d\theta^2}=-\frac{WL}{2}\cos\theta+kL^2(\cos^2\theta-\sin^2\theta).$$

When $\theta=0$ is substituted in this expression, it becomes

$$\frac{d^2V}{d\theta^2}=-\frac{WL}{2}+kL^2=-\frac{100(4)}{2}+25(4)^2=+200,$$

indicating that V has a minimum value and the equilibrium position is stable. When $\theta=60°$ is substituted in the expression, it becomes

$$\frac{d^2V}{d\theta^2}=-\left[\frac{100(4)}{2}\right](0.500)+25(4)^2[(0.500)^2-(0.866)^2]$$
$$=-300,$$

which indicates that $\theta=60°$ is an unstable equilibrium position.

Since the bar is not stable when $\theta=60°$, it is seen that if θ is a little more than 60°, the bar will move to a horizontal position, $\theta=90°$, where it must again be in equilibrium. The vertical force on the block at A results from a *rigid* stop; therefore the potential function will have an infinite discontinuity at $\theta=90°$ when this force is included. If the stop is replaced by a spring having a modulus k_1, however, the potential function, Eq. (a), becomes

$$V=\frac{WL\cos\theta}{2}+\frac{kL^2\sin^2\theta}{2}+\frac{k_1L^2\cos^2\theta}{2}$$

for values of θ more than 90°. When numerical values are substituted in the two expressions for the potential function, they become

$$V=200\cos\theta+200\sin^2\theta,\qquad 0<\theta<90°,$$

and

$$V=200\cos\theta+200\sin^2\theta+8k_1\cos^2\theta,\qquad \theta>90°.$$

The value of V is plotted against θ for three different values of k_1 in Fig. 7-20(b). The graph shows that V is minimum at $\theta=0$ and maximum at $\theta=60°$. A second minimum is indicated for θ slightly more than 90°, depending on the value of k_1. A spring with an infinite stiffness, representing a rigid stop, would give a stable equilibrium position at $\theta=90°$.

A discontinuous potential function occurs whenever the position variable is not free to assume all possible values. The limiting position will be given by the constraints and will be an equilibrium position (also stable) whenever the potential function decreases as the variable approaches the limiting value. When the position variable is limited by constraints, the possibility of equilibrium at the extreme positions should be investigated. Such a situation will always occur when only one value of the variable makes the derivative of the potential function vanish and when this position is found to be unstable.

PROBLEMS

7-32 An 80-lb homogeneous cone is fastened to a 100-lb homogeneous hemisphere as indicated in Fig. P7-32. Is the position of the bodies stable as shown?

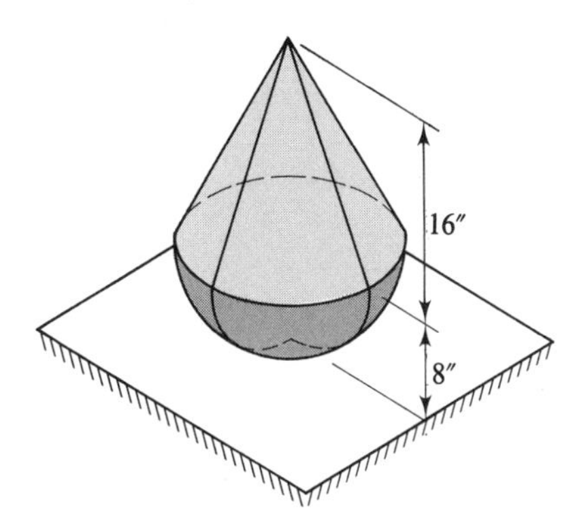

Figure P7-32

7-33 Bodies *A*, *B*, and *C* in Fig. P7-33 are equally spaced around the rim of the wheel and have the weights indicated. Determine the angle(s) line *OA* will make with the horizontal when the system is in equilibrium and indicate whether or not the system is in stable equilibrium.

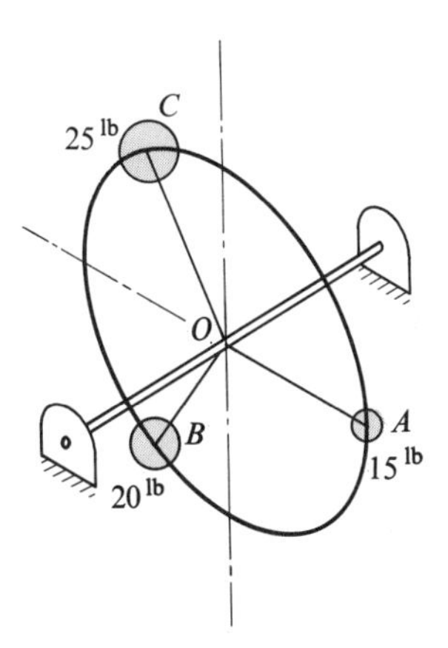

Figure P7-33

7-34 Two weights of 20 and 40 lb are attached to the rim of an otherwise balanced wheel as shown on Fig. P7-34. The wheel is supported on a smooth axle. A third weight of 20 lb is suspended from a cord wrapped around a drum which is attached to the wheel. Determine any value(s) of the angle β for which the system is in equilibrium and indicate the stability of the equilibrium for each value.

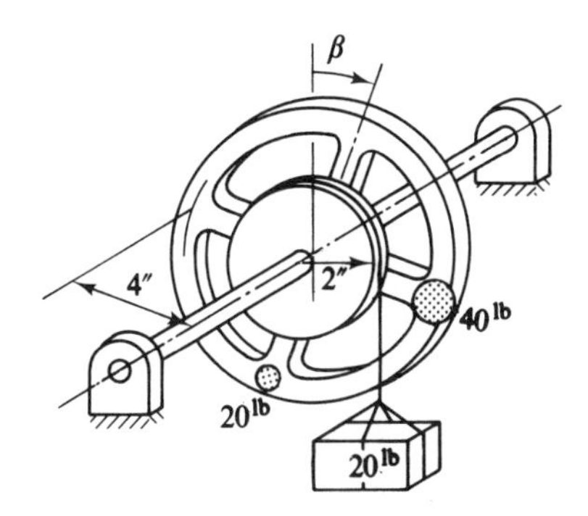

Figure P7-34

***7-35** Check the stability of equilibrium for Problem 7-14.

7-36 Check the stability of equilibrium for Problem 7-16.

7-37 Check the stability of equilibrium for Problem 7-19.

7-38 Check the stability of equilibrium for Problem 7-24.

7-39 Check the stability of equilibrium for Problem 7-30.

7-40 Bar *AB* in Fig. P7-40 has a weight *W* and is held in a vertical position by two springs, each of which has a modulus of *k*. Determine the minimum value of *k* for which the equilibrium will be stable.

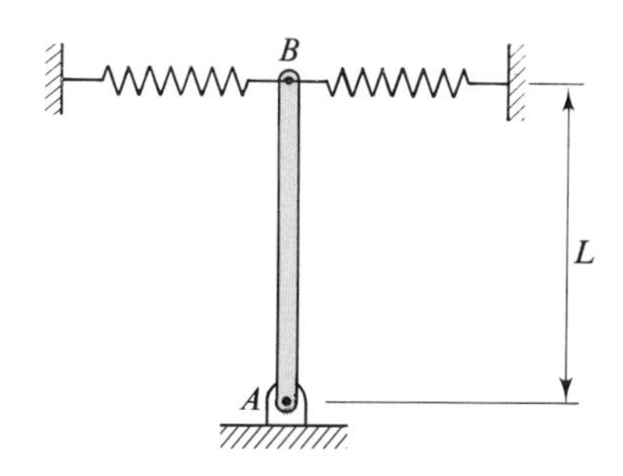

Figure P7-40

7-41 The homogeneous bar AB in Fig. P7-41 has a weight W and is connected to weight Q by a cord which passes over two small pulleys. Show that the system will be in stable equilibrium if Q is more than $W/4$ and $\theta = 0°$ or if Q is less than $3W/4$ and $\theta = 180°$.

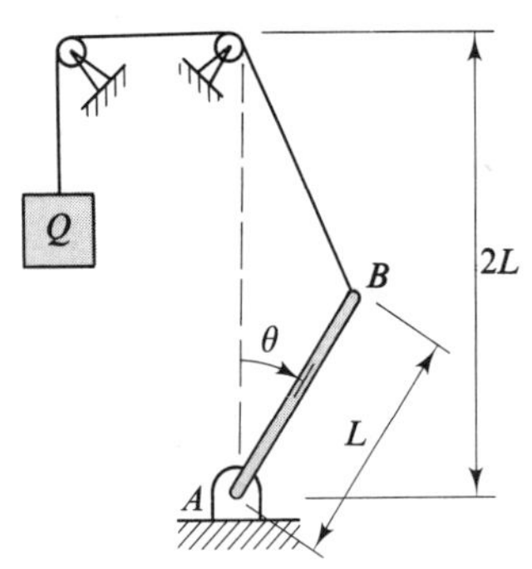

Figure P7-41

7-42 In Problem 7-41 the weight of Q is equal to $W/2$. Determine any values of θ for which the system will be in equilibrium and indicate whether the equilibrium is stable or unstable for each value.

7-43 Each of the members AB, BC, EF, and FG of the mechanism in Fig. P7-43 has a length of $2b$ ($b = 20$ in.). The small connecting blocks slide in smooth slots. The spring modulus is 10 lb per in., and the spring is unstressed when $\theta = 60°$. Show that when $P = 1000$ lb the mechanism will be in equilibrium when $\theta = 46.3°$ and when $\theta = 27.2°$. Determine which, if either, of the positions will result in stable equilibrium.

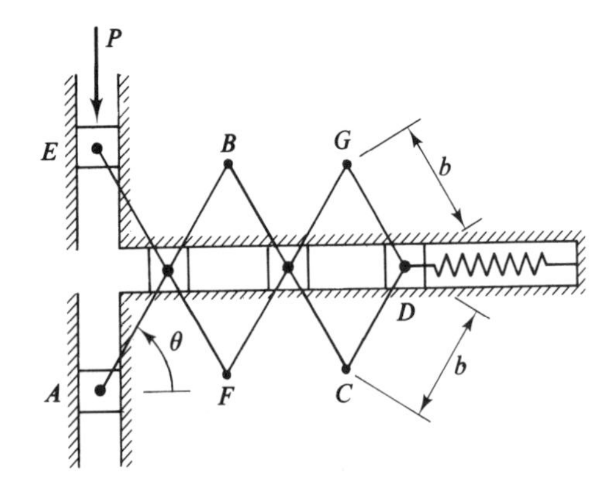

Figure P7-43

8 Kinematics

PART A—KINEMATICS OF PARTICLES

8-1 INTRODUCTION

Kinematics is that branch of mechanics which deals with the motion of particles, lines, and bodies without consideration of the forces required to produce or maintain the motion. A working knowledge of the relationships between position, time, velocity, acceleration, displacement, and distance traveled for particles, lines, and bodies is essential to the study of the effects of unbalanced force systems on bodies. Kinetics is concerned with the force systems that produce accelerated motion of bodies, the inertial properties of the bodies, and the resulting motion of the bodies.

Kinematic quantities, such as position and velocity, are expressed with respect to a system of reference axes. When axes fixed on the earth are used as a reference, the motion is called *absolute motion*, as distinguished from relative motion, which is measured with respect to moving axes. Strictly speaking, absolute motion is motion measured relative to a coordinate system fixed in space, a Newtonian reference frame, and since the earth is moving, motion measured with respect to axes fixed on the earth is actually relative motion. For elementary mechanics, however, the motion of the earth can be neglected, and in this text *motion measured relative to a coordinate system fixed on the earth is defined as absolute motion.*

The term *particle* is used for any body whose size can be neglected without introducing appreciable errors when studying or describing its motion. In planetary motion, the earth is considered as a particle, whereas the balance wheel of a watch must be treated as a body in an analysis of the motion of the watch mechanism.

The word *linear,* as in linear velocity or linear displacement, is used to describe the motion of particles or points. Angular motion is restricted to lines and bodies because two lines are necessary to define an angle.

Particles are said to have *rectilinear motion* when they move along straight lines and *curvilinear motion* when they travel on curved paths. *Uniform motion of a particle is defined as a motion such that equal distances are traversed during equal intervals of time, regardless of how small the time intervals may be.*

8-2 REVIEW OF VECTOR CALCULUS

Vector calculus is useful in kinematics because many of the quantities studied are vectors. In scalar calculus, only changes in the magnitude of a variable are involved, whereas in vector calculus, changes in both magnitude and direction of variables must be considered.

The following material is a brief review of some aspects of vector calculus. A more complete discussion can be obtained from most calculus texts or from books on vector analysis.

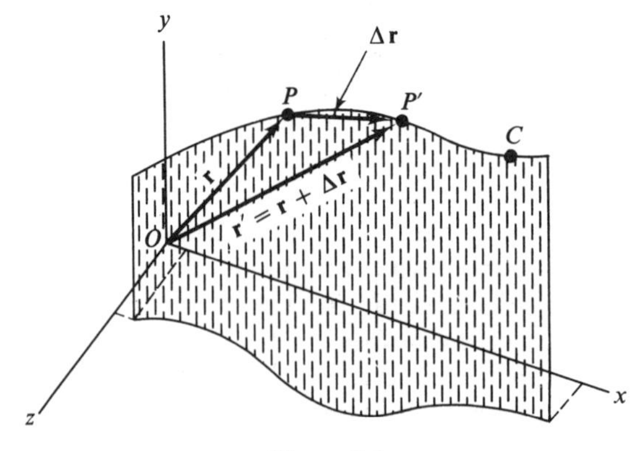

Figure 8-1

Figure 8-1 represents a particle moving along a curved path. Point O is the origin of the fixed coordinate system, xyz. The position vector $\mathbf{r}$ from O to P, expressed as a function of time, locates the particle at any instant. As the particle moves from point P to point P' the change of $\mathbf{r}$ is

$$\Delta\mathbf{r} = \mathbf{r}' - \mathbf{r}.$$

If $\Delta\mathbf{r}$ occurs during a time interval Δt, the ratio $\Delta\mathbf{r}/\Delta t$ is the average time rate of change of position, and the limit as Δt approaches zero is defined as the derivative of the position vector $\mathbf{r}$ with respect to the time t. Thus

$$\frac{d\mathbf{r}}{dt} = \dot{\mathbf{r}} = \lim_{\Delta t \to 0} \frac{\Delta\mathbf{r}}{\Delta t}.$$ [1]

Figure 8-1 shows that $\Delta\mathbf{r}$ is a vector and indicates the change of both magnitude and direction of $\mathbf{r}$. The vector $\mathbf{r}$ is not limited to position vectors but might be any vector which varies with t, and t can be any scalar variable.

To evaluate the derivative of a vector, changes of both magnitude and direction must be considered. If the vector is expressed in terms of unit vectors in fixed directions, the unit vectors can be treated as constants during the differentiation. Consider the function

$$\mathbf{r} = x\mathbf{i} + y\mathbf{j} + z\mathbf{k},$$

where x, y, and z are scalar functions of time and $\mathbf{i}$, $\mathbf{j}$, and $\mathbf{k}$ are unit vectors in the fixed x, y, and z directions. Each term is a product of a variable and a constant, and the derivative with respect to time is

$$\frac{d\mathbf{r}}{dt} = \dot{\mathbf{r}} = \dot{x}\mathbf{i} + \dot{y}\mathbf{j} + \dot{z}\mathbf{k}.$$

If the unit vector or vectors used to express $\mathbf{r}$ change in direction, the differentiation procedure must be modified to account for this change, as explained in Art. 8-8.

The derivative of the sum of two vectors is equal to the sum of the derivatives since vectors obey the distributive law. Thus if $\mathbf{A} = \mathbf{B} + \mathbf{C}$,

$$\frac{d}{dt}\mathbf{A} = \frac{d}{dt}(\mathbf{B} + \mathbf{C}) = \frac{d\mathbf{B}}{dt} + \frac{d\mathbf{C}}{dt}.$$

The following relationships can also be demonstrated by writing the derivative as a limit:

$$\frac{d}{dt}(\mathbf{A} \cdot \mathbf{B}) = \mathbf{A} \cdot \frac{d\mathbf{B}}{dt} + \frac{d\mathbf{A}}{dt} \cdot \mathbf{B} = \mathbf{A} \cdot \dot{\mathbf{B}} + \dot{\mathbf{A}} \cdot \mathbf{B},$$

[1] A single dot above a variable indicates the derivative of the variable with respect to time, and two dots, $\ddot{r}$, indicate the second derivative. The dot notation is used only for differentiation with respect to time and was originally adopted by Newton.

and

$$\frac{d}{dt}(\mathbf{A}\times\mathbf{B})=\mathbf{A}\times\frac{d\mathbf{B}}{dt}+\frac{d\mathbf{A}}{dt}\times\mathbf{B}=\mathbf{A}\times\dot{\mathbf{B}}+\dot{\mathbf{A}}\times\mathbf{B}.$$

It is essential that the order of the factors in each term be maintained when differentiating a vector product.

It was shown that the derivative of a vector with respect to a scalar, such as time, is a vector. Sometimes it is desirable to reverse the process, that is, to integrate a vector. Thus, if

$$\mathbf{A}=\frac{d\mathbf{B}}{dt},$$

the inverse expression would be

$$\mathbf{B}=\int \mathbf{A}\,dt+\mathbf{C},$$

where $\mathbf{C}$ is a vector constant of integration, which is constant in both magnitude and direction. One of the most useful methods of carrying out the integration is by expressing the integrand in terms of constant unit vectors, since such constants are not changed by either integration or differentiation. Thus if

$$\mathbf{A}=A_x\mathbf{i}+A_y\mathbf{j}+A_z\mathbf{k},$$

the integral becomes

$$\begin{aligned}\mathbf{B}&=\int \mathbf{A}\,dt=\int (A_x\mathbf{i}+A_y\mathbf{j}+A_z\mathbf{k})\,dt+\mathbf{C}\\&=\mathbf{i}\int A_x\,dt+\mathbf{j}\int A_y\,dt+\mathbf{k}\int A_z\,dt+\mathbf{C}.\end{aligned}$$

Additional concepts involving the calculus of vectors will be introduced as they are needed.

8-3 BASIC DEFINITIONS

The *instantaneous position* of a particle moving along some path can be specified by the position vector $\mathbf{r}$ from the origin of a fixed coordinate system to the particle. The position of the particle can also be specified by stating the coordinates of the particle or by giving the distance s, measured along the given path, from a fixed point on the path to the particle. It is assumed that all of these quantities are expressed as functions of time.

A consistent and clearly defined sign convention is essential if the algebraic solution is to be given a correct physical interpretation. Positive directions are those of the coordinate axes when given. If axes are not specified, positive directions that are consistent with given data should be selected and shown with problem solutions.

The linear displacement, $\mathbf{q}$, *of a particle during an interval of time is defined as the change of position of the particle during the time interval.* Thus if the particle moves from position P to position P' of Fig. 8-1 during a certain time interval, its displacement is

$$\mathbf{q} = \Delta\mathbf{r} = \mathbf{r}' - \mathbf{r} \tag{8-1}$$

and is the vector from P to P'. If the particle travels from P to C and then back to P', its displacement, $\mathbf{q}$, is still the vector $\Delta\mathbf{r}$ from P to P' and depends only on the initial and final positions of the particle.

The total distance traveled, represented by Q, *is the total accumulated length of path traversed* and thus depends on the shape of the path and whether or not the direction of travel is reversed as well as on the initial and final positions of the particle. From this definition it is seen that the total distance traveled will always be equal to or greater than q, the magnitude of the displacement. Only when the path is a straight line and the motion does not reverse will Q be equal to q.

The displacement $\mathbf{q}$ is a vector quantity and is completely specified only when both the magnitude and direction are given. The total distance Q is a scalar quantity having only magnitude. Linear displacement and total distance traveled represent changes which take place during a time interval, whereas the position, $\mathbf{r}$, is a continuous vector function of time with a definite value for each instant of time. All three quantities have fundamental dimensions of length and common units are feet, inches, meters, and similar units.

The linear velocity $\mathbf{v}$ *of a particle is defined as the time rate of change of the position of the particle.* If the particle in Fig. 8-1 moves from position P to position P' in time Δt, the average time rate of change of position, the average velocity, is $\Delta\mathbf{r}/\Delta t$, and the instantaneous velocity at position P is

$$\mathbf{v} = \lim_{\Delta t \to 0} \frac{\Delta\mathbf{r}}{\Delta t} = \frac{d\mathbf{r}}{dt} = \dot{\mathbf{r}}. \tag{8-2}$$

As the time interval is decreased point P' will approach point P and the displacement $\Delta\mathbf{r}$ will approach the tangent to the curve at P. Therefore *the velocity of a point is a vector and is always tangent to the path along which the point moves.* The magnitude of the velocity, v, is frequently called the *speed.* Equation (8-2) indicates that linear velocity has dimensions of length divided by time, LT^{-1}, and common units are feet per second (fps), meters per second (m/s), knots (one nautical mile, 6080 ft, per hour), and similar units.

The linear acceleration $\mathbf{a}$ *of a particle is defined as the time rate of change of the linear velocity of the particle.* At point P of Fig. 8-2 the velocity is $\mathbf{v}$, and at point P' it is $\mathbf{v}+\Delta\mathbf{v}$. The average time rate of change of linear velocity is $\Delta\mathbf{v}/\Delta t$, and the instantaneous acceleration at position P is

$$\mathbf{a} = \lim_{\Delta t \to 0} \frac{\Delta\mathbf{v}}{\Delta t} = \frac{d\mathbf{v}}{dt} = \frac{d^2\mathbf{r}}{dt^2} = \ddot{\mathbf{r}}. \tag{8-3}$$

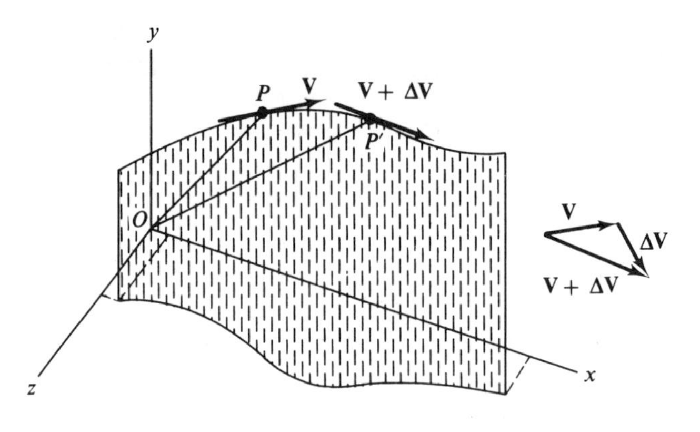

Figure 8-2

Linear acceleration is a vector quantity in the direction of the change of velocity. It should be noted that, although velocity is tangent to the path, the velocity may change in both magnitude and direction, and, in general, the acceleration is not tangent to the path. Equation (8-3) shows that the dimensions of linear acceleration are length divided by time squared, LT^{-2}. Common units are feet per second per second (fps^2), meters per second per second (m/s^2), and similar units. Accelerations are sometimes expressed in gs, where g is the acceleration of gravity (approximately 32.2 fps^2 or 9.81 m/s^2) at the surface of the earth. Thus an acceleration of 5g would be 161 fps^2 or 49.05 m/s^2.

A line has *angular motion* when the angle it makes with a fixed reference line changes. The line may turn about a fixed point on the line or line extended, as in the case of the hand of a clock, or it may be that no point on the line remains fixed, as in the case of a spoke of a rolling wheel. Particles are considered dimensionless, and any angular motion they might have cannot be measured or described; therefore, *angular motion is a property which is restricted to lines and bodies.*

In the following discussion, it is assumed that the line AB in Fig. 8-3 moves only in the xy plane. Noncoplanar motion of a line is discussed in Part B of this chapter. The angular position of line AB at any instant is defined by the angle θ *from* the x axis *to* the line. The angular position function θ, as used here, is a scalar function of time. The net change of angular position of a line during a given time interval is called the angular displacement of the line and is indicated by the symbol $\boldsymbol{\phi}$. Similarly, the symbol Φ is used to designate the total angle turned through during a time interval and will be more than ϕ whenever the direction of rotation changes.

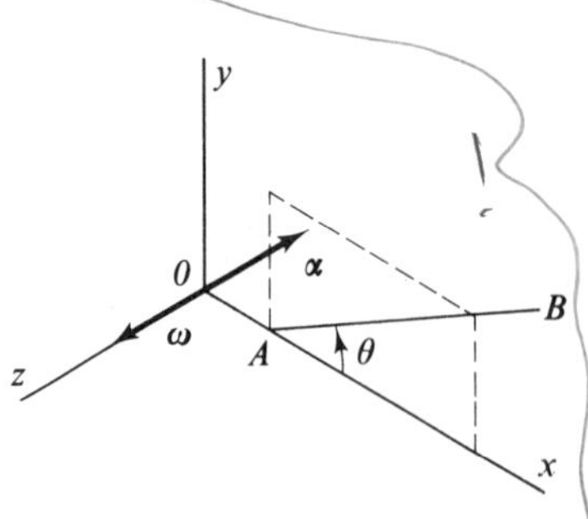

Figure 8-3

The angular velocity, $\boldsymbol{\omega}$, *of a line is defined as the time rate of change of the angular position of the line.* Angular velocity is a free vector quantity which can be represented by a vector perpendicular to the plane of motion of the line in a direction determined by the right-hand screw rule. If line AB in Fig. 8-3 is turning so that θ is increasing, the angular velocity vector is in the z direction as shown. When a line remains in one plane the angular velocity vector is always perpendicular to the plane and it is convenient to indicate the direction of the angular velocity with a clockwise or counterclockwise arrow. For coplanar motion of a line, the magnitude of the angular velocity is

$$\omega = \lim_{\Delta t \to 0} \frac{\Delta\theta}{\Delta t} = \frac{d\theta}{dt} = \dot{\theta}. \tag{8-4}$$

Angular velocity has the dimensions of an angle per unit of time, that is, T^{-1}, since an angle is dimensionless. Common units for angular velocity are radians per second (rad per sec), revolutions per minute (rpm), etc.

The angular acceleration $\boldsymbol{\alpha}$, *of a line is defined as the time rate of change of the angular velocity of the line.* Since angular velocity is a vector quantity, it can change in either magnitude or direction or both. Since only coplanar angular motion is considered in this article, the angular velocity can change only in magnitude and sense. The magnitude of the angular acceleration of a line is

$$\alpha = \lim_{\Delta t \to 0} \frac{\Delta\omega}{\Delta t} = \frac{d\omega}{dt} = \frac{d^2\theta}{dt^2} = \ddot{\theta}. \tag{8-5}$$

The angular acceleration of a line can be represented by a free vector, and for coplanar motion of a line, the vector is perpendicular to the plane of motion with the sense of the vector determined by the right-hand screw rule. If line AB of Fig. 8-3 is rotating counterclockwise with decreasing angular velocity, the angular acceleration is as shown on the figure. For coplanar motion it is convenient to use a clockwise or counterclockwise arrow to represent angular acceleration.

Angular acceleration has dimensions of an angle divided by a time squared, T^{-2}. Common units are radians per second per second (rad per sec^2), revolutions per second per second (rev per sec^2), and similar units.

When angular positions of moving lines are shown on a sketch the direction of the arrow from the fixed reference line is used to indicate the positive direction. Similarly, the coordinate axes are used to indicate positive linear directions. When positive directions are not specified they can frequently be determined from the given data. Where insufficient data permit a choice of positive directions, the assumed directions should be clearly indicated.

8-4 RECTILINEAR MOTION

Rectilinear motion of a particle has been defined as motion along a straight line. When the definitions developed in Art. 8-3 are applied to rectilinear motion it is seen that the velocity of the particle, which is always tangent to the path, will always be along the straight line and thus can change only in magnitude and sense. Since the velocity can change only in magnitude, the acceleration (the time rate of change of velocity) must also be along the straight line. If $\mathbf{i}$ is a unit vector along the path, the following equations are obtained from the definitions in Art. 8-3.

The position $\mathbf{r}$ (of the point) is

$$\mathbf{r} = s\mathbf{i},$$

where s is the distance *from* a fixed point on the path *to* the moving point. The velocity is

$$\mathbf{v} = \frac{d\mathbf{r}}{dt} = \frac{ds}{dt}\mathbf{i} = v\mathbf{i}$$

from which

$$v = \frac{ds}{dt}. \tag{a}$$

Similarly, the acceleration of the point is

$$\mathbf{a} = \frac{d\mathbf{v}}{dt} = \frac{dv}{dt}\mathbf{i} = a\mathbf{i};$$

that is,

$$a = \frac{dv}{dt} = \frac{d^2s}{dt^2}. \tag{b}$$

If v is expressed as a function of s, the acceleration can also be expressed (for rectilinear motion) as

$$a = \frac{dv}{ds}\frac{ds}{dt} = v\frac{dv}{ds}. \tag{c}$$

Note that Eqs. (a)–(c) are scalar equations which give the magnitudes of the velocities and accelerations. The directions are indicated by the unit vector along the path and the algebraic sign of the velocity or acceleration. If the magnitude of the velocity is decreasing, the acceleration is sometimes called a *deceleration*.

When some of the kinematic quantities (position, displacement, velocity, acceleration, and time) are known and others are to be determined Eqs. (8-2) and (8-3), or some of their other forms, can be used to obtain the desired quantities by differentiation or integration. The three scalar equations most useful for rectilinear motion are

$$v = \frac{ds}{dt} \quad \text{or} \quad v\,dt = ds,$$

$$a = \frac{dv}{dt} \quad \text{or} \quad a\,dt = dv,$$

$$a = v\frac{dv}{ds} \quad \text{or} \quad a\,ds = v\,dv.$$

The particular equation or form to use will depend on the given data and the desired results.

The similarity of Eqs. (a) and (8-4) and of (b) and (8-5) indicates a direct analogy between the position, velocity, and acceleration for rectilinear motion of a point and for coplanar angular motion of a line. Algebraic expressions for angular quantities can be differentiated and integrated in the same manner as the analogous expressions for linear quantities. It is often useful to relate the linear position of a point to the angular position of a line by means of geometry and obtain expressions relating velocities and accelerations by differentiation. The following examples illustrate some of the principles just discussed.

EXAMPLE 8-1

The velocity of a particle moving along a horizontal straight line is given by the expression

$$\dot{\mathbf{r}} = \mathbf{v} = (6t^2 - 10t)\mathbf{i},$$

where t is time in seconds and r is measured in feet. The particle is moving to the left when t is 2 sec. Determine

(a) The acceleration of the particle when $t = 2$ sec.
(b) The displacement of the particle during the time interval from $t = 1$ sec to $t = 3$ sec.
(c) The total distance traveled by the particle during the interval from $t = 1$ sec to $t = 3$ sec.

SOLUTION

(a) When $t = 2$ sec

$$\mathbf{v} = [6(2)^2 - 10(2)]\mathbf{i} = +4\mathbf{i}\text{ fps},$$

and the particle is moving to the left; therefore the unit vector **i** must be directed to the left. The acceleration of the particle can be obtained by differentiating the expression for the velocity. Thus

$$\mathbf{a} = \dot{\mathbf{v}} = \frac{d}{dt}(6t^2 - 10t)\mathbf{i} = (12t - 10)\mathbf{i},$$

and when $t = 2$ sec

$$\mathbf{a} = [12(2) - 10]\mathbf{i} = 14\mathbf{i}\text{ fps}^2$$

$$= \underline{14\text{ fps}^2\text{ to the left.}} \qquad \text{Ans.}$$

(b) Since the velocity is the derivative of the position function, Eq. (8-2), the position function can be determined by integrating the expression for the velocity. Thus

$$\mathbf{v} = \frac{d\mathbf{r}}{dt} \quad \text{and} \quad d\mathbf{r} = \mathbf{v}\, dt$$

from which

$$\mathbf{r} = \int \mathbf{v}\, dt.$$

In this example,

$$\mathbf{r} = \int (6t^2 - 10t)\mathbf{i}\, dt = \mathbf{i}\int (6t^2 - 10t)\, dt = \mathbf{i}(2t^3 - 5t^2) + \mathbf{C},$$

where $\mathbf{C}$ is a constant of integration. The location of the origin is not specified and thus may be selected at any convenient point on the line. If the origin is located at the position of the particle when $t = 0$, the constant $\mathbf{C}$ is zero.

When $t = 1$ sec

$$\mathbf{r}_1 = -3\mathbf{i},$$

and when $t = 3$ sec

$$\mathbf{r}_3 = +9\mathbf{i}.$$

The displacement during the time interval is

$$\begin{aligned} \mathbf{q} = \mathbf{r}_3 - \mathbf{r}_1 = 9\mathbf{i} - (-3\mathbf{i}) &= 12\mathbf{i} \text{ ft} \\ &= \underline{12 \text{ ft to the left.}} \qquad \text{Ans.} \end{aligned}$$

(c) For rectilinear motion, the magnitude of the displacement and the total distance traveled will be the same unless the particle reverses its velocity. In this example, the velocity is zero when $t = 0$ and when $t = 1.667$ sec. During the interval from $t = 1$ sec to $t = 1.667$ sec, the magnitude of the velocity is negative; that is, the particle is moving to the right, and from $t = 1.667$ sec to $t = 3$ sec, the velocity is to the left. When $t = 1.667$ sec, the position of the particle is

$$\mathbf{r}_{1.667} = \mathbf{i}[2(1.667)^3 - 5(1.667)^2] = -4.63\mathbf{i}.$$

The total distance traveled is the sum of the distances traveled in the two directions; that is,

$$\begin{aligned} Q &= |\mathbf{r}_{1.667} - \mathbf{r}_1| + |\mathbf{r}_3 - \mathbf{r}_{1.667}| \\ &= |-4.63\mathbf{i} - (-3\mathbf{i})| + |9\mathbf{i} - (-4.63\mathbf{i})| \\ &= 1.63 + 13.63 = \underline{15.26 \text{ ft.}} \qquad \text{Ans.} \end{aligned}$$

The quantity 1.63 ft is the distance traveled to the right during the first two-thirds of a second, and 13.63 ft is the distance traveled to the left after the velocity reverses.

EXAMPLE 8-2

The magnitude of the acceleration of a point moving along a horizontal straight line varies according to the equation $a = 12s^{1/2}$, where a is in feet per second per second and s is the distance of the point from the origin in feet. When the time t is 2 sec, the point is 16 ft to the right of the origin and has a velocity of 32 fps to the right and an acceleration of 48 fps^2 to the right. Determine the velocity and acceleration of the point when the time is 3 sec.

SOLUTION

The positive direction is obtained by observing that s must be positive in order to have the acceleration be real (not imaginary), and the problem specifies that the acceleration is real (48 fps^2) when $s = 16$ ft to the right. Therefore the positive direction is to the right.

When the acceleration is expressed as a function of the position s, it is convenient to write the acceleration in the form

$$a = v\frac{dv}{ds}$$

and

$$v\,dv = a\,ds = 12s^{1/2}\,ds$$

from which

$$\int v\,dv = \int 12s^{1/2}\,ds$$

or

$$\frac{v^2}{2} = 8s^{3/2} + C_1.$$

When $s = +16$ ft, $v = +32$ fps; therefore

$$\frac{32^2}{2} = 8(16)^{3/2} + C_1$$

and

$$C_1 = 0$$

from which

$$v^2 = 16s^{3/2}.$$

The velocity is the derivative of the position function. That is,

$$v = 4s^{3/4} = \frac{ds}{dt}$$

from which

$$4\,dt = s^{-3/4}\,ds$$

and

$$4t = 4s^{1/4} + C_2.$$

When $t = 2$ sec, $s = 16$ ft; therefore

$$4(2) = 4(16)^{1/4} + C_2 = 8 + C_2$$

and $C_2 = 0$. The equation for t thus reduces to

$$t = s^{1/4} \quad \text{or} \quad s = t^4$$

from which

$$v = \frac{ds}{dt} = 4t^3$$

and

$$a = \frac{dv}{dt} = 12t^2.$$

When $t = 3$ sec these expressions give

$$v = 4(3)^3 = 108 \text{ fps}$$
$$\mathbf{v} = \underline{108 \text{ fps to the right.}} \qquad \text{Ans.}$$
$$a = 12(3)^2 = 108 \text{ fps}^2$$
$$\mathbf{a} = \underline{108 \text{ fps}^2 \text{ to the right.}} \qquad \text{Ans.}$$

EXAMPLE 8-3

The acceleration of a particle with rectilinear motion is given by the equation $\mathbf{a} = -k\mathbf{v}$, where a is in meters per second per second, $\mathbf{v}$ is the velocity in meters per second, and k is a constant. When the time t is zero the position and velocity of the particle are $\mathbf{s}_0$ and $\mathbf{v}_0$, respectively.

(a) Derive expressions for the position, velocity, and acceleration of the particle in terms of the time t.
(b) If $\mathbf{s}_0 = 20$ m to the left, $\mathbf{v}_0 = 10$ m/s to the right, and $k = 0.2 \text{ sec}^{-1}$, determine the value of t when $\mathbf{s} = 0$ and the corresponding values of $\mathbf{a}$ and $\mathbf{v}$.

SOLUTION

(a) The positive direction is assumed to be to the right since it is not specified in the problem statement and $\mathbf{i}$ is a unit vector to the right. The magnitude of the acceleration is

$$a = \frac{dv}{dt} = -kv$$

or

$$k\,dt = -\frac{dv}{v}$$

from which

$$kt = -\ln v + C_1.$$

When $t = 0$, $v = v_0$; therefore

$$0 = -\ln v_0 + C_1$$

or

$$C_1 = \ln v_0,$$

and the preceding expression becomes

$$kt = -\ln v + \ln v_0 = \ln \frac{v_0}{v},$$

which can be written as

$$e^{kt} = \frac{v_0}{v}$$

or

$$v = v_0 e^{-kt}.$$

Since $v = ds/dt$, this equation can be written as

$$ds = v_0 e^{-kt}\,dt$$

or

$$s = v_0 \int e^{-kt}\,dt = -\frac{v_0}{k} e^{-kt} + C_2.$$

When $t = 0$, $s = s_0$ and C_2 is found to be

$$C_2 = s_0 + \frac{v_0}{k} e^0 = s_0 + \frac{v_0}{k}.$$

Thus the position function is

$$s = s_0 + \frac{v_0}{k}(1 - e^{-kt}).$$

The acceleration is, from the problem statement,

$$a = -kv = -kv_0 e^{-kt}.$$

The results are

$$\underline{\mathbf{s} = \mathbf{s}_0 + \frac{\mathbf{v}_0}{k}(1 - e^{-kt})}, \qquad \text{Ans.}$$

$$\mathbf{v} = \underline{\mathbf{v}_0 e^{-kt}}, \qquad \text{Ans.}$$

and

$$\mathbf{a} = \underline{-\mathbf{v}_0 k e^{-kt}}. \qquad \text{Ans.}$$

(b) When numerical values for s_0, v_0, and k are substituted into the expression for s it becomes

$$\mathbf{s} = -20\mathbf{i} + \frac{10}{0.2}(1 - e^{-0.2t})\mathbf{i},$$

and when $s = 0$ this equation gives

$$1 - e^{-0.2t} = \frac{20(0.2)}{10} = 0.4$$

$$e^{-0.2t} = 1 - 0.4 = 0.6 = e^{-0.511}$$

and

$$\underline{t = \frac{0.511}{0.2} = 2.555} \qquad \text{Ans.}$$

The velocity is

$$\mathbf{v} = 10\mathbf{i}e^{-0.2t} = \underline{6\mathbf{i}\,\text{m/s}}, \qquad \text{Ans.}$$

and the acceleration is

$$\mathbf{a} = -k\mathbf{v} = -0.2(6\mathbf{i}) = \underline{-1.2\mathbf{i}\ \text{m/s}^2}. \qquad \text{Ans.}$$

EXAMPLE 8-4

A spotlight rotates in a horizontal plane with a constant angular velocity ω as shown in Fig. 8-4. The spot of the light, P, moves along the wall as indicated.

(a) Derive expressions for the velocity and acceleration of point P.
(b) The distance b is 100 ft, and the angular velocity of the beam of

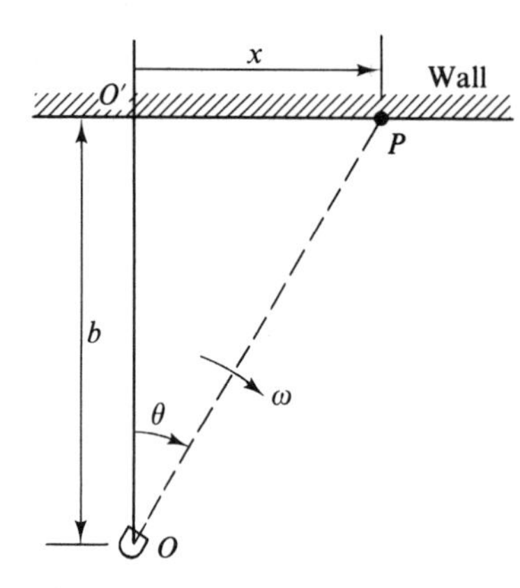

Figure 8-4

light is 5 rpm clockwise. Determine the velocity and acceleration of P when $\theta = 20°$.

SOLUTION

(a) Point P has rectilinear motion, and its position, measured from O', is

$$\mathbf{O'P} = \mathbf{x} = x\mathbf{i} = \mathbf{i}b \tan\theta,$$

where x is positive to the right and θ is positive clockwise. The magnitude of the velocity of P is

$$v_P = \frac{dx}{dt} = b \sec^2\theta \frac{d\theta}{dt} = b\omega \sec^2\theta,$$

and the velocity of P is

$$\mathbf{v}_P = \underline{\mathbf{i}b\omega \sec^2\theta.} \qquad \text{Ans.}$$

The magnitude of the acceleration of P is

$$\begin{aligned} a_P = \frac{dv_P}{dt} &= b\left[\frac{d\omega}{dt}\sec^2\theta + \omega 2 \sec\theta(\sec\theta \tan\theta)\frac{d\theta}{dt}\right] \\ &= b[\alpha \sec^2\theta + 2\omega^2 \sec^2\theta \tan\theta]. \end{aligned}$$

Since ω is constant, α must be zero, and the acceleration is

$$\mathbf{a}_P = \underline{\mathbf{i}2b\omega^2 \sec^2\theta \tan\theta.} \qquad \text{Ans.}$$

(b) The angular velocity of the light beam is

$$\begin{aligned} \omega &= 5 \text{ rpm} \\ &= 5\frac{\text{rev}}{\text{min}}\left(\frac{1 \text{ min}}{60 \text{ sec}}\right)\left(\frac{2\pi \text{ rad}}{\text{rev}}\right) \\ &= \frac{\pi}{6} \text{ rad per sec.} \end{aligned}$$

When numerical values for b, ω, and θ are substituted into the expressions for the velocity and acceleration of P they become

$$\begin{aligned} \mathbf{v}_P &= \mathbf{i}(100)\left(\frac{\pi}{6}\right)\sec^2 20° \\ &= \underline{59.3\mathbf{i} \text{ fps}} \\ &= \underline{59.3 \text{ fps to the right.}} \end{aligned} \qquad \text{Ans.}$$

and

$$\begin{aligned} \mathbf{a}_P &= \mathbf{i}(2)(100)\left(\frac{\pi}{6}\right)^2(\sec^2 20°)(\tan 20°) \\ &= \underline{22.6\mathbf{i} \text{ fps}^2} \\ &= \underline{22.6 \text{ fps}^2 \text{ to the right.}} \end{aligned} \qquad \text{Ans.}$$

PROBLEMS

8-1 The position of a particle moving along the x axis is given by the equation

$$\mathbf{r} = (2t^3 - 15t^2 + 24t - 6)\mathbf{i},$$

where r is measured in feet and t is in seconds. The particle is 5 ft to the left of the origin at $t = 1$ sec. Determine
(a) The velocity at $t = 2$ sec.
(b) The acceleration at $t = 2$ sec.
(c) The total distance traveled during the interval from $t = 0$ to $t = 4$ sec.

8-2 In Problem 8-1 determine
(a) The average velocity during the interval $t = 0$ to $t = 1$ sec.
(b) The average acceleration during the interval $t = 0$ to $t = 1$ sec.
(c) The displacement during the interval $t = 0$ to $t = 4$ sec.

***8-3** The position of a particle with rectilinear motion is given by the equation

$$\mathbf{r} = \mathbf{i}4\sin^2(0.2t),$$

where r is in meters and t is in seconds. The quantity 0.2 is given in radians per second. The positive direction is to the right. Determine
(a) The velocity when $t = 3$ sec.
(b) The acceleration when $t = 3$ sec.
(c) The displacement during the interval $t = 0$ to $t = 3$ sec.

8-4 A line rotates in a vertical plane according to the equation $\theta = 2t^3 - 12t^2$, where θ gives the angular position of the line in radians and t is time in seconds. The line is turning clockwise when $t = 1$ sec. Determine
(a) The value of t when the line changes direction of rotation.
(b) The angular displacement of the line during the interval $t = 1$ to $t = 5$ sec.
(c) The total angle swept through by the line during the interval $t = 1$ to $t = 5$ sec.

8-5 A line rotates in a horizontal plane in accordance with the equation $\theta = t^3 - 2t^2 - 2$, where θ gives the angular position of the line in radians when t is the time in seconds; the line is turning clockwise when $t = 2$ sec. Determine
(a) The angular acceleration of the line when $t = 2$ sec.
(b) The total angle turned through by the line during the time interval $t = 0$ to $t = 3$ sec.

8-6 The velocity of a particle is given by the equation

$$\mathbf{v} = (5t^2 - 8t)\mathbf{i},$$

where v is in feet per second and t is in seconds. The particle is located at $\mathbf{r} = 3\mathbf{i}$ and moving to the left when $t = 1$ sec. Determine
(a) The displacement of the particle during the interval $t = 0$ to $t = 3$ sec.
(b) The total distance traveled by the particle during the interval $t = 0$ to $t = 3$ sec.
(c) The acceleration of the particle when its velocity is zero.

8-7 A particle starts from rest 2 ft to the left of the origin and moves along a horizontal straight line. The magnitude of the velocity is proportional to the square of the time after starting. During the period from $t = 1$ to $t = 3$ sec the particle has a displacement of 39 ft to the left. When $t = 2$ sec determine
(a) The position of the particle.
(b) The velocity of the particle.
(c) The acceleration of the particle.

***8-8** The magnitude of the velocity of a particle moving along a horizontal straight line is

$$v = kt^2 + 12,$$

where v is in centimeters per second, t is in seconds, and k is a constant (positive to the right). When $t = 3$ s the acceleration is 12 cm/s^2 to the left. Determine the total distance traveled during the time interval $t = 1$ to $t = 6$ s.

8-9 The velocity of a particle with rectilinear motion is given by the equation

$$\mathbf{v} = (3t^2 - kt)\mathbf{i},$$

where v is in feet per second when the time t is in seconds. The position vector of the particle from the origin is $\mathbf{x} = -8\mathbf{i}$ ft when $t = 0$ and $-16\mathbf{i}$ ft when $t = 4$ sec. Determine
(a) The acceleration of the particle when $t = 2$ sec.
(b) The total distance traveled by the particle during the time interval $t = 1$ to $t = 5$ sec.

8-10 The angular velocity of a line rotating in a vertical plane varies according to the equation $\omega = 3t^2 - 5t$, where ω is in radians per second and t is the time in seconds. The line is rotating counterclockwise when $t = 2$ sec. Determine
(a) The angular acceleration of the line when $t = 2$ sec.
(b) The angular displacement of the line during the interval $t = 1$ to $t = 3$ sec.

8-11 The angular velocity of a line varies according to the equation $\omega = Kt^2 + 12$, where ω is the magnitude of the angular velocity in radians per second, t is time in seconds, and K is a constant. When $t = 0$ the angular position of the line is 4 rad clockwise and the angular velocity is 12 rad per sec counterclockwise. When $t = 3$ sec the angular position is 14 rad counterclockwise. Determine
(a) The angular acceleration of the line when $t = 4$ sec.
(b) The total angle turned through by the line during the interval $t = 1$ to $t = 5$ sec.

***8-12** The magnitude of the acceleration of a particle with rectilinear motion is

$$a = 6t - 10,$$

where a is in meters per second per second and t is in seconds. When $t = 0$ the acceleration is 10 m/s^2 to the right and the velocity is 5 m/s to the left. When $t = 2$ s the particle is 10 m to the right of the origin. Locate the particle when $t = 4$ s.

8-13 The magnitude of the acceleration of a particle with rectilinear motion is

$$a = 12t - 8,$$

where a is in feet per second per second and t is in seconds. When $t = 0$ the acceleration of the particle is 8 fps^2 to the left. During the period from $t = 1$ to $t = 4$ sec the displacement of the particle is 30 ft to the right. Determine the velocity of the particle at the beginning and at the end of the displacement period.

8-14 The acceleration of a particle moving along the x axis is

$$\mathbf{a} = \mathbf{i}Ae^{pt},$$

where a is in inches per second per second, t is in seconds, $A = 2$ ips^2, and $p = 0.5(\text{sec}^{-1})$. When $t = 0$ the particle is at the origin and has a velocity of $-2\mathbf{i}$ ips (to the left). Determine the position, velocity, and acceleration of the particle when $t = 2$ sec.

8-15 In Problem 8-14 determine the total distance the particle travels during the interval $t = 1$ to $t = 4$ sec.

8-16 The angular acceleration of a line rotating in a horizontal plane varies according to the expression $\alpha = 6t - 12$, where α is measured in radians per second per second and t is in seconds. The line is turning clockwise when $t =$ 1 sec. The angular velocity is zero, and $\theta = 2$ rad counterclockwise when $t = 0$. Determine the total angle turned through during the time interval $t = 0$ to $t = 5$ sec.

8-17 The angular acceleration of a line rotating in a plane is given by the equation $\alpha = Kt^2 - 6$, where α is in radians per second per second when t is in seconds and where K is a constant. When $t = 0$ the angular acceleration is 6 rad per sec^2 counterclockwise, and the angular velocity is 10 rad per sec clockwise. When $t =$ 1 sec the angular velocity is 8 rad per sec clockwise, and the angular position is 5 rad clockwise. Determine the angular position and acceleration of the line when $t = 2$ sec.

8-18 The magnitude of the velocity of a particle with rectilinear motion is given by the equation $v = ks^2$, where v is in feet per second, s is the distance (in feet) of the particle from an origin on the path, and $k = -2 \text{ ft}^{-1} \text{sec}^{-1}$. The particle is 10 ft to the right of the origin at $t = 0$. When $t = 2$ sec determine
(a) The position of the particle.
(b) The velocity of the particle.
(c) The acceleration of the particle.

***8-19** The velocity of a particle moving along the x axis is given by the equation

$$\mathbf{v} = \mathbf{i}v_0 \cos\left(\frac{x}{b}\right),$$

where $v_0 = 10$ m/s and $b = 5$ m. Determine the velocity and acceleration when $x = 5$ m.

***8-20** The particle in Problem 8-19 is at the origin ($x = 0$) when $t = 0$. Determine the value of t when $\mathbf{x} = 5\mathbf{i}$ m.

***8-21** A French artillerist, Captain A. Le Duc, developed an empirical formula for the speed of a projectile in terms of the distance traveled by the projectile in the bore. For a 155-mm gun, this equation is $v = 1010s/(1.78 + s)$, where s is the distance traveled in the bore in meters and v is the speed of the projectile in meters per second. The projectile travels 4.70 m in the bore. Determine
(a) The muzzle velocity of the projectile.
(b) The magnitude of the acceleration of the projectile when $s = 1.3$ m.

8-22 The acceleration of a particle moving along the x axis is given by the equation $\mathbf{a} = 3x^{1/2}\mathbf{i}$, where a is in feet per second per second and x is the distance (in feet) of the particle from the origin. The particle starts from rest at the origin at $t = 0$. Determine the position, velocity, and acceleration of the particle when $t = 3$ sec.

8-23 The acceleration of a particle with rectilinear motion is given by the equation $a\mathbf{i} = -(k/x)\mathbf{i}$, where a is acceleration in inches per second per second, x is the position of the particle in inches, and k is a constant. The velocity is zero when $x = x_0$.
(a) Derive an expression for the velocity in terms of the position x.
(b) Determine the velocity when $x = x_0/2$ and $k = 18\ \text{in}^2/\text{sec}^2$.

8-24 In Problem 8-23 assume that the velocity is $v_0\mathbf{i}$ at position $x_0\mathbf{i}$. Determine the value of x when the particle comes to rest and the corresponding acceleration.

8-25 The magnitude of the angular acceleration of a line moving in a plane varies in accordance with the expression $\alpha = -K\theta^2$, where α is in radians per second per second, θ is the angular position of the line in radians (clockwise is positive), and K is a constant. The angular velocity of the line changes from 6 rad per sec clockwise to 3 rad per sec clockwise while θ changes from 0 to 3 rad clockwise.
(a) Determine the maximum (positive) value of θ.
(b) Determine the corresponding value of α.

8-26 The acceleration of a particle with rectilinear motion is $\mathbf{a} = -0.05v^2\mathbf{i}$, where a is in feet per second per second and v is the velocity in feet per second. The particle starts at the origin with a velocity of $50\mathbf{i}$ fps at $t = 0$. Locate the particle when $t = 5$ sec and calculate the corresponding velocity and acceleration.

8-27 In Problem 8-26 determine the time interval for the particle to travel
(a) The first 20 ft.
(b) The second 20 ft.

***8-28** The acceleration of a particle with rectilinear motion is given by the equation $\mathbf{a} = (k/v)\mathbf{i}$, where a is in meters per second per second, v is velocity in meters per second, and k is a constant. The particle is at the origin with a velocity of $v_0\mathbf{i}$ at $t = 0$.
(a) Derive an equation for the position of the particle as a function of t.
(b) What are the fundamental dimensions of k?

8-29 The angular acceleration of a line moving in a plane is given by the equation $\alpha = -k\omega^2$, where α is in radians per second per second, ω is the angular velocity of the line in radians per second, and k is a constant. Clockwise is positive. When $t = 0$, $\omega = +5$ rad per sec, and when $t = 3$ sec, $\omega = +2$ rad per sec. Determine the angular displacement of the line during the interval $t = 0$ to $t = 3$ sec.

8-30 The spotlight in Example 8-4 is used to follow a point P moving along the wall with a constant linear velocity of 50 fps to the right. The light is 200 ft from the wall. Determine the angular velocity and acceleration of the light beam when $\theta = 30°$.

8-31 Point C of the mechanism in Fig. P8-31

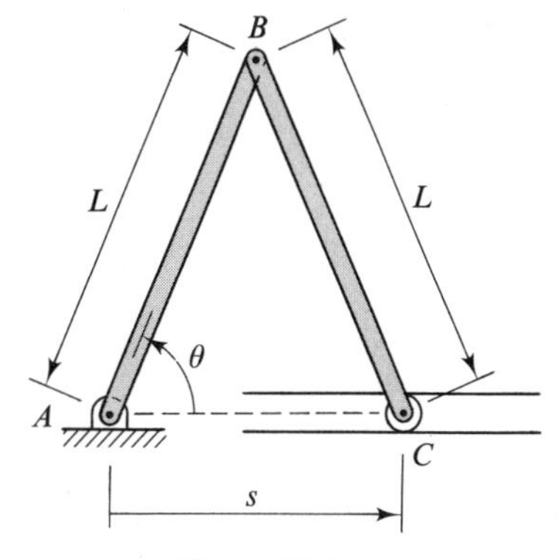

Figure P8-31

moves along the horizontal line through AC. Set up an equation relating s and θ and differentiate this expression to relate the linear velocity and acceleration of C to the angular velocity and acceleration of line AB.

8-32 Members AB and BC of Problem 8-31 each have a length of 10 in. When $\theta = 60°$ point C has a velocity of 30 ips and an acceleration of 20 ips^2, both to the right. Determine the angular velocity and acceleration of member AB for this position.

8-33 Two small vessels are traveling along perpendicular courses at constant speeds as shown in Fig. P8-33. Determine the angular velocity of the line of sight between the ships when OA is 500 ft, OB is 1200 ft, and the speed of each ship is 20 fps.

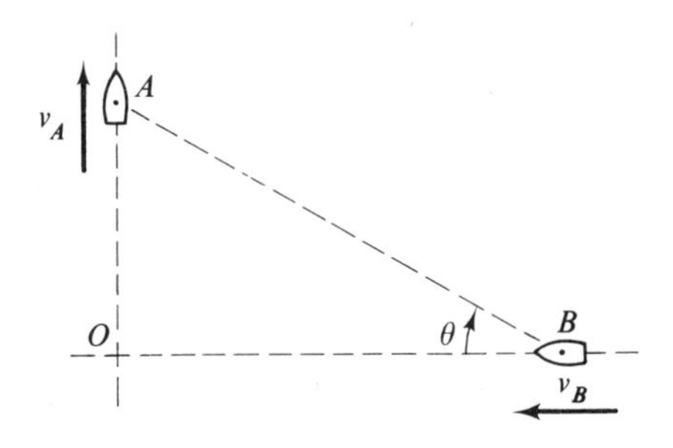

Figure P8-33

8-34 A mechanism causes the spotlight of Example 8-4 to oscillate in accordance with the equation

$$\theta = \theta_0 \sin \omega_0 t,$$

where $\theta_0 = \pi/6$ rad $= 30°$ and $\omega_0 = 10$ rad per sec. The distance b is 500 ft. Determine the magnitudes of the velocity and acceleration of point P when

(a) $\omega_0 t = 0$.

(b) $\omega_0 t = 90°$.

8-35 Solve Problem 8-34 when $\omega_0 t = 45°$.

8-5 MOTION CURVES

For many problems a semigraphical or pictorial procedure utilizing motion curves provides an excellent method of solution and leads to ready visualization of the motion. Motion curves are diagrams showing the relationships between various kinematical quantities. For rectilinear motion the position, velocity, or acceleration may be plotted as ordinates against the corresponding time as the abscissa. In some cases the velocity may be plotted against the position (this is known as a phase-plane plot), or the acceleration may be plotted against the position or velocity. One of the most useful diagrams for many problems is the v-t (velocity-time) diagram. An example of a v-t diagram is shown in Fig. 8-5.

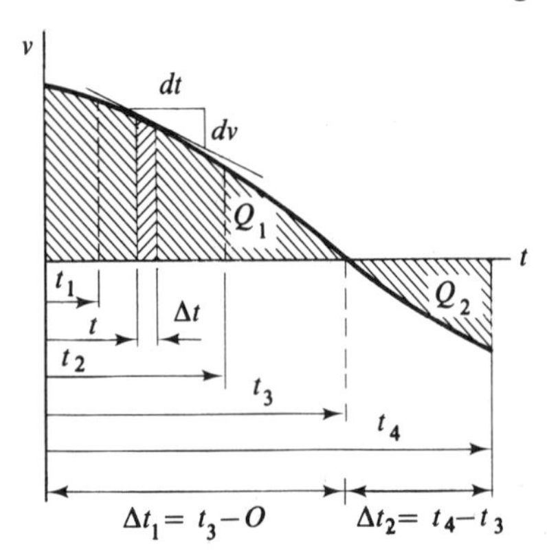

Figure 8-5

The slope of the curve corresponding to any time t represents the change in the magnitude of the velocity dv divided by time interval dt, which, from Eq. (8-3), is the magnitude of the acceleration.

The area under the curve between any two ordinates at t_1 and t_2 of Fig. 8-5 is the definite integral of $v\,dt$. Since $v = ds/dt$, the expression $v\,dt$ is equal to ds, and the integral is

$$\int_{t_1}^{t_2} v\,dt = \int_{s_1}^{s_2} ds = s_2 - s_1 = \Delta s = q,$$

where s_1 and s_2 are the values of s corresponding to t_1 and t_2, respectively. Thus the area under the v-t curve between the ordinates at t_1 and t_2 is the change of s (the magnitude of the displacement) during the corresponding time interval. A reversal of the velocity occurs at $t = t_3$ when the curve crosses the time axis. The area Q_1 represents the distance traveled during the interval Δt_1 from $t = 0$ to $t = t_3$ (prior to the reversal of the velocity). The area Q_2 represents the distance traveled in the opposite direction during the time interval Δt_2 from $t = t_3$ to $t = t_4$. The sum of the two areas represents the total distance traveled from $t = 0$ to $t = t_4$, and the difference is the net distance (the magnitude of the displacement) during the interval.

EXAMPLE 8-5

Draw a-t, v-t, and s-t curves for Example 8-1.

SOLUTION

The three diagrams are shown in Fig. 8-6. The equations for the curves were obtained from Example 8-1.

Recall that a positive sign indicates a vector quantity to the left. The a-t diagram is a straight line, the v-t diagram is a parabola, and the s-t curve is a cubic.

Since the slope of the v-t curve is the acceleration, the time at which the a-t curve crosses the t axis ($t = 0.833$ sec) should correspond to a zero slope on the v-t curve. Also the times at which the v-t curve crosses the t axis ($t = 0$, 1.667 sec) correspond to points of zero slope of the s-t curve.

The area A under the v-t curve represents the distance the particle travels to the right during the interval from $t = 1.000$ to $t = 1.667$ sec. Thus A $= 4.63 - 3.00 = 1.63$ ft. Similarly, the area B represents the distance the particle travels to the left during the period from $t = 1.667$ to $t = 3.00$ sec; that is, $B = 9.00 + 4.63 = 13.63$ ft. The sum of areas A and B is the total distance traveled during the interval, and the difference is the corresponding displacement.

The v-t diagram is probably the most useful diagram in helping to visualize the motion. Its slope gives the acceleration, its area gives the displacement and total distance traveled, and its ordinate gives the velocity.

The v-t diagram is particularly useful when the acceleration is constant for a period of time since the diagram is a straight line and the slope and areas can be computed directly from the diagram. This procedure is illustrated in Examples 8-6 and 8-7.

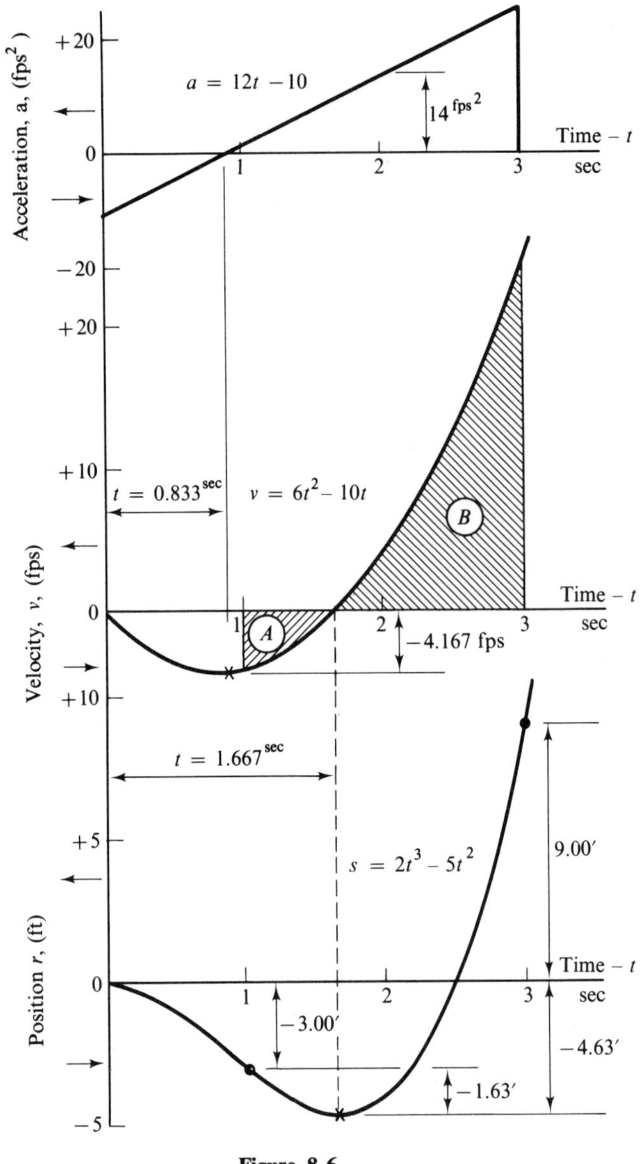

Figure 8-6

EXAMPLE 8-6

A particle moves along a straight line with a constant acceleration. The velocity of the particle changes from 10 fps to the right to 25 fps to the left during a time interval of 7.0 sec. Determine

(a) The acceleration.
(b) The total distance traveled.
(c) The displacement of the particle during the 7-sec interval.

SOLUTION

Since the acceleration is known to be constant, the slope of the v-t diagram must also be constant, and the v-t diagram is a straight line. Initial and final values of v are plotted, and the diagram is drawn as shown in Fig. 8-7.

(a) Since the velocity changes from right to left, the acceleration (constant) must be to the left. The magnitude of the acceleration is given by the slope of the v-t curve and is

$$a = \frac{v_f - v_i}{\Delta t} = \frac{-25 - (+10)}{7.0} = -5$$

and

$$\mathbf{a} = \underline{5 \text{ fps}^2 \text{ to the left.}} \qquad \text{Ans.}$$

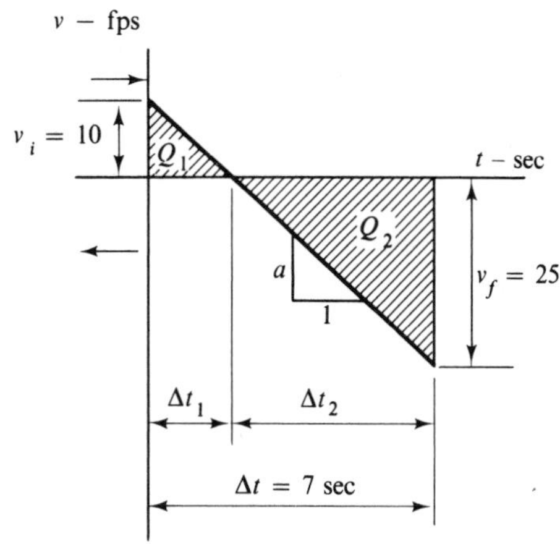

Figure 8-7

(b) The time Δt_1 that has elapsed when the motion reverses direction is, from the slope of the line,

$$\Delta t_1 = \frac{v_i}{a} = \frac{10}{5} = 2 \text{ sec.}$$

The area Q_1 which represents the distance moved to the right during Δt_1 is

$$Q_1 = \tfrac{1}{2}v_i(\Delta t_1) = \tfrac{1}{2}(10)(2) = 10 \text{ ft.}$$

The time interval Δt_2 is

$$\Delta t_2 = \Delta t - \Delta t_1 = 7 - 2 = 5 \text{ sec},$$

and the area Q_2, which represents the distance moved to the left during Δt_2, is

$$Q_2 = \tfrac{1}{2}v_f(\Delta t_2) = \tfrac{1}{2}(25)(5) = 62.5 \text{ ft.}$$

The total distance traveled is

$$Q = Q_1 + Q_2 = 10 + 62.5 = \underline{72.5 \text{ ft.}} \qquad \text{Ans.}$$

(c) Since the area Q_2 is greater than Q_1, the displacement during the 7-sec interval must be to the left. Its magnitude is

$$q = Q_2 - Q_1 = 62.5 - 10 = 52.5 \text{ ft},$$

and the displacement is

$$\mathbf{q} = \underline{52.5 \text{ ft to the left.}} \qquad \text{Ans.}$$

Alternate solution

(a) The following analytical solution is included for comparison. Assume that $t = 0$ and that the particle is at the origin at the beginning of the 7-sec time interval and that the unit vector $\mathbf{i}$ is to the right. Since the acceleration is constant, it can be written as

$$\mathbf{a} = \frac{d\mathbf{v}}{dt} = a\mathbf{i}$$

from which

$$d\mathbf{v} = a\mathbf{i}\,dt$$

and

$$\mathbf{v} = at\mathbf{i} + \mathbf{C}_1.$$

When $t = 0$, $\mathbf{v} = 10\mathbf{i}$ and $\mathbf{C}_1 = 10\mathbf{i}$. Also when $t = 7$ sec, $\mathbf{v} = -25\mathbf{i}$, and when these values are substituted into the equation for $\mathbf{v}$ it becomes

$$-25\mathbf{i} = a7\mathbf{i} + 10\mathbf{i}$$

from which

$$a = -5$$

and

$$\mathbf{a} = \underline{-5\mathbf{i}\text{ fps}^2} = \underline{5\text{ fps}^2\text{ to the left.}} \qquad \text{Ans.}$$

(c) The equation for $\mathbf{v}$ is

$$\mathbf{v} = -5t\mathbf{i} + 10\mathbf{i} = \frac{d\mathbf{r}}{dt}$$

from which

$$\mathbf{r} = \frac{-5t^2}{2}\mathbf{i} + 10t\mathbf{i} + \mathbf{C}_2.$$

Since $\mathbf{r} = 0$ when $t = 0$, $\mathbf{C}_2 = 0$, and

$$\mathbf{r} = (-2.5t^2 + 10t)\mathbf{i}.$$

When $t = 7$ sec, the preceding equation gives $\mathbf{r} = -52.5\mathbf{i}$ ft, and the displacement from $t = 0$ to $t = 7$ sec is

$$\begin{aligned}\mathbf{q} &= \mathbf{r}_7 - \mathbf{r}_0 = -52.5\mathbf{i} - 0 \\ &= \underline{-52.5\mathbf{i}\text{ ft}} = \underline{52.5\text{ ft to the left.}} \qquad \text{Ans.}\end{aligned}$$

(b) The time or times when $\mathbf{v} = 0$ must be obtained to determine the total distance traveled. Since $\mathbf{v} = (-5t + 10)\mathbf{i}$, the velocity will be zero at $t = 2$ sec, and the velocity will be to the right from $t = 0$ to $t = 2$ sec and the left after $t = 2$ sec. When $t = 2$ sec

$$\mathbf{r}_2 = [-2.5(2)^2 + 10(2)]\mathbf{i} = 10\mathbf{i},$$

and the total distance traveled from $t = 0$ to $t = 7$ sec is

$$\begin{aligned}Q &= |\mathbf{r}_2 - \mathbf{r}_0| + |\mathbf{r}_7 - \mathbf{r}_2| \\ &= |10\mathbf{i} - 0| + |-52.5\mathbf{i} - 10\mathbf{i}| \\ &= 10 + 62.5 = \underline{72.5\text{ ft.}} \qquad \text{Ans.}\end{aligned}$$

EXAMPLE 8-7

A particle starts from the origin with a velocity of 2 m/s to the right and moves along a straight line with an acceleration of 4 m/s^2 to the right for 2.5 s. The acceleration then changes to 6 m/s^2 to the left for 3 s, at which time the acceleration becomes 1 m/s^2 to the right for 4 s. Determine

(a) The total distance traveled during the 9.5-s interval.
(b) The displacement of the particle during the first 5.5-s interval.

SOLUTION

Since each of the accelerations is constant during the specified time periods, the v-t diagram provides a convenient solution; see Fig. 8-8. When $t=0$ the velocity is 2 m/s to the right and the acceleration is 4 m/s^2 to the right, which indicates an increase in the magnitude of the velocity. The magnitude of the velocity at $t=2.5$ s can be obtained from the slope of the line; thus

$$a_1 = \text{slope} = \frac{v_{2.5} - v_0}{\Delta t_1} = 4$$

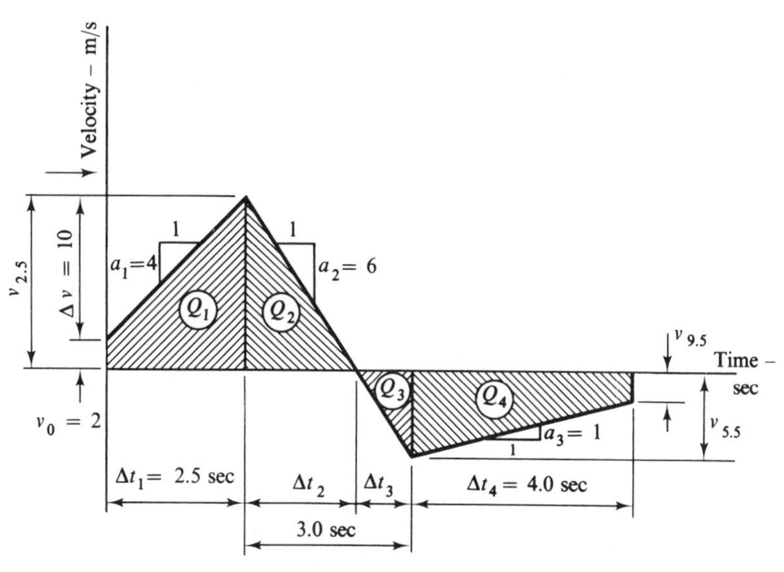

Figure 8-8

from which

$$v_{2.5} = 2 + 4(2.5) = 12 \text{ m/s},$$

and it is represented by the ordinate which is above the t axis and thus the velocity is to the right. During the time interval $\Delta t_2 + \Delta t_3 = 3$ sec the acceleration is 6 m/s^2 to the left; hence the velocity is decreasing. The time Δt_2 is obtained as follows:

$$a_2 = \frac{v_{2.5}}{\Delta t_2}$$

from which

$$\Delta t_2 = \frac{v_{2.5}}{a_2} = \frac{12}{6} = 2 \text{ s},$$

and, from the diagram,

$$\Delta t_3 = 3 - \Delta t_2 = 3 - 2 = 1 \text{ s}.$$

The velocity at the end of 5.5 s can be obtained from the diagram as

$$v_{5.5} = a_2(\Delta t_3) = 6(1) = 6 \text{ m/s}$$

and is represented by an ordinate below the t axis, which indicates a velocity to the left. During the last time interval ($\Delta t_4 = 4$ s) the acceleration is to the right, and the magnitude of the velocity at the end of this interval is

$$v_{9.5} = v_{5.5} - a_3(\Delta t_4) = 6 - 1(4)$$
$$= 2 \text{ m/s}.$$

The ordinate for $v_{9.5}$ is still below the t axis and thus represents a velocity to the left.

(a) The total distance traveled is, from the areas under the diagram,

$$Q = Q_1 + Q_2 + Q_3 + Q_4$$
$$= \frac{(2+12)}{2}(2.5) + \frac{12}{2}(2) + \frac{6}{2}(1) + \frac{(6+2)}{2}(4)$$
$$= 17.5 + 12 + 3 + 16$$
$$= \underline{48.5 \text{ m.}} \qquad \text{Ans.}$$

(b) The magnitude of the displacement during the first 5.5 s is

$$q = Q_1 + Q_2 - Q_3$$
$$= 17.5 + 12 - 3 = 26.5 \text{ m},$$

and the displacement is

$$\mathbf{q} = \underline{26.5 \text{ m to the right.}} \qquad \text{Ans.}$$

PROBLEMS

8-36 Using a v-t diagram for rectilinear motion with constant acceleration, prove that

$$v_f = v_i + a(\Delta t)$$

and that

$$q = v_i(\Delta t) + \frac{a(\Delta t)^2}{2}.$$

8-37 Using a v-t diagram, prove that

$$v_f^2 = v_i^2 + 2aq$$

for rectilinear motion with constant acceleration.

8-38 A particle with rectilinear motion has a constant acceleration, and the velocity reverses direction during a certain time interval. Derive an equation relating the magnitudes of the initial and final velocities, the acceleration, and the total distance traveled by the particle. Why is this equation valid *only* when the velocity reverses?

8-39 A particle has a constant acceleration of 5 fps^2 to the left. During a 7-sec interval the particle has a displacement to the right, and the total distance traveled is twice the magnitude of the displacement. Determine the initial and final velocities of the particle.

***8-40** A particle in a mechanism is to travel along a straight line with a constant acceleration. The particle is to have a displacement of 8.50 cm to the left during an interval of 0.50 s and have a velocity of 30 cm/s to the left at the end of the period. Determine the acceleration of the particle.

8-41 A particle which has a constant acceleration of 6 ips^2 downward is given an initial velocity of 30 ips upward. Determine the time required for the particle to travel a total distance of 90 in. and the velocity of the particle at the end of the period.

8-42 Solve Problem 8-41 if the constant acceleration is upward. All other values are unchanged.

8-43 The acceleration of a particle moving along a horizontal path is constant and equal to 4 fps^2 to the left. During a certain time interval, the particle has a displacement of 54 ft to the right and travels a total distance of 90 ft. Determine the initial and final velocities of the particle and the time interval.

8-44 A particle has a rectilinear motion with constant acceleration to the right. During a 5-sec interval the particle has a displacement of 25 ft to the right while traveling a total distance of 65 ft. Determine the velocities at the beginning and end of the time interval and the acceleration during the interval.

8-45 A particle moves along a horizontal straight line with a constant acceleration. The velocity of the particle is 100 fps to the right at the start of a 6-sec interval. During the interval the particle travels a total distance of 260 ft. Determine the acceleration and the final velocity of the particle.

8-46 Solve Problem 8-45 if the displacement (instead of the total distance traveled) is
(a) 260 ft to the right.
(b) 260 ft to the left.

8-47 Determine the acceleration and final velocity of the particle in Problem 8-45 which will result in the minimum total distance traveled in 6 sec.

8-48 A point having a vertical rectilinear motion starts from rest and is given a constant acceleration of 40 ips^2 upward for 0.30 sec, after which it continues to move for 0.20 sec at constant velocity.
(a) What constant downward acceleration must now be given to the point so that it will reach a maximum height of 6.4 in. above the initial position?
(b) Determine the distance traveled by the point during the first second if the final acceleration is held constant until the end of the first second.

8-49 A skier on a mountain slope travels 60 ft during a 2-sec interval and another 60 ft during the next second. The acceleration was constant during the 3-sec period. Determine the velocity at the end of the 3-sec period.

8-50 A particle starts from rest at the origin and moves to the right with a constant acceleration for 4 sec. The acceleration then becomes 12 fps^2 to the left for a second time interval. The particle travels a total distance of 276 ft and is 24 ft to the left of the origin at the end of the total time interval. Determine
(a) The acceleration during the first 4-sec interval.
(b) The distance traveled during the first 4-sec interval.
(c) The total time interval.

8-51 A car on a straight road can accelerate at 20 fps^2 and decelerate at 30 fps^2. Determine the minimum time required for the car to travel 450 ft from rest if it comes to rest at the end of the 450-ft course. What maximum speed is attained?

8-52 A particle starts from rest and moves along a straight line for Δt_1 sec with a constant acceleration of 4 fps^2 to the right. The acceleration is changed to 10 fps^2 to the left for the next 3 sec, after which the velocity becomes constant for a third time interval. During the total time interval the particle has a displacement of 25 ft to the right and travels a total distance of 115 ft. Determine the total time of travel.

8-53 A particle with rectilinear motion with an initial velocity to the right moves 144 ft to the right during the first 4-sec interval and 56 ft to the right during the following second. Determine the constant acceleration required for these conditions.

8-54 The velocity of a particle moving along a straight line changes uniformly from 0 to 32 fps to the right while traveling 64 ft. The acceleration changes to a different value (constant), and the particle travels 130 ft during the next 5 sec. After the 5-sec interval the acceleration is changed to 10 fps^2 to the left until the particle has traveled a total distance of 300 ft. Determine the total traveling time.

8-55 A particle moves along a horizontal line; it starts from point A with a velocity of 4 fps to the right and an acceleration of 3 fps^2 to the right and travels for 4 sec. Then the acceleration changes to 8 fps^2 to the left for 2.5 sec. At the

end of this time, the particle travels for 5 sec with a constant velocity.
(a) Determine the position of the particle with reference to point A at the end of the last 5-sec interval.
(b) Determine the total distance traveled.

8-56 A man jumped from a stationary balloon at an elevation of 5000 ft above the ground. He waited 10 sec and then pulled his parachute cord. The parachute decelerated him at an average rate of 20 fps^2 until a velocity of 22 fps was reached. He then descended to the ground at a constant velocity of 22 fps. How long did it take the man to descend to the ground? Neglect the effect of air friction during the first 10 sec and assume the acceleration of gravity to be 32.2 fps^2.

8-57 A particle moving along a straight line has a constant acceleration of 8 fps^2 to the right for 5 sec and zero acceleration for the next second. During the 6-sec interval the particle traveled a total distance of 70 ft. Determine the initial velocity of the particle and the resulting displacement during the 6-sec interval.

8-58 Starting with a velocity of 18 fps to the right, a particle moves along a straight line with an acceleration of 3 fps^2 to the left for 8 sec. The acceleration then becomes zero for Δt sec, after which the velocity changes uniformly until it is 8 fps to the right. The total distance traveled is 109 ft, and the linear displacement is 31 ft to the right. Determine the time interval during which the particle traveled with constant speed.

8-59 A particle starts from rest with a constant acceleration of 10 fps^2 to the right for a certain time interval. The acceleration then becomes 20 fps^2 to the left for a second time interval. The sum of the two intervals is 30 sec, and the particle is at the starting point at the end of the second interval. Determine
(a) The total distance traveled.
(b) The maximum speed of the particle.
(c) The average velocity for the 30-sec period.
(d) The average speed for the 30-sec period.

8-6 RELATIVE MOTION—MOTION OF SEVERAL PARTICLES

When two or more particles are in motion simultaneously their motions may be entirely independent. Consider, for example, two trains moving along parallel tracks. The position, velocity, and acceleration of one train does not affect the position, velocity, or acceleration of the second train, and the motions are independent. If the particles are connected by cords, rods, springs, or other constraints, however, the motion of one of the bodies will usually affect the motion of each of the other bodies. Several examples of such motion are illustrated in Fig. 8-9 in which each particle or body has rectilinear motion and the particles are connected by flexible inextensible cables. In Fig. 8-9(a) if body A moves downward a distance Δs_A, body B must move the same distance to the right, and the position coordinates of the two bodies are related by the length of the cord connecting them. The positions of bodies C and D in Fig. 8-9(b) are also related by the length of the cord but the relationship is a little less direct. Figures 8-9(c) and 8-9(d) are other examples in which the position of one of the bodies is related to the positions of the other bodies by the connecting cord or cable.

Relationships between velocities and accelerations of the connected bodies can be developed by writing one or more equations

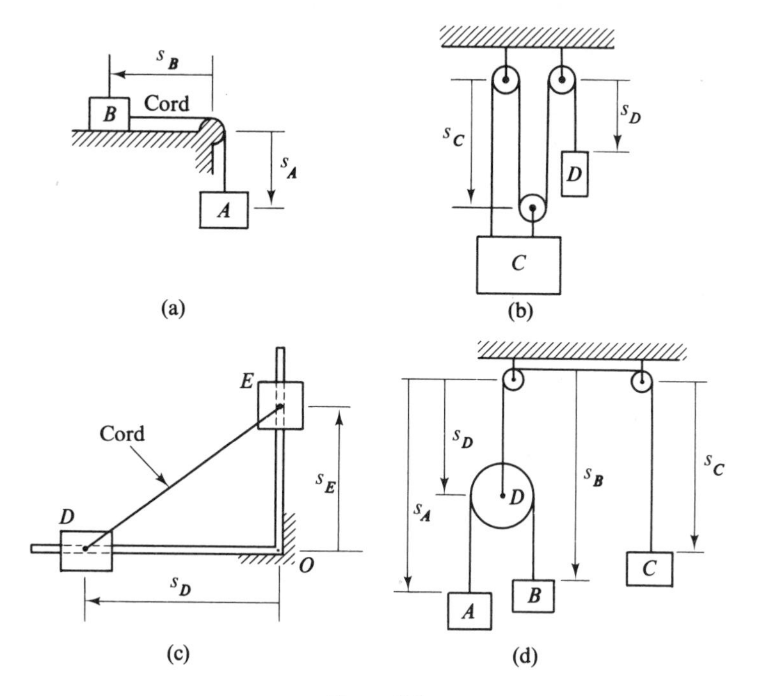

Figure 8-9

relating the positions of the bodies and differentiating with respect to time. A well-defined sign convention is essential if results are to be interpreted correctly. The coordinate giving the position of each of the bodies in Fig. 8-9 is shown on each sketch with an arrow to show the positive direction. In these examples the position of each particle is measured from a fixed reference rather than from a moving origin. Thus the position of A in Fig. 8-9(d) is measured from the fixed pulleys rather than from the moving pulley D.

In Art. 8-1 reference was made to both absolute and relative motion. When the motion is specified with reference to a system of axes fixed in space the motion is called *absolute motion.* When the position, velocity, and acceleration are measured with respect to a moving origin, the motion is called *relative motion.*

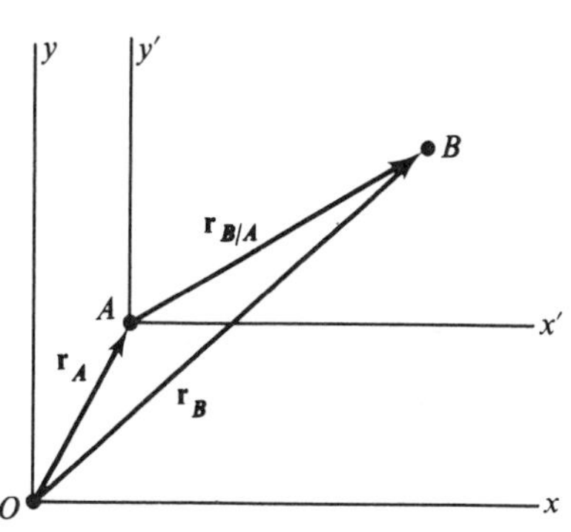

Figure 8-10

Figure 8-10 represents two particles moving in the fixed xy plane. The absolute positions of A and B are shown as $\mathbf{r}_A$ and $\mathbf{r}_B$, respectively. The position of B can also be measured from the $x'y'$ axes which move with point A. The vector $\mathbf{r}_{B/A}$ is defined as the position of B with respect to a system of axes moving with point A. When x' and y' remain parallel to the fixed axes it is convenient to call the position of B relative to the $x'y'$ axes the position of B relative to A. Thus the motion of B measured with respect to the $x'y'$ system is the motion of B relative to A.

A vector equation for the position of B relative to A can be written from Fig. 8-10. That is,

$$\mathbf{r}_B = \mathbf{r}_A + \mathbf{r}_{B/A} \quad \text{or} \quad \mathbf{r}_{B/A} = \mathbf{r}_B - \mathbf{r}_A.$$

This equation can be differentiated to give

$$\mathbf{v}_B = \mathbf{v}_A + \mathbf{v}_{B/A} \quad \text{or} \quad \mathbf{v}_{B/A} = \mathbf{v}_B - \mathbf{v}_A$$

and

$$\mathbf{a}_B = \mathbf{a}_A + \mathbf{a}_{B/A} \quad \text{or} \quad \mathbf{a}_{B/A} = \mathbf{a}_B - \mathbf{a}_A.$$

Note that $\mathbf{r}_B$, $\mathbf{r}_A$, and the other quantities are vectors, and when they are differentiated changes in both magnitude and direction must be included as indicated in Art. 8-2. Such differentiation will be discussed in more detail in the following articles.

The following examples illustrate the kinematic relationships for two or more bodies. The procedure of writing one or more equations relating either absolute or relative position functions and differentiating to obtain velocity and acceleration relationships provides a fundamental means of obtaining such relationships and often gives an independent check of other solutions.

EXAMPLE 8-8

Derive equations relating velocities and accelerations of bodies C and D in Fig. 8-9(b).

SOLUTION

The length of the cable connecting bodies C and D is

$$L = 3s_C + s_D + \sum L_p, \tag{a}$$

where L_p represents the length of the rope wrapped around each of the pulleys. Note that $\sum L_p$ is constant; therefore Eq. (a) can be written

$$3s_C + s_D = L - \sum L_p = L', \tag{b}$$

where L' is a constant. When Eq. (b) is differentiated with respect to time it becomes

$$3\frac{ds_C}{dt} + \frac{ds_D}{dt} = 0$$

or

$$\underline{3v_C + v_D = 0.} \qquad \text{Ans.}$$

This equation indicates that if v_C is positive (downward), v_D will be negative, that is, upward, and have a magnitude equal to $3v_C$.

A second differentiation gives

$$3\frac{dv_C}{dt} + \frac{dv_D}{dt} = 0$$

or

$$\underline{3a_C + a_D = 0.} \qquad \text{Ans.}$$

EXAMPLE 8-9

Particles D and E in Fig. 8-9(c) are connected by a 5-ft inextensible cord. When s_E is 3 ft (upward), particle E has a velocity of 8 fps upward and an acceleration of 4 fps^2 downward. Determine the corresponding velocity and acceleration of D.

SOLUTION

The positive directions for the position functions are selected as shown, and the equation relating the positions is

$$s_D^2 + s_E^2 = 5^2,$$

and when s_E is 3 ft, s_D is 4 ft.

The velocity equation can be obtained by differentiation as

$$2s_D\frac{ds_D}{dt} + 2s_E\frac{ds_E}{dt} = 0$$

or

$$s_D v_D + s_E v_E = 0. \qquad \text{(a)}$$

The velocity of D can be obtained by substituting numerical values for s_D, s_E, and v_E; thus

$$4v_D + 3(8) = 0$$

or

$$v_D = -6$$

and

$$\mathbf{v}_D = \underline{6 \text{ fps to the right.}} \qquad \text{Ans.}$$

Equation (a) can be differentiated to give the following result:

$$\frac{ds_D}{dt}(v_D) + s_D\left(\frac{dv_D}{dt}\right) + \frac{ds_E}{dt}(v_E) + s_E\left(\frac{dv_E}{dt}\right) = 0$$

or

$$v_D^2 + s_D a_D + v_E^2 + s_E a_E = 0.$$

The acceleration of D is obtained by direct substitution. Thus

$$(-6)^2 + 4a_D + (8)^2 + 3(-4) = 0$$

from which

$$4a_D = 12 - 36 - 64 = -88,$$

$$a_D = -22,$$

and

$$\mathbf{a}_D = \underline{22 \text{ fps}^2 \text{ to the right.}} \qquad \text{Ans.}$$

EXAMPLE 8-10

Three bodies A, B, and C are connected by cords and pulleys as shown in Fig. 8-9(d). The three bodies are released from rest at the same elevation. Body A has a constant acceleration of 7 ips^2 upward, and C has a constant acceleration of 2 ips^2 downward. Determine the time required for the vertical distance between A and B to become 20 in.

SOLUTION

The length of cord wrapped around the pulleys and the horizontal distance between the upper pulleys are all constant and can be neglected when determining velocities and accelerations. The length of the cord between A and B is (downward is positive)

$$L_1 = (s_A - s_D) + (s_B - s_D)$$
$$= s_A + s_B - 2s_D.$$

The length of the other cord is

$$L_2 = s_C + s_D.$$

The position s_D can be eliminated from these two equations to give

$$s_A + s_B - 2(L_2 - s_C) = L_1$$

or

$$s_A + s_B + 2s_C = L_1 + 2L_2.$$

When the preceding expression is differentiated with respect to time it becomes

$$\frac{ds_A}{dt} + \frac{ds_B}{dt} + 2\frac{ds_C}{dt} = 0$$

or

$$v_A + v_B + 2v_C = 0.$$

A second differentiation gives

$$a_A + a_B + 2a_C = 0,$$

and when the given data are substituted into this equation it becomes

$$-7 + a_B + 2(+2) = 0$$

or

$$a_B = 7 - 4 = 3 \text{ fps}^2 \text{ (downward)}.$$

Velocity-time diagrams for the three bodies are drawn in Fig. 8-11 with the accelerations and velocities indicated. Since the displacement of A is upward and that of B is downward, the distance between A and B is equal to the sum of the magnitudes of the displacements:

$$|q_A| + |q_B| = \frac{7(\Delta t)^2}{2} + \frac{3(\Delta t)^2}{2} = 20,$$

and the time interval is found to be

$$(\Delta t)^2 = 4$$

or

$$\Delta t = \underline{2 \text{ sec.}} \qquad \text{Ans.}$$

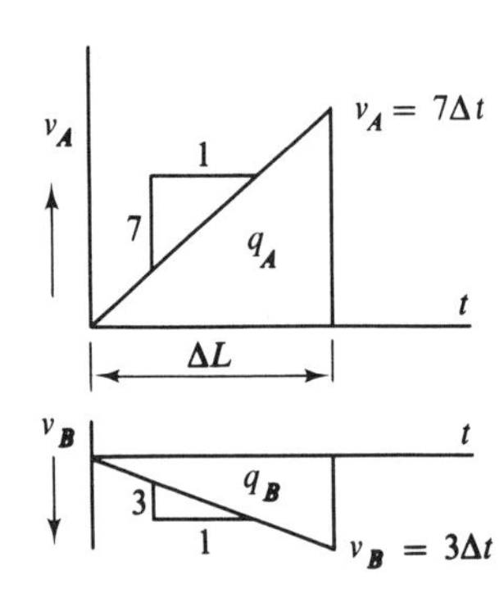

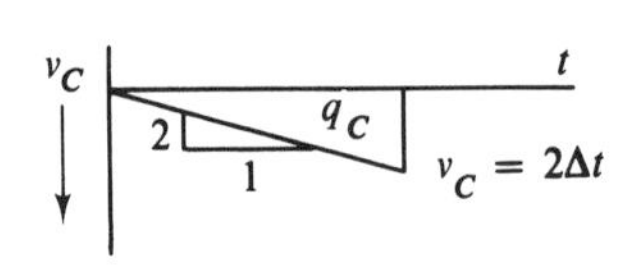

Figure 8-11

EXAMPLE 8-11

A ball is projected vertically upward from the surface of the earth with an initial velocity of 100 fps. Two seconds later a second ball is thrown upward from the same point with a velocity of 80 fps. How far above the earth will the two balls meet? Neglect air resistance and assume that the acceleration of gravity is 32.2 fps^2.

SOLUTION

The acceleration of ball 1 is constant, and an equation for the height of ball 1 above the starting point can be obtained as follows (upward is assumed to be positive):

$$a_1 = \frac{dv_1}{dt} = -32.2$$

$$dv_1 = -32.2\, dt$$

$$v_1 = \int -32.2\, dt = -32.2t + C_1.$$

Since $v_1 = +100$ fps when $t = 0$ the value of C_1 is 100 and the velocity is

$$v_1 = \frac{dy_1}{dt} = -32.2t + 100,$$

where y_1 is the height of the ball. A second integration gives the position, y_1; that is

$$y_1 = \int (-32.2t + 100)\, dt = -16.1t^2 + 100t + C_2.$$

When $t = 0$, $y_1 = 0$; therefore $C_2 = 0$, and the position of ball 1 is

$$y_1 = -16.1t^2 + 100t. \tag{a}$$

Ball 2 also has a constant acceleration of 32.2 fps^2 downward, and when $t = 2$ sec it is at the origin ($y_2 = 0$) with a velocity of 80 fps upward. The equations describing the acceleration and velocity of ball 2 are

$$a_2 = \frac{dv_2}{dt} = -32.2$$

$$v_2 = -32.2t + C_3.$$

Since $v_2 = +80$ when $t = 2$, $C_3 = 80 + 2(32.2) = 144.4$, and v_2 is

$$v_2 = -32.2t + 144.4 = \frac{dy_2}{dt}.$$

The position of ball 2 is

$$y_2 = \int (-32.2t + 144.4)\, dt$$

$$= -16.1t^2 + 144.4t + C_4.$$

Since $y_2 = 0$ when $t = 2$, the value of C_4 is

$$C_4 = 16.1(2)^2 - 144.4(2) = -224.4$$

or

$$y_2 = 16.1t^2 + 144.4t - 224.4 \tag{b}$$

The two balls will meet when $y_1 = y_2$, that is, when

$$-16.1t^2 + 100t = -16.1t^2 + 144.4t - 224.4,$$

which gives

$$t = 5.05 \text{ sec}$$

The value of y when the balls meet, from either Eq. (a) or (b), is

$$y = -16.1(5.05)^2 + 100(5.05)$$
$$= \underline{94.2 \text{ ft.}} \qquad \text{Ans.}$$

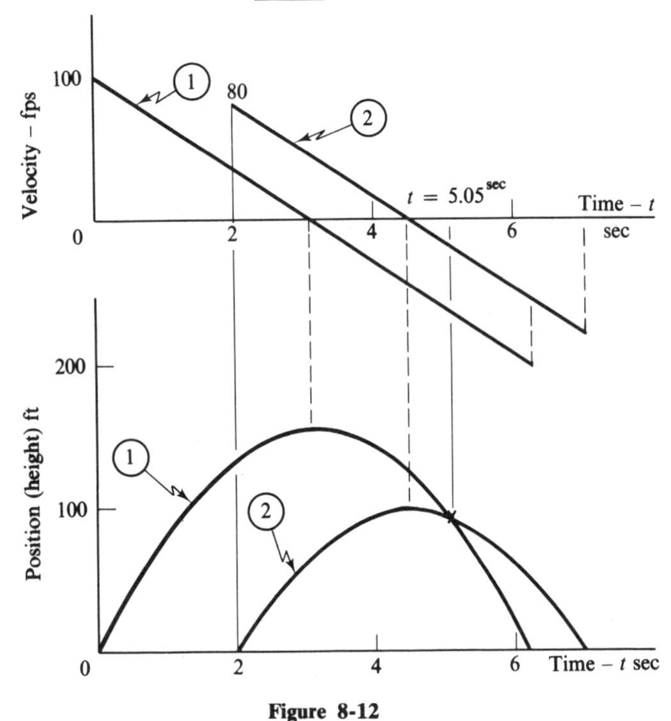

Figure 8-12

Figure 8-12 shows the v-t and s-t diagrams for the two motions. Note that both balls are moving downward when they pass.

EXAMPLE 8-12

Figure 8-13 represents a 1-mile-wide river which has an average velocity of 4 mph south. A motor boat is used to go from A to C. The speed of the boat in still water is 10 mph. In what direction should the boat head in order to go directly to point C, and how much time will be required for the trip?

SOLUTION

Axes are selected as shown on the figure, and angles are measured from the x axis. If the boat is to travel directly from A to C, its absolute velocity must be in the direction from A to C; that is, the angle of the absolute velocity will be $\tan^{-1} 0.5$ from the x axis. The velocity of the boat relative to the water will be 10 mph in a direction to be determined. Note that the speed with which the boat passes a floating leaf or stick, the relative velocity, will be the same whether the water is moving or not. The relative velocity equation for the boat and water is

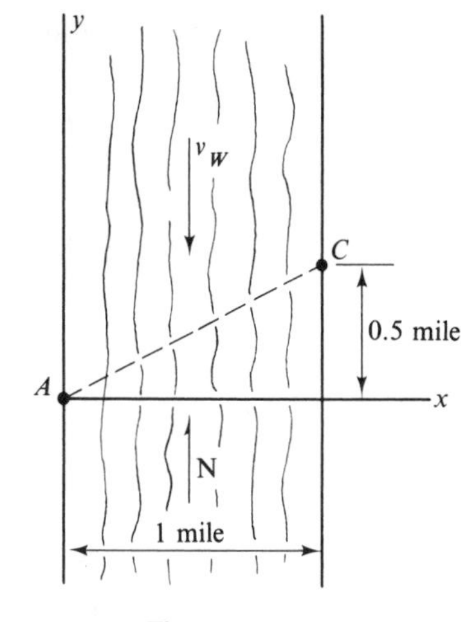

Figure 8-13

$$\mathbf{v}_B = \mathbf{v}_W + \mathbf{v}_{B/W}.$$

This equation is illustrated in Fig. 8-14. The angle θ_2 indicates the direction of the velocity of the boat relative to the water, the direction the boat should head, and v_B is the magnitude of the velocity of the boat. These two quantities can be obtained either from the oblique triangle or by writing the equation in terms of the unit vectors $\mathbf{i}$ and $\mathbf{j}$. When the second approach is used the equation becomes

$$v_B\left(\frac{2\mathbf{i}+\mathbf{j}}{\sqrt{5}}\right) = -4\mathbf{j} + 10(\mathbf{i}\cos\theta_2 + \mathbf{j}\sin\theta_2)$$

from which

$$\mathbf{i}: \quad v_B\frac{2}{\sqrt{5}} = 10\cos\theta_2, \tag{a}$$

$$\mathbf{j}: \quad v_B\frac{1}{\sqrt{5}} = -4 + 10\sin\theta_2. \tag{b}$$

When Eq. (b) is divided by Eq. (a) the result is

$$\frac{1}{2} = \frac{-4+10\sin\theta_2}{10\cos\theta_2}$$

or

$$10\cos\theta_2 = -8 + 20\sin\theta_2 = 10\sqrt{1-\sin^2\theta_2},$$

which gives

$$100(1-\sin^2\theta_2) = 64 - 320\sin\theta_2 + 400\sin^2\theta_2.$$

This equation gives two values of $\sin\theta_2$ as

$$\sin\theta_2 = 0.738 \quad \text{or} \quad -0.0976$$

and

$$\theta_2 = 47.5° \quad \text{or} \quad -5.60°.$$

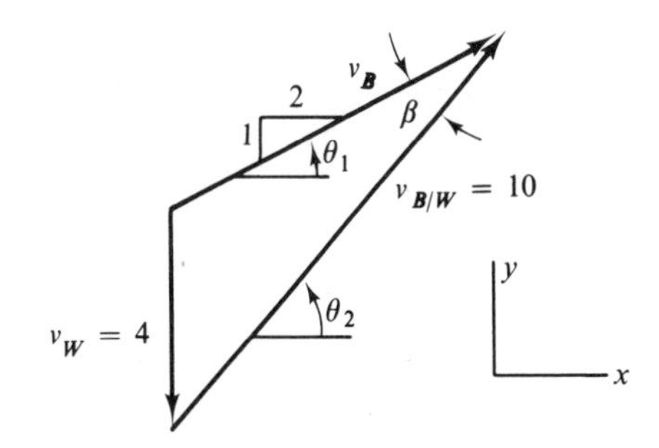

Figure 8-14

The first value of θ_2 gives the direction the boat should head to go from A to C, and the second gives the direction of the relative velocity if the boat is to travel from C to A (the heading would be south of due west). The magnitude of the absolute velocity of the boat, when $\theta_2 = 47.5°$ is substituted into Eq. (a), is

$$v_B = \frac{\sqrt{5}}{2}(10 \cos 47.5°) = 7.55 \text{ mph}.$$

The boat must travel $(1^2 + 0.5^2)^{0.5} = 1.118$ miles from A to C at a speed of 7.55 mph. Therefore the time required is

$$\Delta t = \frac{\Delta s}{v} = \frac{1.118}{7.55} = 0.1481 \text{ hr} = \underline{8.88 \text{ min.}} \qquad \text{Ans.}$$

The boat heading should be

$$\theta_2 = 47.5° \text{ north of due east}$$
$$= \underline{42.5° \text{ east of due north (compass heading)}}. \qquad \text{Ans.}$$

EXAMPLE 8-13

The two cars A and B in Fig. 8-15 are approaching an overpass intersection. A has a constant velocity of 45 mph east, and B has a constant velocity of 60 mph north. At a given instant A is 500 ft west of the intersection, and B is 300 ft south. Neglect the vertical distance between the roadways. Determine
(a) The initial position of A relative to B.
(b) The velocity of A relative to B.
(c) The minimum distance between A and B.

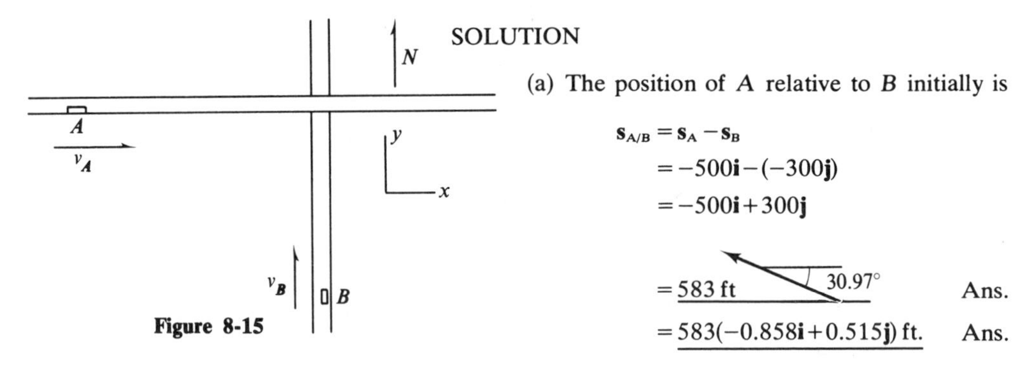

Figure 8-15

SOLUTION

(a) The position of A relative to B initially is

$$\mathbf{s}_{A/B} = \mathbf{s}_A - \mathbf{s}_B$$
$$= -500\mathbf{i} - (-300\mathbf{j})$$
$$= -500\mathbf{i} + 300\mathbf{j}$$

$$= \underline{583 \text{ ft}} \quad 30.97° \qquad \text{Ans.}$$
$$= \underline{583(-0.858\mathbf{i} + 0.515\mathbf{j}) \text{ ft.}} \qquad \text{Ans.}$$

(b) The velocity of A relative to B is

$$\mathbf{v}_{A/B} = \mathbf{v}_A - \mathbf{v}_B$$
$$= 45\mathbf{i} - 60\mathbf{j}$$

$$= \underline{75 \text{ mph}} \quad 53.15° \qquad \text{Ans.}$$
$$= \underline{75(0.6\mathbf{i} - 0.8\mathbf{j}) \text{ mph.}} \qquad \text{Ans.}$$

(c) At any time t (in seconds) after the initial positions the position of A relative to B is

$$\mathbf{s}_{A/B} = (\mathbf{s}_{A/B})_0 + \Delta(\mathbf{s}_{A/B})$$
$$= (\mathbf{s}_{A/B})_0 + (\mathbf{v}_{A/B})t$$

since $\mathbf{v}_{A/B}$ is constant. The relative velocity should be converted to feet per second since $\mathbf{s}_{A/B}$ is in feet:

$$v_{A/B} = \left(\frac{75 \text{ miles}}{\text{hr}}\right)\left(\frac{5280 \text{ ft}}{\text{mile}}\right)\left(\frac{\text{hr}}{3600 \text{ sec}}\right)$$
$$= 110 \text{ fps}.$$

Note that 60 mph is the same as 88 fps. When numerical values are substituted into the relative position equation it becomes

$$\mathbf{s}_{A/B} = -500\mathbf{i} + 300\mathbf{j} + 110t(0.6\mathbf{i} - 0.8\mathbf{j})$$
$$= (-500 + 66t)\mathbf{i} + (300 - 88t)\mathbf{j}.$$

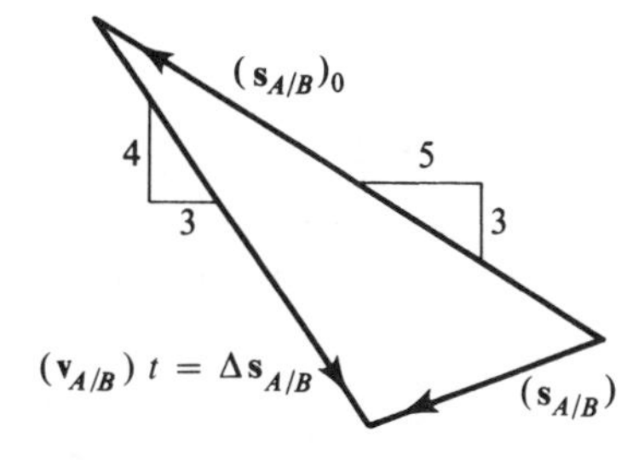

Figure 8-16

The value of t which will make the magnitude of $\mathbf{s}_{A/B}$ a minimum can be determined by writing an expression for the magnitude of $\mathbf{s}_{A/B}$ as a function of t and setting the derivative with respect to t equal to zero. A more convenient solution can be obtained from a graphical representation of the equation for $\mathbf{s}_{A/B}$ as shown in Fig. 8-16. Since $\mathbf{s}_{A/B}$ is equal to the vector sum of $(\mathbf{s}_{A/B})_0$ and $(\mathbf{v}_{A/B})t$, it will be minimum when $\mathbf{s}_{A/B}$ is perpendicular to $\mathbf{v}_{A/B}$, that is, when $\mathbf{s}_{A/B} = s_{A/B}(-0.8\mathbf{i} - 0.6\mathbf{j})$. The vector equation becomes

$$s_{A/B}(-0.8\mathbf{i} - 0.6\mathbf{j}) = -500\mathbf{i} + 300\mathbf{j} + 66t\mathbf{i} - 88t\mathbf{j},$$

and the component equations are

$$\mathbf{i}: \quad -0.8s_{A/B} = -500 + 66t$$
$$\mathbf{j}: \quad -0.6s_{A/B} = 300 - 88t.$$

The simultaneous solution of these equations gives

$$s_{A/B} = \underline{220 \text{ ft.}} \qquad \text{Ans.}$$

Note that if either car A or B has an acceleration, the velocity of A relative to B will not be constant in either magnitude or direction and the vector diagram in Fig. 8-16 will not be valid.

PROBLEMS

***8-60** Body A in Fig. P8-60 has a constant velocity of 0.5 m/s to the right. Determine the displacement of body B relative to body A during a 3-s interval.

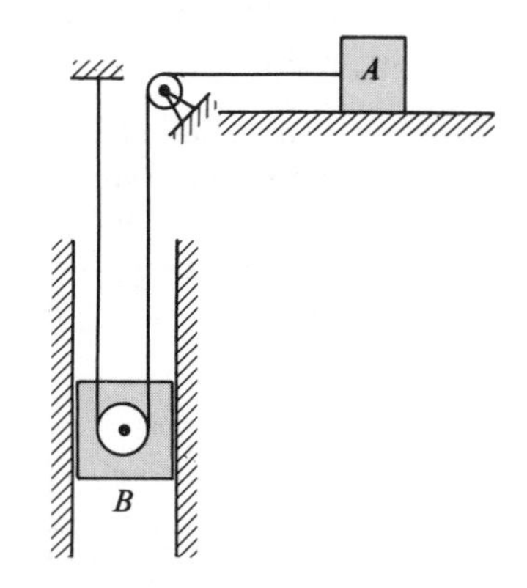

Figure P8-60

8-61 Bodies A and B in Fig. P8-61 are connected by a continuous cord as shown. Body A has a velocity of v_A to the right. Determine
(a) The velocity of B relative to A.
(b) The velocity of point C on the cord relative to A.

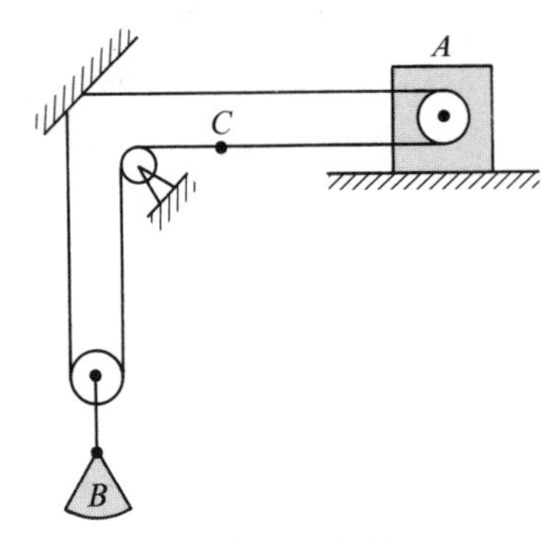

Figure P8-61

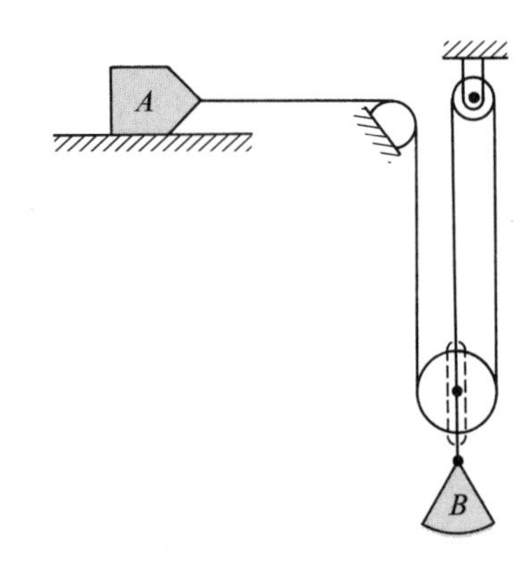

Figure P8-63

8-62 Determine the acceleration of body A in Fig. P8-62 which will result in an acceleration of B relative to A of 12 ips^2 to the right.

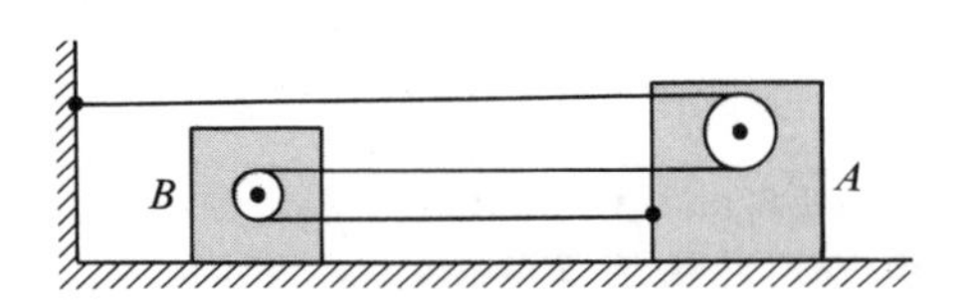

Figure P8-62

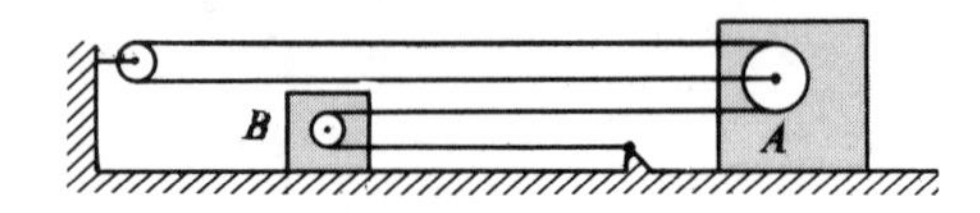

Figure P8-64

8-63 Block A in Fig. P8-63 has a velocity of 6 ips to the right when $t=0$ and a constant acceleration of 5 ips^2 to the left. The cord connecting the bodies remains taut at all times. Determine the total distance traveled by block B during the interval $t=0$ to $t=3$ sec.

***8-64** The two bodies in Fig. P8-64 are initially at rest. Determine the constant acceleration of body A which will cause body B to have a displacement of 200 mm to the right during the first 4 s of motion.

8-65 The system of bodies in Fig. P8-65 starts from rest in the position shown. Determine the constant acceleration of C which will cause the right side of body A to slide to the right side of body B in 4 sec.

8-66 Body A in Fig. P8-66 slides along a fixed horizontal rod, and body B moves in the vertical slot. The cord from body A passes over a *small* peg on B and is tied at C. Derive an equation relating the velocities of A and B. Assume that the cord remains taut.

8-67 Bodies A and B in Fig. P8-67 are connected by a cord which passes over two *small* pulleys.
(a) Write an equation relating s_A and s_B.
(b) Determine the velocity of A in terms of b, s_B, and the velocity of B.

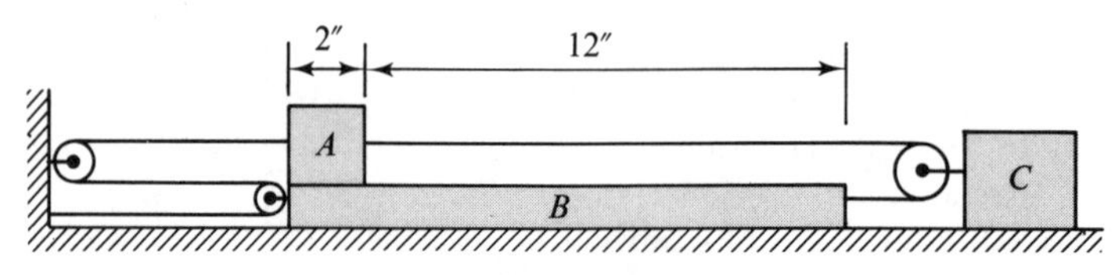

Figure P8-65

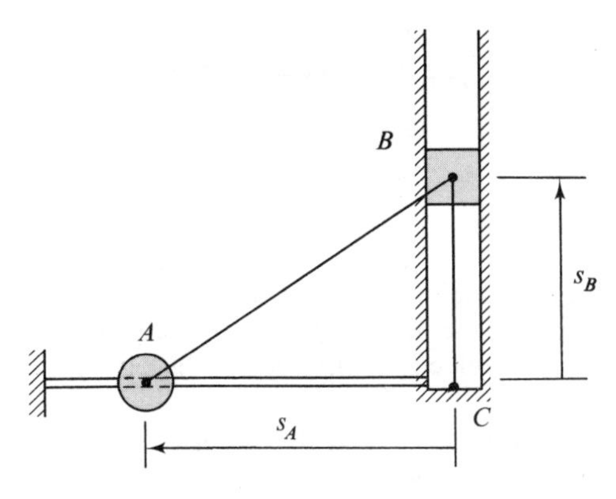

Figure P8-66

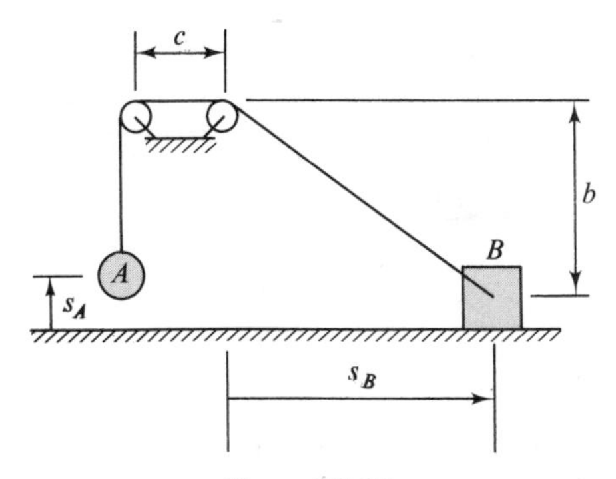

Figure P8-67

8-68 In Problem 8-67 $b = 20$ in. and $c = 10$ in. Body B has a constant velocity of 10 ips to the right. Determine
(a) The velocity of A when $s_B = 0$.
(b) The velocity of A when $s_B = 15$ in.
(c) The acceleration of A when $s_B = 0$.

8-69 Bodies A and B in Fig. P8-69 are connected by a flexible cord which passes over three *small* pulleys.
(a) Write an equation relating s_A and s_B.
(b) Determine the velocity of B in terms of b, s_A, and the velocity of A.

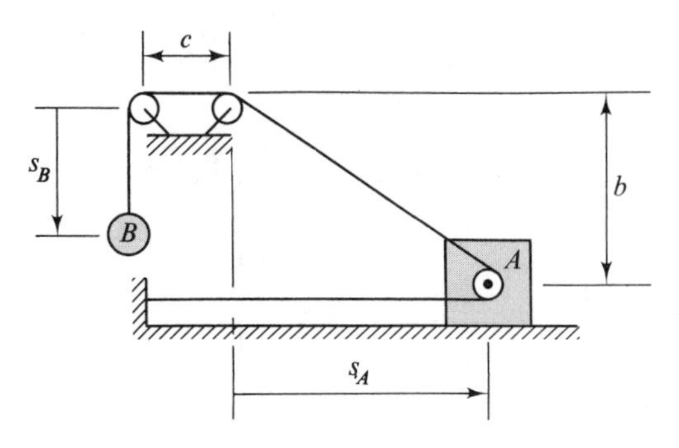

Figure P8-69

8-70 In Problem 8-69 body A has a constant velocity of 10 ips to the right, $b = 8$ in., and $c = 3$ in. Determine
(a) The velocity of B when $s_A = 0$.
(b) The velocity of B when $s_A = 15$ in.

8-71 Bodies A and B in Fig. P8-71 slide along the horizontal rod and are connected by a cord which passes over the small peg at O. Body A has a constant velocity of 10 ips to the right. When A is directly below O body B is 36 in. to the left of A. Determine
(a) The displacement of B corresponding to a displacement of A of 8 in. to the right from a position below O.
(b) The velocity of B when $s_A = 8$ in. to the right.

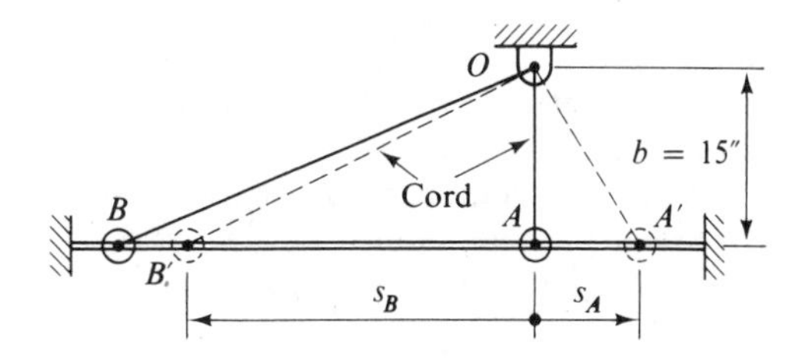

Figure P8-71

8-72 Bodies A and B in Fig. P8-72 slide along fixed rods as shown and are connected by an

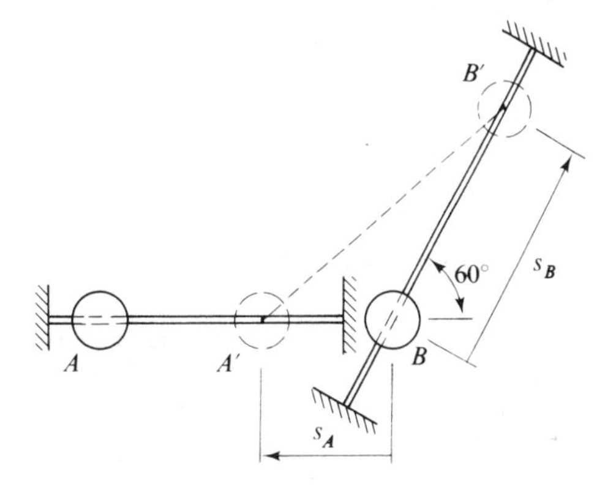

Figure P8-72

inextensible cable. Derive an equation relating the velocities of A and B in terms of their positions.

8-73 The cord connecting A and B in Problem 8-72 is 20 in. long, and B has a constant velocity of 10 ips up and to the right. Determine the acceleration of B when $s_A = s_B$.

8-74 Block C in Fig. P8-74 has a constant acceleration of 1.0 ips^2 and an initial velocity of 2.0 ips, both to the right. Body B has a constant velocity of 3.0 ips to the right. The cord remains taut at all times. Determine the velocity of A when it first hits one of the stops on B.

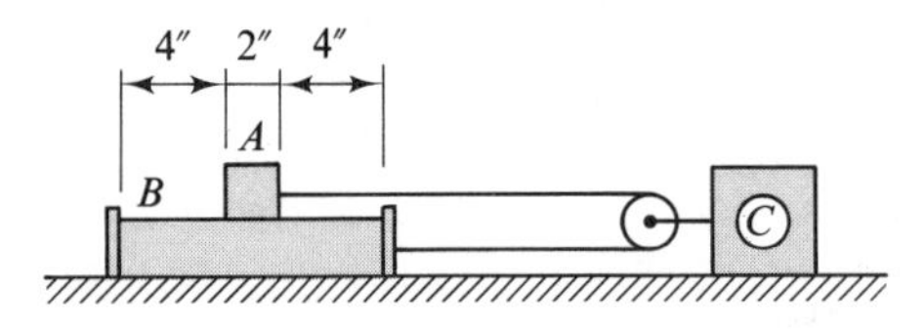

Figure P8-74

8-75 Bodies A, B, and C in Fig. P8-75 start from rest at a common elevation. Body A has a constant acceleration of 2 ips^2 downward. Body B has a zero acceleration for the first 3 sec, after which its acceleration becomes 3 ips^2 upward. Determine the velocity of C after it has traveled a distance of 12 in.

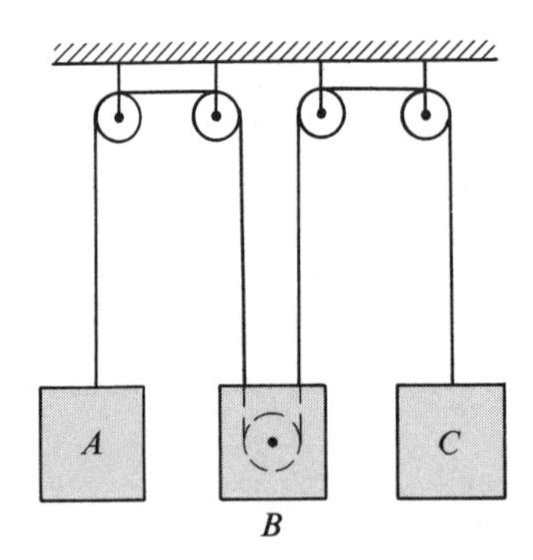

Figure P8-75

***8-76** Blocks A, B, and C in Fig. P8-76 are connected by an inextensible cord as indicated. At a given instant the velocity of C is 90 mm/s to the right, and the velocity of A relative to B is 60 mm/s to the right. Determine the velocity of C relative to A.

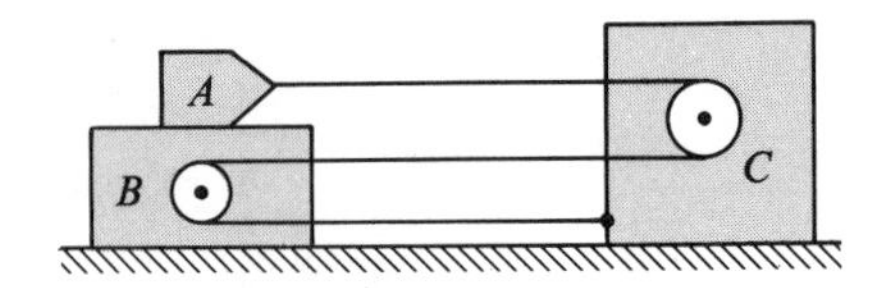

Figure P8-76

***8-77** Bodies A, B, and C in Fig. P8-77 move along parallel paths on a horizontal plane. Body C has a velocity of 0.10 m/s to the right. Determine

(a) The velocity of B if A is fixed.

(b) The velocity of A if the velocity of B is 0.1 m/s to the right.

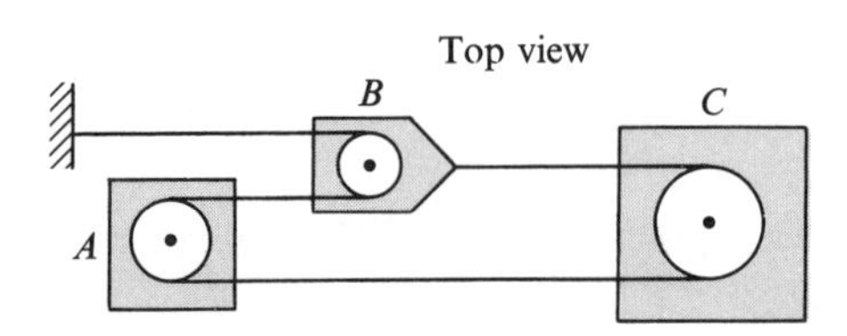

Figure P8-77

8-78 The system of pulleys and cords in Fig. P8-78 support bodies A, B, and C. The system starts from rest in the indicated position. Body A has a constant acceleration of 5 ips^2 downward, and B has a constant acceleration of 5 ips^2

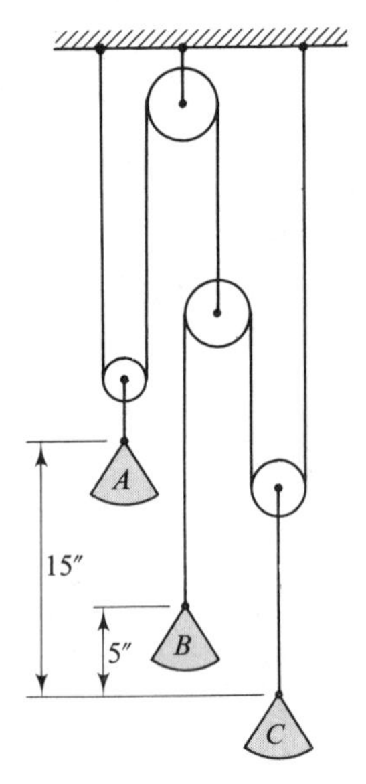

Figure P8-78

upward. Determine the velocity of each body when C first passes either A or B.

8-79 A 64-ft truck and trailer unit has a uniform speed of 60 mph on an interstate highway. The truck is overtaken by a 20-ft passenger car with a uniform speed of 75 mph. The car pulls into the passing lane when it is 100 ft behind the truck and enters the right-hand lane when it is 100 ft ahead of the truck. Determine the total distance the car travels during the passing maneuver.

8-80 Particles A and B move along adjacent parallel paths. Initially, A is 100 ft to the left of B, and each particle has a velocity of 5 fps to the right. The velocity of A remains constant, and the acceleration of B is 2 fps^2 to the left. Determine the total distance traveled by B when the particles pass each other.

8-81 The passenger train A and the freight train B in Fig. P8-81 both start from rest in the position shown and travel to the right. Train A has a constant acceleration of 2 fps^2 to the right until it reaches a maximum speed of 90 mph.

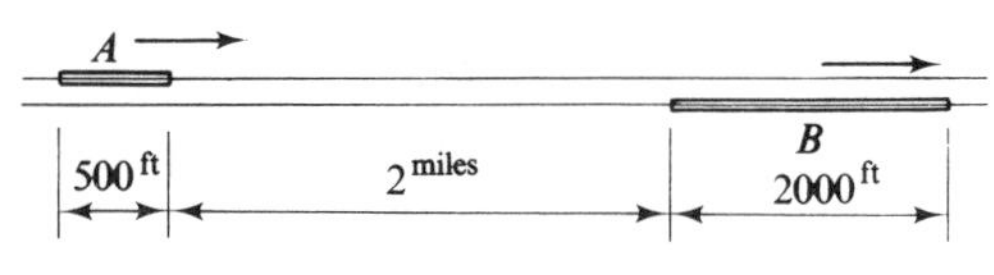

Figure P8-81

The freight has a constant acceleration of 0.5 fps^2 until it attains a maximum speed of 60 mph. Determine
(a) The distance the passenger locomotive travels before overtaking the caboose of the freight.
(b) The time required for the passenger locomotive to pass the freight locomotive.

8-82 A freight elevator is moving upward at a constant rate of 16 fps when it passes a passenger elevator discharging passengers. Two seconds after the freight elevator passes, the passenger elevator starts with a constant acceleration of 12 fps^2 upward. When the velocity of the passenger elevator is 48 fps, the acceleration becomes zero. Determine
(a) The time required by the passenger elevator to overtake the freight elevator.
(b) The distance the passenger elevator travels in overtaking the freight elevator.

8-83 Cars A and B start from rest at a starting line and move along parallel paths. Car A has a constant acceleration of 10 fps^2 to the right for 180 ft, after which it is brought to rest with a constant acceleration of 6 fps^2. Car B has a constant acceleration of 6 fps^2 to the right for 10 sec, after which its velocity remains constant. Determine
(a) The maximum distance A leads B.
(b) The distance A has traveled when the cars have the same velocity.

8-84 Two cars start from rest at the same location and race along a straight track. Car A accelerates at 6.6 fps^2 to a speed of 90 mph and then runs at this constant speed. Car B accelerates at 4.4 fps^2 to a speed of 96 mph and then runs at a constant speed.
(a) Which car will win a 3-mile race and by what distance?
(b) What will be the maximum lead of A over B?
(c) How far will the cars have traveled when B passes A?

8-85 A stone is thrown downward into an abandoned mine shaft with a velocity of 40 fps. Four seconds later the sound of the stone striking water is heard. If sound travels at 1120 fps, determine the distance from the top of the mine shaft to the water surface in the shaft.

8-86 A man is riding in an open hoist (elevator) which is moving upward with a constant velocity of 10 fps. He tosses a stone upward from shoulder height, 5 ft above the floor of the hoist, with a velocity of 20 fps relative to the hoist. Determine
(a) The velocity of the stone when it hits the floor of the hoist.
(b) The maximum height of the stone above the floor of the hoist.
(c) The total vertical distance traveled by the stone before hitting the floor of the hoist.

8-87 Solve Problem 8-86 if the hoist is moving downward with a velocity of 10 fps.

8-88 A stream is 100 ft wide and flows south with an average velocity of 2.0 fps. A man swims

45° north of east and crosses the stream in 30 sec. Determine

(a) The displacement of the man during the 30-sec interval.

(b) The velocity of the man relative to the water.

8-89 An airplane makes a round trip from A to B on each of three days. Point B is 100 miles north of A, and the airspeed of the plane is 120 mph. On the first day there is no wind, on the second day there is a steady 40-mph wind from the north, and on the third day there is a steady 40-mph wind from the west. State which trip will require the least time and which the most.

8-90 A pilot flying a light airplane with an airspeed of 100 mph plans to make a trip to a point due east of his starting point. Upon inquiring at the weather station he learns that he will have a 10-mph head wind at an elevation of 3500 ft and that the wind speed is 20 mph from 20° east of north at an altitude of 5500 ft. At which altitude should he fly to make the best ground speed and what will this ground speed be?

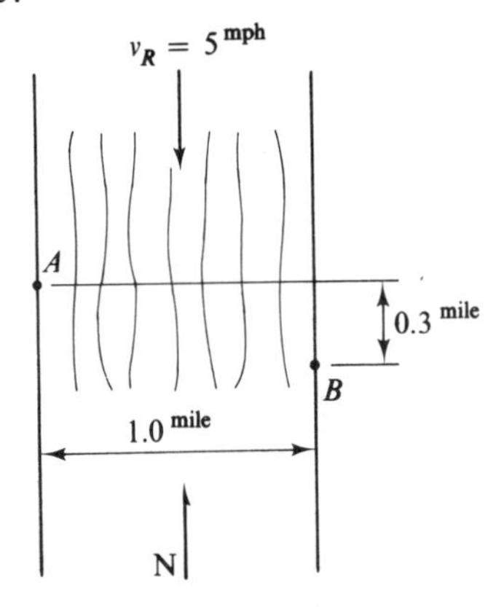

Figure P8-91

8-91 A motor boat is to be used to go from A to B in Fig. P8-91. The speed of the boat in a lake is 10 mph. Determine the least time required for the boat to travel from A to B and the direction the boat should head.

8-92 The boat in Problem 8-91 heads due east for 5 min. What should the heading be for the rest of the trip if the boat is to dock at B?

8-93 Cars A and B in Fig. P8-93 are approaching a grade separation with constant velocities as indicated. A has a speed of 50 mph and is 1.0 mile from the crossover at a given instant. The speed of B is 75 mph, and it is 0.5 mile from the intersection at the same given instant. Determine the location of each car when the distance between them is minimum.

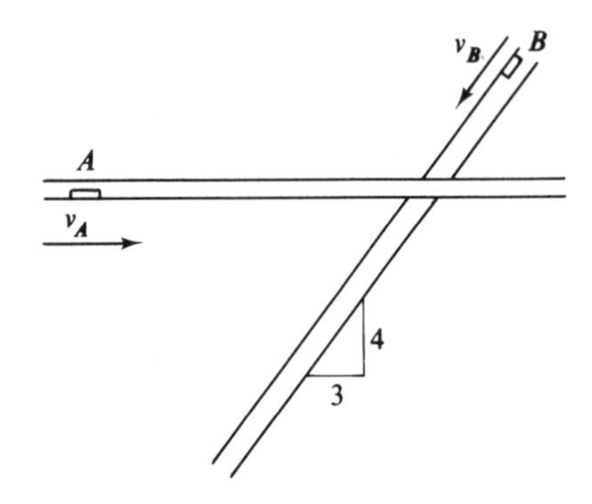

Figure P8-93

8-94 In Example 8-13 at a given instant car A is 500 ft west of the intersection and B is 200 ft south. The velocity of A is 60 fps (constant) to the east. Car B has a constant acceleration of 4 fps^2 north and has a velocity of 30 fps north at the initial instant. Determine the minimum distance between the two cars.

8-7 CURVILINEAR MOTION—RECTANGULAR COMPONENTS

In the preceding articles the particle was assumed to move along a straight line, and the velocity and acceleration were directed along the straight line and could vary only in magnitude (not in direction).

When the particle moves along a curved path these quantities in general vary in both magnitude and direction. It was shown in Art. 8-3 that the velocity is always tangent to the path and thus will change in direction, and perhaps in magnitude, as the particle moves along the path. The acceleration of the particle is not tangent to the path for curvilinear motion. This point will be discussed in more detail in Art. 8-8. Several different coordinate systems can be used to describe the kinematics of curvilinear motion of a particle. The use of rectangular components is discussed in this article.

The position vector from a fixed origin to a point in space can be written in terms of the constant unit vectors **i**, **j**, and **k** (parallel to the fixed x, y, and z axes, respectively) as

$$\mathbf{r} = x\mathbf{i} + y\mathbf{j} + z\mathbf{k}, \tag{8-6}$$

where the rectangular coordinates (x, y, and z) of the point or particle may vary with time. The displacement of the particle during a time interval is

$$\mathbf{q} = \Delta\mathbf{r} = \mathbf{r}_f - \mathbf{r}_i,$$

where $\mathbf{r}_i$ and $\mathbf{r}_f$ are the initial and final positions of the particle.

The velocity of the particle at any instant was defined in Art. 8-3 as the time rate of change of position of the particle. From Eq. (8-6) the velocity is

$$\begin{aligned} \mathbf{v} &= \frac{d\mathbf{r}}{dt} = \dot{\mathbf{r}} = \dot{x}\mathbf{i} + \dot{y}\mathbf{j} + \dot{z}\mathbf{k} \\ &= v_x\mathbf{i} + v_y\mathbf{j} + v_z\mathbf{k}, \end{aligned} \tag{8-7}$$

where $v_x = \dot{x} = dx/dt$ is the magnitude of the component of the velocity in the x direction with similar expressions for v_y and v_z. Note that the derivative of the product $x\mathbf{i}$ is $\dot{x}\mathbf{i} + x\dot{\mathbf{i}} = \dot{x}\mathbf{i}$ since **i** is a constant in both magnitude and direction for fixed axes.

The linear acceleration of the particle at any instant of time is defined as the time rate of change of the velocity of the particle, and from Eq. (8-7) the acceleration is

$$\begin{aligned} \mathbf{a} &= \dot{\mathbf{v}} = \ddot{x}\mathbf{i} + \ddot{y}\mathbf{j} + \ddot{z}\mathbf{k} \\ &= a_x\mathbf{i} + a_y\mathbf{j} + a_z\mathbf{k} \end{aligned} \tag{8-8}$$

in which a_x, a_y, and a_z are the components of the acceleration in the x, y, and z directions, respectively.

Equations (8-6)–(8-8) indicate that curvilinear motion can be considered to be the vector sum of three simultaneous rectilinear motions along the three coordinate axes. This analysis is particularly useful for analyzing motions in which the motion (acceleration and velocity) along each of the coordinate axes is independent.

The motion of a projectile in flight is an example of curvilinear motion. When air resistance is neglected the only force acting on the projectile is its weight[2]; therefore the horizontal component of the acceleration is zero, and the vertical component is the acceleration due to the weight of the body, approximately 32.2 fps^2 or 9.81 m/s^2 directed vertically downward. Later in this article it will be shown that a projectile follows a parabolic path when air resistance is neglected. Air resistance is a variable force which changes in both magnitude and direction. The magnitude varies with the velocity and with the density of the air, and the direction is opposite the direction of the velocity. No simple solution is available for such problems, but approximate solutions have been obtained.[3]

Since the horizontal and vertical components of the acceleration of a projectile are known, the resulting motion is the superposition of two rectilinear motions, and rectangular coordinates provide a convenient method of analysis. The accelerations of the component motions are constant (one is zero); therefore the v-t diagrams are straight lines and can often lead to a convenient solution of the resulting motion. The differential equations of the two simultaneous motions can also be used to analyze the motion. Both procedures are discussed in the following examples.

EXAMPLE 8-14

The velocity (in inches per second) of a particle moving in the xy plane is given by

$$\mathbf{v} = 4(t-1)\mathbf{i} + 3(t^2 - 2t + 1)\mathbf{j},$$

where t is time in seconds. The particle is at $\mathbf{r} = (2\mathbf{i} - \mathbf{j})$ in. when $t = 0$. Determine

(a) The equation of the path of the particle.
(b) The position, velocity, and acceleration of the particle when $t =$ 3 sec.

SOLUTION

(a) Axes and unit vectors are selected as shown in Fig. 8-17(a). The velocity is

$$\mathbf{v} = \dot{\mathbf{r}} = 4(t-1)\mathbf{i} + 3(t^2 - 2t + 1)\mathbf{j} \tag{a}$$

from which

$$\mathbf{r} = \int [4(t-1)\mathbf{i} + 3(t^2 - 2t + 1)\mathbf{j}]\, dt$$
$$= (2t^2 - 4t)\mathbf{i} + (t^3 - 3t^2 + 3t)\mathbf{j} + \mathbf{C},$$

[2] A rocket or missile will also be subject to the rocket thrust, and for long-range firings, the curvature of the earth, the rotation of the earth, and sometimes the attraction of the moon or sun may affect the trajectory. Such effects are neglected in this discussion.

[3] D. T. Greenwood, *Principles of Dynamics* (Englewood Cliffs, N.J.: Prentice-Hall, Inc., 1965), pp. 74–76; S. W. Groesberg, *Advanced Mechanics* (New York: John Wiley & Sons, Inc., 1968), pp. 41, 42.

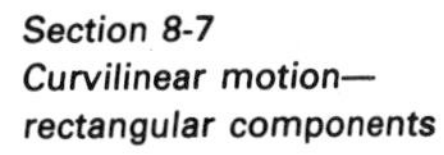

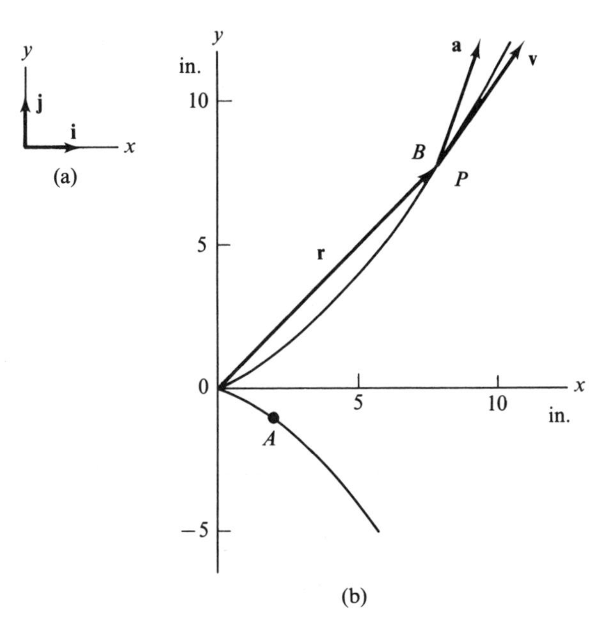

Figure 8-17

where **C** is a constant (vector) of integration. When $t = 0$, $\mathbf{r} = 2\mathbf{i} - \mathbf{j}$; therefore

$$\mathbf{C} = 2\mathbf{i} - \mathbf{j},$$

and the position vector is

$$\begin{aligned} \mathbf{r} &= (2t^2 - 4t + 2)\mathbf{i} + (t^3 - 3t^2 + 3t - 1)\mathbf{j} \\ &= 2(t-1)^2\mathbf{i} + (t-1)^3\mathbf{j} \\ &= x\mathbf{i} + y\mathbf{j}. \end{aligned} \tag{b}$$

That is,

$$x = 2(t-1)^2 \quad \text{and} \quad y = (t-1)^3.$$

These are the equations of the path in parametric form. The time t can be eliminated as follows:

$$(t-1) = \left(\frac{x}{2}\right)^{1/2} = y^{1/3},$$

which reduces to

$$\underline{x^3 = 8y^2.} \qquad \text{Ans.}$$

The curve corresponding to this equation is shown in Fig. 8-17(b).

(b) The acceleration of the particle is obtained by differentiating the velocity expression, Eq. (a), with respect to time, giving

$$\mathbf{a} = \dot{\mathbf{v}} = 4\mathbf{i} + (6t - 6)\mathbf{j}. \tag{c}$$

The position, velocity, and acceleration when $t = 3$ sec are obtained by direct substitution into Eqs. (b), (a), and (c), respectively. Thus

$$\mathbf{r}_{3\text{ sec}} = 2(3-1)^2\mathbf{i} + (3-1)^3\mathbf{j}$$
$$= \underline{(8\mathbf{i} + 8\mathbf{j})\text{ in.}} \qquad \text{Ans.}$$
$$\mathbf{v}_{3\text{ sec}} = 4(3-1)\mathbf{i} + 3(9-6+1)\mathbf{j}$$
$$= \underline{(8\mathbf{i} + 12\mathbf{j})\text{ ips,}} \qquad \text{Ans.}$$

and

$$\mathbf{a}_{3\text{ sec}} = 4\mathbf{i} + (18-6)\mathbf{j}$$
$$= \underline{(4\mathbf{i} + 12\mathbf{j})\text{ ips}^2.} \qquad \text{Ans.}$$

The position of the particle when $t = 3$ sec is shown as point P in Fig. 8-17(b) together with the corresponding velocity and acceleration vectors. Note that the velocity is tangent to the path but that the acceleration is not tangent to the path. As t increases from zero the particle moves along the curved path from A to O and toward B.

EXAMPLE 8-15

A ball is thrown with a velocity of 75 fps with a slope of 4 vertical to 3 horizontal at a building with a flat roof. The building is 45 ft high, and the man throwing the ball is 100 ft from the front wall and is standing in a depression so that the ball leaves his hand at the same elevation as the base of the building. Neglect air resistance.

(a) Will the ball strike the building?
(b) If the ball hits the building, determine where it will hit and with what velocity.

SOLUTION

Figure 8-18(a) shows three possible trajectories of the ball resulting from different magnitudes of the initial velocity. The proper trajectory can be selected by determining the height of the ball when it has traveled 100 ft horizontally. Figure 8-18(b) is drawn on the assumption that the ball will land on the roof, that is, that the ball will be more than 45 ft above the ground when it has traveled 100 ft horizontally. The initial velocity of the ball is

$$\mathbf{v}_i = 75\text{ fps}\ \angle^{4}_{3} = 0.6(75)\overrightarrow{} + 0.8(75)\uparrow$$
$$= (\overrightarrow{45} + \overset{\uparrow}{60})\text{ fps.}$$

Figure 8-18(c) shows the horizontal component of the velocity plotted against time, and Fig. 8-18(d) shows the vertical component of the velocity plotted against time. Although the horizontal distances on the v-t diagrams and on the sketch of the path do not represent the same variable, it is helpful if scales are selected so that points on the v-t diagrams line up vertically with the corresponding points on the sketch

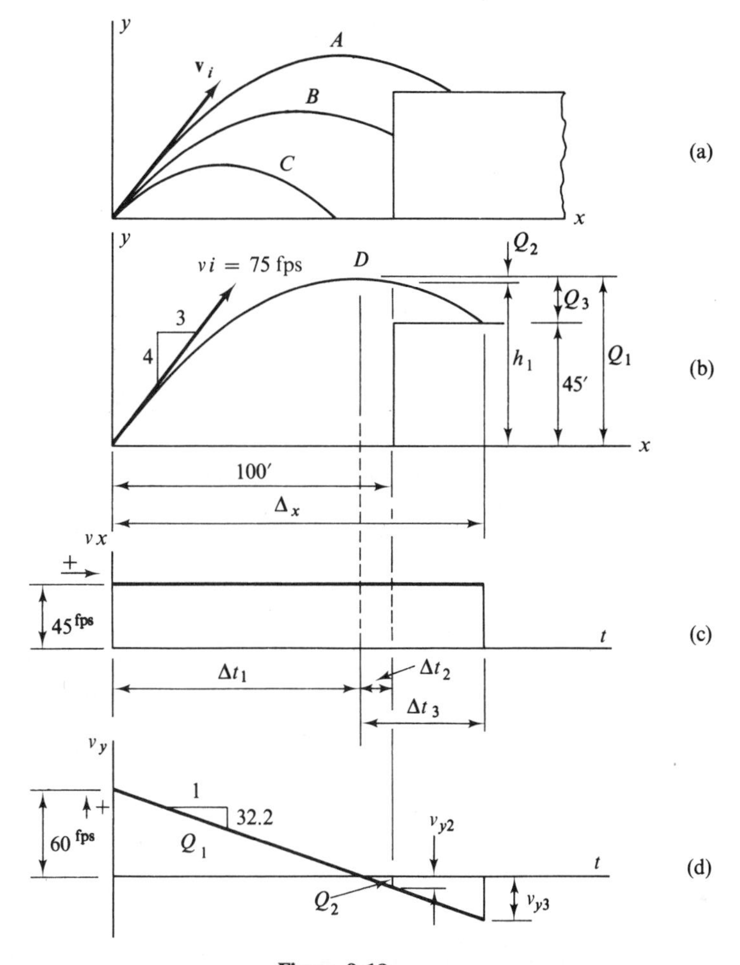

Figure 8-18

of the path. For example, when the ball is at the top of its trajectory, $v_y = 0$ is the corresponding point on the v_y-t diagram. Since a_x is zero, the v_x-t diagram is a horizontal straight line. The vertical component of the acceleration has a constant value of 32.2 fps² downward; thus, the v_y-t diagram in Fig. 8-18(d) is a straight line with a slope of 32.2. The area under the v_x-t diagram represents the horizontal distance traveled during any time interval. Therefore, when the ball has traveled 100 ft horizontally

$$\Delta x = 100 = v_x(\Delta t_1 + \Delta t_2) = 45(\Delta t_1 + \Delta t_2)$$

and

$$\Delta t_1 + \Delta t_2 = \frac{100}{45} = 2.222 \text{ sec.} \tag{a}$$

The time Δt_1 required for the ball to reach its maximum height at D

can be determined from the slope of the v_y-t diagram. Thus

$$\frac{v_{yi}}{\Delta t_1} = a_y = 32.2 = \frac{60}{\Delta t_1}$$

and

$$\Delta t_1 = \frac{60}{32.2} = 1.863 \text{ sec.}$$

From Eq. (a) the value of Δt_2 is

$$\Delta t_2 = 2.222 - \Delta t_1 = 2.222 - 1.863 = 0.359 \text{ sec.}$$

The velocity v_{y2} can be determined from the slope of the v_y-t curve and is

$$v_{y2} = a_y(\Delta t_2) = 32.2(0.359) = 11.56 \text{ fps},$$

$$\mathbf{v}_{y2} = 11.56 \text{ fps downward.}$$

The difference between areas Q_1 and Q_2 represents the height h_1 of the ball above the ground when it reaches the wall; that is,

$$h_1 = Q_1 - Q_2 = \frac{60(1.863)}{2} - \frac{11.56(0.359)}{2}$$

$$= 55.89 - 2.07 = 53.8 \text{ ft.}$$

Since the value of h_1 is more than the height of the building, the ball will fall on the roof of the building. The vertical distance, Q_3, which the ball must fall from the top of its path before landing on the roof is represented by the corresponding area under the v_y-t diagram. That is,

$$Q_3 = Q_1 - 45 = 55.89 - 45 = 10.89 \text{ ft},$$

which is equal to the area of the corresponding triangle. Thus

$$\frac{v_{y3}(\Delta t_3)}{2} = 10.89.$$

The slope of the v_y-t diagram gives

$$a_y = \frac{v_{y3}}{\Delta t_3} = 32.2,$$

and the simultaneous solution of these two equations gives

$$\Delta t_3 = 0.822 \text{ sec}$$

and

$$v_{y3} = 26.5 \text{ fps.}$$

The horizontal distance traveled by the ball, Δx, when it is 45 ft above the ground is, from the area under the v_x-t diagram,

$$\Delta x = v_x(\Delta t_1 + \Delta t_3) = 45(1.863 + 0.822)$$

$$= 120.8 \text{ ft.}$$

Therefore the ball will land on the roof of the building 20.8 ft from the left edge.

The velocity of the ball as it strikes the roof is the vector sum of the component velocities and is

$$\mathbf{v}_f = \mathbf{v}_x + \mathbf{v}_y$$

$$= (45\mathbf{i} - 26.5\mathbf{j}) \text{ fps} = 52.2 \text{ fps} \quad 30.6^\circ \qquad \text{Ans.}$$

ALTERNATE SOLUTION

In some instances, it may be desirable to solve the differential equations of motion instead of using v-t diagrams for projectile motion. The position vector is

$$\mathbf{r} = x\mathbf{i} + y\mathbf{j},$$

and the equation for the acceleration is

$$\ddot{\mathbf{r}} = -g\mathbf{j}. \tag{b}$$

Integration of Eq. (b) gives the velocity as

$$\dot{\mathbf{r}} = -gt\mathbf{j} + \mathbf{C}_1,$$

where $\mathbf{C}_1$ is a constant of integration. If $\mathbf{u}$ is the velocity of the projectile when t is zero, $\mathbf{C}_1$ must be equal to $\mathbf{u}$, and the equation for the velocity becomes

$$\dot{\mathbf{r}} = -gt\mathbf{j} + \mathbf{u} = u_x\mathbf{i} + (u_y - gt)\mathbf{j},$$

where u_x and u_y are the components of the initial velocity. The velocity equation can be integrated to give the position. Thus,

$$\mathbf{r} = u_x t\mathbf{i} + \left(u_y t - \frac{gt^2}{2}\right)\mathbf{j} + \mathbf{C}_2.$$

The constant $\mathbf{C}_2$ gives the position of the projectile when $t = 0$, and if the origin is selected at this position, this constant is zero and the position equation becomes

$$\mathbf{r} = (x\mathbf{i} + y\mathbf{j}) = u_x t\mathbf{i} + \left(u_y t - \frac{gt^2}{2}\right)\mathbf{j}.$$

The parametric equations of the path are

$$x = u_x t \quad \text{and} \quad y = u_y t - \frac{gt^2}{2}. \tag{c}$$

When t is eliminated from these two expressions, the equation of the path in cartesian coordinates becomes

$$y = \frac{u_y}{u_x}x - \frac{gx^2}{2u_x^2}$$

or

$$2u_x^2 y = 2u_x u_y x - gx^2, \tag{d}$$

which is the equation of a parabola.

When numerical data are substituted into Eq. (d) it becomes

$$y = \frac{60}{45}x - \frac{32.2x^2}{2(45)^2}$$

$$= 1.333x - 0.00795x^2. \qquad \text{(e)}$$

The height of the ball when it reaches the building is the value of y when $x = 100$ ft, that is,

$$y = h_1 = 1.333(100) - 0.00795(100)^2$$

$$= 53.8 \text{ ft},$$

which shows that the ball passes above the front of the building. When $y = 45$ ft (the height of the building) the resulting quadratic equation gives two values of x, which are

$$x = 46.8 \text{ ft} \quad \text{and} \quad 120.9 \text{ ft}.$$

The second value locates the point where the ball lands on the roof.

The value of t when $x = 120.9$ ft can be obtained from Eq. (c) as

$$t = \frac{x}{u_x} = \frac{120.9}{45} = 2.69 \text{ sec}.$$

The velocity of the ball as it strikes the roof is obtained by substituting numerical values in the expression for $\dot{\mathbf{r}}$, that is,

$$\dot{\mathbf{r}} = u_x\mathbf{i} + (u_y - gt)\mathbf{j}$$

$$= 45\mathbf{i} + [60 - 32.2(2.69)]\mathbf{j} = 45\mathbf{i} - 26.6\mathbf{j}$$

$$= \underline{52.3(0.861\mathbf{i} - 0.509\mathbf{j}) \text{ fps}.} \qquad \text{Ans.}$$

PROBLEMS

Note: Neglect air resistance in all projectile problems unless otherwise stated.

8-95 The position of a point moving in the xy plane is given by

$$\mathbf{r} = 2t^3\mathbf{i} + 5t^2\mathbf{j},$$

where r is in inches when t is time in seconds. Determine
(a) The displacement during the interval $t = 1$ to $t = 3$ sec.
(b) The average velocity during the interval $t = 1$ to $t = 3$ sec.
(c) The velocity at $t = 2$ sec.
(d) The acceleration at $t = 2$ sec.

***8-96** The position of a particle moving in the xy plane is given by

$$\mathbf{r} = Bt^2\mathbf{i} + \left(\frac{C}{t^2}\right)\mathbf{j},$$

where $B = 2$ m/s^2, $C = 4$ m · s^2, and t is time in seconds. Determine
(a) The displacement during the interval $t = 1$ to $t = 2$ s.
(b) The velocity at $t = 2$ s.
(c) The acceleration at $t = 2$ s.

8-97 The position vector for a particle is

$$\mathbf{r} = 2t\mathbf{i} + (2t^2 - 3)\mathbf{j},$$

where r is measured in feet when the time, t, is in seconds. Determine
(a) The displacement of the particle during the time interval from $t = 1$ to $t = 3$ sec.
(b) The velocity of the particle when $t = 1$ sec.

8-98 The position of a particle is given by

$$\mathbf{r} = (t^2 - 2)\mathbf{i} + t^3\mathbf{j} + e^{-t}\mathbf{k},$$

where r is in inches and t is in seconds.
(a) Locate the particle when $t = 1$ sec.
(b) Determine the velocity of the particle when $t = 1$ sec.
(c) Determine the acceleration of the particle when $t = 1$ sec.

8-99 A point moves along the curve $y^2 = 25x$ (x and y in feet) in such a manner that the y component of the position vector at any time t is $5t^2\mathbf{j}$. Determine the acceleration of the point when $t = 2$ sec.

***8-100** The x coordinate of a particle moving along the path $by = x^2$ is given by $x = b^{nt}$. The constants are $b = 4$ m and $n = 0.2\ \text{s}^{-1}$. Determine the acceleration of the particle when $t = 2$ s.

8-101 The velocity of a particle moving in the xy plane is given by the expression $\mathbf{v} = (4t - 1)\mathbf{i} + 2\mathbf{j}$, where v is in feet per second when t is in seconds. The particle is located at $\mathbf{r} = (3\mathbf{i} + 4\mathbf{j})$ ft when $t = 1$ sec. Determine the equation of the path in terms of x and y.

8-102 The acceleration of a particle moving in the xy plane is given by

$$\mathbf{a} = -\omega^2(\mathbf{i}b \cos \omega t + \mathbf{j}c \sin \omega t),$$

where ω, b, and c are constants. When the time $t = 0$ the position and velocity of the particle are $\mathbf{r} = 2b\mathbf{i}$ and $\dot{\mathbf{r}} = c\omega\mathbf{j}$, respectively.

(a) Determine the equation of the path of the particle.

(b) The values of the constants are $b = 2c = 6$ in. and $\omega = (\pi/2)$ rad per sec. Calculate the position and velocity of the particle at $t = 3$ sec.

8-103 Block A in Fig. P8-103 slides in the vertical slot in member BC as BC moves along the x axis. The curved guide passes through a hole in A and thus causes A to move vertically as BC moves horizontally. The value of b in the equation is 2 in. Determine the velocity and acceleration of A when $\mathbf{x} = 5\mathbf{i}$ in., $\dot{\mathbf{x}} = 3\mathbf{i}$ ips, and $\ddot{\mathbf{x}} = 0$.

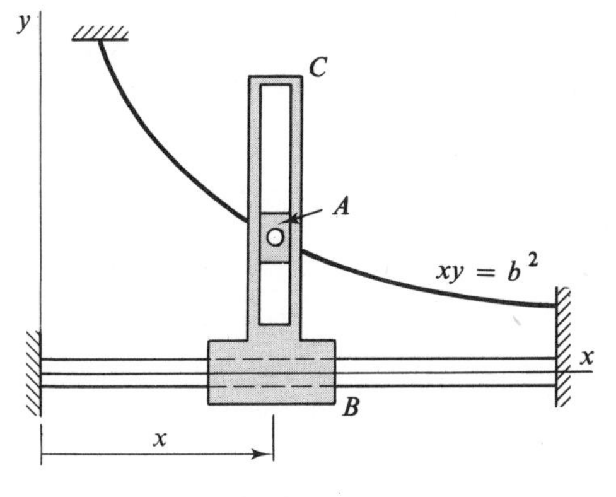

Figure P8-103

8-104 Determine the acceleration of BC in Problem 8-103 which will result in a zero vertical component of the acceleration of A. Other data for Problem 8-103 are unchanged.

***8-105** Roller A in Fig. P8-105 slides along the fixed curved track as it moves in the vertical slot in member BC. Member BC moves horizontally along the fixed track as shown.

(a) Derive equations for the velocity and acceleration of A in terms of b, x, $\dot{x}$, and $\ddot{x}$.

(b) Determine the velocity and acceleration of A when $b = 10$ cm, $\mathbf{x} = 4\mathbf{i}$ cm, $\dot{\mathbf{x}} = 10\mathbf{i}$ cm/s, and $\ddot{\mathbf{x}} = -8\mathbf{i}$ cm/s^2.

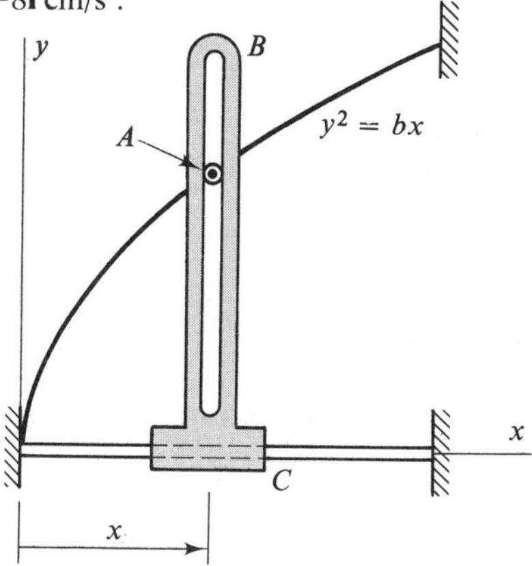

Figure P8-105

8-106 The right triangular plate in Fig. P8-106 moves in a vertical plane, with vertices B and C moving in horizontal and vertical grooves as shown. Determine, in terms of b, θ, $\dot{\theta}$, and $\ddot{\theta}$, the velocity and acceleration of point A.

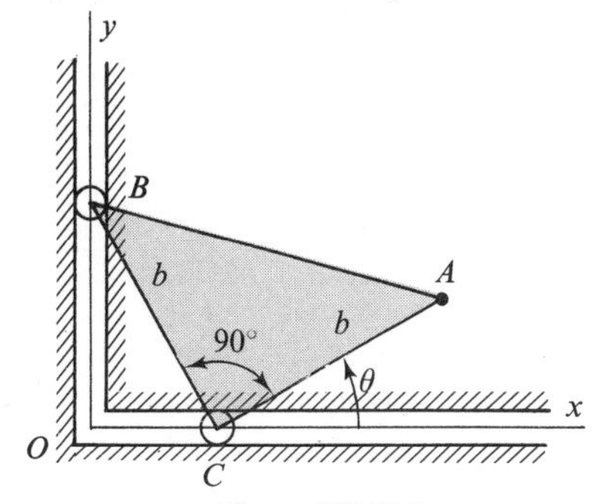

Figure P8-106

8-107 Block B in Fig. P8-107 slides along the fixed curve whose equation is $x^2 = by$ and at the same time slides in a slot in member AC. The value of b is 1 in., and when $x = 2.0$ in. the horizontal component of the velocity of B is 2.0 ips to the right and the horizontal component of its acceleration is zero. Determine the angular velocity and acceleration of AB at this instant.

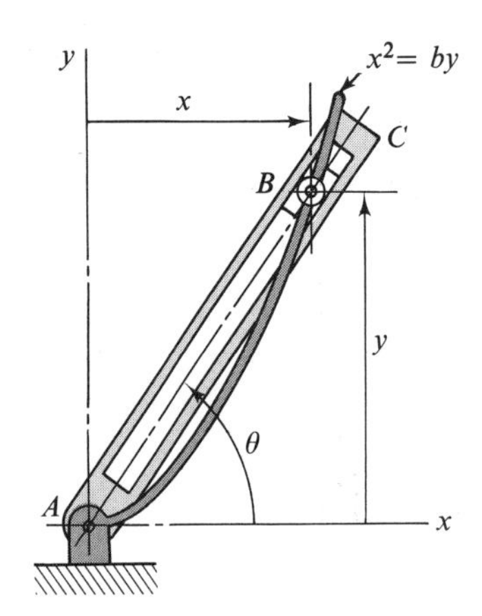

Figure P8-107

8-108 The triangular block ABC in Fig. P8-108 moves in a vertical plane with its sides AB and BC in contact with fixed points D and E, respectively. The angular velocity of the triangle is 5.0 rad per sec counterclockwise when $\tan\theta = 0.75$. Determine the velocity of point B when in this position.

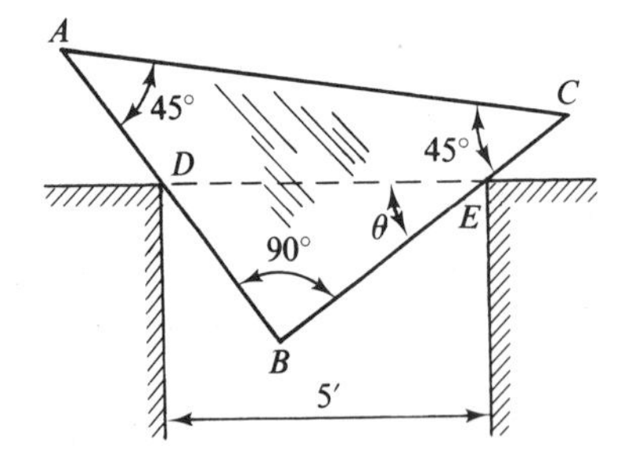

Figure P8-108

8-109 Bar AB in Fig. P8-109 moves with end A in a fixed horizontal slot while the slot in the member slides along the fixed pin C. When $\theta = 60°$, the velocity of A is $\mathbf{v}_A = -16\mathbf{i}$ ips. Determine, for this position,

(a) The angular velocity of the bar.

(b) The velocity of point B.

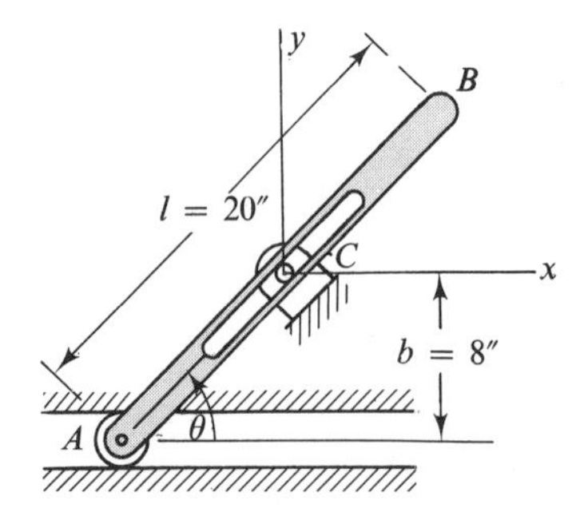

Figure P8-109

8-110 Determine the acceleration of pin A in Problem 8-109 which will result in zero angular acceleration of AB with the conditions given.

8-111 Determine the angular acceleration of the bar in Problem 8-109 which will result in zero acceleration of pin A.

8-112 In Fig. P8-112, AB is a slot fixed in the xy plane. Member C slides horizontally along the x axis and contains a curved slot whose equation is $bx_1 = y_1^2$, where the x_1y_1 axes move with C. The roller P extends through both slots. The position of O_1 from the xy axes is x_2 as shown. Prove that the velocity of P is given by

$$\dot{\mathbf{r}} = (\mathbf{i} - \mathbf{j})\dot{x}_2\left[9 - \left(\frac{4x_2}{b}\right)\right]^{-0.5}.$$

8-113 A ball is thrown horizontally from a 240-ft building with an initial velocity of 80 fps.

(a) Determine the distance from the bottom of the building to the point where the ball hits the ground.

(b) What is the height of the ball from the ground when it is 250 ft from the building?

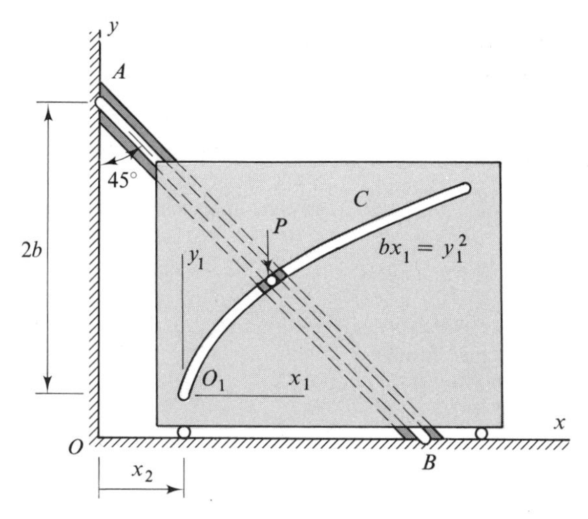

Figure P8-112

8-114 A firing table for a 105-mm howitzer gives a range of 5200 yd for an elevation of 443 mil with an initial velocity of 875 fps. Given that 1600 mil = 90°, determine the reduction in range caused by air resistance.

8-115 A ball is thrown at a building which is 200 ft from the thrower. The initial speed of the ball is 100 fps, and the ball strikes the side of the building 3.0 sec after being thrown. Determine the height (above the initial position) at which the ball strikes the building.

8-116 An airplane at 20,000-ft altitude (see Fig. P8-116) is flying along a course which passes over an antiaircraft gun. The speed of the plane is 800 fps. The gun angle is 60°, and the muzzle velocity of the projectile is 2000 fps. Determine the angle, β, of the line of sight at which the projectile should be fired if it is to hit the plane during the upward flight of the projectile.

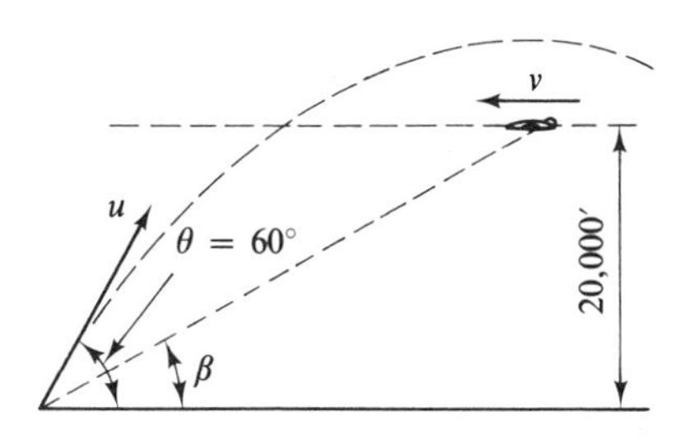

Figure P8-116

8-117 A baseball is hit 3 ft above the ground and just clears the 3-ft center-field fence 400 ft away. The initial speed of the ball is 150 fps. Determine the angle of the initial velocity of the ball (two values) and its maximum elevation corresponding to each angle.

8-118 A batter hit a baseball 4 ft above the ground toward the 10-ft-high center-field fence 455 ft away. The ball left the bat with a velocity of 125 fps with a slope of 3 vertical to 4 horizontal. The center fielder can catch a ball 9 ft above the ground. Did the batter fly out, hit a double against the fence above the fielder's glove, or hit a home run? Show all calculations.

***8-119** A mortar is fired with a muzzle velocity of 250 m/s. A radar range station indicates that the projectile has an elevation of 200 m 23 s after firing. Determine the horizontal distance traveled by the projectile during the 23-s interval and the speed of the projectile at the end of the interval.

***8-120** A low-speed projectile with a muzzle velocity of 100 m/s is fired at a target which is 600 m horizontally and 150 m vertically from the gun. Determine the angle of elevation of the gun necessary for the projectile to hit the target.

8-121 A three-stage rocket fired from Cape Kennedy was known to have reached a speed of 10,000 fps with an unknown slope at an altitude of 150,000 ft when the third stage burned out. The FM telemeter signal faded out exactly 600 sec after burnout as the rocket fell below the eastern horizon. Assume that $g = 30.0\ \text{fps}^2$ (an average value) and neglect curvature and rotation of the earth. Calculate

(a) The maximum altitude attained by the rocket.

(b) The horizontal distance traveled.

8-122 A golf ball is driven from a tee which is 50 ft above a level fairway. The maximum height of the ball in flight above the tee is 70 ft, and the horizontal distance traveled by the ball before it strikes the fairway is 700 ft. Determine the velocity of the ball as it hits the fairway.

8-123 The stream from a fire hose is directed toward a building as shown in Fig. P8-123 and is to hit point P. Show that the minimum required nozzle velocity will occur when θ is given by the equation $\tan 2\theta = -x/y$.

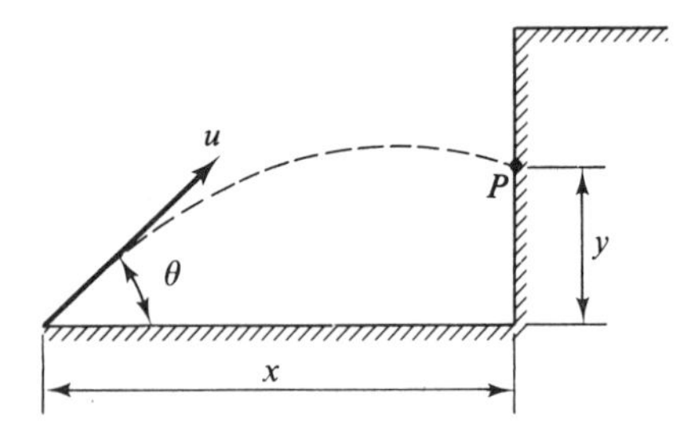

Figure P8-123

8-124 Determine the minimum required nozzle velocity in Problem 8-123 if the stream is to hit point P when $x = 80$ ft and $y = 50$ ft.

8-125 A stream of water falls on a sloping plane as shown in Fig. P8-125. Show that the distance q where the water strikes the plane will be greatest when the angle θ satisfies the equation $\tan 2\theta = -\cot \beta$.

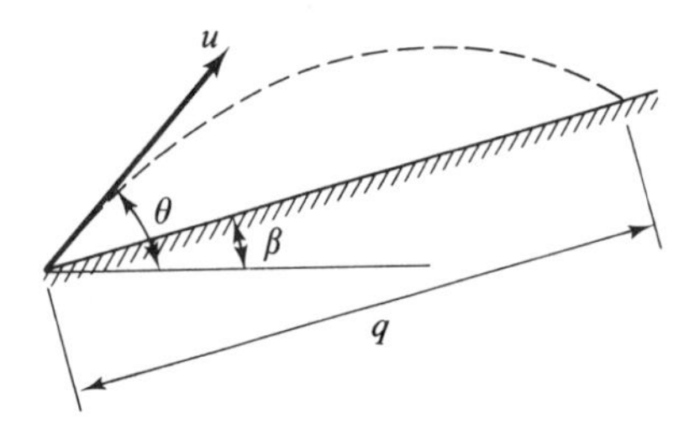

Figure P8-125

8-126 Determine the minimum speed a jeep must have at A in Fig. P8-126 to clear the stream and land on the opposite bank at B.

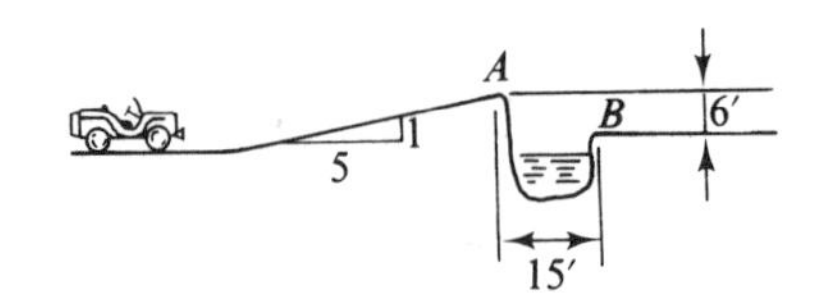

Figure P8-126

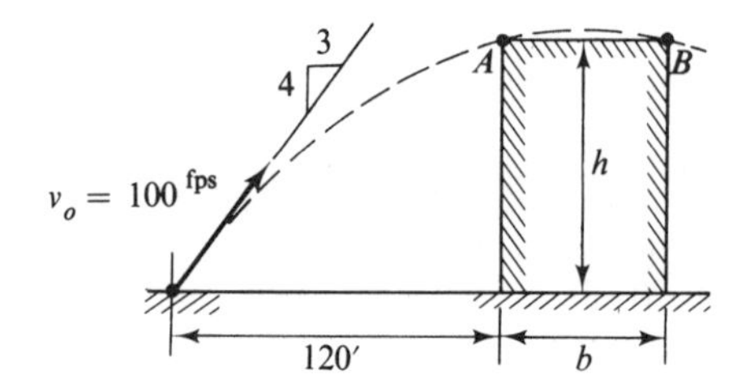

Figure P8-127

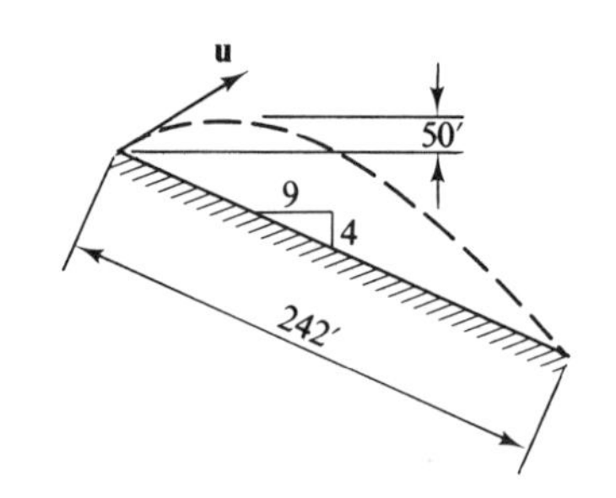

Figure P8-128

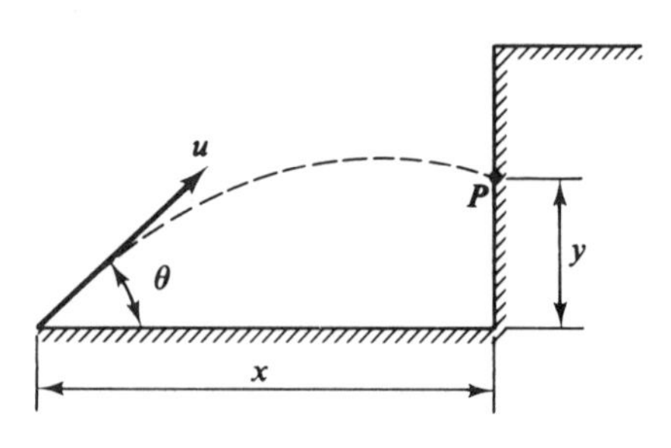

Figure P8-129

8-127 A ball is thrown with an initial velocity of 100 fps as shown in Fig. P8-127 and just clears the edges A and B of a building 120 ft away. Determine the height, h, and the width, b, of the building.

8-128 Determine the initial velocity u of a ball thrown down the slope as shown in Fig. P8-128.

8-129 A fire hose is directed at a building from a point x ft away as shown in Fig. P8-129. The nozzle velocity is u fps.
(a) Show that the maximum value of y will result when $\tan \theta = u^2/gx$.
(b) Determine the velocity of the water as it hits the building at the maximum height when $u = 100$ fps and $x = 120$ ft.

8-8 CURVILINEAR MOTION—NORMAL AND TANGENTIAL COMPONENTS

Curvilinear motion was analyzed by means of rectangular components in Art. 8-7, and in certain types of problems this analysis provides the simplest and most direct solution. Rectangular components of velocity and acceleration, however, fail to relate the components directly to the path being traversed, and this may be a disadvantage for some problems. For certain problems in curvilinear motion, particularly circular motion of a particle, tangential and normal components of velocity and acceleration may be more convenient than rectangular components.

The position of a particle which moves along a curved path is completely specified when the path is known and when the distance, measured along the path, from some fixed point on the path to the particle is known as a function of time. The position of particle P in Fig. 8-19(a) as it moves along the plane curve is given by the distance s measured along the curve from the fixed point A.

It is convenient to introduce unit vectors in the tangential and normal directions, as shown in Fig. 8-19(a). The unit vector $\mathbf{e}_n$ is along

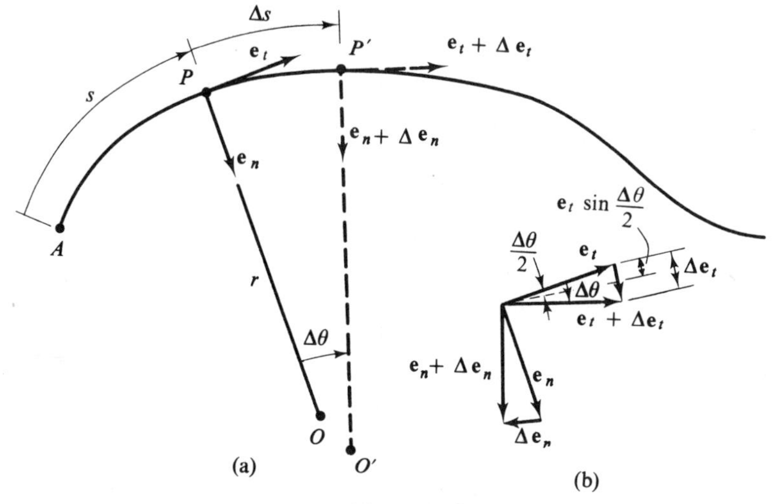

Figure 8-19

the principal radius of curvature[4] at the point, and the positive direction is from the curve toward the center of curvature at O. The unit vector $\mathbf{e}_t$ is tangent to the path at the point, and the positive direction is in the direction of increasing s as shown. *Since these unit vectors will change in direction, they have time derivatives different from zero.*

The velocity of a particle is always tangent to the path as shown in Art. 8-3. The magnitude of the velocity can be obtained by analyzing Fig. 8-19(a). As the particle moves from P to P' it travels a distance Δs in a time interval Δt and has an average speed of $\Delta s/\Delta t$. The magnitude of the instantaneous velocity is the limit of this ratio as Δt approaches zero; that is,

$$v = \lim_{\Delta t \to 0} \frac{\Delta s}{\Delta t} = \frac{ds}{dt} = \dot{s}$$

and

$$\mathbf{v} = \dot{s}\mathbf{e}_t. \tag{8-9}$$

The acceleration of the particle is defined as the time rate of change of the velocity. The velocity from Eq. (8-9) is the product of two factors, and its derivative is

$$\mathbf{a} = \dot{\mathbf{v}} = \ddot{s}\mathbf{e}_t + \dot{s}\dot{\mathbf{e}}_t.$$

The quantity $\dot{\mathbf{e}}_t$ is the time rate of change of $\mathbf{e}_t$ and can be evaluated as indicated in Fig. 8-19(b). Since $\mathbf{e}_t$ has a constant magnitude (unity), it can change only in direction. The magnitude of $\Delta\mathbf{e}_t$ is

$$\Delta e_t = 2\left(e_t \sin\frac{\Delta\theta}{2}\right) = 2(1)\sin\frac{\Delta\theta}{2},$$

and as $\Delta\theta$ becomes very small $\sin(\Delta\theta/2)$ is approximately equal to $\Delta\theta/2$ and Δe_t becomes

$$\Delta e_t = 2\left(\frac{\Delta\theta}{2}\right) = \Delta\theta.$$

The direction of $\Delta\mathbf{e}_t$ is perpendicular to $\mathbf{e}_t$, that is, in the direction of $\mathbf{e}_n$, and the time derivative of $\mathbf{e}_t$ is

$$\dot{\mathbf{e}}_t = \lim_{\Delta t \to 0} \frac{\Delta\mathbf{e}_t}{\Delta t} = \lim_{\Delta t \to 0} \frac{\Delta\theta}{\Delta t}\mathbf{e}_n = \dot{\theta}\mathbf{e}_n. \tag{8-10}$$

Note that $\dot{\theta}$ must be given in radians per unit of time.

Although it is not needed for this analysis, it is useful to note that $\mathbf{e}_n$ can be differentiated in the same manner to give

$$\dot{\mathbf{e}}_n = -\dot{\theta}\mathbf{e}_t \tag{8-10a}$$

since the direction of $\Delta\mathbf{e}_n$ is opposite that of $\mathbf{e}_t$.

[4] The following discussion is limited to coplanar curvilinear motion, but the same reasoning applies for noncoplanar motion, provided that the principal radius of curvature is used.

The linear acceleration of the particle thus becomes

$$\mathbf{a} = \dot{\mathbf{v}} = \ddot{s}\mathbf{e}_t + \dot{s}\dot{\theta}\mathbf{e}_n. \tag{8-11}$$

It is sometimes helpful to relate angular and linear quantities in the preceding equations. Points O and O' in Fig. 8-19(a) are the centers of curvature of the curve at P and P' respectively, and r is the radius of curvature at P. It can be observed that as $\Delta\theta$ approaches zero, O' will approach O and ds will be equal to $r\,d\theta$,[5] where $d\theta$ is in radians.

When this quantity is divided by dt, it becomes

$$v = \frac{ds}{dt} = r\frac{d\theta}{dt} = r\dot{\theta} = \dot{s}, \tag{8-12}$$

and Eq. (8-11) can be written as

$$\mathbf{a} = \ddot{s}\mathbf{e}_t + \frac{\dot{s}^2}{r}\mathbf{e}_n = \ddot{s}\mathbf{e}_t + r\dot{\theta}^2\mathbf{e}_n = \mathbf{a}_t + \mathbf{a}_n, \tag{8-13}$$

where $\mathbf{a}_t$ and $\mathbf{a}_n$ are the tangential and normal components, respectively, of the acceleration of the particle.

It should be noted that the velocity is always tangent to the path and thus has no normal component. The velocity of the point has a magnitude $\dot{s}$ in the direction $\mathbf{e}_t$—see Eq. (8-9). Note that the magnitude of the tangential component of the acceleration, $\ddot{s}$, results from a change in the magnitude of v. The normal component of the acceleration is due to the change of the direction of the velocity as indicated by $\dot{\mathbf{e}}_t$. The normal component is always directed from the particle toward the center of curvature of the path and is zero only when the velocity is zero or is not changing in direction (rectilinear motion). The tangential component is zero when the speed is constant.

For the special case of motion on a circular path, r is constant and from Eqs. (8-12) and (8-5)

$$\ddot{s} = r\ddot{\theta},$$

which makes it possible to write Eq. (8-13) as

$$\mathbf{a} = r\ddot{\theta}\mathbf{e}_t + r\dot{\theta}^2\mathbf{e}_n.$$

Sometimes it is convenient to change the independent variable in the expression for a_t in Eq. (8-13) from t to s. This can be accomplished by writing

$$a_t = \ddot{s} = \dot{v} = \frac{dv}{dt} = \frac{dv}{ds}\frac{ds}{dt} = v\frac{dv}{ds},$$

[5] The radius of curvature is

$$r = \frac{[1 + (dy/dx)^2]^{3/2}}{d^2y/dx^2}$$

as derived in any standard calculus text.

or in vector form

$$\mathbf{a}_t = v\frac{dv}{ds}\mathbf{e}_t.$$

The following examples illustrate the use of the normal and tangential components of acceleration.

EXAMPLE 8-16

The position of a particle moving on a circular path with a radius of 32 ft varies according to the law $s = 3t^2 + 4t$, where s is the distance in feet from a fixed point to the particle measured along the path and t is the time in seconds. The line from the center of the path to the particle is turning counterclockwise when $t = 1$ sec, and the particle is at the top of the path when $t = 2$ sec. Determine the acceleration of the particle when $t = 2$ sec.

SOLUTION

The positive direction of motion along the path can be determined from the value of the velocity when $t = 1$ sec. The magnitude of the velocity is

$$v = \frac{ds}{dt} = 6t + 4.$$

Thus v is positive when $t = 1$ sec and P is moving counterclockwise along the path; therefore counterclockwise is the positive direction. The magnitude of the normal component of the acceleration is v^2/r, and when $t = 2$ sec the magnitude of the velocity is

$$v = 6(2) + 4 = 16 \text{ fps}.$$

The normal component of the acceleration is directed from the particle toward the center of the circular path, which is downward when $t = 2$ sec. Therefore

$$\mathbf{a}_n = \frac{v^2}{r}\mathbf{e}_n = \frac{16^2}{32}\mathbf{e}_n = 8 \text{ fps}^2 \downarrow.$$

The magnitude of the tangential component of the acceleration is

$$a_t = \frac{dv}{dt} = 6 \text{ fps}^2,$$

which is positive for all values of t and shows that the velocity is increasing in magnitude and the direction of the tangential component of acceleration is the same as the direction of the velocity, that is, to the left when $t = 2$ sec. The acceleration when $t = 2$ sec is

$$\mathbf{a} = \mathbf{a}_n + \mathbf{a}_t = \overset{\downarrow}{8} + \overset{\leftarrow}{6} = 10 \text{ fps}^2 \; \swarrow_{6}^{8}.$$ Ans.

EXAMPLE 8-17

Example 8-15 analyzed the motion of a ball which was thrown toward a building. Determine the normal and tangential components of its acceleration and the radius of curvature of the trajectory
(a) When the ball is at the top of the trajectory.
(b) Just before the ball hits the roof.

SOLUTION

(a) At the top of the path the velocity is

$$\mathbf{v} = 45\mathbf{i}\text{ fps},$$

and the acceleration is

$$\mathbf{a} = -32.2\mathbf{j}\text{fps}^2;$$

therefore

$$a_t = \underline{0} \qquad \text{Ans.}$$

and

$$\mathbf{a}_n = \underline{-32.2\mathbf{j}\text{ fps}^2}. \qquad \text{Ans.}$$

The magnitude of the normal component of the acceleration [see Eq. (8-13)] is

$$a_n = \frac{\dot{s}^2}{r} = \frac{v^2}{r} = 32.2$$

from which

$$r = \frac{(45)^2}{32.2} = \underline{62.9\text{ ft.}} \qquad \text{Ans.}$$

(b) Just before the ball hits the roof its velocity (see Example 8-15) is

$$\mathbf{v} = 45\mathbf{i} - 26.5\mathbf{j}$$

or

$$v\mathbf{e}_t = 52.2(0.862\mathbf{i} - 0.508\mathbf{j})\text{ fps},$$

and the acceleration is

$$\mathbf{a} = -32.2\mathbf{j}\text{ fps}^2.$$

The acceleration can be resolved into n and t components by various methods. For example,

$$\begin{aligned}\mathbf{a}_t &= (\mathbf{a}\cdot\mathbf{e}_t)\mathbf{e}_t \\ &= (-32.2\mathbf{j})\cdot(0.862\mathbf{i} - 0.508\mathbf{j})(0.862\mathbf{i} - 0.508\mathbf{j}) \\ &= \underline{(14.10\mathbf{i} - 8.31\mathbf{j})\text{ fps}^2} \\ &= \underline{16.38\mathbf{e}_t\text{ fps}^2}. \qquad \text{Ans.}\end{aligned}$$

and

$$\begin{aligned}\mathbf{a}_n &= \mathbf{a} - \mathbf{a}_t = -32.2\mathbf{j} - (14.10\mathbf{i} - 8.31\mathbf{j}) \\ &= \underline{(-14.10\mathbf{i} - 23.89\mathbf{j})\text{ fps}^2} \\ &= \underline{27.7\mathbf{e}_n\text{ fps}^2}. \qquad \text{Ans.}\end{aligned}$$

The radius curvature of the trajectory just before impact is

$$r = \frac{v^2}{a_n} = \frac{(52.2)^2}{27.7}$$

$$= \underline{98.4 \text{ ft.}} \qquad \text{Ans.}$$

PROBLEMS

8-130 A particle travels in a vertical plane along a 10-ft-diameter circular path. The distance along the path from the bottom of the circle to the particle is $s = t^2 - 3t$, where s is in feet and t is in seconds (counterclockwise motion is positive). Determine the velocity and acceleration of the particle when $t = 3$ sec.

8-131 The velocity of a particle P moving along a circular path varies according to the expression $v = 2t^3 - 6$, where v is the magnitude of the velocity in feet per second and t is time in seconds. The particle is moving downward in the position shown in Fig. P8-131 when $t = 2$ sec. Determine the acceleration of the particle at this instant.

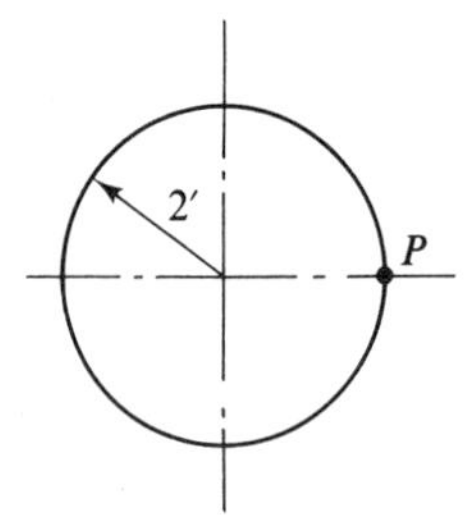

Figure P8-131

***8-132** The particle P in Fig. P8-132 has a tangential acceleration whose magnitude is given by the expression $a_t = 12t - 8$, where a_t is in meters per second per second and t is in seconds. Line OP starts from rest with a clockwise angular acceleration when t is zero. When $t = 2$ s the particle is at point A. Determine the linear acceleration of P when $t = 2$ s.

8-133 The velocity of a point moving along a 40-ft-diameter circular track is given by the equation $v^2 = 16s$, where v is in feet per second and s is the distance in feet from the origin to the point (measured along the track). Determine the magnitude of the acceleration of the point when $s = 4$ ft.

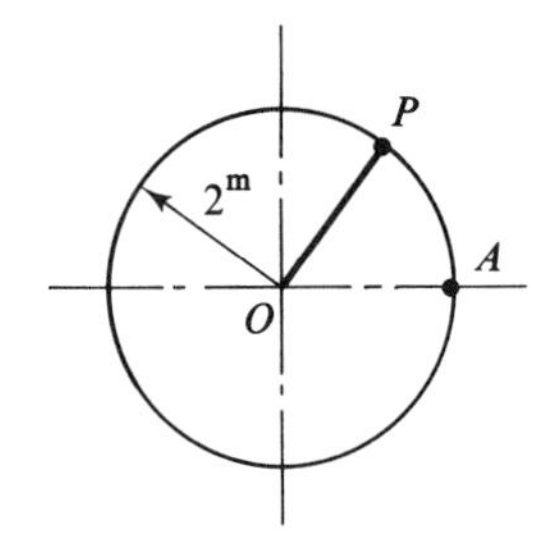

Figure P8-132

8-134 Point A moves on a circle with decreasing magnitude of velocity. At a given instant, the point has the acceleration and position shown in Fig. P8-134. Determine the velocity of point A.

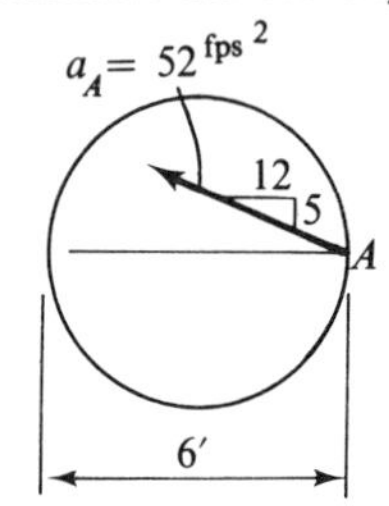

Figure P8-134

8-135 A small cart starts from rest at point A in Fig. P8-135 and moves counterclockwise along the circular track. The speed of the cart increases uniformly at the rate of 25 ips^2. Determine the velocity of the cart when the magnitude of the acceleration is 65 ips^2.

***8-136** The magnitude of the velocity of a particle changes uniformly from 14 m/s downward at A to 4 m/s upward at B (see Fig. P8-136) as the particle moves along the path from A to B. Determine the acceleration of the particle when it is at point C.

8-137 The body in Fig. P8-137 rotates about point O. When line OA is in the position shown; point A has a speed of 8.0 fps and an acceleration of 34.0 fps^2 downward. Determine the angular acceleration of line OA.

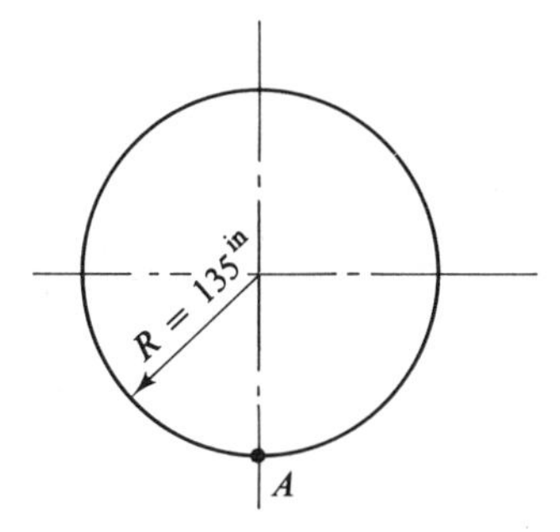

Figure P8-135

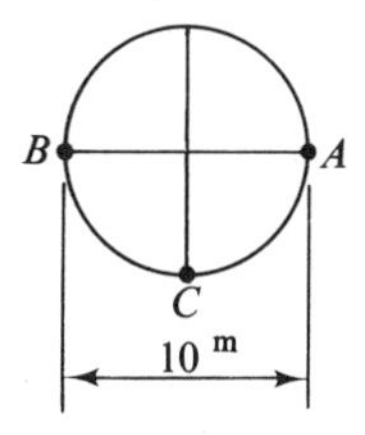

Figure P8-136

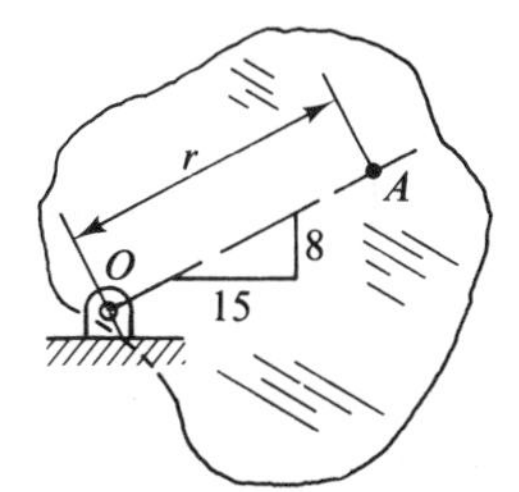

Figure P8-137

8-138 The position of a point moving in the xy plane is given by the equation $\mathbf{r}=2t^3\mathbf{i}+5t^2\mathbf{j}$, where r and t are in inches and seconds, respectively. Determine, for $t=2$ sec,
(a) The tangential component of the acceleration of the point.
(b) The normal component of the acceleration of the point.
(c) The radius of curvature of the path.

***8-139** The position of a particle is given by

$$\mathbf{r}=2t^2\mathbf{i}+\left(\frac{4}{t^2}\right)\mathbf{j},$$

where r and t are in meters and seconds, respectively. Determine, when $t=1$ s,
(a) The magnitudes of the normal and tangential components of the acceleration of the particle.
(b) The radius of curvature of the path.

8-140 A ball is thrown from the top of a 240-ft building with a horizontal velocity of 80 fps. Determine the radius of curvature of the trajectory of the ball when the ball is 250 ft from the building.

8-141 A particle moves along the curve $y=0.2x^2$, where x and y are in feet. When the particle is at the point $x=2.4$ ft, the acceleration is $-125\mathbf{i}$ fps^2 and the speed of the particle is decreasing. Determine the tangential and normal components of the acceleration of the particle and show them on a sketch.

8-142 Solve Problem 8-141 if the acceleration at $x=2.4$ ft is $75\mathbf{j}$ fps^2.

***8-143** A particle starts from rest at $x=0$ and moves to the right along the curve $y=(36/\pi)\cos(\pi x/24)$, where x and y are in meters. The speed of the particle increases uniformly with time. When $x=4$ m the acceleration of the particle is $-16\mathbf{j}$ m/s^2. Determine the elapsed time.

8-144 In Problem 8-98 the position of a particle was given as

$$\mathbf{r}=(t^2-2)\mathbf{i}+t^3\mathbf{j}+e^{-t}\mathbf{k},$$

where r and t are in inches and seconds. Determine, for $t=1$ sec,
(a) The tangential and normal components of the acceleration of the particle.
(b) The position vector from the origin to the center of curvature of the path.

8-9 CURVILINEAR MOTION—RADIAL AND TRANSVERSE COMPONENTS

When a particle travels on a curved path in a plane, it is sometimes convenient to express its position in polar coordinates. The radial distance r from a fixed point and the angular position θ from a fixed reference line, as shown in Fig. 8-20(a), expressed as functions of time, completely determine the location of the particle. When using polar coordinates, the velocity and acceleration are expressed in terms of radial and transverse components. The radial component is along the radius vector to the particle, and the transverse component is perpendicular to the radius vector. Figure 8-20(a) shows unit vectors $\mathbf{e}_r$ and $\mathbf{e}_\theta$ in the positive directions of increasing r and θ.

The position vector for the particle P of Fig. 8-20(a) can be written as

$$\mathbf{r} = r\mathbf{e}_r, \tag{8-14}$$

where r is the radial polar coordinate.

The velocity of the particle is defined as the time rate of change of position of the particle. The unit vectors $\mathbf{e}_r$ and $\mathbf{e}_\theta$ are differentiated in the same manner as were $\mathbf{e}_n$ and $\mathbf{e}_t$ in the preceding section. From Fig. 8-20(b), it is seen that

$$\dot{\mathbf{e}}_r = \lim_{\Delta t \to 0} \frac{(\Delta\theta)\mathbf{e}_\theta}{\Delta t} = \dot{\theta}\mathbf{e}_\theta, \tag{8-15}$$

and in the same manner

$$\dot{\mathbf{e}}_\theta = -\dot{\theta}\mathbf{e}_r. \tag{8-15a}$$

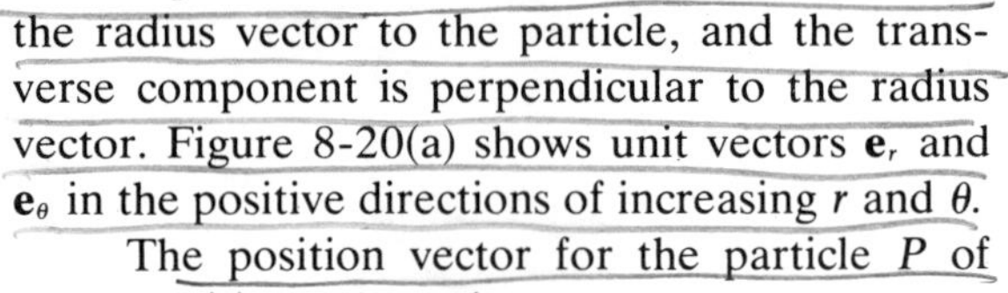

(a)

(b)

Figure 8-20

The linear velocity of the particle P, as obtained by differentiation of Eq. (8-14), is

$$\mathbf{v} = \dot{\mathbf{r}} = \dot{r}\mathbf{e}_r + r\dot{\mathbf{e}}_r = \dot{r}\mathbf{e}_r + r\dot{\theta}\mathbf{e}_\theta = \mathbf{v}_R + \mathbf{v}_T, \tag{8-16}$$

where $\mathbf{v}_R$ and $\mathbf{v}_T$ are the radial and transverse components, respectively, of the velocity.

The linear acceleration of P is the time rate of change of velocity and, from Eq. (8-16), is

$$\left.\begin{aligned} \mathbf{a} = \ddot{\mathbf{r}} &= \ddot{r}\mathbf{e}_r + \dot{r}\dot{\mathbf{e}}_r + \dot{r}\dot{\theta}\mathbf{e}_\theta + r\ddot{\theta}\mathbf{e}_\theta + r\dot{\theta}\dot{\mathbf{e}}_\theta \\ &= \ddot{r}\mathbf{e}_r + \dot{r}\dot{\theta}\mathbf{e}_\theta + \dot{r}\dot{\theta}\mathbf{e}_\theta + r\ddot{\theta}\mathbf{e}_\theta - r\dot{\theta}^2\mathbf{e}_r \\ &= (\ddot{r} - r\dot{\theta}^2)\mathbf{e}_r + (r\ddot{\theta} + 2\dot{r}\dot{\theta})\mathbf{e}_\theta \\ &= (\ddot{r} - r\dot{\theta}^2)\mathbf{e}_r + \frac{1}{r}\frac{d}{dt}(r^2\dot{\theta})\mathbf{e}_\theta = \mathbf{a}_R + \mathbf{a}_T, \end{aligned}\right\} \tag{8-17}$$

where $\mathbf{a}_R$ and $\mathbf{a}_T$ are the radial and transverse components, respectively, of the acceleration. The equivalence of the two forms for $\mathbf{a}_T$ can be verified by expanding the second form. The second form is often useful when it is necessary to integrate the tangential component of the acceleration.

The following example illustrates the use of polar coordinates for finding the velocity and acceleration of a particle moving on a curved path in a plane.

EXAMPLE 8-18

Pin A in Fig. 8-21 moves along the curved fixed slot and also along the slot in member OB as OB rotates about O. The equation of the curved slot is $r = b + c \cos \theta$.

(a) Derive expressions for the velocity and acceleration of A in terms of the angular position, velocity, and acceleration of OB.

(b) Determine the velocity and acceleration of A when $b = 3$ cm, $c = 2$ cm, $\theta = 90°$, $\dot{\theta} = 5$ rad/s counterclockwise, and $\ddot{\theta} = 20$ rad/s^2 clockwise.

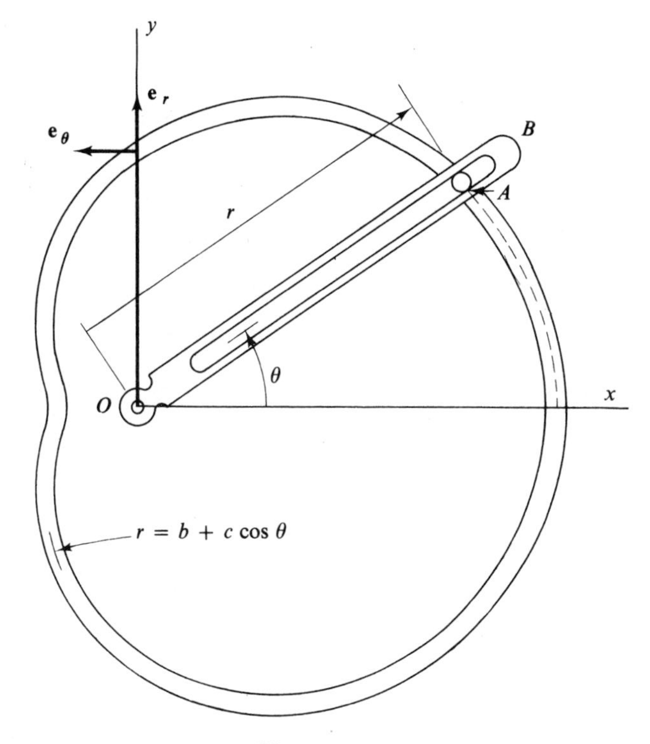

Figure 8-21

SOLUTION

(a) The position of point A, in polar coordinates, is

$$\mathbf{r} = r\mathbf{e}_r = (b + c \cos\theta)\mathbf{e}_r.$$

The velocity and acceleration of A can be obtained by successive differentiations of the position function; thus the velocity is

$$\underline{\mathbf{v} = \dot{\mathbf{r}} = -c \sin\theta\,\dot{\theta}\mathbf{e}_r + (b + c\cos\theta)\dot{\theta}\mathbf{e}_\theta.} \qquad \text{Ans.}$$

A second differentiation will give the acceleration of A. This acceleration can also be obtained by differentiating the expression for the magnitude of r and substituting into Eq. (8-17) as follows:

$$r = b + c\cos\theta$$

$$\dot{r} = -c\sin\theta\,\dot{\theta}$$

$$\ddot{r} = -c\cos\theta\,\dot{\theta}^2 - c\sin\theta\,\ddot{\theta}$$

from which

$$\begin{aligned}\mathbf{a} &= (\ddot{r} - r\dot{\theta}^2)\mathbf{e}_r + (r\ddot{\theta} + 2\dot{r}\dot{\theta})\mathbf{e}_\theta \\ &= [-c\cos\theta\,\dot{\theta}^2 - c\sin\theta\,\ddot{\theta} - (b + c\cos\theta)\,\dot{\theta}^2]\mathbf{e}_r \\ &\quad + [(b + c\cos\theta)\,\ddot{\theta} + 2(-c\sin\theta\,\dot{\theta})\,\dot{\theta}]\mathbf{e}_\theta \\ &= [-(b + 2c\cos\theta)\,\dot{\theta}^2 - c\sin\theta\,\ddot{\theta}]\mathbf{e}_r \\ &\quad + [(b + c\cos\theta)\,\ddot{\theta} - 2c\sin\theta\,\dot{\theta}^2]\mathbf{e}_\theta \qquad \text{Ans.}\end{aligned}$$

(b) From the given data $b = 3$ cm, $c = 2$ cm, $\theta = 90°$, $\dot{\theta} = 5$ rad/s, and $\ddot{\theta} = -20$ rad/s^2, which give

$$r = 3 - 2(\cos 90°) = 3 \text{ cm},$$

$$\dot{r} = -2(\sin 90°)(5) = -10 \text{ cm/s},$$

and

$$\ddot{r} = -2(\cos 90°)(5)^2 - 2(\sin 90°)(-20) = +40 \text{ cm/s}^2.$$

The directions of $\mathbf{e}_r$ and $\mathbf{e}_\theta$ for $\theta = 90°$ are shown on Fig. 8-21. The velocity of A can be obtained by substituting numerical data into either Eq. (8-16) or in the result of part (a), either of which gives

$$\begin{aligned}\mathbf{v} &= \dot{r}\mathbf{e}_r + r\dot{\theta}\mathbf{e}_\theta \\ &= -10\mathbf{e}_r + 3(5)\mathbf{e}_\theta \\ &= -10\mathbf{e}_r + 15\mathbf{e}_\theta = 18.03 \text{ cm/s} \angle 33.7° \qquad \text{Ans.}\end{aligned}$$

Similarly, the acceleration of A is

$$\begin{aligned}\mathbf{a} &= (\ddot{r} - r\dot{\theta}^2)\mathbf{e}_r + (r\ddot{\theta} + 2\dot{r}\dot{\theta})\mathbf{e}_\theta \\ &= [40 - 3(5)^2]\mathbf{e}_r + [3(-20) + 2(-10)(5)]\mathbf{e}_\theta \\ &= -35\mathbf{e}_r - 160\mathbf{e}_\theta = 163.8 \text{ cm/s}^2 \angle 12.34° \qquad \text{Ans.}\end{aligned}$$

The velocity and acceleration are shown on Figs. 8-22(a) and 8-22(b), respectively.

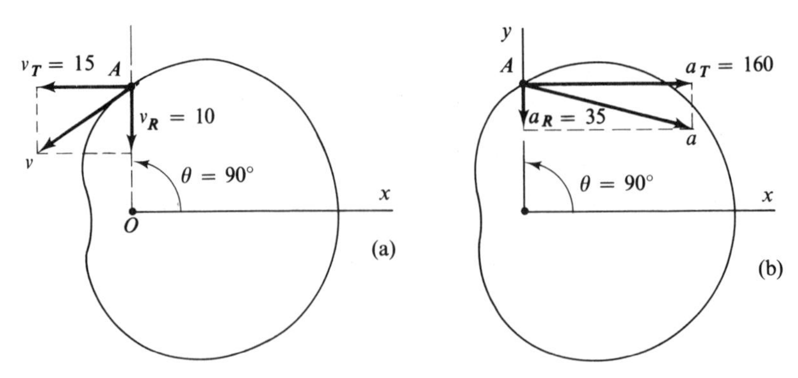

Figure 8-22

PROBLEMS

8-145 Blocks A and B in Fig. P8-145 are pinned together, and A slides along the fixed curved rod, while B slides along member OC. Rod OC rotates about point O, and the equation of the curved rod is

$$r = 10 \sin 2\theta \qquad (r \text{ is in inches}).$$

(a) When $\theta = 45°$, $\dot{\theta} = 10$ rad per sec counterclockwise. Determine the velocity of block A.
(b) When $\theta = 60°$ the speed of A is 100 ips. Determine the angular velocity of OC.

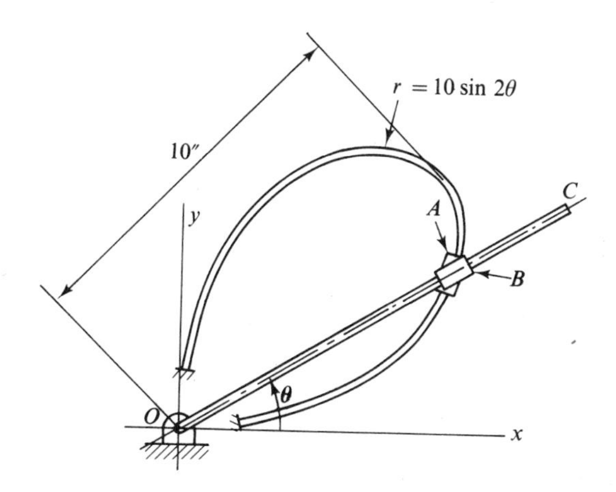

Figure P8-145

8-146 Determine the radial and transverse components of the acceleration of block A in Problem 8-145 when $\theta = 30°$, $\dot{\theta} = 10$ rad per sec counterclockwise, and $\ddot{\theta} = 0$.

***8-147** In Example 8-18 let $b = c = 5$ cm. When $\theta = 45°$ the radial component of the velocity of A is 10 cm/s from O toward B. Determine the velocity of A.

***8-148** In addition to the data in Problem 8-147 the radial component of the acceleration of A is 150 cm/s^2 from B toward O. Determine the acceleration of A.

8-149 Particle A in Fig. P8-149 travels counterclockwise along the circular path with a constant speed v.

(a) Determine the radial and transverse components of the acceleration of A when $h = 0.8b$.
(b) Show that the result from part (a) is the same as the *normal* component of the acceleration of A.

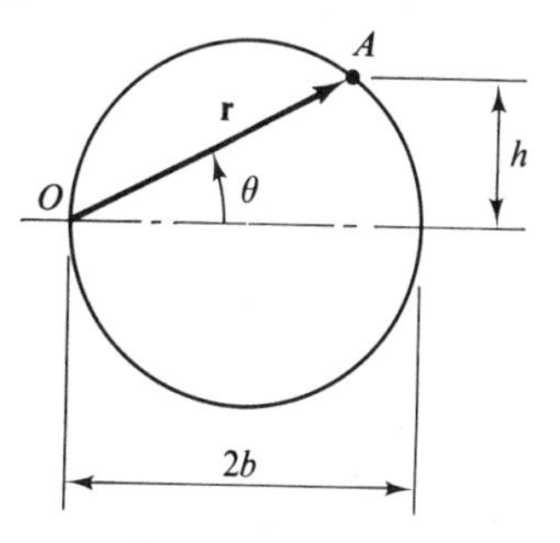

Figure P8-149

8-150 The circular table in Fig. P8-150 rotates in a horizontal plane, while the ball B slides along a slot in the table. At a given instant, the ball is 14 in. from O and is moving outward at the rate of 6 ips, while the plate has an angular velocity of 2 rad per sec and an angular acceleration of 8 rad per sec^2, both clockwise. Determine

(a) The magnitude of the acceleration of B if $\ddot{r} = 0$.

(b) The value of $\ddot{r}$ which will result in the minimum value of a_B.

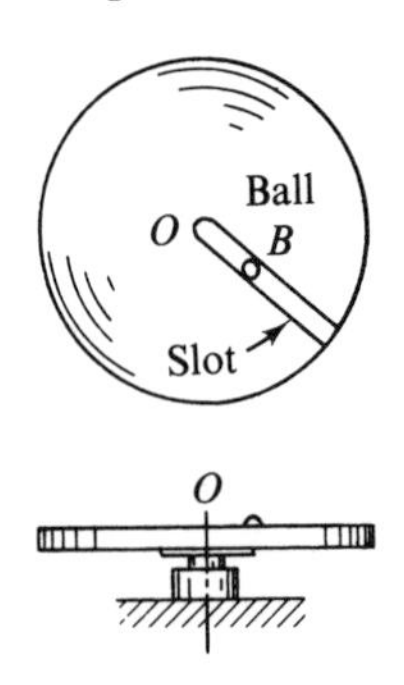

Figure P8-150

***8-151** The ring in Fig. P8-151 slides along the rotating rod OB. In the position shown $r = 0.3$ m, $\tan\theta = \frac{4}{3}$, r is increasing at the rate of 0.5 m/s, and $\dot{r}$ is decreasing at 0.6 m/s^2. Member AB has an angular velocity of 2 rad/s clockwise and an angular acceleration of 5 rad/s^2 clockwise.

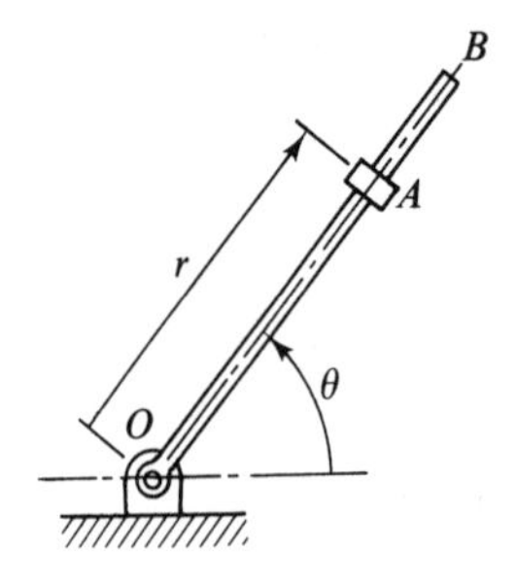

Figure P8-151

(a) Set up an expression for the vertical component of the position of A and determine the vertical component of the acceleration of A by differentiation.

(b) Determine the radial and transverse components of the acceleration of A and verify the answer to part (a).

8-152 The equation of the spiral in Fig. P8-152 is $(r/b)^2 = \theta$, where r and b are in inches and θ is in radians. The angle θ is equal to $6t$, where t is time in seconds. Point P is the intersection of line OA and the spiral. When $b = 2$ in. and $t = \pi/12$ sec, determine

(a) The velocity of P.

(b) The acceleration of P.

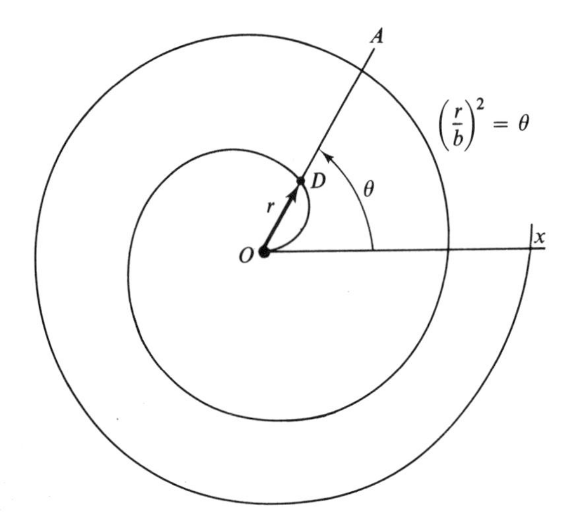

Figure P8-152

8-153 Solve Problem 8-152 when $b = 2$ in. and $t = 5\pi/12$ sec.

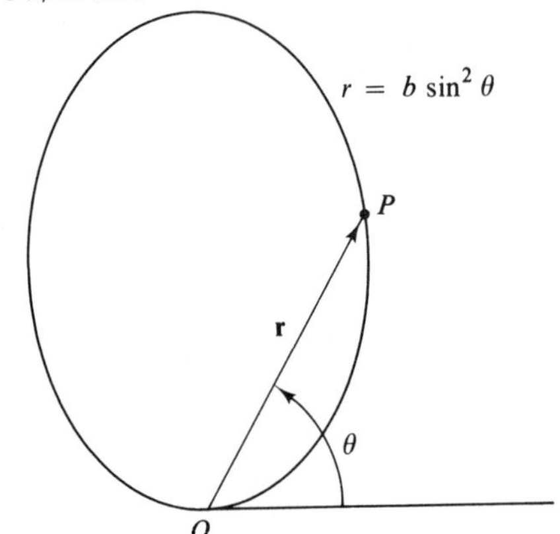

Figure P8-156

8-154 A particle travels along the cardioid $r = b(1-\cos\theta)$, where r and b are in inches. The distance r increases uniformly at 3 ips for $10° < \theta < 170°$. The distance b is 6 in. Determine the value of θ, in the specified range, for which the velocity of the particle will be horizontal and the corresponding velocity.

8-155 Determine the value of θ, in the specified range, in Problem 8-154 for which the velocity of the particle will be vertical and the corresponding acceleration of the particle.

8-156 Point P in Fig. P8-156 moves along the curve whose equation, in polar coordinates, is $r = b\sin^2\theta$, where $b = 2$ in. The radius vector has a constant angular velocity of 5 rad per sec counterclockwise. Determine the velocity and acceleration of P when $\theta = 45°$.

8-10 KINEMATICS OF A PARTICLE USING CYLINDRICAL COORDINATES

The analysis of particle motion using polar coordinates as explained in Art. 8-9 can be extended to three dimensions by the addition of a z coordinate and a unit vector $\mathbf{e}_z$. The added unit vector is constant in direction (parallel to the z axis) as well as in magnitude. The positive directions are in the directions of increasing r, θ, and z, as shown by the unit vectors of Fig. 8-23. It should be noted that $\mathbf{e}_r$, $\mathbf{e}_\theta$, and $\mathbf{e}_z$ define a right-hand system of coordinates.

The position vector from the origin to any particle P is, as shown in Fig. 8-23,

$$\mathbf{r}_1 = r\mathbf{e}_r + z\mathbf{e}_z. \tag{8-18}$$

The velocity of P is defined as the time rate of change of position. The derivative of $\mathbf{e}_r$ is given by Eq. (8-15) of Art. 8-9. Since $\mathbf{e}_z$ is constant in both magnitude and direction, its derivative is zero. The time derivative of Eq. (8-18) is

$$\begin{aligned}\mathbf{v} = \dot{\mathbf{r}}_1 &= \dot{r}\mathbf{e}_r + r\dot{\theta}\mathbf{e}_\theta + \dot{z}\mathbf{e}_z \\ &= \mathbf{v}_R + \mathbf{v}_T + \mathbf{v}_z,\end{aligned} \tag{8-19}$$

where $\mathbf{v}_R$, $\mathbf{v}_T$, and $\mathbf{v}_z$ are the velocity components in the radial, transverse, and z directions, respectively.

In the same manner, differentiation of Eq. (8-19) gives

$$\begin{aligned}\mathbf{a} = \ddot{\mathbf{r}}_1 &= (\ddot{r} - r\dot{\theta}^2)\mathbf{e}_r + (r\ddot{\theta} + 2\dot{\theta}\dot{r})\mathbf{e}_\theta + \ddot{z}\mathbf{e}_z \\ &= \mathbf{a}_R + \mathbf{a}_T + \mathbf{a}_z,\end{aligned} \tag{8-20}$$

where $\mathbf{a}_R$, $\mathbf{a}_T$, and $\mathbf{a}_z$ are the corresponding acceleration components.

The following example illustrates the use of cylindrical coordinates for problems in kinematics.

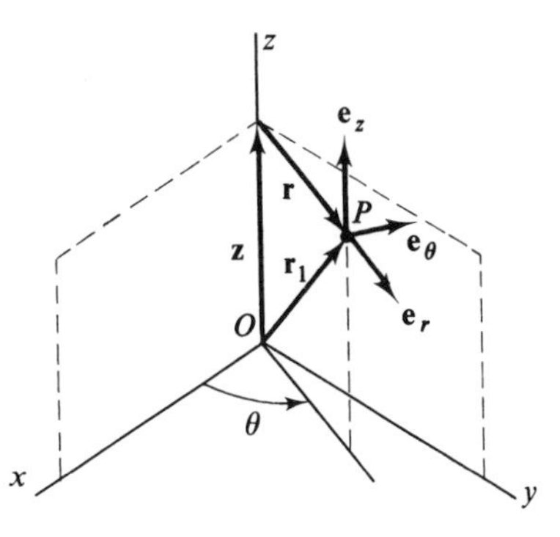

Figure 8-23

EXAMPLE 8-19

A T-29 airplane is cruising at 180 knots east in straight and level flight. Each propeller has a diameter of 13 ft and rotates at 1035 rpm clockwise as viewed from the rear. Determine the velocity and acceleration of the particle at the tip of a propeller when the tip is in the farthest south position. *Note:* A knot is one nautical mile (6080 ft) per hour.

SOLUTION

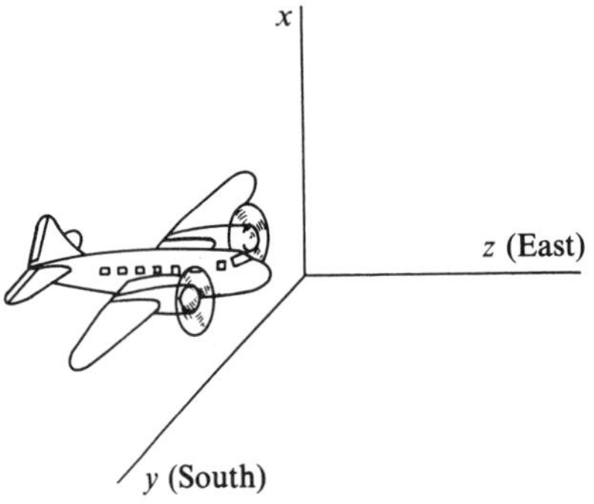

x

z (East)

y (South)

Figure 8-24

A set of axes is selected as shown in Fig. 8-24 with the z axis to the east (along the line of flight), the x axis vertical, and the y axis to the south. For the specified position of the propeller tip $\mathbf{e}_r = \mathbf{e}_y$ and $\mathbf{e}_\theta = -\mathbf{e}_x$. From the problem statement, the z component of the velocity of the tip is

$$\dot{z} = 180 \text{ knots} = \frac{180(6080)}{3600} = 304 \text{ fps}$$

and

$$\dot{\mathbf{z}} = 304\mathbf{e}_z \text{ fps (east).}$$

The magnitude of the transverse component of the tip velocity is

$$v_T = r\dot{\theta} = \frac{(13/2)(1035)(2\pi)}{60} = 705 \text{ fps}$$

and the radial component is zero, since the radius vector has a constant magnitude. The velocity of the tip of the propeller is

$$\mathbf{v} = \mathbf{v}_T + \mathbf{v}_R + \mathbf{v}_z = 705\mathbf{e}_\theta + 304\mathbf{e}_z,$$

and when the propeller is horizontal the velocity of the south tip is

$$\mathbf{v} = -705\mathbf{e}_x + 304\mathbf{e}_z$$

$$= \underline{768(-0.918\mathbf{e}_x + 0.396\mathbf{e}_z) \text{ fps.}} \qquad \text{Ans.}$$

From the statement of the problem, $\ddot{z}$, $\dot{r}$, $\ddot{r}$, and $\ddot{\theta}$ are all zero; therefore, the acceleration of the tip of the propeller is

$$\mathbf{a} = (\ddot{r} - r\dot{\theta}^2)\mathbf{e}_r + (r\ddot{\theta} + 2\dot{r}\dot{\theta})\mathbf{e}_\theta + \ddot{z}\mathbf{e}_z$$

$$= -r\dot{\theta}^2\mathbf{e}_r = -6.5\left[\frac{1035(2\pi)}{60}\right]^2\mathbf{e}_r$$

$$= -76{,}400\mathbf{e}_r$$

and when the propeller is horizontal the acceleration of the south tip is

$$\mathbf{a} = \underline{-76{,}400\mathbf{e}_y \text{ fps}^2 \text{ (to the north).}} \qquad \text{Ans.}$$

PROBLEMS

8-157 A flexible hose is wrapped around the cylinder in Fig. P8-157 to form a helix. The mean radius of the helix is R, and the pitch is p. A small ball P is forced to slide along the helix by the vertical roller AB which rotates about the center of the cylinder. The angular position of the ball in front of the roller is given by the angle θ, where θ is a function of time.
(a) Write an equation for the position of P using cylindrical coordinates.
(b) Derive expressions for the velocity and acceleration of the ball P.
(c) Given $R = 6$ in., $p = 3$ in., and $\theta = ct^2$, where $c = 0.2$ rad per sec^2, determine the velocity and acceleration of P when $t = 5$ sec.

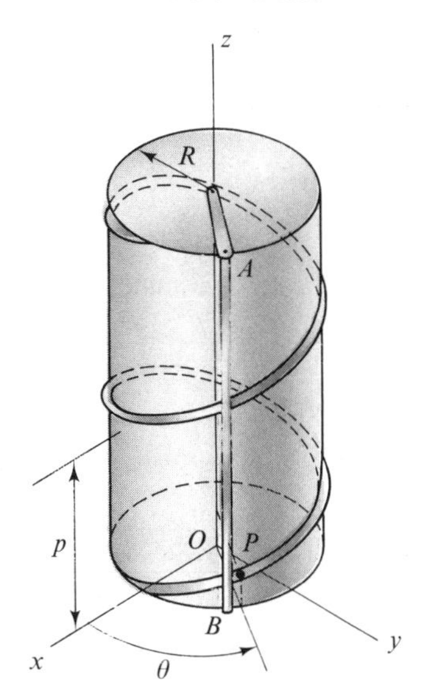

Figure P8-157

8-158 Particle P in Fig. P8-158 is constrained to move along the spiral path on the inside of the cone. The particle makes one revolution about the z axis in 2 sec during which time it moves downward 5 in. The particle starts from point A. Determine the velocity and acceleration of the particle at point B after one complete revolution. (Note that $\dot{\theta}$ and $\dot{r}$ are both constant.)

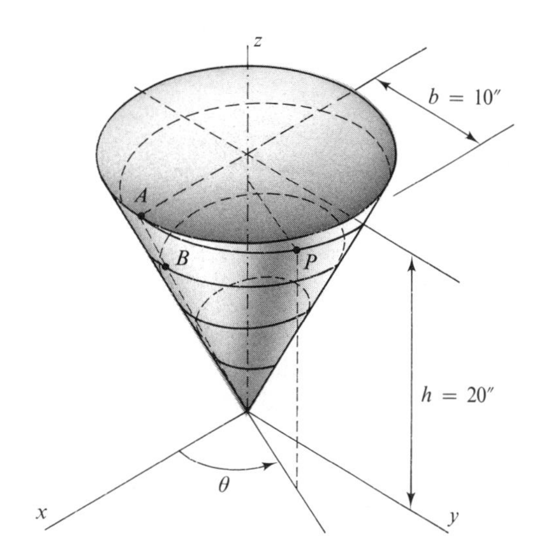

Figure P8-158

8-159 Solve Problem 8-158 if $\ddot{\theta} = \pi$ rad per sec^2. The dimensions of the path are the same as in Problem 8-158. The particle starts from rest at point A.

8-160 The equation of the paraboloid of revolution in Fig. P8-160 is

$$h(x^2 + y^2) = hr^2 = b^2 z.$$

The particle moves along a path on the surface

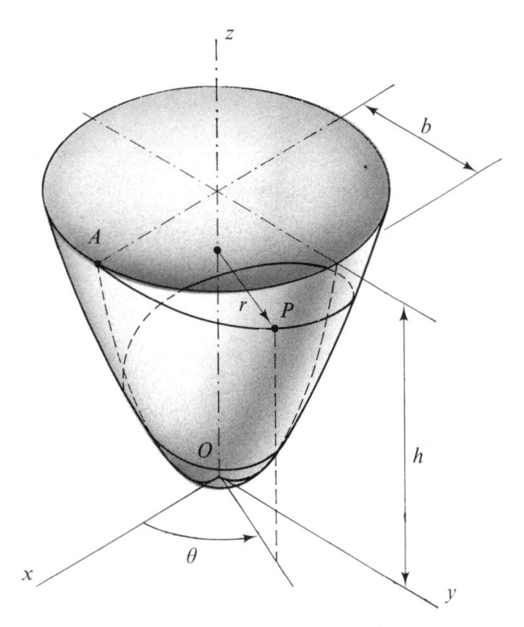

Figure P8-160

in accordance with

$$\theta = c_1 t$$

$$r = b - c_2 t,$$

where c_1 and c_2 are constants.

(a) Derive an expression for the velocity of P.

(b) Let $h = 20$ in., $b = 10$ in., $c_1 = 0.2\pi$ rad per sec, and $c_2 = 0.5$ ips. Determine the velocity of P when $t = 0$ sec and when $t = 10$ sec.

8-161 Determine the acceleration of the point P in Problem 8-160(b) when

(a) $t = 0$ sec.

(b) $t = 10$ sec.

8-11 CLOSURE

The kinematic quantities used to describe the motion of a particle are time, position (including displacement and total distance traveled), velocity, and acceleration. Similar quantities are used to describe the angular motion of a line. The problems of kinematics consist of determining one or more of the preceding quantities from the given data. The various kinematic quantities are related by differential equations. The relationships have been developed using cartesian, normal and tangential, radial and transverse, and cylindrical coordinates or components. Parametric equations are frequently convenient. Motion curves may be used to supplement or replace the analytical solution. Since many of the quantities involved are vectors, it is frequently necessary to differentiate and integrate vector quantities.

REVIEW PROBLEMS

8-162 A particle starts from rest from a point 6 ft to the right of the origin when $t = 0$. The point moves along a horizontal straight line, and the magnitude of its acceleration is directly proportional to the time after starting. The displacement of the particle from $t = 0$ to $t = 2$ sec is 2 ft to the left. Determine the velocity of the particle when it passes the origin.

***8-163** A particle starts from rest and moves along a straight line with an acceleration of 5 m/s^2 to the right for 12 sec. The acceleration is then changed to a different constant value such that the displacement for the entire period is 180 m to the right and the total distance traveled is 780 m. Determine

(a) The acceleration during the second time interval.

(b) The total time interval.

8-164 Body BC in Fig. P8-164 slides in the fixed horizontal slot as shown. The arm OA has an angular velocity of $\dot{\theta}$ and an angular acceleration of $\ddot{\theta}$. Write an equation for the position of B and differentiate it to determine its velocity and acceleration in terms of θ, $\dot{\theta}$, and $\ddot{\theta}$.

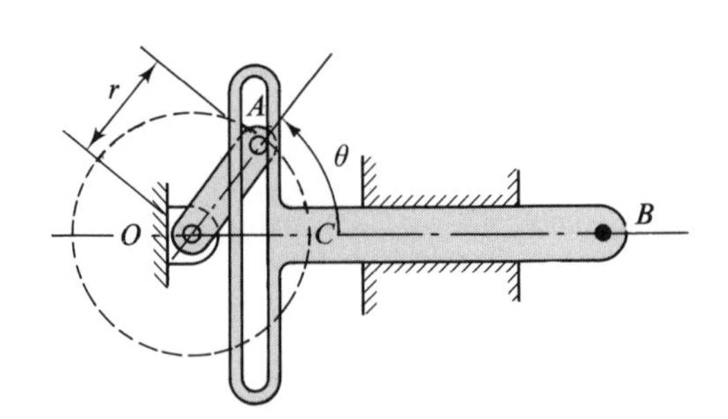

Figure P8-164

8-165 The 12-ft bar AB in Fig. P8-165 moves in a vertical plane with its ends in the guides as shown. When $\tan\theta = \frac{3}{4}$ the bar has an angular velocity of 3 rad per sec clockwise and an angular acceleration of 2 rad per sec^2 counterclockwise. Write an equation for the position of the center, C, of the bar at any instant in terms of θ and differentiate to determine the acceleration of C when $\tan\theta = \frac{3}{4}$.

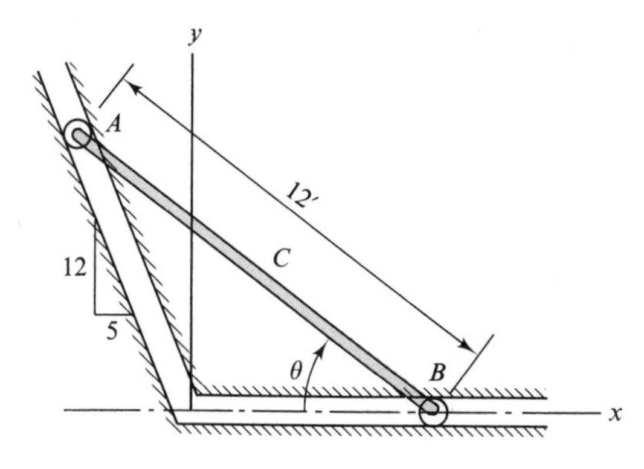

Figure P8-165

8-166 A stone is thrown with an initial velocity of 80 fps in the direction shown in Fig. P8-166 and falls on the inclined plane. Determine
(a) The distance from A to B.
(b) The maximum height z of the stone above the inclined plane.

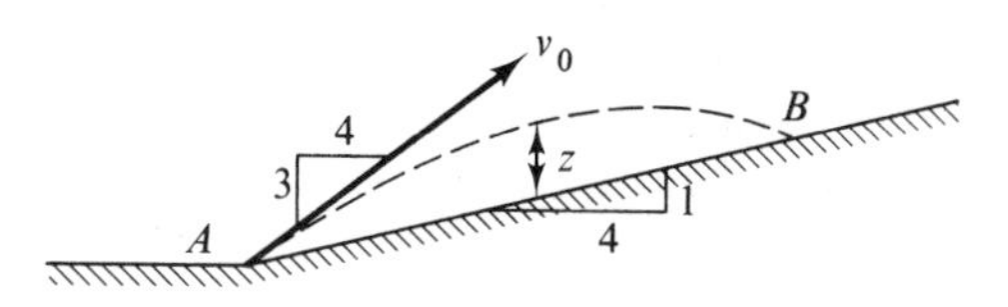

Figure P8-166

***8-167** A point moves along the circular path in Fig. P8-167. When the point is at position A the angular velocity of line OA is 2 rad/s clockwise. The angular velocity of OA changes uniformly in such a manner than the angular displacement is 6 rad clockwise, while the line turns through a total angle of 10 rad. Determine the acceleration of the point 3 s after leaving position A.

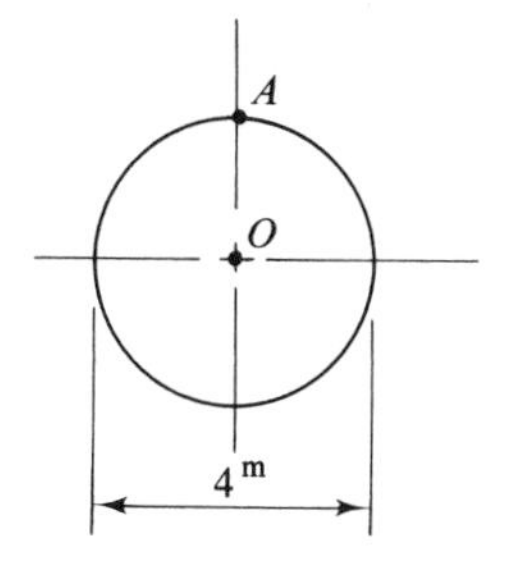

Figure P8-167

***8-168** A particle travels on an elliptical path given by the equation $x^2+4y^2=16$, where x and y are in centimeters. At the point where $y=2$ cm, the particle has a velocity of 6 cm/s to the right, and the velocity is increasing at the rate of 2 cm/s^2. Determine the acceleration of the point when $y=2$ cm.

8-169 A particle moves along the curve $y=0.2x^2$, where x and y are in feet. When the particle is at the point $x=2.4$ ft, the acceleration is $-125\mathbf{i}$ fps^2 and the speed is decreasing. Determine the values of $\dot{s}\mathbf{e}_t$ and $\ddot{s}\mathbf{e}_t$ and show them on a sketch of the curve.

8-170 The velocity of a particle moving along a horizontal path is given by the expression $\mathbf{v}=(s^2+9)\mathbf{i}$, where v is in feet per second and positive is to the right. When $t=0$, $s=3$ ft. Determine s as a function of the time t.

8-171 Rod AB in Fig. P8-171 rotates about an axis at A and is connected to E by a flexible and inextensible cable. As AB turns from the horizontal position through the angle θ, body E moves upward a distance s.
(a) Derive an expression relating s and θ.
(b) Differentiate the equation from part (a) to determine the velocity of E in terms of the angular velocity of AB and the positions of the bodies.
(c) When the angle θ is 60°, the angular velocity of AB is 0.5 rad per sec clockwise and the acceleration of E is 10 fps^2 upward. Determine the angular acceleration of AB for this position.

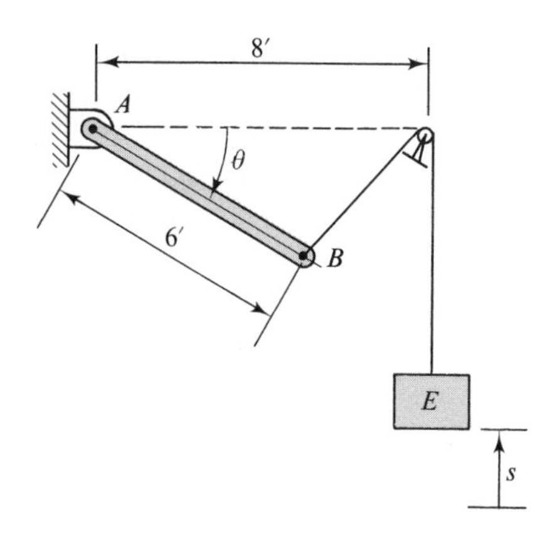

Figure P8-171

PART B—RIGID BODY KINEMATICS

8-12 DEFINITIONS

The motion of points or particles, and of lines connecting them, was analyzed in the preceding articles of this chapter. When the dimensions of a body are small compared with the dimensions of the path of motion of the body, it is usually permissible to treat the body as a particle in kinetics. When the dimensions of the body are not negligible compared to the dimensions of the path it usually is necessary to investigate the relationships between the positions, velocities, and accelerations of various points and lines of the body. A rigid body was defined in Art. 1-2 as one in which the particles remain at fixed distances from each other. Any rigid body motion may be classified as one of the following types.

Translation: When a rigid body moves in such a manner that all lines in the body remain parallel to their original positions, the body has a motion of translation. For such motion, the angular velocity and angular acceleration of all lines in the body must be zero. If the mass center of the body remains in one plane, the motion is defined as *coplanar translation.* Most bodies having a motion of translation will be found to have coplanar translation. Some examples of translation are (1) the piston rod of a locomotive running along a straight level track and also the parallel rod connecting the drivers of a locomotive, (2) the bumper of an automobile traveling in a straight line along a road, and (3) a sliding door whose rollers run on straight tracks.

From the definition of translation, it is apparent that all points of the body must have the same displacement during any time interval and consequently must have identical velocities and identical accelerations at any instant. In other words, *the motion of any particle of a rigid body having a motion of translation determines the motion of every particle of the body.*

Rotation: A rigid body has a motion of rotation if one line in the body or the body extended is fixed and all particles in the body not on the fixed line travel in circular paths with centers on the fixed axis. The fixed line is called the *axis of rotation,* and each particle not on the axis moves in a plane perpendicular to the axis of rotation. Sometimes this motion is called *pure rotation,* to distinguish it from motion in which no line is fixed but in which the body turns or rotates about a moving line as in plane motion (to be defined later).

The rotor of a stationary steam turbine, a water wheel, the hands of a clock, and the crankshaft of an automobile when the car is standing still and the motor is running are examples of rotating rigid bodies. If the car is traveling along a straight road, the crankshaft does not have rotation because the points travel along helical paths instead of circular paths.

Plane motion: When a rigid body moves in such a manner that every particle of the body remains a constant distance from a fixed reference plane, the body has plane motion. Thus, all particles in a rigid body with plane motion move in the same plane or in parallel planes. Rotation is a special case of plane motion, and coplanar translation is also plane motion. The plane in which the mass center moves is called the plane of motion of the body. In general, plane motion may be thought of as a combination of rotation and translation. Two common examples of plane motion are the connecting rod of a stationary engine and the wheel of an automobile running straight along a road.

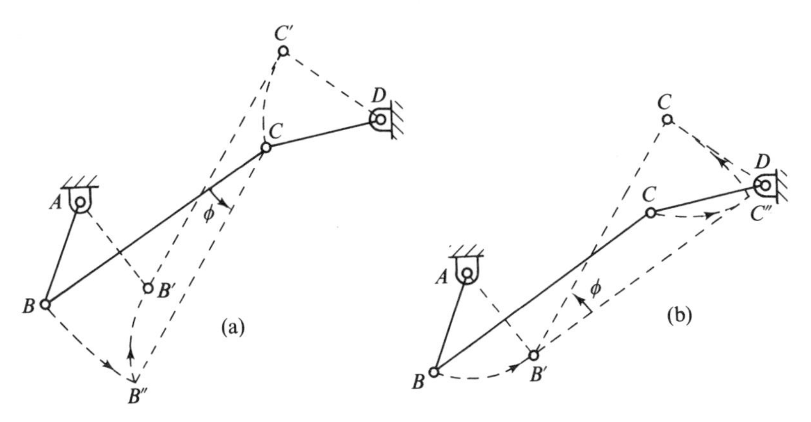

Figure 8-25

Plane motion can be analyzed by treating it as a combination of a rotation of the body about some convenient reference axis in the body, or the body extended, and a translation of the body in which all points have the same motion as the reference point. Member BC of the link mechanism in Fig. 8-25(a) has plane motion. As arm AB rotates to position AB', member DC rotates to DC' and BC moves to $B'C'$. The motion of BC can be visualized as a rotation about an axis at C to position $B''C$ followed by a curvilinear translation to $B'C'$ with all particles of BC moving along curved paths similar to CC' from $B''C$ to position $B'C'$. The motion can also be thought of as a curvilinear translation of BC to position $B'C''$ in Fig. 8-25(b) followed by a rotation about B' to position $B'C'$. *The angular displacement of BC when moving from BC to B′C′ is ϕ counterclockwise whether the rotation is considered as occurring about an axis at C or one at B or about any other axis perpendicular to the plane of motion.* Furthermore, it is immaterial whether the rotation or the translation is considered as occurring first.

Motion about a fixed point: When a body moves in such a manner that one point, O, remains fixed in space, it can be shown that the resulting displacement is a rotation about some axis through O with an angular velocity $\boldsymbol{\omega}$.[6] A general infinitesimal rotation associated with the angular velocity $\boldsymbol{\omega}$ can be considered to be successive rotations about a set of x, y, and z axes with angular velocities $\boldsymbol{\omega}_x = \omega_x \mathbf{i}$, $\boldsymbol{\omega}_y = \omega_y \mathbf{j}$, and $\boldsymbol{\omega}_z = \omega_z \mathbf{k}$. Infinitesimal angular rotations can be shown to be vectors; therefore, the angular velocity $\boldsymbol{\omega}$ is the sum of the three component angular velocities.

General motion. Euler's theorem for motion of a rigid body about a fixed point can be extended to describe the general motion of a rigid body. The general theorem, known as *Chasles' theorem,*[7] states that any displacement can be analyzed as a combination of a translation and a rotation.

When a body has plane motion, all lines in the body parallel to the plane of motion have common angular velocities and accelerations,

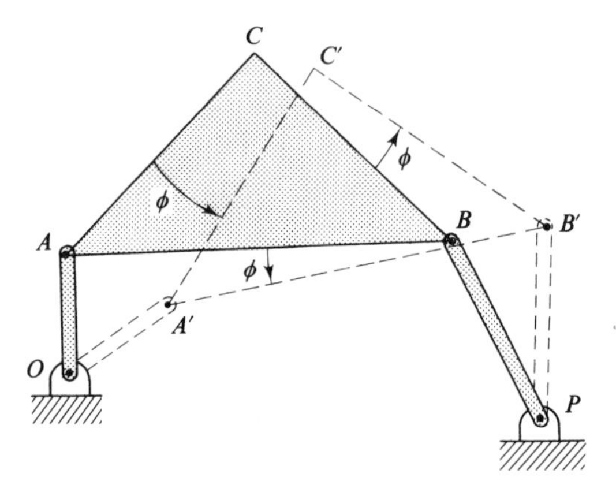

Figure 8-26

and the values are referred to as the angular velocity or angular acceleration of the body. The plate ABC in Fig. 8-26 has plane motion, and as link OA moves to OA' member PB will move to PB' and ABC will move to $A'B'C'$. Line AB on plate ABC will turn through the angle ϕ, and lines BC and AC must also turn through the same angle ϕ since the angles ABC and BAC must remain constant on a rigid body. Thus it is seen that all lines in body ABC will turn through the same angle in the same time interval and thus will have the same angular velocity at any instant. This same reasoning leads to the conclusion that all lines in the body (parallel to the plane of motion) must have a common angular acceleration at any instant.

---

[6] This statement is sometimes called *Euler's theorem;* see Herbert Goldstein, *Classical Mechanics* (Reading, Mass.: Addison-Wesley Publishing Company, Inc., 1950), p. 118.

[7] Goldstein, *Ibid.*, p. 124.

The motion of various points and lines of bodies in a mechanism can be related by deriving expressions relating the linear and angular positions (from geometry) and differentiating these expressions to obtain velocities and accelerations as illustrated in some of the preceding articles. The equations of relative motion, derived in Art. 8-6, can often be applied to rigid body problems in kinematics to obtain more direct solutions. Familiarity with various methods is important if the most direct solution is to be selected.

The notation for absolute and relative motion of positions, velocities, and accelerations of points was discussed in Art. 8-6. The symbol $\mathbf{v}_A$ represents the velocity of point A measured with respect to a fixed set of axes (the earth is assumed to be fixed for most problems). The quantity $\mathbf{v}_{A/B}$ is the velocity of point A measured with respect to a set of translating (nonrotating) axes whose origin moves with point B. Motion with respect to rotating axes is analyzed in Art. 8-21.

Angular motion of a line or body can also be expressed as either an absolute or a relative quantity. The absolute angular motion of a line is defined as the angular motion of the line measured with respect to a set of nonrotating axes. Motion of a line measured with respect to rotating axes is called the *relative angular motion of the line.* In this text, all values of linear and angular quantities are considered absolute unless otherwise specified.

8-13 ROTATION OF A RIGID BODY

The velocity and acceleration of any point on a rotating rigid body can be expressed in terms of the angular motion of the body. Figure 8-27 represents a rigid body which is rotating about the fixed z axis with an angular velocity $\boldsymbol{\omega}=\dot{\boldsymbol{\theta}}$ and an angular acceleration $\boldsymbol{\alpha}=\ddot{\boldsymbol{\theta}}$ as

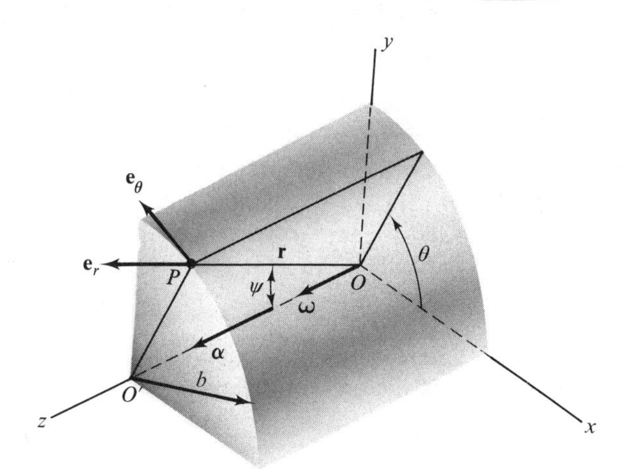

Figure 8-27

shown. The point P moves along a circular path of radius b, and the position of P from an arbitrary origin on the axis of rotation is

$$\mathbf{r} = r\mathbf{e}_r.$$

The velocity of P is tangent to the path and from Art. 8-8 is

$$\mathbf{v}_P = \dot{\mathbf{r}} = b\omega\mathbf{e}_\theta = r(\sin\psi)\omega\mathbf{e}_\theta = \boldsymbol{\omega}\times\mathbf{r}. \tag{8-21}$$

Note that the order of the terms in the vector product must be as indicated in order to give the correct direction for the velocity. The derivative of the vector $\mathbf{r}$ is given in Eq. (8-21) only when the magnitude of $\mathbf{r}$ is constant as when O and P are two points in a rigid body.

The acceleration of P can be determined by differentiating Eq. (8-21) and is

$$\begin{aligned}\mathbf{a}_P = \dot{\mathbf{v}}_P &= \dot{\boldsymbol{\omega}}\times\mathbf{r} + \boldsymbol{\omega}\times\dot{\mathbf{r}}\\ &= \boldsymbol{\alpha}\times\mathbf{r} + \boldsymbol{\omega}\times(\boldsymbol{\omega}\times\mathbf{r}).\end{aligned} \tag{a}$$

The first term on the right can be written as

$$\boldsymbol{\alpha}\times\mathbf{r} = \alpha r\sin\psi\mathbf{e}_\theta = \alpha b\mathbf{e}_\theta = \mathbf{a}_t.$$

Since $\boldsymbol{\omega}$ and $\dot{\mathbf{r}}$ are perpendicular, the magnitude of the second term of Eq. (a) is

$$|\boldsymbol{\omega}\times\dot{\mathbf{r}}| = \omega v_P = b\omega^2,$$

and its direction, perpendicular to $\boldsymbol{\omega}$ and $\mathbf{e}_\theta$, is from P toward O' and is thus seen to be the normal acceleration of P. That is,

$$\mathbf{a}_P = \boldsymbol{\alpha}\times\mathbf{r} + \boldsymbol{\omega}\times(\boldsymbol{\omega}\times\mathbf{r}). \tag{8-22}$$

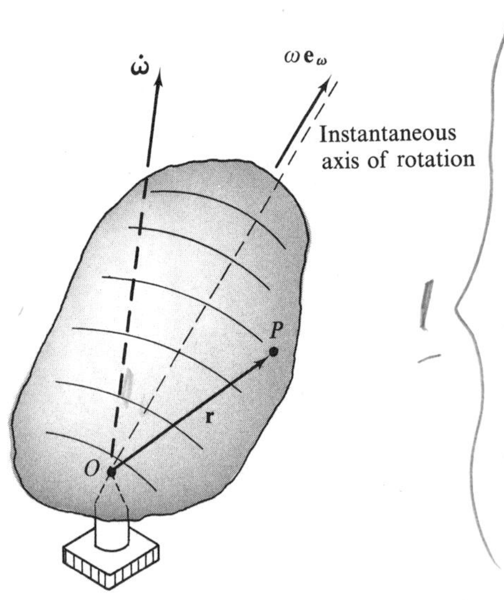

Figure 8-28

Euler's theorem states that rotation of a body about a fixed *point* is also rotation about a line in the body through the fixed point. The angular velocity of the body lies along the instantaneous axis of rotation as shown in Fig. 8-28. The velocity of any point P can be determined by means of Eq. (8-21).

Since the direction of the instantaneous axis of rotation varies with time, the angular velocity will, in general, vary in both magnitude and direction. The angular velocity can be expressed as

$$\boldsymbol{\omega} = \omega\mathbf{e}_1,$$

where $\mathbf{e}_1$ is a unit vector along the instantaneous axis of rotation. The angular acceleration of the body is

$$\begin{aligned}\dot{\boldsymbol{\omega}} &= \dot{\omega}\mathbf{e}_1 + \omega\dot{\mathbf{e}}_1\\ &= \dot{\omega}\mathbf{e}_1 + \omega(\boldsymbol{\omega}_2\times\mathbf{e}_1),\end{aligned}$$

where $\boldsymbol{\omega}_2$ is the angular velocity of the axis of rotation. The first term of this expression is the component of the angular acceleration due to a change in magnitude of the angular velocity, and the second term is the component resulting from a change in direction of the angular velocity. These two components are mutually perpendicular, and since the first component is parallel to the angular velocity, the resultant angular acceleration will be parallel to the angular velocity only when the angular velocity does not change direction. Equation (8-22) can be used to calculate the acceleration of point P in Fig. 8-28, but since $\boldsymbol{\alpha}$ and $\boldsymbol{\omega}$ are not necessarily parallel, the quantity $\boldsymbol{\alpha} \times \mathbf{r}$ is not necessarily parallel to $\boldsymbol{\omega} \times \mathbf{r}$, and the acceleration component given by $\boldsymbol{\alpha} \times \mathbf{r}$ may not be tangent to the path of the particle.

8-14 RELATIVE VELOCITY—NONROTATING AXES

The concept of the motion of point B in Fig. 8-9 with respect to a set of nonrotating axes whose origin moved with point A was discussed in Art. 8-6. The equation relating the velocities of points A and B is

$$\mathbf{v}_B = \mathbf{v}_A + \mathbf{v}_{B/A}. \tag{8-23}$$

Equation (8-23) can be rearranged to give

$$\mathbf{v}_{B/A} = \mathbf{v}_B - \mathbf{v}_A.$$

A helpful interpretation of the last equation can be obtained by considering points A and B in Fig. 8-29, which have velocities $\mathbf{v}_A$ and $\mathbf{v}_B$. If a velocity $-\mathbf{v}_A$ is added to the velocities of each of the particles, the resultant velocity of particle A will be zero and the resulting velocity of B will be $\mathbf{v}_{B/A}$. In other words, the velocity of B relative to A is the velocity B would appear to have to an observer moving with A (having the same velocity as A) but not rotating. The concept of thinking of the motion of B with A fixed will often aid in determining correctly the direction of the velocity of B relative to A.

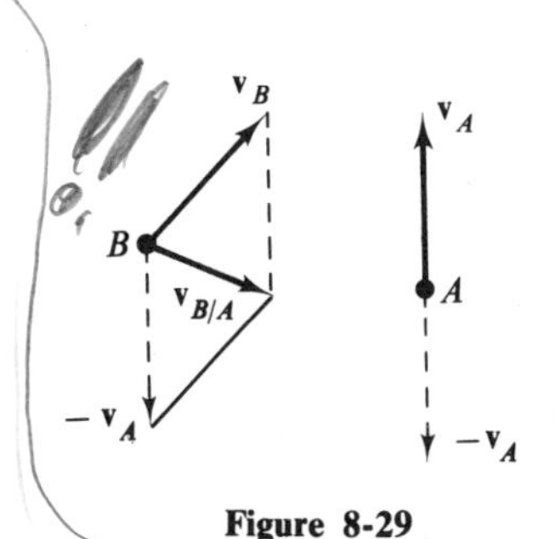

Figure 8-29

When points A and B in Eq. (8-23) are two points on a rigid body the motion of B relative to A must be along a spherical path with its center at A and with a radius equal to the distance from A to B since the two points must remain the same distance apart. In this case the velocity of B relative to A will be given by Eq. (8-21), and the direction of the relative velocity will be perpendicular to the line AB. The velocity of B is thus

$$\mathbf{v}_B = \mathbf{v}_A + \boldsymbol{\omega} \times \mathbf{r}, \tag{8-23a}$$

where $\boldsymbol{\omega}$ is the angular velocity of the member containing A and B and $\mathbf{r}$ is the position of B relative to A.

When Eq. (8-23) is applied to two points which are not on the same rigid body the motion of B relative to A will usually not be on a spherical surface, and the relative velocity can be more difficult to analyze. The use of Eq. (8-23) for plane motion and for three-dimensional motion will be discussed in Art. 8-15 and 8-17.

8-15 RELATIVE VELOCITY—PLANE MOTION

Many of the kinematic problems encountered in engineering practice involve bodies which have plane motion. When points A and B in Eq. (8-23) are points on a rigid body with plane motion the motion of B relative to A will be along a circular path with its center at A, and all velocity vectors will lie in a common plane. The vector equation can be resolved into two scalar equations, which in turn can be solved for two unknown quantities. Note that the angular velocity vector is always perpendicular to the plane of motion and thus can change only in magnitude.

An important case of plane motion occurs when a wheel rolls without slipping along a plane surface. The center of the wheel has rectilinear motion parallel to the plane. Figure 8-30 shows a wheel in its initial position and after it has rolled a distance s_O to the right. The line OA on the wheel turns clockwise through an angle θ to the position $O'A'$, and if there is no slipping, the arc $A'B = r\theta$ must be the same as the distance $AB = s_O$; that is,

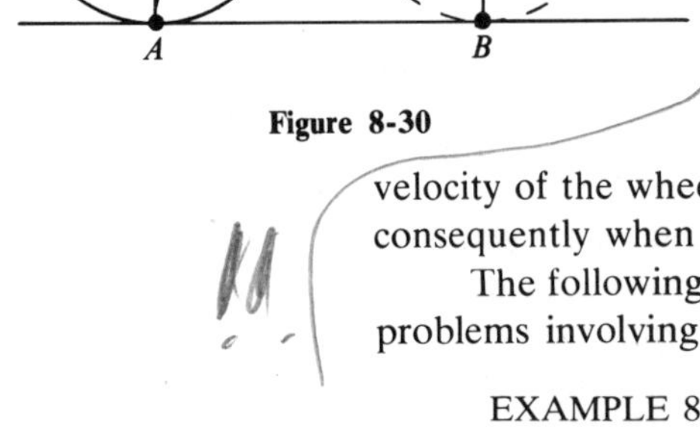

$$s_O = r\theta. \qquad \text{(a)}$$

Figure 8-30

The magnitude of the velocity of point O is

$$v_O = \frac{d}{dt}(r\theta) = r\omega, \qquad \text{(b)}$$

where $\omega = d\theta/dt$ is the magnitude of the angular velocity of the wheel. Note that when s_O is to the right θ is clockwise; consequently when v_O is to the right ω must be clockwise.

The following examples illustrate the application of Eq. (8-23) to problems involving plane motion.

EXAMPLE 8-20

The wheel in Fig. 8-30 rolls without slipping along the horizontal plane. The center of the wheel has a constant velocity of v_O to the right. Determine the velocity of point A' on the rim of the wheel when the angle θ is 150°.

SOLUTION

The velocity of the center of the wheel is

$$\mathbf{v}_O = v_O \mathbf{i},$$

and the angular velocity of the wheel, from Eq. (b), is

$$\boldsymbol{\omega} = \frac{v_O}{r} \circlearrowright = -\frac{v_O}{r}\mathbf{k}.$$

The velocity of A' can be determined by means of the relative velocity equation for points A' and O' since the velocity of O' is known. The equation is

$$\mathbf{v}_{A'} = \mathbf{v}_{O'} + \mathbf{v}_{A'/O'}$$
$$= v_{O'}\mathbf{i} + \boldsymbol{\omega} \times \mathbf{r}_{O'A'}.$$

Note that when θ is 150° point A' will be above and to the left of O' and the vector $r_{O'A'}$ will be $r(-0.500\mathbf{i}+0.866\mathbf{j})$. Thus the equation becomes

$$\mathbf{v}_{A'} = v_{O'}\mathbf{i} - \frac{v_O}{r}\mathbf{k} \times r(-0.500\mathbf{i}+0.866\mathbf{j})$$
$$= v_O\mathbf{i} + 0.500 v_O \mathbf{j} + 0.866 v_O \mathbf{i}$$
$$= v_O(1.866\mathbf{i} + 0.500\mathbf{j})$$
$$= 1.932 v_O \;\angle 15^\circ \qquad \text{Ans.}$$

EXAMPLE 8-21

The mechanism in Fig. 8-31 moves in a vertical plane, and member CD, when in the position shown, has an angular velocity of 8 rad per sec clockwise. Determine
(a) The angular velocities of members AB and BC.
(b) The velocity of point M, the midpoint of member BC.

SOLUTION

(a) Members AB and CD each have a motion of rotation, and pins B and C move in circular paths about centers at A and D, respectively. Pins B and C are also points on the rigid body BC, which has plane motion, and thus the relative motion of B with respect to C (or of C with respect to B) must be along a circular path whose radius is the distance from B to C. The absolute and relative velocities of points B and C can be expressed in terms of the angular velocities of the members and the radius vectors by means of Eq. (8-21). Unknown angular velocities are assumed to be counterclockwise (in this example), and a negative result will indicate that the assumed direction was incorrect. The angular velocities are

$$\boldsymbol{\omega}_{AB} = \omega_{AB}\mathbf{k}, \qquad \boldsymbol{\omega}_{BC} = \omega_{BC}\mathbf{k},$$
$$\boldsymbol{\omega}_{CD} = -8\mathbf{k} \text{ rad per sec},$$

and the radius vectors are

$$\mathbf{r}_{AB} = 3\mathbf{j} \text{ in.}, \qquad \mathbf{r}_{CB} = (-12\mathbf{i}+5\mathbf{j}) \text{ in.}, \qquad \mathbf{r}_{DC} = (-3\mathbf{i}-4\mathbf{j}) \text{ in.}$$

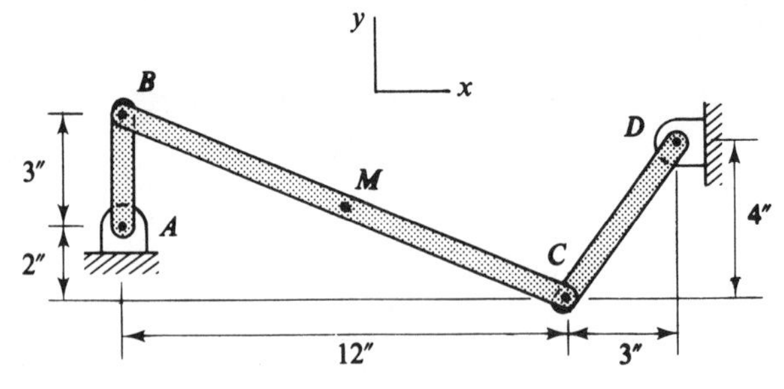

Figure 8-31

The relative velocity equation for points B and C is

$$\mathbf{v}_B = \mathbf{v}_C + \mathbf{v}_{B/C}$$

or

$$\boldsymbol{\omega}_{AB} \times \mathbf{r}_{AB} = \boldsymbol{\omega}_{CD} \times \mathbf{r}_{DC} + \boldsymbol{\omega}_{BC} \times \mathbf{r}_{CB}$$

which becomes

$$\omega_{AB}\mathbf{k}\times 3\mathbf{j} = -8\mathbf{k}\times(-3\mathbf{i}-4\mathbf{j}) + \omega_{BC}\mathbf{k}\times(-12\mathbf{i}+5\mathbf{j}). \qquad \text{(a)}$$

Note that the radius vector is the vector *from* the fixed point (A or D) or the reference point (point C for the velocity of B with respect to C) *to* the moving point. When Eq. (a) is expanded it becomes

$$-3\omega_{AB}\mathbf{i} = +24\mathbf{j} - 32\mathbf{i} - 12\omega_{BC}\mathbf{j} - 5\omega_{BC}\mathbf{i}$$

or

$$-3\omega_{AB}\mathbf{i} = \mathbf{i}(-32 - 5\omega_{BC}) + \mathbf{j}(24 - 12\omega_{BC}).$$

If two vectors are equal, their components (the coefficients of the unit vectors) must be equal. Thus the coplanar vector equation can be reduced to the following two scalar equations:

$$\mathbf{i}: \quad -3\omega_{AB} = -32 - 5\omega_{BC},$$

$$\mathbf{j}: \quad 0 = +24 - 12\omega_{BC}.$$

The simultaneous solution of these equations gives

$$\omega_{BC} = 2.00 \text{ rad per sec}; \qquad \boldsymbol{\omega}_{BC} = \underline{2.00\mathbf{k} \text{ rad per sec},} \quad \text{Ans.}$$

$$\omega_{AB} = 14.00 \text{ rad per sec}; \qquad \boldsymbol{\omega}_{AB} = \underline{14.00\mathbf{k} \text{ rad per sec}.} \quad \text{Ans.}$$

(b) The velocity of point M can be obtained by writing a relative velocity equation between M and either B or C. If point B is used for the reference point, the relative velocity equation becomes

$$\begin{aligned}\mathbf{v}_M &= \mathbf{v}_B + \mathbf{v}_{M/B} \\ &= \boldsymbol{\omega}_{AB} \times \mathbf{r}_{AB} + \boldsymbol{\omega}_{BC} \times \mathbf{r}_{BM} \\ &= 14.00\mathbf{k}\times 3\mathbf{j} + 2.00\mathbf{k}\times(6\mathbf{i} - 2.5\mathbf{j}) \\ &= -42.0\mathbf{i} + 12.0\mathbf{j} + 5.0\mathbf{i} \\ &= -37.0\mathbf{i} + 12.0\mathbf{j} \\ &= \underline{38.9 \text{ ips} \nwarrow 17.97^\circ.} \qquad \text{Ans.}\end{aligned}$$

ALTERNATE SOLUTION

The solution can also be accomplished by a semigraphical procedure. The relative velocity equation for points B and C in algebraic and graphical form is

$$\mathbf{v}_B = \mathbf{v}_C + \mathbf{v}_{B/C}.$$

$v_B = 3\omega_{AB}$ $\qquad v_C = 8\,(5)$, $\omega_{CD} = 8$ $\qquad v_{B/C} = 13\omega_{BC}$

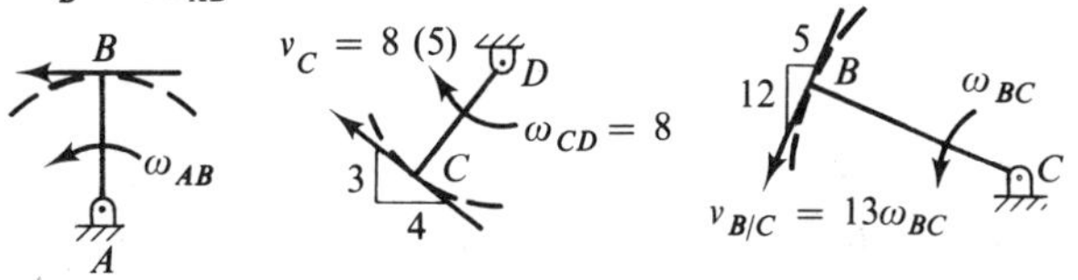

These sketches show the slope and sense of each of the velocity terms beneath the corresponding terms of the vector equation. Since B is traveling along a circular path about A, its velocity must be tangent to the path, and its sense must agree with the assumed angular velocity of AB. If ω_{AB} is found to have a negative sign, the assumed sense is wrong. In the diagram showing the velocity of B relative to C, point C is not actually fixed, but it is convenient to show it fixed temporarily to indicate the correspondence of the direction of $\boldsymbol{\omega}_{BC}$ and $\mathbf{v}_{B/C}$. The vector equation contains two unknown quantities, ω_{AB} and ω_{BC} (or v_B and $v_{B/C}$), and the two unknowns can be obtained by equating the horizontal and vertical components of the various terms. The sum of the horizontal components (assume positive to the right) gives

$$(\xrightarrow{+})\qquad -3\omega_{AB} = -\tfrac{4}{5}(8)(5) - \tfrac{5}{13}(13\omega_{BC}) = -32 - 5\omega_{BC}.$$

In a similar manner the sum of the vertical components of the velocities gives

$$(\uparrow +)\qquad 0 = \tfrac{3}{5}(8)(5) - \tfrac{12}{13}(13\omega_{BC}) = 24 - 12\omega_{BC}.$$

The solution of these two algebraic equations gives

$$\omega_{BC} = 2 \text{ rad per sec},\quad \omega_{AB} = 14 \text{ rad per sec}.$$

Both quantities have positive signs, indicating that the assumed directions were correct. Therefore

$\boldsymbol{\omega}_{BC} = \underline{2.00 \text{ rad per sec counterclockwise,}}$ Ans.

$\boldsymbol{\omega}_{AB} = \underline{14.00 \text{ rad per sec counterclockwise.}}$ Ans.

The velocity of M can be obtained by writing an equation for the velocity of M relative to either B or C. The equation relating v_M and v_B is

$$\mathbf{v}_M = \mathbf{v}_B + \mathbf{v}_{M/B}$$

$v_B = 3\,(14) = 42$, $\omega = 14$ $\qquad v_{M/B} = 6.5\,(2) = 13$, $\omega = 2$

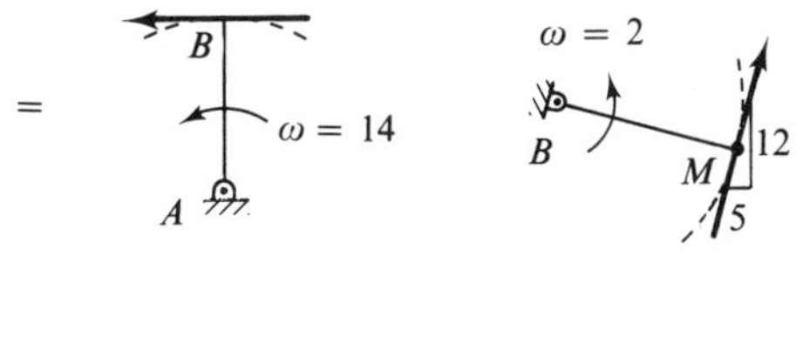

$=$ 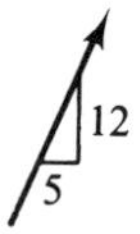

$= \overleftarrow{42} + 13 = \overleftarrow{42} + \overrightarrow{5} + \overset{\uparrow}{12}$

$= \overleftarrow{37} + \overset{\uparrow}{12} = 38.9 \text{ ips}$ 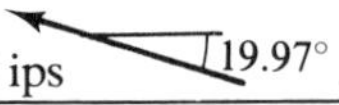

. Ans.

EXAMPLE 8-22

Blocks D and E in Fig. 8-32(a) are pinned together. Block D slides in a slot in the rotating member OA, and E slides in the fixed vertical slot. Determine the velocity of E as a function of θ and $\dot{\theta}$ by

(a) Differentiating the position vector of E.
(b) Using the equation of relative velocity.

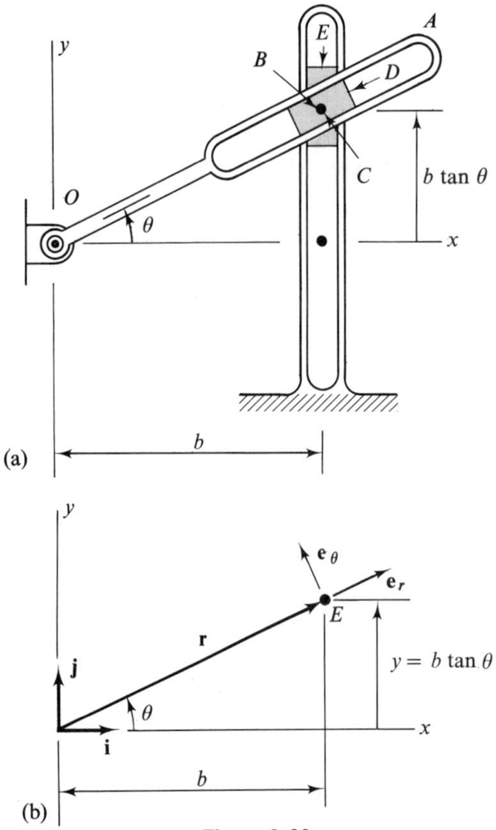

Figure 8-32

SOLUTION

(a) The position vector for block E [see Fig. 8-32(b)] can be written using either polar or rectangular coordinates as

$$\mathbf{r} = r\mathbf{e}_r = b\mathbf{i} + b \tan\theta\,\mathbf{j}. \qquad \text{(a)}$$

Since both r and $\mathbf{e}_r$ vary with time, the second (rectangular form of Eq. (a) is more convenient to use. The velocity of E is

$$\mathbf{v}_E = \dot{\mathbf{r}} = b \sec^2\theta\,\dot{\theta}\,\mathbf{j}. \qquad \text{Ans.}$$

(b) Point B is on block E at the center of the pin connecting the two blocks, and C is a point on the rotating arm OA, which is coincident with B in the position shown. A relative velocity equation can be written between points B and C, and diagrams representing each of these terms are drawn below each term of the equation. An explanation of these diagrams is given below.

$$\mathbf{v}_B \quad = \quad \mathbf{v}_C \quad + \quad \mathbf{v}_{B/C}$$

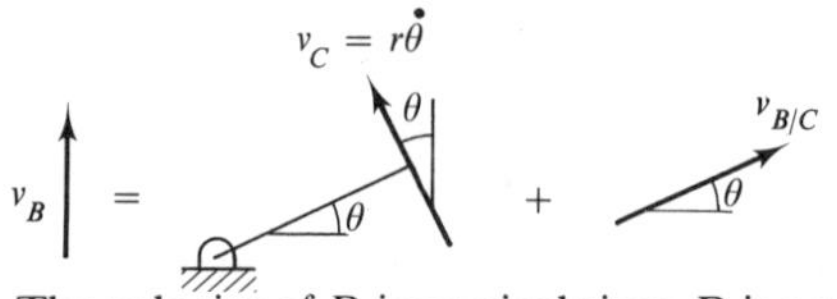

The velocity of B is vertical since B is on block E, which is constrained to move in the vertical slot. Point C is on the rotating member and its velocity is perpendicular to OA with a magnitude of $r\theta$. The velocity of B relative to C is the velocity B would appear to have to an observer at point C, and this would be along OA,[8] when B and C

[8] It should be noted that, although the relative velocity of B with respect to C is along OA, the relative velocity may be changing in both magnitude and direction, and the relative acceleration of the two points will not, in general, be along the member OA.

are coincident. In these diagrams there are two unknowns, the magnitudes v_B and $v_{B/C}$. Note that $r = b \sec\theta$. The value of v_B can be determined by summing velocity components in the direction perpendicular to member OA:

$$v_B \cos\theta = b \sec\theta\, \dot{\theta}$$

$$v_B = b \sec^2\theta\, \dot{\theta}$$

$$\mathbf{v}_E = \mathbf{v}_B = b \sec^2\theta\, \dot{\theta} \uparrow. \qquad \text{Ans.}$$

The velocity of B relative to C ($v_{B/C}$) can be determined by summing velocity components in the direction along OA but is not needed for this problem.

PROBLEMS

8-172 The angular velocity of member AB in Fig. P8-172 is 9 rad-per sec clockwise in the position shown. Determine
(a) The angular velocities of members CE and CD.
(b) The velocity of point E.

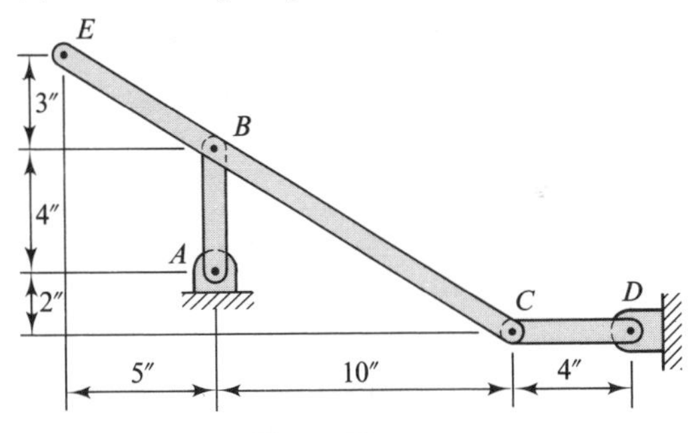

Figure P8-172

8-173 The angular velocity of member AB in Fig. P8-173 is 21 rad per sec counterclockwise in the position shown. Determine
(a) The angular velocities of members BC and CD.
(b) The velocity of point E.

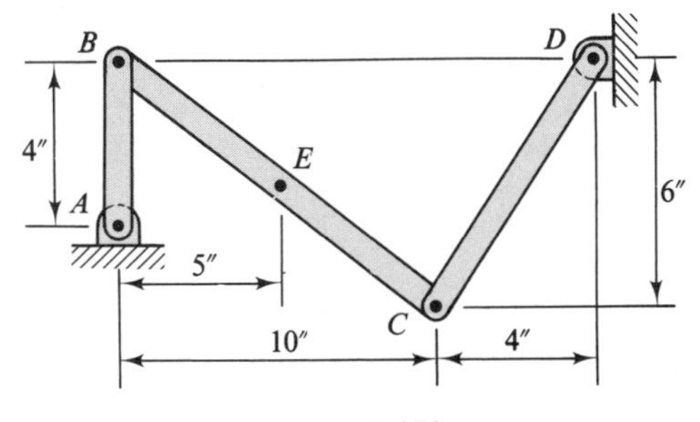

Figure P8-173

8-174 A simplified version of the mechanism for operating the hydraulic dump truck illustrated in Fig. P8-174(a) is shown in Fig. P8-174(b). When the truck body is in the position shown, the angular velocity of link A is 0.035 rad per sec clockwise. Determine the angular velocity of the truck body for this position.

(a)
Courtesy Hercules Steel Products Corp., Galion, Ohio.

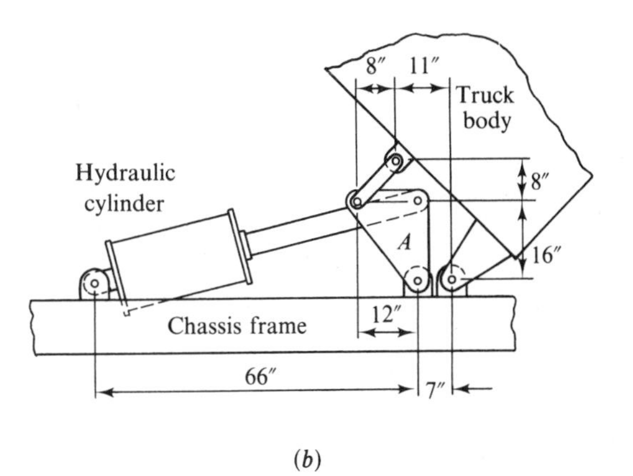

(b)
Figure P8-174

8-175 Determine the velocity of the piston relative to the cylinder in Problem 8-174. All data in the original problem apply to this problem.

8-176 The crank, connecting rod, and piston arrangement for a large reciprocating gas engine is shown in Fig. P8-176. The angular velocity of the crank shaft is 300 rpm clockwise. Determine the velocity of the piston for the position indicated.

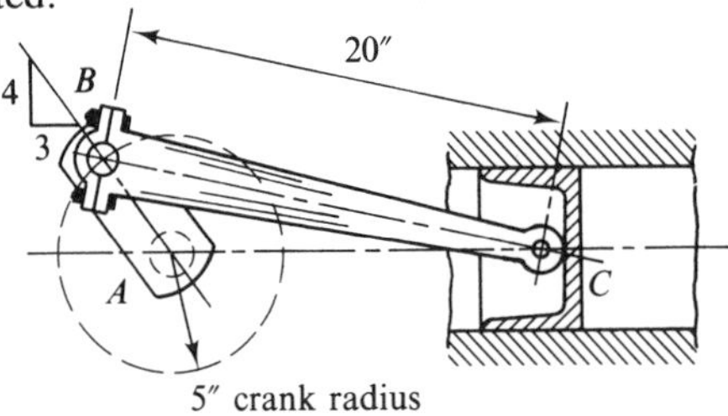

Figure P8-176

***8-177** Roller C of member CD in Fig. P8-177 has a velocity of 0.30 m/s upward along the slot. Determine

(a) The angular velocities of members AB and CD.

(b) The velocity of point D.

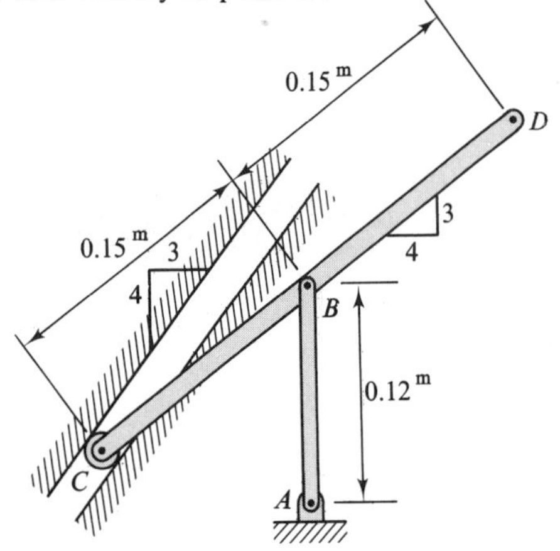

Figure P8-177

***8-178** The angular velocity of member BCD in Fig. P8-178 is 0.50 rad/s clockwise. Determine the velocity of point D.

8-179 The wheel in Fig. P8-179 rolls without slipping along an elevated track with an angular velocity of 4 rad per sec clockwise. Determine the linear velocities of points B, C, D, and E.

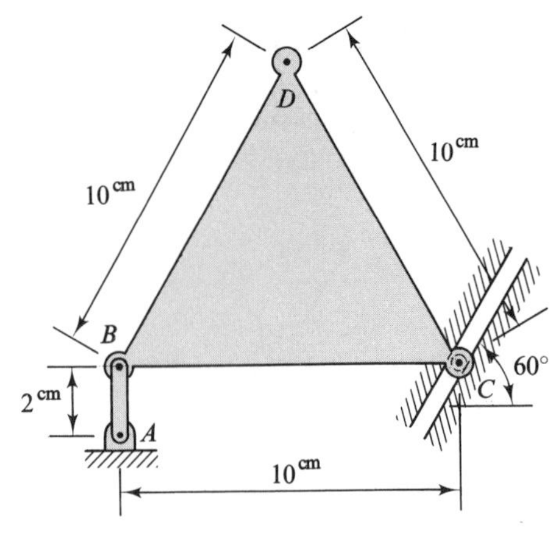

Figure P8-178

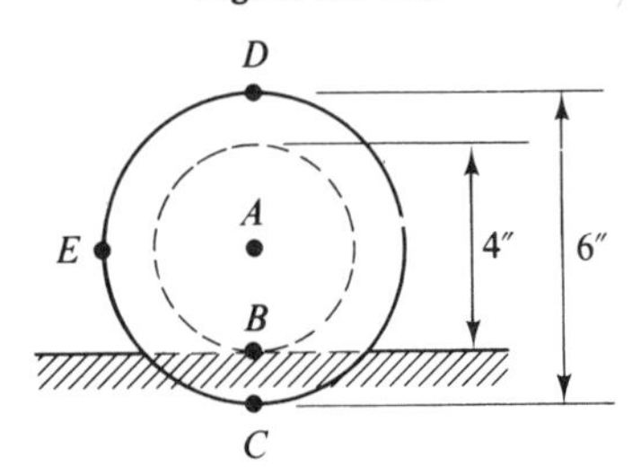

Figure P8-179

8-180 Solve Problem 8-179 if the wheel rolls and slips on the track. The velocity of point A is 6.0 ips to the left, and the angular velocity of the wheel is 4.0 rad per sec clockwise.

8-181 The wheel in Fig. P8-181 rolls without slipping along the curved track. As the center moves from O to O' the radius of the wheel will turn through an angle θ as shown. Derive the relationship between θ and ϕ and show that the speed of the center of the wheel is equal to $r\dot{\theta}$.

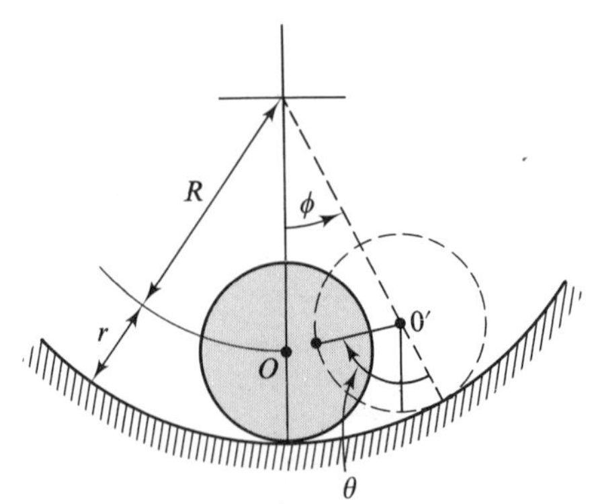

Figure P8-181

8-182 The gear in Fig. P8-182 rotates about a fixed axis at C with an angular velocity of 2.0 rad per sec clockwise in the position shown. The rack D slides along the horizontal plane, and block A slides in a slot in rack D. Determine the velocity of block A relative to the rack.

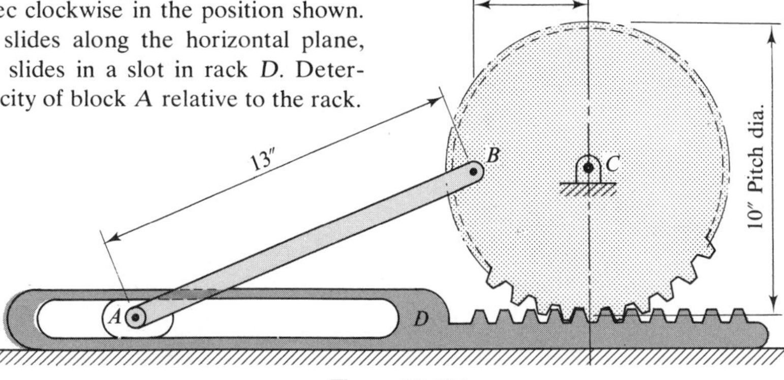

Figure P8-182

8-183 The crank OA in Fig. P8-183 has a constant clockwise angular velocity of 6.25 rad per sec. When the mechanism is in the position shown, determine

(a) The angular velocity of member AB.

(b) The linear velocity of the point on AB which is coincident with the roller at C.

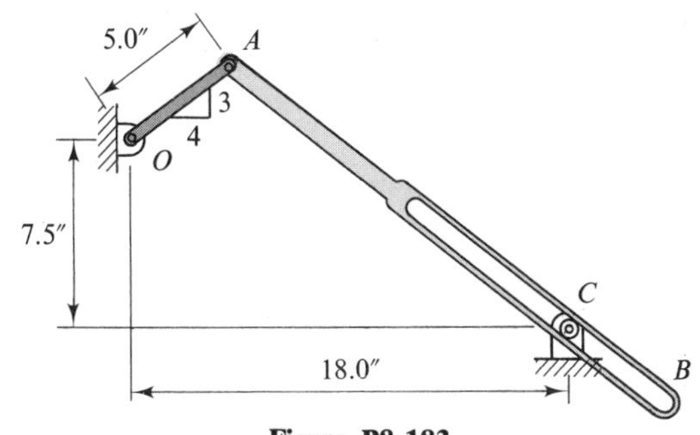

Figure P8-183

8-184 Member OA in Fig. P8-184 has an angular velocity of 2 rad per sec clockwise when in the position shown. Determine

(a) The angular velocity of body C.

(b) The velocity of point B.

(c) The velocity of AB relative to C.

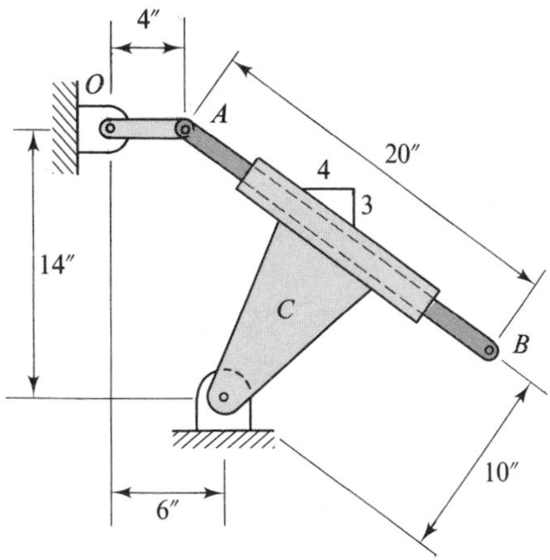

Figure P8-184

8-185 Roller A on member AB in Fig. P8-185 slides along the horizontal slot, while the member causes the wheel to rotate. There is no slipping between AB and the wheel. Derive expressions relating

(a) The velocity of A and the angular velocity of the wheel.

(b) The velocity of A and the angular velocity of AB.

(c) The angular velocities of the wheel and of AB.

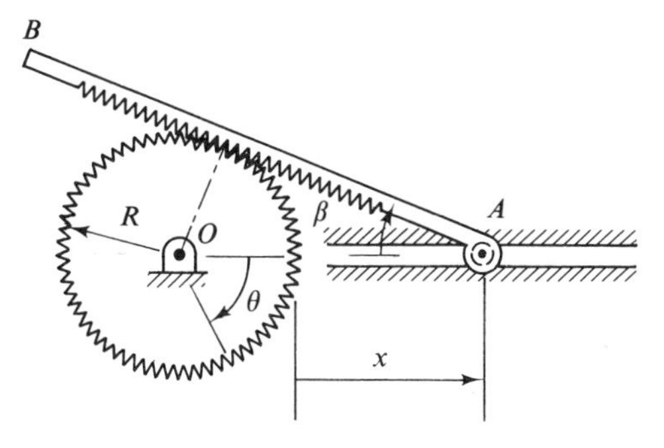

Figure P8-185

8-186 Member OP in Fig. P8-186 has an angular velocity of 5.0 rad per sec clockwise in the position shown. Determine the velocity of point Q.

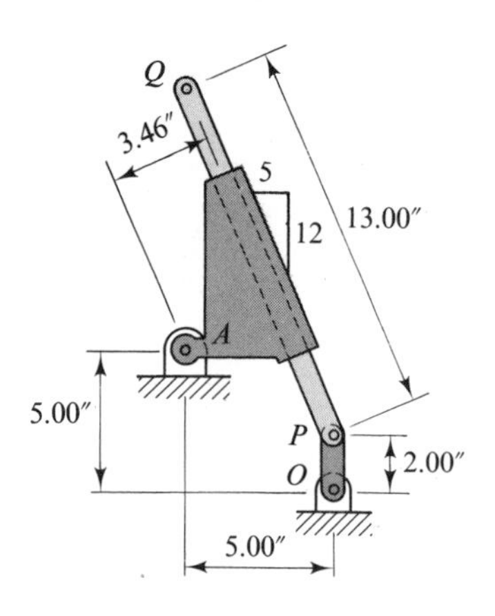

Figure P8-186

8-16 INSTANTANEOUS CENTERS FOR PLANE MOTION

Frequently, it is necessary to determine the linear velocity of some point on a rigid body or the angular velocity of a line in a rigid body when the body has plane motion. These velocities can usually be obtained from Eq. (8-23) for relative velocities. The relative velocity equation is

$$\mathbf{v}_B = \mathbf{v}_A + \mathbf{v}_{B/A},$$

where B and A are any two points on the rigid body. If a point A with zero velocity can be located, the velocity of B [see Eq. (8-21)] becomes

$$\mathbf{v}_B = \mathbf{v}_{B/A} = \boldsymbol{\omega} \times \mathbf{r}, \tag{8-24}$$

where $\mathbf{r}$ is the position vector of B from the point of zero velocity and $\boldsymbol{\omega}$ is the angular velocity of the body. When A and B are in any plane parallel to the plane of motion, the two vectors $\boldsymbol{\omega}$ and $\mathbf{r}$ in Eq. (8-24) are perpendicular, and the equation can be written in scalar form as

$$v_B = \omega r. \tag{8-25}$$

Either Eq. (8-24) or (8-25) provides a convenient method for determining the velocity of any point B on a rigid body having plane motion provided a point (on the body, or the body extended) whose velocity is zero at the instant in question can be readily located. For kinematics, a body, such as the rod CD in Fig. 8-33, can be considered to be extended or enlarged in any manner, provided the enlarged body remains a rigid body. The concept of a body extended is needed

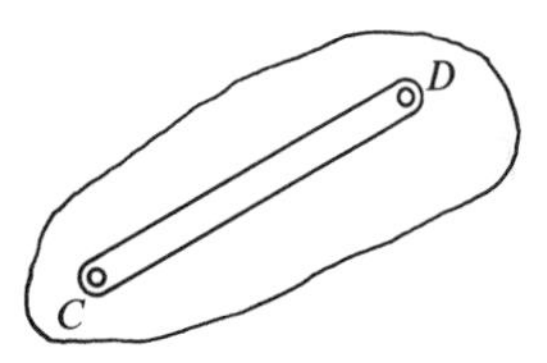

Figure 8-33

because the point with zero velocity is not necessarily a point on the actual body.

When a rigid body has plane motion, the axis in the body, or the body extended, *whose particles,* at any instant, *have zero velocity is defined as the instantaneous axis of zero velocity* or the instantaneous axis of the body. The instantaneous axis of zero velocity is always perpendicular to the plane of motion of the body. The point of intersection of the instantaneous axis and the plane of motion is defined as the *instantaneous center* of zero velocity. The velocity of any point of the body not on the instantaneous axis is perpendicular to the line from the instantaneous axis of the body to the point. Thus the velocities (but not the accelerations) of points on a body with plane motion can be obtained by considering the body to be rotating about the instantaneous axis of zero velocity at any instant. At the next instant, the instantaneous center is usually another point in the rigid body or body extended. The instantaneous axis is not a fixed axis in the body, nor is it a fixed axis in space.

The point in the plane of motion of the body, or body extended, which at any instant has zero velocity can be readily located if the directions of the velocities of any two points in the plane of motion of the body are known, provided the velocities are not parallel. The rigid body in Fig. 8-34 has plane motion, and the directions of the absolute velocities of two points A and B are as shown. Since at the instant in question the velocity of A is perpendicular to the line from the instantaneous center to A, the instantaneous center must be somewhere on the line ac through A perpendicular to $\mathbf{v}_A$. Likewise, the instantaneous center must be on the line bd through B and perpendicular to $\mathbf{v}_B$. Consequently, the instantaneous center of zero velocity is the intersection of these lines at point O. The magnitude of the angular velocity of the body is

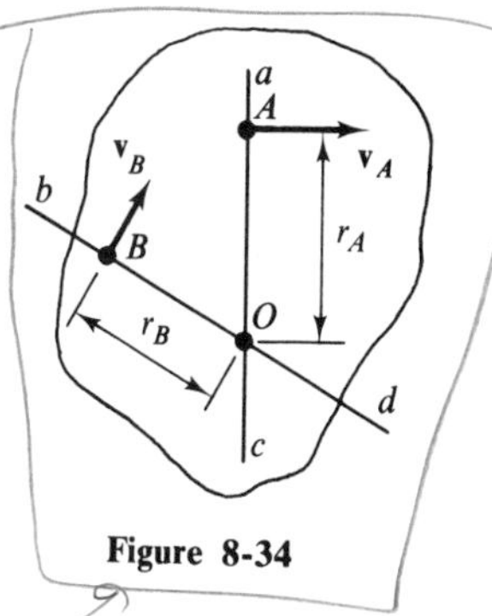

Figure 8-34

$$\omega = \frac{v_A}{r_A} = \frac{v_B}{r_B}. \tag{a}$$

Thus if the directions of two nonparallel velocities in the plane of motion and the magnitude of one of them are known, the location of the instantaneous center and the angular velocity of the body can be obtained, and consequently the linear velocity of any other point on the body can be determined.

When the velocities of two points in the plane of motion of a rigid body with plane motion are parallel, equal in magnitude, and in the same direction, the instantaneous center is at infinity, the angular velocity of the body is zero, and all points on the body have the same velocity. When the velocities of the two points are parallel and unequal in magnitude, lines drawn through the two points perpendicular to their velocities will be collinear, and the instantaneous center of the body will be on this line. If the sense and magnitudes of the velocities are known, the instantaneous center can be located by using Eq. (a). If the velocities of C and A in Fig. 8-35 are to the right and if v_C is less than v_A, the instantaneous center will be a distance r_C below C on ab. The distance r_C can be obtained from Eq. (a) as follows:

$$\omega = \frac{v_C}{r_C} = \frac{v_A}{r_A} = \frac{v_A}{d + r_C},$$

and if v_A, v_C, and d are known, r_C can be determined.

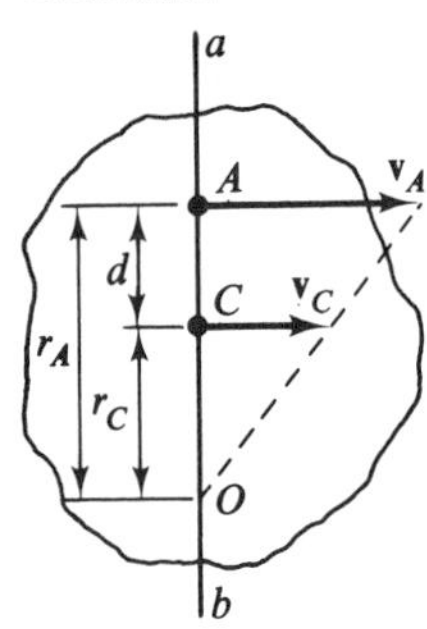

Figure 8-35

The instantaneous center of zero velocity of a rigid body can also be located if the velocity of one point on the body and the angular velocity of the body are known for the instant in question. If the velocity of point A on the rigid body in Fig. 8-36 has a known value to the right and if the angular velocity of the body has a known value counterclockwise, the distance from A to the instantaneous center is

$$r_A = \frac{v_A}{\omega}$$

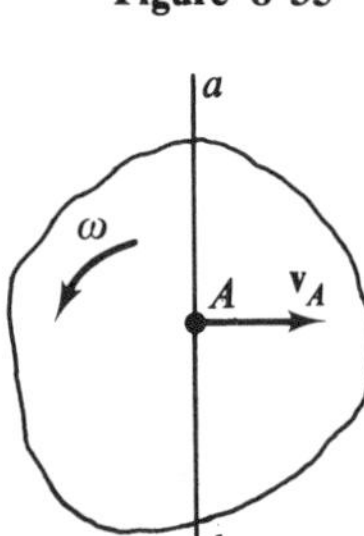

Figure 8-36

from Eq. (a). The instantaneous center must be on the line ab, and if the velocity of A is to be to the right when the angular velocity of the body is counterclockwise, the instantaneous center must be above A.

Once the instantaneous center of zero velocity is located, the magnitude of the velocity of any particle of the body at that instant is given by Eq. (a), where ω is the angular velocity of any line in the body in, or parallel to, the plane of motion and r is the distance from the instantaneous axis to the particle whose velocity is to be found. The value of ω can be obtained from Eq. (a) if it is not given. The direction of the velocity is obtained by inspection, since it is perpendicular to the line from O to the point and its sense is consistent with the sense of ω.

The instantaneous center of zero velocity is, in general, not a point of zero acceleration. No convenient method is available for locating the point of zero acceleration. The principles of relative motion as developed in Art. 8-14 for velocities will be extended in Art. 8-18 to apply to the accelerations of particles of a rigid body. It should be emphasized that *the instantaneous center of zero velocity cannot be used to obtain accelerations because the instantaneous center does not, in general, have zero acceleration.*

The instantaneous center of zero velocity as presented in this article is the instantaneous center of a body or member with respect to the earth. For use in courses in kinematics, this concept can be readily extended to determine the instantaneous center of one moving member of a mechanism relative to another moving member of the mechanism. The instantaneous center of two bodies with plane motion is the point common to the two bodies, or bodies extended, which has the same linear velocity in each body. In the previous discussion no mention was made of the second body, but if the instantaneous center of zero velocity is desired, the second body is understood to be the earth. For example, when a wheel rolls without slipping on a fixed surface, the point on the body which is in contact with the fixed surface (the earth) is the instantaneous center of zero velocity.

EXAMPLE 8-23

The wheel shown in Fig. 8-37 rolls without slipping on a horizontal elevated track with an angular velocity of 5 rad per sec clockwise. Locate the instantaneous center of zero velocity and determine the velocities of points A, D, and E.

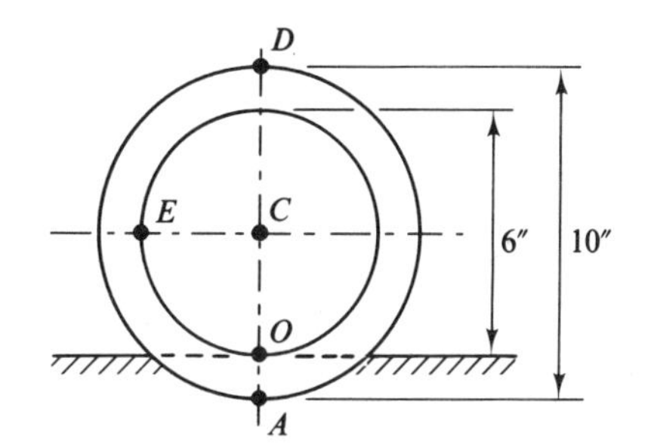

Figure 8-37

SOLUTION

The magnitude of the velocity of C, as determined by Example 8-21, is

$$v_C = r\omega = 3(5) = 15$$

and

$$\mathbf{v}_C = 15 \text{ ips} \rightarrow.$$

The instantaneous center of zero velocity must therefore be on a vertical line through C. Furthermore, from Eq. (8-25),

$$v_C = r_C\omega,$$

where r_C is the distance from the instantaneous center to C. Thus, $r_C = 3$ in., and point O is the instantaneous center of zero velocity. This result verifies the last statement preceding this example.

Again from Eq. (8-25),

$$v_A = r_A\omega = 2(5) = 10 \text{ ips}$$

and

$$\mathbf{v}_A = \underline{10 \text{ ips} \leftarrow.} \qquad \text{Ans.}$$

In a similar manner,

$$v_E = r_E\omega = 3\sqrt{2}(5) = 21.2 \text{ ips}$$

and

$$\mathbf{v}_E = \underline{21.2 \text{ ips} \angle 45^\circ} \qquad \text{Ans.}$$

Also

$$v_D = r_D\omega = 8(5) = 40 \text{ ips}$$

and

$$\mathbf{v}_D = \underline{40 \text{ ips} \rightarrow.} \qquad \text{Ans.}$$

EXAMPLE 8-24

The angular velocity of member AB of the link mechanism in Fig. 8-38(a) is 8 rad per sec clockwise. Determine, for the given position,
(a) The location of the instantaneous center of zero velocity of member BC.
(b) The angular velocity of BC.
(c) The angular velocity of CD.
(d) The velocity of point E, the midpoint of BC.

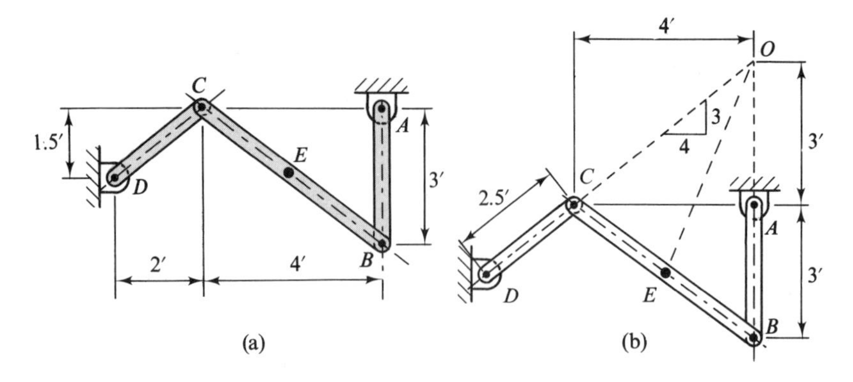

Figure 8-38

SOLUTION

(a) For the position indicated, point B has a horizontal velocity; therefore, the instantaneous center of member BC is on a vertical line through B. Point C has a velocity perpendicular to DC, and the instantaneous center of BC is also along line CD. These lines intersect at O, the instantaneous center of member BC, located as shown in Fig. 8-38(b).

(b) The magnitude of the velocity of point B is

$$v_B = r_{AB}\omega_{AB} = 3(8) = 24 \text{ fps}$$

and

$$\mathbf{v}_B = 24 \text{ fps} \leftarrow.$$

From Eq. (a),

$$\omega_{OB} = \frac{v_B}{r_{OB}} = \frac{24}{6} = 4 \text{ rad per sec}$$

and

$$\boldsymbol{\omega}_{BC} = \boldsymbol{\omega}_{OB} = \boldsymbol{\omega}_{OC} = \underline{4 \text{ rad per sec} \circlearrowright}. \qquad \text{Ans.}$$

(c) From Eq. (8-25),

$$v_C = \omega_{OC} r_{OC} = 4(5) = 20 \text{ fps}$$

and

$$\mathbf{v}_C = 20 \text{ fps} \quad (4\text{ up}, 3\text{ left})$$

Finally,

$$\omega_{CD} = \frac{v_C}{r_{CD}} = \frac{20}{2.5} = 8 \text{ rad per sec}$$

and

$$\boldsymbol{\omega}_{CD} = \underline{8 \text{ rad per sec} \circlearrowleft}. \qquad \text{Ans.}$$

(d) Point E is 2.0 ft to the left of AB and 1.5 ft below AC; therefore,

$$r_{OE} = \sqrt{(2.0)^2 + (3.0 + 1.5)^2} = 4.92 \text{ ft}.$$

The magnitude of the velocity of point E is

$$v_E = r_{OE}\omega_{BC} = 4.92(4.0) = 19.68 \text{ fps},$$

and the velocity of E is

$$\mathbf{v}_E = 19.68 \text{ fps} \; \nwarrow 23.96^\circ. \qquad \text{Ans.}$$

PROBLEMS

***8-187** Member AB in Fig. P8-187 has an angular velocity of 12 rad/s clockwise in the position shown. Determine the angular velocities of members BC and CD.

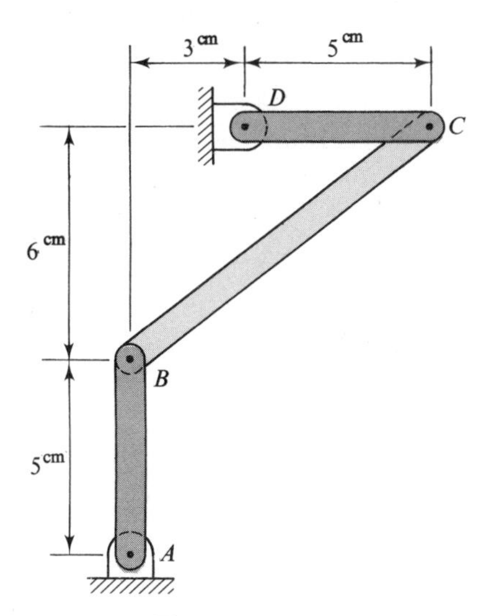

Figure P8-187

8-188 Member AB in Fig. P8-188 has an angular velocity of 6 rad per sec counterclockwise in the position shown. Determine the angular velocities of members BC and CD.

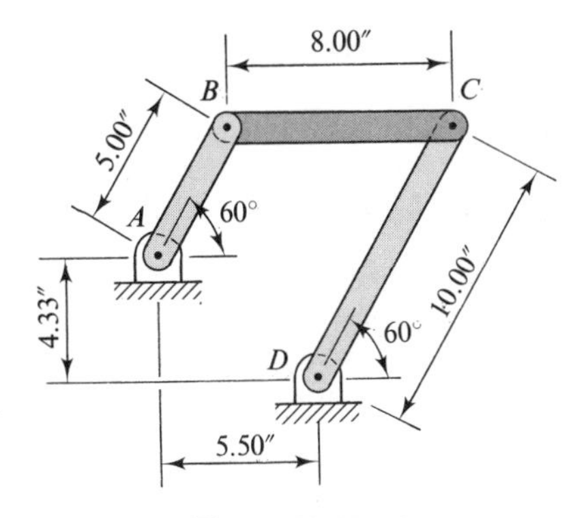

Figure P8-188

***8-189** Member AB in Fig. P8-189 has a constant angular velocity of 15 rad/s clockwise. Determine

(a) The angular velocity of member CD.

(b) The velocity of E, the midpoint of member BC.

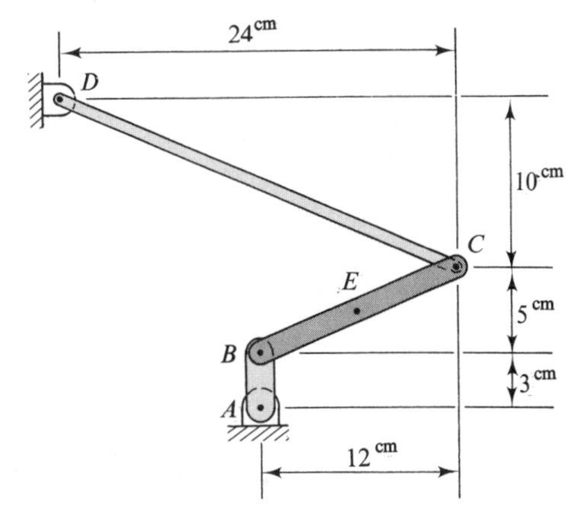

Figure P8-189

8-190 Member AB in Fig. P8-190 has an angular velocity of 4 rad per sec counterclockwise in the position shown. Determine

(a) The velocity of point C.

(b) The velocity of point E.

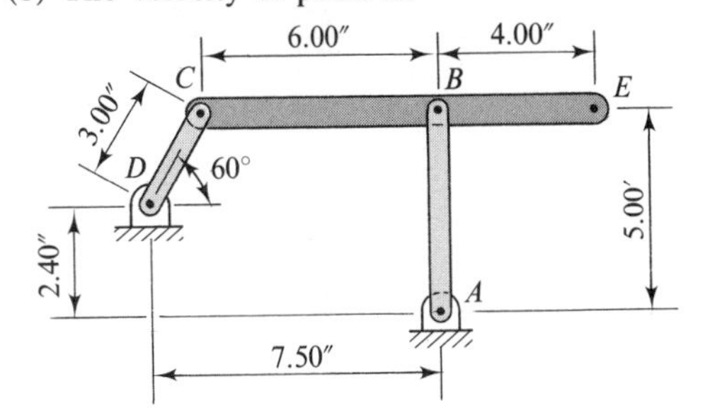

Figure P8-190

8-191 Determine the angular velocity of member *CD* in Problem 8-190 for which the speed of *E* will be 30 ips. (The angular velocity of *AB* will not be the same as for Problem 8-190.)

8-192 Member *AB* in Fig. P8-192 has an angular velocity of 3.0 rad per sec clockwise in the position shown. Determine
(a) The velocity of point *B*.
(b) The angular velocity of the crank *OA*.

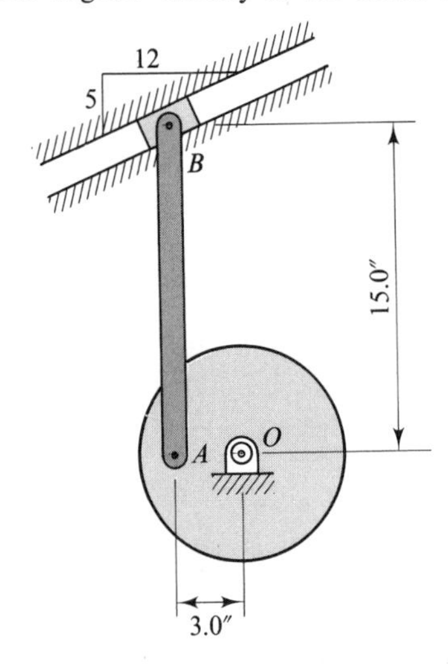

Figure P8-192

8-193 The velocity of point *A* on member *AC* in Fig. P8-193 is 75 ips up the slot in the position shown. Determine the velocity of point *C* and the angular velocity of member *BD*.

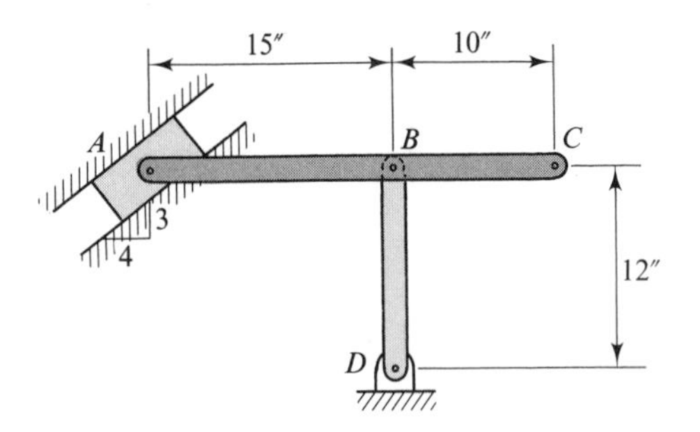

Figure P8-193

8-194 Plate *ABCD* in Fig. P8-194 is pinned to *OA* at *A*, and the roller at *C* slides in the horizontal slot. In the position shown the velocity of *C* is 35 ips to the right. Determine the velocities of points *A*, *B*, and *D*.

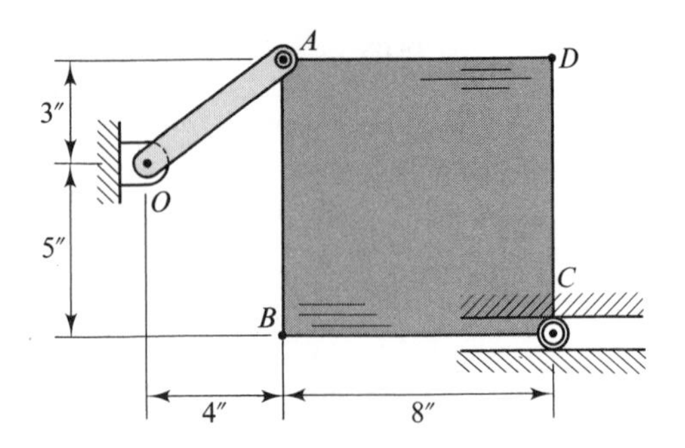

Figure P8-194

8-195 The angular velocity of member *C* in Fig. P8-195 is 0.80 rad per sec counterclockwise in the position shown. Determine the velocities of points *A* and *B*.

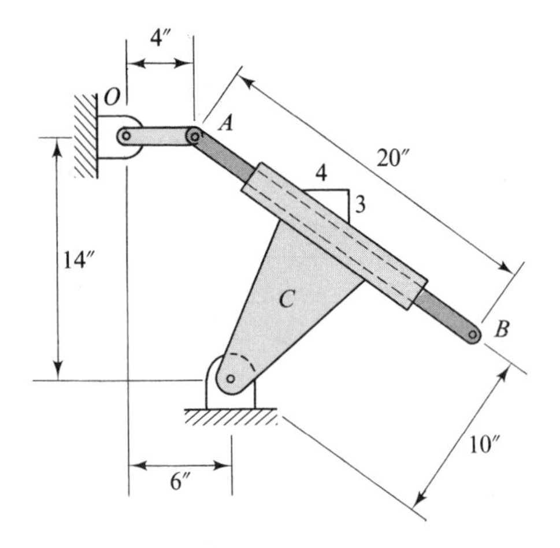

Figure P8-195

8-196 The angular velocity of member *A* in Fig. P8-196 is 2.0 rad per sec clockwise. Determine the velocities of points *P* and *Q*.

***8-197** The wheel in Fig. P8-197 rolls on the hub without slipping along the horizontal plane. The velocity of point *C* is 2.0 m/s to the right. Determine the velocities of points *A* and *M*.

***8-198** The wheel in Problem 8-197 rolls and slips along the plane in such a manner that the velocity of *C* is 1.5 m/s to the right and the

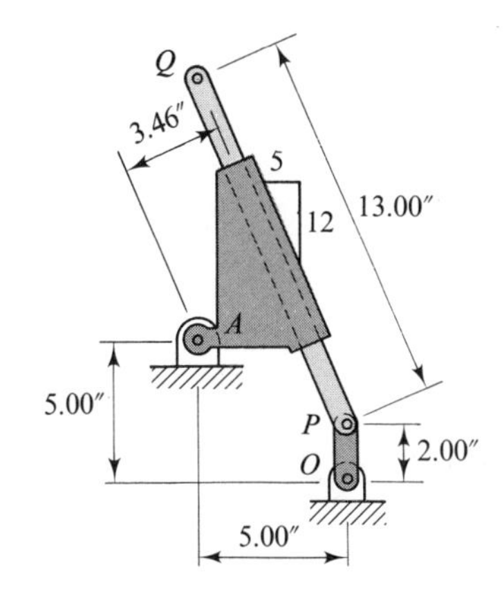

Figure P8-196

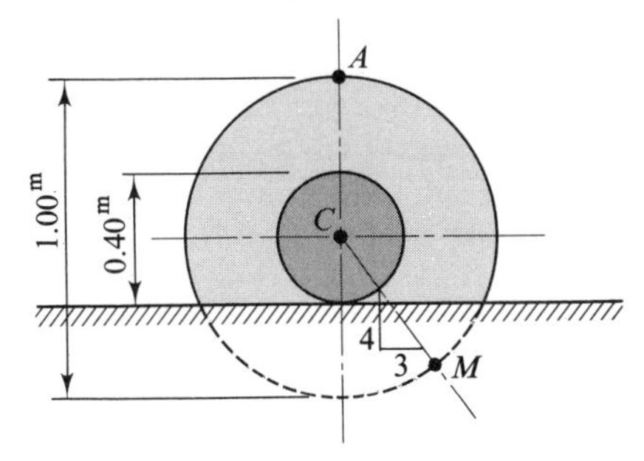

Figure P8-197

angular velocity of the wheel is 5 rad/s counterclockwise. Determine the velocities of points A and M.

***8-199** Point A on the wheel in Fig. P8-199 has a velocity of 2.0 m/s in the direction shown on the figure. Determine the angular velocity of the wheel if it remains in contact with the plane.

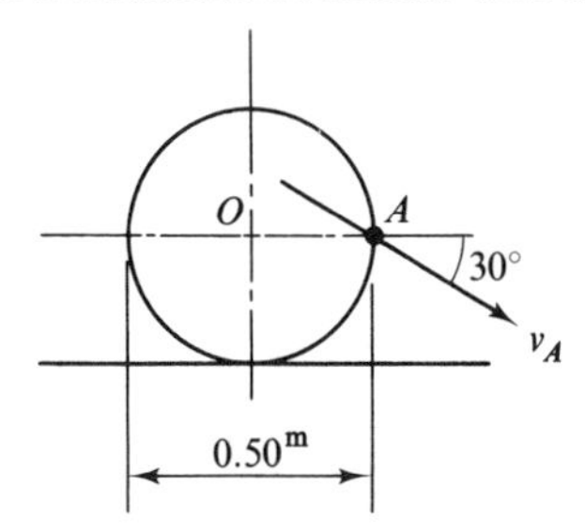

Figure P8-199

***8-200** The planetary gear in Fig. P8-200 has an outer gear A, an inner gear B, and a spider on which are mounted three small planet gears D. The pitch diameters are 30 cm for A, 20 cm for B, and 5 cm for D. The angular velocity of A is zero, and the angular velocity of C is 10 rpm clockwise. Determine

(a) The angular velocity of B.

(b) The velocity of point P on D.

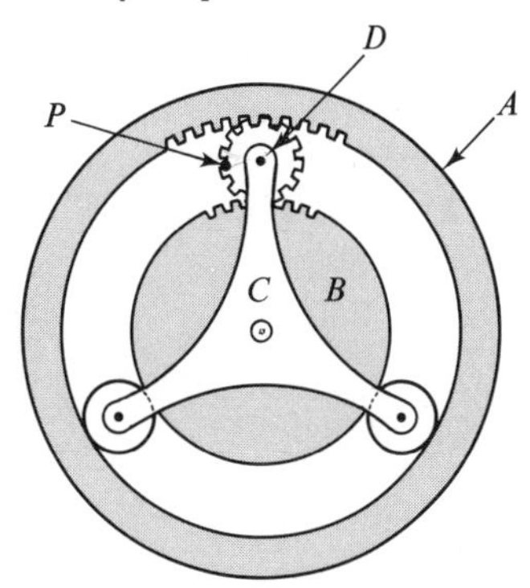

Figure P8-200

***8-201** In Problem 8-200 the inner gear B is stationary, and the spider C has an angular velocity of 10 rpm clockwise. Determine

(a) The angular velocity of A.

(b) The velocity of point P on D.

***8-202** In Problem 8-200 the spider C is stationary, and the inner gear B has an angular velocity of 10 rpm clockwise. Determine

(a) The angular velocity of A.

(b) The velocity of point P on D.

***8-203** The bar AB in Fig. P8-203 moves with its ends in contact with the horizontal and vertical walls. When $\theta = 60°$ the velocity of A is 50 cm/s downward. Determine the velocities of points B and G for this position.

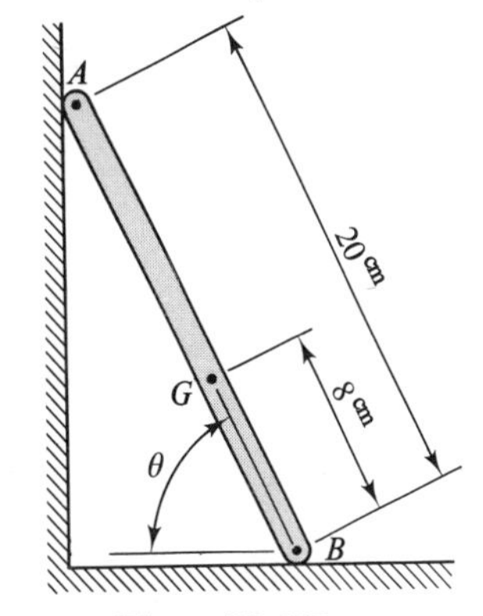

Figure P8-203

8-204 Body ABC in Fig. P8-204 is a rigid body with plane motion. In the position shown the velocity of C is 42 ips downward. Determine the velocities of points A and B.

8-205 The angular velocity of member ABC in Fig. P8-205 is 2.0 rad per sec counterclockwise in the position shown. Determine the velocities of points A, B, and C.

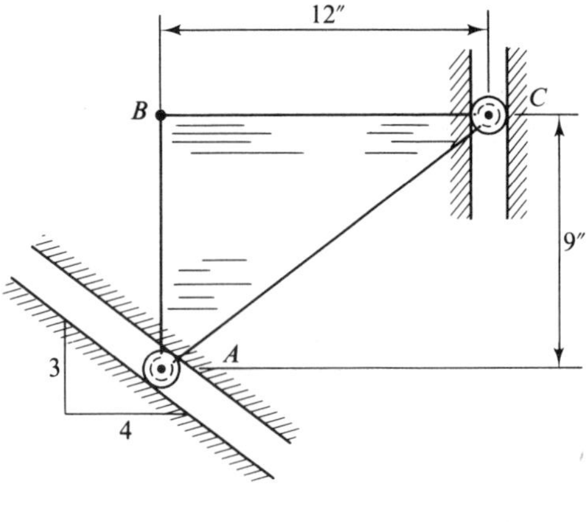

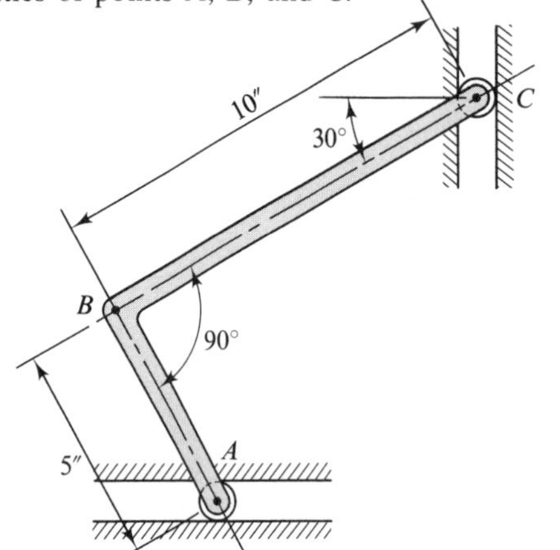

Figure P8-205

Figure P8-204

***8-206** The wheel in Fig. P8-206 rolls without slipping, and point O has a constant velocity of 40 cm/s to the left. Determine the linear velocity of B for the position shown.

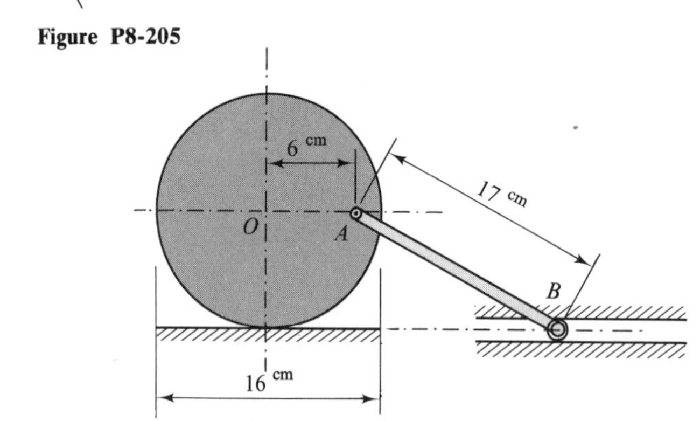

Figure P8-206

***8-207** Solve Problem 8-206 when point A is directly above point O.

***8-208** Solve Problem 8-206 when point A is to the left of point O.

***8-209** The two wheels in Fig. P8-209 roll along the horizontal plane without slipping and are connected by the rod AB. The velocity of point P is 12 cm/s to the right. Determine the angular velocity of AB and the linear velocity of Q (a) When A is to the right of P and B is to the right of Q as shown.

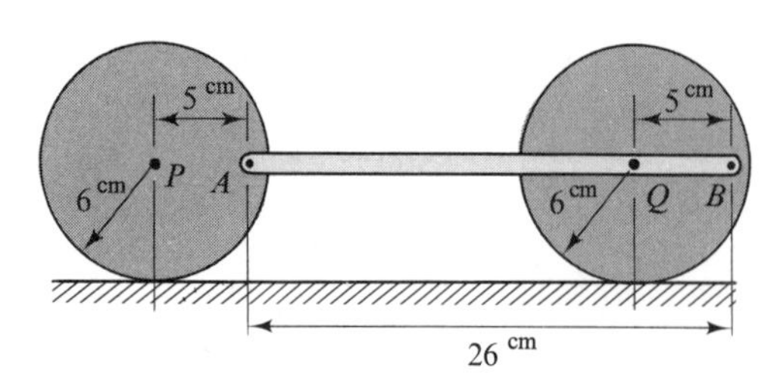

Figure P8-209

(b) When A is to the right of P and B is to the left of Q.

***8-210** Solve Problem 8-209 when the bodies are arranged as shown in Fig. P8-210.

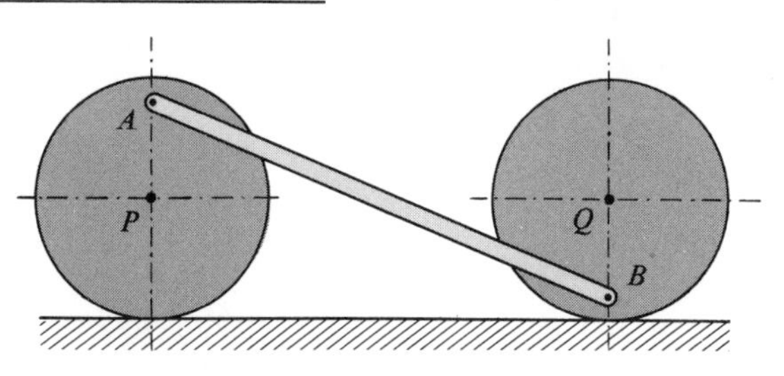

Figure P8-210

8-211 Gear A in Fig. P8-211 rotates with an angular velocity of 2 rad per sec clockwise. Gear B rolls without slipping on gear A. Arm C connects the gears as indicated and turns with an angular velocity of 3 rad per sec clockwise. Determine
(a) The angular velocity of gear B.
(b) The linear velocity of point D on gear B.

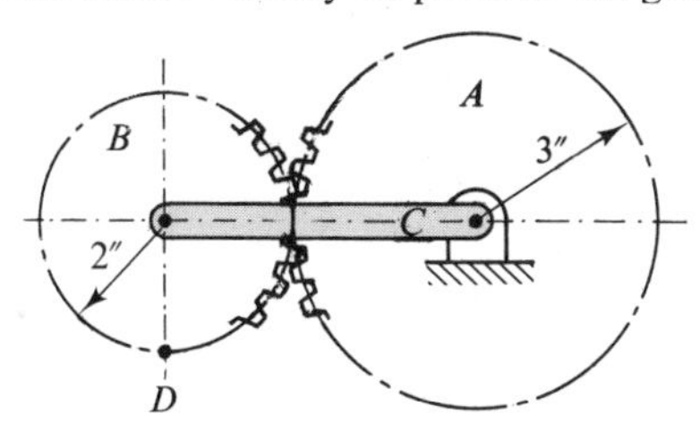

Figure P8-211

***8-212** Gears A and B in Fig. P8-212 are connected by smooth pins to the horizontal arm C at E and D. Arm C rotates with a constant angular velocity of 3 rad/s clockwise. Gear B rolls without slipping on either the fixed track or gear A. Determine the velocity of point P on gear A in the position shown.

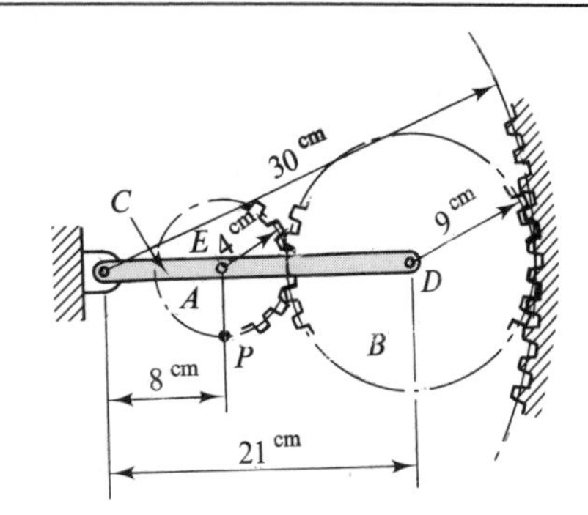

Figure P8-212

***8-213** Gears D and E in Fig. P8-213 are pinned to body ABC at B and C. Member ABC rotates about A with an angular velocity of 2 rad/s clockwise. Gear D rolls without slipping on either gear E or the fixed surface. Determine the velocity of point P on gear E.

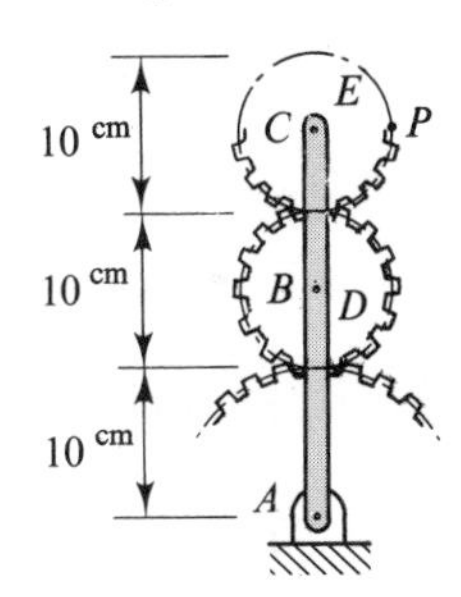

Figure P8-213

8-214 The three rigid bodies in Fig. P8-214 all move in the same plane. Member AE has an angular velocity of 5 rad per sec counterclockwise in the indicated position. Determine the linear velocities of points B and F.

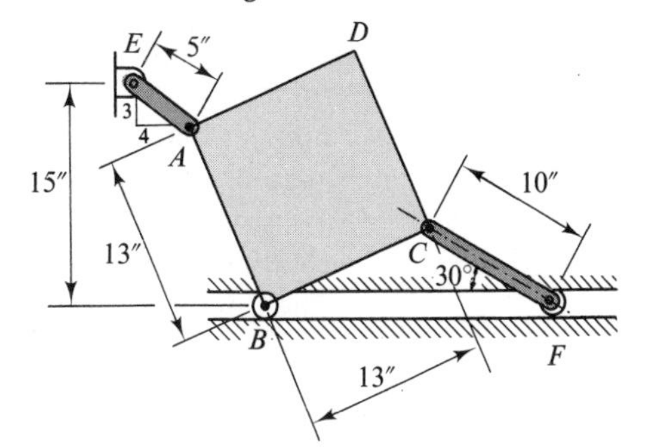

Figure P8-214

***8-215** Member ABC in Fig. P8-215 is pinned at C to bar CD. The blocks A, B, and D move in fixed slots. In the position shown A has a velocity of 10 cm/s to the right. Determine the velocities of points B and D for this position.

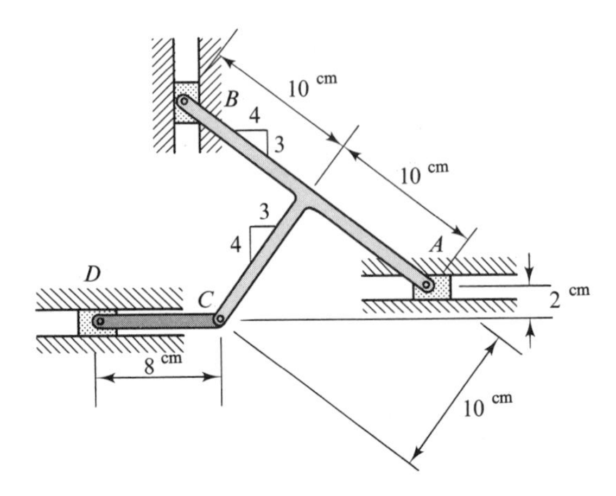

Figure P8-215

8-17 RELATIVE VELOCITY—THREE-DIMENSIONAL MOTION

As indicated in Art. 8-14, the equation

$$\mathbf{v}_B = \mathbf{v}_A + \mathbf{v}_{B/A}$$

is valid for any two points A and B provided nonrotating reference axes are used to describe the three terms. When points A and B are points on a rigid body with three-dimensional motion, the motion of B relative to A will be tangent to a spherical surface with its center at A and with a radius equal to the distance from A to B. The velocity of B relative to A can be expressed as

$$\mathbf{v}_{B/A} = \boldsymbol{\omega} \times \mathbf{r},$$

where $\boldsymbol{\omega}$ is the absolute angular velocity of the body and $\mathbf{r}$ is the vector *from A to B* as shown in Art. 8-13. Even though the velocities of points A and B are known, if the system is not fully constrained, the angular velocity $\boldsymbol{\omega}$ cannot be uniquely determined. An example of such a case is indicated by Fig. 8-39 in which member AB is connected to a rotating wheel and to a sliding collar at B by ball and socket joints. The angular velocity, $\boldsymbol{\omega}$, of AB can be resolved into two components, $\boldsymbol{\omega}_t$ parallel to AB and $\boldsymbol{\omega}_n$ normal to AB. The component $\boldsymbol{\omega}_t$ represents the *spin* of the member about its axis, and its magnitude will have no effect on the motion of points A and B. If one of the ball and socket joints is replaced by a clevis as shown in Fig. 8-41 (p. 432) or if the angular motion of AB is constrained by other means, the quantity $\boldsymbol{\omega}_t$, as well and $\boldsymbol{\omega}_n$, will have a definite value and $\boldsymbol{\omega}$ is therefore determinate. The angular velocity can be expressed as

$$\boldsymbol{\omega} = \omega_x \mathbf{i} + \omega_y \mathbf{j} + \omega_z \mathbf{k},$$

and when $\boldsymbol{\omega}$ is indeterminate it is sometimes convenient to assume one of the components to be equal to zero and solve for the remaining components. If the resulting angular velocity is resolved into $\boldsymbol{\omega}_t$ and $\boldsymbol{\omega}_n$ components, the $\boldsymbol{\omega}_n$ component will be unique, but the value of $\boldsymbol{\omega}_t$ will depend on which of the axial components (ω_x, ω_y, or ω_z) was assumed to be zero or some other arbitrary value. The following examples will demonstrate these concepts.

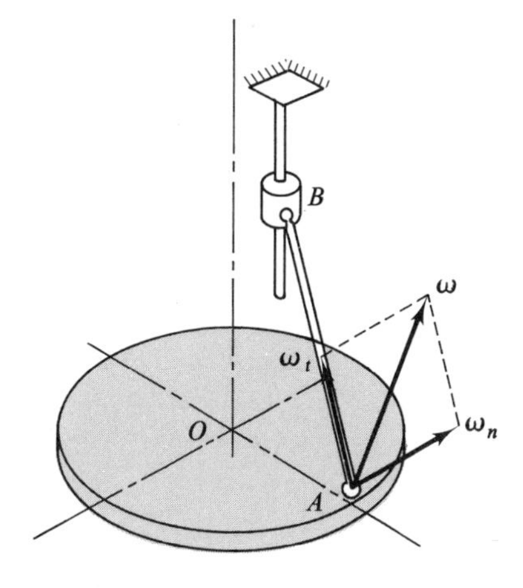

Figure 8-39

EXAMPLE 8-25

Collars A and B in Fig. 8-40 slide along fixed rods. The rod C is connected to A by a clevis and to B by a ball and socket joint. The velocity of A is $v_A \mathbf{i}$. Determine the velocity of B and the angular velocity of member C.

SOLUTION

The system is fully constrained; therefore the angular velocity of C can be uniquely determined. Rod C will have an angular velocity component due to the rotation of the collar and clevis pin and a component due to its rotation about the clevis pin. These components are shown in Fig. 8-41(a), and the angular velocity is

$$\begin{aligned}\boldsymbol{\omega} &= \boldsymbol{\omega}_1 + \boldsymbol{\omega}_2 \\ &= \omega_1 \mathbf{i} + \omega_2 \mathbf{e}_p \\ &= \omega_1 \mathbf{i} + \omega_2 \frac{d\mathbf{j} + c\mathbf{k}}{\sqrt{c^2 + d^2}}.\end{aligned} \tag{a}$$

The unit vector $\mathbf{e}_p$ (parallel to the pin) can often be visualized more readily by drawing a sketch looking at the end of the collar as in Fig. 8-41(b). Since $\mathbf{e}_p$ is perpendicular to both member C and the rod along which A slides, it can also be calculated as

$$\mathbf{e}_p = \frac{\mathbf{r}_{AB} \times \mathbf{r}_{DE}}{|\mathbf{r}_{AB} \times \mathbf{r}_{DE}|}.$$

The relative velocity equation is

$$\mathbf{v}_B = \mathbf{v}_A + \mathbf{v}_{B/A} = \mathbf{v}_A + \boldsymbol{\omega} \times \mathbf{r}_{B/A},$$

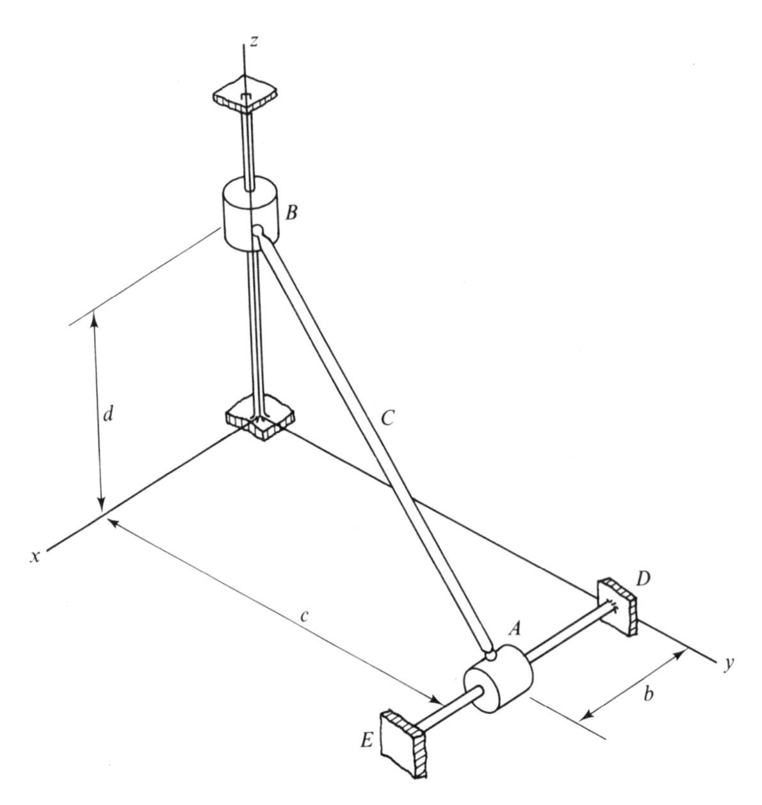

Figure 8-40

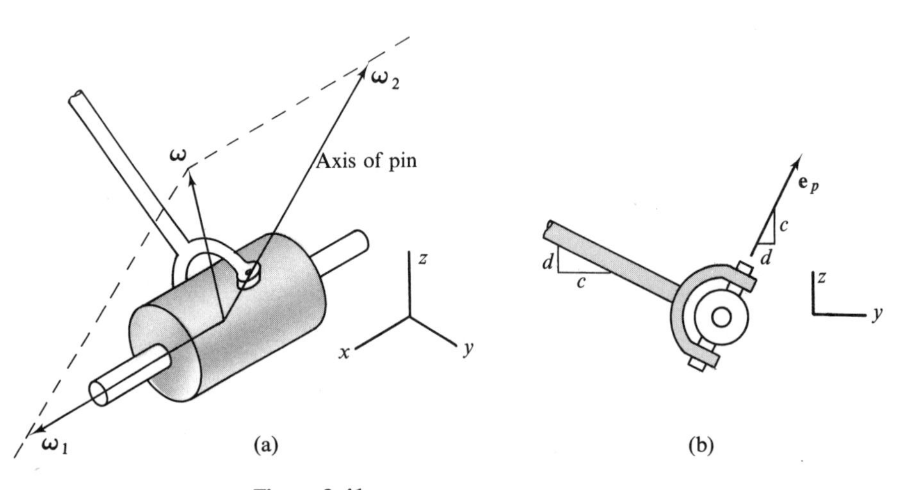

Figure 8-41

where $\mathbf{r}_{B/A} = -b\mathbf{i} - c\mathbf{j} + d\mathbf{k}$ is the position of B relative to A. The velocity equation can be expanded as

$$v_B\mathbf{k} = v_A\mathbf{i} + \begin{vmatrix} \mathbf{i} & \mathbf{j} & \mathbf{k} \\ \omega_1 & \dfrac{d\omega_2}{\sqrt{c^2+d^2}} & \dfrac{c\omega_2}{\sqrt{c^2+d^2}} \\ -b & -c & d \end{vmatrix}$$

$$= \mathbf{i}\left(v_A + \frac{d^2+c^2}{\sqrt{c^2+d^2}}\,\omega_2\right) + \mathbf{j}\left(\frac{-bc\omega_2}{\sqrt{c^2+d^2}} - d\omega_1\right)$$

$$+ \mathbf{k}\left(-c\omega_1 + \frac{bd\omega_2}{\sqrt{c^2+d^2}}\right).$$

The component equations are

$$\sqrt{c^2+d^2}\,\omega_2 = -v_A, \tag{b}$$

$$d\omega_1 + \frac{bc}{\sqrt{c^2+d^2}}\,\omega_2 = 0, \tag{c}$$

$$-c\omega_1 + \frac{bd}{\sqrt{c^2+d^2}}\,\omega_2 = v_B. \tag{d}$$

Equations (b) and (c) give

$$\omega_2 = \frac{-v_A}{\sqrt{c^2+d^2}},$$

$$\omega_1 = \frac{bcv_A}{d(c^2+d^2)},$$

and when these values are substituted into Eq. (a) the angular velocity of C is

$$\underline{\boldsymbol{\omega} = \frac{v_A}{d(c^2+d^2)}(bc\mathbf{i} - d^2\mathbf{j} - cd\mathbf{k}).} \qquad \text{Ans.}$$

The velocity of B is obtained by substituting values of ω_1 and ω_2 into Eq. (d), thus

$$v_B = \frac{-bc^2v_A}{d(c^2+d^2)} + \frac{bd}{\sqrt{c^2+d^2}}\left(\frac{-v_A}{\sqrt{c^2+d^2}}\right),$$

which reduces to

$$v_B = -\frac{b}{d}\,v_A$$

and

$$\underline{\mathbf{v}_B = -\frac{b}{d}\,v_A\mathbf{k}.} \qquad \text{Ans.}$$

When numerical data are given, numbers can be substituted for symbols and thus simplify the calculations, but this makes it impossible to use dimensional homogeneity to check the algebraic manipulations.

EXAMPLE 8-26

Solve Example 8-25 if the clevis at A is replaced a ball and socket joint.

SOLUTION

When the clevis at A is replaced by a ball and socket the system is no longer completely constrained and the angular velocity of C cannot be uniquely determined. The velocity of B, however, can be uniquely determined, and an angular velocity of C can be obtained by introducing an arbitrary constraint on the system. The problem will be solved twice, first by introducing the constraint $\omega_x = 0$ and second by using the constraint ω_t (the spin velocity) $= 0$.

The relative velocity equation for points A and B is

$$\mathbf{v}_B = \mathbf{v}_A + \mathbf{v}_{B/A},$$

$$v_B\mathbf{k} = v_A\mathbf{i} + \boldsymbol{\omega} \times \mathbf{r}_{B/A},$$

where

$$\boldsymbol{\omega} = \omega_x\mathbf{i} + \omega_y\mathbf{j} + \omega_z\mathbf{k}.$$

Thus the velocity equation becomes

$$v_B\mathbf{k} = v_A\mathbf{i} + \begin{vmatrix} \mathbf{i} & \mathbf{j} & \mathbf{k} \\ \omega_x & \omega_y & \omega_z \\ -b & -c & d \end{vmatrix}$$

$$= v_A\mathbf{i} + \mathbf{i}(d\omega_y + c\omega_z) - \mathbf{j}(d\omega_x + b\omega_z) + \mathbf{k}(-c\omega_x + b\omega_y),$$

which is equivalent to the following three algebraic equations:

$$\mathbf{i}: \quad +d\omega_y + c\omega_z = -v_A, \tag{a}$$

$$\mathbf{j}: \quad -d\omega_x - b\omega_z = 0 \tag{b}$$

$$\mathbf{k}: \quad -c\omega_x + b\omega_y = v_B. \tag{c}$$

Since there are four unknown quantities, ω_x, ω_y, ω_z, and v_B, and only three equations, not all the unknowns can be determined. When $\boldsymbol{\omega}$ is indeterminate a possible solution can be obtained by assuming a value for one of the components and solving for the other two. From the assumption that $\omega_x = 0$

$$\omega_z = -\frac{d}{b}\omega_x = 0,$$

$$\omega_y = \frac{-v_A}{d},$$

and

$$\boldsymbol{\omega} = \frac{-v_A}{d}\mathbf{j}. \qquad \text{Ans.}$$

From Eq. (c)

$$v_B = b\omega_y = -\frac{b}{d} v_A$$

and

$$\mathbf{v}_B = -\frac{b}{d} v_A \mathbf{k}. \qquad \text{Ans.}$$

It is seen that an initial assumption of *either* $\omega_x = 0$ or $\omega_y = 0$ yields the same axial components for $\boldsymbol{\omega}$. Assuming $\omega_y = 0$ will yield a different set of axial components for $\boldsymbol{\omega}$ and will also increase the complexity of the calculations for $\boldsymbol{\omega}$. All three of these assumptions, however, result in the same answer for $\mathbf{v}_B$.

The angular velocity of C can be resolved into normal and tangential components as follows. The component parallel to C is

$$\boldsymbol{\omega}_t = (\boldsymbol{\omega} \cdot \mathbf{e}_t)\mathbf{e}_t$$

$$= \frac{-v_A}{d} \mathbf{j} \cdot \left(\frac{-b\mathbf{i} - c\mathbf{j} + d\mathbf{k}}{\sqrt{b^2 + c^2 + d^2}} \right) \left(\frac{-b\mathbf{i} - c\mathbf{j} + d\mathbf{k}}{\sqrt{b^2 + c^2 + d^2}} \right)$$

$$= \frac{v_A c}{d(b^2 + c^2 + d^2)} (-b\mathbf{i} - c\mathbf{j} + d\mathbf{k}).$$

Note that $\mathbf{e}_t$ is a unit vector parallel to AB. The component of the angular velocity normal to AB is

$$\boldsymbol{\omega}_n = \boldsymbol{\omega} - \boldsymbol{\omega}_t$$

$$= -\frac{v_A}{d} \mathbf{j} - \frac{v_A c}{d(b^2 + c^2 + d^2)} (-b\mathbf{i} - c\mathbf{j} + d\mathbf{k})$$

$$= \frac{v_A}{d(b^2 + c^2 + d^2)} [bc\mathbf{i} - (b^2 + d^2)\mathbf{j} - cd\mathbf{k}].$$

The magnitude of the tangential component of the angular velocity (the spin velocity) is $\boldsymbol{\omega} \cdot \mathbf{e}_t$, and when this quantity is set equal to zero it provides a fourth equation,

$$\omega_t = (\omega_x \mathbf{i} + \omega_y \mathbf{j} + \omega_z \mathbf{k}) \cdot \left(\frac{-b\mathbf{i} - c\mathbf{j} + d\mathbf{k}}{\sqrt{b^2 + c^2 + d^2}} \right) = 0$$

or

$$-b\omega_x - c\omega_y + d\omega_z = 0. \qquad \text{(d)}$$

There are now four equations (a), (b), (c), and (d) which can be solved simultaneously for the four unknowns ω_x, ω_y, ω_z, and v_B.

From Eq. (a),

$$\omega_y = -\frac{v_A}{d} - \frac{c\omega_z}{d}, \qquad \text{(e)}$$

and from Eq. (b),

$$\omega_x = -\frac{b\omega_z}{d}. \tag{f}$$

Equations (e) and (f) can be substituted into Eq. (d) to give

$$\frac{b^2\omega_z}{d} + \frac{cv_A}{d} + \frac{c^2\omega_z}{d} + d\omega_z = 0$$

from which

$$\omega_z = -\frac{cv_A}{b^2 + c^2 + d^2}.$$

The values for ω_x and ω_y can be determined by substituting into Eq. (e) and (f), and the final result is

$$\underline{\boldsymbol{\omega} = \frac{v_A}{d(b^2 + c^2 + d^2)}[bc\mathbf{i} - (b^2 + d^2)\mathbf{j} - cd\mathbf{k}].} \qquad \text{Ans.}$$

This result is the same as the value of $\boldsymbol{\omega}_n$ obtained in the first solution, which serves to confirm the fact that $\boldsymbol{\omega}_n$ is independent of the assumptions made in obtaining a value for the indeterminate angular velocity of the bar.

The velocity of B is obtained by substituting into Eq. (c); that is

$$\begin{aligned} v_B &= -c\omega_x + b\omega_y \\ &= \frac{v_A}{d(b^2 + c^2 + d^2)}(-bc^2 - b^3 - bd^2) \\ &= -\frac{b}{d}v_A \end{aligned}$$

and

$$\underline{\mathbf{v}_B = -\frac{b}{d}v_A\mathbf{k}.} \qquad \text{Ans.}$$

The procedures used in the preceding examples for solving the relative velocity equation

$$\mathbf{v}_B = \mathbf{v}_A + \mathbf{v}_{B/A}$$

can be summarized as follows:

1. If the system is fully constrained, the relative velocity equation together with the equations of constraint give the same number of algebraic equations as the number of unknowns, and $\boldsymbol{\omega}$ can be uniquely determined (as in Example 8-25).

2. If the system is not fully constrained, the number of algebraic equations is less than the number of unknowns. Additional arbitrary constraints (which do not violate any existing constraints) must be introduced to solve for $\boldsymbol{\omega}$, which will then not be unique (as in Example 8-26).

PROBLEMS

***8-216** The bent rod CDE is welded to rod AB in Fig. P8-216, and point C slides on the horizontal xy plane. Rod AB is connected to the two collars by means of ball and socket joints. Determine the velocity of C and the angular velocity of member ABC in terms of v_A.

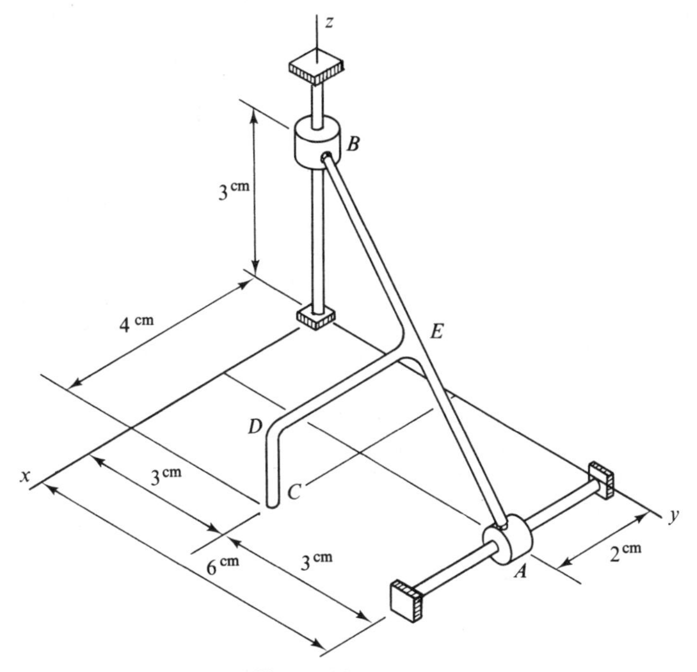

Figure P8-216

8-217 The crank OA in Fig. P8-217 rotates in the yz plane about an axis through O with a constant angular velocity of $5\mathbf{i}$ rad per sec. Link AB has a ball and socket joint at each end.

(a) Determine the velocity of the collar at B.
(b) Determine the component of the angular velocity of AB which is perpendicular to AB.

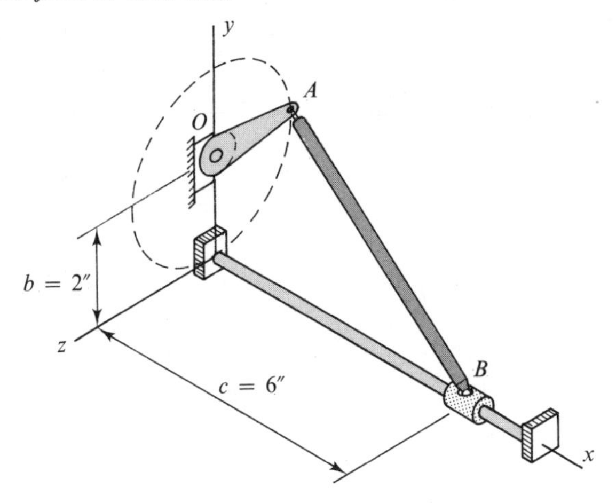

Figure P8-217

8-218 Determine the angular velocity of member AB in Problem 8-217 if it is connected to the collar at B by a clevis as shown in Fig. P8-218.

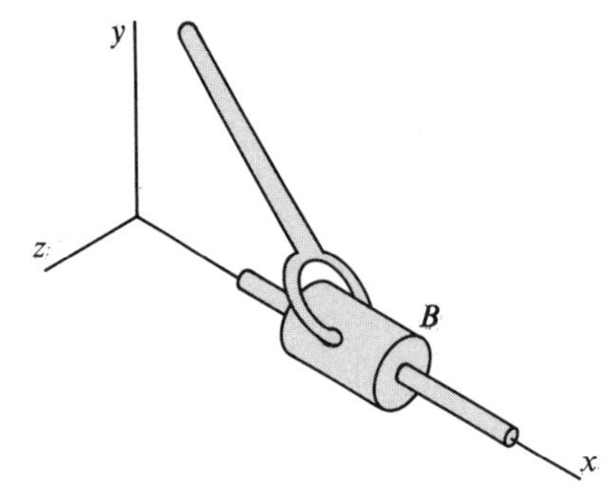

Figure P8-218

8-219 Figure P8-219 represents a pair of bevel gears A and B. Gear A rotates in a horizontal plane about a vertical axis at O. Gear B rotates around the shaft OC, which in turn rotates about the vertical axis at O. The angular velocity of the shaft OC is $10\mathbf{j}$ rad per sec, and gear A is stationary. Determine

(a) The angular velocity of gear B.

(b) The velocity of point D at the top of the gear.

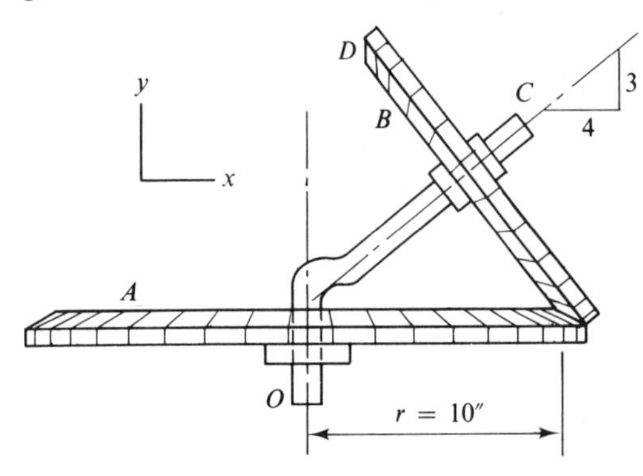

Figure P8-219

8-220 Solve Problem 8-219 if gear A has an angular velocity of $4\mathbf{j}$ rad per sec. All other data are the same as for Problem 8-219.

8-221 Disk D in Fig. P8-221 rotates about an axis at O with an angular velocity of $\omega_1\mathbf{k}$. The collar B slides along the inclined rod which is parallel to the yz plane. Member AB is

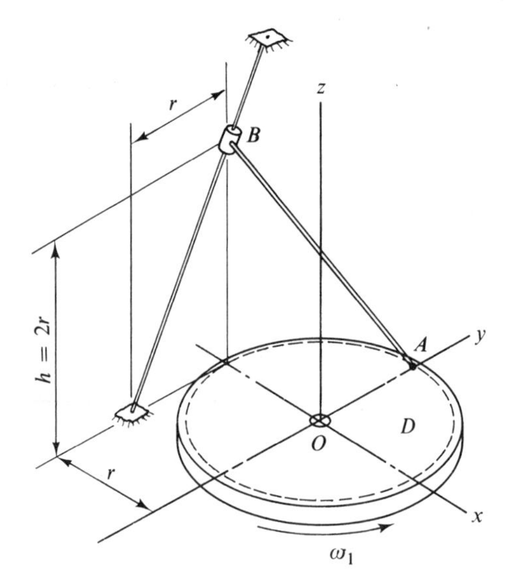

Figure P8-221

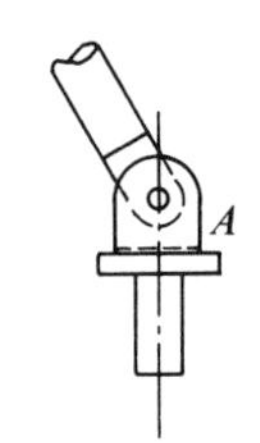

Figure P8-222

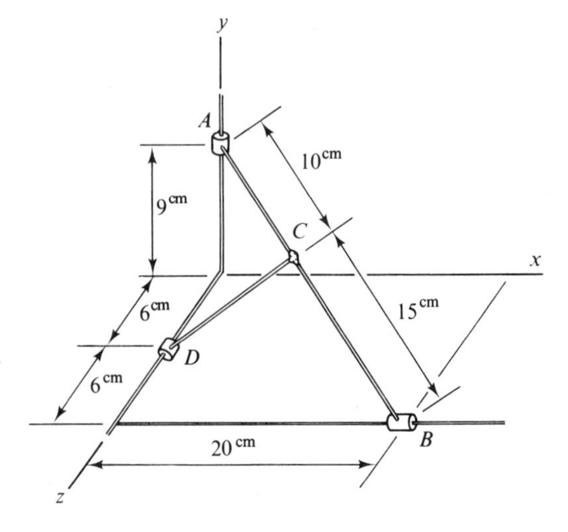

Figure P8-223

connected to the disk and to the collar by ball and socket joints. Determine
(a) The velocity of collar B.
(b) The component of the angular velocity of AB which is normal to AB.

8-222 The ball and socket at A in Problem 8-221 is replaced by the joint shown in Fig. P8-222 in which the vertical member rotates in a hole in the disk and is connected to AB by a horizontal pin. Determine the angular velocity of member AB.

***8-223** Cylinders A, B, and D in Fig. P8-223 slide along smooth rods as shown. The rigid member AB is connected to A and B with ball and socket joints. Member CD is connected to point C on member AB and to the cylinder D by ball and socket joints. The velocity of A is 12 cm/s downward in the position shown. Determine the velocities of points B and D.

8-18 RELATIVE ACCELERATION—NONROTATING AXES

In Art 8-6 equations were developed relating the absolute motion of a moving point to its motion measured relative to a set of nonrotating axes at a second moving point and the absolute motion of A. The equation relating the accelerations is

$$\mathbf{a}_B = \mathbf{a}_A + \mathbf{a}_{B/A}, \tag{8-26}$$

where $\mathbf{a}_B$ and $\mathbf{a}_A$ are the absolute acceleration of points B and A, respectively, and $\mathbf{a}_{B/A}$ is the acceleration of B relative to a set of nonrotating axes attached to A. The acceleration of B relative to A is the acceleration B would appear to have to an observer moving with point A (but not rotating).

When A and B are two points on a rigid body, the motion of B relative to A must be along a path on a spherical surface with its center at A and with a radius equal to the distance from A to B since the two points must remain the same distance apart. The acceleration of B relative to A is given by Eq. (8-22) as

$$\begin{aligned}\mathbf{a}_{B/A} &= \boldsymbol{\alpha}\times\mathbf{r}+\boldsymbol{\omega}\times(\boldsymbol{\omega}\times\mathbf{r})\\ &= \boldsymbol{\alpha}\times\mathbf{r}+\boldsymbol{\omega}\times\mathbf{v}_{B/A},\end{aligned}$$

where $\boldsymbol{\alpha}$ and $\boldsymbol{\omega}$ are the absolute angular acceleration and velocity of the rigid body and $\mathbf{r}$ is the position vector *from A to B*. When the body has plane motion the angular acceleration and angular velocity vectors are parallel. Thus for plane motion $\boldsymbol{\alpha}\times\mathbf{r}$ is parallel to $\boldsymbol{\omega}\times\mathbf{r}$ and is tangent to the path of B relative to A. The term $\boldsymbol{\omega}\times\mathbf{v}_{B/A}$ is normal to the relative velocity and thus is normal to the path. When the body containing A and B has three-dimensional motion the vectors representing $\boldsymbol{\alpha}$ and $\boldsymbol{\omega}$ are usually not parallel, and the quantity $\boldsymbol{\alpha}\times\mathbf{r}$ will not, in general, be parallel to $\boldsymbol{\omega}\times\mathbf{r}$ and thus will not be the tangential component of the acceleration of B relative to A.

If Eq. (8-26) is applied to two points not on the same rigid body, the motion of B relative to A will not, in general, be along a circular path, and the relative acceleration is usually more difficult to analyze. Application of Eq. (8-26) to problems involving plane motion and three-dimensional motion are discussed in the following two articles.

8-19 RELATIVE ACCELERATION—PLANE MOTION

In Art. 8-18 it was pointed out that when a rigid body has plane motion the angular velocity and angular acceleration vectors are parallel and can change in magnitude but not in direction. When the position vector $\mathbf{r}$ from A to B is parallel to the plane of motion it is normal to $\boldsymbol{\alpha}$ and $\boldsymbol{\omega}$ and Eq. (8-26) can be written as

$$\begin{aligned}\mathbf{a}_{B/A} &= \boldsymbol{\alpha}\times\mathbf{r}+\boldsymbol{\omega}\times(\boldsymbol{\omega}\times\mathbf{r})\\ &= \boldsymbol{\alpha}\times\mathbf{r}-\omega^2\mathbf{r}\\ &= \alpha r\mathbf{e}_t+\omega^2 r\mathbf{e}_n,\end{aligned}$$

where $\mathbf{e}_t$ is a unit vector tangent to the path of B relative to A and $\mathbf{e}_n$ is a unit vector normal to the path and directed from B to A. Note that $\boldsymbol{\alpha}$ and $\boldsymbol{\omega}$ are absolute angular quantities (measured with respect to nonrotating axes) and that $\mathbf{r}$ is the position vector of B relative to A.

In Art. 8-15 it was noted that a wheel which rolls without slipping along a fixed plane surface is an important case of plane motion of a rigid body. It was shown that the center O of the wheel has rectilinear motion and that the magnitudes of the position and velocity of point O are related to the magnitudes of the angular position and velocity of the wheel by the expressions [see Fig. 8-30]

$$s_O = r\theta \tag{a}$$

$$v_O = r\dot{\theta} = r\omega \tag{b}$$

in which s_O and v_O are to the right when θ and ω are clockwise. The acceleration of point O can be obtained by differentiating Eq. (b) and is

$$a_O = r\dot{\omega} = r\alpha. \tag{c}$$

Note that the acceleration of O will be to the right when $\dot{\omega}=\alpha$ is clockwise and that r is the distance from the fixed plane to the center of the wheel. These three expressions are valid when the wheel rolls without slipping along a fixed plane. If the wheel rolls without slipping along a curved surface, a similar proof shows that the quantity $r\alpha$ is the magnitude of the tangential component of the acceleration of the center of the wheel. The normal component of the acceleration of the center of the wheel will depend on the velocity of the center and the radius of curvature of the path. If the wheel rolls without slipping along a moving surface, for example, a drum rolling on the bed of a moving truck, the preceding expressions give the velocity and acceleration of the center of the wheel *relative* to the moving surface. None of the preceding expressions are valid if the wheel rolls and slips on a fixed or moving surface.

The following examples illustrate the use of Eq. (8-26) for problems involving plane motion.

EXAMPLE 8-27

The large unbalanced wheel in Fig. 8-42 rolls without slipping along the horizontal plane. In the position shown the center O has a velocity of 4.0 m/s to the right and an acceleration of 10.0 m/s^2 to the left. Determine the velocity and acceleration of the mass center, G.

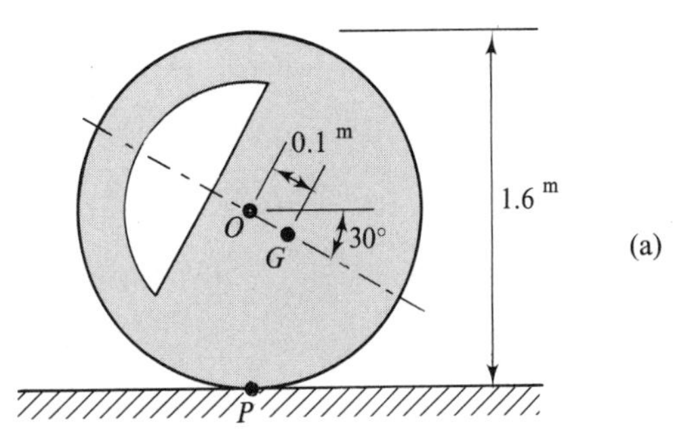

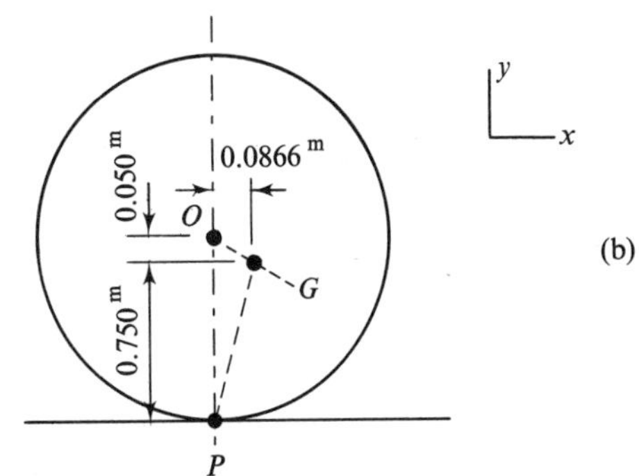

Figure 8-42

SOLUTION

The wheel rolls without slipping along a fixed horizontal plane; consequently the angular velocity and acceleration of the wheel are

$$\omega = \frac{v_O}{r} = \frac{4.0}{0.8} = 5.0 \text{ rad/s} \circlearrowright,$$

$$\alpha = \frac{a_O}{r} = \frac{10.0}{0.8} = 12.5 \text{ rad/} \circlearrowleft \text{s}^2.$$

The instantaneous center of zero velocity of the wheel is at point P, and the velocity of G is

$$\mathbf{v}_G = r_{PG}\omega\mathbf{e}_t$$
$$= [(0.75)^2 + (0.0866)^2]^{1/2}(5.0)\mathbf{e}_t$$
$$= 3.77 \text{ m/s} \searrow 6.59° \quad \text{Ans.}$$

The acceleration of G is related to the acceleration of O by

$$\mathbf{a}_G = \mathbf{a}_O + \mathbf{a}_{G/O}$$

$$= \overleftarrow{10} + \quad O,\ \alpha = 12.5,\ \omega = 5,\ 30°,\ G,\ 0.1(12.5),\ 0.1(5)^2$$

$$= -10\mathbf{i} + 1.25(\mathbf{i}\sin 30° + \mathbf{j}\cos 30°) + 2.5(-\mathbf{i}\cos 30° + \mathbf{j}\sin 30°)$$
$$= -10\mathbf{i} + 0.625\mathbf{i} + 1.082\mathbf{j} - 2.165\mathbf{i} + 1.25\mathbf{j} = -11.54\mathbf{i} + 2.33\mathbf{j}$$
$$= 11.77 \text{ m/s}^2 \nwarrow 11.40° \quad \text{Ans.}$$

EXAMPLE 8-28

The links of the mechanism in Fig. 8-43 move in a vertical plane. When they are in the position shown, CD has an angular velocity of 8 rad per sec clockwise and AB has an angular acceleration of 30 rad per sec^2 clockwise, see Example 8-21. Determine the acceleration of point C for the indicated position.

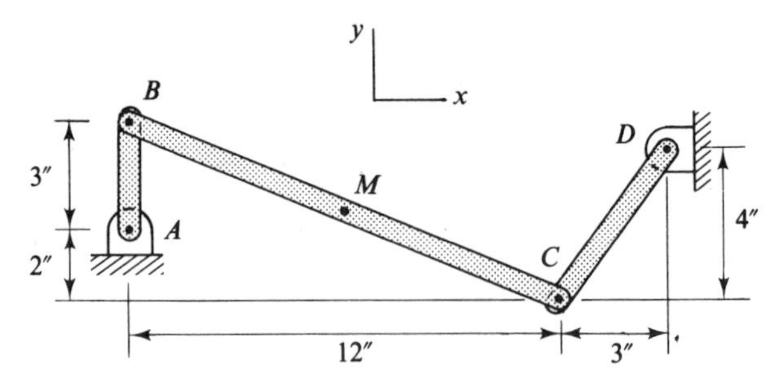

Figure 8-43

SOLUTION

The angular velocities of members AB and BC were found in Example 8-21 to be

$$\boldsymbol{\omega}_{AB} = 14.00 \text{ rad per sec clockwise}$$

and

$$\boldsymbol{\omega}_{BC} = 2.00 \text{ rad per sec counterclockwise.}$$

The angular velocities can also be determined by use of instantaneous centers. In any case they must be computed before the relative acceleration equation can be solved for the angular accelerations. A relative acceleration equation can be written for any two points. Whenever possible, two points on a rigid body should be selected since the path of the relative motion is a circle. In addition, the points selected should be points which move along known paths and whose accelerations can be expressed in terms of known or desired values. For this example points B and C are chosen since they are points on member BC and since they move along circular paths. The relative acceleration equation is

$$\mathbf{a}_C \quad = \quad \mathbf{a}_B \quad + \quad \mathbf{a}_{C/B}$$

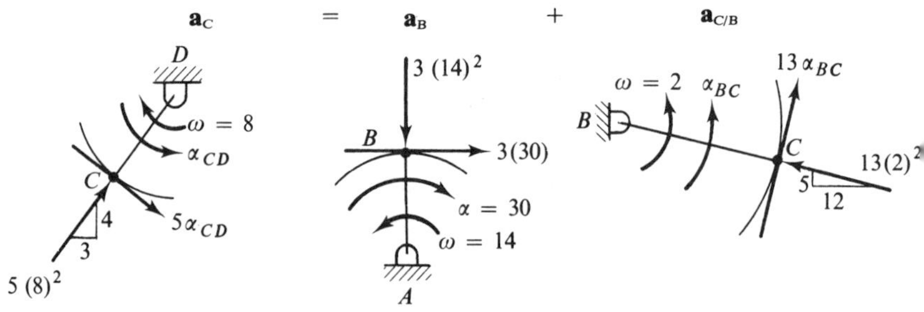

The diagrams below the equation represent the normal and tangential components of the accelerations and provide a graphical representation of the vector equation. All directions are known and there are two unknown magnitudes, α_{CD} and α_{BC}. This coplanar vector equation is equivalent to two scalar equations representing the horizontal and vertical components of the various terms. Thus

$$\xrightarrow{+} \qquad \frac{4}{5}(5)\alpha_{CD} + \frac{3}{5}(5)(8^2) = 3(30) + \frac{5}{13}(13)\alpha_{BC} - \frac{12}{13}(13)(2^2)$$

$$+\uparrow \qquad \frac{-3}{5}(5)\alpha_{CD} + \frac{4}{5}(5)(8^2) = -3(14^2) + \frac{12}{13}(13)\alpha_{BC} + \frac{5}{13}(13)(2^2),$$

which become

$$4\alpha_{CD} + 192 = 90 + 5\alpha_{BC} - 48 \qquad -3\alpha_{CD} + 256 = -588 + 12\alpha_{BC} + 20$$

or

$$5\alpha_{BC} - 4\alpha_{CD} = 150$$

$$12\alpha_{BC} + 3\alpha_{CD} = 824.$$

The solution of these simultaneous equations gives

$$\alpha_{BC} = 59.5 \text{ rad per sec}^2$$

and

$$\alpha_{CD} = 36.8 \text{ rad per sec}^2.$$

Both of these results are positive, which indicates that the assumed directions (both counterclockwise) were correct. The acceleration of C is

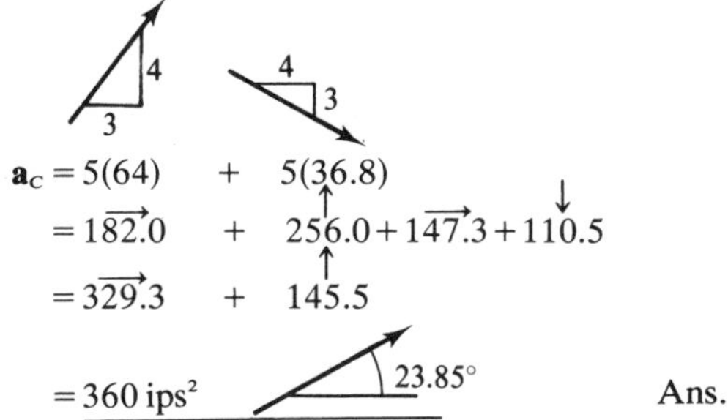

Ans.

ALTERNATE SOLUTION

The solution can also be obtained using vector notation. When the axes are selected as shown on the figure the angular velocities and accelerations are

$$\boldsymbol{\omega}_{AB} = +14\mathbf{k}, \qquad \boldsymbol{\omega}_{BC} = 2\mathbf{k}, \qquad \boldsymbol{\omega}_{CD} = -8\mathbf{k},$$

$$\boldsymbol{\alpha}_{AB} = -30\mathbf{k}, \qquad \boldsymbol{\alpha}_{BC} = \alpha_{BC}\mathbf{k}, \qquad \boldsymbol{\alpha}_{CD} = \alpha_{CD}\mathbf{k}.$$

Note that $\boldsymbol{\omega}_{AB}$ and $\boldsymbol{\omega}_{BC}$ were obtained from Example 8-21. The radius vectors are

$$\mathbf{r}_{AB} = 3\mathbf{j}, \qquad \mathbf{r}_{BC} = 12\mathbf{i} - 5\mathbf{j}, \qquad \mathbf{r}_{CD} = -3\mathbf{i} - 4\mathbf{j}.$$

In each case the radius vector is the vector from the reference (or fixed) point to the corresponding moving point. The acceleration equation is

$$\mathbf{a}_C = \mathbf{a}_B + \mathbf{a}_{C/B}. \tag{a}$$

The acceleration of point C is

$$\begin{aligned}
\mathbf{a}_C &= \boldsymbol{\alpha}_{CD} \times \mathbf{r}_{CD} + \boldsymbol{\omega}_{CD} \times (\boldsymbol{\omega}_{CD} \times \mathbf{r}_{CD}) \\
&= \boldsymbol{\alpha}_{CD} \times \mathbf{r}_{CD} - \omega_{CD}^2 \mathbf{r}_{CD} \\
&= \alpha_{CD}\mathbf{k} \times (-3\mathbf{i} - 4\mathbf{j}) - (8)^2(-3\mathbf{i} - 4\mathbf{j}) \\
&= -3\alpha_{CD}\mathbf{j} + 4\alpha_{CD}\mathbf{i} + 192\mathbf{i} + 256\mathbf{j} \\
&= (4\alpha_{CD} + 192)\mathbf{i} + (-3\alpha_{CD} + 256)\mathbf{j}.
\end{aligned}$$

Similarly,

$$\begin{aligned}
\mathbf{a}_B &= \boldsymbol{\alpha}_{AB} \times \mathbf{r}_{AB} - \omega_{AB}^2 \mathbf{r}_{AB} \\
&= -30\mathbf{k} \times 3\mathbf{j} - (14)^2 3\mathbf{j} \\
&= +90\mathbf{i} - 588\mathbf{j}
\end{aligned}$$

and

$$\begin{aligned}\mathbf{a}_{C/B} &= \boldsymbol{\alpha}_{BC} \times \mathbf{r}_{BC} - (\omega_{BC})^2 \mathbf{r}_{BC} \\ &= \alpha_{BC}\mathbf{k} \times (12\mathbf{i} - 5\mathbf{j}) - (2)^2(12\mathbf{i} - 5\mathbf{j}) \\ &= 12\alpha_{BC}\mathbf{j} + 5\alpha_{BC}\mathbf{i} - 48\mathbf{i} + 20\mathbf{j} \\ &= (5\alpha_{BC} - 48)\mathbf{i} + (12\alpha_{BC} + 20)\mathbf{j}.\end{aligned}$$

When these expressions are substituted into Eq. (a) it becomes

$$(4\alpha_{CD} + 192)\mathbf{i} + (-3\alpha_{CD} + 256)\mathbf{j} = 90\mathbf{i} - 588\mathbf{j} + (5\alpha_{BC} - 48)\mathbf{i} + (12\alpha_{BC} + 20)\mathbf{j}.$$

When the coefficients of the unit vectors on the left-hand side of the equation are equated to the corresponding coefficients of the right-hand side the vector equation gives

$$\begin{aligned}\mathbf{i}&: \quad 4\alpha_{CD} + 192 = 90 + 5\alpha_{BC} - 48, \\ \mathbf{j}&: \quad -3\alpha_{CD} + 256 = -588 + 12\alpha_{BC} + 20,\end{aligned}$$

which, of course, is the same as the result obtained by means of the "pictorial" analysis.

EXAMPLE 8-29

The bottom of the 25-ft ladder in Fig. 8-44 slides along the floor with a velocity of 6 fps to the left and an acceleration of 8 fps^2 to the left when $\theta = \tan^{-1}\frac{24}{7}$ (as shown). The top of the ladder remains in contact with the vertical wall. Determine, for this position, the velocity and acceleration of the mass center G.

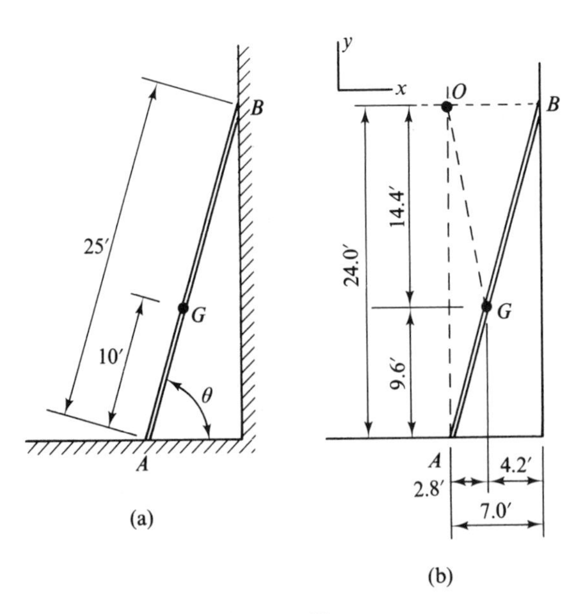

Figure 8-44

SOLUTION

The velocity of G can be obtained by applying the equations of relative velocity, as discussed in Art. 8-15, or by means of instantaneous centers. The instantaneous center is at point O of Fig. 8-44(b), and the magnitude of the angular velocity is

$$\omega = \frac{V_A}{OA} = \frac{6}{24} = 0.25 \text{ rad per sec}$$

and the angular velocity is

$$\boldsymbol{\omega} = 0.25 \text{ rad per sec clockwise.}$$

The velocity of G has a magnitude of

$$v_G = OG(\omega) = [(14.4)^2 + (2.8)^2]^{1/2}(0.25) = 3.67 \text{ fps},$$

and its direction is perpendicular to OG; thus

$\mathbf{v}_G = 3.67$ fps 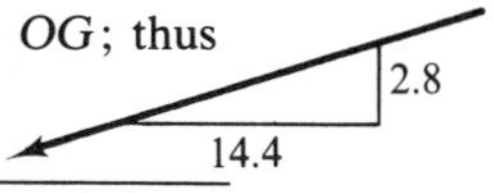

Ans.

Since the acceleration of point A is given, points A and G are selected for the relative acceleration equation. That is,

$$\mathbf{a}_G = \mathbf{a}_A + \mathbf{a}_{G/A}$$

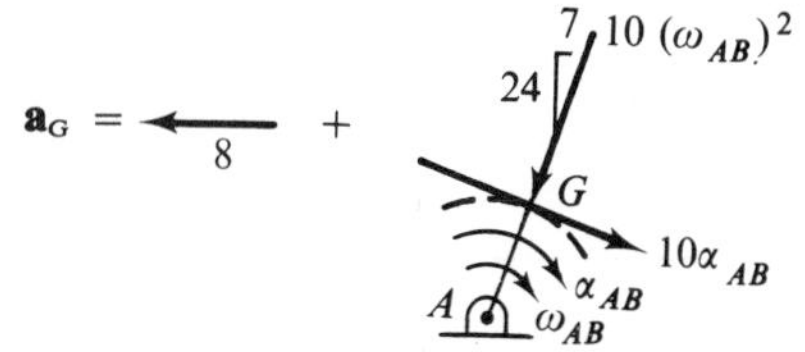

and the terms are represented by the sketches just below the equation. Both the magnitude and direction of $\mathbf{a}_G$ are unknown, and the magnitude of α_{AB} (assumed to be clockwise) is also unknown. The angular velocity of the ladder ($\omega = 0.25$ rad per sec) was determined by means of instantaneous centers. The acceleration equation can be written in terms of $\mathbf{i}$ and $\mathbf{j}$ as

$$\mathbf{a}_G = a_{Gx}\mathbf{i} + a_{Gy}\mathbf{j}$$
$$= -8\mathbf{i} + 10(0.25)^2\left(\frac{-7\mathbf{i} - 24\mathbf{j}}{25}\right) + 10\alpha\left(\frac{24\mathbf{i} - 7\mathbf{j}}{25}\right).$$

This vector equation contains three unknowns and is equivalent to two scalar equations; therefore it cannot be solved without more information. Since point B moves vertically, the direction of its acceleration must be vertical, and a relative acceleration equation for points A and B will contain only two unknown magnitudes. The equation is

$$\mathbf{a}_B = \mathbf{a}_A + \mathbf{a}_{B/A},$$

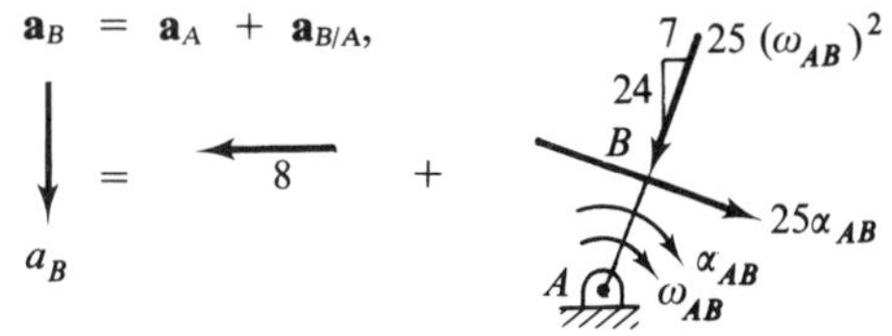

which can be written in terms of **i** and **j** as

$$-a_B\mathbf{j} = -8\mathbf{i} + 25(0.25)^2\left[\frac{-7\mathbf{i}-24\mathbf{j}}{25}\right] + 25\alpha\left[\frac{24\mathbf{i}-7\mathbf{j}}{25}\right].$$

This equation contains two unknown magnitudes, a_B and α, which can be determined by solving the two scalar equations representing the coefficients of **i** and **j**. The coefficients of **i** give

$$0 = -8 + 25(0.25)^2\left(\frac{-7}{25}\right) + 25\alpha\left(\frac{24}{25}\right)$$

from which

$$\alpha = 0.352 \text{ rad per sec}^2,$$

and the positive sign indicates that the assumed direction of α is correct; therefore

$$\boldsymbol{\alpha} = 0.352 \text{ rad per sec}^2 \text{ clockwise}$$
$$= -0.352\mathbf{k} \text{ rad per sec}^2.$$

When this value of α is substituted into Eq. (a) it becomes

$$\mathbf{a}_G = -8\mathbf{i} + 10(0.25)^2\left(\frac{-7\mathbf{i}-24\mathbf{j}}{25}\right) + 10(0.352)\left(\frac{24\mathbf{i}-7\mathbf{j}}{25}\right)$$
$$= -8\mathbf{i} - 0.175\mathbf{i} - 0.600\mathbf{j} + 3.375\mathbf{i} - 0.984\mathbf{j}$$
$$= -4.800\mathbf{i} - 1.584\mathbf{j}$$
$$= 5.05 \text{ fps}^2 \; \swarrow 18.27^\circ$$

Ans.

PROBLEMS

8-224 The wheel in Fig. P8-224 rolls without slipping along the fixed surface. The velocity and acceleration of the center O are 20 ips to the left and 30 ips^2 to the right for the position shown. Determine the velocity and acceleration of point A for this position.

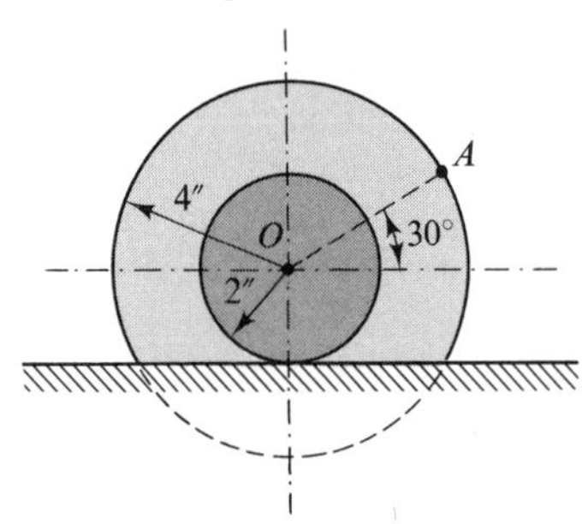

Figure P8-224

***8-225** Disk A in Fig. P8-225 rotates about a fixed axis through O. Disk B is connected to A by a smooth pin. When in the position shown, A has an angular velocity of 4 rad per sec counterclockwise and an angular acceleration of 6 rad per sec^2 clockwise, and B has an angular velocity of 12 rad per sec and an angular acceleration of 20 rad per sec^2, both counterclockwise. Determine the acceleration of point Q on disk B.

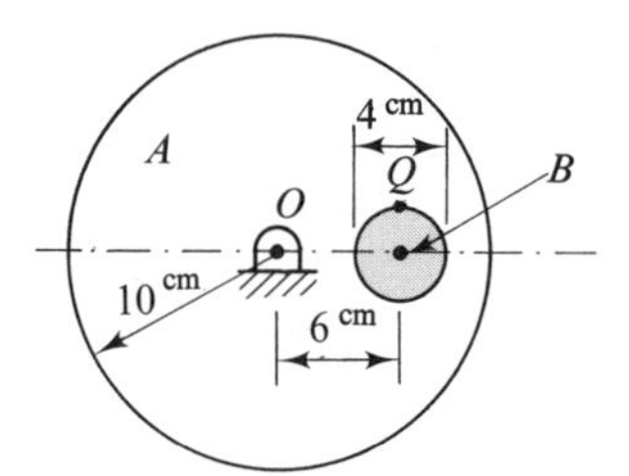

Figure P8-225

8-226 When a wheel rolls along a circular surface, the center of the wheel also moves along a circular path as indicated in Fig. P8-226. Determine the relationships between the angular velocity and acceleration of the wheel and the linear velocity and acceleration of the center O when the wheel rolls without slipping. *Hint:* Write an expression for the position of O in terms of the angular position of the wheel and differentiate to determine the magnitude of the velocity and of the tangential acceleration of O.]

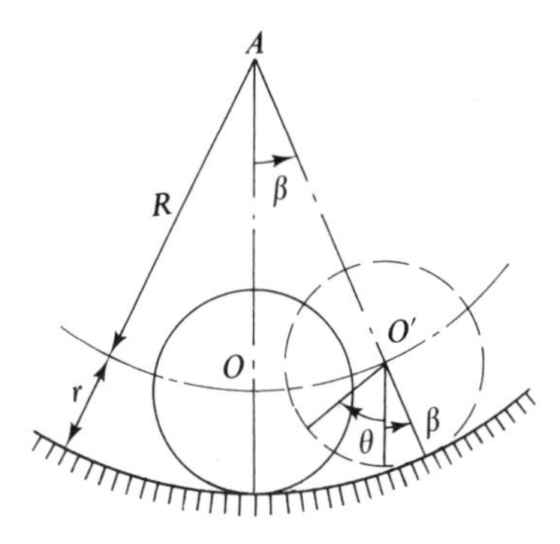

Figure P8-226

8-227 The wheel in Fig. P8-227 rolls without slipping with decreasing angular velocity along the horizontal plane. When in the position shown, point B has an acceleration of 5 ips^2 to the right. Determine the velocity and acceleration of point O.

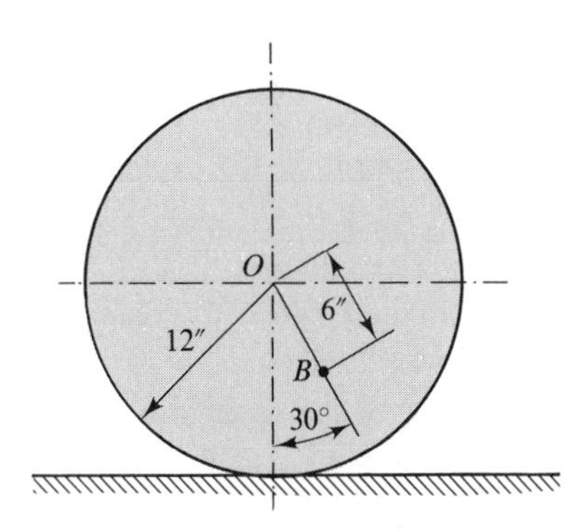

Figure P8-227

***8-228** Wheel A in Fig. P8-228 rolls along the horizontal plane without slipping and is connected to body B by a flexible cord. In the position shown, point C has a velocity of 8 cm/s and an acceleration of 20 cm/s^2, both to the left. Determine the acceleration of body B with respect to point D.

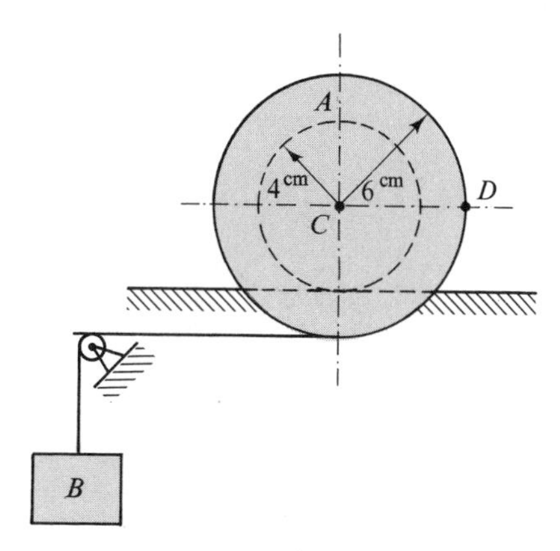

Figure P8-228

***8-229** Crank A in Fig. P8-229 is connected to member C by the connecting rod B. Crank A has a constant angular velocity of 10 rad/s clockwise. Determine the angular accelerations of members B and C when in the position shown.

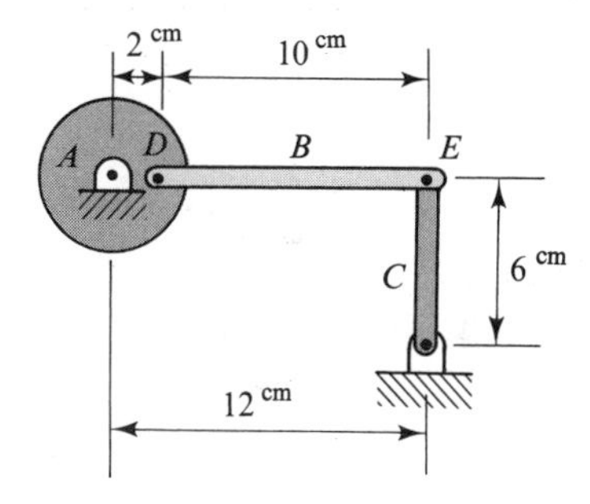

Figure P8-229

8-230 In the position shown in Fig. P8-230 member CD has an angular velocity of 2 rad per sec counterclockwise and an angular acceleration of 2 rad per sec^2 clockwise. Determine the acceleration of point B.

8-231 Member AB in Fig. P8-231 has an angular velocity of 4 rad per sec clockwise and an angular acceleration of 6 rad per sec^2 counterclockwise in the indicated position. Determine the acceleration of point C when in this position.

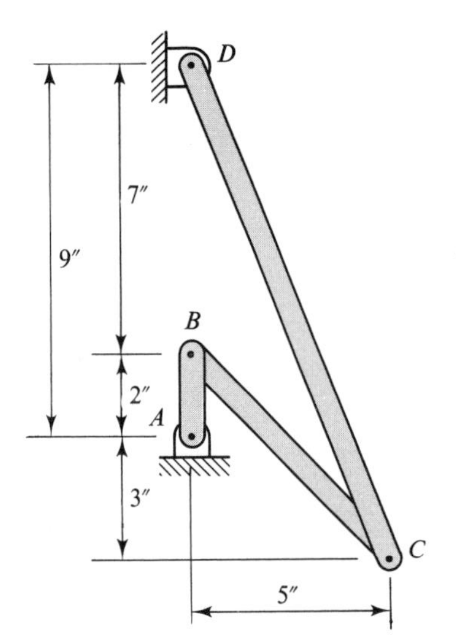

Figure P8-230

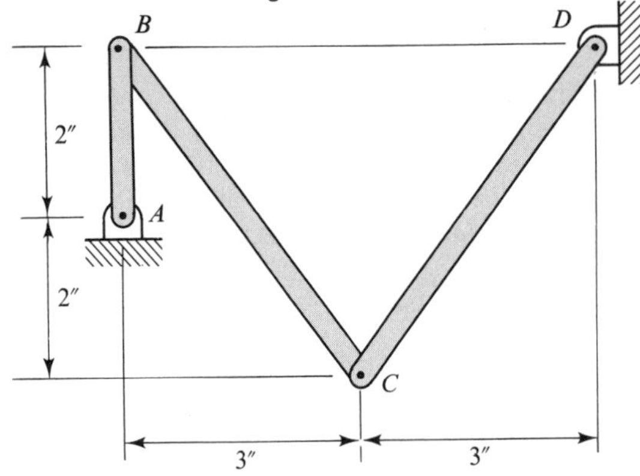

Figure P8-231

8-232 Member AB in Fig. P8-232 has an angular velocity of 3 rad per sec and an angular acceleration of 10 rad per sec^2, both counterclockwise, when in the indicated position. Determine the angular acceleration of member BC.

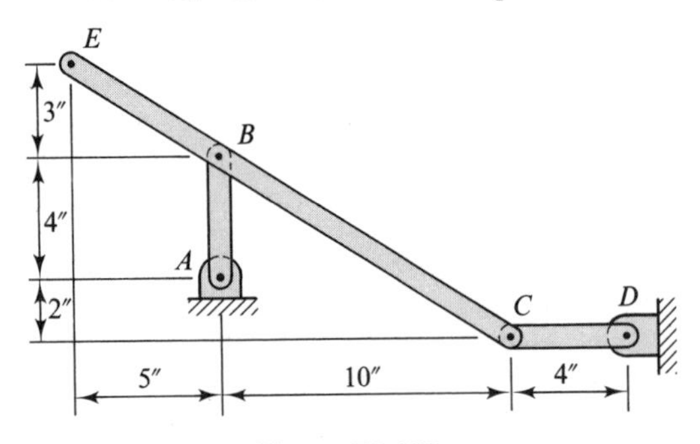

Figure P8-232

8-233 Bar CD in Fig. P8-233 has an angular velocity of 1.0 rad per sec and an angular acceleration of 2.0 rad per sec^2, both clockwise, at the instant shown. Determine the angular acceleration of bar BC for this position.

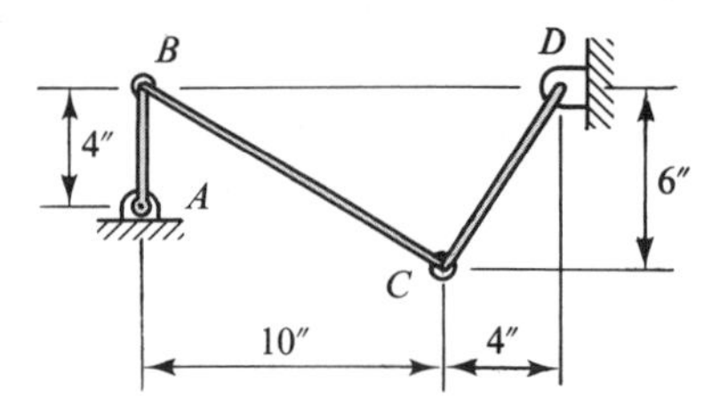

Figure P8-233

8-234 Arm AB in Fig. P8-234 has an angular velocity of 2 rad per sec clockwise and an angular acceleration of 10 rad per sec^2 counterclockwise when in the position shown. Determine the velocity and acceleration of block C.

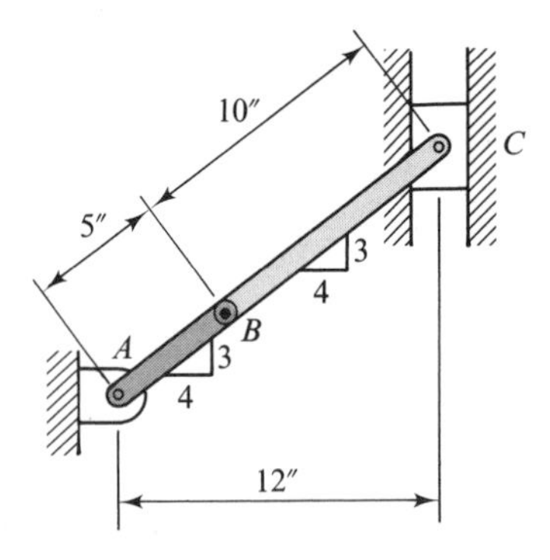

Figure P8-234

8-235 Determine the angular acceleration of member BC in Fig. P8-235 which will result in a zero acceleration of point C when C has a velocity of 28 ips upward.

8-236 Roller C of the mechanism in Fig. P8-236 is to have a velocity of 42 ips to the right and be slowing down at the rate of 100 ips^2 in the position shown. Determine the required angular velocity and acceleration of member OA.

8-237 The block in Fig. P8-237 has a clockwise angular velocity of 5 rad per sec and a counterclockwise angular acceleration of 20 rad per

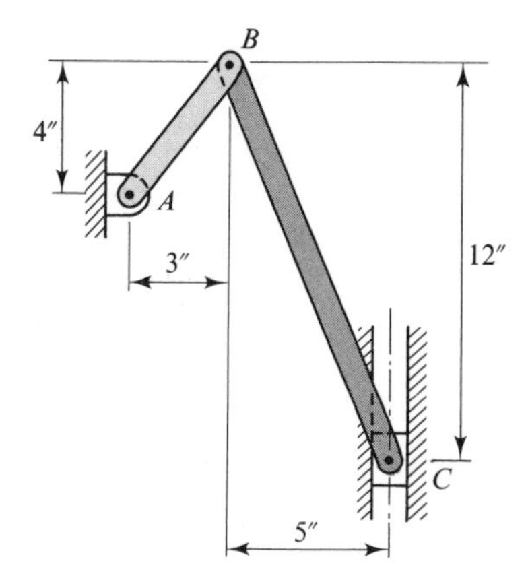

Figure P8-235

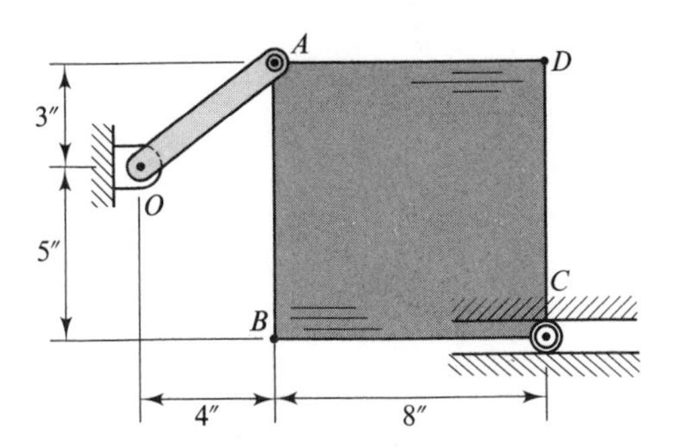

Figure P8-236

sec² in the indicated position. Determine the acceleration of point D.

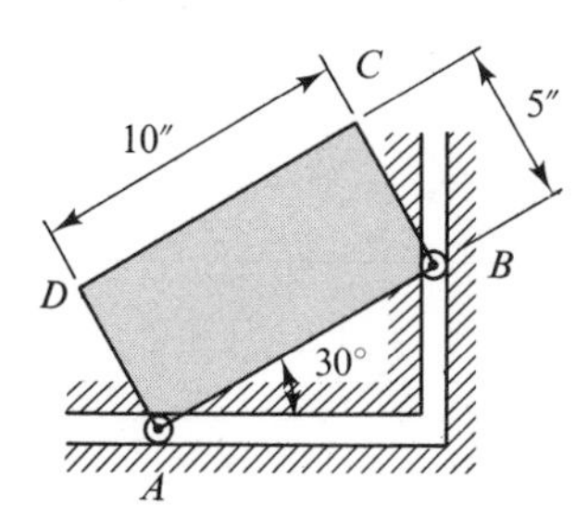

Figure P8-237

8-238 The velocity of point A in Fig. P8-238 is 10.5 ips down the slot and its acceleration is 100 ips² up the slot when in the position shown. Determine the acceleration of point C.

***8-239** Roller A in Fig. P8-239 has a velocity of 0.70 m/s downward and an acceleration of 1.50 m/s² upward in the position shown. Deter-

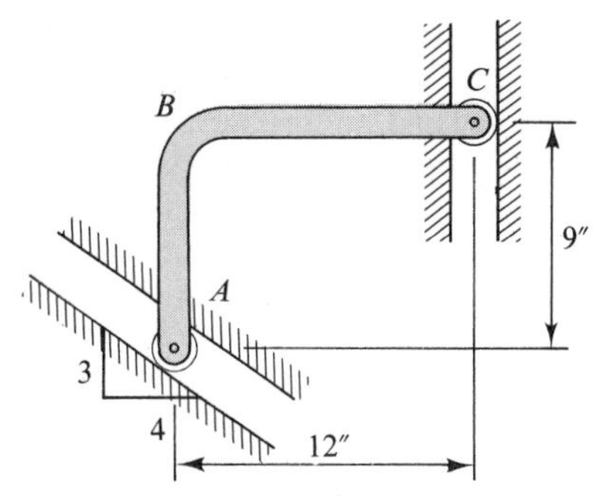

Figure P8-238

mine the angular velocity and acceleration of member AB for this position.

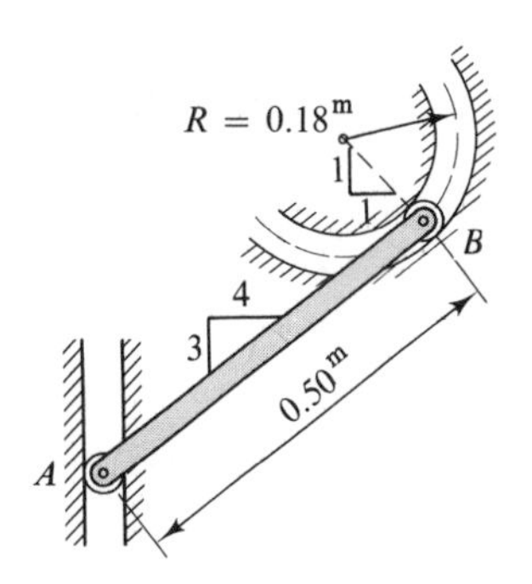

Figure P8-239

8-240 The disk in Fig. P8-240 rolls without slipping along the circular track. The disk has a constant clockwise angular velocity of 30 rpm.
(a) Determine the time required for the center of the disk to make one complete revolution about the point B.
(b) Determine the acceleration of point A when the disk is in the position shown.
(c) If, in addition to the given angular velocity, the disk has a counterclockwise angular acceleration of 4 rad per sec², determine the acceleration of point A.

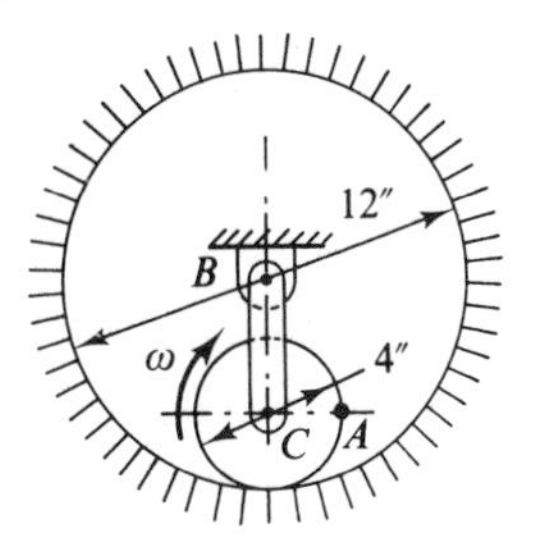

Figure P8-240

8-241 In the planetary gear assembly in Fig. P8-241 the inner gear A has an angular velocity of 2 rad per sec and an angular acceleration of 4 rad per sec², both counterclockwise, and the outer gear is clamped. The planet gears B, C, and D roll between the inner and outer gears. Determine the velocity and acceleration of point P on gear B.

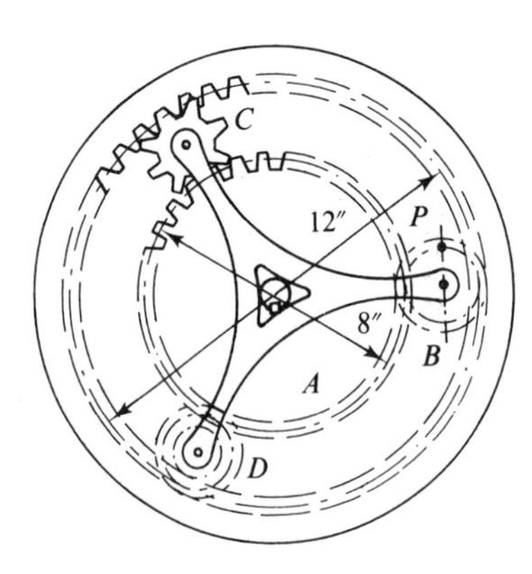

Figure P8-241

8-242 The wheel in Fig. P8-242 rolls without slipping, and when in the indicated position it has a clockwise angular velocity of 2 rad per sec and a counterclockwise angular acceleration of 6 rad per sec². Determine the angular accelerations of AB and BC.

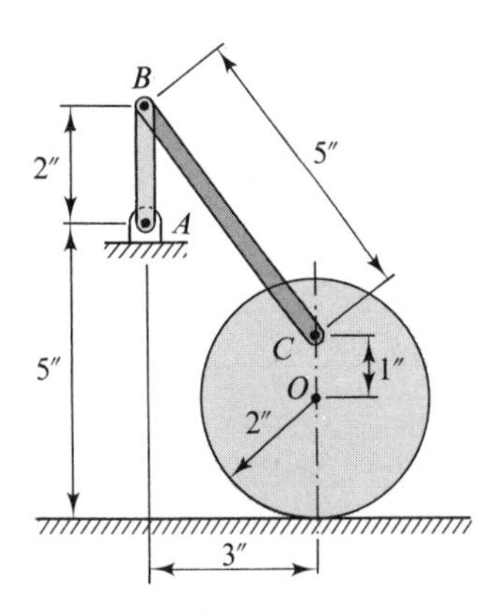

Figure P8-242

8-243 The wheel in Fig. P8-243 rolls without slipping along the curved surface, and in the position shown it has an angular velocity of 3 rad per sec and an angular acceleration of 5 rad per sec², both clockwise. Determine

(a) The angular acceleration of member AB.

(b) The acceleration of slider B.

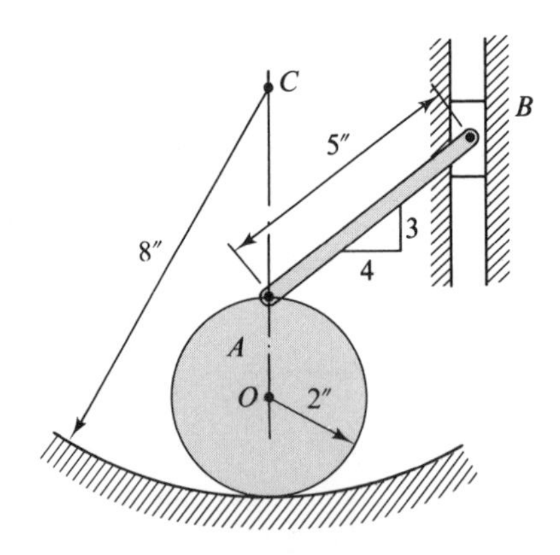

Figure P8-243

***8-244** Block D in Fig. P8-244 has a velocity of 16 cm/s to the left and an acceleration of 30 cm/s² to the right in the indicated position. Determine the velocity and acceleration of block A for this position.

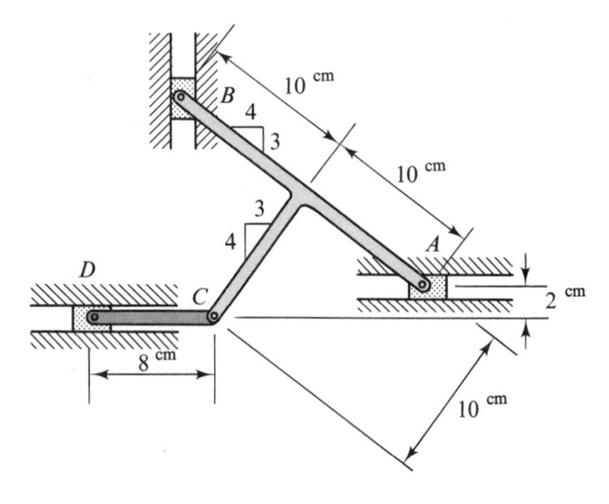

Figure P8-244

8-20 RELATIVE ACCELERATION—THREE-DIMENSIONAL MOTION

Equation (8-26) of Art. 8-18 was applied to rigid bodies with plane motion in Art. 8-19. The equation is also valid for rigid bodies which have three-dimensional motion provided nonrotating reference axes are used. The equation is

$$\mathbf{a}_B = \mathbf{a}_A + \mathbf{a}_{B/A},$$

and when A and B are points on a rigid body the relative acceleration term is

$$\mathbf{a}_{B/A} = \dot{\boldsymbol{\omega}} \times \mathbf{r} + \boldsymbol{\omega} \times (\boldsymbol{\omega} \times \mathbf{r}), \tag{a}$$

where $\dot{\boldsymbol{\omega}}$ is the absolute angular acceleration of the body, $\boldsymbol{\omega}$ is the absolute angular velocity of the body, and $\mathbf{r}$ is the vector *from A to B* (the position of B relative to A). As noted in Art. 8-18, $\dot{\boldsymbol{\omega}}$ or $\boldsymbol{\alpha}$ is usually not parallel to $\boldsymbol{\omega}$ for three-dimensional motion.

Figure 8-45 represents a member AB connected by ball and socket joints to a rotating disk and to a sliding collar. The angular velocity is shown resolved into two components, one tangent to AB and one normal to AB. Since the angular velocity can change in both magnitude and direction, the angular acceleration will, in general, not be parallel to the angular velocity. The angular acceleration can also be resolved into two components tangent and normal to member AB, that is,

$$\dot{\boldsymbol{\omega}} = \dot{\boldsymbol{\omega}}_t + \dot{\boldsymbol{\omega}}_n,$$

and the quantity $\dot{\boldsymbol{\omega}} \times \mathbf{r}$ becomes

$$\dot{\boldsymbol{\omega}} \times \mathbf{r} = (\dot{\boldsymbol{\omega}}_t + \dot{\boldsymbol{\omega}}_n) \times \mathbf{r} = \dot{\boldsymbol{\omega}}_n \times \mathbf{r}$$

since $\dot{\boldsymbol{\omega}}_t$ is parallel to $\mathbf{r}$. This result indicates that $\dot{\boldsymbol{\omega}} \times \mathbf{r}$ is independent of $\dot{\boldsymbol{\omega}}_t$; that is $\dot{\boldsymbol{\omega}}_t$ can have any value without affecting Eq. (a).

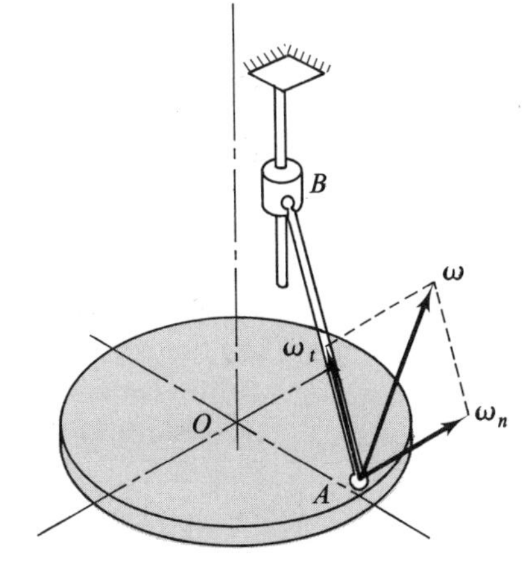

Figure 8-45

It was shown in Art. 8-17 that the value of $\boldsymbol{\omega}_t$ was indeterminate for this system and could have any assigned magnitude. When $\boldsymbol{\omega}_t$ is made equal to zero the value of $\boldsymbol{\omega}$ becomes $\boldsymbol{\omega}_n$ and is perpendicular to $\mathbf{r}$. In this case, the right-hand term in Eq. (a) becomes

$$\boldsymbol{\omega} \times (\boldsymbol{\omega} \times \mathbf{r}) = \boldsymbol{\omega}_n \times (\boldsymbol{\omega}_n \times \mathbf{r}) = -\omega_n^2 \mathbf{r}.$$

Note that this result is true only when $\boldsymbol{\omega}$ and $\mathbf{r}$ are normal to each other.

If the ball and socket joint at B in Fig. 8-45 is replaced by the clevis in Fig. 8-46, the angular velocity and acceleration will not be indeterminate. The angular velocity can be expressed as the vector sum

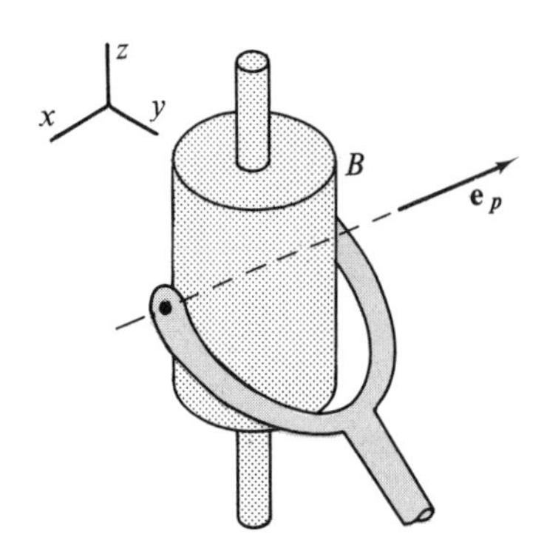

Figure 8-46

of the angular velocity of the collar and the angular velocity of member AB around the pin. Thus

$$\boldsymbol{\omega} = \omega_1\mathbf{k} + \omega_2\mathbf{e}_p,$$

where ω_1 is the magnitude of the angular velocity of the collar, ω_2 is the magnitude of the angular velocity of AB around the pin, and $\mathbf{e}_p$ is a unit vector parallel to the pin. The pin must be perpendicular to the shaft on which the collar slides and to the member AB; therefore

$$\mathbf{e}_p = \frac{\mathbf{k}\times\mathbf{r}_{AB}}{|\mathbf{k}\times\mathbf{r}_{AB}|}.$$

The angular acceleration of AB is

$$\boldsymbol{\alpha} = \dot{\boldsymbol{\omega}} = \dot{\omega}_1\mathbf{k} + \omega_1\dot{\mathbf{k}} + \dot{\omega}_2\mathbf{e}_p + \omega_2\dot{\mathbf{e}}_p.$$

The vector $\mathbf{k}$ is constant; therefore $\dot{\mathbf{k}} = 0$. Since $\mathbf{e}_p$ is parallel to the pin, it must rotate with the pin so that

$$\dot{\mathbf{e}}_p = \boldsymbol{\omega}_{\text{pin}}\times\mathbf{e}_p = \omega_1\mathbf{k}\times\mathbf{e}_p$$

where

$$\boldsymbol{\omega}_{\text{pin}} = \boldsymbol{\omega}_{\text{collar}} = \omega_1\mathbf{k}.$$

Thus

$$\boldsymbol{\alpha} = \dot{\omega}_1\mathbf{k} + \dot{\omega}_2\mathbf{e}_p + \omega_2(\omega_1\mathbf{k}\times\mathbf{e}_p). \qquad \text{(b)}$$

The first term in Eq. (b) represents the rate of change of magnitude of $\boldsymbol{\omega}_1$, the second term is the rate of change of magnitude of $\boldsymbol{\omega}_2$, and the third term is the change of $\boldsymbol{\omega}_2$ resulting from the change of direction of the pin (or of the collar). The following examples illustrate the application of Eq. (8-26).

EXAMPLE 8-30

Collars A and B of the mechanism in Fig. 8-47 slide along fixed rods and are connected to member C by a clevis at A and by a ball and socket at B. The clevis at A is illustrated in Fig. 8-41 (which is repeated here for convenience). In the position shown $c = d = 2b$ and collar A has a velocity of $v_A\mathbf{i}$ and an acceleration of $a_A\mathbf{i}$. Determine the acceleration of B and the angular acceleration of member C.

SOLUTION

The angular velocity of C is the sum of the angular velocity, $\boldsymbol{\omega}_1$, of the collar and the angular velocity, $\boldsymbol{\omega}_2$, of C about the clevis pin. The unit

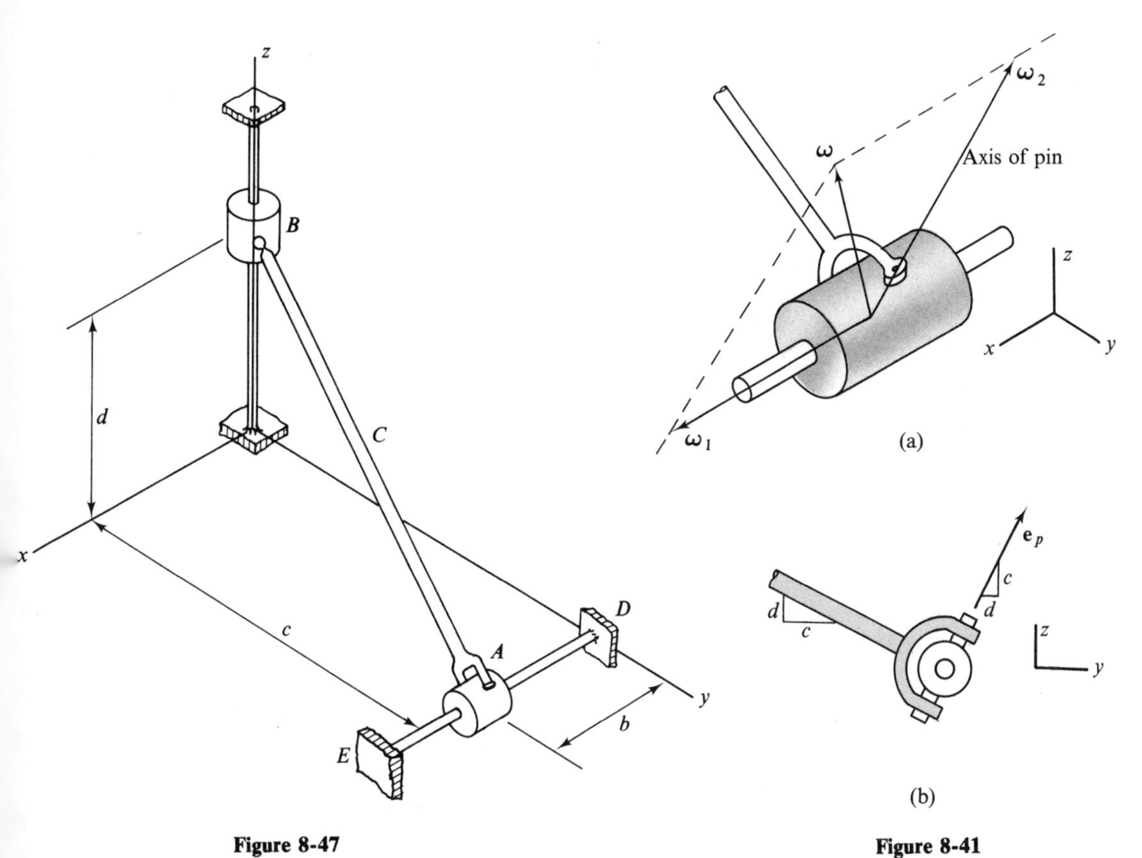

Figure 8-47

Figure 8-41

vector along the pin is

$$\mathbf{e}_p = \frac{\mathbf{r}_{AB} \times \mathbf{i}}{|\mathbf{r}_{AB} \times \mathbf{i}|} = \frac{(-b\mathbf{i} - 2b\mathbf{j} + 2b\mathbf{k}) \times \mathbf{i}}{|(-b\mathbf{i} - 2b\mathbf{j} + 2b\mathbf{k}) \times \mathbf{i}|}$$

$$= \frac{\mathbf{j} + \mathbf{k}}{\sqrt{2}},$$

and the angular velocity of C is

$$\left.\begin{aligned} \boldsymbol{\omega} &= \omega_1 \mathbf{i} + \omega_2 \mathbf{e}_p \\ &= \omega_1 \mathbf{i} + \frac{\omega_2}{\sqrt{2}}(\mathbf{j} + \mathbf{k}). \end{aligned}\right\} \qquad \text{(a)}$$

The relative velocity equation of B with respect to A is

$$v_B \mathbf{k} = v_A \mathbf{i} + \boldsymbol{\omega} \times \mathbf{r}$$

$$= v_A \mathbf{i} + b \begin{vmatrix} \mathbf{i} & \mathbf{j} & \mathbf{k} \\ \omega_1 & \dfrac{\omega_2}{\sqrt{2}} & \dfrac{\omega_2}{\sqrt{2}} \\ -1 & -2 & +2 \end{vmatrix}$$

$$= v_A \mathbf{i} + b\left[\mathbf{i}(2)\sqrt{2}\,\omega_2 + \mathbf{j}\left(-\frac{\sqrt{2}}{2}\omega_2 - 2\omega_1\right) + \mathbf{k}\left(-2\omega_1 + \frac{\sqrt{2}}{2}\omega_2\right)\right]$$

from which

$$\mathbf{i}: \quad 2\sqrt{2}\,\omega_2 = -\frac{v_A}{b},$$

$$\mathbf{j}: \quad -2\omega_1 - \frac{\sqrt{2}}{2}\,\omega_2 = 0,$$

$$\mathbf{k}: \quad -2\omega_1 + \frac{\sqrt{2}}{2}\,\omega_2 = \frac{v_B}{b},$$

which in turn gives

$$\omega_2 = -\frac{\sqrt{2}}{4}\frac{v_A}{b},$$

$$\omega_1 = \frac{1}{8}\frac{v_A}{b}.$$

The angular velocity of AB is

$$\boldsymbol{\omega} = \frac{v_A}{8b}(\mathbf{i} - 2\mathbf{j} - 2\mathbf{k}).$$

The acceleration of B is

$$\begin{aligned}\mathbf{a}_B &= \mathbf{a}_A + \mathbf{a}_{B/A}\\ &= \mathbf{a}_A + \boldsymbol{\alpha}\times\mathbf{r} + \boldsymbol{\omega}\times(\boldsymbol{\omega}\times\mathbf{r}).\end{aligned} \tag{b}$$

The angular velocity as given in Eq. (a) is

$$\boldsymbol{\omega} = \omega_1\mathbf{i} + \omega_2\mathbf{e}_p,$$

and the angular acceleration is

$$\boldsymbol{\alpha} = \dot{\boldsymbol{\omega}} = \dot{\omega}_1\mathbf{i} + \omega_1\dot{\mathbf{i}} + \dot{\omega}_2\mathbf{e}_p + \omega_2\dot{\mathbf{e}}_p.$$

Since $\mathbf{i}$ is constant, $\dot{\mathbf{i}}$ is zero. The quantity $\mathbf{e}_p$ is a unit vector which can change in direction but not in magnitude. The angular velocity of $\mathbf{e}_p$ is the angular velocity of the collar, that is, $\omega_1\mathbf{i}$; therefore

$$\begin{aligned}\dot{\mathbf{e}}_p &= \omega_1\mathbf{i}\times\mathbf{e}_p\\ &= \omega_1\mathbf{i}\times\left(\frac{\mathbf{j}+\mathbf{k}}{\sqrt{2}}\right)\\ &= \frac{\omega_1}{\sqrt{2}}(-\mathbf{j}+\mathbf{k}).\end{aligned}$$

The angular acceleration thus becomes

$$\begin{aligned}\boldsymbol{\alpha} &= \dot{\omega}_1\mathbf{i} + \frac{\dot{\omega}_2}{\sqrt{2}}(\mathbf{j}+\mathbf{k}) + \frac{\omega_2\omega_1}{\sqrt{2}}(-\mathbf{j}+\mathbf{k})\\ &= \dot{\omega}_1\mathbf{i} + \frac{\dot{\omega}_2}{\sqrt{2}}(\mathbf{j}+\mathbf{k}) - \frac{\sqrt{2}}{32}\left(\frac{v_A}{b}\right)^2\frac{1}{\sqrt{2}}(-\mathbf{j}+\mathbf{k})\\ &= \dot{\omega}_1\mathbf{i} + \left[\frac{\dot{\omega}_2}{\sqrt{2}} + \frac{1}{32}\left(\frac{v_A}{b}\right)^2\right]\mathbf{j} + \left[\frac{\dot{\omega}_2}{\sqrt{2}} - \frac{1}{32}\left(\frac{v_A}{b}\right)^2\right]\mathbf{k},\end{aligned} \tag{c}$$

and the quantity $\boldsymbol{\alpha}\times\mathbf{r}$ is

$$\boldsymbol{\alpha}\times\mathbf{r}=b\begin{vmatrix}\mathbf{i} & \mathbf{j} & \mathbf{k}\\ \dot{\omega}_1 & \dfrac{\dot{\omega}_2}{\sqrt{2}}+\dfrac{1}{32}\left(\dfrac{v_A}{b}\right)^2 & \dfrac{\dot{\omega}_2}{\sqrt{2}}-\dfrac{1}{32}\left(\dfrac{v_A}{b}\right)^2\\ -1 & -2 & 2\end{vmatrix}$$

$$=b\left\{2\sqrt{2}\,\dot{\omega}_2\mathbf{i}+\left[-2\dot{\omega}_1-\frac{\sqrt{2}}{2}\,\dot{\omega}_2+\frac{1}{32}\left(\frac{v_A}{b}\right)^2\right]\mathbf{j}\right.$$

$$\left.+\left[-2\dot{\omega}_1+\frac{\sqrt{2}}{2}\,\dot{\omega}_2+\frac{1}{32}\left(\frac{v_A}{b}\right)^2\right]\mathbf{k}\right\}.$$

The quantity $\boldsymbol{\omega}\times\mathbf{r}$ is

$$\boldsymbol{\omega}\times\mathbf{r}=\frac{v_A b}{8b}\begin{vmatrix}\mathbf{i} & \mathbf{j} & \mathbf{k}\\ 1 & -2 & -2\\ -1 & -2 & 2\end{vmatrix}$$

$$=\frac{v_A}{8}[\mathbf{i}(-4-4)+\mathbf{j}(2-2)+\mathbf{k}(-2-2)]$$

$$=\frac{v_A}{2}(-2\mathbf{i}-\mathbf{k}),$$

and the vector triple product becomes

$$\boldsymbol{\omega}\times(\boldsymbol{\omega}\times\mathbf{r})=\frac{v_A}{8b}\frac{v_A}{2}\begin{vmatrix}\mathbf{i} & \mathbf{j} & \mathbf{k}\\ 1 & -2 & -2\\ -2 & 0 & -1\end{vmatrix}$$

$$=\frac{v_A^2}{16b}(2\mathbf{i}+5\mathbf{j}-4\mathbf{k}).$$

The preceding values can be substituted in Eq. (b) to give

$$a_B\mathbf{k}=a_A\mathbf{i}+b\left\{2\sqrt{2}\,\dot{\omega}_2\mathbf{i}+\left[-2\dot{\omega}_1-\frac{\sqrt{2}}{2}\,\dot{\omega}_2+\frac{1}{32}\left(\frac{v_A}{b}\right)^2\right]\mathbf{j}\right.$$

$$\left.+\left[-2\dot{\omega}_1+\frac{\sqrt{2}}{2}\,\dot{\omega}_2+\frac{1}{32}\left(\frac{v_A}{b}\right)^2\right]\mathbf{k}\right\}+\frac{(v_A)^2}{16b}(2\mathbf{i}+5\mathbf{j}-4\mathbf{k}),$$

which yields the following three scalar equations:

$$\mathbf{i}:\quad 2\sqrt{2}\,\dot{\omega}_2=-\frac{1}{8}\left(\frac{v_A}{b}\right)^2-\frac{a_A}{b},$$

$$\mathbf{j}:\quad -2\dot{\omega}_1-\frac{\sqrt{2}}{2}\,\dot{\omega}_2=-\frac{11}{32}\left(\frac{v_A}{b}\right)^2,$$

$$\mathbf{k}:\quad -2\dot{\omega}_1+\frac{\sqrt{2}}{2}\,\dot{\omega}_2=\frac{7}{32}\left(\frac{v_A}{b}\right)^2+\frac{a_B}{b}.$$

These three equations contain three unknown quantities, $\dot{\omega}_1$, $\dot{\omega}_2$, and

a_B. The simultaneous solution of these equations gives

$$\dot{\omega}_1 = \frac{1}{8}\frac{a_A}{b} + \frac{3}{16}\left(\frac{v_A}{b}\right)^2,$$

$$\dot{\omega}_2 = -\frac{\sqrt{2}}{4}\frac{a_A}{b} - \frac{\sqrt{2}}{32}\left(\frac{v_A}{b}\right)^2,$$

and

$$a_B = -\frac{a_A}{2} - \frac{5}{8}\frac{v_A^2}{b}.$$

The angular acceleration of C is obtained by substituting these values into Eq. (c); that is

$$\boldsymbol{\alpha} = \frac{a_A}{8b}(\mathbf{i} - 2\mathbf{j} - 2\mathbf{k}) + \frac{1}{16}\left(\frac{v_A}{b}\right)^2(3\mathbf{i} - \mathbf{k}). \qquad \text{Ans.}$$

The acceleration of B is

$$\mathbf{a}_B = a_B\mathbf{k} = -\left(\frac{a_A}{2} + \frac{5v_A^2}{8b}\right)\mathbf{k}. \qquad \text{Ans.}$$

EXAMPLE 8-31

Solve Example 8-30 if the clevis at A is replaced by a ball and socket joint.

SOLUTION

The component of the angular velocity of the bar parallel to the bar, $\boldsymbol{\omega}_t$, is indeterminate when both ends are connected to ball and socket joints as noted in Art. 8-17. The same reasoning indicates that the component of the angular acceleration parallel to the bar is indeterminate. Since the direction of $\boldsymbol{\omega}_t$ is changing, the value assumed for the magnitude of $\boldsymbol{\omega}_t$ will be a factor in determining the value of $\boldsymbol{\alpha}_n$ which is normal to the bar. For convenience it is assumed that $\boldsymbol{\omega}_t$ is zero. A different assumed value for $\boldsymbol{\omega}_t$ would result in a different value for $\boldsymbol{\alpha}_n$, but it would not affect the acceleration of B. The relative velocity equation for points A and B is

$$\mathbf{v}_B = \mathbf{v}_A + \mathbf{v}_{B/A} = \mathbf{v}_A + \boldsymbol{\omega} \times \mathbf{r},$$

$$v_B\mathbf{k} = v_A\mathbf{i} + (\omega_x\mathbf{i} + \omega_y\mathbf{j} + \omega_z\mathbf{k}) \times (-b\mathbf{i} - 2b\mathbf{j} + 2b\mathbf{k}),$$

where $\boldsymbol{\omega} = \omega_x\mathbf{i} + \omega_y\mathbf{j} + \omega_z\mathbf{k}$ is the angular velocity of AB. When this equation is expanded it becomes

$$v_B\mathbf{k} = v_A\mathbf{i} + b\begin{vmatrix} \mathbf{i} & \mathbf{j} & \mathbf{k} \\ \omega_x & \omega_y & \omega_z \\ -1 & -2 & 2 \end{vmatrix}$$

$$= v_A\mathbf{i} + b[\mathbf{i}(2\omega_y + 2\omega_z) + \mathbf{j}(-\omega_z - 2\omega_x) + \mathbf{k}(-2\omega_x + \omega_y)],$$

and when the coefficients of **i**, **j**, and **k** are equated the equation gives

$$\mathbf{i}: \quad 2\omega_y + 2\omega_z = -\frac{v_A}{b} \qquad \text{(a)}$$

$$\mathbf{j}: \quad -2\omega_x - \omega_z = 0 \tag{b}$$

$$\mathbf{k}: \quad -2\omega_x + \omega_y = \frac{v_B}{b} \tag{c}$$

The requirement that $\omega_t = \boldsymbol{\omega}\cdot(\mathbf{r}/r) = 0$ gives

$$-\omega_x - 2\omega_y + 2\omega_z = 0. \tag{d}$$

When Eq. (b) is multiplied by $-\frac{1}{2}$ and added to the sum of Eq. (a) and (d) the result is

$$2\omega_z + \frac{1}{2}\omega_z + 2\omega_z = -\frac{v_A}{b}$$

from which

$$\omega_z = -\frac{2v_A}{9b}.$$

From Eq. (b),

$$\omega_x = -\frac{1}{2}\omega_z = \frac{v_A}{9b},$$

and from Eq. (a) or (d),

$$\omega_y = -\frac{5v_A}{18b}.$$

The angular velocity is

$$\boldsymbol{\omega} = \boldsymbol{\omega}_t + \boldsymbol{\omega}_n = 0 + \boldsymbol{\omega}_n$$

$$= \frac{v_A}{18b}(2\mathbf{i} - 5\mathbf{j} - 4\mathbf{k}) \text{ rad per sec.}$$

This result can also be obtained from Example 8-25 by substituting $c = d = 2b$.

The relative acceleration equation for points A and B is

$$\mathbf{a}_B = \mathbf{a}_A + \mathbf{a}_{B/A}$$

$$= \mathbf{a}_A + \boldsymbol{\alpha}\times\mathbf{r} + \boldsymbol{\omega}\times(\boldsymbol{\omega}\times\mathbf{r}),$$

where $\mathbf{r}$ is the position of B relative to A and $\boldsymbol{\alpha} = \alpha_x\mathbf{i} + \alpha_y\mathbf{j} + \alpha_z\mathbf{k}$ is the angular acceleration of AB. Since $\boldsymbol{\omega}_t$ was set equal to zero, the quantity $\boldsymbol{\omega}\times(\boldsymbol{\omega}\times\mathbf{r})$ is equal to $-\omega^2\mathbf{r}$. The magnitude of $\boldsymbol{\omega}$ is

$$\omega_n = \frac{v_A}{18b}\sqrt{2^2 + 5^2 + 4^2} = \frac{\sqrt{45}\,v_A}{18b},$$

and the vector triple product becomes

$$\boldsymbol{\omega}\times(\boldsymbol{\omega}\times\mathbf{r}) = -\frac{45}{(18)^2}\left(\frac{v_A}{b}\right)^2 b(-\mathbf{i} - 2\mathbf{j} + 2\mathbf{k})$$

$$= -\frac{5v_A^2}{36b}(-\mathbf{i} - 2\mathbf{j} + 2\mathbf{k}).$$

The relative acceleration equation, in terms of **i**, **j**, and **k**, is

$$a_B\mathbf{k}=a_A\mathbf{i}+b\begin{vmatrix}\mathbf{i} & \mathbf{j} & \mathbf{k}\\ \alpha_x & \alpha_y & \alpha_z\\ -1 & -2 & 2\end{vmatrix}-\frac{5v_A^2}{36b}(-\mathbf{i}-2\mathbf{j}+2\mathbf{k})$$

$$=a_A\mathbf{i}+b[\mathbf{i}(2\alpha_y+2\alpha_z)+\mathbf{j}(-\alpha_z-2\alpha_x)+\mathbf{k}(-2\alpha_x+\alpha_y)]$$

$$+\frac{5v_A^2}{36b}(+\mathbf{i}+2\mathbf{j}-2\mathbf{k}),$$

which reduces to the following three equations:

$$\mathbf{i}:\quad +2\alpha_y+2\alpha_z=-\frac{5}{36}\left(\frac{v_A}{b}\right)^2-\frac{a_A}{b} \tag{e}$$

$$\mathbf{j}:\quad -2\alpha_x-\alpha_z=-\frac{10}{36}\left(\frac{v_A}{b}\right)^2 \tag{f}$$

$$\mathbf{k}:\quad -2\alpha_x+\alpha_y=+\frac{10}{36}\left(\frac{v_A}{b}\right)^2+\frac{a_B}{b}. \tag{g}$$

The requirement that $\boldsymbol{\alpha}_t$ be zero is satisfied by $\boldsymbol{\alpha}\cdot\mathbf{r}=0$; that is

$$b(-\alpha_x-2\alpha_y+2\alpha_z)=0. \tag{h}$$

The components α_x, α_y, and α_z can be eliminated by adding one-half of Eq. (e) to Eq. (f) and subtracting Eq. (g), which gives

$$0=-\frac{5}{72}\left(\frac{v_A}{b}\right)^2-\frac{a_A}{2b}-\frac{10}{36}\left(\frac{v_A}{b}\right)^2-\frac{10}{36}\left(\frac{v_A}{b}\right)^2-\frac{a_B}{b}$$

from which

$$\frac{a_B}{b}=-\frac{a_A}{2b}-\frac{5}{8}\left(\frac{v_A}{b}\right)^2,$$

and the acceleration of B is

$$\mathbf{a}_B=-\left(\frac{a_A}{2}+\frac{5v_A^2}{8b}\right)\mathbf{k}. \qquad \text{Ans.}$$

The values of α_x, α_y, and α_z, obtained by solving Eq. (e), (f), and (h) simultaneously, are

$$\alpha_x=\frac{5}{36}\left(\frac{v_A}{b}\right)^2+\frac{1}{9}\left(\frac{a_A}{b}\right),$$

$$\alpha_y=-\frac{5}{72}\left(\frac{v_A}{b}\right)^2-\frac{5}{18}\left(\frac{a_A}{b}\right),$$

$$\alpha_z=-\frac{2}{9}\left(\frac{a_A}{b}\right),$$

and the component of the angular acceleration normal to AB is

$$\boldsymbol{\alpha}_n=\frac{5}{72}\left(\frac{v_A}{b}\right)^2(2\mathbf{i}-\mathbf{j})+\frac{1}{18}\left(\frac{a_A}{b}\right)(2\mathbf{i}-5\mathbf{j}-4\mathbf{k}). \qquad \text{Ans.}$$

PROBLEMS

8-245 Crank OA in Fig. P8-245 rotates in the yz plane with a constant angular velocity of $6\mathbf{i}$ rad per sec. Member AB has a ball and socket joint at each end. Determine the acceleration of collar B.

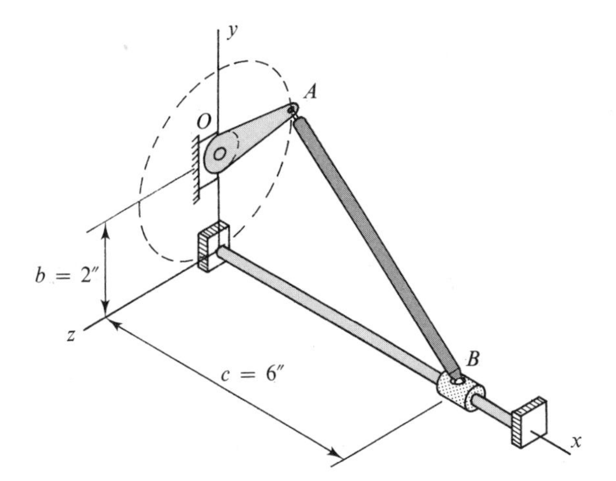

Figure P8-245

8-246 Assume the component of the angular velocity of AB in Problem 8-245 parallel to AB to be zero and calculate the component of the angular acceleration of AB normal to AB.

8-247 The ball and socket at B in Problem 8-245 is replaced by a clevis; see Fig. P8-218. Determine the angular acceleration of member AB.

8-248 Figure P8-248 represents a pair of bevel gears A and B. Gear A rotates in a horizontal plane about a vertical axis at O. Gear B rotates around the shaft OC, which in turn rotates about the vertical axis at O (see Problem 8-219). The shaft OC has a constant angular velocity of $\omega_2\mathbf{j}$, and gear A is stationary. Determine, for the position shown,
(a) The angular acceleration of gear B.
(b) The linear acceleration of point D on gear B.

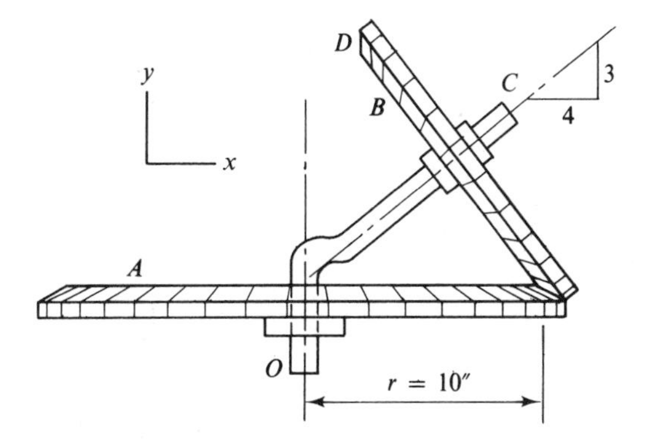

Figure P8-248

8-249 Determine the angular acceleration of gear B in Problem 8-248 when gear A has an angular velocity of $30\mathbf{j}$ rad per sec and an angular acceleration of $-20\mathbf{j}$ rad per sec^2 and member OC has an angular velocity of $10\mathbf{j}$ rad per sec and an angular acceleration of $25\mathbf{j}$ rad per sec^2.

8-250 Disk D in Fig. P8-250 rotates about an axis at O with a constant angular velocity of $\omega_1\mathbf{k}$ rad per sec. The collar B slides along the inclined rod which is parallel to the yz plane. The link AB is connected to the disk and collar by ball and socket joints. Determine the acceleration of collar B.

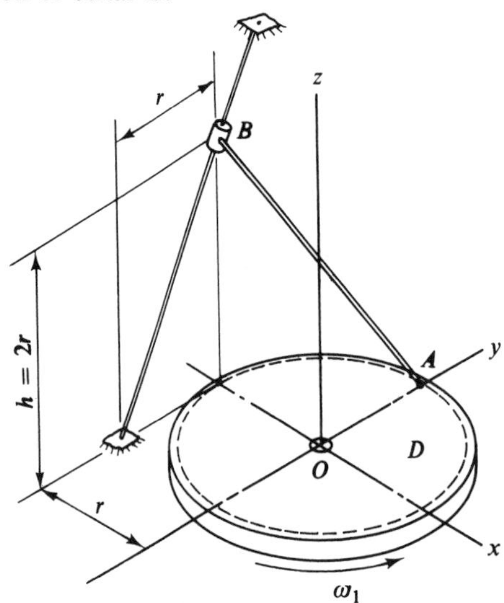

Figure P8-250

8-251 The ball and socket joint at A in Problem 8-250 is replaced by the joint shown in Fig. P8-251 in which the vertical member rotates in a hole in the disk D and is connected to AB by a horizontal pin at A. Determine the angular acceleration of member AB.

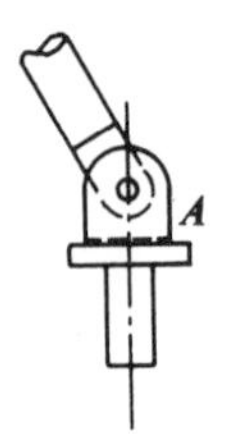

Figure P8-251

***8-252** Rod CDE in Fig. P8-252 is welded to AB, and point C slides on the horizontal xy plane. Rod AB is connected to two collars by means of ball and socket joints. In the position shown the velocity of A is $27\mathbf{i}$ cm/s and its acceleration is zero. Determine

(a) The angular acceleration of ABC.

(b) The linear acceleration of point C.

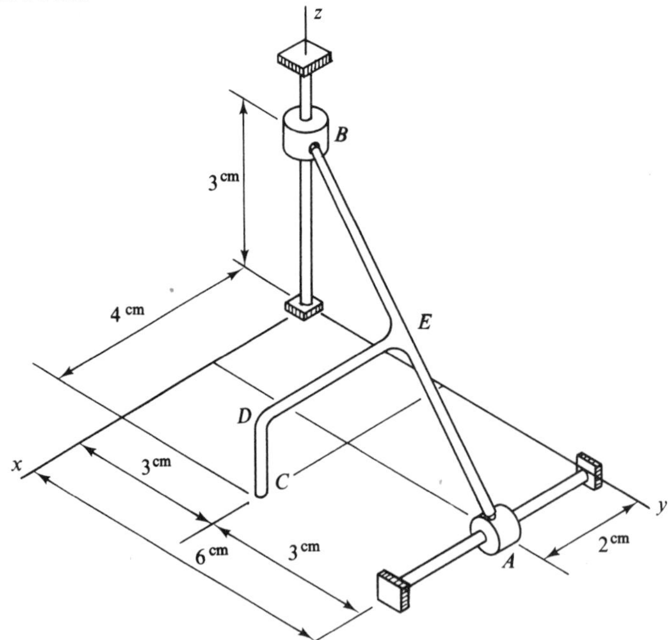

Figure P8-252

8-21 RELATIVE MOTION WITH RESPECT TO ROTATING AXES

Equations relating absolute and relative positions, velocities, and accelerations of two points were developed in Art. 8-6. Application of these equations to problems in which the moving system of coordinates did not rotate was discussed in some detail in Art. 8-14 through 8-20. The equations were particularly useful when the two points were on a rigid body. In some instances it is convenient to use axes that rotate with the body. Moments and products of inertia are usually functions of time unless the moving axes are fixed in the body. Rotating axes may be advantageous when the two points are not on the same rigid body as when a point on one body slides along a path on the second body.

Consider the XYZ system of coordinates in Fig. 8-48 to be fixed in space so that the unit vectors $\mathbf{i}'$, $\mathbf{j}'$, and $\mathbf{k}'$ are constant in direction.

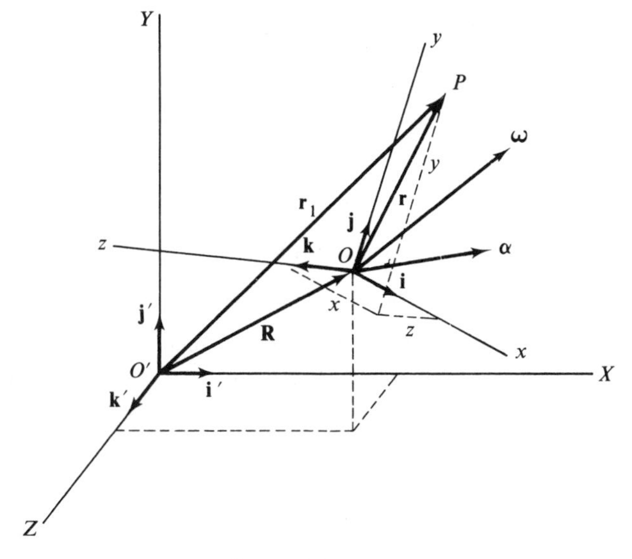

Figure 8-48

The origin of the xyz coordinates moves with point O, and at the same time the axes have an angular velocity $\boldsymbol{\omega}$ and an angular acceleration $\boldsymbol{\alpha}=\dot{\boldsymbol{\omega}}$. The quantities $\boldsymbol{\omega}$ and $\boldsymbol{\alpha}$ are measured with respect to the fixed XYZ axes and are thus absolute quantities. In general, $\boldsymbol{\omega}$ and $\boldsymbol{\alpha}$ will not be collinear unless the xyz axes have plane motion. Note that the unit vectors $\mathbf{i}$, $\mathbf{j}$, and $\mathbf{k}$ are not constant since they rotate with the xyz axes. The time rate of change of these unit vectors will be needed in the following development and can be determined from the principles developed in Art. 8-13.

Consider $\mathbf{i}$ to be the position vector for the tip of the unit vector along the x axis. The velocity of the tip relative to a set of nonrotating axes at O (not shown) is

$$\dot{\mathbf{i}} = \boldsymbol{\omega} \times \mathbf{i},$$

Similarly,

$$\dot{\mathbf{j}} = \boldsymbol{\omega} \times \mathbf{j} \quad \text{and} \quad \dot{\mathbf{k}} = \boldsymbol{\omega} \times \mathbf{k}.$$

The position of point P in Fig. 8-48 is

$$\mathbf{r}_1 = \mathbf{R} + \mathbf{r},$$

where

$$\mathbf{r} = x\mathbf{i} + y\mathbf{j} + z\mathbf{k} \tag{a}$$

is the position of P relative to the moving axes (as seen by an observer moving with the xyz reference system). The velocity of P is

$$\mathbf{v}_P = \dot{\mathbf{r}}_1 = \dot{\mathbf{R}} + \dot{\mathbf{r}},$$

and the value of $\dot{\mathbf{r}}$, from Eq. (a), is

$$\begin{aligned}
\dot{\mathbf{r}} &= (\dot{x}\mathbf{i} + \dot{y}\mathbf{j} + \dot{z}\mathbf{k}) + (x\dot{\mathbf{i}} + y\dot{\mathbf{j}} + z\dot{\mathbf{k}}) \\
&= (\dot{x}\mathbf{i} + \dot{y}\mathbf{j} + \dot{z}\mathbf{k}) + (x\boldsymbol{\omega} \times \mathbf{i} + y\boldsymbol{\omega} \times \mathbf{j} + z\boldsymbol{\omega} \times \mathbf{k}) \\
&= (\dot{x}\mathbf{i} + \dot{y}\mathbf{j} + \dot{z}\mathbf{k}) + \boldsymbol{\omega} \times (x\mathbf{i} + y\mathbf{j} + z\mathbf{k}) \\
&= \mathbf{v}_{P/xyz} + \boldsymbol{\omega} \times \mathbf{r}
\end{aligned} \tag{b}$$

in which $\mathbf{v}_{P/xyz} = (\dot{x}\mathbf{i} + \dot{y}\mathbf{j} + \dot{z}\mathbf{k})$ is the velocity of P measured relative to the moving references axes. Thus the velocity of P is

$$\mathbf{v}_P = \dot{\mathbf{R}} + \mathbf{v}_{P/xyz} + \boldsymbol{\omega} \times \mathbf{r}. \tag{8-27}$$

If P' is a point fixed to the rotating axes which is coincident with P at the instant under consideration, the quantity $\boldsymbol{\omega} \times \mathbf{r}$ is the velocity of P' relative to a set of nonrotating axes at O (the velocity of P' relative to O). The velocity of P is seen to consist of three terms:

$\dot{\mathbf{R}}$: the velocity of the moving origin.
$\mathbf{v}_{P/xyz} = (\dot{x}\mathbf{i} + \dot{y}\mathbf{j} + \dot{x}\mathbf{k})$: the velocity of P relative to the moving axes.
$\boldsymbol{\omega} \times \mathbf{r}$: the velocity of P' relative to O.

The sum of the second and third terms on the right-hand side of Eq. (8-27) is the velocity of P relative to a set of nonrotating axes at O, and in Art. 8-14 it was designated as $\mathbf{v}_{P/O}$; that is

$$\mathbf{v}_{P/O} = \mathbf{v}_{P/xyz} + \boldsymbol{\omega} \times \mathbf{r}$$

and

$$\mathbf{v}_P = \mathbf{v}_O + \mathbf{v}_{P/O}.$$

The following example illustrates the use of Eq. (8-27) with various sets of moving axes.

EXAMPLE 8-32

The small disk D in Fig. 8-49 is pinned to the larger disk C at point A. Disk C rotates about a vertical axis at O' with an absolute angular velocity of $5\mathbf{k}$ rad per sec, and D has an absolute angular velocity of $8\mathbf{k}$ rad per sec. Determine the velocity of point P on disk D.

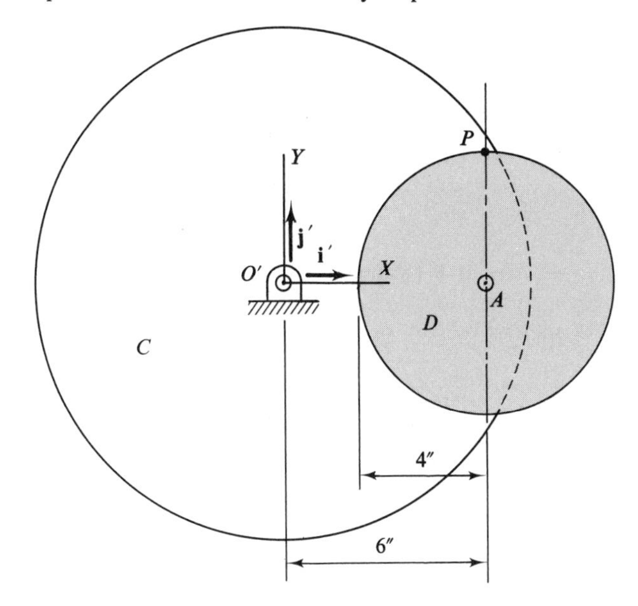

Figure 8-49

SOLUTION

Four different choices of moving axes xyz are used to demonstrate the various terms in Eq. (8-27). The x and y axes are chosen parallel to the X and Y axes for each application; consequently $\mathbf{i}=\mathbf{i}'$, $\mathbf{j}=\mathbf{j}'$, and $\mathbf{k}=\mathbf{k}'$.

(a) *Nonrotating axes, origin at* A. The velocity equation is

$$\mathbf{v}_P = \mathbf{v}_O + \mathbf{v}_{P/xyz} + \boldsymbol{\omega}\times\mathbf{r},$$

where

$$\mathbf{v}_O = \mathbf{v}_A = \boldsymbol{\omega}_C \times \mathbf{O'A} = 5\mathbf{k}\times 6\mathbf{i} = 30\mathbf{j} \text{ ips}$$

and

$$\mathbf{v}_{P/xyz} = \boldsymbol{\omega}_D \times \mathbf{AP} = 8\mathbf{k}\times 4\mathbf{j} = -32\mathbf{i} \text{ ips}.$$

Since the xy axes are nonrotating, $\boldsymbol{\omega}=\boldsymbol{\omega}_{xyz}=0$, and the third term in the equation is zero. Therefore

$$\mathbf{v}_P = (30\mathbf{j} - 32\mathbf{i}) \text{ ips} = 43.9 \text{ ips} \;\; \angle 43.2^\circ \qquad \text{Ans.}$$

(b) *Axes rotate with body* D, *origin at* A. The angular velocity of the xyz axes is $8\mathbf{k}$ rad per sec. Since P is a point fixed on the rotating axes, $\mathbf{v}_{P/xyz}$ must be zero.

The velocity equation becomes

$$\mathbf{v}_P = \mathbf{v}_O + \mathbf{v}_{P/xyz} + \boldsymbol{\omega} \times \mathbf{r}$$
$$= \mathbf{v}_A + 0 + 8\mathbf{k} \times 4\mathbf{j}$$
$$= \underline{(30\mathbf{j} - 32\mathbf{i}) \text{ ips as before.}} \qquad \text{Ans.}$$

Note that when P is fixed on the moving axes the velocity of P relative to the axes is zero.

(c) *Axes rotate with body C, origin at P.* The velocity of the origin is the velocity of the point on C which is coincident with P at the instant under consideration; thus

$$\mathbf{v}_O = \boldsymbol{\omega}_C \times \mathbf{r} = 5\mathbf{k} \times (6\mathbf{i} + 4\mathbf{j})$$
$$= (-20\mathbf{i} + 30\mathbf{j}) \text{ ips.}$$

The velocity of P relative to the rotating axes is the velocity it would *appear* to have to an observer who was moving (rotating) with the axes. The angular velocity of disk D relative to the xyz axes is

$$\boldsymbol{\omega}_{D/xyz} = \boldsymbol{\omega}_D - \boldsymbol{\omega}_{xyz} = 8\mathbf{k} - 5\mathbf{k} = 3\mathbf{k} \text{ rad per sec,}$$

and the relative velocity of P is

$$\mathbf{v}_{P/xyz} = \boldsymbol{\omega}_{D/xyz} \times \mathbf{r}_{AP} = 3\mathbf{k} \times 4\mathbf{j} = -12\mathbf{i} \text{ ips.}$$

Note that the path of P relative to the rotating axes is a circular path which moves with the axes. The third term in the velocity equation is zero since P is at the origin and the value of $\mathbf{r}$ (the position of P from the origin) is zero. Thus the velocity of P is

$$\mathbf{v}_P = \mathbf{v}_O + \mathbf{v}_{P/xyz} + \boldsymbol{\omega} \times \mathbf{r}$$
$$= (-20\mathbf{i} + 30\mathbf{j}) + (-12\mathbf{i}) + 5\mathbf{k} \times 0$$
$$= \underline{(-32\mathbf{i} + 30\mathbf{j}) \text{ ips as before.}} \qquad \text{Ans.}$$

(d) *Axes rotate with body C, origin at A.* The angular velocity of the rotating axes is $5\mathbf{k}$ rad per sec, and the angular velocity of body D relative to the rotating axes is $3\mathbf{k}$ rad per sec. The velocity of P is

$$\mathbf{v}_P = \mathbf{v}_O + \mathbf{v}_{P/xyz} + \boldsymbol{\omega} \times \mathbf{r}$$
$$= 5\mathbf{k} \times 6\mathbf{i} + 3\mathbf{k} \times 4\mathbf{j} + 5\mathbf{k} \times 4\mathbf{j}$$
$$= 30\mathbf{j} - 12\mathbf{i} - 20\mathbf{i}$$
$$= \underline{(-32\mathbf{i} + 30\mathbf{j}) \text{ ips.}} \qquad \text{Ans.}$$

The result must, of course, be the same for any choice of rotating or nonrotating axes and for any origin. In this particular example each set of axes gave results with about the same effort. It is important, however, that the effects of the different sets be completely understood.

The acceleration of P can be obtained by differentiating Eq. (8-27) with respect to time. The acceleration is

$$\mathbf{a}_P = \dot{\mathbf{v}}_P = \ddot{\mathbf{R}} + \dot{\mathbf{v}}_{P/xyz} + \dot{\boldsymbol{\omega}} \times \mathbf{r} + \boldsymbol{\omega} \times \dot{\mathbf{r}}. \tag{c}$$

Section 8-21
Relative motion with respect to rotating axes

The quantity $\mathbf{v}_{P/xyz}=(\dot{x}\mathbf{i}+\dot{y}\mathbf{j}+\dot{z}\mathbf{k})$ can be differentiated to give

$$\begin{aligned}\dot{\mathbf{v}}_{P/xyz}&=(\ddot{x}\mathbf{i}+\ddot{y}\mathbf{j}+\ddot{z}\mathbf{k})+(\dot{x}\dot{\mathbf{i}}+\dot{y}\dot{\mathbf{j}}+\dot{z}\dot{\mathbf{k}})\\&=(\ddot{x}\mathbf{i}+\ddot{y}\mathbf{j}+\ddot{z}\mathbf{k})+(\dot{x}\boldsymbol{\omega}\times\mathbf{i}+\dot{y}\boldsymbol{\omega}\times\mathbf{j}+\dot{z}\boldsymbol{\omega}\times\mathbf{k})\\&=\mathbf{a}_{P/xyz}+\boldsymbol{\omega}\times\mathbf{v}_{P/xyz},\end{aligned}$$

where $\mathbf{a}_{P/xyz}$ is the acceleration of P relative to the moving xyz coordinate system. It is the acceleration P would *appear* to have to an observer who was rotating with the xyz axes.

The quantity $\dot{\mathbf{r}}$ from Eq. (b) is substituted into the last term of Eq. (c) to give

$$\begin{aligned}\boldsymbol{\omega}\times\dot{\mathbf{r}}&=\boldsymbol{\omega}\times(\mathbf{v}_{P/xyz}+\boldsymbol{\omega}\times\mathbf{r})\\&=\boldsymbol{\omega}\times\mathbf{v}_{P/xyz}+\boldsymbol{\omega}\times(\boldsymbol{\omega}\times\mathbf{r}).\end{aligned}$$

The acceleration of P thus becomes

$$\begin{aligned}\mathbf{a}_P&=\ddot{\mathbf{R}}+\mathbf{a}_{P/xyz}+\boldsymbol{\omega}\times\mathbf{v}_{P/xyz}+\dot{\boldsymbol{\omega}}\times\mathbf{r}+\boldsymbol{\omega}\times\mathbf{v}_{P/xyz}+\boldsymbol{\omega}\times(\boldsymbol{\omega}\times\mathbf{r})\\&=\ddot{\mathbf{R}}+\dot{\boldsymbol{\omega}}\times\mathbf{r}+\boldsymbol{\omega}\times(\boldsymbol{\omega}\times\mathbf{r})+\mathbf{a}_{P/xyz}+2\boldsymbol{\omega}\times\mathbf{v}_{P/xyz}.\end{aligned}\tag{8-28}$$

The sum of the first three terms on the right-hand side of Eq. (8-28) gives the acceleration of a point P', fixed in the xyz coordinate system, which is coincident with particle P at the instant under consideration. Thus the acceleration of P can be written as

$$\mathbf{a}_P=\mathbf{a}_{P'}+\mathbf{a}_{P/xyz}+2\boldsymbol{\omega}\times\mathbf{v}_{P/xyz}.\tag{8-28a}$$

The quantity $\mathbf{a}_{P/xyz}$ has already been defined as the acceleration of P measured relative to the moving (xyz) coordinate system. Any angular velocity or acceleration used to compute this quantity must be measured relative to the moving axes.

The last term of Eq. (8-28) is the Coriolis[9] or supplementary acceleration. This component is normal to the plane containing the vectors $\boldsymbol{\omega}$ and $\mathbf{v}_{P/xyz}$ as determined by the right-hand rule for vector products. The sense of the Coriolis component can be visualized by noting the motion of the tip of the vector $\mathbf{v}_{P/xyz}$ as it is rotated in the direction indicated by $\boldsymbol{\omega}$.

If a rigid body is attached to the moving axes and if points O and P are points on the rigid body (P and P' will be identical), the last two terms of Eq. (8-28) will be zero since both $\mathbf{a}_{P/xyz}$ and $\mathbf{v}_{P/xyz}$ are zero. The second and third terms of Eq. (8-28) are the acceleration of P' relative to nonrotating axes at O.

It is sometimes convenient to express the angular velocity of one rigid body relative to a second rigid body as in the discussion of the

[9] Named after G. C. Coriolis (1792–1843), a French scientist, who was the first to publish a discussion of this acceleration.

angular velocity of a member connected to a rotating collar by a clevis in Art. 8-17. As another example, assume that the wheel B in Fig. 8-50(a) rotates about a horizontal axis along line OP on the disk A as A rotates about the z axis. The absolute angular velocity of A is $\boldsymbol{\omega}_A$, and wheel B has an angular velocity relative to disk A of $\boldsymbol{\omega}_{B/A}$ along the axis OP. The absolute angular velocity of B is

$$\boldsymbol{\omega}_B = \boldsymbol{\omega}_A + \boldsymbol{\omega}_{B/A}$$

$$= \boldsymbol{\omega}_A + \omega_{B/A}\mathbf{e}_{OP},$$

where $\mathbf{e}_{OP}$ is a unit vector parallel to line OP, the axis of the relative angular velocity. The angular velocity of B is shown in Fig. 8-50(b).

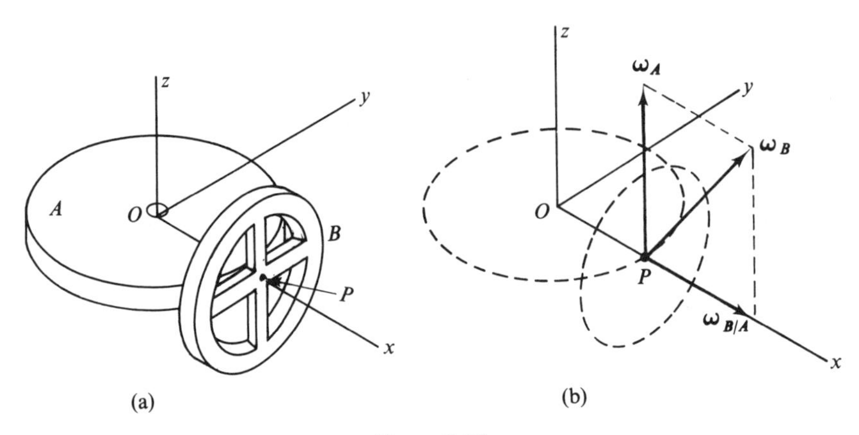

Figure 8-50

The angular acceleration of B is obtained by differentiating the angular velocity; that is,

$$\boldsymbol{\alpha}_B = \dot{\boldsymbol{\omega}}_B = \dot{\boldsymbol{\omega}}_A + \dot{\omega}_{B/A}\mathbf{e}_{OP} + \omega_{B/A}\dot{\mathbf{e}}_{OP}$$

$$= \dot{\boldsymbol{\omega}}_A + \dot{\omega}_{B/A}\mathbf{e}_{OP} + \omega_{B/A}\boldsymbol{\omega}_A \times \mathbf{e}_{OP}$$

$$= \boldsymbol{\alpha}_B + \boldsymbol{\alpha}_{B/A} + \boldsymbol{\omega}_A \times \boldsymbol{\omega}_{B/A}.$$

The third term in the preceding equation results from the change in direction of the relative angular velocity vector. This term will be zero only when $\boldsymbol{\omega}_A$ or $\boldsymbol{\omega}_{B/A}$ is zero or when $\boldsymbol{\omega}_A$ and $\boldsymbol{\omega}_{B/A}$ are parallel.

The determination of accelerations using rotating axes, using nonrotating axes, and by differentiation is discussed in the following examples. Link mechanisms with no sliding connections can usually be solved most directly by using nonrotating axes as in Art. 8-19 and 8-20. Mechanisms with sliding connections can often be solved more readily using rotating axes and Eq. (8-28); see Example 8-34. Many problems, such as Example 8-33, can be solved most directly by differentiating the position function to determine the velocity and acceleration. Most problems can be solved by more than one method, and personal preference dictates the choice. Solution by two different methods provides an excellent check on the work.

EXAMPLE 8-33

Blocks D and E in Fig. 8-51 are pinned together, and D slides along the slot in member OA, while E slides in the vertical slot. The angular velocity of OA is $\omega\mathbf{k}$, and its angular acceleration is $\dot{\omega}\mathbf{k}$ when the angle θ is as shown. Determine the acceleration of block E for this position.

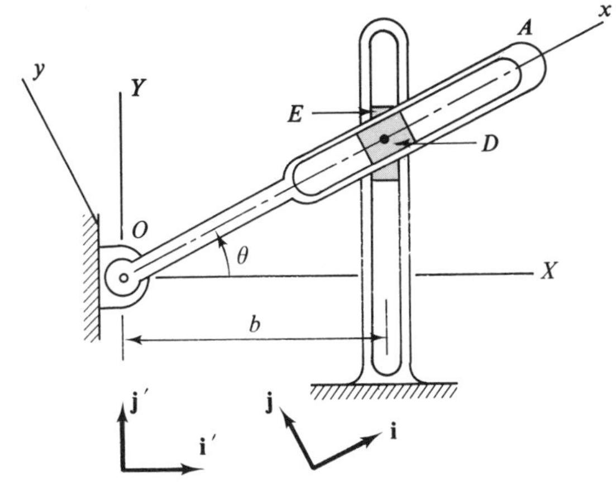

Figure 8-51

SOLUTION

Since block D slides along member OA, a set of rotating, xyz, axes are selected as shown. The fixed, XYZ, axes and the rotating xyz axes both have their origins at O in this example. The angular velocity and acceleration of the rotating axes are the same as for member OA. The unit vectors $\mathbf{i}'$ and $\mathbf{j}'$ are parallel to X and Y and $\mathbf{i}$ and $\mathbf{j}$ are parallel to x and y, respectively. Equation (8-28) is

$$\mathbf{a}_E = \ddot{\mathbf{R}} + \dot{\boldsymbol{\omega}} \times \mathbf{r} + \boldsymbol{\omega} \times (\boldsymbol{\omega} \times \mathbf{r}) + \mathbf{a}_{P/xyz} + 2\boldsymbol{\omega} \times \mathbf{v}_{P/xyz}.$$

The following values are available from the given data and the selection of axes:

$$\mathbf{R} = \dot{\mathbf{R}} = \ddot{\mathbf{R}} = 0, \qquad \mathbf{r} = r\mathbf{i}, \qquad \dot{\boldsymbol{\omega}} = \dot{\omega}\mathbf{k}, \qquad \mathbf{v}_{P/xyz} = \dot{r}\mathbf{i},$$

and

$$\mathbf{a}_{P/xyz} = \ddot{r}\mathbf{i}.$$

Note that r is the distance from O to D. Since E moves along the vertical slot, its acceleration is $\mathbf{a}_E = a_E\mathbf{j}'$. When these values are substituted into the acceleration equation it becomes

$$\begin{aligned} a_E\mathbf{j}' &= 0 + \dot{\omega}\mathbf{k} \times r\mathbf{i} + \omega\mathbf{k} \times (\omega\mathbf{k} \times r\mathbf{i}) + \ddot{r}\mathbf{i} + 2\omega\mathbf{k} \times \dot{r}\mathbf{i} \\ &= \dot{\omega}r\mathbf{j} - \omega^2 r\mathbf{i} + \ddot{r}\mathbf{i} + 2\omega\dot{r}\mathbf{j}. \qquad \text{(a)} \end{aligned}$$

The quantity $r = b \sec\theta$ can be determined for any angle θ, but the equation still contains three unknown magnitudes: a_E, $\dot{r}$, and $\ddot{r}$. The value of $\dot{r}$ can be determined from Eq. (8-27); thus

$$\mathbf{v}_E = \dot{\mathbf{R}} + \mathbf{v}_{E/xyz} + \boldsymbol{\omega} \times \mathbf{r}$$

or

$$v_E \mathbf{j}' = 0 + \dot{r}\mathbf{i} + \omega r \mathbf{j}.$$

Since $\mathbf{v}_E$ is vertical, the horizontal component of $\mathbf{v}_E$ must be zero; that is,

$$\mathbf{v}_E \cdot \mathbf{i}' = 0. \tag{b}$$

Note that

$$\mathbf{i}' = \mathbf{i} \cos\theta - \mathbf{j} \sin\theta;$$

therefore Eq. (b) becomes

$$(\dot{r}\mathbf{i} + \omega r \mathbf{j}) \cdot (\mathbf{i}\cos\theta - \mathbf{j}\sin\theta) = \dot{r}\cos\theta - \omega r \sin\theta = 0$$

from which

$$\dot{r} = \omega r \tan\theta.$$

When this quantity is substituted in Eq. (a) it becomes

$$a_E \mathbf{j}' = \dot{\omega} r \mathbf{j} - \omega^2 r \mathbf{i} + \ddot{r}\mathbf{i} + 2\omega^2 r \tan\theta \mathbf{j}$$
$$= (\ddot{r} - r\omega^2)\mathbf{i} + r(\dot{\omega} + 2\omega^2 \tan\theta)\mathbf{j} \tag{c}$$

in which the two unknowns are a_E and $\ddot{r}$. The first term on the right in this equation can be eliminated by replacing $\mathbf{j}'$ by its equivalent $(\mathbf{i}\sin\theta + \mathbf{j}\cos\theta)$ and taking the scalar or dot product of the equation with $\mathbf{j}$. That is,

$$a_E(\mathbf{i}\sin\theta + \mathbf{j}\cos\theta) \cdot \mathbf{j} = (\ddot{r} - r\omega^2)\mathbf{i} \cdot \mathbf{j} + r(\dot{\omega} + 2\omega^2 \tan\theta)\mathbf{j} \cdot \mathbf{j}$$

or

$$a_E \cos\theta = 0 + r(\dot{\omega} + 2\omega^2 \tan\theta)$$

from which

$$\mathbf{a}_E = r \sec\theta(\dot{\omega} + 2\omega^2 \tan\theta)\mathbf{j}'$$
$$= \underline{b \sec^2\theta(\dot{\omega} + 2\omega^2 \tan\theta)\mathbf{j}'}. \qquad \text{Ans.}$$

As indicated in the paragraph preceding this example, this particular problem could be solved more readily by deriving an expression relating the position of D to the angular position of OA and differentiating twice. The purpose of this example is to illustrate the application of relative acceleration with respect to rotating axes.

EXAMPLE 8-34

Block D in Fig. 8-52 is pinned to member CD and slides along member OAB. The mechanism moves in a common plane and in the position shown $\boldsymbol{\omega}_{CD} = -4\mathbf{k}$ rad per sec, and $\dot{\boldsymbol{\omega}}_{CD} = -10\mathbf{k}$ rad per sec^2. Determine the angular acceleration of member OAB.

SOLUTION

Since block D slides along member AB, the motion of D relative to axes which rotate with OAB will be rectilinear and should provide a

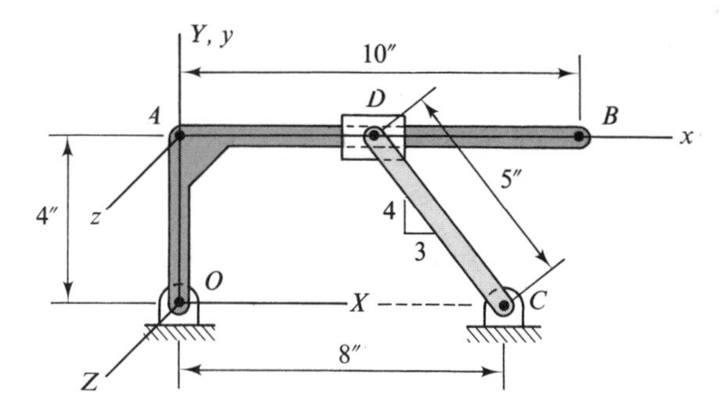

Figure 8-52

direct solution. The origin of the moving axes can be selected at any convenient point, such as O, A, B, or the point on AB which is coincident with D. Point A will be selected as the origin for this example in which case $\mathbf{R}$ and $\mathbf{r}$ in Eq. (8-28) are

$$\mathbf{R} = \mathbf{R}_{OA} = 4\mathbf{j} \quad \text{and} \quad \mathbf{r} = \mathbf{r}_{AD} = 5\mathbf{i}.$$

The relative velocity $\mathbf{v}_{D/xyz}$ and the angular velocity of the xyz axes are needed for the relative acceleration equation and can be obtained from the relative velocity equation; thus

$$\mathbf{v}_D = \dot{\mathbf{R}} + \boldsymbol{\omega} \times \mathbf{r} + \mathbf{v}_{D/xyz}$$

in which $\boldsymbol{\omega}$ is the angular velocity of the xyz axes. The velocity of D is

$$\mathbf{v}_D = \boldsymbol{\omega}_{CD} \times \mathbf{r}_{CD} = -4\mathbf{k} \times (-3\mathbf{i} + 4\mathbf{j}) = 16\mathbf{i} + 12\mathbf{j},$$

the quantity $\dot{\mathbf{R}}$ is

$$\dot{\mathbf{R}} = \boldsymbol{\omega} \times \mathbf{R} = \omega\mathbf{k} \times (4\mathbf{j}) = -4\omega\mathbf{i},$$

the vector $\boldsymbol{\omega} \times \mathbf{r}$ is

$$\boldsymbol{\omega} \times \mathbf{r} = \omega\mathbf{k} \times (5\mathbf{i}) = 5\omega\mathbf{j},$$

and

$$\mathbf{v}_{D/xyz} = v_{D/xyz}\mathbf{i}$$

from which

$$16\mathbf{i} + 12\mathbf{j} = -4\omega\mathbf{i} + 5\omega\mathbf{j} + v_{D/xyz}\mathbf{i}$$

This vector equation is equivalent to

$$16 = -4\omega + v_{D/xyz}$$

and

$$12 = 5\omega$$

from which

$$\boldsymbol{\omega} = 2.4\mathbf{k} \quad \text{and} \quad \mathbf{v}_{D/xyz} = 25.6\mathbf{i}.$$

The acceleration of D is

$$\begin{aligned}\mathbf{a}_D &= \dot{\boldsymbol{\omega}}_{CD} \times \mathbf{r}_{CD} + \boldsymbol{\omega}_{CD} \times (\boldsymbol{\omega}_{CD} \times \mathbf{r}_{CD}) \\ &= -10\mathbf{k} \times (-3\mathbf{i} + 4\mathbf{j}) + (-4\mathbf{k}) \times [(-4\mathbf{k}) \times (-3\mathbf{i} + 4\mathbf{j})] \\ &= 88\mathbf{i} - 34\mathbf{j}.\end{aligned}$$

The acceleration of D can also be expressed as

$$\mathbf{a}_D = \ddot{\mathbf{R}} + \dot{\boldsymbol{\omega}} \times \mathbf{r} + \boldsymbol{\omega} \times (\boldsymbol{\omega} \times \mathbf{r}) + \mathbf{a}_{D/xyz} + 2\boldsymbol{\omega} \times \mathbf{v}_{D/xyz},$$

where $\boldsymbol{\omega}$ and $\dot{\boldsymbol{\omega}}$ refer to the rotating axes. The vector $\ddot{\mathbf{R}}$ is

$$\ddot{\mathbf{R}} = \dot{\omega}\mathbf{k} \times (4\mathbf{j}) + 2.4\mathbf{k} \times (2.4\mathbf{k} \times 4\mathbf{j}) = -4\dot{\omega}\mathbf{i} - 23.04\mathbf{j}.$$

When numerical data are substituted for $\boldsymbol{\omega}$, $\mathbf{r}$, and $\mathbf{v}_{D/xyz}$ the acceleration equation becomes

$$\begin{aligned} 88\mathbf{i} - 34\mathbf{j} &= (-4\dot{\omega}\mathbf{i} - 23.04\mathbf{j}) + \dot{\omega}\mathbf{k} \times 5\mathbf{i} + 2.4\mathbf{k} \times (2.4\mathbf{k} \times 5\mathbf{i}) \\ &\quad + a_{D/xyz}\mathbf{i} + 2(2.4\mathbf{k} \times 25.6\mathbf{i}) \\ &= -4\dot{\omega}\mathbf{i} - 23.04\mathbf{j} + 5\dot{\omega}\mathbf{j} - 28.8\mathbf{i} + a_{D/xyz}\mathbf{i} + 122.9\mathbf{j}. \end{aligned}$$

This vector equation is equivalent to

$$88\mathbf{i} = (-4\dot{\omega} - 28.8 + a_{D/xyz})\mathbf{i}$$

and

$$-34\mathbf{j} = (-23.04 + 5\dot{\omega} + 122.9)\mathbf{j}$$

from which

$$\dot{\omega} = -26.8 \text{ rad per sec}^2$$

or

$$\dot{\boldsymbol{\omega}} = \boldsymbol{\alpha}_{OAB} = \underline{-26.8\mathbf{k} \text{ rad per sec}^2 \text{ (clockwise).}} \qquad \text{Ans.}$$

The first equation, in terms of the $\mathbf{i}$ components, is not used for this problem.

EXAMPLE 8-35

The helicopter in Fig. 8-53 is hovering above a point on the ground so that point O' is stationary. The body A has an angular velocity of

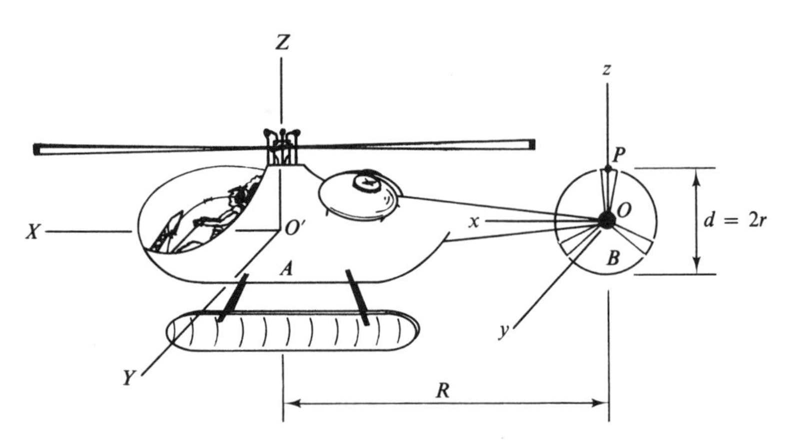

Figure 8-53

$\boldsymbol{\omega}_A = \omega_A\mathbf{k}'$ which is slowing down with an angular acceleration $\dot{\boldsymbol{\omega}}_A = \dot{\omega}_A\mathbf{k}'$. Note that the numerical value of $\dot{\omega}_A$ will be negative. The tail rotor B has a constant angular velocity relative to A and in the position shown $\boldsymbol{\omega}_{B/A} = \omega_{B/A}\mathbf{j}$. Determine

(a) The angular acceleration of the tail rotor.
(b) The acceleration of the point P at the top of the tail rotor.

SOLUTION

(a) The XYZ axes ($\mathbf{i}'$, $\mathbf{j}'$, $\mathbf{k}'$ unit vectors) are fixed, and the moving xyz axes ($\mathbf{i}$, $\mathbf{j}$, $\mathbf{k}$ unit vectors) at O move with the body A. The angular velocity of the xyz axes therefore is $\boldsymbol{\omega}_A$. The angular velocity of B is

$$\begin{aligned}\boldsymbol{\omega}_B &= \boldsymbol{\omega}_A + \boldsymbol{\omega}_{B/A}\\ &= \omega_A \mathbf{k}' + \omega_{B/A}\mathbf{j},\end{aligned}$$

and the angular acceleration of B is

$$\boldsymbol{\alpha}_B = \dot{\boldsymbol{\omega}}_B = \dot{\omega}_A \mathbf{k}' + \omega_A \dot{\mathbf{k}}' + \dot{\omega}_{B/A}\mathbf{j} + \omega_{B/A}\dot{\mathbf{j}}.$$

The factor $\dot{\mathbf{k}}'$ is zero since $\mathbf{k}'$ does not change direction. The quantity $\dot{\omega}_{B/A}$ is equal to zero, and $\dot{\mathbf{j}} = \boldsymbol{\omega}_A \times \mathbf{j}$; therefore

$$\begin{aligned}\boldsymbol{\alpha}_B &= \dot{\omega}_A \mathbf{k}' + 0 + 0 + \omega_{B/A}\mathbf{k}' \times \mathbf{j}\\ &= \underline{\dot{\omega}_A \mathbf{k}' - \omega_{B/A}\omega_A \mathbf{i}'}. \qquad \text{Ans.}\end{aligned}$$

Note that $\mathbf{i} = \mathbf{i}'$ at the instant under consideration.

(b) The acceleration of P, from Eq. (8-28), is

$$\mathbf{a}_P = \ddot{\mathbf{R}} + \dot{\boldsymbol{\omega}}_A \times \mathbf{r} + \boldsymbol{\omega}_A \times (\boldsymbol{\omega}_A \times \mathbf{r}) + \mathbf{a}_{P/xyz} + 2\boldsymbol{\omega}_A \times \mathbf{v}_{P/xyz}.$$

The same axes are used as for part (a), and the following quantities can be written for this choice of axes. The vector $\ddot{\mathbf{R}}$ is the acceleration of O which moves in a horizontal circle around O'; that is

$$\begin{aligned}\ddot{\mathbf{R}} &= \dot{\omega}_A \mathbf{k} \times (-R\mathbf{i}) + \omega_A \mathbf{k} \times [\omega_A \mathbf{k} \times (-R\mathbf{i})]\\ &= -R\dot{\omega}_A \mathbf{j} + R\omega_A^2 \mathbf{i}.\end{aligned}$$

The vector $\mathbf{r}$ is $r\mathbf{k}$; therefore

$$\dot{\boldsymbol{\omega}}_A \times \mathbf{r} + \boldsymbol{\omega}_A \times (\boldsymbol{\omega}_A \times \mathbf{r}) = 0$$

since $\mathbf{r}$ is parallel to both $\boldsymbol{\omega}_A$ and $\dot{\boldsymbol{\omega}}_A$. Point A moves on a circle of radius r relative to the xyz axes; therefore

$$\mathbf{v}_{P/xyz} = \boldsymbol{\omega}_{B/A} \times \mathbf{r} = r\omega_{B/A}\mathbf{i}$$

and

$$\begin{aligned}\mathbf{a}_{P/xyz} &= \boldsymbol{\alpha}_{B/A} \times \mathbf{r} + \boldsymbol{\omega}_{B/A} \times (\boldsymbol{\omega}_{B/A} \times \mathbf{r})\\ &= 0 - r(\omega_{B/A})^2 \mathbf{k}.\end{aligned}$$

The acceleration of P thus becomes

$$\begin{aligned}\mathbf{a}_P &= -R\dot{\omega}_A \mathbf{j} + R\omega_A^2 \mathbf{i} + 0 + 0 - r(\omega_{B/A})^2 \mathbf{k} + 2\omega_A \mathbf{k} \times r\omega_{B/A}\mathbf{i}\\ &= -R\dot{\omega}_A \mathbf{j} + R\omega_A^2 \mathbf{i} - r(\omega_{B/A})^2 \mathbf{k} + 2r\omega_A \omega_{B/A}\mathbf{j}\\ &= \underline{R\omega_A^2 \mathbf{i} + (2r\omega_A\omega_{B/A} - R\dot{\omega}_A)\mathbf{j} - r(\omega_{B/A})^2 \mathbf{k}} \qquad \text{Ans.}\end{aligned}$$

A check on this solution can be made by selecting xyz axes at O which do not rotate or by letting the xyz axes rotate with the rotor B.

PROBLEMS

8-253 Rods *AB* and *CD* in Fig. P8-253 rotate in vertical planes and are connected by a slider which is pinned to *CD* and slides along *AB*. In the position shown, member *CD* has an angular velocity of 3 rad per sec clockwise and an angular acceleration of 5 rad per sec^2 counterclockwise. Determine the angular velocity and acceleration of member *AB*.

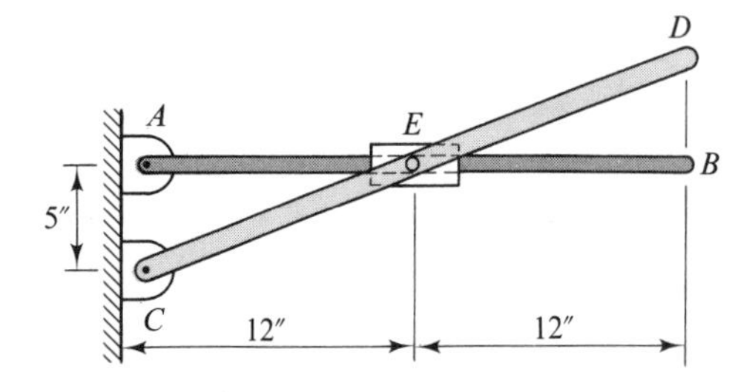

Figure P8-253

8-254 Solve Problem 8-253 when the slider is pinned to *AB* and slides along *CD* as in Fig. P8-254. All other data are the same as in Problem 8-253.

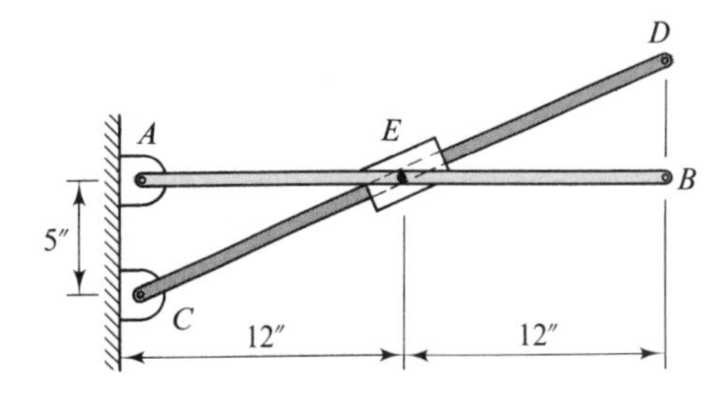

Figure P8-254

***8-255** Bar *AB* in Fig. P8-255 is pinned to block *A*, which slides along the horizontal slot, while the bar slides past the fixed pin *C*. Block *A* has a velocity of 26 cm/s to the right and an acceleration of 65 cm/s^2 to the left when in the position shown. Determine the angular velocity and acceleration of the bar. Use axes which rotate with *AB* and whose origin is at the moving point *A*.

***8-256** Solve Problem 8-255 using axes which rotate with *AB* and whose origin is at *C*.

8-257 Member *AB* in Fig. P8-257 is pinned to crank *OA* and slides in block *C*. Block *C* is free

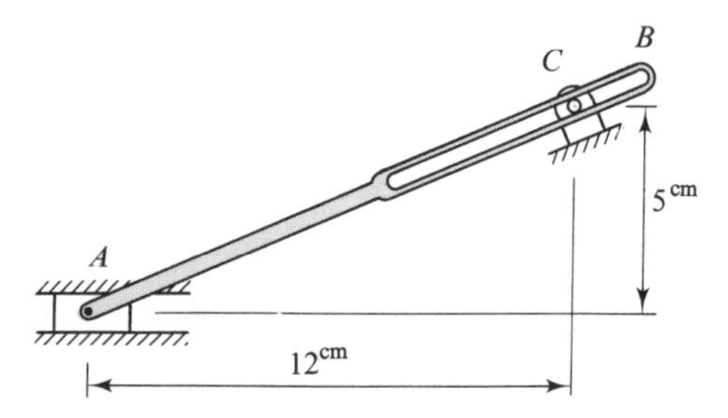

Figure P8-255

to rotate. Crank *OA* has an angular velocity of 4 rad per sec counterclockwise and an angular acceleration of 10 rad per sec^2 clockwise in the indicated position. Determine the angular velocity and acceleration of member *AB* using axes which rotate with *AB* and whose origin is at
(a) Point *A* (moving).
(b) Point *C* (fixed).

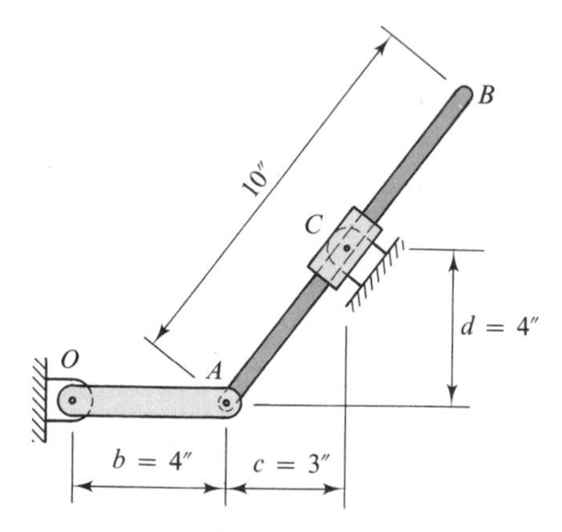

Figure P8-257

8-258 Pin *P* on crank *B* in Fig. P8-258 slides in the slot in body *A*. In the position shown, *B* has an angular velocity of 8 rad per sec clockwise and an angular acceleration of 10 rad per sec^2 counterclockwise. Determine the angular acceleration of body *A* and the acceleration of *P* relative to body *A*.

8-259 A 20-ft ladder leans against a wall as shown in Fig. P8-259, and a cat is on the ladder 6 ft from end *B*. For the given position, the velocity of end *B* is 6 fps downward and the acceleration of *B* is 7 fps^2 downward. The cat has a velocity relative to the ladder of 9 fps along the ladder toward *B* and an acceleration relative to the ladder of 2 fps^2 along the ladder

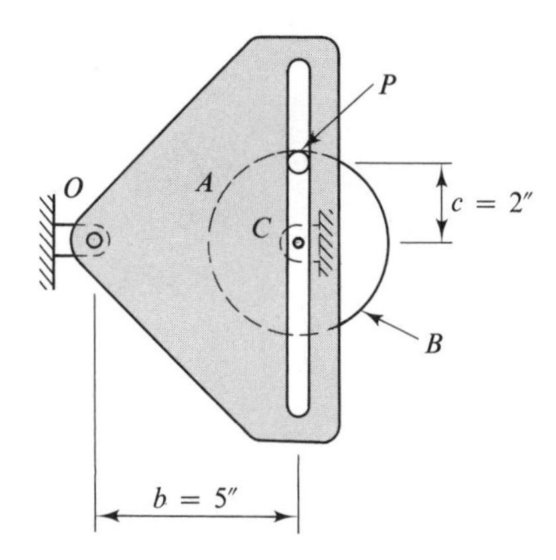

Figure P8-258

toward A. Determine the acceleration of the cat. Both ends of the ladder remain in contact with the surfaces.

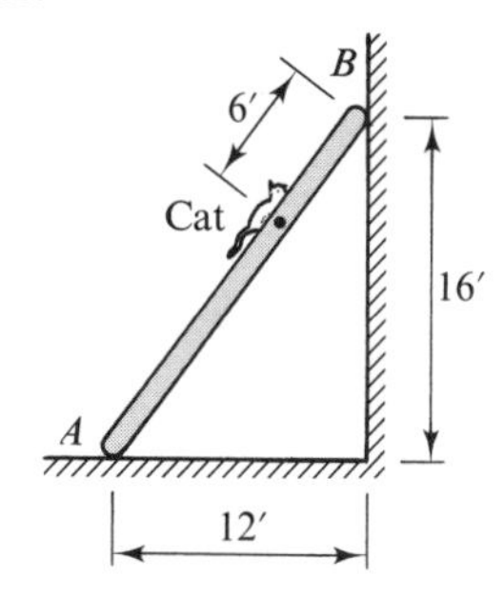

Figure P8-259

8-260 Particle *P* in Fig. P8-260 travels along a straight rod, *AB*, which is attached to disk *C* that rotates at a constant angular velocity of 2**k** rad per sec. The *xyz* axes rotate with the disk. At the instant the particle crosses the y axes, its velocity and acceleration relative to the rod are observed to be 25 fps toward *B* and 50 fps² toward *A*. Determine the velocity and acceleration of particle *P*.

8-261 In the mechanism in Fig. P8-261, the angular velocity and acceleration of rod *AB* are 6 rad per sec clockwise and 8 rad per sec² counterclockwise. Determine the angular velocity and acceleration of rod *BD*.

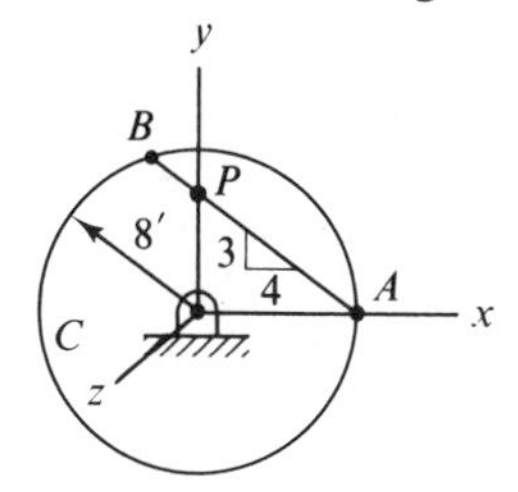

Figure P8-260

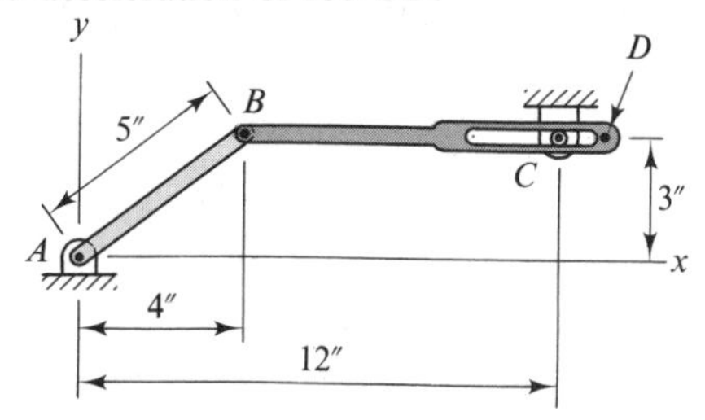

Figure P8-261

8-262 Member *OA* of Fig. P8-262 has an angular velocity of 2 rad per sec clockwise and an angular acceleration of 8 rad per sec² counterclockwise in the position shown. Determine (a) The angular acceleration of member *C*. (b) The linear acceleration of point *B*.

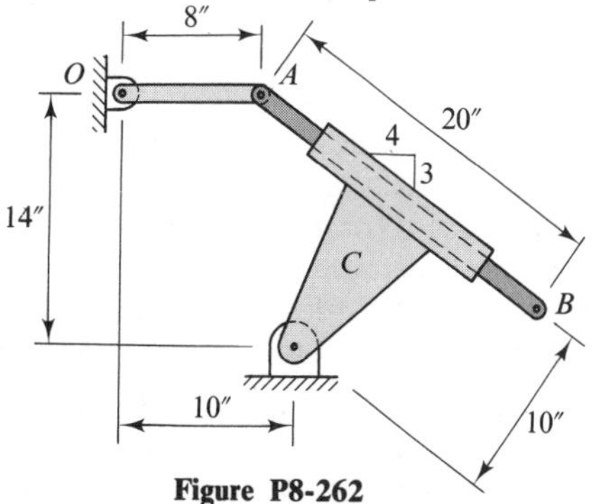

Figure P8-262

8-263 Pin *E* on member *AB* in Fig. P8-263 slides in the slot in member *CD*. In the position shown, block *A* has a velocity of 16 ips to the left and an acceleration of 56 ips² to the right. Determine the angular acceleration of member *CD*.

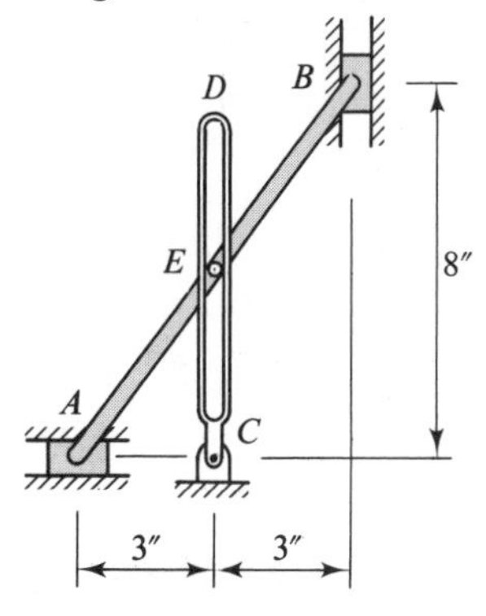

Figure P8-263

8-264 Block A in Fig. P8-264 has a velocity of 16 ips to the left and an acceleration of 56 ips² to the right in the position shown. Determine the angular acceleration of member *CD*.

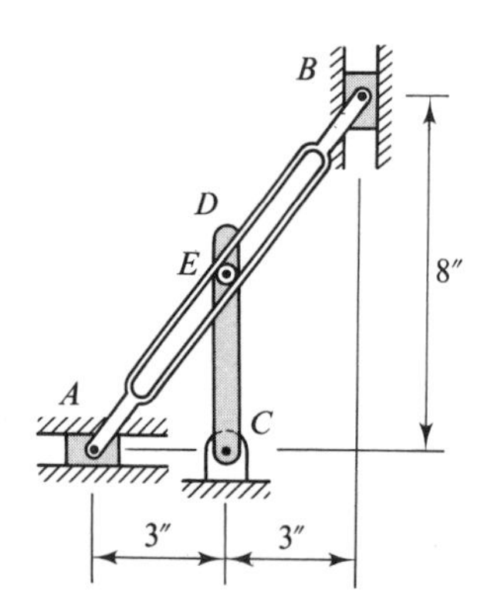

Figure P8-264

8-265 Wheel *C* in Fig. P8-265 rolls without slipping along a horizontal plane. Crank *B* rotates about a fixed axis at *O*, and pin *P* on the crank slides along the slot in the wheel. The crank has a constant angular velocity of ω_B rad per sec counterclockwise. Determine the angular velocity and acceleration of *C* and the acceleration of *P* relative to *C* for the position shown.

***8-266** Wheel *A* in Fig. P8-266 rolls without slipping along the horizontal plane. The wheel is driven by the pin *P* on the crank *OP*. The pin slides along the slot in the wheel. In the position shown, *OP* has an angular velocity of 8 rad/s clockwise and an angular acceleration of 10 rad/s² counterclockwise. Determine the velocity and acceleration of the center of the wheel.

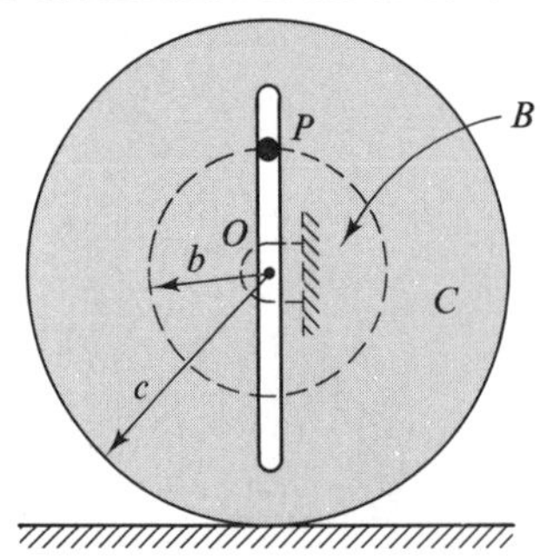

Figure P8-265

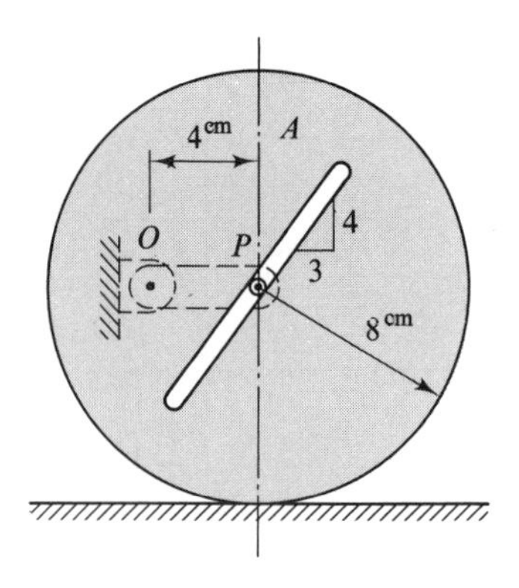

Figure P8-266

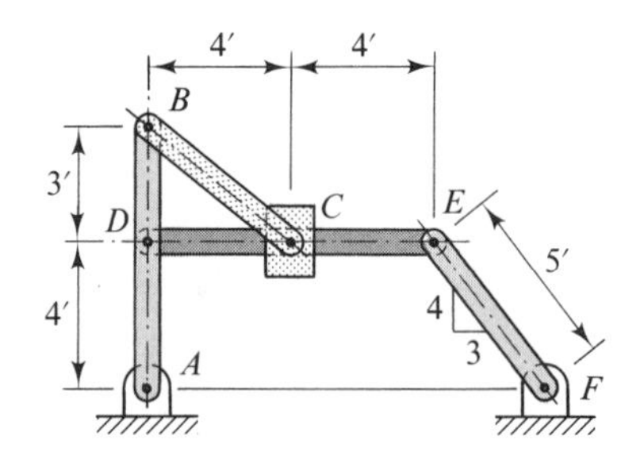

Figure P8-267

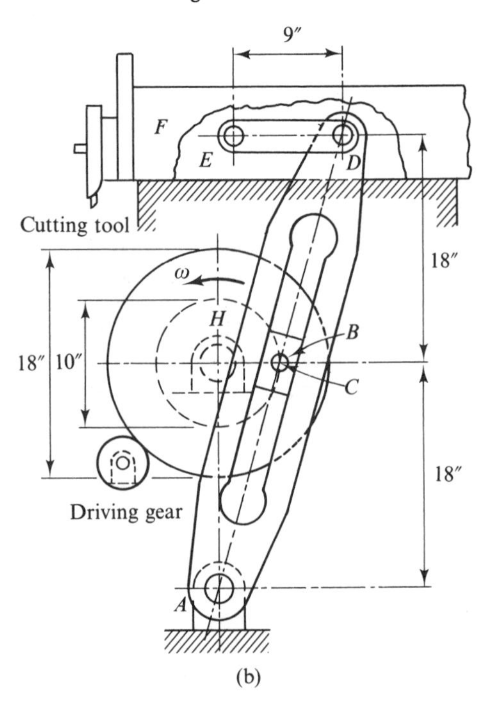

Figure P8-268

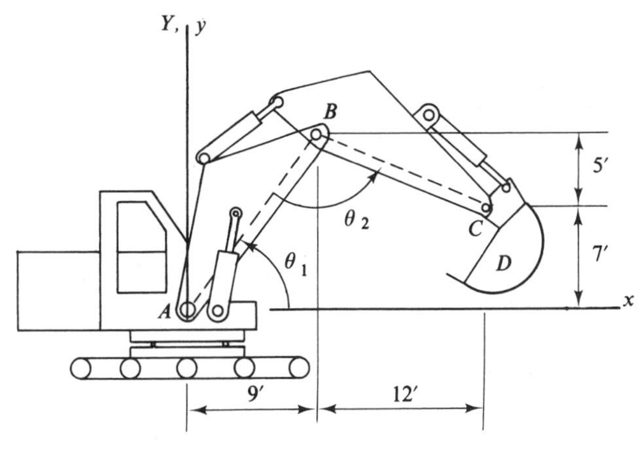

Figure P8-269

8-267 The mechanism in Fig. P8-267, consisting of rigid members AB, BC, DE, and EF, is pinned at A, B, D, E, and F and moves in the vertical plane. End C of member BC is pinned to a block which slides on member DE. For the given position $\boldsymbol{\omega}_{AB} = 2$ rad per sec clockwise and $\boldsymbol{\alpha}_{AB} = 3$ rad per sec^2 counterclockwise. Determine the acceleration of the block at C for this position.

8-268 Figure P8-268(a) is a photograph of the rocker arm and bull gear for the quick-return mechanism of a shaper. The drawing in Fig. P8-268(b) shows the mechanism connected to the ram, F. The bull gear H operates with a constant angular velocity of 45 rpm counterclockwise. When the rocker arm, AD, is in the position shown, determine

(a) The velocity of the ram, F.
(b) The angular acceleration of AD.
(c) The acceleration of the ram, F.

8-269 Figure P8-269 shows a "back-hoe" used by the construction industry. When in the indicated position, the entire machine is rotating about the Y axis with a constant angular velocity of $1.2\mathbf{j}$ rad per sec. Determine the velocity and acceleration of point C when θ_1 is increasing at the constant rate of 0.8 rad per sec and θ_2 is constant.

8-270 Solve Problem 8-269 if θ_1 is constant and θ_2 is increasing at the constant rate of 1.5 rad per sec.

8-271 A tube is formed into a circular ring and mounted on a vertical shaft AB as shown in Fig. P8-271. The ring rotates about AB with a constant angular velocity $\boldsymbol{\omega}_0$. A small ball P slides inside the tube from C to D with a constant speed v relative to the tube. Determine the velocity and acceleration of P when it reaches point D and the ring is in the yz plane.

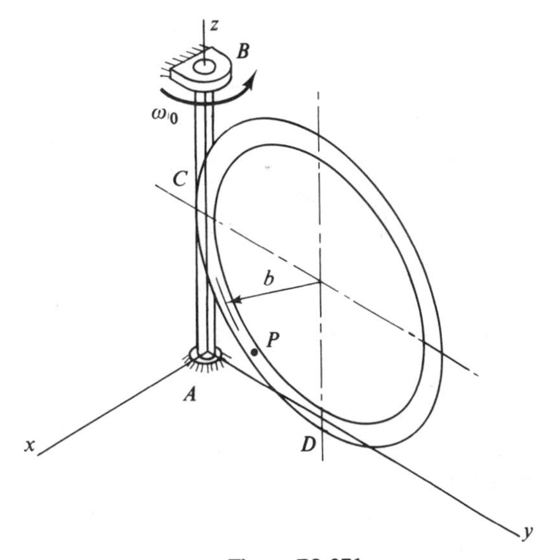

Figure P8-271

8-272 An amusement park ride called the "Rolla-Plane" is illustrated in Fig. P8-272. Member AB rotates about a horizontal axle at O. The cars C and D are mounted in bearings at the ends of A and B and can rotate with an angular velocity of $\boldsymbol{\omega}_1$ relative to member AB. Assume that the axle at O is fixed in the horizontal direction, while AB has a constant angular velocity of 1.0 rad per sec counterclockwise and car C has a constant relative angular velocity of 3.0 rad per sec clockwise looking from G to H. Determine the velocity and acceleration of the passenger's head P when the angle θ is zero and the passenger is vertical as shown.

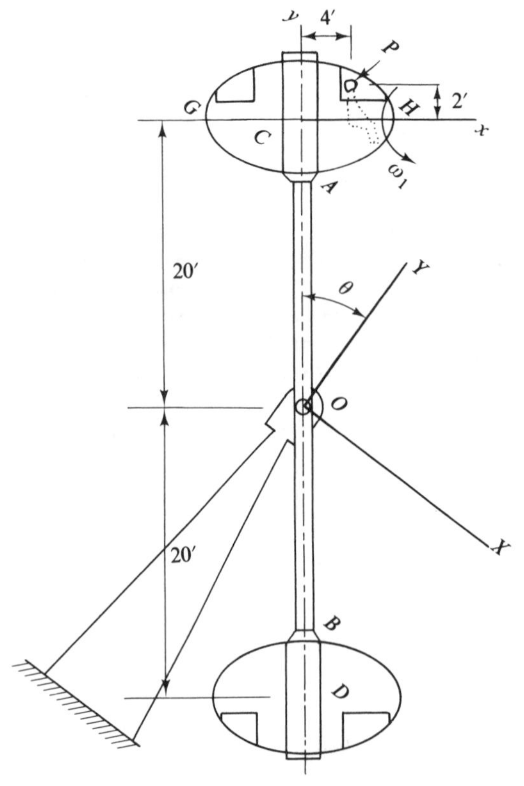

Figure P8-272

8-273 Disk A of Fig. P8-273 rotates about the Y axis, and rod BC is connected to A by a horizontal shaft which rotates with A. In the position shown, disk A has an angular acceleration of $3\mathbf{j}'$ rad per sec^2 and an angular velocity of $-5\mathbf{j}'$ rad per sec, and the rod has an angular acceleration of $-4\mathbf{i}$ rad per sec^2 and an angular velocity of $6\mathbf{i}$ rad per sec, both relative to the disk A. Determine

(a) The linear velocity of point B.

(b) The linear acceleration of point B.

(c) The angular acceleration of BC.

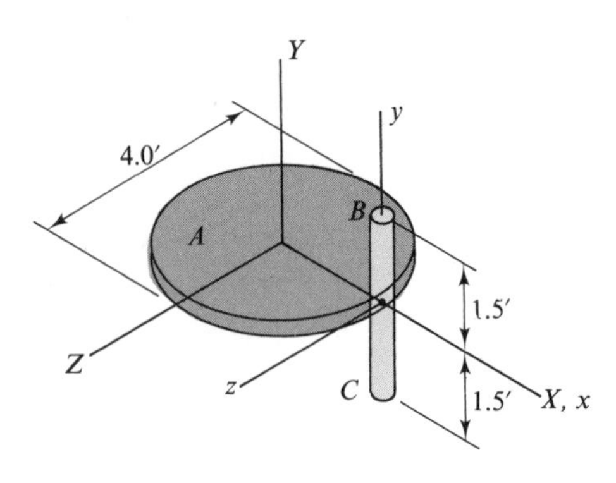

Figure P8-273

8-274 Disk W in Fig. P8-274 slides along the rod OD and also rotates around the rod as OD rotates in the xy plane around the z axis. When in the position shown, the angular velocity of OD is $2\mathbf{k}$ rad per sec and its angular acceleration is $3\mathbf{k}$ rad per sec^2; the angular velocity and acceleration of W relative of OD are $4\mathbf{j}$ rad per sec and $5\mathbf{j}$ rad per sec^2, respectively; and the disk is sliding along the rod from O to D with a constant speed of 6 ips. Determine

(a) The angular acceleration of W.

(b) The velocity of point A on the disk.

(c) The acceleration of point A.

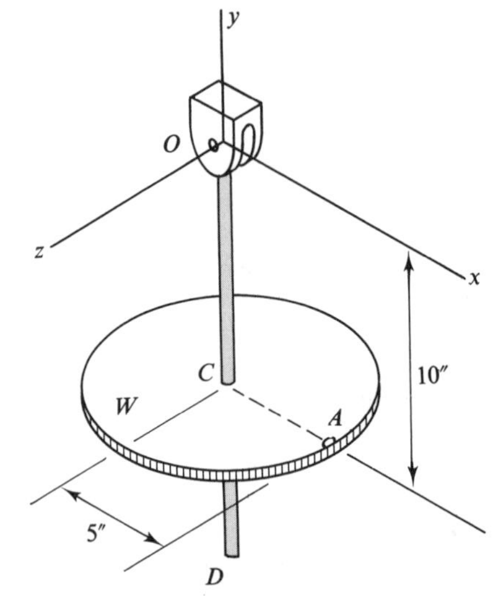

Figure P8-274

8-275 The wheel in Fig. P8-275 rotates about the shaft *AB* as member *OAB* rotates in the *xy* plane about the *z* axis. The ring *R* slides along a spoke on the wheel. When the system is in the position shown *OAB* has an angular velocity of $2\mathbf{k}$ rad per sec and an angular acceleration of $5\mathbf{k}$ rad per sec^2, and the wheel has an angular velocity of $6\mathbf{i}$ rad per sec and an angular acceleration of $8\mathbf{i}$ rad per sec^2, both relative to member *OAB*. At the same instant the ring *R* has a velocity of $10\mathbf{j}$ ips and an acceleration of $-7\mathbf{j}$ ips^2, both relative to the wheel. Determine
(a) The angular acceleration of the wheel.
(b) The acceleration of *R*.

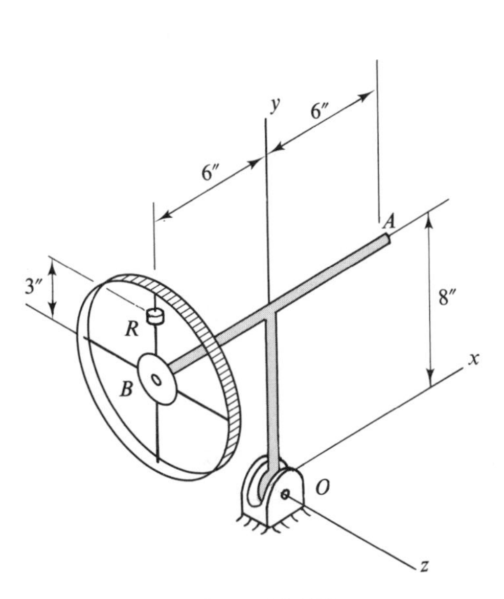

Figure P8-275

8-276 Body *A* in Fig. P8-276 rotates about a fixed vertical axis at *O*. Body *B* is connected to bodies *A* and *C* by horizontal pins. When in the position shown, *A* has an angular velocity of $4\mathbf{j}$ rad per sec and an angular acceleration of $3\mathbf{j}$ rad per sec^2; *B* has an angular velocity of $2\mathbf{k}$ rad per sec and an angular acceleration of $6\mathbf{k}$ rad per sec^2, both with respect to body *A*; and *C* has a constant angular velocity of $-2\mathbf{k}$ rad per sec with respect to *B*. Determine
(a) The angular acceleration of body *B*.
(b) The acceleration of point *D* on body *C*.

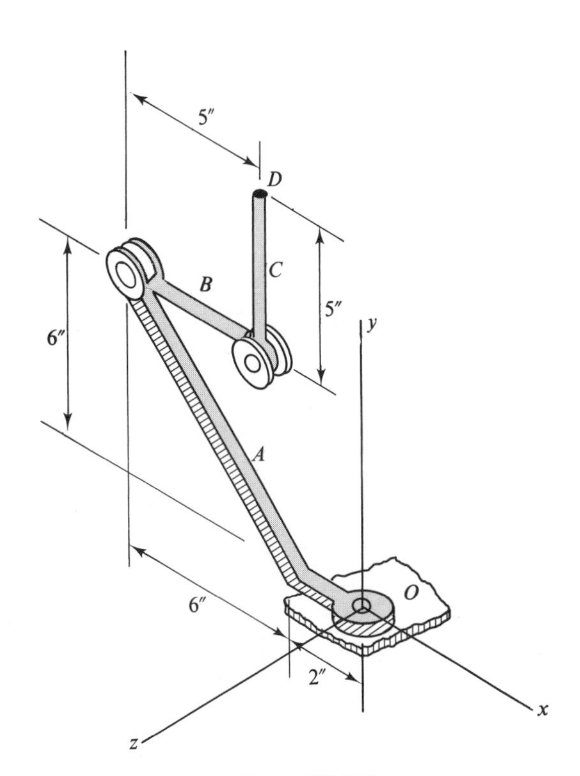

Figure P8-276

8-277 The turntable *T* in Fig. P8-277 rotates about the y axis. The rod *OA* is connected to *T* by a horizontal pin at *O*, and *R* is a ring which slides along OA. In the position shown, T has an angular velocity of $2\mathbf{j}$ rad per sec and an angular acceleration of $4\mathbf{j}$ rad per sec^2; OA has an angular velocity of $3\mathbf{k}$ rad per sec and an angular acceleration of $-5\mathbf{k}$ rad per sec^2, both relative to the turntable; and the ring has a relative velocity along the rod of 10 ips toward A and a relative acceleration of 15 ips^2 toward O. Determine the acceleration of the ring in the indicated position.

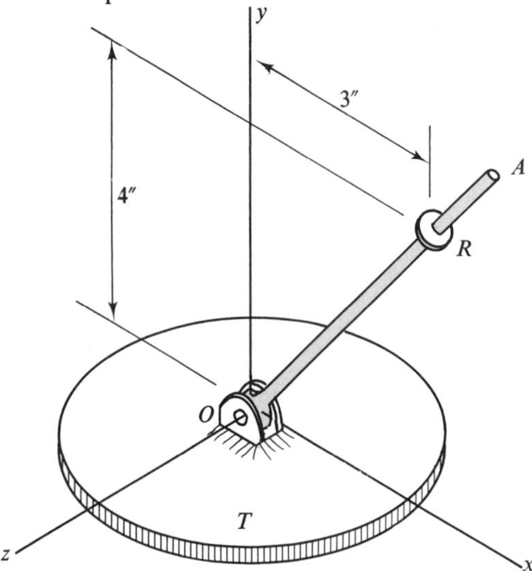

Figure P8-277

8-278 Member AB of Fig. P8-278 rotates in a horizontal plane around the z axis with a constant angular velocity of ω_0. The collar D slides along AB. Member E is connected to point C and to the collar by ball and socket joints. Determine the velocity and acceleration of the collar D when in the indicated position.

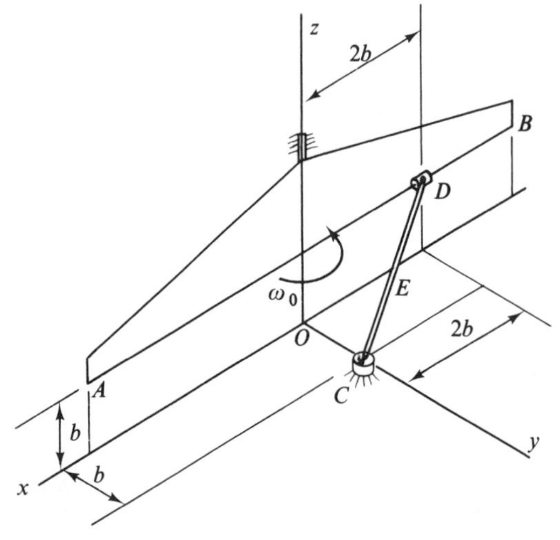

Figure P8-278

8-279 Disk A in Fig. P8-279 rotates about the z axis with a constant angular velocity of $\boldsymbol{\omega}_0$ rad per sec as shown. The rod BC extends through a small hole in A and is pivoted about a ball and socket joint at B. Point P is the point on BC which is passing through the plate in the position shown. Determine the velocity and acceleration of point P.

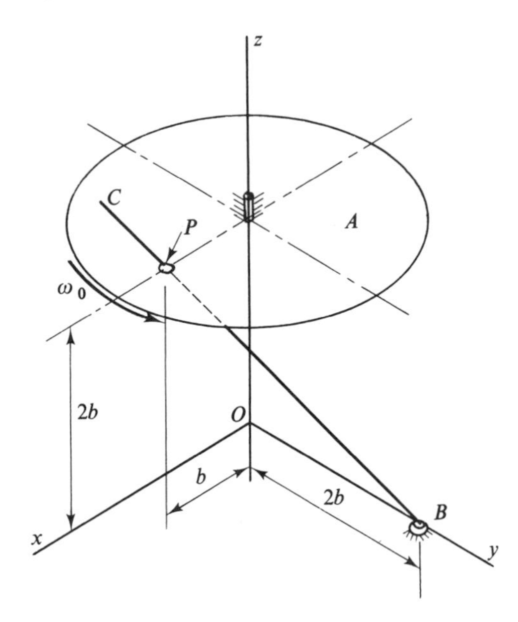

Figure P8-279

8-280 Crank OA in Fig. P8-280 (p. 479) rotates about the z axis (in a horizontal plane) with a constant angular velocity of $\boldsymbol{\omega}_0$ rad per sec. The slender rod AB is connected to the crank by a ball and socket joint and passes through a small hole in the fixed plate Q. Point P on AB is coincident with the hole in the plate in the position shown. Determine the velocity and acceleration of P for this position.

8-281 The bent rod $OABC$ in Fig. P8-281 oscillates about the z axis, and when in the position shown it has an angular velocity of 3 rad per sec and no angular acceleration.

Member D is connected to O' and to the sliding collar P by ball and socket joints. Determine the velocity and acceleration of P for the given position.

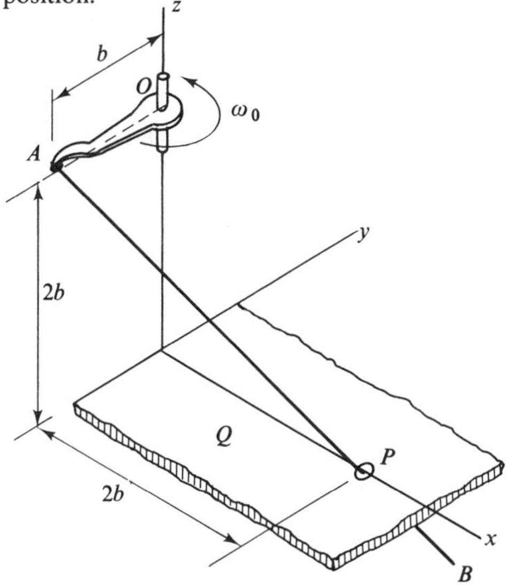

Figure P8-280

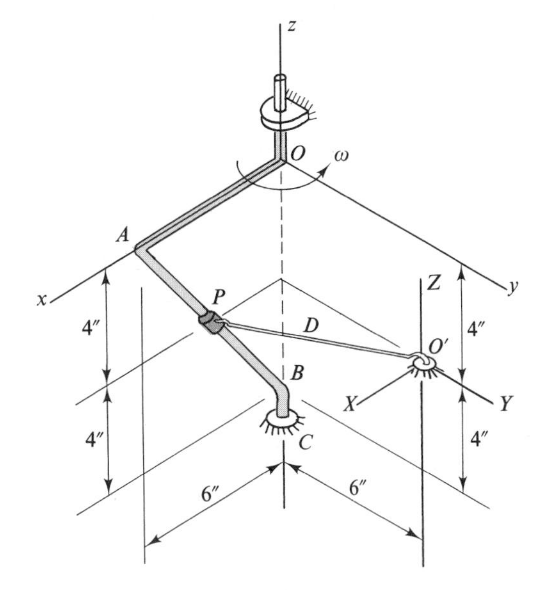

Figure P8-281

8-22 CLOSURE

The position, velocity, and acceleration of a particle can be expressed in terms of the motion of the particle relative to a set of moving axes and the motion of the axes. When the relative motion is expressed with respect to a set of translating (nonrotating) axes, the relative velocity and acceleration are given by Eq. (8-21) and (8-22). When the relative motion is expressed with respect to a set of rotating axes, Eq. (8-27) and (8-28) give the velocity and acceleration of the moving point. Note that Eq. (8-21) and (8-22) apply to two points on a *rigid body*, while Eq. (8-27) and (8-28) apply to any two points and are *not* restricted to rigid bodies.

The *velocities*, but not the accelerations, of points and lines on rigid bodies which have plane motion can be obtained by means of instantaneous centers.

The kinematic relationships between positions, velocities, and accelerations of points and lines can also be obtained by writing an expression for the position of the point or line and differentiating, as was discussed in Part A of this chapter. This mathematical approach to the problem provides an independent check on the solution by relative motion and sometimes leads to a more direct solution to the problem.

PROBLEMS

8-282 In the linkage of Fig. P8-282 the angular velocity of AB is 3 rad per sec clockwise. Determine the velocities of points C and P.

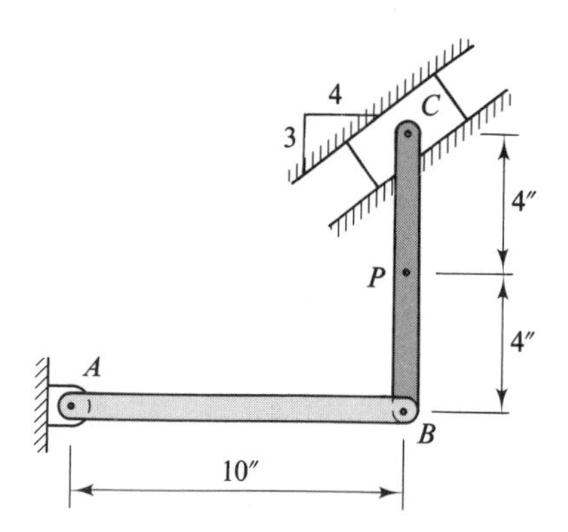

Figure P8-282

***8-283** Link AB in Fig. P8-283 has an angular velocity of 6 rad/s clockwise. Determine, for this position,

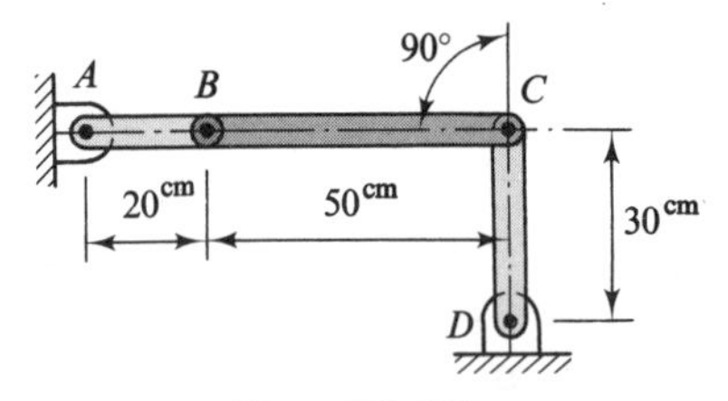

Figure P8-283

(a) The location of the instantaneous center of zero velocity of link BC.
(b) The angular velocity of link BC.

8-284 The two meshed gears A and B in Fig. P8-284 are rotating as indicated. When body A is in the given position, its angular velocity is 40 rpm clockwise and its angular acceleration is 8 rad per sec² counterclockwise. Determine the acceleration of point D with respect to point C.

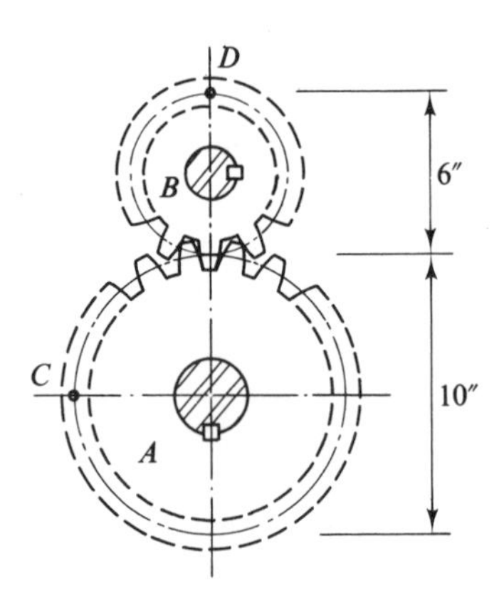

Figure P8-284

8-285 The six rods in Fig. P8-285 are welded to form a rigid structure which rotates about AC with a constant angular velocity. At the instant shown the x component of the velocity of D is 80 ips to the right. Determine
(a) The angular velocity of the structure.
(b) The acceleration of point D.

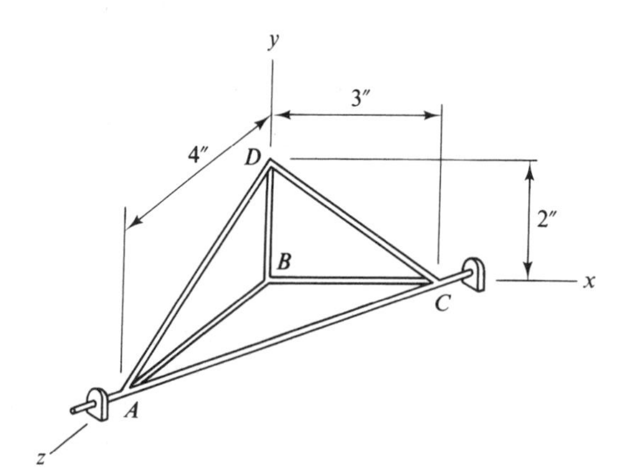

Figure P8-285

8-286 The block in Fig. P8-286 rotates about an axle along AB. In the position shown the block has an angular velocity of 5 rad per sec and an angular acceleration of 10 rad per sec^2, both clockwise when looking from B to A. Determine the velocity and acceleration of point C.

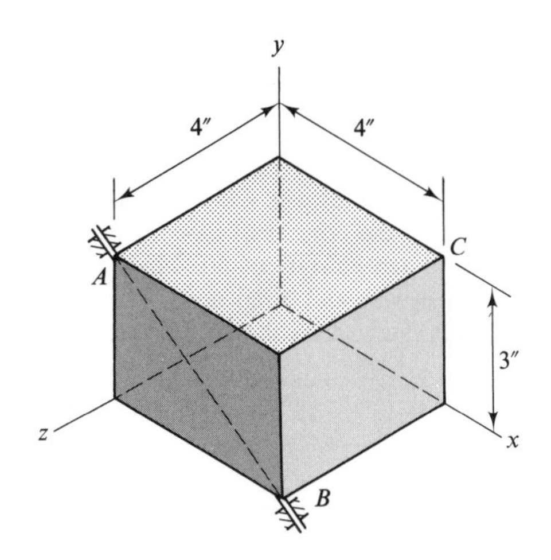

Figure P8-286

8-287 Block A in Fig. P8-287 slides along the horizontal slot as member AB slides past the fixed pin at C. Write an equation relating the position of A to the angular position of AB and, by differentiation, derive equations for the velocity and acceleration of A in terms of θ and its derivatives.

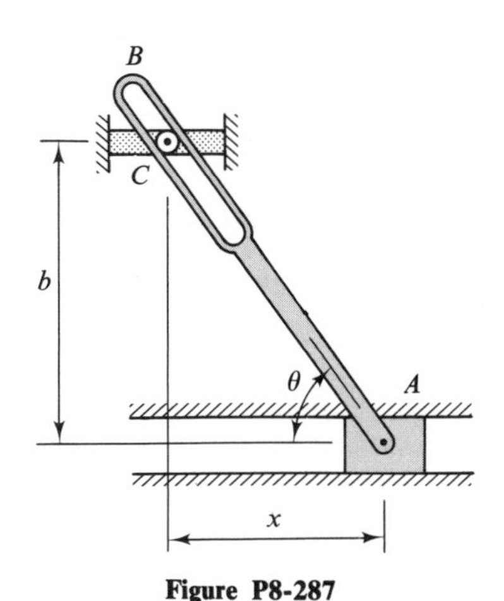

Figure P8-287

8-288 The crank in Fig. P8-288 rotates with a constant angular velocity of 10 rad per sec clockwise. Determine the velocity and acceleration of point C when in the position shown.

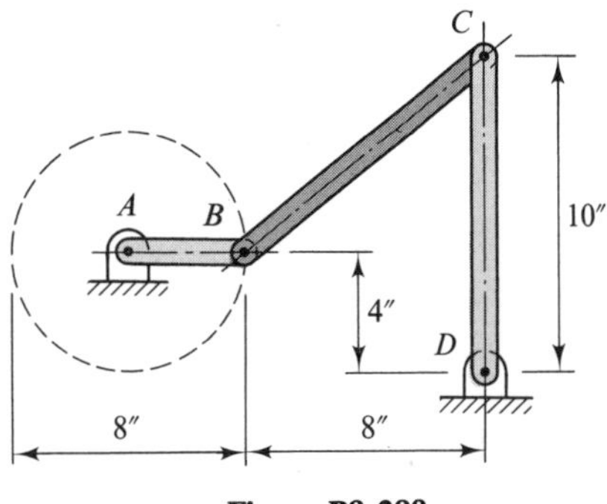

Figure P8-288

8-289 The angular velocity of BC of Fig. P8-289 is 2 rad per sec counterclockwise, and the angular acceleration of BC is 4 rad per sec^2 clockwise when in the indicated position. Determine the acceleration of block C.

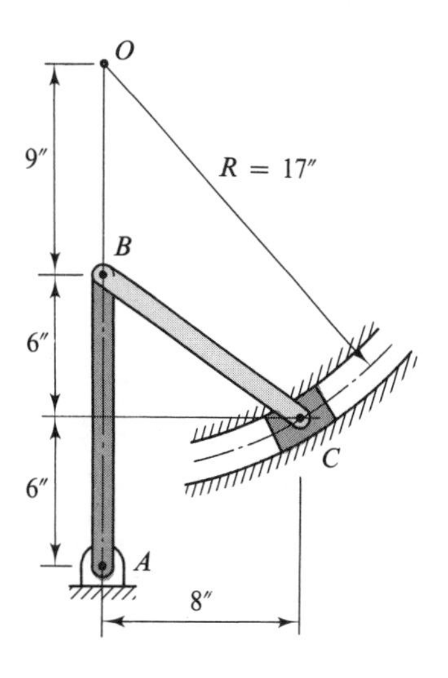

Figure P8-289

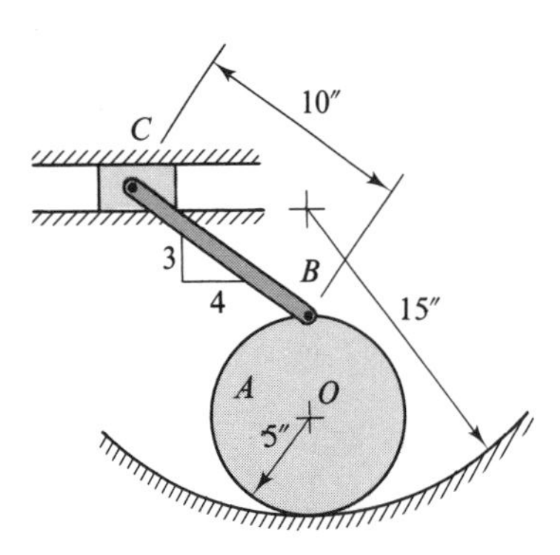

Figure P8-290

8-290 Block C of Fig. P8-290 has a velocity of 20 ips to the right and an acceleration of 30 ips^2 to the left when in the indicated position. The wheel rolls without slipping. Determine the angular acceleration of the wheel.

***8-291** The wheel in Fig. P8-291 rolls along the horizontal plane without slipping and is connected to B by a flexible cord which is wrapped around a drum on the wheel. When the bodies are in the position shown the acceleration of point A relative to point B is 50 cm/s^2 to the right and the speed of B is decreasing.
(a) Determine the velocity and acceleration of point C.
(b) During a certain time interval C is displaced 3 cm to the left. Determine the displacement of B during this time interval.

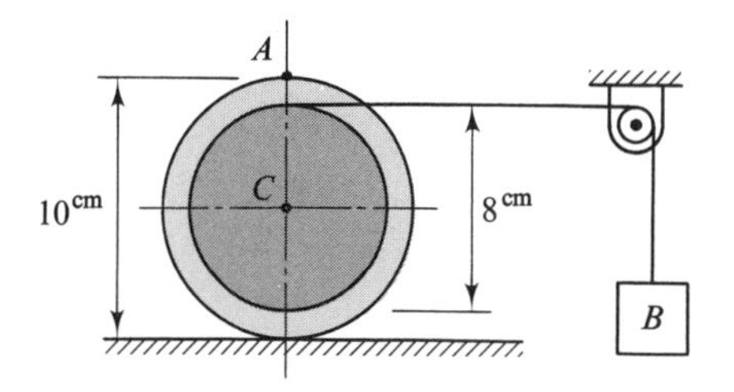

Figure P8-291

8-292 Rod BC in Fig. P8-292 is pinned to the wheel at B in such a manner that the rim of the wheel does not strike the rod. Block C slides in the horizontal slot, and the wheel rolls without slipping. When the assembly is in the position shown the acceleration of point O is 20 ips^2 to the right and the angular velocity of the wheel is 3 rad per sec clockwise. Determine

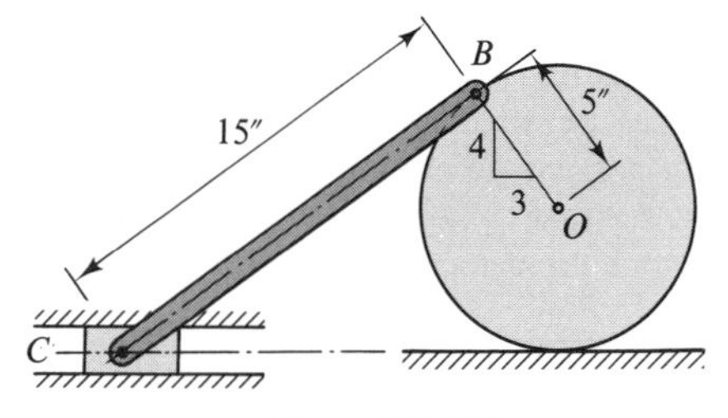

Figure P8-292

(a) The velocity of C.
(b) The acceleration of C.

8-293 Wheel D of Fig. P8-293 rolls without slipping along the horizontal plane. The angular velocity and acceleration of the wheel are 6 rad per sec clockwise and 5 rad per sec^2 clockwise, respectively. Determine the angular acceleration of BC.

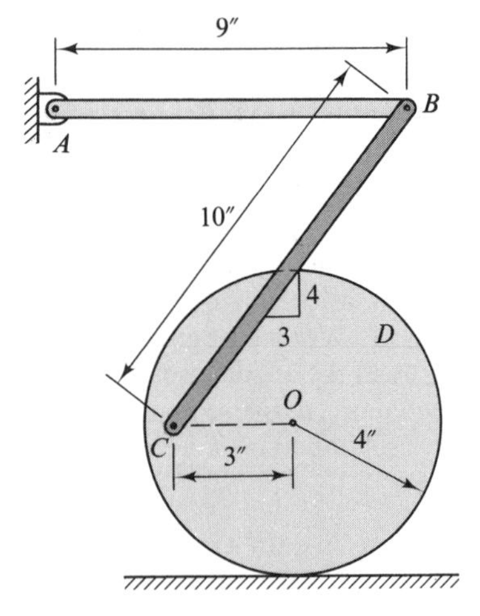

Figure P8-293

***8-294** Crank OA in Fig. P8-294 rotates with a constant angular velocity of 10 rad/s clockwise. Determine the angular velocity and acceleration of member AB when in the position shown.

8-295 Wheel C of Fig. P8-295 rolls without slipping along the horizontal plane. The crank B oscillates about an axis at O and drives the wheel by means of a pin moving in the slot in the wheel. The two lengths are $b = 5$ in. and $c = 10$ in. In the position shown, the angular velocity of B is 5 rad per sec clockwise, and its angular acceleration is zero. Determine

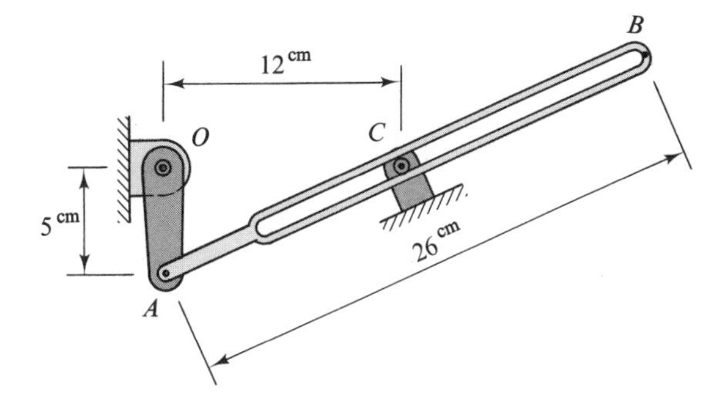

Figure P8-294

(a) The angular acceleration of C.
(b) The acceleration of P relative to the slot in C.

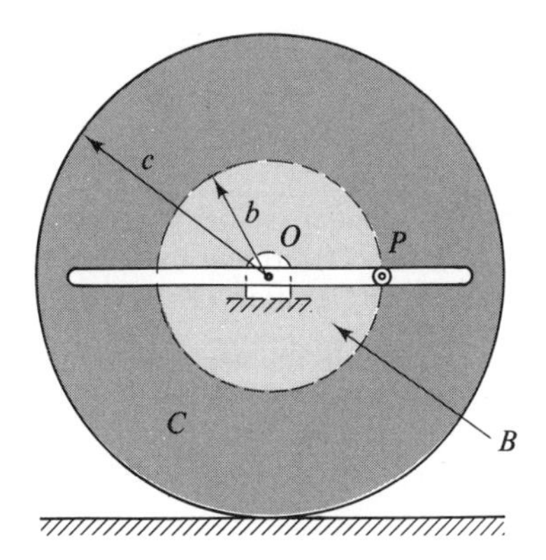

Figure P8-295

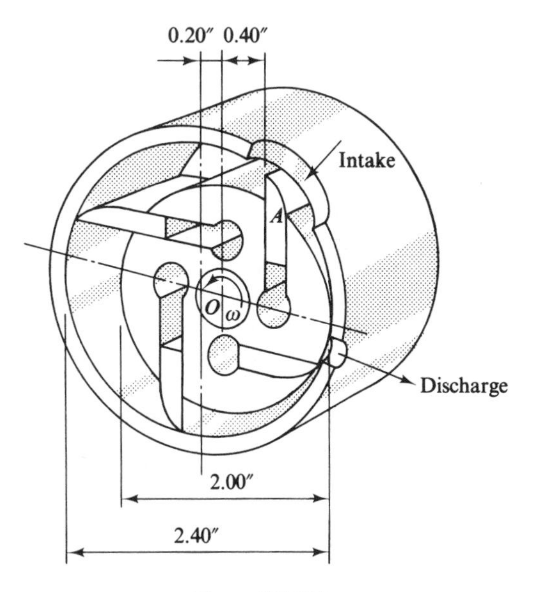

Figure P8-296

8-296 A vane-type pump or compressor is illustrated in Fig. P8-296. The smaller circular disk with slots rotates about a fixed axis at O with a constant angular velocity of 1700 rpm. The vanes slide in the slots, and the outer ends of the vanes maintain contact with the large stationary cylinder. Determine the velocity and acceleration of point A on the vertical vane when it is in the indicated position.

8-297 In the mechanism in Fig. P8-297 block B has a velocity of 10 ips down the slot and acceleration of 20 ips^2 up the slot when in the indicated position. Determine
(a) The angular acceleration of member CD.
(b) The acceleration of pin E relative to member CD.

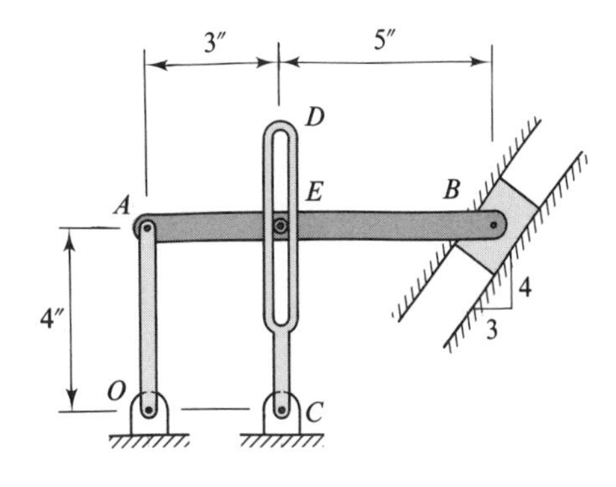

Figure P8-297

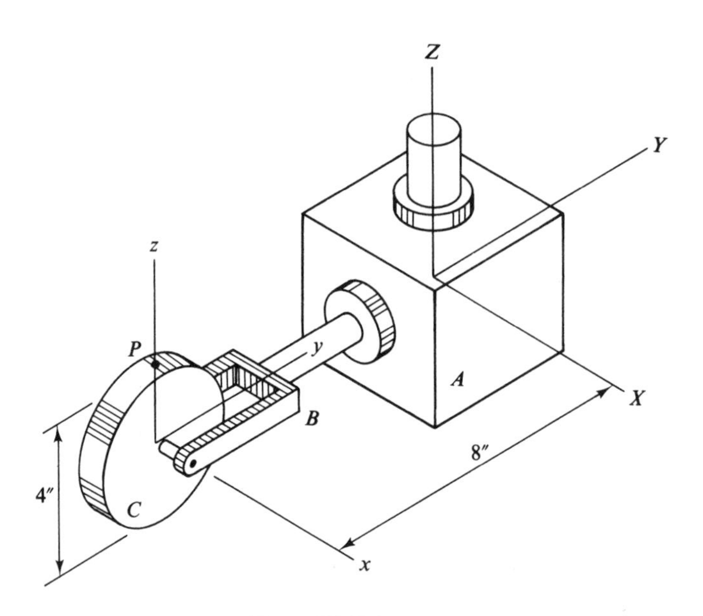

Figure P8-298

8-298 Body A in Fig. P8-298 rotates about the Z axis with a constant angular velocity of $3\mathbf{k}$ rad per sec. The forked member B rotates in bearings in A with a constant relative angular velocity of 5 rad per sec clockwise looking from B toward A. The disk C rotates in the fork with a constant angular velocity relative to B of $10\mathbf{i}$ rad per sec. Determine the absolute angular velocity and acceleration of the disk C.

8-299 Determine the velocity and acceleration of point P at the top of disk C in Problem 8-298.

9

Kinetics–Force, Mass, and Acceleration

PART A–KINETICS OF PARTICLES

9-1 INTRODUCTION

Kinetics is the study of the relationship between the resultant force system and the associated motion of bodies that are not in equilibrium. The parallelogram law for the addition of forces and methods for determining the resultant of force systems were studied in Chapters 1 and 2. When the resultant of a force system acting on a body is zero, the body is in equilibrium. Equations of equilibrium were studied in Chapter 4 using the free-body diagram concept as an aid in the visualization of forces. Kinematics, the geometric study of motion without considering the associated forces, was presented in Chapter 8 as a preliminary to the study of the general kinetics problem. Equations relating the unbalanced force systems, the inertial properties (mass and mass moment of inertia), and the motion of bodies not in equilibrium will be developed in this and succeeding chapters.

The foundation for the classical study of kinetics is based on the laws of motion first presented in a formal manner in 1687 by Sir Isaac Newton (1642–1727). Although called Newton's laws of motion, it should be noted that Newton drew upon the scientific achievements of others over a historical period ranging from the builders of the pyramids to Newton's contemporaries. Perhaps the most significant discoveries were made by Galileo Galilei (1564–1642), who performed experiments on falling bodies and pendulums and disproved earlier false theories about dynamics. Significant developments in science pertinent to kinetics have been achieved since Newton's time, and these developments will certainly continue in the future, but the laws of motion stated by Newton are still recognized as the basic laws of engineering kinetics.

The motion of bodies acted upon by unbalanced force systems can be analyzed by several methods. The most useful method for any particular problem depends on the forces acting (constant or variable) and the results to be obtained (reactions, accelerations, velocities, displacements, and so on). The analysis of kinetics problems by force, mass, and acceleration, by the principle of work and energy, and by the impulse-momentum principle is discussed in the following chapters. Although these principles appear to differ in form, they are all immediately derivable from Newton's laws.

9-2 NEWTON'S LAWS OF MOTION

Newton's laws can be stated as follows:

1. If the resultant force acting on a particle is zero, the particle will remain at rest or move with a constant velocity (constant speed and direction).
2. If the resultant force acting on a particle is not zero, the particle will be accelerated in the direction of the force, and the magnitude of the acceleration will be directly proportional to the force and inversely proportional to the mass of the particle.
3. The force exerted by one particle on a second particle is equal in magnitude and opposite in direction to the force exerted by the second particle on the first.

These three laws were developed from a study of planetary motion, that is, the motion of particles. As stated, Newton's laws apply only to the motion of particles. Bodies are made up of particles, however, and these laws have been extended to apply to bodies.

The first law indicates that a particle has some property called *inertia* and that as a result a force is required to produce a change in the motion of the particle.

The second law is a quantitative statement of the relationship which must exist between the force acting on the particle, the inertia (mass) of the particle, and the resulting acceleration.

A mathematical statement of Newton's second law is

$$\mathbf{a} = K\frac{\mathbf{F}}{m}, \tag{9-1}$$

where $\mathbf{a}$ is the absolute acceleration, $\mathbf{F}$ is the resultant force, m is the mass of the particle, and K is a dimensionless constant. If units of acceleration, force, and mass are chosen arbitrarily, the constant K must be determined experimentally. It is convenient to have the constant K be unity, which is possible if the units of either acceleration, force, or mass are selected in terms of the other two quantities. Only those cases in which K is unity are considered here, so that Eq. (9-1) becomes

$$\mathbf{a} = \frac{\mathbf{F}}{m}. \tag{9-1a}$$

A suitable frame of reference must be used for the measurement of absolute accelerations in order to satisfy Newton's second law. Such a reference frame is called a *newtonian* or *inertial reference frame* and is ideally fixed in space. Suppose that an inertial reference frame does exist, and as a result of a force $\mathbf{F}$, an observer in that frame measures an absolute acceleration $\mathbf{a}_P$ satisfying the second law,

$$\mathbf{a}_P = \frac{\mathbf{F}}{m}. \tag{a}$$

Let $\mathbf{a}_{P/M}$ be the acceleration observed in a moving frame due to the same force, and let $\mathbf{a}_M$ be the acceleration of the moving frame itself, so that

$$\mathbf{a}_P = \mathbf{a}_M + \mathbf{a}_{P/M}.$$

When this expression is substituted into Eq. (a) it becomes

$$\mathbf{a}_M + \mathbf{a}_{P/M} = \frac{\mathbf{F}}{m}$$

or

$$\mathbf{a}_{P/M} = \frac{\mathbf{F}}{m} - \mathbf{a}_M. \tag{b}$$

The acceleration $\mathbf{a}_{P/M}$ *observed* in the moving frame satisfies the second law only when $\mathbf{a}_M$ is equal to zero, that is, when the velocity $\mathbf{v}_M$ is a constant, where $\mathbf{v}_M$ is the velocity of the moving frame. Thus, a true *inertial reference frame* is one that is fixed in space, or translates with constant velocity and has no rotation. In most engineering applications, however, motion is measured relative to a reference frame fixed on the rotating earth, since the error introduced by the $\mathbf{a}_M$ term in Eq. (b) is negligible. (For example, the maximum error in calculating the acceleration of a freely falling particle near the earth's surface is approximately 0.34 percent when the rotation of the earth about its axis is neglected.)

Newton's third law makes it possible to extend the second law for particles to bodies, since bodies are made up of a system of particles and the forces between the particles always occur in collinear pairs which balance each other.

A fourth law formulated by Newton is Newton's law of universal gravitation, which states that the force of attraction **F** between two masses m_1 and m_2 separated by a distance r has the magnitude

$$F = \frac{Gm_1m_2}{r^2}.$$

G is a universal gravitational constant, which has an experimentally determined value of

$$G = 3.438(10)^{-8}\,\frac{\text{ft}^4}{\text{lb-sec}^4} = 6.670(10)^{-11}\,\text{m}^3/(\text{kg}\cdot\text{s}^2).$$

9-3 UNITS AND DIMENSIONS

Physical quantities have both *qualitative measures* called *dimensions* and *quantitative* measures called *units*. The application of Newton's second law requires the use of a homogeneous set of dimensions. Since the second law relates the three quantities of force, mass, and acceleration, the dimensions of only two of the quantities can be specified. The dimensions of the third quantity are then derived from Eq. (9-1a). It is commonly agreed that the dimensions of length (L) and time (T) are fundamental, so that the dimensions of acceleration are specified (LT^{-2}). One now has the choice of specifying either the mass (m) or the force (F) as a fundamental dimension.

When length, time, and mass are selected as fundamental dimensions the dimensions of force are derived from Eq. (9-1a). A *unit force* can then be defined as the force necessary to produce a unit acceleration of a particle having a unit mass. Systems of units in which the dimensions of mass, length, and time are fundamental and the dimension of force is the derived quantity are called *absolute systems* of units. The *International System* of Units (SI) is an absolute system in which acceleration is measured in meters per second per second (m/s^2) and mass is measured in kilograms (kg). The derived unit of force is then the force which will produce an acceleration of 1 m/s^2 when applied to a mass of 1 kg and is called a *newton* (N). One newton is the same as 1 m-kg/s^2.

When length, time, and force are selected as fundamental dimensions the dimension of mass is derived from Eq. (9-1a). A *unit mass* can thus be defined as the mass of a particle which will have a unit acceleration when acted upon by a unit force. Systems of units in which the dimensions of force, length, and time are fundamental and the dimension of mass is the derived quantity are called *gravitational systems* of units. In gravitational systems, the unit of force is defined in terms of the gravitational attraction of the earth for a standard mass. If the location of the standard mass is changed, the unit of force is changed due to variations in the earth's gravitational field. Thus, the term *gravitational* implies that the units change when the standard mass is moved in the gravitational field, whereas the term *absolute* implies that the units are independent of the position of the standard mass.

In the *British gravitational system* of units, acceleration is measured in feet per second per second and force in pounds. The derived unit of mass will be 1 lb divided by 1 fps^2, or 1 lb-sec^2-per ft. The unit of mass of 1 lb-sec^2-per ft is frequently called a *slug* or a *g-pound*. Both the *SI* and *British gravitational systems* of units will be used in this text and are summarized in Table 9-1. When the British gravitational system is used, the unit of mass will be called a slug.

The mass of a body in British gravitational units can be derived from Eq. (9-1a) if the acceleration produced by some known force can be determined. Experiments show that if the attraction of the earth for the body (the *weight* of the body) is the only force acting on the body, the acceleration will be approximately $g = 32.2$ fps^2 downward. When the y axis is vertical, Newton's second law is

$$\mathbf{a} = \frac{\mathbf{F}}{m},$$

and when values are substituted in this expression it becomes

$$-g\mathbf{j} = \frac{-W\mathbf{j}}{m}$$

from which

$$m = \frac{W}{g}. \tag{9-2}$$

In the *British gravitational system* of units, the weight is measured in pounds, and the acceleration of gravity is approximately 32.2 fps^2. In

TABLE 9-1 SYSTEMS OF UNITS

Dimension	*British Gravitational Units*	*x Factor =*	*International System of Units*
length	foot (ft)	0.3048	meter (m)
time	second (sec)	1.000	second (s)
force	pound (lb)	4.448	newton (N)
mass	slug (lb-sec^2/ft)	14.59	kilogram (kg)

this text the acceleration of gravity on or near the surface of the earth will be taken as 32.2 fps^2 or 9.81 m/s^2.

9-4 EQUATIONS OF MOTION OF A PARTICLE

When a force system acts on a particle, the system must be concurrent (a particle has no dimensions), and the resultant is a single force acting through the particle. Newton's second law of motion for a particle can be expressed mathematically as

$$\mathbf{R} = m\mathbf{a} \tag{9-3}$$

where $\mathbf{R}$ is the resultant force, m is the mass of the particle, and $\mathbf{a}$ is its acceleration. Equation (9-3) is a vector equation of motion for particles which expresses the equality of the resultant force vector $\mathbf{R}$ and the $m\mathbf{a}$ vector as shown in Fig. 9-1. The $m\mathbf{a}$ vector, which is the effect of the resultant force, is frequently called the *effective* force of the particle because of this equality.

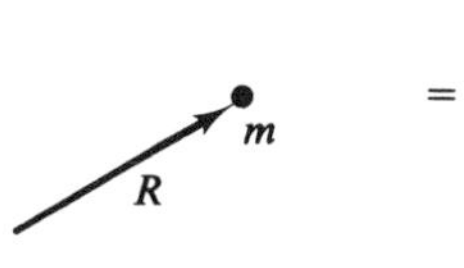

=

m ma

Figure 9-1

In the solution of kinetics problems (as in kinematic analysis) it is generally necessary to resolve the vector equation into a set of component equations suitable to the particular problem being analyzed. Resolving both sides of Eq. (9-3) into rectangular components, the vector equation of motion can be written in component form as

$$R_x\mathbf{i} + R_y\mathbf{j} + R_z\mathbf{k} = \sum F_x\mathbf{i} + \sum F_y\mathbf{j} + \sum F_z\mathbf{k} = m(a_x\mathbf{i} + a_y\mathbf{j} + a_z\mathbf{k}) \tag{9-4a}$$

or in scalar form as

$$R_x = \sum F_x = ma_x, \qquad R_y = \sum F_y = ma_y, \qquad R_z = \sum F_z = ma_z. \tag{9-4b}$$

When using these equations, a sign convention should be established for the forces and acceleration. A convenient sign convention is to show a set of axes and consider that the signs of all force and acceleration components are the same as the signs of the corresponding axes. The direction of unknown forces should be assumed and shown on the free-body diagram. Unknown accelerations should generally be assumed in the directions of the positive axes.

The kinematics of a problem, discussed in Chapter 8, generally indicate which components are most convenient for the kinetic analysis. As an example, normal and tangential components are particularly useful in problems where the particle is constrained to move in a circular path. In this case, when there are forces perpendicular to the plane of the circle, the resultant of these forces must be zero since there is no acceleration component in this direction.

9-5 MOTION OF THE MASS CENTER OF A SYSTEM OF PARTICLES

The motion of the mass center of a system of particles can be obtained from Newton's equation of motion,

$$\mathbf{R} = m\mathbf{a},$$

for a single particle. The system of particles may be considered either as discrete particles, the particles of matter which make up a solid body, rigid or nonrigid, or the individual particles of a gas or fluid. Consider the collection of particles indicated by Fig. 9-2(a). The ith particle has a mass m_i and a position vector from the origin of coordinates $\mathbf{r}_i$. As shown in Fig. 9-2(b), each particle may be subjected to an external force $\mathbf{F}_i$ plus internal forces from the other particles of the system shown as $\mathbf{F}_{i_1}$, $\mathbf{F}_{i_2}$, $\mathbf{F}_{i_3}$, $\mathbf{F}_{ij}, \cdots$. These internal forces arise from such sources as magnetic fields, electrical charges, and molecular and elastic forces. Note that, according to Newton's third law, on mass m_j there is a force $\mathbf{F}_{ji}$ exerted by the mass m_i which is equal in magnitude to and collinear with $\mathbf{F}_{ij}$ but in an opposite direction so that

$$\mathbf{F}_{ij} + \mathbf{F}_{ji} = 0.$$

According to Newton's second law, the equation of motion for each particle is

$$\mathbf{F}_i + \sum_{j=1}^{n} \mathbf{F}_{ij} = m_i \mathbf{a}_i, \tag{9-5}$$

where n is the number of particles in the system. An equation of motion for the entire system may be obtained by adding all the equations of motion for the individual particles as follows:

$$\sum_{i=1}^{n} \mathbf{F}_i + \sum_{i=1}^{n}\left(\sum_{j=1}^{n} \mathbf{F}_{ij}\right) = \sum_{i=1}^{n} m_i \mathbf{a}_i.$$

The second term is the sum of all internal forces of the system which

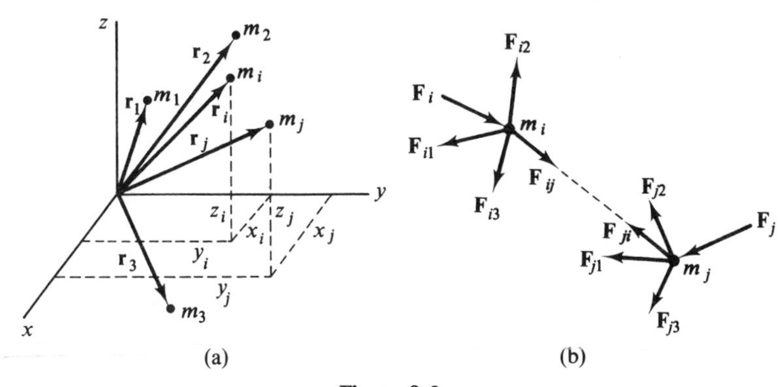

Figure 9-2

must be zero, since they occur in equal, opposite, collinear pairs.[1] If **R** represents the resultant external force acting upon the system, this equation may be written as

$$\mathbf{R} = \sum_{i=1}^{n} \mathbf{F}_i = \sum_{i=1}^{n} m_i \mathbf{a}_i. \tag{9-6}$$

Thus the *resultant* (magnitude, direction, and a point on the line of action) *of the external forces acting on a system of particles is equal to the resultant of the* **ma** *vectors* (effective forces) *of the particles of the system.* The preceding statement, a corollary of Newton's second and third laws, is called *d'Alembert's principle*, in honor of Jean LeRond d'Alembert (1717–1783), a French mathematician, who was the first to publish it.

The position of the mass center of a system of particles can be located by Eq. (3-3a). Thus, for the system of particles in Fig. 9-2(a),

$$m\mathbf{r}_G = \sum_{i=1}^{n} m_i \mathbf{r}_i. \tag{9-7}$$

Equation (9-7) can be differentiated with respect to time to give

$$m\dot{\mathbf{r}}_G = m\mathbf{v}_G = \sum_{i=1}^{n} m_i \dot{\mathbf{r}}_i = \sum_{i=1}^{n} m_i \mathbf{v}_i \tag{9-8}$$

and

$$m\ddot{\mathbf{r}}_G = m\mathbf{a}_G = \sum_{i=1}^{n} m_i \ddot{\mathbf{r}}_i = \sum_{i=1}^{n} m_i \mathbf{a}_i. \tag{9-9}$$

Equations (9-6) and (9-9) can be combined to give

$$\mathbf{R} = \sum \mathbf{F} = m\mathbf{a}_G, \tag{9-10}$$

which can also be written in vector component or scalar form similar to Eq. (9-4). The G subscript should be retained (for example, $\mathbf{a}_{Gx}$) to emphasize that $\mathbf{a}_G$ is the acceleration of the mass center. Equation (9-10) expresses the relationship between the resultant of the external forces acting on any system of particles and the *acceleration of the mass center* of the system regardless of the type of motion. This relationship is commonly called the *principle of motion of the mass center* of a system of particles. Equation (9-10) for a system of particles is the same as the equation of motion for a single particle located at the mass center of the system and having a mass equal to the total mass of the system. Thus, any body can be considered to be a particle in applying the equations of motion just developed. One very important difference should be noted, however. With a single particle, the line of action of the resultant of the force system acting on the particle passes through the particle, since the force system is concurrent. *With a system of particles, the external force system is not necessarily concurrent, and the*

[1] The $\mathbf{F}_{ii}$ terms will be zero, since a particle cannot exert a force upon itself.

resultant usually does not pass through the mass center of the system of particles. To locate the resultant force acting on a body or system of particles, it is necessary to determine the moment of the force system with respect to one or more axes. Methods of determining these moments are developed in the articles for translation, rotation, and plane motion of rigid bodies. If Eq. (9-3) or (9-10) does not supply sufficient information to solve the problem, the body cannot be treated as a particle, and moment equations (which take into account the dimensions of the body and its angular motion) must be used.

In some instances, bodies with negligible mass, such as pulleys and rollers, may be involved in dynamics problems. Force and moment equations for such bodies can be written as in statics since terms involving accelerations will also involve zero mass.

9-6 PROCEDURE FOR THE SOLUTION OF PROBLEMS IN KINETICS

Problems can usually be analyzed most readily and with less chance of error if a definite procedure is followed. The following steps have been found to be helpful in problems in kinetics and will be emphasized in the succeeding articles:

1. Determine carefully what data are given and what is required in the problem.

2. Draw a free-body diagram of one of the particles or bodies involved. (It is usually desirable to include only one body in a free-body diagram when using the method of force, mass, and acceleration, instead of combining several bodies as was frequently done in statics.)

3. Decide what type of motion is involved (particle motion, translation, rotation, or plane motion of a rigid body), select suitable component directions for the problem, and write the general equations of motion for the particle or body being considered.

4. From the free-body diagram (step 2) and the general equations of motion (step 3), determine the number of unknown quantities involved (such as forces, acceleration components, and distances). A list of the unknowns may prove helpful.

5. (a) If the number of equations in step 3 is equal to (or greater than) the number of unknowns, proceed to step 8.

(b) If step 4 indicates more unknowns than the number of equations from step 3, additional equations must be obtained as indicated in step 6.[2]

[2] Special cases may arise in which certain unknowns (but never all of them) may be evaluated even though there are more unknowns than independent equations.

6. Obtain additional equations in one or more of the following ways:

(a) By using the kinematic relationships between the motions of the various particles, lines, or bodies.

(b) By drawing other free-body diagrams and repeating step 3.

(c) By using the equations of friction if sliding or impending motion exists between two rough surfaces.

7. When as many independent equations are available as the total number of unknown quantities, the analysis is complete.

8. Substitute specific values in the equations of motion from steps 3 and 6 (if more than one free body is necessary). When two or more free-body diagrams are used, the directions of the reference axes (positive directions) must be consistent. Finally, solve for the desired unknowns as determined in step 1.

The above procedure will be used in this text primarily to determine unknown quantities (such as the forces acting on a particle or its acceleration) at a particular instant. Kinetics can also be used to calculate these quantities as a function of time or position along the path of motion, or even to calculate the path of motion itself if it is unknown. In the latter case, Newton's laws of motion are used to derive one or more *differential equations of motion* which must be integrated to obtain equations for the position of the body as a function of time or some other parameter.

Deriving the correct differential equations of motion is an essential step in the calculation of the path of motion of a body. Since the engineer is the one most familiar with the particular problem being studied, he is usually the best qualified person to derive the differential equations of motion. The resulting differential equations of motion are often quite difficult to solve, and an analytical solution in terms of elementary algebraic functions might not be possible. In such cases, engineering solutions can usually be obtained by numerically integrating the differential equations of motion. Digital or analog computers are often necessary, and the services of trained computer programmers and mathematicians might be required. The most essential feature of such an analysis, however, is a properly derived set of differential equations of motion which, together with the appropriate *initial conditions*, accurately represent the physical problem. As the term implies, initial conditions are the conditions which are necessary to describe mathematically the position and velocity of the particle at the beginning of the analysis (usually at $t=0$).

It is beyond the scope of this text to give a detailed presentation of the calculation of paths of motion including the solution of various types of differential equations; consequently, the application of Newton's laws in deriving the differential equations of motion will be emphasized. It will frequently be found that a first integration of the differential equations can be obtained by using some of the techniques presented in Chapter 8. As will be shown later, a first integration of the differential equations can also frequently be obtained by applications of the work-energy principle (Chapter 10) or the impulse-momentum principle (Chapter 11).

EXAMPLE 9-1

An airplane pilot completes a dive by pulling out in a circular path in the vertical plane. The speed of the airplane at the bottom of the path is 300 mph. The pilot weighs 170 lb, and the maximum acceleration he can withstand is 8 g. Determine the minimum radius of the circle and the *apparent* weight of the pilot at the bottom of the path.

SOLUTION

The trajectory of the airplane is shown in Fig. 9-3(a). At the time of interest the trajectory is a circular path so that the indicated n and t directions will provide suitable coordinates for the problem.

The minimum radius of the circle can be determined from kinematics. The magnitude of the normal acceleration at the bottom of the path is

$$a_n = \frac{v^2}{r}, \qquad \text{(a)}$$

where

$$v = 300 \text{ mph} = 300\left(\frac{5280}{3600}\right) = 440 \text{ fps}$$

and

$$a_n = 8g = 8(32.2) = 258 \text{ fps}^2.$$

Substituting the above values, Eq. (a) becomes

$$258 = \frac{(440)^2}{r},$$

from which

$$r = \underline{750 \text{ ft.}} \qquad \text{Ans.}$$

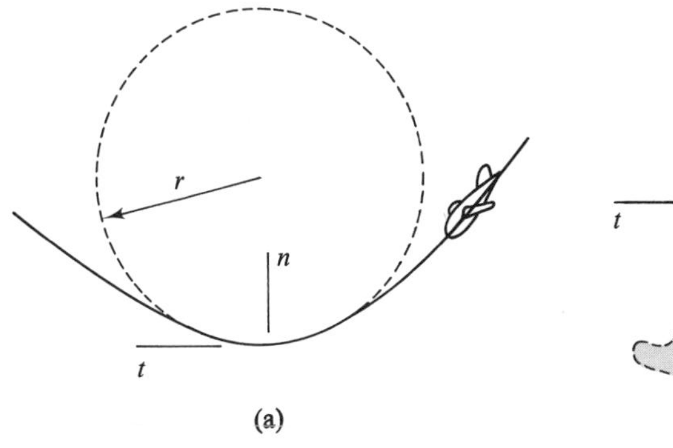

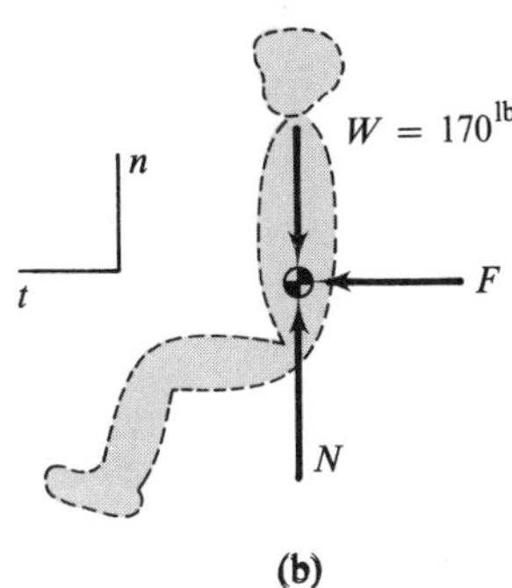

Figure 9-3

A free-body diagram of the pilot is shown in Fig. 9-3(b). The force **N** is the pilot's apparent weight, i.e., the force that would be recorded on a spring scale located in the seat cushion. The equation of motion is

$$(N - W)\mathbf{e}_n + F\mathbf{e}_t = m(a_n\mathbf{e}_n + a_t\mathbf{e}_t)$$

or in scalar form

$$\mathbf{e}_n: \quad N - W = \frac{W}{g}(8g), \tag{b}$$

$$\mathbf{e}_t: \quad F = \frac{W}{g}a_t \tag{c}$$

The force N can be calculated from Eq. (b) as

$$N = 9W$$

or

$$N = 9(170) = \underline{1530\text{ lb.}}$$

Ans.

Equation (c), which relates the force and acceleration in the tangential direction, is not required in the solution for N, and it should be noted that N is independent of the tangential acceleration.

EXAMPLE 9-2

The small body A in Fig. 9-4(a) has a mass of 0.15 kg and swings in a horizontal plane on the end of cord AB. The line AO has a constant angular speed of 25 rpm. Determine the tension in the cord and the angle ϕ.

SOLUTION

The free-body diagram of body A is shown in Fig. 9-4(b). The circular path of the body suggests that normal and tangential components

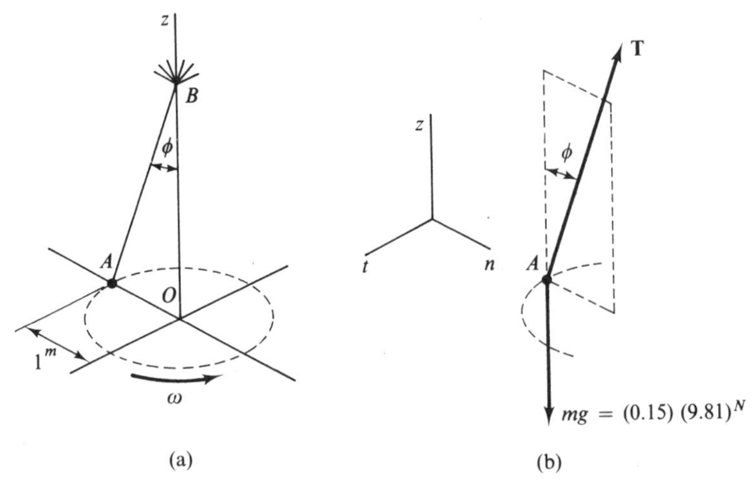

Figure 9-4

would be appropriate for this problem. The positive directions of the axes are shown in Fig. 9-4(b).

The force **T** in the free-body diagram has two unknown characteristics, the magnitude T and the angle ϕ. The equation of motion is

$$\mathbf{R} = T(\cos\phi\mathbf{e}_z + \sin\phi\mathbf{e}_n) - mg\mathbf{e}_z = m(a_n\mathbf{e}_n + a_t\mathbf{e}_t + a_z\mathbf{e}_z).$$

There are no forces in the t direction, and

$$a_t = 0$$

since the angular speed ω of line OA is constant. The vector equation of motion thus gives two scalar equations:

$$\mathbf{e}_n: \quad T\sin\phi = ma_n,$$

$$\mathbf{e}_z: \quad T\cos\phi - mg = ma_z.$$

The body moves in a horizontal plane so that

$$a_z = 0,$$

and from kinematics

$$a_n = r\omega^2 = 1.00\left[\frac{25(2\pi)}{60}\right]^2 = 6.85 \text{ m/s}^2.$$

Substituting values for m, g, and the acceleration components into the scalar equation gives

$$T\sin\phi = 0.15(6.85) = 1.028$$

and

$$T\cos\phi - 0.15(9.81) = T\cos\phi - 1.472 = 0$$

from which

$$\underline{T = 1.795\text{N},} \qquad \underline{\phi = 34.9°.} \qquad \text{Ans.}$$

EXAMPLE 9-3

Block A in Fig. 9-5(a), (p. 499), weighs 10.00 lb, and block B weighs 6.44 lb. The blocks are connected by a flexible inextensible cord which passes over a smooth pulley at C. The weights of the cord and the pulley can be neglected. The coefficient of friction between the plane and block A is 0.30. The blocks are initially at rest. Determine the tension in the cord and the velocity of A after it has traveled 5.00 ft.

SOLUTION

The velocity of A can be determined from the principles of kinematics once the acceleration is known. A free-body diagram of A is shown in Fig. 9-5(b). The frictional force **F** must be in a direction opposite to the direction of the velocity. Since the blocks are released from rest, the velocity a short time after release will be in the same direction as the acceleration. The direction of the acceleration is unknown and must be assumed. In this case, the acceleration is assumed to be up the plane, which means that the velocity will also be up the plane. Therefore, the frictional force will be down the plane.

Coordinate axes are chosen parallel and perpendicular to the plane because block A slides along the plane, thus, $\mathbf{a}_{Ay} = 0$ and $\mathbf{a}_{Ax} = \mathbf{a}_A$. The equation of motion for body A is

$$\mathbf{R} = \sum \mathbf{F} = m\mathbf{a}_A$$

or

$$T\mathbf{i} - F\mathbf{i} + N\mathbf{j} + 5.00(0.8\mathbf{i} - 0.6\mathbf{j}) + 10.00(-0.6\mathbf{i} - 0.8\mathbf{j}) = \frac{10.00}{32.2} a_A \mathbf{i}.$$

This vector equation of motion for body A gives the two scalar equations

$$\mathbf{i}: \quad T - F + 4.00 - 6.00 = 0.310 a_A \tag{a}$$

and

$$\mathbf{j}: \quad N - 3.00 - 8.00 = 0. \tag{b}$$

Equations (a) and (b) contain four unknowns, T, F, a_A, and N. An additional equation is available from friction since the block is assumed to slide on the plane, and the magnitude of the friction force is therefore equal to the limiting value

$$F = \mu N = 0.30N. \tag{c}$$

This still leaves four unknowns and three equations.

Another equation can be obtained by drawing a free-body diagram of body B [Fig. 9-5(c)], and applying Newton's second law:

$$\mathbf{R} = \sum \mathbf{F} = m\mathbf{a}_B.$$

Note that $\mathbf{a}_B = \mathbf{a}_{By}$ since block B has motion only in the y direction. It should also be noted that the positive y direction is down for the free-body diagram of block B. The logic for this selection will be explained later.

The cord tension $\mathbf{T}$ has the same magnitude in both free-body diagrams. This fact can be demonstrated by drawing a free-body diagram of the pulley as in Fig. 9-5(d). Since the pulley has negligible mass, the equations of equilibrium are valid, and from the equation $\sum M_O = 0$ the two tensions are equal. (It will be shown later that when the pulley mass is not zero, the two tensions are not equal.) The equation of motion for block B thus becomes

$$-T\mathbf{j} + 6.44\mathbf{j} = \frac{6.44}{32.2} a_B \mathbf{j}$$

or

$$-T + 6.44 = 0.20 a_B \tag{d}$$

Equation (d) gives a fourth equation but introduces a fifth unknown, a_B. Since the cord is inextensible, however, kinematics gives

$$a_B = a_A \tag{e}$$

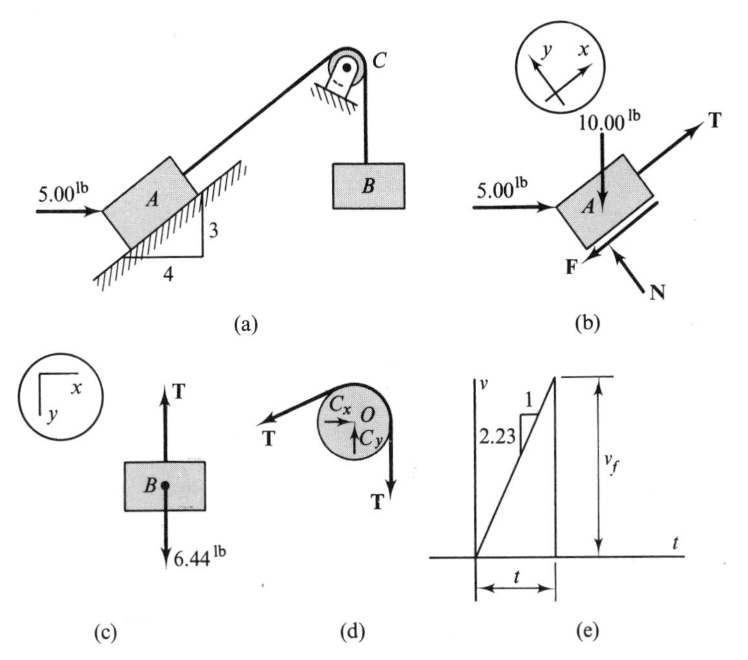

Figure 9-5

for the positive directions shown in Fig. 9.5(b) and 9.5(c). Consistent positive directions must be selected for these accelerations to avoid the use of a negative sign when they are equated. If $\mathbf{a}_A$ were assumed to be down the plane, $\mathbf{a}_B$ should be assumed upward to be consistent. Note that the friction force **F** should also be reversed in this case.

There are now five equations (a), (b), (c), (d), and (e) and five unknowns. The magnitude

$$N = 11 \text{ lb}$$

is obtained from Eq. (b) and substituted into Eq. (c) so that

$$F = 0.3N = 0.3(11) = 3.30 \text{ lb}.$$

When the value of F is substituted in Eq. (a), it becomes

$$T - 3.30 + 4.00 - 6.00 = 0.310a_A,$$

and Eq. (d) becomes

$$T + 6.44 = 0.20a_B = 0.20a_A.$$

The above two equations are solved simultaneously to give

$$\underline{T = 5.99 \text{ lb}}, \qquad \underline{\mathbf{a}_A = 2.23\mathbf{i} \text{ fps}^2}. \qquad \text{Ans.}$$

A v-t diagram for block A is shown in Fig. 9-5(e). The area of the triangle is 5.0 ft, and the slope is 2.23 fps^2. Thus,

$$\text{area} = \frac{v_f t}{2} = 5.0$$

and

$$\text{slope} = \frac{v_f}{t} = 2.23,$$

from which

$$\mathbf{v}_f = \underline{4.72\mathbf{i}\ \text{fps.}}$$

Ans.

If the acceleration had been assumed in the wrong direction, the value of a_A would have had a negative sign, indicating that the resulting velocity and hence the frictional force were assumed to be in the wrong direction. As a result it would be necessary to solve the problem again with the frictional force in the correct direction.

Another method of solution is to assume that the bodies are in equilibrium and determine the frictional force necessary to maintain equilibrium. If the friction thus determined is less than the limiting value (μN), the system will remain at rest. If the frictional force for equilibrium is more than the limiting value, the system will be accelerated in a direction opposite to the direction of the friction (since the blocks are initially at rest), and the direction of the velocity and acceleration will then be known.

If the force system acting on block A produces a reversal of velocity during the time interval, the direction of the frictional force will also reverse, and the acceleration will change in magnitude though not necessarily in direction.

EXAMPLE 9-4

The 2.00-lb body A and 4.00-lb body B in Fig. 9-6(a) are initially at rest. The coefficients of friction are $\mu_A = 0.70$ between bodies A and B, and $\mu_B = 0.20$ between body B and the horizontal surface. Determine the accelerations of the two bodies due to the 5.00-lb force applied to body A.

SOLUTION

Three different combinations of motion are possible for the system shown in Fig. 9-6(a):

(a) Body A accelerates, and body B remains at rest.

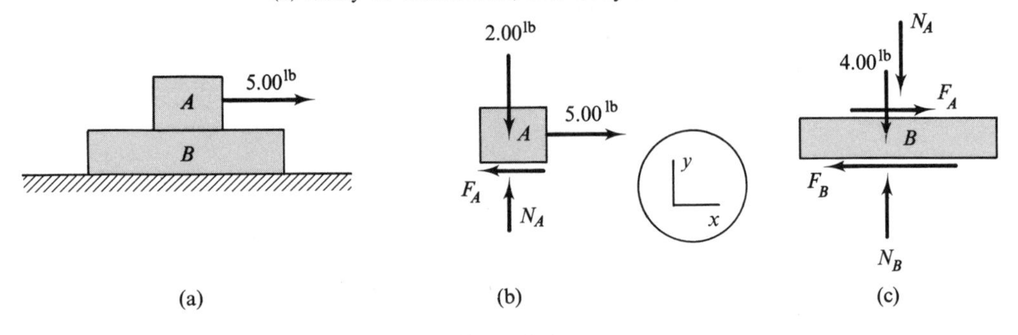

Figure 9-6

(b) Bodies A and B both move with the same acceleration.
(c) Bodies A and B both move but with different accelerations.

Free-body diagrams of bodies A and B are shown in Fig. 9-6(b) and 9-6(c). Note that the friction forces must oppose any slipping or tendency to slip. The equations of motion for bodies A and B are, respectively,

$$\mathbf{R}_A = (5.00 - F_A)\mathbf{i} + (N_A - 2.00)\mathbf{j} = \frac{2.00}{32.2} a_A \mathbf{i},$$

$$\mathbf{R}_B = (F_A - F_B)\mathbf{i} + (N_B - N_A - 4.00)\mathbf{j} = \frac{4.00}{32.2} a_B \mathbf{i}.$$

The scalar equations for body A are

$$\mathbf{i}: \quad 5.00 - F_A = \frac{2.00}{32.2} a_A, \qquad \text{(a)}$$

$$\mathbf{j}: \quad N_A - 2.00 = 0, \qquad \text{(b)}$$

and for body B the equations are

$$\mathbf{i}: \quad F_A - F_B = \frac{4.00}{32.2} a_B, \qquad \text{(c)}$$

$$\mathbf{j}: \quad N_B - N_A - 4.00 = 0. \qquad \text{(d)}$$

The four equations (a), (b), (c), and (d) contain six unknowns, N_A, N_B, F_A, F_B, a_A, and a_B. The force N_A as determined from Eq. (b) is

$$N_A = 2.00 \text{ lb},$$

and the force N_B, determined from Eq. (d), is

$$N_B = N_A + 4.00 = 2.00 + 4.00 = 6.00 \text{ lb}.$$

Two equations, (a) and (c), and four unknowns, F_A, F_B, a_A, and a_B, remain. The two additional equations necessary to complete the solution are determined by considering the three possible combinations of motion.

(a) Assume that body A accelerates and that body B remains at rest. In this case the two additional equations are

$$a_B = 0$$

and

$$F_A = F'_A = \mu_A N_A = 0.70(2.00) = 1.400 \text{ lb}.$$

In addition, since body B remains at rest, the friction force F_B must be less than its limiting value F'_B, where

$$F'_B = \mu_B N_B = (0.20)(6.00) = 1.200 \text{ lb}.$$

When $a_B = 0$ is substituted into Eq. (c) it becomes

$$F_A - F_B = 0$$

or

$$F_B = F_A = 1.400 \text{ lb}.$$

Since F_B is greater than F'_B, the assumption that body B remains at rest is invalid.

(b) Assume that bodies A and B move with the same acceleration. In this case

$$a_A = a_B$$

and

$$F_B = F'_B = 1.200 \text{ lb}$$

since B slides on the horizontal surface. In addition, since there is no relative motion between A and B, the friction force F_A must be less than its limiting value; that is,

$$F_A < F'_A$$

or

$$F_A < 1.400 \text{ lb}.$$

Equations (a) and (c) now become

$$5.00 - F_A = \frac{2.00}{32.2} a_A = \frac{2.00}{32.2} a_B$$

and

$$F_A - F'_B = F_A - 1.20 = \frac{4.00}{32.2} a_B.$$

The above two equations can be solved simultaneously to give

$$a_B = a_A = 20.4 \text{ fps}^2$$

and

$$F_A = 3.73 \text{ lb}.$$

Since F_A is greater than F'_A, the assumption that bodies A and B move with the same acceleration is invalid.

(c) Assume that bodies A and B both move but with different accelerations. In this case

$$F_A = F'_A = 1.400 \text{ lb}$$

and

$$F_B = F'_B = 1.200 \text{ lb},$$

and in addition a_A must be greater than a_B. Substituting the above values into Eq.(a) and (c) yields

$$5.00 - F'_A = 5.00 - 1.400 = \frac{2.00}{32.2} a_A,$$

$$F'_A - F'_B = 1.400 - 1.200 = \frac{4.00}{32.2} a_B$$

from which

$$a_A = 58.0 \text{ fps}^2, \qquad a_B = 1.610 \text{ fps}^2$$

and

$$\mathbf{a}_A = \underline{58.0\mathbf{i} \text{ fps}^2}, \qquad \mathbf{a}_B = \underline{1.610\mathbf{i} \text{ fps}^2}. \qquad \text{Ans.}$$

Since $\mathbf{a}_A$ is greater than $\mathbf{a}_B$, block A will slip to the right on B, and the limiting force will oppose motion as shown.

PROBLEMS

Note: In the following problems unless otherwise specified, all cords, ropes, and cables are assumed to be flexible, inextensible, and of negligible weight. Pulleys are assumed to have negligible mass, and pins, axles, and pegs are assumed to be smooth unless otherwise stated.

9-1 A spring-operated platform scale is placed on the floor of an elevator. Neglect the weight of the platform scale. When the elevator is at rest, a man stands on the scale, which indicates 180 lb. Determine
(a) The acceleration of the elevator when the scale indicates a load of 225 lb.
(b) The load indicated by the scale when the elevator has an acceleration of 15 fps^2 downward.

9-2 An astronaut weighs 170 lb on the surface of the earth. Assume the earth is a sphere with a 4000-mile radius. Determine
(a) The astronaut's mass and weight when he occupies a space station which has a circular orbit 500 miles above the surface of the earth.
(b) The satellite speed required for the astronaut to experience a weightless sensation.

9-3 The 50-lb block A in Fig. P9-3 is initially at rest on a horizontal surface. The coefficients of static and kinetic friction between the block and surface are, respectively $\mu_s = 0.40$ and $\mu_k = 0.30$. The applied force **P** is slowly increased from zero until the block moves. Determine the minimum magnitude of the force **P** required to move the block and the resulting acceleration.

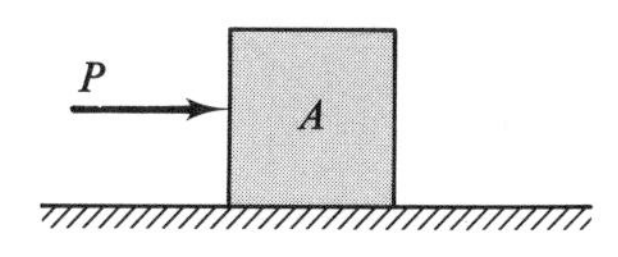

Figure P9-3

9-4 Block A in Fig. P9-4 weighs 75 lb, and the coefficient of friction between A and the plane is 0.50. Determine the acceleration of A under the action of the 25-lb force when the velocity of A is (a) 8 fps up the plane, (b) 15 fps down the plane, (c) zero.

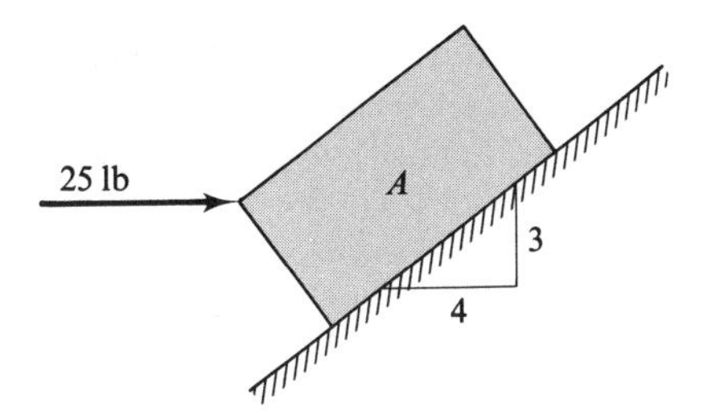

Figure P9-4

9-5 A 7-lb box is given a velocity of 10 fps up a plane which is inclined 30° to the horizontal. The coefficient of friction between the box and the plane is 0.60. Determine the velocity of the box (a) after 0.25 sec, (b) after 0.50 sec.

9-6 Solve Problem 9-5 if the coefficient of friction is 0.40.

9-7. Body A in Fig. P9-7 weighs W lb and is initially moving to the right under the action of the force **P**, which has a constant magnitude. The coefficient of friction between A and the horizontal plane is μ. Determine
(a) The angle θ for maximum acceleration.
(b) The magnitude of the resulting acceleration.

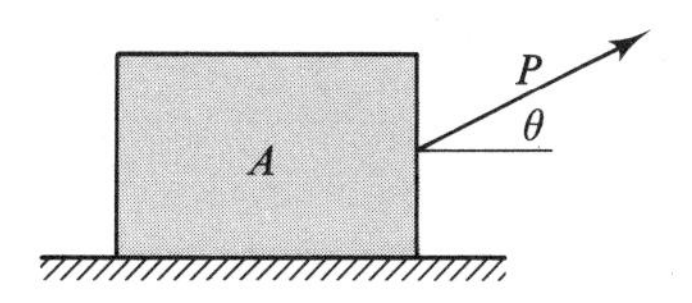

Figure P9-7

9-8 Solve Problem 9-7 if body A is initially moving to the left.

***9-9** Block A in Fig. P9-9 has a mass of 5 kg, and block B has a mass of 2 kg. The coefficient of friction between A and the plane is 0.4. The blocks are initially at rest. Determine the acceleration of B and the tension in the cord.

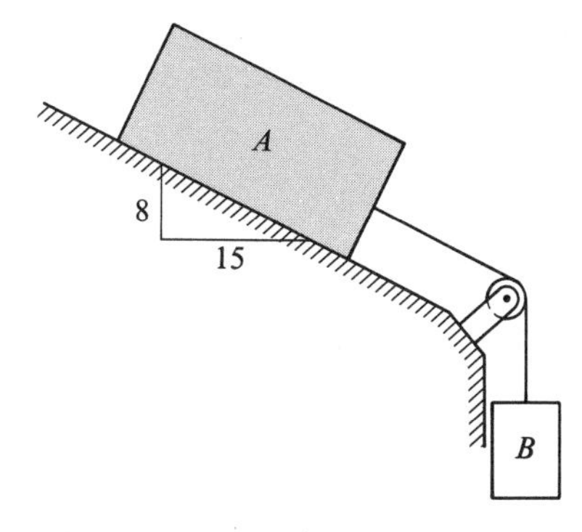

Figure P9-9

***9-10** The masses of blocks A, B, and C in Fig. P9-10 are 15, 25, and 10 kg, respectively, and the system is initially at rest. The coefficient of friction between B and the horizontal surface is 0.10. Determine
(a) The tension in each cord.
(b) The acceleration of body B.

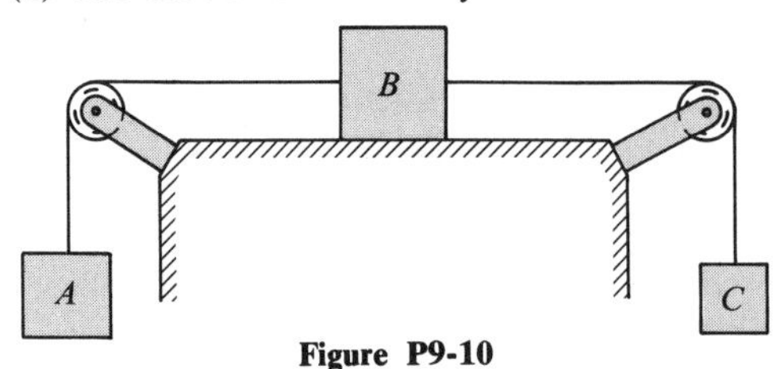

Figure P9-10

9-11 The three bodies A, B, and C in Fig. P9-11 weigh 20, 25, and 15 lb, respectively. The horizontal plane is smooth. Determine
(a) The acceleration of body C.
(b) The tension in the cord connecting B and C.
(c) The tension in the cord connecting A and B.

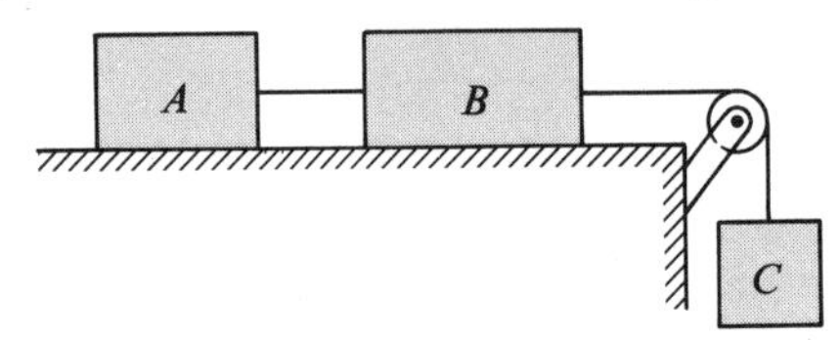

Figure P9-11

9-12 Solve Problem 9-11 if the coefficient of friction between the horizontal plane and bodies A and B is 0.2. The blocks start from rest.

9-13 Bodies A and B in Fig. P9-13 weigh 50 and 30 lb, respectively. The coefficient of friction between A and the inclined plane is 0.3. Body A has an initial velocity of 10 fps up the plane. Determine
(a) The acceleration of block A.
(b) The tension in the cord.

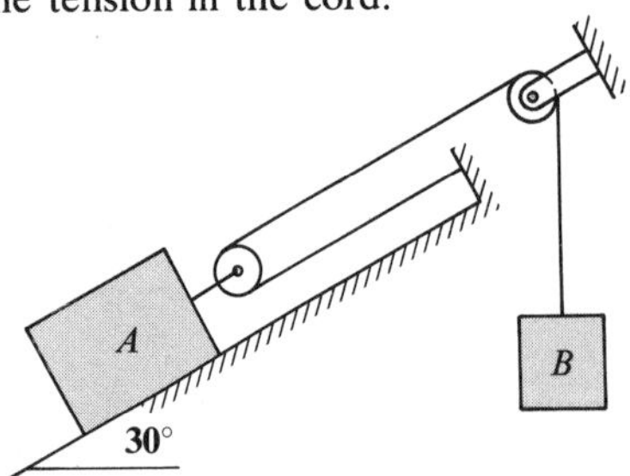

Figure P9-13

9-14 Body A in Fig. P9-13 has a velocity of 12 fps down the plane. Determine the weight of body B required to produce an acceleration of body A of 8.05 fps^2 up the plane. All other data are the same as in Problem 9-13.

***9-15** Block A in Fig. P9-15 has a mass of 2.6 kg, and block B has a mass of 1.0 kg. The coefficient of friction between A and the plane is 0.25. Block A has an initial velocity of 3 m/s down the plane. Determine
(a) The acceleration of block A.
(b) The tension in the cable.

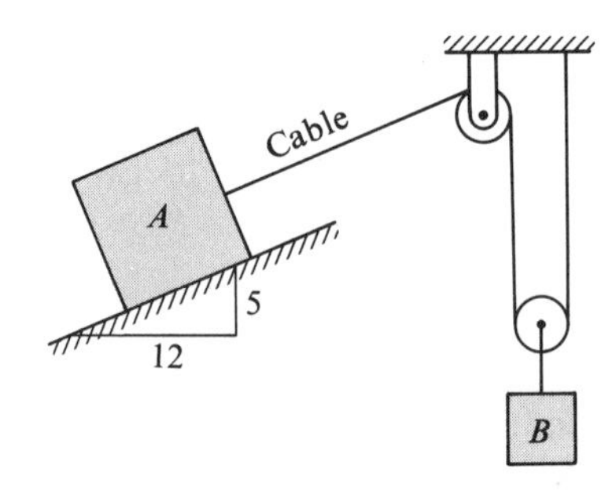

Figure P9-15

9-16 The two blocks A and B in Fig. P9-16 weigh 15 and 10 lb, respectively. The coefficient of friction is 0.50 between A and the plane and 0.20 between B and the plane. Determine the magnitude of the resultant force between the blocks after they have been released from rest.

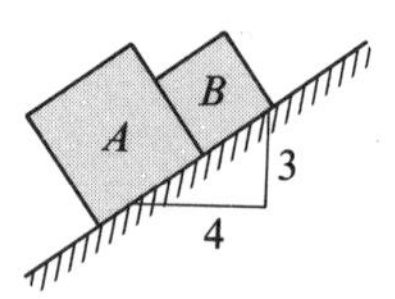

Figure P9-16

9-17 In Fig. P9-17 body A weighs 15 lb and body B weighs 35 lb. The bodies are initially at rest on the smooth horizontal plane. The coefficient of friction between A and B is 0.25. Determine the acceleration of each of the bodies if the magnitude of the force P is (a) 10 lb, (b) 20 lb.

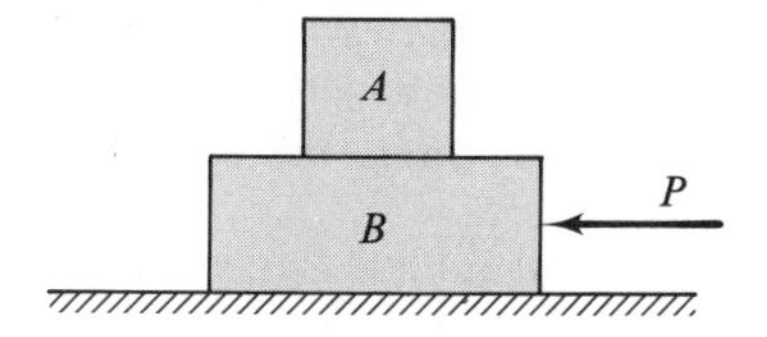

Figure P9-17

9-18 Blocks A and B in Fig. P9-18 are released from rest in the position shown. The weight of B is twice the weight of A. The coefficient of static friction for all surfaces is 0.40, and the coefficient of kinetic friction is 0.30. Determine the acceleration of block A.

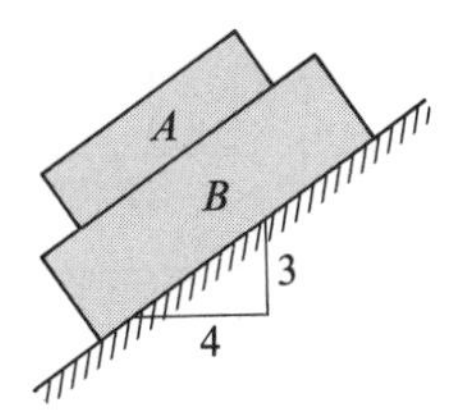

Figure P9-18

9-19 Block A in Fig. P9-19 weighs W lb and slides along a vertical circular track 10 ft in diameter. The coefficient of friction between A and the track is 0.30. When in the position shown, block A has a velocity of 10 fps down and to the left. Determine the acceleration of A for this position.

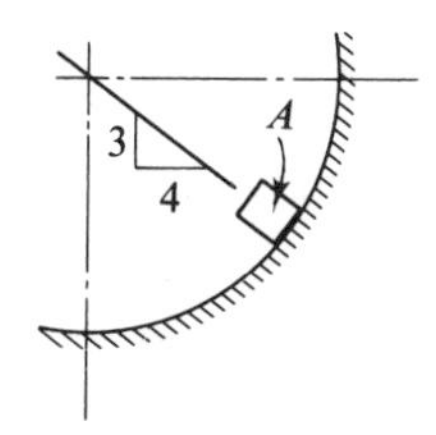

Figure P9-19

9-20 Particle A in Fig. P9-20 weighs 5.0 lb and swings in a vertical plane at the end of an 8-ft cord. When the angle θ is 30°, the magnitude of the velocity of A is 4 fps. Determine the tension in the cord and the angular acceleration of the cord for this position.

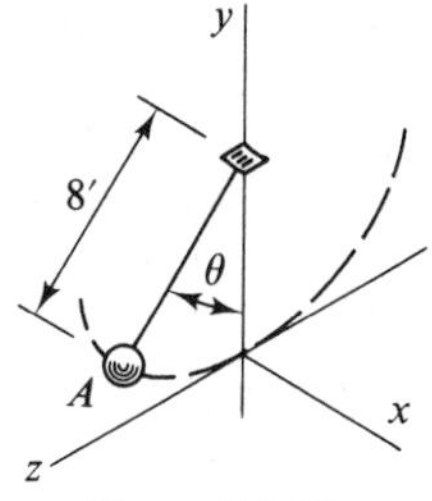

Figure P9-20

9-21 Particle A of mass m travels on a circular path, as shown in Fig. P9-21, with a speed of 14 fps. Determine the radius r of the path.

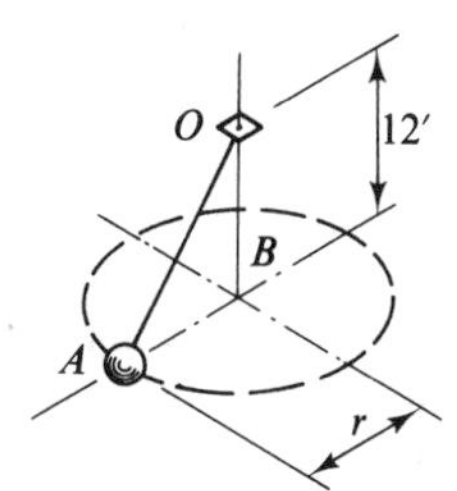

Figure P9-21

9-22 In Fig. P9-21, particle A is connected to point B by a 9-ft cord and to point O by a 15-ft cord.

(a) Derive an expression for the tension in AB in terms of the mass of the body and the angular velocity of AB.

(b) At what angular velocity will the tension in the cord AB become zero?

(c) As the angular velocity is decreased, the angle AOB will also decrease. Determine the angular velocity at which the angle will become zero.

***9-23** Body A in Fig. P9-23 has a mass of 9 kg and rests on a smooth plane. The force P applied to body A is increased from zero at a very slow rate. Determine the maximum force P that can be applied without causing the smooth 6-kg cylinder to leave body A at point C.

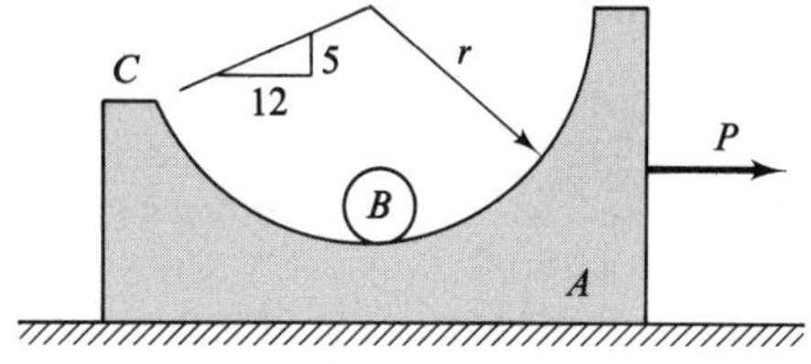

Figure P9-23

***9-24** A racing car travels around a curve at a constant speed of 250 km/hr. The radius of curvature of the track is 500 m, measured horizontally, and the super elevation of the track, angle θ in Fig. P9-24, is 30°. Determine the minimum coefficient of friction between the track and tires if there is to be no tendency for the car to slip.

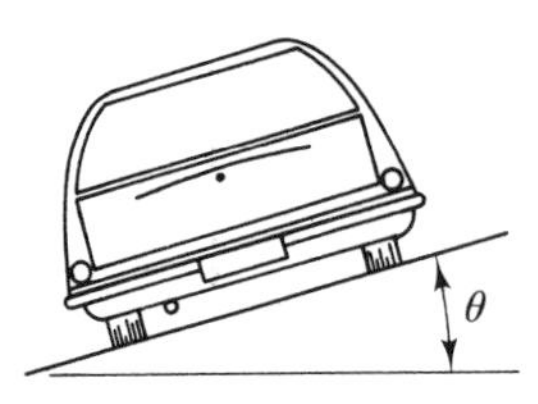

Figure P9-24

9-25 The turntable in Fig. P9-25 rotates in a horizontal plane with a constant angular acceleration of 3.0 rad per sec² counterclockwise when looking downward. The small block A on the disk weighs 2.0 lb. Determine the magnitude of the frictional force of the disk on the block when the angular velocity of the disk is 2.0 rad per sec counterclockwise. Assume that the block does not slip on the disk.

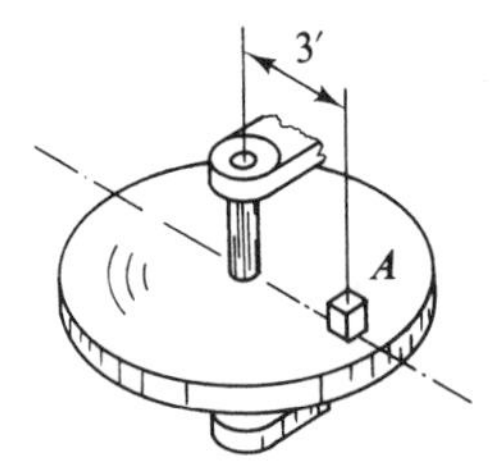

Figure P9-25

9-26 The coefficient of friction between the block and disk in Problem 9-25 is 0.60. Determine the angular velocity of the disk when the block starts to slip. The angular acceleration of the disk is the same as in Problem 9-25.

9-27 A cardboard box weighing 50 lb comes out the end of a chute with a horizontal velocity of 15 fps and lands without bouncing on the platform of a 100-lb pushcart as shown in Fig. P9-27. The coefficient of friction between the cart and the box is 0.4. The inertia and the rolling resistance of the cart wheels can be neglected. Assume the box comes to rest relative to the cart before impacting the end of the cart. Determine

(a) The final velocity of the box and cart after the box has ceased to slide with respect to the cart.

(b) The distance the cart slides along the top of the cart.

(c) The total distance the cart moves before the box and cart have the same velocity.

Determine the time required for block A to travel the length of block B and the corresponding displacement of block B.

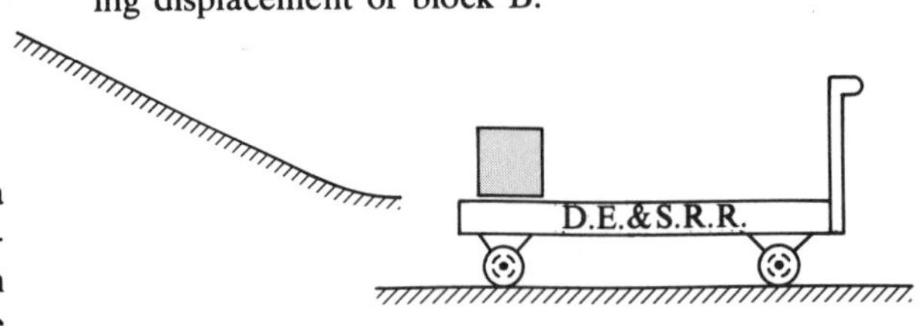

Figure P9-27

9-28 Block A in Fig. P9-28 weighs 10.00 lb and is 5.00 in. long. Block B weighs 20.00 lb and is 15.0 in. long. Block B is on a smooth plane, and the coefficient of friction between A and B is 0.40. Both blocks are initially at rest when the 8.00-lb force is applied to block A.

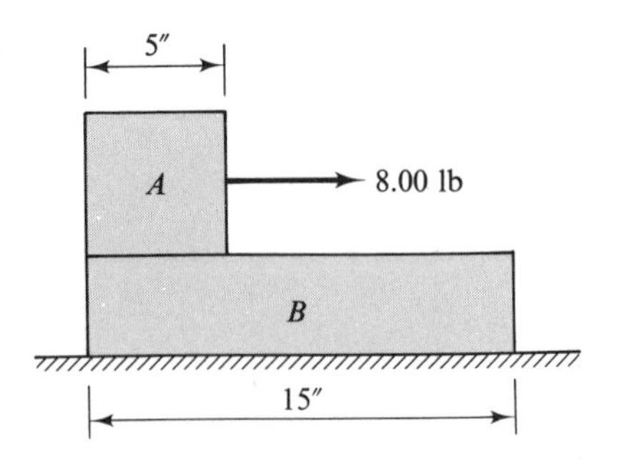

Figure P9-28

9-29 Bodies A, B, and C in Fig. P9-29 weigh 40, 30, and 20 lb, respectively. The system is released from rest. Determine the acceleration of body A and the tension in the cord connecting bodies B and C.

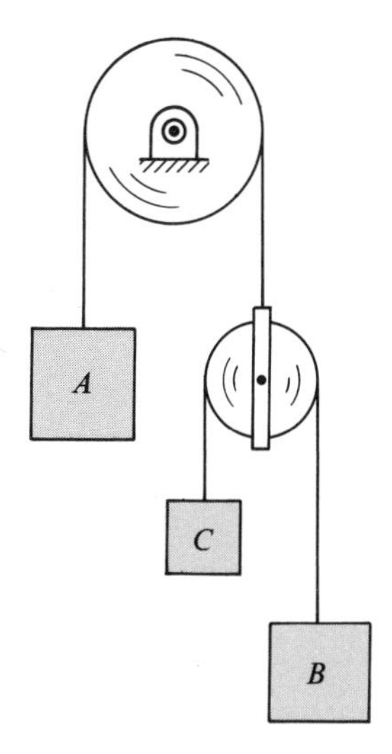

Figure P9-29

9-30 A *top view* of three bodies is shown in Fig. P9-30 as they slide on a smooth horizontal plane. Bodies A, B, and C weigh, respectively, 32.2, 64.4, and 96.6 lb. Determine the accelerations of bodies A, B, and C.

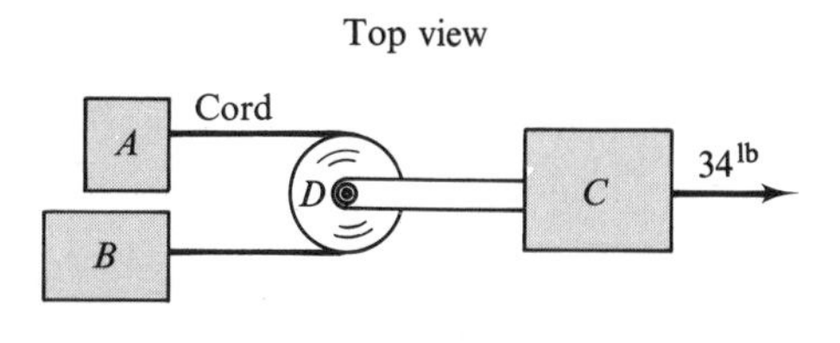

Figure P9-30

9-31 In Fig. P9-31, body A weighs 5.0 lb and body B weighs 10.0 lb. The coefficients of friction are 0.4 between A and B and 0.2 between B and the horizontal surface. The weight of body C can be neglected. Determine

(a) The maximum value of P for which there is no motion.

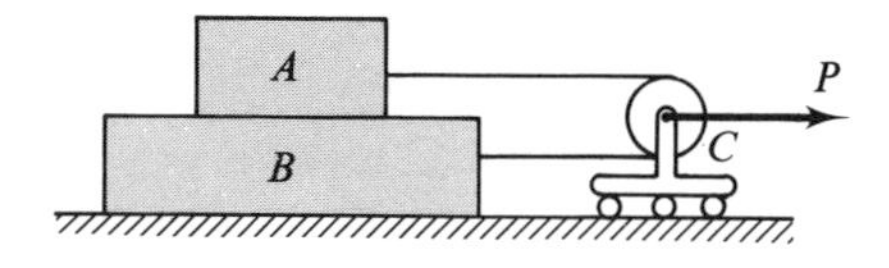

Figure P9-31

(b) The minimum value of P for which A slides on B.

9-32 Ring B in Fig. P9-32 slides along rod OA, which rotates in the vertical plane. In the position shown, rod OA has an angular velocity of 5 rad per sec counterclockwise and an angular acceleration of 3 rad per sec^2 clockwise. The distance r is 1 ft and is increasing at a rate of 2 fps. The coefficient of friction between the ring and rod is 0.2. Determine the acceleration of the ring relative to the rod.

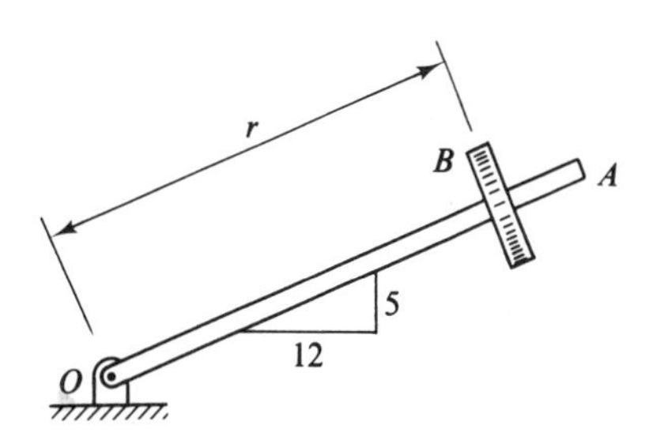

Figure P9-32

9-33 The cone in Fig. P9-33 has a central angle θ and rotates about the vertical z axis. A small body which has a mass m is connected to the vertex of the cone by a cord of length L and rotates with the cone. Derive, in terms of the length L, the angle θ, the mass m, and the angular velocity of the cone, an expression for

(a) The tension in the cord.

(b) The reaction of the cone on the particle.

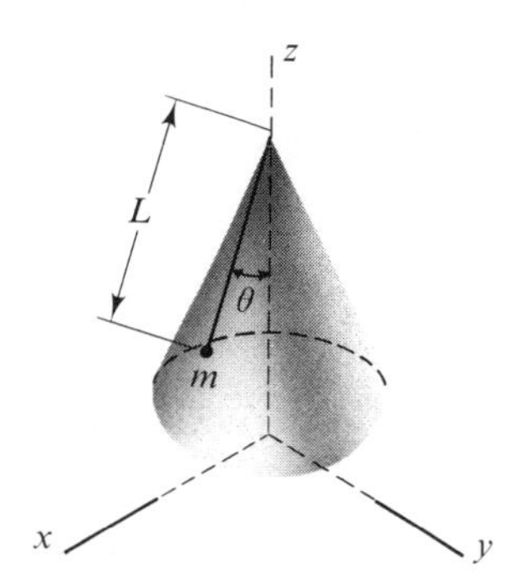

Figure P9-33

9-34 The 4-in. bar AB in Fig. P9-34 rotates in the horizontal (xy) plane around the z axis with an angular speed of ω rad per sec. Particle P weighs 2 lb and is suspended from AB by two 7-in. cords.
(a) Determine the angular speed of AB and the tension in each cord when the distance $d = 6$ in.
(b) As ω decreases the distance d will also decrease. Determine the value of ω for which d approaches zero.

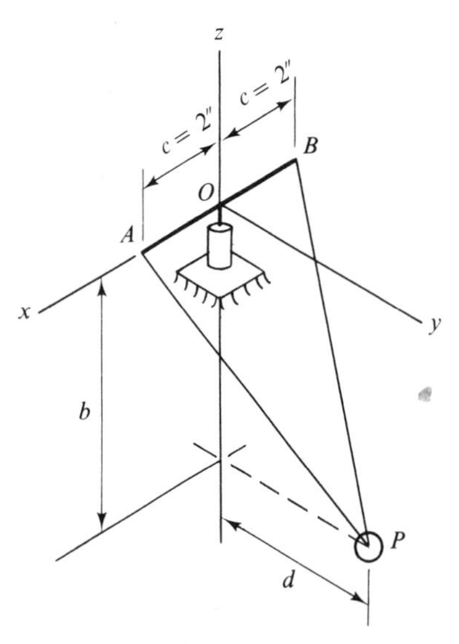

Figure P9-34

9-35 The 8.05-lb body A in Fig. P9-35 is connected to the 16.1-lb body B by a 7.0-ft cord which passes over the smooth peg at C. The horizontal slot is smooth. When the distance x is 4.0 ft the velocity of B is 5.0 fps to the right and the force P is 10.0 lb to the right. Determine the acceleration of B and the tension in the cord.

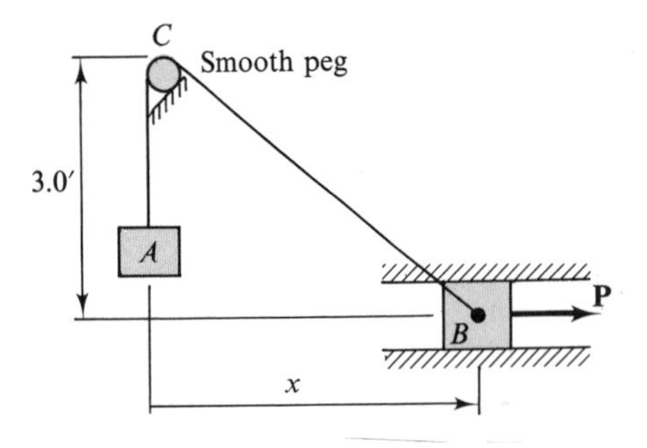

Figure P9-35

9-36 Bodies A and B in Fig. P9-36 weigh 10 lb each. The cable attached to body B passes around a smooth peg inside body A. In the position shown, A has a velocity of 6 fps to the right on the smooth horizontal surface. Determine
(a) The tension in the cord.
(b) The acceleration of A.
(c) The acceleration of B.

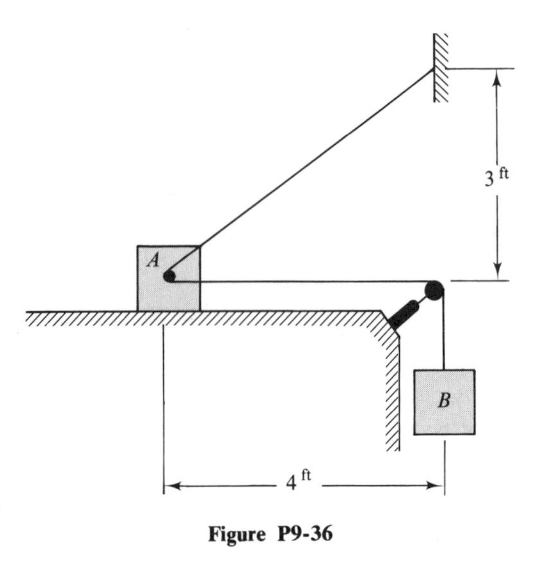

Figure P9-36

9-37 In Fig. P9-37, body A is connected to body B by a cord which goes through smooth rollers at C. Body A weighs 15 lb and body B weighs 12 lb. The coefficient of friction between body A and the horizontal surface is 0.6. At the instant shown, body B has a velocity of 6 fps up and to the right. If the cord connecting body A to the wall is cut at this instant, determine the acceleration of body A.

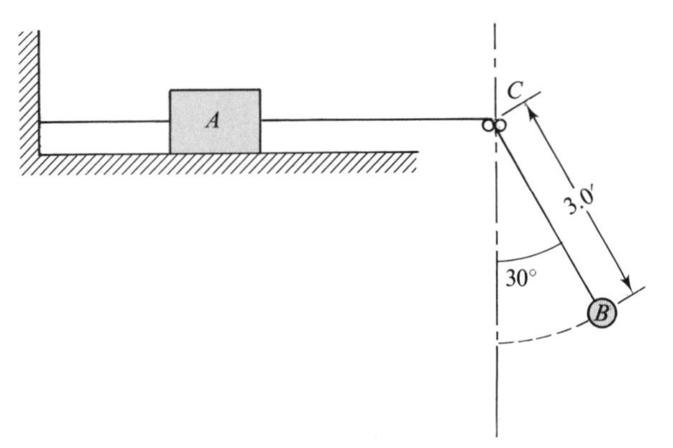

Figure P9-37

9-7 REVERSED EFFECTIVE FORCES

It was first noted by Jean LeRond d'Alembert (1717–1783) that dynamics problems may be treated by the methods of statics if Newton's second law of motion is written as

$$\mathbf{F}+(-m\mathbf{a})=0,$$

where the $(-m\mathbf{a})$ term is known as the *reversed effective force.* A corollary of d'Alembert's principle [see Art. 9-5] can be stated as follows: *If a force equal in magnitude to and collinear with, but opposite in sense to, the resultant of the* $m\mathbf{a}$ *quantities of the particles is added to the external forces acting on the particles, the resulting force system will be in equilibrium.* Although equivalent to forces, reversed effective forces are not true forces since they do not represent the action of one body upon another.

The concept of reversed effective forces does not present any new information and is therefore only an alternative to the method previously presented in this chapter. It is important to decide which method will be used early in the solution of any problem since if the two methods are mixed in a single solution, the $m\mathbf{a}$ terms are often used once as *reversed effective forces* and again as *effective forces,* resulting in an incorrect solution. When applying d'Alembert's principle, it is desirable to show the reversed effective forces as dashed lines to distinguish them from true forces.

The following example illustrates the use of reversed effective forces.

EXAMPLE 9-5

A small body weighing 2.0 lb is suspended from the top of a bus by a 52-in. cord. As the bus is accelerated to the left the body swings toward

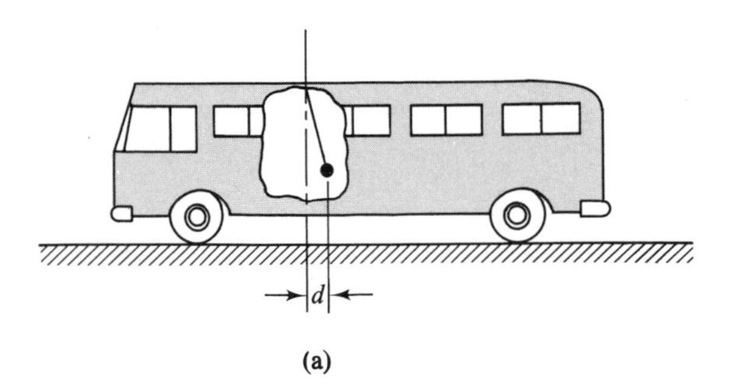

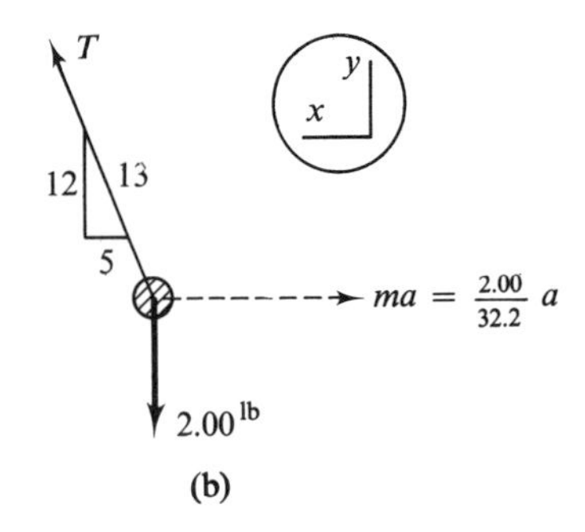

Figure 9-7

the rear of the bus as shown in Fig. 9-7(a). Assume that the acceleration is constant and that the pendulum does not oscillate. When the distance d is 20.0 in., determine (a) the acceleration of the bus and (b) the magnitude of the tension in the cord.

SOLUTION

A free-body diagram of the body is illustrated in Fig. 9-7(b). The 2.00-lb weight **W** and tension **T** are the external forces acting on the body. The reversed effective force is shown as a dashed line to distinguish it from the true forces. It should be noted that since the acceleration is known to be toward the left, the reversed effective force is known to act toward the right. The equilibrium equation for the body is

$$\mathbf{R}=0$$

$$T\left(\frac{5}{13}\mathbf{i}+\frac{12}{13}\mathbf{j}\right)-2.00\mathbf{j}-\frac{2.00}{32.2}a\mathbf{i}=0.$$

The associated scalar equations are

$$\mathbf{i}:\quad \frac{5}{13}T-\frac{2.00}{32.2}a=0 \qquad \text{(a)}$$

$$\mathbf{j}:\quad \frac{12}{13}T-2.00=0. \qquad \text{(b)}$$

The tension T, from Eq. (b), is

$$\underline{T=2.17\text{ lb}}, \qquad \text{Ans.}$$

and when this value is substituted into Eq. (a) the value of a becomes

$$a=13.42$$

and

$$\underline{\mathbf{a}=13.42\mathbf{i}\text{ fps}^2}. \qquad \text{Ans.}$$

PROBLEMS

***9-38** Solve Problem 9-9 by the reversed effective-force method.

9-39 Solve Problem 9-20 by the reversed effective-force method.

***9-40** Solve Problem 9-23 by the reversed effective-force method.

***9-41** Solve Problem 9-24 by the reversed effective-force method.

9-42 Solve Problem 9-25 by the reversed effective-force method.

9-43 A package rests on the front seat of an automobile as shown in Fig. P9-43. The coefficient of friction between the package and the seat cushion is 0.6. When stopping at a traffic signal, at what deceleration will the package begin to slide from the seat?

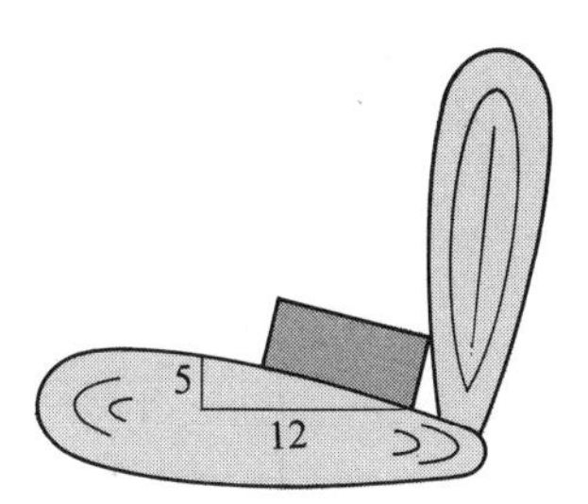

Figure P9-43

9-44 In Fig. P9-44 the homogeneous 5-lb ball *C* is suspended from the 20-lb body *B* by a cord attached at *A*. The motion of the system is such that the angle θ remains constant at 30° when the velocity of *B* is to the left. The coefficient of friction between *B* and the plane is 0.4. Determine the magnitude of the force **P**.

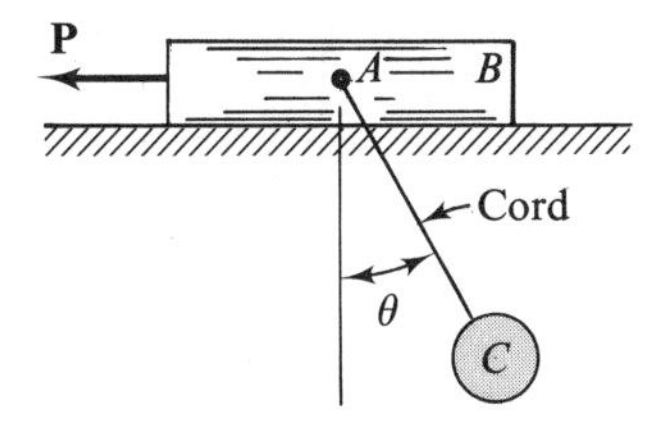

Figure P9-44

9-45 The shaft in Fig. P9-45 rotates in bearings *A* and *B* at a speed of 1200 rpm. The weights at *C* and *D* are 2 and 5 lb, respectively. Neglect the effects of gravity. Determine the forces exerted on the shaft by the bearings when in the position shown.

9-46 The 5-in. radius disk in Fig. P9-46 spins about the vertical axis at a constant speed. The pendulum arms are 10 in. long. Determine the rate of spin when the angle θ is 30°.

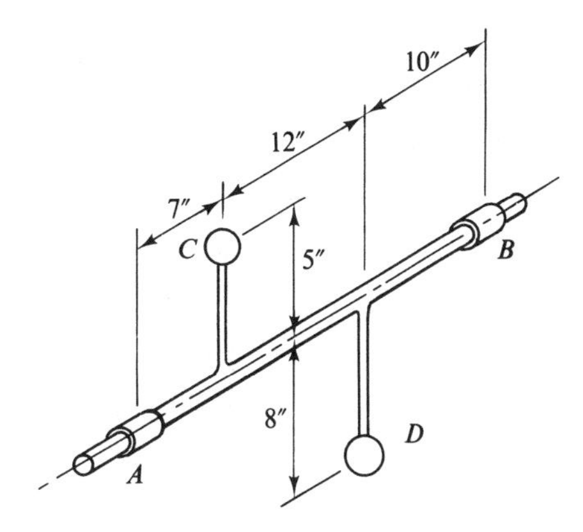

Figure P9-45

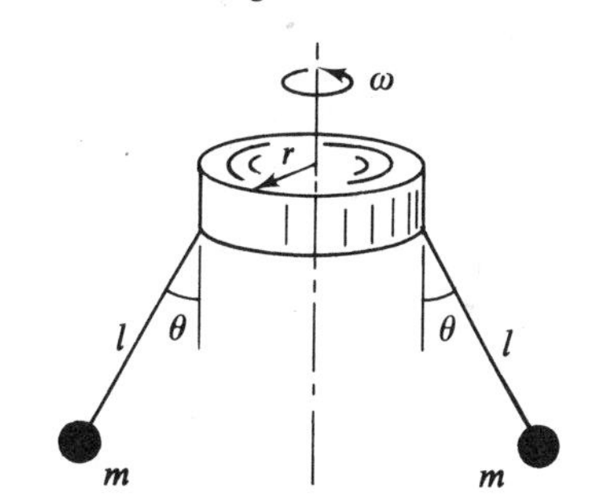

Figure P9-46

***9-47** In Fig. P9-47, blocks *A* and *B* have masses of 10 and 40 kg, respectively. Determine the acceleration of *A* and the tension in the cord if the fixed drum is smooth.

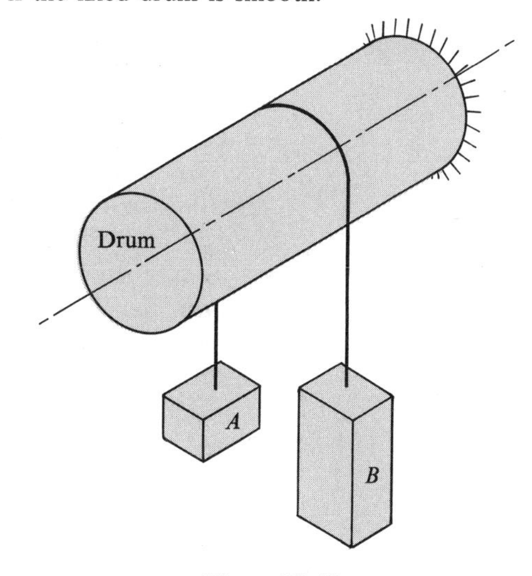

Figure P9-47

***9-48** Solve Problem 9-47 if the coefficient of friction between the cord and the fixed drum is 0.15 and the bodies are released from rest. [*Hint:* Review belt friction in Art. 5-6 if necessary.]

9-49 Derive an equation relating the large and small tensions in the flat belt in Fig. P9-49 when slipping impends between the belt and rotating pulley. The belt weighs w lb per ft of length, and the pulley has a constant angular velocity of ω rad per sec. Neglect the force that the earth exerts on the belt (the weight). (Review belt-friction equations for stationary belts or band brakes in Art. 5-6 if necessary.)

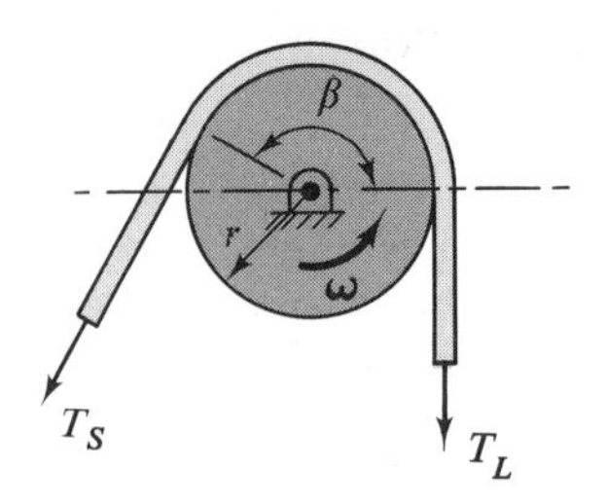

Figure P9-49

PART B–KINETICS OF RIGID BODIES

9-8 INTRODUCTION

The application of Newton's laws of motion to the kinetics of rigid bodies will be developed in Part B of this chapter. Kinetics has been defined as the relationship between the resultant force system and the associated motion of bodies that are not in equilibrium. When the size of a body can be neglected in kinetic analysis, the body can be treated as a *particle* with its mass concentrated at a single point, and the line of action of the resultant force must pass through the particle. Consequently, a single vector equation obtained from Newton's second law, $\mathbf{R} = m\mathbf{a}$, is sufficient to describe the relation of the unbalanced force system acting on a particle to the inertia or mass of the particle and the resulting acceleration.

It was shown in Art. 9-5 that the relationship between the resultant of the external forces acting on any *system of particles* and the acceleration of the mass center of the system is also governed by a single vector equation, $\mathbf{R} = m\mathbf{a}_G$. For a system of particles, however, the external force system is not necessarily concurrent, and the resultant usually does not pass through the mass center of the system. In addition, the equation of motion for the mass center of a system of particles gives no information regarding the motion of the individual particles *relative* to the mass center.

A *rigid body* is a system of particles in which all particles remain at fixed distances from each other. The remainder of this chapter will be devoted to the effects of the points of application of forces on the motion of a rigid body relative to its center of mass, as well as on the motion of the mass center itself. It should be obvious that additional equations of motion will be required. These equations can be obtained from moment equations which take into account the point of application of forces, the dimensions of the body, and its angular motion. Methods of determining these equations will be developed for translation, rotation, and plane motion of rigid bodies in the following articles.

9-9 TRANSLATION OF A RIGID BODY

When every line in a rigid body remains parallel to its initial position the motion of the body is defined as *translation*. If a rigid body has translation, all particles of the body have the same acceleration.

Figure 9-8 represents a rigid body which has a motion of translation so that each particle of the body has an acceleration **a** as indicated. The resultant of all external forces acting on the body is shown as **R**. From d'Alembert's principle [Eq. (9-6)] the resultant of the external forces **R** must be equal to the sum of the $m_i\mathbf{a}_i$ values for any system of particles which includes rigid bodies. Let the body be made up of n particles and m_i be the mass of the ith particle. The force **R** is

$$\mathbf{R}=\sum_{i=1}^{n} m_i\,\mathbf{a}_i=\left(\sum_{i=1}^{n} m_i\right)\mathbf{a}=m\mathbf{a},$$

where m is the mass of the body and **a** is the acceleration of each point in the body, including the mass center.

Let Q be any point on the line of action of the resultant of the

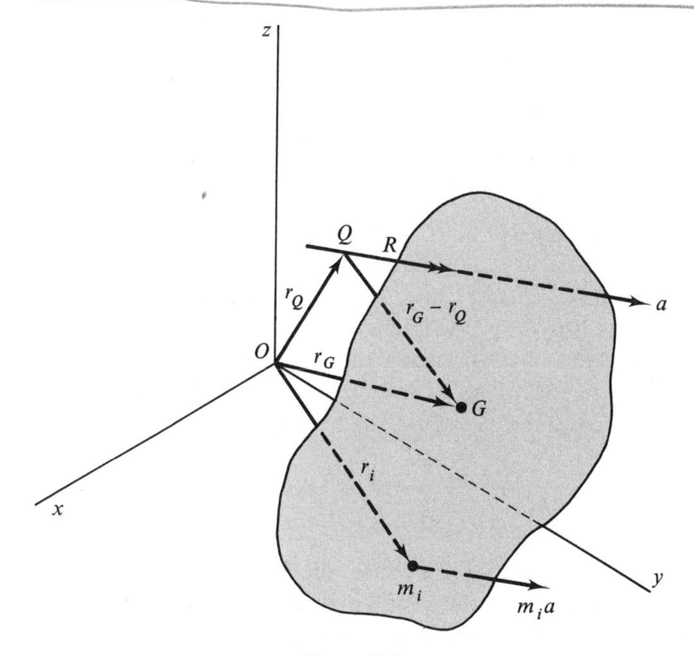

Figure 9-8

external forces, **R**. The moment of **R** with respect to the origin O is

$$\mathbf{M}_O = \mathbf{r}_Q \times \mathbf{R} = \mathbf{r}_Q \times m\mathbf{a}. \tag{a}$$

The moment with respect to O must also be equal to the sum of the moments of the $m_i\mathbf{a}_i$ values; that is,

$$\mathbf{M}_O = \sum_{i=1}^{n} (\mathbf{r}_i \times m_i\mathbf{a}_i) = \sum_{i=1}^{n} (\mathbf{r}_i \times m_i\mathbf{a})$$

$$= \sum_{i=1}^{n} (m_i\mathbf{r}_i) \times \mathbf{a} = m\mathbf{r}_G \times \mathbf{a}. \tag{b}$$

When the expressions for the moment $\mathbf{M}_O$ in Eq. (a) and (b) are equated, the result is

$$\mathbf{r}_Q \times m\mathbf{a} = m\mathbf{r}_G \times \mathbf{a},$$

which reduces to

$$m(\mathbf{r}_G - \mathbf{r}_Q) \times \mathbf{a} = 0.$$

The last equation is satisfied if $\mathbf{r}_Q = \mathbf{r}_G$, in which case the point Q is at the mass center G, and **R** passes through G. The equation is also satisfied provided that $\mathbf{r}_Q - \mathbf{r}_G$ is parallel to **a**. However, from Eq. (9-10), **R** has the same sense and slope as **a**, which means that if $\mathbf{r}_G - \mathbf{r}_Q$ is parallel to **a**, **R** must pass through G. Consequently, *if a rigid body has a motion of translation, the resultant of the external forces applied to the body must pass through the mass center of the body and the resultant moment of the external forces about the mass center must be zero.* The equations of motion for translation are

$$\sum \mathbf{F} = m\mathbf{a}, \quad \sum \mathbf{M}_G = 0, \tag{9-11a}$$

which are equivalent to

$$\begin{matrix} \sum F_x = ma_x, & \sum F_y = ma_y, & \sum F_z = ma_z, \\ \sum M_{Gx} = 0, & \sum M_{Gy} = 0, & \sum M_{Gz} = 0. \end{matrix} \tag{9-11b}$$

In many problems, a body may have coplanar translation, and if the external forces are all in the plane of motion, only the force components in the plane of motion and the moment equation about an axis perpendicular to the plane of motion are relevant. If the mass center moves in the xy plane, Eq. (9-11b) becomes

$$\sum F_x\mathbf{i} + \sum F_y\mathbf{j} = m(a_x\mathbf{i} + a_y\mathbf{j}), \quad \sum M_{Gz}\mathbf{k} = 0. \tag{9-12}$$

When a translating body has rectilinear motion, it is usually desirable to select either the x or y axis parallel to the acceleration, in which case the other acceleration component is zero. When a body has plane curvilinear translation, it is frequently convenient to use Eq.

(9-12) with the x and y axes replaced by n and t axes in the directions of the normal and tangential components of acceleration. The following examples illustrate the principles just discussed.

EXAMPLE 9-6

A man wants to slide the homogeneous 100-lb box in Fig. 9-9(a) across the floor by pushing on it with a force **P** as indicated. The coefficient of friction between the box and the floor is 0.20. Can an acceleration of 8.0 fps^2 to the right be produced without tipping the box? If so, determine the magnitude of the force **P**.

SOLUTION

Figure 9-9(b) is a free-body diagram of the box. Since the body has a motion of translation, the equations of motion are

$$\sum F_x\mathbf{i}+\sum F_y\mathbf{j}=m(a_{Gx}\mathbf{i}+a_{Gy}\mathbf{j}),$$
$$\sum M_{Gz}\mathbf{k}=\mathbf{0}.$$

Assume that the box will slide without tipping. A check for the validity of this assumption can be obtained by determining the distance x; if x is less than 1.5 ft, the box will not tip. The force **P** is large enough to cause the box to slip; therefore, the friction will equal the limiting value. That is,

$$F=F'=\mu N=0.20N. \qquad \text{(a)}$$

The equations of motion for the axes shown on the free-body diagram are

$$P\mathbf{i}-F\mathbf{i}+N\mathbf{j}-100\mathbf{j}=\frac{100}{32.2}(8.0\mathbf{i}+0\mathbf{j})=24.8\mathbf{i}$$

and

$$\sum M_{Gz}\mathbf{k}=xN\mathbf{k}-2P\mathbf{k}-2.5F\mathbf{k}=0.$$

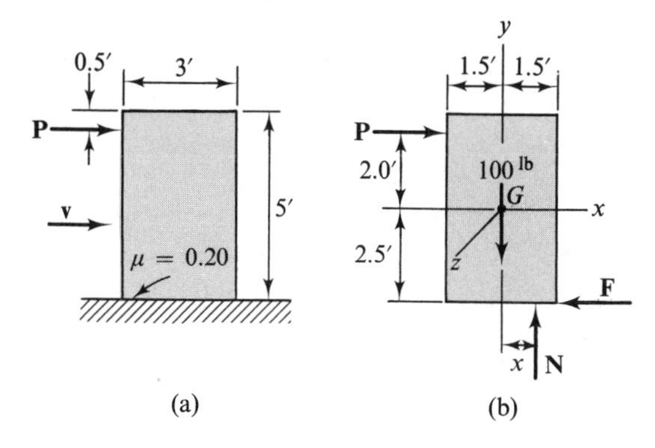

Figure 9-9

The scalar equations are

$$\mathbf{i}:\quad P-F=24.8, \tag{b}$$

$$\mathbf{j}:\quad N-100=0, \tag{c}$$

$$\mathbf{k}:\quad xN-2P-2.5F=0. \tag{d}$$

Equations (a)–(d) can be used to determine three unknown forces and one unknown distance; thus all unknown values can be determined. The value for N is determined from Eq. (c) and substituted into Eq. (a) and (b). The results are

$$\mathbf{N}=100\mathbf{j}\,\text{lb},\qquad \mathbf{F}=-20\mathbf{i}\,\text{lb},\qquad \mathbf{P}=\underline{44.8\mathbf{i}\,\text{lb}.}\quad \text{(Ans.)}$$

Substituting these values into Eq. (d), the value for x is found to be

$$x=1.397\text{ ft}.$$

Since x is less than 1.5 ft, tipping does not impend. If x were equal to 1.5 ft, the box would be on the verge of tipping. If x were more than 1.5 ft, the box would tip over before **P** increased to 44.8 lb, and the solution would not be valid.

EXAMPLE 9-7

The assembly in Fig. 9-10(a) has translation with the acceleration of the particles parallel to the x axis. The homogeneous 1600-lb body A is supported above body B by bearings C and D on a smooth rod. Cord EF, which is parallel to the xz plane, is subjected to a tensile force of 2000 lb. Determine the bearing reactions at C and D on body A and the acceleration of the assembly.

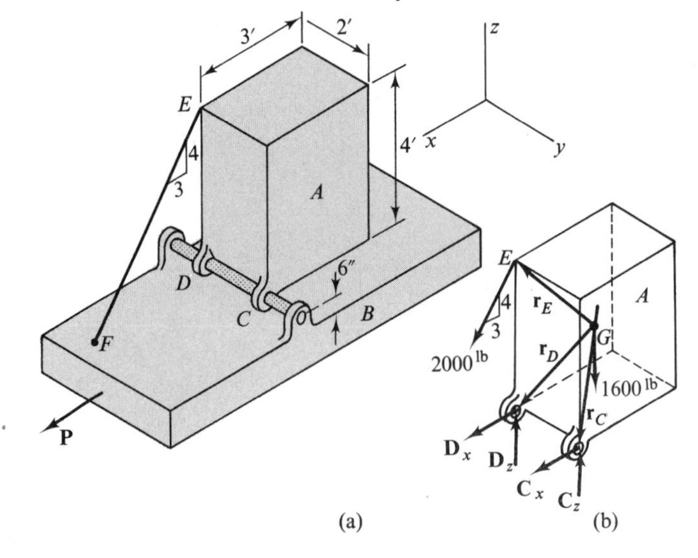

Figure 9-10

SOLUTION

The weight of A and the cord tension are given. The reactions at C and D on A and the acceleration of A are required.

Figure 9-10(b) is a free-body diagram of body A. Since the supporting rod is smooth, there are no components of force in the y direction at C and D. Body A has translation, but not all forces are in one plane; therefore Eq. (9-11) are required. The equations are

$$\sum \mathbf{F} = m\mathbf{a}, \qquad \sum \mathbf{M}_G = 0.$$

There are five unknown quantities on the free-body diagram and in the equation of motion, namely $\mathbf{C}_x$, $\mathbf{C}_z$, $\mathbf{D}_x$, $\mathbf{D}_z$, and $\mathbf{a} = a_x\mathbf{i}$.

The two vector equations are, in general, equivalent to six algebraic equations. Since there are no forces or accelerations in the y direction, however, the component equation $\sum \mathbf{F}_y = m\mathbf{a}_y$ will not contribute any useful information; thus only five equations are available to determine the five unknowns.

The force in the cable is parallel to the xz plane, and it can be written as

$$\mathbf{T} = 2000(0.6\mathbf{i} - 0.8\mathbf{k}) = (1200\mathbf{i} - 1600\mathbf{k})\text{ lb}.$$

The positive directions are indicated by the coordinate axes. The force equation gives

$$\sum \mathbf{F} = \mathbf{T} + \mathbf{C} + \mathbf{D} + \mathbf{W} = \text{m}(\text{a}_x)\mathbf{i},$$

which becomes

$$(1200\mathbf{i} - 1600\mathbf{k}) + (C_x\mathbf{i} + C_z\mathbf{k}) + (D_x\mathbf{i} + D_z\mathbf{k}) + (-1600\mathbf{k}) = \left(\frac{1600}{32.2}\right)(a_x\mathbf{i}).$$

In component form, the equation becomes

$$1200 + C_x + D_x = 49.7a_x \tag{a}$$

and

$$-1600 + C_z + D_z - 1600 = 0. \tag{b}$$

To write the moment equation of motion in vector form, the position vectors from G to points on the action lines of forces $\mathbf{C}$, $\mathbf{D}$, and $\mathbf{T}$ are required. Points C, D, and E are convenient and will be used. Since the block is homogeneous, G lies at the geometrical center of the block, and the position vectors are

$$\mathbf{r}_C = (1.5\mathbf{i} + \mathbf{j} - 2\mathbf{k})\text{ ft}, \qquad \mathbf{r}_D = (1.5\mathbf{i} - \mathbf{j} - 2\mathbf{k})\text{ ft},$$

$$\mathbf{r}_E = (1.5\mathbf{i} - \mathbf{j} + 2\mathbf{k})\text{ ft}.$$

The moment equation is

$$\sum \mathbf{M}_G = \mathbf{r}_C \times \mathbf{F}_C + \mathbf{r}_D \times \mathbf{F}_D + \mathbf{r}_E \times \mathbf{T} = 0,$$

or

$$(1.5\mathbf{i} + \mathbf{j} - 2\mathbf{k}) \times (C_x\mathbf{i} + C_z\mathbf{k}) + (1.5\mathbf{i} - \mathbf{j} - 2\mathbf{k}) + (D_x\mathbf{i} + D_z\mathbf{k})$$
$$+ (1.5\mathbf{i} - \mathbf{j} + 2\mathbf{k}) \times (1200\mathbf{i} - 1600\mathbf{k}) = 0,$$

which expands to

$$[C_z\mathbf{i}-(2C_x+1.5C_z)\mathbf{j}-C_x\mathbf{k}]+[-D_z\mathbf{i}-(2D_x+1.5D_z)\mathbf{j}+D_x\mathbf{k}]$$
$$+[1600\mathbf{i}+(2400+2400)\mathbf{j}+1200\mathbf{k}]=0.$$

Since the **i**, **j**, and **k** components must each be zero, this is equivalent to the following three scalar equations:

$$\mathbf{i}:\quad C_z-D_z+1600=0, \tag{c}$$

$$\mathbf{j}:\quad -2(C_x+D_x)-1.5(C_z+D_z)+4800=0, \tag{d}$$

$$\mathbf{k}:\quad -C_x+D_x+1200=0. \tag{e}$$

Equations (b) and (c) may be solved simultaneously for the values of C_z and D_z:

$$C_z=+800, \qquad D_z=+2400.$$

Since, from Eq. (b), $C_z+D_z=3200$, Eq. (d) may be written as

$$-2(C_x+D_x)-1.5(3200)+4800=0$$

or

$$C_x+D_x=0. \tag{f}$$

The simultaneous solution of Eq. (e) and (f) gives

$$C_x=+600, \qquad D_x=-600.$$

The bearing reactions on body A, are, therefore,

$$\mathbf{F}_C=(600\mathbf{i}+800\mathbf{k})\text{ lb}=\underline{1000(0.600\mathbf{i}+0.800\mathbf{k})\text{ lb}} \qquad \text{Ans.}$$

and

$$\mathbf{F}_D=(-600\mathbf{i}+2400\mathbf{k})\text{ lb}=\underline{2470(-0.243\mathbf{i}+0.970\mathbf{k})\text{ lb.}} \qquad \text{Ans.}$$

The acceleration can be found from Eq. (a) as

$$1200+600-600=49.7a_x \quad\text{or}\quad a_x=24.15$$

from which

$$\mathbf{a}=\underline{24.2\mathbf{i}\text{ fps}^2.} \qquad \text{Ans.}$$

PROBLEMS

9-50 The resultant of the force system acting on body E in Fig. P9-50 passes through the mass center G. The resultant moment of external forces about point H is found to be zero. What conclusion can be drawn from these facts concerning the motion of the body and the acceleration of point P?

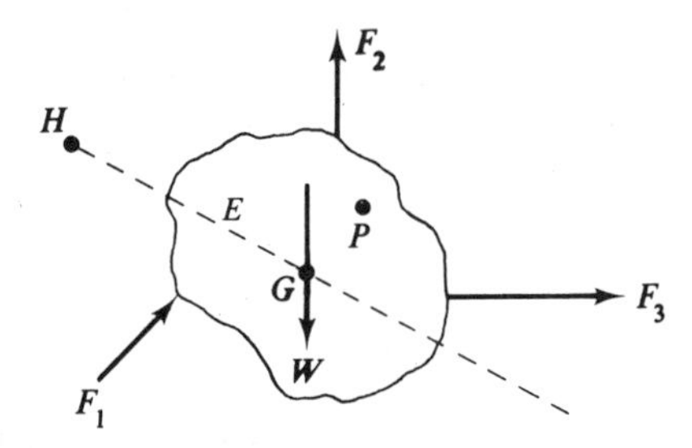

Figure P9-50

***9-51** The 90,000-kg jet airliner of Fig. P9-51 is being stopped by brake action only. The engines are at idle and producing negligible thrust, and the velocity of the aircraft is too slow to produce measurable aerodynamic lift or drag.

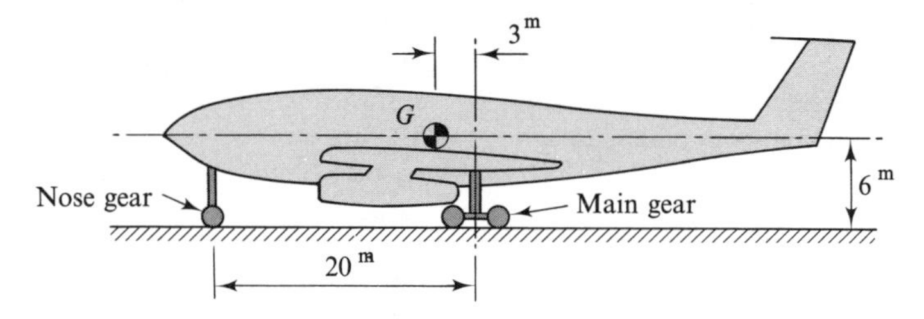

Figure P9-51

Only the main landing gear is equipped with brakes, and the coefficient of friction between the tires and runway is 0.7. If under these conditions the brakes are applied hard enough that the main wheels are on the verge of slipping, determine the vertical force in the nose wheel strut.

9-52 The uniform bar *AB* in Fig. P9-52 weighs 10.00 lb and is connected to two collars of negligible weight. The collars slide freely on parallel shafts which are in a vertical plane. Determine

(a) The acceleration of the bar.
(b) The reactions at *A* and *B*.

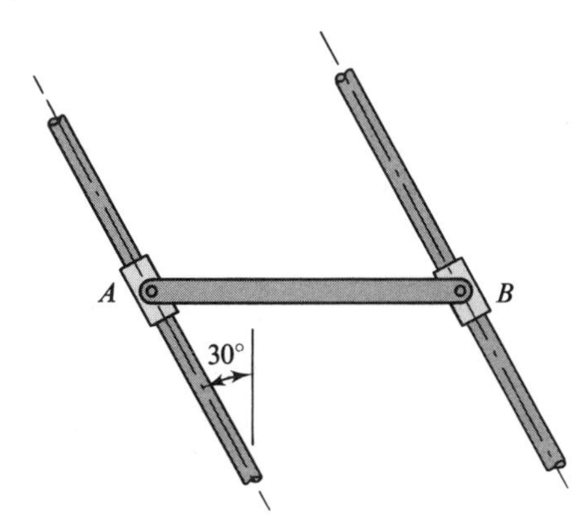

Figure P9-52

9-53 The homogeneous bar in Fig. P9-53 weighs 90.0 lb and slides on the smooth horizontal plane. It is connected to the 50.0-lb body *B* by a cord which passes over a smooth peg at *C*. Determine the force *P* which will result in translation of body *A*.

9-54 The crate in Fig. P9-54 weighs 100 lb and is lowered by three vertical cables connected at points *A*, *B*, and *C*. A constant tension of 30.0 lb is maintained in cable *A*. If the crate is to have translation, determine

(a) The tension required in the other two cables.
(b) The acceleration of the crate.

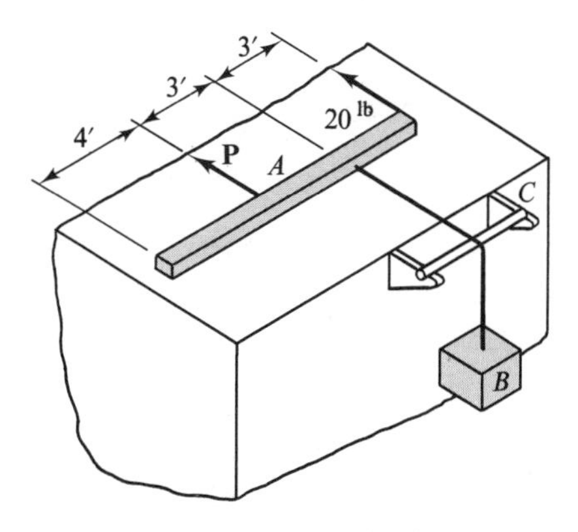

Figure P9-53

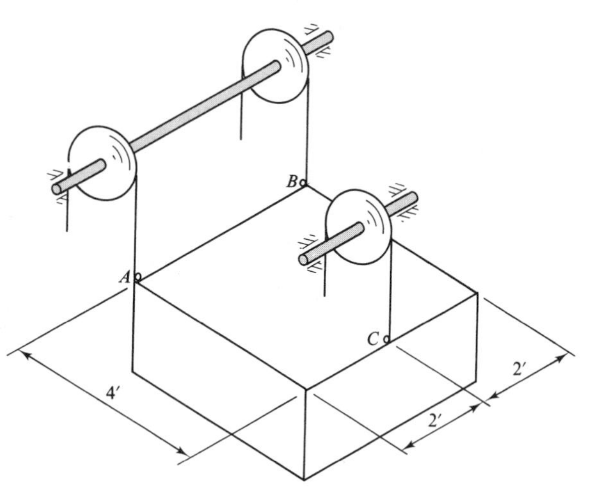

Figure P9-54

9-55 Determine the minimum accelerations of the blocks in Fig. P9-55 if A is not to move relative to B. The coefficient of friction between A and B is 0.20, and the horizontal plane is smooth.

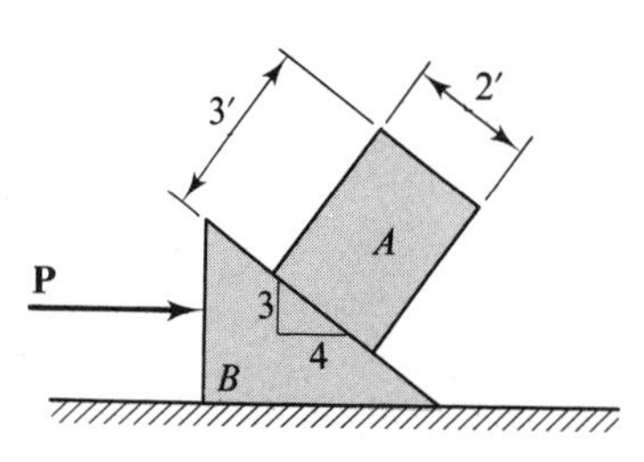

Figure P9-55

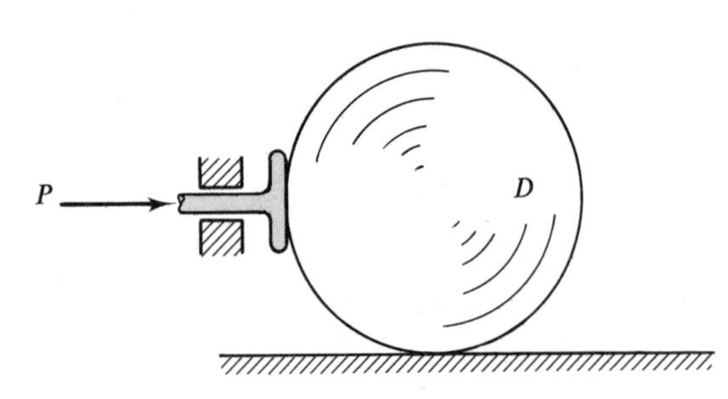

Figure P9-56

9-56 The homogeneous cylindrical drum D in Fig. P9-56 is to be pushed along the horizontal surface by the bumper B. The drum weighs 42.0 lb, and the coefficients of static and kinetic friction between the drum and all surfaces are, respectively, $\mu_s = 0.60$ and $\mu_k = 0.50$. The drum is to have translation. Determine

(a) The minimum force P required.

(b) The resulting acceleration of the drum.

9-57 Bar BC of Fig. P9-57 weighs 161 lb and has a radius of gyration of mass with respect to a horizontal axis through G of 3.0 ft. The bar is supported by flexible cords AB and CD which rotate in the same vertical plane. The speed of the mass center, when in the position shown, is 12 fps. Determine the tension in cable AB for this position.

9-58 The homogeneous body in Fig. P9-58 weighs 0.50 lb-per ft^3 and is supported by three 6-ft flexible cords attached to the body at A, B, and C. The body moves parallel to the yz plane and, in the position shown, has a velocity of 12.0 fps. Determine the tension in each of the cords and the acceleration of the body.

9-59 The 32.2-lb homogeneous rod BC is connected to the 96.6-lb frame D in Fig. P9-59 by the pin at C and the cord AB. Determine the maximum magnitude of P for which no relative motion will take place between BC and D.

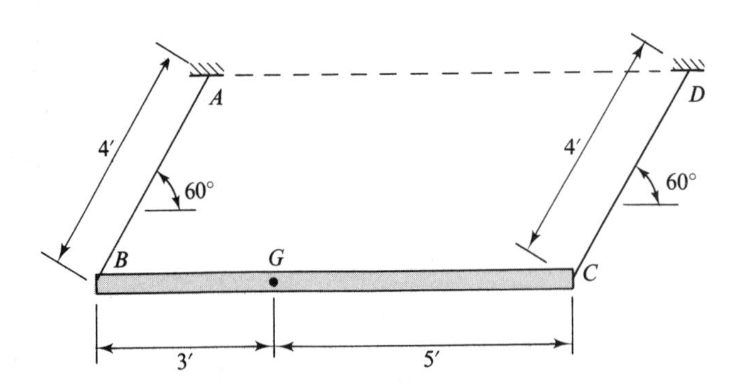

Figure P9-57

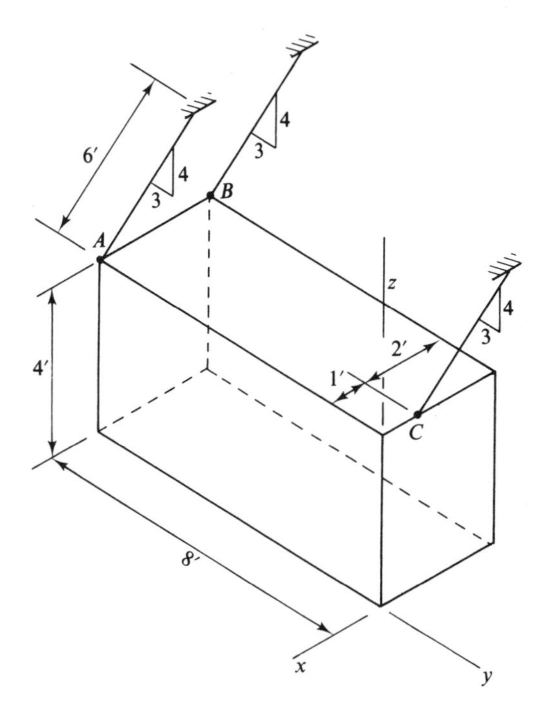

Figure P9-58

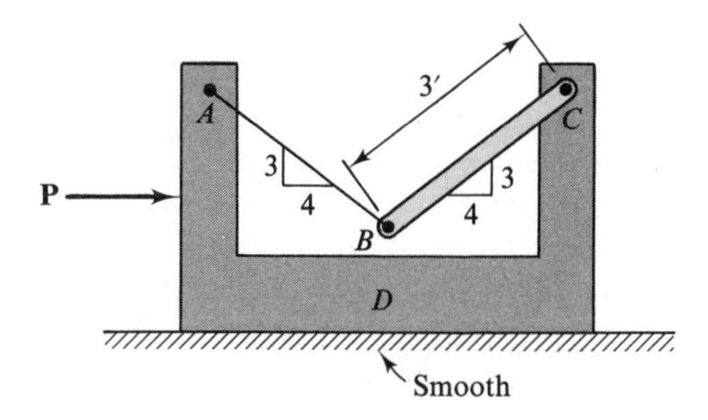

Figure P9-59

9-60 The sliding door in Fig. P9-60 weighs 900 lb. The mass and friction of the rollers may be considered negligible. Counterweight B weighs 100 lb and is connected to the door by a cable. Determine the acceleration of the door and the reactions of the track on the rollers.

9-61 Blocks A and B in Fig. P9-61 are connected by cable CD. Both blocks are symmetrical about the plane of motion, and cable CD and the force P are in the plane of motion.

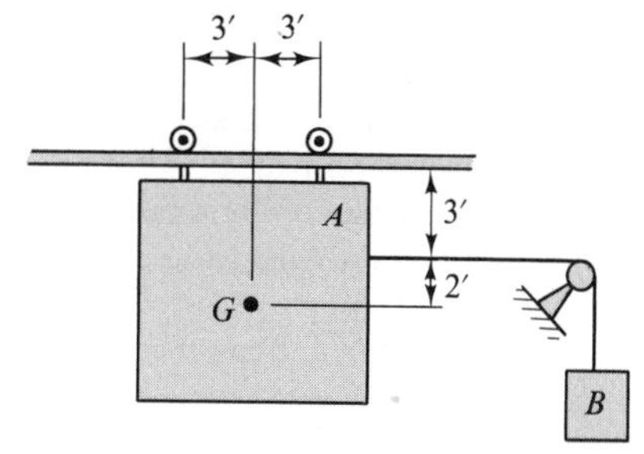

Figure P9-60

Block A weighs 6.0 lb, and block B weighs 5.0 lb. Both blocks are to move in translation. Determine the maximum allowable force P.

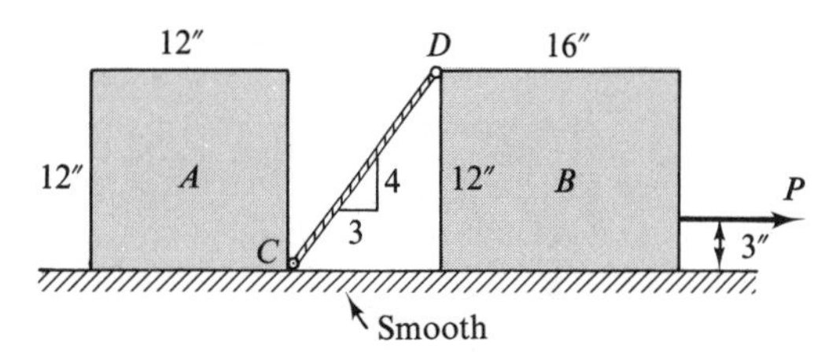

Figure P9-61

9-62 A long slender rod is connected to a block by a smooth pin at A in Fig. P9-62. The coefficient of friction between the block and the slot is 0.5. When the block is moving down the slot, determine the acceleration of the system and the corresponding angle θ between the rod and the vertical.

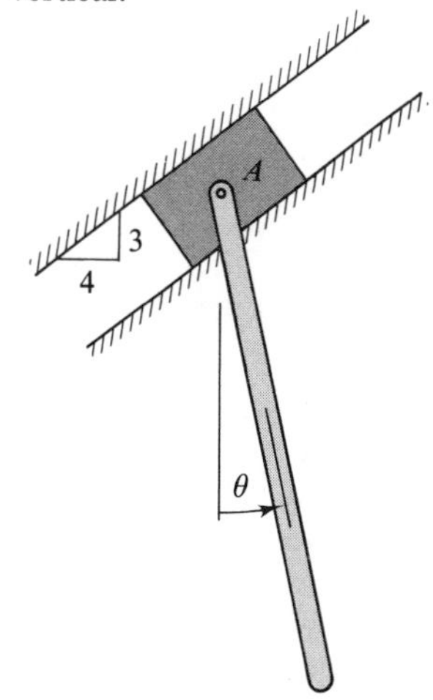

Figure P9-62

9-63 The uniform bars AB and CD in Fig. P9-63 weigh 64.4 and 128.8 lb, respectively. The 48.3-lb ball E is suspended by a cord as shown. The acceleration of F is gradually increased from zero to a constant value of 12 fps^2 to the right. For this acceleration, determine
(a) The angle θ.
(b) The components of the pin reaction at B on CD.

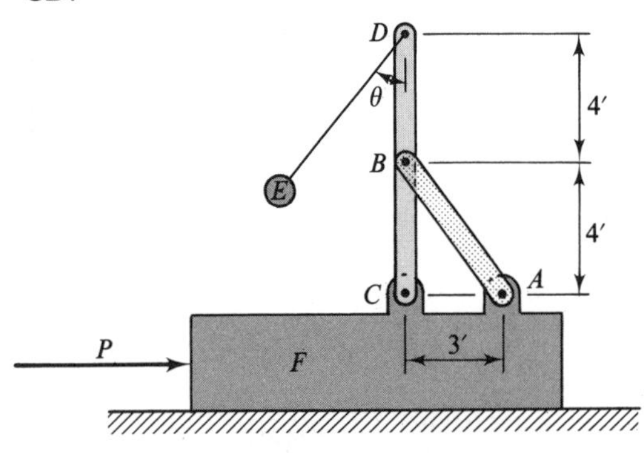

Figure P9-63

9-64 The homogeneous bar AB in Fig. P9-64 weighs 5 lb and is connected to the 25-lb block C by means of a smooth pin at B. Determine the components of the reaction at B on bar AB.

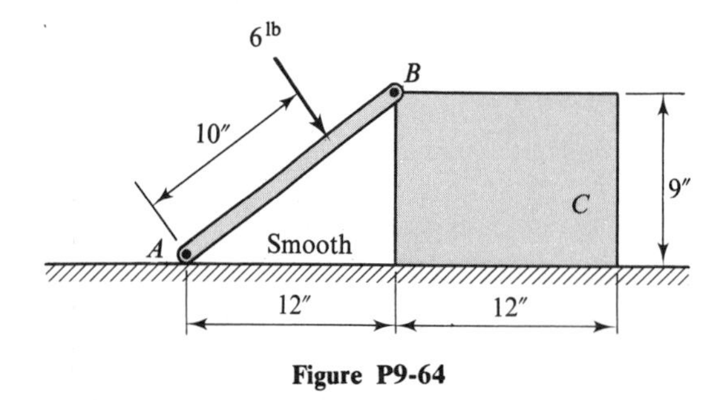

Figure P9-64

***9-65** Wheel C in Fig. P9-65 rolls along the plane without slipping. The homogeneous 8-kg bar AB is pinned to body C at A. The coefficient of friction between the bar at B and the plane is 0.6. The angular velocity and acceleration of wheel C at this instant are 4 rad per sec clockwise and 2 rad per sec^2 counterclockwise, respectively. Determine, for this instant, the components of the pin reaction at A on body AB.

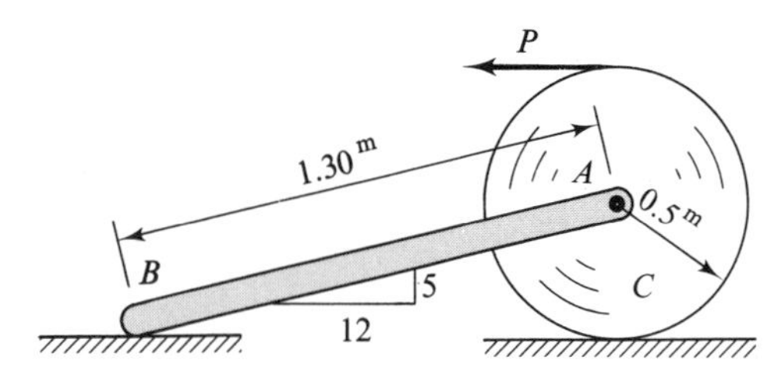

Figure P9-65

***9-66** The crate on the truck in Fig. P9-66 has a mass of m kg and is restrained from tipping by the cable. The coefficient of friction between the crate and truck is 0.30. Determine the maximum acceleration of the truck to the left if the crate is to remain upright. Would such an acceleration be reasonable?

***9-67** The truck in Problem 9-66 has a velocity of 50 km/hr to the left when the brakes are applied. Determine the shortest stopping distance if the crate is not to tip or slide into the truck cab.

9-68 The homogeneous wedge-shaped block in Fig. P9-68 has a mass of m units and is mounted on the frame E by means of smooth bearings at A and B and balls at C and D. A small part of the wedge is cut away at the bearings to eliminate axial thrust along the pins. The assembly is given an acceleration $a\mathbf{i}$ in the x direction. Determine the reactions of body E on the block.

9-69 The homogeneous bar AB in Fig. P9-69 is 25 in. long and weighs 32.2 lb. The carriage E weighs 60 lb and slides along a smooth horizontal plane. AB is connected to E by a ball and socket at B, a smooth surface at A, and a flexible cord from O to C. Point C is 10 in. from A (15 in. from B). The force $P = 30$ lb results in translation parallel to the xy plane. Determine all unknown forces acting on AB.

9-70 In Fig. P9-70 a 25-lb container of nitroglycerine is being moved in transporter B, which weighs 50 lb. The explosive is suspended inside the transporter by wires at C, D, and E and can be moved in translation only. The breaking strength of the wires is 90 lb. Determine the maximum force $\mathbf{P}$ that can be applied to the transporter.

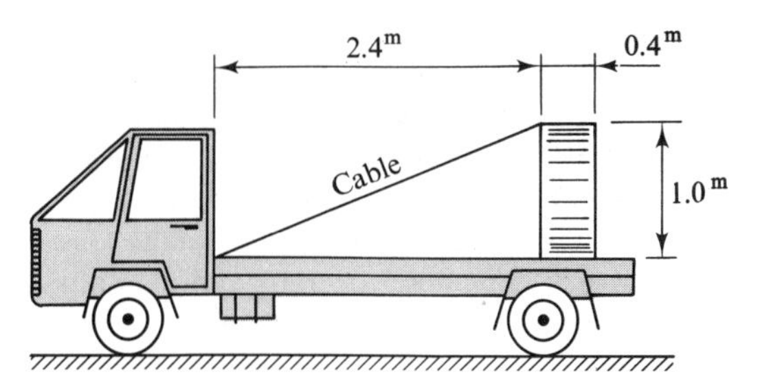

Figure P9-66

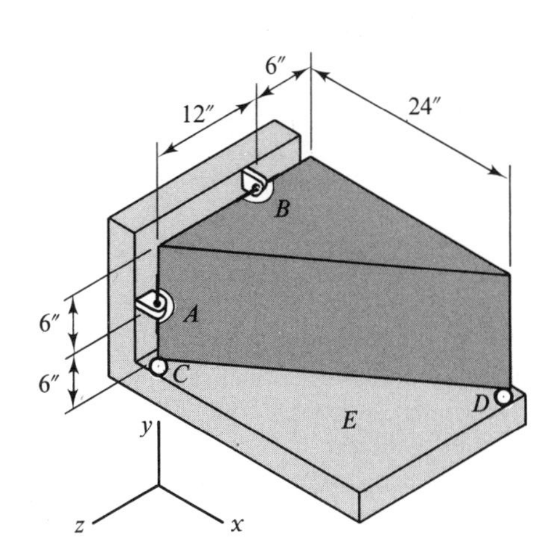

Figure P9-68

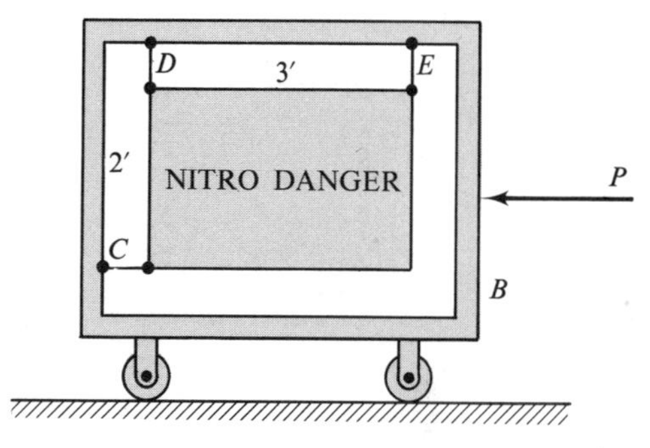

Figure P9-70

9-71 In Fig. P9-71, block A weighs 10 lb, and block B weighs 15 lb. The constant force P has the maximum possible magnitude without causing A to tip.

(a) Determine the force P.

(b) If B has an initial velocity of 32.2 fps to the left, determine its velocity after moving a total distance of 50 ft.

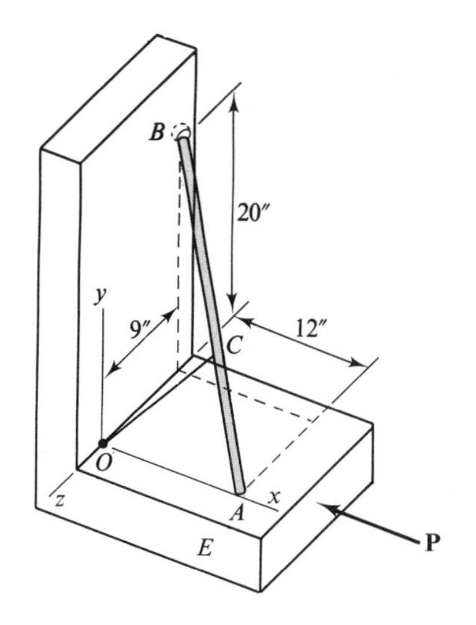

Figure P9-69

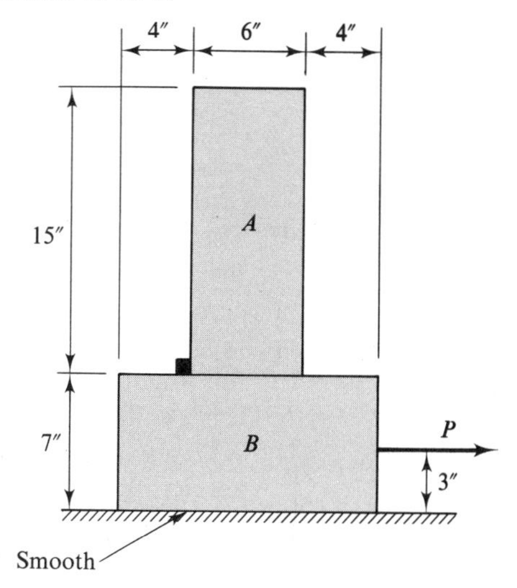

Figure P9-71

9-72 The 500-lb homogeneous block A in Fig. P9-72 is placed on a frame as shown. The support at C is smooth, and the coefficient of friction at D is 0.20. Members EF and GH rotate at a constant angular speed of 2 rad per sec. Determine the components of the reactions at C and D on A when the system is in the configuration shown.

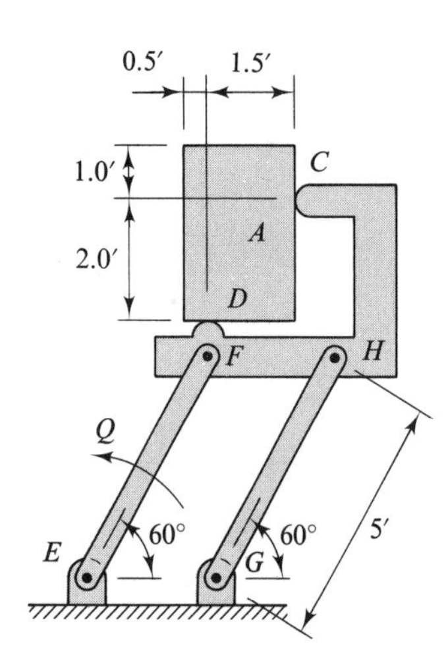

Figure P9-72

***9-73** Member BC in Fig. P9-73 has a mass of 4 kg, and A has a mass of 2 kg. The surface between A and BC is smooth. The system is held in equilibrium by the four cords. The angle θ is 30° when the horizontal cord at B is cut. Show that the tension in the cord D will become zero and determine the acceleration of A just after the cord is cut.

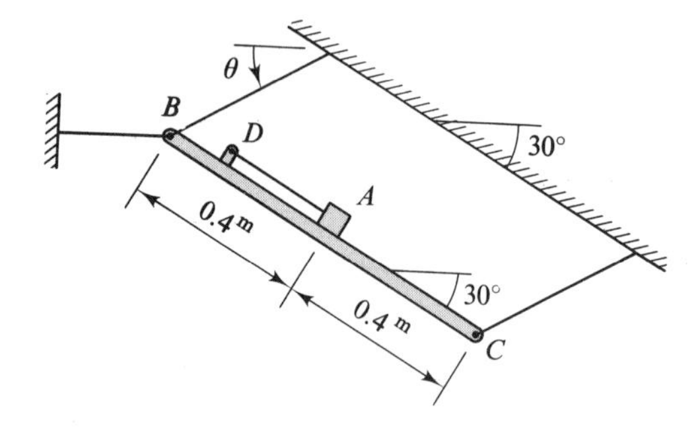

Figure P9-73

***9-74** Cord D in Problem 9-73 is removed, and the smooth surface between A and BC is replaced by a rough one. Determine the minimum coefficient of friction if A is not to slip on BC either before or after the horizontal cord is cut.

9-10 ROTATION OF A RIGID BODY

When a rigid body has a motion of rotation, all particles of the body travel in circular paths about the fixed axis of rotation. Figure 9-11(a) shows a rigid body rotating about the fixed Z axis. The XYZ coordinate system is fixed in space with its origin O at the intersection of the axis of rotation and the plane in which the mass center moves. The xyz coordinate system also has its origin at point O and is fixed in the body and rotates with it. The z and Z axes are collinear, and for convenience the x and y axes are so oriented that the x axis passes through G, the mass center of the body. The angular velocity $\boldsymbol{\omega}$ and the angular acceleration $\boldsymbol{\alpha}$ will lie along the z axis, and $\boldsymbol{\omega} = \omega\mathbf{k}$ and $\boldsymbol{\alpha} = \alpha\mathbf{k}$. The mass center travels on a circular path of radius r_G in the XY plane with a normal acceleration of $-r_G\omega^2\mathbf{i}$ and a tangential acceleration of $r_G\alpha\mathbf{j}$. Equation (9-10), when applied to a rotating body, becomes

$$\mathbf{R} = \sum \mathbf{F} = m\mathbf{a}_G = m(a_{Gx}\mathbf{i} + a_{Gy}\mathbf{j}) \qquad \text{(9-13a)}$$

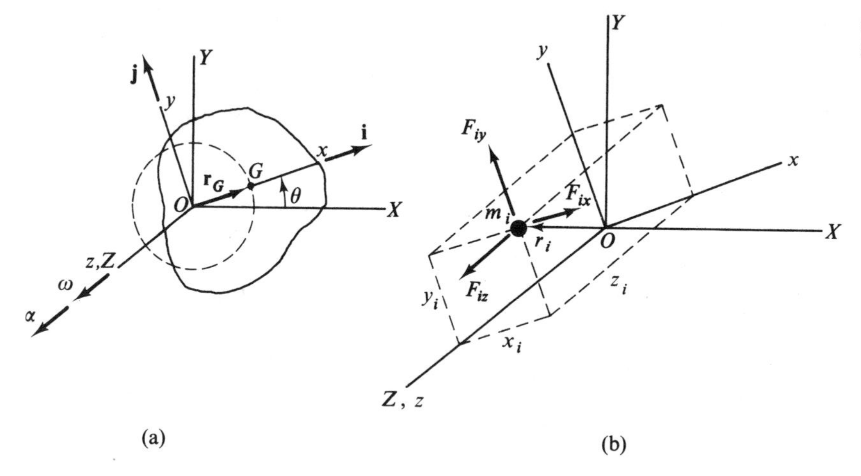

Figure 9-11

or in scalar form

$$
\begin{aligned}
\mathbf{i}:&\quad \sum F_x = ma_{Gx} = m(-r_G\omega^2),\\
\mathbf{j}:&\quad \sum F_y = ma_{Gy} = m(r_G\alpha),\\
\mathbf{k}:&\quad \sum F_z = 0.
\end{aligned}
\qquad (9\text{-}13b)
$$

Equations (9-13a) and (9-13b) were derived with the x axis through G. Sometimes, particularly for composite bodies, it is not convenient or necessary to locate the mass center of the body and place the x axis through it. The only requirement is that the directions of the components of $\mathbf{a}_G$ (or of $m\mathbf{a}_G$) for each component part of the body be consistent with the axes used to express the components of the forces.

The moment equation of motion for rotation can be developed by consideration of a particle of the rotating body as shown in Fig. 9-11(b). The position of the particle m_i relative to the x, y, z axes is $\mathbf{r}_i = x_i\mathbf{i} + y_i\mathbf{j} + z_i\mathbf{k}$. Since m_i is a particle of the rigid body, its acceleration is, from Eq. (8-22),

$$
\begin{aligned}
\mathbf{a}_i &= \boldsymbol{\alpha}\times\mathbf{r}_i + \boldsymbol{\omega}\times(\boldsymbol{\omega}\times\mathbf{r}_i)\\
&= (\alpha\mathbf{k})\times(x_i\mathbf{i} + y_i\mathbf{j} + z_i\mathbf{k}) + (\omega\mathbf{k})\times[\omega\mathbf{k}\times(x_i\mathbf{i} + y_i\mathbf{j} + z_i\mathbf{k})]\\
&= -(x_i\omega^2 + y_i\alpha)\mathbf{i} + (x_i\alpha - y_i\omega^2)\mathbf{j}.
\end{aligned}
$$

From Newton's second law, the resultant force (both internal and external) acting on the particle is

$$\mathbf{R}_i = m_i\mathbf{a}_i,$$

and the moment of the resultant force about point O is

$$\mathbf{M}_{Oi} = \mathbf{r}_i \times \mathbf{R}_i = \mathbf{r}_i \times m_i \mathbf{a}_i.$$

When $\mathbf{r}_i$ and $\mathbf{a}_i$ are expressed in component form the moment becomes

$$\mathbf{M}_{Oi} = (x_i\mathbf{i} + y_i\mathbf{j} + z_i\mathbf{k}) \times m_i[-(x_i\omega^2 + y_i\alpha)\mathbf{i} + (x_i\alpha - y_i\omega^2)\mathbf{j}]$$

$$= \mathbf{i}m_i(y_iz_i\omega^2 - x_iz_i\alpha) - \mathbf{j}m_i(x_iz_i\omega^2 + y_iz_i\alpha) + \mathbf{k}m_i(x_i^2 + y_i^2)\alpha.$$

When the moments about O of all the forces applied to all the particles of a body are added together, the moments of the internal forces are eliminated since they always occur in balancing pairs, and thus the sum of the moments about O of the external forces becomes

$$\sum \mathbf{M}_O = \sum_{i=1}^{n} \mathbf{M}_{Oi} = \mathbf{i} \sum_{i=1}^{n} m_i(y_iz_i\omega^2 - x_iz_i\alpha)$$

$$-\mathbf{j} \sum_{i=1}^{n} m_i(x_iz_i\omega^2 + y_iz_i\alpha) + \mathbf{k} \sum_{i=1}^{n} m_i(x_i^2 + y_i^2)\alpha.$$

For a rigid body with a continuous mass distribution, the particle mass m_i can be replaced by the mass dm of a small element of the body and the summation $\sum_{i=1}^{n}$ can be replaced by the integration $\int_m$, and the moment becomes

$$\sum \mathbf{M}_O = \int_m d\mathbf{M}_O = \mathbf{i}\left(\omega^2 \int_m yz\,dm - \alpha \int_m xz\,dm\right)$$

$$-\mathbf{j}\left(\omega^2 \int_m xz\,dm + \alpha \int_m yz\,dm\right) + \mathbf{k}\alpha \int_m (x^2 + y^2)\,dm.$$

The integrals represent products and moments of inertia of the mass of the body with respect to the xyz coordinate system which is fixed in the body. Using the notation developed in Chapter 6, the moment equation becomes

$$\sum \mathbf{M}_O = (\omega^2 I_{yz} - \alpha I_{xz})\mathbf{i} - (\omega^2 I_{xz} + \alpha I_{yz})\mathbf{j} + \alpha I_z \mathbf{k}, \qquad \text{(9-14a)}$$

which is equivalent to

$$\begin{aligned} \mathbf{i}:&\quad \sum \mathbf{M}_{Ox} = (\omega^2 I_{yz} - \alpha I_{xz}), \\ \mathbf{j}:&\quad \sum \mathbf{M}_{Oy} = -(\omega^2 I_{xz} + \alpha I_{yz}), \qquad \text{(9-14b)} \\ \mathbf{k}:&\quad \sum \mathbf{M}_{OZ} = \alpha I_z. \end{aligned}$$

Figure 9-11(a) was drawn with the x axis through the mass center. This limitation, however, was not involved in deriving Eq. (9-14), and the only restrictions on the coordinate system are that it be a right-hand system and that the z axis be the axis of rotation.

If the axis of rotation, the z axis, is parallel to an axis of revolution of a homogeneous body (the axis of revolution will pass through G) and if the xy plane is the plane of motion of the body, the products of inertia I_{xz} and I_{yz} will be zero. They are zero because $I_{xz'}$ and $I_{y'z'}$—where y' and z' are axes through G parallel to y and z, respectively—are zero due to symmetry, and the transfer terms are zero because z_G is zero.

If the xz plane is a plane of symmetry, $I_{yz}=0$ as shown in Chapter 6, and if the yz plane is a plane of symmetry, $I_{xz}=0$.

If the force system and body are both symmetrical about the xy plane, the equations for $\sum M_{Ox}$, $\sum M_{Oy}$, and $\sum F_z$ provide no useful information (both sides of the equations vanish), and Eq. (9-13b) and (9-14b) become

$$\sum F_x = ma_{Gx},\quad \sum F_y = ma_{Gy},\quad \sum M_{Oz} = I_z\alpha. \tag{9-15}$$

When using Eq. (9-13), (9-14), and (9-15), a consistent sign convention is essential. In all derivations, the positive directions were the same as the positive directions of the coordinate axes. The vector representations for positive $\boldsymbol{\omega}$ and $\boldsymbol{\alpha}$ have the same direction as the z axis.

Because the normal and tangential components of the acceleration of the mass center of a rotating rigid body are generally the most convenient components to determine, it is usually easier to determine the normal and tangential components of the pin or axle reaction on the body than to determine the horizontal and vertical components. It is often convenient to show normal and tangential components of the pin reaction on the free-body diagram. In the preceding derivation, the x axis is the normal direction for the acceleration of the mass center, and the y axis is the tangential direction.

In Art. 9-8 it was shown that, for a body with translation, the resultant external force acts through the mass center of the body. When a rigid body has rotation, however, the resultant external force does not pass through the mass center. If the mass center is on the axis of rotation, the resultant of the external force system is a couple because $\mathbf{R}$ is zero from Eq. (9-13) and $\mathbf{M}$ is different from zero in Eq. (9-14) unless α and ω are zero.

The location of a point on the action line of the resultant external force, when it is a force, can be determined in the following manner. For simplicity, consider a body which is symmetrical about the plane of motion (the xy plane) and assume that the forces are also symmetrical with respect to this plane. Figure 9-12 represents the plane of motion of the mass center of the body, and the components of the resultant external force (including the reaction at O) are shown as $\mathbf{R}_x$ and $\mathbf{R}_y$ at point P. The moment of the resultant force about O is

$$\mathbf{M}_O = (q\mathbf{i})\times(R_x\mathbf{i}+R_y\mathbf{j}) = qR_y\mathbf{k}.$$

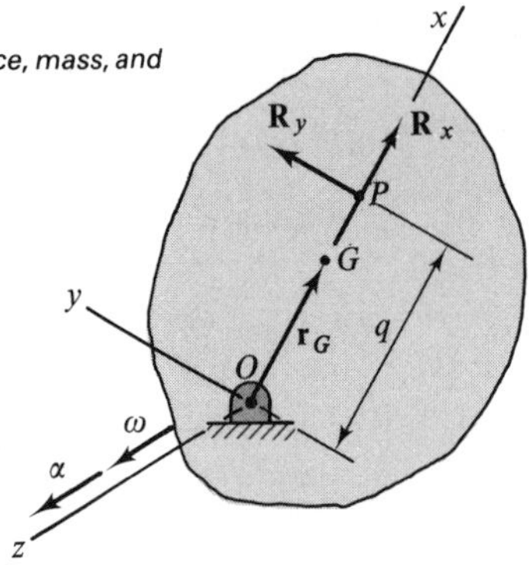

Figure 9-12

From Eq. (9-13b), $R_y = mr_G\alpha$, and from Eq. (9-14b), $M_O = \alpha I_z$; therefore,

$$\alpha I_z = qmr_G\alpha$$

and

$$q = \frac{I_z}{mr_G}.$$

The parallel-axis theorem for I_z gives

$$I_z = (I_G)_z + mr_G^2 = mk_G^2 + mr_G^2,$$

where k_G is the radius of gyration of mass of the body about an axis through G parallel to the z axis. When this expression is substituted into the equation for q it becomes

$$q = \frac{mk_G^2 + mr_G^2}{mr_G} = \frac{k_G^2}{r_G} + r_G.$$

This equation shows that q is always more than r_G.

The point P at which the resultant of the external forces acting on a rotating rigid body (symmetrical about the *xy* plane) *intersects the x axis is called the center of percussion of the body.* Since the moment of the external forces about the center of percussion is zero, a force applied to the body through the center of percussion will not affect the tangential component of the reaction of the pin or axle at the axis of rotation. With ballistic pendulums, impact-testing machines, and similar apparatus, the load is usually applied to the body through the center of percussion to eliminate the bearing reaction due to the applied load. In baseball, the bat will sting the batter's hands unless the ball is hit at the center of percussion of the swinging bat.

EXAMPLE 9-8

The unbalanced wheel A in Fig. 9-13(a) weighs 64.4 lb and has a radius of gyration of mass relative to the horizontal axis of rotation

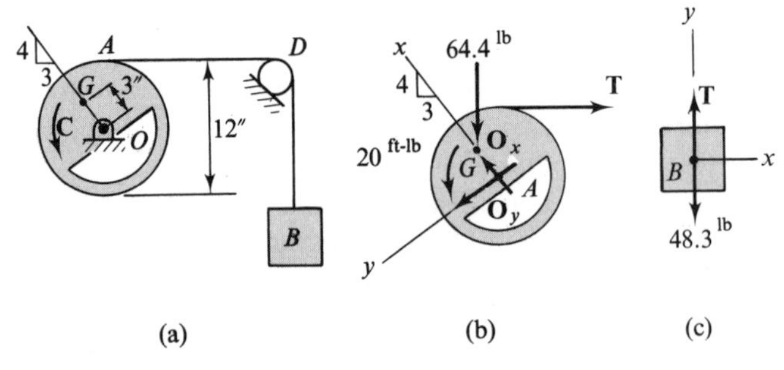

Figure 9-13

through O of 0.40 ft. The wheel is symmetrical about the plane of motion. The block B weighs 48.3 lb and is fastened to A by an inextensible cord which passes over the smooth drum at D and is wrapped around A. The moment of the couple **C** applied to A is 20.0 ft-lb counterclockwise. When A is in the position indicated, it has an angular velocity of 6.0 rad per sec clockwise. Determine the normal and tangential components of the reaction at O on body A. Neglect bearing friction at O.

SOLUTION

The free-body diagram of body A is shown in Fig. 9-13(b). The reaction at O is resolved into normal and tangential components along the x and y axes, respectively. Since the body is symmetrical about the plane of motion, Eq. (9-15) applies, and the three equations of motion for body A are

$$\sum F_x = ma_{Gx} = -mr_G\omega^2,$$

$$O_x - 0.60T - 0.80(64.4) = -2.0(\tfrac{3}{12})(-6.0)^2; \tag{a}$$

$$\sum F_y = ma_{Gy} = mr_G\alpha,$$

$$O_y - 0.80T + 0.60(64.4) = 2.0(\tfrac{3}{12})\alpha; \tag{b}$$

$$\sum M_{Oz} = I_z\alpha = mk_z^2\alpha,$$

$$20.0 - \tfrac{6}{12}T + 0.60(\tfrac{3}{12})(64.4) = 2.0(0.40)^2\alpha. \tag{c}$$

The three equations of motion for body A contain four unknowns, O_x, O_y, T, and α. An additional equation can be obtained from the free-body diagram of body B in Fig. 9-13(c). The equation of motion in the y direction is

$$\sum F_y = ma_y,$$

$$T - 48.3 = 1.5a_{By}. \tag{d}$$

This equation contains no additional unknown forces, and the unknown acceleration a_{By} can be expressed in terms of the angular acceleration of A. Noting the direction of the positive y axis for body B and the direction of $\boldsymbol{\alpha}$ for body A, this kinematical relationship becomes

$$a_{By} = \tfrac{6}{12}\alpha. \tag{e}$$

There are now five equations and five unknowns. When Eq. (e) is substituted into Eq. (d), Eq. (d) and (c) can be solved simultaneously for T, which is

$$T = 54.2 \text{ lb},$$

and for α, which is

$$\alpha = 7.93 \text{ rad per sec}^2.$$

These results can then be substituted into the first two equations to determine the components of the reactions. The results are

$$\mathbf{O}_x = \underline{66.1\mathbf{i}\text{ lb}} \qquad \text{Ans.}$$

and

$$\mathbf{O}_y = \underline{8.7\mathbf{j}\text{ lb.}} \qquad \text{Ans.}$$

EXAMPLE 9-9

The 36-lb thin, homogeneous, right triangular plate A in Fig. 9-14(a) rotates about the z axis with an angular velocity of $4\mathbf{k}$ rad per sec and an angular acceleration of $6\mathbf{k}$ rad per sec^2 in the position shown. The plate is mounted on a smooth, horizontal circular rod and supported by smooth bearings at B and C. Determine the torque $\mathbf{T}$ applied to the rod and the bearing reactions on the rod when the plate is in the given position. Neglect the weight of the circular rod.

SOLUTION

The angular velocity and acceleration of the plate, together with its weight and dimensions, are given. The couple $\mathbf{T}$ and the bearing reactions are required.

A free-body diagram of the plate and rod is shown in Fig. 9-14(b). Since the bearing rod is smooth, there are no z components of force at B and C.

Body A has rotation, but neither the body nor the force system is symmetrical with respect to the plane of motion (the xy plane), and although Eq. (9-15) are valid for this problem, they do not provide enough information to solve the problem. The general force and moment equations, Eq. (9-13) and (9-14), can be used to solve this problem. The equations are

$$\sum \mathbf{F} = m\mathbf{a}_G$$

and

$$\sum \mathbf{M}_O = (\omega^2 I_{yz} - \alpha I_{xz})\mathbf{i} - (\omega^2 I_{xz} + \alpha I_{yz})\mathbf{j} + \alpha I_z \mathbf{k}.$$

These equations involve four unknown force components and an unknown moment $\mathbf{T}$. The acceleration of G can be obtained from the given data and the principles of kinematics. The moment and products

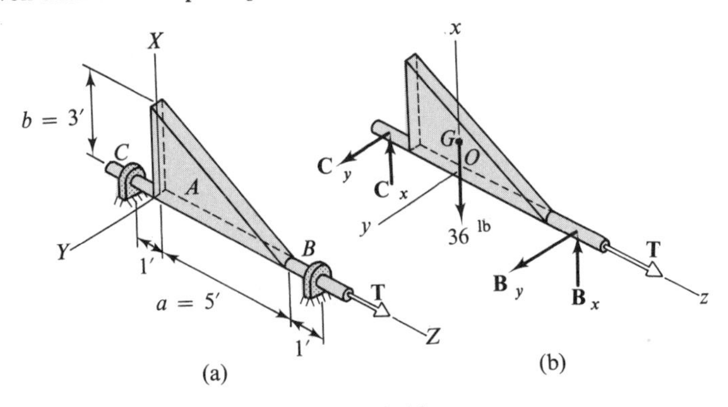

Figure 9-14

of inertia can be determined from the data, either by integration or from the table on p. 296 or Appendix B.

The two vector equations of motion are equivalent to six component equations. As previously noted, however, there are no forces in the z direction, and the force equation

$$\sum F_z = ma_{Gz}$$

will not provide any useful information; therefore, the two general motion equations are equivalent to only five scalar equations for this problem. Since there are only five unknown force and couple components, the analysis is complete.

The values of the moments and products of inertia are obtained from the parallel-axis theorem. Thus (see Chapter 6 or Appendix B)

$$\begin{aligned} I_z &= I_{zG} + mx_G^2 = m(2b^2+3c^2)/36 + md^2 \\ &= \frac{36}{32.2}\left[\frac{2(3)^2+3(0)^2}{36}\right] + \frac{36}{32.2}(1)^2 \\ &= 1.677 \text{ slug-ft}^2, \end{aligned}$$

$$\begin{aligned} I_{xz} &= (I_{xz})_G + mx_G z_G = -mab/36 + m\left(\frac{a}{3}\right)\left(\frac{c}{2}\right) \\ &= -\frac{36}{32.2}\left[\frac{5(3)}{36}\right] + 0 \\ &= -0.466 \text{ slug-ft}^2. \end{aligned}$$

Since the xz plane is a plane of symmetry, I_{yz} is zero.

The acceleration of G is

$$\begin{aligned} \mathbf{a}_G &= -r_G\omega^2\mathbf{i} + r_G\alpha\mathbf{j} = -(1)(4)^2\mathbf{i} + (1)(6)\mathbf{j} \\ &= (-16\mathbf{i} + 6\mathbf{j}) \text{ fps}^2. \end{aligned}$$

Equation (9-13a) becomes

$$\sum \mathbf{F} = m\mathbf{a}_G$$

$$B_x\mathbf{i} + B_y\mathbf{j} + C_x\mathbf{i} + C_y\mathbf{j} - 36\mathbf{i} = \frac{36}{32.2}(-16\mathbf{i} + 6\mathbf{j}),$$

which is equivalent to

$$\mathbf{i}: \quad B_x + C_x = 36 - \frac{36}{32.2}(16) = 18.11 \qquad \text{(a)}$$

and

$$\mathbf{j}: \quad B_y + C_y = 6.71. \qquad \text{(b)}$$

Equation (9-14a) for moments becomes

$$\sum \mathbf{M}_O = \mathbf{r}_B \times \mathbf{F}_B + \mathbf{r}_C \times \mathbf{F}_C + \mathbf{r}_G \times \mathbf{W} + \mathbf{T} = -\alpha I_{xz}\mathbf{i} - \omega^2 I_{xz}\mathbf{j} + \alpha I_z\mathbf{k},$$

or

$$\begin{aligned} &(4.33\mathbf{k}) \times (B_x\mathbf{i} + B_y\mathbf{j}) + (-2.67\mathbf{k}) \times (C_x\mathbf{i} + C_y\mathbf{j}) + (1\mathbf{i}) \times (-36\mathbf{i}) + T\mathbf{k} \\ &\qquad = -6(-0.466)\mathbf{i} - (4)^2(-0.466)\mathbf{j} + 6(1.677)\mathbf{k}, \end{aligned}$$

which, when expanded, is equivalent to

$$\mathbf{i}: \quad -4.33B_y + 2.67C_y = 2.80, \tag{c}$$

$$\mathbf{j}: \quad 4.33B_x - 2.67C_x = 7.46, \tag{d}$$

$$\mathbf{k}: \quad T = 10.06. \tag{e}$$

Equations (a) and (d) can be solved simultaneously for B_x and C_x, Eq. (b) and (c) can be solved for B_y and C_y, and Eq. (e) gives T. The components are

$$B_x = 7.97 \text{ lb}, \qquad B_y = 2.16 \text{ lb},$$

$$C_x = 10.15 \text{ lb}, \qquad C_y = 4.55 \text{ lb},$$

and

$$T = 10.06 \text{ ft-lb}.$$

The couple and bearing reactions are

$$\mathbf{T} = \underline{10.06\mathbf{k} \text{ ft-lb}}, \qquad \text{Ans.}$$

$$\mathbf{F}_B = 7.97\mathbf{i} + 2.16\mathbf{j} = \underline{8.25(0.966\mathbf{i} + 0.262\mathbf{j}) \text{ lb}}, \qquad \text{Ans.}$$

and

$$\mathbf{F}_C = 10.15\mathbf{i} + 4.55\mathbf{j} = \underline{11.12(0.913\mathbf{i} + 0.409\mathbf{j}) \text{ lb}}. \qquad \text{Ans.}$$

PROBLEMS

9-75 The slender, homogeneous rod in Fig. P9-75 is 3 ft long and weighs 2 lb. In the configuration shown, the rod rotates about point A with a counterclockwise angular velocity of 5 rad per sec. Determine
(a) The angular acceleration of the rod.
(b) The normal and tangential components of the reaction at A.

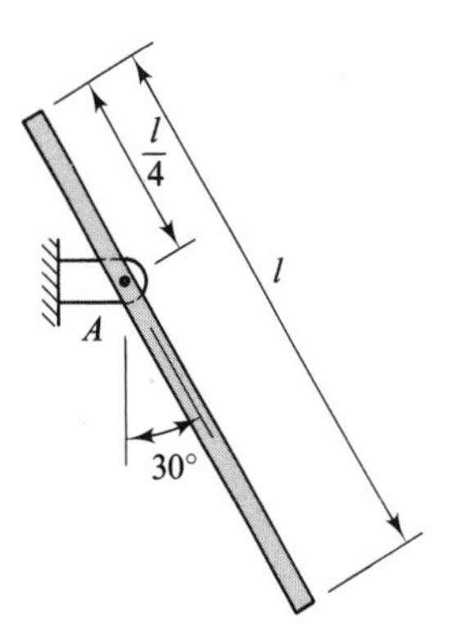

Figure P9-75

9-76 In Fig. P9-76, a 50-lb homogeneous drum is at rest against a small step at A. In an effort to push the drum over the step, a force of 20 lb is applied as shown. Determine
(a) The initial angular acceleration of the drum.
(b) The corresponding normal and tangential components of the reaction at A.

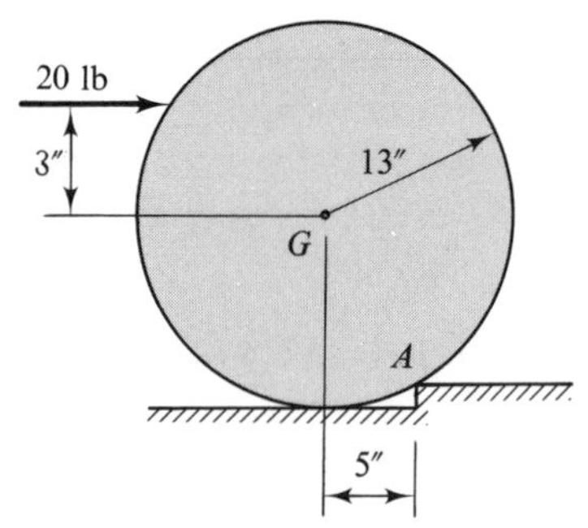

Figure P9-76

***9-77** The solid homogeneous cylinder in Fig. P9-77 has a mass of 30 kg and is rotating at 1200 rpm clockwise about a fixed horizontal axis through O. The coefficient of kinetic friction between the brake and the cylinder is 0.20. If the tension in the spring when the brake is applied is 100 N, determine the time required for the cylinder to stop rotating. Neglect the thickness of the vertical members.

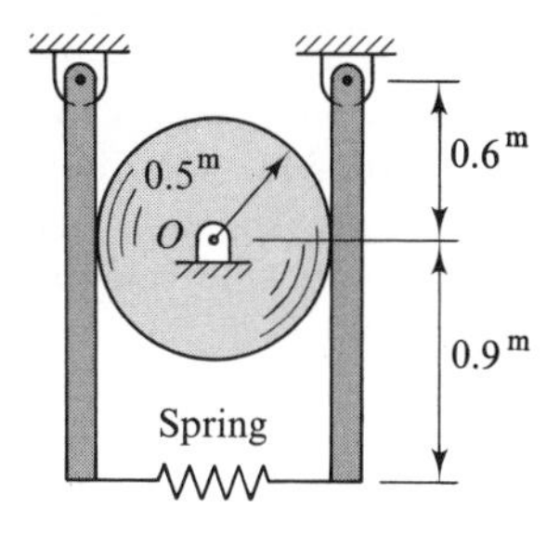

Figure P9-77

9-78 Body A in Fig. P9-78 weighs 32.2 lb and is rotating about a horizontal axis at O with an angular velocity of 3.0 rad per sec counterclockwise when in the position indicated. The coefficient of friction between the brake drum and the brake BCD is 0.50, and body A has a radius of gyration of mass with respect to a horizontal axis through G of 6.0 in. Assume symmetry about the plane of motion. Determine the angular acceleration of A.

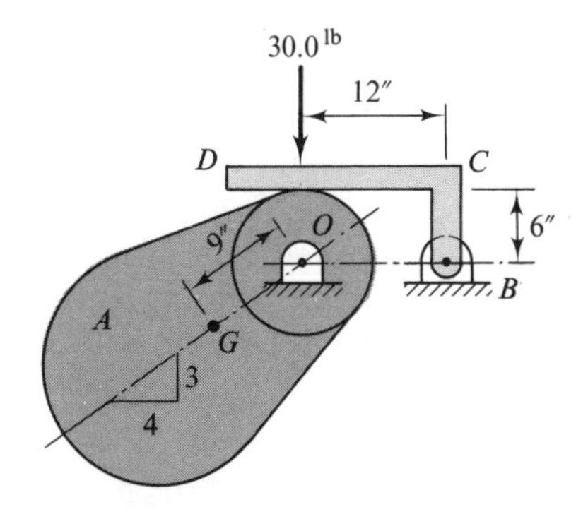

Figure P9-78

9-79 The chain-drive system in Fig. P9-79 consists of a 4-lb sprocket A and a 1-lb sprocket B. Sprocket A is driven by a 30-in-lb torque as indicated. Determine the angular acceleration of sprocket B.

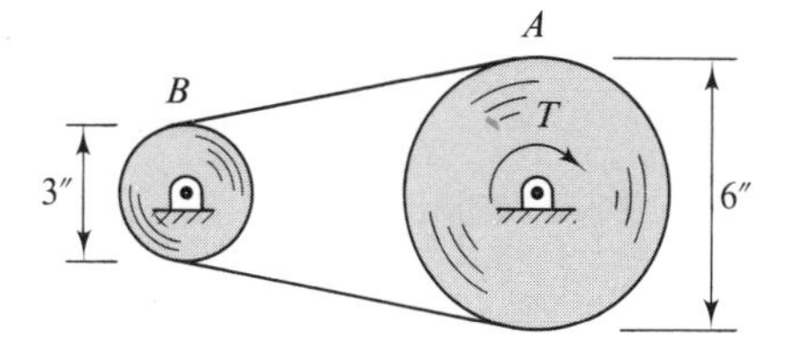

Figure P9-79

9-80 The pinion, gear, and flywheel in Fig. P9-80 have moments of inertia about their axes of rotation of 0.030, 0.050, and 0.750 slug-ft^2, respectively. A torque of 120 ft-lb is applied to the pinion in the direction indicated. Determine the angular acceleration of the flywheel. Neglect bearing friction.

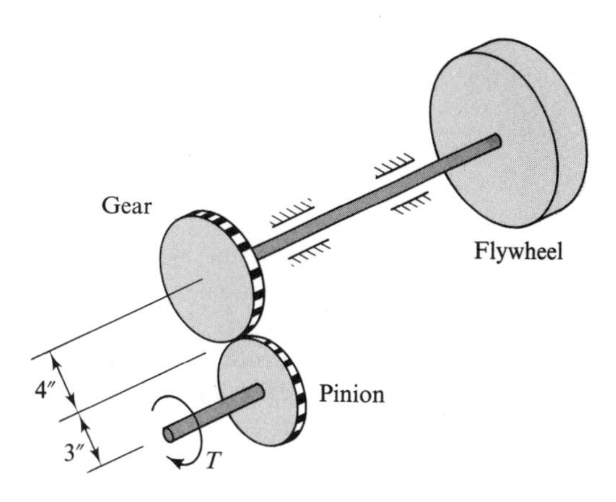

Figure P9-80

***9-81** The 5-kg block in Fig. P9-81 slides on a smooth horizontal surface. The pulley has a mass moment of inertia of 0.5 kg · m^2 about its axis of rotation, and there is no slipping between the pulley and the cable. Determine the angular acceleration of the pulley and the forces exerted by the cable on the blocks.

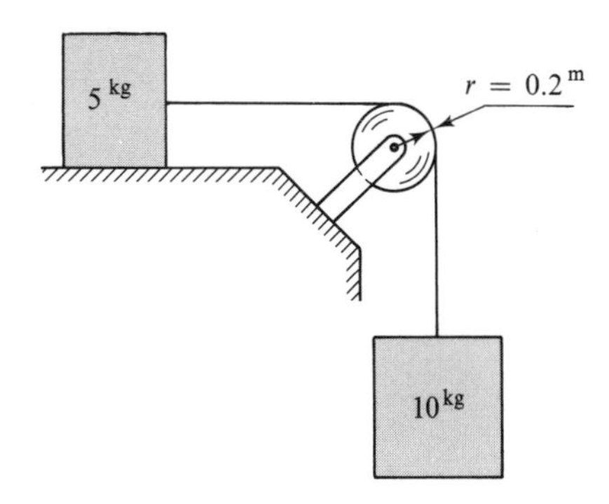

Figure P9-81

9-82 The 12-lb rigid bar in Fig. P9-82 rotates in the vertical plane about the fixed point O. In the position shown, the mass center G of the bar has a tangential acceleration of 7 ft per sec^2 down. Determine the radius of gyration of mass of the bar with respect to G.

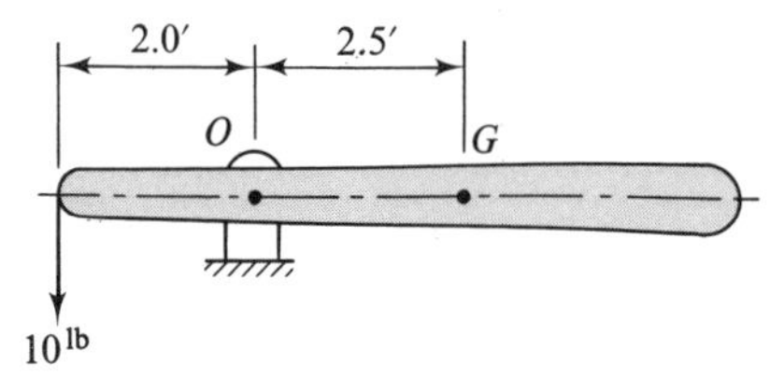

Figure P9-82

9-83 The slender homogeneous rod in Fig. P9-83 is supported by a cord at A and a horizontal pin at B. The cord is suddenly cut. Determine, for the instant of release,

(a) The location of pin B which will result in the maximum angular acceleration of the rod.

(b) The resulting angular acceleration of the rod.

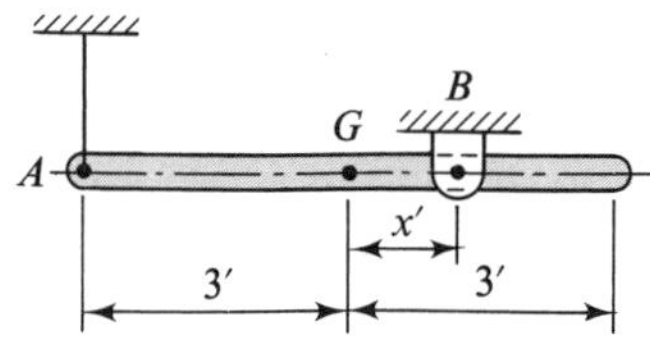

Figure P9-83

***9-84** The uniform homogeneous rocker arm ACE in Fig. P9-84 rotates about a smooth horizontal axis through C with an angular velocity of 8 rad/s counterclockwise and an angular acceleration of 24 rads/s^2 clockwise in the position shown. Part AC has a mass of 20 kg, and part CE has a mass of 30 kg. Determine, for this position, the force P and the components of the pin reaction at C.

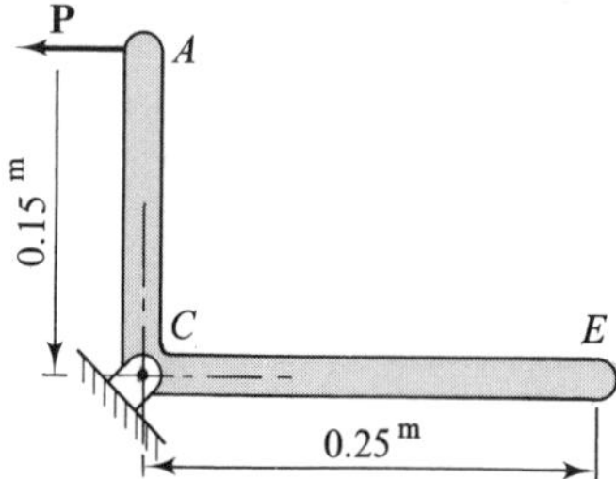

Figure P9-84

9-85 The 15-lb bar in Fig. P9-85 has a radius of gyration of mass with respect to a horizontal axis through G of 3 in. and rotates about a horizontal axis at A. The 5-lb body C is restrained from sliding in the smooth slot by the cord CD. When AB is in the position shown, the tension in the cord is 4 lb. Determine the normal component of the reaction at A on the bar and the angular acceleration of AB.

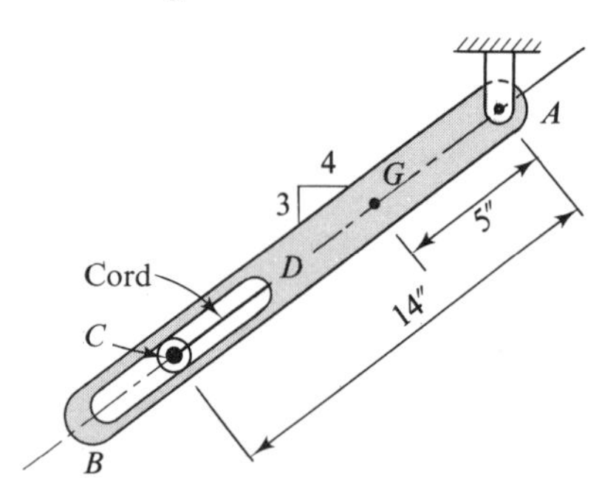

Figure P9-85

9-86 Figure P9-86 represents the pendulum of an impact-testing machine used to measure the strength of the test specimen used as a cantilever beam. The rod AB weighs 20.0 lb, and each of the cylindrical disks weighs 60.0 lb. Determine the distance the striking edge should be placed from the center of the disks to eliminate any horizontal bearing reaction at the instant of impact.

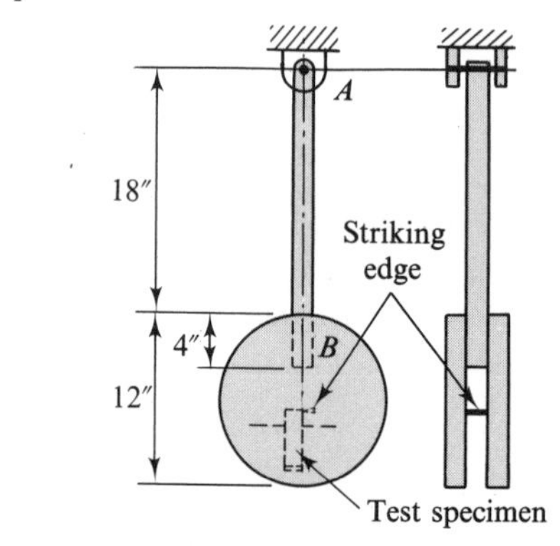

Figure P9-86

9-87 The homogeneous cylinder in Fig. P9-87 has an initial angular velocity of 800 rpm counterclockwise. The coefficient of friction between the cylinder and the horizontal and vertical surfaces is 0.2. Determine the time required for the cylinder to come to rest.

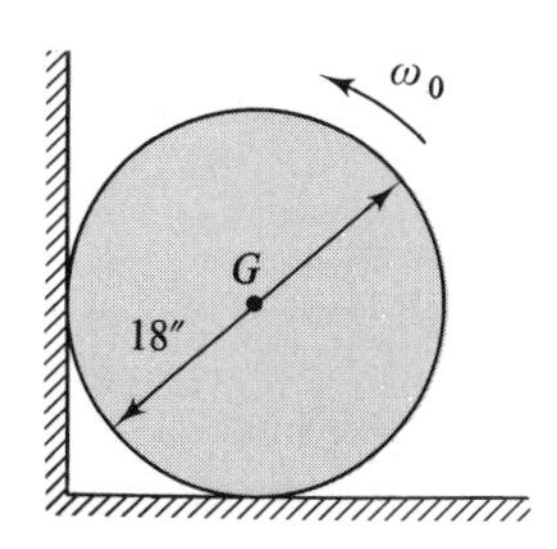

Figure P9-87

*9-88 The assembly in Fig. P9-88 rotates about a fixed horizontal axis through A. The solid homogeneous disk has a mass of 30 kg and is connected to the rigid A-frame ABC by the cord and a pin at C. When the assembly is in the position shown, it has an angular velocity of 2 rad per sec clockwise. The weight of the A-frame can be neglected. Determine the tension in the cord.

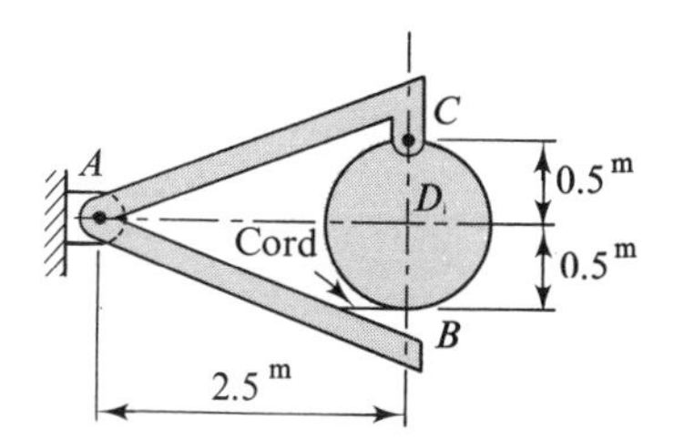

Figure P9-88

9-89 The center O of the homogeneous cylinder in Fig. P9-89 is restrained by a smooth slot. The cylinder is at rest when it is lowered onto the belt, which has a constant velocity of 20 fps as shown. The coefficient of friction between the belt and the cylinder is 0.4. Determine
(a) The maximum angular velocity of the cylinder.
(b) The distance traveled by the belt during the time required for the cylinder to reach its maximum angular velocity.

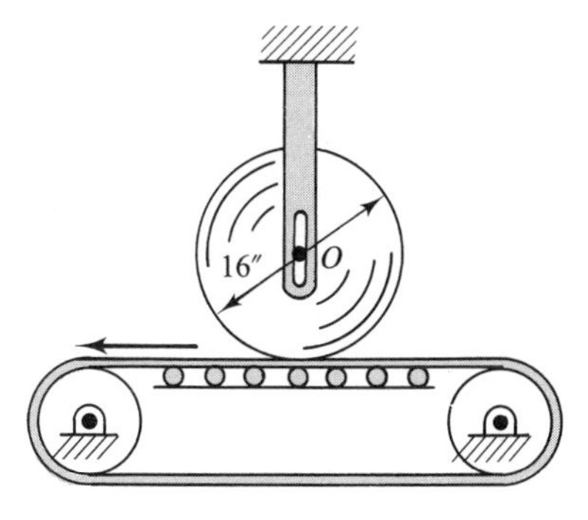

Figure P9-89

9-90 The homogeneous cylinder A weighs 161 lb and rotates about a vertical axis as shown in Fig. P9-90. A small disk B weighing 16.1 lb, which may be considered a particle, is placed on the cylinder. The coefficient of static friction between the disk and cylinder is 0.50. The system is at rest when the 63-lb force is applied. Determine the angular velocity of the cylinder when the disk starts to slip.

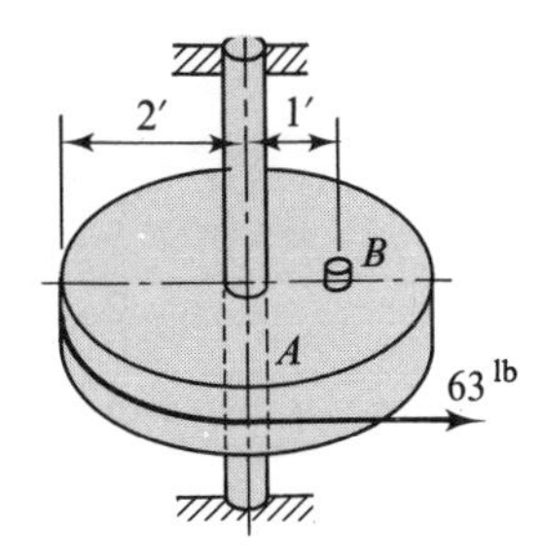

Figure P9-90

9-91 Cylinder A in Fig. P9-91 weighs 8 lb and has an initial angular velocity of 300 rpm clockwise. The 2-lb cylinder B is initially at rest when

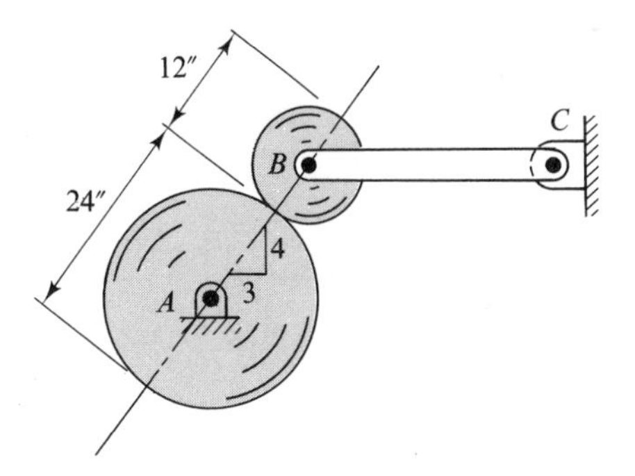

Figure P9-91

it is lowered onto cylinder A by the arm BC. The weight of arm BC is negligible, and the coefficient of friction between the cylinders is 0.2. Determine the final angular velocity of the cylinders and the contact time required to reach those velocities.

***9-92** The 20-kg cage A in Fig. P9-92 is raised and lowered by a cable and drum mounted on gear B. The drum and gear B have a mass of 30 kg and a radius of gyration of mass of 0.40 m with respect to the axis of rotation. The 15-kg gear C has a radius of gyration of mass of 0.20 m with respect to its axis of rotation. Determine the acceleration of A when the 120-N force is applied to C as shown.

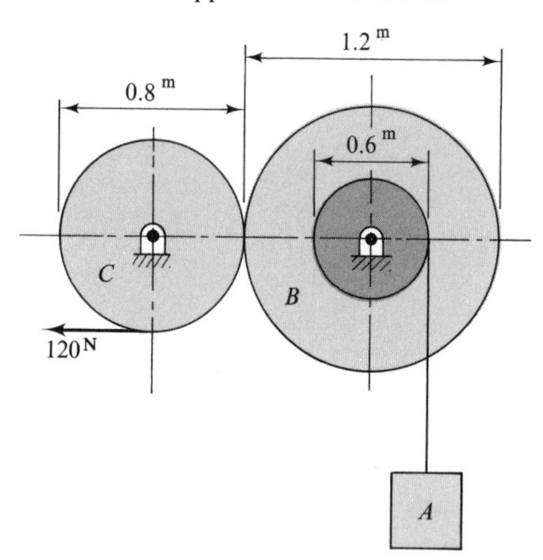

Figure P9-92

9-93 The unbalanced 32.2-lb cylinder in Fig. P9-93 has a radius of gyration of mass about a horizontal axis through G of 0.70 ft and rotates about a parallel axis through E. In the position shown, the velocity of O is 2.0 fps downward. Determine

(a) The acceleration of point O.

(b) The normal component (along line EG) of the pin reaction at E on the cylinder.

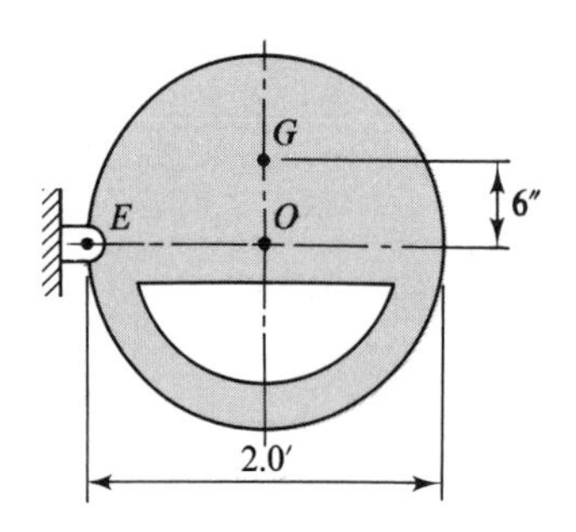

Figure P9-93

9-94 The symmetrical body A in Fig. P9-94 weighs 644 lb, and body B weighs 64.4 lb. Body B has a velocity of 16 fps upward at the beginning of a certain time interval. During the time interval, body B travels 104 ft and has a displacement of 40 ft downward. Determine the radius of gyration of mass of body A with respect to its axis of rotation.

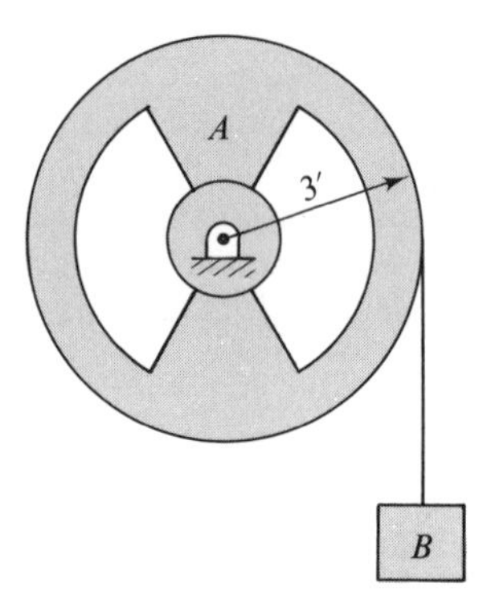

Figure P9-94

9-95 The drum in Fig. P9-95 weighs 322 lb and has a radius of gyration of mass with respect to a horizontal axis through G of 1.50 ft. Body A weighs 64.4 lb and has a velocity of 10.0 fps upward in the position indicated. Determine

(a) The acceleration of body A.

(b) The reaction of the axle on the drum.

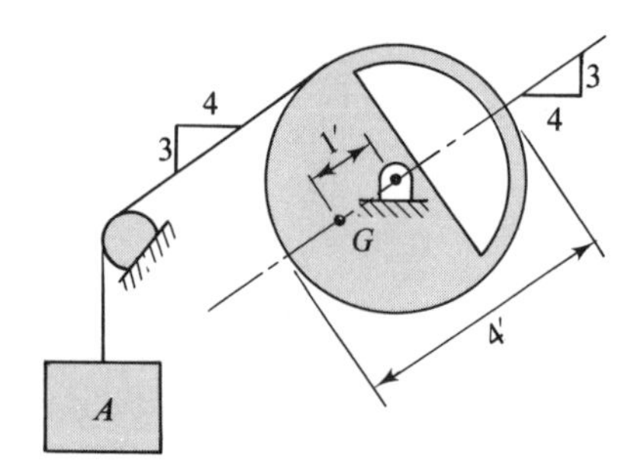

Figure P9-95

9-96 Body A in Fig. P9-96 weighs 161 lb and has a radius of gyration of mass with respect to a horizontal axis through G of 3.0 ft. The homogeneous body B weighs 95 lb, and the coefficient of friction between A and B is 0.50. Body A has an angular velocity of 3.0 rad per sec clockwise in the position shown. Determine (for this position)

(a) The angular acceleration of body A.
(b) The reaction at O on body A.

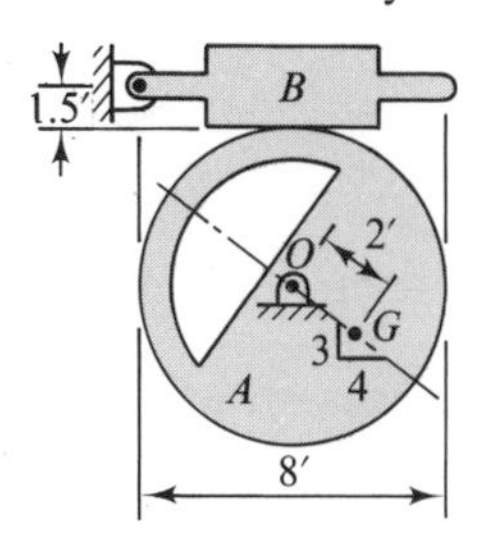

Figure P9-96

9-97 The thin rod AB in Fig. P9-97 slides along the inside of the smooth fixed cylinder. The rod weighs 2 lb and in the configuration shown has an angular velocity of 5.0 rad per sec clockwise. Determine the angular acceleration of the rod and the reactions at A and B.

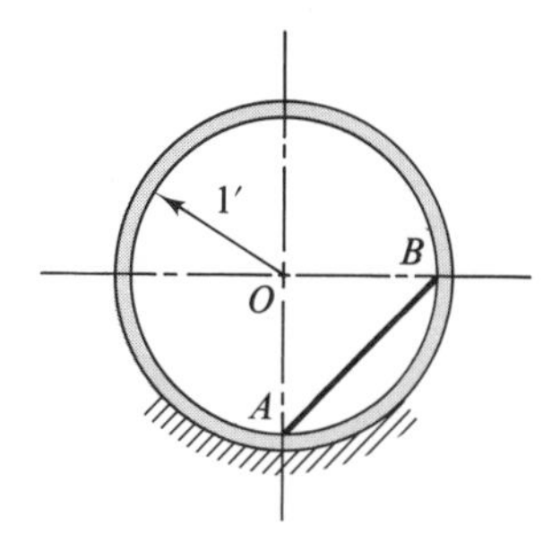

Figure P9-97

9-98 The slender homogeneous bar AB in Fig. P9-98 has a mass m and a length $\sqrt{2}l$ and is supported by three cords as shown. Determine the tensions in cords OA and OB just after cord C is cut.

9-99 The slender homogeneous bar in Fig. P9-99 weighs 20 lb and is held in equilibrium by three cords as shown. Determine the tensions in cords A and B just after cord C is cut.

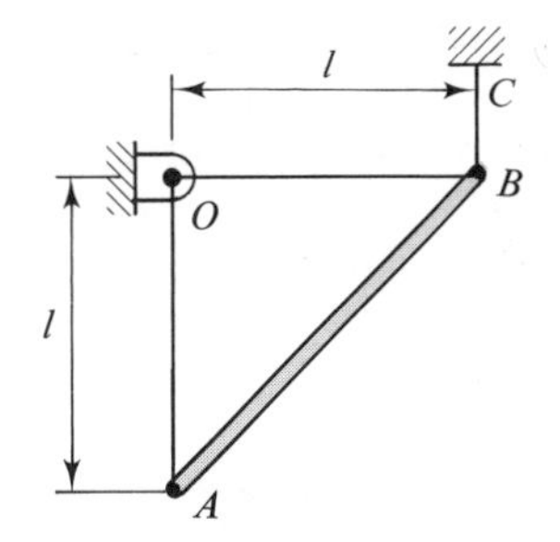

Figure P9-98

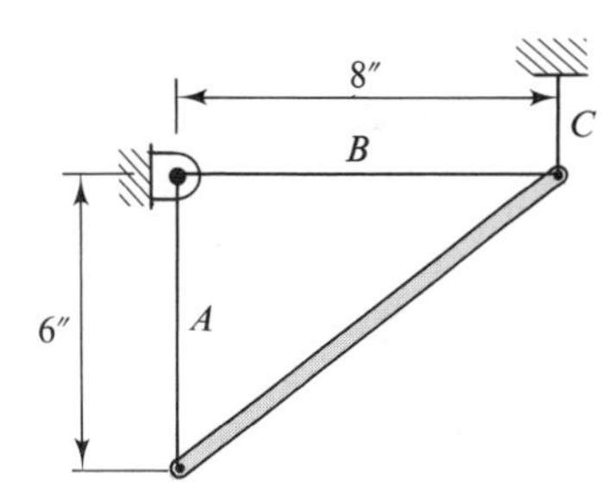

Figure P9-99

9-100 The system in Fig. P9-100 consists of three slender homogeneous rods pinned at points A, B, and C to form an equilateral triangle. Each rod is 21 in. long and weighs 8 lb. Determine the reaction on the frame at A at the instant the system is released from rest with member AC aligned with the vertical direction.

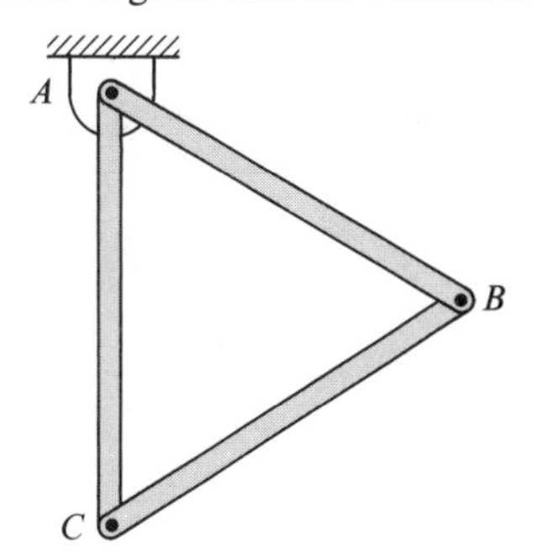

Figure P9-100

9-101 A torque **T** of 24 in-lb applied as shown in Fig. P9-101 produces an angular acceleration of disk A of 4 rad per sec^2. Disk A and its power shaft have a weight of 18 lb and a radius of gyration of mass of 2.5 in. with respect to the geometric axis through its mass center. Disk B

weighs 30 lb and has a radius of gyration of mass with respect to the vertical axis through its mass center of 4.0 in. Assume that the thrust bearing is frictionless and that A does not slip on B. Determine the radial distance x from the center of B to the point of contact of A on B.

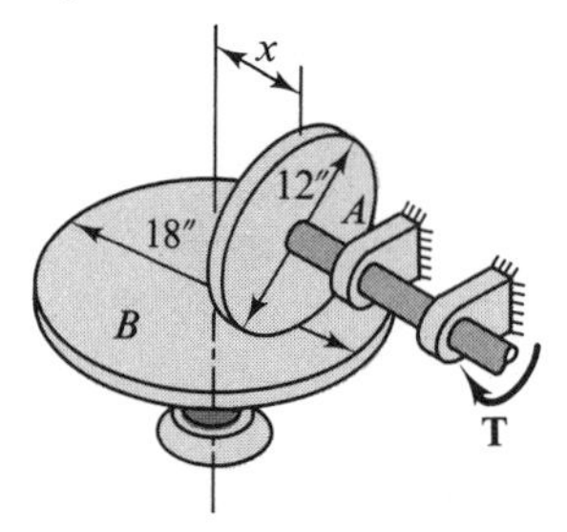

Figure P9-101

9-102 The system in Fig. P9-102 consists of a 9-lb slender homogeneous rod and a 25-lb homogeneous disk. The disk rotates in smooth bearings at A and B, and the rod rotates in a smooth ball and socket joint at O. The disk and rod are rigidly connected at C. A 24-in-lb torque **T** is applied to the disk as shown. Determine the angular acceleration of the system when in the indicated configuration.

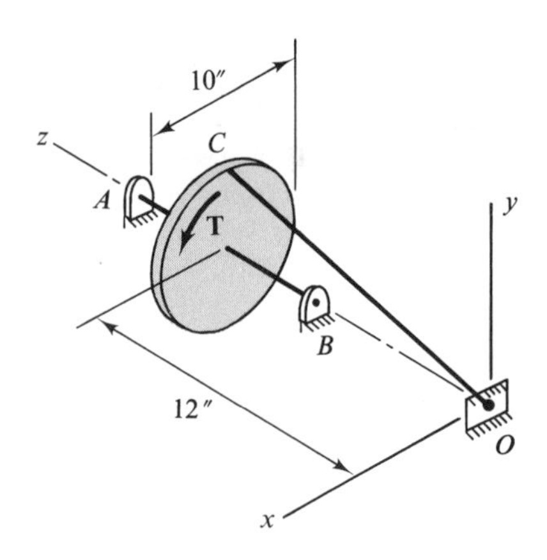

Figure P9-102

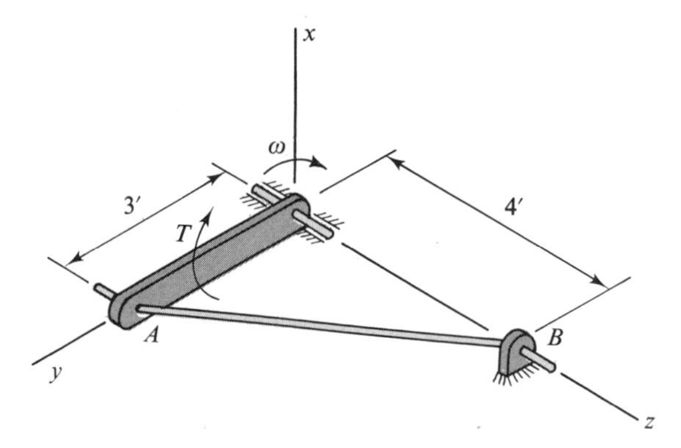

Figure P9-103

9-103 Rod AB in Fig. P9-103 weighs 20 lb and rotates with its ends in horizontal smooth bearings. The torque **T** causes member OA to rotate about the z axis with a constant angular velocity of 10 rad per sec. Determine the bearing reactions on AB at A and B when in the position shown.

9-104 The system in Fig. P9-104 consists of two slender vertical bars BC and AD connected by the slender horizontal rod AB. The vertical bars weigh 4 lb each and rotate in smooth bearings at C and D. The horizontal rod weighs 6 lb and is rigidly connected to the vertical bars. The 20-lb force applied to the horizontal rod at P lies in the xy plane. In the configuration shown, the angular velocity of the system is 10**k** rad per sec. Determine

(a) The angular acceleration of the system.

(b) The bearing reactions at C and D.

***9-105** The eccentric circular cone in Fig. P9-105 has a mass of 600 kg and is mounted on a horizontal shaft whose mass can be neglected.

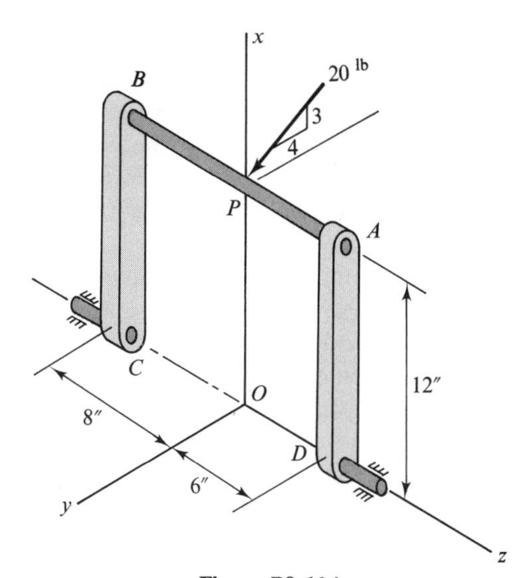

Figure P9-104

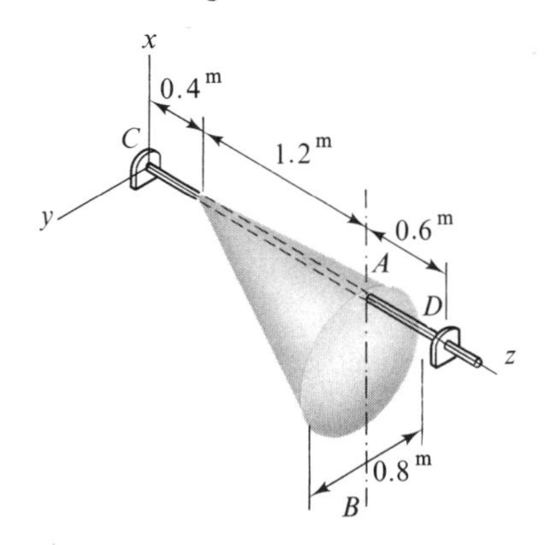

Figure P9-105

The cone has an angular velocity of 5**k** rad/s and an angular acceleration of −15**k** rad/s² when in the position shown. Determine the bearing reactions at C and D and the couple which must be applied to the shaft.

9-106 Figure P9-106 represents a slender rod A and a small ball B, each weighing 0.60 lb. The bodies rotate about the vertical shaft and are supported by a smooth step or thrust bearing at D (a bearing which can exert a force in the z direction) and by the cord C. Determine the tension in the cord and the bearing reaction when the angular velocity of the system is 20**k** rad per sec.

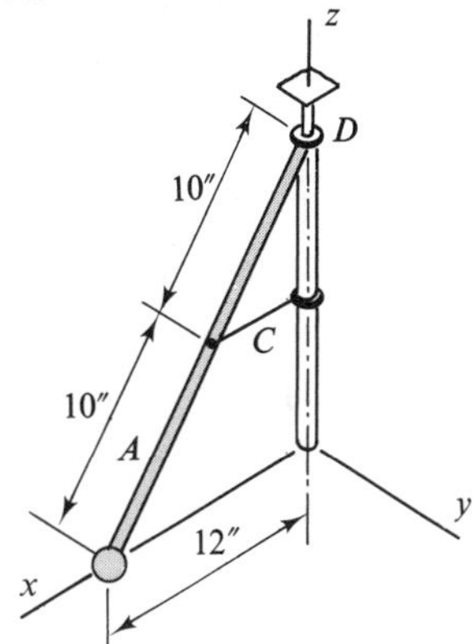

Figure P9-106

9-107 Rod AB and body C in Fig. P9-107 each have a mass m. The system is released from rest when $\theta = \theta_0$. Determine the distance x which will result in the maximum angular acceleration of rod AB and the resulting angular acceleration when $\theta_0 = 30°$

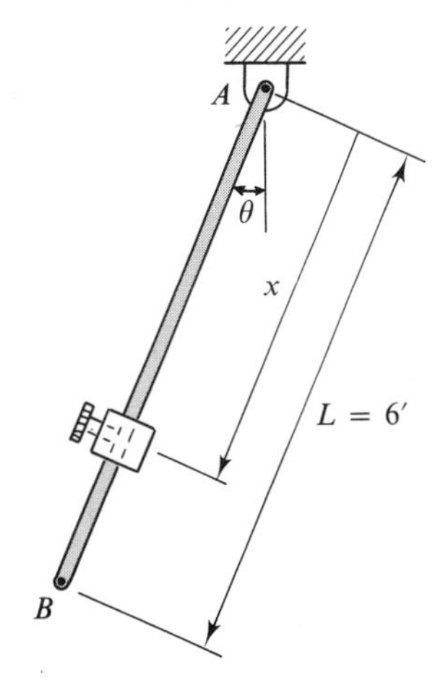

Figure P9-107

9-11 PLANE MOTION OF A RIGID BODY

When every particle of a moving rigid body remains a constant distance from a fixed reference plane, the body has plane motion. All lines in the body in, or parallel to, the plane of motion (the plane in which the mass center moves) have the same angular velocity and the same angular acceleration at any instant. The particles of a rigid body having plane motion all travel along plane curves, and the relations between the angular motion of the body and the accelerations of various particles of the body can be obtained from kinematics as indicated in Chapter 8.

The general equations of plane motion can be conveniently developed by selecting an XYZ coordinate system fixed in space and an xyz coordinate system fixed in, and moving with, the body with its origin at the mass center. The XY and xy planes are placed in the plane of motion as shown in Fig. 9-15. Vectors representing the angular velocity and the angular acceleration of the body will be parallel to the Z axis. The xyz coordinate system moves with the body, and in general the x and y axes will not be parallel to the X and Y axes. Since $\mathbf{R}_G$ is the position vector for the mass center, G, the equation of motion of the mass center is [see Eq. (9-10)]

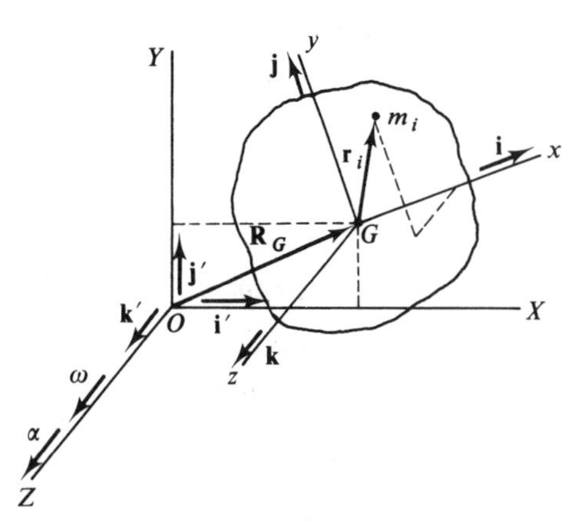

Figure 9-15

$$\sum \mathbf{F} = m\mathbf{a}_G = m\ddot{\mathbf{R}}_G \tag{9-16a}$$

or

$$\begin{aligned} \sum F_x &= ma_{Gx} = m\ddot{R}_{Gx}, \\ \sum F_y &= ma_{Gy} = m\ddot{R}_{Gy}, \\ \sum F_z &= ma_{Gz} = 0. \end{aligned} \tag{9-16b}$$

The angular velocity and angular acceleration of the x and y axes, as well as of any other line in the body parallel to the plane of motion, are $\omega\mathbf{k}$ and $\alpha\mathbf{k}$, respectively. The acceleration of any particle m_i in Fig. 9-15 can be expressed in terms of its position in the moving coordinate system (see Art. 8-21). The position of the particle relative to the xyz axes is $\mathbf{r}_i = x_i\mathbf{i} + y_i\mathbf{j} + z_i\mathbf{k}$, and since x, y, and z are constant for any given particle, $\mathbf{v}_{P/xyz}$ and $\mathbf{a}_{P/xyz}$ are zero. Thus, the acceleration of m_i, from Eq. (8-28), is

$$\begin{aligned} \mathbf{a}_i &= \mathbf{a}_G + \boldsymbol{\alpha}\times\mathbf{r}_i + \boldsymbol{\omega}\times(\boldsymbol{\omega}\times\mathbf{r}_i) \\ &= \mathbf{a}_G + (\alpha\mathbf{k})\times(x_i\mathbf{i} + y_i\mathbf{j} + z_i\mathbf{k}) + (\omega\mathbf{k})\times[(\omega\mathbf{k})\times(x_i\mathbf{i} + y_i\mathbf{j} + z_i\mathbf{k})] \\ &= \mathbf{a}_G - (\alpha y_i + \omega^2 x_i)\mathbf{i} + (\alpha x_i - \omega^2 y_i)\mathbf{j}. \end{aligned}$$

The resultant force, both internal and external forces, applied to the particle is

$$\mathbf{F}_i = m_i\mathbf{a}_i,$$

and the moment of this resultant, with respect to the mass center of the body, is

$$\mathbf{M}_{Gi} = \mathbf{r}_i \times \mathbf{F}_i = \mathbf{r}_i \times m_i \mathbf{a}_i$$

$$= (x_i\mathbf{i} + y_i\mathbf{j} + z_i\mathbf{k}) \times m_i[\mathbf{a}_G - (\alpha y_i + \omega^2 x_i)\mathbf{i} + (\alpha x_i - \omega^2 y_i)\mathbf{j}],$$

which, when expanded, becomes

$$\mathbf{M}_{Gi} = \mathbf{r}_i \times m_i \mathbf{a}_G + m_i(\omega^2 y_i z_i - \alpha x_i z_i)\mathbf{i}$$

$$- m_i(\omega^2 x_i z_i + \alpha y_i z_i)\mathbf{j} + m_i \alpha (x_i^2 + y_i^2)\mathbf{k}.$$

The resultant moment of the external force system with respect to the mass center is

$$\mathbf{M}_G = \sum_{i=1}^{n} \mathbf{M}_{Gi}$$

since the internal forces are eliminated in the summation. As in Art. 9-9, the mass of the particle m_i can be replaced by the mass of the element dm, and the summation can be replaced by integration for a rigid body with a continuous mass distribution. The resultant moment of the external force system about the center of mass then becomes

$$\mathbf{M}_G = \left(\int_m \mathbf{r}\, dm\right) \times \mathbf{a}_G + \omega^2 \mathbf{i} \int_m yz\, dm - \alpha \mathbf{i} \int_m xz\, dm$$

$$- \omega^2 \mathbf{j} \int_m xz\, dm - \alpha \mathbf{j} \int_m yz\, dm + \alpha \mathbf{k} \int_m (x^2 + y^2)\, dm.$$

The quantity $\int_m \mathbf{r}\, dm$ must be zero since $\mathbf{r}$ is measured from G [see Eq. (3-3b)]. The other integrals are moments and products of inertia with respect to the xyz coordinate system, and the moment equation reduces to

$$\mathbf{M}_G = (\omega^2 I_{yz} - \alpha I_{xz})\mathbf{i} - (\omega^2 I_{xz} + \alpha I_{yz})\mathbf{j} + \alpha I_z \mathbf{k}. \tag{9-17a}$$

Equation (9-17a) yields the three scalar equations:

$$\begin{aligned} \mathbf{i}&: \quad M_{Gx} = \omega^2 I_{yz} - \alpha I_{xz}, \\ \mathbf{j}&: \quad M_{Gy} = -\omega^2 I_{xz} - \alpha I_{yz}, \\ \mathbf{k}&: \quad M_{Gz} = \alpha I_z. \end{aligned} \tag{9-17b}$$

It should be noted that in the preceding equations the moments and products of inertia are computed with respect to axes which have an origin at the mass center. Equations (9-17) are identical to Eq. (9-14) except that the moment center is on the axis of rotation for Eq. (9-14) and at the mass center for Eq. (9-17). Hence, plane motion of a rigid body can be analyzed in two steps: (1) the motion of the mass center and (2) the rotation of the body with respect to an axis through the mass center. The motion of the mass center can be analyzed by Newton's second law just as if the rigid body were a mass particle, and the rotation of the body about an axis through the mass center is treated as if the body were rotating about a fixed axis through G.

In many engineering problems involving plane motion, the body has an axis of revolution (the axis of revolution will pass through the mass center of a homogeneous body) or is symmetrical with respect to the plane of motion (the xy plane). In either case, I_{xz} and I_{yz} are zero, and Eq. (9-17b) reduce to

$$M_{Gx} = 0, \quad M_{Gy} = 0, \quad M_{Gz} = \alpha I_z. \tag{9-18}$$

Furthermore, many plane motion problems are essentially two-dimensional in that the body is symmetrical with respect to the plane of motion and the applied forces can be assumed to be in this plane, in which case Eq. (9-16) and Eq. (9-18) reduce to

$$\sum F_x = ma_{Gx}, \quad \sum F_y = ma_{Gy}, \quad \sum M_{Gz} = I_z\alpha. \tag{9-19}$$

The other three equations yield no useful information.

In the preceding analysis, the origin of the moving coordinate system was placed at the mass center because the equations developed in this manner are in general the most useful. There are some problems, however, for which other points within the body may be used for the origin of the moving coordinate system and result in a considerable saving of effort.

Figure 9-16 shows a rigid body with plane motion similar to Fig. 9-15 except that the origin of the moving coordinate system is at some point A which does not coincide with the mass center G but does lie in the plane of motion of the mass center. The development of the moment equation of the body with respect to A proceeds as before, so that

$$\mathbf{M}_{Ai} = \mathbf{r}_i \times m_i \mathbf{a}_A + m_i(\omega^2 y_i z_i - \alpha x_i z_i)\mathbf{i} - m_i(\omega^2 x_i z_i + \alpha y_i z_i)\mathbf{j} + m_i\alpha(x_i^2 + y_i^2)\mathbf{k}.$$

From Fig. 9-16,

$$\mathbf{r}_i = \mathbf{r}_G + \boldsymbol{\rho}_i,$$

and the first term in the moment equation can be written as

$$\mathbf{r}_i \times m_i \mathbf{a}_A = \mathbf{r}_G \times m_i \mathbf{a}_A + \boldsymbol{\rho}_i \times m_i \mathbf{a}_A.$$

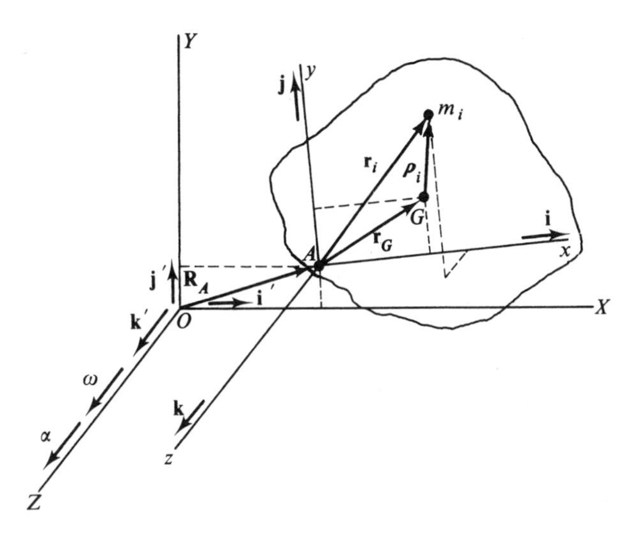

Figure 9-16

The summation of moments about A gives

$$\mathbf{M}_A = \sum_{i=1}^{n} \mathbf{M}_{Ai} = \sum_{i=1}^{n} \mathbf{r}_G \times m_i \mathbf{a}_A + \sum_{i=1}^{n} \boldsymbol{\rho}_i \times m_i \mathbf{a}_A$$

$$+\sum_{i=1}^{n} m_i(\omega^2 y_i z_i - \alpha x_i z_i)\,\mathbf{i}$$

$$-\sum_{i=1}^{n} m_i(\omega^2 x_i z_i + \alpha y_i z_i)\mathbf{j} + \sum_{i=1}^{n} m_i \alpha (x_i^2 + y_i^2)\mathbf{k}.$$

Upon replacing the summation by integration, the first term of the preceding equation becomes

$$\int_m \mathbf{r}_G \times \mathbf{a}_A \, dm = \mathbf{r}_G \times \mathbf{a}_A \int_m dm = (\mathbf{r}_G \times \mathbf{a}_A)m.$$

Similarly, the second term becomes

$$\int_m \boldsymbol{\rho} \times \mathbf{a}_A \, dm = \int_m \boldsymbol{\rho} \, dm \times \mathbf{a}_A,$$

which must be zero as in the previous discussion, since $\int \boldsymbol{\rho}\, dm$ is the first moment of the mass with respect to G. The last three terms produce moment of inertia terms similar to those in Eq. (9-17a), and the moment of the external forces about point A becomes

$$\mathbf{M}_A = (\mathbf{r}_G \times \mathbf{a}_A)m + (\omega^2 I_{yz} - \alpha I_{xz})\mathbf{i} - (\omega^2 I_{xz} + \alpha I_{yz})\mathbf{j}$$
$$- \alpha I_z \mathbf{k}. \tag{9-20}$$

Note that in Eq. (9-20) the moment of inertia terms are computed with respect to axes which have an origin at A, whereas in Eq. (9-17) the origin of the axes is at G. Equation (9-20) reduces to the form of Eq. (9-17a) if the first term on the right is zero and this will occur if

1. $\mathbf{r}_G = 0$, in which case A coincides with G.
2. $\mathbf{a}_A = 0$, which corresponds to the rotation about a fixed axis (see Art. 9-9).
3. $\mathbf{r}_G \times \mathbf{a}_A = 0$, which implies that $\mathbf{r}_G$ and $\mathbf{a}_A$ are parallel.

The third case provides the only new information. Equation (9-20) reduces to the same form as Eq. (9-17a), provided the acceleration of the moment center is along the line AG. For example, when a wheel is rolling without slipping along a fixed plane, the acceleration of the point of contact (the instantaneous center of zero velocity) is normal to the plane, and if the mass center is on the line through the point of contact normal to the plane, Eq. (9-17) can be applied with respect to an axis through the instantaneous center. It should be noted, however, that this procedure is valid only when the acceleration of the moment axis is toward or away from the mass center.

When any of the equations of motion for plane motion is used, a consistent sign convention must be followed. In the derivations, the positive directions for velocities, accelerations, and forces were the positive directions of the coordinate axes. Any given quantity in the opposite direction must be used as a negative number in the equations of motion. When two or more bodies are involved in a single problem, different orientations of the axes and different sign conventions can be used for different bodies, provided related kinematical equations are properly written.

In problems involving wheels, cylinders, and similar bodies rolling on a plane, the data may not always indicate whether or not the body slides as it rolls along the plane. There are three possibilities for problems of this type: (1) The problem may state that the wheel *rolls without slipping* or that *slipping impends.* (2) The problem may state that the body *rolls and slips.* (3) The problem may not specify which of these conditions exists. *The method of solution for each of these situations is discussed assuming that the body and force system are symmetrical about the plane of motion and that the mass center is at the geometrical center of the wheel.* A similar analysis applies if the mass center of the wheel is not at the geometrical center except that the kinematic relationships are somewhat more involved. Note that Eq. (9-19) are the only equations needed for the problem as stated.

1. When the problem states that the body *rolls without slipping* the acceleration of the mass center is $a_G = r\alpha$—see Art. 8-19. This equation of kinematics can be used with the equations of motion to solve the problem. The maximum friction ($F' = \mu N$) is usually *not* developed in this type of problem. The friction is a force unknown in magnitude and direction and can be determined from the equations of

motion. A special case of this type of problem is illustrated in Example 9-10 in which *slipping impends*. In this case, both $F = F' = \mu N$ and $a_G = r\alpha$ are valid.

2. When the problem states that the body *rolls and slips*, the friction force must equal its limiting value of μN. The direction of the velocity of the point of contact must be determined from the velocity data in order to show the friction with the correct sense (opposite to this velocity). *Note that* a_G *does not equal* $r\alpha$.

3. When the problem does not state that one of these conditions exists (that is, no slipping, slipping impends, or rolls and slips), one of the conditions (usually no slipping) must be assumed, and a check made to indicate the validity of the assumption. The following steps are suggested:

a. Assume that no slipping occurs (write "assume no slip" as part of the solution) and use $a_G = r\alpha$.
b. Solve the equations of motion for the friction and normal forces. This friction is the force which must be developed to prevent slipping.
c. Compute the maximum friction available from $F' = \mu N$.
d. Check the assumption; that is, compare F and F'.
   If $F < F'$, assumption is correct and $a_G = r\alpha$.
   If $F = F'$, slipping impends, assumption is correct, and $a_G = r\alpha$.
   If $F > F'$ the body rolls and slips and a_G *is not equal to* $r\alpha$.
e. If the check indicates that the body rolls and slips, the preceding solution is not valid. The friction force is equal to F' in the direction of F found in step b. Values of a_G and α can be determined from the equations of motion.

EXAMPLE 9-10

Body A in Fig. 9-17(a) is a 100-lb solid homogeneous cylinder with a narrow slot cut in it as indicated by the dashed circle. The effect of the slot on the moment of inertia of the cylinder can be neglected. The coefficient of friction between the cylinder and plane is 0.40. Block B is connected to A by the flexible cord which passes over the smooth drum and is wrapped around the cylinder in the slot. Determine the maximum weight B can have if the cylinder is to roll without slipping along the inclined plane. The point on A in contact with the plane has impending motion down the plane.

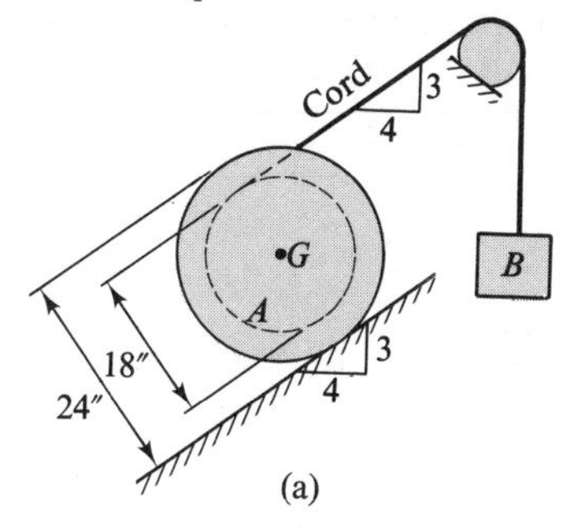

Figure 9-17

SOLUTION

The free body diagrams for the two bodies are shown in Fig. 9-17(b) and (c). For impending motion, the frictional force $\mathbf{F}$ opposes the motion or impending motion of the point of contact; therefore, the friction on A is up the plane. Body A has plane motion and body B has translation. The general equations of motion for A are

$$\sum F_x = ma_{Gx}, \qquad \sum F_y = ma_{Gy}, \qquad \sum M_{Gz} = I_z\alpha,$$

Figure 9-17 *(Continued)*

and the equation of motion for B is

$$\sum F_y = ma_{By}.$$

The x, y, and z axes for body A are selected parallel and perpendicular to the plane as shown, and a_{Gy} is zero. The axes for Fig. 9-17(b) are selected as shown so that a positive angular acceleration, $\alpha\mathbf{k}$, will result in a positive acceleration of G, $a_{Gx}\mathbf{i}$. In the two free-body diagrams there are four unknown forces ($\mathbf{W}_B$, $\mathbf{T}$, $\mathbf{F}$, and $\mathbf{N}$), and the four motion equations contain three unknown accelerations (a_{Gx}, α, and a_{By}), making seven unknowns and four equations of motion. Three additional relationships must be obtained from friction and kinematics.

Since slipping impends, the magnitude of the frictional force is

$$F = F' = 0.40N.$$

The wheel does not slip; therefore, from kinematics, the acceleration of G is

$$a_{Gx}\mathbf{i} = r\alpha\mathbf{i} = \alpha\mathbf{i}.$$

Note: $\boldsymbol{\alpha} = \alpha\mathbf{k}$ is assumed clockwise and $(\mathbf{a}_G)_x$ is assumed up the plane. The x component of the acceleration of E (equal in magnitude to the acceleration of B) can be expressed in terms of the angular acceleration of A by relative motion. Thus, with $\boldsymbol{\alpha}$ assumed to be clockwise,

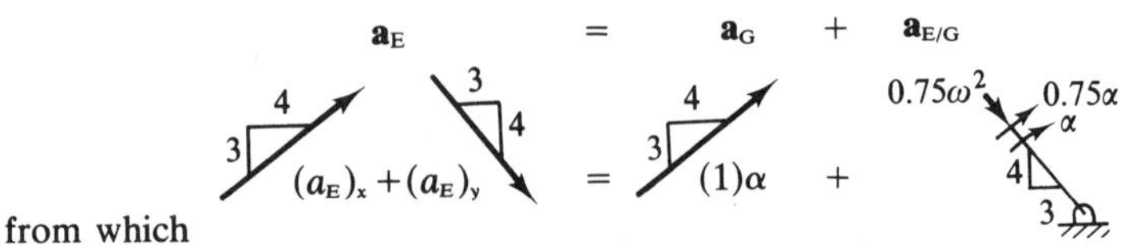

from which

$$(a_E)_x = (a_B)_y = +1.00\alpha + 0.75\alpha = +1.75\alpha,$$

where $(\mathbf{a}_E)_x$ is up the plane and $(\mathbf{a}_B)_y$ is downward.

When specific values are substituted in the equations of motion for A, they become

$$\sum F_x = T + F - \tfrac{3}{5}(100) = \frac{100}{32.2}\, a_{Gx} = \frac{100}{32.2}\,\alpha,$$

$$\sum F_y = \tfrac{4}{5}(100) - N = \frac{100}{32.2}\, a_{Gy} = 0,$$

and

$$\sum M_{Gz} = 0.75T - 1.00F$$

$$= \frac{1}{2}\frac{100}{32.2}(1.00)^2\alpha.$$

Likewise, for B,

$$\sum F_y = W_B - T = \frac{W_B}{32.2}\, a_{By}$$

$$= \frac{W_B}{32.2}\, 1.75\alpha,$$

and from friction,

$$F = 0.40N.$$

Solution of these equations gives

$$W_B = \underline{313\ \text{lb.}} \qquad \text{Ans.}$$

EXAMPLE 9-11

The wheel in Fig. 9-18(a) weighs 96.6 lb and rolls along the horizontal plane. The coefficient of friction between the wheel and plane is 0.20. The radius of gyration of the mass of the wheel with respect to its geometric axis is 1.30 ft. Determine the acceleration of the center of the wheel and the angular acceleration of the wheel.

SOLUTION

The free-body diagram of the wheel is drawn in Fig. 9-18(b). *Assume that the wheel does not slip* since the statement of the problem did not specify either slipping or no slipping. The positive axes are shown on the diagram, and the accelerations are assumed to be in the positive

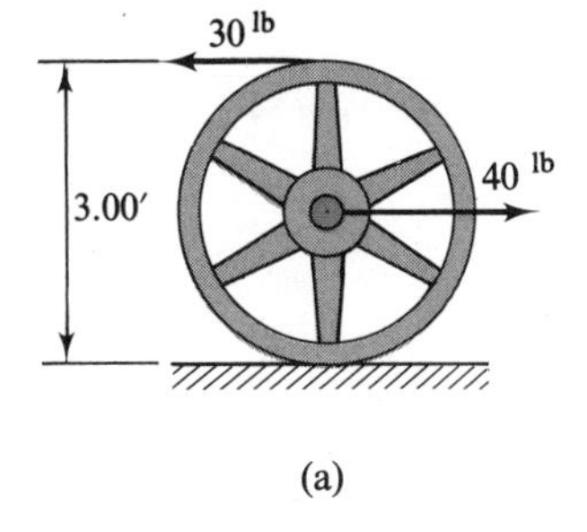

(a)

Figure 9-18

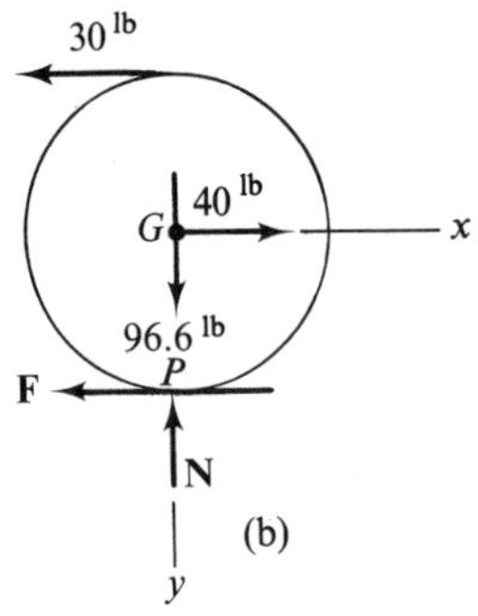

Figure 9-18 (*Continued*)

directions of the axes. The equations of motion are

$$\sum F_x = 40 - 30 - F = ma_{Gx} = 3.0a_{Gx}, \tag{a}$$

$$\sum F_y = 96.6 - N = ma_{Gy} = 0, \tag{b}$$

and

$$\sum M_{Gz} = 1.5F - 1.5(30) = I_{Gz}\alpha = 3.0(1.30)^2\alpha. \tag{c}$$

Since rolling without slipping is assumed, the accelerations are related by

$$a_{Gx} = r\alpha = 1.5\alpha.$$

Note: This equation is valid only for the given orientation of the axes. If y had been chosen upward and z to the front, a_{Gx} would equal $-ra$ for a right-hand system of axes.

Simultaneous solution of these equations gives

$$F = 21.42 \text{ lb} \qquad \mathbf{F} = -21.42\mathbf{i} \text{ lb},$$

$$N = 96.6 \text{ lb}, \qquad \mathbf{N} = -96.6\mathbf{j} \text{ lb},$$

$$\alpha = -2.54 \text{ rad per sec}^2, \qquad \boldsymbol{\alpha} = -2.54\mathbf{k} \text{ rad per sec}^2,$$

$$a_G = 1.5(-2.54) = -3.81 \text{ fps}^2, \qquad \mathbf{a}_G = -3.81\mathbf{i} \text{ fps}^2.$$

The force $F = 21.42$ lb is the friction required to prevent slipping. The maximum friction which can be developed is

$$F' = \mu N = 0.20(96.6) = 19.32 \text{ lb}.$$

Since F' is less than F, the assumption that the wheel does not slip is wrong, and the solution is incorrect. The value obtained for N is correct, however, since the equation for N did not involve the assumption. When $F = F' = 19.32$ lb is substituted into Eq. (a) and (c) they become

$$\sum F_x = 40 - 30.0 - 19.32 = 3.00a_{Gx}$$

and

$$\sum M_{Gz} = 1.50(19.32) - 1.50(30) = 5.07\alpha.$$

Solution of these equations gives

$$a_{Gx} = -3.11 \text{ fps}^2, \qquad \underline{\mathbf{a}_{Gz} = -3.11\mathbf{i} \text{ fps}^2,} \qquad \text{Ans.}$$

and

$$\alpha = -3.16 \text{ rad per sec}^2, \qquad \underline{\boldsymbol{\alpha} = -3.16\mathbf{k} \text{ rad per sec}^2.} \qquad \text{Ans.}$$

EXAMPLE 9-12

Bar AB in Fig. 9-19(a) is 5.0 m long and has a mass of 40 kg. The ends of the rod are connected to rollers of negligible mass which move in the smooth slots as indicated. When the bar is in the position shown, the

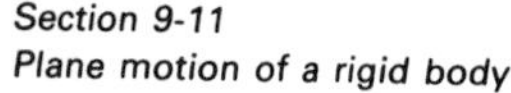

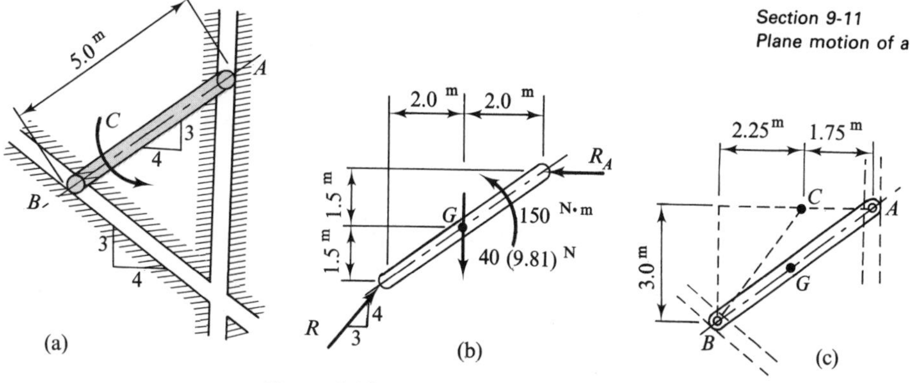

Figure 9-19

moment of the couple C is 150 N · m counterclockwise, and the velocity of A is 3.5 m/s down the slot. Determine
(a) the angular acceleration of AB.
(b) The reactions on the bar at A and B.

SOLUTION

Figure 9-19(b) is a free-body diagram of the bar. The body is symmetrical about the plane of motion and the forces are all in this plane; therefore, Eq. (9-19) are the only motion equations which provide useful information. The motion equations are

$$\sum F_x = ma_{Gx}, \qquad \sum F_y = ma_{Gy}, \qquad \sum M_{Gz} = I_{Gz}\alpha.$$

The free-body diagram has two unknown forces, and the three equations of motion contain two unknown linear acceleration components and the unknown angular acceleration of the rod. It is therefore necessary to obtain two additional equations relating some of the five unknown quantities before the problem can be solved.

The ends of the bar are constrained to move along the slots, and thus they have rectilinear motion with accelerations and velocities along the slots. With this information, it is possible to express the acceleration of the mass center in terms of the angular velocity and angular acceleration of the bar. The angular velocity of the bar can be obtained from the instantaneous center as shown in Fig. 9-19(c), and the acceleration terms can be related by means of relative acceleration equations. The velocity of A is 3.5 m/s downward, and from Fig. 9-19(c) it is also equal to 1.75ω; thus

$$1.75\omega = 3.5$$

from which $\omega = 2$ rad/s and $\boldsymbol{\omega} = 2$ rad/s↻. The angular acceleration is assumed to be clockwise, and the accelerations of A and B are related by the equation

$$\mathbf{a}_B \quad = \mathbf{a}_A \quad + \quad \mathbf{a}_{B/A}.$$

$$a_B \ (\text{4–3 slope}) \quad = a_A \ \uparrow \quad + \quad 5\alpha, \ 5\omega^2 \ (\text{4–3 slope}), \ \alpha$$

When horizontal components are equated the equation becomes

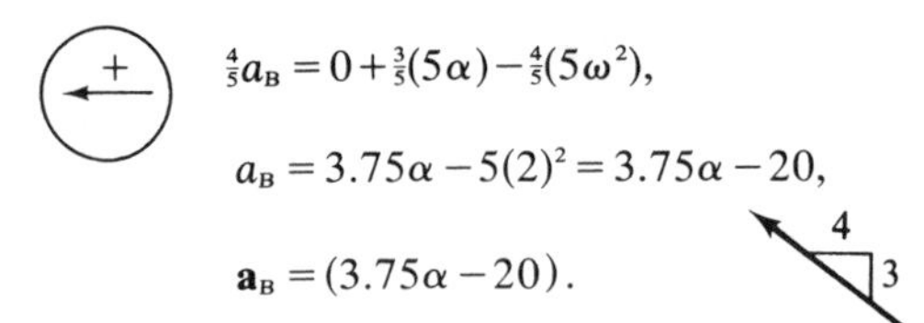

$$\tfrac{4}{5}a_B = 0 + \tfrac{3}{5}(5\alpha) - \tfrac{4}{5}(5\omega^2),$$

$$a_B = 3.75\alpha - 5(2)^2 = 3.75\alpha - 20,$$

and

$$\mathbf{a}_B = (3.75\alpha - 20).$$

The relative acceleration equation for points G and B is

$$\mathbf{a}_G = \mathbf{a}_B + \mathbf{a}_{G/B}$$

$2.5\,\omega^2 = 10$, $2.5\,\alpha$

$$= (3.75\alpha - 20) +$$

$$= \overleftarrow{3.0\alpha} + \overset{\uparrow}{2.25\alpha} + \overrightarrow{16} + \overset{\downarrow}{12} + \overrightarrow{1.5\alpha} + \overset{\downarrow}{2\alpha} + \overleftarrow{8} + \overset{\downarrow}{6}$$

$$= (\overleftarrow{1.5\alpha - 8}) + (\overset{\uparrow}{0.25\alpha - 18}).$$

When this expression for $\mathbf{a}_G$ is substituted into the force equations of motion they become

$$\sum F_x = ma_{Gx},$$

$$R_A - 0.6R_B = 40(1.5\alpha - 8) = 60\alpha - 320; \qquad \text{(a)}$$

$$\sum F_y = ma_{Gy},$$

$$0.8R_B - 40(9.81) = 40(0.25\alpha - 18) = 10\alpha - 720. \qquad \text{(b)}$$

The moment equation can be written conveniently by resolving $\mathbf{R}_B$ into components and applying the principle of moments; thus

$$\sum M_{Gz} = I_{Gz}\alpha.$$

$$0.8R_B(2.0) - 0.6R_B(1.5) - R_A(1.5) - 150 = \tfrac{1}{12}(40)(5)^2\alpha$$

or

$$0.7R_B - 1.5R_A = 83.3\alpha + 150. \qquad \text{(c)}$$

When Eq. (a)–(c) are solved simultaneously the results are

$$R_A = -408\text{N}$$

$$R_B = -380\text{N}$$

$$\alpha = 2.34 \text{ rad/s}^2.$$

The negative signs indicate that $\mathbf{R}_A$ and $\mathbf{R}_B$ were assumed to be In the wrong direction. The results are

$$\mathbf{R}_A = 408\text{N} \rightarrow, \qquad \text{Ans.}$$

$$\mathbf{R}_B = 380\text{N} \ (3,4 \text{ slope}), \qquad \text{Ans.}$$

$$\boldsymbol{\alpha} = 2.34 \text{ rad/s}^2 \circlearrowright. \qquad \text{Ans.}$$

EXAMPLE 9-13

Two 2-ft-diameter homogeneous disks weighing 96.6 lb each are connected by a 4-ft rod of negligible weight. Welded to this rod are two homogeneous 1- by 2-ft rectangular plates as shown in Fig. 9-20(a). Each plate weighs 48.3 lb. The assembly rolls without slipping along a straight path on a horizontal plane. When in the position indicated, the angular velocity is 10 rad per sec as shown, and the magnitude of the couple **C** is such that slipping impends. The coefficient of friction between the wheels and plane is 0.25. Determine the angular acceleration and the couple **C**.

SOLUTION

Figure 9-20(b) is a free-body diagram of the assembly showing the orientation of the moving *xyz* coordinate system with its origin at the mass center, *G*. The friction forces are assumed to have moments about the *z* axis which oppose **C**. Due to symmetry, *G* lies midway between the disks on the center line of the rod, and the total weight of the assembly is

$$W = 2(96.6) + 2(48.3) = 289.8 \text{ lb}.$$

The body is not symmetrical about the plane of motion, and the forces do not all lie in this plane; therefore, the general moment equation, [Eq. (9-17a)] will be needed, and it will be necessary to obtain values for I_{xz}, I_{yz}, and I_z.

The body is symmetrical with respect to the *yz* plane; therefore, I_{xz} must be zero. The *xz* plane is a plane of symmetry for the two disks, so I_{yz} for the disks is also zero. The product of inertia of each plate with respect to planes of symmetry is zero (see Art. 6-10 or Appendix B), but the transfer terms are different from zero, and I_{yz} becomes

$$I_{yz} = \sum [(I_{yz})_G + m y_G z_G]$$

$$= 0 + \frac{48.3}{32.2}\left(-\frac{1}{2}\right)(1) + 0 + \frac{48.3}{32.2}\left(\frac{1}{2}\right)(-1)$$

$$= -1.500 \text{ slug-ft}^2.$$

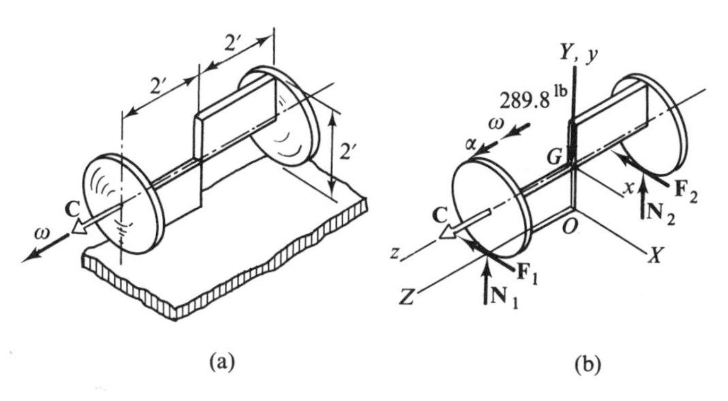

Figure 9-20

The value of I_z is also obtained by the parallel axis theorem (see the table in Art. 6-14 or Appendix B).

The moment of inertia is

$$I_z = \sum (I_z)_{\text{disks}} + \sum [(I_z)_G + md^2]_{\text{plates}}$$

$$= 2\left[\frac{1}{2}\frac{96.6}{32.2}(1.00)^2\right] + 2\left[\frac{1}{12}\frac{48.3}{32.2}(1.00)^2 + \frac{48.3}{32.2}(0.50)^2\right]$$

$$= 4.00 \text{ slug-ft}^2.$$

The force equation of motion is

$$\sum \mathbf{F} = m\mathbf{a}_G,$$

which, from Fig. 9-20(b), becomes

$$-F_1\mathbf{i} - F_2\mathbf{i} + N_1\mathbf{j} + N_2\mathbf{j} - 289.8\mathbf{j} = \frac{289.8}{32.2}\mathbf{a}_G.$$

The moments of the forces about G can be obtained by vector multiplication, but in this example the moments can be determined more readily by inspection of Fig. 9-20(b). The equation is

$$\sum \mathbf{M}_G = (\omega^2 I_{yz} - \alpha I_{xz})\mathbf{i} - (\omega^2 I_{xz} + \alpha I_{yz})\mathbf{j} + \alpha I_z \mathbf{k},$$

which becomes

$$(2N_2 - 2N_1)\mathbf{i} + (2F_2 - 2F_1)\mathbf{j} + (C - 1F_1 - 1F_2)\mathbf{k}$$
$$= (10)^2(-1.500)\mathbf{i} - \alpha(-1.500)\mathbf{j} + \alpha(4.00)\mathbf{k}.$$

The two general equations of motion are equivalent to five algebraic equations (there are no forces or acceleration component in the z direction). There are five unknown forces and couples in addition to the linear acceleration of G and the angular acceleration of the body. The problem stated that the wheels roll without slipping; therefore,

$$\mathbf{a}_G = r\alpha(-\mathbf{i}) = -\alpha\mathbf{i}.$$

The algebraic equations of motion are

$$-F_1 - F_2 = 9a_G = -9\alpha, \tag{a}$$

$$N_1 + N_2 - 289.8 = 0, \tag{b}$$

$$2N_2 - 2N_1 = -150, \tag{c}$$

$$2F_2 - 2F_1 = 1.500\alpha, \tag{d}$$

and

$$C - F_1 - F_2 = 4.00\alpha \tag{e}$$

These five equations contain six unknowns, and one additional equation can be obtained from friction, according to the statement of the problem. As C is increased, F_1 and F_2 will also increase until one of them is equal to the limiting value. Since there are two possibilities, an assumption must be made and checked. If F_1 is assumed to equal μN_1, a simultaneous solution of the equations gives

$$N_1 = 182.4 \text{ lb},$$

$$N_2 = 107.4 \text{ lb},$$

$$F_1 = \mu N_1 = 0.25(182.4) = 45.6 \text{ lb},$$

and

$$F_2 = 53.9 \text{ lb}.$$

Since F_2 is greater than μN_2, which is 0.25(107.4) or 26.8 lb, the right-hand wheel will slip first, and the problem must be solved assuming that F_2 is equal to μN_2. This solution gives

$$N_1 = 182.4 \text{ lb}, \qquad N_2 = 107.4 \text{ lb},$$

$$F_2 = \mu N_2 = 26.85 \text{ lb}$$

$$F_1 = 22.72 \text{ lb},$$

$$\alpha = 5.51 \text{ rad/sec}^2,$$

and

$$C = 71.6 \text{ ft-lb}.$$

Since F_1 is less than μN_1, this is the correct assumption, and slipping impends at the right wheel. The results are

$$\underline{\alpha = 5.51\mathbf{k} \text{ rad per sec}^2}, \qquad \text{Ans.}$$

and

$$\underline{\mathbf{C} = 71.6\mathbf{k} \text{ ft-lb.}} \qquad \text{Ans.}$$

PROBLEMS

9-108 The 16.1-lb wheel in Fig. P9-108 is a homogeneous cylinder with a narrow slot in which the cord is wrapped. Neglect the effect of the slot on the moment of inertia of the mass of the wheel. The wheel rolls without slipping on the horizontal plane. Determine
(a) The acceleration of G.
(b) The coefficient of friction required between the wheel and the plane to prevent slipping.

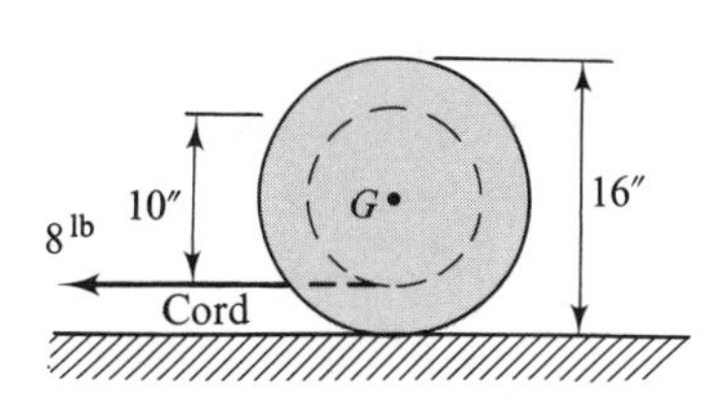

Figure P9-108

***9-109** Figure P9-109 represents a 5-kg solid homogeneous cylinder. Determine
(a) The least coefficient of friction between the cylinder and the plane for which the cylinder will roll down the plane without slipping starting from rest.
(b) The acceleration of the center of the cylinder when it rolls without slipping.

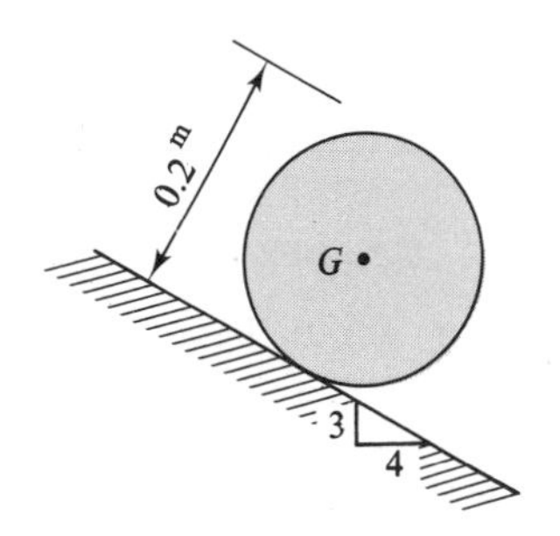

Figure P9-109

9-110 The 9-lb homogeneous sphere in Fig. P9-110 has a narrow slot cut in it for the cord. Neglect the effect of the slot on the moment of inertia of mass of the sphere, and assume that the cord acts in the vertical plane of symmetry. The sphere is released from rest with G 80 in. above the floor. Determine

(a) The angular acceleration of the sphere.

(b) The velocity of G immediately prior to impact with the floor.

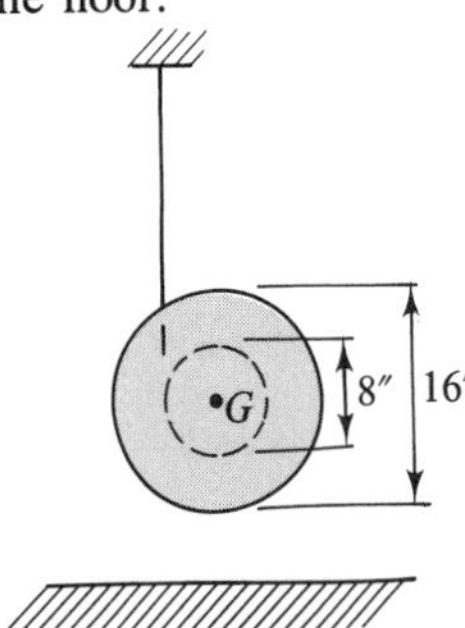

Figure P9-110

9-111 The 32.2-lb homogeneous sphere in Fig. P9-111 has a narrow slot cut in it for the cord. Neglect the effect of the slot on the moment of inertia of mass of the sphere, and assume that both cords act in a vertical plane of symmetry. Determine the force **P** when the angular acceleration of the sphere is 12 rad per sec² counterclockwise.

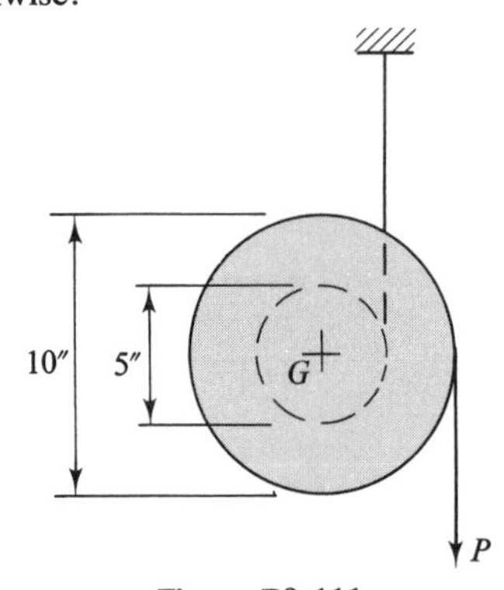

Figure P9-111

9-112 Cylinder A in Fig. P9-112 weighs 32.2 lb and rolls along the inclined plane without slipping. Neglect the effect of the slot on the moment of inertia of the disk. Body B weighs 16.1 lb. Determine the angular acceleration of the disk.

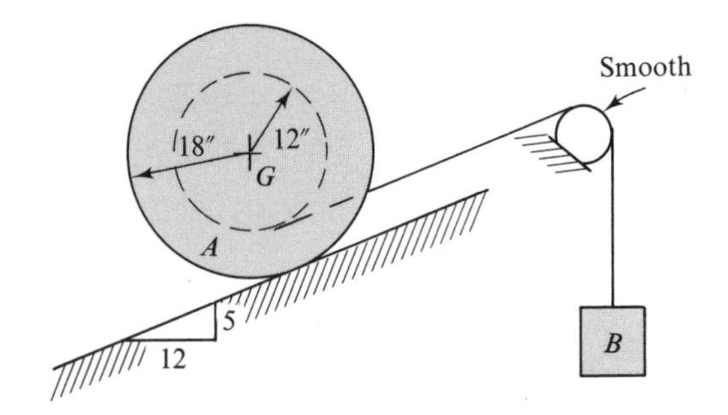

Figure P9-112

9-113 Wheels A and B in Fig. P9-113 are identical 16.1-lb solid homogeneous cylinders with smooth axles at their geometric centers. The 32.2-lb carriage C, with mass center at G, is supported on the axles of A and B, and the wheels roll without slipping on the inclined plane. Determine the acceleration of the carriage C.

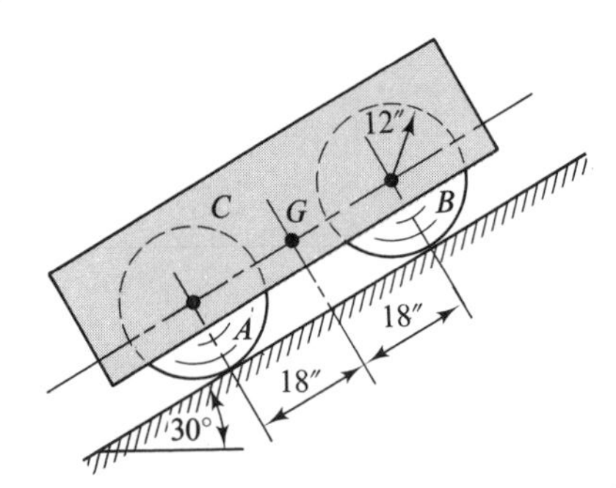

Figure P9-113

9-114 In Fig. P9-114 the homogeneous cylinder A is mounted on smooth bearings inside box B and connected by a cable to body C. Bodies A, B, and C weigh 50, 20, and 70 lb, respectively. All surfaces are smooth. Determine the acceleration of bodies B and C.

***9-115** The solid homogeneous cylinder in Fig. P9-115 has a mass of 13 kg. The force P is applied to a cord wrapped around a thin slot (neglect the effect of the slot on the moment of inertia of the cylinder). The coefficient of friction between the cylinder and the inclined plane is 0.30. Determine the angular acceleration of

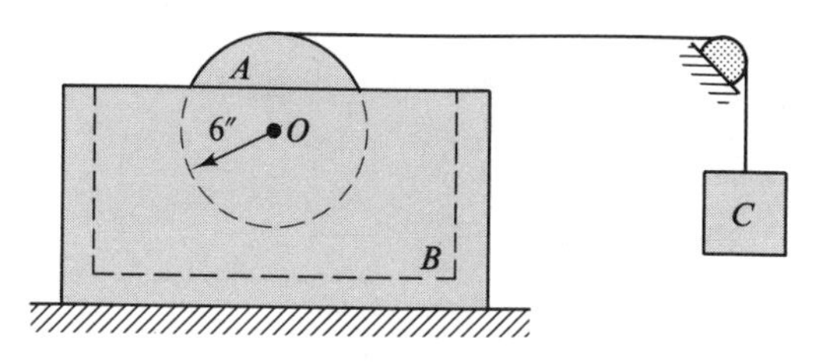

Figure P9-114

the cylinder and the linear acceleration of G when
(a) $P = 24$ N.
(b) $P = 60$ N.

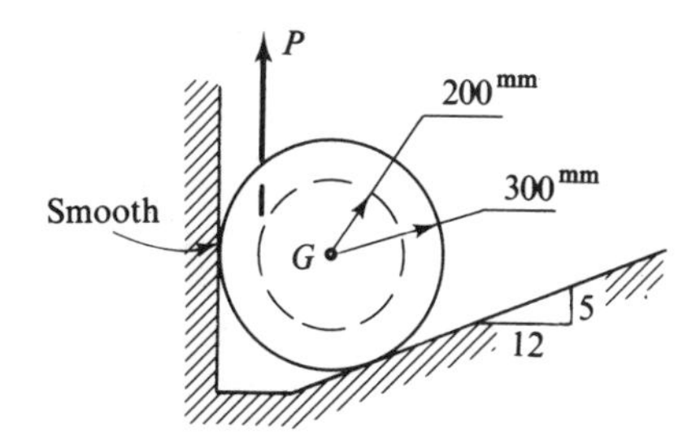

Figure P9-115

***9-116** The coefficient of friction between the cylinder and the inclined plane in Problem 9-115 is increased to 0.60. Determine the angular acceleration of the cylinder and the linear acceleration of G when
(a) $P = 60$ N.
(b) $P = 75$ N.

9-117 The 12.88-lb homogeneous disk A in Fig. P9-117 is connected by a rope to the 32.2-lb homogeneous cylinder B. Determine the angular acceleration of body B.

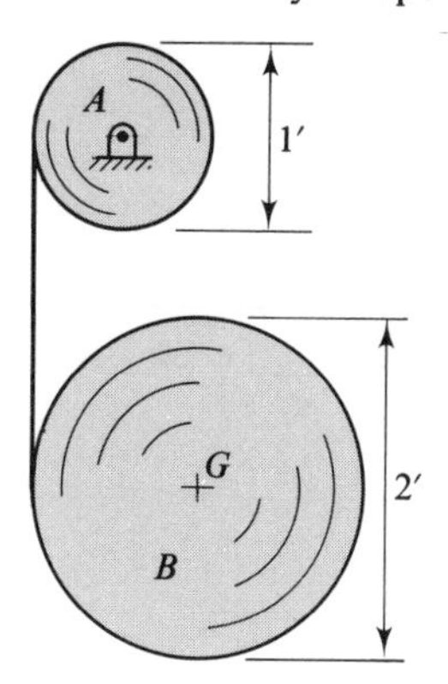

Figure P9-117

***9-118** Body A in Fig. P9-118 is a 60-kg homogeneous cylinder with a narrow slot cut for the flexible cable. Neglect the effect of the slot on the moment of inertia. When in the position shown, the acceleration of the mass center of the cylinder is 2 m/s^2 upward. Determine the mass of body B.

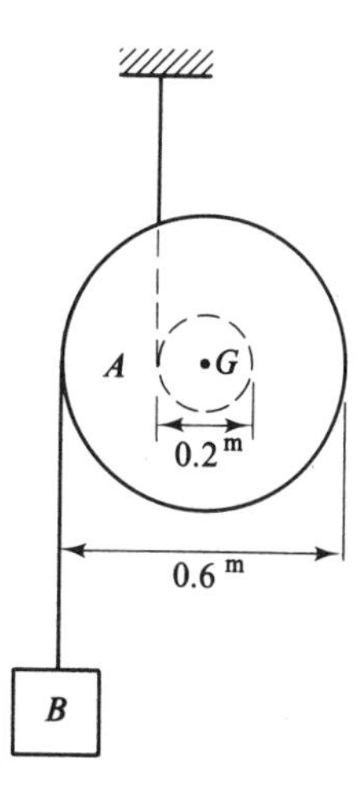

Figure P9-118

9-119 In Fig. P9-119 the homogeneous rod CD is suspended from homogeneous cylinders A and B by cables which are free to unwind from the cylinders without slipping. The weights of cylinders A and B and the rod are 10, 6, and 30 lb, respectively. The system is released from rest in the configuration shown. Determine the angular acceleration of the rod.

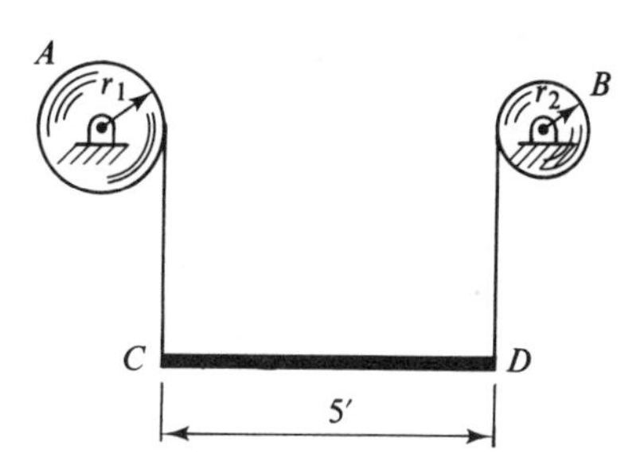

Figure P9-119

9-120 Body A in Fig. P9-120 rolls without slipping, weighs 322 lb, and has a radius of gyration of mass with respect to G of 2 ft. The angular acceleration of A is 10 rad per sec^2 clockwise, and the velocity of G is 8 fps to the right. Body B weighs 96.6 lb and is acted on by a horizontal force **P** of proper magnitude to keep the cord supporting B vertical. Determine

(a) The force **P**.

(b) The couple **C**.

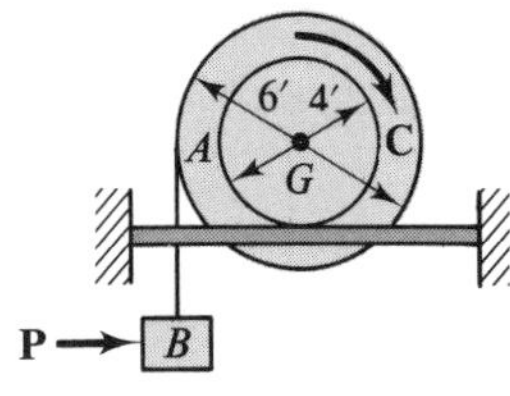

Figure P9-120

9-121 The solid homogeneous sphere in Fig. P9-121 weighs 64.4 lb and rolls without slipping on the horizontal plane. Determine the distance x for which the frictional force will be zero and the corresponding angular acceleration of the sphere.

***9-122** The coefficient of friction between the homogeneous 145-kg cylinder in Fig. P9-122 and the horizontal plane is 0.30. Neglect the effect of the slot on the moment of inertia of the cylinder. The cylinder is acted upon by a 700-N force as shown. Determine

(a) The frictional force on the cylinder.

(b) The acceleration of the mass center of the cylinder.

(c) The angular acceleration of the cylinder.

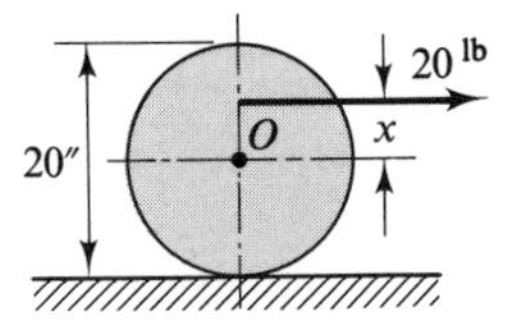

Figure P9-121

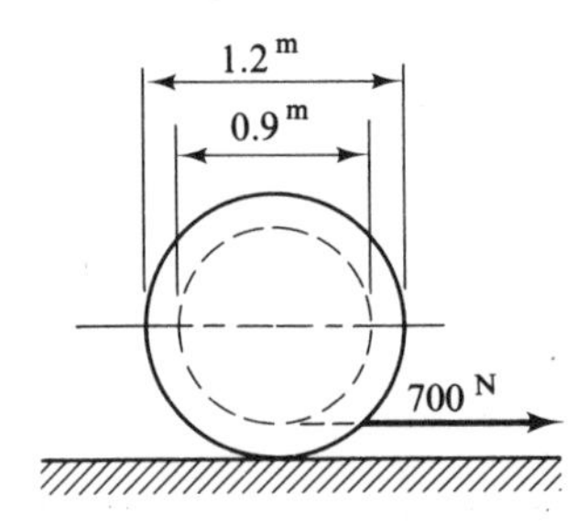

Figure P9-122

9-123 The 4-lb homogeneous rod in Fig. P9-123 is supported on a horizontal surface by friction pads at A and B. The coefficient of friction between the pads and the surface is 0.5. The rod is at rest, and the force P is 2 lb as shown. At the instant the cord is cut, determine

(a) The angular acceleration of the rod.

(b) The acceleration of the mass center G.

9-124 Solve Problem 9-123 when the force P is 8 lb as shown.

9-125 The homogeneous rod AB in Fig. P9-125 is held in equilibrium by the smooth horizontal plane and the cord BC. Determine the acceleration of the mass center of the rod and the angular acceleration of the rod at the instant the cord is cut.

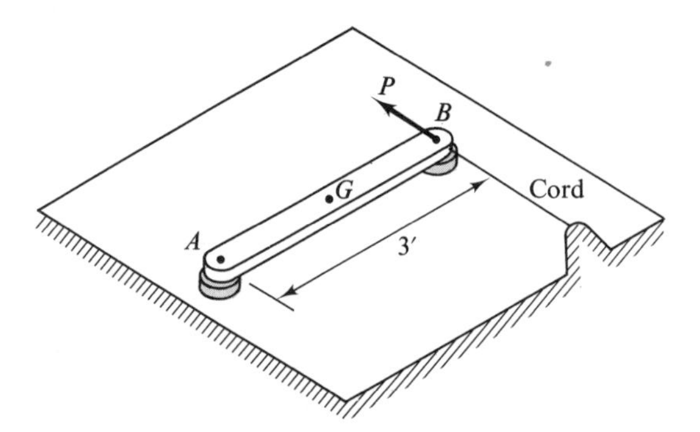

Figure P9-123

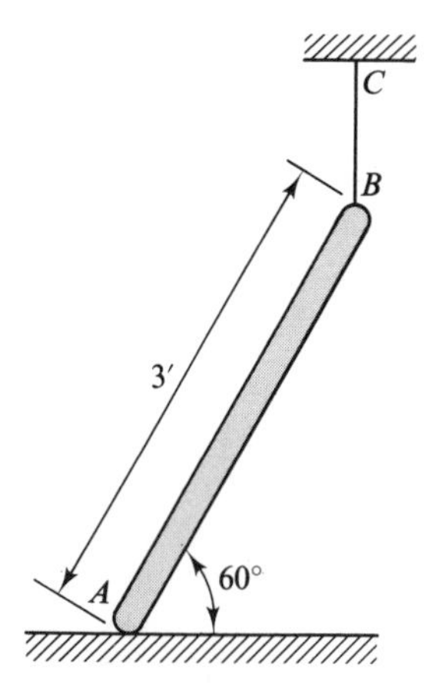

Figure P9-125

9-126 Solve Problem 9-125 if the coefficient of friction between the rod and the horizontal plane is 0.6.

9-127 Solve Problem 9-125 if the coefficient of friction between the rod and the horizontal plane is 0.2.

9-128 The wheel in Fig. P9-128 weighs 322 lb and has a radius of gyration of mass of 1.6 ft with respect to a horizontal axis through *O*. The wheel rolls and slips on the horizontal plane. In the position shown, the velocity of *O* is 4.0 fps to the right, the angular velocity is 3.0 rad per sec clockwise, and the angular acceleration is 2.0 rad per sec^2 clockwise. The coefficient of friction between the wheel and the plane is 0.20. Determine

(a) The couple *C*.
(b) The acceleration of point *O*.

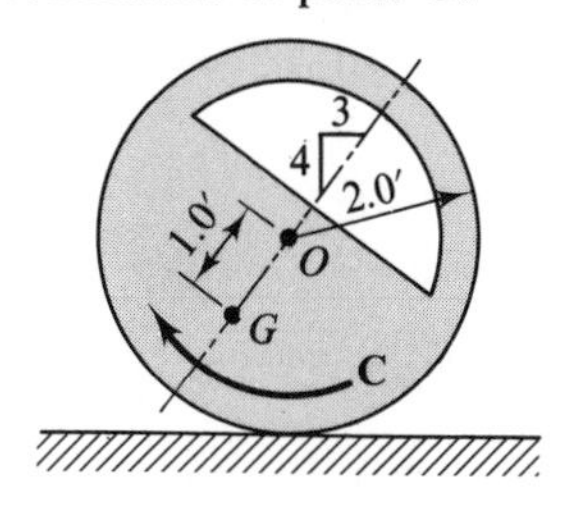

Figure P9-128

***9-129** The 150-kg wheel in Fig. P9-129 has a radius of gyration of mass with respect to an axis through the mass center *G* of 0.30 m. A couple **T** of 300 N · m acts on the wheel. The coefficient of static friction between the horizontal surface and the wheel is 0.40. Determine the magnitude of the maximum angular velocity the wheel can have in the position shown if the wheel does not slip.

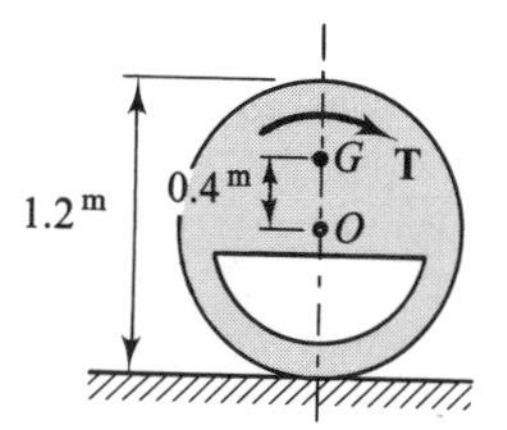

Figure P9-129

9-130 A homogeneous sphere with a 9-in. radius weighs *W* lb and is placed on an inclined plane with a slope of 3 vertical to 4 horizontal. The coefficient of friction between the sphere and plane is 0.20. If the sphere is released from rest, determine the distance the center of the sphere will move in 2.0 sec.

9-131 The homogeneous cylinder *C* in Fig. P9-131 is initially at rest when it is placed on a belt moving with a constant velocity of 40 fps to the left. The coefficient of friction between the belt and the cylinder is 0.5. Determine the angular velocity of the cylinder when it strikes the wall.

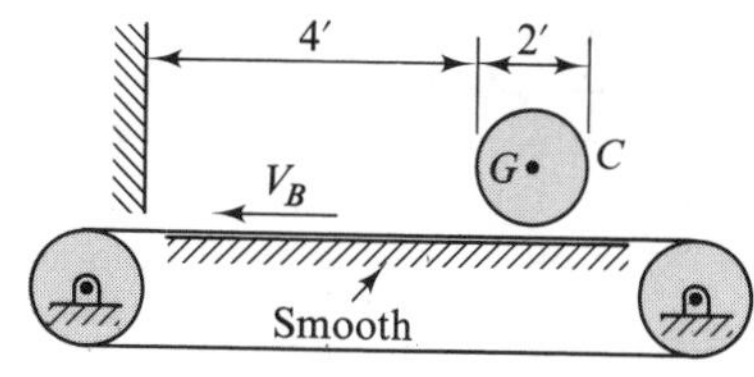

Figure P9-131

9-132 Solve Problem 9-131 if the belt moves with a constant velocity of 30 fps to the left.

***9-133** The homogeneous cylinder in Fig. P9-133 (p. 558) has a mass of 45 kg and is initially at rest on cart *B*. A force **P** is applied to the cart, giving it an acceleration of 5 m/s^2 to the right. The cylinder does not slip on the cart. Determine.

(a) The acceleration of the mass center of the cylinder.
(b) The friction force acting on the cylinder.

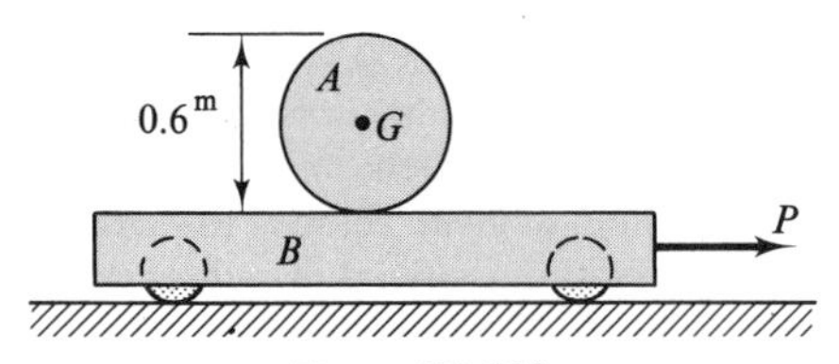

Figure P9-133

9-134 Block A and homogeneous cylinder B in Fig. P9-134 weigh 32.2 lb each. The inclined plane is smooth, and the coefficient of friction between A and B is 0.5. The bodies are released from rest when the 20-lb force is applied to A. Determine the acceleration of body A and the angular acceleration of cylinder B.

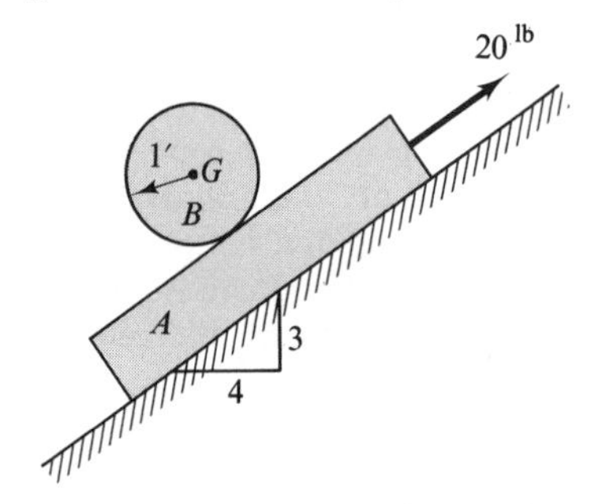

Figure P9-134

9-135 Body A in Fig. P9-135 weighs 128.8 lb, and the coefficient of friction between it and the plane is 0.2. Body B weighs 644 lb, has a radius of gyration of mass of 1.5 ft with respect to a horizontal axis through G, and rolls without slipping. The moment of the couple **C** is 100 ft-lb. Determine the acceleration of body A when it is being pulled to the right by body B.

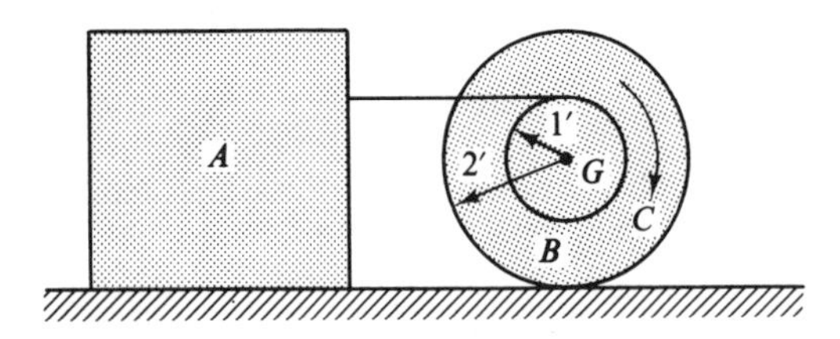

Figure P9-135

9-136 The 7-lb homogeneous rod AB in Fig. P9-136 is supported by two thin inextensible wires AD and BC. At the instant wire BC is cut, determine

(a) The angular acceleration of the rod.
(b) The tension in wire AD.

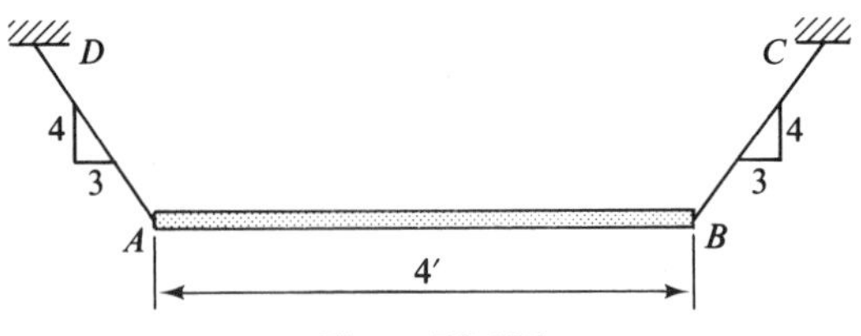

Figure P9-136

9-137 The slotted bar AB in Fig. P9-137 has a velocity of 4.0 fps to the left in the position shown when the cord breaks. The radius of gyration of the mass of the bar about its mass center is 12 in. The coefficient of friction between the bar and the pin at O is 0.50. Determine the acceleration of the mass center of the bar just after the cord breaks.

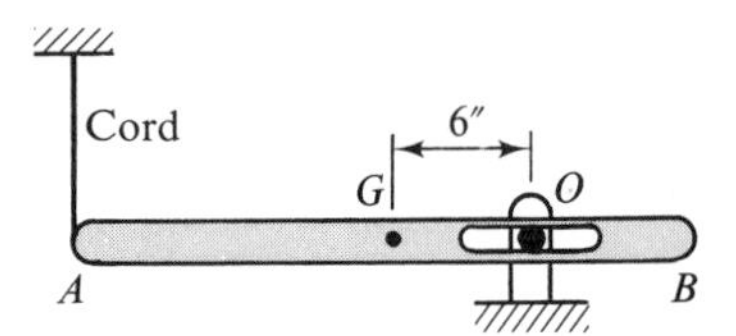

Figure P9-137

9-138 The homogeneous rod AB in Fig. P9-138 is held in equilibrium on the smooth corner C by cord AD. Determine the angular acceleration of the rod at the instant the cord is cut.

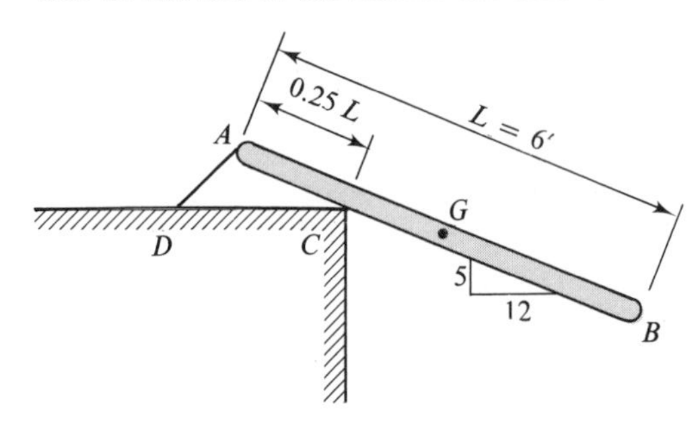

Figure P9-138

9-139 The uniform slender bar BC in Fig. P9-139 weighs 193.2 lb. It has an angular velocity of 2 rad per sec and an angular acceleration of 6 rad per sec^2, both clockwise, in this position. Determine the components of all unknown forces acting on BC.

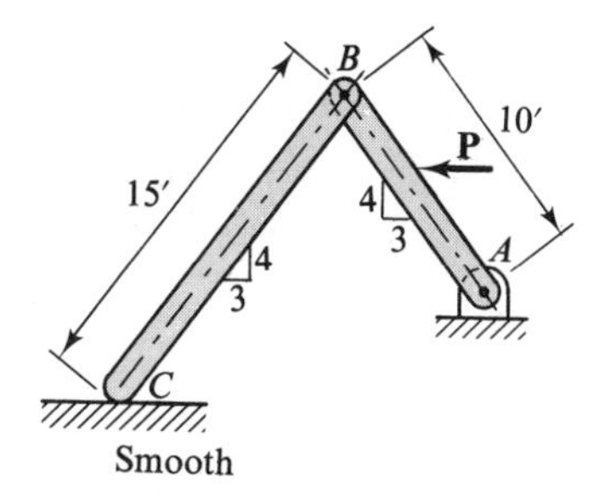

Figure P9-139

*9-140 The velocity of point B on the 40-kg homogeneous rod in Fig. P9-140 is 3 m/s down. End B of the rod is smooth, and the coefficient of friction between end A and the horizontal plane is 0.5. Determine the angular acceleration of the rod.

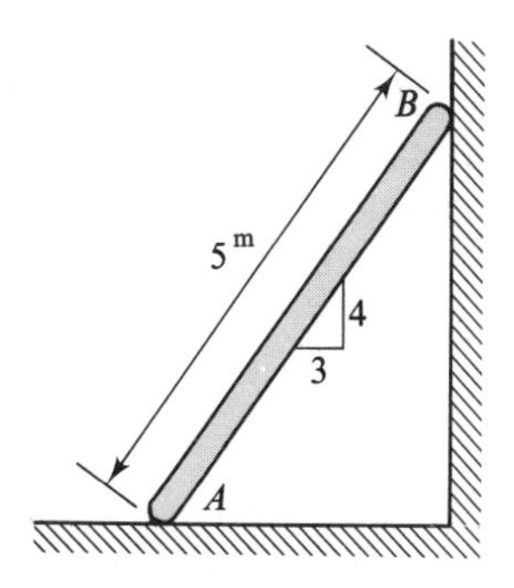

Figure P9-140

9-141 The 10-ft bar AB in Fig. P9-141 weighs 322 lb and has a radius of gyration of mass of 3 ft with respect to a horizontal axis through G. The angular velocity of the bar is 2 rad per sec clockwise in the position shown. Determine the couple **C** that will cause the bar to have an angular acceleration of 5 rad per sec^2 counterclockwise. Neglect the weights of the smooth guide blocks at A and B.

9-142 The homogeneous cylinder A in Fig. P9-142 weighs 193.2 lb and in the position shown is acted on by the torque $\mathbf{T} = 40$ ft-lb counterclockwise. The wheel has a constant angular velocity of 8 rad per sec counterclockwise. Rods BC and DE each weigh 64.4 lb. Determine the reaction of the pin at B on member BC.

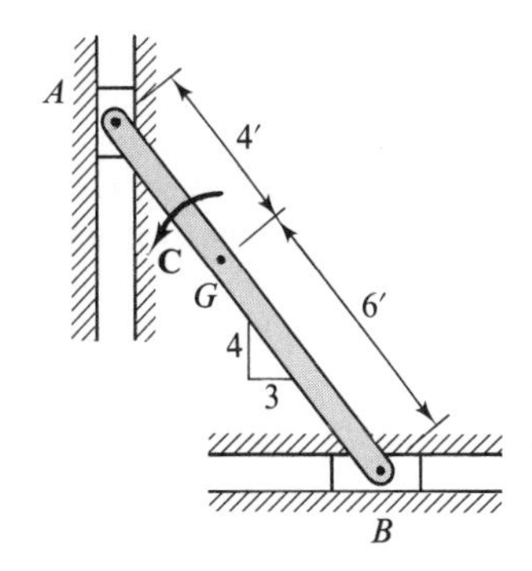

Figure P9-141

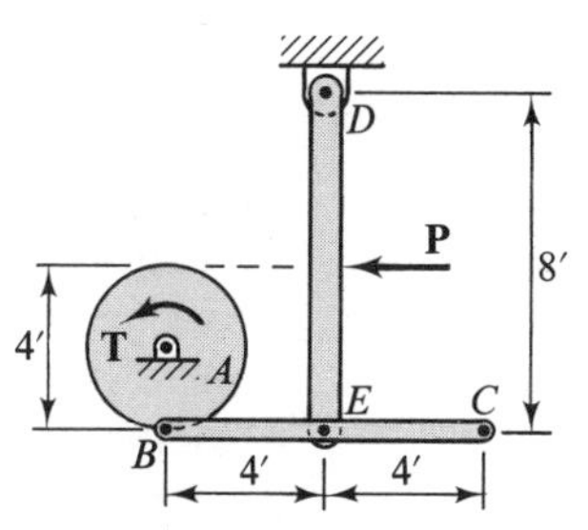

Figure P9-142

9-143 In Fig. P9-143, body A is a homogeneous cylinder weighing 64.4 lb, and B is a block weighing 8.05 lb. The flexible cord connecting A and B passes over two small pulleys whose weights and diameters may be neglected. When the angle θ is 90°, the velocity of A is 10 fps to the right. Body A rolls along the plane without slipping. Determine all unknown forces acting on A when $\theta = 90°$.

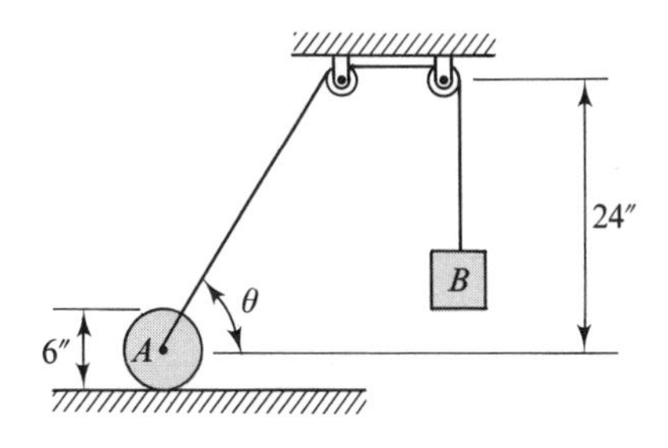

Figure P9-143

9-144 The unbalanced wheel in Fig. P9-144 weighs 64.4 lb and rolls without slipping along the inclined plane. The center of the wheel C has a velocity of 2.0 fps and an acceleration of 10.0 fps^2, both up the plane, when in the position shown. Determine the radius of gyration of the mass of the wheel with respect to a horizontal axis through G.

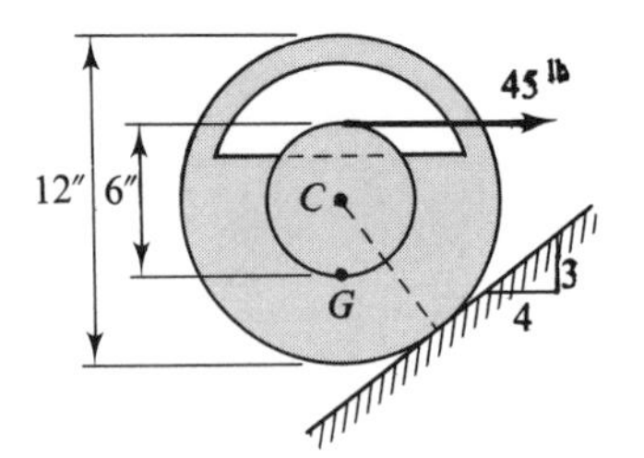

Figure P9-144

9-12 REVERSED EFFECTIVE FORCES AND COUPLES

A corollary of d'Alembert's principle for a system of particles was presented in Art. 9-7. The corollary states the following: *If a force equal in magnitude to and collinear with, but opposite in sense to, the resultant of the* m**a** *quantities of the particles is added to the external forces acting on the system, the resulting force system will be in equilibrium.* d'Alembert's reasoning can be extended to rigid bodies. Thus, by d'Alembert's principle, the equation of motion of the mass center of a rigid body becomes

$$\mathbf{F}+(-m\mathbf{a}_G)=0. \tag{9-21}$$

When the body has translation, the resultant external force passes through the mass center so the $m\mathbf{a}_G$ vector also must pass through the mass center. For translation, therefore, the problem may be solved by the methods of statics if the reversed effective force $(-m\mathbf{a}_G)$ is placed at the mass center in the free-body diagram. The equations $\sum \mathbf{F}=0$ and $\sum \mathbf{M}=0$ can be applied using all forces on the diagram including the reversed effective force $(-m\mathbf{a}_G)$. This method offers the advantage that the moment equation can be written with respect to any point or axis, thereby making it possible to select moment centers which eliminate the greatest number of unknowns from the moment equation.

The application of *d'Alembert's principle* becomes somewhat more involved when the body has angular motion. For example, if the equations of motion for a rigid body with plane motion are examined—see Eq. (9-16) and (9-17)—it is evident that by *d'Alembert's principle* they become

$$\mathbf{F}+(-m\mathbf{a}_G)=0$$

and

$$\sum \mathbf{M}_G+[-(\omega^2 I_{yz}-\alpha I_{xz})\mathbf{i}+(\omega^2 I_{xz}+\alpha I_{yz})\mathbf{j}-\alpha I_z\mathbf{k}]=0, \tag{9-22}$$

where the second term of the last equation represents the *reversed effective couple.* Reversed effective couples are not true couples any more than reversed effective forces are true forces. By placing the reversed effective force ($-m\mathbf{a}_G$) at the mass center and the reversed effective couple anywhere on the body, the problem may be treated as one of statics. Care must be taken to make sure that the moments and products of inertia are computed with respect to axes through the mass center of the body. Once the reversed effective forces and couples are shown on the free-body diagram, the equations of motion are simply $\sum \mathbf{F}=0$ and $\sum \mathbf{M}=0$, where all forces and moments, including the reversed effective forces and couples, are used. Any convenient point may be used as a moment center.

When solving a problem involving a rigid body rotating about a fixed axis, it may be more convenient to use Eq. (9-14) to compute the moment of the reversed effective couple. The xyz axes are used to compute moments and products of inertia. If this is done, the reversed effective force ($-m\mathbf{a}_G$) must be placed through the origin used to determine the reversed effective couple.

As indicated in Art. 9-7, the concept of reversed effective forces and couples does not present any new information and is therefore only an alternative to the methods previously presented. It is important to decide which method will be used early in the solution of any problem since if the two methods are mixed in a single solution, the $m\mathbf{a}$ terms are often used once as reversed effective forces and again as acceleration terms, resulting in incorrect results. When applying d'Alembert's principle, it is desirable to show the reversed effective forces and couples as dashed lines to distinguish them from true forces and couples.

Reversed effective forces are helpful in the study of the banking or superelevation of highway and railroad curves. *Superelevation* is defined as the difference in elevation between the outer and inner edges of a roadway or between the outer and inner rails of railroad tracks. The analysis is similar for highway and railroad curves; hence, the following discussion of superelevation of highways is also applicable to railroads.

The tendency for an automobile to slide away from the center of curvature of a curve or to overturn while traveling on a curve can be overcome by banking the highway. The car in Fig. 9-21(a) (p. 562) is traveling along a horizontal highway curve which has a radius of R ft. The speed of the car is v fps. The free-body diagram of the car is shown in Fig. 9-21(b). The mass of the car is obviously not symmetrical with respect to the plane of motion, the horizontal plane in which G moves, and therefore the reversed effective force should not pass through G. The exact location of the reversed effective force would be difficult to determine, however, and it is found that the error introduced by placing the force through G is comparatively small; therefore, the

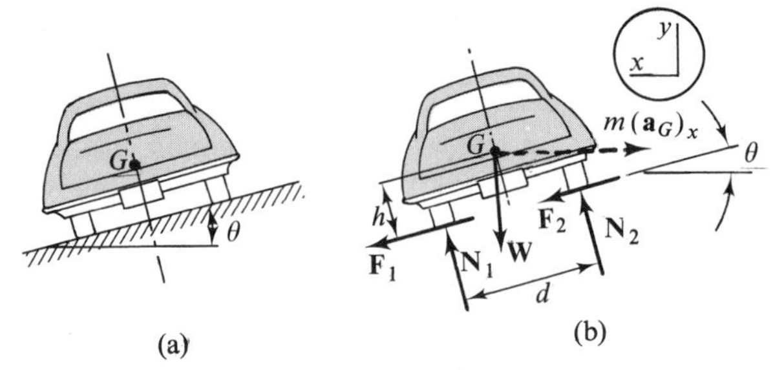

Figure 9-21

reversed effective force is shown acting through G. Moreover, the position of the reversed effective force does not affect the force equations and is important only in an investigation of the tendency of the car to overturn. The ideal angle of bank, θ, is one which will eliminate the frictional forces $\mathbf{F}_1$ and $\mathbf{F}_2$. This ideal angle can be determined from the following equations of equilibrium when F_1 and F_2 are set equal to zero:

$$\sum F_x = N_1 \sin\theta + N_2 \sin\theta - m(a_G)_x = 0,$$

$$\sum F_y = N_1 \cos\theta + N_2 \cos\theta - W = 0.$$

The value of $(a_G)_x$ is v^2/R. Therefore,

$$\sin\theta(N_1 + N_2) = \frac{W}{g}\frac{v^2}{R} \quad \text{and} \quad \cos\theta(N_1 + N_2) = W$$

from which

$$\tan\theta = \frac{v^2}{gR}. \tag{a}$$

The angle θ is seen to be a function of the speed of the car and the radius of the curve.

When the speed of a car is greater than the value of v, from Eq. (a) the car will tend to slide outward along the radius of curvature, and when the speed is less than the value from Eq. (a), the car will tend to slide in toward the center of curvature. This tendency for slower-moving vehicles to slide toward the center of the curve limits the permissible amount of superelevation because the amount of friction that can be developed is rather small when the highway is icy.

When the car is on the verge of overturning, the reaction N_1 of Fig. 9-21(b) will be zero. The overturning velocity can be shown to be a function of d, h, θ, and R by use of the equations of equilibrium for the free-body diagram of Fig. 9-21(b) when the car is assumed not to slip. Whether the car will slip or overturn first will depend on the coefficient of friction as well as the quantities d, h, θ, and R.

The following example illustrates the use of reversed effective forces and couples.

EXAMPLE 9-14

The uniform slender bar BE in Fig. 9-22(a) weighs 64.4 lb and is fastened to the vertical shaft by a smooth pin at E and the cord CD. The mechanism rotates about the vertical axis with a constant angular velocity of 50 rpm. Determine the force of the cord and the components of the pin reaction at E on BE.

SOLUTION

The xyz coordinate system is attached to the body and rotates with it so that the body is always in the xz plane. The angular-velocity is $\omega\mathbf{k} = 50(2\pi/60)\mathbf{k}$ rad per sec. The angular acceleration is zero, and the acceleration of the mass center is

$$\mathbf{a}_G = -x_G\omega^2\mathbf{i} = -1.8\left(50\frac{2\pi}{60}\right)^2\mathbf{i} = -49.35\mathbf{i}\ \text{fps}^2.$$

Therefore, the reversed effective force is

$$-m\mathbf{a}_G = \frac{64.4}{32.2}(49.35)\mathbf{i} = 98.7\mathbf{i}\ \text{lb}.$$

The bar is not symmetrical with respect to the plane of motion, the xy plane; therefore, Eq. (9-14a) must be used to determine the *reversed*

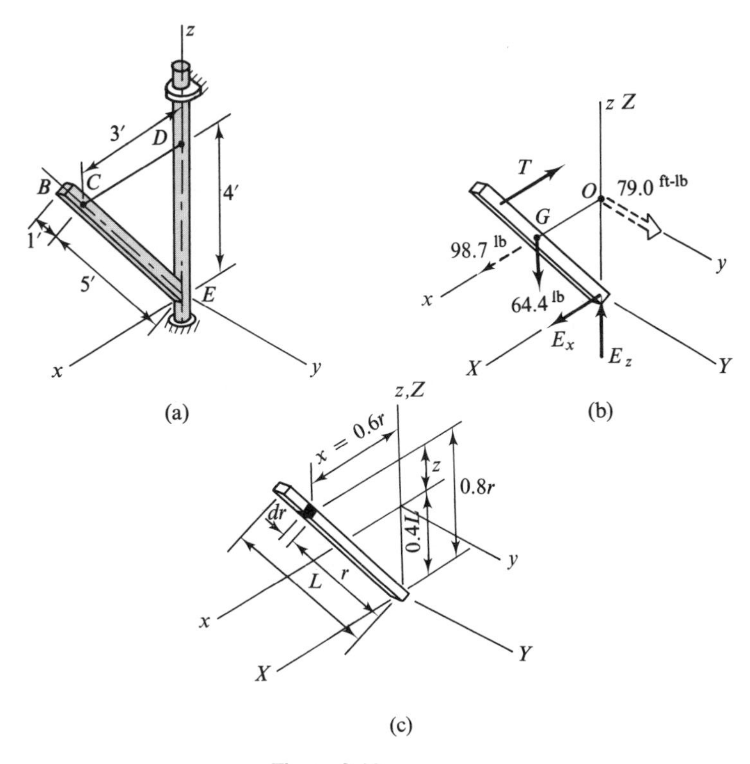

Figure 9-22

effective couple. Note that the reversed effective force must act through the origin at O if Eq. (9-14a) is used to compute the moment of the couple. The moment and products of inertia are obtained using Fig. 9-22(c). The mass of the element is

$$dm = \rho A\, dr,$$

and the mass of the bar is

$$m = \rho AL = 2.00 \text{ slugs}.$$

The moment of inertia of the element about the z axis is

$$dI_z = x^2\, dm = (0.6r)^2 \rho A\, dr,$$

and the resultant moment of inertia is

$$I_z = \rho A(0.36)\int_0^L r^2\, dr = 0.12\rho AL^3 = 0.12mL^2$$

$$= 0.12(2.00)(6.0)^2 = 8.64 \text{ slug-ft}^2.$$

The bar is symmetrical about the xz plane; therefore, I_{yz} is zero. The product of inertia of the element about the xy and yz planes is

$$dI_{xz} = xz\, dm = (0.6r)(0.8r - 0.4L)\rho A\, dr,$$

and the resultant product of inertia is

$$I_{xz} = 0.6\rho A\int_0^L (0.8r^2 - 0.4Lr)\, dr$$

$$= 0.04\rho AL^3 = 0.04mL^2 = 0.04(2.00)(6.0)^2$$

$$= 2.88 \text{ slug-ft}^2.$$

From Eq. (9-14a), the reversed effective couple is

$$\mathbf{C}_R = 0\mathbf{i} + \left(50\frac{2\pi}{60}\right)^2 (2.88)\mathbf{j} - 0\mathbf{k}$$

$$= 79.0\mathbf{j} \text{ ft-lb}.$$

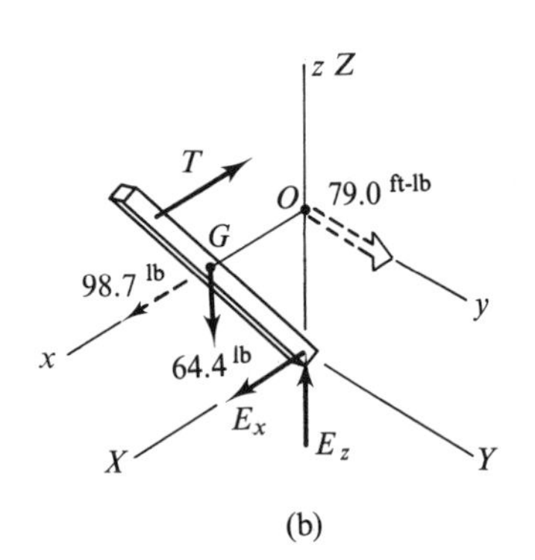

(b)

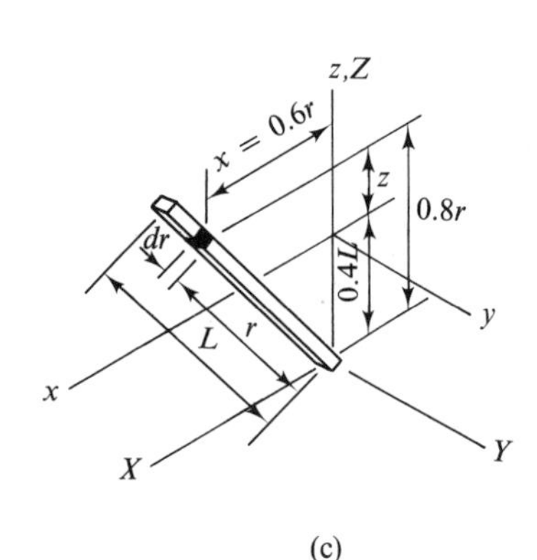

(c)

Figure 9-22(b) and (c)

The reversed effective force and couple are shown by dashed lines on Fig. 9-22(b). The equations of equilibrium are

$$\sum F_z = E_z - 64.4 = 0$$

from which

$$E_z = 64.4 \text{ lb}$$

and

$$\mathbf{E}_z = \underline{64.4\mathbf{k} \text{ lb};} \qquad \text{Ans.}$$

$$\sum M_y = -4T + 0.8(3)(98.7) + 0.6(3)64.4 + 79.0 = 0$$

from which

$$T = 107.9 \text{ lb}$$

and

$$\mathbf{T} = \underline{-107.9\mathbf{i} \text{ lb};} \qquad \text{Ans.}$$

and

$$\sum F_x = E_x + 98.7 - 107.9 = 0$$

from which

$$E_x = 9.2 \text{ lb}$$

and

$$\mathbf{E}_x = \underline{9.2\mathbf{i} \text{ lb.}} \qquad \text{Ans.}$$

PROBLEMS

9-145 Solve Problem 9-62 by the reversed effective-force method.

9-146 Solve Problem 9-71 by the reversed effective-force method.

9-147 Solve Problem 9-72 by the reversed effective-force method.

***9-148** Solve Problem 9-88 by the reversed effective-force method.

9-149 Solve Problem 9-93 by the reversed effective-force method.

9-150 Solve Problem 9-128 by the reversed effective-force method.

9-151 Solve Problem 9-141 by the reversed effective-force method.

9-152 Block A in Fig. P9-152 weighs 128.8 lb, slides in a smooth track, and is connected by a smooth pin D to the uniform bar B, which weighs 64.4 lb. In the position shown, the acceleration of the mass center of B is 15 fps^2 at a slope of 4 horizontally to the left and 3 vertically upward, and the angular acceleration of B is 8 rad per sec^2 clockwise. Determine the force P and the couple C which act on bar B.

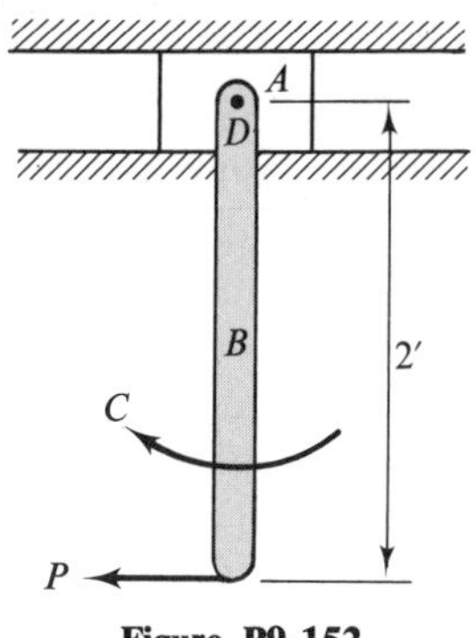

Figure P9-152

9-153 The rigid frame $ABCD$ in Fig. P9-153 slides along the horizontal plane with an acceleration of 6 fps^2 to the right. The solid homogeneous cylinder F weighs 161 lb and is fastened to the frame at E with a smooth pin. When in the position shown, the cylinder has an angular velocity of 5 rad per sec and an angular acceleration of 10 rad per sec^2, both clockwise. Determine the force Q and the components of the force on the body at E.

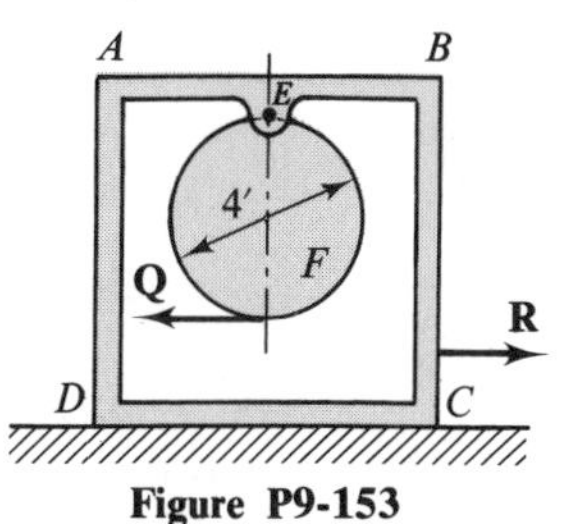

Figure P9-153

***9-154** Figure P9-154 represents a 30-kg solid homogeneous cylinder and two 15-kg slender rods. The cylinder rolls without slipping on the horizontal surface. Determine the acceleration of the center of the cylinder.

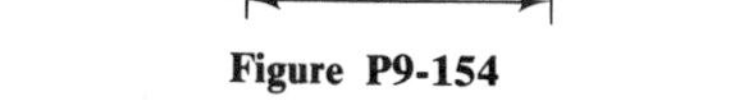

Figure P9-154

9-155 The 64.4-lb homogeneous bar AB in Fig. P9-155 (p. 566) is pinned at A and rests against the 96.6-lb body D at B. Determine the maximum weight body C can have without causing AB to rotate. All surfaces of contact are smooth, and it is assumed that body D will not tip.

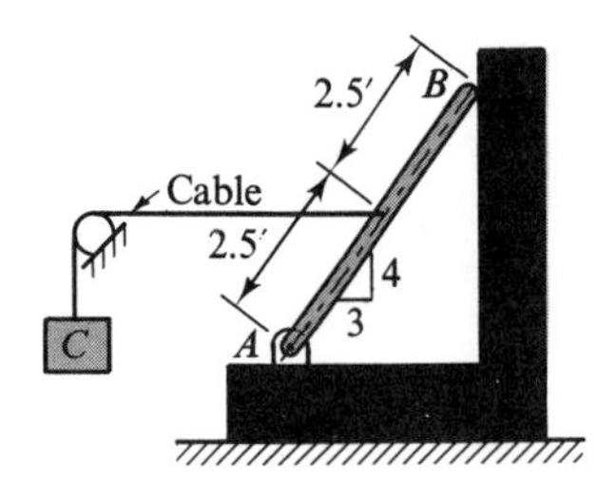

9-156 The homogeneous disk A and pulley B in Fig. P9-156 are supported by and keyed to a shaft which is mounted in bearings at C and D. The disk weighs 50 lb, body W weighs 20 lb, and the weight of the pulley and shaft may be neglected. In the position shown, the magnitude of the angular velocity of A is 2.0 rad per sec. Determine the components of the bearing reactions on the shaft at C and D.

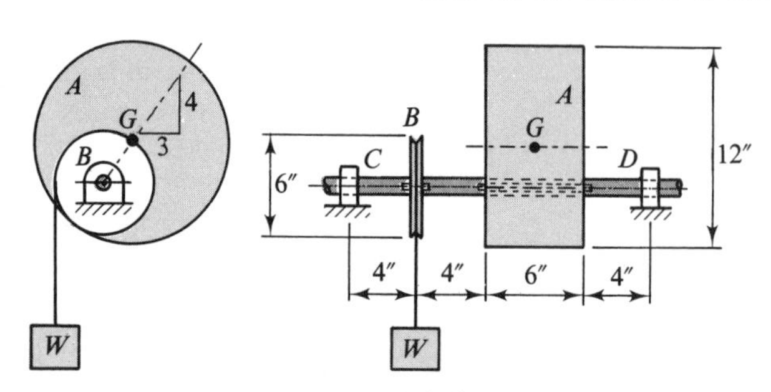

Figure P9-156

9-13 CLOSURE

An orderly method of procedure is a definite aid in solving problems in kinetics. The main essentials of the systematic procedure of Art. 9-6 include (1) a complete free-body diagram; (2) a careful analysis of the unknown forces, accelerations, and so on; (3) the timely use of kinematic and friction relations to reduce the number of unknowns involved; and (4) the use of the appropriate equations of motion.

When a rigid body has pure rotation, it is usually desirable to resolve the pin reaction into components in the directions of the normal and tangential components of the acceleration of the mass center of the body. Moments are usually computed with respect to an axis through the mass center when the body has either translation or plane motion and with respect to the axis of rotation when the body has rotation. When the reversed effective-force method is used, moments can be taken with respect to any axis. Whether or not to use reversed effective forces is a matter of individual preference.

Instantaneous values of forces and accelerations are obtained by the methods of force, mass, and acceleration. Velocities, displacements, and distances traveled can be determined by means of v-t or ω-t diagrams when accelerations are constant. If some of the forces acting on a body are variable, the acceleration is variable, and the differential equations of kinematics must be used to determine velocities, displacements, and so on. In this case, the principle of work and energy or of impulse and momentum as discussed in the next two chapters may provide a more direct solution.

PROBLEMS

Note: Unless otherwise specified, all cords, ropes, and cables are assumed to be flexible, inextensible, and of negligible weight. All pins and axles are assumed to be smooth unless otherwise stated.

9-157 Body *A* in Fig. P9-157 weighs 32.2 lb, and body *B* weighs 48.3 lb. The coefficient of friction between *A* and *B* is 0.40, and the plane is smooth. Determine the maximum horizontal force which can be applied at *C* without causing *A* to tip or slide on *B*
(a) If the force is to the left.
(b) If the force is to the right.

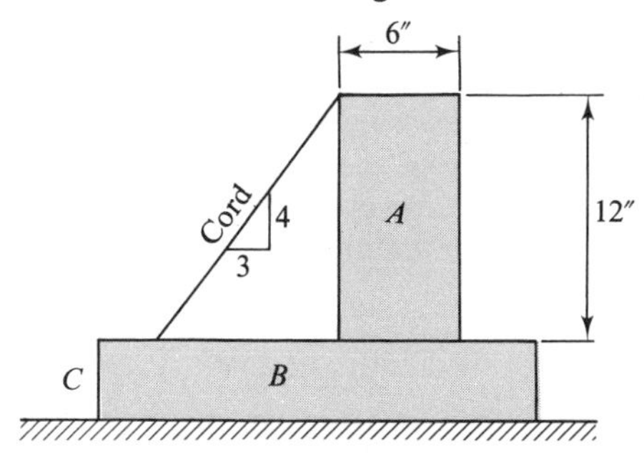

Figure P9-157

9-158 Body *A* in Fig. P9-158 weighs 20.0 lb, and body *B* weighs 10.0 lb. The coefficient of friction is 0.20 at *C*, 0.50 at *D*, and zero at *E*. The bodies are initially at rest when the force $P = 8.00$ lb is applied. Determine the acceleration of *A* with respect to *B* immediately after the force is applied.

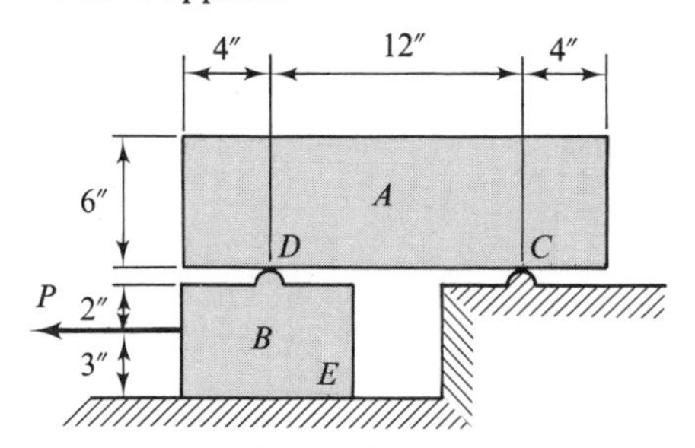

Figure P9-158

9-159 The 64.4-lb homogeneous triangular block *B* in Fig. P9-159 is connected to frame *A* by a smooth pin at *O*. Determine the maximum acceleration up the plane that can be given frame *A* without causing *B* to rotate.

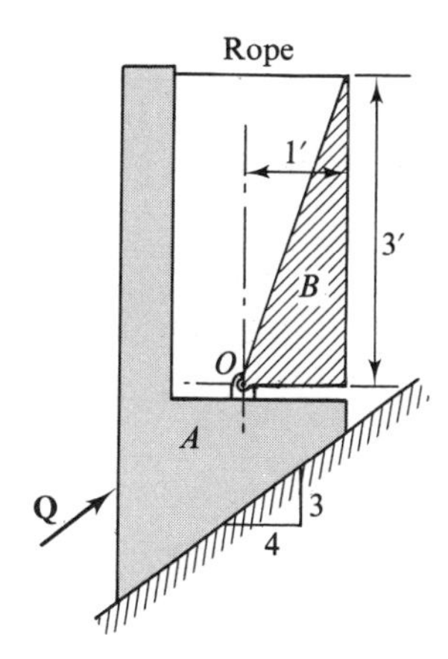

Figure P9-159

***9-160** The two slender homogeneous bars in Fig. P9-160 are rigidly connected at *B* and rotate in a vertical plane. The mass of bar *AB* is 45 kg, and the mass of bar *CD* is 15 kg. In the position shown, the angular velocity is 3.0 rad/s clockwise. Determine
(a) The angular acceleration of the bars.
(b) The reaction of the pin at *A* on *AB*.

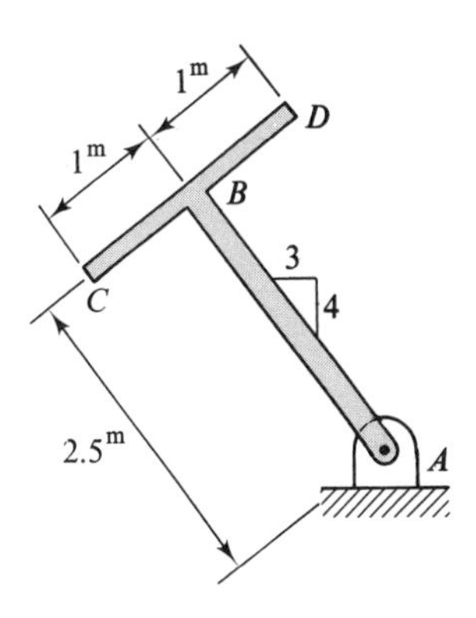

Figure P9-160

9-161 Body A in Fig. P9-161 is a 96.6-lb slender homogeneous bar which is rigidly connected to the 193.2-lb homogeneous half-cylinder B. Bodies A and B rotate about a horizontal axis at O. Block C is attached to a rope which wraps around B. In the position shown, the angular velocity and acceleration of A and B are 3.0 rad per sec and 2.0 rad per sec^2, both clockwise. Determine
(a) The weight of block C.
(b) The magnitude of the bending moment at the junction of A and B.

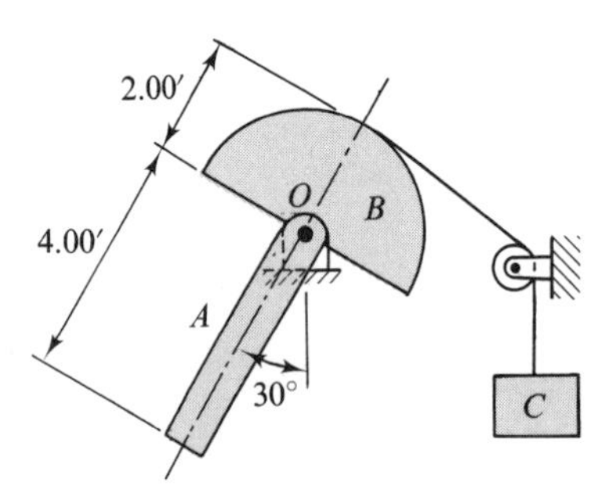

Figure P9-161

9-162 The homogeneous body A in Fig. P9-162 rolls and slips on the plane. It has a radius of gyration of mass with respect to a horizontal axis through G of 2.0 ft. The weight of A is 644 lb. The coefficient of friction between A and the plane is 0.2. The initial velocity of G is 4.5 fps down the plane, and the cable is taut. Determine the acceleration of point G.

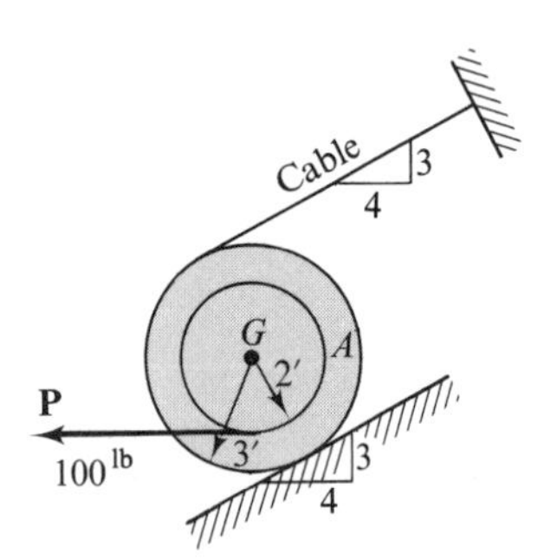

Figure P9-162

***9-163** The homogeneous cylinder A in Fig. P9-163 has a mass of 60 kg, and the slender homogeneous bar BC has a mass of 15 kg. The bar is welded to the cylinder in its plane of symmetry. The assembly moves on a smooth horizontal surface. In the position shown, the body has an angular velocity of 2.0 rad/s clockwise. Determine the acceleration of point C at the instant shown.

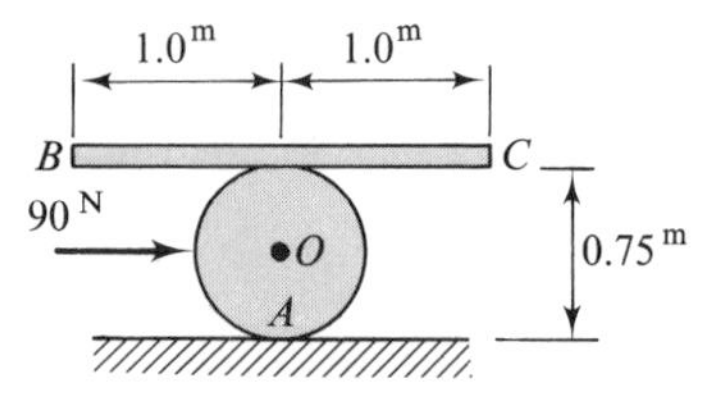

Figure P9-163

9-164 Determine the components of the force exerted by the piston pin A on the connecting rod AB of the reciprocating engine in Fig. P9-164 for the crank position of 90°. The crank has a constant angular velocity of 100 rad per sec clockwise. The connecting rod weighs 8.05 lb and has a centroidal radius of gyration of 2.77 in. The pressure of the expanding gases on the 7.00-lb piston at this position is 85.0 psi. Neglect friction between the piston and the cylinder.

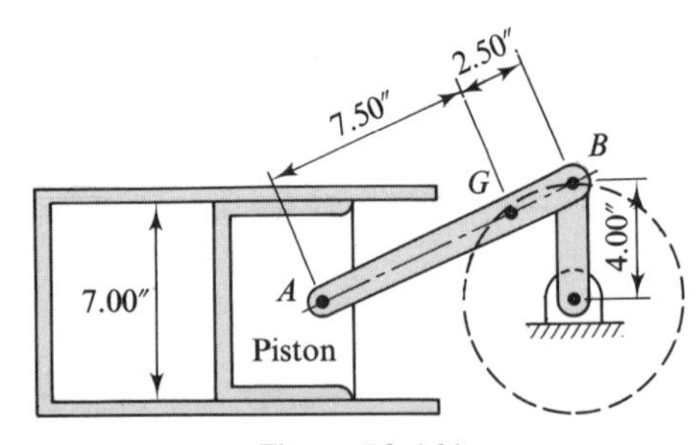

Figure P9-164

9-165 Body A in Fig. P9-165 weighs 128.8 lb (including brackets) and is pinned at O to the 64.4-lb solid homogeneous drum C. The cord which unwinds from C passes over a smooth peg as indicated. Determine the acceleration of the 96.6-lb block B.

9-166 The assembly in Fig. P9-166 consists of a 32.2-lb solid homogeneous cylinder C rigidly connected to the 64.4-lb slender homogeneous bar AB. The assembly rotates in a vertical plane about the fixed axis at A and, in the position shown, has an angular velocity of 4 rad per sec

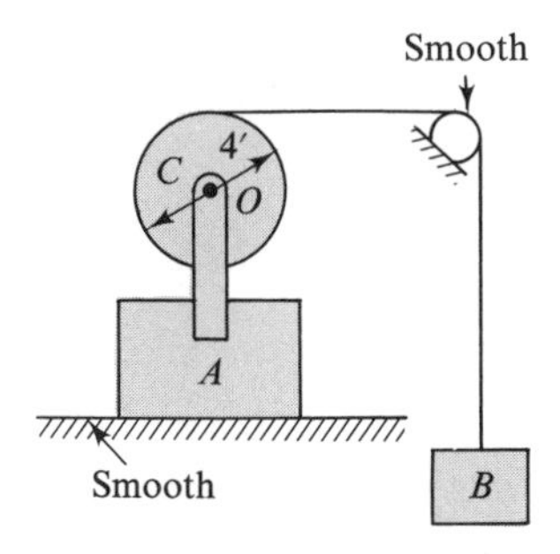

Figure P9-165

clockwise. The magnitude of the moment of the couple **T** is 100 ft-lb. Determine, for this position,

(a) The angular acceleration of the assembly.
(b) The bearing reaction at A on the bar.

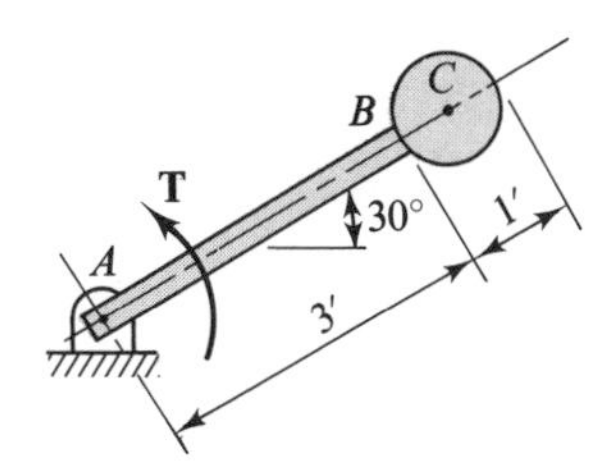

Figure P9-166

9-167 Both rods in Fig. P9-167 move in the vertical plane. When in the position shown, rod AB has an angular velocity of 2 rad per sec clockwise and an angular acceleration of 3 rad per sec^2 counterclockwise, and the homogeneous 64.4-lb rod BC has an angular velocity of 3 rad per sec clockwise. Determine

(a) The angular acceleration of rod BC.
(b) The pin reaction on rod BC.

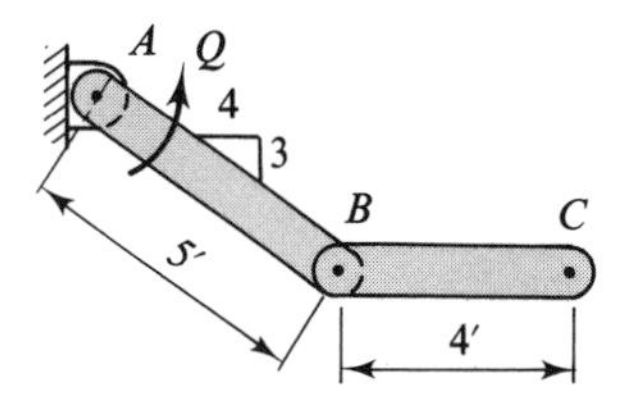

Figure P9-167

9-168 The 50-lb homogeneous cylinder in Fig. P9-168 is connected to the wall by cord AB and is acted upon by the vertical force **P**. The cord and the force **P** both lie in the plane of motion, and the effect of the slot can be neglected. The coefficient of friction between the cylinder and the floor is 0.20. Determine the acceleration of G and the tension in the cord when the force P is (a) 4.0 lb, (b) 10.0 lb.

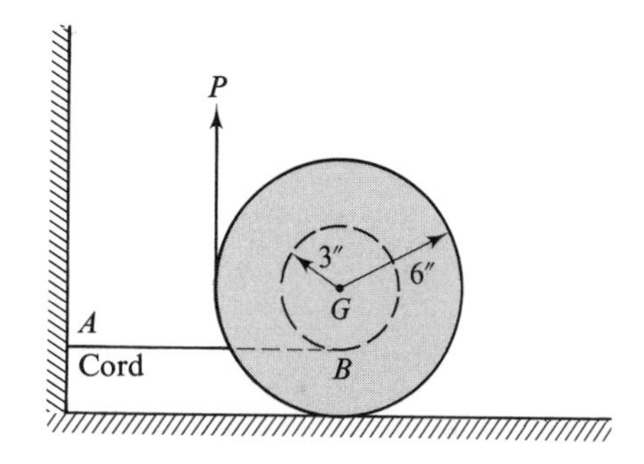

Figure P9-168

9-169 Solve Problem 9-168 for the case when the force P is 20.0 lb.

10 Kinetics–Work and Energy

PART A–PARTICLES

10-1 INTRODUCTION

In Chapter 9, Newton's laws of motion relating force, mass, and acceleration were used in the solution of problems in kinetics. Newton's laws are concerned with the forces acting on a particle or body at any instant and the resulting acceleration. If the position or velocity is required as a function of time, the acceleration obtained by the methods of Chapter 9 can be integrated as discussed in Chapter 8. This integration is usually straightforward if the acceleration is found to be constant or a function of the time. When the acceleration is obtained as a function of the position of the particle or body the problem can frequently be solved more readily by applying the principle of work and kinetic energy, which equates the work done on a particle or body to the change in kinetic energy of the particle or body. When any of the forces acting on a particle vary with the position of the particle the method of work and energy is particularly effective.

10-2 WORK DONE BY A FORCE OR A SYSTEM OF FORCES

The term *work*, as used in mechanics, should not be confused with such common terms as yardwork, roadwork, and housework. The definition of *work* and its calculation are presented in Art. 7-3. This material is included here as a review and for a ready reference.

The work done by a constant force **F** during a displacement **q** is defined as the product of the rectangular component of the force in the direction of the displacement and the magnitude of the displacement. If the angle between **F** and **q** is θ, the work done can be expressed as

$$U = Fq \cos \theta, \tag{10-1}$$

where U is the work done. Figure 10-1 shows that the work done by the force **F** during the displacement **q** as the block moves from position A to position B is the product of the magnitude of the rectangular component of **F** in the direction of the displacement, $F \cos \theta$, and the magnitude of the displacement. Thus U is equal to $(F \cos \theta)q$. The work done may also be thought of as the product of the magnitude of **F** and the magnitude of the component of the displacement in the direction of **F**; that is, $U = F(q \cos \theta)$.

Since the definition of the work done by the force **F** during the displacement **q** is the same as the definition of the scalar or dot product of the vectors **F** and **q**, it is convenient to express the work done as

$$U = \mathbf{F} \cdot \mathbf{q}. \tag{10-1a}$$

Work is the scalar product of two vectors; therefore it is a scalar quantity having only magnitude and algebraic sign. The quantities **F** and **q** can be expressed in vector notation as

$$\mathbf{F} = (F_x \mathbf{i} + F_y \mathbf{j} + F_z \mathbf{k})$$

and

$$\mathbf{q} = (q_x \mathbf{i} + q_y \mathbf{j} + q_z \mathbf{k}).$$

The work done by **F** can be obtained by means of the dot product which determines both the magnitude and algebraic sign; thus,

$$U = \mathbf{F} \cdot \mathbf{q} = F_x q_x + F_y q_y + F_z q_z.$$

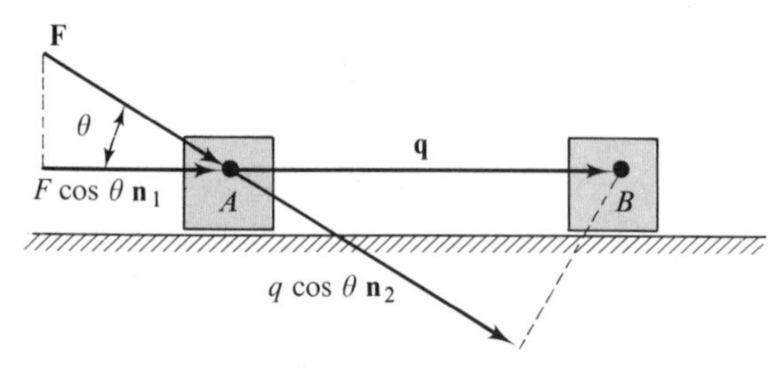

Figure 10-1

From the definition of work [see Eq. (10-1)] it may be noted that the work done by a force is positive when the displacement and the component of the force along the displacement have the same sense ($0 \le \theta < 90°$). When the displacement and the component of the force in the direction of the displacement are in opposite directions ($90° < \theta \le 180°$), the force does negative work. When $\theta = 90°$, the force has no component along the displacement and the work done by the force is zero. Since work is the product of force and displacement, it has dimensions of FL and common units are ft-lb, in-lb, and N · m (Newton·meter or joule).

Note that, by definition, the work done by a *constant* force does not depend on the path traveled by the point of application of the force during the displacement. Thus, the work done by the constant force **F** in Fig. 10-2 is the same if it is moved from position A to position B along path a, along path b, or along the straight line from A to B. Since the weight of a particle, W, is constant, the work done by the weight is Wh, where h is the vertical distance the particle moves. This work is positive when the particle moves downward and negative when it moves upward.

When the force varies either in magnitude or direction, as in Fig. 10-3, Eq. (10-1) is valid only for an infinitesimal change in position, and the work done by the force as it moves from A to B is

$$U = \int_{\mathbf{r}_A}^{\mathbf{r}_B} \mathbf{F} \cdot d\mathbf{r} = \int_{s_A}^{s_B} F \cos \theta \, ds. \tag{10-2}$$

Note that ds is the magnitude of the infinitesimal displacement $d\mathbf{r}$ and that θ is the angle between **F** and $d\mathbf{r}$. Equation (10-2) is valid for any general force and displacement, and it can be used to compute the work done if relationships between F, θ, and s can be established. Equation (10-2) can be written in a form which is sometimes more convenient to use; that is,

$$U = \int \mathbf{F} \cdot d\mathbf{r} = \int (F_x\mathbf{i} + F_y\mathbf{j} + F_z\mathbf{k}) \cdot (dx\mathbf{i} + dy\mathbf{j} + dz\mathbf{k})$$

$$= \int F_x \, dx + \int F_y \, dy + \int F_z \, dz. \tag{10-2a}$$

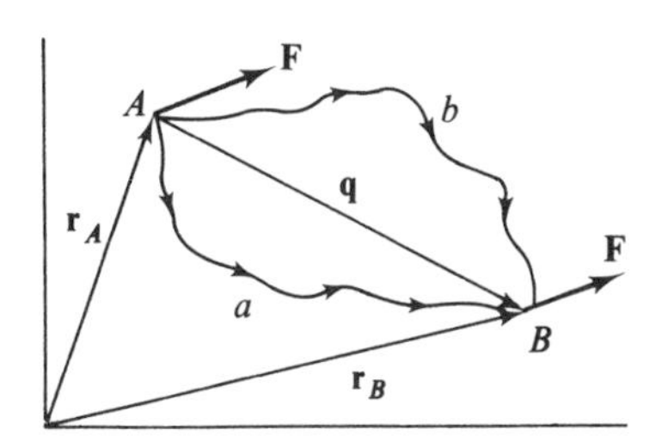

Figure 10-2

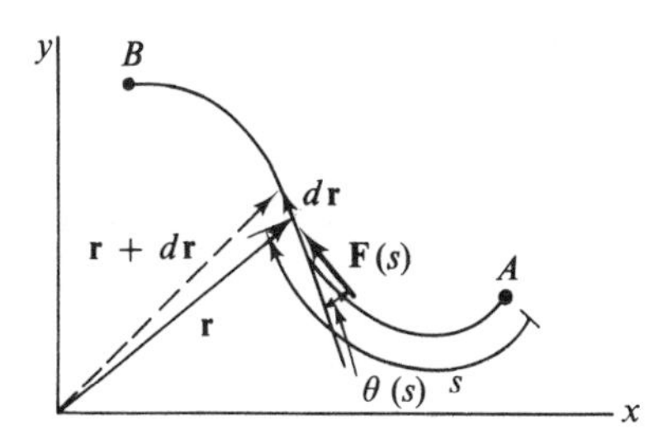

Figure 10-3

Although $\mathbf{F}$ and $d\mathbf{r}$ are vector quantities, the dot product is a scalar quantity, and ordinary scalar integration can be used to determine the work done.

Sometimes, a graphical or semigraphical method for determining the work done by a force is useful, especially when the mathematical relation between F, θ, and s in Eq. (10-2) is either unknown or cumbersome. When the component of the force in the direction of the displacement, $F\cos\theta$, is plotted against the position function s as indicated in Fig. 10-4, the area under the curve represents the work done—see Eq. (10-2). The indicator card used to determine the work done by the steam on the piston of a steam engine is an example of an F-s diagram.

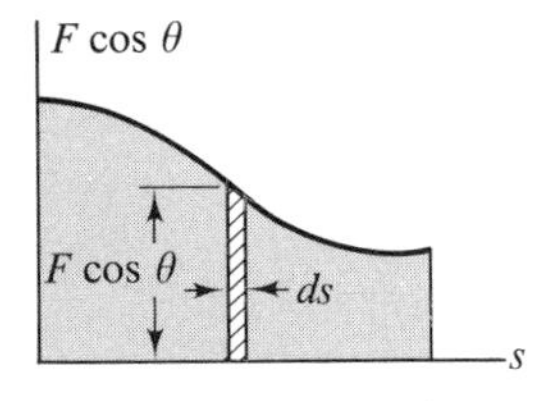

Figure 10-4

A spring is a body which deforms when acted on by a force. The relationship of the force and deformation depends on the material used and on the dimensions and shape of the spring (helical, spiral, conical, cantilever, or other). Only springs which exhibit a linear relationship between force and displacement or deformation are considered in this section. In this case, the magnitude, F, of the force is given by the expression

$$F = ks,$$

where s is the deformation of the spring from its unloaded position and k is a constant known as the *modulus* of the spring. The force-displacement diagram for a spring is shown in Fig. 10-5. As the spring is stretched an amount s from its unstretched length l_o, the force increases from zero to $F = ks$. The spring modulus, k, is the slope of the F-s diagram and has dimensions of force divided by length, often expressed as pounds per foot, pounds per inch, or newtons per meter.

The work done in stretching the spring in Fig. 10-5 from an initial length $l_o + s_1$ to a final length $l_o + s_2$ can be determined from Eq. (10-2a). Since the force and displacement are in the same direction,

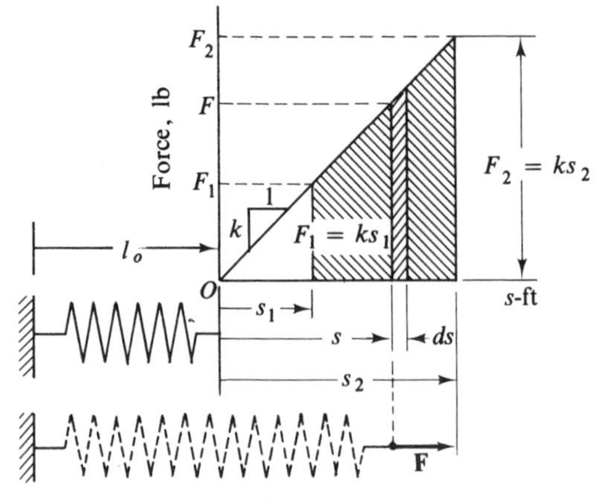

Figure 10-5

the angle θ is zero and the equation becomes

$$U = \int_{s_1}^{s_2} F \cos \theta \, ds = \int_{s_1}^{s_2} ks \, ds = \frac{k(s_2^2 - s_1^2)}{2}.$$

The force **F** is the force exerted by a body on the spring. The spring must exert the same force on the body except in the opposite direction. Consequently,

$$U_{\text{on body}} = -U_{\text{on spring}} = -\int_{s_1}^{s_2} ks \, ds.$$

The preceding expression can also be obtained as the area under the F-s diagram between s_1 and s_2, the shaded trapezoidal area. This area is

$$U = \text{area} = \frac{F_1 + F_2}{2}(s_2 - s_1).$$

When $F_1 = ks_1$ and $F_2 = ks_2$ are substituted into the expression, it becomes

$$U = \frac{ks_1 + ks_2}{2}(s_2 - s_1) = \frac{k(s_2^2 - s_1^2)}{2}$$

as before. When the spring is initially unstretched, the trapezoid becomes a triangle and the work done on the spring is $ks_2^2/2$.

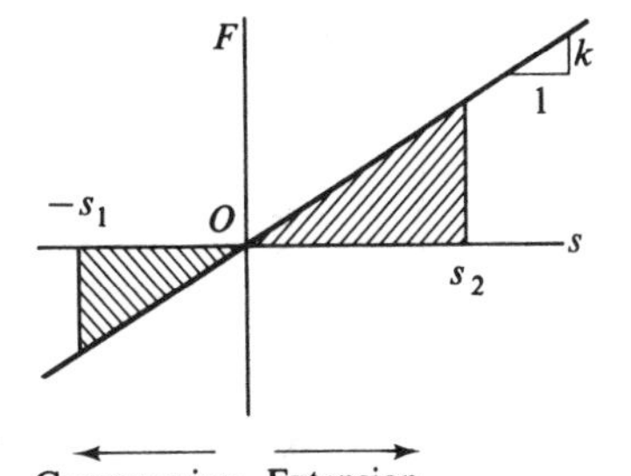

Figure 10-6

It should be noted that as a spring is being deformed (either stretched or compressed) the force on the spring and the displacement are in the same direction and the work done *on the spring* is positive. If a spring is initially deformed and released gradually, the force and displacement are in opposite directions and the work done *on the spring* is negative. The force of the spring on a body or particle is opposite the force of the body on the spring, but both the particle and the end of the spring have the same displacement; therefore, the work done *by a particle on a spring* and the work done *by the spring on the particle* will have the same magnitudes but opposite signs.

When a spring is constructed so that it can act in either tension or compression, such as an automobile suspension spring, it is usually assumed that the spring modulus is the same in tension and compression. The force-displacement diagram will be similar to Fig. 10-6, and it is apparent that, during a change of length from tension to compression (or the reverse), the force will change direction while the displacement will continue in the same direction. Consequently, the area on one side of the s axis will represent positive work and that on the other side will give negative work.

When two bodies are connected by a flexible inextensible cable the resultant work done on the bodies by the force in the cable is zero. The two forces at the ends of the cable have the same magnitude (the mass of the cable is assumed to be negligible), and if the cable is inextensible, the components of the displacements of the two ends in the direction of the forces must have the same magnitude; therefore the resultant work done by the cable is zero. The following examples demonstrate the calculation of the work done by various force systems.

EXAMPLE 10-1

The coefficient of friction between the 100-lb body A in Fig. 10-7(a) and the inclined plane is 0.20. The spring has a modulus of 20 lb per ft and is stretched 1 ft in the position shown. Determine the total work done on body A while it is moving 5 ft down the plane.

SOLUTION

A free-body diagram of body A is shown in Fig. 10-7(b). There are four forces acting on the body, and three will do work as body A moves 5 ft down the plane. The body has no acceleration normal to the plane; therefore $N = W \cos\theta = 100(\frac{4}{5}) = 80$ lb. The frictional force opposes the motion and thus acts up the plane, and since slipping occurs, $F = \mu N = 0.20(80) = 16$ lb. The work done by the friction is negative since it is opposite the displacement; thus

$$U_F = -F\,\Delta s = -16(5) = -80 \text{ ft-lb.}$$

The work done by the weight is positive since the vertical component of the displacement is in the direction of the weight; that is,

$$U_W = W(\Delta s_y) = 100[(\tfrac{3}{5})5] = 300 \text{ ft-lb.}$$

The work done by the spring force is negative since it is always directed up the plane and the movement is down the plane. The spring force is equal to the modulus (20 lb per ft) multiplied by the stretch, s, of the spring. The work may be computed by drawing an F-s diagram or making use of an integral. Since the initial stretch is 1 ft and the final stretch is 6 ft, the work done by the spring is

$$U_{\text{spring}} = -\int_1^6 20s\,ds = -10s^2\Big]_1^6 = -350 \text{ ft-lb.}$$

The total work is

$$U = -80 + 300 - 350 = \underline{-130 \text{ ft-lb.}} \qquad \text{Ans.}$$

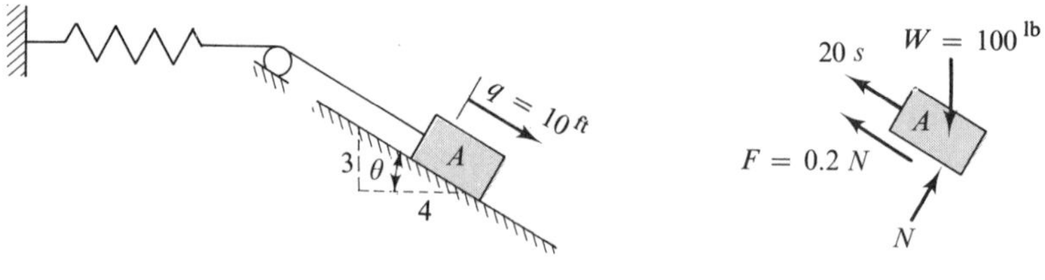

Figure 10-7

EXAMPLE 10-2

The 200-lb body A in Fig. 10-8(a) slides in the smooth slot and is connected to the 300-lb body B by an inextensible cord. The spring has a modulus of 20 lb per ft and is arranged so that it can act in either compression or tension. The unstressed length of the spring is 7 ft, and the spring is initially perpendicular to the slot. Determine the work done by the forces acting on the two bodies as A moves 12 ft down the slot from the position shown. The slope and magnitude and the 100-lb force remain constant throughout the motion.

SOLUTION

A free-body diagram of the system in an intermediate position is drawn in Fig. 10-8(b). The force in the cord is an internal force of the system, and since the cord is inextensible, the work done by the cord is zero. There are six forces on the free-body diagram, and two of these forces, **N** and **R**, are normal to their corresponding displacements; therefore they do no work. The resultant work on the system is thus

$$U = U_{300\,\text{lb}} + U_{200\,\text{lb}} + U_{100\,\text{lb}} + U_{\text{spring}}.$$

The work done by the 300-lb force is

$$U_{300\,\text{lb}} = 300(12) = 3600 \text{ ft-lb}.$$

The work done by the 200-lb force is

$$U_{200\,\text{lb}} = 200(\tfrac{4}{5})(12) = 1920 \text{ ft-lb}.$$

The work done by the 100-lb force is

$$U_{100\,\text{lb}} = \mathbf{F}\cdot\mathbf{q} = 100(0.6\mathbf{i}+0.8\mathbf{j})\cdot 12(0.6\mathbf{i}-0.8\mathbf{j})$$

$$= 1200(0.36-0.64) = -336 \text{ ft-lb}.$$

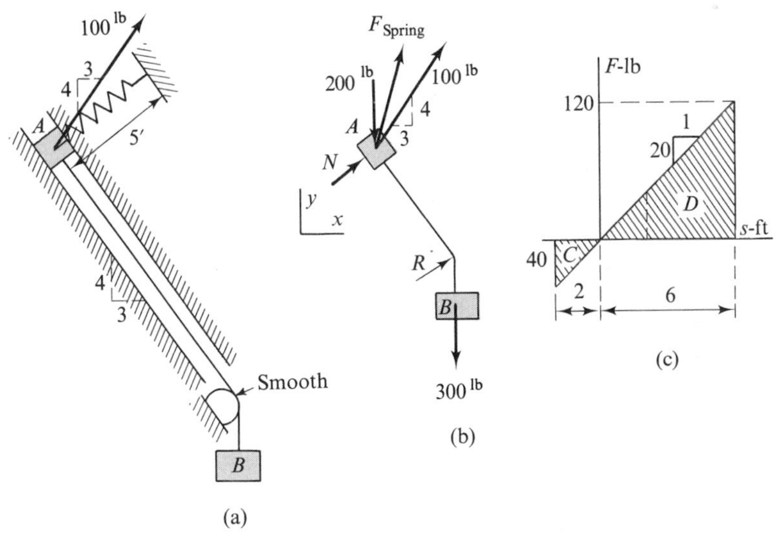

Figure 10-8

The work done by the spring can be computed from the F-s diagram in Fig. 10-8(c). As body A moves 12 ft down the slot the length of the spring changes from 5 to 13 ft. Since the spring has an unstressed length of 7 ft, it is initially compressed 2 ft and exerts a 40-lb compressive force on A. In the final position, the spring is stretched 6 ft and exerts a tensile force of 120 lb on A. The area C of the F-s diagram represents negative work done by A on the spring since the spring is in compression but is increasing in length; however, area C represents positive work of the spring on body A since the component of the force on body A along the slot and the displacement of body A are both down the slot. In a similar manner, area D represents negative work done by the spring on body A. The work done by the spring on A is

$$U_{\text{spring}} = C - D = \frac{40(2)}{2} - \frac{120(6)}{2} = -320 \text{ ft-lb}.$$

The total work done on the two bodies is

$$U = 3600 + 1920 - 336 - 320 = \underline{4864 \text{ ft-lb.}}$$

Ans.

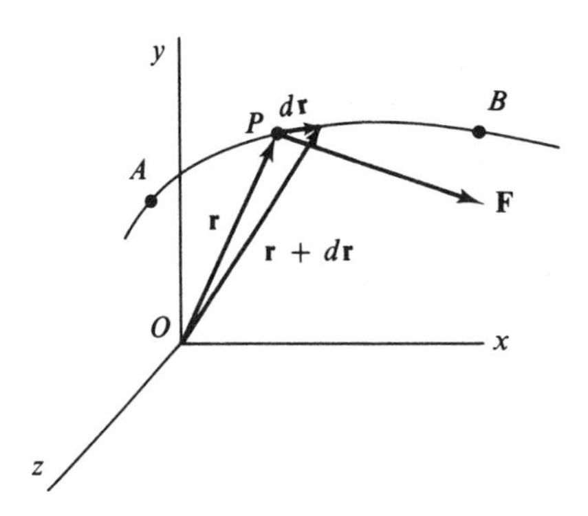

Figure 10-9

10-3 PRINCIPLE OF WORK AND KINETIC ENERGY

The particle P in Fig. 10-9 has a mass m and travels along the path from A to B. Let F be the resultant force acting on the particle, and let the axes shown be a newtonian set of axes. The work done by the force as it moves through a differential displacement, $d\mathbf{r}$, is

$$dU = \mathbf{F}\cdot d\mathbf{r} = m\ddot{\mathbf{r}}\cdot\dot{\mathbf{r}}\,dt = \tfrac{1}{2}m\frac{d}{dt}(\dot{\mathbf{r}}\cdot\dot{\mathbf{r}})\,dt = \tfrac{1}{2}m\,d(v^2),$$

where $v = |\dot{\mathbf{r}}|$. The work done in traveling from the initial position, A, to the final position, B, is

$$U = \tfrac{1}{2}m\int_{v_i}^{v_f} d(v^2) = \tfrac{1}{2}m(v_f^2 - v_i^2),$$

where the subscripts i and f indicate initial and final velocities. The term $\frac{1}{2}mv^2$ is called the *kinetic energy* of the particle and is denoted by the symbol T. The kinetic energy of the particle is its capacity to do work by virtue of its velocity. The above equation can be written

$$U = \Delta T = T_f - T_i. \tag{10-3}$$

This equation leads to the *principle of work and kinetic energy*, which can be stated as follows: *The work done on a particle* (by all forces acting on the particle) *is equal to the change of the kinetic energy of the particle.*

EXAMPLE 10-3

The 20-lb body in Fig. 10-10(a) is constrained to slide along the smooth rod BC which lies in a vertical plane. When the body is in position A its velocity is 30 fps to the right, and the spring (modulus of 40 lb per ft) is stretched 6 ft. Determine the speed of the body when it reaches the position A'.

SOLUTION

A free-body diagram of the body is drawn in an intermediate position in Fig. 10-10(b). The equation relating the work and kinetic energy is

$$U = \Delta T$$
$$= \tfrac{1}{2}m(v_f^2 - v_i^2).$$

Since the force N is the reaction of the smooth rod on the particle and is always normal to the motion of the particle, N does no work. The length of the spring changes from 21 to 17 ft. Since it is initially stretched 6 ft, it is stretched 2 ft in the final position. The spring is under tension; therefore the component of the spring force along the path is always in the direction of the movement of the body, and the spring will do positive work on the body. This work on the body can be evaluated from an integral as follows:

$$U_{\text{spring}} = -\int_{s_i}^{s_f} ks\,ds = -\int_6^2 40s\,ds = -20s^2\Big]_6^2$$
$$= 640 \text{ ft-lb}.$$

The work done by the weight on the body is

$$U_{20\text{lb}} = W(\Delta s_y) = -(20)(6) = -120 \text{ ft-lb}.$$

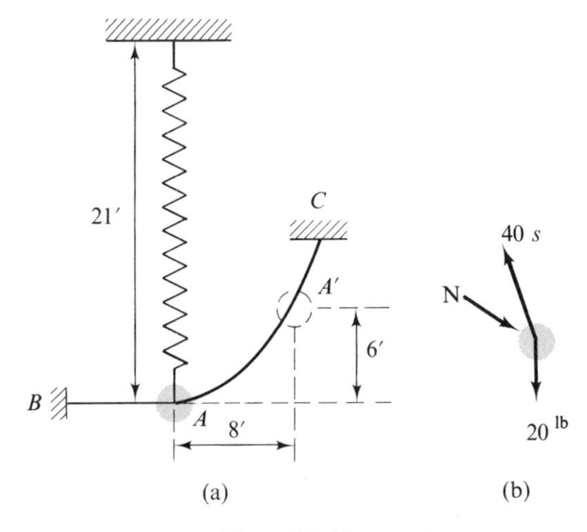

Figure 10-10

The total work done on the body is

$$U = U_{\text{spring}} + U_{20\text{lb}}$$
$$= 640 - 120 = 520 \text{ ft-lb}.$$

The work and kinetic energy equation becomes

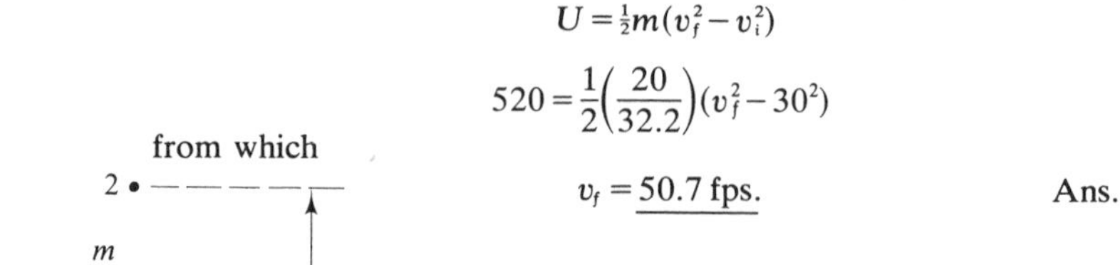

$$U = \tfrac{1}{2}m(v_f^2 - v_i^2)$$
$$520 = \frac{1}{2}\left(\frac{20}{32.2}\right)(v_f^2 - 30^2)$$

from which

$$v_f = \underline{50.7 \text{ fps.}} \qquad \text{Ans.}$$

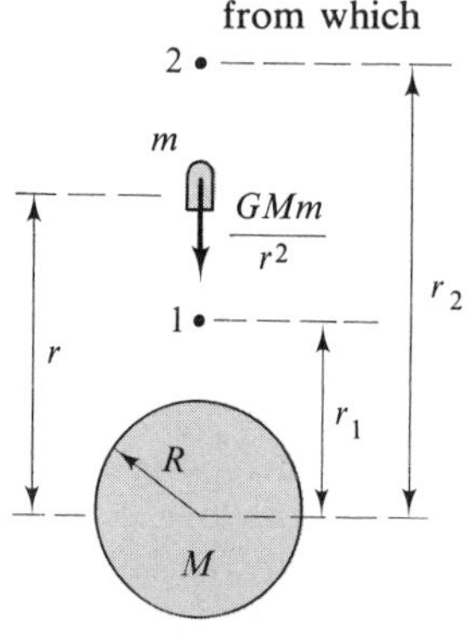

Figure 10-11

EXAMPLE 10-4

A rocket is fired vertically upward, and after burnout at position 1 it has a mass m and a vertical velocity v_1. From position 1 to position 2, as shown in Fig. 10-11, the only force acting on it is due to the inverse square law of gravitation, GMm/r^2, where G is the universal gravitation constant, M is the mass of the earth, and r is the distance from the center of the earth to the rocket. Determine the speed of the body when it reaches position 2.

SOLUTION

Since the gravitational force is a function of the position of the rocket, the principle of work and kinetic energy is used:

$$U = \tfrac{1}{2}m(v_f^2 - v_i^2).$$

The gravitational force is downward, and the movement of the mass m is upward; therefore, the work done by gravity is negative. Hence

$$U = -\int_{r_1}^{r_2} \frac{GMm}{r^2}\,dr = \frac{GMm}{r}\Bigg]_{r_1}^{r_2} = GMm\left(\frac{1}{r_2} - \frac{1}{r_1}\right)$$
$$= GMm\frac{r_1 - r_2}{r_1 r_2} = -GMm\frac{r_2 - r_1}{r_1 r_2}$$

and

$$\Delta T = \tfrac{1}{2}m(v_2^2 - v_1^2) = U = -GMm\left(\frac{r_2 - r_1}{r_1 r_2}\right)$$

from which

$$v_2^2 = v_1^2 - 2GM\left(\frac{r_2 - r_1}{r_1 r_2}\right)$$

or

$$v_2 = \underline{\left[v_1^2 - 2GM\left(\frac{r_2 - r_1}{r_1 r_2}\right)\right]^{1/2}}. \qquad \text{Ans.}$$

PROBLEMS

Note: Unless otherwise specified, the mass of each spring and cord may be neglected. All cords are to be considered flexible and inextensible; all pins, pulleys, and fixed drums are assumed to be smooth.

10-1 A stone weighing 0.25 lb is thrown with a velocity of 100 fps at an angle of 60° above the horizontal. By making use of the principle of work and kinetic energy, determine the maximum height it will rise above the point of throwing.

10-2 The force P acting on the 40-lb block A in Fig. P10-2 has a constant direction, as shown, and a magnitude that varies according to the equation $P = 6s^2$ lb, where s gives the position of A in feet. When $s = 2$ ft body A has a velocity of 4 fps to the right. Determine the velocity of A when $s = 5$ ft. The plane is smooth.

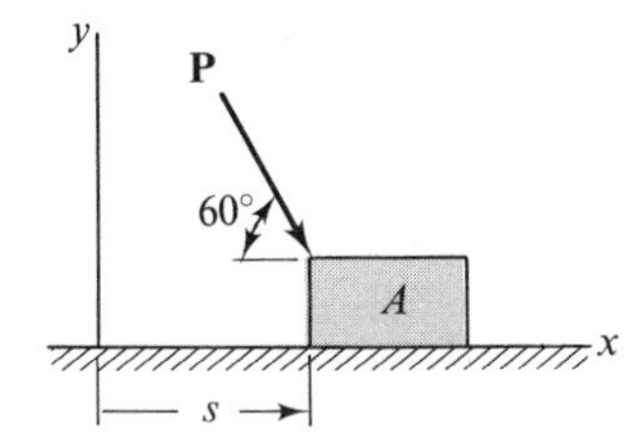

Figure P10-2

10-3 Solve Problem 10-2 if the coefficient of friction between body A and the horizontal plane is 0.10.

***10-4** Block A in Fig. P10-4 has a mass of 2.0 kg and has a velocity of 5 m/s up the plane. The coefficient of friction between the block and surface is 0.20. There is a gradual transition from the horizontal to the inclined surface. Locate the position where the block will finally come to rest.

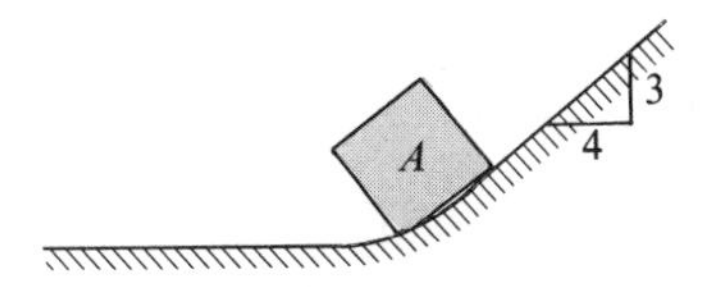

Figure P10-4

10-5 The 10-lb block A in Fig. P10-5 is projected from the 10-ft tube by a spring whose unstressed length is 2.5 ft. The modulus of the spring is 400 lb per in. The friction between the block and tube is 2.0 lb. Determine the velocity A as it leaves the tube.

10-6 The 64.4-lb collar shown in Fig. P10-6 is released from rest in position A where the spring is unstretched. Determine the modulus of

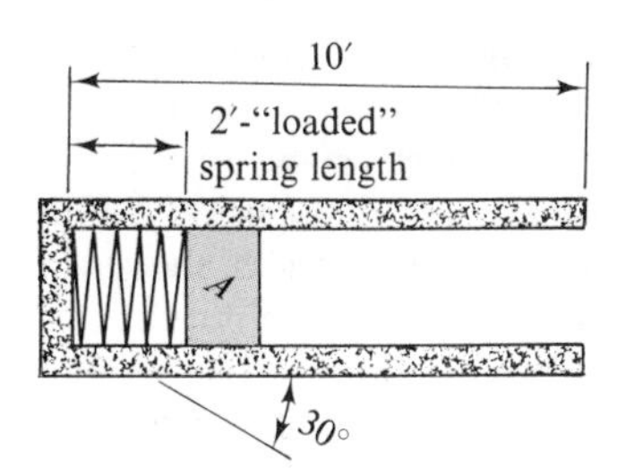

Figure P10-5

the spring if the velocity of the collar is 6 fps after it slides downward 5 in. The rod is smooth, and the 50-lb force maintains the same slope throughout the motion.

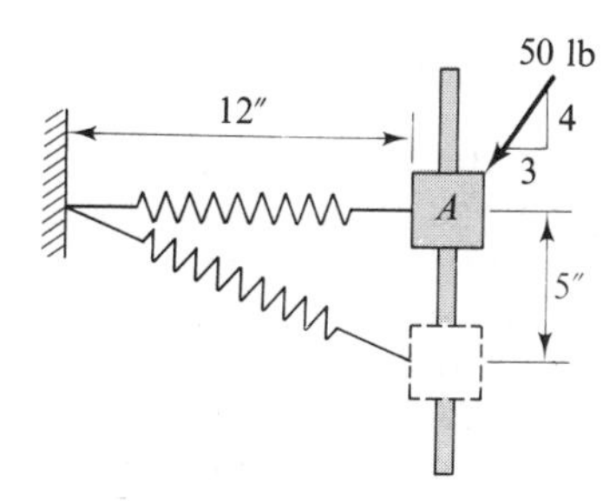

Figure P10-6

10-7 The 3.22-lb particle A in Fig. P10-7 is moving on a smooth horizontal surface and is connected to point O by an elastic cord which has a modulus of 18.7 lb per ft. In the position shown, the elastic cord is stretched 1 ft, and the velocity of A is as indicated. Determine the magnitude of the velocity of A when it reaches its maximum distance from point O of 10 ft.

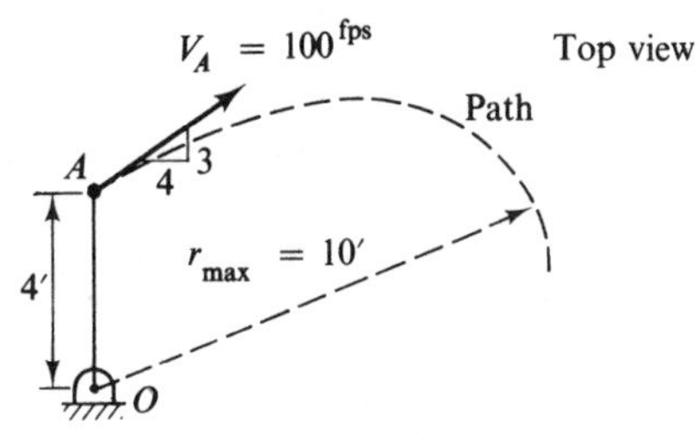

Figure P10-7

10-8 The smooth rod in Fig. P10-8 has a negligible mass and rotates about a vertical axis through O with a clockwise angular velocity of 6 rad per sec. Block B weighs 32.2 lb and is held by a string in contact with the uncompressed spring. The modulus of the spring is 27 lb per ft. The string breaks, and in the resulting motion the spring is compressed a maximum of 1 ft. Determine the speed of the block when the spring reaches its maximum compression.

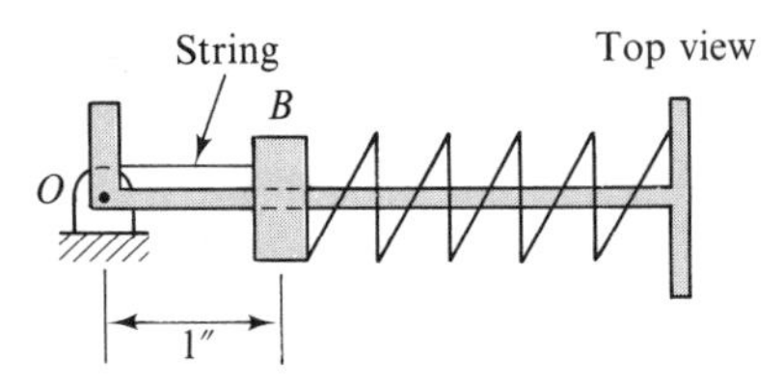

Figure P10-8

10-9 The 64.4-lb body A in Fig. P10-9 slides in the smooth slot. In the position shown, the spring is compressed 1 ft, and body A has a velocity of 10 fps up the slot. The velocity of the body is reduced to 8 fps up the slot after moving 5 ft. The 130-lb force maintains a constant slope and magnitude during the motion. The spring is attached to the block and to point C. Determine the modulus of the spring.

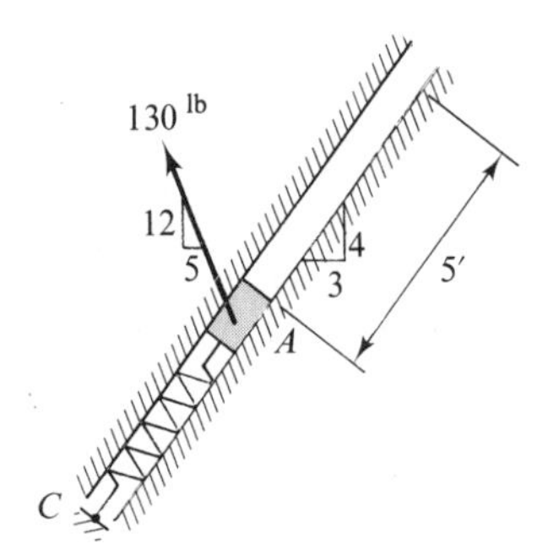

Figure P10-9

***10-10** The 20-kg block A in Fig. P10-10 slides along the smooth slot. Spring B has a modulus of 1000 N/m, and spring C has a modulus of 500 N/m. In the position shown, spring B is stretched 0.80 m, spring C is unstressed, and the block has a velocity of 10 m/s down the slot. Determine the velocity of A when it reaches position D.

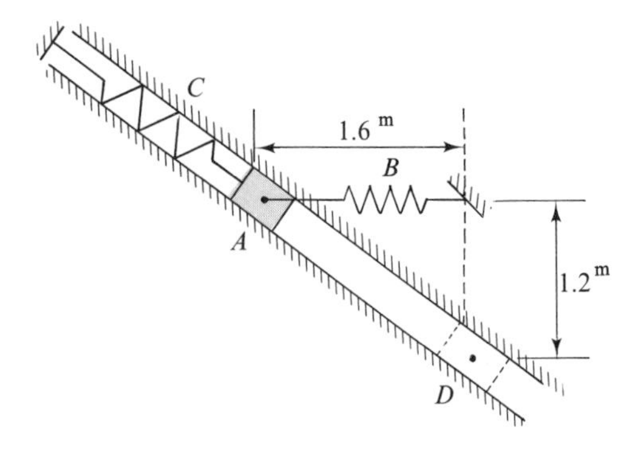

Figure P10-10

10-11 The 10-lb body A in Fig. P10-11 is 6 in. from the spring when its velocity is 6 fps down the plane. The modulus of the spring is 10 lb per inch, and the coefficient of friction between A and the inclined plane is 0.20. Determine the distance traveled down the plane by the body before coming to rest.

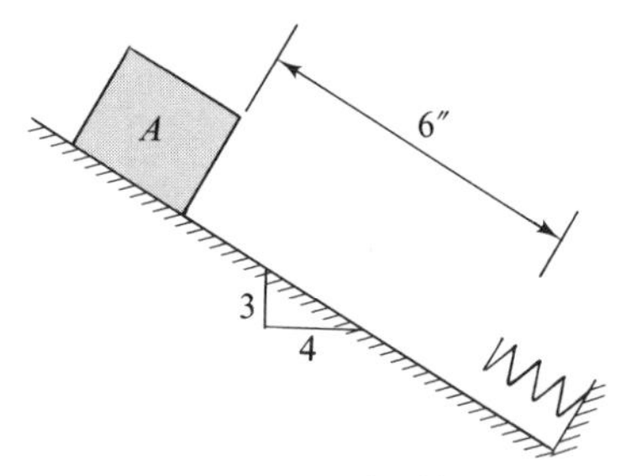

Figure P10-11

10-12 The 32.2-lb block W in Fig. P10-12 is released from rest in the position shown when the spring is stretched 1 ft. The spring constant is 4 lb per ft, and the coefficient of friction between the block and the inclined plane is 0.20. Determine

(a) The expression for the velocity of the block in terms of its displacement, x, during its motion down the plane.

(b) The maximum compression of the spring.

***10-13** The 30-kg block shown in Fig. P10-13 slides on a smooth plane and is connected to a spring with a modulus of 150 N/m. When in the position shown, the block has a velocity of 10 m/s up the plane, and the spring is stretched

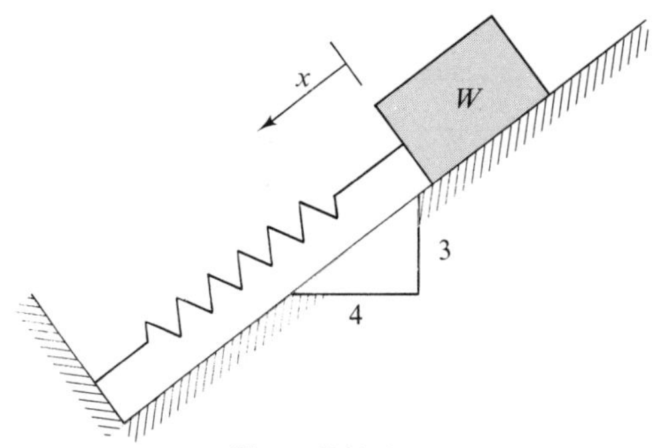

Figure P10-12

0.30 m. Determine the velocity of the block after it moves

(a) Two meters up the plane.

(b) Four meters up the plane.

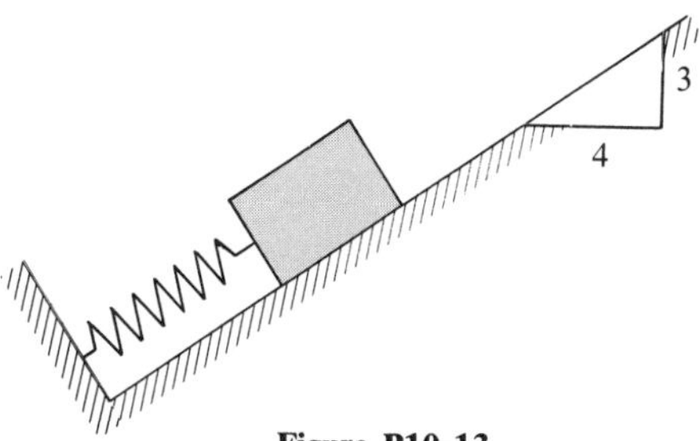

Figure P10-13

*__10-14__ The block in Fig. P10-14 has a mass of 2 kg and slides along the rough ($\mu = 0.20$) plane. Spring A has a modulus of 15 N/m, and spring B has a modulus of 60 N/m. When the block has a velocity of 0.60 m/s to the right spring A is compressed 0.40 m and spring B is stretched 0.50 m. Determine the velocity of the block after it has moved 0.40 m.

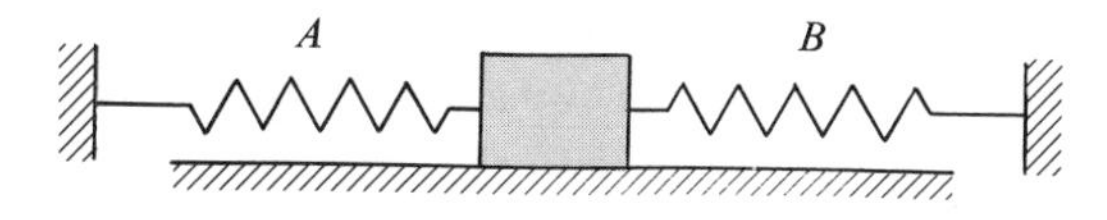

Figure P10-14

10-15 The body in Fig. P10-15 weighs 32.2 lb and is constrained to slide along the smooth rod which lies in a vertical plane. The tension in the nonlinear spring is given by the equation $F = 9s^2$, where F is in pounds and s is the distance in

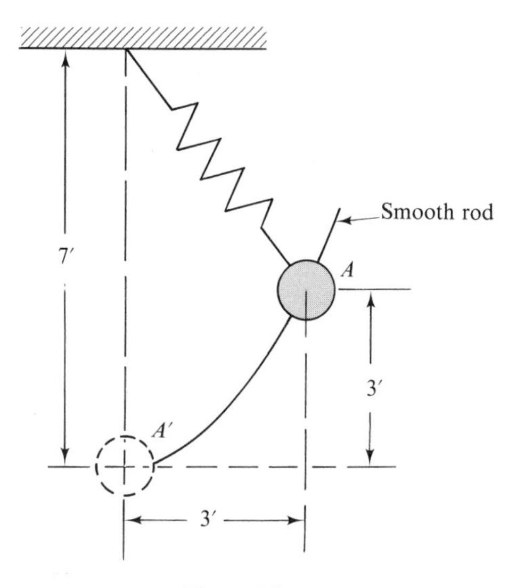

Figure P10-15

feet the spring is stretched. When the body is in position A it has a velocity of 5.0 fps down along the rod, and the spring is stretched 1.0 ft. Determine the velocity of the body when it reaches position A'.

10-16 A pendulum which weighs 3.86 lb and has a cord length of 10 in. is released from rest when the cord is horizontal as shown in Fig. P10-16. When the mass reaches the bottom of its path it strikes a spring with a modulus of 100 lb per in. Determine

(a) The maximum kinetic energy of the mass.

(b) The maximum compression of the spring.

(c) The tension in the cord just before the mass contacts the spring.

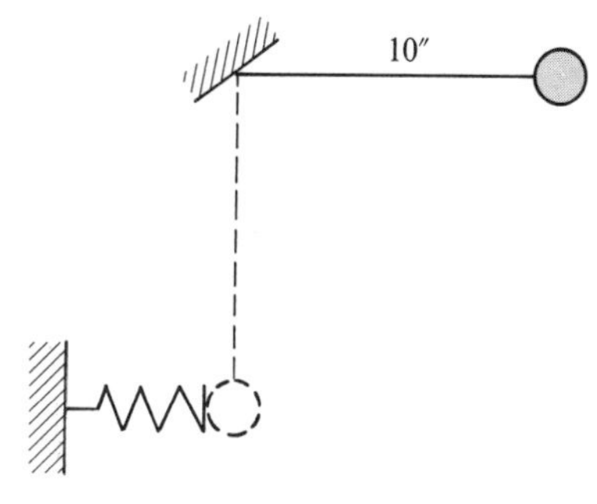

Figure P10-16

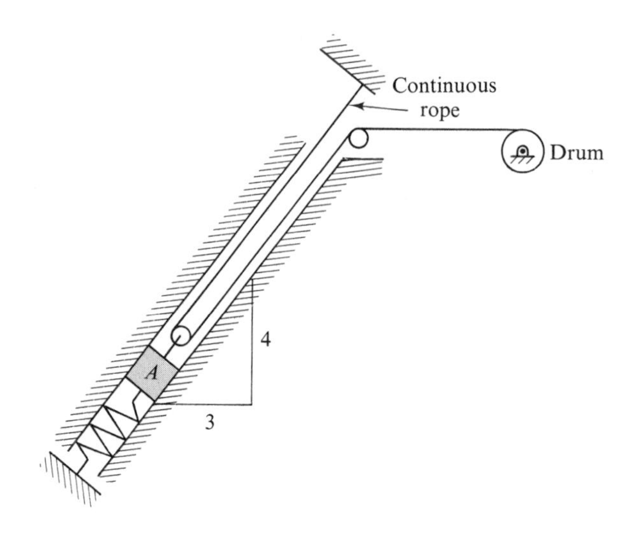

Figure P10-17

10-17 Body A in Fig. P10-17 weighs 25 lb, and the modulus of the spring is 10 lb per ft. Initially the system is in equilibrium, and the tension in the rope is zero. The drum starts to rotate and produces a constant tension in the rope of 50 lb. Determine the velocity of A after it has moved 5.0 ft up the smooth slot.

10-18 The particle C in Fig. P10-18 weighs 40 lb and is constrained to move along a smooth circular path which lies in a vertical plane. The particle is released from rest. The spring tension is 25 lb when in the position shown. Determine the modulus of the spring if the velocity of C is 8.0 fps to the left as it passes through point B.

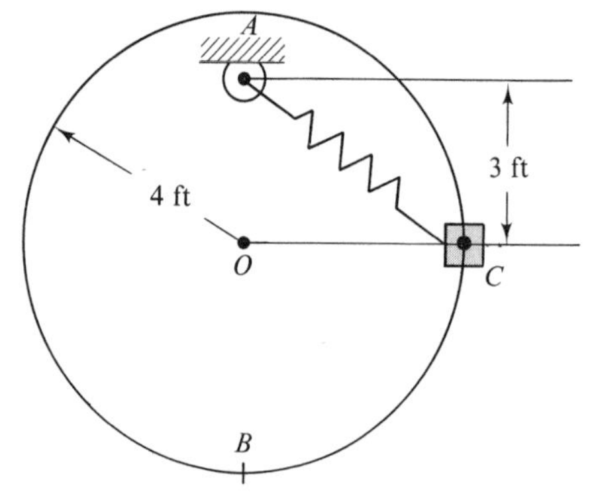

Figure P10-18

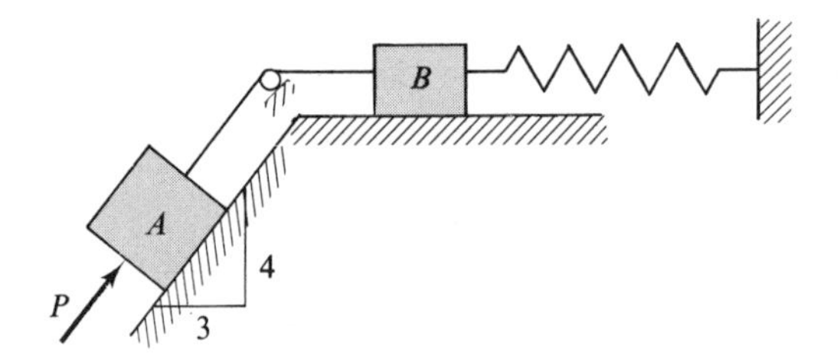

Figure P10-19

***10-19** The 50-kg block A and the 20-kg block B in Fig. P10-19 are held in equilibrium by the force P and the spring. The spring has a modulus of 150 N/m and has a tension of 50 N in the position shown. The coefficient of friction between the blocks and the planes is 0.10. The force **P** is removed, allowing the blocks to move. Determine the velocity of A after it has moved 2.0 m.

10-20 The 322-lb body A in Fig. P10-20 slides on the smooth horizontal surface, and body B weighs 128.8 lb. In the position shown, the spring is stretched 1 ft, and body A has a velocity of 10 fps to the right. When body A has moved 6 ft its velocity is 5 fps to the right. Determine the modulus of the spring.

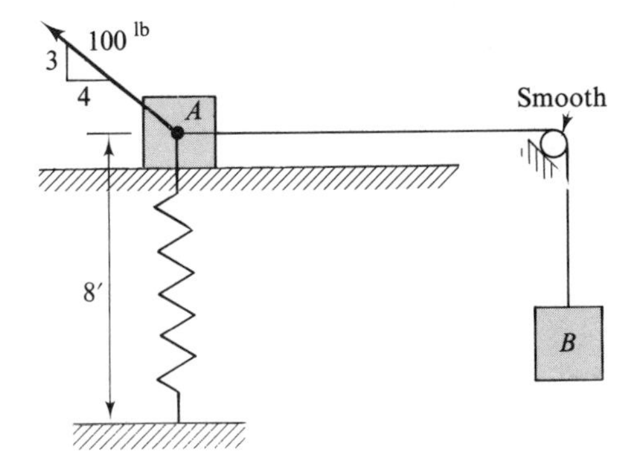

Figure P10-20

10-21 The 322-lb body A in Fig. P10-21 slides on the smooth horizontal surface, and the 64.4-lb body B is suspended from a pulley of negligible weight. The spring modulus is 200 lb per ft. In the position shown, body A has a velocity of 10 fps to the right, and the spring is stretched 2 ft. Determine the velocity of A when it has moved 5 ft to the right.

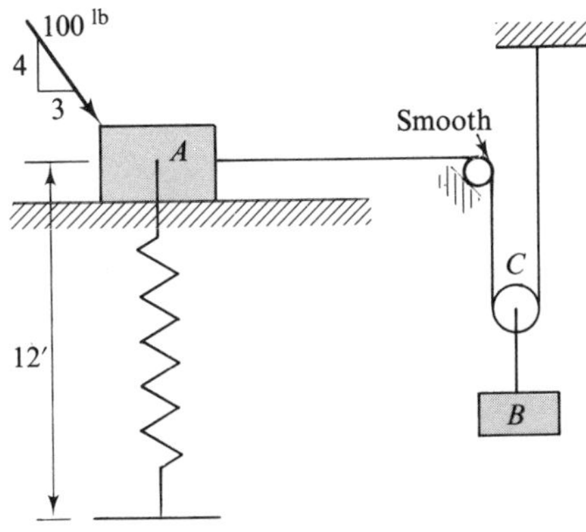

Figure P10-21

10-22 In Fig. P10-22 block A weighs 64.4 lb, block B weighs 32.2 lb, and the weight of the pulley is negligible. The modulus of spring S_1 is 50 lb per ft. When in the position shown, the tension in S_1 is 50 lb, the tension in S_2 is 300 lb, and the velocity of A is 10 fps to the right along the smooth plane. Determine the modulus of spring S_2 if body A comes to rest after moving 4 ft.

***10-23** The 50-kg body A in Fig. P10-23 has a velocity of 3 m/s to the right when the spring

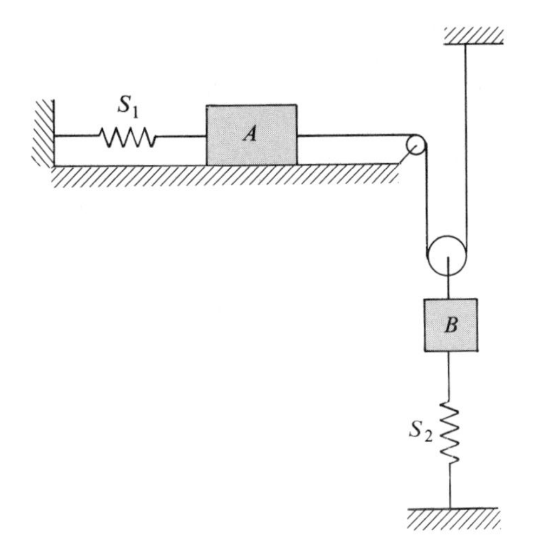

Figure P10-22

(modulus of 50 N/m) is stretched 0.3 m. The coefficient of friction between A and the horizontal plane is 0.3. Body B has a mass of 30 kg. Determine the distance A will have moved when its velocity is 1 m/s to the right.

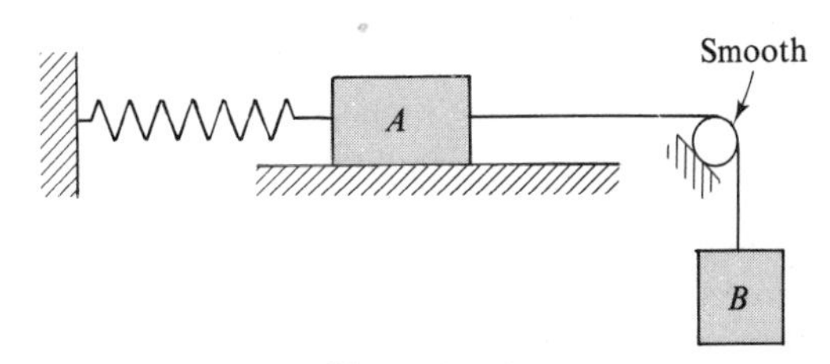

Figure P10-23

10-24 The 644-lb body A in Fig. P10-24 slides on a smooth horizontal plane. In the position shown, the 128.8-lb body B has a velocity of 8 fps downward, and the spring is stretched 2 ft. The modulus of the spring is 20 lb per ft. Determine the velocity of A when it passes under the smooth drum at C.

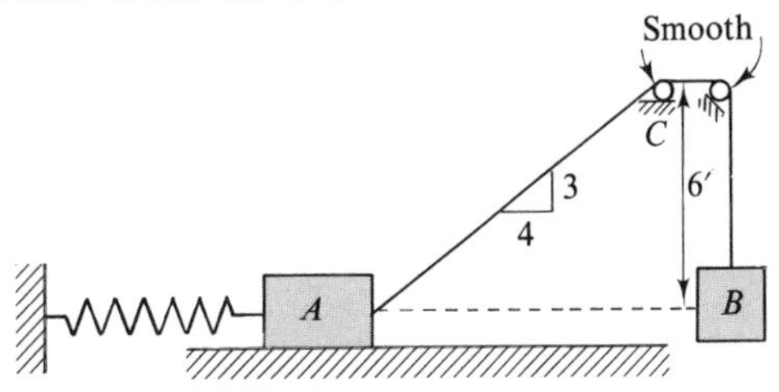

Figure P10-24

10-25 The 20-lb block A in Fig. P10-25 is connected to the 30-lb block B by the 20-in. bar of negligible weight. The blocks slide in smooth slots and are released from rest in the position shown. The spring has a modulus of 2 lb per in. and is stretched 6 in. in the indicated position. Determine the velocity of B when it has moved downward 4 in.

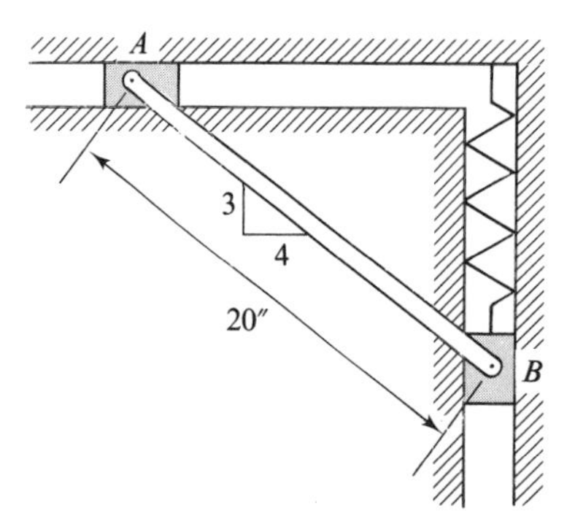

Figure P10-25

10-26 In Fig. P10-26, the two blocks A and B, each weighing 40 lb and connected by the rigid 25-in. bar of negligible mass, slide in smooth slots. In the position shown block A has velocity of 8 ips down the slot, and the spring (modulus of 16 lb per in.) is compressed 5 in. The constant force F has a magnitude of 65 lb and remains at the indicated slope throughout the motion. Determine the velocity of block A when it has moved 15 in. downward to point A'.

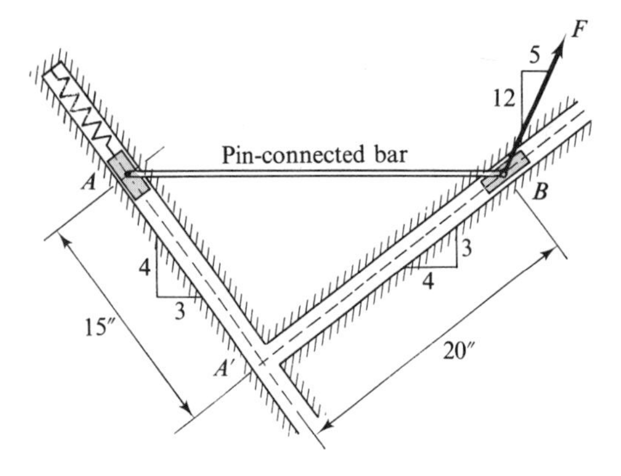

Figure P10-26

***10-27** In Fig. P10-27 body A has a mass of 15 kg, body B has a mass of 7 kg, and the modulus of the spring is 8.0 N/m. The plane is smooth. In the position shown, body A is at rest, and the tension in the spring is 60 N. Determine

(a) The velocity of A as it passes under the small peg at C.

(b) The tension in the cord as A passes under C.

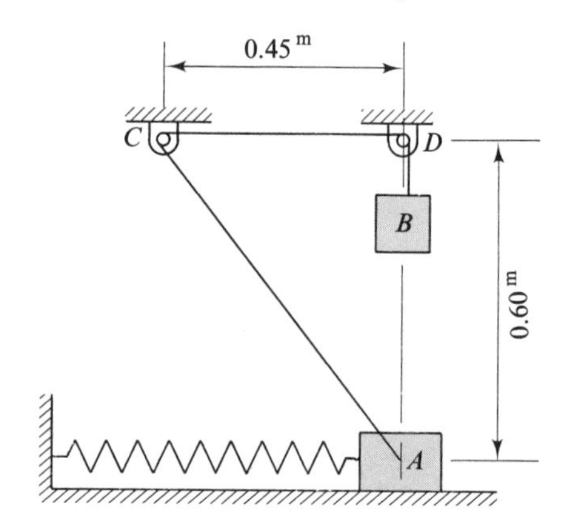

Figure P10-27

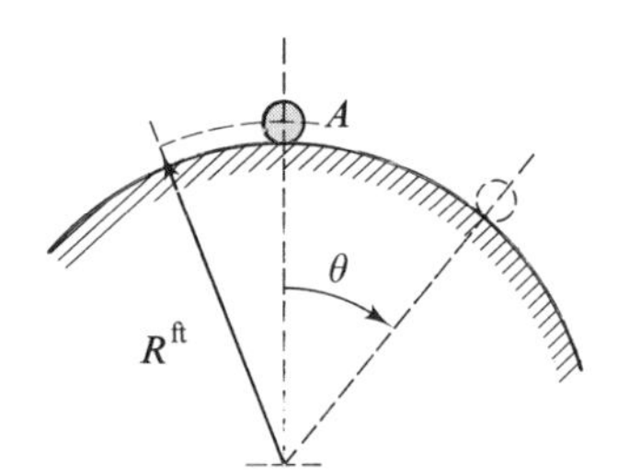

Figure P10-28

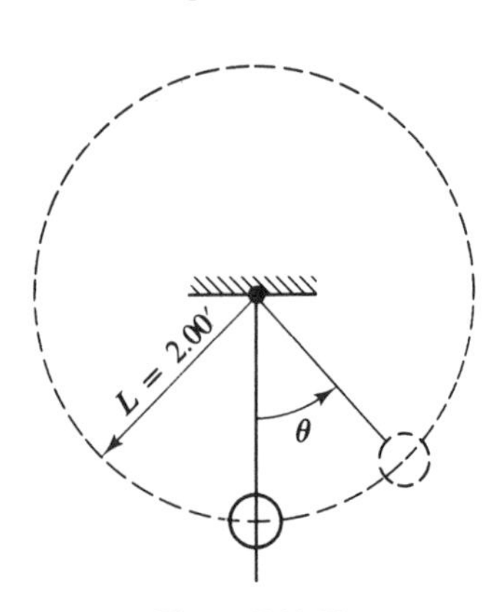

Figure P10-29

10-28 A particle of weight W starts from rest in position A in Fig. P10-28 and slides on the smooth cylindrical surface. Determine
(a) The velocity of the particle as a function of θ.
(b) The value of the angle θ when the particle leaves the surface.

10-29 The small body in Fig. P10-29 weighs 3.0 lb and swings in a vertical plane at the end of a 2.00-ft cord. When the angle θ is 20° the velocity of the body is 9.00 fps. Determine
(a) The tension in the cord when $\theta = 60°$.
(b) The tension in the cord when $\theta = 0°$.

10-30 Determine the minimum speed that the body in Problem 10-29 must have when $\theta = 0$ if it is to travel in a full circle.

10-31 The speed of the body in Problem 10-29 is $2\sqrt{gL}$ when $\theta = 0$. Determine the angle θ when the cord becomes slack.

10-4 POTENTIAL ENERGY—CONSERVATION OF ENERGY

The potential energy of a body or system of bodies is the capacity of the body to do work as a result of the relative positions of the various particles of the system. If a spring is compressed, the material of the spring is stressed and energy is stored in the spring. The stresses in the spring tend to restore the spring to its original unstressed length, and the force exerted by the spring can do work as the spring returns to its original length. As long as the stresses in the spring remain below the elastic limit[1] for the material of the spring, the *elastic potential energy* or strain energy stored in the spring is equal to the work done on the spring in compressing it. Steam compressed in a boiler has potential energy due to its pressure, and this energy is transformed into work as the steam drives the piston. Steam also has thermal energy due to the temperature difference between the steam and the atmosphere. These two forms of stored energy in steam are closely related and are usually combined in describing the energy of the steam.

When a body is raised above any reference or datum plane, it is able to do work as it returns to the reference plane, as a result of the attraction of the earth for the body. The hammer of a pile driver is an example of such a body. Many pile drivers do work as a result of the attraction of the earth for the hammer. The *gravitational potential energy* is not stored in the hammer alone but rather in the system of bodies consisting of the hammer and the earth.

The work done by the weight $\mathbf{W}$ of particle P as it moves from position P_1 to position P_2, as shown in Fig. 10-12, (p. 588) is

$$U = -W(y_2 - y_1) = Wy_1 - Wy_2. \qquad (10\text{-}4)$$

When the work done by a force depends only on the initial and final positions of the particle and is independent of the path the force is called a conservative force. The force W in Fig. 10-12 is an example of

[1] The greatest stress which can be developed in a material without a permanent deformation remaining upon complete release of the stress is defined as the *elastic limit* of the material.

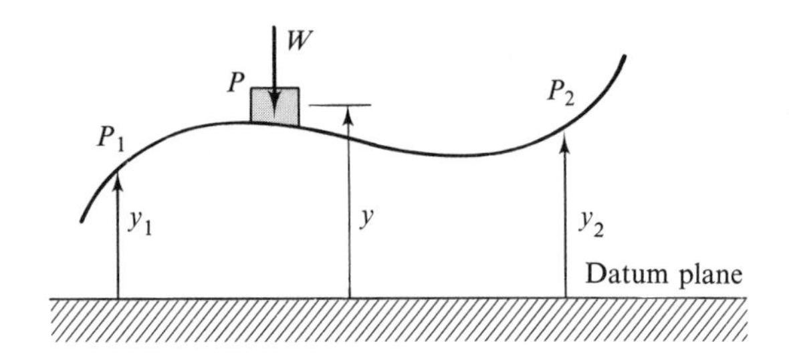

Figure 10-12

a conservative force. The scalar function Wy, which is equal to the work done by the force W in moving from the displaced position to the datum plane, is called the potential energy, V, of the gravity force with respect to the datum plane.

Equation (10-4) may be written in the form

$$U = V_1 - V_2 = -(V_2 - V_1) = -\Delta V. \tag{10-4a}$$

From this equation it is observed that the work done by a conservative force is equal to the negative of its change in potential energy. Since only the change in the potential energy is involved in Eq. (10-4a), the choice of a datum plane is arbitrary because the potential energy varies only by a constant for different datum planes, and the *change* of the potential energies in moving from the initial to the final positions is always the same.

Body A of weight W, shown in Fig. 10-13(a), slides on a smooth horizontal plane and is attached to a spring having a modulus k and an unstretched length L_O. A free-body diagram of A is shown in Fig. 10-13(b). The potential energy of the spring force, ks, acting on body A is equal to the work done by the spring on body A as it moves from a stretched position (length $= L_O + s$) to a datum or reference position (length $= L_O$). The spring force on body A is to the left, and the movement is also to the left; therefore the work done during the displacement **s**, which is the potential energy, must be positive. Thus

$$V = -\int_s^0 ks\,ds = -\frac{k}{2}\left[s^2\right]_s^0 = \frac{ks^2}{2}, \tag{10-5}$$

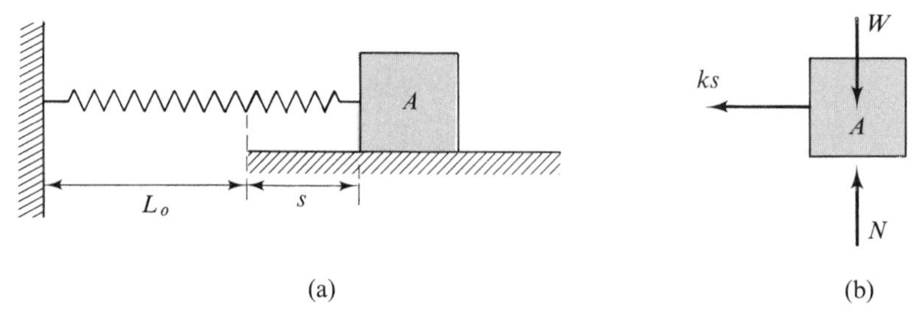

Figure 10-13

which is the potential energy of the spring force when the spring is stretched (or compressed) a distance s.

In the discussion of Fig. 10-12 the weight of the body was considered to be constant. When bodies move in space the variation of the gravity force with the distance from the center of the earth must be considered. A body of mass m is shown in Fig. 10-14 at a distance r from the center of the earth. The gravity force, GMm/r^2, from Newton's law of universal gravitation, is also shown. The potential energy of this force is the work done by it as it moves from the position shown to the reference position, which, for convenience, is selected at $r=\infty$. Since the force on the body is always toward the earth and the movement is away from the earth, the work done by the force on the body is negative. The potential energy is

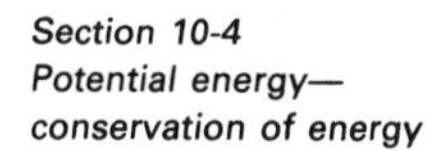
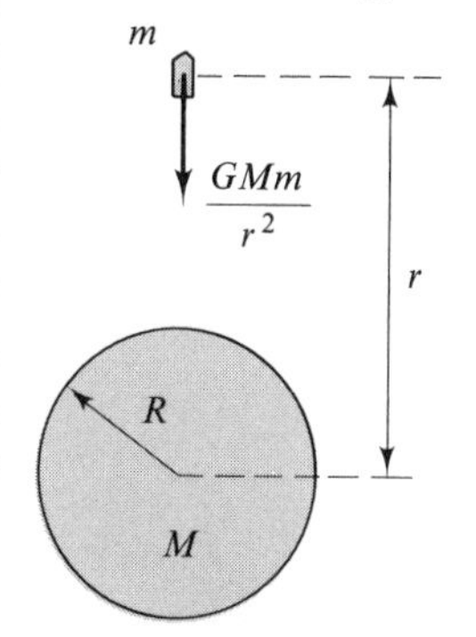

Figure 10-14

$$V=-\int_r^\infty \frac{GMm}{r^2}\,dr=GMm\left[\frac{1}{r}\right]_r^\infty=-\frac{GMm}{r}. \tag{10-6}$$

When a particle is acted upon by conservative forces, such as the weight and spring forces, the work and kinetic energy relationship as given by Eq. (10-3),

$$U=\Delta T=T_f-T_i,$$

can be written in a form involving potential energy. Since the work done by conservative forces is equal to the negative of the change in potential energy of these forces, this equation becomes

$$-(V_f-V_i)=T_f-T_i,$$

where V_f and V_i are the potential energies in the final and initial positions. This equation can be rewritten as

$$V_i+T_i=V_f+T_f, \tag{10-7}$$

which states that the sum of the initial potential and kinetic energies must equal the sum of the final potential and kinetic energies. This statement is known as the principle of conservation of mechanical energy.

EXAMPLE 10-5

The small 2-lb collar of Fig. 10-15 is attached to a spring which has a modulus of 4 lb per ft and an unstretched length of 3 ft. The collar slides along the fixed smooth circular wire which is in a vertical plane, and when in the position A the collar has a velocity of 3 fps to the

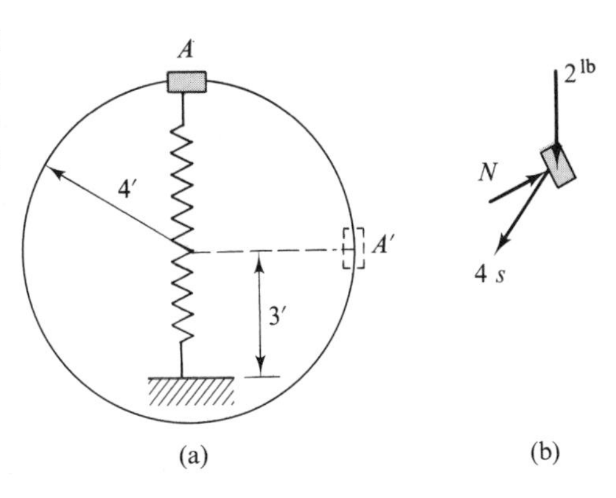

Figure 10-15

right. Determine the velocity of the collar when it reaches the position A′.

SOLUTION

A free-body diagram of the particle in an intermediate position is shown in Fig. 10-15(b). The weight and spring forces are conservative, and the normal force, N, does no work and so will not enter into the conservation of energy equation which will be used in solving this problem. The datum plane for the weight is chosen as the horizontal diameter of the circle, and the unstretched length of the spring ($s = 0$) is used as the reference position for the spring.

The conservation of energy equation is

$$V_i + T_i = V_f + T_f. \qquad \text{(a)}$$

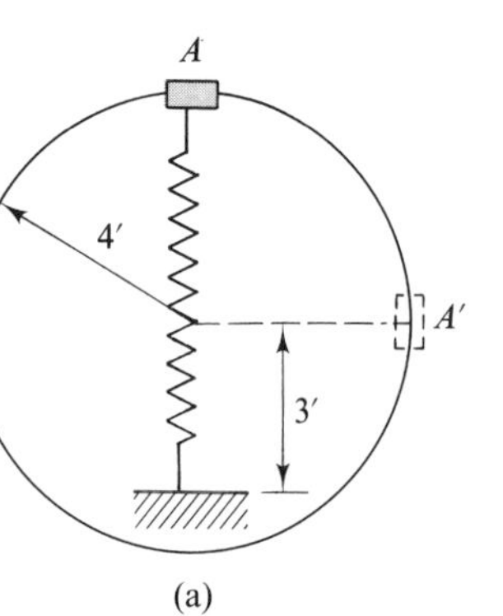

(a)

2 lb
N
4 s

(b)

Figure 10-15

The initial stretch of the spring is $s_i = 4$ ft, and the final stretch is $s_f = 2$ ft. Thus

$$V_i = \frac{1}{2}ks_i^2 + Wh_i = \frac{1}{2}(4)(4)^2 + 2(4) = 40 \text{ ft-lb},$$

$$T_i = \frac{1}{2}mv_i^2 = \frac{1}{2}\left(\frac{2}{32.2}\right)(3)^2 = 0.28 \text{ ft-lb},$$

$$V_f = \frac{1}{2}ks_f^2 + Wh_f = \frac{1}{2}(4)(2)^2 + 0 = 8 \text{ ft-lb},$$

$$T_f = \frac{1}{2}mv_f^2 = \frac{1}{2}\left(\frac{2}{32.2}\right)v_f^2 = 0.0311v_f^2.$$

When these values are substituted into Eq. (a) it becomes

$$40 + 0.28 = 8 + 0.0311v_f^2,$$

which gives

$$v_f = 32.2 \text{ fps}$$

and

$$\mathbf{v}_f = \underline{32.2 \text{ fps downward.}} \qquad \text{Ans.}$$

PROBLEMS

10-32 The small 6.00-lb body in Fig. P10-32 swings in a vertical plane at the end of a 10-ft cord. The angular velocity of the cord is 0.50 rad per sec clockwise in the position shown.
(a) Use the principle of conservation of mechanical energy to determine the velocity of the body when in its lowest position.
(b) Determine the tension in the cord when the body is in its lowest position.

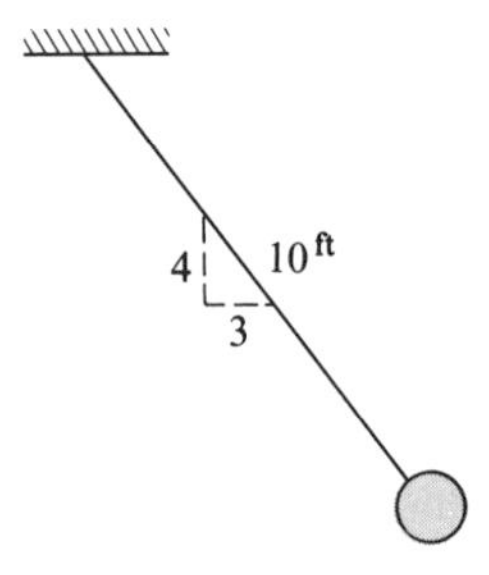

Figure P10-32

***10-33** The 150-kg block in Fig. P10-33 slides down the smooth plane. In the position shown the block has a velocity of 3 m/s down the plane and the spring (modulus = 25 N/m) is stretched

0.3 m. Use the conservation of mechanical energy equation to determine the velocity of the block after it has moved 1.5 m down the plane.

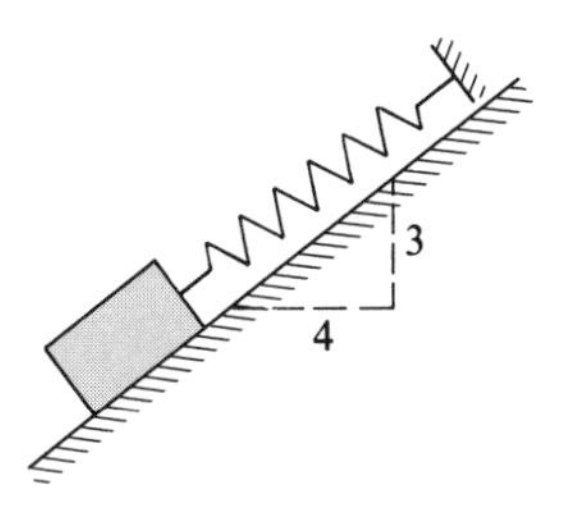

Figure P10-33

10-34 The 128.8-lb body in Fig. P10-34 slides along the smooth rod BC and is connected to springs S_1 and S_2. The modulus of S_1 is 40 lb per ft and its unstretched length is 7 ft. The modulus of S_2 is 20 lb per ft, and its unstretched length is 8 ft. The body has a velocity of 10 fps down the rod when in position A. Determine its velocity when it reaches A'.

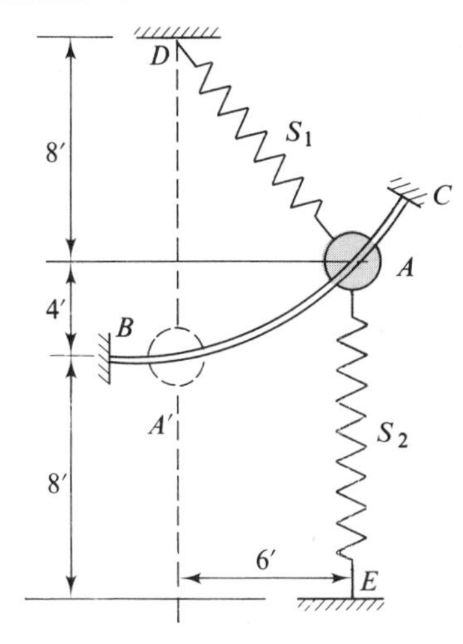

Figure P10-34

10-35 The 12.88-lb body A in Fig. P10-35 slides in the smooth slot and is attached to a spring having a modulus of 4.0 lb per ft. When the spring is compressed 2 ft body A has a velocity of 3 fps down the slot. The 13-lb force is constant throughout the motion. Determine the velocity of A after it has moved 10 ft down the slot.

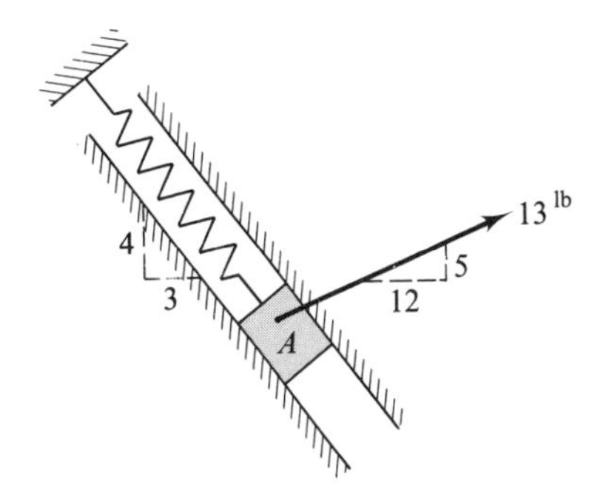

Figure P10-35

10-36 A satellite of mass m describes an orbit about the earth as shown in Fig. P10-36. When the satellite is at position A it has a speed of v_1. Use the principle of conservation of mechanical energy to show that the speed at position B is

$$v_2 = \left[v_1^2 + \frac{2gR^2(r_1 - r_2)}{r_1 r_2} \right]^{1/2}$$

where g is the acceleration of gravity at the earth's surface.

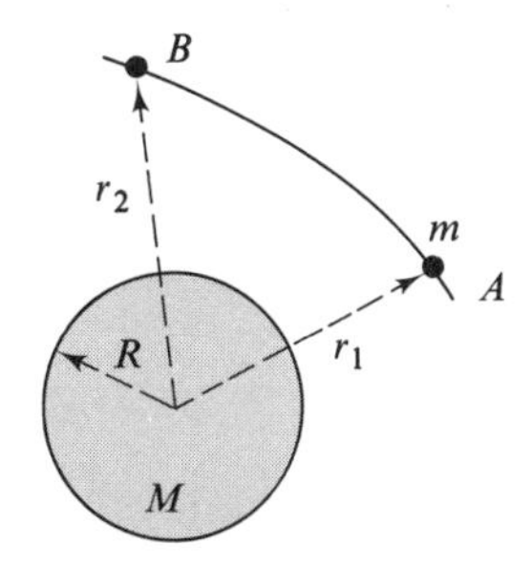

Figure P10-36

***10-37** Solve Problem 10-27(a) by using the conservation of mechanical energy equation.

10-38 Solve Problem 10-28 by using the conservation of mechanical energy equation.

10-5 CONSERVATIVE FORCES

As indicated in Art. 10-4, if the work done by a force as it moves from an initial to a final position is independent of the path traveled, the force is called a conservative force. When the potential energy of a conservative force can be expressed as a function of its position, that is, when $V = V(x, y, z)$, the differential of the potential energy is

$$dV = \frac{\partial V}{\partial x}\,dx + \frac{\partial V}{\partial y}\,dy + \frac{\partial V}{\partial z}\,dz.$$

The work done by the force F during a displacement $d\mathbf{r} = \mathbf{i}\,dx + \mathbf{j}\,dy + \mathbf{k}\,dz$ is

$$dU = \mathbf{F}\cdot d\mathbf{r} = F_x\,dx + F_y\,dy + F_z\,dz.$$

From Eq. (10-4a) $dU = -dV$; therefore

$$F_x\,dx + F_y\,dy + F_z\,dz = -\frac{\partial V}{\partial x}\,dx - \frac{\partial V}{\partial y}\,dy - \frac{\partial V}{\partial z}\,dz$$

or

$$\left(F_x + \frac{\partial V}{\partial x}\right)dx + \left(F_y + \frac{\partial V}{\partial y}\right)dy + \left(F_z + \frac{\partial V}{\partial z}\right)dz = 0.$$

Since the force is conservative and the work done is independent of the path, the preceding equation must be satisfied for any arbitrary choice of dx, dy, and dz. This requires that the coefficients of the differentials must be zero. Hence

$$F_x = -\frac{\partial V}{\partial x}, \qquad F_y = -\frac{\partial V}{\partial y}, \qquad F_z = -\frac{\partial V}{\partial z}. \tag{10-8}$$

These equations state that if a force is conservative, its components are given by the negative partial derivatives of its potential function as indicated by Eq. (10-8).

When the partial derivatives of F_x and F_y are taken with respect to y and x, respectively, they become

$$\frac{\partial F_x}{\partial y} = -\frac{\partial^2 V}{\partial y\,\partial x} \quad \text{and} \quad \frac{\partial F_y}{\partial x} = -\frac{\partial^2 V}{\partial x\,\partial y}.$$

Since the order of the partial derivatives is immaterial, it follows that

$$\frac{\partial F_x}{\partial y} = \frac{\partial F_y}{\partial x}. \tag{10-9a}$$

In a similar manner it can be shown that

$$\frac{\partial F_y}{\partial z} = \frac{\partial F_z}{\partial y} \tag{10-9b}$$

and

$$\frac{\partial F_z}{\partial x} = \frac{\partial F_x}{\partial z}. \tag{10-9c}$$

These three equations must be satisfied by any force which is conservative. Thus Eq. (10-9) provide a convenient method to determine whether or not a force is conservative.

EXAMPLE 10-6

Show that the force $\mathbf{F} = x^2\mathbf{i} + y^2\mathbf{j} + z^2\mathbf{k}$ is conservative and determine the potential function of the force.

SOLUTION

Since $F_x = x^2$, $\partial F_x/\partial y = \partial F_x/\partial z = 0$. Similarly, $\partial F_y/\partial x = \partial F_y/\partial z = 0$ and $\partial F_z/\partial x = \partial F_z/\partial y = 0$. Consequently, Eq. (10-9a, b, c) are satisfied, and the force is conservative.

From Eq. (10-8),

$$F_x = x^2 = -\frac{\partial V}{\partial x}$$

from which

$$V = -\frac{x^3}{3} + f_1(y, z),$$

where $f_1(y, z)$ is a function of y and z but not of x. In a similar manner,

$$F_y = y^2 = -\frac{\partial V}{\partial y} \quad \text{and} \quad F_z = z^2 = -\frac{\partial V}{\partial z},$$

which give

$$V = -\frac{y^3}{3} + f_2(x, z)$$

and

$$V = -\frac{z^3}{3} + f_3(x, y).$$

It can be shown by direct substitution in these three equations for V that

$$f_1(y, z) = -\frac{y^3 + z^3}{3} + C_1,$$

$$f_2(x, z) = -\frac{x^3 + z^3}{3} + C_2,$$

and

$$f_3(x, y) = -\frac{x^3 + y^3}{3} + C_3,$$

and thus the potential energy function can be written as

$$V = -\frac{x^3 + y^3 + z^3}{3} + C, \qquad \text{Ans.}$$

where C is an arbitrary constant which can be determined if the value of the potential energy is known for any specified set of values of x, y, and z.

PROBLEMS

10-39 The force $\mathbf{F} = [xy\mathbf{i} + (x^2/2)\mathbf{j}]$ lb acts on a particle which moves in the xy plane.

(a) Show that the force is conservative.
(b) Determine the potential energy function.

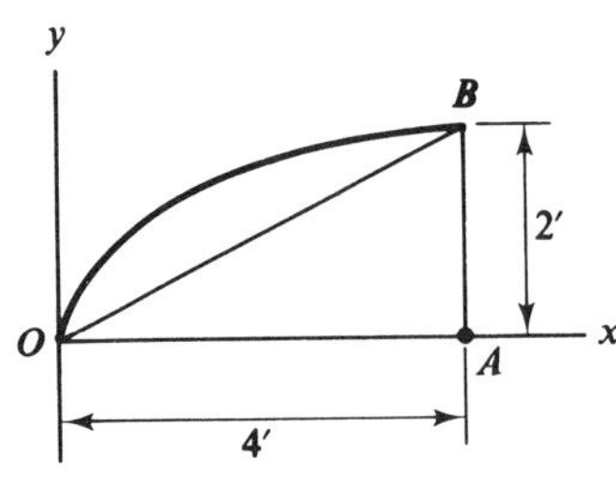

Figure P10-40

10-40 Determine the work done by the force of Problem 10-39 as it moves from O to B along the paths shown in Fig. P10-40
(a) From O to A to B.
(b) From O to B along the straight line OB.
(c) From O to B along the parabola $y^2 = x$.

10-41 Show that the force $\mathbf{F} = (x^2\mathbf{i} + xy\mathbf{j})$ lb is nonconservative and determine the work done by the force in moving from O to B along the straight line and along the parabola $y^2 = x$ as shown in Fig. P10-40.

10-42 Which of the following forces are conservative?
(a) $\mathbf{F} = (2x\mathbf{i} + 3xy\mathbf{j} - 6\mathbf{k})$ lb.
(b) $\mathbf{F} = (x\mathbf{i} + y\mathbf{j} + z\mathbf{k})$ lb.
(c) $\mathbf{F} = (4\mathbf{i} + 3y\mathbf{j} + xy\mathbf{k})$ lb.
Determine the potential energy function of the forces which are conservative.

PART B–RIGID BODIES

10-6 WORK OF COUPLES AND FORCES ON RIGID BODIES

The work done by a force was shown to be given by the dot product of two vectors, force and linear displacement. The work done by a couple can be obtained by applying this definition to each force of the couple.

Consider the work done by the couple in Fig. 10-16(a) as it rotates through the infinitesimal angle $d\theta$. The work done by the two forces is

$$U = \mathbf{P}_1 \cdot d\mathbf{r}_1 + \mathbf{P}_2 \cdot d\mathbf{r}_2,$$

where $\mathbf{r}_1$ and $\mathbf{r}_2$ are the position vectors to forces $\mathbf{P}_1$ and $\mathbf{P}_2$, respectively, and $d\mathbf{r}_1$ and $d\mathbf{r}_2$ are the displacements of the two forces. Since $\mathbf{P}_1$ and $d\mathbf{r}_1$ are in the same direction, $\mathbf{P}_1$ does positive work, while $\mathbf{P}_2$

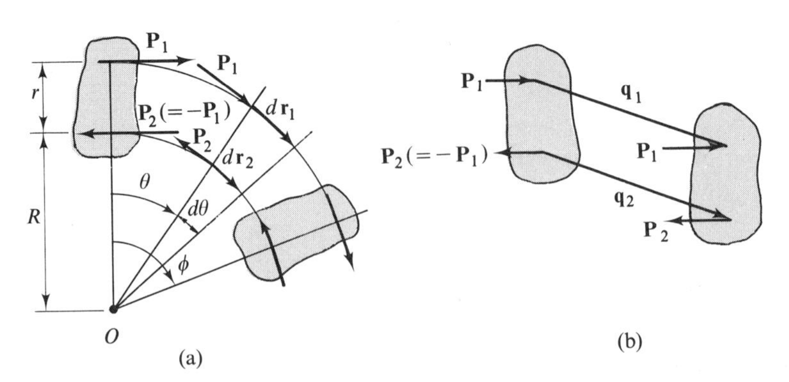

Figure 10-16

does negative work by the same reasoning. Thus the work can be written as

$$dU = P_1\,dr_1 - P_2\,dr_2.$$

From the definition of a couple, P_1 is equal to P_2, and from the geometry of the figure,

$$dr_1 = (R + r)\,d\theta \quad \text{and} \quad dr_2 = R\,d\theta.$$

Therefore, the work of the couple is

$$dU = P_1(R + r)\,d\theta - P_1 R\,d\theta = P_1 r\,d\theta = T\,d\theta.$$

where T is the moment of the couple.

The work done during a finite angular displacement can be obtained by integration of the expression

$$U = \int_{\theta_i}^{\theta_f} T\,d\theta, \qquad (10\text{-}10)$$

and if T is constant, the work is

$$U = T(\theta_f - \theta_i) = T\,\Delta\theta,$$

where $\Delta\theta$ is the change of θ, that is, the angular displacement of the couple. The work is positive if the angular displacement is in the same direction as the sense of rotation of the couple and negative if the displacement is in the opposite direction.

If the body is translated as shown in Fig. 10-16(b), the work done by the couple is

$$U = \mathbf{P}_1 \cdot \mathbf{q}_1 + \mathbf{P}_2 \cdot \mathbf{q}_2 = \mathbf{P}_1 \cdot \mathbf{q}_1 - \mathbf{P}_1 \cdot \mathbf{q}_1 = 0$$

since $\mathbf{P}_2 = -\mathbf{P}_1$ and $\mathbf{q}_2 = \mathbf{q}_1$. Thus a couple does no work due to a translation of the body on which it acts. When a couple acts on a body which is simultaneously translated and rotated the couple does work only as a result of the rotation.

When the magnitude of T is a variable, Eq. (10-10) must be integrated to obtain the work done. Moment-angular position diagrams (similar to F-s diagrams) are frequently useful for problems involving variable couples, particularly when the T-θ diagram is a straight line. Linear variations occur for some torsion springs (such as a watch spring), in which case the moment and angular position are related by

$$T = k\theta,$$

where k is the torsional spring constant and is often expressed in such units as foot-pounds per radian, inch-ounces per radian, or newton meters per radian.

The work done on a *rigid* body by a system of external forces and couples is the algebraic sum of works done by the individual forces and couples. The internal forces of a body occur in equal and opposite collinear pairs, and if the body is rigid, the component of the displacement of each force of the pair in the direction of the forces must be equal in magnitude; therefore the resultant work done by the internal forces is zero. *If the body is not rigid, this statement is not valid since work will be done by the internal forces.* When two or more rigid bodies are connected by smooth pins or by flexible, inextensible cables the resultant work done by the connecting members of the system is also zero. The two forces at the ends of a cable have the same magnitude (the mass of the cable is assumed to be negligible), and if the cable is inextensible, the components of the displacements of the two ends in the direction of the forces must have the same magnitudes; therefore, the resultant work done by the cable must be zero. When two bodies are connected by an *extensible* cable the bodies will do work on the cable when its length is changed, and the cable will do an equal amount of work on the bodies with an opposite sign. Thus if the cable is stretched, the bodies will do positive work on the cable, while the cable will do negative work on the bodies. The following examples demonstrate the calculation of the work done by various forces and force systems.

EXAMPLE 10-7

The bar AB in Fig. 10-17(a) weighs 20 lb and is connected to blocks at the ends which slide in smooth horizontal and vertical slots as indicated. The spring OA is arranged so that it can act in either tension or compression. It has a modulus of 5.0 lb per in. and an unstressed length of 7 in. Determine the work done by all forces acting on the bar as it moves to the horizontal position $A'B'$ as indicated by dashed lines. The weights of the blocks at A and B may be neglected.

SOLUTION

A free-body diagram of the bar in an intermediate position is shown in Fig. 10-17(b). There are four forces and a couple acting on the body. Two of the forces, $\mathbf{R}_1$ and $\mathbf{R}_2$, are normal to their corresponding displacements, and as a result they do no work. The resultant work on the bar is thus

$$U = U_{\text{spring}} + U_{\text{weight}} + U_{\text{couple}}.$$

An F-s diagram for the spring is shown in Fig. 10-17(c). The length of the spring changes 8 in., from 5 to 13 in. The spring is initially compressed 2 in. and exerts a 10-lb force on the block A. In the final position, the spring is stretched 6 in. and has a tensile force of 30 lb. The area C represents negative work *on the spring* since the spring is in compression but is increasing in length. The area C represents positive work on block A since the vertical component of the force on A and the displacement are in the same (downward) direction. In a similar manner, the area D represents negative work of the spring on block A.

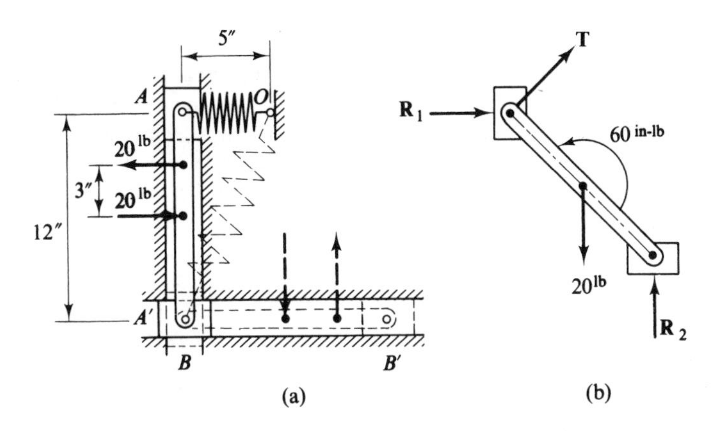

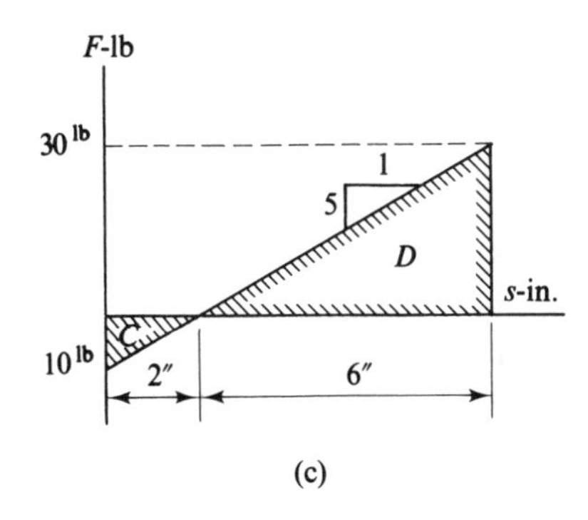

Figure 10-17

The work of the spring on AB is

$$U_{\text{spring}} = \frac{10(2)}{2} - \frac{30(6)}{2} = -80 \text{ in-lb on } AB.$$

The work done by the weight is

$$U_{\text{weight}} = 20(6) = 120 \text{ in-lb on } AB,$$

and the work done by the couple is

$$U_{\text{couple}} = 60\frac{\pi}{2} = 94.2 \text{ in-lb on } AB.$$

The total work done on AB is

$$U = -80 + 120 + 94.2 = \underline{134.2 \text{ in-lb.}} \qquad \text{Ans.}$$

EXAMPLE 10-8

The 20-lb wheel A in Fig. 10-18(a) is connected to the 50-lb block B by means of an inextensible cord of negligible weight which passes over a smooth peg at C. The wheel rolls on plane E without slipping, and the coefficient of friction between B and plane H is 0.30. Determine the work done on the system of bodies (A and B) while body A moves 10 ft up the plane E.

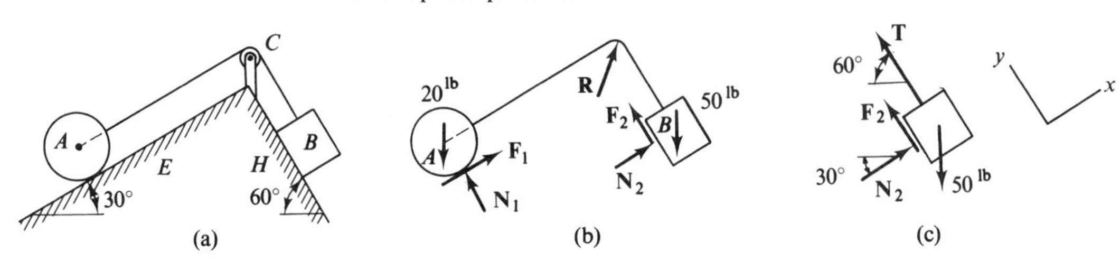

Figure 10-18

SOLUTION

The free-body diagram of the system of bodies is drawn in Fig. 10-18(b). The forces $\mathbf{N}_1$, $\mathbf{N}_2$, and $\mathbf{R}$ are perpendicular to the motion of their points of application, and so they do no work. Since the wheel rolls without slipping, the frictional force $\mathbf{F}_1$ also does no work because the particle of the wheel on which $\mathbf{F}_1$ acts has no velocity and hence no displacement while the force is acting on it.[2] Thus the two weights and the frictional force $\mathbf{F}_2$ are the only forces that do work on the system of bodies.

As body A rolls 10 ft up to the right on plane E, block B slides 10 ft down to the right on plane H. From the free-body diagram of block B in Fig. 10-18(c)

$$\sum \mathbf{F}_x = 0,$$

since there is no acceleration in the x direction. Therefore,

$$[N_2 - 0.50(50)]\mathbf{i} = 0$$

or

$$\mathbf{N}_2 = 25\mathbf{i} \text{ lb},$$

and

$$\mathbf{F}_2 = \mu N_2 \mathbf{j} = 0.30(25)\mathbf{j} = 7.5\mathbf{j} \text{ lb}.$$

---

[2] Forces $\mathbf{N}_1$ and $\mathbf{F}_1$ do no work because they are the components of the force exerted by one particle (or, more exactly, one group of adjacent particles) of the plane on one particle of the periphery of the wheel at an instant when neither particle is moving. At the next instant, the force will be exerted by another particle of the plane on an adjacent particle of the wheel. The work done is zero because none of the forces moves. Each acts for an instant as a holding force and becomes zero as a new force exerted by the next particle becomes the holding force or reaction of the plane on the wheel. If a wheel rolls and slips on a fixed plane or rolls without slipping on a moving plane, the frictional force will do work.

The work done by $\mathbf{F}_2$ is

$$U_{F_2} = \mathbf{F}_2 \cdot \mathbf{q}_B = (7.5\mathbf{j}) \cdot (-10\mathbf{j}) = -75 \text{ ft-lb.}$$

The work done by the weight of A is

$$U_{20\text{lb}} = (-10.0\mathbf{i} - 17.32\mathbf{j}) \cdot (10\mathbf{i}) = -100 \text{ ft-lb.}$$

The work done by the weight of B is

$$U_{50\text{lb}} = (-25.0\mathbf{i} - 43.3\mathbf{j}) \cdot (-10\mathbf{j}) = 433 \text{ ft-lb.}$$

The total work done on the two bodies is

$$U = -75 - 100 + 433 = \underline{258 \text{ ft-lb.}} \qquad \text{Ans.}$$

10-7 KINETIC ENERGY OF A RIGID BODY—PLANE MOTION

The kinetic energy of any body, whether rigid or not, is the sum of the kinetic energies of the separate particles. If the body is not rigid, there is no simple equation relating the velocities of the different particles. The velocities of the various particles in rigid bodies, however, are directly related to each other and to the angular motion of the body. An expression for the kinetic energy of a rigid body which has plane motion is developed here. An expression for the kinetic energy of a rigid body with general motion will be developed in Art. 11-9 in terms of the angular momentum of the body. Consider the body in Fig. 10-19, which has plane motion parallel to the XY plane. Point A is any point fixed in the body and is the origin of the moving xyz axes which move with the body. The xy plane is parallel to the XY plane. The velocity of any particle of mass dm of the body is (see the figure)

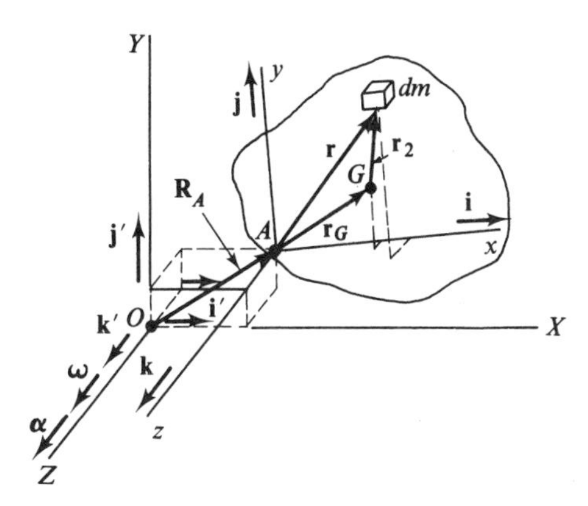

Figure 10-19

$$\mathbf{v} = \dot{\mathbf{R}}_A + \dot{\mathbf{r}} = \dot{\mathbf{R}}_A + \boldsymbol{\omega} \times \mathbf{r},$$

where $\boldsymbol{\omega}$ is the angular velocity of the body. The kinetic energy of the particle is

$$dT = \frac{dm}{2} v^2 = \frac{dm}{2} \mathbf{v} \cdot \mathbf{v}$$

$$= \frac{dm}{2} [\dot{\mathbf{R}}_A + \boldsymbol{\omega} \times \mathbf{r}] \cdot [\dot{\mathbf{R}}_A + \boldsymbol{\omega} \times \mathbf{r}]$$

$$= \frac{dm}{2} \dot{R}_A^2 + dm\dot{\mathbf{R}}_A \cdot (\boldsymbol{\omega} \times \mathbf{r}) + \frac{dm}{2} (\boldsymbol{\omega} \times \mathbf{r}) \cdot (\boldsymbol{\omega} \times \mathbf{r}).$$

The body has plane motion; therefore, $\boldsymbol{\omega} = \omega\mathbf{k}$ and $\mathbf{r} = x\mathbf{i} + y\mathbf{j} + z\mathbf{k}$, from which

$$\boldsymbol{\omega} \times \mathbf{r} = \omega(x\mathbf{j} - y\mathbf{i})$$

and

$$(\boldsymbol{\omega} \times \mathbf{r}) \cdot (\boldsymbol{\omega} \times \mathbf{r}) = \omega^2(x^2 + y^2).$$

From Fig. 10-19, $\mathbf{r} = \mathbf{r}_G + \mathbf{r}_2$, and the kinetic energy of the particle can be written

$$dT = \frac{dm}{2}\dot{R}_A^2 + dm\dot{\mathbf{R}}_A \cdot (\boldsymbol{\omega} \times \mathbf{r}_G) + dm\dot{\mathbf{R}}_A \cdot (\boldsymbol{\omega} \times \mathbf{r}_2) + \frac{dm}{2}\omega^2(x^2 + y^2).$$

Since $\dot{\mathbf{R}}_A$, $\mathbf{r}_G$, and $\boldsymbol{\omega}$ do not vary from point to point on the body, the kinetic energy of the body can be determined by integration over the mass as

$$T = \frac{\dot{R}_A^2}{2}\int_m dm + \dot{\mathbf{R}}_A \cdot (\boldsymbol{\omega} \times \mathbf{r}_G)\int_m dm + \dot{\mathbf{R}}_A \cdot \left(\boldsymbol{\omega} \times \int_m \mathbf{r}_2\, dm\right)$$

$$+ \frac{\omega^2}{2}\int_m (x^2 + y^2)\, dm$$

$$= \frac{m\dot{R}_A^2}{2} + m\dot{\mathbf{R}}_A \cdot (\boldsymbol{\omega} \times \mathbf{r}_G) + 0 + \frac{I_z\omega^2}{2}. \qquad (10\text{-}11)$$

Note that the quantity $\int_m \mathbf{r}_2\, dm$ is zero because $\mathbf{r}_2\, dm$ is the first moment of mass with respect to the mass center (see Chapter 3).

Since A is any point in the body, Eq. (10-11) can be simplified by the proper choice of point A. If A is selected at the mass center, $\mathbf{r}_G$ becomes zero, and the equation reduces to

$$T = \frac{mv_G^2}{2} + \frac{I_z\omega^2}{2}, \qquad (10\text{-}11a)$$

where $\dot{R}_A$ is v_G and the z axis passes through G.

If point A is a point of zero velocity (the instantaneous axis of zero velocity or a point on the axis of rotation for a body which has rotation), $\dot{R}_A$ is zero, and the equation becomes

$$T = \frac{I_z\omega^2}{2}, \qquad (10\text{-}11b)$$

where the z axis is the instantaneous axis of zero velocity or the fixed axis of rotation for bodies having rotation.

When the body has translation, $\boldsymbol{\omega}$ is zero, and Eq. (10-11) reduces to

$$T = \frac{mv^2}{2}, \qquad (10\text{-}11c)$$

where v is the magnitude of the velocity of any point on the body since all points have the same velocity.

Since the kinetic energy of a rigid body is a positive scalar quantity, the kinetic energy of a system of bodies is the sum of the kinetic energies of the bodies of the system.

EXAMPLE 10-9

The slender homogeneous bar AC in Fig. 10-20 weighs 25 lb, and body B weighs 100 lb. Bar AC rotates about a horizontal axis at A, and B is pinned to AC at C. The radius of gyration of the mass of B with respect to a horizontal axis through its mass center G is 1.50 ft. In the position shown, AC has an angular velocity of 2.0 rad per sec clockwise, and B has an angular velocity of 5.0 rad per sec clockwise. Determine the kinetic energy of the system.

SOLUTION

The bar rotates about a fixed axis at A, and body B has plane motion. The kinetic energy of the system is the sum of the kinetic energies of the two bodies; thus

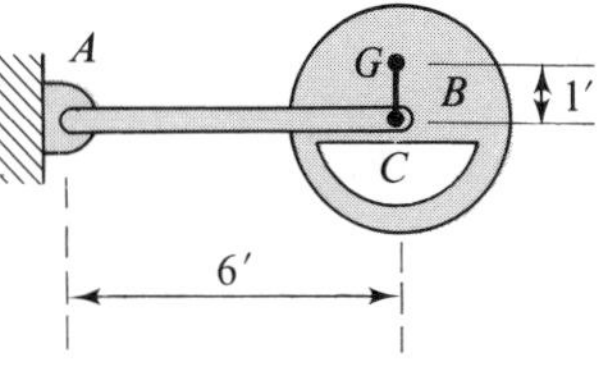

Figure 10-20

$$T = T_{AC} + T_B.$$

The velocity of G can be determined by writing a relative velocity equation for points G and C; that is,

$$\mathbf{v}_G = \mathbf{v}_C \quad + \quad \mathbf{v}_{G/C}$$

$\omega = 2$ 6 (2) + 1 (5) $\omega = 5$

$$= \overset{\downarrow}{12} + \overset{\rightarrow}{5} \quad = 13 \text{ fps} \; \overset{5}{\underset{12}{\searrow}}.$$

The moments of inertia of mass are determined by the procedures discussed in Chapter 6. The kinetic energy of AC, from Eq. (10-11b), is

$$T_{AC} = \frac{1}{2} I_A \omega_{AC}^2$$

$$= \frac{1}{2}\left(\frac{1}{3} mL^2\right)\omega_{AC}^2$$

$$= \frac{1}{2}\left[\frac{1}{3}\left(\frac{25}{32.2}\right)(6)^2\right](2)^2$$

$$= 18.63 \text{ ft-lb.}$$

The kinetic energy of body B, from Eq. (10-11a), is

$$T_B = \frac{1}{2} m_B v_G^2 + \frac{1}{2} I_G \omega_B^2$$

$$= \frac{1}{2} m_B v_G^2 + \frac{1}{2}(m_B k_G^2)\omega_B^2$$

$$= \frac{1}{2}\left(\frac{100}{32.2}\right)(13)^3 + \frac{1}{2}\left[\left(\frac{100}{32.2}\right)(1.5)^2\right](5)^2$$

$$= 350 \text{ ft-lb.}$$

The kinetic energy of the system is

$$T = 18.63 + 350$$
$$= \underline{369 \text{ ft-lb.}} \qquad \text{Ans.}$$

10-8 PRINCIPLE OF WORK AND KINETIC ENERGY—PLANE MOTION

The principle of work and kinetic energy, as applied to particles, was developed in Art. 10-3. The principle states that the work done on a particle (by all forces acting on the particle) is equal to the change of kinetic energy of the particle; that is,

$$U = \Delta T = T_f - T_i.$$

The principle of work and kinetic energy can be extended to include bodies and systems of bodies, since bodies are composed of particles. The total work done by all the forces acting on the particles of a body is equal to the sum of the changes of the kinetic energy of the particles, which is the same as the change of the kinetic energy of the body. The internal forces between the particles can do work as well as the external forces due to outside bodies. If the body is rigid, however, the resultant work done by the internal forces is zero, as was indicated in Art. 10-6. Furthermore, if two or more bodies are connected by inextensible cords or cables, the net work done by the connecting members is zero, and the work done by the external forces on the entire system is equal to the change of the kinetic energy of the system of rigid bodies. Thus the principle of work and kinetic energy *for a system of rigid bodies* is expressed

$$U_{\text{ext}} = \Delta T = T_f - T_i. \tag{10-12}$$

The general equation can be written in the following special forms for the particular type of motion indicated:

translation: $U = \frac{1}{2}m(v_f^2 - v_i^2)$, (10-12a)

rotation: $U = \frac{1}{2}I_z(\omega_f^2 - \omega_i^2)$, (10-12b)

plane motion: $U = \frac{1}{2}m[(v_G)_f^2 - (v_G)_i^2] + \frac{1}{2}I_z(\omega_f^2 - \omega_i^2)$. (10-12c)

Note that the z axis used to calculate I_z is the axis of rotation in Eq. (10-12b) and the axis through the mass center perpendicular to the plane of motion in Eq. (10-12c). The work and kinetic energy equation for plane motion can also be written in terms of the moment of inertia of mass with respect to the instantaneous axis of zero velocity. Since the moment of inertia is not necessarily constant with respect to the axis through the instantaneous center, the equation for plane motion is

$$U = \tfrac{1}{2}(I_O)_f\omega_f^2 - \tfrac{1}{2}(I_O)_i\omega_i^2, \tag{10-12d}$$

where $(I_O)_i$ and $(I_O)_f$ are the moments of inertia of the mass of the body with respect to the initial and final instantaneous axes of zero velocity.

Equation (10-12) is not limited to any specific types of motion and can be used when Eq. (10-12a)–(10-12d) do not apply. Since work and energy are scalar quantities, free-body diagrams involving two or more bodies connected by smooth pins, flexible cables, and so on, can be used without encountering the difficulties which arise with vector quantities when working with force, mass, and acceleration.

EXAMPLE 10-10

The unbalanced wheel A in Fig. 10-21(a) weighs 100 lb and has a radius of gyration of mass with respect to a horizontal axis through O of 0.80 ft. The wheel rolls along the horizontal plane without slipping and is connected to the 50.0-lb body B by means of an inextensible cord which passes over the smooth pulley at C and is wrapped around A. Neglect the mass of the pulley at C. The modulus of the spring S is 20.0 lb per ft, and its mass may be neglected. The spring is attached to the wheel at O. When the mass center G of the wheel is in the position indicated on the diagram, the tension in the spring is 50.0 lb, and the velocity of O is 4.0 fps to the left. Determine the velocity of O when the wheel has rolled 90° counterclockwise.

SOLUTION

Figure 10-21(b) is a free-body diagram of the system of bodies. The equation relating the work and kinetic energy of this system is

$$U = T_f - T_i$$

$$U_{100} + U_{50} + U_P = \tfrac{1}{2}m_A[(v_G)_f^2 - (v_G)_i^2] + \tfrac{1}{2}(I_z)_A[(\omega_A)_f^2 - (\omega_A)_i^2] + \tfrac{1}{2}m_B[(v_B)_f^2 - (v_B)_i^2].$$

The forces **F**, **N**, and **Q** do no work, as was shown in Art. 10-6. The

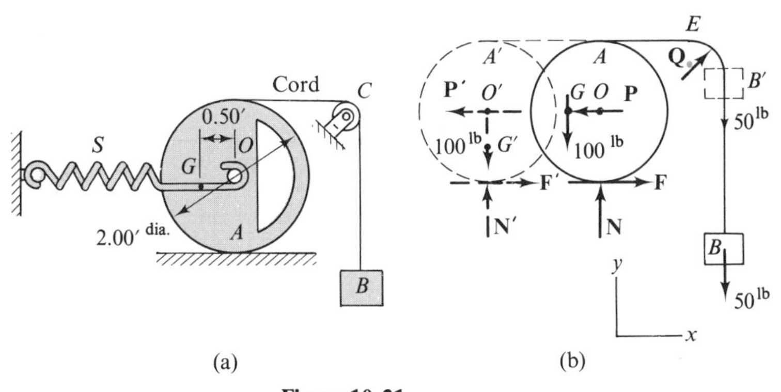

Figure 10-21

final positions of the two bodies are indicated by dashed lines. The displacement of point O is

$$\mathbf{q}_O = -r(\Delta\theta)\mathbf{i} = -1\,\frac{\pi}{2}\,\mathbf{i} = -1.571\mathbf{i}\text{ ft.}$$

The displacement of body B has the same magnitude as the displacement of point E on the cord, which can be obtained by relative motion. Thus,

$$\begin{aligned}\mathbf{q}_E &= \mathbf{q}_O + \mathbf{q}_{E/O} = \mathbf{q}_O - r(\Delta\theta)\mathbf{i}\\ &= -1.571\mathbf{i} - \left(\frac{\pi}{2}\right)\mathbf{i}\\ &= -3.142\mathbf{i}\text{ ft}\end{aligned}$$

and

$$\mathbf{q}_B = 3.14\mathbf{j}\text{ ft.}$$

The displacement of the mass center G has both a horizontal and vertical component, but since the force at G (the weight) is vertical, only the vertical component of the displacement of G (0.50 ft downward) is needed. The work done on the system by the two weights is

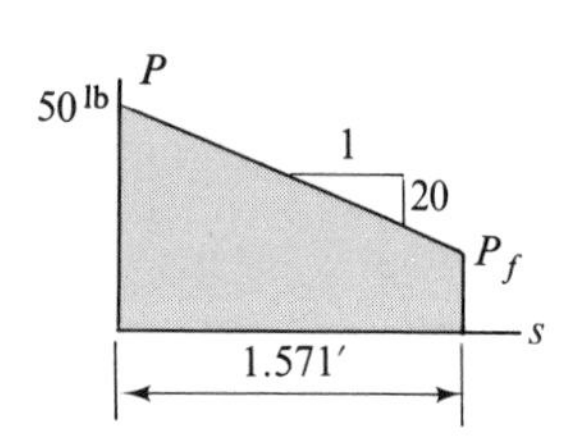

Figure 10-22

$$U_{100} = F(q_G)_y = 100(0.50) = 50.0\text{ ft-lb},$$

$$U_{50} = Fq_B = -50(3.142) = -157.1\text{ ft-lb}.$$

The force of the spring on the body varies as the wheel rolls, and the work done can be determined by integration or from the F-s diagram for the spring (Fig. 10-22). The final force is determined from the slope of the diagram as follows:

$$k = 20 = \frac{50.0 - P_f}{1.571}$$

from which

$$P_f = 50.0 - 20(1.571) = 18.58\text{ lb.}$$

The work done by the spring is proportional to the area under the diagram and is

$$U_P = \frac{50.0 + 18.58}{2}(1.571) = 53.9\text{ ft-lb.}$$

The total work on the system of bodies is

$$U = 50 - 157.1 + 53.9 = -53.2\text{ ft-lb.}$$

Equations relating the various velocities can be determined from kinematics using the instantaneous center of zero velocity of A. Thus,

$$(\omega_A)_i = \frac{(v_O)_i}{r} = \frac{4.00}{1.00} = 4.00\text{ rad per sec},$$

$$(v_G)_i = (1.00^2 + 0.50^2)^{1/2}(\omega_A)_i = 1.118(4.00) = 4.47\text{ fps},$$

$$(v_B)_i = 2.00(\omega_A)_i = (2.00)(4.00) = 8.00\text{ fps.}$$

Similarly,

$$(v_G)_f = 0.50(\omega_A)_f, \qquad (v_B)_f = 2.00(\omega_A)_f, \qquad (v_O)_f = 1.00(\omega_A)_f.$$

The moment of inertia $(I_z)_A$, where the z axis passes through G, as obtained by the parallel-axis theorem, is

$$(I_z)_A = I_O - md^2$$

$$= \frac{100}{32.2}(0.80)^2 - \frac{100}{32.2}(0.50)^2 = 1.211 \text{ slug-ft}^2.$$

When these values are substituted in the work and kinetic energy equation, it becomes

$$-53.2 = \frac{1}{2}\frac{100}{32.2}[0.50^2(\omega_A)_f^2 - 4.47^2] + \frac{1}{2}(1.211)[(\omega_A)_f^2 - 4.00^2]$$

$$+\frac{1}{2}\frac{50}{32.2}[2.00^2(\omega_A)_f^2 - 8.00^2]$$

from which

$$(\omega_A)_f = 3.01 \text{ rad per sec}$$

and

$$(\boldsymbol{\omega}_A)_f = 3.01\mathbf{k} \text{ rad per sec.}$$

The magnitude of the final velocity of O is

$$(v_O)_f = 1.00(\omega_A)_f = 3.01 \text{ fps}$$

and

$$(\mathbf{v}_O)_f = \underline{-3.01\mathbf{i} \text{ fps.}} \qquad \text{Ans.}$$

Since $(\omega_A)_f$ is obtained as a square root, it can be either positive or negative, and the direction is determined from the statement of the problem. In this example, the wheel is turning counterclockwise initially, and the velocity is to be determined after it turns 90° counterclockwise; therefore, the final angular velocity must also be counterclockwise.

PROBLEMS

Note: Unless otherwise specified, the mass of all springs and cords may be neglected. All cords are to be considered flexible and inextensible, and all pins are smooth. Where cords or cables are shown tangent to a circular body, it is assumed that they are wrapped around it and that they wind up on the body or unwind from it as the body moves unless the statement of the problem indicates otherwise.

10-43 The angular velocity of the 3.86-lb homogeneous bar in Fig. P10-43 is 10 rad per sec clockwise in the vertical position. After the bar rotates 90° clockwise, it strikes the two springs and continues to rotate until the spring S_1 is compressed 0.20 in. The modulus of S_1 is twice that of S_2, and the springs are unstressed when the bar first strikes them. Determine the modulus of S_1.

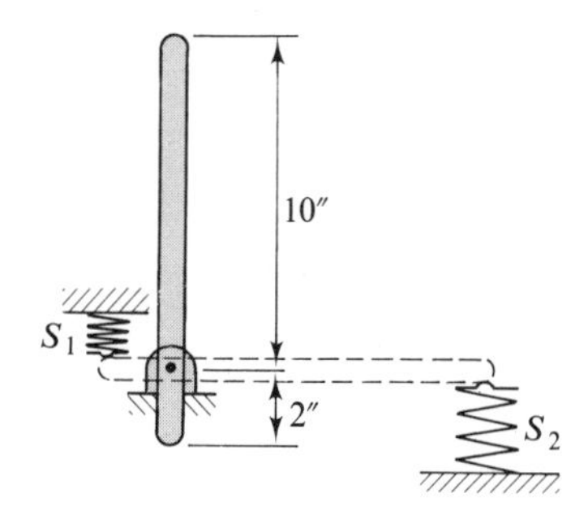

Figure P10-43

***10-44** The 50-kg homogeneous bar AB in Fig. P10-44 has an angular velocity of 4 rad/s clockwise in the position shown, and the spring is stretched 0.3 m in this position. The modulus of the spring is 25 N/m. Will the bar reach the horizontal position? If it becomes horizontal, what will be its angular velocity?

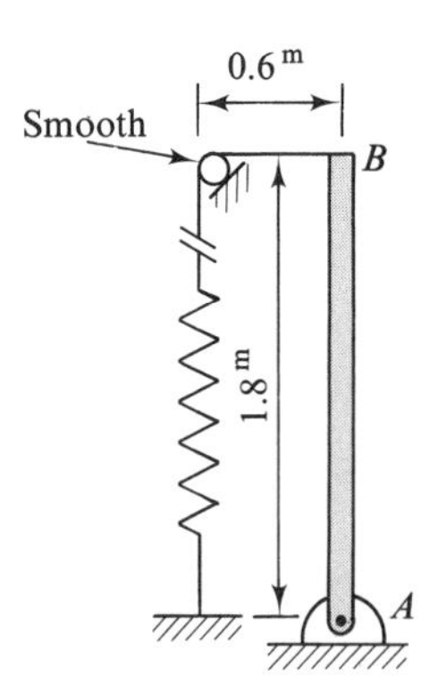

Figure P10-44

10-45 In Fig. P10-45, block A weighs 32.2 lb, block B weighs 96.6 lb, and the unbalanced drum weighs 64.4 lb. The radius of gyration of the mass of the drum with respect to the axis of rotation is 1.50 ft. In the position shown, the angular velocity of the drum is 4.0 rad per sec counterclockwise. Determine the angular velocity of the drum the first time G is directly below the axis of rotation. Assume that the cable does not slip on the pulley.

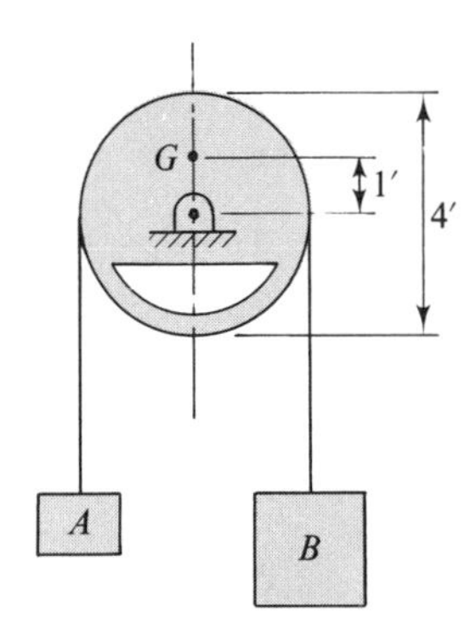

Figure P10-45

10-46 The homogeneous disk B in Fig. P10-46 weighs 322 lb and has a 4-ft-diameter hub of negligible mass. The 128.8-lb body C has a velocity of 20 fps downward in the position shown. Determine the spring modulus required to make the velocity of C 10 fps downward after it has fallen 4 ft. The spring has an initial tension of 200 lb.

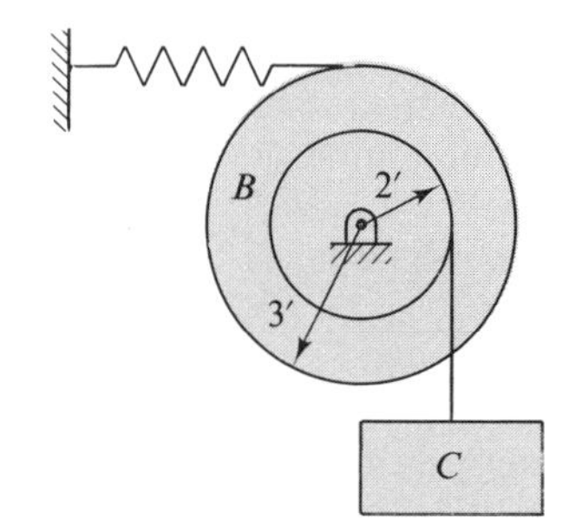

Figure P10-46

10-47 In Fig. P10-47, the homogeneous bars AB ($m = 6$ slugs) and BC ($m = 3$ slugs) are welded at B and rotate about a horizontal axis at A. When AB is horizontal, its angular velocity is 2.0 rad per sec clockwise, and the spring force is 20 lb. After the bars have rotated 90° clockwise, their angular velocity is 3.0 rad per sec clockwise. Determine the spring modulus.

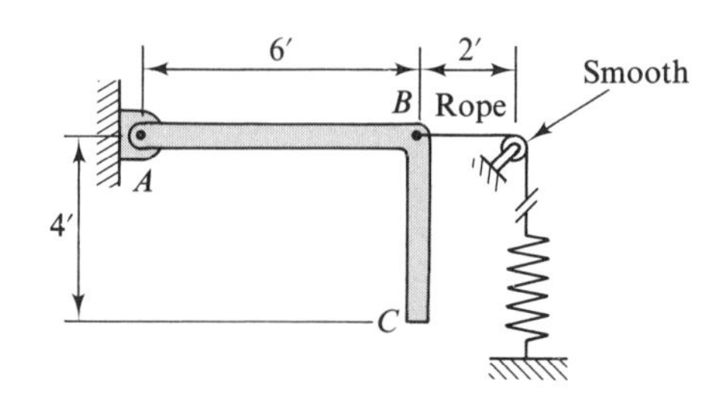

Figure P10-47

10-48 The pulley in Fig. P10-48 weighs 96.6 lb and has a radius of gyration of mass with respect to its axis of rotation of 0.60 ft. Body A weighs 32.2 lb, and the modulus of the spring is 10 lb per ft. When A has a velocity of 8.0 fps downward, the spring is stretched 6.0 in. Determine the displacement of A while its velocity is changed to 2.0 fps downward.

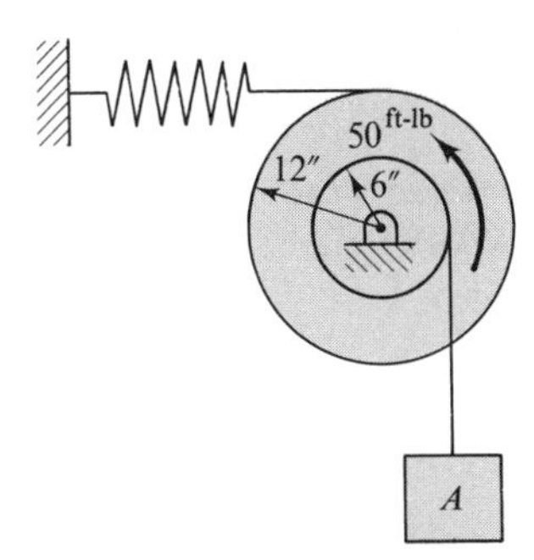

Figure P10-48

10-49 The 322-lb symmetrical body A in Fig. P10-49 has a radius of gyration of mass of 2 ft with respect to its axis of rotation. The modulus of the spring is 15 lb per in. The 64.4-lb body B has a velocity of 12 fps downward when the spring tension is 300 lb. Determine the distance B will have moved downward when its velocity is 3 fps downward.

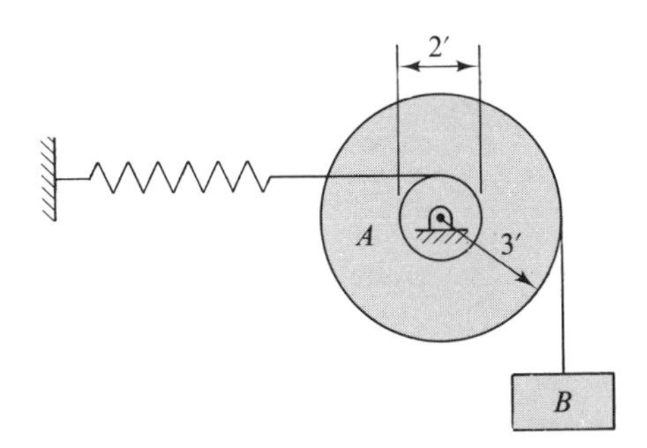

Figure P10-49

10-50 Body A in Fig. P10-50 weighs 64.4 lb and has a radius of gyration of mass with respect to its horizontal axis of rotation of 0.25 ft. The modulus of the spring is 40 lb per ft. The angular velocity of the body is 3 rad per sec clockwise, and the tension in the spring is 2.0 lb in the position shown. Determine the vertical component of the pin reaction on body A when it has rotated 90° clockwise.

10-51 Body A in Fig. P10-51 weighs 60 lb, has a radius of gyration of mass with respect to a horizontal axis through G of 0.60 ft, and rolls without slipping along the inclined plane. The velocity of body B changes from 15 to 6 fps

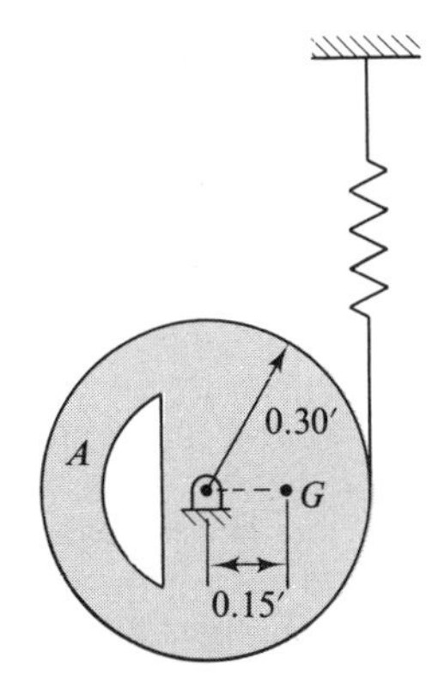

Figure P10-50

(both upward) while moving upward 6 ft. Determine the weight of B.

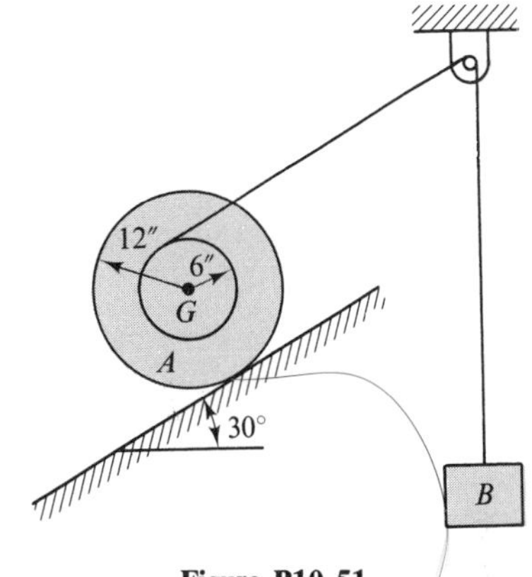

Figure P10-51

*__10-52__ The 300-kg wheel in Fig. P10-52 rolls without slipping on the horizontal plane and has a radius of gyration of mass with respect to a horizontal axis through G of 0.4 m. The spring, with a modulus of 50 N/m, is stretched 0.6 m in

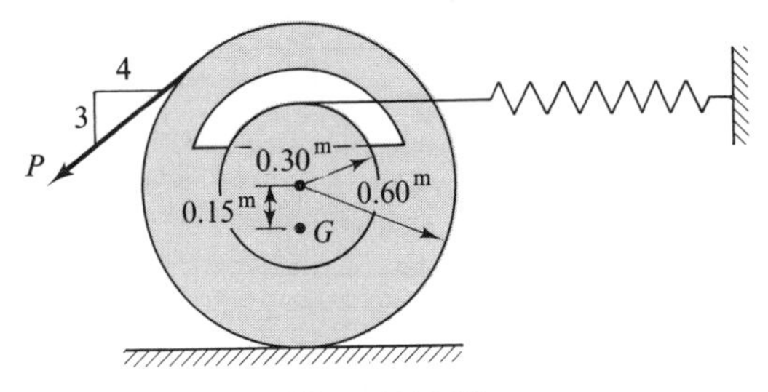

Figure P10-52

the position shown, and the angular velocity of the wheel is 5 rad/s counterclockwise. The constant force **P** is applied at the same slope throughout the motion. Determine the force **P** if the angular velocity of the wheel is 4 rad/s counterclockwise after it rolls 90° counterclockwise.

10-53 The unbalanced wheel in Fig. P10-53 weighs 644 lb and has a radius of gyration of mass with respect to a horizontal axis through G of 1.5 ft. The wheel rolls without slipping, and the modulus of the spring is 20 lb per ft. In the position shown the spring is stretched 1 ft, and the angular velocity of the wheel is 4 rad per sec counterclockwise. Determine if it is possible for the wheel to roll through a 90° counterclockwise angle before it stops. The 100-lb force is applied at the same slope throughout the motion.

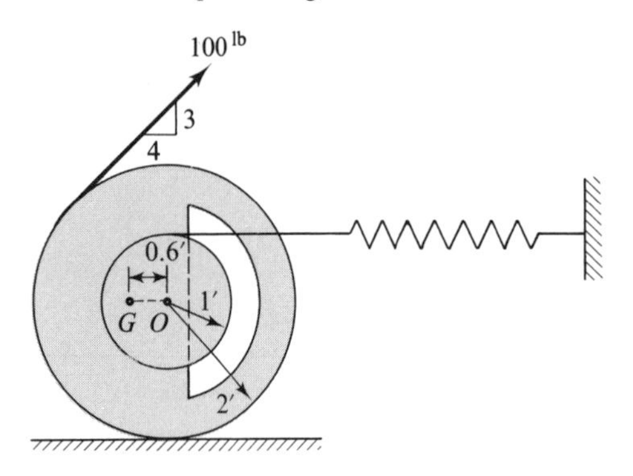

Figure P10-53

10-54 The eccentric wheel in Fig. P10-54 weighs 100 lb, has a radius of gyration of mass with respect to a horizontal axis through G of 0.40 ft, and rolls without slipping along the horizontal plane. The slope of the 30-lb force remains constant. The angular velocity of A changes from 8 to 4 rad per sec, both clockwise, while the wheel rolls 180° clockwise from the position shown. Determine the weight of body B.

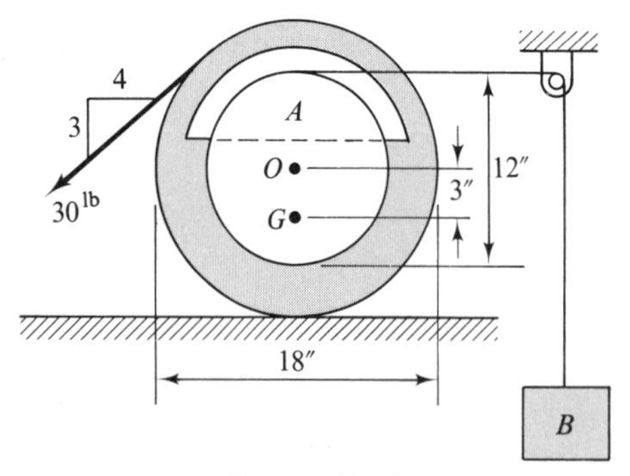

Figure P10-54

10-55 The wheel in Fig. P10-55 weighs 322 lb and has a radius of gyration of mass with respect to an axis through G of 2.0 ft. Body A weighs 257.6 lb, and the modulus of the spring is 40 lb per ft. Initially the spring is stretched 1.0 ft, and the angular velocity of the wheel is 4 rad per sec counterclockwise. Determine the angular velocity of the wheel after it has rotated 90° counterclockwise.

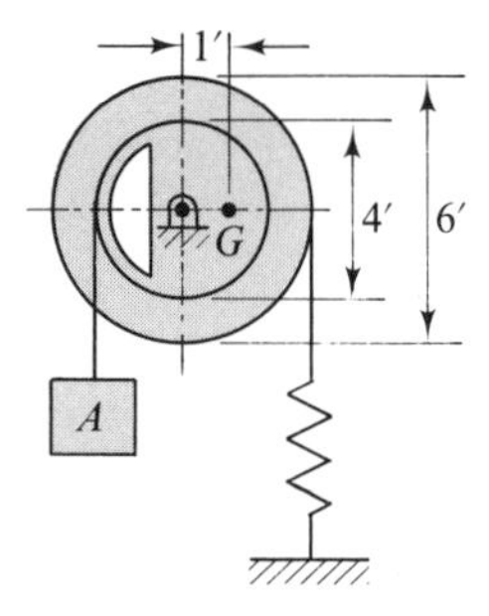

Figure P10-55

10-56 The coefficient of friction between the 32.2-lb body A in Fig. P10-56 and the plane is 0.50. The 64.4-lb wheel B rolls without slipping and has a radius of gyration of mass with respect to a horizontal axis through G of 2.0 in. The spring modulus is 1.0 lb per in. In the position shown, the velocity of A is 4.0 fps to the left, and the spring is unstretched. Determine the force P which will cause the velocity of A to increase to 5.0 fps to the left while A moves 1.0 ft.

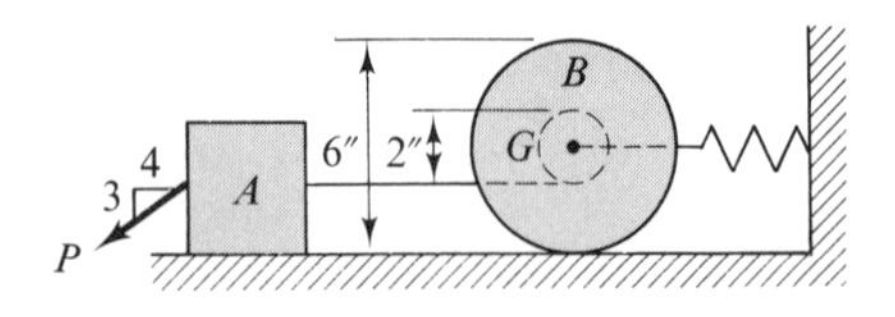

Figure P10-56

***10-57** The 150-kg homogeneous cylinder A in Fig. P10-57 rolls without slipping on the horizontal plane. In the position shown, the 30-kg body B has a velocity of 5 m/s downward, and the spring is stretched 0.6 m. The spring modulus is 50 N/m. Determine the angular velocity of A when it passes directly under the smooth peg C.

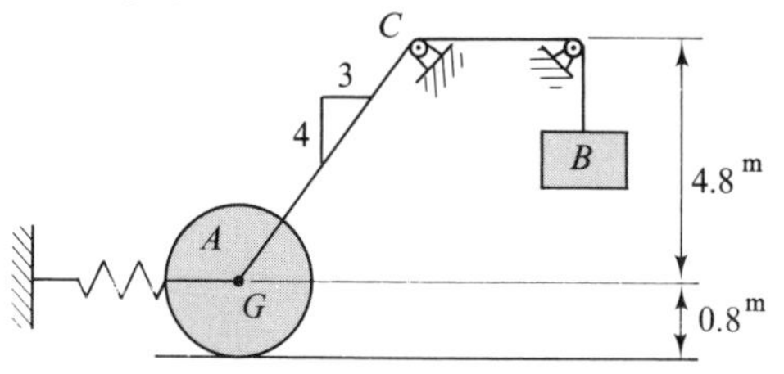

Figure P10-57

10-58 The 644-lb symmetrical body A in Fig. P10-58 rolls without slipping and has a radius of gyration of mass of 2 ft with respect to a horizontal axis through G. The homogeneous disk B weighs 128.8 lb, and body C weighs 64.4 lb. The spring (modulus of 20 lb per ft) is stretched 2 ft when body A has an angular velocity of 6 rad per sec clockwise. Determine the angular velocity of A after G has moved 3 ft to the right.

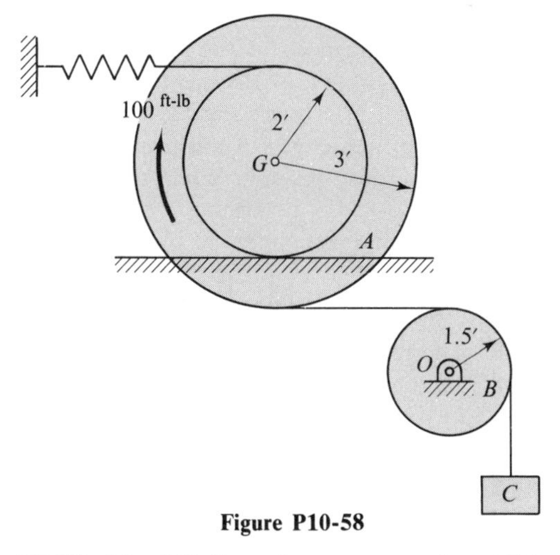

Figure P10-58

10-59 The 161-lb nonhomogeneous bar AB in Fig. P10-59 has its mass center at G and is pinned at the ends to guide blocks, of negligible weight, which slide in smooth slots. The radius of gyration of the mass of AB with respect to a horizontal axis through G is 4 ft. In the position shown the bar has an angular velocity of 3.0 rad per sec counterclockwise, and the spring is compressed 2 ft. Determine the modulus of the spring which will reduce the angular velocity of the bar to 1.5 rad per sec when it reaches the horizontal position.

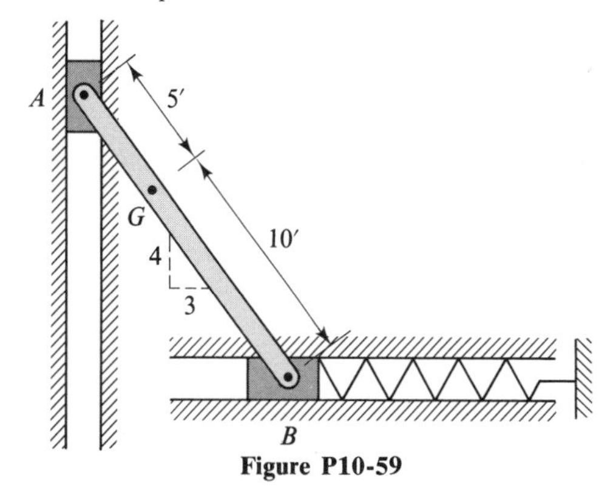

Figure P10-59

10-60 The 322-lb nonhomogeneous bar in Fig. P10-60 has a radius of gyration of mass of 4 ft with respect to a horizontal axis through the mass center G. In the position shown the angular velocity of the bar is 4 rad per sec clockwise, and the spring is compressed 2 ft. Determine the modulus of the spring if the angular velocity of the bar is 3 rad per sec clockwise when it reaches the horizontal position. Neglect the mass of the smooth guide blocks at A and B.

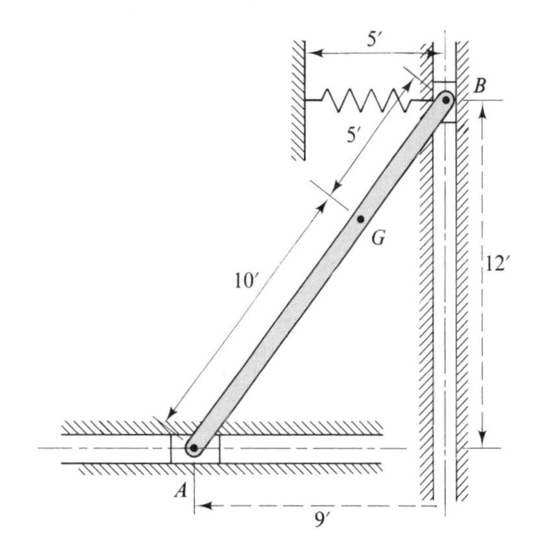

Figure P10-60

*10-61 The semicircular body A in Fig. P10-61 has a mass of 17 kg and rotates about a fixed horizontal axis at O. The homogeneous rod BC has a mass of 3 kg and moves in a vertical plane. When line OB is vertical as shown, the angular velocity of A is 7.0 rad/s clockwise. Determine the angular velocity of A after it has rotated 90° clockwise to the position shown by dashed lines.

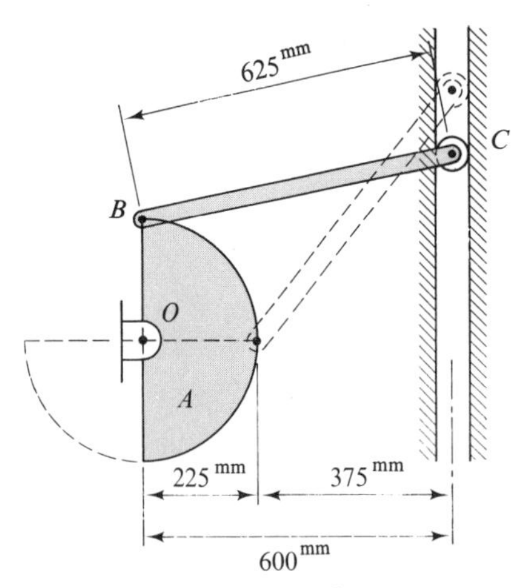

Figure P10-61

*10-62 If body A in Fig. P10-61 has a mass of 10 kg, BC has a mass of 3 kg, and the system is at rest initially (when B is directly above O), will A start rotating clockwise or counterclockwise? Determine the angular velocity of A when OB is horizontal after starting from rest.

10-63 In Fig. P10-63, A weighs 128.8 lb, B weighs 966 lb, and D weighs 322 lb. Body B is a solid homogeneous cylinder, and body D is a solid homogeneous sphere which rolls without slipping. The couple $\mathbf{C}_v$ on body B varies according to the equation $C_v = 30\theta^2 + 90$, where θ, the angular position of B, is in radians and C_v is in foot-pounds. The positive direction is clockwise for C_v and θ. If θ is zero when the velocity of A is 20 fps downward, determine the distance A moves while its velocity is changing to 10 fps downward.

*10-64 The homogeneous 150-kg rod in Fig. P10-64 is released from rest in the position shown. The initial tension in the spring is 500 N. The rod attains an angular velocity of 2.0 rad/s clockwise when it becomes horizontal. Determine the spring constant. Neglect the weights of blocks A and B.

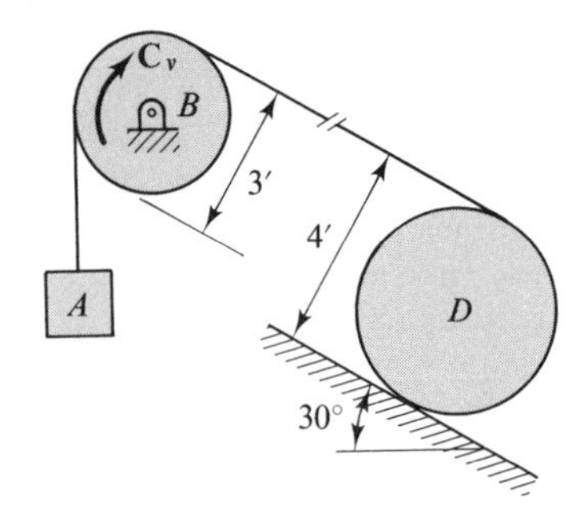

Figure P10-63

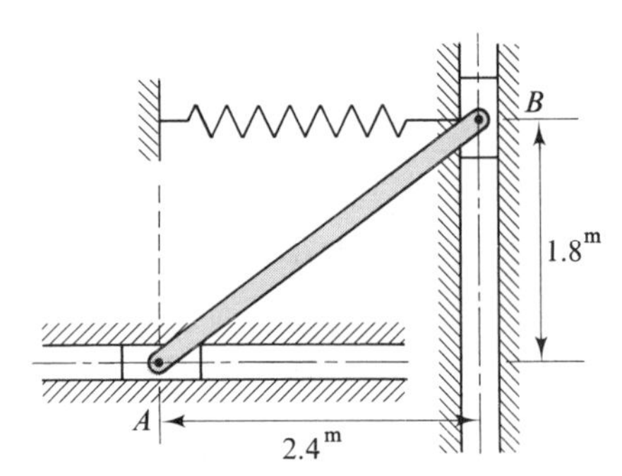

Figure P10-64

10-65 The spring in Fig. P10-65 has a modulus of 10 lb per ft and is stretched 2 ft in the position shown. The homogeneous 96.6-lb bar has an angular velocity of 3 rad per sec clockwise in the vertical position. When the bar has rotated 90° clockwise, determine

(a) The horizontal component of the reaction at O on the bar.

(b) The angular acceleration of the bar.

10-66 The homogeneous 10-ft bar AB in Fig. P10-66 has a mass of 6 slugs and is connected to body C by a flexible cable. Ends A and B move

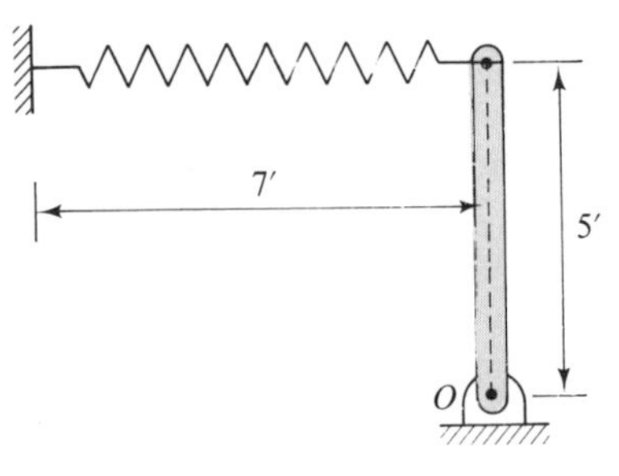

Figure P10-65

in smooth slots as shown. The system is released from rest when AB is vertical. When C has fallen 8 ft to C' its velocity is 15 fps. Determine the mass of C.

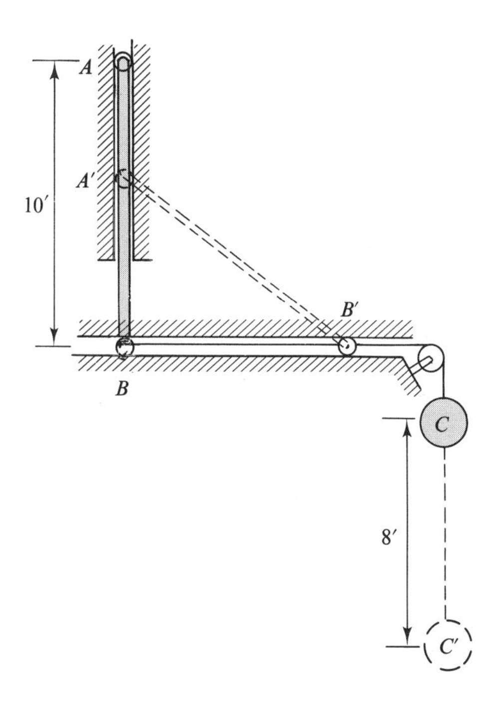

Figure P10-66

10-67 The 64.4-lb body A in Fig. P10-67 rolls without slipping and has a radius of gyration of mass of 2 in. with respect to a horizontal axis through G. The homogeneous disk B weighs 32.2 lb, and body C weighs 12.88 lb. The modulus of the spring is 1.0 lb per in., and the spring is stretched 6 in. when the velocity of C is 1.0 fps downward. Determine the angular velocity of A after C has moved 12 in. downward.

10-68 The 322-lb nonhomogeneous bar AB in Fig. P10-68 has a radius of gyration of mass of 4 ft with respect to a horizontal axis through the mass center G. In the position shown the angular velocity of the bar is 4 rad per sec clockwise, and the spring (modulus of 40 lb per ft) is stretched 1 ft. Determine the angular velocity of the bar when it passes through the horizontal position. Neglect the weights of the smooth guide blocks A and B.

***10-69** The homogeneous 30-kg bar AB in Fig. P10-69 is at rest in the position shown. The homogeneous bar BC has a mass of 90 kg, and the mass of the roller at C may be neglected. The modulus of the spring is 300 N/m, and its unstretched length is 3 m. Determine the angular velocity of AB after it has rotated 90° counterclockwise.

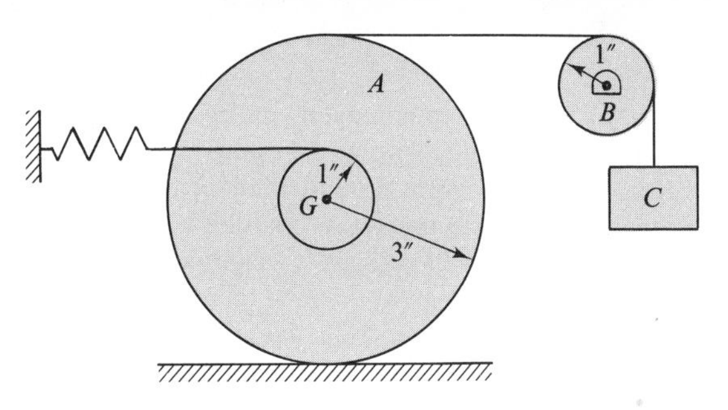

Figure P10-67

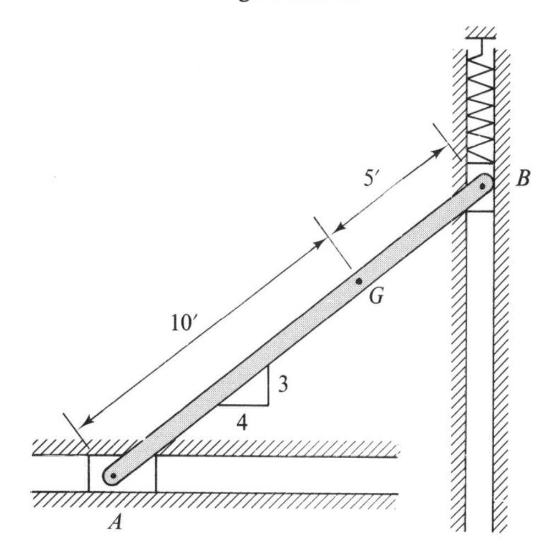

Figure P10-68

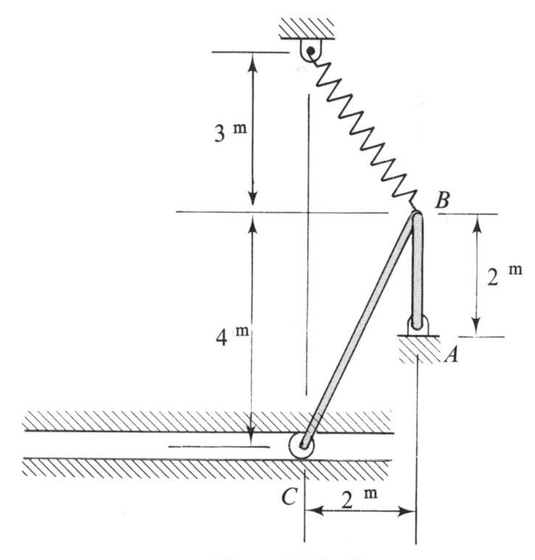

Figure P10-69

10-70 The 322-lb homogeneous cylinder A in Fig. P10-70 rolls without slipping. The cord from the spring wraps around a hub of negligible mass. The 96.6-lb homogeneous slender rod BC has a small roller of negligible weight at C. Body A has an initial angular velocity of 1 rad per sec counterclockwise, and the spring is initially stretched 2 ft. Determine the spring constant required for the entire assembly to come to rest after body A rotates 90° counterclockwise.

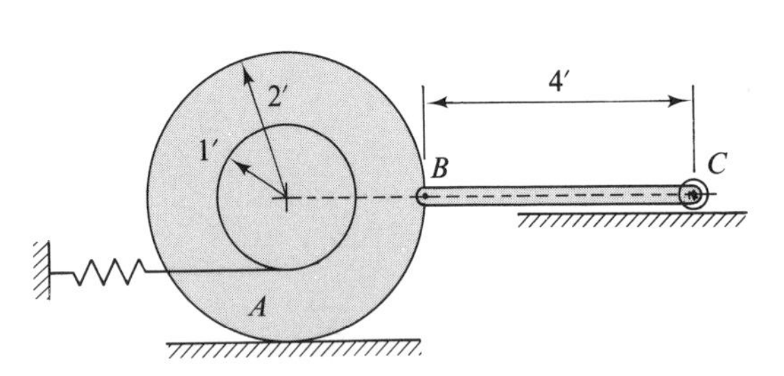

Figure P10-70

10-71 The 96.6-lb cylindrical body in Fig. P10-71 has a radius of gyration of mass with respect to a horizontal axis through the mass center G of 2.5 in. and rolls without slipping. The body has an angular velocity of 1 rad per sec clockwise in the position shown, and the spring is stretched 1 in. Determine the modulus of the spring if the angular velocity of the body is 3 rad per sec clockwise after it has turned 90° clockwise. The 20-lb force remains at the same slope throughout the motion.

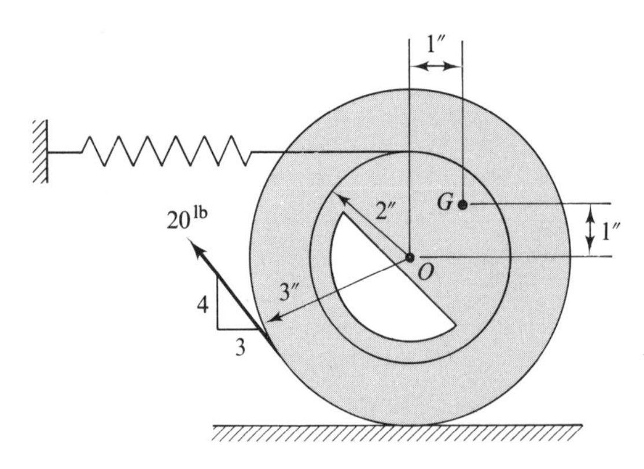

Figure P10-71

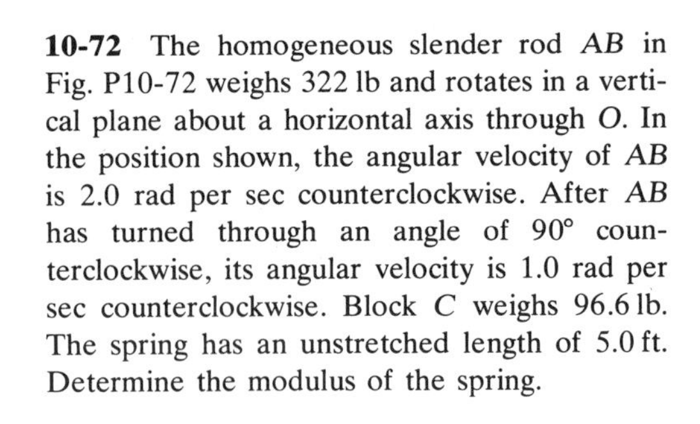

10-72 The homogeneous slender rod AB in Fig. P10-72 weighs 322 lb and rotates in a vertical plane about a horizontal axis through O. In the position shown, the angular velocity of AB is 2.0 rad per sec counterclockwise. After AB has turned through an angle of 90° counterclockwise, its angular velocity is 1.0 rad per sec counterclockwise. Block C weighs 96.6 lb. The spring has an unstretched length of 5.0 ft. Determine the modulus of the spring.

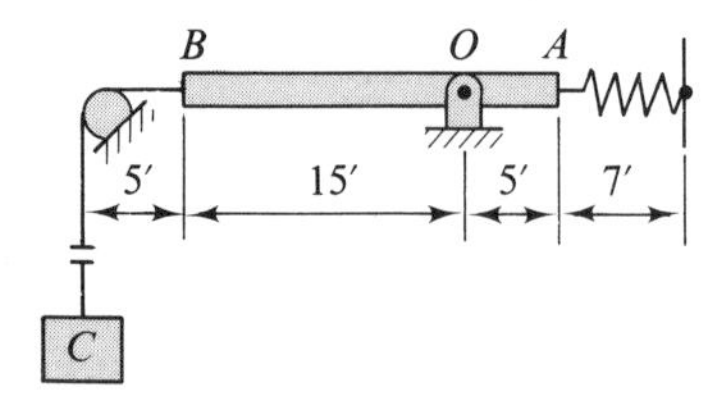

Figure P10-72

10-73 The 161-lb unbalanced wheel A in Fig. P10-73 rolls without slipping on the circular surface and has a radius of gyration of mass with respect to a horizontal axis through G of 1.5 ft. The weight of B can be neglected. The spring has a modulus of 100 lb per ft, has an unstretched length of 4.0 ft, and is attached to member B and to the support at D. The wheel has an angular velocity of 3.0 rad per sec clockwise when in the top position. Determine the angular velocity of A after it has rolled to position A' (shown by dashed lines).

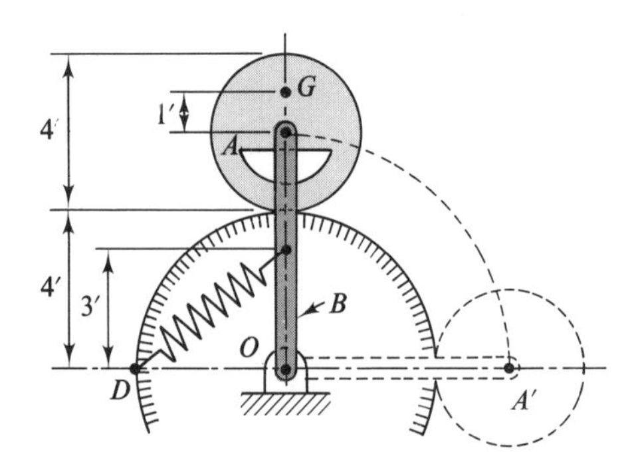

Figure P10-73

*10-74 When the 150-kg homogeneous rod AB is in the position shown in Fig. P10-74 the *tension* in the spring is 225 N. The weights of the blocks at A and B and friction can be neglected. The bar is released from rest in the indicated position. When the bar becomes vertical its angular velocity is 1.0 rad/s counterclockwise. Determine the spring modulus.

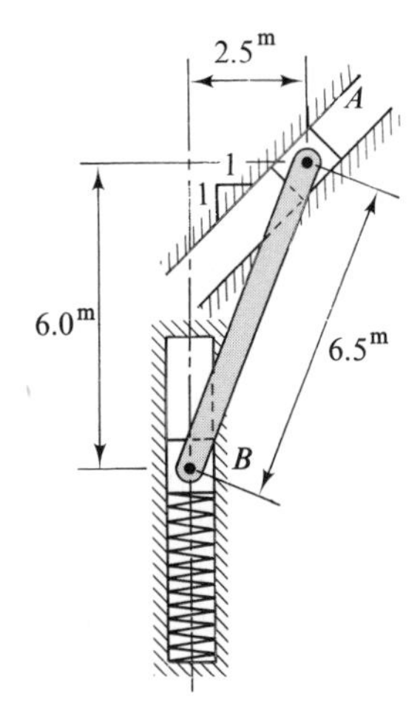

Figure P10-74

10-75 The slender homogeneous bars AB and BC in Fig. P10-75 weigh 64.4 and 128.8 lb, respectively. A torsion spring with a modulus of 120 ft-lb per rad is attached to bar AB and the support at A. The system is at rest, and the torque of the spring on AB is zero in the position shown. The end C of the bar BC is displaced slightly to the right and released. Determine the angular velocity of AB when it has rotated through 120°.

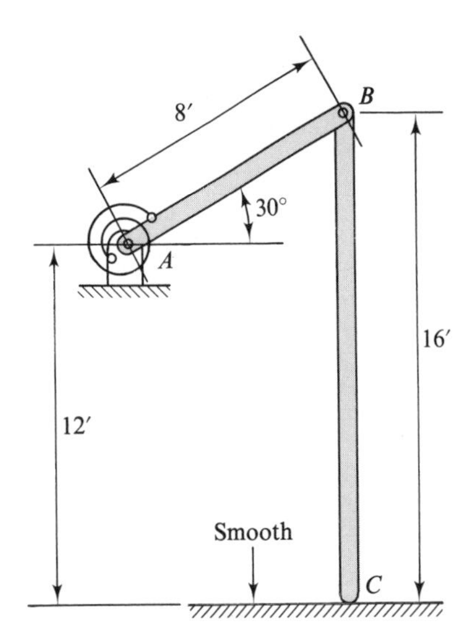

Figure P10-75

10-76 In Fig. P10-76, collars A and B each weigh 16.1 lb, and the weight of the bar can be neglected. The 25-in. bar is fastened to A and B by ball and socket joints, and length b is 9 in. The system starts from rest when $x = 0$, and a force $5\mathbf{i}$ lb is applied to B. Determine the velocity of B when $\mathbf{x} = 20\mathbf{i}$ in. Neglect friction.

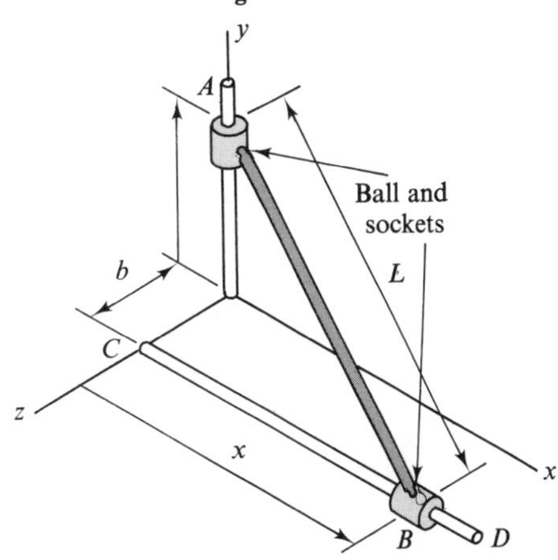

Figure P10-76

10-77 The homogeneous semicircular body in Fig. P10-77 has a uniform thickness and weighs 32 lb. The uniform bar AB weighs 20 lb and is pinned to C. Body C rolls without slipping, and the surface at A is smooth. The system is at rest in the position shown. When C has rotated 90° clockwise it has an angular velocity of 2.0 rad per sec clockwise. Determine the constant force $\mathbf{P}$.

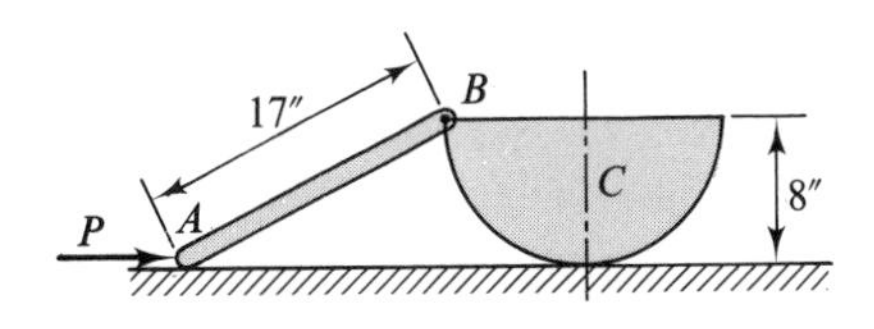

Figure P10-77

*10-78 The homogeneous body in Fig. P10-78 is released from rest in the position shown. Determine the velocity of A just before AB becomes horizontal.

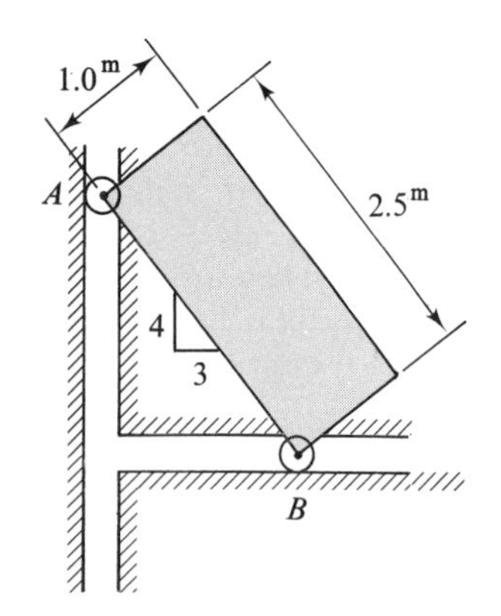

Figure P10-78

10-9 POWER AND EFFICIENCY

Power, as used in mechanics, is defined as *the time rate of doing work*. From the definition of power, its dimensions are force multiplied by length and divided by time, FL/T, and common units are foot-pounds per second, foot-pounds per minute, horsepower (hp), watts, kilowatts (kW), British thermal units per hour, and so on. The magnitudes of the various units have the relationships indicated in the following equations: 1 hp = 550 ft-lb per sec = 33,000 ft-lb per min = 746 W = 0.746 kW. By definition, 1 British thermal unit (Btu) is equal to 778 ft-lb of energy.

Power is used in rating machines. It is usually more important to know the amount of work the machine can do in a second, hour, or day than merely to know that the machine can do so much work. A large pump and a small pump may both be capable of filling a tank with water from a reservoir, but the time required may differ considerably for the two pumps, and thus their rates of doing the work (power) will not be the same.

Since power is the time rate of doing work, the power delivered to a body by a force $\mathbf{F}$ acting on the body which has a displacement $d\mathbf{r}$ during a time dt is

$$P = \frac{dU}{dt} = \frac{\mathbf{F}\cdot d\mathbf{r}}{dt} = \mathbf{F}\cdot\mathbf{v}. \tag{10-13a}$$

The power delivered by a couple with a moment $\mathbf{T}$ acting on a body which has an angular displacement $d\theta$ in the same plane as the couple during a time dt is

$$P = \frac{dU}{dt} = \frac{T\,d\theta}{dt} = T\omega.$$

This expression is equivalent to

$$P = \mathbf{T}\cdot\boldsymbol{\omega}. \tag{10-13b}$$

The units used in expressing power can also be used with an interval of time to express work or energy. Thus a horsepower-hour (hp-hr) is the amount of work done in an hour by a force working at the rate of one horsepower. That is,

$$1 \text{ hp-hr} = \left(33{,}000\,\frac{\text{ft-lb-hr}}{\text{min}}\right)\left(60\,\frac{\text{min}}{\text{hr}}\right) = 1{,}980{,}000 \text{ ft-lb} = 2545 \text{ Btu}.$$

Similarly, 1 kilowatt-hour (kWh) = 2,655,000 ft-lb = 3413 Btu.

The term *efficiency* can be used to denote a number of different quantities. When referring to machines, *efficiency is defined as the ratio of the energy output of the machine per unit time to the energy supplied to the machine in the same time interval.* Efficiency can be further qualified by calling it mechanical efficiency, electrical efficiency, hydraulic efficiency, overall efficiency, and so on. The mechanical efficiency of a machine is

$$\eta_{\text{mech}} = \frac{\text{power output}}{\text{power input}}.$$

It is impossible to eliminate all frictional forces between the moving parts of a machine. Therefore, some of the energy supplied to the machine must be utilized in overcoming the frictional resistance; hence, the power output is less than the energy supplied per unit of time. The kinetic energy of the water as it leaves a water wheel is another example of energy which is not used by the machine to do work. These are only two of many examples which could be cited indicating that the efficiency of any engine is always less than unity, or 100%. Some machines for transforming electric or hydraulic energy to mechanical energy, such as motors and turbines, have efficiencies which may run as high as 90% or more when operating under ideal conditions.

10-10 DISSIPATION OF MECHANICAL ENERGY

Frictional forces are usually undesirable in machines, since they reduce the efficiency of the machines by dissipating some of the energy in unusable forms which are manifested by an increase of temperature, the production of sound, and other phenomena. In some instances, however, the transformation of mechanical energy to thermal energy by means of frictional forces may be desirable, as in the use of brakes to stop an automobile. The kinetic energy of the automobile is dissipated (transformed into thermal energy) through the action of the brake lining sliding on the brake drum. The energy which is transferred to the brake lining by the frictional forces, and which is indicated by an increase in the temperature of the lining, is eventually dissipated into the atmosphere. Frictional forces are also utilized for transmitting

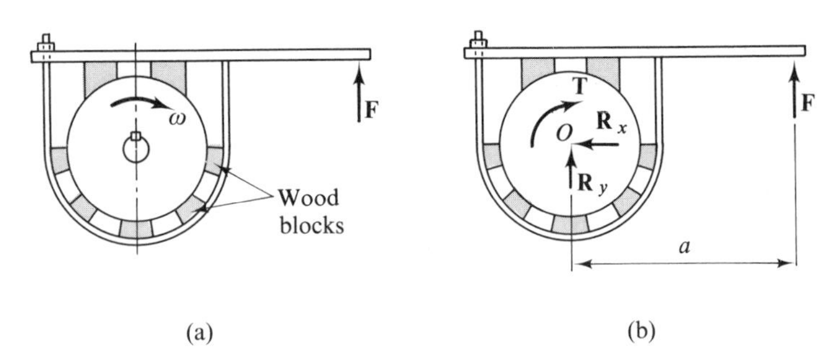

Figure 10-23

energy in such apparatus as clutches and belt drives, but the dissipation of energy in such applications is usually small.

A *dynamometer* is a machine or instrument for measuring the force or power developed by an engine. The efficiency can be determined when the power developed is known. The Prony brake illustrated in Fig. 10-23(a) is an example of an absorption-type dynamometer in which the energy transmitted to the drum by the shaft is transformed into thermal energy by frictional forces and dissipated into the atmosphere or into cooling water. The brake drum rotates in a clockwise direction due to the turning moment transmitted by the shaft. When the turning moment is balanced by the frictional moment of the brake on the drum, the angular velocity of the drum remains constant. The work done on the drum by the torque in the shaft is obtained as the product of the turning moment and the angle through which the drum rotates. The power supplied to the drum (and dissipated by the friction) is given by Eq. (10-13b) as

$$P = \mathbf{T} \cdot \boldsymbol{\omega},$$

where P is the power, $\mathbf{T}$ is the turning moment, and $\boldsymbol{\omega}$ is the angular velocity of the drum and shaft.

The turning moment of the shaft on the drum in Fig. 10-23(a) can be obtained from the free-body diagram of the brake and drum in Fig. 10-23(b). The reaction of the shaft on the drum is resolved into a pair of component forces, $\mathbf{R}_x$ and $\mathbf{R}_y$, and a couple, $\mathbf{T}$. The system is in equilibrium; therefore,

$$\sum M_O = 0$$

from which

$$T = Fa,$$

where F is the net load at the end of the brake arm. The expression just developed is not valid for all brake arrangements, and a free-body diagram of either the brake, the drum, or both together should be drawn to help determine the turning moment.

PROBLEMS

10-79 A jet of water flows from a nozzle with a velocity of 200 fps. The diameter of the jet is 3.00 in., and the water weighs 62.4 lb per ft³. Determine
(a) The kinetic energy per pound of water in the jet.
(b) The kinetic energy of the water which flows from the nozzle in 1 sec.
(c) The horsepower of the water in the jet.

***10-80** Determine the torque that must be developed by a motor revolving at 2000 rpm to generate 2.5 kW.

10-81 A 165-lb man walks up a flight of stairs 10.5 ft high in 10 sec. Determine the average horsepower exerted by the man.

10-82 The pumps at the Grand Coulee Dam raise the water to be used in irrigation a distance of 280 ft. Each pump is operated by a 65,000-hp motor and has a capacity of 700,000 gal per min. Water weighs 8.34 lb per gal. Determine the overall efficiency of the pumping system.

10-83 A hoist lifts a load of 3000 lb at a speed of 10 fps when 50 kW is supplied to the motor. Determine the overall efficiency of the motor-hoist unit.

10-84 (a) A 2000-lb load is being lifted with a constant velocity of 10 fps. Determine the power required.
(b) Determine the power required if the 2000-lb load has a velocity of 10 fps upward and is accelerated 4 fps² upward.

***10-85** The moment **M** in Fig. P10-85 causes the drum to have an angular velocity of 150 rpm clockwise. The coefficient of friction between the belt and drum is 0.30. Determine the power dissipated by the brake.

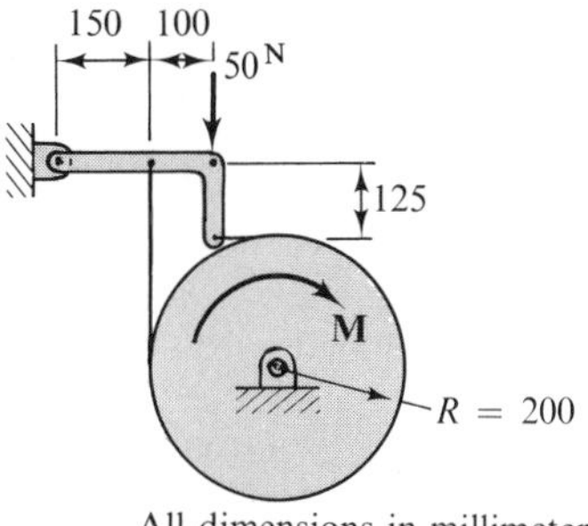

Figure P10-85

10-86 The Prony brake in Fig. P10-86 is used to measure the power developed by a small impulse-type water wheel. The couple **C** transmitted to the drum by the shaft causes the brake drum to have a clockwise angular velocity of 50 rad per sec. The net load **Q** (neglecting the weight of the brake arm) is 12.5 lb upward. The power in the jet of water driving the wheel is 2.40 hp before it strikes the wheel. Determine
(a) The power dissipated by the Prony brake.
(b) The efficiency of the water wheel.

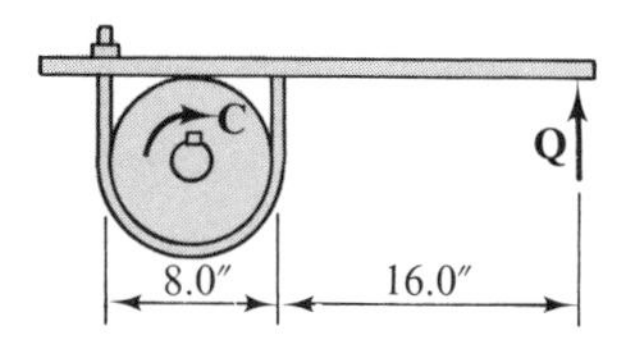

Figure P10-86

10-87 The couple **C** acting on the drum in Fig. P10-87 causes it to rotate with a constant angular velocity of 24 rad per sec clockwise. The coefficient of friction between the brake and drum is 0.50. Determine the power dissipated by the brake.

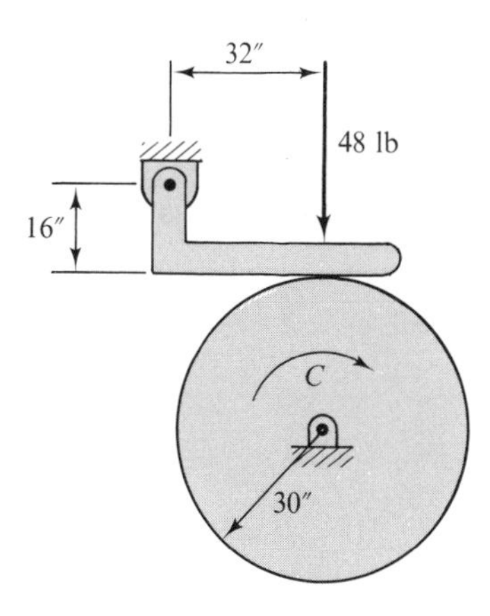

Figure P10-87

10-88 The brake mechanism in Fig. P10-88 is used to lower body A with a constant velocity of 3.0 m/s. The coefficient of friction between the brake and drum D is 0.40. Neglect the weight of the brake. Determine
(a) The mass of body A.
(b) The power dissipated by the brake.

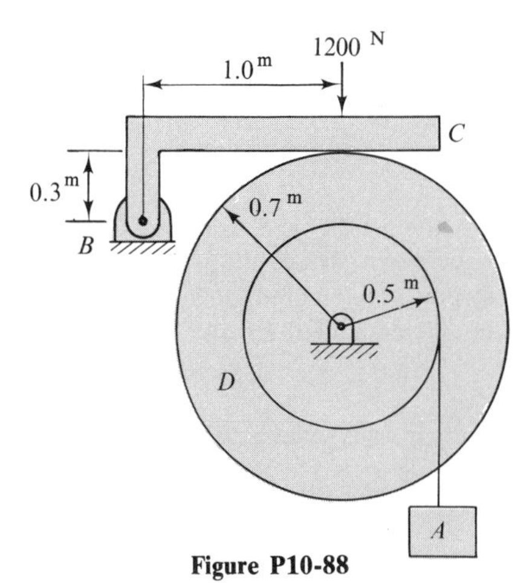

Figure P10-88

10-89 The drum in Fig. P10-89 has a constant angular velocity of $\boldsymbol{\omega}$ rad per sec clockwise, and the coefficient of friction between the drum and brake band is μ. Derive an expression for the power dissipated by the brake in terms of the force **Q**, the angular velocity of the drum, the coefficient of friction between the brake and drum, and the dimensions of the brake.

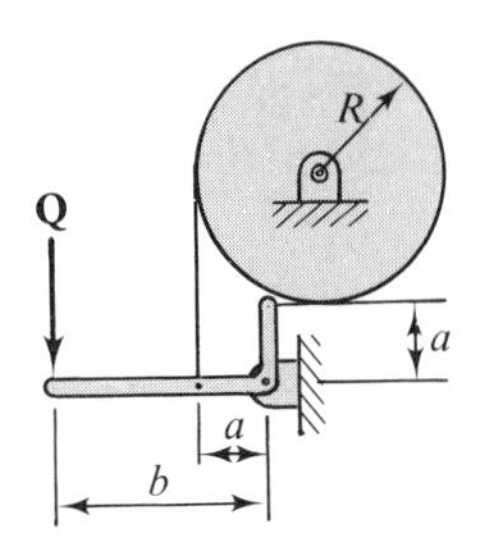

Figure P10-89

10-90 The electric generator in Fig. P10-90 is used as a dynamometer to measure the power output of a motor. The frame of the generator (the field) is mounted on smooth bearings and is restrained from turning by the two forces **Q** and **Q′**, which constitute a couple. The armature of the generator is driven by the motor. As the load on the generator is increased, the forces **Q** and **Q′** also increase. When the motor is driving the generator with an angular velocity of 1700 rpm counterclockwise, the magnitudes Q and Q' are 9.75 lb each, and 8.50 kW is being supplied to the motor. Determine the efficiency of the motor.

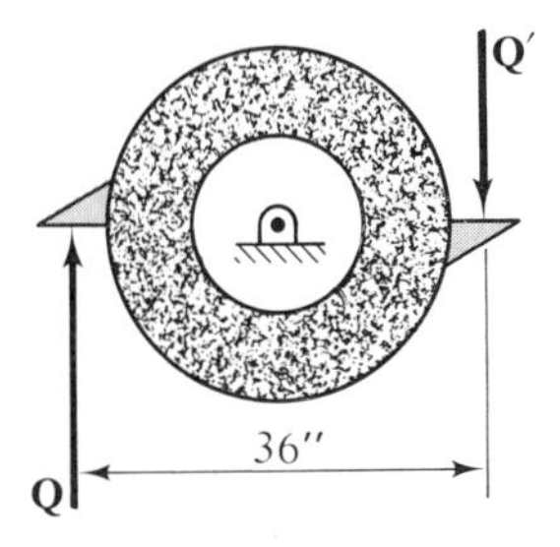

Figure P10-90

10-91 A truck-and-trailer unit which weighs 10 tons travels along a highway at 45 mph. The wind and frictional resistance is 50 lb per ton. The efficiency of the power transmission from the engine to the wheels is 65%. Determine the power which must be developed by the engine while moving up a 4% grade (4-ft rise per 100 ft of horizontal distance).

10-11 CLOSURE

The principle of work and kinetic energy equates the work done by all the forces acting on a body or system of bodies to the change in the kinetic energy of the bodies. Work is either a positive or a negative scalar product of force and displacement, whereas kinetic energy is a positive scalar property of a body. The work done by the internal forces of a rigid body is zero, but for a nonrigid body the internal forces do work. When a system of rigid bodies is connected by inextensible cords, smooth pins, and similar devices, no work is done on the system by the connections, and one free-body diagram of the system of bodies is normally used for the work and energy method. It should be remembered that it is usually desirable to draw separate free-body diagrams of each body when using the method of force, mass, and acceleration.

The method of work and kinetic energy is particularly useful for problems involving variable forces whose magnitudes are functions of the position of the body, such as the force exerted by a spring. When all the forces acting on a body are constant, the work and kinetic energy method frequently gives a more direct solution for an unknown velocity or displacement than the method of force, mass, and acceleration combined with a v-t diagram or other kinematic relations.

The method of work and kinetic energy cannot be used to determine forces which do no work on a body. When the method is used to obtain the magnitude of a variable force which does work on a body, the position average value of the force is usually obtained rather than an instantaneous value. Accelerations and time intervals are other quantities which cannot be obtained directly from the principle of work and kinetic energy.

Sometimes it is convenient to use both the methods of work and energy and of force, mass, and acceleration in the same problem. In some examples, the acceleration of the mass center of a body is a function of the angular velocity of the body or the linear velocity of a connected body, and it may be easier to determine the unknown velocity by work and energy than by using force, mass, and acceleration and the differential equations of kinematics. The velocity can be substituted in the equations of motion relating force, mass, and acceleration. The following problems are examples in which the two methods can be used to supplement each other.

PROBLEMS

10-92 The ball in Fig. P10-92 is released from rest in position A. Determine the angle the cord makes with the vertical when the cord tension is 2.5 times the weight of the body.

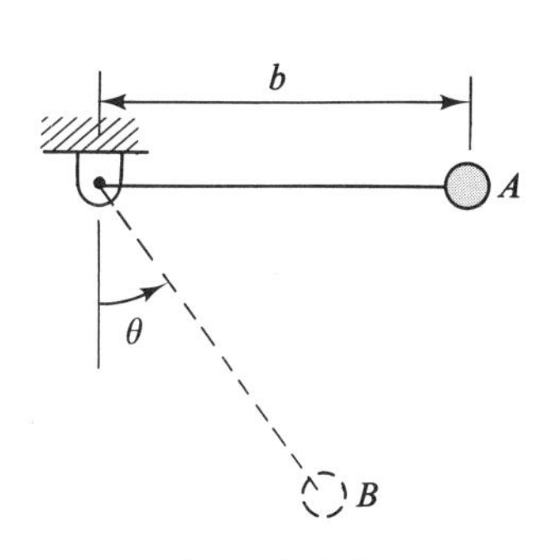

Figure P10-92

*10-93 The ball in Fig. P10-93 is given a velocity of 5.0 m/s to the right when $\theta = 0$.
(a) Determine the maximum height the ball will reach on a circular path.
(b) Will the body leave the circular path? If so, determine the velocity of the ball as it leaves the path.

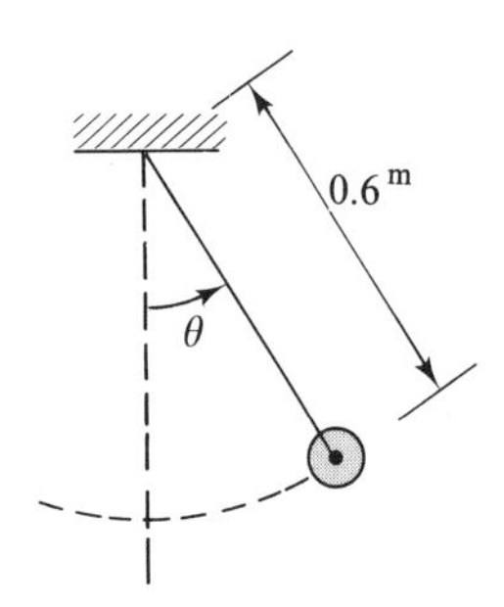

Figure P10-93

10-94 Body A in Fig. P10-94 weighs 2 lb, and B weighs 5 lb. Determine the maximum angle θ from which A can be released from rest without causing B to leave the floor while A swings through a vertical arc.

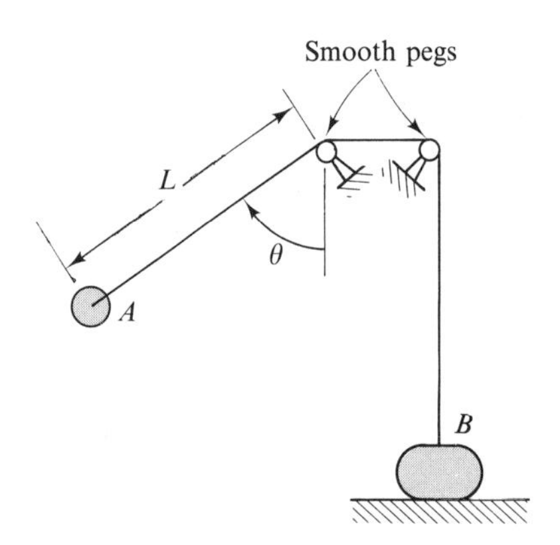

Figure P10-94

10-95 A small body P slides along the smooth vertical track in Fig. P10-95. Determine
(a) The velocity of the particle at point B if it leaves the track at point C.
(b) Where and with what velocity the body will fall back on the track.

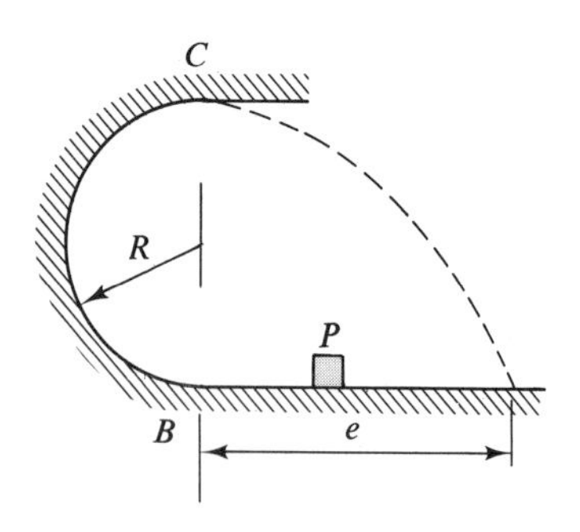

Figure P10-95

10-96 Body A in Fig. P10-96 is a solid homogeneous cylinder which weighs 64.4 lb and rolls along the horizontal plane without slipping. The cylinder is connected to the 32.2-lb block B by an inextensible cord which passes over the smooth pegs C and D. The cylinder is released from rest when $s = 6$ ft. Determine all unknown forces acting on the cylinder when
(a) $s = 4$ ft.
(b) $s = 0$.

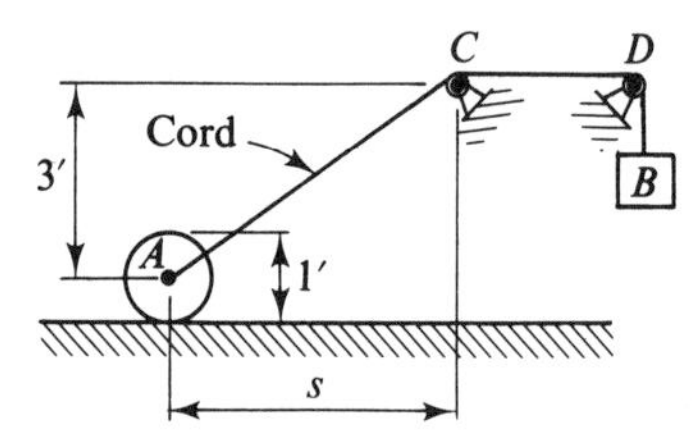

Figure P10-96

10-97 Body A in Fig. P10-97 is a solid homogeneous sphere which rolls on the fixed cylindrical surface as shown. The coefficient of friction between the ball and the cylinder is 0.50. The ball starts from rest at the top of the cylinder. At what angle θ will the ball start to slip on the cylinder?

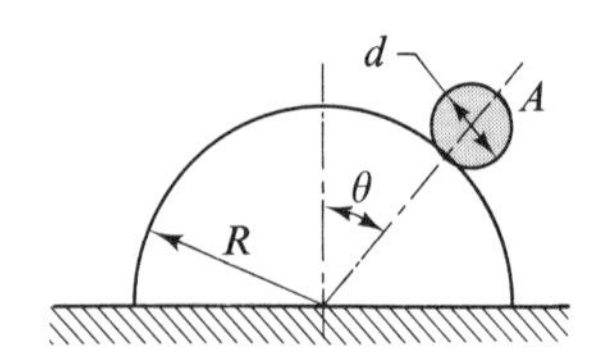

Figure P10-97

*10-98 Bar AB in Fig. P10-98 has a mass of 20 kg and is connected to small (negligible mass) blocks at its ends which slide in smooth guides.

The 10-kg body C is connected to block B by a flexible inextensible cord. The system is released from rest when AB is vertical. Determine
(a) The tension in the cord just after the system is released.
(b) The velocity of C after it has moved 1.6 m downward.

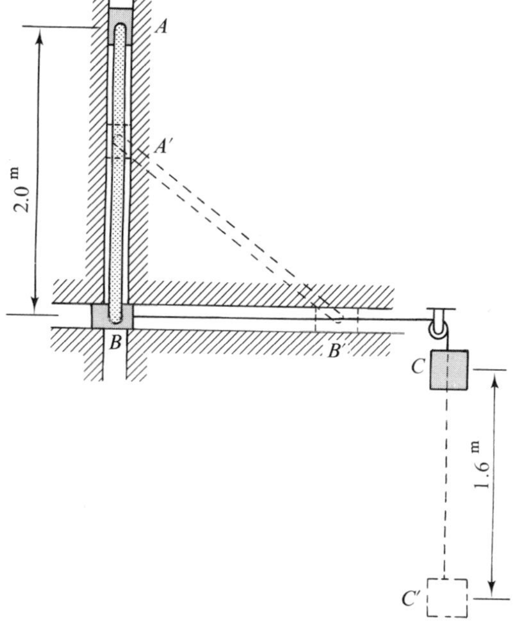

Figure P10-98

*10-99 Determine the reactions on the ends of the bar in Problem 10-98 at the instant the bar becomes horizontal.

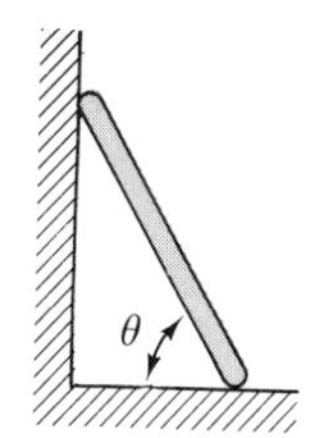

Figure P10-100

10-100 The homogeneous slender rod in Fig. P10-100 moves with its ends in contact with the two smooth surfaces as indicated. The rod is at rest in the vertical position when the lower end is pulled out slightly to start the bar moving. Determine the angle θ when the upper end leaves the vertical wall.

10-101 The upper end of the rod in Problem 10-100 is constrained by a roller in a smooth slot to move along a vertical line as shown in Fig. P10-101. Determine the angle θ at which the lower end will leave the floor. Motion is started in the same manner as in Problem 10-100.

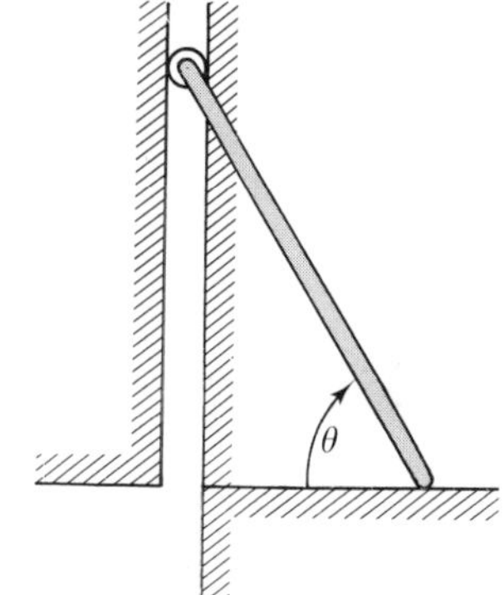

Figure P10-101

11 Kinetics–Impulse and Momentum

PART A–PARTICLES

11-1 INTRODUCTION

The principles of force, mass, and acceleration were developed from Newton's laws of motion and their application to kinetics problems was illustrated in Chapter 9. In Chapter 10, the principle of work and kinetic energy was derived from Newton's laws of motion and applied to the analysis of problems in kinetics. The principles of impulse and momentum, which also are based on Newton's laws of motion, will be developed and their use explained in this chapter. Since each of the three methods has certain advantages for some types of problems, it is desirable to understand all three methods.

When an acceleration or a force at an instant is to be obtained, the method of force, mass, and acceleration is usually the most direct. When one or more of the forces involved are variables and can be expressed as functions of the position of the body, the method of work and kinetic energy is generally best. If one or more of the forces involved are variables and can be expressed as functions of time, the principles of impulse and momentum usually provide the most direct solution. The principles of impulse and momentum are particularly effective for problems involving an impact or collision between two bodies, for problems dealing with fluids, and for problems in which the mass varies with time, such as rockets and hoisting cables. Many problems can be solved readily by more than one of the three methods. In such cases, solutions by two methods provide an excellent means of checking the result.

11-2 PRINCIPLE OF LINEAR IMPULSE AND LINEAR MOMENTUM

The linear impulse $\mathbf{I}_L$ *of a force* $\mathbf{F}$ *during a time interval from* t_i *to* t_f is defined *as the definite integral of the product of the force and the time dt from* t_i *to* t_f; that is,

$$\mathbf{I}_L = \int_{t_i}^{t_f} \mathbf{F}\, dt. \tag{11-1}$$

The linear impulse is a vector and has fundamental dimensions of FT with common units of pound-seconds and newton-seconds. If the force $\mathbf{F}$ is constant in direction, the linear impulse has the same line of action as the force. When the force varies in direction, the component form of Eq. (11-1) can be used to obtain the components of the linear impulse vector for the time interval involved. The x component is

$$(\mathbf{I}_L)_x = \int_{t_i}^{t_f} \mathbf{F}_x\, dt$$

with similar expressions for the y and z components. The component form is convenient to indicate the relationship between the force and time graphically. In Fig. 11-1, the magnitudes of the x component of the force are plotted against the corresponding values of time. The shaded area under the curve of Fig. 11-1 equals the magnitude of the x component of the linear impulse of the force $\mathbf{F}$ during the time interval from t_i to t_f.

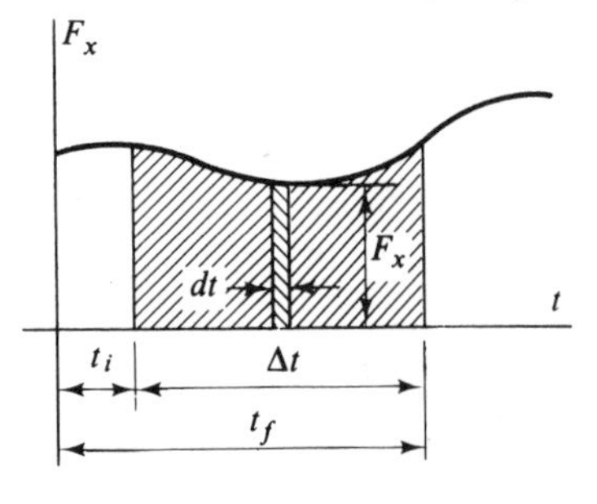

Figure 11-1

The linear impulse in the x direction can also be written

$$(\mathbf{I}_L)_x = (\mathbf{F}_x)_{avg}\, \Delta t,$$

where $(\mathbf{F}_x)_{avg}$ is the time average value of $\mathbf{F}_x$. It should be noted that the time average value of the force is usually not the same as the distance average value of the force used in determining the work done by a force. Problem 11-6 illustrates the preceding statement.

In the preceding chapter, it was shown that a force does work only when it has a component in the direction of its displacement. Every force that acts during a time interval does, however, have a linear impulse whether it does any work or not.

The linear impulse of a system of forces during a time interval is the vector sum of the linear impulses of the separate forces of the system during that time interval.

The linear momentum of a particle of mass m at any instant is defined as the product of the mass of the particle and its velocity at that instant; that is,

$$\mathbf{G} = m\mathbf{v}, \tag{11-2}$$

where $\mathbf{G}$ is the linear momentum of the particle.

The linear momentum of a particle is a vector quantity with the same direction as the velocity of the particle. The vector representing the momentum of a particle has a line of action which passes through the particle, and thus it is a localized vector. The dimensions of linear momentum are obtained as follows:

$$mv = \frac{FT^2}{L}\left(\frac{L}{T}\right) = FT.$$

Thus the dimensions, and consequently the common units, of linear momentum and linear impulse are the same.

The linear momentum of a system of n particles at any instant is the vector sum of the linear momentums of the particles of the system at that instant. This sum, as given by Eq. (9-8), is

$$\mathbf{G} = \sum_{i=1}^{n} m_i \mathbf{v}_i = m\mathbf{v}_G, \tag{11-3}$$

where m is the total mass of the system of particles and $\mathbf{v}_G$ is the velocity of the mass center of the system of particles.

Equation (11-3) gives the magnitude and direction of the linear momentum of any system of particles. The linear-momentum vector usually does not pass through the mass center of the system of particles. The position of the resultant linear momentum vector of a rigid body can be obtained by the principle of moments as explained in Art. 11-9.

From the principle of motion of the mass center for a system of particles having a total mass m the equation of motion given by Eq. (9-10) may be written as

$$\mathbf{R} = m\mathbf{a}_G = m\frac{d}{dt}(\mathbf{v}_G),$$

where $\mathbf{R}$ is the resultant of all the external forces acting on the system of particles. This equation may be written as

$$\mathbf{R}\,dt = m\,d\mathbf{v}_G. \tag{11-4}$$

The definite integral of Eq. (11-4) is

$$\left.\begin{aligned}\int_{t_i}^{t_f} \mathbf{R}\,dt = m\int_{\mathbf{v}_i}^{\mathbf{v}_f} d\mathbf{v}_G = m(\mathbf{v}_{Gf} - \mathbf{v}_{Gi}) \\ \text{or} \qquad \mathbf{I}_L = \mathbf{G}_f - \mathbf{G}_i,\end{aligned}\right\} \tag{11-5}$$

where the subscripts i and f signify initial and final values of time and velocity.

The principle of linear impulse and linear momentum as expressed mathematically in Eq.(11-5) can be stated in words as follows: *The linear impulse of a force system acting on any system of particles during a time interval is equal to the change in the linear momentum of the system of particles during that time interval.*

Equation (11-4) with m constant can also be written as

$$\mathbf{R} = \frac{d}{dt}(m\mathbf{v}_G), \tag{11-6}$$

which indicates that the resultant force acting on a system of particles is equal to the time rate of change of the linear momentum of the system of particles. Newton's second law of motion *for a particle* is sometimes stated in this form. Although Eq. (11-6) is generally not directly applicable to a system of particles that is gaining or losing material, there are instances for which it can be correctly used for variable mass problems. If the particles being added to the system have no velocity before the addition takes place or if the velocity of the particles being ejected from the system is zero after being ejected, then Eq. (11-6) can be applied.

Equations (11-5) are vector equations and can be used in component form. However, a consistent sign convention must be used for both impulses (or forces) and momentums (or velocities). Positive directions should be assumed if not given, and errors will frequently be avoided if positive axes are shown on the sketch.

The principle of linear impulse and linear momentum applies to any system of particles including a system of rigid bodies. Because of the vector quantities involved, however, it is usually better to draw separate free-body diagrams of each rigid body than to use combinations of bodies, as is commonly done when using the principle of work and kinetic energy.

The following examples illustrate the application of the principle of linear impulse and momentum.

EXAMPLE 11-1

The coefficient of friction between the 200-lb body A in Fig. 11-2(a) and the horizontal plane is 0.40. Body B weighs 60 lb. The magnitude of the variable force $\mathbf{P}_v$ is given by the equation $P_v = 3t^2$, where P_v is in

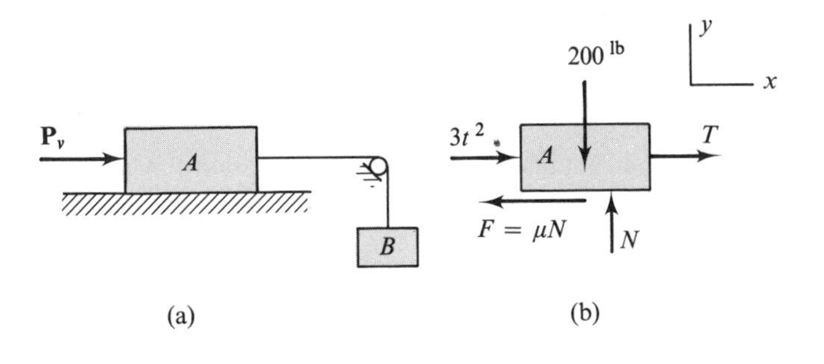

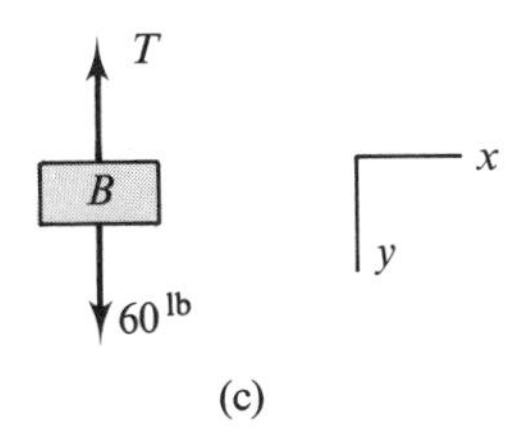

Figure 11-2

pounds when t is in seconds. When $t=0$ the velocity of A is 5.0 fps to the right. Determine the velocity of A when $t=5$ sec.

SOLUTION

Free-body diagrams of bodies A and B are shown in Fig. 11-2(b) and (c). Positive axes are included with each diagram. Body A is initially moving to the right; therefore the friction will be to the left and will be equal to the limiting value ($F'=\mu N$). The magnitudes of the velocities of A and B are the same, and their final velocities are $\mathbf{v}_{Af}=v_f\mathbf{i}$ and $\mathbf{v}_{Bf}=v_f\mathbf{j}$. There are three unknown quantities, N, T, and v_f, and there are two component linear impulse and linear momentum equations available for body A and one for body B. The component (scalar) equation for body A in the y direction is

$$(I_L)_y = m_A(v_{Ayf} - v_{Ayi})$$

or

$$\int_0^5 (N-200)\,dt = \frac{200}{32.2}(0-0)$$

from which

$$N = 200\text{ lb}.$$

The component equation in the x direction for body A is

$$(I_L)_x = m_A(v_{Axf} - v_{Axi}),$$

$$\int_0^5 (3t^2 + T - \mu N)\,dt = \frac{200}{32.2}(v_f - 5),$$

$$t^3\Big]_0^5 + \int_0^5 T\,dt - 0.4(200)t\Big]_0^5 = \frac{200}{32.2}(v_f - 5),$$

which reduces to

$$\int_0^5 T\,dt - 275 = \frac{200}{32.2}(v_f - 5). \tag{a}$$

The linear impulse and momentum equation for body B is

$$(I_L)_y = m_B(v_{Bf} - v_{Bi}),$$

$$\int_0^5 (60 - T)\,dt = \frac{60}{32.2}(v_f - 5),$$

which reduces to

$$300 - \int_0^5 T\,dt = \frac{60}{32.2}(v_f - 5). \tag{b}$$

The simultaneous solution of Eq. (a) and (b) gives

$$v_f = 8.10 \text{ fps}$$

and

$$\mathbf{v}_{Af} = \underline{8.10 \text{ fps} \rightarrow}. \qquad \text{Ans.}$$

The tension T is a variable, and this solution does not provide a specific value of T. The integral $\int_0^5 T\,dt$ is equal to the time average value of T multiplied by the time interval, and thus an *average* value for T can be obtained. This solution assumed that the friction force on A was constant, which requires that the velocity not become zero. This assumption can be checked by determining the minimum velocity which will occur when the acceleration is zero. If the velocity of A is to the right when the acceleration is zero, the assumption is correct. If the minimum velocity is found to be to the left, the assumption that the friction is constant (equal to μN) is not valid, and the problem would have to be solved using this information. For this problem the acceleration is zero when $t = 2.58$ sec and the minimum velocity is 0.733 fps to the right. Thus the solution is valid. It is suggested that this minimum velocity, and the time at which it occurs, be checked by the reader.

EXAMPLE 11-2

The coefficient of friction between the 200-lb block of Fig. 11-3(a) and the plane is 0.50. The magnitude of the variable force $\mathbf{P}_v$ is given by the expression $P_v = 10t + 100$, where P_v is in pounds when t is in seconds.

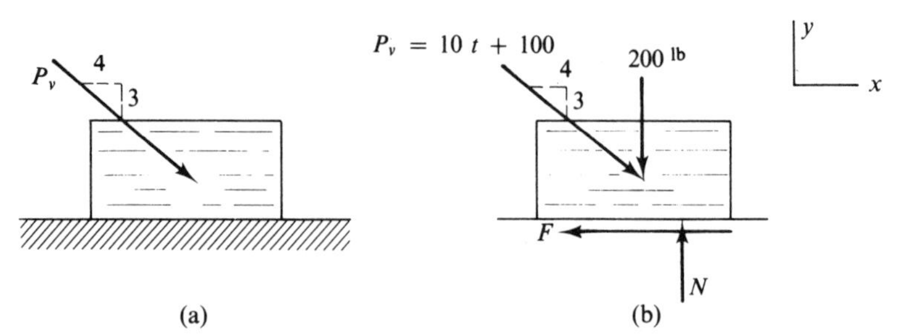

Figure 11-3

The block is at rest when $t=0$. Determine the velocity of the body when $t=15$ sec.

SOLUTION

The free-body diagram of the body is shown in Fig. 11-3(b). The impulse and momentum equation in the y direction is

$$(I_L)_y = m(v_{yf} - v_{yi}),$$

$$\int_0^5 \left(N - 200 - \frac{3}{5} P_v\right) dt = \frac{200}{32.2}(0-0)$$

or

$$N = 200 + 0.6(10t + 100)$$

$$= (260 + 6t) \text{ lb}.$$

If the horizontal component of P_v is greater than the limiting value of F, the block will start to slide. When $t=0$, $N=260$ lb and $\mu N = 130$ lb. The horizontal component of P_v is $0.8(100) = 80$ lb when $t=0$; therefore the block will not slip. The value of t when motion starts can be obtained by equating the horizontal component of $\mathbf{P}_v$ to the limiting value of F, that is,

$$0.8(10t + 100) = \mu N = 0.5(260 + 6t)$$

from which

$$t = 10 \text{ sec}.$$

The velocity will be zero from $t=0$ to $t=10$ sec. For t greater than 10 sec the force F will be

$$F' = \mu N = 0.5(260 + 6t) = 130 + 3t.$$

The linear impulse and momentum equation for the interval from 10 sec$<t<$15 sec is

$$(I_L)_x = m(v_{xf} - v_{xi}),$$

$$\int_{10}^{15} (0.8P_v - F')\, dt = m(v_f - 0),$$

$$\int_{10}^{15} [0.8(10t + 100) - (130 + 3t)]\, dt = \frac{200}{32.2}(v_f - 0),$$

which gives

$$\left.\frac{5t^2}{2} - 50t\right]_{10}^{15} = \frac{200}{32.2}(v_f - 0),$$

$$(562.5 - 750) - (250 - 500) = \frac{200}{32.2} v_f,$$

$$v_f = 10.06 \text{ fps}$$

and

$$\mathbf{v}_f = \underline{10.06 \text{ fps.} \rightarrow}. \text{ Ans.}$$

The principle of linear impulse and linear momentum is particularly useful in determining the force developed when a jet of water or other fluid is deflected by a blade. Equation (11-5) can be written as

$$\mathbf{R}\,\Delta t = m(\mathbf{v}'' - \mathbf{v}'),$$

where m is the mass of fluid deflected during the time interval Δt. The force $\mathbf{R}$ is the resultant force acting on the fluid, and $\mathbf{v}'$ and $\mathbf{v}''$ are the velocities of the fluid before and after it is deflected by the blade.

The mass of fluid striking the blade in time Δt depends on the cross-sectional area and the velocity of the jet, the density (mass per unit volume) of the fluid, the velocity of the blade, and whether a single blade or a series of blades mounted on the periphery of a wheel is involved. If the jet strikes a stationary blade, as shown in Fig. 11-4(a), the mass of fluid striking the blade in a time Δt is $A_J v_J'(\Delta t)\rho$,

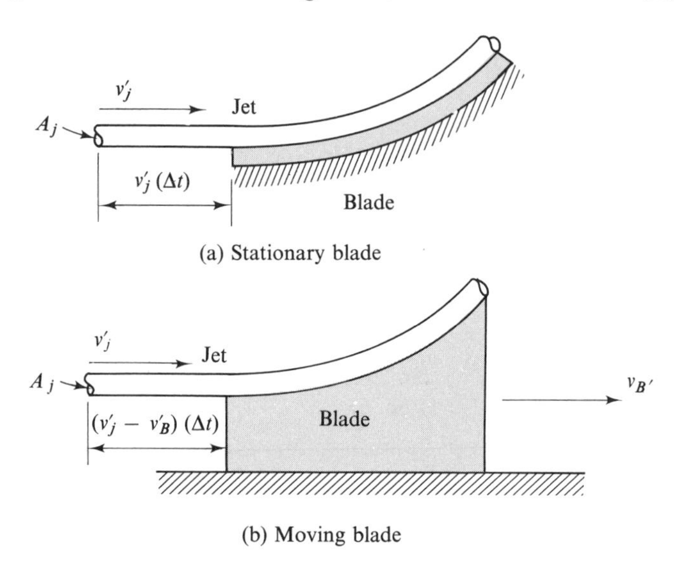

(a) Stationary blade

(b) Moving blade

Figure 11-4

where A_J is the area of the jet, v_J' is the magnitude of its velocity, and ρ is the density of the fluid. If the blade is moving in the same direction as the jet with a speed v_B [see Fig. 11-4(b)], the length of the stream which strikes the blade in a time Δt is $(v_J' - v_B')\,\Delta t$, and the mass striking the blade in the time Δt is $A_J(v_J' - v_B')(\Delta t)\rho$. When a jet of fluid strikes a series of blades on a wheel (such as the runner of an impulse turbine), all the fluid issuing from the nozzle strikes the blades, and the mass of fluid deflected during the time interval Δt is $A_J v_J'(\Delta t)\rho$.

The velocity of the jet as it leaves the blade can be determined from the equation of relative velocity. If the blade is smooth, the force of the blade on the fluid is always normal to the surface and hence normal to the relative velocity of the fluid with respect to the blade. This normal force produces only a normal acceleration of the fluid, and therefore the magnitude of the relative velocity of the fluid leaving the blade is the same as the magnitude of the relative velocity with which the fluid strikes the blade. The direction of the final velocity of fluid relative to the blade is tangent to the blade. The angle through which the relative velocity is changed is called the *deflection angle* of the blade. The final absolute velocity of the water as it leaves the blade can be determined from

$$\mathbf{v}'_{J/B} = \mathbf{v}'_J - \mathbf{v}'_B \quad \text{and} \quad \mathbf{v}''_J = \mathbf{v}''_B + \mathbf{v}''_{J/B},$$

where

$$v''_{J/B} = v'_{J/B}.$$

If the surface of the blade is not smooth, the reaction of the blade on the water will have a component tangent to the surface which will reduce the magnitude of the final relative velocity of the jet with respect to the blade. Thus the final relative velocity might be reduced to 80 or 90% of the initial relative velocity. In a well-designed impulse turbine, the friction between the blade and the jet is usually quite small, and the turbine efficiency may be as much as 90%.

The following problem demonstrates the application of the principle of linear impulse and momentum to a jet.

EXAMPLE 11-3

A jet of water with a cross-sectional area of 2.0 in^2. has a velocity of 180 fps to the right when it strikes the single curved blade in Fig. 11-5(a). The velocity of the blade is 80 fps to the right, and friction between the blade and water is negligible. The jet remains in a horizontal plane. The water weighs 62.4 lb per ft^3. Determine (a) the force of the water on the blade and (b) the power developed.

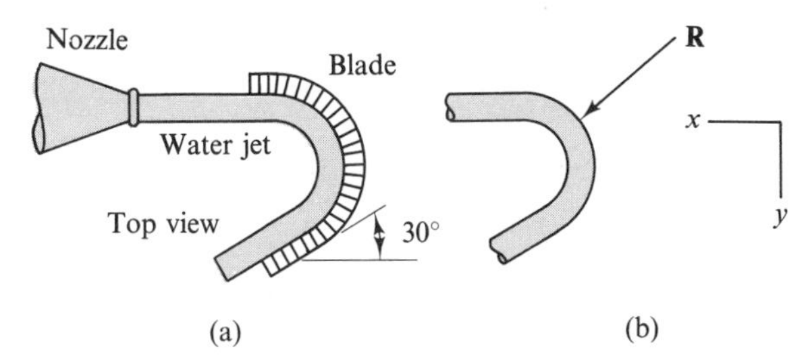

Figure 11-5

SOLUTION

(a) The free-body diagram (top view) of the water which strikes the blade during the time interval Δt is shown in Fig. 11-5(b). In this example, the weight of the fluid is perpendicular to the plane in which the jet moves, but in any case it is customary to neglect the weight of the fluid in comparison with the other forces acting on the jet. The equation of impulse and momentum is

$$\mathbf{R}\,\Delta t = m(\mathbf{v}'' - \mathbf{v}').$$

The mass of water which strikes the blade in time Δt is

$$m = A_J(v_J - v_B)\,\Delta t\rho = \frac{2.0}{144}(180-80)\,\Delta t\frac{62.4}{32.2} = 2.69\,\Delta t.$$

The velocity of the jet relative to the blade before impact is

$$\mathbf{v}'_{J/B} = -180\mathbf{i} - (-80\mathbf{i}) = -100\mathbf{i}\text{ fps}.$$

Since the blade is smooth, the final relative velocity is

$$\mathbf{v}''_{J/B} = 100(0.866\mathbf{i} + 0.500\mathbf{j})\text{ fps},$$

and the final absolute velocity of the jet is

$$\mathbf{v}''_J = \mathbf{v}''_B + \mathbf{v}''_{J/B} = -80\mathbf{i} + 100(0.866\mathbf{i} + 0.500\mathbf{j})$$

$$= (6.6\mathbf{i} + 50.0\mathbf{j})\text{ fps}.$$

When numerical data are substituted into the equation of impulse and momentum, it becomes

$$\mathbf{R}\,(\Delta t) = 2.69\,(\Delta t)[(6.6\mathbf{i} + 50\mathbf{j}) - (-180\mathbf{i})]$$

from which

$$\mathbf{R} = (502\mathbf{i} + 134.5\mathbf{j})\text{ lb on the water.}$$

The resultant force *on the blade* is

$$\mathbf{R} = -502\mathbf{i} - 134.5\mathbf{j} = \underline{520(-0.965\mathbf{i} - 0.259\mathbf{j})\text{ lb.}} \qquad \text{Ans.}$$

The initial and final absolute velocities of the water used in the equations of impulse and momentum can be replaced by the corresponding velocities of the water relative to the blade because the difference of the two absolute velocities is the same as the difference of the two relative velocities. By the use of relative velocities, the impulse and momentum equation is

$$\mathbf{R}\,(\Delta t) = m(\mathbf{v}''_{J/B} - \mathbf{v}'_{J/B})$$

$$= 2.69(\Delta t)[(86.6\mathbf{i} + 50\mathbf{j}) - (-100\mathbf{i})]$$

and

$$\mathbf{R} = (502\mathbf{i} + 134.5\mathbf{j})\text{ lb on the jet.}$$

This is the same result as that obtained using the absolute velocities.

(b) The reaction on the blade moves in the x direction with the blade; therefore, $\mathbf{R}_y$ does no work. The power delivered to the blade is, from Eq. (10-13a), the product of the force exerted on the blade in the direction of the velocity multiplied by the velocity of the force (and blade); that is,

$$P = R_x v_B = 502(80) = 40{,}200\text{ ft-lb per sec}$$

$$= \frac{40{,}200}{550} = \underline{73.0\text{ hp.}} \qquad \text{Ans.}$$

PROBLEMS

Note: In the following problems, unless otherwise specified, all cords, ropes, and cables are assumed to be flexible, inextensible, and of negligible weight. Pulleys are assumed to have negligible mass, and pins, axles, and pegs are assumed to be smooth unless otherwise stated.

11-1 The coefficient of friction between the 100-lb block A in Fig. P11-1 and the horizontal surface is 0.2. The force $\mathbf{P}=3t^2\mathbf{i}$ is in pounds when t is in seconds. Determine the resultant linear impulse on the block from $t=0$ to $t=5$ sec if the body moves to the right during the entire time interval.

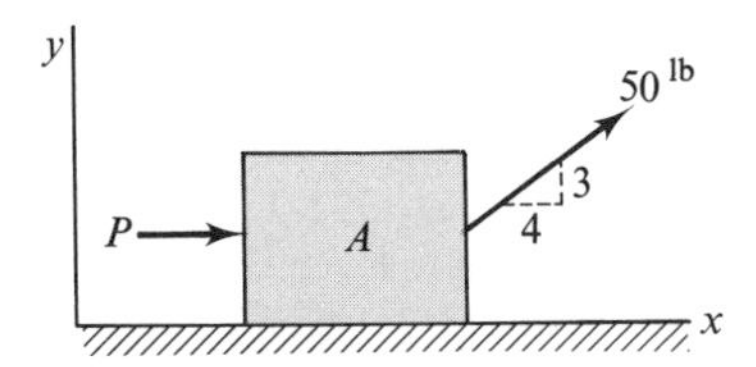

Figure P11-1

***11-2** Block A in Fig. P11-2 moves along the smooth plane. Determine the linear impulse of the force **P** in the direction parallel to the inclined plane from $t=0$ to $t=17$ s when the magnitude of **P** is (a) 13N, (b) $(2t+20)$N, where t is in seconds.

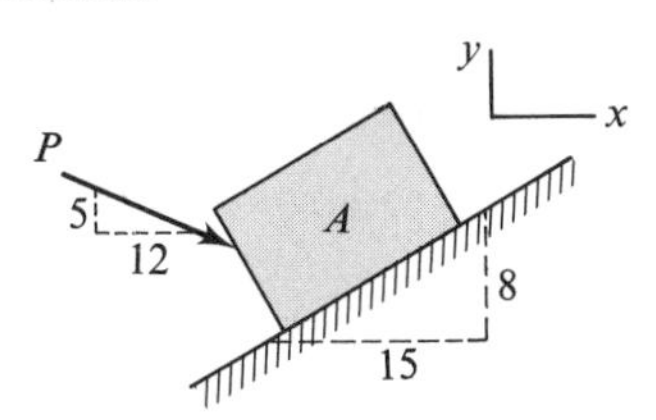

Figure P11-2

***11-3** Solve Problem 11-2 if the magnitude of the force **P** is given by the force time diagram in Fig. P11-3.

11-4 The coefficient of friction between the 25-lb body A in Fig. P11-4 and the plane is 0.2. The force $\mathbf{P}=(50+30t)\mathbf{i}$ is in pounds when t is in seconds. The body is at rest when $t=0$. Determine the velocity of the body when $t=2$ sec.

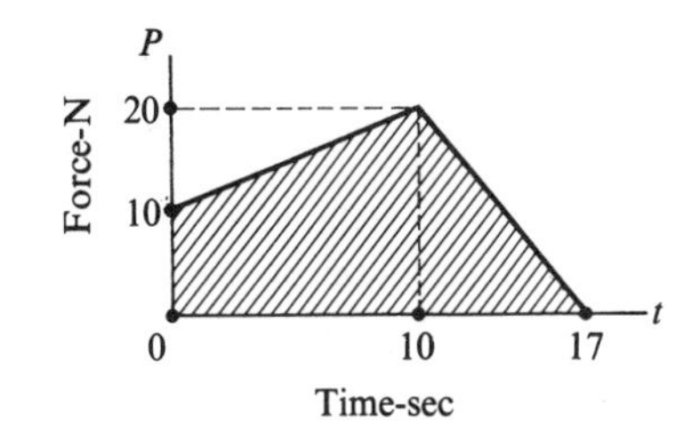

Figure P11-3

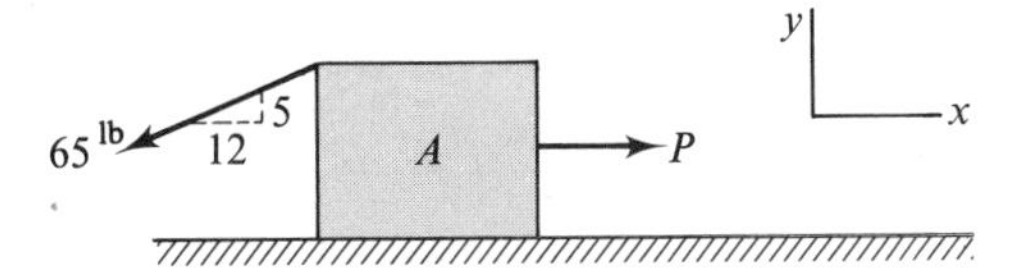

Figure P11-4

11-5 The 250-lb block in Fig. P11-5 is moved to the right along the rough plane ($\mu=0.2$) for 5 sec by the 90-lb force. Determine the angle θ (between 0 and 90°) for the maximum resultant linear impulse and the magnitude and direction of this impulse.

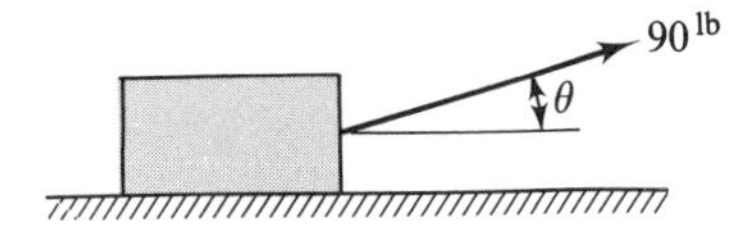

Figure P11-5

11-6 The 64.4-lb block in Fig. P11-6 moves on a smooth horizontal plane and is acted on by the force **F**, which varies according to the equation $F=6t^2$, where F is in pounds when t is in seconds. The block starts from rest at the origin when $t=0$.

(a) Determine the time average value of the force **F** during the interval from $t=0$ to $t=3$ sec.

(b) Derive the expression for the position s of the block as a function of time.

(c) Determine the work done by the force **F** during the time interval from $t=0$ to $t=3$ sec, using the expression $U=\int F\,ds$.

(d) Determine the position average value of the force **F** from the expression $U=(F_{avg})(\Delta s)$.

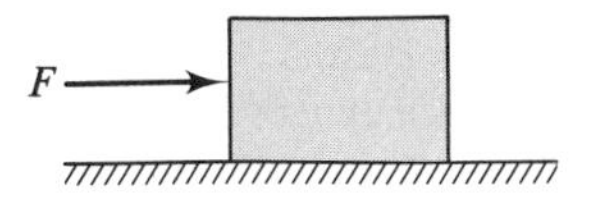

Figure P11-6

***11-7** The three particles in Fig. P11-7 are located in the xy plane. The mass of each particle in kilograms and the velocity in meters per second are indicated on the diagram. Determine
(a) The linear momentum of the system.
(b) The velocity of the mass center.

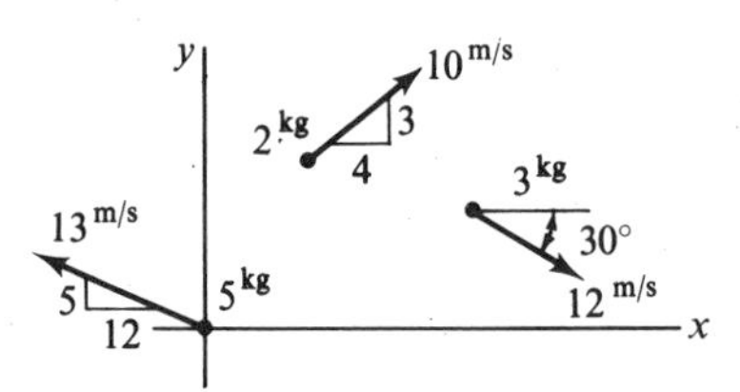

Figure P11-7

***11-8** Body A in Fig. P11-8 has a mass of 2 kg, and bodies B and C have masses of 3 and 5 kg, respectively. Determine the linear momentum of the system when A has a velocity of 10 m/s downward.

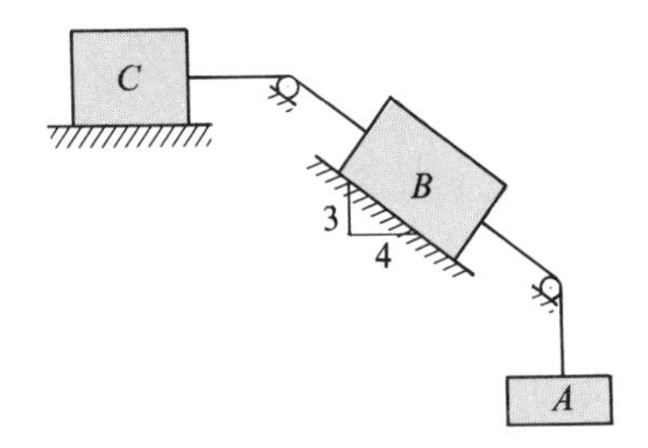

Figure P11-8

11-9 The coefficient of friction between the 100-lb block in Fig. P11-9 and the plane is 0.4. Determine the time elapsed for the 120-lb force **P** to change the velocity of the block from 0 to 64.4 fps to the right.

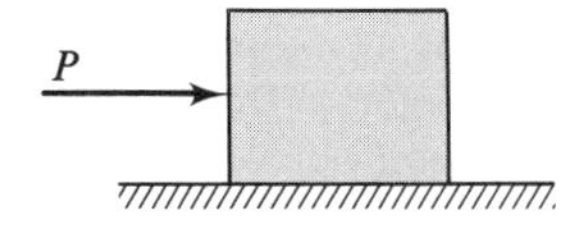

Figure P11-9

11-10 Solve Problem 11-9 if the magnitude of the force is $P = 80t$, where P is in pounds when t is in seconds, and the velocity of the block is zero when $t = 0$.

11-11 The 10-lb body A in Fig. P11-11 is at rest against the stop on the smooth plane when $t = 0$. The magnitude of the force **P** varies as indicated in the diagram. Determine the velocity of A when $t = 15$ sec.

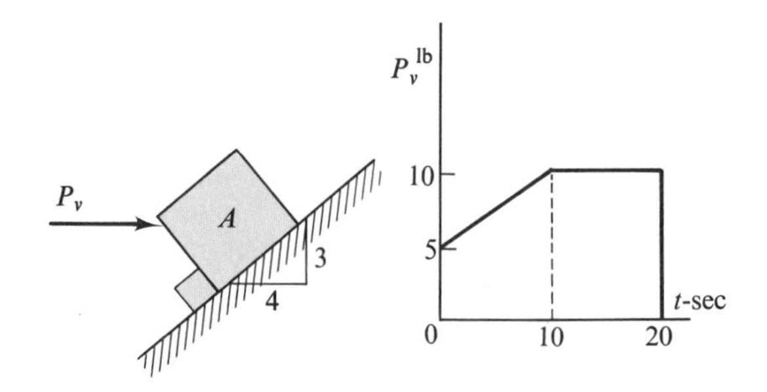

Figure P11-11

11-12 The coefficient of friction between the 322-lb body A in Fig. P11-12 and the plane is 0.2. Body B weighs 64.4 lb. The magnitude of the force **P** in pounds is given by the expression $P = 150 + 30t^2$, where t is in seconds. When $t = 0$, A has a velocity of 2 fps to the right. Determine the velocity of A when $t = 3$ sec.

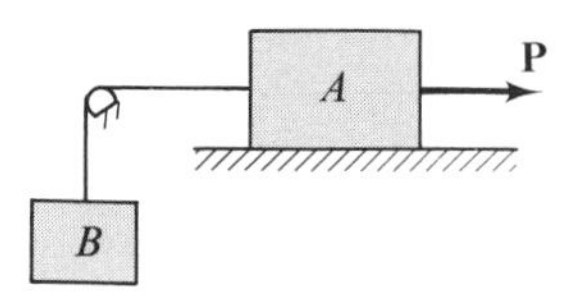

Figure P11-12

***11-13** The 5-kg body in Fig. P11-13 slides on a smooth horizontal surface. The magnitude of the force **P** is given by $P = 3t^2 + 20$, where P is in newtons when t is in seconds, and the force maintains the constant direction indicated in the plane of motion. When $t = 0$ the velocity of the mass center G is $\mathbf{v}_G = (-6\mathbf{i} + 4\mathbf{j})$ m/s. Determine the velocity of G when $t = 8$ s.

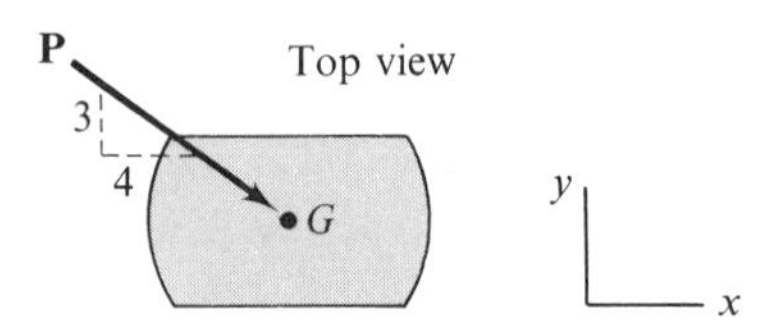

Figure P11-13

*11-14 The magnitude of the force **P** in Problem 11-13 varies as shown in Fig. P11-14, while its direction remains constant. The mass and initial velocity of the body are the same as in Problem 11-13. Determine the velocity of G when $t = 8$ s.

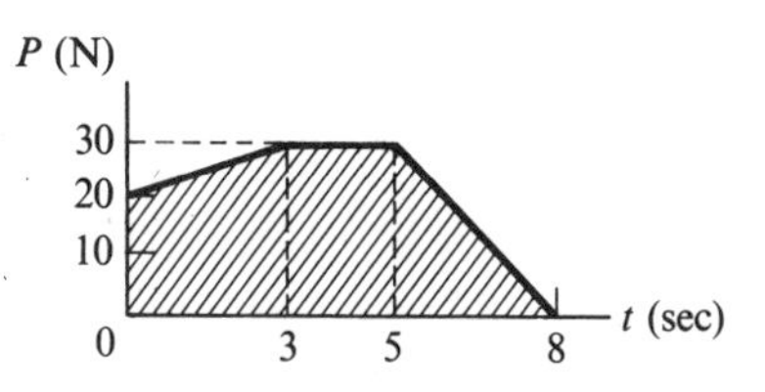

Figure P11-14

*11-15 The masses of bodies A and B in Fig. P11-15 are 2 and 4 kg, respectively. The coefficient of friction is 0.2 between body A and the horizontal plane and 0.4 between body B and the inclined plane. Body B has an initial velocity of 2 m/s down the plane. Determine the velocity of A 3 s later.

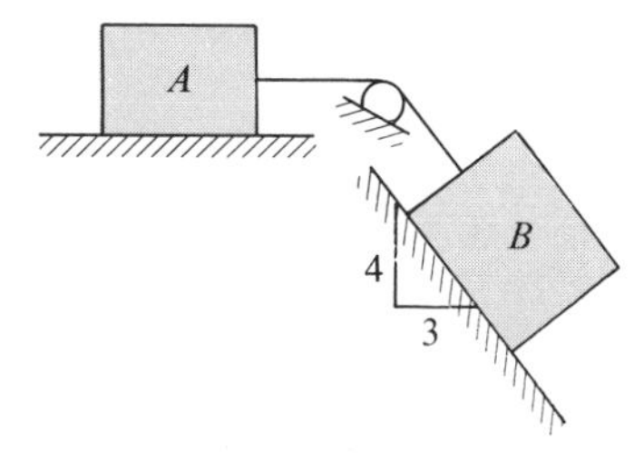

Figure P11-15

*11-16 A box having a mass of 10 kg is given a velocity of 5 m/s down an unloading chute that makes an angle of 20° with the horizontal. The box comes to rest in 3 s. Determine the coefficient of friction between the box and the chute.

11-17 The coefficient of friction between the block A in Fig. P11-17 and the horizontal plane is 0.4. Determine the minimum value of the force **P**, in terms of the weight W of the block, required to give the block a velocity of 12 fps in 2 sec starting from rest.

11-18 Block A in Fig. P11-18 weighs 40 lb and slides along the rough plane ($\mu = 0.25$). The magnitude of the force **P** in pounds is given by the equation $P = 2t$, where t is in seconds. Body A has a velocity of 15 fps to the left when $t = 0$.

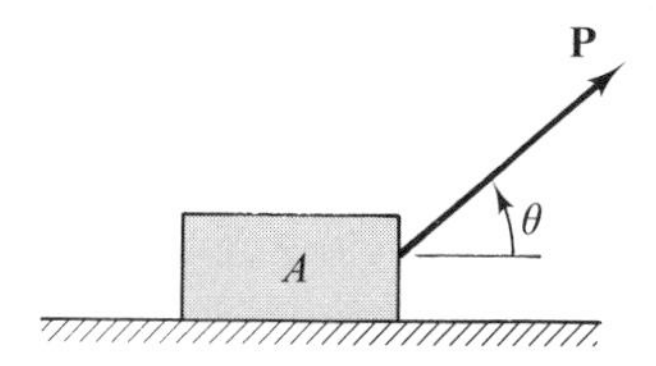

Figure P11-17

Determine

(a) The time t when the block comes to rest.

(b) The velocity of A when $t = 8$ sec.

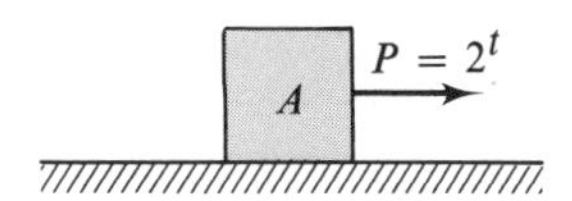

Figure P11-18

11-19 The force **P** acting on the 10-lb body in Fig. P11-19 varies as shown. The coefficient of friction between the body and the plane is 0.40, and the body is at rest when $t = 0$. Determine

(a) The maximum velocity of the body.

(b) The velocity of the body when $t = 3$ sec.

(c) The velocity of the body when $t = 5$ sec.

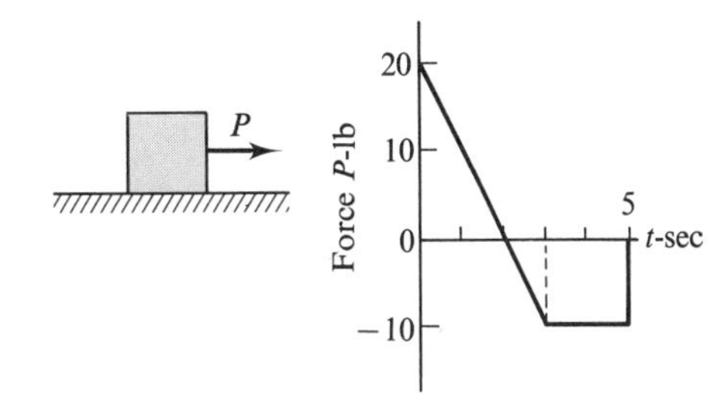

Figure P11-19

11-20 The 6.50-lb block A in Fig. P11-20 rests at the bottom of the smooth inclined plane. The magnitude of the force **P** is given by the equation $P = 20(1 - e^{-0.1t})$, where P is in pounds when t is in seconds. Determine the velocity of A when $t = 4$ sec.

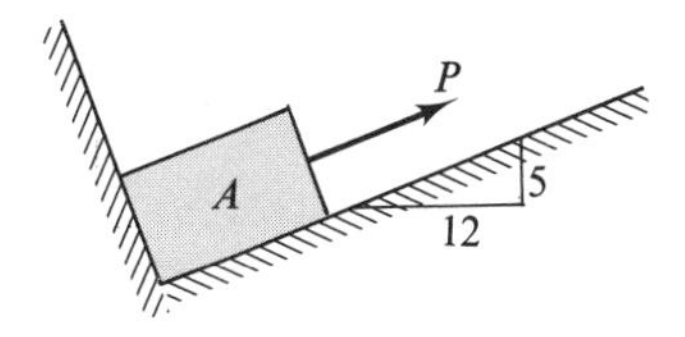

Figure P11-20

11-21 Determine the time t in Problem 11-20 at which the velocity of body A will be 40 fps up the plane.

11-22 The 257.6-lb block A in Fig. P11-22 is connected to the 96.6-lb block B by a flexible wire. The coefficient of friction is 0.20 between A and the plane and 0.25 between B and the plane. Determine the velocity of A 3 sec after it starts from rest.

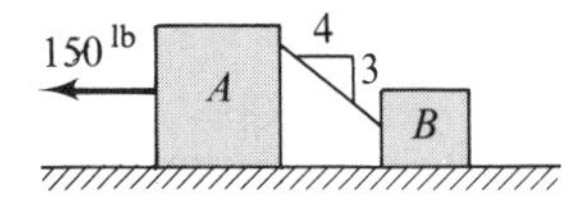

Figure P11-22

11-23 The 5.0-lb body A in Fig. P11-23 is pulled along the plane by a flexible cord which passes over a drum mounted in the body. The coefficient of friction is 0.40 between the cord and the drum and between the body and the plane. The velocity of the body is 5.0 fps to the right when $t = 0$. *Note:* The body does not stop during the 2-sec interval. Determine

(a) The velocity of A when $t = 1$ sec.
(b) The velocity of A when $t = 2$ sec.
(c) The maximum velocity of A.

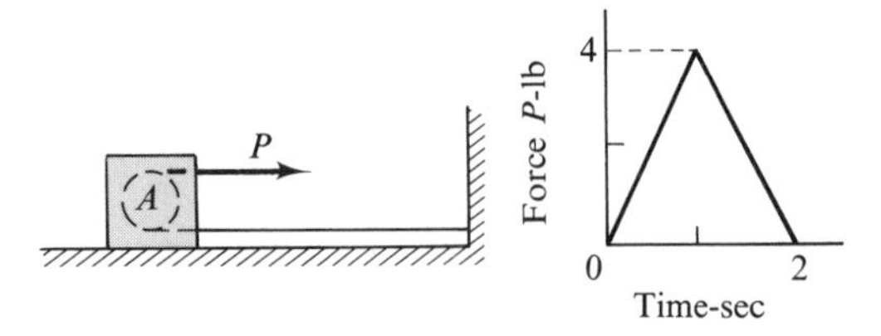

Figure P11-23

11-24 Body A in Fig. P11-24 weighs 10 lb, and B weighs 5 lb. The bodies are connected by the cord which passes over smooth pulleys at A and C. The coefficient of friction between B and the plane is 0.20. The magnitude of the force $\mathbf{P}$ is indicated by the diagram. When $t = 0$ the velocity of B is 30 fps to the right.

(a) Does the velocity of B become zero during the interval $0 < t < 4$ sec?
(b) Determine the velocity of B when $t = 4$ sec.

11-25 Solve Problem 11-24 if the pulley on body A is locked and the coefficient of friction between the pulley and cord is 0.20. All other data are unchanged.

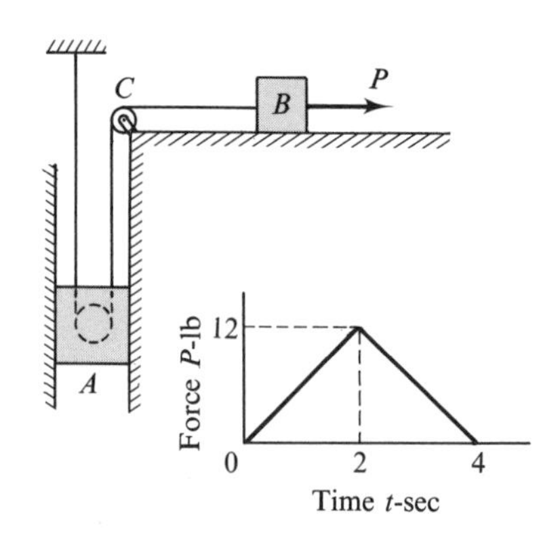

Figure P11-24

11-26 Bodies A, B, and C in Fig. P11-26 have weights of 3, 8, and 5 lb, respectively. The planes on which the bodies slide and the pulleys are smooth. The magnitude of the force $\mathbf{P}$ varies according to the equation $P = 0.5 + 1.2t$, where P is in pounds when t is in seconds. The system is at rest when $t = 0$. Determine the velocity of each body when $t = 3$ sec.

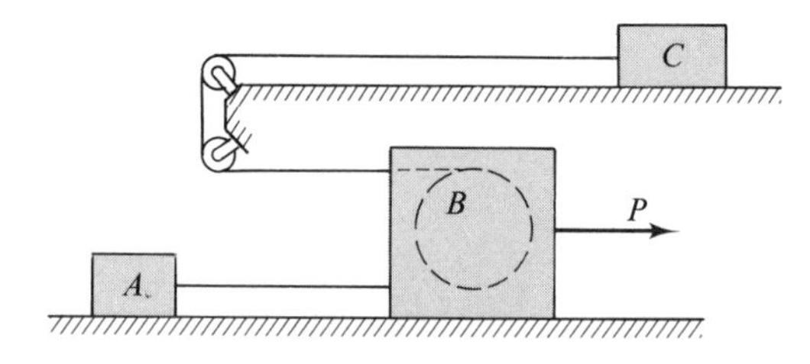

Figure P11-26

11-27 Bodies A, B, and C in Fig. P11-27 weigh 60, 30, and 10 lb, respectively. The coefficient of friction is 0.50 between A and B and 0.10 between B and the plane. The magnitude of the force $\mathbf{P}$ is equal to $0.6t^2$, where P is in pounds when t is in seconds. The system is at rest when $t = 0$. Determine

(a) The value of t when motion impends.
(b) The velocity of A when $t = 10$ sec.

11-28 A jet of water 2 in. in diameter impinges on a fixed blade as shown in Fig. P11-28. The velocity of the jet is 60 fps to the right. Determine the force exerted on the smooth blade by the water.

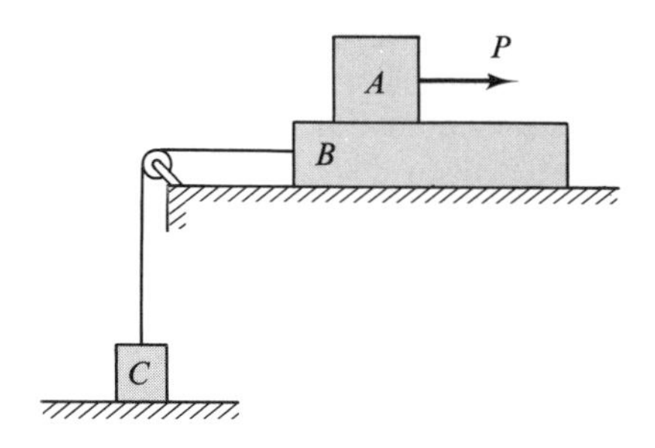

Figure P11-27

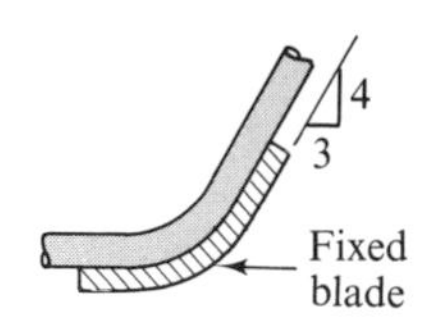

Figure P11-28

11-29 A jet of water having a cross-sectional area of a in^2. strikes a fixed blade and is deflected 90° as shown in Fig. P11-29. The forces required to hold the blade in equilibrium are shown on the figure. The speed of the jet as it strikes the blade is 100 fps. Determine the percentage reduction of the jet speed due to blade friction.

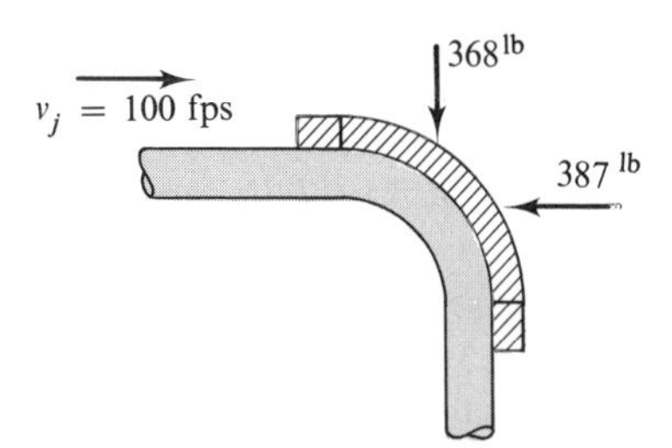

Figure P11-29

11-30 A jet of water 2 in. in diameter strikes the vane shown in Fig. P11-30. The spring has a modulus of 400 lb per ft. Determine the speed of the jet which will cause the spring to be compressed 1 ft.

11-31 The jet of water in Fig. P11-31 has a 0.2-ft^2 cross-sectional area, leaves the nozzle at 200 fps, and strikes the curved blade as shown. The blade has a constant velocity of 50 fps to the right. Determine

(a) The horsepower delivered to the blade by the water.

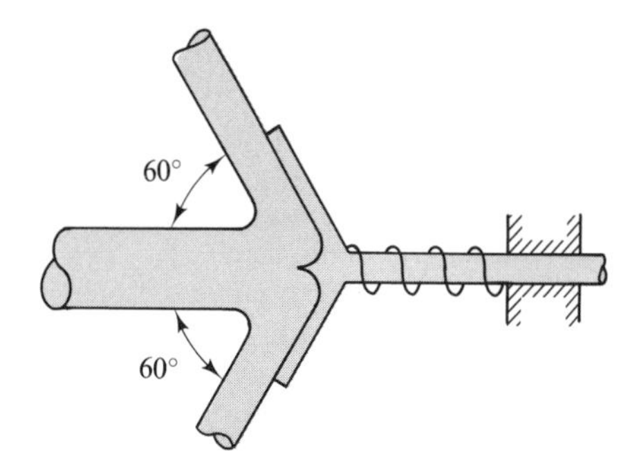

Figure P11-30

(b) The velocity of the water as it leaves the blade.

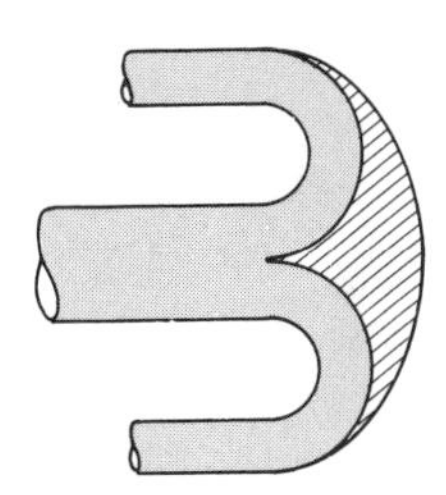

Figure P11-31

11-32 A jet of water having a cross-sectional area of 3 in.^{2} and a velocity of 170 fps to the right strikes a single curved blade and is divided into two equal streams as shown in Fig. P11-32. The blade has a constant velocity of 50 fps to the right, and the friction on the blade reduces the magnitude of the relative velocity of the water with respect to the blade by 6%. Determine

(a) The force exerted on the blade by the water.

(b) The horsepower delivered to the blade by the water.

(c) The kinetic energy remaining in each pound of water as it leaves the blade.

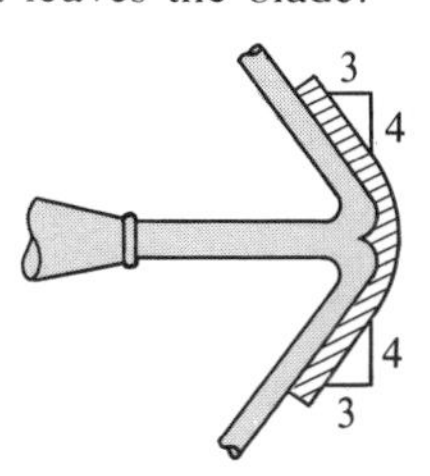

Figure P11-32

11-33 A jet of water flows from a 2-in.-diameter pipe with a velocity of 100 fps and strikes a smooth plunger as indicated in Fig. P11-33. Determine

(a) The force which must be applied to the plunger to keep it stationary.

(b) The force required to move the plunger 2 ft to the left in 0.05 sec with a constant velocity.

(c) The work done in moving the plunger against the water as required in part (b).

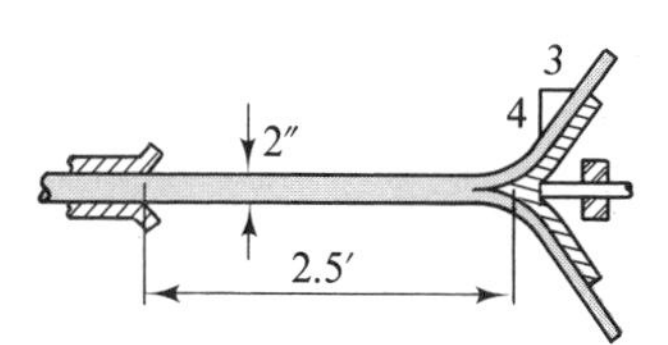

Figure P11-33

11-34 A 4.0-in.-diameter jet of water impinges on the single vane shown in Fig. P11-34. The velocities of the jet and vane are 72 and 20 fps, respectively, both to the right. Determine the angle ϕ for which 25 hp will be developed by the vane.

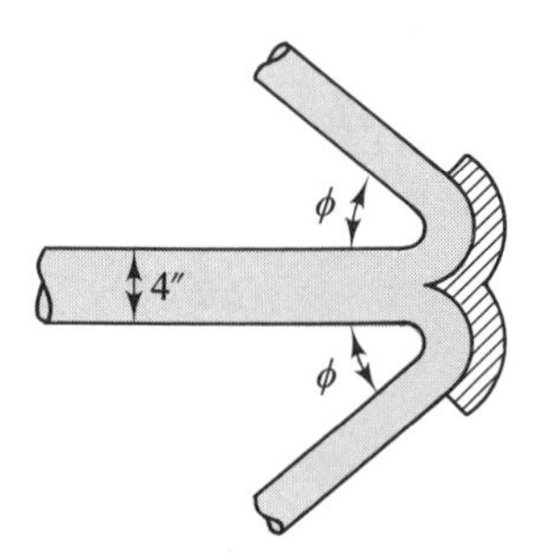

Figure P11-34

11-3 CONSERVATION OF LINEAR MOMENTUM

In Art 11-2, it was shown that the linear impulse of a force system acting on any system of particles during a time interval is equal to the change in the linear momentum of the system of particles during that time interval. *The principle of conservation of linear momentum states that when the linear impulse exerted on a system of particles is zero the change in the linear momentum of the system of particles is zero, and therefore the linear momentum of the system is conserved.* Furthermore, when there is no linear impulse on a system of particles in a given direction the linear momentum is unchanged in that direction even though the total linear momentum may not be constant.

When two particles or bodies, A and B, collide, the linear impulse of A on B is equal in magnitude and opposite in sense to the linear impulse of B on A. Thus, if no other forces are acting on the two bodies in a given direction, the linear momentum of the system composed of the two bodies is conserved in that direction during the impact. The two blocks A and B in Fig. 11-6 are moving along the same path on a smooth plane when they collide. Since there are no horizontal forces external to the system composed of blocks A and B acting on the blocks during the collision, the linear momentum of the system in the

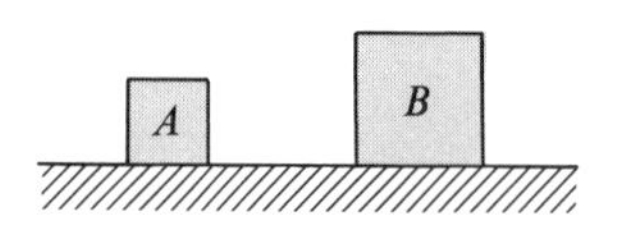

Figure 11-6

horizontal direction is constant. That is,

$$m_A(\mathbf{v}_A)_i + m_B(\mathbf{v}_B)_i = m_A(\mathbf{v}_A)_f + m_B(\mathbf{v}_B)_f, \qquad (11\text{-}7)$$

where the subscripts i and f denote initial and final velocities, respectively. When Eq. (11-7) is used, a positive direction must be selected, and all velocities in the opposite direction must be treated as negative quantities.

When the linear impulse in a given direction is not zero but is known to be relatively small, it can frequently be neglected in order to obtain an approximate solution sufficiently accurate for many purposes. If the linear momentum of the system composed of A and B in Fig. 11-6 is large compared to the linear impulse of the frictional force, Eq. (11-7) is approximately true even though friction exists between the plane and the blocks. The frictional forces could not exceed μN, and the time of impact would be small; therefore, the impulse of the friction on the blocks *during the impact period* would not change the large linear momentum of the blocks materially.

Conservation of linear momentum should not be confused with conservation of kinetic energy, which was discussed in Art. 10-4. The linear momentum of a body or system of bodies is conserved when the linear impulse acting on the body is zero. The kinetic energy of a body is conserved when no work is done on it. It is possible to have the linear momentum of two bodies conserved even though most or all of the kinetic energy is dissipated by producing deformation of the bodies, increasing the temperature, and producing sound and other vibrations. On the other hand, when an elastic steel ball strikes and rebounds from a large steel plate, most of the kinetic energy is conserved, while the linear momentum of the ball may change by a complete reversal in sense with little change in magnitude.

EXAMPLE 11-4

A 10.00-lb block is suspended by a long cord. The block is at rest when a 0.050-lb bullet traveling horizontally to the left strikes the block and is embedded in it. The impact causes the block to swing upward 0.50 ft measured vertically from its lowest position. Determine

(a) The velocity of the bullet just before it strikes the block.
(b) The loss of kinetic energy of the system during impact.

SOLUTION

(a) The velocity of the block and bullet immediately after impact can be found by the principle of work and kinetic energy. Since no dimensions are given, the block is treated as a particle. There are only two forces acting on the block and bullet after impact as indicated in the free-body diagram in Fig. 11-7, and the force **P** does no work because there is no displacement in the direction of **P**. When values are

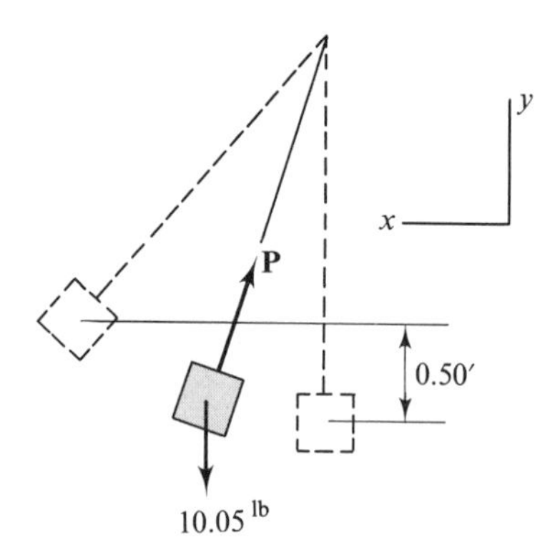

Figure 11-7

substituted into

$$U = \tfrac{1}{2}m[v_f^2 - v_i^2],$$

the result is

$$-10.05(0.50) = \frac{1}{2}\frac{10.05}{32.2}[0 - v^2]$$

from which

$$\mathbf{v} = 5.67\mathbf{i} \text{ fps},$$

where $\mathbf{v}$ is the velocity of the block and bullet immediately after impact. During the time of the collision, no forces external to the block and bullet are acting in the horizontal direction. Consequently, the horizontal linear momentum is conserved. The positive x axis has been selected to the left for this example, and the conservation of momentum principle gives

$$\frac{10.00}{32.2}(0) + \frac{0.05}{32.2}\mathbf{v}_B = \frac{10.05}{32.2}(5.67\mathbf{i})$$

from which

$$\mathbf{v}_B = \underline{1140\mathbf{i} \text{ fps}}, \qquad \text{Ans.}$$

where $\mathbf{v}_B$ is the velocity of the bullet before impact.

(b) The kinetic energy of the system before impact is

$$\frac{1}{2}\left(\frac{0.05}{32.2}\right)(1140)^2 = 1009 \text{ ft-lb}.$$

The kinetic energy of the system after impact is

$$\frac{1}{2}\left(\frac{10.05}{32.2}\right)(5.67)^2 = 5.03 \text{ ft-lb}.$$

The loss of kinetic energy during the collision is $1009 - 5.03 = 1004$ ft-lb, or 99.5%. It should be noted that even though the momentum is conserved during impact, nearly all of the kinetic energy is dissipated as heat and internal work in deforming the block and the bullet.

EXAMPLE 11-5

Crate A in Fig. 11-8(a) slides down the chute with a velocity of 10 m/s and falls on the stationary cart B. The mass of A is 5 kg, and the mass of B is 15 kg. The coefficient of friction between A and B is 0.40. The crate slides along the cart with no rebound. Neglect the mass of the wheels and bearing friction.

(a) Determine the velocity of the cart after the bodies attain a common velocity.
(b) Will the crate hit the end of the cart?
(c) If so, determine the impulse of the cart on the crate.

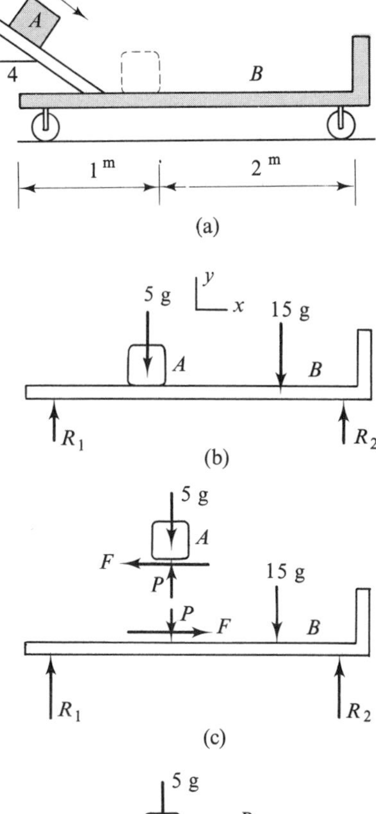

Figure 11-8

SOLUTION

(a) Figure 11-8(b) is a free-body diagram of the system. There are no horizontal forces external to the system; therefore the linear momentum is conserved in the horizontal direction:

$$\sum mv_{xi} = \sum mv_{xf},$$

$$m_A v_{Axi} + m_B v_{Bxi} = m_A v_{Axf} + m_B v_{Bxf} = (m_A + m_B) v_{xf},$$

$$5(\tfrac{4}{5})10 + 0 = (5 + 15) v_{xf}$$

from which

$$\mathbf{v}_{xf} = \underline{2.0 \text{ m/s} \rightarrow}. \qquad \text{Ans.}$$

(b) Various procedures are available for answering part (b). For example, assume that A strikes B after sliding 2 m relative to B. The velocity of A will be more than the velocity of B if the assumption is correct. The linear momentum of the system will be conserved, but the friction force between A and B will do work on the system. From the free-body diagram of body A in Fig. 11-8(c) the normal force P is 5g N, and the friction (as long as sliding occurs) is $\mu P = 0.4(5g) = 2g$ N. The work done on the two bodies during the relative motion is

$$U = -F\,\Delta s_A + F\,\Delta s_B = -F(\Delta s_A - \Delta s_B),$$

and since A slides 2 m farther than B, the work done is

$$U = -F(\Delta s_{A/B}) = -2g(2) = -4g \text{ N·m}.$$

The work and energy principle gives

$$U = \Delta T$$

$$-4g = \tfrac{1}{2}m_A(v_{A2}^2 - v_{A1}^2) + \tfrac{1}{2}m_B(v_{B2}^2 - v_{B1}^2),$$

where the subscript 1 refers to velocities when slipping starts and 2 refers to velocities just before impact. Note that some of the kinetic energy of A was lost when A first hit the cart and that its velocity changed from 10 m/s down the chute to 8 m/s horizontally. The work and energy equation becomes

$$-4g = \tfrac{1}{2}(5)(v_{A2}^2 - 8^2) + \tfrac{1}{2}(15)(v_{B2}^2 - 0). \qquad \text{(a)}$$

The conservation of linear momentum equation from 1 to 2 gives

$$m_A v_{A1} + m_B v_{B1} = m_A v_{A2} + m_B v_{B2}$$

or

$$5(8) + 0 = 5v_{A2} + 15v_{B2}. \qquad \text{(b)}$$

Equation (a) becomes

$$v_{A2}^2 + 3v_{B2}^2 = -1.6g + 64, \qquad \text{(c)}$$

and from Eq. (b),

$$v_{A2} = 8 - 3v_{B2}. \qquad \text{(d)}$$

When Eq. (d) is substituted into Eq. (c) it becomes

$$12v_{B2}^2 - 48v_{B2} + 1.6(9.81) = 0$$

from which

$$v_{B2} = 3.64 \text{ m/s} \quad \text{or} \quad 0.36 \text{ m/s},$$

and from Eq. (d),

$$v_{A2} = -2.92 \text{ m/s} \quad \text{or} \quad 6.92 \text{ m/s}.$$

The first pair of values ($v_{B2} = 3.64$ m/s and $v_{A2} = -2.92$ m/s) does not satisfy the assumption that $v_A > v_B$. The second set of values, however, does satisfy the assumption; therefore the crate will strike the end of the cart.

(c) The free-body diagram of the crate during the impact period is shown in Fig. 11-8(d). Since the duration of the impact is short, the impulse of the friction force is neglected as compared with the impulse of the force R. The impulse-momentum equation for A is

$$-\int R\,dt = m(v_{Af} - v_{A2}) = 5(2 - 6.92)$$

or

$$\int \mathbf{R}\,dt = \underline{24.6 \text{ N·s} \leftarrow \text{on A.}} \qquad \text{Ans.}$$

PROBLEMS

***11-35** The 10-kg body A in Fig. P11-35 slides on a smooth horizontal surface with a velocity of 8 m/s to the right and collides with the 5-kg body B, which has a velocity of 6 m/s to the left just before impact. The two bodies lock together on impact.
(a) Determine their common velocity after impact.
(b) If the time of impact is 0.1 s, determine the time average value of the force exerted by body A on B.

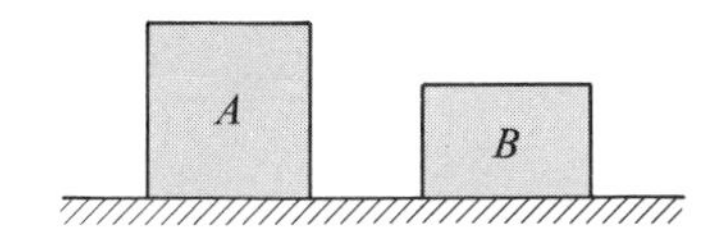

Figure P11-35

11-36 A 150-lb man running with a velocity of 12 fps to the right jumps on a 300-lb cart moving with a velocity of 4 fps to the right. Determine
(a) The common velocity of the man and cart.
(b) The loss of kinetic energy during the impact.

11-37 A 400-lb body has a velocity of 800 fps in free flight when an explosion causes a separation into two parts. Part A weighs 300 lb, and B weighs 100 lb. Both parts continue in motion along the original line of flight. After the explosion B is moving 120 fps faster than A. Determine
(a) The speed of A.
(b) The linear impulse on B caused by the explosion.

***11-38** The block in Fig. P11-38 has a mass of M kg and is supported by an inextensible cord L m long. The block is at rest when a bullet of mass m kg and traveling horizontally to the right strikes the block and is embedded in it. The impact causes the block to move so that the cord makes a maximum angle θ with its initial position. Derive, as a function of θ, L, M, and m, a relation for the velocity of the bullet just before impact.

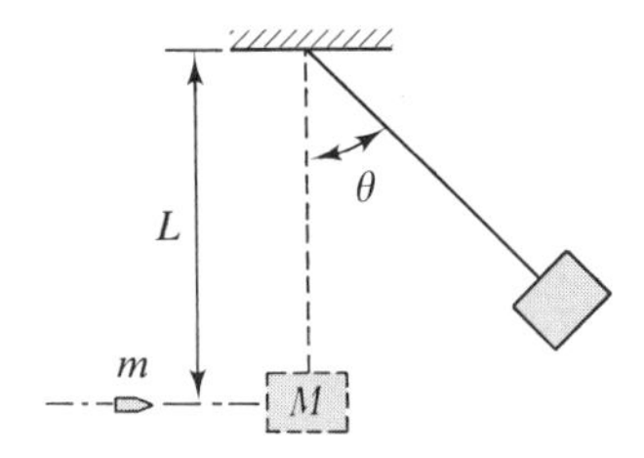

Figure P11-38

***11-39** In Fig. P11-39 block A has a mass of 9 kg and is moving toward the left with a velocity of 1.5 m/s just before it is struck by the bullet B, which has a mass of 50 g and a velocity of 800 m/s to the right. The bullet is embedded in the block. The coefficient of friction between body A and the plane is 0.2. How long will the block and bullet continue to move after the impact?

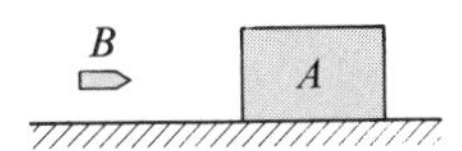

Figure P11-39

11-40 The 4.00-lb body A in Fig. P11-40 is suspended by means of an inextensible cord. The body is swinging to the left through its lowest position with a velocity of 6 fps when a bullet weighing 0.01 lb having a velocity of 1500 fps as shown is embedded in the block. How far, measured vertically, will body A rise above its lowest position?

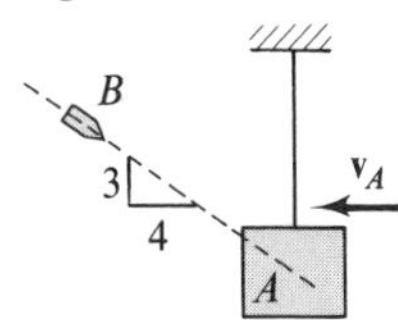

Figure P11-40

11-41 The 26-lb block A in Fig. P11-41 rests upon a smooth plane as shown. A bullet weighing 0.1 lb is fired horizontally with a velocity of 1300 fps and is embedded in the block. Determine the maximum distance the block will move up the plane.

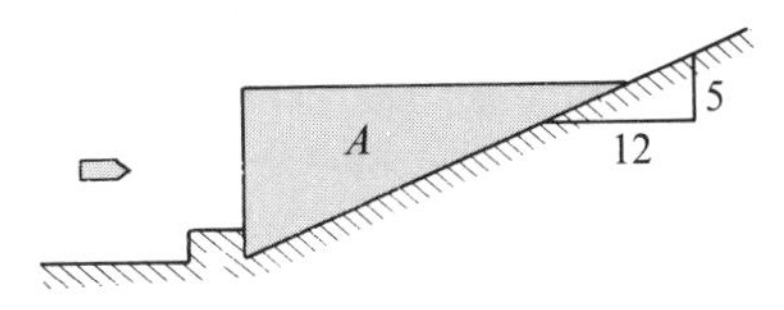

Figure P11-41

11-42 In Fig. P11-42, the projectile weighs 60 lb and has a velocity of 1500 fps as shown as it leaves the gun. The gun and carriage weigh 3000 lb. Neglecting the mass of the gas, determine

(a) The velocity of the gun and carriage just after the projectile leaves the gun.

(b) The required modulus of the spring if it is not to be deflected more than 18 in. when the gun and carriage hit it. Assume that the spring is initially uncompressed. Neglect the mass of the spring and the small rollers.

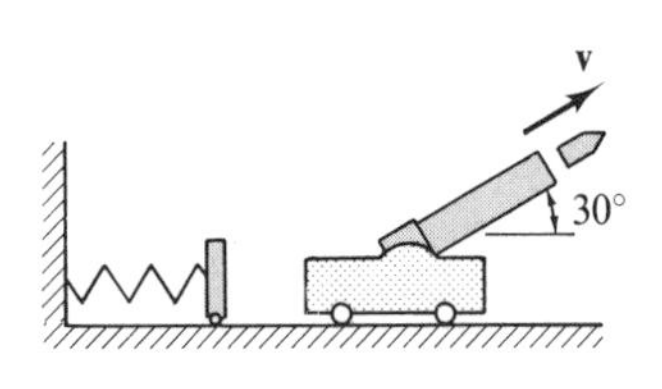

Figure P11-42

*__11-43__ A body having a mass of 200 kg is traveling in free flight with a velocity of 600 m/s as shown in Fig. P11-43(a). An instant later an explosion separates the body into three parts *A*, *B*, and *C* with masses of 45, 25, and 130 kg, respectively. The velocity of *A* and the directions of the velocities of *B* and *C* are indicated in Fig. P11-43(b). Determine the magnitudes of the velocities of *B* and *C*.

11-44 The 100-lb body *A* in Fig. P11-44 drops 4.5 ft from rest on the 50-lb plate *B* and remains in contact with it. Before contact body *B* is at rest on the spring, which has a modulus of 200 lb-per ft. Determine the maximum downward displacement of body *B*.

11-45 A falling weight of 400 lb is used to drive a 600-lb pile into the ground. Determine the penetration of the pile when the weight is dropped 10 ft. Assume an average resisting force of 15,000 lb and no rebound of the weight after impact.

11-46 A frame of weight $6W$ is at rest on a smooth plane in the position shown in Fig. P11-46. A particle of weight W is attached to the end of rod AB, which is pinned to the frame at A. Neglect the mass of the rod. The rod is released from rest in the horizontal position. Determine the velocity of the frame when the rod is vertical.

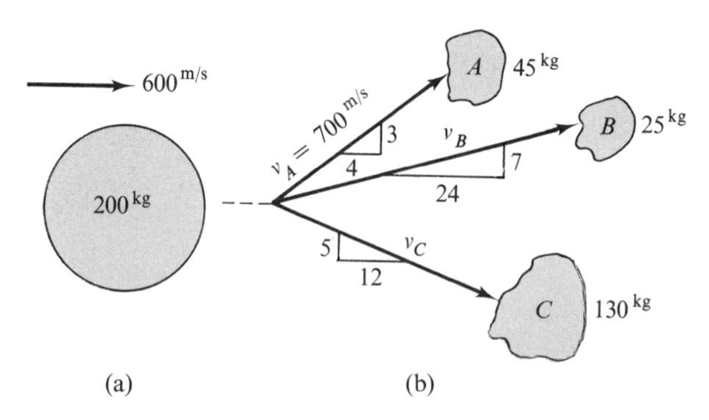

Figure P11-43

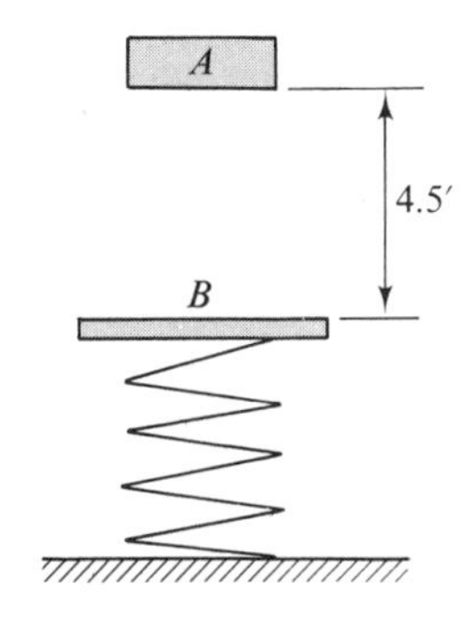

Figure P11-44

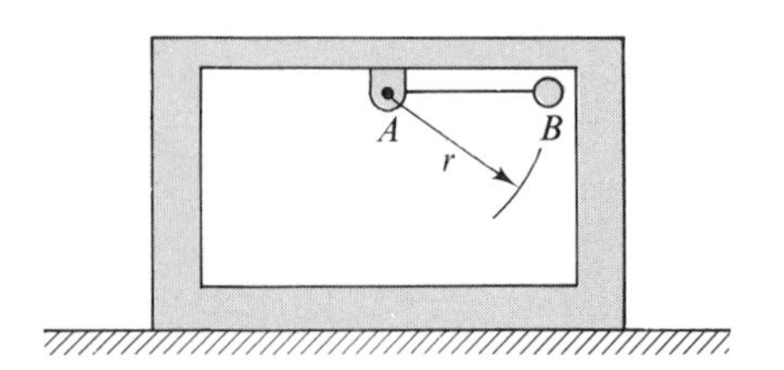

Figure P11-46

11-47 An open cart of mass M_0 is rolling with a velocity v_0 to the right along a smooth surface. Rain begins to fall with a velocity **u** in the direction shown in Fig. P11-47. The rate (mass per unit of time) at which the mass of water accumulates in the cart is r. Derive an expression for the velocity of the cart in terms of the given data and time.

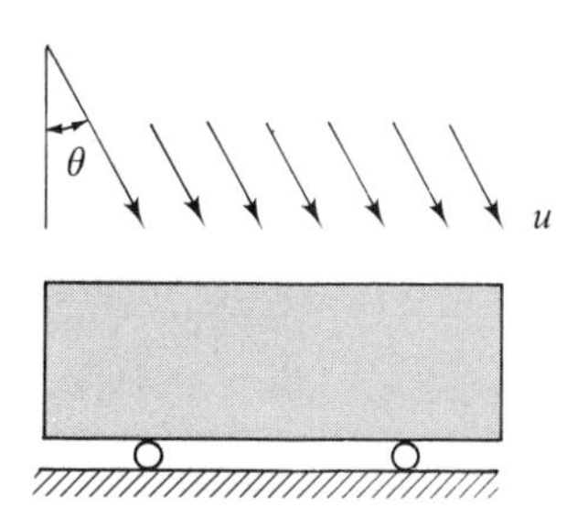

Figure P11-47

***11-48** The 1.0-kg block A in Fig. P11-48 slides on a smooth horizontal plane with a velocity of $30\mathbf{i}$ m/s when the bullet B strikes and is embedded in the block. The mass of the bullet is 10 g, and it has a velocity of 500 m/s parallel to the yz plane as shown. Determine the final velocity of A.

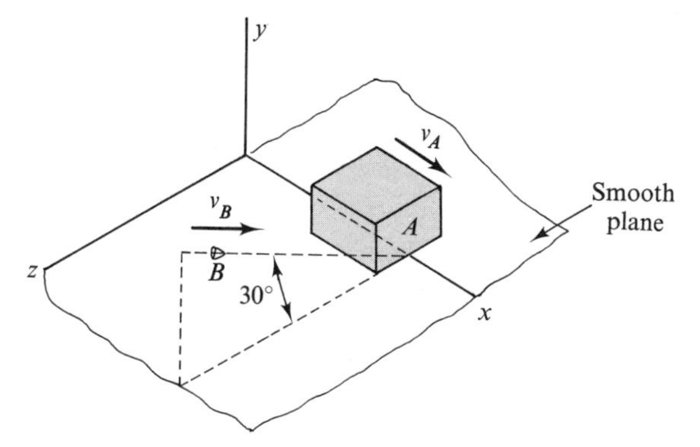

Figure P11-48

***11-49** Block A in Fig. P11-49 has a mass of 2.0 kg and slides on the smooth inclined plane. Body B has a mass of 1.0 kg and falls on block A and sticks to A. Just before the impact A has a horizontal velocity of 30 m/s as shown, and B has a vertical velocity of 40 m/s downward. Determine

(a) The final velocity of body B.

(b) The average force of A on B during the impact assuming the duration of the impact to be 0.10 s.

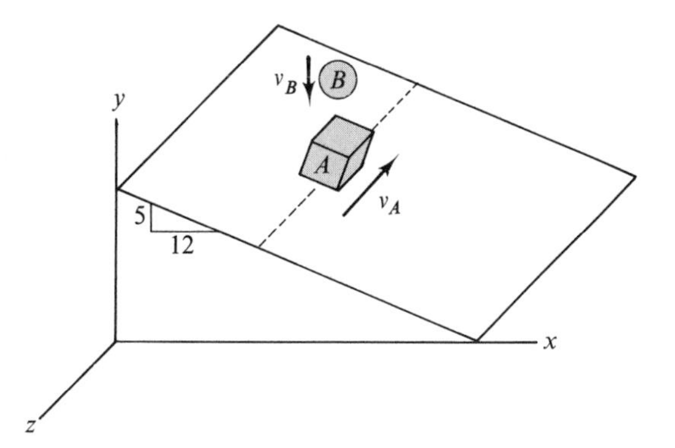

Figure P11-49

11-4 ELASTIC IMPACT

A collision between two bodies which occurs in a very small interval of time and where relatively large reaction forces exist is called *impact.* When two elastic bodies collide, they are compressed until their mass centers attain a common velocity, and then they move apart as the forces within the bodies act to restore the bodies to their original shapes. There has been considerable study of the forces acting during impact and of the resulting stresses and strains in the bodies. The primary concern in elementary engineering mechanics, however, is to obtain relations between the velocities of the bodies before and after impact. For this purpose, some additional definitions will be helpful.

When two bodies collide the straight line normal to the striking surfaces and passing through the point of contact is called the line of impact. When contact occurs over an area instead of at a point, the line of impact is defined as the line of action of the resultant normal force exerted by the bodies on each other. When the centers of gravity of the two colliding bodies are on this line the impact is defined as *central impact. Eccentric impact* occurs when the center of gravity of one or of both bodies is not on the line of impact. When the velocities of the points of contact of the two colliding bodies are along the line of impact *the impact is direct.* Direct impact implies a head-on collision as distinguished from the case of one body striking another body with a glancing blow, which is defined as *oblique impact.* Figure 11-9 shows top views of various types of impact of smooth circular disks sliding on horizontal surfaces and rods rotating on the same surfaces about vertical axes. Only central impact, either direct or oblique, can occur between particles. Eccentric impact is discussed in Art. 11-12.

When two bodies collide, the principle of conservation of linear momentum can frequently be applied for one or more directions to give one or more relations between the velocities of the bodies before and after impact. The velocities of the bodies after impact also depend

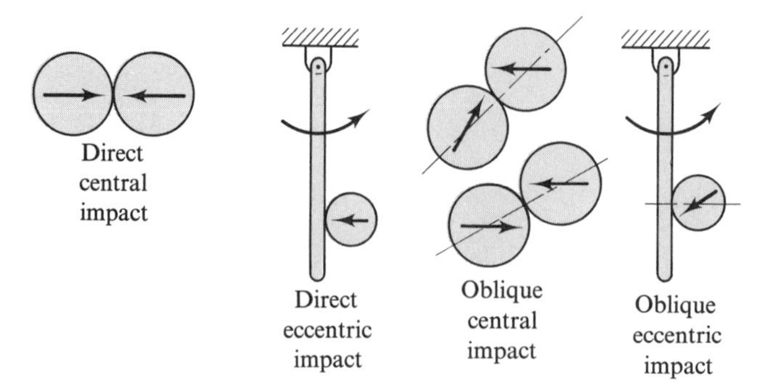

Figure 11-9

on the material properties of the bodies. Experiments indicate that the magnitude of the relative velocity of departure of two spheres which collide with direct central impact depends primarily on their relative velocity of approach and the material properties of the spheres. The preceding statement is not valid if the forces of impact are so large that the bodies are excessively deformed or shattered. *The ratio of the magnitude of the relative velocity of departure of the points of contact of two bodies that collide with direct impact to their relative velocity of approach is defined as the coefficient of restitution for the two bodies.* This ratio is a measure of the elastic properties of the bodies and must be determined experimentally. In mathematical form,

$$e = -\frac{(v_{A/B})_f}{(v_{A/B})_i} = -\frac{(v_A)_f - (v_B)_f}{(v_A)_i - (v_B)_i}, \tag{11-8}$$

where e is the coefficient of restitution, A and B are the points of contact of the two colliding bodies, and i and f indicate initial and final velocities, respectively. The relative velocities of approach and departure are always of opposite sense; therefore, the negative sign before the fraction must be included if the value of e is to be positive.

For perfectly elastic bodies, the coefficient of restitution is approximately one, and for inelastic bodies which remain together after impact the coefficient is zero. A collision in which the final relative velocity is zero is often called plastic impact. If the coefficient of restitution is unity, the kinetic energy of the system is conserved during impact (see Problem 11-68), but since some energy is always used in producing sound, vibrations of the bodies, permanent deformations of the bodies, and possibly in other ways, the coefficient is always less than one.[1] The value of e for any two bodies is therefore between the limits of zero and unity.

The table, following, of approximate values of coefficients of restitution for direct, central impact of spheres of various materials gives an indication of the range of values.

APPROXIMATE COEFFICIENTS OF RESTITUTION

Materials	e
Glass on glass	0.93–0.95
Ivory on ivory	0.88–0.89
Steel on steel	0.5–0.8
Cast iron on cast iron	0.4–0.7
Lead on lead	0.12–0.18
Iron on lead	0.11–0.15
Cork on cork	0.5–0.6
Wood on wood	0.4–0.6
Clay on clay (moist)	0
Putty on putty (moist)	0

[1] See Werner Goldsmith, "The Coefficient of Restitution," *Bulletin of the Mechanics Division of the American Society for Engineering Education* (May 1952). Vol. 2, no. 2, p. 10.

When the impact of two bodies is oblique the components of the velocities of the points of contact normal to the surface of contact are used in Eq. (11-8), and the coefficient of restitution is assumed to be the same as for direct impact. The coefficient of restitution for nonspherical bodies depends on the shape of the bodies and on the positions of the bodies relative to the line of impact as well as on the material of the bodies. For oblique impact of *smooth* bodies, the components of the velocities tangent to the contacting surfaces are not altered by the impact because there is no linear impulse on either body in the tangential direction.

EXAMPLE 11-6

Two smooth 4-oz pucks are sliding on smooth ice when they collide as indicated in the top view in Fig. 11-10(a). The velocity of A before impact was 40 fps and the velocity of B before impact was 50 fps in the directions indicated on the figure. The coefficient of restitution for the bodies is 0.56. Determine the velocity of each puck after impact.

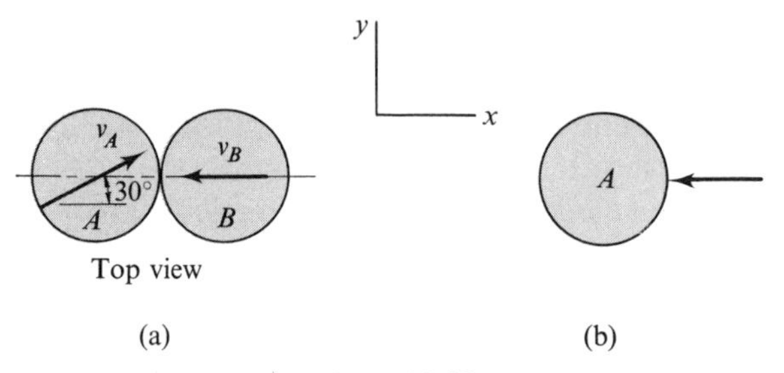

Figure 11-10

SOLUTION

The line of impact is in the x direction and the x components of the velocities of the pucks are changed. The principle of conservation of linear momentum for this direction, with positive to the right, gives

$$[m_A(v_A)_x + m_B(v_B)_x]_i = [m_A(v_A)_x + m_B(v_B)_x]_f,$$

which, in scalar form, becomes

$$\frac{4}{16g}(40)(0.866) + \frac{4}{16g}(-50) = \frac{4}{16g}(v_A)_x + \frac{4}{16g}(v_B)_x, \tag{a}$$

where $\mathbf{v}_A$ and $\mathbf{v}_B$ are the final velocities of the pucks. The x components of the final velocities are both assumed to be to the right. Since the impact is oblique, the components of the velocities *along the line of impact* must be used in Eq. (11-8), and, for this position, the equation becomes

$$e = 0.56 = -\frac{(v_A)_x - (v_B)_x}{40(0.866) - (-50)}. \tag{b}$$

Equations (a) and (b) can be solved simultaneously to give

$$(v_A)_x = -31.4 \text{ fps}, \qquad (v_B)_x = 16.02 \text{ fps}.$$

The free-body diagram of puck A in Fig. 11-10(b) shows that there is no impulse on the body in the y direction; therefore the linear momentum of A is conserved in the y direction, and the y component of its velocity does not change. A similar analysis shows that the impulse on puck B in the y direction is also zero and that the y component of the velocity of B does not change during the impact. Therefore, the final velocity of puck A is

$$\mathbf{v}_A = -31.4\mathbf{i} + 0.5(40)\mathbf{j} = \underline{37.2 \text{ fps } 32.5^\circ}, \quad \text{Ans.}$$

and the final velocity of B is

$$\mathbf{v}_B = 16.02\mathbf{i} = \underline{16.02 \text{ fps} \rightarrow}. \quad \text{Ans.}$$

PROBLEMS

***11-50** Two bodies are moving on a smooth horizontal plane when they collide with direct central impact. Before impact the velocity of the 40-kg body A is 20 m/s to the right, and the 20-kg body B has a velocity of 10 m/s to the left. The coefficient of restitution is 0.6. Determine
(a) The velocity of each body after impact.
(b) The time average value of the force exerted by body A on body B if the contact time is 0.05 s.

11-51 The 40-lb body A in Fig. P11-51 has a velocity of 20 fps to the right when it collides with the 100-lb body B, which is at rest before impact. The spring (modulus 40 lb-per ft) is unstressed before impact. The coefficient of restitution is 0.4. Determine the maximum compression of the spring if the horizontal plane is smooth.

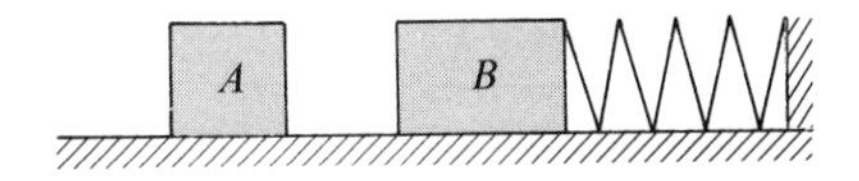

Figure P11-51

11-52 Body A in Fig. P11-51 weighs 32.2 lb, and body B weighs 64.4 lb, and the modulus of the spring is 40 lb per ft. Body A has a velocity of 15 fps to the right when it collides with body B, which is at rest before impact. Determine the coefficient of restitution if the maximum compression of the spring is 2 ft. The spring is initially uncompressed.

11-53 Body A in Fig. P11-53 weighs 4.0 lb and is released from rest in the position shown. Body A strikes the 2.0-lb body B with direct central impact. Body B is at rest on the smooth horizontal plane before impact. The coefficient of restitution is 0.5. Determine
(a) The velocity of A after impact.
(b) The loss of kinetic energy of body A during impact.

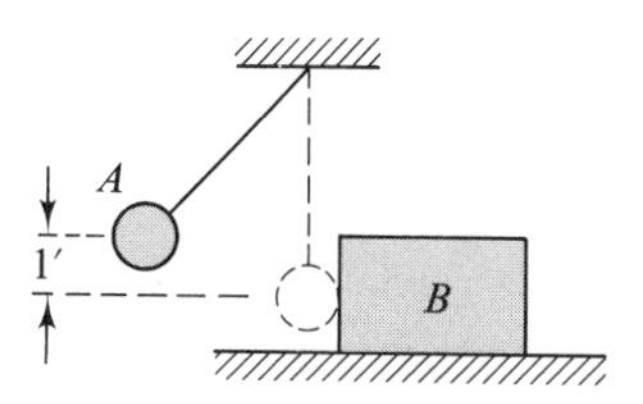

Figure P11-53

11-54 Block A in Fig. P11-54 weighs 50 lb and has a velocity of 30 fps to the left when it

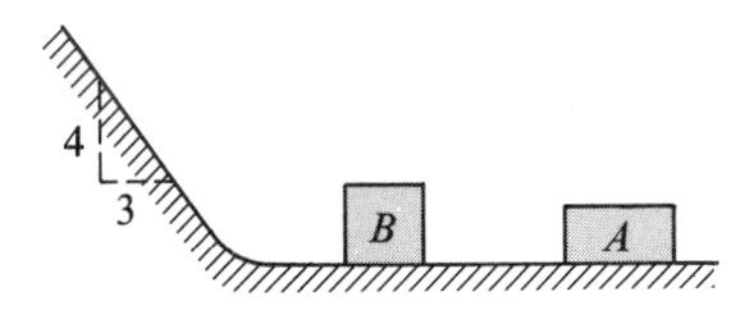

Figure P11-54

strikes the 70-lb body B, which is at rest. The coefficient of restitution between the bodies is 0.6. The horizontal plane is smooth, but the inclined plane and body B have a coefficient of friction of 0.3. Determine the distance body B will travel up the inclined plane before coming to rest.

11-55 The 5-lb smooth disk A and the 15-lb smooth disk B in Fig. P11-55 slide on a horizontal surface and collide with oblique central impact. Before impact the velocities of A and B are 10 and 13 fps, as shown. The coefficient of restitution is 0.6. Determine the velocity of body A after impact.

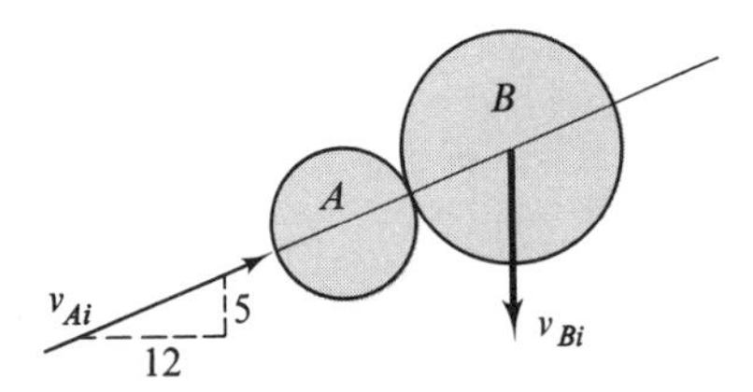

Figure P11-55

11-56 Disks A and B in Fig. P11-56 weigh 25 and 15 lb, respectively, and slide on a smooth horizontal plane. Just before colliding with oblique central impact, the velocities of A and B are 39 and 20 fps, as shown. The striking surfaces are smooth, and the coefficient of restitution is 0.4. Determine the velocity of A after impact.

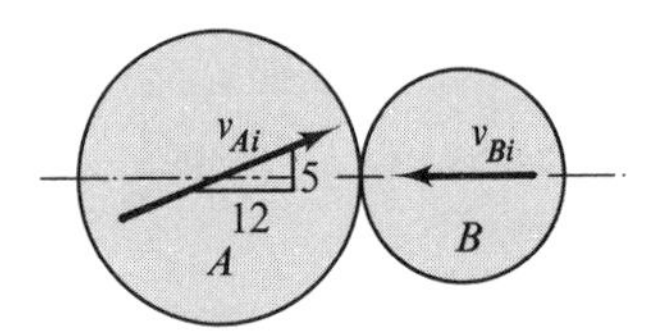

Figure P11-56

11-57 The 5-lb smooth disk A in Fig. P11-57 and the 20-lb smooth disk B slide on a horizontal plane when they collide with oblique central impact. Before impact the velocities are as shown. The coefficient of restitution is 0.5. Determine the kinetic energy of A after impact.

11-58 Disks A and B in Fig. P11-58(a) slide on a smooth horizontal surface and collide with

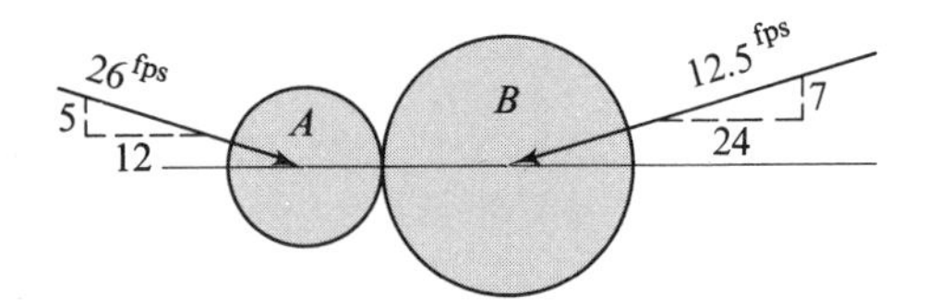

Figure P11-57

direct oblique impact. Disk A weighs 1 lb and has a velocity of 10 fps to the right before impact. Disk B weighs 2 lb and has a velocity of 12 fps upward before impact. The coefficient of restitution for bodies A and B is 0.80. Determine the velocities of A and B after impact if the bodies are in the position shown in Fig. P11-58(b) just before impact.

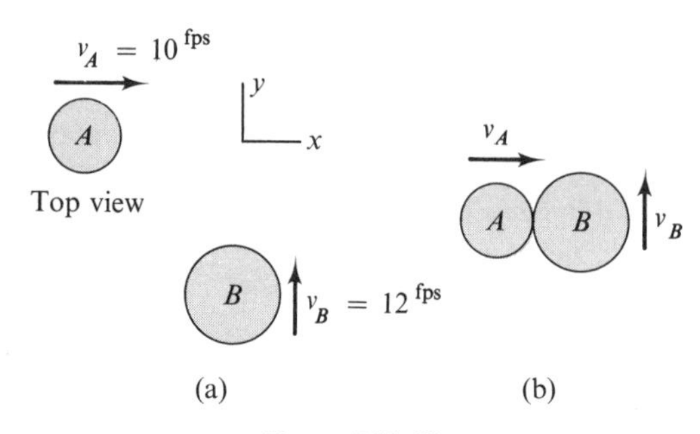

Figure P11-58

11-59 Solve Problem 11-58 if disks A and B are in the position shown in Fig. P11-59 at the instant of impact.

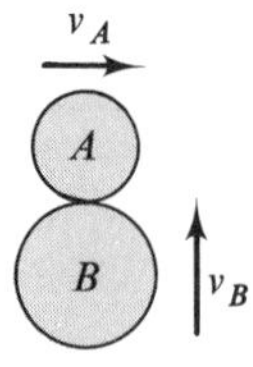

Figure P11-59

11-60 Solve Problem 11-58 if disks A and B are in the position shown in Fig. P11-60 just before impact.

11-61 Two spheres A and B in Fig. P11-61 are dropped simultaneously from a height h above a horizontal plane. The coefficients of restitution are e_1 between A and the plane and e_2 between B and the plane. Show that the

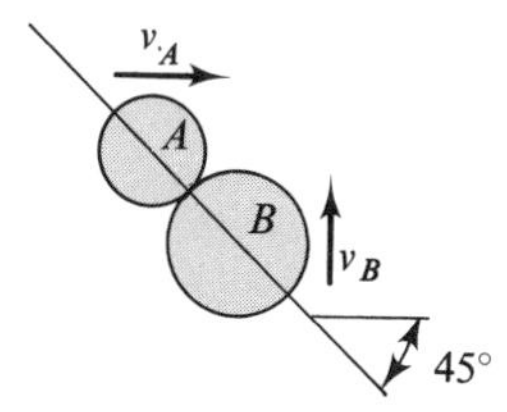

Figure P11-60

velocity of A with respect to B immediately after impact is $(e_1 - e_2)\sqrt{2gh}$ upward.

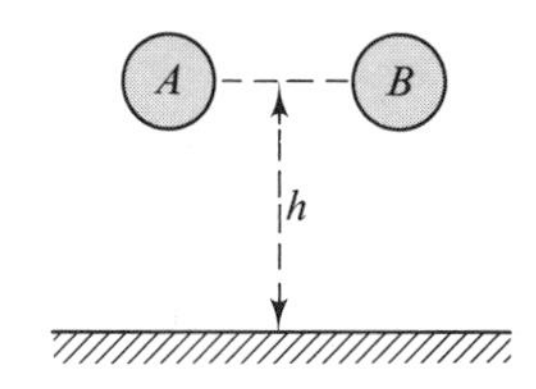

Figure P11-61

***11-62** A ball rebounds from a smooth inclined plane as shown in Fig. P11-62. The ball has a velocity of 20 m/s downward just before it strikes the plane, and the coefficient of restitution is 0.8. Determine the velocity just after the impact.

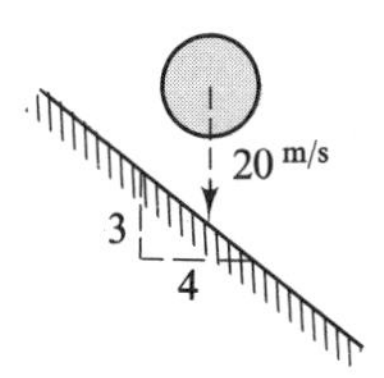

Figure P11-62

***11-63** The 2-kg ball in Fig. P11-63 strikes the smooth horizontal surface with a velocity of 65 m/s. The coefficient of restitution is 0.6. Determine

(a) The velocity of the ball after impact.

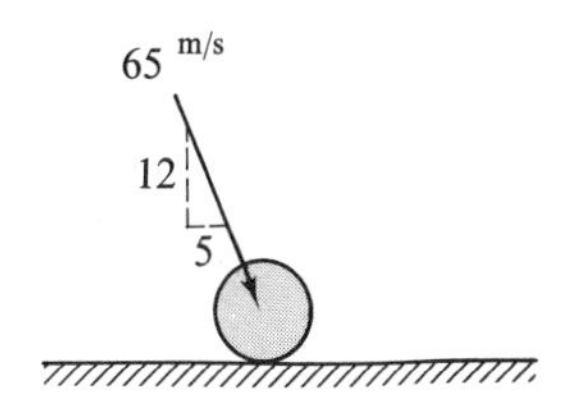

Figure P11-63

(b) The maximum height above the surface that the ball will rise.

(c) The time average value of the force exerted by the surface on the ball if the contact time is 0.02 s.

11-64 A ball hits a smooth horizontal plane with a velocity of v_i and rebounds with a velocity of v_f as shown in Fig. P11-64. The coefficient of restitution is e. Show that $\tan\theta_f = e\tan\theta_i$.

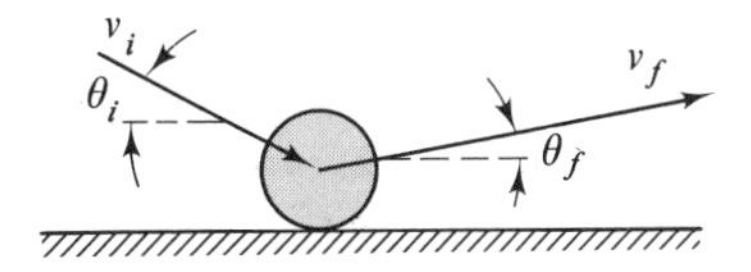

Figure P11-64

11-65 A 0.322-lb ball hits a rough horizontal plane with a velocity of 60 fps as shown in Fig. P11-65 and rebounds at the indicated angle. The coefficient of restitution is 0.8. Determine the linear impulse of the frictional force on the ball.

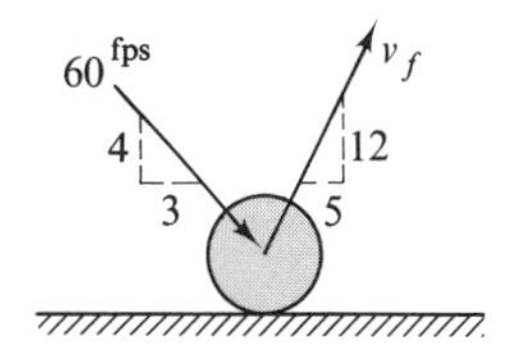

Figure P11-65

***11-66** Each of the balls A and B in Fig. P11-66 has a mass of 1.0 kg. Ball B is suspended at rest from a vertical inextensible cord

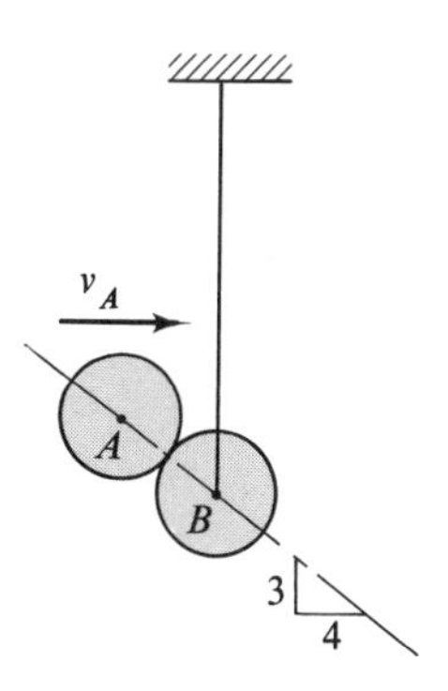

Figure P11-66

when it is struck by A, which has an initial velocity of 25 m/s to the right. The coefficient of restitution for the balls is 0.80. Determine the velocity of each ball after impact.

11-67 The two disks A and B in Fig. P11-67 have identical masses and slide on a smooth

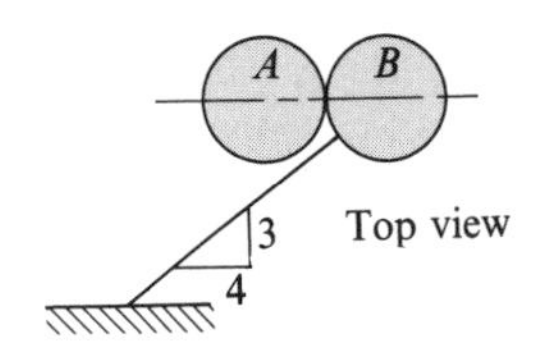

Figure P11-67

horizontal plane. Body B is connected to the wall by an extensible cord. The coefficient of restitution for disks A and B is 0.80. Initially the velocity of A is 25 fps to the right, and B is at rest. Determine the velocity of each body after impact.

11-68 The two bodies A and B in Fig. P11-68 are moving to the right along the smooth horizontal plane when they collide with direct central impact. Determine the percentage loss of kinetic energy of the two bodies during impact in terms of two masses, the initial velocities, and the coefficient of restitution.

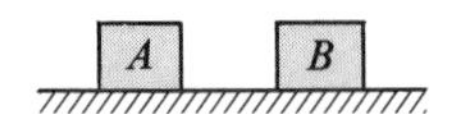

Figure P11-68

11-5 ANGULAR MOMENTUM OF A PARTICLE—CONSERVATION OF ANGULAR MOMENTUM

The linear momentum $\mathbf{G}$ of a particle having a mass m and a velocity $\mathbf{v}$ is $\mathbf{G} = m\mathbf{v}$. The vector $\mathbf{G}$ is localized and passes through the particle as shown in Fig. 11-11. Let the fixed point O be the origin of a Newtonian reference system $Oxyz$. The *angular momentum* of the particle with respect to the fixed point O is the moment of the linear momentum with respect to O. Using the symbol $\mathbf{H}_O$ for angular

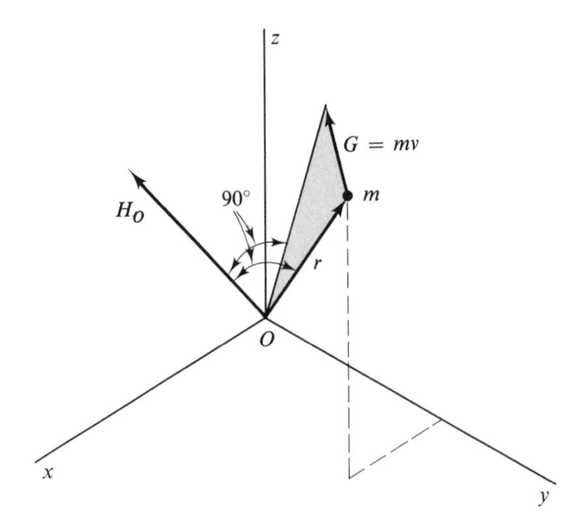

Figure 11-11

momentum, it can be written as

$$\mathbf{H}_O = \mathbf{r} \times m\mathbf{v}, \tag{11-9}$$

where $\mathbf{r}$ is the position vector from O to the particle. The angular momentum vector $\mathbf{H}_O$ is perpendicular to the plane formed by the position vector $\mathbf{r}$ and the velocity vector $\mathbf{v}$ as shown in Fig. 11-11.

When Eq. (11-9) is differentiated with respect to time, the time rate of change of the angular momentum is

$$\dot{\mathbf{H}}_O = \dot{\mathbf{r}} \times m\mathbf{v} + \mathbf{r} \times m\dot{\mathbf{v}}$$

or

$$\dot{\mathbf{H}}_O = \mathbf{v} \times m\mathbf{v} + \mathbf{r} \times m\mathbf{a}.$$

The first term on the right is zero, and the second term is the same as $\mathbf{r} \times \mathbf{F}$, where $\mathbf{F}$ is the resultant force acting on the particle. The time rate of change of the angular momentum can thus be written as

$$\dot{\mathbf{H}}_O = \mathbf{M}_O, \tag{11-10}$$

where $\mathbf{M}_O$ is the moment about point O of the resultant force acting on the particle. Eq. (11-10) is a vector equation and can be resolved into component scalar equations as

$$\begin{aligned} \dot{H}_x &= M_x, \\ \dot{H}_y &= M_y, \\ \dot{H}_z &= M_z, \end{aligned} \tag{11-10a}$$

where H_x, H_y, H_z and M_x, M_y, M_z are the angular momentum and the moment of the resultant force, respectively, about fixed x, y, and z axes.

Equations (11-10) and (11-10a) are the mathematical form of the *principle of angular momentum for particles, which states that the time rate of change of the angular momentum of a particle with respect to a fixed point O (or a fixed axis) is equal to the moment about point O (or about the fixed axis) of the resultant force acting on the particle.*

When the moment $\mathbf{M}_O$ can be expressed as a function of time in a Newtonian reference system, Eq. (11-10) can be integrated from some initial time t_i to some final time t_f to give

$$\Delta\mathbf{H}_O = (\mathbf{H}_O)_f - (\mathbf{H}_O)_i = \int_{t_i}^{t_f} \mathbf{M}_O \, dt, \tag{11-11}$$

where $\int_{t_i}^{t_f} M_O \, dt$ is the *angular impulse*, with respect to O, of the resultant force acting on the particle. Equation (11-11) is the mathematical form of the *principle of angular impulse and angular momentum for a particle and states that during the time interval t_i to t_f the change in the angular momentum about any fixed point O is equal to the angular impulse about point O acting on the particle.* Equation (11-11) can be resolved into three component expressions in the same manner as that explained for Eq. (11-10).

If $\mathbf{M}_O$ is zero, Eq. (11-10) becomes

$$\dot{\mathbf{H}}_O = 0.$$

Since the derivative of $\mathbf{H}_O$ is zero, $\mathbf{H}_O$ must be constant during the time interval that $\mathbf{M}_O$ is zero, and

$$(\mathbf{H}_O)_f = (\mathbf{H}_O)_i.$$

The *principle of conservation of angular momentum for a particle states that if the moment about any fixed point O of the resultant force acting on a particle is zero, the angular momentum of the particle about that point is conserved.*

Equation (11-10) is a vector equation, and although the total moment $\mathbf{M}_O$ may not be zero, it frequently occurs that a component of $\mathbf{M}_O$ is zero along some fixed axis through point O. In such a case, although the total angular momentum $\mathbf{H}_O$ is not conserved, the component of $\mathbf{H}_O$ in the direction of zero moment is conserved. Such a case is illustrated in Example 11-8.

EXAMPLE 11-7

The bob of a simple pendulum has a mass m and is supported by a cord of length L which swings in a vertical plane. Use the principle of angular momentum to determine the differential equation of motion.

SOLUTION

A free-body diagram of the bob swinging in the xz plane is shown in Fig. 11-12. There are two forces acting on the bob, the weight and the tension in the cord. Since the cord passes through O, the moment of the tensile force with respect to O is zero; therefore the resultant moment of all forces about O is the moment of the weight. That is,

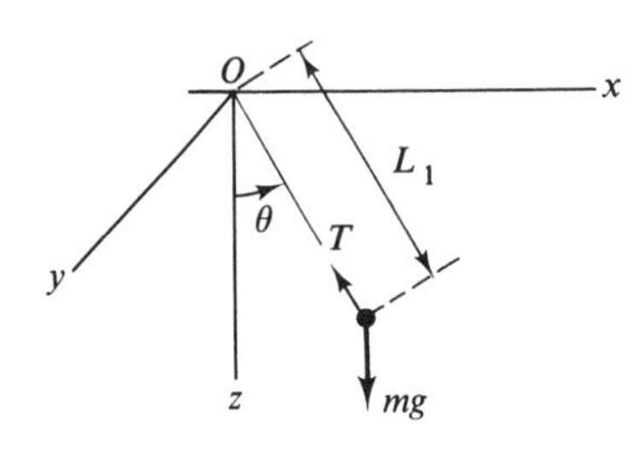

Figure 11-12

$$\mathbf{M}_O = -mgL \sin\theta\, \mathbf{j}.$$

The magnitude of the linear momentum of the pendulum bob is $mL\dot{\theta}$, and the angular momentum is the moment of the linear momentum; hence

$$\mathbf{H}_O = L(mL\dot{\theta})\mathbf{j}.$$

When $\mathbf{H}_O$ is differentiated with respect to time, the principle of angular momentum gives

$$mL^2\ddot{\theta}\mathbf{j} = -mgL \sin\theta\, \mathbf{j}$$

or

$$\underline{\ddot{\theta} + \frac{g}{L}\sin\theta = 0.} \qquad \text{Ans.}$$

EXAMPLE 11-8

The spherical pendulum in Fig. 11-13(a) is given an initial velocity of v_1 parallel to the y axis when the angle θ is 30°. If θ is to increase to a maximum value of 45°, determine the required value of v_1.

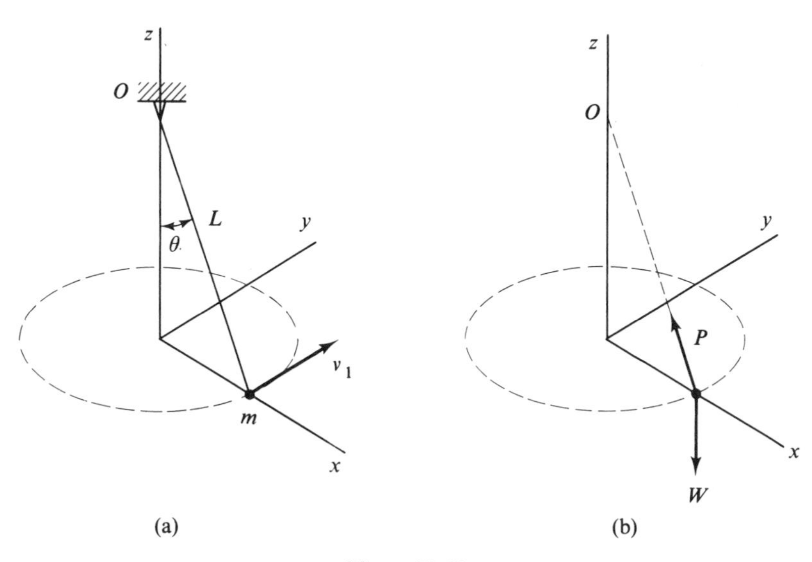

Figure 11-13

SOLUTION

A free-body diagram of the particle m is shown in Fig. 11-13(b). The moment of the forces acting on the particle about the fixed pivot O is

$$\mathbf{M}_O = (\mathbf{L}\times\mathbf{P}) + (\mathbf{L}\times\mathbf{W}),$$

where $\mathbf{L}$ is the vector from O to m. The moment due to $\mathbf{P}$ is zero since $\mathbf{P}$ is always directed through O. The moment due to $\mathbf{W}$ is always perpendicular to the plane formed by $\mathbf{L}$ and $\mathbf{W}$ and consequently has no component in the z direction; therefore the component of the angular momentum along the z axis is conserved.

Since the initial velocity $\mathbf{v}_1$ is parallel to the xy plane, the initial angular momentum of the particle about the fixed z axis is

$$(\mathbf{H}_z)_1 = [(\mathbf{L}\times m\mathbf{v}_1)\cdot\mathbf{k}]\mathbf{k} = Lmv_1 \sin 30°\mathbf{k}$$

and

$$(H_z)_1 = Lmv_1 \sin 30°.$$

Although the trajectory of the particle is unknown, when θ reaches a maximum value the velocity v_2 must again be parallel to the xy plane; therefore

$$(\mathbf{H}_z)_2 = [(\mathbf{L}\times m\mathbf{v}_2)\cdot\mathbf{k}]\mathbf{k} = Lmv_2 \sin 45°\mathbf{k}$$

and

$$(H_z)_2 = Lmv_2 \sin 45°.$$

Since the z component of the angular momentum is conserved, $(H_z)_1$ must equal $(H_z)_2$, that is,

$$Lmv_1 \sin 30° = Lmv_2 \sin 45°$$

from which

$$v_2 = v_1\left(\frac{\sin 30°}{\sin 45°}\right). \tag{a}$$

A second equation relating v_2 and v_1 can be obtained from the work-energy principle; that is,

$$U_{1-2} = T_2 - T_1$$

or

$$-W(L\cos 30° - L\cos 45°) = \frac{1}{2}\frac{W}{g}v_2^2 - \frac{1}{2}\frac{W}{g}v_1^2.$$

The last equation can be rearranged to the form

$$v_2^2 = v_1^2 - 2gL(\cos 30° - \cos 45°). \tag{b}$$

Substitution of Eq. (a) into Eq. (b) gives

$$v_1^2\left(\frac{\sin 30°}{\sin 45°}\right)^2 = v_1^2 - 2gL(\cos 30° - \cos 45°)$$

or

$$v_1^2 = \frac{2gL(\cos 30° - \cos 45°)}{1 - (\sin 30°/\sin 45°)^2} = 0.318(2gL).$$

When $L = 3$ ft is substituted into the last equation the solution is obtained as

$$v_1^2 = 61.4$$

or

$$\underline{v_1 = 7.84 \text{ fps.}} \qquad \text{Ans.}$$

PROBLEMS

11-69 (a) Determine the angular momentum of a 1.61-lb particle with respect to the origin O of a fixed coordinate system when the particle is passing through the point $(4\mathbf{i} + 10\mathbf{j} - 10\mathbf{k})$ ft with a velocity of $(2\mathbf{i} - 6\mathbf{j} + 10\mathbf{k})$ fps.
(b) Solve part (a) if the particle has a mass of 0.5 kg, the coordinates of the point are (4, 10, −5) m, and $\mathbf{v} = (5\mathbf{i} + 2\mathbf{j} - 7\mathbf{k})$ m/s.

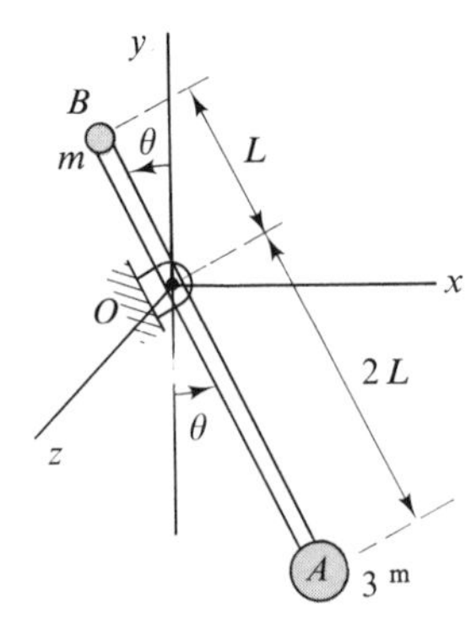

Figure P11-70

11-70 The particles A and B shown in Fig. P11-70 have masses of $3m$ and m, respectively, and are joined by a rigid bar, the mass of which may be neglected. The assembly rotates in the vertical xy plane about a horizontal axis through O. Determine
(a) The angular momentum of the two particles with respect to O.
(b) The differential equation of motion by using the principle of angular momentum.

11-71 Particles A and B shown in Fig. P11-71 have masses of $2m$ and m, respectively, and are joined by rigid bars, the masses of which may be neglected. The assembly rotates in the vertical xy plane. Determine
(a) The angular momentum of the two particles with respect to O.
(b) The differential equation of motion by using Eq. (11-10).

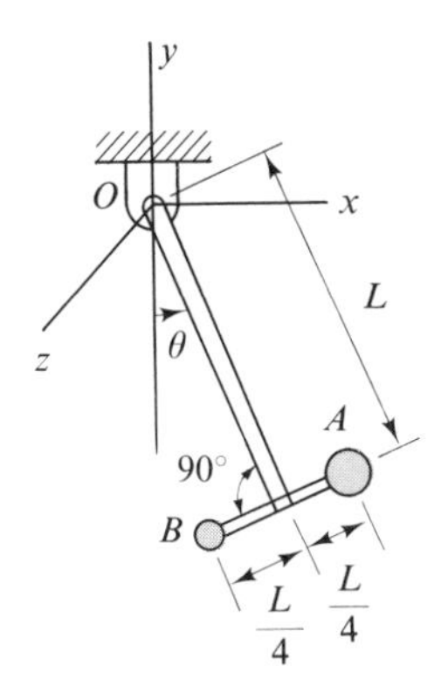

Figure P11-71

11-72 The smooth rod of negligible mass in Fig. P11-72 rotates in a horizontal plane about a vertical axis at O with an angular velocity of 3 rad per sec clockwise. The particle A of mass m can slide along the rod and is held in position by the string. When in the position shown, the string breaks, and the mass is free to slide along the rod. Determine
(a) The velocity of A just before it hits the stop.
(b) The velocity of A just after it hits the stop (assume no rebound).
(c) The impulse of the stop on A during the impact.

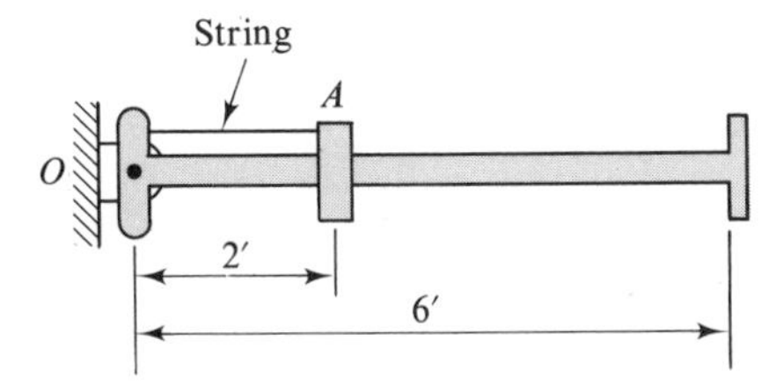

Figure P11-72

11-73 A smooth thin rod of negligible mass rotates in a horizontal plane about a vertical axis through O as shown in Fig. P11-73 with a clockwise angular velocity of 6 rad per sec. A 32.2-lb disk is held by a string and is in contact with an unstressed spring. The string then breaks. Determine the value of the spring constant if the maximum compression of the spring is 1 ft.

11-74 The 3.22-lb particle A in Fig. P11-74 slides on the smooth horizontal plane and is

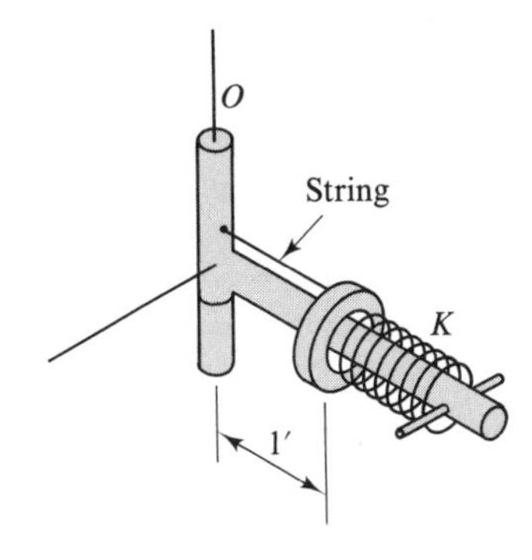

Figure P11-73

connected to the spring by a cord which passes through a smooth hole in the plane. In the position shown, A has a velocity of 100 fps to the right as indicated, there is no slack in the cord, and the spring is unstretched. During the motion the particle reaches a maximum distance of 10 ft from the hole. Determine the spring constant.

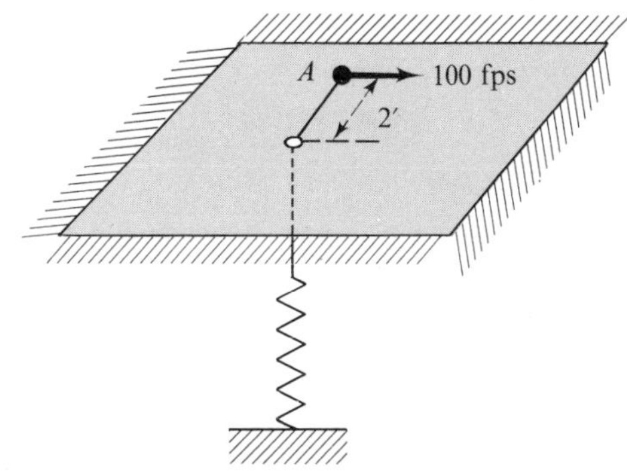

Figure P11-74

***11-75** The 0.1-kg particle A in Fig. P11-75 slides on a smooth horizontal plane and is attached to the fixed point O by means of an elastic cord having a modulus of 20 N/m and an

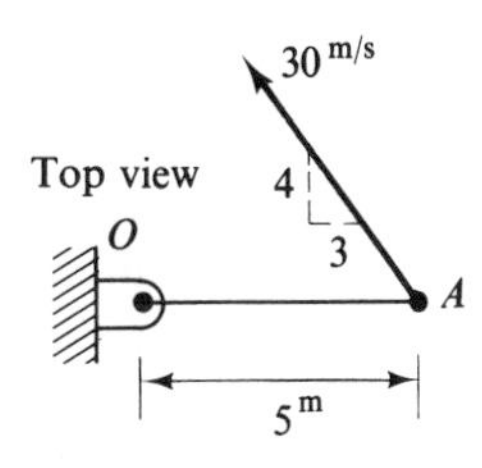

Figure P11-75

unstretched length of 3 m. When A is in the position shown its velocity is 30 m/s as indicated. Determine

(a) The speed of the particle when the cord becomes slack.

(b) The minimum distance from O to the particle A during the resulting motion.

11-76 The 6.44-lb particle A in Fig. P11-76 is connected to the fixed point O by a spring with a modulus of 20 lb per ft. In the position shown, the spring is stretched 2 ft, and A has the velocity indicated. Determine the maximum distance from O to A during the resulting motion.

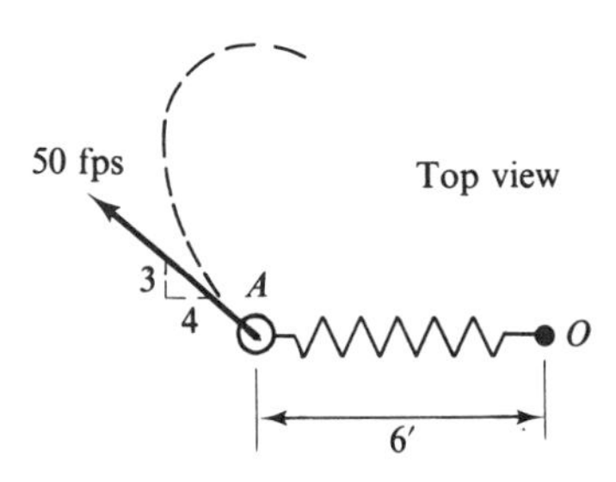

Figure P11-76

11-77 A particle weighing 6.44 lb moves on a smooth horizontal plane as shown in Fig. P11-77. Throughout the motion a force of magnitude $F = 4r^3$ acts on the particle and is always directed toward the fixed point O. The force is in pounds when r (the distance of the particle from the fixed point) is in feet. Determine the velocity of the particle at point A if the particle reaches a maximum distance from O of 8 ft.

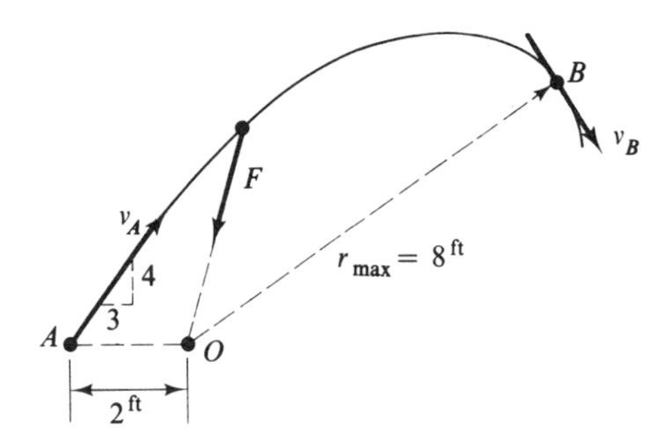

Figure P11-77

11-78 A particle travels on the smooth inner surface of a hollow cone with a vertical axis of revolution. When in the position shown in Fig. P11-78, the mass is given an initial velocity in a *horizontal* plane. Determine the radius of the path when the velocity of the mass is again horizontal if the initial speed is

(a) 3 fps.

(b) 30 fps.

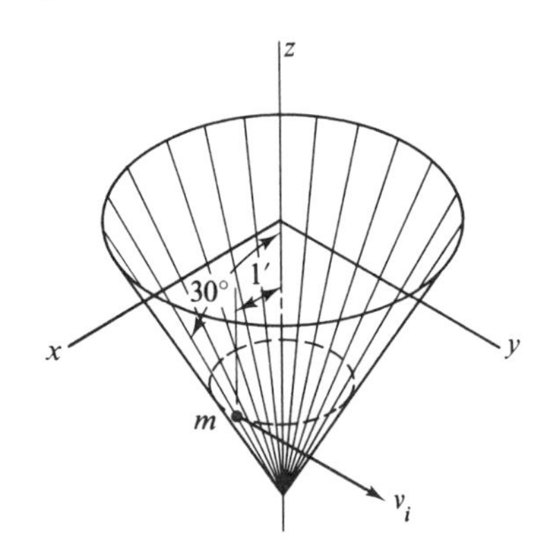

Figure P11-78

11-6 CENTRAL-FORCE MOTION

A central force is defined as a force with a line of action that always passes through a fixed point, known as the center of force. Some common examples of central-force systems include the motion of the planets about the sun, the motion of the moon (as well as artificial satellites) about the earth, the motion of an electron about the nucleus of an atom, and the motion of an α particle in the vicinity of a nuclear charge.

Consider the motion of particle P of mass m in Fig. 11-14 under the action of the central force $\mathbf{F}$ always directed through the fixed point

O. From Eq. (11-10),

$$\dot{\mathbf{H}}_O = \mathbf{M}_O = 0$$

because $\mathbf{F}$ has no moment about O. Since $\dot{\mathbf{H}}_O$ is zero, the integral of $\dot{\mathbf{H}}_O$ must be a constant; that is

$$\mathbf{H}_O = \mathbf{C}.$$

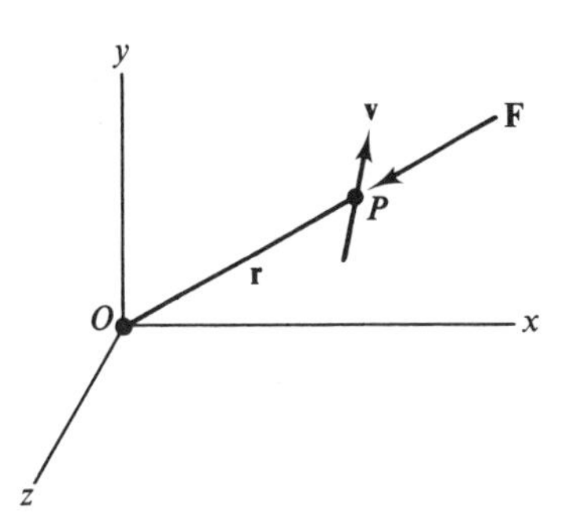

Figure 11-14

The angular momentum of a particle about a fixed point was defined as the moment of the linear momentum ($m\mathbf{v}$) about the point; therefore,

$$\mathbf{H}_O = \mathbf{r} \times (m\mathbf{v}) = \mathbf{C}.$$

Since $\mathbf{H}_O$ is a constant in both magnitude and direction and, from the definition of a cross product, $\mathbf{H}_O$ is perpendicular to the plane of $\mathbf{r}$ and $\mathbf{v}$, it follows that $\mathbf{r}$ and $\mathbf{v}$ must remain in the same plane. Therefore *central-force motion is motion in a plane.* The plane is determined by the center of force, the position of the particle, and its velocity in that position.

Newton's law of universal gravitation states that the force of attraction, $\mathbf{F}$, between two masses M and m separated by a distance r has the magnitude

$$F = \frac{GMm}{r^2},$$

where G is the universal gravitational constant. The experimentally determined value of G is

$$G = 3.438(10)^{-8} \frac{\text{lb-ft}^2}{\text{slug}^2}.$$

This unit could also be written as

$$\frac{\text{ft}^3}{\text{slug-sec}^2} \quad \text{or} \quad \frac{\text{ft}^4}{\text{lb-sec}^4}.$$

In SI units the constant G is

$$G = 6.670(10)^{-11}\ \text{m}^3/\text{kg} \cdot \text{s}^2.$$

If the mass of one of the two bodies is much greater than that of the other body, the larger mass can be considered fixed in space because the smaller mass would have a negligible effect upon the motion of the heavier body. The masses of the sun, earth, and moon are given for reference in this regard. They are

$$M_{\text{sun}} = 1.97(10)^{29}\ \text{slugs} = 2.87(10)^{30}\ \text{kg},$$

$$M_{\text{earth}} = 4.09(10)^{23}\ \text{slugs} = 5.97(10)^{24}\ \text{kg},$$

and

$$M_{\text{moon}} = 5.05(10)^{21}\ \text{slugs} = 7.37(10)^{22}\ \text{kg}.$$

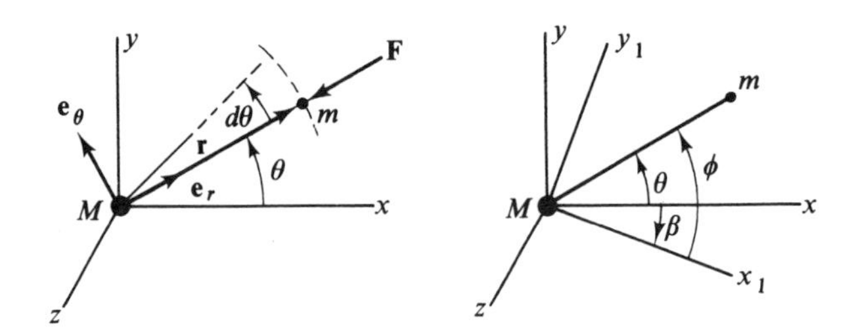

Figure 11-15

Figure 11-15(a) shows a large mass M, a much smaller mass m moving in the xy plane, and the force $\mathbf{F}$ exerted on m by M. Using polar coordinates and considering M fixed in space, the equations of motion of m in component form are, from Eq. (8-17),

$$\sum F_R = ma_R,$$

$$-\frac{GMm}{r^2} = m(\ddot{r} - r\dot{\theta}^2); \tag{a}$$

and

$$\sum F_T = ma_T,$$

$$0 = \frac{m}{r}\frac{d}{dt}(r^2\dot{\theta}). \tag{b}$$

When Eq. (b) is multiplied by r/m and integrated, the result is

$$r^2\dot{\theta} = \text{a constant} = h. \tag{c}$$

Note that the area swept out by the radius vector as it turns through an angle $d\theta$ in time dt is $dA = r(r\,d\theta)/2$; therefore,

$$2\frac{dA}{dt} = r^2\frac{d\theta}{dt} = \text{a constant} = h. \tag{11-12}$$

The quantity $dA/dt = h/2$ is called the *areal speed*, and, as shown here, it is a constant for any central-force system. Equation (11-12) expresses Kepler's second law of planetary motion which states that *the radius vector describes equal areas in equal intervals of time.*

The equation of the path of motion can be obtained by eliminating the parameter t from Eq. (a) and (c) as follows: Let $u = 1/r$, from which $du = -dr/r^2$ and, with Eq. (11-12),

$$\frac{dr}{dt} = \frac{dr}{d\theta}\frac{d\theta}{dt} = \frac{h}{r^2}\frac{dr}{d\theta} = -h\frac{du}{d\theta}.$$

Also

$$\frac{d^2r}{dt^2} = \frac{d}{dt}\left(-h\frac{du}{d\theta}\right) = -h\frac{d^2u}{d\theta^2}\frac{d\theta}{dt} = -h\frac{d^2u}{d\theta^2}\frac{h}{r^2} = -h^2u^2\frac{d^2u}{d\theta^2}. \tag{d}$$

Substitution of Eq (11-12), of Eq. (d), and of u for $1/r$ into Eq. (a) gives

$$-GMmu^2 = m\left[-h^2u^2\frac{d^2u}{d\theta^2} - \frac{1}{u}(hu^2)^2\right]$$

or

$$\frac{d^2u}{d\theta^2} + u = \frac{GM}{h^2}. \tag{e}$$

Equation (e) is an ordinary, second-order differential equation with constant coefficients and with the right-hand side different from zero. It can be solved by several methods; probably the most common is to assume a solution and verify it by substitution in the differential equation. Since u and its second derivative with respect to θ are involved, a function of θ with a second derivative equal to the negative of the original function plus or minus a constant is desired. Some exponential and trigonometric functions have this property, and the following assumed solution will be used:

$$u = C\cos(\theta + \beta) + D.$$

Substitution into Eq. (e) gives

$$D = \frac{GM}{h^2},$$

and the general solution of the differential equation is

$$u = C\cos(\theta + \beta) + \frac{GM}{h^2} = \frac{1}{r}, \tag{f}$$

where C and β are constants of integration with values determined by the initial conditions. The value of h is given by Eq. (c). By measuring the polar angle from an x_1 axis as shown in Fig. 11-15(b) instead of from the x axis, the angle $\theta + \beta$ can be replaced by an angle ϕ, and Eq. (f) becomes

$$r = \frac{h^2/GM}{1 + (h^2C/GM)\cos\phi}. \tag{11-13a}$$

Equation (11-13a) is the equation of a conic section in polar form with the origin at one focus. For the simplest case with C equal to zero, r is a constant and the curve is a circle. A conic section is defined as the path of a point P (Fig. 11-16) which moves in such a manner that

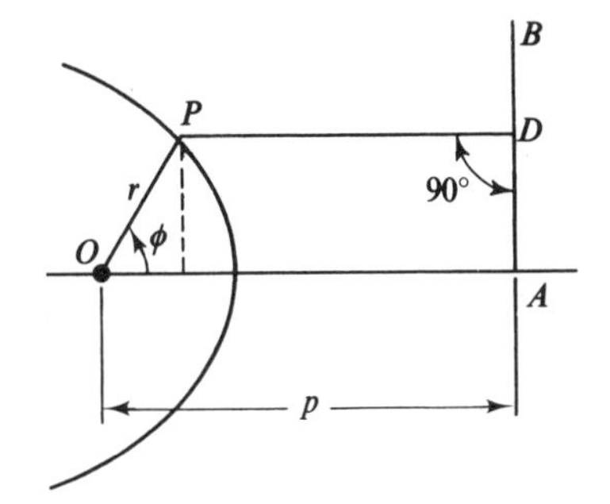

Figure 11-16

$$\frac{OP}{PD} = \text{a constant} = e,$$

where O is a fixed point, AB is a fixed line, OP is the distance from O to P, PD is the perpendicular distance from P to the fixed line, and e is the eccentricity. This relation can be written as

$$OP = r = e(PD) = e(p - r\cos\phi)$$

from which

$$r = \frac{ep}{1 + e\cos\phi}. \tag{11-13b}$$

Equation (11-13b) is the same as Eq. (11-13a) when the eccentricity of the conic section is

$$e = \frac{h^2C}{GM} \tag{g}$$

and

$$p = \frac{1}{C}.$$

Therefore

$$ep = \frac{h^2}{GM}.$$

Equation (11-13a) or (11-13b) can now be written as

$$r = \frac{h^2/GM}{1 + e\cos\phi}. \tag{11-14a}$$

Also Eq. (f) can be written as

$$\frac{1}{r} = C\cos\phi + \frac{GM}{h^2}, \tag{11-14b}$$

which is the equation of the path of the particle in polar form.

The preceding analysis proves Kepler's first law, which states that *the trajectory or orbit of a particle moving under the action of an inverse-square central force is a conic section with the center of force at one of the foci.* When e is less than 1, the locus is a closed path (an ellipse or circle when $e = 0$) since the denominator of Eq. (11-14a) is never zero for any value of ϕ. For values of $e \geq 1$ the path is not closed because, for certain values of ϕ, the denominator becomes zero and r becomes infinite. For $e = 1$ the curve is a parabola, and for $e > 1$ it is a hyperbola.[2]

For planetary motion about the sun and for artificial earth satellites, the orbits are ellipses and e is less than 1. The minimum distance from the center of attraction (at one focus) to the particle, corresponding to $\phi = 0$, is called perigee (r_p), and the maximum distance (for an elliptical orbit), occurring when $\phi = \pi$, is called apogee (r_a). When $\phi = 0$ in Fig. 11-16 the velocity of the particle is perpendicular to the radius vector, and from Eq. (c)

$$r^2\dot{\theta} = r^2\dot{\phi} = r_p v_p = h.$$

[2] See the discussion of conic sections in any standard text on analytical geometry.

This equation makes it possible to compute the constant h if the velocity and distance at perigee are known.

For an elliptical orbit, Eq. (11-14a) shows that the perigee and apogee distances are

$$\left.\begin{aligned} r_p &= \frac{h^2/GM}{1+e} \\ \text{and} \qquad r_a &= \frac{h^2/GM}{1-e}. \end{aligned}\right\} \tag{h}$$

These two equations can be solved for e by eliminating h^2/GM, and the result is

$$e = \frac{r_a - r_p}{r_a + r_p}. \tag{i}$$

The semimajor axis of the ellipse is

$$a = \frac{r_p + r_a}{2},$$

which, from Eq. (h), becomes

$$a = \frac{h^2}{2GM}\left(\frac{1}{1+e} + \frac{1}{1-e}\right) = \frac{h^2}{GM}\left(\frac{1}{1-e^2}\right). \tag{j}$$

The semiminor axis is

$$b = a(1-e^2)^{1/2} = \frac{h^2}{GM}(1-e^2)^{-1/2}.$$

The area of an ellipse is

$$A = \pi ab = \pi a^2(1-e^2)^{1/2}.$$

The areal speed, $dA/dt = h/2$, from Eq. (11-12) gives $A = (h/2)T$, where T is the period (the time for one orbit) and

$$T = \frac{2A}{h} = \frac{2\pi a^2}{h}(1-e^2)^{1/2}.$$

The value of $1-e^2$ can be obtained from Eq. (j), and the period becomes

$$\left.\begin{aligned} T &= \frac{2\pi a^2}{h}\left(\frac{h^2}{GMa}\right)^{1/2} = \frac{2\pi a^{3/2}}{(GM)^{1/2}} \\ \text{or} \qquad \frac{T^2}{a^3} &= \frac{4\pi^2}{GM}. \end{aligned}\right\} \tag{11-15}$$

Note that the period is independent of the mass of the orbiting body. Equation (11-15) demonstrates Kepler's third law[3] of planetary motion; namely, *the squares of the periods of the planets are proportional to the cubes of the semimajor axes of their orbits.*

The orbit of the moon around the earth is very nearly an ellipse (the sun perturbs it slightly) with $e = 0.056$. Its perigee and apogee distances are 222,000 and 253,000 miles, respectively. Some data on the planets are shown in the accompanying table.

PLANETARY DATA

Planet	*e*	*Mean Distance from Sun in Astronomical Units**	*Time for One Revolution About the Sun in Days* (d) *or Years* (yr)†
Mercury	0.206	0.387	88 d
Venus	0.007	0.723	225 d
Earth	0.017	1.000	365 d
Mars	0.093	1.524	687 d
Jupiter	0.048	5.203	11.9 yr
Saturn	0.056	9.539	29.5 yr

* An astronomical unit equals the mean distance from the earth to the sun which is approximately $9.26(10^7)$ miles or $1.49(10^8)$ km.

† The days and years used are the conventional time periods, or solar time, as distinguished from sidereal, or star time, measured by the rotation of the earth in relation to the stars. A sidereal day equals 23 hr, 56 min, and 4.091 sec of mean solar time.

When a satellite is "launched" with a velocity v_p parallel to the earth's surface at a distance r_p from the center of the earth the constant C can be determined by means of Eq. (11-14b). The angle ϕ is zero, and the value of C is

$$C = \frac{1}{r_p} - \frac{GM}{h^2} = \frac{1}{r_p} - \frac{GM}{r_p^2 v_p^2}.$$

When the value of C is substituted into Eq. (g) the eccentricity becomes

$$e = \frac{h^2}{GM}\left(\frac{1}{r_p} - \frac{GM}{h^2}\right)$$

$$= \frac{r_p v_p^2}{GM} - 1. \tag{k}$$

The constant GM for earth satellites can be determined from the force of gravity acting on a particle of mass m located at the surface of

[3] The first two laws were published in 1609 and the third in 1619. Newton derived the law of gravitational attraction from Kepler's laws.

the earth. By the law of universal gravitation,

$$\frac{GMm}{R^2} = mg,$$

where R, the radius of the earth, is 3960 miles or 6370 km; therefore

$$GM = gR^2$$
$$= 32.2(3960)^2(5280)^2 = 1.408(10)^{16}\ \text{ft}^3/\text{sec}^2.$$

Other values of gR^2 that are commonly used are

$$gR^2 = 1.239(10)^{12}\ \text{miles}^3/\text{hr}^2$$
$$= 3.98(10)^{13}\ \text{m}^3/\text{s}^2$$
$$= 5.16(10)^{12}\ \text{km}^3/\text{hr}^2.$$

Equation (k) can be used to determine the minimum velocity the satellite must have to escape from the earth's gravitational field. The minimum velocity will occur when the satellite is on a parabolic path, that is, when $e = 1$. When $e = 1$ is substituted into Eq. (k) the escape velocity is found to be

$$v_e = \sqrt{\frac{2GM}{r_p}} = \sqrt{\frac{2gR^2}{r_p}}.$$

When the eccentricity is zero the particle will have a circular orbit, and the corresponding velocity, from Eq. (k), is

$$v_{\text{cir}} = \sqrt{\frac{GM}{r_p}} = \sqrt{\frac{gR^2}{r_p}}.$$

The two preceding equations show that

$$v_e = \sqrt{2}v_{\text{cir}}.$$

EXAMPLE 11-9

An earth satellite is launched with a velocity of 18,000 mph parallel to the surface of the earth at an altitude of 540 miles. Determine the maximum altitude reached by the satellite.

SOLUTION

Figure 11-17 shows the satellite P' moving from perigee to apogee. The angular momentum at perigee must equal the angular momentum at apogee; that is,

$$mv_pr_p = mv_ar_a$$

or

$$v_a = \frac{r_p}{r_a}v_p. \tag{a}$$

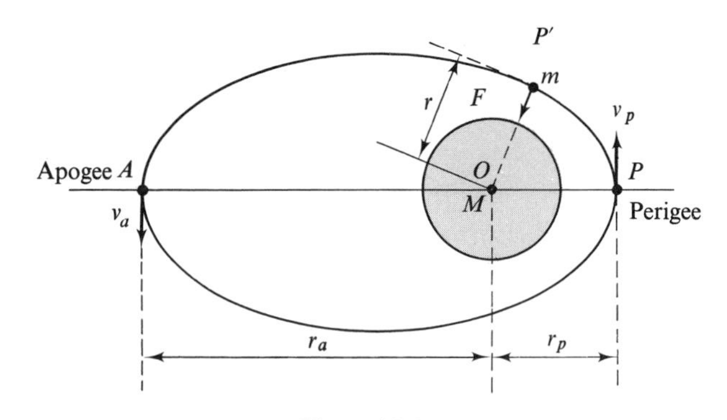

Figure 11-17

The principle of work and energy gives

$$U_{p\to a} = T_a - T_p$$

or

$$-\int_{r_p}^{r_a} \frac{GMm}{r^2}\,dr = \frac{1}{2}\,m(v_a^2 - v_p^2)$$

from which

$$GMm\left(\frac{1}{r}\right)_{r_p}^{r_a} = GMm\left(\frac{r_p - r_a}{r_p r_a}\right) = \frac{1}{2}\,m(v_a^2 - v_p^2). \tag{b}$$

When Eq. (a) is substituted into Eq. (b) the result is

$$GM\left(\frac{r_p - r_a}{r_p r_a}\right) = \frac{1}{2}\left(\frac{r_p^2}{r_a^2}\,v_p^2 - v_p^2\right) = \frac{v_p^2}{2}\,\frac{r_p^2 - r_a^2}{r_a^2},$$

which can be solved for r_a as

$$r_a = \frac{v_p^2 r_p^2}{2GM - v_p^2 r_p}. \tag{c}$$

The radius of the earth is 3960 miles; therefore

$$r_p = 3960 + 540 = 4500 \text{ miles.}$$

The velocity v_p is given as 18,000 mph, and the value of MG is $1.239(10)^{12}$ miles^3/hr^2 in consistent units. When these values are substituted into Eq. (c) the radius at apogee is

$$r_a = \frac{(18{,}000)^2(4500)^2}{2(1.239)(10)^{12} - (18{,}000)^2(4500)}$$

$$= 6430 \text{ miles.}$$

The altitude at apogee is

altitude = 6430 − 3960 = 2470 miles above the earth's surface.

Ans.

EXAMPLE 11-10

The satellite of Fig. 11-18 is launched with a velocity of 21,000 mph parallel to the surface of the earth at an altitude of 600 miles.

Determine

(a) The eccentricity of the orbit.
(b) The equation of the orbit.
(c) The time for one orbit.

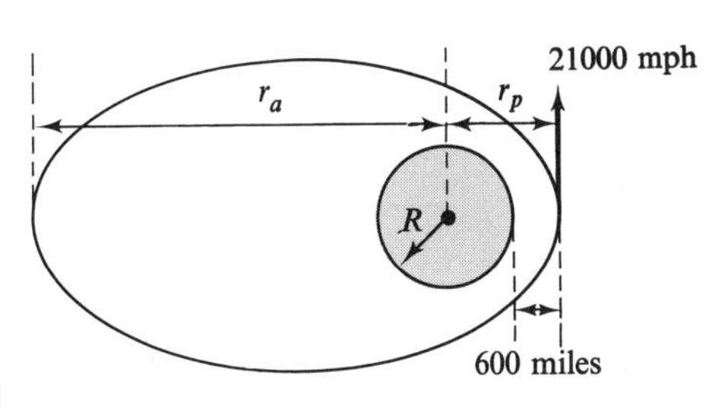

Figure 11-18

SOLUTION

(a) The initial distance of the satellite from the center of the earth, r_p, is

$$r_p = 3960 + 600 = 4560 \text{ miles.}$$

The eccentricity, from Eq. (k), is

$$e = \frac{r_p v_p^2}{GM} - 1$$

$$= \frac{4560(21{,}000)^2}{1.239(10)^{12}} - 1 = \underline{0.623.} \qquad \text{Ans.}$$

(b) The equation of the orbit can be expressed in a variety of forms. If Eq. (11-14a) is used, the value of h^2/GM must be determined. Since h is twice the constant areal speed, its value is

$$h = v_p r_p,$$

and the numerator of Eq. (11-14a) is

$$\frac{h^2}{GM} = \frac{v_p^2 r_p^2}{GM}$$

$$= \frac{(21{,}000)^2(4560)^2}{1.239(10)^{12}}$$

$$= 7400 \text{ miles.}$$

The equation of the orbit is

$$r = \frac{h^2/GM}{1 + e\cos\phi}$$

$$= \frac{7400}{1 + 0.623\cos\phi} \qquad \text{Ans.}$$

(c) The period can be obtained from Eq. (11-15) after the semimajor axis, a, has been obtained. Equation (j) for the semimajor axis gives

$$a = \frac{h^2}{GM}\left(\frac{1}{1-e^2}\right)$$

$$= 7400\left(\frac{1}{1-0.623^2}\right)$$

$$= 12{,}094 \text{ miles,}$$

and the period is

$$T = 2\pi\left(\frac{a^3}{GM}\right)^{1/2}$$

$$= 2\pi\left[\frac{12{,}094^3}{1.239(10)^{12}}\right]^{1/2}$$

$$= \underline{7.51 \text{ hr.}} \qquad \text{Ans.}$$

PROBLEMS

11-79 A satellite is launched with a velocity of 20,000 mph parallel to the earth's surface at an altitude of 300 miles. Determine
(a) The altitude at apogee.
(b) The distance from the center of the earth to the center of the elliptical path.

***11-80** A satellite is launched with a velocity of 34,000 km/hr parallel to the surface of the earth at an altitude of 650 km. Determine
(a) The altitude at apogee.
(b) The speed of the satellite when it is at the extremity of the minor axis. (*Note:* From analytical geometry $a^2 = b^2 + c^2$, where a is the semimajor axis, b is the semiminor axis, and c is the distance from the center of the ellipse to the center of the earth).

11-81 At burnout the satellite in Fig. P11-81 reached an altitude of 1200 miles with a velocity of 18,000 mph, making an angle of 80° with the vertical. Determine the perigee and apogee distances.

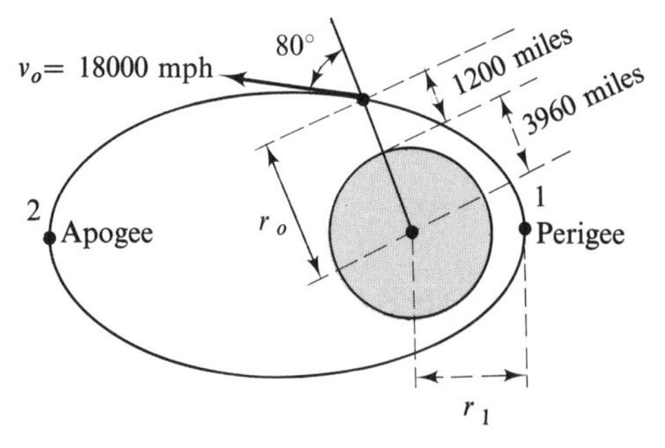

Figure P11-81

11-82 (a) A satellite is traveling in a circular orbit at an altitude of 1000 miles above the earth. Determine the velocity of the satellite.
(b) The velocity of the satellite of part (a) is increased to 19,000 mph by firing booster rockets. Determine the apogee distance for the resulting elliptical orbit.

11-83 A satellite is to be launched with a velocity parallel to the earth's surface at an altitude of 300 miles. Determine
(a) The escape velocity.
(b) The circular orbit velocity.

***11-84** The perigee and apogee distances of a satellite are 7200 and 19,000 km, respectively. Determine
(a) The eccentricity of the orbit.
(b) The velocity at perigee.

11-85 A satellite is launched with a velocity of 21,000 mph parallel to the earth's surface at an altitude of 600 miles. Determine
(a) The equation of the orbit.
(b) The apogee distance.

***11-86** A satellite has an elliptical orbit ($e = 0.2$) about the earth with a minimum distance from the center of the earth of 9600 km. Determine
(a) The distance from the center of the earth to the apogee.
(b) The period of the orbit.

11-87 A particle of mass m is subjected to a central force and travels on a path defined by the expression $r = \theta^2$. Develop the expression for the central force as a function of h and r.

11-88 The central force **F** of Fig. 11-15(a) is $-(km/r^3)\mathbf{e}_r$. For this value of **F**, determine the polar equation of the path of the particle of mass m.

***11-89** A satellite is launched with a velocity of 30,000 km/hr *parallel to the earth* at a distance of 7250 km from the center of the earth.

Determine
(a) The equation of the orbit.
(b) The eccentricity of the orbit.

11-90 The mass of the earth is approximately 81.3 times the mass of the moon. The radius of the moon is approximately 1080 miles. Determine the acceleration of gravity on the moon's surface.

11-91 Figure P11-91 represents the elliptical path of a satellite with one focus at the center of the earth, O, and the other focus at F. An ellipse has the property that the sum of the distances from any point on the ellipse to the two foci is equal to the length of the major axis; that is, $r_p + r_a = 2a$. Show that the length of the semiminor axis, b, is $b = \sqrt{r_p r_a}$, where r_p and r_a are the perigee and apogee distances.

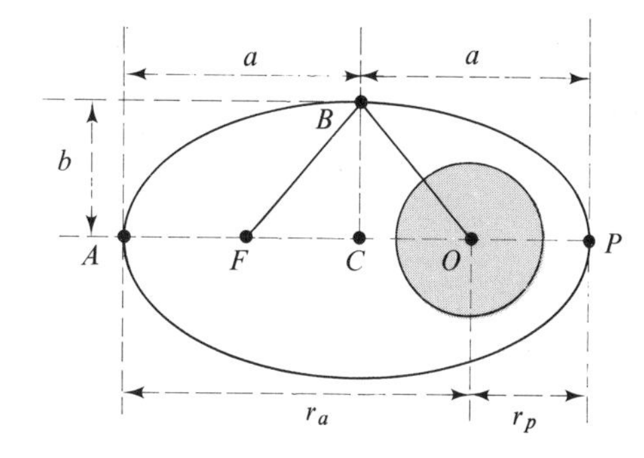

Figure P11-91

11-92 A satellite of the earth has an apogee distance of 10,000 miles and a perigee distance of 5000 miles. Determine
(a) The period of the orbit.
(b) The semiminor axis by using the relationship developed in Problem 11-91.

11-7 SYSTEMS WITH VARIABLE MASS

In previous articles, the mass of the system considered remained constant. There are, however, many dynamics problems such as those involving rockets, hoisting cables, and high-speed particles in which the mass of the system changes with time or in which some parts of the system gain or lose mass from other parts of the system. Variable-mass problems can be divided into two[4] types:

1. Problems in which the mass of the system varies because the number of individual particles, each of constant mass, in the system varies.
2. Problems in which masses of the individual particles change with time, while the number of particles remains constant.

The first variable-mass system is commonly encountered in rocket problems where the system to be studied is defined as the rocket shell plus the unburned fuel.

The second type occurs when the velocities of individual particles of the system become so large that the relativistic changes of mass with velocities must be considered. Relativistic changes in mass are not normally important unless the velocities involved are a significant fraction of the speed of light. Such velocities are attained by charged

[4] There could conceivably be a combination of these two types, but such a combination will not be considered.

particles in cyclotrons, X-ray machines, and television picture tubes. In this text, only variable-mass problems of the first type will be treated.[5]

The general procedure to be developed will apply to any body which is either acquiring or expelling mass. The following assumptions are made to simplify the derivation of the differential equation of motion and the subsequent analysis:

1. The motion of the system is along a straight line; consequently, the resultant of the external forces applied to the system is along the path of motion.
2. The acceleration of gravity, g, is constant.

The first assumption excludes any lateral maneuver of the system but permits investigation of the fundamental equations relating acceleration, velocity, and distance traveled along the path. These equations are dependent on the variation of mass of the body. Since the acceleration due to gravity is inversely proportional to the square of the distance from the center of the earth, the second assumption is accurate to within 5% for altitudes less than about 100 miles above the earth.

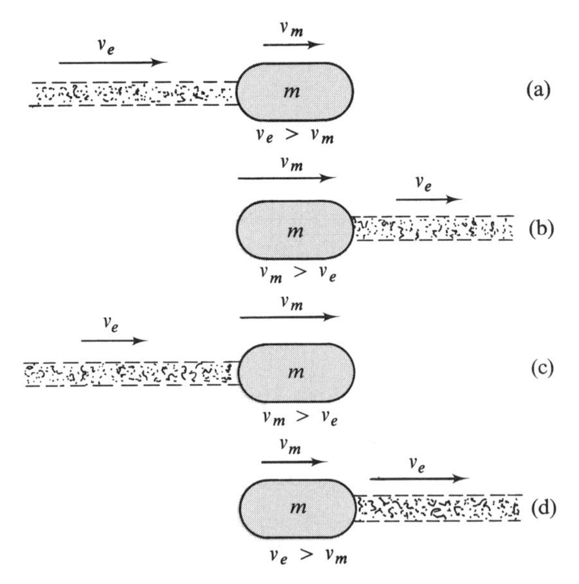

Figure 11-19

Four possible situations in which a large mass m is gaining or losing mass from a stream of particles are illustrated in Fig. 11-19. All velocities are assumed to be to the right, and any velocity to the left

[5] Those interested in relativistic mass changes are referred to books on the special theory of relativity, such as M. Russell Wehr and James A. Richards, *Physics of the Atom* (Reading, Mass.: Addison-Wesley Publishing Company, Inc., 1960), Chap. 5.

will be indicated by a minus sign. Figures 11-19(a) and 11-19(b) represent situations in which the large mass m is being increased, first because the stream of particles (with velocity $\mathbf{v}_e$) is to the left of m and $\mathbf{v}_e$ is greater than $\mathbf{v}_m$ and second because the stream is to the right of m and is being overtaken by m ($\mathbf{v}_m$ is greater than $\mathbf{v}_e$). Figures 11-19(c) and 11-19(d) represent cases in which mass is being expelled from m and would occur when m was being accelerated by firing a rocket engine [Fig. 11-19(c)] or decelerated by firing a retrorocket [Fig. 11-19(d)].

The equation of motion will be derived for the situation shown in Fig. 11-19(a) but will be valid for any of the four cases when consistent signs are used for velocities and rate of change of mass. Figure 11-20(a) shows the system at time t when the mass m has a velocity $\mathbf{v}_m$ and the increment of mass Δm has a velocity of $\mathbf{v}_e$. The forces $\mathbf{T}_1$ and $\mathbf{T}_2$ are the thrust or mutual reaction between m and Δm, and $\mathbf{R}$ is the resultant of the external forces (such as weight and aerodynamic drag) acting on the system. The diagram in Fig. 11-20(b) illustrates the system at time $(t+\Delta t)$ when Δm has joined m with a common velocity of $\mathbf{v}_m+\Delta\mathbf{v}_m$. The linear momentum of the system at time t is

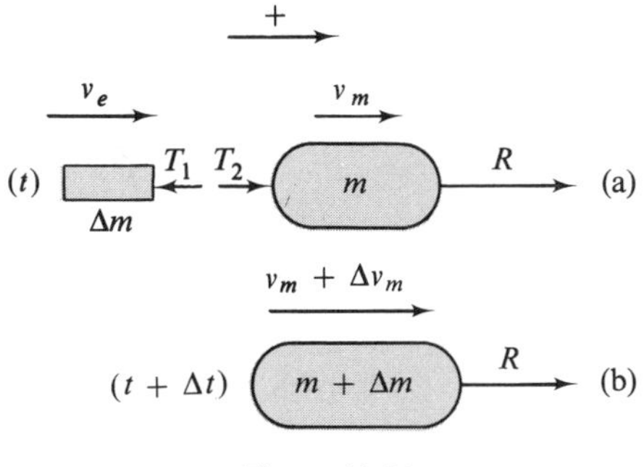

Figure 11-20

$$\mathbf{G}(t) = m\mathbf{v}_m + \Delta m\mathbf{v}_e,$$

and the momentum at $t+\Delta t$ is

$$\mathbf{G}(t+\Delta t) = (m+\Delta m)(\mathbf{v}_m + \Delta\mathbf{v}_m).$$

The impulse of the external forces (note that $\mathbf{T}_1$ and $\mathbf{T}_2$ are internal to the system) is equal to the change of the momentum of the system; that is,

$$\begin{aligned}\mathbf{R}\,\Delta t &= \mathbf{G}(t+\Delta t) - \mathbf{G}(t)\\ &= (m\mathbf{v}_m + m\,\Delta\mathbf{v}_m + \Delta m\mathbf{v}_m + \Delta m\,\Delta\mathbf{v}_m) - (m\mathbf{v}_m + \Delta m\mathbf{v}_e)\\ &= m\,\Delta\mathbf{v}_m - \Delta m(\mathbf{v}_e - \mathbf{v}_m) + \Delta m\,\Delta\mathbf{v}_m.\end{aligned}$$

As the time interval approaches zero the incremental quantities Δt, Δm, and $\Delta\mathbf{v}_m$ become the differential quantities dt, dm, and $d\mathbf{v}_m$, respectively, and the equation can be written as

$$\mathbf{R}\,dt = m\,d\mathbf{v}_m - dm(\mathbf{v}_e - \mathbf{v}_m) + dm\,d\mathbf{v}_e.$$

The last term in this expression is a second-order differential and can be neglected. When the equation is divided by dt it becomes

$$\begin{aligned}\mathbf{R} &= m\frac{d\mathbf{v}_m}{dt} - \frac{dm}{dt}(\mathbf{v}_e - \mathbf{v}_m)\\ &= m\dot{\mathbf{v}}_m - \dot{m}(\mathbf{v}_e - \mathbf{v}_m) = m\dot{\mathbf{v}}_m - \dot{m}\mathbf{v}_{e/m}. \qquad (11\text{-}16)\end{aligned}$$

If v_m is greater than v_e (the large mass is overtaking Δm), the equation becomes

$$\mathbf{R} = m\dot{\mathbf{v}}_m + \dot{m}(\mathbf{v}_m - \mathbf{v}_e) = m\dot{\mathbf{v}}_m + \dot{m}\mathbf{v}_{m/e}. \quad (11\text{-}16a)$$

When Δm is being expelled as in Fig. 11-19(c) or (d), $\dot{m}$ will be negative in Eq. (11-16) and (11-16a).

Note that in general $\mathbf{v}_e$ is not zero, although it may be in special cases, for example, when Δm is at rest before it is "picked up." If Newton's second law of motion is written in the form

$$\mathbf{R} = \frac{d}{dt}(m\mathbf{v}_m) = \dot{m}\mathbf{v}_m + m\dot{\mathbf{v}}_m, \quad (11\text{-}17)$$

it is seen that this expression agrees with Eq. (11-16) or (11-16a) *only* when $\mathbf{v}_e$ is zero, that is, when the initial velocity of the acquired mass or the final velocity of the expelled mass is zero.

The thrust T can be obtained by applying the impulse-momentum equation to the mass Δm in Fig. 11-20. When $\mathbf{i}$ is used as the unit vector to the right the impulse-momentum equation is

$$-T_1\,\Delta t\mathbf{i} = \Delta m(v_m + \Delta v_m - v_e)\mathbf{i}$$

from which

$$-T_1 = \lim_{\Delta t \to 0} \frac{\Delta m}{\Delta t}(v_m - v_e) = \dot{m}v_{m/e} = -\dot{m}v_{e/m}$$

when the higher-order term $\Delta m(\Delta v_m)$ is omitted. Therefore,

$$T_1 = \dot{m}v_{e/m},$$

where T_1 is the magnitude of the thrust force, and since $\dot{m}$ and $v_{e/m}$ are both positive when v_e is greater than v_m, T_1 is positive and is in the direction shown (to the left) in Fig. 11-20(a). The vector expression for the thrust force is $\mathbf{T}_1 = -\dot{m}\mathbf{v}_{e/m}$. The force $\mathbf{T}_2$, which has the same magnitude as $\mathbf{T}_1$ but acts to the right, is

$$\mathbf{T}_2 = \dot{m}v_{e/m}\mathbf{i} = \dot{m}\mathbf{v}_{e/m}. \quad (a)$$

The last equation can be combined with Eq. (11-16) to give

$$\mathbf{R} = m\dot{\mathbf{v}}_m - \mathbf{T}_2$$

or

$$m\dot{\mathbf{v}}_m = \mathbf{R} + \mathbf{T}_2. \quad (11\text{-}16b)$$

Again it should be noted that when mass is being expelled from m, as in a rocket, the factor $\dot{m}$ is negative in the preceding equations.

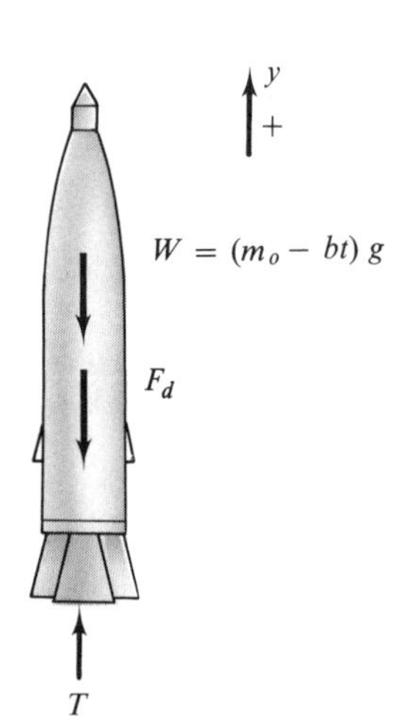

Figure 11-21

Consider the rocket in Fig. 11-21 with an

initial mass m_0 which burns fuel at a constant rate b; that is, $\dot{m} = -b$. The velocity of the burned gasses with respect to the rocket is $\mathbf{v}_{e/m}$ and is assumed to be constant. The mass of the rocket at time t is

$$m = m_0 - bt,$$

and the magnitude of the thrust is

$$T = \dot{m}v_{e/m} = bv_{e/m}.$$

Note that $\dot{m} = -b$ and that $\mathbf{v}_{e/m}$ is downward; therefore the force $\mathbf{T}_2$ is opposite $\mathbf{v}_{e/m}$; that is $\mathbf{T}_2$ acts upward on the rocket.

The external forces in the vertical direction are the weight $W = mg = (m_0 - bt)g$ and the aerodynamic drag force F_d, both downward. Equation (11-16b) becomes

$$(m_0 - bt)\dot{v}_m = -(m_0 - bt)g - F_d + T \tag{11-18a}$$

$$= -(m_0 - bt)g - F_d + bv_{e/m}. \tag{11-18b}$$

When any rocket is fired vertically, subject to the assumptions listed, Eq. (11-18a) can be applied if the thrust is known, or Eq. (11-18b) can be used if the relative speed of the exhaust gases is known. If the rocket travels horizontally along a track, the weight and track reaction will be normal to the velocity and thus will be omitted from the equation. The drag force will be zero in the absence of frictional and aerodynamic forces or may have a magnitude kv^n in the presence of an atmosphere where n varies with the speed of the rocket but is generally taken to be 1 or 2. During the analysis, it is usually helpful to use symbols for the constants as well as the variables. Numerical values can be substituted when the formula has been developed to yield the desired results.

EXAMPLE 11-11

The single-stage rocket in Fig. 11-21 has an initial weight of 8050 lb and is fired vertically from rest from the surface of the earth. The rocket contains 6440 lb of fuel, which is consumed in 80 sec. The speed of the exhaust gases relative to the rocket is 5000 fps. At the time of burnout, determine

(a) The acceleration of the rocket in g units.
(b) The velocity of the rocket in miles per hour.
(c) The altitude of the rocket in miles.

Neglect drag forces.

SOLUTION

(a) When air resistance is neglected, the differential equation of the motion, Eq. (11-18b), becomes

$$(m_0 - bt)\dot{v}_m = -(m_0 - bt)g + bv_{e/m}$$

from which

$$\frac{dv}{dt}=\frac{bv_{e/m}}{(m_0-bt)}-g. \tag{a}$$

From the statement of the problem, $m_0=8050/32.2=250$ slugs and $b=(6440/32.2)/80=2.5$ slugs per sec. Thus, at burnout, when $t=80$ sec, Eq. (a) gives

$$\frac{dv}{dt}=a=\frac{(2.5)(5000)}{250-(2.5)(80)}-g=250-g$$

or

$$\mathbf{a}=\left[\left(\frac{250}{32.2}\right)g-g\right]\mathbf{j}=\underline{6.76g\uparrow}. \qquad \text{Ans.}$$

(b) When $t=0$, $v=0$, and Eq. (a) can be integrated to obtain the velocity. Thus

$$\int_0^v dv=v_{e/m}\int_0^t\frac{b\,dt}{m_0-bt}-g\int_0^t dt$$

and

$$\Big[v\Big]_0^v=v_{e/m}\Big[-\ln(m_0-bt)\Big]_0^t-\Big[gt\Big]_0^t$$

from which

$$v=-v_{e/m}\ln\left(\frac{m_0-bt}{m_0}\right)-gt. \tag{b}$$

When $t=80$ sec, the speed of the rocket is

$$v=-(5000)\ln\left[\frac{250-(2.5)(80)}{250}\right]-32.2(80)=5471\text{ fps}$$

and

$$\mathbf{v}=5471\left(\frac{3600}{5280}\right)\mathbf{j}=\underline{3730\text{ mph}\uparrow}. \qquad \text{Ans.}$$

(c) When v in Eq. (b) is replaced by dy/dt, it can be integrated between limits $y=0$ when $t=0$ and $y=y$ when $t=t$. That is,

$$\int_0^y dy=-v_{e/m}\int_0^t\ln\left(\frac{m_0-bt}{m_0}\right)dt-g\int_0^t t\,dt.$$

The first integral to the right of the equal sign can be expressed in the form $\int\ln u\,du$, which is equal to $u(\ln u-1)$, if dt is multiplied by $-b/m_0$ and the coefficient of the integral is multiplied by $-m_0/b$. The equation becomes

$$\int_0^y dy=\frac{m_0v_{e/m}}{b}\int_0^t\left[\ln\left(\frac{m_0-bt}{m_0}\right)\right]\left(\frac{-b}{m_0}dt\right)-g\int_0^t t\,dt.$$

This equation can be integrated to give

$$\Big[y\Big]_0^y=\frac{m_0v_{e/m}}{b}\left\{\left(\frac{m_0-bt}{m_0}\right)\left[\ln\left(\frac{m_0-bt}{m_0}\right)-1\right]\right\}_0^t-g\left[\frac{t^2}{2}\right]_0^t$$

or

$$y=\frac{m_0 v_{e/m}}{b}\left\{\left(\frac{m_0-bt}{m_0}\right)\left[\ln\left(\frac{m_0-bt}{m_0}\right)-1\right]+1\right\}-\frac{1}{2}gt^2.$$

The value of y, when $t=80$ sec, is

$$y=\frac{(250)(5000)}{(2.5)}\left\{\left(\frac{250-(2.5)(80)}{250}\right)\left[\ln\left(\frac{250-(2.5)(80)}{250}\right)-1\right]+1\right\}$$

$$-\frac{32.2}{2}(80)^2=136{,}000\text{ ft}$$

from which

$$\mathbf{y}=(136{,}000/5280)\mathbf{j}=\underline{25.8\text{ miles}\uparrow}. \qquad \text{Ans.}$$

A more complete discussion of the performance of rockets can be found in various advanced texts on rocket performance.[6]

Another example of variable-mass problems arises in the motion of chains or flexible cables. When the velocity of the acquired, or lost, mass is zero before it becomes part of the moving system, or after it has left the system, Eq. (11-17) may be applied directly.

If a chain is allowed to fall link by link through a hole in a support [see Fig. 11-22(a)] or is raised from a fixed plane [see Fig. 11-22(b)], the only external forces acting on the *moving* part of the chain are the weight of the moving part of the chain and any force on the end of the chain away from the support. The floor supports the stationary part of the chain but does not act on the moving links.

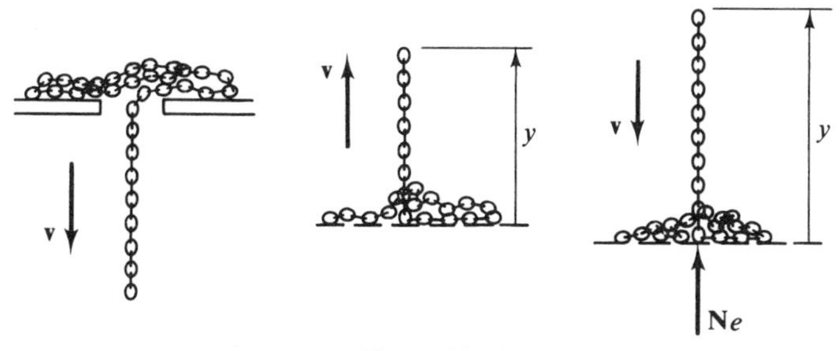

Figure 11-22

When chains are being lowered or dropped onto a fixed plane, as in Fig. 11-22(c), a normal upward force is exerted by the plane on the lowest link of the chain in order to bring it to rest as it strikes the plane. Only that portion of the normal force (the total force includes the weight of the chain on the plane) which changes the velocity of the moving link to zero is shown as N_e. If the chain in Fig. 11-22(c) weighs

---

[6] See, for example, Arthur I. Berman, *The Physical Principles of Astronautics* (New York: John Wiley & Sons, Inc. 1961).

q lb per ft and is moving downward with a speed $\dot{y}$, the mass of the chain brought to rest during a time interval Δt is $(q/g)(\Delta y)$. The velocity of each link changes from $-\dot{y}\mathbf{j}$ to zero during the time Δt. Therefore, the principle of linear impulse and momentum gives

$$(\mathbf{I}_L)_y = N_e\,\Delta t\mathbf{j} = \Delta\mathbf{G} = \frac{q}{g}(\Delta y)[0-(-\dot{y}\mathbf{j})]$$

or

$$N_e = \frac{q}{g}\frac{\Delta y}{\Delta t}\dot{y},$$

which becomes, as Δt approaches zero,

$$N_e = \frac{q}{g}\dot{y}^2.$$

The free-body diagram of the moving portion of the chain being lowered onto a plane must include not only the force (if any) applied to the upper end of the chain and the weight of the vertical portion of the chain but also a vertical force, with a magnitude of $(q/g)\dot{y}^2$, applied to the lower end.

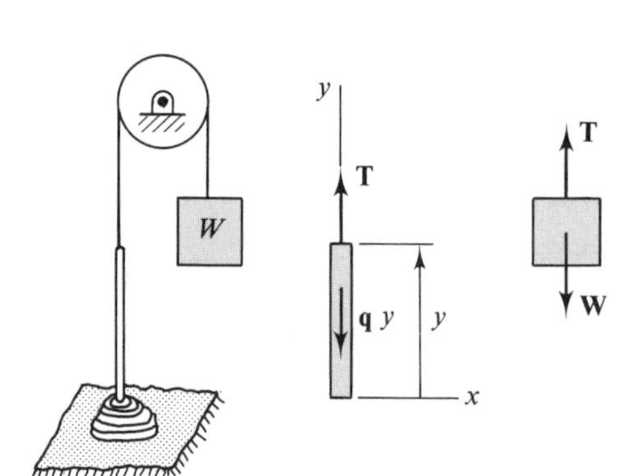

Figure 11-23

EXAMPLE 11-12

A 10-ft cable weighing 4 lb per ft is loosely coiled on the floor. Attached to one end of the cable is a light cord which passes over a smooth pulley to a 16-lb weight as shown in Fig. 11-23(a). A 2-ft length of the cable is initially raised from the floor, and the system is released from rest. Determine the maximum height reached by the upper end of the cable.

SOLUTION

A free-body diagram of the moving portion of the cable is shown in Fig. 11-23(b). It is assumed that y_{max} is less than the length of the cable (10 ft) so that the weight of the moving portion is given by qy. The resultant external force on the body is $(T-qy)\mathbf{j}$. Since the velocity of the part of the cable on the floor is zero before it leaves the floor, Eq. (11-17) applies, and the equation of motion is

$$\mathbf{R} = \frac{d}{dt}(m\mathbf{v}_m),$$

which becomes, in scalar form,

$$T - qy = \frac{d}{dt}\left(\frac{qy}{g}\dot{y}\right). \tag{a}$$

The acceleration of the weight W is $-\ddot{y}\mathbf{j}$, and its equation of motion [see Fig. 11-23(c)] is

$$T - W = \frac{W}{g}(-\ddot{y}) = -\frac{d}{dt}\left(\frac{W}{g}\dot{y}\right). \tag{b}$$

The simultaneous solution of Eq. (a) and (b) gives

$$W - qy = \frac{d}{dt}\left(\frac{qy}{g}\dot{y}\right) + \frac{d}{dt}\left(\frac{W}{g}\dot{y}\right)$$

$$= \frac{1}{g}\frac{d}{dt}[(W + qy)\dot{y}].$$

This differential equation is easier to solve if the right-hand side is rewritten as

$$\frac{du}{dt} = \frac{du}{dy}\frac{dy}{dt} = \dot{y}\frac{du}{dy},$$

where $u = (W - qy)\dot{y}$. Thus,

$$W - qy = \frac{\dot{y}}{g}\frac{d}{dy}[(W + qy)\dot{y}]$$

or

$$(W - qy)\,dy = \left(\frac{\dot{y}}{g}\right)d[(W + qy)\dot{y}]. \tag{c}$$

When Eq. (c) is multiplied by $W + qy$, the right-hand side assumes the form $u\,du$, which is the differential of $u^2/2$. Therefore,

$$(W + qy)(W - qy)\,dy = \frac{1}{g}[(W + qy)\dot{y}]d[(W + qy)\dot{y}]$$

or

$$(W^2 - q^2y^2)\,dy = \frac{1}{2g}\,d[(W + qy)\dot{y}]^2.$$

The integral of this equation is

$$W^2y - \frac{q^2y^3}{3} + C = \frac{[(W + qy)\dot{y}]^2}{2g}. \tag{d}$$

From the given data, when $t = 0$, $y = y_0 = 2$ ft and $\dot{y} = 0$; therefore,

$$W^2y_0 - \frac{q^2y_0^3}{3} + C = 0$$

and

$$C = \frac{q^2y_0^3}{3} - W^2y_0.$$

Consequently Eq. (d) becomes

$$W^2(y - y_0) - \frac{q^2(y^3 - y_0^3)}{3} = \frac{[(W + qy)\dot{y}]^2}{2g}.$$

The maximum height of the cable occurs when $\dot{y} = 0$; therefore,

$$W^2(y_{max} - y_0) - \frac{q^2(y_{max}^3 - y_0^3)}{3} = 0$$

or

$$(y_{max}-y_0)\left[W^2-\frac{q^2(y_{max}^2+y_{max}y_0+y_0^2)}{3}\right]=0.$$

Since $y_{max}-y_0\neq 0$, the second factor must be zero; that is,

$$y_{max}^2+y_{max}y_0+y_0^2-\frac{3W^2}{q^2}=0.$$

When the values $y_o=2$ ft, $W=16$ lb, and $q=4$ lb per ft are substituted, this equation becomes

$$y_{max}^2+2y_{max}-44=0$$

from which

$$y_{max}=5.708 \quad \text{or} \quad -7.708.$$

The negative root has no meaning for this problem; therefore the solution is

$$y_{max}=\underline{5.71\text{ ft.}} \qquad \text{Ans.}$$

Since y_{max} is less than 10 ft, the assumption that some of the cable remains on the floor is justified. If y_{max} had been more than 10 ft, it would be necessary to determine the velocity when $y=10$ ft after which the acceleration would be constant and the added distance the cable would rise could be determined directly.

PROBLEMS

11-93 Determine the weight of the fuel consumed in 2 min by a rocket engine producing a thrust of 1,400,000 lb if the exhaust gas has a velocity of 7000 fps relative to the rocket.

***11-94** A booster rocket is used to give additional starting power to an airplane. The fuel is burned at a rate of 4 kg/s and ejected with a relative velocity of 1500 m/s. Determine the additional starting thrust generated by the rocket.

11-95 A fueled rocket with a total mass of 50 slugs, including 40 slugs of fuel, is fired vertically. The fuel is consumed at the rate of 2 slugs per sec and is ejected with a relative velocity of 6000 fps. Neglect the effect of air friction. Determine
(a) The acceleration of the rocket, expressed in terms of g, immediately after ignition.
(b) The velocity of the rocket at time of burnout.

11-96 The two-stage rocket shown in Fig. P11-96 is fired vertically. The shell of rocket A weighs 100 lb and contains 600 lb fuel. The shell of rocket B weighs 50 lb and contains 450 lb of fuel. The payload C weighs 50 lb. When rocket A expends its fuel, shell A is released, and rocket B is immediately fired. Rocket A consumes its fuel at the rate of 20 lb per sec, and rocket B burns its fuel at the rate of 15 lb per sec. The relative velocity of the exhaust for both rockets is 7000 fps. Neglect air friction. Determine

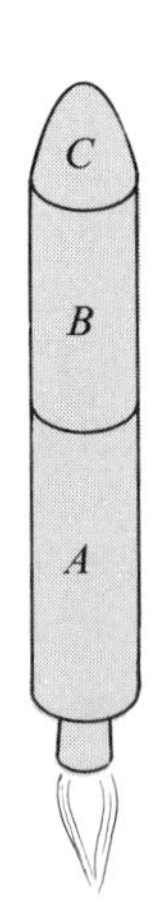

Figure P11-96

(a) The speed when shell A is released.
(b) The maximum speed attained by the payload C.

11-97 A small CO_2 rocket travels along a horizontal frictionless track. The initial and final weights of the rocket are 12 and 4 lb, respectively. The CO_2 is discharged at the uniform rate of 2 lb per sec with a velocity of 500 fps relative to the rocket. The rocket is subjected to a drag force $F_D = -kv$, where $k = 0.02$ lb-sec per ft and v is the velocity in feet per second. Determine the maximum velocity attained by the rocket. The rocket starts from rest.

11-98 Determine the distance traveled by the rocket in Problem 11-97 during the time required for the velocity to change from zero to its maximum value.

11-99 A fueled rocket with a total mass of 40 slugs, including 30 slugs of fuel, approaches the moon along a flight path which is perpendicular to the moon's orbital path. The speed of the rocket is 8000 fps when a retrorocket is fired to bring the payload to rest as it touches the moon's surface. If the moon's gravitational attraction is one-sixth that of the earth's, what must be the burning time of the rocket, and at what altitude above the moon's surface must it be fired? The velocity of the exhaust gases with respect to the rocket is 7000 fps. Assume that the attraction of the moon for the rocket does not vary with the distance from the rocket to the moon.

11-100 One end of a pile of a long, uniform chain falls through a hole in its support and pulls the remaining links after it in a steady flow. Assume that the links which are initially at rest acquire the velocity of the falling chain without frictional resistance or interference from the adjacent links. Determine the velocity of the moving links after 10 ft of the chain has fallen through the hole.

***11-101** A body with a mass of 1 kg is dropped from rest and accumulates mass at a rate of 0.1 kg/m of travel. The mass accumulated was at rest before becoming attached to the body. Determine the velocity of the body after it has dropped 10 m. Neglect air resistance.

11-102 A uniform chain 16 ft long and weighing 0.1 lb per ft is coiled neatly on a table. A 21-ft cord is fastened to one end of the chain and passes over a small pulley 18 ft above the coil. The other end of the cord is connected to a 2-lb body A. Determine the velocity of body A just before it hits the table if it is released from rest 15 ft above the table.

11-103 Body A weighs 4 lb and is attached to an 18-ft uniform chain which weighs 0.1 lb per ft. The chain is coiled on the floor. Body A is projected upward from floor level with an initial velocity of 40 fps. Determine the maximum height to which body A will rise.

11-104 Solve Problem 11-103 if the initial velocity of body A is 60 fps upward.

***11-105** A portion of the uniform chain in Fig. P11-105 is coiled on a table, and the other portion passes over a small pulley and hangs down on the other side a distance x. When $x = 2$ m, the chain is released from rest. Determine the relationship between the velocity of the chain at some subsequent time and the length x.

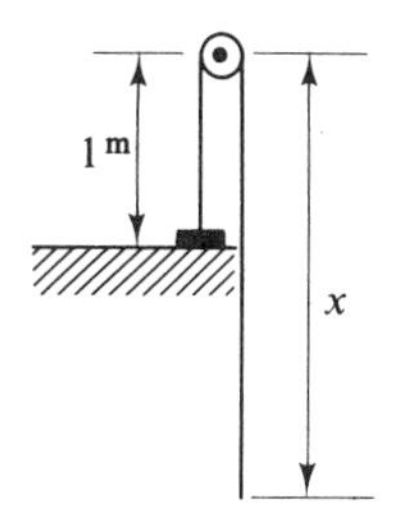

Figure P11-105

11-106 A uniform chain 20 ft long and weighing 0.1 lb per ft is partly coiled on the floor. A cord is fastened to the upper end of the chain and passes over a small pulley 12 ft above the coiled chain. The other end of the cord is connected to body A, which weighs 0.5 lb. The body is released from rest when 10 ft of the chain is suspended in air. Determine the velocity of body A as a function of the length of the chain suspended above the floor.

11-8 CLOSURE

The principle of linear impulse and linear momentum equates the linear impulse of the forces acting on a body or system of particles to the change of linear momentum of the body or system of particles. The principle of angular impulse and angular momentum equates the angular impulse of the forces acting on a body or system of particles with respect to a point fixed in space to the change of angular momentum of the particle with respect to the corresponding point. When either the linear or angular impulse is zero, the corresponding linear or angular momentum is unchanged or conserved.

The principle of angular momentum equates the time rate of change of angular momentum of a particle, with respect to a point fixed in space, to the sum of the moments of the external forces acting on the particle with respect to the point. When the resultant moment is zero with respect to any fixed point the angular momentum is constant or conserved with respect to the point.

The concepts of linear and angular impulse and momentum are particularly useful in the analysis of variable-mass problems and of problems involving elastic impact and central-force motion. Since impulse and momentum are vector quantities, it is usually desirable to cut the cords, pull the pins, and perform other acts of liberation so as to work with free-body diagrams of separate bodies.

The coefficient of restitution is the ratio of the magnitude of the relative velocity of departure of two colliding bodies to the magnitude of their relative velocity of approach.

Three methods of solving problems in kinetics have been developed in Chapters 9, 10, and 11. The method of force, mass, and acceleration can be used to obtain instantaneous values of forces and acceleration. The principles of force, mass, and acceleration, together with the equations of kinematics, can also be used when a time interval, a distance traveled, and/or a change in velocity are involved.

The principle of work and kinetic energy equates work done by the forces acting on a body or system of bodies to the change in kinetic energy of the system. Thus it is the logical method when the problem involves forces, distances, and velocities. It is particularly useful when the forces and acceleration of the body vary with the position of the body, for example, a body which is acted on by a spring or a body swinging on the end of a cord.

The method of impulse and momentum is concerned with forces, time intervals, and changes in velocity. It can be applied to problems involving these quantities for either constant or variable forces, although it is particularly useful when an applied force is specified as a function of time. This method is the only one whose application to fluid flow and variable-mass problems was discussed, although the method of force, mass, and acceleration can be used for such problems with necessary modifications.

Neither the method of impulse and momentum nor that of work and kinetic energy gives instantaneous values of acceleration, and the method of work and energy does not give even an average value of forces which do no work. In any problem, it is important that a careful analysis be made to ascertain the given data and the unknown quantities to be obtained before a method of solution is selected. Two of the methods may sometimes be used to supplement each other. For example, in a ballistic pendulum, the principles of work and kinetic energy and of conservation of linear momentum can both be used to good advantage. When the problem can be solved completely by two different methods, they provide a reliable check on the solution.

PROBLEMS

11-107 The coefficient of friction between the 200-lb block in Fig. P11-107 and the horizontal surface is 0.4. The force $20t$ is in pounds when t is in seconds. When $t = 0$ the block has a velocity of 40 fps to the left. Determine
(a) The time interval that the block will *not* be moving.
(b) The velocity of the block when $t = 6$ sec.

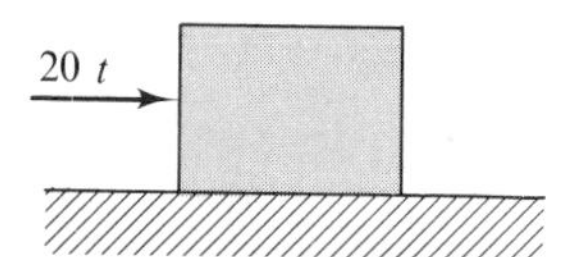

Figure P11-107

11-108 The 12.88-lb body A in Fig. P11-108 is connected to the 6.44-lb body B by a rope passing over the two pulleys C and D of negligible mass. The force $2.0t$ is in pounds when t is in seconds. When $t = 0$, B has a velocity of 10 fps upward. Determine the velocity of body B when $t = 5$ sec.

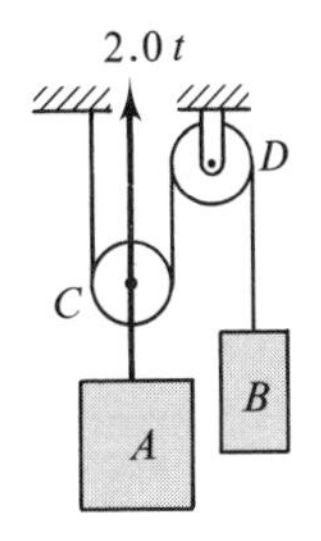

Figure P11-108

11-109 A bead of mass m, attached to an inextensible cord of negligible mass, is revolving on a smooth horizontal table as shown in Fig. P11-109. The cord passes through a hole in the table and is being pulled downward at a constant rate u by the force P. At the instant shown, the bead is at a distance R from the hole, and the magnitude of the component of the velocity of the bead perpendicular to the cord is v. Determine
(a) The kinetic energy of the bead in the position shown.
(b) The work done by the force P in decreasing the distance of the bead from the hole to $R/2$.

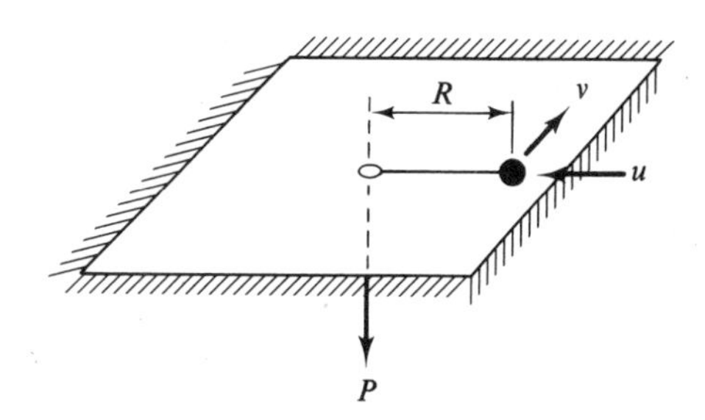

Figure P11-109

11-110 The 9.66-lb body A in Fig. P11-110 slides on a smooth horizontal surface and has a velocity of 20 fps to the right when it strikes the 6.44-lb body B, which is at rest. The coefficient of restitution is 0.5. After impact body B moves up the circular surface. Determine the reaction of the curved surface on body B just as it starts to move up the circular surface (position B in the diagram).

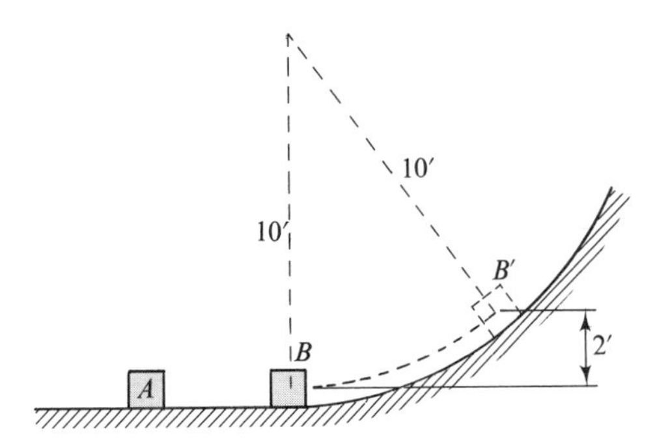

Figure P11-110

11-111 Using the data given in Problem 11-110, determine the reaction of the curved surface on body B when it has moved to position B'.

11-112 Particle A weighs 16.1 lb and moves on a smooth horizontal surface as shown in Fig. P11-112. Throughout the motion a force $F = kr^3$, where F is in pounds when r is in feet, acts on the particle and is always directed toward the fixed point O. When the particle is at its maximum distance from O the magnitude of its velocity is 20 fps. Determine the value of k.

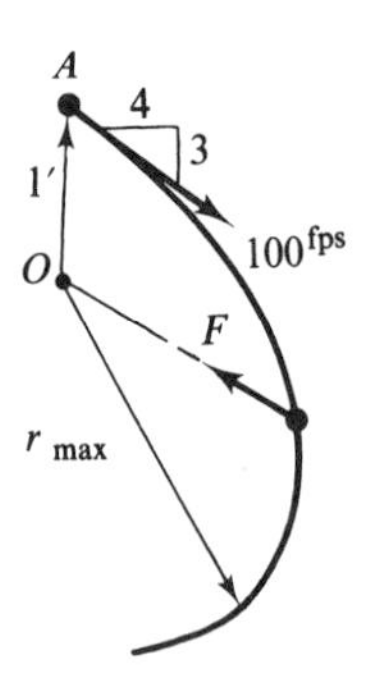

Figure P11-112

***11-113** The two-stage rocket shown in Fig. P11-113 is fired vertically. The shell of each rocket, A and B, has a mass of 50 kg, and each rocket contains 500 kg of fuel. The payload C has a mass of 20 kg. When rocket A consumes its fuel, shell A is released and rocket B is immediately fired. Both rockets burn fuel at a rate of 25 kg/s, and the relative velocity of efflux for both rockets is 2100 m/s. Neglect air friction. Determine

(a) The thrust immediately after rocket A is fired.

(b) The velocity just after rocket A has consumed its fuel.

(c) The maximum speed attained by the payload C.

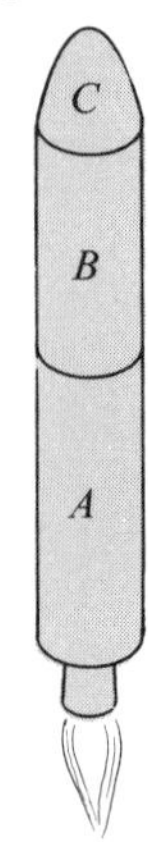

Figure P11-113

11-114 A snow plow with a rotary blade [Fig. P11-114(a)] is moving with a constant speed of 15 fps and throwing snow with a velocity of 30 fps relative to the plow at an angle of 30° as shown in Fig. P11-114(b). If 300 lb of snow per second is thrown by the plow, determine the magnitude of the forces F and R which must be exerted due to the snow removal.

11-115 A satellite is launched with a velocity of 20,000 mph parallel to the earth's surface at an altitude of 500 miles. Determine

(a) The eccentricity of the orbit.

(b) The equation of the orbit.

(c) The period of the orbit.

11-116 The 8.05-lb body A in Fig. P11-116(a) is moving to the right on a smooth horizontal plane with a velocity of 20 fps when it strikes the dynamic force gage B. The contact force between the gage and body A is shown in Fig. P11-116(b). Determine

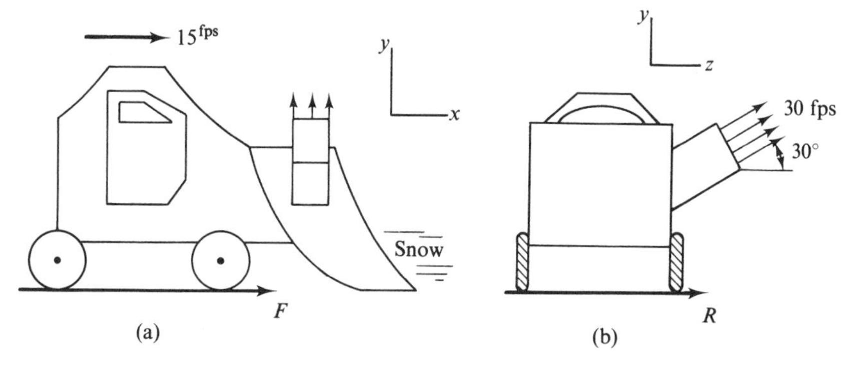

Figure P11-114

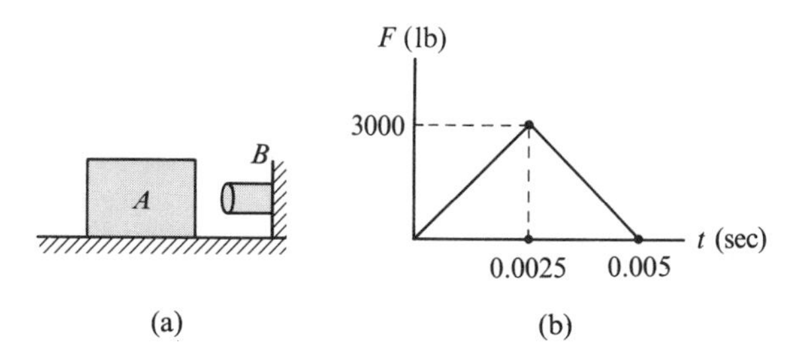

Figure P11-116

(a) The velocity of body A after impact.
(b) The coefficient of restitution between A and the gage.

***11-117** Body A in Fig. P11-117 has a mass of 20 kg, and the cart B has a mass of 80 kg. The coefficient of friction between A and B is 0.20, and the coefficient of restitution is zero. The bodies are at rest when the force P (magnitude = 300 N) is applied. Determine
(a) The velocity of B when it has moved 4.5 m to the left.
(b) The velocity of A when it has moved 4.5 m to the left.
(c) The velocity of A just after it collides with the right end of B.

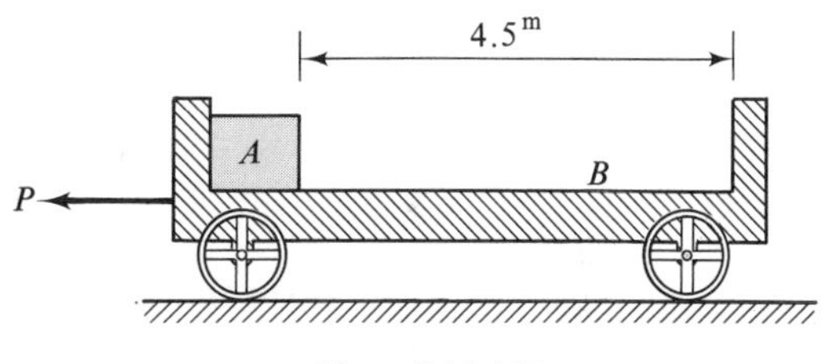

Figure P11-117

(d) The work done by the friction force on each body while A is sliding on B.
(e) The energy dissipated by the plastic collision.

PART B–RIGID BODIES

11-9 ANGULAR MOMENTUM OF A RIGID BODY OR SYSTEM OF PARTICLES

The angular momentum of a particle with respect to a fixed point was defined in Art. 11-5. In general, *the angular momentum of a*

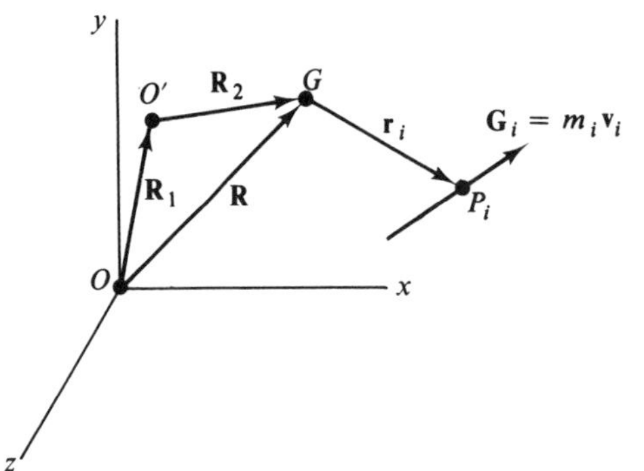

Figure 11-24

particle with respect to any point, fixed or moving, is the moment of the linear momentum of the particle with respect to that point. The angular momentum of a system of particles about any point is the sum of the angular momenta of the individual particles. Figure 11-24 represents a system of n particles (only the ith particle P_i is shown on the diagram) where G is the mass center of the particles. The XYZ axes are fixed in space, and O' is any point in space. The linear momentum of P_i is $m_i\mathbf{v}_i$ as shown, and the angular momentum of P_i with respect to point O' is the vector

$$(\mathbf{H}_{O'})_i = (\mathbf{R}_2 + \mathbf{r}_i) \times \mathbf{G}_i = (\mathbf{R}_2 + \mathbf{r}_i) \times m_i\mathbf{v}_i. \qquad \text{(a)}$$

The velocity of P_i in terms of its position vector is

$$\mathbf{v}_i = \frac{d}{dt}(\mathbf{R} + \mathbf{r}_i) = \dot{\mathbf{R}} + \dot{\mathbf{r}}_i.$$

The angular momentum of the system of particles with respect to O' is

$$\mathbf{H}_{O'} = \sum_{i=1}^{n} (\mathbf{R}_2 + \mathbf{r}_i) \times m_i(\dot{\mathbf{R}} + \dot{\mathbf{r}}_i). \qquad \text{(11-19)}$$

When O' is moved to G the vector $\mathbf{R}_2$ becomes zero, and the angular momentum of the system of particles with respect to G becomes

$$\mathbf{H}_G = \sum_{i=1}^{n} \mathbf{r}_i \times m_i(\dot{\mathbf{R}} + \dot{\mathbf{r}}_i)$$

$$= \left(\sum_{i=1}^{n} m_i\mathbf{r}_i\right) \times \dot{\mathbf{R}} + \sum_{i=1}^{n} m_i\mathbf{r}_i \times \dot{\mathbf{r}}_i.$$

The first term in the preceding equation is zero since the summation is the first moment of the mass with respect to its mass center. The vector $\dot{\mathbf{r}}_i$ is the velocity of the ith particle relative to G. The angular momentum of the system with respect to the mass center thus becomes

$$\mathbf{H}_G = \sum_{i=1}^{n} m_i\mathbf{r}_i \times \dot{\mathbf{r}}_i. \qquad \text{(b)}$$

When O' is selected as the fixed point O, $\mathbf{R}_2$ is equal to $\mathbf{R}$, and Eq. (11-19) becomes

$$\mathbf{H}_O = \sum_{i=1}^{n} (\mathbf{R} + \mathbf{r}_i) \times m_i(\dot{\mathbf{R}} + \dot{\mathbf{r}}_i)$$

$$= \mathbf{R} \times \dot{\mathbf{R}} \sum_{i=1}^{n} m_i + \mathbf{R} \times \frac{d}{dt}\left(\sum_{i=1}^{n} m_i\mathbf{r}_i\right) + \left(\sum_{i=1}^{n} \mathbf{r}_i m_i\right) \times \dot{\mathbf{R}}$$

$$+ \sum_{i=1}^{n} m_i\mathbf{r}_i \times \dot{\mathbf{r}}_i.$$

The summation $\sum_{i=1}^{n} m_i\mathbf{r}_i$ is zero in the second and third terms of the preceding equation, and when Eq. (b) is substituted for the last term the equation becomes

$$\mathbf{H}_O = \mathbf{R}\times\dot{\mathbf{R}}m + \mathbf{H}_G = \mathbf{R}\times m\mathbf{v}_G + \mathbf{H}_G. \tag{c}$$

Equation (c) demonstrates that the angular momentum of any system of particles with respect to a fixed point O is the sum of the moment of the linear momentum of the mass of the system considered concentrated at the mass center and moving with the velocity of the mass center added to the angular momentum of the particles of the system with respect to the mass center.

Equation (11-19) was developed for any system of particles. When the particles constitute a rigid body the velocities of the various particles can be expressed in terms of the angular velocity of the body and the linear velocity of the mass center. Figure 11-25 represents a rigid body in which P is a particle of mass dm and G is the mass center of the body. The XYZ axes are fixed in space, and the xyz coordinate system is fixed in the body with its origin at G.

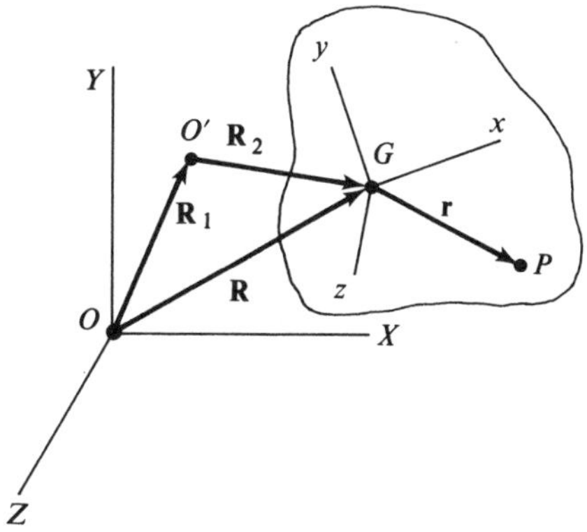

Figure 11-25

From Eq. (8-27) the absolute velocity of particle P is

$$\mathbf{v}_P = \dot{\mathbf{R}} + \mathbf{v}_{P/xyz} + \boldsymbol{\omega}\times\mathbf{r}$$

where $\mathbf{v}_{P/xyz}$ is the velocity of P relative to the moving axes and is zero in this case since P does not move relative to the xyz axes and $\boldsymbol{\omega}$ is the angular velocity of the body (the xyz axes). The angular momentum of the particle P with respect to point O' (either fixed or moving) is

$$\begin{aligned} d\mathbf{H}_{O'} &= (\mathbf{R}_2+\mathbf{r})\times(\mathbf{v}_P\,dm) \\ &= (\mathbf{R}_2+\mathbf{r})\times(\dot{\mathbf{R}}+\boldsymbol{\omega}\times\mathbf{r})\,dm \\ &= [\mathbf{R}_2\times\dot{\mathbf{R}}+\mathbf{R}_2\times(\boldsymbol{\omega}\times\mathbf{r})+\mathbf{r}\times\dot{\mathbf{R}}+\mathbf{r}\times(\boldsymbol{\omega}\times\mathbf{r})]\,dm. \end{aligned}$$

The angular momentum of the rigid body can be obtained by integrating the preceding expression. The quantities $\mathbf{R}_2$, $\dot{\mathbf{R}}$, and $\boldsymbol{\omega}$ are constant for all elements of mass and can thus be removed from the integrals. The angular momentum of the body is

$$\mathbf{H}_{O'} = \mathbf{R}_2\times\dot{\mathbf{R}}\int_m dm + \mathbf{R}_2\times\left(\boldsymbol{\omega}\times\int_m \mathbf{r}\,dm\right) + \left(\int_m \mathbf{r}\,dm\right)\times\dot{\mathbf{R}} + \int_m \mathbf{r}\times(\boldsymbol{\omega}\times\mathbf{r})\,dm.$$

The quantity $\int_m \mathbf{r}\,dm$ is zero since $\mathbf{r}$ is measured from the mass center of the body and $\mathbf{H}_{O'}$ reduces to

$$\mathbf{H}_{O'} = \mathbf{R}_2 \times \dot{\mathbf{R}} m + \int_m \mathbf{r} \times (\boldsymbol{\omega} \times \mathbf{r})\, dm$$

$$= \mathbf{R}_2 \times m\mathbf{v}_G + \int_m \mathbf{r} \times (\boldsymbol{\omega} \times \mathbf{r})\, dm \tag{11-20}$$

Equation (11-20) gives the angular momentum of a rigid body at any instant with respect to any point O'. If O' is moved to O, $\mathbf{R}_2$ becomes $\mathbf{R}$, and Eq. (11-20) becomes

$$\mathbf{H}_O = \mathbf{R} \times (m\mathbf{v}_G) + \int_m \mathbf{r} \times (\boldsymbol{\omega} \times \mathbf{r})\, dm. \tag{11-21}$$

If O' is moved to G, $\mathbf{R}_2$ becomes zero, and Eq. (11-20), for moments with respect to the mass center, is

$$\mathbf{H}_G = \int_m \mathbf{r} \times (\boldsymbol{\omega} \times \mathbf{r})\, dm. \tag{11-22}$$

Although Eq. (11-20)–(11-22) provide concise expressions for the moment of momentum of a rigid body, they are more useful when the integral is expressed in terms of moments and products of inertia. The quantity $\mathbf{r}$ can be written as

$$\mathbf{r} = x\mathbf{i} + y\mathbf{j} + z\mathbf{k}$$

and $\boldsymbol{\omega}$ as

$$\boldsymbol{\omega} = \omega_x \mathbf{i} + \omega_y \mathbf{j} + \omega_z \mathbf{k}$$

in which case $\boldsymbol{\omega} \times \mathbf{r}$ becomes

$$\boldsymbol{\omega} \times \mathbf{r} = (\omega_y z - \omega_z y)\mathbf{i} + (\omega_z x - \omega_x z)\mathbf{j} + (\omega_x y - \omega_y x)\mathbf{k}.$$

The vector triple product $\mathbf{r} \times (\boldsymbol{\omega} \times \mathbf{r})$ reduces to

$$\begin{aligned}\mathbf{r} \times (\boldsymbol{\omega} \times \mathbf{r}) &= [\omega_x(y^2+z^2) - \omega_y xy - \omega_z xz]\mathbf{i} \\ &\quad + [-\omega_x xy + \omega_y(z^2+x^2) - \omega_z yz]\mathbf{j} \\ &\quad + [-\omega_x xz - \omega_y yz + \omega_z(x^2+y^2)]\mathbf{k}.\end{aligned}$$

Equation (11-22) for $\mathbf{H}_G$ can now be written as

$$\begin{aligned}\mathbf{H}_G &= \left[\omega_x \int_m (y^2+z^2)\, dm - \omega_y \int_m xy\, dm - \omega_z \int_m xz\, dm\right]\mathbf{i} \\ &\quad + \left[-\omega_x \int_m xy\, dm + \omega_y \int_m (z^2+x^2)\, dm - \omega_z \int_m yz\, dm\right]\mathbf{j} \\ &\quad + \left[-\omega_x \int_m xz\, dm - \omega_y \int_m yz\, dm + \omega_z \int_m (x^2+y^2)\, dm\right]\mathbf{k} \\ &= (\omega_x I_x - \omega_y I_{xy} - \omega_z I_{xz})\mathbf{i} + (-\omega_x I_{xy} + \omega_y I_y - \omega_z I_{yz})\mathbf{j} \\ &\quad + (-\omega_x I_{xz} - \omega_y I_{yz} + \omega_z I_z)\mathbf{k}.\end{aligned} \tag{11-22a}$$

Note that Eq. (11-21) can be written as

$$\mathbf{H}_O = \mathbf{R} \times m\mathbf{v}_G + \mathbf{H}_G, \tag{11-21a}$$

where $\mathbf{H}_G$ is given by Eq. (11-22a).

Equation (11-22a) for $\mathbf{H}_G$, in terms of the moments and products of inertia of the body and the rectangular components of the angular velocity of the body for general motion, appears rather complicated. For most practical cases, however, many of the terms are zero. Some important special cases are given here for ready reference.

For translation $\boldsymbol{\omega} = 0$, and

$$\mathbf{H}_G = 0. \tag{11-23a}$$

When a body has plane motion parallel to the xy plane (note that this includes rotation about an axis parallel to the z axis) ω_x and ω_y are both zero, and Eq. (11-22a) becomes

$$\mathbf{H}_G = \omega_z(-I_{xz}\mathbf{i} - I_{yz}\mathbf{j} + I_z\mathbf{k}). \tag{11-23b}$$

This equation will be further simplified if the xy plane is a plane of symmetry or if the z axis is an axis of symmetry since the product of inertia terms will be zero. In this case, the angular momentum about G becomes

$$\mathbf{H}_G = I_z\omega_z\mathbf{k}. \tag{11-23c}$$

Note that Eq. (11-23c) applies to bodies which have rotation about an axis parallel to the z axis or plane motion parallel to the xy plane provided the z axis is an axis of symmetry or the xy plane is a plane of symmetry.

When a body rotates about a fixed (Z) axis, ω_x and ω_y are both zero. If point O is selected as the intersection of the Z axis and the plane of motion, the angular momentum of the body with respect to O becomes, from Eq. (11-21a) and (11-22a),

$$\begin{aligned}
\mathbf{H}_O &= \mathbf{R} \times m\mathbf{v}_G + \mathbf{H}_G \\
&= \mathbf{R} \times m(\omega_z\mathbf{k} \times \mathbf{R}) + \omega_z(-I_{xz}\mathbf{i} - I_{yz}\mathbf{j} + I_z\mathbf{k}) \\
&= \omega_z[-I_{xz}\mathbf{i} - I_{yz}\mathbf{j} + (mR^2 + I_z)\mathbf{k}] \\
&= \omega_z(-I_{xz}\mathbf{i} - I_{yz}\mathbf{j} + I_Z\mathbf{k}),
\end{aligned} \tag{11-23d}$$

where I_{xz} and I_{yz} refer to axes through G and I_Z is the moment of inertia with respect to the fixed axis of rotation. Note that $\mathbf{R}$ and $\mathbf{k}$ are mutually perpendicular so that $\mathbf{R} \times (\mathbf{k} \times \mathbf{R}) = R^2\mathbf{k}$.

If the body is symmetrical with respect to the plane of motion or the z axis, the product of inertia terms are zero, and the value of $\mathbf{H}_O$ reduces to

$$\left.\begin{aligned}
&\mathbf{H}_O = (mR^2 + I_z)\omega_z\mathbf{k} \\
\text{or}\quad & \\
&\mathbf{H}_{AR} = I_{AR}\omega\mathbf{k}
\end{aligned}\right\} \tag{11-23e}$$

where the subscript AR refers to the axis of rotation.

The angular momentum of a particle is defined [see Eq. (11-9)] as the moment of its linear momentum. The angular momentum of a system of particles, rigid or nonrigid, with respect to any point is the sum of the moments of the linear momenta of all the particles of the system with respect to that same point as shown by Eq. (11-19). The angular momentum of a rigid body, however, is equal to the moment of the linear momentum of the body only when certain conditions of symmetry are met, as shown in the next paragraph.

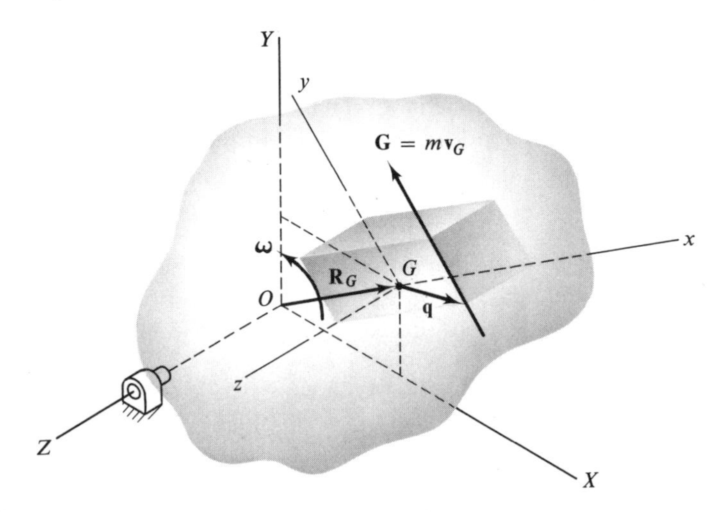

Figure 11-26

The rigid body of Fig. 11-26 is rotating about the Z axis through O parallel to the moving z axis. The xyz axes are fixed in the body and oriented as shown. The linear momentum of the body is

$$\mathbf{G} = m\mathbf{v}_G = m\omega R_G\mathbf{j},$$

and the position vector from the mass center to the intersection of **G** and the xz plane is

$$\mathbf{q} = x\mathbf{i} + z\mathbf{k}.$$

The moment of **G** with respect to the mass center is

$$\begin{aligned}\mathbf{H}'_G &= \mathbf{q}\times\mathbf{G} = (x\mathbf{i}+z\mathbf{k})\times(m\omega R_G\mathbf{j})\\ &= xm\omega R_G\mathbf{k} - zm\omega R_G\mathbf{i}.\end{aligned}$$

From Eq. (11-23b), the angular momentum about the mass center for this example is

$$\mathbf{H}_G = \omega(-I_{xz}\mathbf{i} - I_{yz}\mathbf{j} + I_z\mathbf{k}).$$

Since $\mathbf{H}'_G$ contains no coefficient of **j**, these quantities can be equal only when I_{yz} is zero.

When the xy plane in Fig. 11-26 is a plane of symmetry, the expression for $\mathbf{H}_G$ reduces to $\omega I_z\mathbf{k}$. In this case, $\mathbf{H}'_G$ will be the same as $\mathbf{H}_G$ if $\mathbf{G}$ is in the xy plane (that is, if z is zero), and the value of x, the distance from the mass center to the action line of $\mathbf{G}$, is

$$x=\frac{\omega I_z}{m\omega R_G}=\frac{mk_G^2}{mR_G}=\frac{k_G^2}{R_G},$$

where R_G is the distance from O to G. The distance from the axis of rotation to the linear momentum vector is

$$\frac{k_G^2}{R_G}+R_G,$$

which is identical to the value obtained in Art. 9-10 for the distance to the center of percussion. Consequently *the linear momentum vector acts through the center of percussion of a rotating rigid body that is symmetrical with respect to the plane of motion.*

An expression for the kinetic energy of a rigid body with plane motion was developed in Art. 10-6. By using Eq. (11-22) for the angular momentum of a rigid body with respect to its mass center, a concise expression for the kinetic energy of a rigid body with general motion can be developed.

As previously shown, the velocity of any particle P of the rigid body in Fig. 11-25 is

$$\mathbf{v}_P=\mathbf{v}_G+\boldsymbol{\omega}\times\mathbf{R},$$

and from Art. 10-3, the kinetic energy of the particle P is

$$\begin{aligned} dT &= dm\frac{(\mathbf{v}_P)^2}{2}=\frac{dm}{2}\mathbf{v}_P\cdot\mathbf{v}_P \\ &= \frac{dm}{2}(\mathbf{v}_G+\boldsymbol{\omega}\times\mathbf{r})\cdot(\mathbf{v}_G+\boldsymbol{\omega}\times\mathbf{r}) \\ &= \frac{dm}{2}[v_G^2+2\mathbf{v}_G\cdot(\boldsymbol{\omega}\times\mathbf{r})+(\boldsymbol{\omega}\times\mathbf{r})\cdot(\boldsymbol{\omega}\times\mathbf{r})]. \end{aligned}$$

Since $\mathbf{v}_G$ and $\boldsymbol{\omega}$ do not vary over the body, the kinetic energy of the entire body can be written

$$\begin{aligned} T &= \frac{v_G^2}{2}\int_m dm+\mathbf{v}_G\cdot\left(\boldsymbol{\omega}\times\int_m \mathbf{r}\,dm\right) \\ &\quad +\frac{1}{2}\int_m[(\boldsymbol{\omega}\times\mathbf{r})\cdot(\boldsymbol{\omega}\times\mathbf{r})]\,dm. \end{aligned} \tag{d}$$

The term $\int_m \mathbf{r}\,dm$ is zero, since $\mathbf{r}$ is measured from the mass center. The identity

$$(\mathbf{a}\times\mathbf{b})\cdot\mathbf{c}=\mathbf{a}\cdot(\mathbf{b}\times\mathbf{c})$$

can be proved by writing the three vectors in terms of $\mathbf{i}$, $\mathbf{j}$, and $\mathbf{k}$

components and expanding (see any text on vector analysis). When **a** is replaced by **ω**, **b** by **r**, and **c** by **ω×r**, the identity shows that the last integral of Eq. (d) can be written as

$$\frac{1}{2}\boldsymbol{\omega}\cdot\int_m [\mathbf{r}\times(\boldsymbol{\omega}\times\mathbf{r})]\,dm=\frac{1}{2}\boldsymbol{\omega}\cdot\mathbf{H}_G$$

from Eq. (11-22). With these results, Eq. (d) reduces to

$$T=\tfrac{1}{2}mv_G^2+\tfrac{1}{2}\boldsymbol{\omega}\cdot\mathbf{H}_G.$$

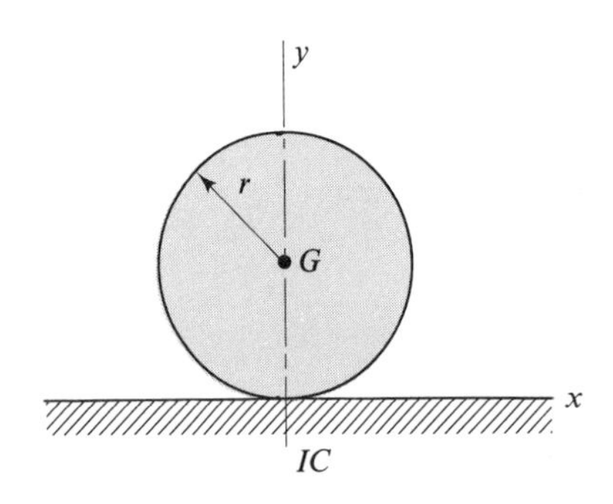

Figure 11-27

EXAMPLE 11-13

The homogeneous cylinder of mass m shown in Fig. 11-27 rolls without slipping on a horizontal plane with a counterclockwise angular velocity of ω rad per sec. Show that the angular momentum of the cylinder with respect to the instantaneous center, IC, is

$$\mathbf{H}_{IC}=I_{IC}\boldsymbol{\omega}=I_{IC}\omega\mathbf{k}.$$

SOLUTION

Since the cylinder rolls without slipping,

$$\mathbf{v}_G=-r\omega\mathbf{i}.$$

From Eq. (11-23c),

$$\mathbf{H}_G=I_G\omega\mathbf{k},$$

and from Eq. (11-21a),

$$\mathbf{H}_{IC}=\mathbf{R}\times m\mathbf{v}_G+\mathbf{H}_G;$$

therefore

$$\begin{aligned}\mathbf{H}_{IC}&=r\mathbf{j}\times m\mathbf{v}_G+\mathbf{H}_G\\&=r\mathbf{j}\times m(-r\omega\mathbf{i})+I_G\omega\mathbf{k}\\&=mr^2\omega\mathbf{k}+I_G\,\omega\mathbf{k}\\&=(mr^2+I_G)\omega\mathbf{k}\\&=\underline{I_{IC}\omega\mathbf{k}}.\end{aligned}$$

Ans.

PROBLEMS

11-118 The homogeneous uniform 96.6-lb bar OA in Fig. P11-118 and the homogeneous 161-lb sphere are welded together to form a rigid body that is symmetrical with respect to the xy plane. The body has an angular velocity of 5 rad per sec counterclockwise about the z axis in the position shown.
(a) Determine the angular momentum with respect to the origin O.
(b) Locate the linear momentum vector of the body.

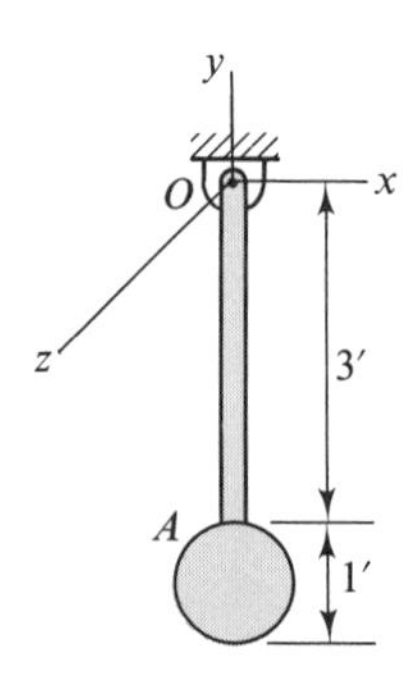

Figure P11-118

11-119 (a) If point A is at the top of the cylinder of Example 11-13, determine the angular momentum of the cylinder with respect to A. (b) Is the answer to part (a) equal to $I_A\omega$?

***11-120** The homogeneous bar OB in Fig. P11-120 has a mass of 10 kg and is attached by a pin to the homogeneous 30-kg cylinder C. The cylinder rolls without slipping on the circular track. The system is symmetrical with respect to the xy plane. Determine the angular momentum of the system with respect to point O when the bar has an angular velocity of 2 rad/s counterclockwise. The cylinder has a diameter of 1 m.

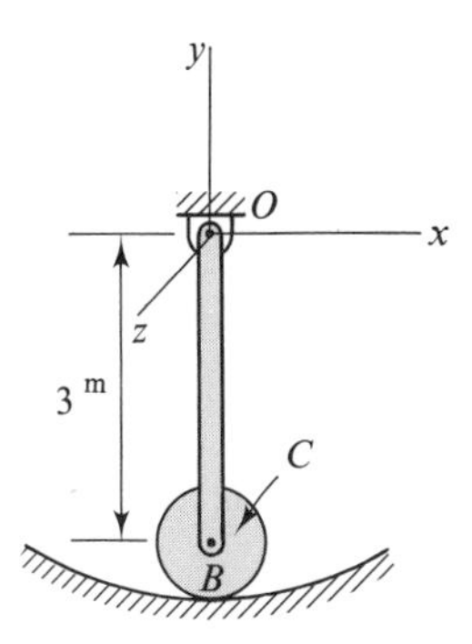

Figure P11-120

11-121 The 128.8-lb homogeneous bar AB in Fig. P11-121 moves on a horizontal xy plane. The bar has an angular velocity of 4 rad per sec clockwise, and the velocity of the mass center is shown on the figure. Determine
(a) The angular momentum of the bar with respect to point C.
(b) The kinetic energy of the bar.

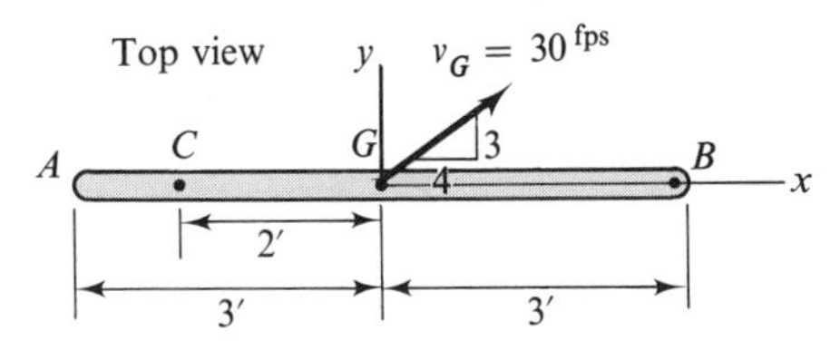

Figure P11-121

***11-122** The radius of gyration of mass of the unbalanced 250-kg body in Fig. P11-122 is 300 mm with respect to an axis through the mass center G perpendicular to the plane of motion. The body rolls without slipping on the horizontal plane and in the position shown has an angular velocity of 2 rad/s clockwise.
(a) Determine the angular momentum of the body with respect to G.
(b) Locate the linear momentum vector with respect to G.

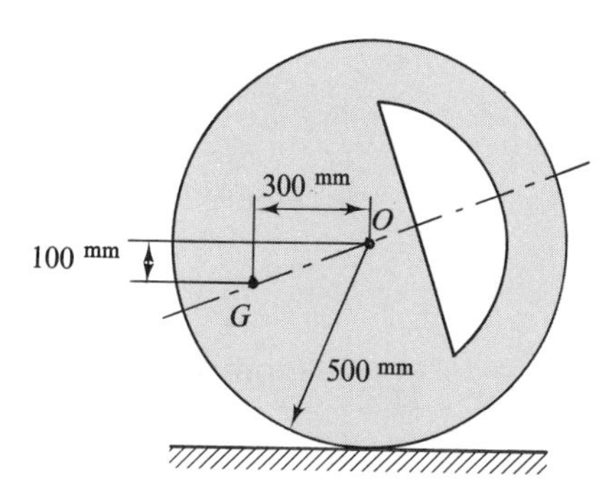

Figure P11-122

11-123 Two 1.61-lb weights are attached to two light rods as shown in Fig. P11-123. In the position shown, the angular velocity of rod OA is $8\mathbf{k}$ rad per sec, and the angular velocity of BC *relative* to OA is $40\mathbf{j}$ rad per sec.
(a) Determine the moment of momentum of the two weights with respect to point O.
(b) Calculate the linear momentum of the system and, if possible, locate the point at which it pierces the yz plane.

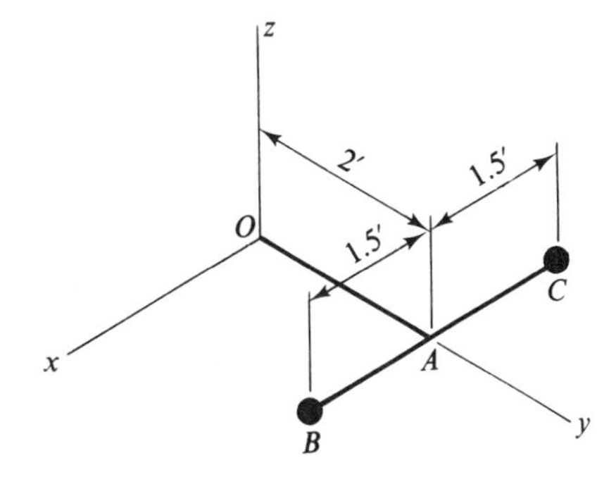

Figure P11-123

11-10 PRINCIPLE OF ANGULAR MOMENTUM

Equation (11-19) is an expression for the angular momentum of a system of particles with respect to any point. A useful relation can be developed for the time rate of change of the moment of momentum by differentiating Eq. (11-19) as follows:

$$\dot{\mathbf{H}}_{O'} = \sum_{i=1}^{n} [(\mathbf{R}_2 + \mathbf{r}_i) \times m_i(\ddot{\mathbf{R}} + \ddot{\mathbf{r}}_i) + (\dot{\mathbf{R}}_2 + \dot{\mathbf{r}}_i) \times m_i(\dot{\mathbf{R}} + \dot{\mathbf{r}}_i)]. \tag{a}$$

When O' is the fixed point O, $\mathbf{R}_2$ becomes $\mathbf{R}$, and

$$(\dot{\mathbf{R}} + \dot{\mathbf{r}}_i) \times (\dot{\mathbf{R}} + \dot{\mathbf{r}}_i) = 0.$$

Therefore Eq. (a) can be written

$$\dot{\mathbf{H}}_O = \sum_{i=1}^{n} [(\mathbf{R} + \mathbf{r}_i) \times m_i(\ddot{\mathbf{R}} + \ddot{\mathbf{r}}_i)]. \tag{b}$$

From Eq. (9-3),

$$\mathbf{F}_i = m_i \mathbf{a}_i = m_i(\ddot{\mathbf{R}} + \ddot{\mathbf{r}}_i),$$

and Eq. (b) reduces to

$$\dot{\mathbf{H}}_O = \sum_{i=1}^{n} [(\mathbf{R} + \mathbf{r}) \times \mathbf{F}_i] = \sum_{i=1}^{n} \mathbf{M}_{O_i} = \mathbf{M}_O, \tag{11-24}$$

where $\mathbf{M}_O$ is the sum of the moments with respect to O of the external forces acting on the system of particles. Equation (11-24) shows that *the time rate of change of the angular momentum of any system of particles, with respect to a point fixed in space, is equal to the moment of the external forces acting on the system about that point.* This relationship is known as the *principle of angular momentum for a fixed point.*

The principle of angular momentum can be derived for moments about the mass center by letting O' in Fig. 11-24 become G. In this case R_2 is zero, and Eq. (a) becomes

$$\dot{\mathbf{H}}_G = \sum_{i=1}^{n} [\mathbf{r}_i \times m_i(\ddot{\mathbf{R}} + \ddot{\mathbf{r}}_i) + \dot{\mathbf{r}}_i \times m_i(\dot{\mathbf{R}} + \dot{\mathbf{r}}_i)]$$

$$= \sum_{i=1}^{n} [(\mathbf{r}_i \times \mathbf{F}_i) + (\dot{\mathbf{r}}_i m_i) \times \dot{\mathbf{R}} + 0] \tag{c}$$

since

$$\dot{\mathbf{r}}_i \times \dot{\mathbf{r}}_i = 0.$$

In Eq. (c),

$$\dot{\mathbf{r}}_i m_i = \frac{d}{dt}(\mathbf{r}_i m_i),$$

and the last term of Eq. (c) can be written

$$\frac{d}{dt}\left[\sum_{i=1}^{n} (\mathbf{r}_i m_i)\right] \times \dot{\mathbf{R}}$$

since the summation does not involve $\dot{\mathbf{R}}$ or d/dt. The vectors $\mathbf{r}_i$ are all measured from the mass center; therefore, the summation is zero, and Eq. (c) becomes

$$\dot{\mathbf{H}}_G = \sum_{i=1}^{n} (\mathbf{r}_i \times \mathbf{F}_i) = \sum_{i=1}^{n} \mathbf{M}_{G_i} = \mathbf{M}_G. \quad (11\text{-}25)$$

When the particles of the system constitute a rigid body Eq. (11-22a) and the various forms of Eq. (11-23) can be used to simplify the calculations.

Equations (11-24) and (11-25) will be utilized to develop closely related concepts of the next three articles that are useful for solving a variety of problems.

11-11 PRINCIPLE OF ANGULAR IMPULSE AND ANGULAR MOMENTUM

In Art. 11-10 the relation $\dot{\mathbf{H}} = \mathbf{M}$ was derived where the reference point was either a fixed point in space or the center of mass of a system of particles. The system of particles can be a rigid body or merely a group of particles. This relation is useful in the form developed in Art. 11-10, and it also leads to the principle of angular impulse and angular momentum. From Eq. (11-24),

$$\dot{\mathbf{H}}_O = \frac{d\mathbf{H}_O}{dt} = \mathbf{M}_O$$

or

$$d\mathbf{H}_O = \mathbf{M}_O \, dt$$

from which

$$\int_{\mathbf{H}_{Oi}}^{\mathbf{H}_{Of}} d\mathbf{H}_O = \int_{t_i}^{t_f} \mathbf{M}_O \, dt$$

and

$$\mathbf{H}_{Of} - \mathbf{H}_{Oi} = \int_{t_i}^{t_f} \mathbf{M}_O \, dt. \quad (11\text{-}26a)$$

The right-hand side of Eq. (11-26a) is defined as the angular impulse of the forces acting on a system of particles with respect to the fixed point O, and the left-hand side is the change in angular momentum of the system of particles with respect to the same point during the time interval from t_i to t_f.

By starting with Eq. (11-25), the same development that gave Eq. (11-26a) gives

$$\mathbf{H}_{Gf} - \mathbf{H}_{Gi} = \int_{t_i}^{t_f} \mathbf{M}_G \, dt, \quad (11\text{-}26b)$$

where G is the center of mass of the system of particles.

Equations (11-26a) and (11-26b) are valid for any system of particles. If the particles are so connected that they constitute a rigid body, Eq. (11-22a) is useful in determining the angular momentum of the body. When the body has plane motion parallel to the xy plane Eq. (11-23b) can be substituted into Eq. (11-26b) to give

$$\int_{t_i}^{t_f} \mathbf{M}_G \, dt = \mathbf{H}_{Gf} - \mathbf{H}_{Gi}$$

$$= (-I_{xz}\mathbf{i} - I_{yz}\mathbf{j} + I_z\mathbf{k})(\omega_{zf} - \omega_{zi}). \quad (11\text{-}27a)$$

If the body is symmetrical with respect to the plane of motion, the product of inertia terms are zero, and the equation reduces to

$$\int_{t_i}^{t_f} \mathbf{M}_G \, dt = I_z\mathbf{k}(\omega_{zf} - \omega_{zi})$$

$$= I_G(\omega_f - \omega_i)\mathbf{k}, \quad (11\text{-}27b)$$

where I_G is the moment of inertia of the body with respect to the axis through G perpendicular to the plane of motion.

When a body rotates about a fixed axis (say the Z axis) Eq. (11-26a) is valid if O is any point on the Z axis. If O is the intersection of the Z axis and the plane of motion, Eq. (11-23d) can be substituted into Eq. (11-26a) to give

$$\int_{t_i}^{t_f} \mathbf{M}_O \, dt = \mathbf{H}_{Of} - \mathbf{H}_{Oi}$$

$$= (-I_{xz}\mathbf{i} - I_{yz}\mathbf{j} + I_Z\mathbf{k})(\omega_{zf} - \omega_{zi}). \quad (11\text{-}28a)$$

When the body is symmetrical with respect to the plane of motion or the z axis the angular momentum is given by Eq. (11-23e), and Eq. (11-26a) reduces to

$$\int_{t_i}^{t_f} \mathbf{M}_{AR} \, dt = I_{AR}(\omega_f - \omega_i)\mathbf{k}. \quad (11\text{-}28b)$$

In addition to the angular impulse and momentum equations, the equations of linear impulse and momentum developed in Art. 11-2 provide additional information which may be needed to solve some problems. The general equation for linear impulse and momentum is

$$\int_{t_i}^{t_f} \mathbf{F} \, dt = \Delta\mathbf{G} = m(\mathbf{v}_{Gf} - \mathbf{v}_{Gi}),$$

which can be resolved into two or three component equations as needed.

When a symmetrical wheel rolls without slipping along a plane surface as in Fig. 11-27 a special angular impulse and momentum relationship can be developed using the instantaneous center of zero velocity. This point moves from point to point during a time interval and thus is neither a *fixed* point nor the mass center. The acceleration of the instantaneous center is perpendicular to the plane on which the wheel rolls and thus is toward the mass center of the symmetrical body.

Equation (9-20), from Chapter 9, is

$$\mathbf{M}_A = (\mathbf{r}_G \times \mathbf{a}_A)m + (\omega^2 I_{yz} - \alpha I_{xz})\mathbf{i}$$
$$-(\omega^2 I_{xz} + \alpha I_{yz})\mathbf{j} + \alpha I_z \mathbf{k},$$

and when point A is the instantaneous center, $\mathbf{r}_G$ is parallel to $\mathbf{a}_A$ and the first term on the right is zero. Since the wheel is assumed to be symmetrical with respect to the plane of motion, the product of inertia terms are zero. Therefore the equation reduces to

$$\mathbf{M}_{IC} = I_{IC}\boldsymbol{\alpha} = I_{IC}\frac{d\boldsymbol{\omega}}{dt}.$$

When this equation is integrated with respect to time it becomes

$$\int_{t_i}^{t_f} \mathbf{M}_{IC}\, dt = I_{IC}\int_{\omega_i}^{\omega_f} d\boldsymbol{\omega} = I_{IC}(\boldsymbol{\omega}_f - \boldsymbol{\omega}_i). \tag{11-28c}$$

This equation indicates that the angular impulse of the external forces acting on the body, with respect to the (moving) instantaneous center, is equal to the change of the angular momentum of the body with respect to this axis. It should be emphasized, however, that this equation is valid *only* for a symmetrical wheel which rolls without slipping along a plane surface.

When the principle of angular impulse and angular momentum is used, it is usually better to draw separate free-body diagrams of each rigid body involved than to use combinations of bodies as is commonly done in using the principle of work and kinetic energy.

EXAMPLE 11-14

The sprinkler in Fig. 11-28(a) consists of a rotating horizontal tube AB with closed ends. The tube contains two orifices located on opposite sides of the tube, each at a distance r from the center of the tube. The volume rate of flow of water is Q ft^3 per sec, the velocity of efflux relative to the orifice is v fps perpendicular to the tube, and the tube has a constant angular velocity of ω rad per sec. Determine the frictional drag moment of the support on the tube.

SOLUTION

The principle of angular impulse and momentum will be applied to the tube and water for a differential time interval, dt. The mass of water flowing through the sprinkler in time dt is

$$dm = \rho Q\, dt,$$

where ρ is the density (mass per unit volume) of the water. The velocity leaving the orifice at

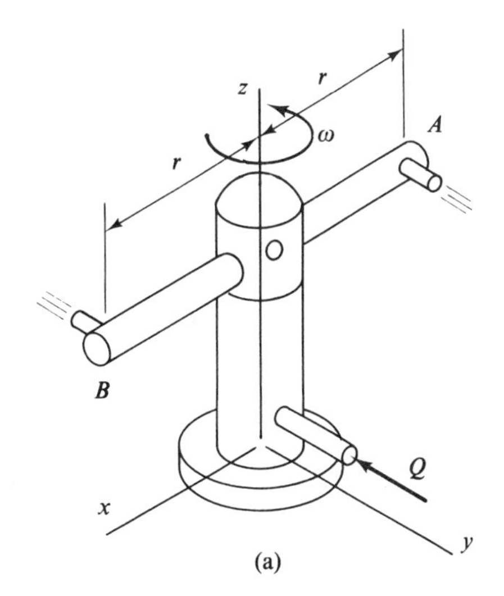

(a)

Figure 11-28(a)

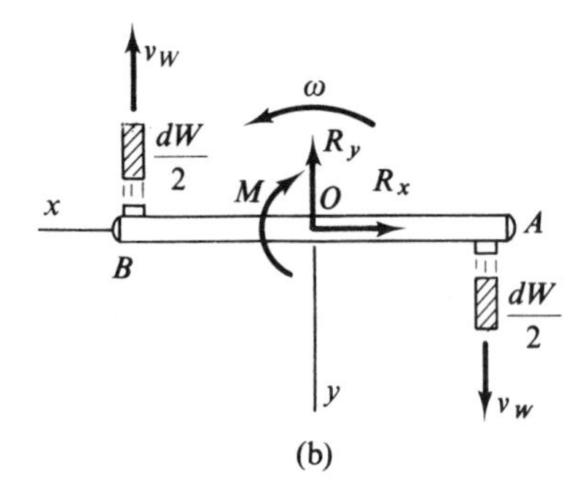

Figure 11-28(b)

A is

$$\mathbf{v}_W = \mathbf{v}_A + \mathbf{v}_{W/A} = -r\omega\mathbf{j} + v\mathbf{j} = (v - r\omega)\mathbf{j}.$$

The velocity of the water leaving the orifice at B has the same magnitude but is opposite in direction. The angular momentum of the tube and of the water in the tube is constant, and the angular momentum of the water entering the tube is zero (about the axis of rotation) so that the only *change* of angular momentum is that due to the momentum of the water leaving the tube. From the free-body diagram of the system in Fig. 11-28(b), the equation of angular impulse and momentum is

$$-M\mathbf{k}\,dt = d\mathbf{H} = -2\left(\frac{dm}{2}v_W\right)r\mathbf{k} = -Q\rho\,dt(v - r\omega)r\mathbf{k}$$

from which

$$\underline{M = Q\rho r(v - r\omega).} \qquad \text{Ans.}$$

EXAMPLE 11-15

The body in Fig. 11-29(a) rotates about a horizontal axis at O and has a mass of 30 kg and a radius of gyration of mass with respect to

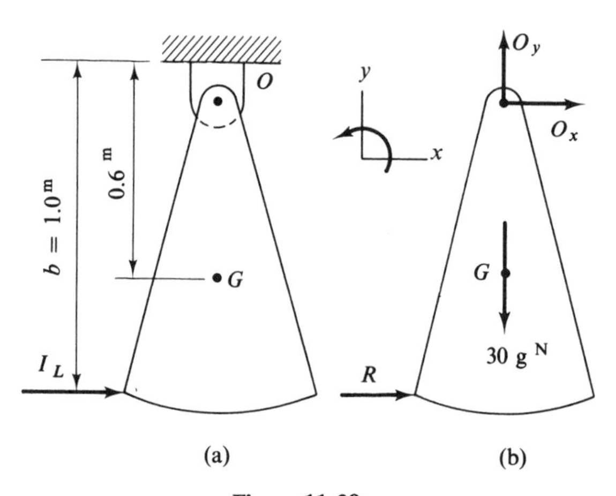

Figure 11-29

the axis of rotation of 0.80 m. The body has an angular velocity of 3.0 rad/s clockwise when an impulsive blow to the right is administered to the lower end of the body and changes the angular velocity of the body to 2.0 rad/s counterclockwise. Determine
(a) The linear impulse of the blow.
(b) The linear impulse of the horizontal component of the pin reaction at O during the time the blow was struck.

SOLUTION

(a) A free-body diagram of the body is shown in Fig. 11-29(b) where the force R is the magnitude of the impulsive blow. The body has rotation about a fixed axis and is assumed to be symmetrical with respect to the plane of motion; thus Eq. (11-28b) is valid. The angular impulse of the impulsive blow is the moment of the linear impulse; thus

$$\int Rb\,dt = I_{AR}(\omega_f - \omega_i) = mk_{AR}^2(\omega_f - \omega_i),$$

$$(1)\quad \int R\,dt = 30(0.8)^2[2-(-3)],$$

$$\int \mathbf{R}\,dt = \underline{96.0\ \text{N}\cdot\text{s}\rightarrow}. \qquad \text{Ans.}$$

(b) The impulse of O_x can be obtained by applying either the principle of linear impulse and momentum or the principle of angular impulse and momentum about the mass center. When linear impulse and momentum is used the equation is

$$\left(\overset{+}{\rightarrow}\right)\quad \sum \int F_x\,dt = m(v_{Gxf} - v_{Gxi}),$$

$$\int R\,dt + \int O_x\,dt = 30[0.6(2.0) - 0.6(-3.0)]$$

from which

$$\int O_x\,dt = 90.0 - 96.0 = -6.0,$$

$$\int \mathbf{O}_x\,dt = \underline{6.0\ \text{N}\cdot\text{s}\leftarrow}. \qquad \text{Ans.}$$

EXAMPLE 11-16

The wheel A of Fig. 11-30(a) weighs 1288 lb, rolls without slipping on the inclined plane, is symmetrical with respect to its geometrical axis, and has a radius of gyration of mass with respect to its geometrical axis of 2.0 ft. The homogeneous cylindrical drum B weighs 322 lb and is connected to A by an inextensible cable which wraps around both bodies. The couple $\mathbf{C}$ equals $(600-3t^2)\mathbf{k}$ ft-lb. When $t=0$, $\boldsymbol{\omega}_B$ is $8\mathbf{k}$ rad per sec. Determine $\boldsymbol{\omega}_B$ when $t=6$ sec.

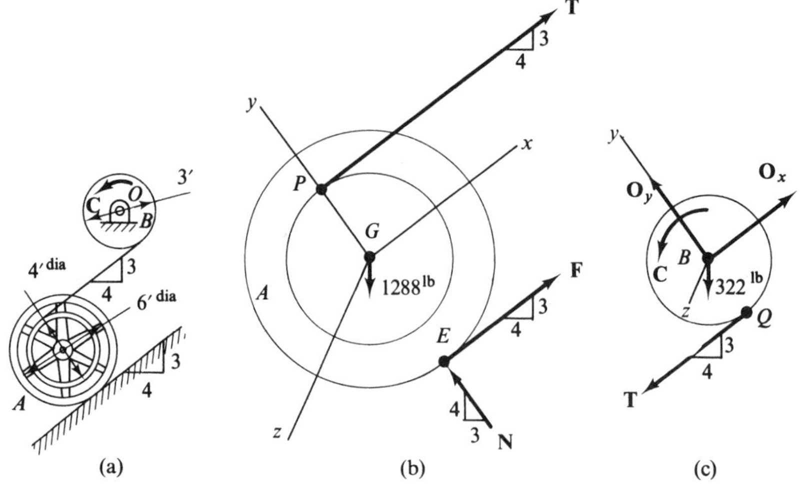

Figure 11-30

SOLUTION

Separate free-body diagrams of A and B are shown in Fig. 11-30(b) and 11-30(c). Body A has plane motion, and B has rotation about the z axis. The equations of linear and angular impulse and momentum for body A are

$$\left.\begin{aligned} \mathbf{I}_{Ly} &= \mathbf{G}_{yf} - \mathbf{G}_{yi} = m(\mathbf{v}_G)_{yf} - m(\mathbf{v}_G)_{yi} = 0, \\ \mathbf{I}_{Lx} &= \mathbf{G}_{xf} - \mathbf{G}_{xi} = m(\mathbf{v}_G)_{xf} - m(\mathbf{v}_G)_{xi}, \\ \text{and} \qquad \int_{t_i}^{t_f} \mathbf{M}_G\, dt &= \mathbf{H}_{Gf} - \mathbf{H}_{Gi} = \omega_{zf} I_z \mathbf{k} - \omega_{zi} I_z \mathbf{k}. \end{aligned}\right\} \quad \text{(a)}$$

Note that $\mathbf{v}_{Gy}$ is zero and $\mathbf{H}_G$ is $I_z\omega_z\mathbf{k}$ due to symmetry—see Eq. (11-23c). The equations of linear and angular impulse and momentum for body B are

$$\left.\begin{aligned} \mathbf{I}_{Lx} &= m(\mathbf{v}_G)_{xf} - m(\mathbf{v}_G)_{xi} = 0, \\ \mathbf{I}_{Ly} &= m(\mathbf{v}_G)_{yf} - m(\mathbf{v}_G)_{yi} = 0, \\ \text{and} \qquad \int_{t_i}^{t_f} \mathbf{M}_O\, dt &= \omega_{zf} I_z \mathbf{k} - \omega_{zi} I_z \mathbf{k}. \end{aligned}\right\} \quad \text{(b)}$$

In Eq. (b), $\mathbf{v}_G$ is zero and $\mathbf{H}_O$ is equal to $\omega_z I_z \mathbf{k}$ from Eq. (11-23c).

There are five unknown forces ($\mathbf{T}$, $\mathbf{F}$, $\mathbf{N}$, $\mathbf{O}_x$, and $\mathbf{O}_y$) and five unknown velocities ($\boldsymbol{\omega}_{Bf}$, $\boldsymbol{\omega}_{Ai}$, $\boldsymbol{\omega}_{Af}$, $\mathbf{v}_{Gi}$, and $\mathbf{v}_{Gf}$). Since only six motion equations are available, four additional equations must be obtained from kinematics. The velocities of points P and Q are equal, since they are joined by an inextensible cable. Point E is the instantaneous center of body A. Consequently, the following kinematic equations can be written:

$$\mathbf{v}_P = \mathbf{v}_Q = -5\omega_A \mathbf{i} = 1.5\omega_B \mathbf{i},$$

where $\boldsymbol{\omega}_A$ and $\boldsymbol{\omega}_B$ are both assumed to be positive (counterclockwise). As a result,

$$\omega_A = -0.3\omega_B.$$

From the given data,

$$\omega_{Ai} = -0.3\omega_{Bi} = -0.3(8) = -2.4 \text{ rad per sec}$$

and

$$\boldsymbol{\omega}_{Ai} = -2.4\mathbf{k} \text{ rad per sec (clockwise).} \qquad \text{(c)}$$

Also

$$v_{Gi} = r\omega_{Ai} = 3(2.4) = 7.2 \text{ fps}$$

and

$$\mathbf{v}_{Gi} = 7.2\mathbf{i} \text{ fps.} \qquad \text{(d)}$$

Similarly,

$$\boldsymbol{\omega}_{Af} = -0.3\omega_{Bf}\mathbf{k} \text{ (clockwise if } \omega_{Bf} \text{ is positive),} \qquad \text{(e)}$$

$$\mathbf{v}_{Gf} = 3(0.3)\omega_{Bf}\mathbf{i} = 0.9\omega_{Bf}\mathbf{i}. \qquad \text{(f)}$$

The first of Eq. (a) and the first two of Eq. (b) involve $\mathbf{N}$, $\mathbf{O}_x$, and $\mathbf{O}_y$, which are not required. When Eq. (c), Eq. (d), Eq. (e), and Eq. (f) are substituted into the other three of Eq. (a) and Eq. (b) they become

$$\mathbf{I}_{Lx} = \mathbf{i}\int_0^6 F\,dt + \mathbf{i}\int_0^6 T\,dt - 0.6(1288)6\mathbf{i}$$

$$= m[\mathbf{v}_{Gf} - \mathbf{v}_{Gi}] = 40[0.9\omega_{Bf} - 7.2]\mathbf{i},$$

$$\int_0^6 \mathbf{M}_G\,dt = \mathbf{k}\int_0^6 (3F)\,dt - \mathbf{k}\int_0^6 (2T)\,dt$$

$$= I_G(\boldsymbol{\omega}_{Af} - \boldsymbol{\omega}_{Bf}] = 40(2)^2[-0.3\omega_{Bf} - (-2.4)]\mathbf{k},$$

and

$$\int_0^6 \mathbf{M}_O\,dt = \mathbf{k}\int_0^6 (600 - 3t^2)\,dt - \tfrac{3}{2}\mathbf{k}\int_0^6 T\,dt$$

$$= \mathrm{I}_z[\boldsymbol{\omega}_{Bf} - \boldsymbol{\omega}_{Bi}] = \tfrac{1}{2}(10)(\tfrac{3}{2})^2[\omega_{Bf} - 8]\mathbf{k}.$$

These three equations reduce to

$$\int_0^6 F\,dt + \int_0^6 T\,dt - 4637 = +36\omega_{Bf} - 288,$$

$$3\int_0^6 F\,dt - 2\int_0^6 T\,dt = -48\omega_{Bf} + 384,$$

and

$$3384 - 1.5\int_0^6 T\,dt = (11.25)\omega_{Bf} - 90.$$

The simultaneous solution of these three equations gives

$$\omega_{Bf} = -5.59$$

and

$$\boldsymbol{\omega}_{Bf} = \underline{-5.59\mathbf{k} \text{ rad per sec.}} \qquad \text{Ans.}$$

Since the mass center of A is at its geometric center, Eq. (11-28c) can be used where point O is at the instantaneous center. The equation would give the same result as the simultaneous solution of the last two equations of Eq. (a).

PROBLEMS

11-124 Equation (b) of Art. 11-9 is

$$\mathbf{H}_G = \sum m_i \mathbf{r}_i \times \dot{\mathbf{r}}_i.$$

Show by differentiating this expression that $\dot{\mathbf{H}}_G = \mathbf{M}_G$, where $\mathbf{M}_G$ is the moment about the mass center of the external forces acting on the system of particles.

11-125 (a) Water enters the horizontal pipe in Fig. P11-125 through a vertical inlet at O and discharges through two short nozzles at a distance L from the fixed axis at O. Determine the magnitude of the moment $\mathbf{M}$ which must be applied to the pipe to maintain a constant angular velocity $\boldsymbol{\omega}$ when the system is discharging Q ft³ per sec of water. The density of water is ρ. The water leaves the nozzles with a velocity $\mathbf{v}_r$ relative to the nozzles.
(b) Determine the value of $\boldsymbol{\omega}$ if the moment M is zero.

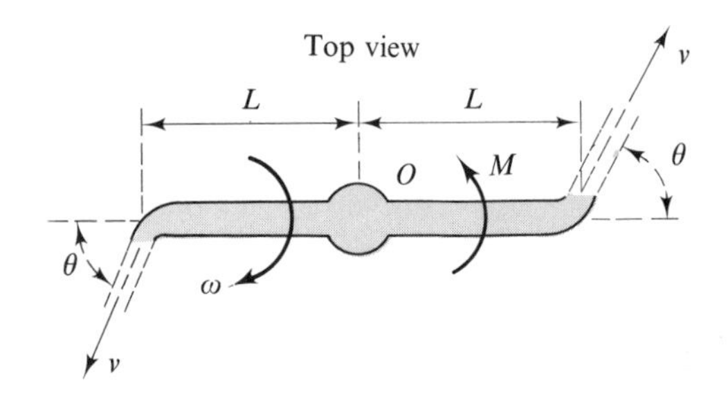

Figure P11-125

11-126 A flywheel is keyed to a shaft 4 in. in diameter, the combined weight being 644 lb. A constant turning moment of 3000 in-lb clockwise is exerted on the shaft causing the angular velocity to change from 60 to 420 rpm, both clockwise, in a time interval of 6 sec. The bearing (journal) friction on the shaft is 6 lb. Determine the radius of gyration of mass of the combined assembly with respect to the axis of rotation.

***11-127** The rotor of a turbine has a mass of 300 kg. When no other forces are acting, bearing friction will reduce the angular velocity from 150 rpm to zero in 2 minutes. The radius of gyration of mass with respect to the axis of rotation is 330 mm. Determine the frictional moment.

11-128 The armature of an electric motor weighs 640 lb, and its radius of gyration of mass with respect to the axis of rotation is 9 in. The radius of the shaft is 4 in. A period of 25 minutes is required for the bearing friction to bring the armature to rest from an angular velocity of 1800 rpm. Determine the coefficient of kinetic friction between the shaft and the bearings.

11-129 The 322-lb homogeneous disk in Fig. P11-129 has an angular velocity of 5 rad per sec counterclockwise when $t = 0$. The force P_v in pounds varies with time as indicated by the diagram. Determine the angular velocity of the disk when $t = 5$ sec.

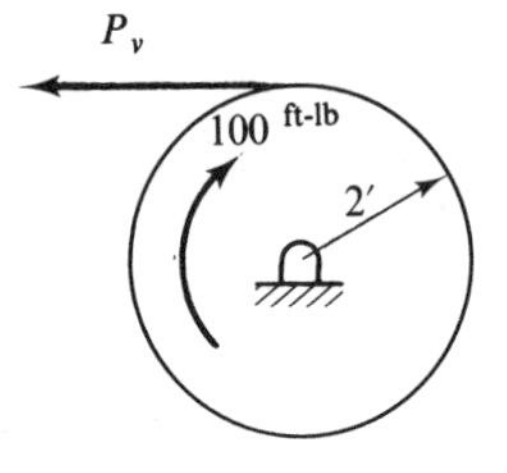

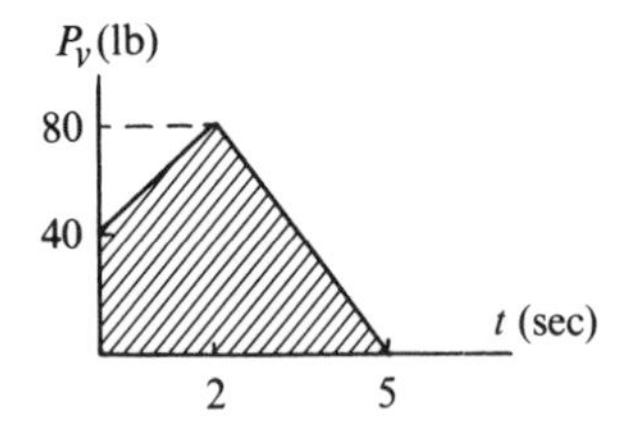

Figure P11-129

11-130 The force P_v in Problem 11-129 varies according to the equation $P_v = 3t^2$, where P_v is in pounds when t is in seconds. Determine the value or values of t when the angular velocity is again 5 rad per sec counterclockwise.

11-131 The 322-lb disk in Fig. P11-131 has a radius of gyration of mass of 2 ft with respect to a vertical axis through the mass center G and slides on a smooth *horizontal* plane. The force F_v in pounds, applied to a rope which wraps

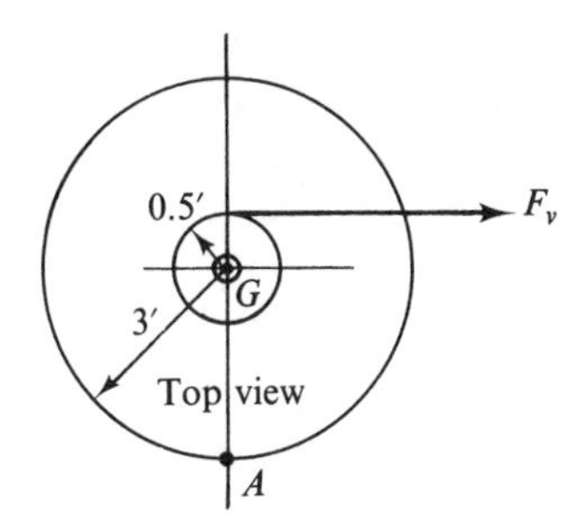

Figure P11-131

around the hub, varies according to the expression $F_v = 12t^2 + 20$, where t is in seconds. When $t = 0$, the velocity of G is 10 fps to the left, and the angular velocity of the disk is 5 rad per sec counterclockwise. The body is in the indicated position when $t = 5$ sec. Determine the velocity of point A at this instant.

***11-132** The force $\mathbf{F}_v$ in Fig. P11-132 is applied to a rope which wraps around a hub with a radius of 200 mm. The disk has a mass of 30 kg. The disk moves in a vertical plane and has a radius of gyration of mass of 0.5 m with respect to a horizontal axis through the mass center perpendicular to the plane of motion. The magnitude of the force $\mathbf{F}_v$ in newtons is given by the expression $F_v = 100 + 30t^2$, where t is in seconds. The system is released from rest when t is zero. When $t = 4$ s, determine
(a) The angular velocity of the disk.
(b) The velocity of point A on the rope.

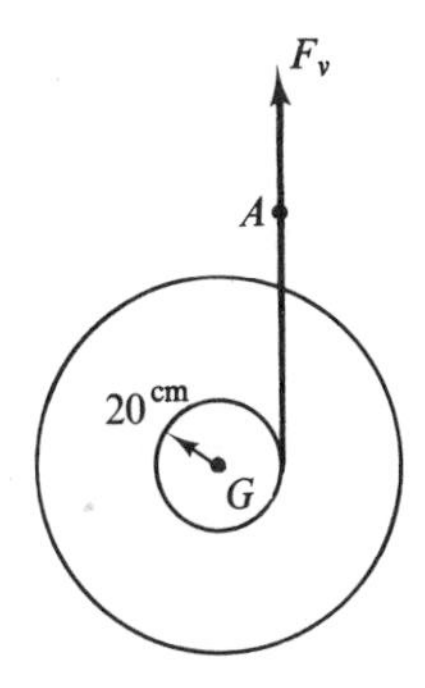

Figure P11-132

11-133 The homogeneous rod AB in Fig. P11-133 weighs 96.6 lb and is at rest in the vertical position when an impulsive blow of 6 lb-sec is applied at the end B of the rod. Determine
(a) The angular velocity of the rod after the impulse has been applied.
(b) The change in the linear momentum of the rod.
(c) The time average force acting on the rod exerted by the horizontal component of the pin reaction at A if the impact time interval is 0.002 sec.

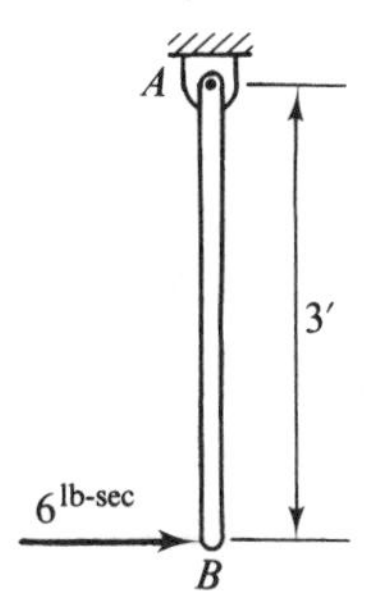

Figure P11-133

11-134 The 6-ft slender homogeneous 60-lb bar AB in Fig. P11-134 is suspended from a smooth pin at A and is at rest when the 10-lb ball C moving with a velocity of 20 fps to the left strikes the bar and rebounds to the right with 36% of its original kinetic energy. Determine
(a) The velocity of C after impact.
(b) The linear impulse of AB on C during the impact.
(c) The angular velocity of the bar after the impact.
(d) The linear impulse of the horizontal component of the pin reaction at A on the bar during the impact period.

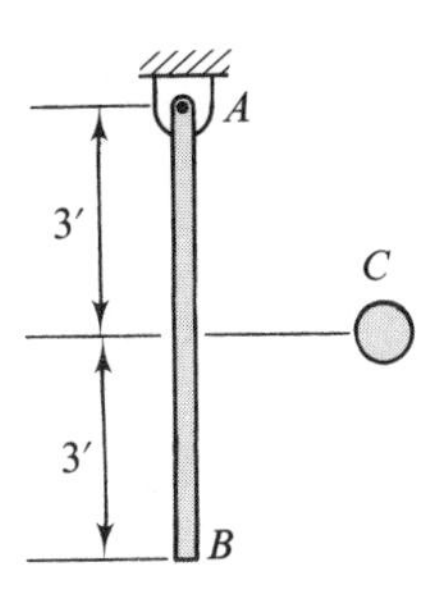

Figure P11-134

*11-135 The wheel in Fig. P11-135 rolls without slipping and has a radius of gyration of mass of 75 mm with respect to a horizontal axis through the mass center G. The magnitude of the variable force $\mathbf{P}_v$ is given by the equation $P_v = 3t^2$, where P_v is in newtons when t is in seconds. The force $\mathbf{P}_v$ is applied to a rope which is wrapped around the hub. During the interval from $t = 2$ to $t = 5$ s the velocity of G changes from 200 mm/s to 1000 mm/s, both to the right. Determine the mass of the body.

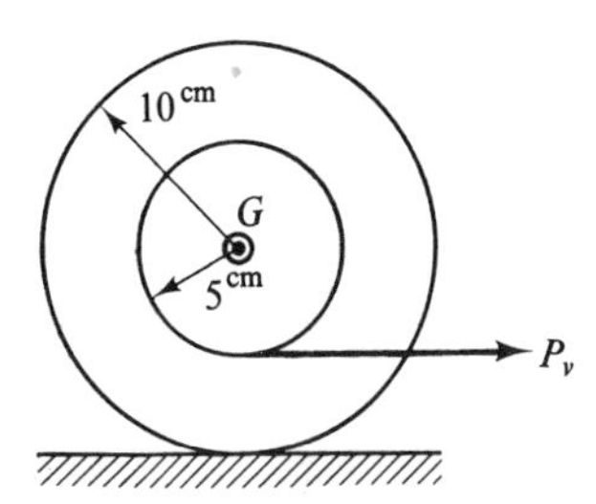

Figure P11-135

11-136 Body A in Fig. P11-136 weighs 16.1 lb and has a radius of gyration of mass with respect to its mass center of 0.20 ft. The variable force P_v is in pounds when t is in seconds. When $t = 0$ the velocity of the mass center is 2.0 fps downward. Determine the velocity of the mass center at $t = 5$ sec.

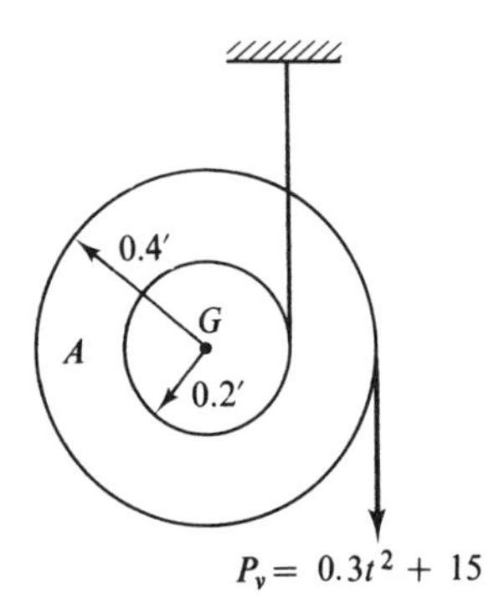

Figure P11-136

11-137 The solid homogeneous cylinder in Fig. P11-137 has a mass m and a radius r and rolls and slips on the horizontal plane. Initially the velocity of the center of the cylinder is v_0 down the plane, and the angular velocity of the cylinder is $4v_0/r$ counterclockwise. The coefficient of friction between the cylinder and plan˜ is 0.1. Determine the time, in terms of v_0, that will elapse before the wheel will roll without slipping.

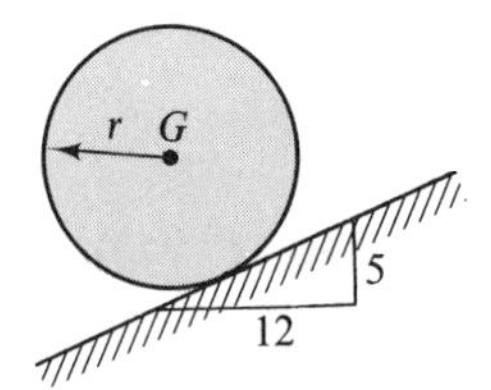

Figure P11-137

11-138 The homogeneous cylinder in Fig. P11-138 has a mass of 5 slugs and a radius of 1 ft. The variable couple $6t^2$ is in foot-pounds when t is in seconds. When $t = 0$, the angular velocity is zero. Determine

(a) The angular velocity when $t = 8$ sec.

(b) The value of t when the angular velocity is again zero.

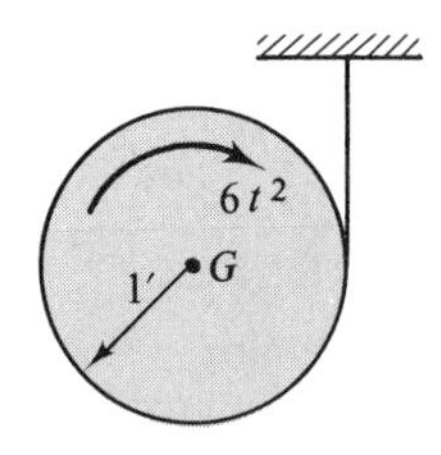

Figure P11-138

11-139 The homogeneous 64.4-lb cylinder in Fig. P11-139 has an angular velocity of 20 rad

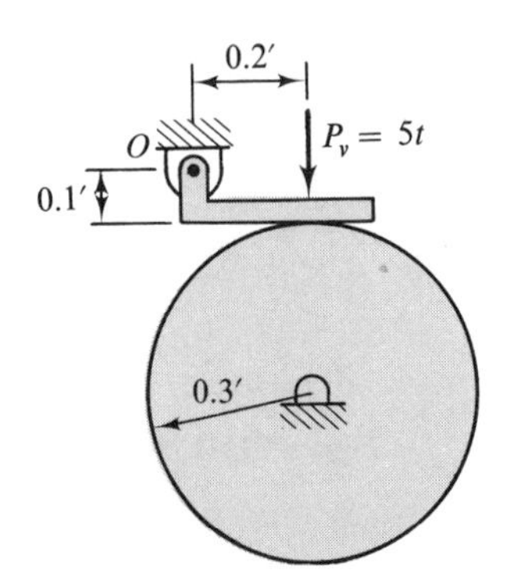

Figure P11-139

per sec clockwise when $t=0$. Determine the time required to reduce the angular velocity to 10 rad per sec clockwise. The force $\mathbf{P}_v$ is in pounds when t is in seconds, and the coefficient of friction between the brake and cylinder is 0.25. Neglect the weight of the brake.

***11-140** The homogeneous cylinder A in Fig. P11-140 has a mass of 80 kg, and the mass of the hub can be neglected. The magnitude of the variable force P_v is in newtons when t is in seconds. The velocity of body B changes from 2 m/s downward when $t=0$ to 4 m/s downward when $t=5$ s. Determine the mass of B.

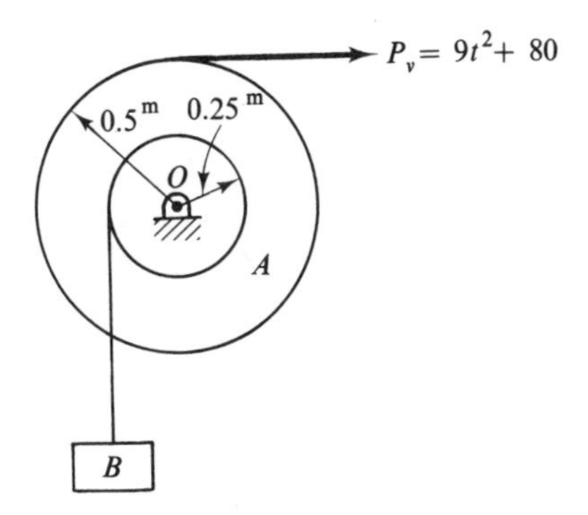

Figure P11-140

11-141 The symmetrical 64.4-lb drum A in Fig. P11-141 has a radius of gyration of mass of 6 in. with respect to its axis of rotation. The variable force $P=12t^3+3t^2$ is in pounds when t is in seconds. The coefficient of friction between the 32.2-lb body B and the horizontal plane is 0.5. When $t=0$, the angular velocity of A is 16 rad per sec counterclockwise. Determine the velocity of B when $t=2$ sec.

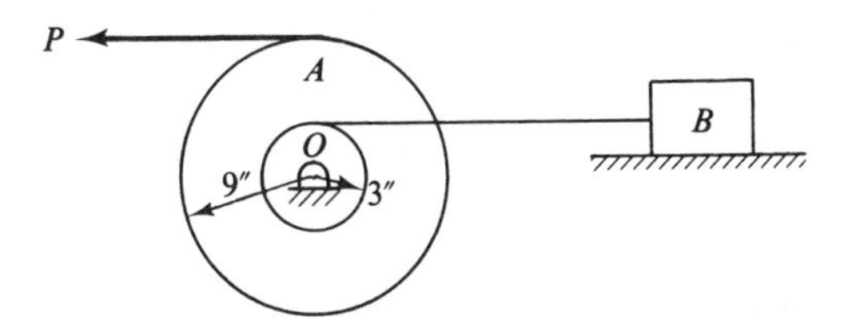

Figure P11-141

11-142 Wheel W in Fig. P11-142 weighs 644 lb, and body D weighs 322 lb. The radius of gyration of mass of W about its axis of rotation is 2.0 ft. The coefficient of friction between W and the brake is 0.50. Neglect the weight of the brake. The moment of the couple C varies according to the equation $C=kt^2$, where C is in foot-pounds when t is in seconds. The velocity of D changes from 20 fps downward when $t=0$ to 14 fps downward when $t=5$ sec. Determine the constant k.

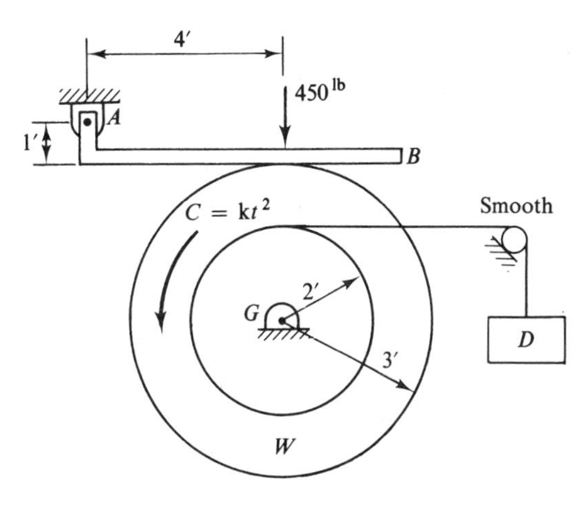

Figure P11-142

11-143 The 30-lb symmetrical body A in Fig. P11-143 has a radius of gyration of mass of 4 in. with respect to the horizontal axis through G. Body A rolls without slipping, and the moment of the variable couple $C=0.3t^2$ is in inch-pounds when t is in seconds. When $t=0$, the 4.0-lb body B has a velocity of 8 ips downward. Determine the angular velocity of A when $t=$ 10 sec.

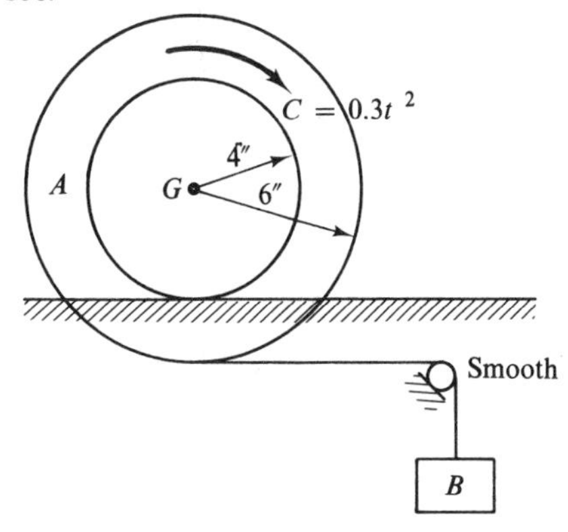

Figure P11-143

11-144 Body A in Fig. P11-144 weighs 30 lb, has a radius of gyration of mass with respect to a horizontal axis through its mass center of 3.0 in., and rolls without slipping. The magnitude of the moment of the variable couple is given by $C = 0.3t^2$, where C is in inch-pounds when t is in seconds. The angular velocity of body A changes from 20 rad per sec clockwise when $t = 0$ to 12 rad per sec counterclockwise when $t = 10$ sec. Determine the weight of body B.

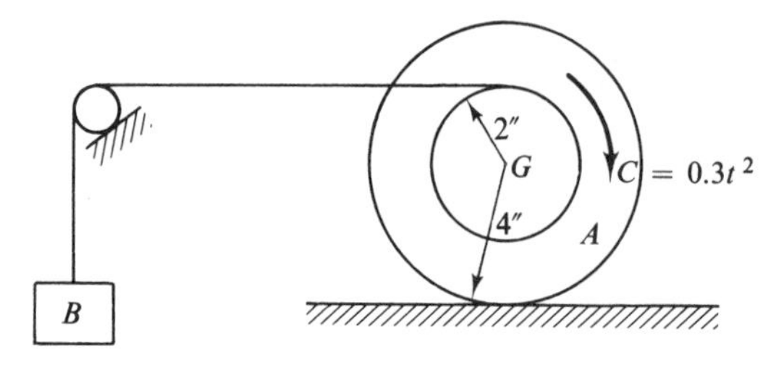

Figure P11-144

11-145 The symmetrical 644-lb body A in Fig. P11-145 rolls without slipping on the horizontal plane and has a radius of gyration of mass of 1.5 ft with respect to a horizontal axis through G. The homogeneous 128.8-lb disk B is acted upon by a variable couple $C_v = 3t^2$, where C_v is in foot-pounds when t is in seconds. The angular velocity of body A is 5 rad per sec clockwise when $t = 0$. Determine the angular velocity of body A 10 sec later.

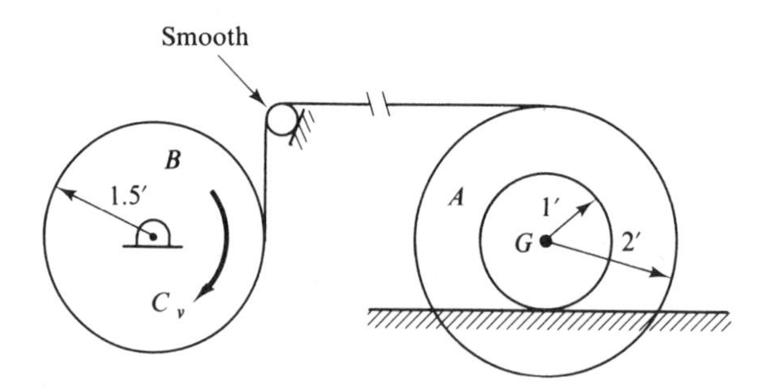

Figure P11-145

11-146 In Fig. P11-146, the homogeneous cylinder H weighs 322 lb and has a narrow slot cut into it for cord B, which has a negligible effect on its moment of inertia. Determine the time required for the center of the cylinder to reach a velocity of 9 fps to the right if it is at rest when $t = 0$. The variable force, of magnitude $P_v = 15t + 60$, is in pounds when t is in seconds. The coefficient of friction between H and the plane is 0.20.

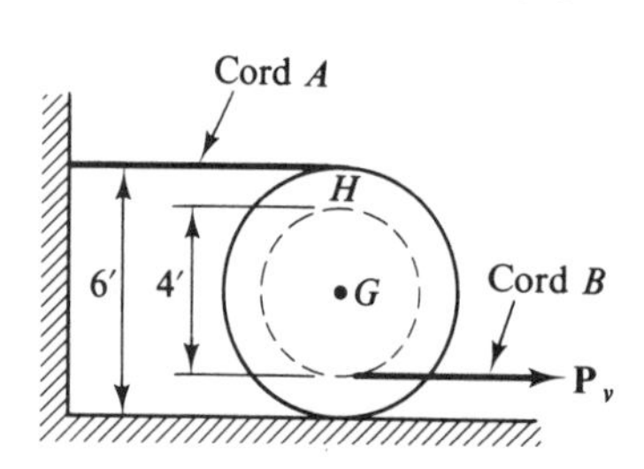

Figure P11-146

11-147 Body A in Fig. P11-147 weighs 322 lb and has a radius of gyration of mass with respect to a horizontal axis through G of 2 ft. Body B weighs 96.6 lb and when $t = 0$ has a velocity of 10 fps downward. The moment of the variable couple is given by the expression $C_v = 60t^2$, where C_v is in foot-pounds when t is in seconds. Determine the velocity of B when $t = 10$ sec.

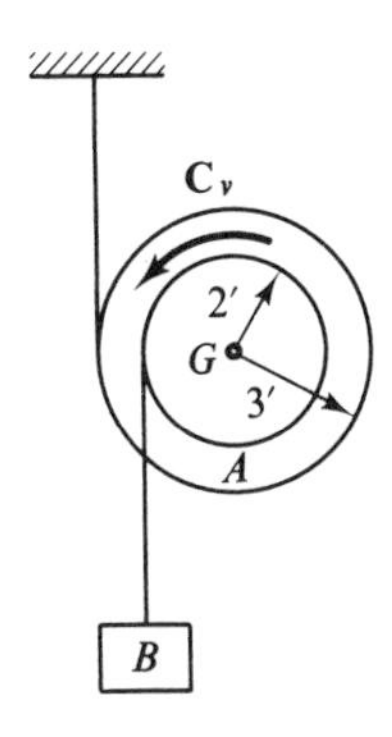

Figure P11-147

11-148 Body A in Fig. P11-148 weighs 60 lb and rolls without slipping along the horizontal plane. Body B is a homogeneous cylinder which weighs 10 lb. The moment of the variable torque $\mathbf{C}_v$ is given by the expression $C_v = 5.0 + 0.3t^2$, where C_v is in inch-pounds when t is in seconds. The angular velocity of A changes from 12 rad per sec counterclockwise when $t = 0$ to 30 rad per sec clockwise when $t = 5$ sec. Determine the radius of gyration of mass of A with respect to the horizontal axis through G.

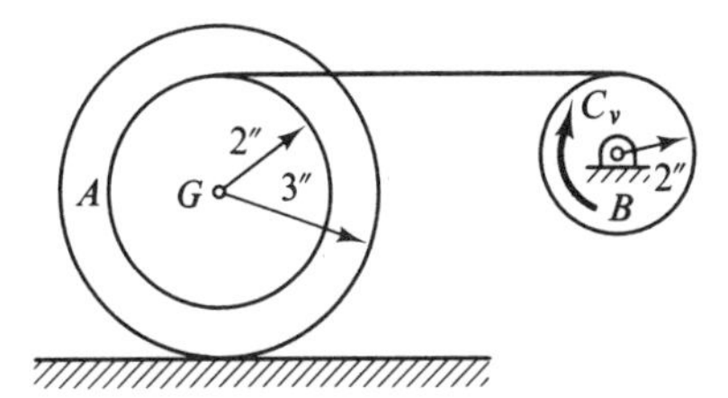

Figure P11-148

11-149 The 322-lb body A in Fig. P11-149 has a radius of gyration of mass with respect to the horizontal axis through the mass center of 1.5 ft. The moment of the variable couple $\mathbf{C}_v$ is equal to $60t^2$ in foot-pounds when t is in seconds. The 64.4-lb body B has a velocity of 6 fps downward when $t = 0$. Determine the angular velocity of body A when $t = 5$ sec.

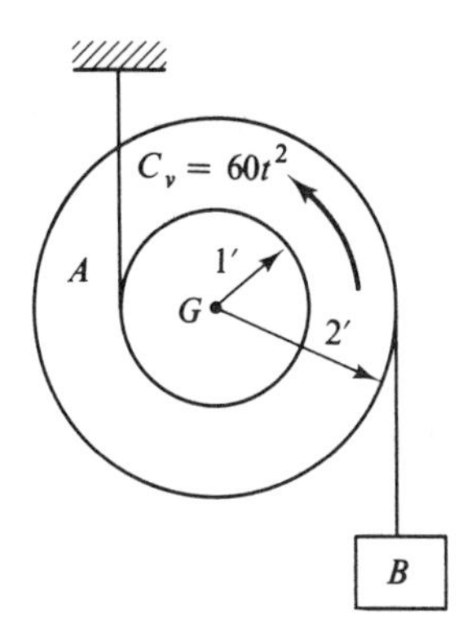

Figure P11-149

11-150 A flexible rope whose weight may be neglected has one end wrapped around the symmetrical pulley A and the other end around the solid homogeneous cylinder B as shown in Fig. P11-150. The pulley has a radius of gyration of mass with respect to a horizontal axis through O of 9 in. and weighs 16.1 lb. The cylinder B weighs 24.15 lb. There is no initial slack in the rope, and the system starts from rest. Determine the time required for the mass center of B to attain a velocity of 50 fps downward.

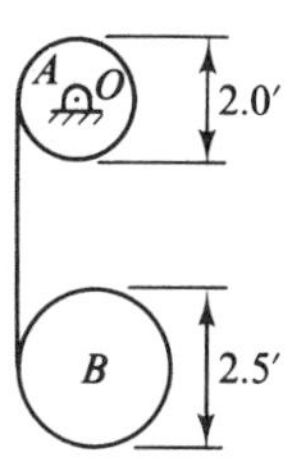

Figure P11-150

***11-151** The wheel in Fig. P11-151(a) has a mass of 30 kg and a radius of gyration of mass with respect to a horizontal axis through G of 20 cm and rolls without slipping on the horizontal track. The force P_v varies in magnitude in the manner indicated in Fig. P11-151(b), and the magnitude of the clockwise torque $\mathbf{T}_V$ varies according to the equation $T_v = 3t^2$, where T_v is in newton·meters when t is in seconds. When $t = 0$, the velocity of G is 60 cm/s to the right. Determine the velocity of G when $t = 5$ s.

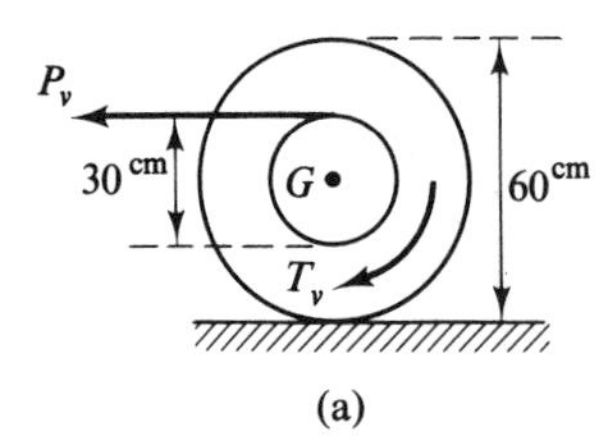

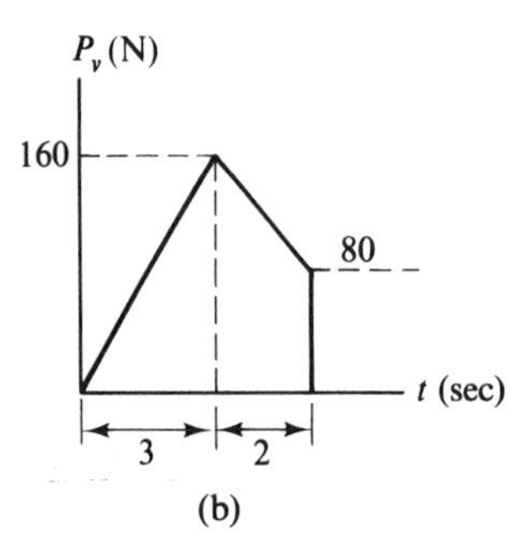

Figure P11-151

***11-152** The homogeneous sphere in Fig. P11-152 has a mass of 7.0 kg. It is held above the horizontal plane and rotates with an angular velocity of 30 rad/s clockwise. The sphere is then lowered to the plane. The coefficient of friction between the sphere and plane is 0.1. Determine

(a) The period of time before the velocity of G becomes constant.

(b) The velocity of G after it becomes constant.

(c) The distance G has moved when the sphere begins to roll without slipping.

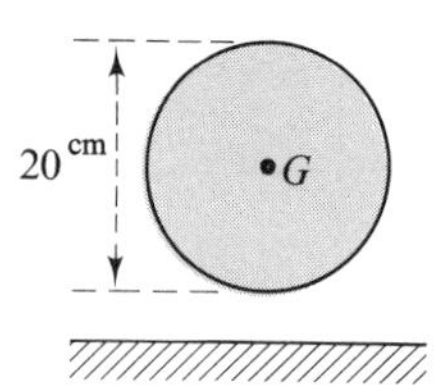

Figure P11-152

***11-153** The 150-kg homogeneous sphere A in Fig. P11-153 and the 40-kg body B are at rest when the 600-N force is applied to body B. Determine the velocity of the center of the sphere 0.5 s after the application of the force. Assume that the sphere does not slip on body B and that the friction between B and the horizontal plane is negligible.

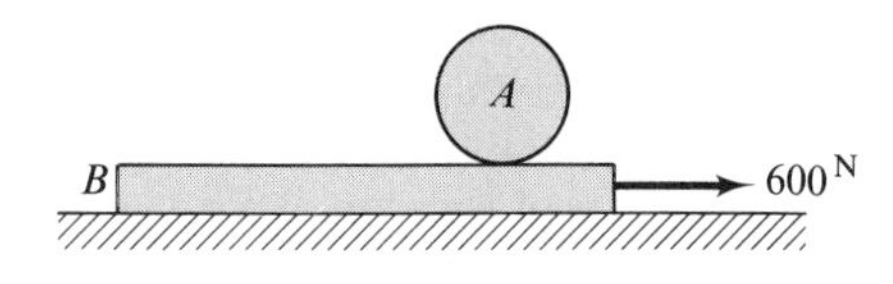

Figure P11-153

11-12 CONSERVATION OF ANGULAR MOMENTUM

It was shown in Art. 11-10 that for any system of particles, rigid or nonrigid, the time rate of change of angular momentum, with respect to either a point fixed in space or the mass center of the system of particles, is equal to the resultant moment of the external forces acting on the system about that same point. When the moment of the external forces acting on a system of particles with respect to either a fixed point or the mass center is zero, the time rate of change of the angular momentum about the corresponding point is zero from Eq. (11-24) or (11-25), and the angular momentum is constant, or in other words, it is conserved. Therefore, *the angular momentum about a fixed point or the mass center is conserved or remains constant when the moment of the external forces about that point is zero.* In mathematical form, this statement is

$$\mathbf{H}_f - \mathbf{H}_i = 0. \tag{11-29}$$

If the resultant moment with respect to any fixed point or the mass center is relatively small and the time interval during which the moment acts is very small, the angular impulse can usually be neglected, and the angular momentum of the system of bodies can be assumed to be constant without introducing an appreciable error. The principle of conservation of angular momentum may be observed by means of slow-motion pictures of fancy dives. When a diver wishes to turn over in the air several times, he doubles up, holding his knees close to his chest. His moment of inertia about an axis through his mass center is small, and his angular velocity is relatively large. Just before the diver goes into the water, he straightens out, increasing his moment of inertia and decreasing his angular velocity so that he has nearly stopped turning when he enters the water.

An important practical use of the conservation of angular momentum occurs in stopping the spin of a space vehicle in flight. The so-called yo-yo despinner used in the Navy's Hugo III photographic-reconnaissance rocket is an example of a device designed to take advantage of the conservation of angular momentum.[7] With this device, the angular velocity of the spinning rocket can be reduced to zero or to any specific amount by transferring the angular momentum of the spinning mass to a pair of small connected masses as they are allowed to unwind from the main body.

When two bodies collide or react in any manner and the angular impulse on the system is zero with respect to some fixed point the angular momentum of the system with respect to the same point is conserved. It should be emphasized, however, that some of the kinetic energy of the system is usually lost during the impact or reaction.

In addition to conservation of angular momentum, it is often useful to write equations indicating that the linear momentum of a body or system of bodies is conserved in one or more directions. If the bodies are elastic and rebound after impact, the coefficient of restitution can be used to relate the relative velocity of departure of the points of contact to their relative velocity of approach (see Art. 11-4). It should be noted that the velocities of the points of contact are, in general, different from the velocities of the mass centers of the bodies. When eccentric impact is involved, the coefficient of restitution relates the *components* of the relative velocities along the line of impact.

EXAMPLE 11-17

The homogeneous 20-kg bar AB in Fig. 11-31(a) is falling with a velocity of 10 m/s downward and no angular velocity when the pin at C (extending from both sides of the bar) strikes the fixed bearings at D

---

[7] J. V. Fedor, "Theory and Design Curves for a Yo-Yo De-Spin Mechanism for Satellites," *Technical Note D*-708 (Washington, D.C.: National Aeronautics and Space Administration, 1961).

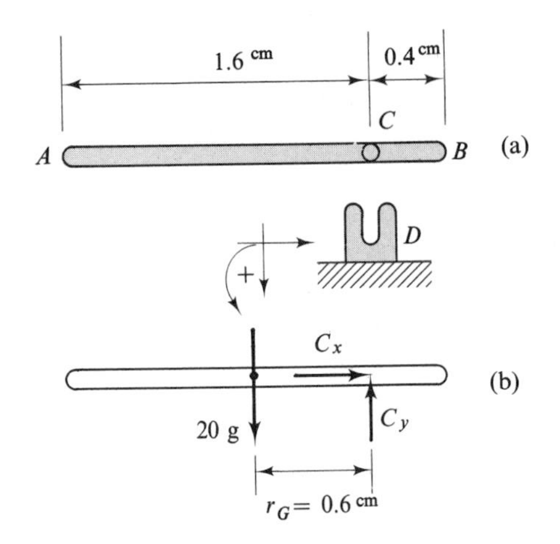

Figure 11-31

with plastic impact (no rebound). Determine
(a) The angular velocity of the bar immediately after impact.
(b) The loss of kinetic energy of the bar during the impact period.
(c) The horizontal component of the reaction of the bearing on the bar immediately after the impact.

SOLUTION

(a) A free-body diagram of the bar during the impact period is shown in Fig. 11-31(b). Since the time of the impact is very small, the angular impulse of the weight with respect to the fixed bearing at C may be neglected during the impact period, and the angular momentum of the body about an axis through C is conserved; that is,

$$\mathbf{H}_{Ci} = \mathbf{H}_{Cf},$$

$$mv_{Gi}r_G = I_{AR}\omega_f,$$

$$20(10)(0.6) = [\tfrac{1}{12}(20)(2)^2 + 20(0.6)^2]\omega_f$$

from which

$$\omega_f = 8.64 \text{ rad/s}$$

and

$$\boldsymbol{\omega}_f = \underline{8.65 \text{ rad/s}} \circlearrowleft. \qquad \text{Ans.}$$

(b) The kinetic energy before impact is

$$T_i = \tfrac{1}{2}mv_{Gi}^2 = \tfrac{1}{2}(20)(10)^2 = 1000 \text{ N·m},$$

and the kinetic energy just after impact is

$$T_f = \tfrac{1}{2}I_{AR}\omega_f^2 = \tfrac{1}{2}[\tfrac{1}{12}(20)(2)^2 + 20(0.6)^2](8.65)^2$$

$$= 519 \text{ N·m}.$$

The loss of kinetic energy is

$$T_{\text{loss}} = T_i - T_f = 1000 - 519$$

$$= \underline{481 \text{ N·m}}. \qquad \text{Ans.}$$

(c) The horizontal component of the reaction can be determined by means of Newton's equation of motion. The horizontal component of the acceleration of the mass center is $r_G\omega^2$ to the right, and the equation of motion in the horizontal direction is

$$\sum F_x = ma_{Gx} = mr_G\omega^2,$$

$$\overset{+}{\rightarrow} C_x = 20(0.6)(8.65)^2 = 899 \text{ N},$$

and

$$\mathbf{C}_x = \underline{899 \text{ N} \rightarrow}. \qquad \text{Ans.}$$

EXAMPLE 11-18

Bar AB in Fig. 11-32(a) weighs 25 lb and spins about its mass center on a smooth horizontal plane with an angular velocity of 5 rad per sec counterclockwise. Disk C weighs 5 lb and has a velocity of 50 fps as shown just before it strikes AB at point D. The coefficient of restitution between AB and C is 0.9. Determine the velocity of C just after impact and the corresponding angular velocity of AB.

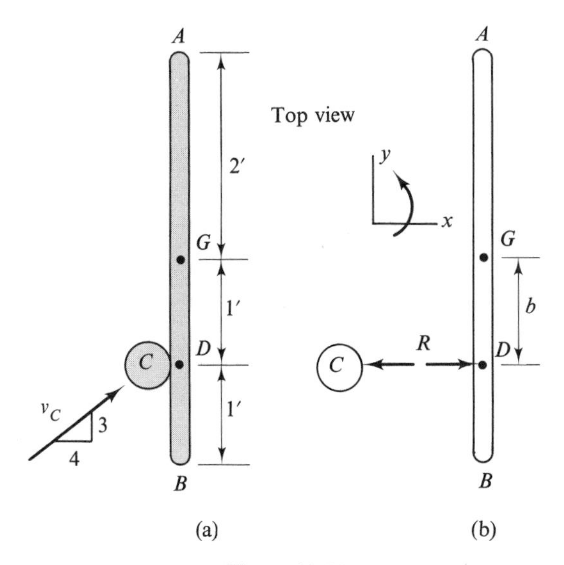

Figure 11-32

SOLUTION

Figure 11-32(b) is a free-body diagram of bar AB and disk C during the impact period, and it is seen that there is no impulse on either body in the y direction. Therefore, for C,

$$v_{Cyf} = v_{Cyi} = 0.6(50) = 30 \text{ fps}$$

and

$$\mathbf{v}_{Cyf} = 30 \text{ fps} \uparrow .$$

When the two bodies are considered as a system there are no external forces in the horizontal plane; therefore both the linear momentum and the angular momentum about any vertical axis must be conserved. All unknown linear and angular velocities are assumed to be in the positive directions as shown on the diagram. The conservation of linear momentum in the x direction gives

$$\sum mv_{xi} = \sum mv_{xf},$$

$$m_{AB}v_{Gxi} + m_C v_{Cxi} = m_{AB}v_{Gxf} + m_C v_{Cxf},$$

$$0 + \frac{5}{g}(40) = \frac{25}{g}v_{Gxf} + \frac{5}{g}v_{Cxf}$$

from which

$$40 = 5v_{Gxf} + v_{Cxf}. \tag{a}$$

The angular momentum of the system about any vertical axis is conserved. When the vertical axis through G is selected, the angular momentum equation becomes

$$\sum \mathbf{H}_{Gi} = \sum \mathbf{H}_{Gf},$$

$$m_C v_{Cxi}(b) + \tfrac{1}{12} m_{AB} L^2 \omega_i = m_C v_{Cxf}(b) + \tfrac{1}{12} m_{AB} L^2 \omega_f,$$

$$\frac{5}{g}(40)(1) + \frac{1}{12}\frac{25}{g}(4)^2(5) = \frac{5}{g} v_{Cxf}(1) + \frac{1}{12}\frac{25}{g}(4)^2 \omega_f$$

from which

$$220 = 3v_{Cxf} + 20\omega_f. \tag{b}$$

The coefficient of restitution relates the relative velocities of approach and departure of the points of contact; that is,

$$e = -\frac{v_{Cxf} - v_{Dxf}}{v_{Cxi} - v_{Dxi}},$$

$$0.9 = -\frac{v_{Cxf} - v_{Dxf}}{40 - v_{Dxi}},$$

or

$$v_{Cxf} - v_{Dxf} = -0.9(40 - v_{Dxi}). \tag{c}$$

Equations (a)–(c) contain five unknown quantities, v_{Gxf}, v_{Cxf}, ω_f, v_{Dxi}, and v_{Dxf}. The relative velocity equation for points D and G on body AB provides additional relationships for the initial and final velocities. The equation, using only the x components, is

$$v_{Dx} = v_{Gx} + v_{(D/G)x} = v_{Gx} + b\omega.$$

The initial velocity of D is

$$v_{Dxi} = v_{Gxi} + b\omega_i$$
$$= 0 + 1(5) = 5 \text{ fps}. \tag{d}$$

The final velocity of D, in the x direction, is

$$v_{Dxf} = v_{Gxf} + 1\omega_f. \tag{e}$$

Equation (d) can be substituted in Eq. (c) and the four equations (a), (b), (c), and (e) rearranged as follows:

$$5v_{Gxf} + v_{Cxf} = 40 \tag{a'}$$
$$\omega_f + 0.15v_{Cxf} = 11 \tag{b'}$$
$$v_{Cxf} - v_{Dxf} = -31.5 \tag{c'}$$
$$-v_{Gxf} - \omega_f + v_{Dxf} = 0 \tag{e'}$$

When Eq. (a′) is multiplied by 0.2, the four equations can be added together to eliminate three of the four unknowns; thus

$$v_{Cxf}(0.2 + 0.15 + 1) = 8 + 11 - 31.5$$

from which

$$v_{Cxf} = -\frac{12.5}{1.35} = -9.26 \text{ fps}$$

and

$$\mathbf{v}_{Cxf} = 9.26 \text{ fps} \leftarrow.$$

The final velocity of C is

$$\mathbf{v}_{Cf} = v_{Cxf}\mathbf{i} + v_{Cyf}\mathbf{j}$$

$$= -9.26\mathbf{i} + 30.0\mathbf{j}$$

$$= \underline{31.4 \text{ fps}} \quad 72.8^\circ \qquad \text{Ans.}$$

The value of v_{Cxf} can be substituted into Eq. (b′) to give

$$\omega_f = 12.39 \text{ rad per sec},$$

$$\boldsymbol{\omega}_f = \underline{12.39 \text{ rad per sec}} \circlearrowright. \qquad \text{Ans.}$$

PROBLEMS

***11-154** A slender homogeneous 2-m rod has a mass of 30 kg and rotates in a horizontal plane about a vertical axis which is 0.30 m from one end of the rod. The rod has an angular velocity of 4 rad/s clockwise when it is struck by a particle having a mass of 1 kg and a velocity of 120 m/s in the same direction as the velocity of the point of contact on the rod. The coefficient of restitution is zero. After impact the angular velocity of the rod is 8 rad/s. Determine the location of the point of impact relative to the axis of rotation.

11-155 The 3.0-ft bar in Fig. P11-155 has a mass of 0.10 slug and is suspended at rest in the vertical position. A ball with a mass of 0.01 slug is thrown horizontally with a velocity of 80 fps and strikes the lower end of the bar. After impact the ball rebounds with a velocity of 20 fps in the opposite direction. Determine
(a) The angular velocity of the bar after collision.
(b) The horizontal linear impulse imparted to the bar by the pivot during the impact period.
(c) The coefficient of restitution between the ball and bar.

11-156 The homogeneous bar in Fig. P11-156 weighs 64.4 lb and is free to rotate in a vertical plane about a smooth horizontal axis at O. The

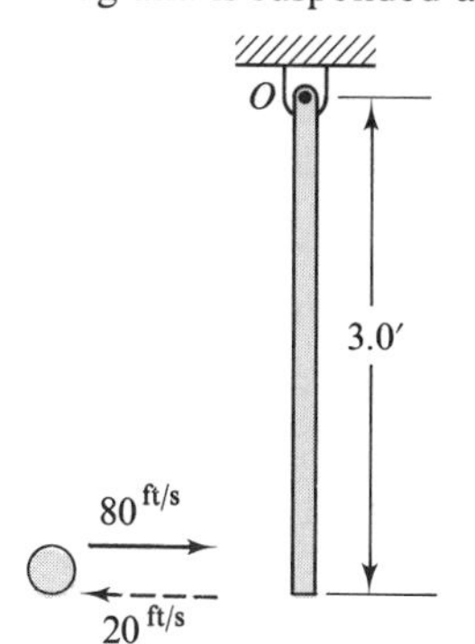

Figure P11-155

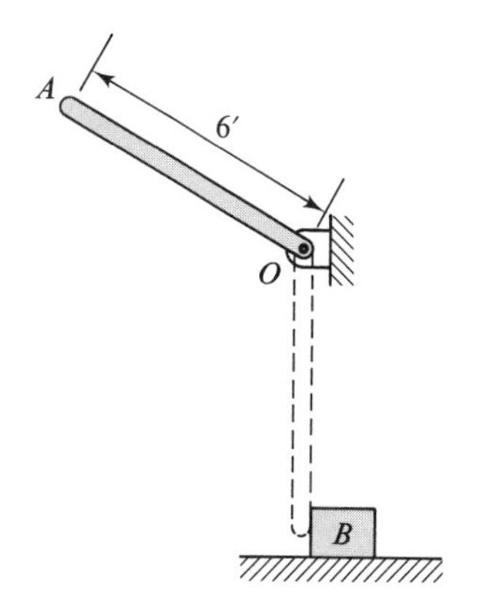

Figure P11-156

bar is released from rest in the OA position. The bar has an angular velocity of 5 rad per sec counterclockwise when it strikes the 16.1-lb block B, which is at rest on the smooth horizontal plane. The coefficient of restitution between the bar and block is 0.5. Determine the velocity of the block immediately after impact.

***11-157** Bar AB in Fig. P11-157 has a mass of 30 kg and rotates about a vertical axis through the end B. The small body C has a mass of 5 kg and has a hole through it so that it can slide on the bar. A cord D keeps C from sliding when the bar rotates. When C is 1 m from B as shown the magnitude of the angular velocity of the bar is 50 rad/s. The cord is cut, and C slides out to the stop. Determine
(a) The final angular velocity of the bar.
(b) The loss of kinetic energy in the system.

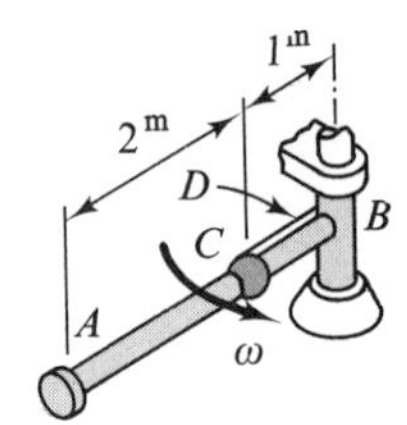

Figure P11-157

11-158 The horizontal turntable in Fig. P11-158 weighs 1000 lb and is mounted on bearings whose friction may be neglected. The radius of gyration of the mass of the turntable with respect to its axis of rotation is 6.0 ft. A 200-lb man stands on the turntable at A. The turntable and the man are initially at rest. The man starts walking along the circular path on the platform with a velocity, relative to the platform, of 6.0 fps. Determine the magnitude of the absolute velocity of the man.

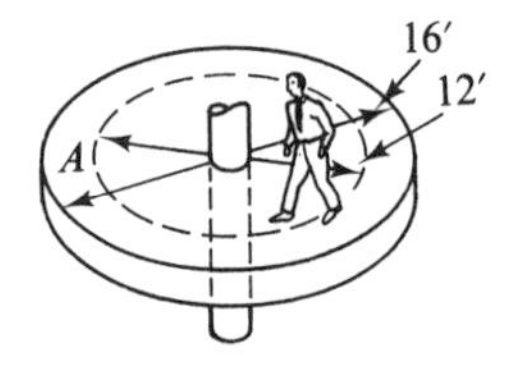

Figure P11-158

11-159 The slender homogeneous bar AB in Fig. P11-159 is rotating about its mass center with an angular velocity ω_0 clockwise when it strikes the rigid stop at D. The coefficient of restitution for AB and D is 0.75. Determine the angular velocity of AB and the linear velocity of G just after impact.

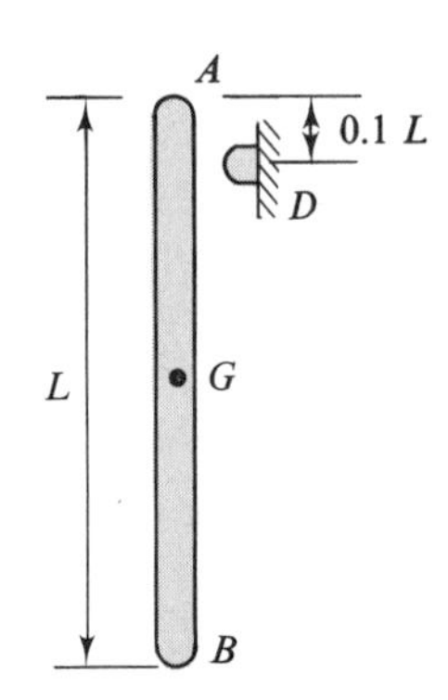

Figure P11-159

11-160 Solve Problem 11-159 if the rod weighs 10 lb and is 5 ft long and if the angular velocity of AB is 20 rad per sec clockwise and the velocity of G is 30 fps to the right just before impact.

11-161 In Fig. P11-161, A is a 4.0-ft-diameter homogeneous disk which weighs 322 lb and is keyed to a vertical shaft rotating at 10 rad per sec as shown. The homogeneous cylinder B is 2 ft in diameter, weighs 322 lb, is not keyed to the shaft, and is not rotating. The mass of the shaft may be neglected. If the cylinder B is allowed to slide down the shaft until it comes in contact with A, determine
(a) The angular velocity of the two bodies after slipping has ceased.
(b) The frictional moments exerted on B by A if slipping occurs for 4 sec.
(c) The percentage loss of kinetic energy of the system.

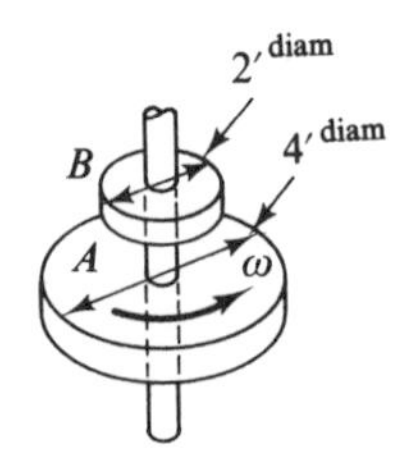

Figure P11-161

11-162 The two homogeneous disks A and B in Fig. P11-162 are free to rotate on the small horizontal shaft. Disk A is 12 in. in diameter and weighs 40 lb; B is 18 in. in diameter and weighs 30 lb. The helical spring connecting the two disks has a torsional modulus of 20 ft-lb per rad; that is, a moment of 20 ft-lb must be applied to B to turn it 1 rad when A is held stationary. Disk A is held stationary, while B is turned two revolutions clockwise looking to the left. The two disks are then released simultaneously. Determine the angular velocity of each disk when the spring is unwound.

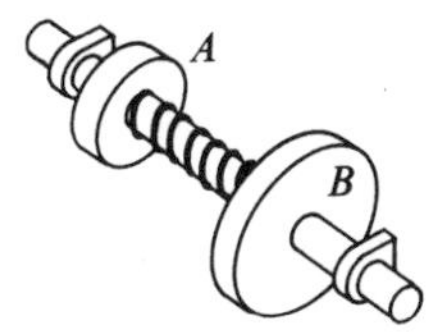

Figure P11-162

***11-163** A 10-kg homogeneous sphere S is fastened to the end of the 20-kg homogeneous rod OA as shown in Fig. P11-163. The 0.04-kg bullet B is fired with a horizontal velocity of 600 m/s. The bullet passes through the center of the sphere and emerges with a horizontal velocity of 100 m/s to the right. If the rod and sphere were stationary before impact, determine their angular velocity after the bullet passes completely through the sphere.

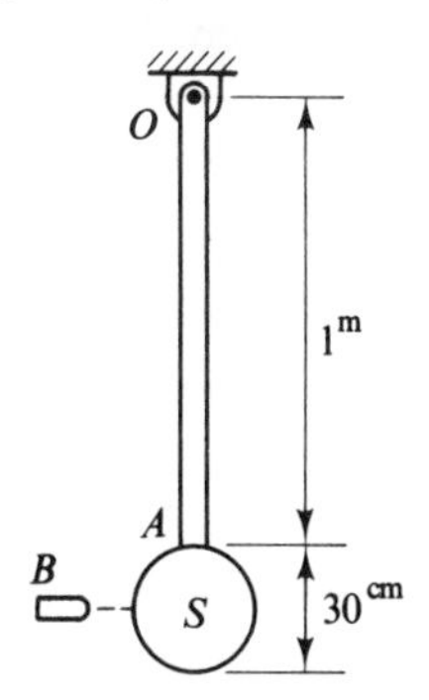

Figure P11-163

11-164 In Fig. P11-164 the smooth homogeneous 60-lb bar which is hanging vertically at rest is struck by the 6.0-lb body A with the velocity shown. If the coefficient of restitution is 0.8, determine the angular velocity of the bar immediately after impact.

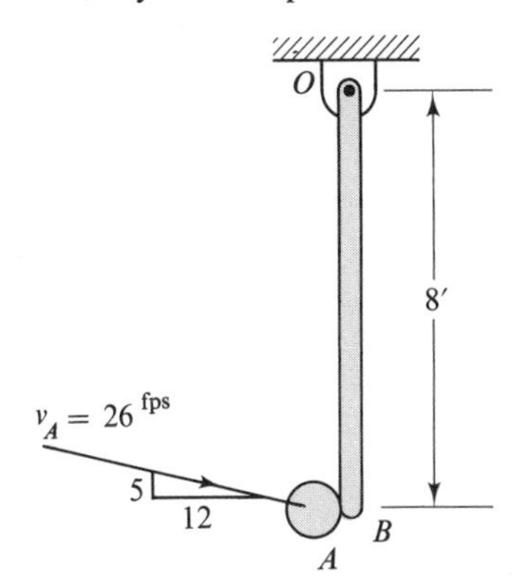

Figure P11-164

11-165 The homogeneous bar AB in Fig. P11-165 weighs 50 lb and rotates in a horizontal plane about a vertical axis at A. Body C weighs 10 lb and has a velocity of 20 fps as shown when it strikes the smooth bar. The bar has an angular velocity of 2 rad per sec counterclockwise just before impact. The coefficient of restitution is 0.4. Determine the velocity of body C immediately after impact.

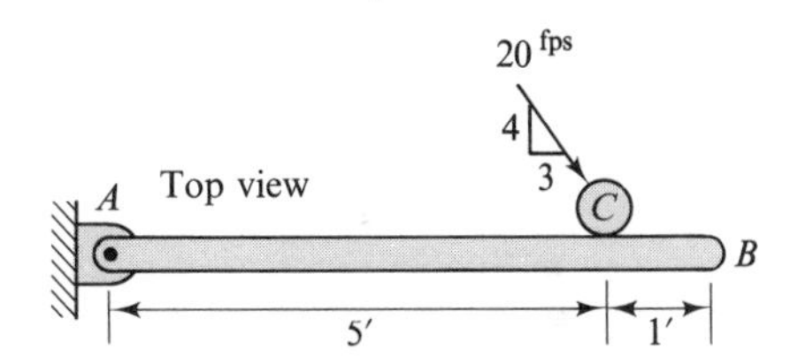

Figure P11-165

***11-166** The homogeneous 2-m bar in Fig. P11-166 (p. 714) has a mass of 3 kg and rotates about a vertical axis through A. The ball D (considered as a particle) has a mass of 5 kg and can slide along the smooth bar. In the position shown the angular velocity of the bar is 4 rad/s, and the ball is held in place by a cord which passes through the smooth bearing. The ball is then pulled in with the cord and held 0.5 m from the axis of rotation. Determine

(a) The angular velocity in the new position.

(b) The tensions in the cord in the initial and final positions.

(c) The work done by the force P.

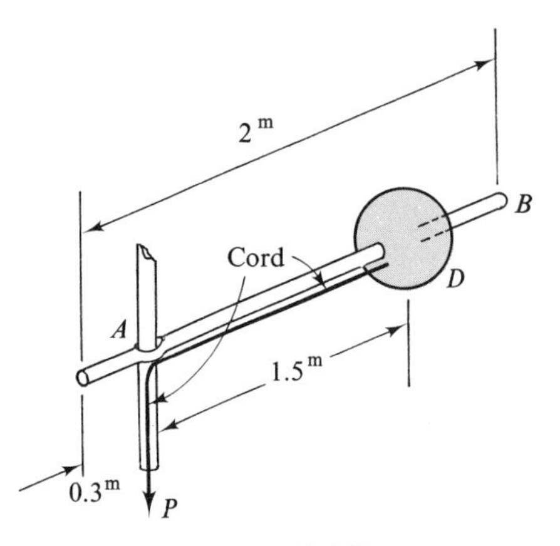

Figure P11-166

*11-167 When the 40-kg block A in Fig. P11-167 is in the position shown, 5 m of cord have been unwound from the drum. The drum is an 80-kg homogeneous cylinder and is initially at rest. Body A is dropped from rest from the indicated position. Determine the velocity of A just before the cord becomes taut and just after the impact when the speed of a point on the rim becomes equal to the speed of A.

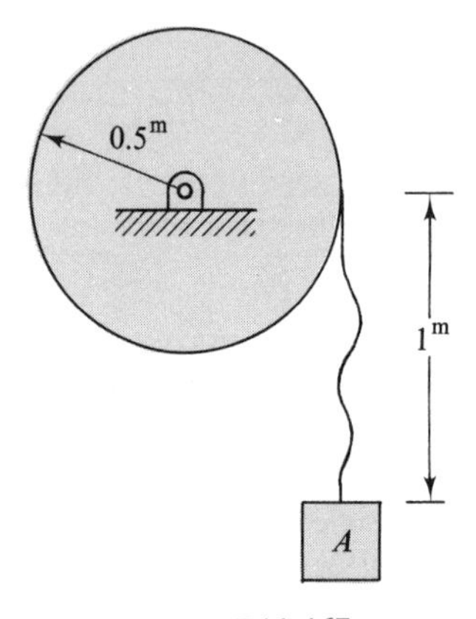

Figure P11-167

11-168 Body A in Fig. P11-168 weighs 5.0 lb, B weighs 8.0 lb, and C is a solid homogeneous disk which weighs 20.0 lb. The two cords fastened to A and B are wrapped around C and are fastened to C. When A and C are released from rest there is 5 ft of slack in the cord between B and C. How far will B be lifted before coming to rest?

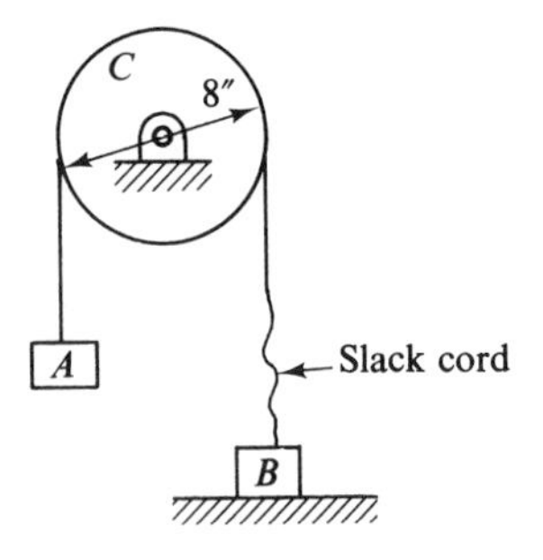

Figure P11-168

*11-169 The 5-kg homogeneous disk D in Fig. P11-169 rotates with a constant angular velocity of 10 rad/s counterclockwise. The 2-kg homogeneous rod BC rests on two smooth supports as shown. Rod BC is pushed to the left so that the small pin A on D strikes the end B of the rod. The coefficient of restitution for A and B is 0.80. Determine the angular velocity of each body just after the impact.

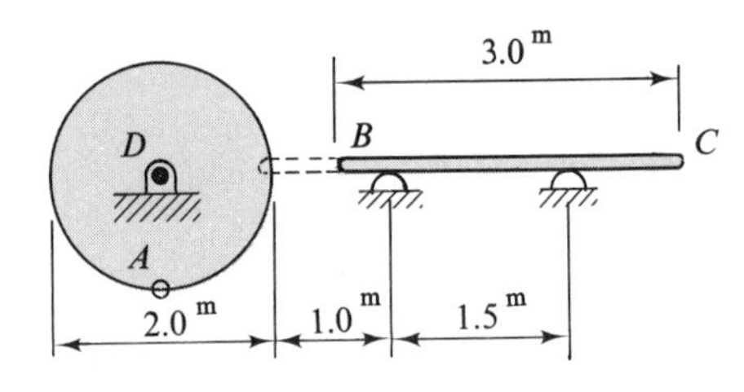

Figure P11-169

11-170 The homogeneous rectangular block A in Fig. P11-170 has a motion of translation with a velocity of 100 ips downward just before it strikes the *smooth* stop at B. The coefficient of restitution between A and B is 0.60. Determine

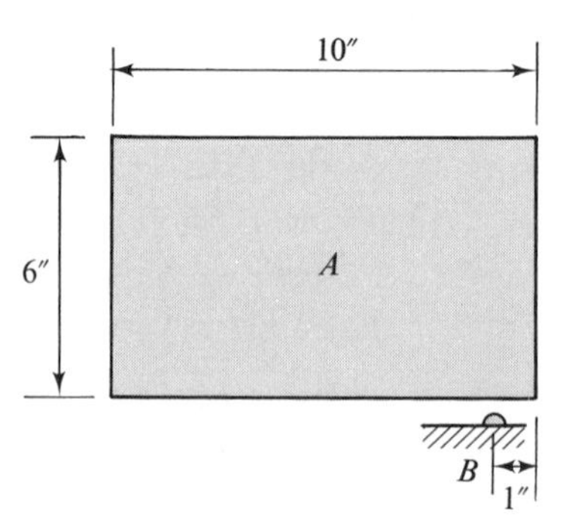

Figure P11-170

the angular velocity of A and the linear velocity of its mass center just after it strikes B.

11-171 Solve Problem 11-170 if block A has a small notch in it at B and there is no rebound after the impact.

11-172 The two identical slender bars in Fig. P11-172 spin on a smooth horizontal plane. Bar CD is pinned at D. Just before ends B and C collide the angular velocities are 30 rad per sec for AB and 5 rad per sec for CD, both counterclockwise, and the linear velocity of G is zero. The coefficient of restitution for B and C is 0.80. Determine the angular velocity of each bar after impact.

***11-173** The three identical bars in Fig. P11-173 each have a mass m and are welded to form a rigid structure which is initially at rest. Body A also has a mass m and has an initial velocity of 10 m/s to the left. The coefficient of restitution is 0.80. Determine the velocity of A and the angular velocity of BCD just after impact.

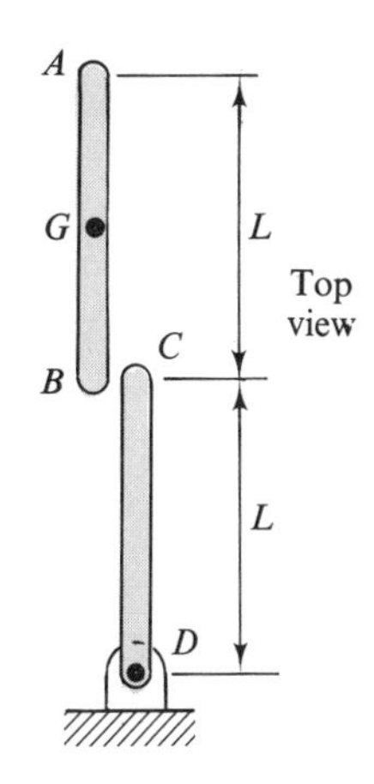

Figure P11-172

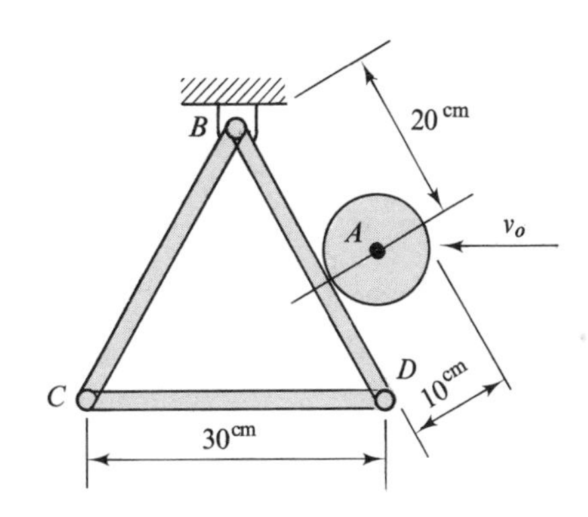

Figure P11-173

11-13 SYMMETRICAL TOPS AND GYROSCOPES

In the entire study of dynamics, perhaps no motion is more interesting than that of a symmetrical top or gyroscope. Gyroscopes have long since passed the stage of being children's toys and now find application in such instruments as automatic pilots, stabilized gun platforms, and inertial guidance systems of aircraft, missiles, and submarines. Gyroscopic effects are noticeable in a variety of instances where a moving piece of machinery contains rotating elements, such as the propeller or jet engine rotor of an airplane, the wheels of motorcycles and bicycles, and spinning bullets.

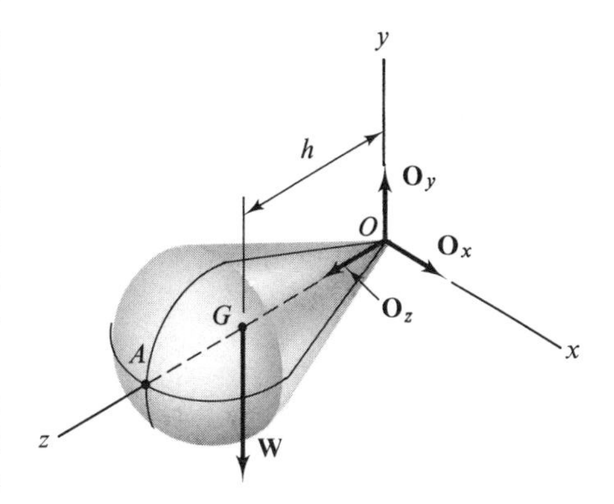

Figure 11-33

The body of Fig. 11-33 is symmetrical with respect to the z axis and supported by a ball and socket joint, with negligible friction, at O. If the body is released from rest from the position shown, it will rotate about the x axis, owing to the moment of the weight about O of

$$\mathbf{M}_O = Wh\mathbf{i}.$$

However, if the body is spinning about the z axis with a large angular velocity $\boldsymbol{\omega} = \omega\mathbf{k}$ when it is released, it will not rotate about the x axis but instead it will rotate or precess around the y axis. In this case, the x axis is called the *moment axis*, the y axis is called the *precession axis*, and the z axis is the *axis of spin. This tendency of a rapidly spinning body to turn about a second axis not parallel to the axis of spin, when acted on by a couple or torque about a third axis, is called gyroscopic action.*

Another important characteristic of a rapidly spinning body is its *rigidity in space* so effectively used for navigation. If the body in Fig. 11-33 is supported by frictionless supports called *gimbals* (see Fig. 11-34) which prevent any external moment from being applied to the body, the axis of spin will remain fixed in space. Both gyroscopic action and rigidity in space can be explained by means of Eq. (11-24),

$$\dot{\mathbf{H}}_O = \mathbf{M}_O,$$

where O is any point fixed in space.

Assume that a second ball and socket joint is placed at A in Fig. 11-33, and that the body is given a constant angular velocity $\boldsymbol{\omega}_z = \omega_z\mathbf{k}$. Since the body is not free to rotate about any axis other than the z axis, $\boldsymbol{\omega}_x$ and $\boldsymbol{\omega}_y$ are zero. Because of symmetry, both I_{xz} and I_{yz} are zero, and the mass center has zero velocity as it is on the axis of rotation. For this condition, Eq. (11-21a) becomes

$$\mathbf{H}_O = \omega_z I_z \mathbf{k}. \tag{a}$$

Gyroscopic action will take place when the support at A is gradually removed, allowing the external moment on the gyroscope with respect to O to become

$$\mathbf{M}_O = Wh\mathbf{i}.$$

Under these starting conditions, Eq. (11-24) becomes

$$\mathbf{M}_O = \dot{\mathbf{H}}_O$$

or

$$Wh\mathbf{i} = \frac{d}{dt}(\omega_z I_z \mathbf{k}) = \omega_z I_z \dot{\mathbf{k}} = \omega_z I_z (\boldsymbol{\omega} \times \mathbf{k}), \tag{b}$$

where $\boldsymbol{\omega}$ is the angular velocity of $\mathbf{k}$, that is, of the z axis. Equation (b) requires that $\boldsymbol{\omega} \times \mathbf{k}$ be in the direction of $\mathbf{i}$. Therefore,

$$\boldsymbol{\omega} = \omega_y \mathbf{j} \quad \text{and} \quad \boldsymbol{\omega} \times \mathbf{k} = \omega_y \mathbf{i}.$$

The value of ω_y can be obtained by substituting the preceding expression in Eq. (b) and is

$$\omega_y = \frac{Wh}{\omega_z I_z}. \tag{11-30}$$

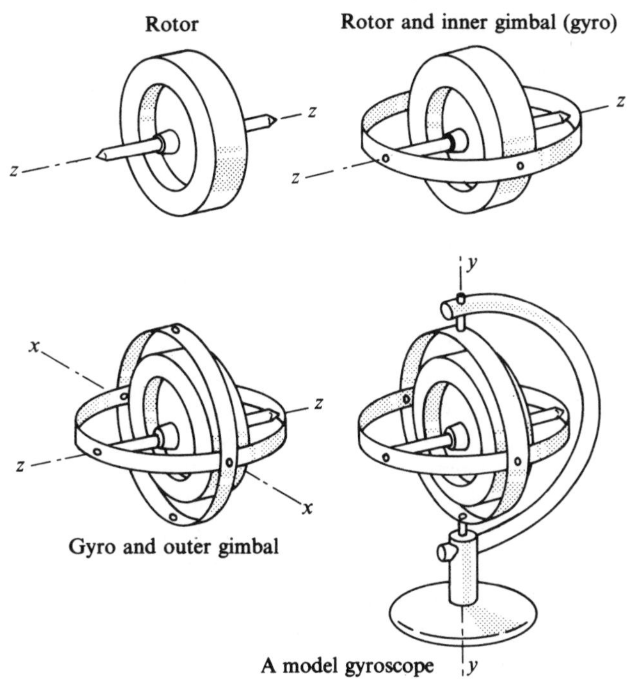

Figure 11-34

Equation (11-30) shows that the initial motion is that motion defined as gyroscopic action. Once the motion starts, Eq. (a) is no longer correct for $\mathbf{H}_O$ because ω_y and v_G are no longer zero. The fact that the motion continues to be gyroscopic action will be demonstrated with a more general case.

When a gyroscope is supported by gimbals (see Fig. 11-34) the external moment about any point will be zero, and the principle of angular momentum gives

$$\dot{\mathbf{H}}_O = 0 \quad \text{or} \quad \mathbf{H}_O = \mathbf{C} = \text{a constant.}$$

The angular momentum will be constant only if the gyroscope continues to spin about the *fixed* z axis and thus it will remain rigid in space.

The two major features of the motion of gyroscopes have been explained for a simple case. A derivation for a more general motion can be developed as follows.

The body A in Fig. 11-35 is symmetrical about the z axis and mounted in a ball and socket joint with negligible friction so that it is free to rotate about any axis through the fixed point O. The XYZ coordinate system is fixed in space, and the xyz coordinate system

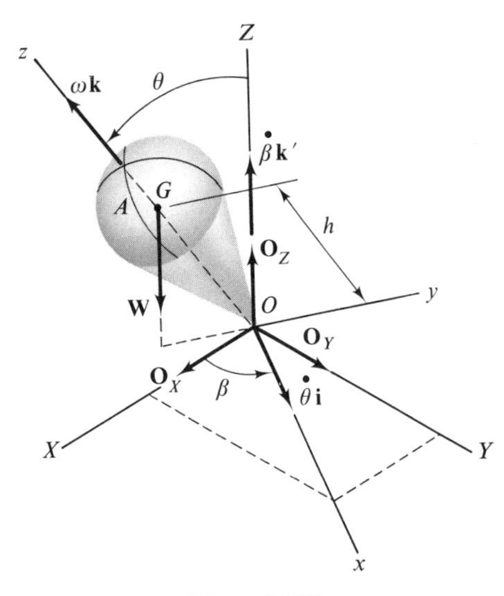

Figure 11-35

rotates about O in such a manner that the z axis is always the axis of symmetry of A and the x axis is always in the XY plane. The fact that the x axis is always in the XY plane in no way restricts or limits the orientation of the z axis. The x axis is retained in the XY plane for simplicity and convenience.

The relative angular velocity of body A with respect to the xyz coordinate system is $\boldsymbol{\omega} = \omega\mathbf{k}$, *and it is commonly called the rate of spin.* The angular velocity of the x axis with respect to the XYZ coordinate system is $\dot{\boldsymbol{\beta}} = \dot{\beta}\mathbf{k}'$ and is the *precessional angular velocity.* The angular velocity $\dot{\theta}\mathbf{i}$ (see Fig. 11-35) is called *nutation.* The general equations of motion of the gyroscope, in which $\omega\mathbf{k}$, $\dot{\beta}\mathbf{k}'$, and $\dot{\theta}\mathbf{i}$ are allowed to vary with time, have thus far defied exact solution even though attempts to solve them have been made for centuries. In almost all instrument applications of gyroscopes, nutation is undesirable, and much effort is expended to eliminate it.

In the following analysis, $\dot{\theta}$ is assumed to be zero and both ω and $\dot{\beta}$ are assumed to be constant. If $\dot{\theta}$ is zero, there will be no nutation, and θ will be constant. Consequently, the analysis is limited to steady-state precession.

The equation of motion for a gyroscope with the specified conditions is derived from the general relation $\dot{\mathbf{H}}_O = \mathbf{M}_O$. Body A is symmetrical with respect to the z axis; therefore,

$$I_{xz} = I_{yz} = 0 \quad \text{and} \quad I_{Gx} = I_{Gy} = I_G.$$

With these simplifications, Eq. (11-21a) for the angular momentum with respect to the fixed point O becomes

$$\begin{aligned} \mathbf{H}_O &= \mathbf{R} \times (m\mathbf{v}_G) + \mathbf{H}_G \\ &= (h\mathbf{k}) \times m[(\dot{\beta}\mathbf{k}') \times (h\mathbf{k})] + \omega_x I_G \mathbf{i} + \omega_y I_G \mathbf{j} + \omega_z I_z \mathbf{k}. \end{aligned} \tag{c}$$

Note that the unit vectors $\mathbf{i}$, $\mathbf{j}$, and $\mathbf{k}$ are in the direction of the x, y, and z axes and that the $\mathbf{i}'$, $\mathbf{j}'$, and $\mathbf{k}'$ unit vectors are in the direction of the X, Y, and Z axes, respectively. The vector $\dot{\beta}\mathbf{k}'$ is perpendicular to the x axis because the x axis is in the XY plane; therefore, the Z axis is in the yz plane. The magnitudes of the components of $\dot{\beta}\mathbf{k}'$ in the y and z directions are $\dot{\beta}\sin\theta$ and $\dot{\beta}\cos\theta$, respectively, and the components of the absolute angular velocity of the body A in the x, y, and z directions are

$$\left.\begin{aligned} \omega_x &= \dot{\theta} = 0, \\ \omega_y &= \dot{\beta}\sin\theta, \\ \text{and} \qquad \omega_z &= \omega + \dot{\beta}\cos\theta. \end{aligned}\right\} \tag{d}$$

When Eq. (c) is expanded, using these values, it becomes

$$\mathbf{H}_O = m\dot{\beta}h^2 \sin\theta\mathbf{j} + \dot{\beta}I_G \sin\theta\mathbf{j} + (\omega + \dot{\beta}\cos\theta)I_z\mathbf{k}. \quad \text{(e)}$$

Since $\dot{\theta}$ is zero, the angular velocity of the rotating coordinate system is $\dot{\beta}\mathbf{k}'$; thus,

$$\dot{\mathbf{j}} = (\dot{\beta}\mathbf{k}') \times \mathbf{j}$$

and

$$\dot{\mathbf{k}} = (\dot{\beta}\mathbf{k}') \times \mathbf{k}.$$

All the quantities in Eq. (e) are constant with respect to time except $\mathbf{j}$ and $\mathbf{k}$, and the time derivative of Eq. (e) gives

$$\dot{\mathbf{H}}_O = (mh^2 + I_G)\dot{\beta}\sin\theta(\dot{\beta}\mathbf{k}' \times \mathbf{j}) + (\omega + \dot{\beta}\cos\theta)I_z(\dot{\beta}\mathbf{k}' \times \mathbf{k}).$$

Since $\mathbf{k}' = \sin\theta\mathbf{j} + \cos\theta\mathbf{k}$, the expression for $\dot{\mathbf{H}}_O$ becomes

$$\begin{aligned}\dot{\mathbf{H}}_O &= (mh^2 + I_G)\dot{\beta}^2 \sin\theta[(\sin\theta\mathbf{j} + \cos\theta\mathbf{k}) \times \mathbf{j}] \\ &\quad + (\omega + \dot{\beta}\cos\theta)I_z\dot{\beta}[(\sin\theta\mathbf{j} + \cos\theta\mathbf{k}) \times \mathbf{k}] \\ &= [(I_z - mh^2 - I_G)\dot{\beta}^2 \sin\theta\cos\theta + \omega I_z\dot{\beta}\sin\theta]\mathbf{i}. \quad \text{(11-31)}\end{aligned}$$

The moment of the external forces about the origin O is (see Fig. 11-35)

$$\mathbf{M}_O = Wh(\sin\theta)\mathbf{i},$$

and from the principle of angular momentum,

$$\mathbf{M}_O = \dot{\mathbf{H}}_O,$$

which becomes

$$Wh\sin\theta\mathbf{i} = [(I_z - mh^2 - I_G)\dot{\beta}^2\cos\theta + \omega I_z\dot{\beta}](\sin\theta)\mathbf{i}$$

and

$$Wh = (I_z - I_y)\dot{\beta}^2\cos\theta + \omega I_z\dot{\beta}. \quad \text{(11-32)}$$

Note that $I_y = I_G + mh^2$.

Equation (11-32) can be used to determine the precessional angular velocity, $\dot{\beta}$, in terms of the magnitude of the spin angular velocity, ω, and the angle θ. Since the equation is a quadratic in $\dot{\beta}$ for values of θ other than 90°, there will, in general, be two distinct values of $\dot{\beta}$. By a more detailed analysis involving stability, it can be shown that, except in some unusual cases, the observed precessional angular velocity will be the smaller value obtained from Eq. (11-32).

For $\theta = 90°$, $\cos\theta$ is zero, and Eq. (11-32) reduces to Eq. (11-30); thus,

$$Wh = \omega I_z\dot{\beta}$$

or

$$\dot{\beta} = \frac{Wh}{\omega I_z}.$$

Equation (11-30) shows that as ω increases the precessional angular velocity decreases. In most instrument applications, it is desirable to have the precessional velocity as small as possible, which requires the use of a large rate of spin, ω. When ω is very large compared to $\dot{\beta}$, Eq. (11-30) approaches the exact solution of Eq. (11-32) for θ other than 90°.

Gyroscopic problems fall generally into two types: (1) The applied forces or moments are given, and either ω or $\dot{\beta}$ is required with the other known. (2) Both ω and $\dot{\beta}$ are given, and either the applied moment or the bearing reactions are required. In all cases, a free-body diagram is a necessary first step in the solution. When the moment is given the following additional steps are suggested:

1. Show the vector representation of $\mathbf{M}_O$ on the free-body diagram.
2. Calculate and show on the free-body diagram the y and z components of $\mathbf{H}_O$. One or both of these components will be in terms of an unknown angular velocity.
3. Determine the direction of rotation of $\dot{\beta}$ or ω by use of the fact that the gyroscope must precess in such a manner that the tip of the $\mathbf{H}_z$ or $\mathbf{H}$ vector moves in the direction of $\mathbf{M}$ when $\mathbf{H}_z$ or $\mathbf{H}$ is drawn with its tail at O.
4. Determine the magnitude of the unknown angular velocity from $\dot{\mathbf{H}}_O = \mathbf{M}_O$ or from Eq. (11-30) or (11-31).

When both angular velocities are given the following additional steps are suggested:

1. Determine $\mathbf{H}_O$ and show its z component on the free-body diagram. Use $\mathbf{H}_G$ when no point is fixed.
2. Determine the sense of $\mathbf{M}_O$ from the direction of precession and determine $\mathbf{M}_O$ from the equation $\dot{\mathbf{H}}_O = \mathbf{M}_O$.
3. If bearing reactions are required, use force-mass-acceleration relations as required.

The following examples illustrate these procedures.

EXAMPLE 11-19

A 6.00-in.-diameter thin homogeneous disk weighing 3.00 lb is mounted on a 9.00-in.-long rod of negligible weight as shown in Fig. 11-36(a). The disk is given a spin angular velocity of $-4800\mathbf{k}$ rpm, and the assembly is placed on the pedestal shown. Determine the steady-state precessional angular velocity of the rod.

SOLUTION

The free-body diagram of the disk and rod is shown in Fig. 11-36(b). The vector representation of $\mathbf{M}_O$ is shown on the free-body diagram in

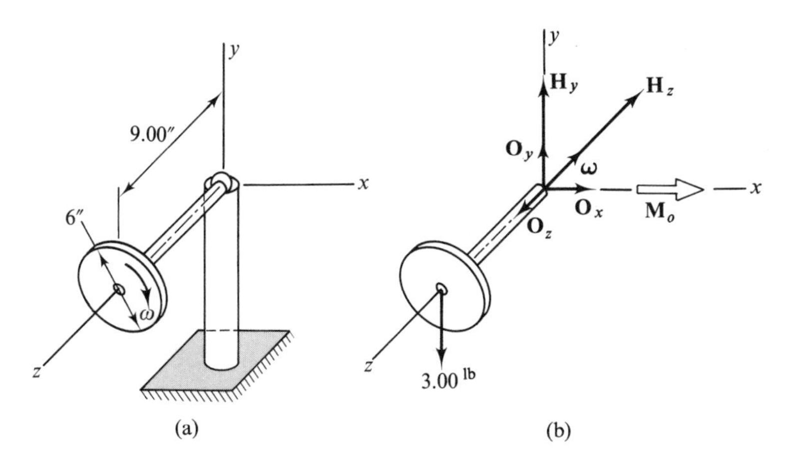

Figure 11-36

Fig. 11-36(b), and its value is

$$\mathbf{M}_O = 9(3.00)\mathbf{i} = 27.0\mathbf{i} \text{ in-lb.}$$

With no nutation, $\mathbf{H}_x = 0$, and $\mathbf{H}_y$ is constant in both magnitude and direction for steady-state precessional motion. The time derivative of $\mathbf{H}_y$ is zero, and thus $\mathbf{H}_y$ is not required for the solution. Therefore $\mathbf{H}_z$ is the only component of $\mathbf{H}_O$ required and is

$$\mathbf{H}_z = \omega_z I_z,$$

where

$$\omega_z = -4800\frac{2\pi}{60}\mathbf{k} = -503\mathbf{k} \text{ rad per sec}$$

and

$$I_z = \frac{1}{2}\frac{3.00}{386}(3)^2 = 0.0350 \text{ lb-sec}^2\text{-in.},$$

thus giving

$$\mathbf{H}_z = -503(0.0350)\mathbf{k} = -17.58\mathbf{k} \text{ lb-sec-in.}$$

The vector representation of $\mathbf{H}_z$ is shown on the free-body diagram. From the free-body diagram, $\mathbf{M}_O$ will move the tip of $\mathbf{H}_z$ clockwise about the y axis looking down on the figure. Since $\dot{\mathbf{k}}$ is equal to $\boldsymbol{\omega}_y \times \mathbf{k}$, the rate of change of $\mathbf{H}$ is

$$\dot{\mathbf{H}}_z = \dot{\mathbf{H}}_O = -17.58\dot{\mathbf{k}} = -17.58(\boldsymbol{\omega}_y \times \mathbf{k})$$
$$= -17.58[(\omega_y \mathbf{j}) \times \mathbf{k}] = -17.58\omega_y \mathbf{i}.$$

From the principle of angular momentum,

$$\dot{\mathbf{H}}_O = \mathbf{M}_O,$$
$$-17.58\omega_y \mathbf{i} = 27.0\mathbf{i}$$

and

$$\omega_y = \dot{\beta} = \frac{-27.0}{17.58} = -1.536$$

or

$$\boldsymbol{\omega}_y = \underline{-1.536\mathbf{j} \text{ rad per sec.}} \qquad \text{Ans.}$$

EXAMPLE 11-20

The body of Example 11-19 is placed in the position shown in Fig. 11-37(a). All data are as given in Example 11-19. Determine the precessional angular velocity and compare it with the solution of Example 11-19.

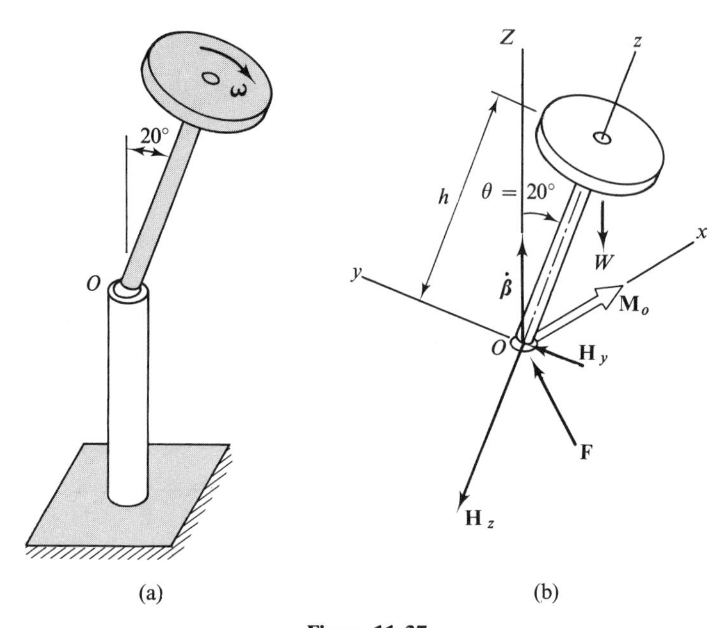

Figure 11-37

SOLUTION

Figure 11-37(b) shows a free-body diagram of the gyroscope. The force **F** is the reaction of the support at O. The vector representation of $\mathbf{M}_O$ is shown on the free-body diagram, and its value is

$$\mathbf{M}_O = Wh \sin \theta \mathbf{i}.$$

The angular velocities are [see Eq. (d)]

$$\boldsymbol{\omega}_x = 0, \qquad \boldsymbol{\omega}_y = \dot{\beta} \sin \theta \mathbf{j}, \qquad \boldsymbol{\omega}_z = (\dot{\beta} \cos \theta - \omega)\mathbf{k},$$

where ω is the angular velocity of spin and $\dot{\beta}$ is assumed to be upward (in the positive Z direction). The angular momentum about O is

$$\mathbf{H}_O = \mathbf{H}_y + \mathbf{H}_z = I_y \dot{\beta} \sin \theta \mathbf{j} + I_z(\dot{\beta} \cos \theta - \omega)\mathbf{k}.$$

The only variables in this equation are **j** and **k**, and their derivatives can be determined as follows:

$$\dot{\boldsymbol{\beta}} = \dot{\beta}\mathbf{k}' = \dot{\beta}(\sin\theta\mathbf{j} + \cos\theta\mathbf{k}),$$

$$\dot{\mathbf{j}} = \dot{\boldsymbol{\beta}} \times \mathbf{j} = -\dot{\beta}\cos\theta\mathbf{i},$$

$$\dot{\mathbf{k}} = \dot{\boldsymbol{\beta}} \times \mathbf{k} = \dot{\beta}\sin\theta\mathbf{i}.$$

The moment of the external forces about O is equal to the time rate of change of the angular momentum about O; therefore,

$$\mathbf{M}_O = \dot{\mathbf{H}}_O$$

or

$$Wh\sin\theta\mathbf{i} = I_y\dot{\beta}\sin\theta(-\dot{\beta}\cos\theta\mathbf{i}) + I_z(\dot{\beta}\cos\theta - \omega)(\dot{\beta}\sin\theta\mathbf{i})$$

from which

$$\dot{\beta}^2\cos\theta(I_z - I_y) - \dot{\beta}\omega I_z = Wh.$$

Numerical values for W and h are given in the problem statement. The value of ω is 503 rad per sec (see Example 11-19). The moments of inertia (see Chapter 6 or Appendix B) are

$$\begin{aligned} I_y &= I_G + mh^2 \\ &= \frac{mR^2}{4} + mh^2 \\ &= \frac{3.00}{386}\left(\frac{3.00^2}{4} + 9.00^2\right) = 0.6470 \text{ lb-sec}^2\text{-in.} \end{aligned}$$

and

$$I_z = 0.0350 \text{ lb-sec}^2\text{-in.} \qquad \text{(from Example 11-19).}$$

When the data are substituted into the quadratic equation for $\dot{\beta}$, it becomes

$$\dot{\beta}^2(\cos 20°)(0.0350 - 0.6470) - \dot{\beta}(503)(0.0350) = 3.00(9.00),$$

which reduces to

$$\dot{\beta}^2 + 30.57\dot{\beta} + 46.95 = 0.$$

The roots of this quadratic equation are

$$\dot{\beta} = -1.622 \quad \text{and} \quad -28.95.$$

The smaller value is the correct solution [see the comment following Eq. (11-32)], and the result is

$$\underline{\dot{\boldsymbol{\beta}} = -1.622\mathbf{k} \text{ rad per sec.}} \qquad \text{Ans.}$$

When this result is compared with the solution to Example 11-19, the error involved in assuming $\dot{\beta}$ to be independent of θ is

$$\text{error} = \frac{1.622 - 1.536}{1.622} = 5.30\%.$$

EXAMPLE 11-21

The rotating element of a jet engine weighs 193.2 lb and spins at 12,000**k** rpm. The radius of gyration of the mass with respect to the axis of rotation is 9.00 in., and the rotor is symmetrical about the axis AB in Fig. 11-38(a). Determine the bearing reactions on the rotor as the plane in which the engine is mounted makes a horizontal turn to the left with a radius of 4000 ft at a velocity of 480 mph.

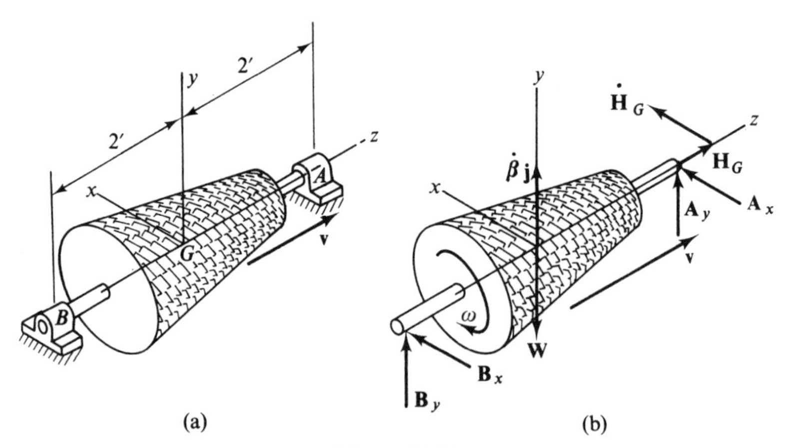

Figure 11-38

SOLUTION

Figure 11-38(b) is a free-body diagram of the rotor. The forward speed of the plane is

$$v = 480 \text{ mph} = 704 \text{ fps}.$$

Therefore, the precessional angular velocity is

$$\dot{\boldsymbol{\beta}} = \dot{\beta}\mathbf{j} = \frac{v}{R}\mathbf{j} = \frac{704}{4000}\mathbf{j} = 0.1760\mathbf{j} \text{ rad per sec},$$

and the acceleration of the mass center is

$$\mathbf{a}_G = \frac{v^2}{R}\mathbf{i} = \left[\frac{(704)^2}{4000}\right]\mathbf{i} = 123.9\mathbf{i} \text{ fps}^2.$$

The rotor is symmetrical about the z axis; therefore, all products of inertia are zero. Also ω_x is zero from the statement of the problem. Therefore the component of the angular momentum along the z axis is

$$\mathbf{H}_z = \omega_z I_z \mathbf{k} = \frac{12{,}000(2\pi)}{60}\,\frac{193.2}{32.2}\left(\frac{9}{12}\right)^2\mathbf{k}$$

$$= 4241\mathbf{k} \text{ ft-lb-sec}.$$

Since $\mathbf{H}_y$ is constant in both magnitude and direction, its derivative will be zero. Consequently,

$$\dot{\mathbf{H}}_G = \dot{\mathbf{H}}_z = 4241\dot{\mathbf{k}} = 4241[(\dot{\beta}\mathbf{j})\times\mathbf{k}] = 4241(0.1760)\mathbf{i}$$

$$= 746\mathbf{i} \text{ ft-lb}.$$

From the free-body diagram in Fig. 11-38(b), the moment of the external forces about G is

$$\sum \mathbf{M}_G = (2\mathbf{k})\times(A_x\mathbf{i}+A_y\mathbf{j})+(-2\mathbf{k})\times(B_x\mathbf{i}+B_y\mathbf{j}),$$

and this moment must be equal to $\dot{\mathbf{H}}_G$. Therefore,

$$746\mathbf{i}=(2A_x-2B_x)\mathbf{j}+(2B_y-2A_y)\mathbf{i}.$$

The component equations from this vector equation are

$$A_x = B_x$$

and

$$B_y - A_y = 373.$$

The force equation of motion $\sum \mathbf{F} = m\mathbf{a}_G$ gives

$$(A_x\mathbf{i}+A_y\mathbf{j})+(B_x\mathbf{i}+B_y\mathbf{j})-193.2\mathbf{j}=\frac{193.2}{32.2}\,123.9\mathbf{i},$$

or in component form,

$$A_x + B_x = 743.4,$$

and

$$A_y + B_y - 193.2 = 0.$$

The simultaneous solution of the four component equations gives

$$A_x = B_x = 371.7 \text{ lb}, \qquad A_y = -89.9 \text{ lb}, \qquad B_y = 283.1 \text{ lb}.$$

The bearing reactions are

$$\mathbf{A} = 371.7\mathbf{i} - 89.9\mathbf{j} = \underline{382(0.972\mathbf{i} - 0.235\mathbf{j}) \text{ lb}} \qquad \text{Ans.}$$

and

$$\mathbf{B} = 371.7\mathbf{i} + 283.1\mathbf{j} = \underline{467(0.796\mathbf{i} + 0.606\mathbf{j}) \text{ lb.}} \qquad \text{Ans.}$$

PROBLEMS

11-174 The homogeneous disk in Fig. P11-174 weighs 6.0 lb and rotates about bearings at the end of the shaft with an angular velocity of 3000 rpm in the direction shown. Body B weighs 10 lb, and the weight of the shaft is negligible. Determine the angular velocity of precession when the distance d is (a) 6 in., (b) 10 in.

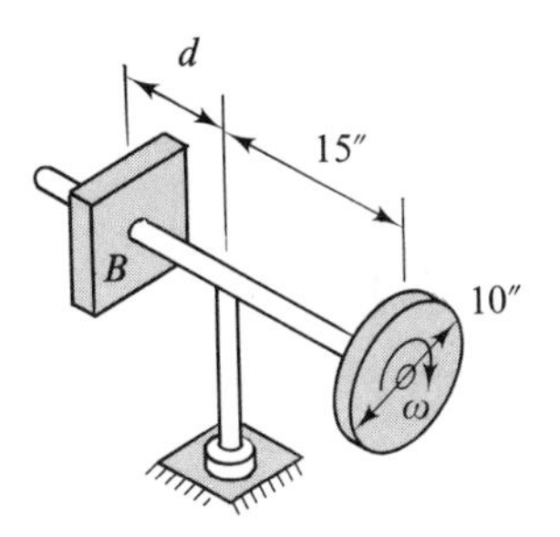

Figure P11-174

***11-175** A homogeneous 20-cm-diameter disk has a mass of 1 kg and is placed on a pedestal so that the assembly precesses in a horizontal plane as shown in Fig. P11-175. If the precessional rate is 20**i** rpm, determine the magnitude and direction of the spin of the disk about its own axle.

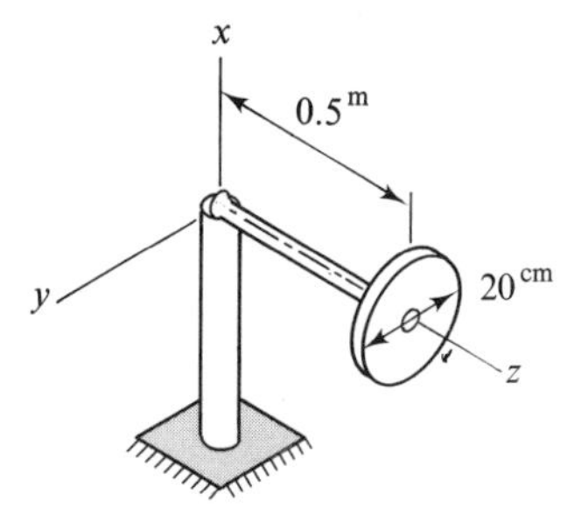

Figure P11-175

11-176 The armature of a motor is mounted on a rotating turntable as shown in Fig. P11-176. The 161-lb armature has a radius of gyration of mass of 0.5 ft with respect to its geometric axis. Assume that the armature is symmetrical with respect to the bearings, which are 2.0 ft between centers. Neglect bearing friction. The angular velocity of the armature is 1200 rpm in the direction indicated, and the angular velocity of the turntable is 4 rad per sec clockwise looking downward. Determine the components of the bearing reactions at A and B on the armature.

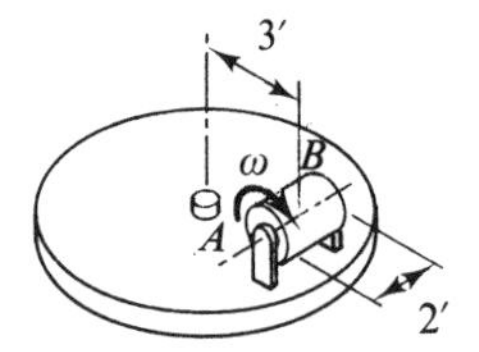

Figure P11-176

11-177 The propeller of a small airplane weighs 25.0 lb and has a radius of gyration of mass of 1.8 ft with respect to its axis of rotation. The plane has a speed of 120 mph when the propeller is rotating at 2000 rpm clockwise when viewed from the rear. The airplane is flying a level course when it makes a turn to the right on a 2000-ft radius. Will the gyroscopic effect of the propeller cause the plane to nose upward or downward? Determine the moment of the propeller on the air frame.

11-178 The rotor of the turbine in Fig. P11-178 may be approximated by a 16.1-lb homogeneous disk with a 6-in. radius. The turbine rotates clockwise at 12,000 rpm as viewed from the positive z axis. Determine the components of the forces exerted by the bearings on axle AB if the angular velocity of the turbine housing is $\omega_x = 2$ rad per sec as shown.

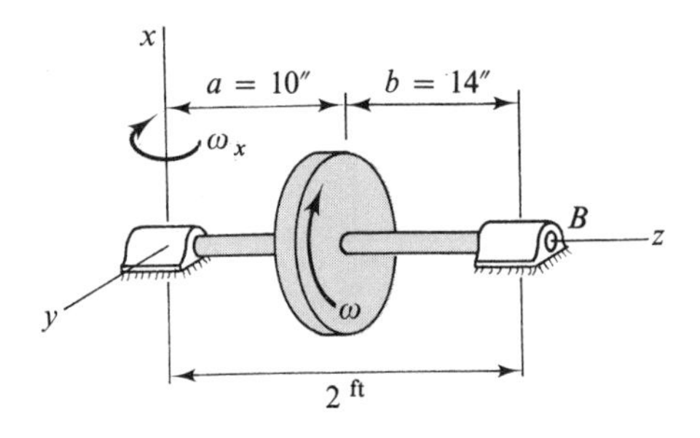

Figure P11-178

11-179 The top in Fig. P11-179 weighs 0.80 lb. The principal radii of the gyration of mass for the origin are 0.60 in. for the axis of symmetry and 2.10 in. for any axis through the origin perpendicular to the axis of symmetry. The top spins about its geometrical axis with an angular velocity of 3600 rpm clockwise, looking down on the top. When the angle θ is 30°, determine the angular velocity of precession of the top about the Z axis.

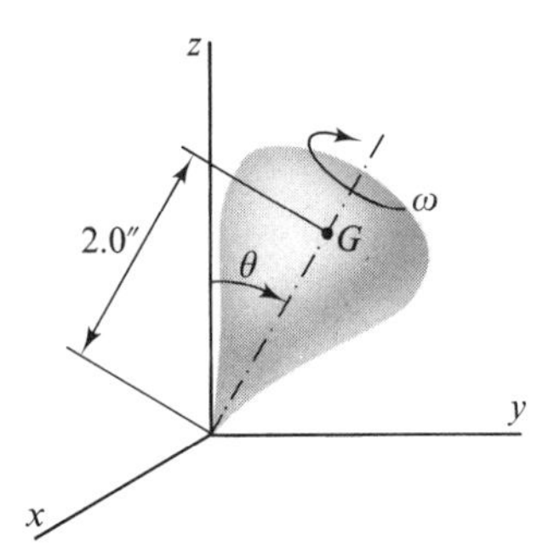

Figure P11-179

11-180 A homogeneous disk of radius r and weight w is traveling at a uniform speed, v, in a circular path of radius R on a horizontal plane. Due to gyroscopic effects, the disk leans at an angle θ, as shown in Fig. P11-180. Prove that the angle θ is given by

$$\tan\theta = \frac{3}{2g}\frac{v^2}{R}.$$

The wheel does not slip on the plane.

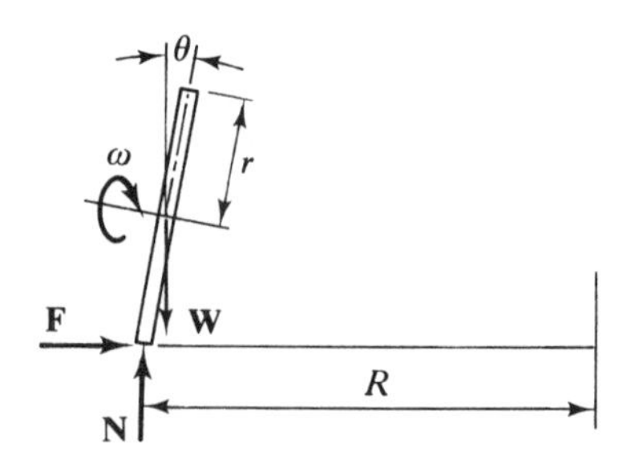

Figure P11-180

11-181 Two identical homogeneous cylinders with small flanges are mounted on a rigid axle as shown in Fig. P11-181. The wheels are set in bearings so each may rotate independently about the axle. The assembly is moving on a circular track on a horizontal plane as shown, and each wheel rolls without slipping. At a sufficiently large angular velocity, $\dot{\beta}$, one of the wheels will be on the verge of lifting off the track.

(a) Which wheel tends to lift off the track?

(b) Prove that the angular velocity for this condition is given by

$$\dot{\beta}^2 = \frac{2gL}{[3r(2R+L)]}$$

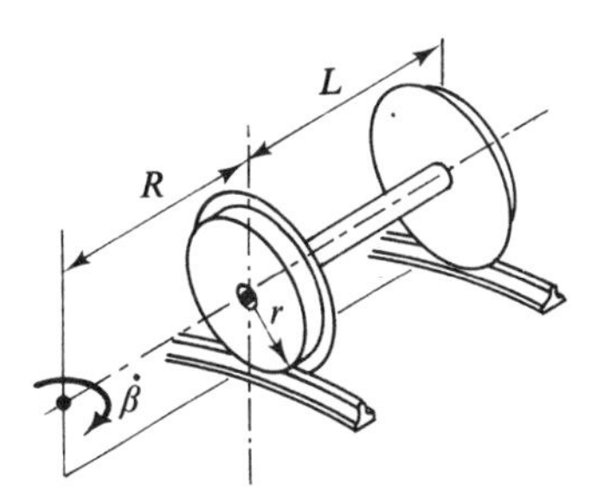

Figure P11-181

***11-182** The 20-cm-diameter homogeneous disk in Fig. P11-182 has a mass of 1.5 kg and is mounted on a 25-cm rod of negligible mass. The disk has a spin angular velocity of 5400**k** rpm. The rod makes an angle of 25° with the vertical axis through O. Determine the steady-state precessional angular velocity about the Z axis.

11-183 The rotor of a free gyroscope (one with two gimbals as in Fig. 11-34) is kept spinning at a high speed. The gyroscope is placed at the equator with its spin axis vertical (normal to the earth's surface) at noon. Determine the position of the spin axis relative to the normal to the earth's surface at 9:00 P.M.

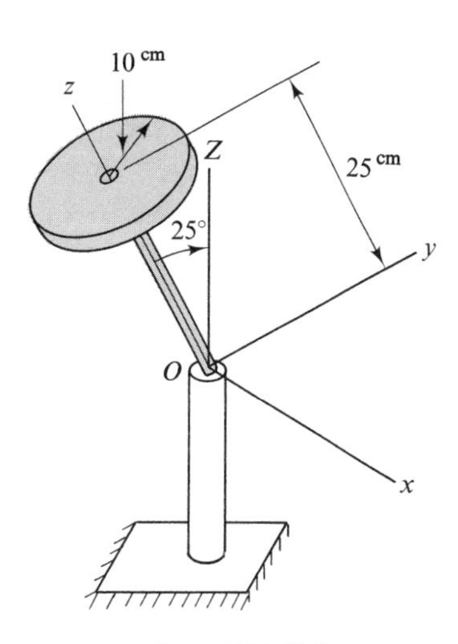

Figure P11-182

11-184 The gyroscope in Problem 11-183 is placed at a point with a latitude of 30° north with its spin axis vertical at noon. Determine the angles the spin axis will make with the vertical and with the north-south meridian at 6:00 P.M.

11-185 The 3.00-lb disk in Fig. P11-185 is mounted on a massless shaft and collar which rotate with a constant angular velocity of 20**j** rad per sec. The disk rolls without slipping on the horizontal surface. The spring has a constant of 10 lb per in and is adjusted so that the shaft remains horizontal. Determine

(a) The normal force between the disk and the plane.

(b) The force in the spring.

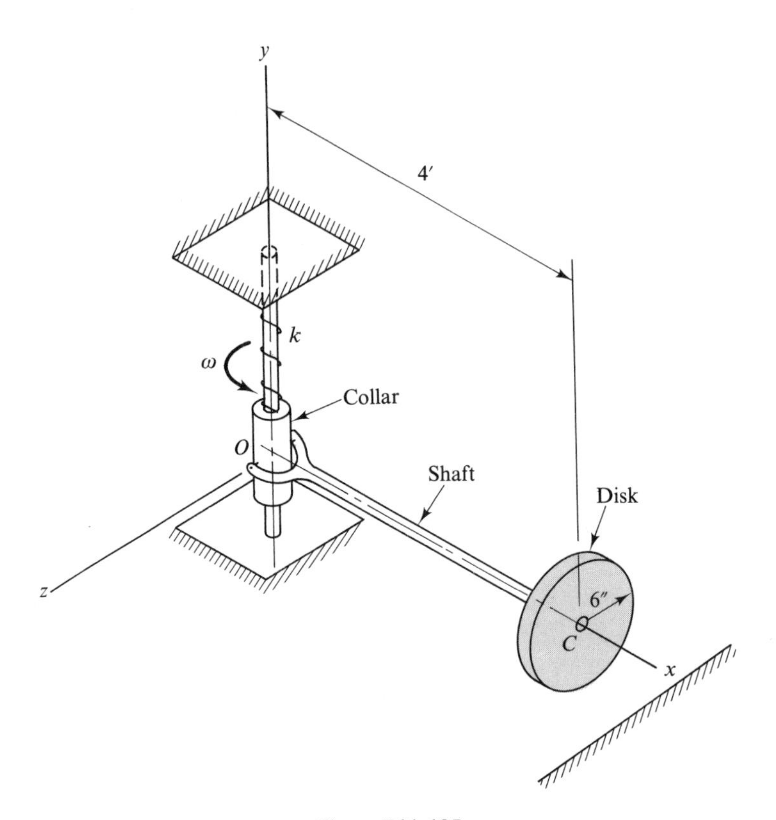

Figure P11-185

***11-186** The thin disk in Fig. P11-186 has a mass of 5.0 kg and is mounted on a rigid shaft with a negligible mass. The shaft is supported by a ball and socket at A and by a sleeve bearing at B. Points A and B are located on the turntable, which rotates about the vertical axis with an angular velocity ω_2 of 30 rpm as shown. The disk spins about the shaft with an angular velocity ω_1 of 600 rpm as indicated. Determine the forces on the shaft at A and B for the given position.

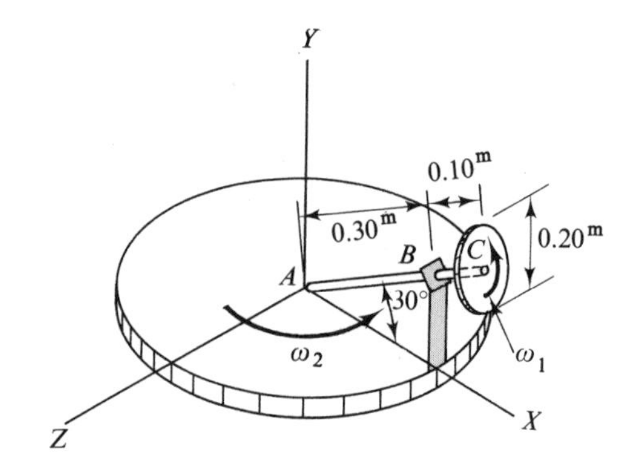

Figure P11-186

12 Mechanical Vibrations

12-1 INTRODUCTION

A *mechanical vibration* is an oscillatory, periodically repeated motion of a particle or body about a position of equilibrium. An engineer is frequently confronted with the problem of mechanical vibrations since they are encountered to some degree in almost all types of machinery and structures. Most vibrations are undesirable in machines and structures because they produce excessive stresses or repeated stresses; cause added wear; increase bearing loads; induce fatigue; create acute passenger discomfort in planes, ships, trains, buses, and automobiles; and absorb energy that could otherwise do useful work. The collapse of the Tacoma Narrows Bridge in 1940 is an example of structural failure due to excessive stresses produced by vibrations. The accuracy of precision instruments, tools, and machines may be impaired by excessive vibrations. Rotating machine parts require careful balancing in order to prevent damage from vibrations. When part of the propeller of an airplane is shot off or breaks off in flight the propeller is no longer symmetrical, and the vibrations from the engine may tear the engine from the plane unless it can be stopped in time. The vibrations produced in an automobile by the engine or by driving on rough roads set up repeated stresses in certain parts that can eventually lead to fatigue failure of the members.

Vibrations are sometimes used to produce desirable effects. For example, vibrators are used to compact concrete in the forms and to separate the grain from the chaff in threshing machines. Instruments which function properly on an airplane with a conventional engine may tend to stick when used in gliders or jet-powered planes because of the lack of vibration. In such instances, a vibrator is sometimes installed on the instrument panel.

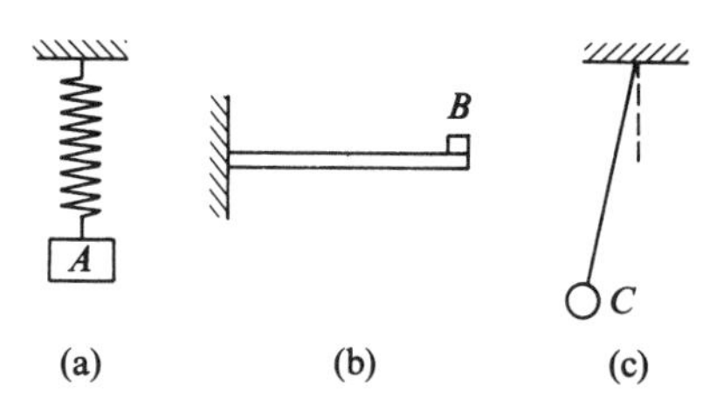

Figure 12-1

When a particle or body that is supported by a system of springs, a shaft, a beam, or any other elastic system is disturbed from its position of equilibrium by the application and sudden removal of an additional force the particle or body will vibrate. Some common examples are (1) the vertical vibratory motion of body A in Fig. 12-1(a) when it is displaced vertically from its equilibrium position on a helical spring and then released; (2) the vertical vibratory motion of body B of Fig. 12-1(b) when it is displaced from its equilibrium position on a flexible springboard (cantilever beam) of negligible weight and released; and (3) the motion of the pendulum bob C in Fig. 12-1(c) as it swings in a vertical plane when supported by a cord of negligible weight.

Figure 12-2(a) shows a weight W suspended from a support by a spring and hanging at its equilibrium position. If the weight is pulled from its equilibrium position by a force $\mathbf{F}$ and released, it will, in the absence of any frictional forces, continue to oscillate about the equilibrium position indefinitely. Figure 12-2(b) shows a graph of the displacement, y, of the weight W from its equilibrium position as a function of time. One of the fundamental properties of mechanical vibrations is that the motion repeats itself at *definite intervals of time.* The *period*, T, of the oscillation is the *minimum amount of elapsed time before the motion starts to repeat itself*. The motion completed in one period is a *cycle.* The *frequency*, f, of the oscillation is the number of cycles which occur in a given unit of time; the customary unit is cycles per second, cps, or hertz, Hz. It should be noted that the frequency is the reciprocal of the period, or, in equation form,

$$f=\frac{1}{T}. \tag{12-1}$$

The *amplitude*, A, of the oscillation is the maximum displacement, either linear or angular, of the body from its equilibrium position.

Mechanical vibrations which are maintained by elastic, and

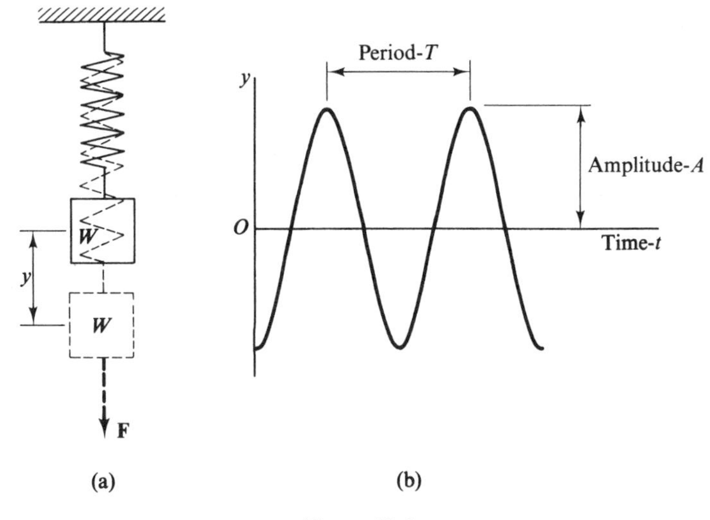

Figure 12-2

sometimes gravity, forces are called *free vibrations.* Once started, a free vibration, frequently called a *natural vibration,* continues to oscillate at its natural frequency. A *forced vibration* is produced and maintained by a periodic exciting force external to the system, and it takes place at the frequency of the exciting force. If the system is nearly frictionless and contains no dissipative elements, the amplitude of the forced oscillation will become very large when the frequency of the exciting force is close to the natural frequency of the system. Thus, although the frequency of a forced vibration is independent of the natural frequency of the elastic system, the resulting amplitude is affected by both frequencies. Vibrations are also classified as *damped* or *undamped.* When friction, air resistance, viscous damping (see Art. 12-5), and all other resisting forces are negligible the vibration is undamped. When any of these effects is appreciable the body has damped vibrations. In practice there is always a frictional damping force which will eventually stop a free vibration, even though it may be neglected for many purposes.

If the motion of a particle or body is constrained so that its position can be completely specified by one coordinate, it is said to have a *single degree of freedom.* If the system can vibrate in two directions or is composed of two particles that can each vibrate independently in one direction, it is said to have *two degrees of freedom* because two coordinates are required to specify the position of the system at any instant. For example, a system composed of a single particle, supported by a spring as indicated in Fig. 12-3(a), which vibrates in a vertical direction only, has one degree of freedom. A system composed of two particles which move vertically and are supported as indicated in Fig. 12-3(b) has two degrees of freedom

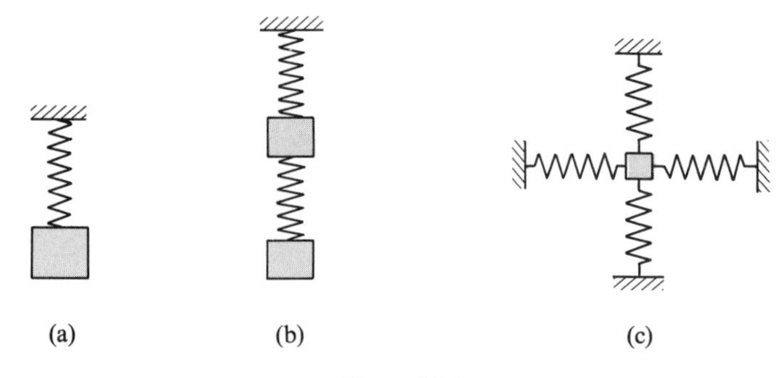

Figure 12-3

because two coordinates are required to locate the particles at any instant. A system composed of one particle supported by four springs and constrained to move in a vertical plane as indicated in Fig. 12-3(c) has two degrees of freedom because two coordinates are required to locate the particle in the vertical plane at any instant. A single rigid body has, in general, six degrees of freedom, since it may have translation in three coordinate directions and it may also rotate about three coordinate axes. The following discussion is concerned only with systems which have a single degree of freedom or which vibrate in only one way at the instant being considered.

12-2 SIMPLE HARMONIC MOTION

A vibration has been defined as an oscillatory, periodically repeated motion. The simplest form of periodic motion is called *simple harmonic motion* and is defined as *any motion which has an acceleration which is proportional to the displacement from a fixed point on the path of the motion and which is directed toward the fixed point.* The motion of many vibratory systems having a single degree of freedom closely approximates simple harmonic motion, and a thorough knowledge of this concept is most helpful when analyzing such systems.

Expressions for velocity and acceleration of a point were obtained in Chapter 8 by differentiating a position function with respect to time. Consider line OB in Fig. 12-4(a), which has a length A and which rotates about O with a constant angular velocity ω. Point P is the projection of point B on the y axis, and the position of P at time t is

$$y = A \cos \omega t. \qquad \text{(a)}$$

Note that each time the angle ωt increases by 2π rad, point P will have moved through one complete cycle, as shown in Fig. 12-4(b).

The velocity of P can be obtained by differentiating Eq. (a) with respect to time; that is,

$$v = \dot{y} = -A\omega \sin \omega t = A\omega \cos\left(\omega t + \frac{\pi}{2}\right). \qquad \text{(b)}$$

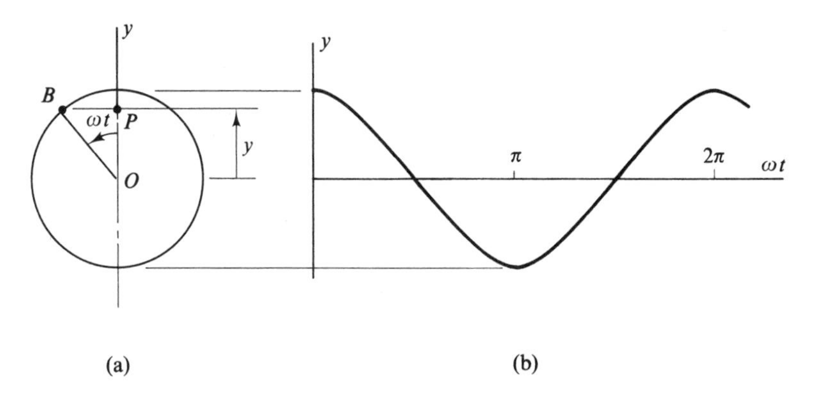

Figure 12-4

In a similar manner the acceleration of P is

$$a = \ddot{y} = -A\omega^2 \cos \omega t = A\omega^2 \cos(\omega t + \pi). \qquad \text{(c)}$$

Note that the only differences in the three equations are that $\dot{y}$ and $\ddot{y}$ are multiplied by ω and ω^2, respectively, and that their angles differ by $\pi/2$ and π from the variable ωt. The angles $\pi/2$ and π are known as *phase angles.* From Eq. (b) and (c) it is seen that the maximum values of the velocity and acceleration are $A\omega$ and $A\omega^2$, respectively. When Eq. (a) is substituted into Eq. (c) the acceleration becomes

$$\ddot{y} = -\omega^2 y \quad \text{or} \quad \ddot{y} + \omega^2 y = 0. \qquad \text{(12-2)}$$

This equation shows that the acceleration of P is proportional to its displacement, y, from the origin O, and the negative sign indicates that the acceleration is opposite the displacement; that is, the acceleration is directed toward O. By definition, therefore, point P executes simple harmonic motion. Any motion described by Eq. (12-2) is simple harmonic motion, and Eq. (12-2) is the *differential equation of motion* for simple harmonic motion.

The factor ω^2 in Eq. (12-2) is the square of the angular velocity of line OB. When line OB has rotated one revolution, 2π rad, point P has completed one cycle of motion. The time, T, required for this motion is the period of the vibratory motion of P, that is,

$$T = \frac{2\pi}{\omega},$$

and the frequency of the vibratory motion is

$$f = \frac{1}{T} = \frac{\omega}{2\pi}.$$

Since ω is closely related to the frequency of vibration, it is called the *circular frequency* of the motion.

12-3 FREE UNDAMPED VIBRATIONS OF A PARTICLE

In many systems whose mass and restoring forces have discrete values the small energy loss due to air resistance, internal friction of springs (hysteresis), or other friction forces can be neglected. Such systems are called *undamped* systems. It should be recognized that, although such systems do not actually exist, the errors introduced by neglecting these losses are small enough that the analysis often gives quite satisfactory engineering results. Experience has shown that the small damping forces which eventually stop a freely vibrating system have little effect on such quantities as frequency and period of vibration.

Figure 12-5(a) represents a particle B with a mass m which is suspended by means of a spring with a modulus k. In the equilibrium position the spring tension will be equal to the weight; that is,

$$ky_{st} = W = mg.$$

If the body is pulled downward a distance y and released, it will oscillate up and down. The free-body diagram in Fig. 12-5(c) shows the forces acting on the body when it is displaced y units downward. The magnitude of the force F is

$$F = k(y_{st} + y) = mg + ky,$$

and the equation of motion for B is

$$m\ddot{y} = \sum F = mg - F = mg - (mg + ky) = -ky$$

from which

$$\ddot{y} = -\frac{k}{m}y. \qquad (12\text{-}2a)$$

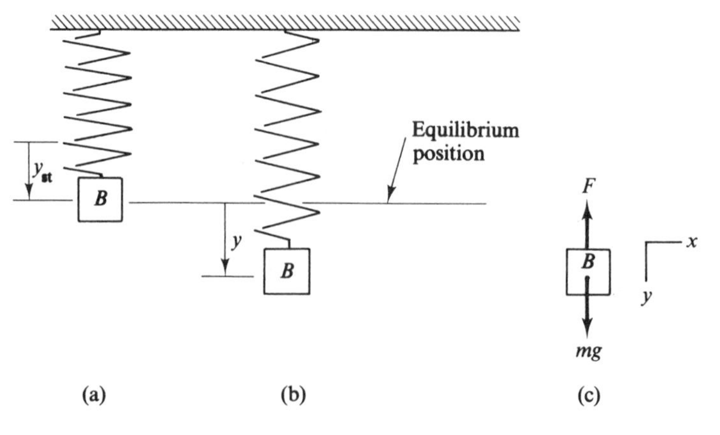

Figure 12-5

This equation shows that the acceleration of B is proportional to its displacement from the equilibrium position and is directed toward that position. Therefore the body has simple harmonic motion. When Eq. (12-2a) is compared with Eq. (12-2) the two expressions are identical if k/m is replaced by ω^2 or ω_n^2. Since f_n, the free undamped frequency of this system, is related to ω_n^2, the subscript n is used for all free undamped systems. The natural frequency of the vibration is

$$f_n = \frac{\omega_n}{2\pi} = \frac{1}{2\pi}\sqrt{\frac{k}{m}}.$$

Equation (12-2) or (12-2a) is a second-order linear differential equation with constant coefficients. Although there are many ways to solve this common differential equation, perhaps the simplest is to assume a solution and verify it by substitution into the differential equation. Note that the second derivative of the function must equal the function multiplied by a constant coefficient in order to satisfy the differential equation. Some exponential and trigonometric functions have this property, and it can be shown by substitution that the following trigonometric function will satisfy the differential equation:

$$y = B \cos \omega_n t + C \sin \omega_n t. \tag{12-3a}$$

The quantities B and C are the constants of integration and can be determined from the initial starting conditions, and ω_n is the square root of k/m.

Equation (12-3a) can also be written as

$$y = B \cos \omega_n t + C \sin \omega_n t = A \cos(\omega_n t - \phi). \tag{12-3b}$$

To show that this is possible Eq. (12-3b) can be expanded to give

$$B \cos \omega_n t + C \sin \omega_n t = A(\cos \omega_n t \cos \phi + \sin \omega_n t \sin \phi)$$

from which

$$(B - A \cos \phi)\cos \omega_n t + (C - A \sin \phi)\sin \omega_n t = 0.$$

If this equation is to be valid for any value of t, the coefficients in parentheses must both be zero. Therefore,

$$B = A \cos \phi \quad \text{and} \quad C = A \sin \phi$$

from which

and

$$\left.\begin{aligned} A &= (B^2 + C^2)^{1/2} \\ \tan \phi &= \frac{C}{B}. \end{aligned}\right\} \tag{12-4}$$

Since the maximum value of $\cos(\omega_n t - \phi)$ is 1, the amplitude of the oscillation is $A = (B^2 + C^2)^{1/2}$.

Equations for two other properties of the oscillation can be derived from Eq. (12-3b). The velocity of the particle is

$$\dot{y} = -\omega_n A \sin(\omega_n t - \phi),$$

and the magnitude of the maximum velocity of the particle is

$$\dot{y}_{max} = \omega_n A = \omega_n \sqrt{B^2 + C^2}. \tag{12-5a}$$

Similarly, the acceleration of the particle is

$$\ddot{y} = -\omega_n^2 A \cos(\omega_n t - \phi),$$

and the magnitude of the maximum acceleration is

$$\ddot{y}_{max} = \omega_n^2 A = \omega_n^2 \sqrt{B^2 + C^2}. \tag{12-5b}$$

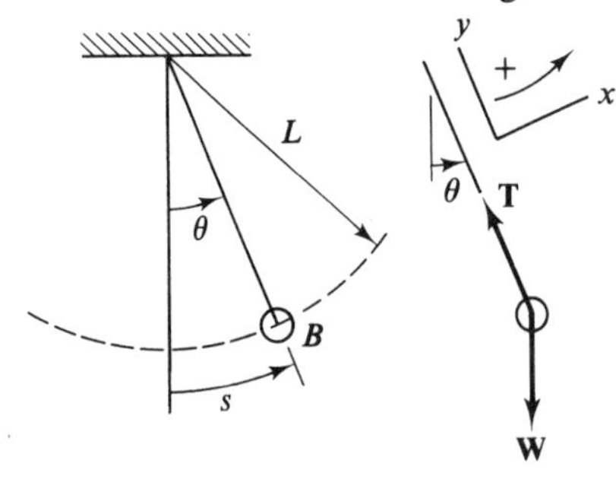

Figure 12-6

When the amplitudes of oscillations are small, it is convenient to set up the equation of motion in terms of inches instead of feet. If the acceleration is in inches per second per second, the mass must be expressed in units of pound-second squared per inch instead of pound-second squared per foot (slugs). If the mass is calculated from the expression W/g, the value of g must be in inches per second per second, approximately 386 ips^2. When using SI units the value of g is 9.81 m/s^2 and all lengths are measured in meters.

In any problem involving free undamped vibrations, a free-body diagram of the vibrating particle should be drawn showing it displaced from its equilibrium position. The acceleration can be determined from the general motion equation in terms of the position of the particle. If the acceleration is of the form

$$a = \ddot{y} = -Ky,$$

where K is any positive constant and y is the position coordinate of the particle measured from its equilibrium position, the resulting motion will be simple harmonic motion. If the acceleration is not equal to $-Ky$, the motion will not be simple harmonic motion, although it may still be an oscillatory motion, and the solution must be obtained from the differential equation.

As an example, consider the oscillations of the simple pendulum in Fig. 12-6. The pendulum, consisting of a particle of weight W suspended by a cord of length L, swings in a vertical plane. The cord is inextensible, and its weight is negligible. The particle is moved out until the cord makes an angle θ_{max} with the vertical and is then released to oscillate in a vertical plane. A free-body diagram of the particle, when it is displaced any angle θ, is shown in Fig. 12-6, and the positive directions are indicated near the diagram. The tangential component of the acceleration can be determined from

$$\sum F_x = ma_x$$

or

$$-W \sin\theta = \frac{W}{g} a_x$$

from which

$$a_x = -g \sin\theta.$$

The tangential acceleration can be expressed in terms of the angular acceleration of the cord, that is, $a_x = L\ddot{\theta}$, and the differential equation becomes

$$L\ddot{\theta} = -g \sin\theta \quad \text{or} \quad \ddot{\theta} = -\frac{g}{L} \sin\theta.$$

When the maximum value of θ is small, $\sin\theta$ is approximately equal to θ, and the equation becomes

$$\ddot{\theta} = -\frac{g}{L}\theta,$$

which indicates that the pendulum has simple harmonic motion. If the angle θ is not small, the resulting motion will not be simple harmonic motion, although it will be an oscillatory motion.

EXAMPLE 12-1

The 7.72-lb body A in Fig. 12-7 moves along a smooth horizontal plane and is restrained by two identical springs each with a modulus of 4 lb-per in. The oscillation is started by moving body A 2 in. to the right of its equilibrium position and giving it a velocity of 30 ips to the left. Determine the period, frequency, amplitude, maximum velocity, and maximum acceleration of the motion. Both springs are in tension at all times.

SOLUTION

Figure 12-7(b) is a free-body diagram of body A displaced a positive distance, x, from the equilibrium position. At the equilibrium position, the tension in each spring must be the same, T_O, and as the mass is displaced to the right a distance x, the tensions in the left and right

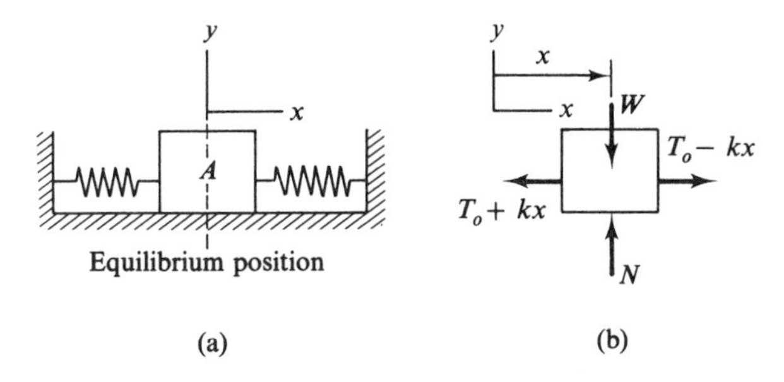

Figure 12-7

springs become T_O+kx and T_O-kx, respectively. The differential equation of motion is obtained from Newton's equation of motion and the free-body diagram of Fig. 12-7(b) as follows:

$$\sum F_x=(T_O-4x)-(T_O+4x)=ma_x=\left(\frac{7.72}{386}\right)\ddot{x}$$

from which

$$\ddot{x}+400x=0.$$

Note that k, x, and m are expressed in terms of pounds, seconds, and inches; therefore, $\ddot{x}$ will be in inches per second per second. The value of ω_n^2 is the coefficient of the displacement, x. Therefore $\omega_n=$ 20 rad per sec, and the period is

$$T=\frac{2\pi}{\omega_n}=\frac{2\pi}{20}=\underline{0.314\text{ sec.}}\qquad\text{Ans.}$$

The frequency is the reciprocal of the period; that is,

$$f=\frac{1}{T}=\frac{1}{0.314}=\underline{3.18\text{ cps}=3.18\text{ Hz.}}\qquad\text{Ans.}$$

The solution to the differential equation of motion [see Eq. (12-3b)] is

$$x=B\cos(20t)+C\sin(20t)=A\cos(20t-\phi).$$

The constants B and C, or A and ϕ, can be obtained from the initial conditions. When $t=0$, $x=+2.0$ and $\dot{x}=-30$. If the second form of the solution is used, the velocity is

$$\dot{x}=-20A\sin(20t-\phi).\qquad\text{(a)}$$

The initial conditions (when $t=0$) give

$$x(0)=+2.0=A\cos(-\phi)=A\cos\phi$$

and

$$\dot{x}(0)=-30=-20A\sin(-\phi)=20A\sin\phi.$$

These equations can be solved simultaneously to give

$$A=2.50\text{ in.},\qquad \phi=\tan^{-1}(-0.750)=-0.644\text{ rad}.$$

The angle ϕ is not needed for this solution, but its value is given to provide a check on the calculations. The amplitude of the motion is

$$A=\underline{2.50\text{ in.}}\qquad\text{Ans.}$$

The maximum velocity and acceleration are

$$v_{max}=A\omega=2.50(20)=\underline{50\text{ ips}}\qquad\text{Ans.}$$

and

$$a_{max}=A\omega^2=2.50(20)^2=\underline{1000\text{ ips}^2.}\qquad\text{Ans.}$$

These values can be obtained from Eq. (a) by differentiation and substitution.

PROBLEMS

12-1 Two bodies with simple harmonic motions have equal maximum velocities, but their frequencies are in the ratio of 1 to 4. Determine the ratio of their amplitudes.

12-2 The position of a particle at any time t is given by the equation $x = 8 \cos \pi t/2$, where x is in inches and t is in seconds. Plot the displacement, velocity, and acceleration versus time curves for one complete cycle.

***12-3** The maximum velocity of a body with simple harmonic motion is 30 mm/s, and its frequency is 60 Hz. Determine the amplitude, maximum acceleration, and period of the motion.

***12-4** Vibration-measuring instruments indicate that a body is vibrating harmonically at a frequency of 480 Hz with a maximum acceleration of 500 cm/s^2. Determine the amplitude of the vibration and the maximum speed of the body.

12-5 A particle moves with simple harmonic motion, and the maximum magnitudes of its velocity and acceleration are 20 ips and 80 ips^2, respectively. Determine the magnitude of the acceleration of the point when the speed is 10 ips.

12-6 A particle moves with simple harmonic motion. The amplitude of the motion is 5 in., and the magnitude of the velocity is 12 ips when the point is displaced 3 in. from the position of maximum velocity. Determine the magnitude of the acceleration of the point for this position.

12-7 A particle has simple harmonic motion. When it is displaced 8 in. from the center of its path, the magnitudes of its velocity and acceleration are 30 ips and 200 ips^2, respectively. Determine
(a) The period of the motion.
(b) The maximum speed of the particle.

***12-8** The 6.0-kg body in Fig. P12-8 is displaced 0.20 m to the right of its equilibrium position and released from rest when $t = 0$. The modulus of the spring is 50 N/m. Determine the frequency of the resulting vibration and the solution of the differential equation.

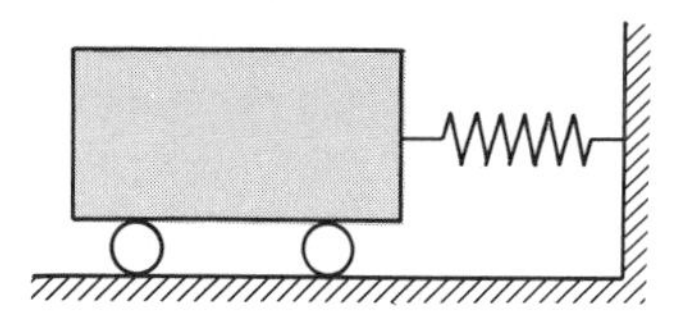

Figure P12-8

12-9 A weight of 25 lb is attached to a vertical spring with a modulus of 40 lb per ft. The weight is released when the spring is unstressed. Determine
(a) The distance the weight falls before coming to rest.
(b) The period of the motion.

12-10 The 48.3-lb block B is supported as indicated in Fig. P12-10 by a spring with a modulus of 8.0 lb per in. The velocity of B is 2.4 fps upward when B is 4.8 in. below its equilibrium position. Determine
(a) The amplitude of the free vibration of B.
(b) The maximum acceleration of B.

Figure P12-10

***12-11** The 8.0-kg ball in Fig. P12-11 swings in a vertical plane on a flexible wire with a length of 2.5 m. When $t = 0$, $\dot{\theta} = 0$ and $\theta = 0.1$ rad. Derive and solve the differential equation of motion of the body. Consider the body to be a particle.

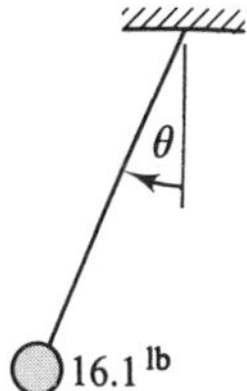

Figure P12-11

12-12 Particle C in Fig. P12-12 weighs W lb and is supported by a 3-ft rod whose weight can be neglected. When the rod is displaced an angle θ of 5° and released from rest, C swings in a vertical plane as a simple pendulum. Determine the magnitude of the maximum velocity of C.

12-13 Block B of mass m is free to slide on the smooth rod CD as indicated in Fig. P12-13.

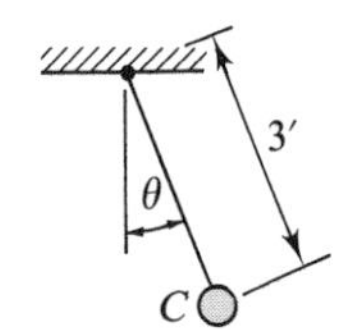

Figure P12-12

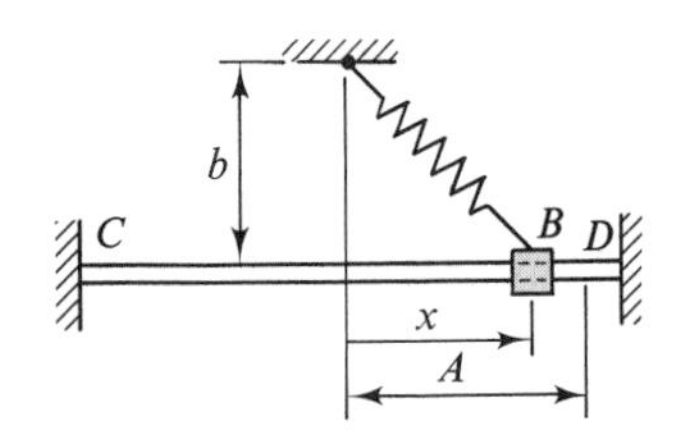

Figure P12-13

The modulus of the spring is k, and its unstretched length is b. The block is moved to the right a distance A and released from rest.
(a) Determine the acceleration of B as a function of the position coordinate x.
(b) Is the motion of B simple harmonic motion?
(c) If so, determine the period of the motion.

12-4 FREE UNDAMPED VIBRATIONS OF RIGID BODIES

Free vibrations of a particle were discussed in Art. 12-3 and the procedures developed there apply to rigid bodies that have translation without any rotation. When a rigid body is supported in such a manner that the body has rotational motion as it vibrates, the period and frequency of the resulting vibratory motion will depend on the moment of inertia of the mass of the body as well as on its mass and the forces that act upon the body.

For rectilinear motion, the criterion for simple harmonic motion is that the acceleration of the particle be of the form

$$a = \ddot{x} = -Kx.$$

An analogous equation for rotational motion in terms of the angular acceleration and position of the body was developed as

$$\alpha = \ddot{\theta} = -K\theta.$$

If the equations of motion for a body reduce to the preceding equation, the body has a *simple harmonic angular motion.* With angular harmonic motion, the angle θ is used in place of a distance x or y, and therefore the *amplitude of the motion is the maximum angular displacement of the body* from its position of equilibrium.

A *compound pendulum* is a rigid body of finite dimensions which oscillates about a fixed horizontal axis through the body. The period of vibration for small oscillations of the compound pendulum in Fig. 12-8(a) can be determined by obtaining an expression for the angular acceleration of the body in terms of its angular position. Figure 12-8(b) is a free-body diagram showing the body displaced through a positive angle θ from its equilibrium position. The moment equation of motion in scalar form is

$$\sum M_O = I_O\ddot{\theta},$$

which becomes

$$-Wr_G \sin\theta = I_O\ddot{\theta} \quad \text{or} \quad \ddot{\theta} = -\frac{Wr_G}{I_O}\sin\theta.$$

If the amplitude of the vibratory motion is small, the value of $\sin\theta$ is approximately equal to the value of θ in radians, and the moment equation becomes

$$\ddot{\theta} = -\frac{Wr_G}{I_O}\theta = -K\theta.$$

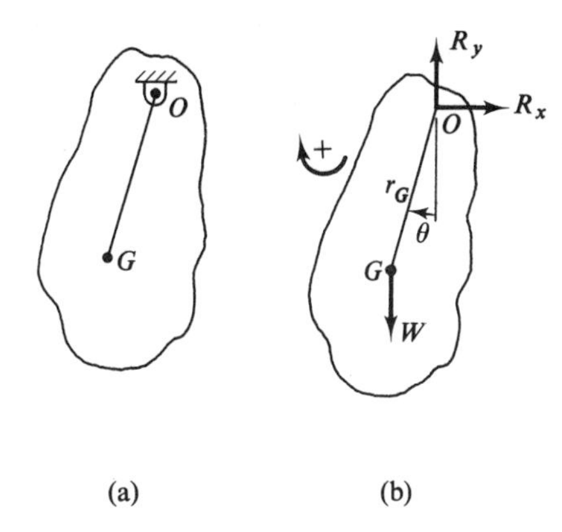

Figure 12-8

The natural circular frequency of the simple harmonic angular motion is

$$\omega_n = \sqrt{K} = \sqrt{\frac{Wr_G}{I_O}},$$

and the period of the vibration is

$$T = \frac{2\pi}{\omega_n} = 2\pi\sqrt{\frac{I_O}{Wr_G}}.$$

A *torsional pendulum* consists of a rigid body supported on a shaft as shown in Fig. 12-9. When the pendulum is twisted through a small angle θ and released, a simple harmonic angular motion results from the stresses developed in the shaft. Assume that a vibratory motion of body B in Fig. 12-9 is started by twisting the assembly through the small angle θ, as shown, and then releasing the pendulum. It is shown in texts on strength of materials that if the proportional limit[1] of the material in a *solid circular shaft* is not exceeded, the moment necessary to twist the shaft is proportional to the angle of twist and can be obtained from

$$T = \frac{JG}{L}\theta = \frac{\pi r^4 G}{2L}\theta = k\theta,$$

where

$$J = \frac{\pi r^4}{2} = \frac{\pi d^4}{32}$$

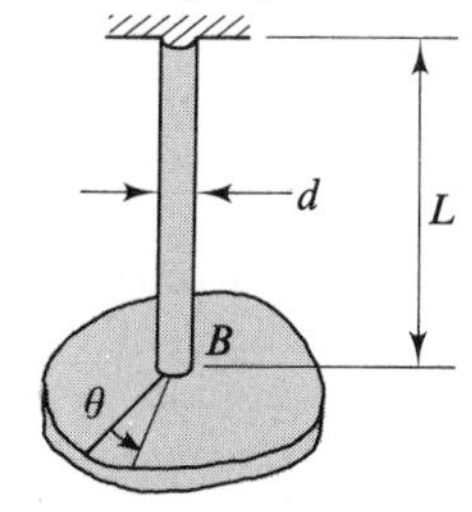

Figure 12-9

is the polar moment of inertia of the cross-sectional area of the shaft, G is the shearing modulus of elasticity of the material, L is the length

[1] The *proportional limit* is the highest unit stress at which strain (deformation) is proportional to the accompanying stress.

of the shaft, and θ is the angle, in radians, through which the shaft is twisted. The moment exerted by the shaft on the body B is equal in magnitude and opposite in sense to the moment, T, required to twist the shaft. The moment exerted by the shaft on B has a sense opposite to the sense of the angular displacement of the lower end of the shaft. The equation of motion for body B, in scalar form, is

$$\sum M_z = I_z\ddot{\theta}$$

or

$$-T = -k\theta = I_z\ddot{\theta}.$$

The angular acceleration is

$$\ddot{\theta} = -\frac{k}{I_z}\theta.$$

This last equation proves that body B has simple harmonic angular motion, and the natural circular frequency of the motion is

$$\omega_n = \sqrt{\frac{k}{I_z}} = \sqrt{\frac{\pi r^4 G}{2LI_z}}.$$

The frequency and the period of the motion are

$$f = \frac{\omega_n}{2\pi} \quad \text{and} \quad T = \frac{2\pi}{\omega_n},$$

respectively.

The moment of inertia of a body can be determined experimentally by suspending it on a shaft as a torsional pendulum or by suspending it from some axis not through the mass center as a compound pendulum and measuring the period of vibration for oscillations of small amplitude.

The solution of problems in free vibrations of rigid bodies is further illustrated by the following examples.

EXAMPLE 12-2

The slender homogeneous rod BC in Fig. 12-10(a) weighs 8 lb, and the small 4.5-lb body E is welded to the rod. The spring has a modulus of 7 lb per in. Rod BC is horizontal when it is in equilibrium. The rod is displaced an angle of 8° clockwise and released from rest. Determine the natural circular frequency and the maximum angular velocity of the vibration.

SOLUTION

Figure 12-10(b) is a free-body diagram of the two bodies displaced an angle θ (positive) from the equilibrium position. The spring force is

$$T = T_O + ky_C,$$

where T_O is the tension in the equilibrium position and y_C is the displacement of the end of the rod. The bar is in equilibrium when

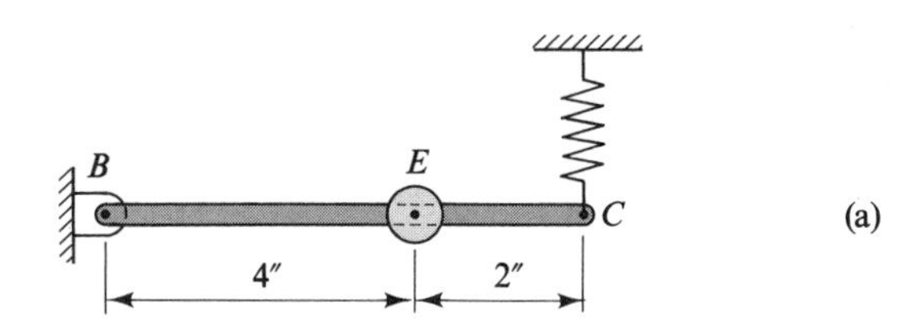

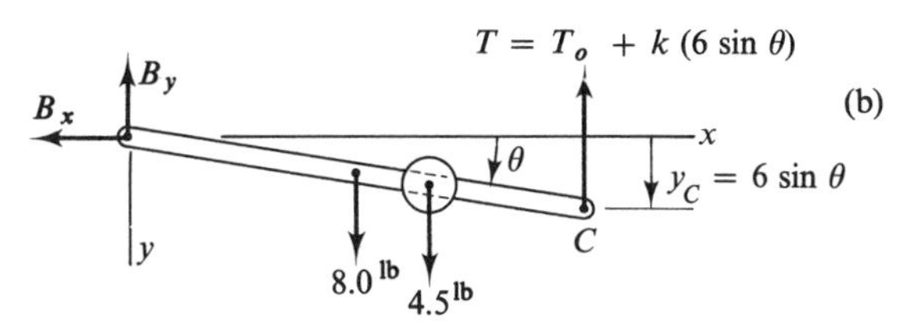

Figure 12-10

$\theta = 0$, and the moment equation about B gives

$$6T_O = 3(8.0) + 4(4.5)$$

from which

$$T_O = 7.0 \text{ lb}.$$

In the displaced position, the moment equation of motion, in scalar form, is

$$\sum M_B = I_B\ddot{\theta}$$

or

$$8.0(3\cos\theta) + 4.5(4\cos\theta) - (T_O + k6\sin\theta)(6\cos\theta) = \left[\frac{1}{3}\frac{8.0}{386}(6)^2 + \frac{4.5}{386}(4)^2\right]\ddot{\theta}.$$

When the angle θ is small, $\cos\theta \approx 1$ and $\sin\theta \approx \theta$ and the equation reduces to

$$24 + 18 - (7 + 42\theta)6 = 0.435\ddot{\theta}$$

or

$$\ddot{\theta} = -579\theta.$$

This is the differential equation of a simple harmonic angular motion with a natural circular frequency of

$$\omega_n = \sqrt{579} = \underline{24.1 \text{ rad per sec.}} \qquad \text{Ans.}$$

The solution to the differential equation is

$$\theta = A\cos(\omega_n t - \phi).$$

In this example, $\theta = 8° = 8\pi/180$ rad when $t = 0$ and $\dot{\theta} = 0$ when $t = 0$. Therefore,

$$\frac{8\pi}{180} = A\cos(-\phi)$$

and

$$0 = \dot{\theta} = -A\omega_n \sin(-\phi).$$

From these equations, $\phi = 0$ and $A = 8\pi/180 = 0.1396$ rad. The solution to the equation is

$$\theta = 0.1396 \cos(24.1t)$$

from which

$$\dot{\theta} = -0.1396(24.1)\sin(24.1t) = -3.36 \sin(24.1t),$$

and the maximum angular velocity of the body is

$$\underline{\dot{\theta}_{max} = 3.36 \text{ rad per sec.}} \qquad \text{Ans.}$$

Note the distinction between $\dot{\theta}_{max}$ and ω_n. These two quantities are sometimes confused since they have the same units.

EXAMPLE 12-3

The 64.4-lb body E in Fig. 12-11(a) is supported by an inextensible flexible belt which is wrapped around a flange on wheel C. The 96.6-lb wheel C is supported by a smooth bearing at O and is connected to a system of springs as indicated. The moduli of the springs are as indicated, and the mass of the springs and of bar AB is negligible. The radius of gyration of the mass of wheel C with respect to the axis of rotation through the mass center at O is 9.6 in. Body E is in the equilibrium position and has a velocity of 2 fps downward when $t = 0$. Determine the differential equation of motion of the system and the solution of the equation.

SOLUTION

Free-body diagrams of AB, C, and E are shown in Fig. 12-11(b). The resulting equations can be written in terms of y_D, y_E, θ_C, or some other

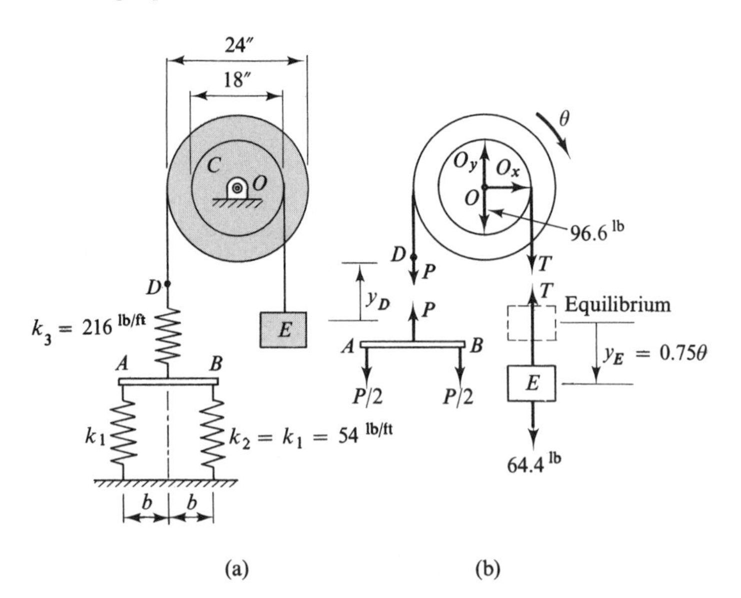

Figure 12-11

variable. The variable θ will be used in this example, in which case $y_E = 0.75\theta$ and $y_D = 1.00\theta$. The angle θ is assumed to be positive when clockwise; therefore, downward is positive for y_E and upward is positive for y_D.

When the system is an equilibrium, $T = 64.4$ lb and $P = P_O = (9/12)64.4 = 48.3$ lb. When the system is given a positive displacement, the force P can be expressed as $P_O + \Delta P$, where ΔP is the change of P due to a change of θ. Let $\Delta_1 = \Delta_2$ be the stretch of springs with moduli k_1 and k_2 and Δ_3 be the stretch of the third spring. The deflections Δ_1 and Δ_2 are equal since AB is loaded symmetrically and since $k_1 = k_2$.

$$y_D = \Delta_1 + \Delta_3 = \frac{\Delta P/2}{k_1} + \frac{\Delta P}{k_3} = \frac{\Delta P}{2(54)} + \frac{\Delta P}{216} = \frac{\Delta P}{72}$$

or

$$\Delta P = 72 y_D = 72\theta$$

and

$$P = P_O + \Delta P = 48.3 + 72\theta.$$

The equations of motion, in scalar form, are as follows: For body C,

$$\sum M_O = I_O \ddot{\theta}$$

or

$$0.75T - 1.00(48.3 + 72\theta) = \frac{96.6}{32.2}\left(\frac{9.6}{12}\right)^2 \ddot{\theta} = 1.92\ddot{\theta},$$

and for body E,

$$\sum F_y = m a_E$$

or

$$64.4 - T = \frac{64.4}{32.2} a_E = 2.00(0.75\ddot{\theta}) = 1.50\ddot{\theta}.$$

The elimination of T in these equations gives

$$-72\theta = 3.045\ddot{\theta}$$

or

$$\underline{\ddot{\theta} = -23.6\theta.} \qquad \text{Ans.}$$

This differential equation of motion demonstrates that the system has simple harmonic motion with a natural circular frequency of

$$\omega_n = \sqrt{23.6} = 4.86 \text{ rad per sec.}$$

The solution of the differential equation can be written as

$$\theta = B \cos \omega_n t + C \sin \omega_n t.$$

From the initial condition $\theta = (\frac{4}{3}) y_E = 0$ when $t = 0$, the value of B must be zero. The angular velocity, with $B = 0$, is

$$\dot{\theta} = C\omega_n \cos \omega_n t,$$

and since $\dot{\theta} = (\frac{4}{3})\dot{y}_E = (\frac{4}{3})2$ when $t = 0$,

$$\frac{8}{3} = C\omega_n$$

and

$$C = 0.548 \text{ rad}.$$

The solution of the differential equation of motion of the system is

$$\theta = \underline{0.548 \sin 4.86t}. \qquad \text{Ans.}$$

PROBLEMS

12-14 Body B in Fig. P12-14 weighs 20 lb, and the moduli of springs C and E are 5 and 15 lb per in., respectively. The speed of B while passing through its equilibrium position is 15 ips. Determine the amplitude and frequency of the vibration of B.

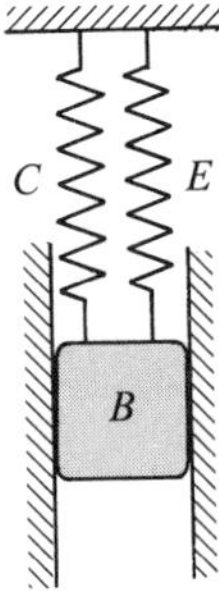

Figure P12-14

12-15 The springs in Problem 12-14 are arranged as shown in Fig. P12-15. All other data are unchanged. Determine the amplitude and frequency of the vibration of B.

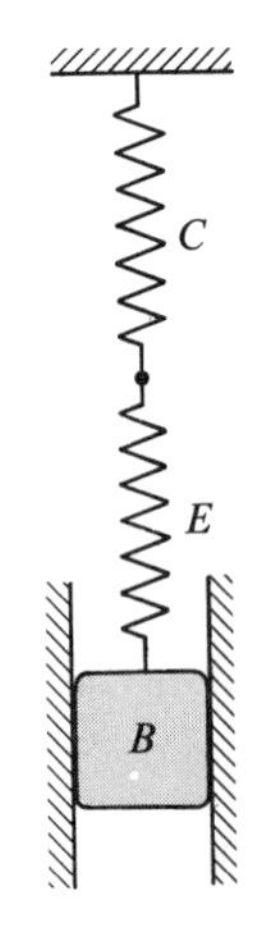

Figure P12-15

12-16 When a small body which weighs W lb is supported by a spring with a modulus of k lb per in. as shown in Fig. P12-16(a), its natural frequency is found to be 2.0 Hz. When an additional spring with a modulus of 5.5 lb per in. is added, as shown in Fig. P12-16(b), the frequency is increased to 3.5 Hz. Determine the weight W and the modulus of the upper spring. Both springs are in tension during the entire vibration.

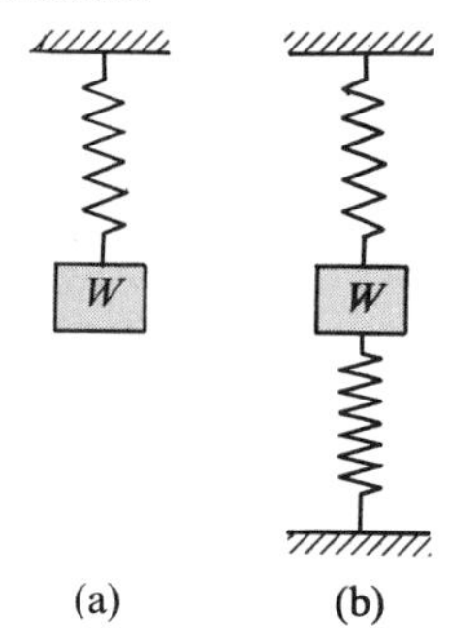

Figure P12-16

***12-17** The 3.0-kg mass C in Fig. P12-17 is supported by the rigid body ABC whose mass is negligible. The modulus of the spring at A is 3600 N/m. The arm BC is horizontal when the system is in equilibrium. Determine the period of the vibration of C when it is displaced slightly from its equilibrium position and released.

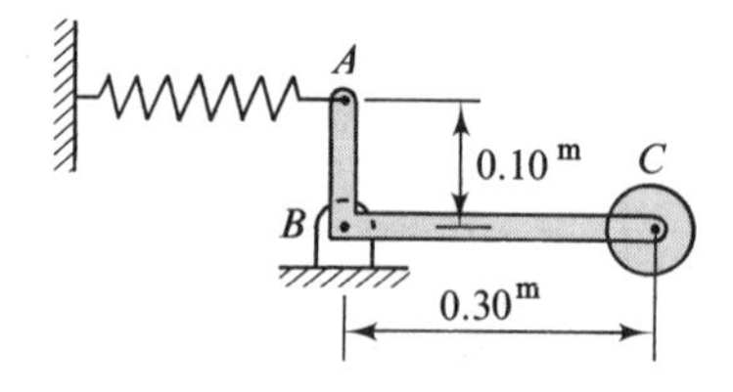

Figure P12-17

***12-18** Determine the distance BC in Problem 12-17 which will result in a frequency of vibration of 1.2 Hz. All other data are unchanged.

12-19 Bar AB in Fig. P12-19 is suspended from a smooth horizontal axis at O. When the spring S_1 at B is the only one acting, the frequency of the free vibrations is f_1 Hz. The modulus of S_1 is 15 lb per in. The natural frequency of the bar is to be doubled by adding a second spring S_2 at A. Both springs are in tension at all times, and the bar is in equilibrium in the horizontal position. Determine the modulus of S_2.

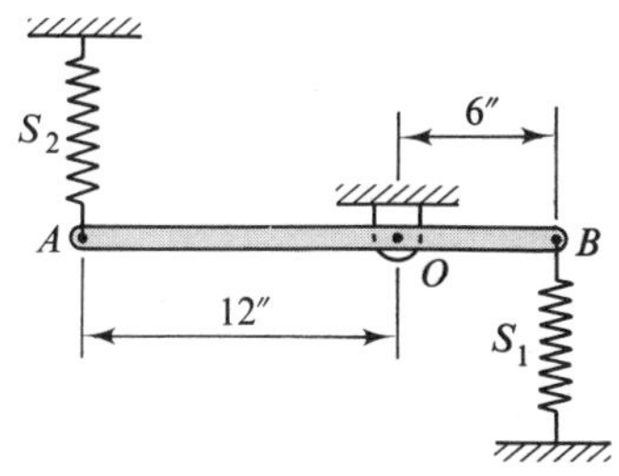

Figure P12-19

12-20 For the system shown in Fig. P12-20 derive and solve the differential equation of motion with the initial conditions $\theta = 0.1$ rad and $\dot{\theta} = 0.6$ rad per sec when $t = 0$. Neglect the weight of the rod.

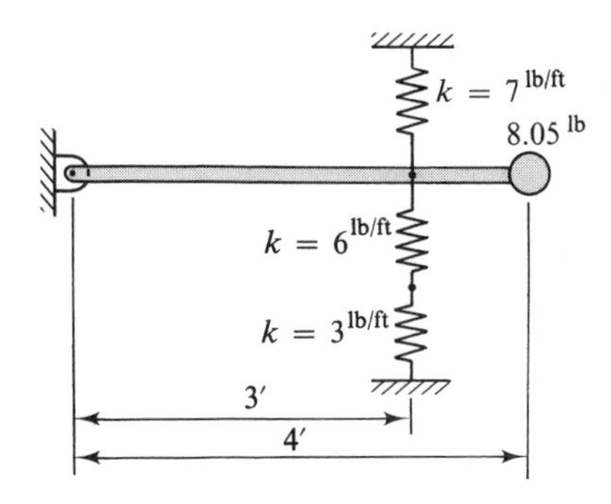

Figure P12-20

12-21 Determine the mass of the uniform bar in Problem 12-20 if the frequency of the free vibration is 0.50 Hz.

***12-22** The radius of gyration of the 8.0-kg body in Fig. P12-22 with respect to an axis through the center of gravity is 0.25 m. When $t = 0$, $\theta = \theta_0$ and $\dot{\theta} = 0$. Derive and solve the differential equation of motion for small values of θ and determine the period of the vibration.

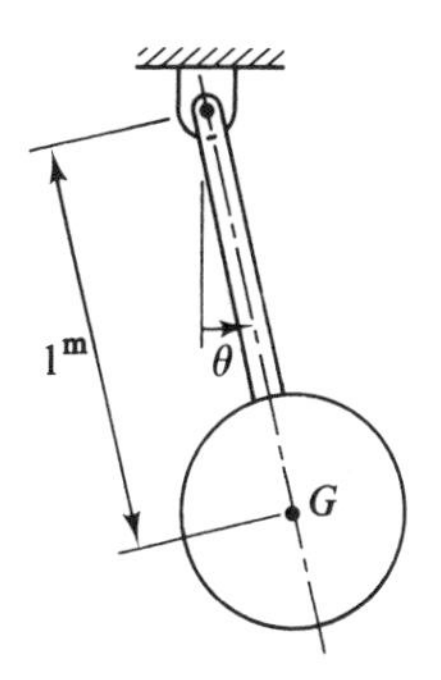

Figure P12-22

12-23 The particle B in Fig. P12-23 is mounted on the rigid rod BC, whose weight may be neglected. The particle B weighs 0.50 lb, and the modulus of each of the springs is 10.0 lb per ft. The springs each have a tension of 2.00 lb when BC is vertical. B is moved 1 in. to the right and released from rest. Determine the frequency of the resulting vibration.

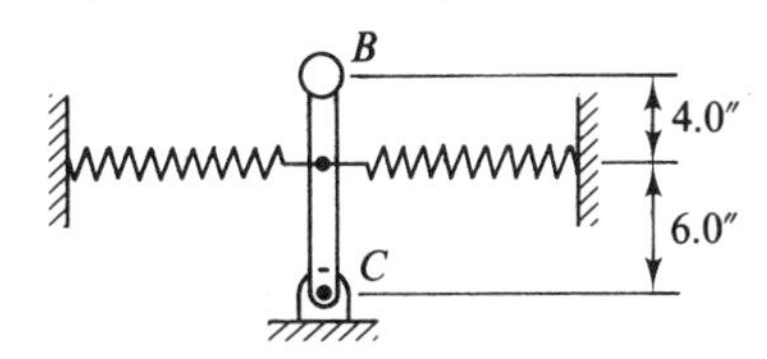

Figure P12-23

12-24 In Problem 12-23 determine the maximum weight of particle B for which member BC will vibrate. All other data are as given in Problem 12-23.

12-25 The solid homogeneous 20-lb sphere in Fig. P12-25 rolls without slipping when displaced from its equilibrium position. Initially each spring has a tension of 50 lb. The center of the sphere is moved 3.0 in. to the right and released from rest. Determine the resulting frequency and the maximum speed of the center of the sphere.

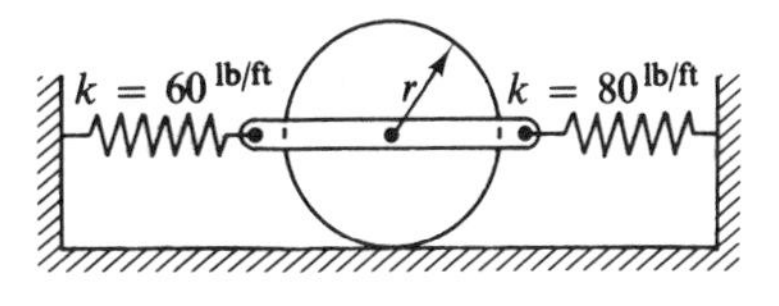

Figure P12-25

12-26 Two identical drums A and B rotate about fixed horizontal axes, as indicated in Fig. P12-26, with constant angular velocities of 60 rpm. The coefficient of kinetic friction between the 20-lb homogeneous plank C and the drums is 0.50. The plank is displaced 3.0 in. to the right from its position of equilibrium and released from rest.
(a) Show that the plank has simple harmonic motion.
(b) Determine the maximum speed of the plank during the resulting motion.

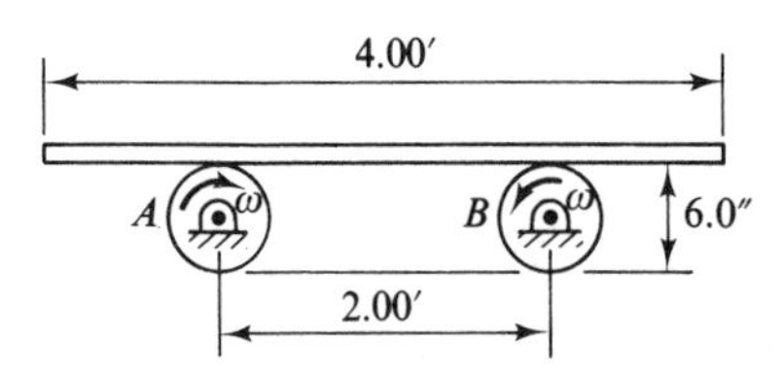

Figure P12-26

12-27 Solve Problem 12-26 if the drums are replaced by V-groove pulleys (see Fig. P12-27) and the plank by a cylindrical rod. The angle ϕ is 60°.

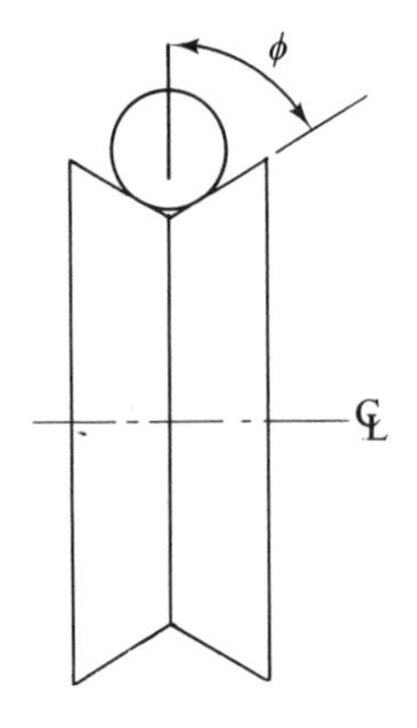

Figure P12-27

12-28 The mechanism in Problem 12-26 is inclined an angle β of 10° as shown in Fig. P12-28. The plank is released from rest in the position indicated. Determine the frequency and amplitude of the resulting motion.

12-29 The solid homogeneous 128.8-lb cylinder in Fig. P12-29 rolls without slipping. Neglect the weight of the hub for the cords which are

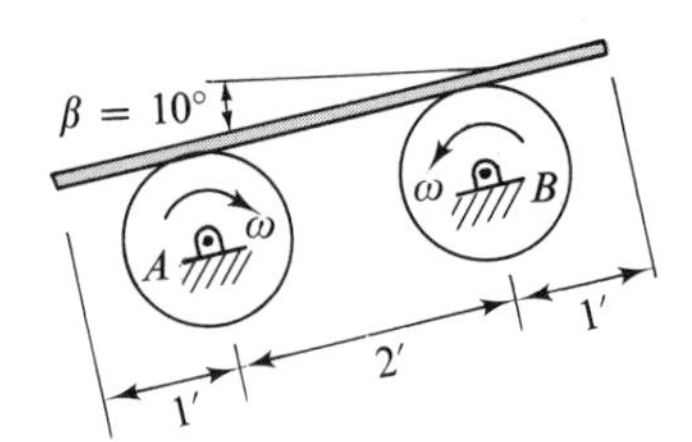

Figure P12-28

attached to the springs. The modulus of spring A is 8 lb per in., and the modulus of B is 2 lb per in. The springs are always in tension. When the cylinder is in equilibrium, a clockwise couple of 60 in-lb is gradually applied to the cylinder and then suddenly released. Determine the amplitude and period of the resulting free vibration.

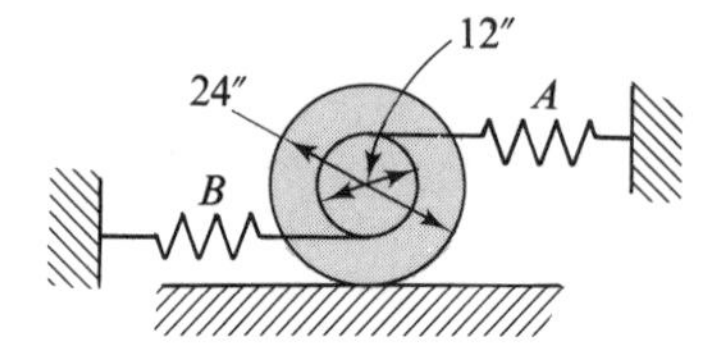

Figure P12-29

***12-30** The 30-kg solid homogeneous cylinder in Fig. P12-30 rolls without slipping along the inclined plane. The wheel is released from rest when $t = 0$ and the spring tension is zero. Derive the differential equation of motion and its solution. Also determine the frequency of the vibration.

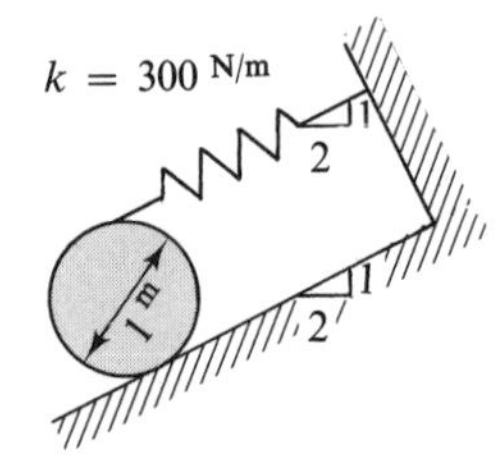

Figure P12-30

***12-31** The body in Fig. P12-31 is suspended from a smooth horizontal bearing. When the body is displaced slightly from the equilibrium position in Fig. P12-31(a) and released, it is found to oscillate 50 times in 68.0 s. The force R

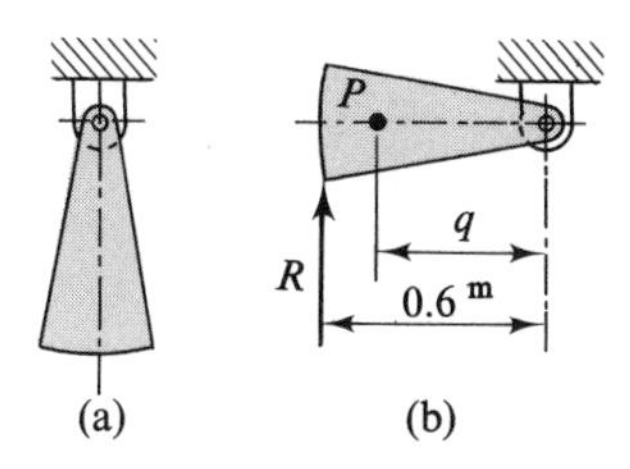

Figure P12-31

required to support the body in the position in Fig. P12-31(b) is 90 N. Determine
(a) The moment of inertia of the mass of the body with respect to its axis of rotation.
(b) The distance q from the axis of rotation to the center of percussion P.

12-32 A motor weighing 100 lb is mounted on two springs as indicated in Fig. P12-32. The springs are deflected 2.00 in. by the weight of the motor. The radius of gyration of the mass of the motor with respect to a horizontal axis through G perpendicular to AB is 4.00 in. Two 25-lb forces, one downward at A and the other upward at B, are applied and suddenly removed. Determine the period and amplitude of the resulting vibrations.

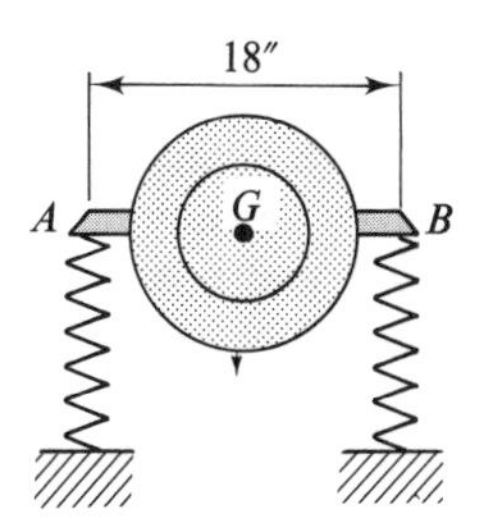

Figure P12-32

***12-33** The two bars in Fig. P12-33 each have a mass $m = 2$ kg and a length $L = 0.50$ m and are welded in the shape shown. The bars are held in equilibrium, with AB vertical, by the spring which has a modulus k. Show that the system will have simple harmonic motion with small angles only when k is more than $3mg/2L$, where g is the acceleration of gravity. Determine the frequency of the vibration when $k = 120$ N/m.

***12-34** Bar AB in Fig. P12-34 has a mass of 2 kg and CD has a mass of 4 kg. Each spring has

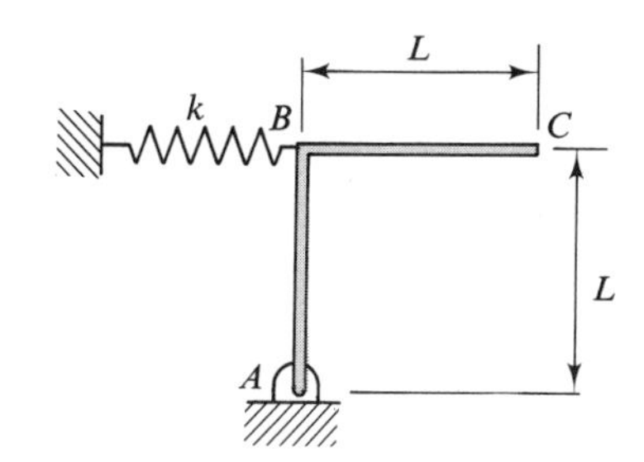

Figure P12-33

a modulus of 300 N/m and is stretched 0.10 m in the position shown. The member oscillates in the vertical plane, and the maximum velocity of D is observed to be 0.50 m/s. Determine the maximum tension in the springs.

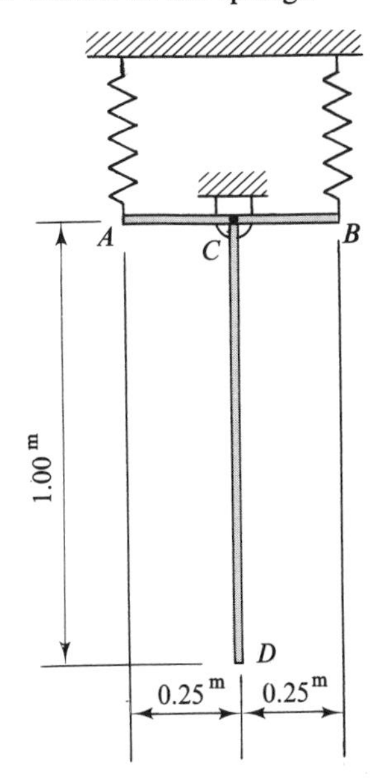

Figure P12-34

***12-35** Solve Problem 12-34 if the body is inverted so that D is above C.

12-36 The solid homogeneous cylinder in Fig. P12-36 (p. 750) has a mass m and rolls along the curved surface without slipping. The cylinder is displaced slightly from its equilibrium position and is allowed to oscillate.
(a) Will the resulting motion be simple harmonic motion for small values of θ?
(b) If so, determine the frequency of the resulting motion.

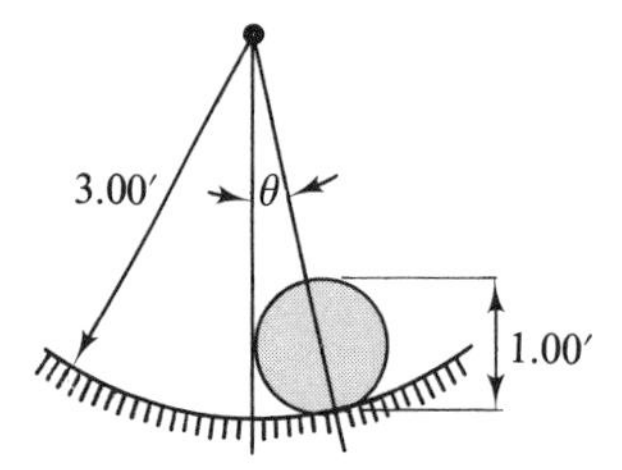

Figure P12-36

12-37 Bars *AB* and *CD* in Fig. P12-37 are made of steel weighing 0.284 lb per in.^{3} The shearing modulus of elasticity of steel is 12,000,000 psi. Neglect the weight of *CD* and assume that bar *AB* is rigid. Bar *AB* has an angular velocity of 0.50 rad per sec when *CD* is not twisted. Determine

(a) The period of oscillation of *AB*.

(b) The maximum angular acceleration of *AB*.

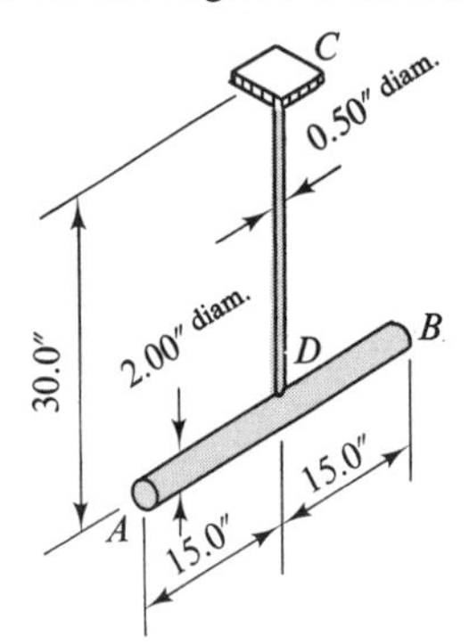

Figure P12-37

***12-38** The solid homogeneous cylindrical disk in Fig. P12-38 has a mass of 1200 kg. The upper rod is steel with a shearing modulus of elasticity of $8.3(10^{10})$ N/m^2. The lower rod is aluminum with a shearing modulus of elasticity of $2.8(10^{10})$ N/m^2. Determine the frequency of the torsional vibration of the disk.

12-39 The torsional pendulum in Fig. P12-39 is to be used as an instrument for measuring increments of time. The pendulum is to have a period of 2.00 sec. The torsional stiffness of the wire is 4.10 in-lb per rad. The circular disk is 1.00 in. thick and is made of material having a density of 15.0 slugs per ft^3. Determine the diameter of the disk.

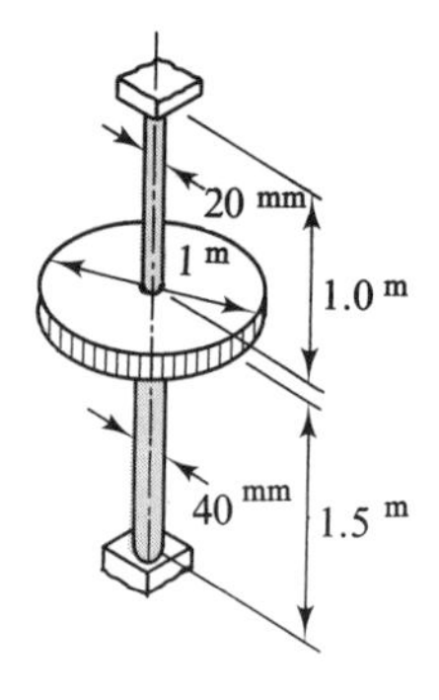

Figure P12-38

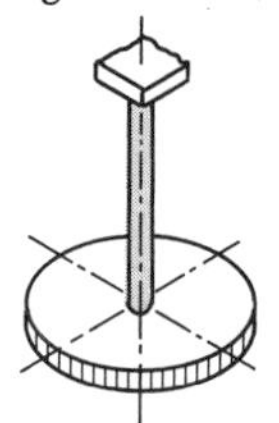

Figure P12-39

12-40 The torsion-bar suspension in Fig. P12-40 has been proposed as a mounting for an automobile wheel. The bar *AB* is assumed to be fixed at *A* and supported by a smooth bearing at *C*. Arm *BD* is fastened to the rod at *B* and carries the wheel bearing at *D*. The wheel weighs 75 lb and has a radius of gyration of mass of 12.0 in. about its geometrical axis. The arm *BD* is 10.0 in. long and is assumed to be rigid. The torsional modulus of *AB* is 20,000 in-lb per rad. Neglect the mass of *AB* and *BD* and assume that *AB* is twisted but does not bend. Determine the natural frequency of vibration of the wheel if

(a) The wheel is supported on its bearing at *D* with the brake released.

(b) The brake is locked so that the wheel must turn with the bearing and arm *BD*.

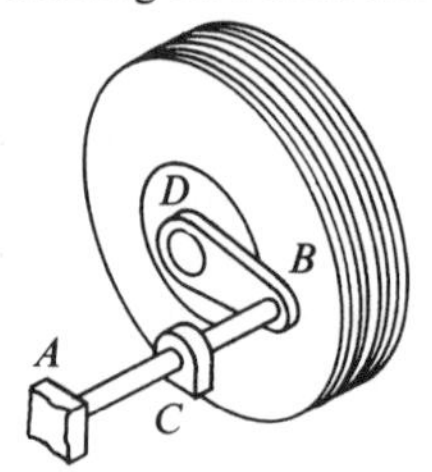

Figure P12-40

12-41 A weight W is suspended as shown in Fig. P12-41 on rigid rods with negligible weight. The system is free to rotate as indicated in the projected sketch. Determine the natural frequency of free vibrations of small oscillations.

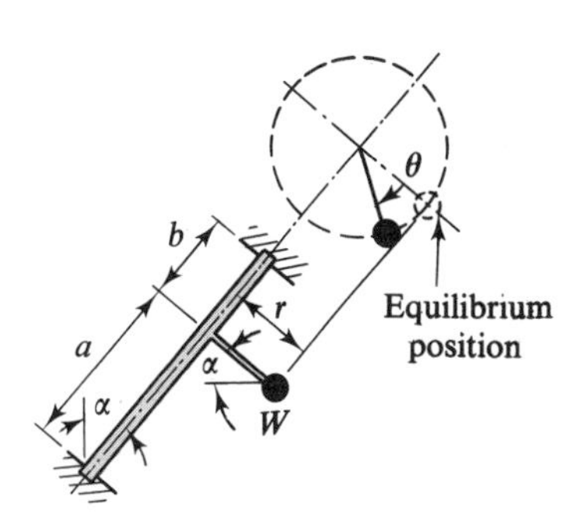

Figure P12-41

12-42 A 30-in. slender rod is bent as shown in Fig. P12-42 and suspended by a pin at O. End B is pulled down 1.0 in. and released from rest. Determine the maximum speed of point A during the resulting motion and the frequency of the motion.

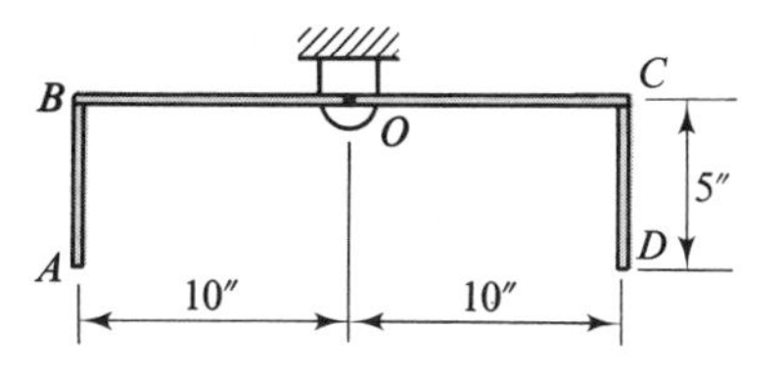

Figure P12-42

12-43 Rod AB and the small body C in Fig. P12-43 each have a mass m. Determine

(a) The distance x at which C should be located in order that the resulting frequency of small vibrations will be maximum.

(b) The frequency resulting from part (a).

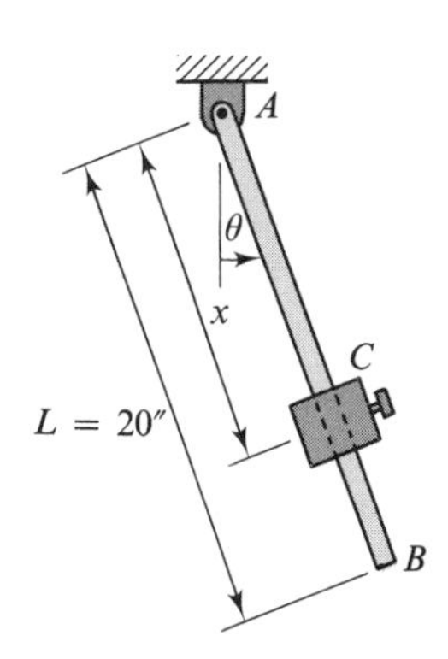

Figure P12-43

12-5 FREE VIBRATIONS WITH VISCOUS DAMPING

A vibrating system contains damping if it possesses elements which remove energy from the system. There are several types of damping: *viscous damping,* which is experienced by bodies moving through fluids with moderate velocities; *coulomb damping,* which arises from the relative motion of dry surfaces; and *structural* or *solid damping,* sometimes called mechanical *hysteresis,* which is caused by the internal friction of the elastic material. Only viscous damping is considered here because viscous dampers or dashpots are often used to control or limit vibrations, and the sliding of well-lubricated surfaces can be closely approximated by assuming viscous damping. Viscous damping results in a direct straightforward solution and is often used as a good approximation for other types of damping.

The free vibrations of systems with viscous damping are transient oscillations that gradually decrease in amplitude and eventually die out altogether. Although this might seem to indicate that the study of free, damped motion would be of little practical interest, it provides a desirable background for the study of forced, damped vibrations.

Body A in Fig. 12-12(a) has a weight W, is supported by a spring of modulus k, and is acted on by a viscous damper. The damper is shown as a dashpot, and c is the *coefficient of viscous damping*. The damping force is proportional to the velocity and its magnitude is $c\dot{y}$. The damping force always opposes the direction of the velocity as shown in the free-body diagram of Fig. 12-12(b).

The positive direction for y, $\dot{y}$, and $\ddot{y}$ is downward, and the free-body diagram is drawn with y and $\dot{y}$ both assumed to be positive. When the body is in equilibrium, F_d is zero and T is equal to W. In the displaced position, the tension is

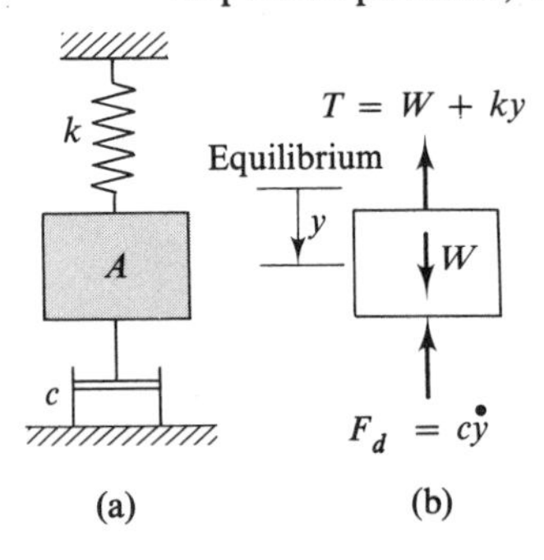

Figure 12-12

$$T = W + ky,$$

and the equation of motion in the y direction is

$$\sum F_y = W - (W + ky) - c\dot{y} = m\ddot{y}$$

or

$$m\ddot{y} + c\dot{y} + ky = 0. \tag{12-6}$$

The solution of any linear, ordinary differential equation with constant coefficients can be obtained by assuming that

$$y = Be^{rt}$$

and determining values of B and r which satisfy the differential equation and the initial or boundary conditions. Substitution into the differential equation gives

$$Be^{rt}[mr^2 + cr + k] = 0.$$

The differential equation will be satisfied if

$$B = 0$$

or if

$$mr^2 + cr + k = 0.$$

The first equation is a trivial solution, and the second equation gives

$$r_{1,2} = \frac{-c \pm \sqrt{c^2 - 4mk}}{2m}. \tag{12-7}$$

The general solution to the differential equation must have two arbitrary constants and can be written as

$$y = Be^{r_1 t} + Ce^{r_2 t}. \tag{12-8}$$

The constants B and C can be determined from the initial conditions for the motion, and the constants r_1 and r_2 are obtained from Eq. (12-7). The behavior of the system depends on the quantity within the radical of Eq. (12-7). The quantity can be zero, positive, or negative and the radical will be zero, real, or imaginary, respectively.

The *critical damping coefficient* is defined as the value of c which will make the radical zero; therefore,

$$c_{cr} = 2\sqrt{km}.$$

The critical damping coefficient represents the minimum amount of damping for nonvibratory motion. The solution to Eq. (12-6) will have three distinct forms, depending on whether c is greater than, equal to, or less than c_{cr}.

Case I: $c > c_{cr}$. In this case, the radical in Eq. (12-7) is real, and both roots are real and negative. The motion is nonvibratory or *aperiodic*, and such a system will return slowly to its equilibrium position with not more than one (usually none) reversal of motion.

Case II: $c = c_{cr}$. In this case, in which the damping is critical, the two roots in Eq. (12-7) are identical and are equal to

$$r = -\frac{c_{cr}}{2m} = -\frac{2\sqrt{km}}{2m} = -\sqrt{\frac{k}{m}} = -\omega_n.$$

In this case, Eq. (12-8) will have only one arbitrary constant, and therefore it is not the general solution. The general solution, when $c = c_{cr}$, is

$$y = (B + Ct)e^{rt} = (B + Ct)e^{-\omega_n t}.$$

This statement can be verified by substitution into Eq. (12-6). The motion given by this equation is also aperiodic, and it is of special importance because it represents the dividing point between aperiodic motion and damped oscillatory motion.

Critical damping is the minimum damping for which a system will return to rest without oscillation. Furthermore, the system will return to rest in the *least possible time* for any specified initial conditions.

Case III: $c < c_{cr}$. In this case, the system is underdamped, the radical in Eq. (12-7) is imaginary, and the roots are

$$r_{1,2} = -\frac{c}{2m} \pm i\sqrt{\frac{k}{m} - \left(\frac{c}{2m}\right)^2} = \alpha \pm i\omega_d,$$

where $i = \sqrt{-1}$, $\alpha = -c/(2m)$, and

$$\omega_d = \sqrt{\frac{k}{m} - \left(\frac{c}{2m}\right)^2}$$

is called the damped natural frequency.

When these values of r_1 and r_2 are substituted into Eq. (12-8) it becomes

$$y = B_1 e^{(\alpha + i\omega_d)t} + C_1 e^{(\alpha - i\omega_d)t} = e^{\alpha t}(B_1 e^{i\omega_d t} + C_1 e^{-i\omega_d t}).$$

By means of a series expansion, it can be shown that

$$e^{\pm i\omega_d t} = \cos \omega_d t \pm i \sin \omega_d t,$$

and after some algebraic reduction, the solution to the differential equation becomes

$$y = e^{\alpha t}(B \cos \omega_d t + C \sin \omega_d t) \tag{12-9a}$$

$$= Ae^{\alpha t} \cos(\omega_d t - \phi), \tag{12-9b}$$

where B and C or A and ϕ are constants which are determined from the initial conditions. This motion, called *time-periodic,* is oscillatory in nature, but the amplitude, $Ae^{\alpha t}$, varies with time. Since the constant $\alpha = -c/(2m)$ is negative, the amplitude will always decrease with time.

Damped free vibration ($c < c_{cr}$) does not repeat itself as does free undamped motion; therefore, it does not have a period or frequency as defined for free undamped vibrations. It is customary, however, to call the time interval between pairs of successive "zeroes or peaks" of the motion the *damped period.* When Eq. (12-9) and (12-3) are compared the damped period and frequency are seen to be

$$T_d = \frac{2\pi}{\omega_d} \quad \text{and} \quad f_d = \frac{\omega_d}{2\pi}.$$

The value of ω_d is

$$\omega_d = \sqrt{\frac{k}{m} - \left(\frac{c}{2m}\right)^2},$$

and it is apparent that ω_d will always be less than $\omega_n = \sqrt{k/m}$ for any positive value of c (less than c_{cr}).

Figure 12-13 shows the response of the system in Fig. 12-12 for

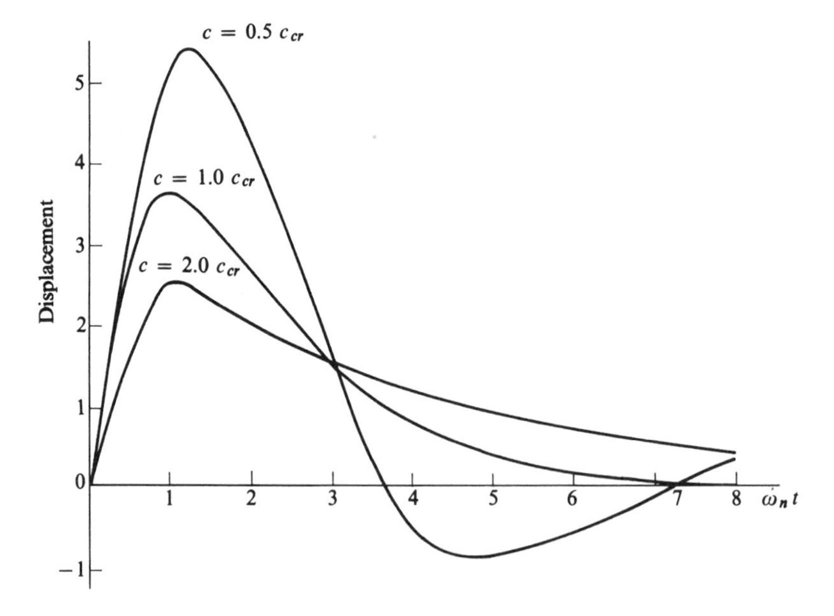

Figure 12-13

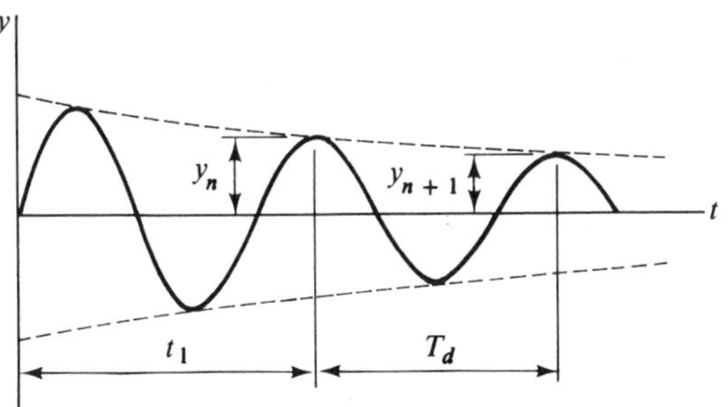

Figure 12-14

the three cases. The motion is started with the same positive velocity and with zero displacement for all three curves.

The amount of damping in a system is indicated by the rate of decay of the oscillation. The rate can be indicated by a quantity called the *logarithmic decrement*, δ, which is defined as the logarithm of the ratio of any two successive positive (or negative) amplitudes. Figure 12-14 represents the response of a system with a relatively small amount of damping. The amplitude at time t_1 is

$$y_n = Ae^{-(c/2m)t_1},$$

and the next amplitude is

$$y_{n+1} = Ae^{-(c/2m)(t_1+T_d)}.$$

The ratio of the two amplitudes is

$$\frac{y_n}{y_{n+1}} = \frac{Ae^{-(c/2m)t_1}}{Ae^{-(c/2m)(t_1+T_d)}} = e^{cT_d/(2m)},$$

and the logarithmic decrement is

$$\delta = \ln e^{cT_d/(2m)} = \frac{cT_d}{2m}. \tag{12-10}$$

The logarithmic decrement may be expressed in various forms since

$$T_d = \frac{2\pi}{\omega_d} = \frac{2\pi}{\sqrt{(k/m)-(c/2m)^2}}.$$

The ratio $\zeta = c/c_{cr}$ is called the *damping factor* and is convenient in expressing many of the preceding equations. When $\sqrt{k/m}$ is replaced by ω_n, $2\sqrt{km}$ by c_{cr}, and c by ζc_{cr}, the following results can be derived:

$$\omega_d = \omega_n\sqrt{1-\zeta^2},$$

$$T_d = \frac{2\pi}{\omega_n\sqrt{1-\zeta^2}},$$

and

$$\delta = \frac{2\pi\zeta}{\sqrt{1-\zeta^2}}.$$

It is suggested that these expressions be verified by substitution.

EXAMPLE 12-4

The 9.66-lb body E in Fig. 12-15(a) is fastened to the rod DF whose weight can be neglected. The spring has a modulus of 9.00 lb per ft, and the coefficient of the dashpot is 2.40 lb-sec per ft. The system is in equilibrium when DF is horizontal. The rod is displaced 0.10 rad clockwise and released from rest when $t=0$. Determine
(a) The equation of motion of the rod.
(b) The frequency of the motion.
(c) The logarithmic decrement of the motion.

SOLUTION

(a) A free-body diagram of the rod, displaced in the positive direction, is shown in Fig. 12-15(b). When the body is in equilibrium, θ and $\dot{\theta}$ are both zero, and $T=T_0=(\frac{4}{5})9.66$. The moment equation for the body with the displacement and velocity indicated in Fig. 12-15(b) is

$$4(9.66)-4F_d-5T=I_O\ddot{\theta}$$

or

$$4(9.66)-4(2.40)(4\dot{\theta})-5\left[\left(\frac{4}{5}\right)9.66+9(5\theta)\right]=\frac{9.66}{32.2}(4.00)^2\ddot{\theta}.$$

This equation can be reduced to

$$4.8\ddot{\theta}+38.4\dot{\theta}+225\theta=0$$

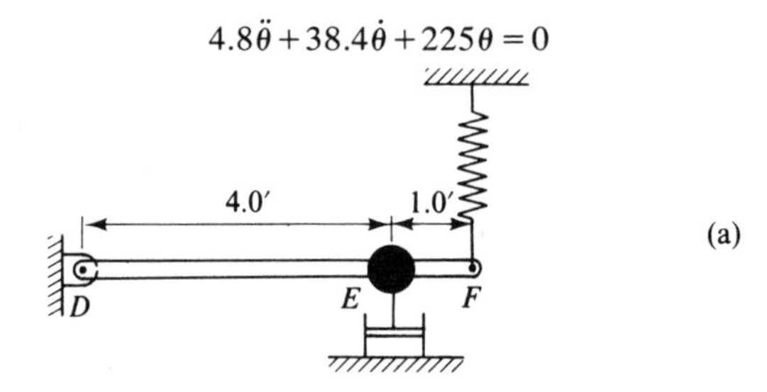

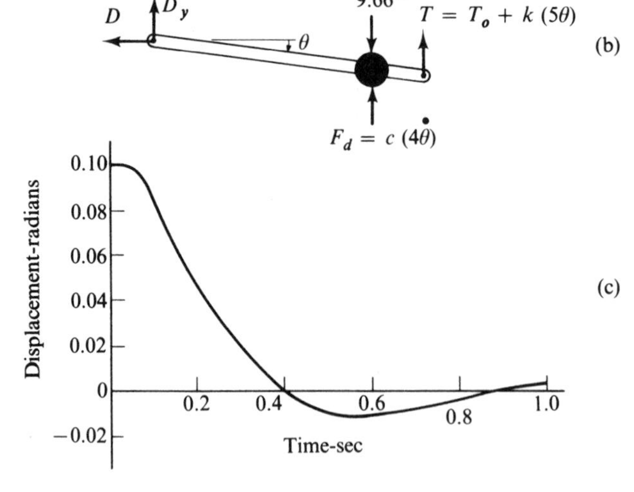

Figure 12-15

or

$$\ddot{\theta}+8.00\dot{\theta}+46.9\theta=0.$$

The general solution can be obtained by substituting $\theta=De^{rt}$ in this equation to give

$$De^{rt}(r^2+8r+46.9)=0 \qquad \text{(a)}$$

from which

$$r_{1,2}=-4.00\pm i5.56=\alpha\pm i\omega_d.$$

The equation of motion is

$$\theta=e^{-4t}(B\cos 5.56t+C\sin 5.56t).$$

When $t=0$, $\theta=0.1$; therefore,

$$B=0.100\text{ rad}.$$

The angular velocity of the rod is

$$\dot{\theta}=e^{-4t}(-4B\cos 5.56t-4C\sin 5.56t-5.56B\sin 5.56t +5.56C\cos 5.56t)$$

and when $t=0$, $\dot{\theta}=0$; therefore,

$$0=-4B+5.56C$$

and

$$C=\frac{4(0.100)}{5.56}=0.0720\text{ rad}.$$

The solution is

$$\underline{\theta=e^{-4.00t}(0.100\cos 5.56t+0.0720\sin 5.56t)}. \qquad \text{Ans.}$$

The graph of this solution is shown in Fig. 12-15(c). For critical damping, the coefficient of r in Eq. (a) should be $\sqrt{4(46.9)}=13.70$. The ratio of 8 to 13.70 is 0.584, which is the damping factor. With this amount of damping, the vibrations are damped out very quickly.

(b) The circular frequency of the motion is

$$\omega_d=5.56\text{ rad per sec},$$

and the frequency is

$$f_d=\frac{\omega_d}{2\pi}=\frac{5.56}{6.28}=\underline{0.884\text{ Hz}}. \qquad \text{Ans.}$$

(c) The logarithmic decrement is

$$\delta=\ln\frac{\theta_0 e^{-4t_1}}{\theta_0 e^{-4(t_1+T_d)}}=4T_d=\frac{4}{f_d}=\frac{4}{0.884}=4.52. \qquad \text{Ans.}$$

In this expression, θ_0 is the maximum value of the quantity in parentheses in the answer to part (a) and occurs at $t=t_1$. Since the values of θ_0 and t_1 both drop out in this evaluation, it is not necessary to calculate either t_1 or θ_0.

PROBLEMS

***12-44** Body A in Fig. P12-44 has a mass of 2.0 kg, and the spring has a modulus of 50 N/m. The system is critically damped by the dashpot. Body A is moved 5.0 cm to the right of the equilibrium position and released with an initial velocity of 50 cm/s to the left. Determine

(a) The value of c_{cr}.

(b) The position and velocity of A when $t = 0.20$ s.

(c) The position and velocity of A when $t = 1.00$ s.

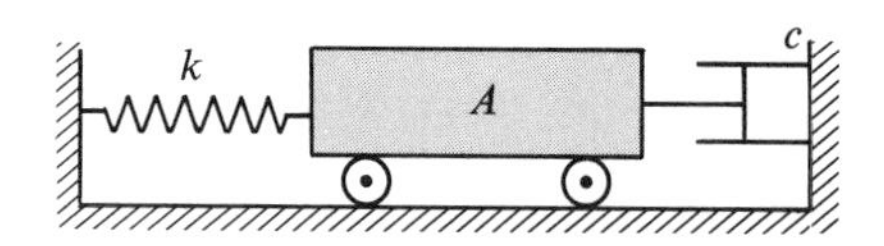

Figure P12-44

***12-45** Solve Problem 12-44, parts (b) and (c), if the damping coefficient is 24 N · s/m.

12-46 Body B in Fig. P12-46 has a mass of 2.0 slugs and the mass of the T-shaped bar can be neglected. The spring modulus is 80 lb per ft, and the damping coefficient of the dashpot is 16 lb-sec per ft. The system is in equilibrium when AB is horizontal. When $b = c = 2$ ft and the system is disturbed from its position of equilibrium, determine

(a) The type of motion which will exist.

(b) The frequency of oscillation (if any exists).

(c) The damping factor.

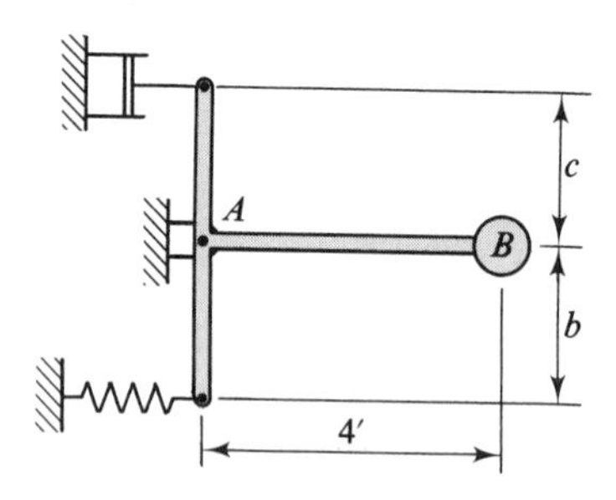

Figure P12-46

***12-47** Body A in Fig. P12-47 has a mass of 5 kg, the spring has a modulus of 300 N/m, and the coefficient of the viscous damper is 180 N · s/m. The mass of the T-shaped member is negligible, and the system is in equilibrium when OA is horizontal. Determine the distance y which will result in critical damping.

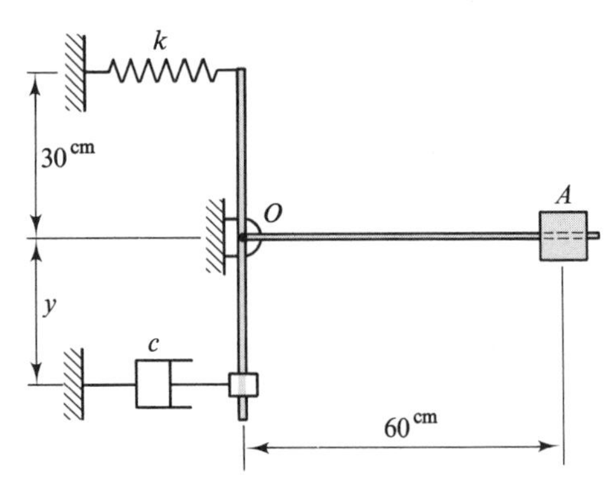

Figure P12-47

***12-48** The damper coefficient in Problem 12-47 is 20 N · s/m, and the distance y is 30 cm. All other data are the same as in Problem 12-47. Body A is pulled 4.0 cm downward from the equilibrium position and released from rest. Determine

(a) The logarithmic decrement of the resulting motion.

(b) The damping factor ζ.

***12-49** Determine the velocity of body A in Problem 12-48 the first time OA is horizontal after being released.

12-50 The body in Fig. P12-50 weighs 96.6 lb, the spring stiffness is 39 lb per ft, and the dashpot coefficient is 12 lb-sec per ft. Determine the critical damping coefficient and the logarithmic decrement.

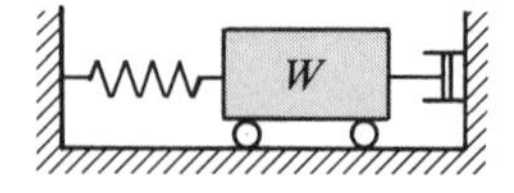

Figure P12-50

12-51 The rigid T-bar of negligible weight in Fig. P12-51 rotates in a vertical plane about a horizontal axis at O. The bar is caused to vibrate by displacing the weight and releasing it from rest. Determine the damped frequency of vibration and the ratio of the positive amplitudes of the third and fourth cycles. What is the ratio of the amplitudes of the first and third cycles?

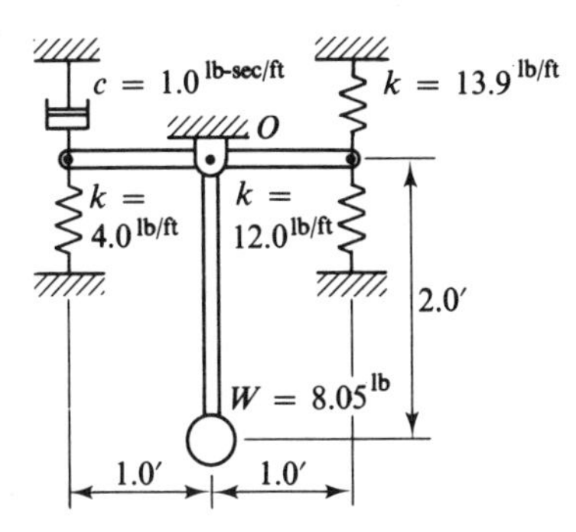

Figure P12-51

12-52 The body in Fig. P12-52 weighs 24.15 lb and is supported by three springs and three dashpots as shown. The spring moduli are $k_1 = k_2 = 10$ lb per ft and $k_3 = 7$ lb per ft. The dashpot coefficients are $c_1 = c_2 = 0.05$ lb-sec per ft and $c_3 = 0.08$ lb-sec per ft. The body is displaced 4 in. downward from equilibrium and released from rest. Determine the number of oscillations which will occur before the amplitude of the vibration is reduced to 20% of its original value.

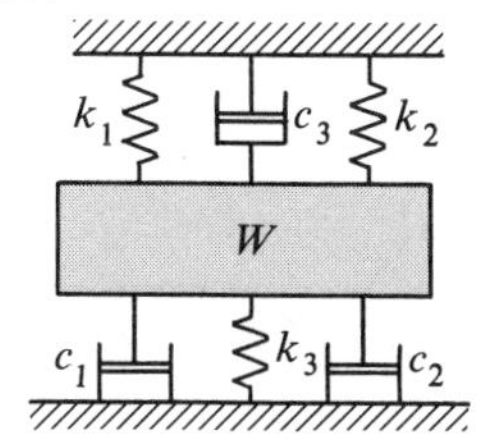

Figure P12-52

12-53 The homogeneous bar in Fig. P12-53 has a mass of 0.8 lb-sec² per ft and is in equilibrium when it is horizontal. The spring modulus is 3.0 lb per ft, and the dashpot coefficient is 1.5 lb-sec per ft. The rod is given an angular displacement of 0.06 rad and an angular velocity of 1.4 rad per sec, both clockwise, when $t = 0$. Solve the differential equation of motion for θ as a function of time and determine the angular position and velocity of the bar when $t = 1.0$ sec.

12-54 Set up the differential equation of motion for the inverted pendulum of Fig. P12-54.

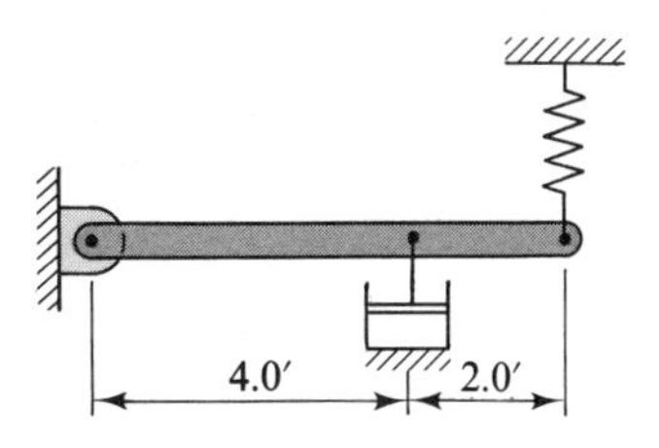

Figure P12-53

The mass of the L-shaped rod can be neglected. Derive an expression for the critical damping coefficient.

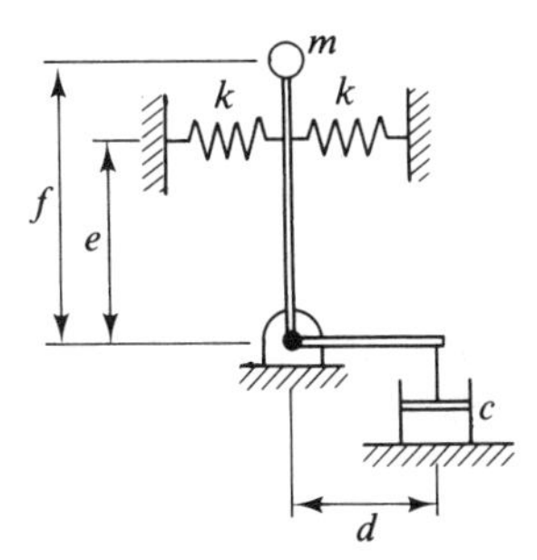

Figure P12-54

12-55 The disk in Fig. P12-55 has a moment of inertia of 0.300 lb-sec²-in. about the z axis and is suspended from a wire as shown. The frequency of torsional vibrations in air (assume zero damping) is 0.50 Hz. When the disk is immersed in a fluid and caused to oscillate by turning the disk 1.00 rad and releasing it, the amplitude of the oscillation is reduced to 0.20 rad after five complete cycles.

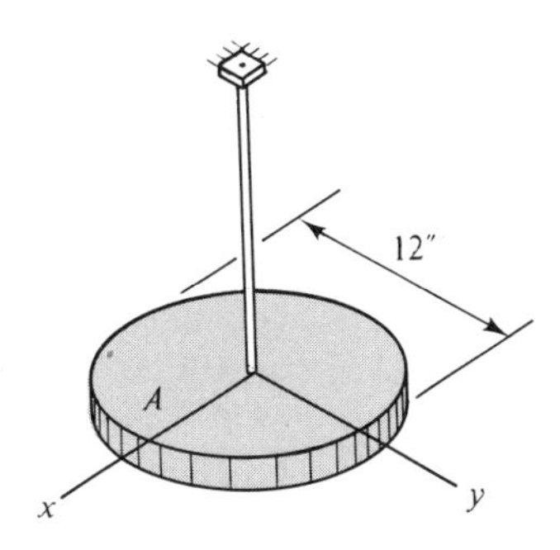

Figure P12-55

(a) Determine the damping coefficient of the disk and fluid.
(b) What would be the value of c if it were assumed that the frequency did not change due to damping?

12-56 Body m in Fig. P12-56 has a mass of 0.025 lb-sec^2 per in., and the weight of the bar can be neglected. During 20 cycles of free vibration of the system, the amplitude of the motion of D decreases from 3.0 to 1.0 in. in 10 sec. Determine
(a) The damping coefficient c.
(b) The spring stiffness k.
(c) The critical damping coefficient c_{cr}.
(d) The logarithmic decrement of the system.

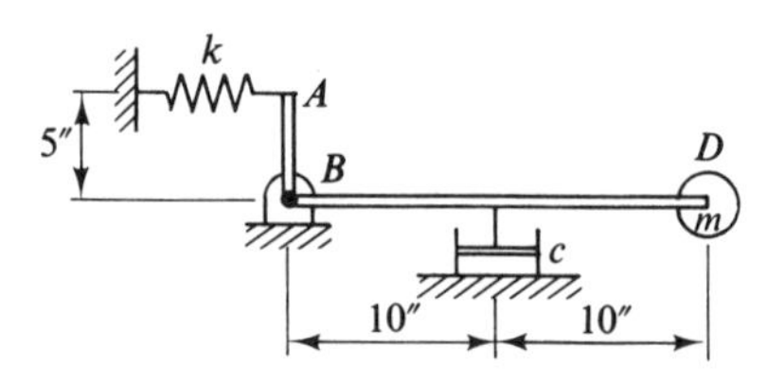

Figure P12-56

12-6 ROTATING VECTORS

Rotating vectors, as indicated in Art. 12-2, can be used to provide a graphical interpretation of some of the properties of simple harmonic motion.

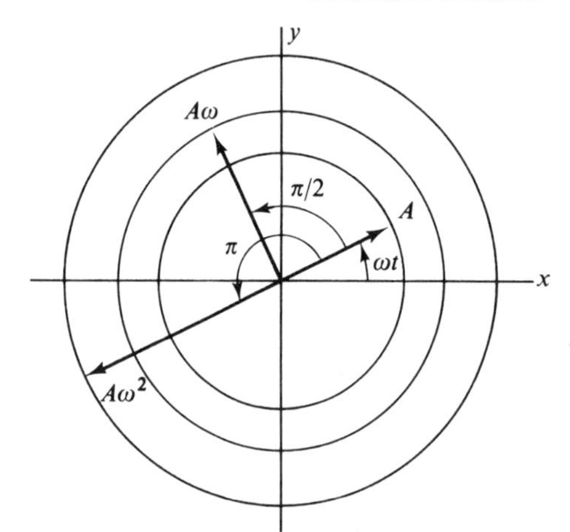

Figure 12-16

A simple harmonic motion can be described by the equation

$$y = A \sin \omega t. \tag{a}$$

If a vector of length A rotates about a fixed point with a constant angular velocity ω as shown in Fig. 12-16, its projection on the y axis is given by Eq. (a). This figure assumes that t was zero when the vector was horizontal.

Differentiation of Eq. (a) with respect to time gives

$$\dot{y} = A\omega \cos \omega t = A\omega \sin\left(\omega t + \frac{\pi}{2}\right)$$

and

$$\ddot{y} = -A\omega^2 \sin \omega t = A\omega^2 \sin(\omega t + \pi).$$

These expressions are similar to Eq. (a) except that the amplitude is multiplied by ω for $\dot{y}$ and by ω^2 for $\ddot{y}$ and the angle ωt is increased by $\pi/2$ for $\dot{y}$ and by π for $\ddot{y}$. Rotating vectors to represent $\dot{y}$ and $\ddot{y}$ are included in Fig. 12-16. The angles $\pi/2$ and π are called *phase angles*. All three vectors rotate with the same angular velocity ω, and the velocity vector is said to *lead* the displacement vector by a phase angle of $\pi/2$ rad. Note that the velocity attains its maximum value one-fourth of a cycle *before* the displacement becomes maximum. Similarly, the acceleration leads the velocity by a phase angle of $\pi/2$ rad and the displacement by π rad.

Figure 12-17 shows the displacement, velocity, and acceleration

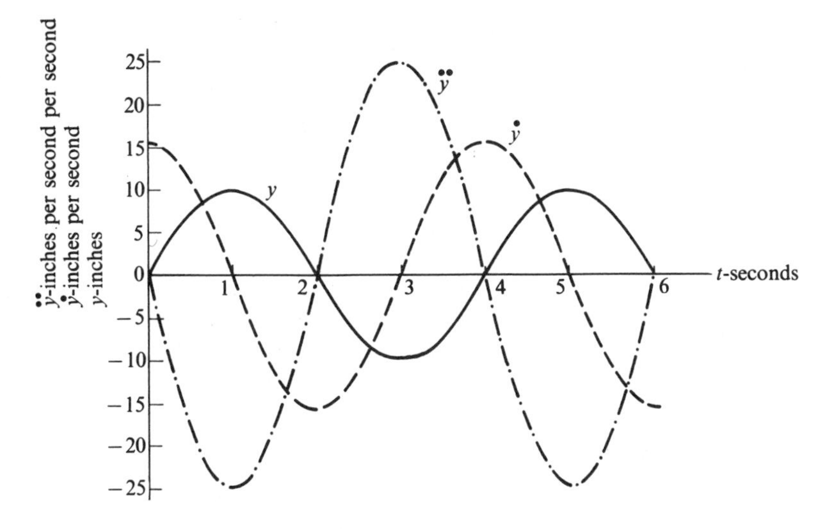

Figure 12-17

of the harmonic motion just discussed when plotted as functions of time. In this figure, A is 10 in. and ω is $\pi/2$ rad per sec. Note that y is maximum at $t=1$ sec, while $\dot{y}$ is maximum at $t=0$, and the acceleration passed through its maximum value 1 sec earlier.

Two harmonic motions with the same frequency but with different phase angles can be combined by vector addition to give the resultant motion. Consider two motions given by

$$y_1 = A_1 \sin(\omega t + \phi_1), \qquad y_2 = A_2 \sin(\omega t + \phi_2).$$

The sum of these motions is

$$\begin{aligned} y_1 + y_2 &= A_1 \sin(\omega t + \phi_1) + A_2 \sin(\omega t + \phi_2) \\ &= A_1(\sin \omega t \cos \phi_1 + \cos \omega t \sin \phi_1) \\ &\quad + A_2(\sin \omega t \cos \phi_2 + \cos \omega t \sin \phi_2) \\ &= (A_1 \cos \phi_1 + A_2 \cos \phi_2)\sin \omega t + (A_1 \sin \phi_1 + A_2 \sin \phi_2)\cos \omega t. \end{aligned}$$

The last expression can also be written as

$$y_1 + y_2 = A \sin(\omega t + \phi),$$

where

$$A = [(A_1 \cos \phi_1 + A_2 \cos \phi_2)^2 + (A_1 \sin \phi_1 + A_2 \sin \phi_2)^2]^{1/2} \tag{b}$$

and

$$\tan \phi = \frac{A_1 \sin \phi_1 + A_2 \sin \phi_2}{A_1 \cos \phi_1 + A_2 \cos \phi_2}. \tag{c}$$

Figure 12-18 shows the rotating vectors $\mathbf{A}_1$ and $\mathbf{A}_2$ together with their resultant $\mathbf{A}$. The sum of the projections of $\mathbf{A}_1$ and $\mathbf{A}_2$ on line OC is $A_1 \cos \phi_1 + A_2 \cos \phi_2$, and the sum of their projections perpendicular

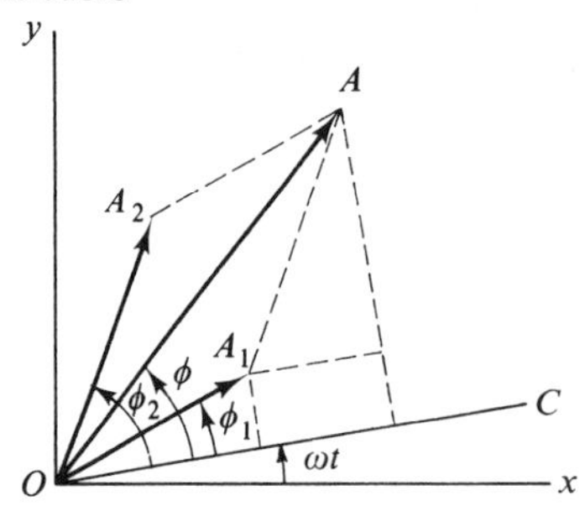

Figure 12-18

to OC is $A_1 \sin \phi_1 + A_2 \sin \phi_2$. Thus the magnitude of vector **A** is

$$A = [(A_1 \cos \phi_1 + A_2 \cos \phi_2)^2 + (A_1 \sin \phi_1 + A_2 \sin \phi_2)^2]^{1/2},$$

which is the same as Eq. (b). From a similar analysis, the angle ϕ in Fig. 12-18 is given by Eq. (c). Therefore, any number of harmonic motions with the same frequency and with various amplitudes and phase angles can be added or subtracted either algebraically or by vector addition of the rotating vectors which represent the motions.

12-7 FORCED VIBRATIONS

When a force which varies periodically is applied to a body mounted on springs or other elastic supports a forced vibration of the body will result. A forced vibration of a body can also be produced by giving the body supporting the vibrating system a periodic motion. For example, if a weight is hung from a spring and the upper end of the spring is moved vertically with a periodic motion, forced vibrations of the weight will be produced. In the last example, the variable force is transmitted through the spring to the weight.

A body subjected to a periodic force and arbitrary initial conditions will have a combination of free and forced vibrations at the start of the motion. In all practical cases, however, the damping forces will eliminate the free vibrations (often called *transients*), and the remaining motion is called the *steady-state* vibration.

The period and frequency of free vibrations depend on the mass of the body, on the stiffness of the elastic support, and on the damping coefficient. The amplitude of free vibrations depends on the starting conditions and in general on the circular frequency. The frequency of steady-state forced vibrations depends on the frequency of the applied load but not on the characteristics of the vibrating body. The amplitude of the steady-state forced vibrations depends on the magnitude and frequency of the applied load and on the frequency of the free vibrations but not on the starting conditions.

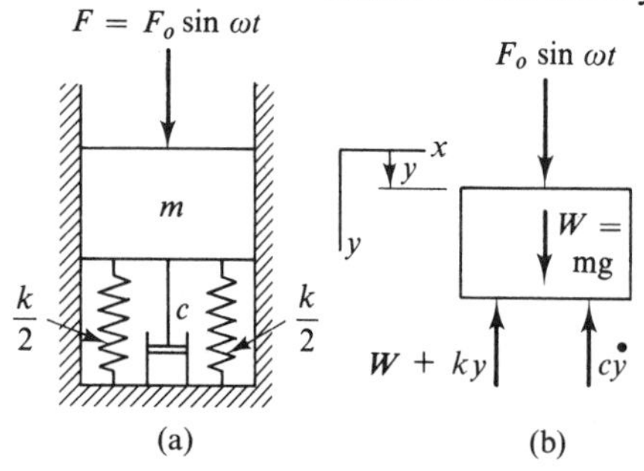

Figure 12-19

Any periodically varying applied force will result in forced vibrations. A common type of variable force is one which can be expressed as a sine or cosine function of time.

Figure 12-19 represents a body of mass m, supported by a spring whose modulus is k and a dashpot whose coefficient is c. The body is acted on by the force $F = F_O \sin \omega t$, where F_O is the amplitude of the variable force and ω is its circular frequency. A free-body diagram of the body displaced a distance y from the equilibrium position is shown in Fig. 12-19(b). The spring force is equal to $W+ky$, where W is the weight of the body and is also the spring force when the body is in equilibrium and the force F is zero. The differential equation of motion, in scalar form, is

$$\sum F_y = m\ddot{y},$$

$$W-(W+ky)-c\dot{y}+F_O \sin \omega t = m\ddot{y}$$

or

$$m\ddot{y}+c\dot{y}+ky = F_O \sin \omega t. \qquad (12\text{-}11)$$

The general solution to Eq. (12-11) is made up of two parts: the complementary function when the right-hand side is zero (see Art. 12-5) and the particular solution which satisfies the equation as written. The complementary function decreases exponentially with time and contains two constants of integration. It is often called the *transient* term. The particular solution is the *steady-state* solution and is usually the one of primary interest. The transient solution has already been discussed in Art. 12-5 and will not be considered further.

The steady-state solution can be obtained by assuming y to be of the form

$$y = A \sin(\omega t+\phi) = B \sin \omega t + C \cos \omega t, \qquad (12\text{-}12)$$

substituting it into Eq. (12-11), and solving for either A and ϕ or B and C. A convenient graphical representation of Eq. (12-11) can be obtained by means of rotating vectors. If the steady-state solution is assumed to be $y = A \sin(\omega t+\phi)$, the various terms of Eq. (12-11) are

$$m\ddot{y} = -Am\omega^2 \sin(\omega t+\phi) = Am\omega^2 \sin(\omega t+\phi+\pi),$$

$$c\dot{y} = Ac\omega \cos(\omega t+\phi) = Ac\omega \sin\left(\omega t+\phi+\frac{\pi}{2}\right),$$

and

$$ky = Ak \sin(\omega t+\phi).$$

Notice that the quantity $m\ddot{\mathbf{y}}$ leads $\mathbf{y}$ by π rad and $c\dot{\mathbf{y}}$ leads $\mathbf{y}$ by $\pi/2$ rad. The vector sum of these three rotating vectors must be equal to $F_O \sin \omega t$ from Eq. (12-11). Figure 12-20(a) shows three vectors with magnitudes $Am\omega^2$, $Ac\omega$, and Ak, which, when added vectorially, must equal F_O. The vector $\mathbf{A}$ is in the direction of $k\mathbf{y}$ (opposite $m\ddot{\mathbf{y}}$) and lags the force $\mathbf{F}$ by the angle ϕ. In this case, ϕ will be negative.

Since the amplitude of the resulting motion occurs in three of the rotating vectors, it is often convenient to divide all amplitudes by A to get Fig. 12-20(b). Note that if $k = m\omega^2$, $(\omega = \omega_n = \sqrt{k/m})$, the applied force will lead the displacement by $\pi/2$ rad. This case is called *resonance*. If c is zero (no damping), the vector polygon becomes a straight line, and ϕ is either zero or π rad, depending on whether k is more or less than $m\omega^2$. When there is no damping and $k = m\omega^2$, the

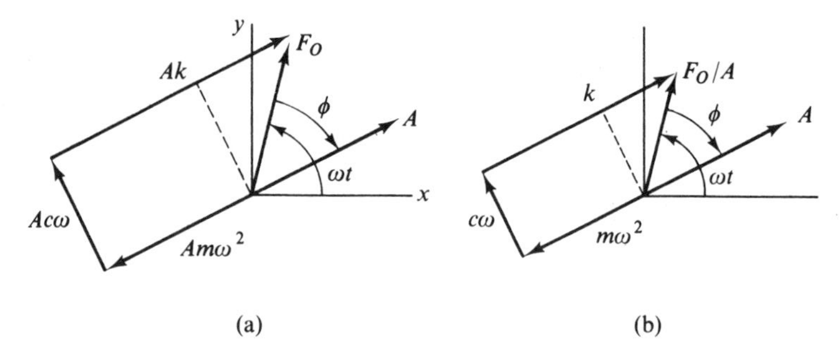

Figure 12-20

steady-state amplitude becomes infinite. The rate at which the amplitude will increase can be determined by obtaining the particular solution to Eq. (12-11) subject to the conditions $c=0$ and $k=m\omega^2$. The solution will be of the form

$$y = A_O t \cos \omega t.$$

This solution can be demonstrated by direct substitution. The buildup of the amplitude may be quite rapid, and unless stops to prevent excessive motion are provided, damage to the equipment will result.

Solutions for the amplitude and phase angle, satisfying Eq. (12-11), can be determined from Fig. 12-20. Thus,

$$\left(\frac{F_O}{A}\right)^2 = (k - m\omega^2)^2 + (c\omega)^2$$

or

$$A = \frac{F_O}{\sqrt{(k-m\omega^2)^2+(c\omega)^2}}. \tag{a}$$

Similarly,

$$\tan\phi = \frac{c\omega}{k-m\omega^2}. \tag{b}$$

When other than rectilinear motion is involved, a differential equation similar to Eq. (12-11) must be derived, and the same analysis can be used to obtain the linear or angular quantities used to derive the equation. When the numerator and denominator of Eq. (a) are both divided by k and when c is replaced by $\zeta c_{cr} = \zeta 2\sqrt{mk}$ and $\sqrt{k/m}$ is replaced by ω_n Eq. (a) becomes

$$A = \frac{F_O/k}{[(1-\omega^2/\omega_n^2)^2+(2\zeta\omega/\omega_n)^2]^{1/2}}. \tag{a'}$$

The ratio of A to F_O/k is called the *magnification factor.* It is the ratio by which the quantity F_O/k, the amount the spring would stretch due to a static load F_O, must be multiplied to give A. The curves in Fig. 12-21 show the variation of the magnification factor with ω/ω_n for various values of ζ.

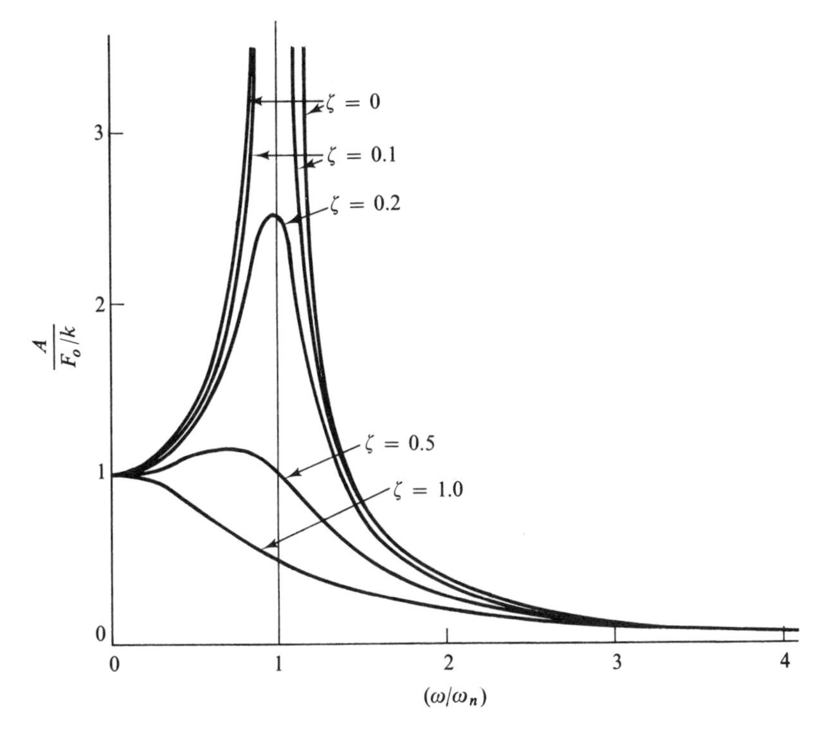

Figure 12-21

A differential equation of motion similar to Eq. (12-11) will result from the rotation of an unbalanced mass or from the harmonic oscillation of a body supporting the spring-mass system as shown in the following examples.

EXAMPLE 12-5

The motor in Fig. 12-22(a) is mounted on two springs, each with a modulus of $k/2 = 60$ lb per in. The dashpot has a coefficient of 0.80 lb-sec per in. The motor, including the unbalanced mass B, weighs 38.6 lb, and the unbalanced body B weighs 1.0 lb and is located 3.0 in. from the center of the shaft.

(a) The motor rotates at 300 rpm. Determine the amplitude and phase angle (relative to the position of B) of the resulting motion.

(b) Determine the maximum and minimum force on the motor exerted by the springs, by the damper, and by the springs and damper together when the motor rotates at 300 rpm.

(c) Determine the resonant speed and the resulting amplitude of the motion.

SOLUTION

(a) Figure 12-22(b) is a free-body diagram of the motor without the weight B, and Fig. 12-22(c) is a free-body diagram of B. The variable y gives the position of the center of the motor shaft measured from the

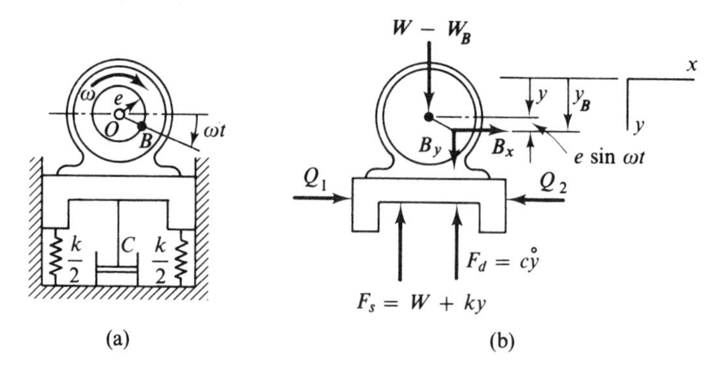

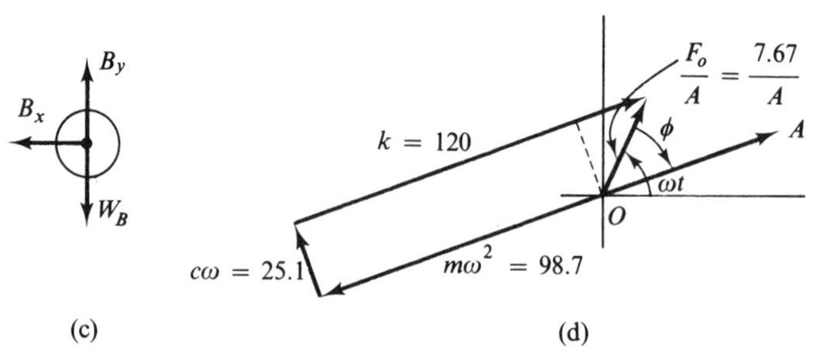

Figure 12-22

equilibrium position, and y_B is the vertical coordinate of B. From the figure,

$$y_B = y + e \sin \omega t;$$

therefore,

$$\ddot{y}_B = \ddot{y} - e\omega^2 \sin \omega t.$$

The equations of motion in the y direction for the two bodies, in scalar form, are

$$(m - m_B)\ddot{y} = (W - W_B) + B_y - (W + ky) - c\dot{y}$$

and

$$m_B(\ddot{y}_B) = m_B(\ddot{y} - e\omega^2 \sin \omega t) = W_B - B_y.$$

When these two equations are added to eliminate B_y, the result is

$$m\ddot{y} - m_B e\omega^2 \sin \omega t = -ky - c\dot{y}$$

or

$$m\ddot{y} + c\dot{y} + ky = m_B e\omega^2 \sin \omega t = F_O \sin \omega t,$$

where

$$F_O = m_B e\omega^2.$$

The equation is the same as Eq. (12-11) and can be solved by means of a vector diagram similar to Fig. 12-20(b). From the data,

$$\omega = \frac{300}{60}(2\pi) = 31.4 \text{ rad per sec},$$

$$m\omega^2 = \frac{38.6}{386}(31.4)^2 = 98.7 \text{ lb per in.},$$

$$c\omega = 0.8(31.4) = 25.1 \text{ lb per in.},$$

and

$$F_O = \frac{1.0}{386}(3.0)(31.4)^2 = 7.67 \text{ lb}.$$

These values are shown on Fig. 12-22(d), and from the right triangle,

$$\left(\frac{7.67}{A}\right)^2 = (120 - 98.7)^2 + (25.1)^2,$$

which gives

$$A = \underline{0.233 \text{ in.}} \qquad \text{Ans.}$$

Also from the diagram,

$$\tan\phi = \frac{25.1}{21.3}$$

from which

$$\phi = 49.7^\circ \text{ (y lags behind } y_B\text{)}. \qquad \text{Ans.}$$

(b) The magnitude of the force exerted by the springs is

$$F_{sp} = W + ky = 38.6 + 120y,$$

and the maximum and minimum values are

$$(F_{sp})_{max} = 38.6 + 120(0.233) = \underline{66.56 \text{ lb}}, \qquad \text{Ans.}$$

$$(F_{sp})_{min} = 38.6 + 120(-0.233) = \underline{10.64 \text{ lb}}. \qquad \text{Ans.}$$

The magnitude of the force exerted by the damper is

$$F_d = c\dot{y} = 0.8\dot{y},$$

and the maximum (and minimum) value of F_d is

$$(F_d)_{max} = 0.8\dot{y}_{max} = 0.8A\omega = 0.8(0.233)(31.4)$$
$$= \underline{5.85 \text{ lb}}. \qquad \text{Ans.}$$

The variable part of the spring force and the damper force are 90° out of phase, as shown by the vector diagram; therefore, the amplitude of the resultant variable force is

$$F_v = [(27.96)^2 + 5.85^2]^{1/2} = 28.6 \text{ lb},$$

and the maximum and minimum resultant force of the spring and damper on the motor is

$$F_{max,min} = 38.6 \pm 28.6 = \underline{67.2 \text{ lb}} \text{ and } \underline{10.0 \text{ lb}}. \qquad \text{Ans.}$$

(c) The resonant speed occurs when $m\omega^2 = k$ or when

$$\omega = \omega_n = \sqrt{\frac{k}{m}} = \sqrt{\frac{120}{0.1}} = 34.64 \text{ rad per sec}$$

$$= \frac{34.64}{2\pi}(60) = \underline{331 \text{ rpm.}} \qquad \text{Ans.}$$

Since

$$F_0 = m_B e\omega^2,$$

the amplitude at resonance is

$$A = \frac{F_0}{c\omega} = \frac{m_B e\omega}{c} = \frac{1.0(3.0)(34.6)}{(386)(0.8)} = \underline{0.337 \text{ in.}} \qquad \text{Ans.}$$

EXAMPLE 12-6

Body D in Fig. 12-23(a) weighs 19.3 lb and is supported by a spring with a modulus of 60 lb per in. Body B at the upper end of the spring is given a vertical harmonic motion by the crank, which has an angular velocity of 40 rad per sec as shown. The length of the crank is 0.50 in.
(a) Determine the amplitude and phase angle of the motion of D when the damping coefficient of the damper is 0.60 lb-sec per in. and when the damper is disconnected.
(b) Determine the range of values of ω (if any) which will limit the amplitude of the motion of D to 0.80 in. when $c = 0$.

SOLUTION

(a) A free-body diagram of body D is shown in Fig. 12-23(b). The downward direction is positive. Note that the tension in the spring is equal to the weight of D plus the spring modulus multiplied by the

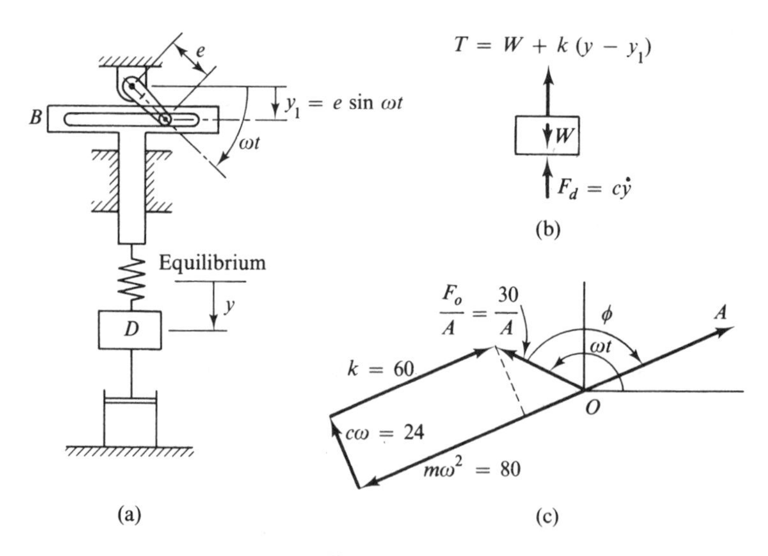

Figure 12-23

amount the spring is stretched. The differential equation of motion in the y direction is

$$m\ddot{y} = W - c\dot{y} - [W + k(y - y_1)]$$

or

$$m\ddot{y} + c\dot{y} + ky = ky_1 = ke \sin \omega t.$$

This is the same as Eq. (12-11) if ke is set equal to F_O.

The steady-state solution is assumed to be of the form

$$y = A \sin(\omega t + \phi),$$

and the vector diagram shown in Fig. 12-23(c) can be drawn when numerical values are substituted for m, c, k, e, and ω. The numerical values for the diagram are

$$m\omega^2 = \frac{19.3}{386}(40)^2 = 80 \text{ lb per in.},$$

$$c\omega = 0.60(40) = 24 \text{ lb per in.},$$

$$k = 60 \text{ lb per in.},$$

and

$$F_O = ek = 0.50(60) = 30 \text{ lb}.$$

From the diagram,

$$\left(\frac{30}{A}\right)^2 = (24)^2 + (80 - 60)^2,$$

which gives

$$A = \underline{0.960 \text{ in.}} \qquad \text{Ans.}$$

The phase angle is

$$\phi = \tan^{-1}\frac{24}{-20} = \underline{129.8°}. \qquad \text{Ans.}$$

This angle indicates that D will reach its lowest position 129.8° after B is at its lowest position. This angle corresponds to a time interval of

$$t = \frac{\phi}{\omega} = \frac{129.8\pi}{180(40)} = 0.0566 \text{ sec}.$$

When c is equal to zero, the polygon in Fig. 12-23(c) is a straight line, and since k is less than $m\omega^2$, y will lag y_1 by 180°. The amplitude of y is obtained from

$$\frac{30}{A} = 80 - 60 = 20$$

or

$$A = \underline{1.500 \text{ in.}} \qquad \text{Ans.}$$

(b) When there is no damping, the value of A will be less than 0.80 in. if the difference between $m\omega^2$ and k is greater than $F_O/A = 30/0.80 = 37.5$. In this case, two possible solutions exist, one with y in phase with y_1 and $k > m\omega^2$, and a second solution with y out of phase with y_1 and $k < m\omega^2$. The first solution will not exist for small amplitudes of A. In this example,

$$60 - 0.05\omega_1^2 = 37.5 \quad \text{and} \quad \omega_1 = 21.2 \text{ rad per sec}$$

or

$$0.05\omega_2^2 - 60 = 37.5 \quad \text{and} \quad \omega_2 = 44.2 \text{ rad per sec.}$$

Thus the amplitude of the motion of D will be less than 0.80 in. if ω is less than 21.2 rad per sec or greater than 44.2 rad per sec.

PROBLEMS

12-57 Body W in Fig. P12-57 weighs 8.05 lb, the modulus k_1 is 20 lb per ft, and k_2 is 16 lb per ft. The displacement of E is given by the expression $y_E = 4 \cos 2t$ where y_E is in feet and t is in seconds. Determine the amplitude of the motion of W.

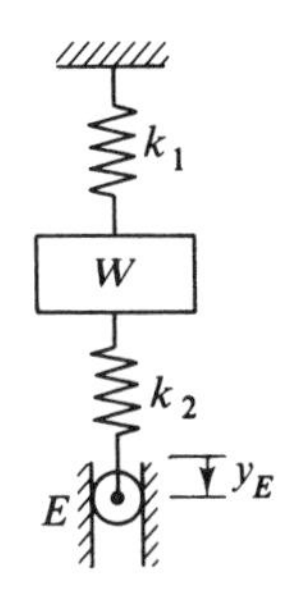

Figure P12-57

12-58 Body W in Fig. P12-58 weighs 60 lb, and the spring moduli are 5 lb per in. for S_1 and 2 lb per in. for S_2. The force F in pounds is given by the expression $F = 2 \cos 2t$, where t is in seconds.
(a) Derive the steady-state solution for the motion of W.
(b) Determine the maximum velocity of W.

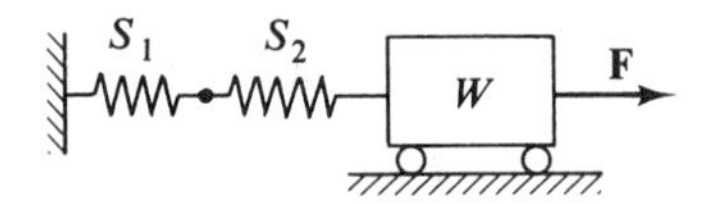

Figure P12-58

12-59 A 50-lb ball is supported as shown in Fig. P12-59. The mass of the bar can be neglected, and the spring modulus is 2 lb per in. Roller E has a harmonic motion given by the expression $x_E = 0.5 \cos 7t$, where x_E is in inches and t is in seconds. Obtain the steady-state solution for the motion of B.

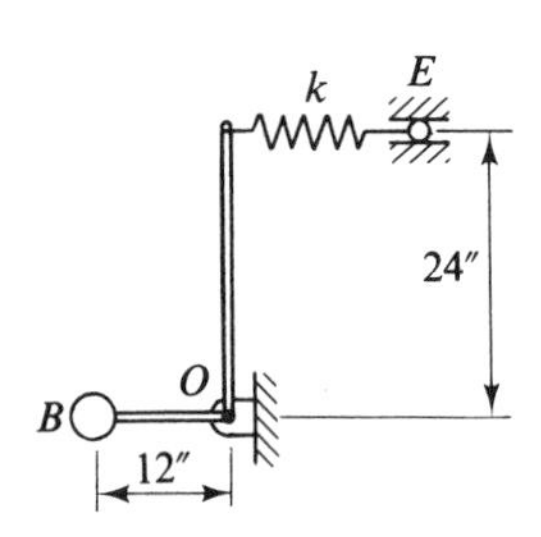

Figure P12-59

12-60 Determine the components of the reaction at O in Problem 12-59 when
(a) Arm OB is horizontal.
(b) Body B is in its lowest position.

12-61 Block E in Fig. P12-61 has a harmonic motion given by the equation $y_E = 0.500 \sin 5t$(y_E is in feet). The modulus of S_1 is 8.0 lb per ft and the modulus of S_2 is 15.0 lb per ft. The weight of the bars supporting the 32.2-lb weight can be neglected. Determine the steady-state solution for the vibrating system. The springs can act in either tension or compression.

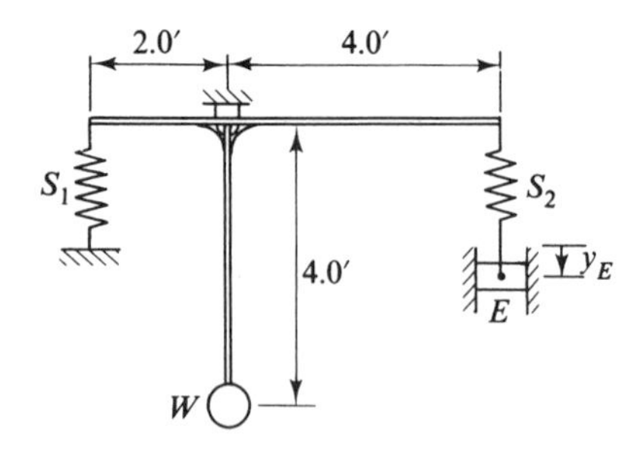

Figure P12-61

12-62 An instrument panel which weighs 20 lb is to be mounted on a machine which vibrates with a frequency of 1200 rpm and an amplitude of 0.25 in. It is necessary to reduce the amplitude of the vibration of the instruments to not more than 0.10 in., and it is proposed to use a spring mounting to accomplish this reduction.

Determine

(a) The range or ranges of spring stiffness which will be satisfactory.

(b) The amplitude of the *variable* part of the force transmitted by the spring when the amplitude of motion of the panel is 0.10 in.

12-63 The curves in Fig. 12-21 indicate the amplitude of motion of a body with forced vibration if the force causing the vibration has a constant amplitude. With an unbalanced rotor as shown in Fig. 12-22, the amplitude of the force producing the vibrations is not constant. Draw a curve, similar to Fig. 12-21, showing the variation of the amplitude of the motion of the body with the speed of the motor when $c = 0$.

12-64 A small variable-speed motor weighing 18.0 lb is mounted on an elastic beam as indicated in Fig. P12-64. The motor rotates with a 2.0-lb eccentric weight 2.0 in. from the center of the shaft. When the motor is not running, the motor and the eccentric weight cause the beam to deflect 0.50 in. Determine

(a) The speed of the system at resonance.

(b) The amplitude of the forced vibrations when the motor is running at 300 rpm.

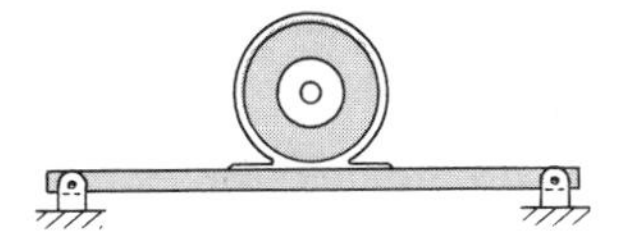

Figure P12-64

12-65 Would it be possible to reduce the amplitude of the forced vibration of the motor in part (b), Problem 12-64, by fastening an additional weight to the motor? If so, what weight should be added to reduce the amplitude of the vibration to 0.500 in.?

***12-66** Block B in Fig. P12-66 is connected to body A by a viscous dashpot. The mass of B is 2.0 kg, and the coefficient of the dashpot is 40 N · s/m. Body A has simple harmonic motion given by the equation $x_1 = e \sin \omega t$, where e is 50 mm and ω is 20 rad/s. Determine the amplitude of the steady-state motion of B and the phase angle by which B leads or lags A.

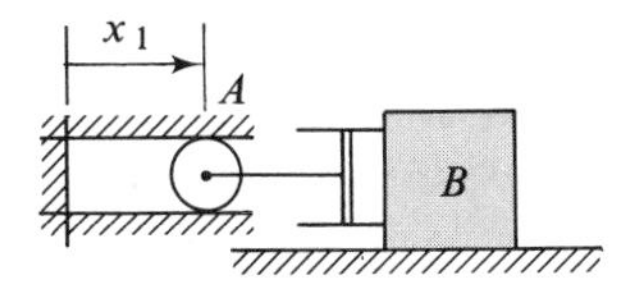

Figure P12-66

12-67 Body A in Fig. P12-67 weighs 5.0 lb, and the viscous damper has a coefficient of 0.10 lb-sec per in. The magnitude of the force F is given by the equation $F = F_o \sin \omega t$, where F_o is 2.0 lb and ω is 10 rad per sec. Determine the amplitude of the motion of A and the maximum force in member B.

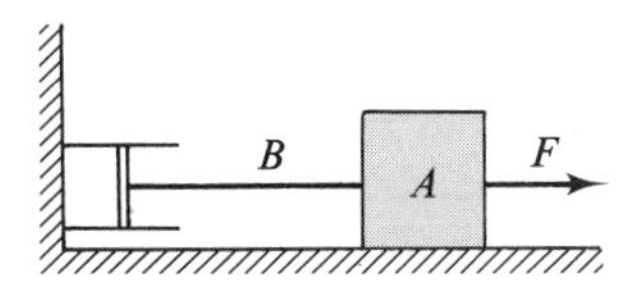

Figure P12-67

12-68 A block which weighs 60 lb is suspended by a spring with a modulus of 48 lb per in. and is subjected to a shaking force $F = 2 \sin 30t$ (F is in pounds and t is in seconds). The block is connected to the floor through a dashpot with a coefficient of 4.0 lb-sec per in. Determine the amplitude of the force transmitted through the dashpot.

12-69 Block A in Fig. P12-69 has simple harmonic motion with an amplitude e and a circular frequency ω. The frequency of the motion is less than the resonant frequency. Body B has a mass m, and the moduli of the spring and damper are k and c, respectively. Determine the amplitude of the motion of B and indicate whether it leads or lags the motion of A.

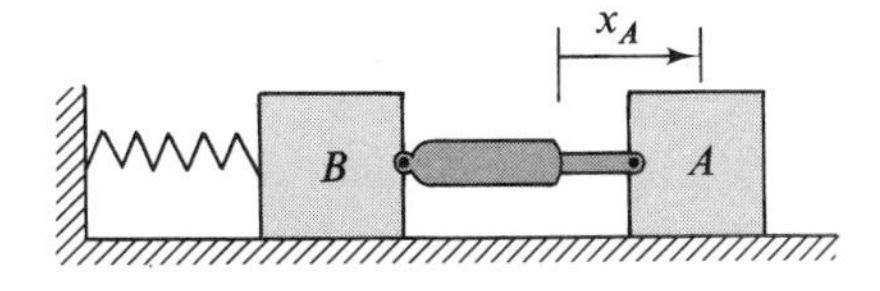

Figure P12-69

12-70 A body weighing 30 lb is mounted on four springs (in parallel) each having a modulus of 2.0 lb per in. The body is acted on by a force with an amplitude of 3.0 lb and a circular frequency of 10 rad per sec. This force produces

excessive vibrations, and three possible remedies are suggested:
(a) Add a weight of 15 lb to the body.
(b) Add a fifth spring with the same modulus.
(c) Place a dashpot with a damping coefficient of 0.90 lb-sec per in. between the body and foundation.
Determine the effect of each of these proposed remedies, applied separately, on the amplitude of the resulting vibrations and on the variable portion of the force transmitted to the foundation.

12-71 Body A in Fig. P12-71 weighs 193 lb, the damping coefficient c is 5.0 lb-sec per in., and the spring modulus k is 40 lb per in. Body B has harmonic motion due to the uniform rotation of the crank. The crank arm e is 2.0 in., and it rotates at 10 rad per sec.
(a) Determine the amplitude of the motion of A and its phase relationship with respect to the motion of B.
(b) (1) Would a small increase in k increase or decrease the amplitude of the motion of A?
(2) Would a small increase in c increase or decrease the amplitude of the motion of A?

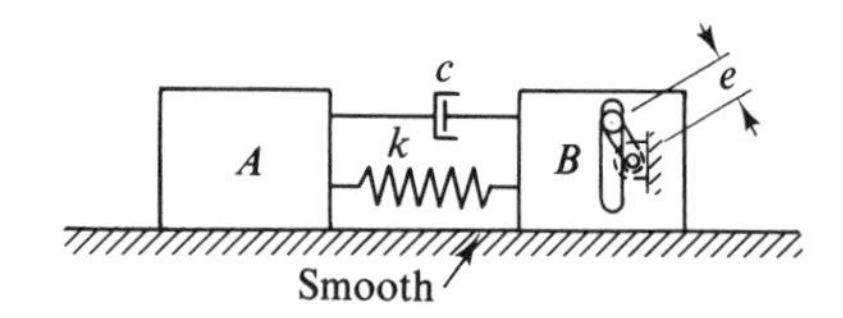

Figure P12-71

*__12-72__ The 10-kg body in Fig. P12-72 is mounted on springs with a modulus of 7000 N/m. The force $F = 40 \sin 30t$ (F in newtons and t in seconds) acts on the body as shown. A dashpot is to be placed between the body and foundation to limit the amplitude of the motion to 1.20 cm. Determine
(a) The coefficient of the dashpot.
(b) The amplitude of the variable portion of the force transmitted to the foundation.

12-73 Solve part (b), Example 12-6, for $A_{max} =$ 0.80 in. and $c = 0.60$ lb-sec per in.

12-74 Solve part (b), Example 12-6, for $A_{max} =$ 0.30 in. and $c = 0$.

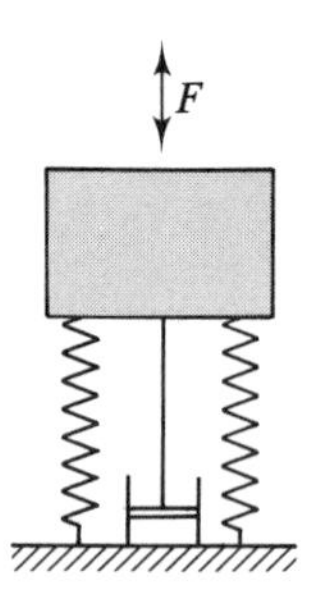

Figure P12-72

12-75 The system in Fig. P12-75 is adjusted to be in equilibrium when AB is horizontal and x_E is zero. Body B weighs 50 lb, the spring modulus is 6.0 lb per in., and the damping coefficient is 1.5 lb-sec per in. The position of block E (in inches) is given by the expression $x_E = 5.0 \sin 5t$. Determine the amplitude of the motion of B and its maximum velocity.

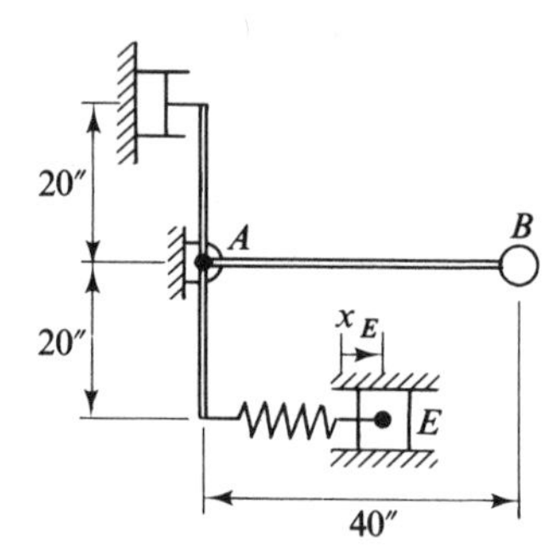

Figure P12-75

*__12-76__ The 30-kg body A of Fig. P12-76 is connected to two springs, each with a modulus of 350 N/m, and a dashpot with a coefficient of 120 N·s/m. The force P (in newtons) acting on A is given by the equation $P = 20.0 \cos 3t$, where t is in seconds. Obtain the steady-state solution for the motion of A and determine the maximum force transmitted to the wall by the springs and dashpot.

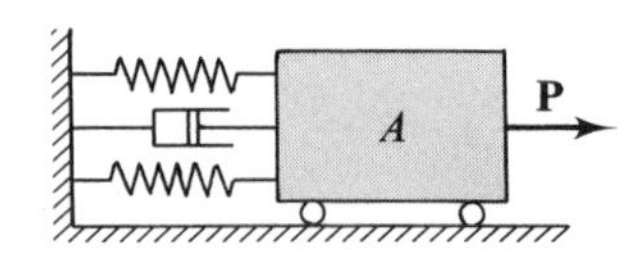

Figure P12-76

12-77 The 386-lb air compressor of Fig. P12-77 is supported by four spring-dashpot mounts, each having a spring modulus of 250 lb per in. and a damping coefficient of 5.0 lb-sec per in. The piston of the compressor weighs 3.86 lb, has a 4.00-in. stroke, and moves in a vertical cylinder. The crankshaft rotates at 40 rad per sec. Assume that the motion of the piston *with respect to the cylinder* can be approximated by

$$y_{P/C} = A \sin \omega t.$$

Neglect the weights of the crank and connecting rod. Determine the amplitude of the vertical vibrations of the compressor.

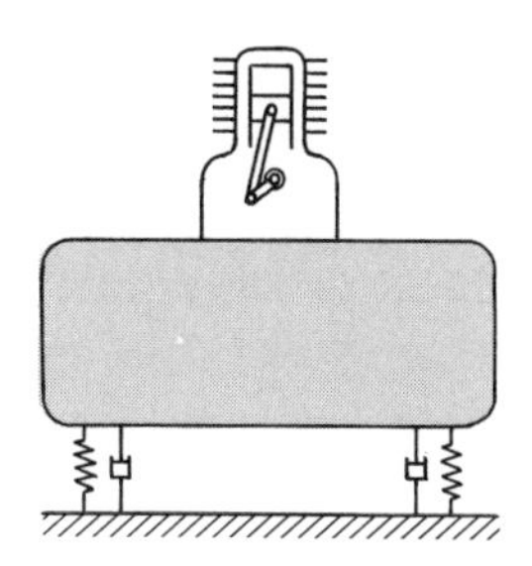

Figure P12-77

12-78 The two disks in Fig. P12-78 are coupled by two springs, each of modulus k, and a damper, coefficient c. The springs are always under tension, and the cable does not slip on either wheel. Each disk has a moment of inertia with respect to its axis of rotation of I. A variable torque causes disk A to have simple harmonic rotation with an amplitude of θ_0 and a circular frequency of ω.

(a) Determine the value of ω for resonance.

(b) Determine the range of values of ω, if any, for which an increase in c will result in a decrease in the amplitude of disk B.

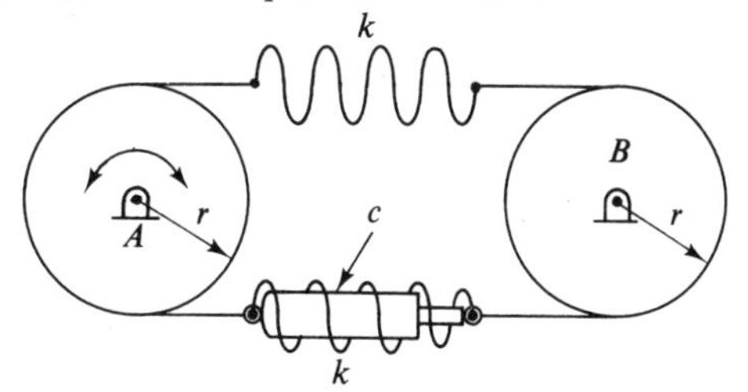

Figure P12-78

Appendix A

International System of Units

The International System of Units, commonly called SI units, was formally adopted in 1960 by the International Organization for Standardization (ISO). The ISO is composed of representatives of all the major industrial countries of the world. SI is a modification of the metric system of units and is under continuous study for necessary refinement and additions.

The seven fundamental or base units defined in this system are length, mass, time, thermodynamic temperature, electric current, luminous intensity, and amount of substance. Mass, length, time, and sometimes temperature are the quantities commonly encountered in engineering mechanics. Since mass instead of force is a base unit or quantity, SI is an absolute system of units in contrast to the conventional gravitational system used by engineers in the United States in which force is a fundamental quantity. Quantities involving units other than those selected as base units are formed as multiples of base units and sometimes given special names. Table A-1 lists the base units and SI symbols together with selected derived units.

TABLE A-1
SI UNITS AND SYMBOLS

	Physical Quantity	*Name of Unit*	*SI Symbol*
Base unit	Length	meter	m
	Mass	kilogram	kg
	Time	second	s
Derived unit	Area	square meter	m^2
	Volume	cubic meter	m^3
	Force	newton	N, $kg \cdot m/s^2$
	Moment of a force	newton · meter	$N \cdot m$
	Work	joule, newton · meter	J, $N \cdot m$
	Power	watt, joule per second	W, J/s, $N \cdot m/s$
	Linear velocity	meter per second	m/s
	Linear acceleration	meter per second squared	m/s^2
	Density, mass per unit volume	kilogram per cubic meter	kg/m^3
	Frequency	hertz, cycles per second	Hz

Prefixes, as listed in Table A-2, are used to describe quantities which are much larger or smaller than the base units or derived unit in order to use numbers which are more easily comprehended. Thus a length of 120 000 meters would be stated as 120 km, and a length of 25 500 000 meters would be 25.5 Mm. It is recommended that prefixes be selected which will result in numerical values between 0.1 and 1000. Use of prefixes hecto, deca, deci, and centi is discouraged unless the recommended prefix results in inconvenient numbers. For example, the volume of a cube which is 35 mm on a side would be either 42 875 mm^3, or 0.000 042 875 m^3. In such a situation the volume

TABLE A-2
SI PREFIXES

Multiplication Factor	*Prefix*	
	Name	*Symbol*
10^{12}	tera	T
10^9	giga	G
10^6	mega	M
10^3	kilo	k
10^2	hecto	h
10	deka	da
10^{-1}	deci	d
10^{-2}	centi	c
10^{-3}	milli	m
10^{-6}	micro	μ
10^{-9}	nano	n
10^{-12}	pico	p
10^{-15}	femto	f
10^{-18}	atto	a

should be expressed as $42.9\ cm^3$. Errors in calculations involving prefixes can be avoided more easily if all quantities are expressed in SI units by replacing prefixes by powers of 10, e.g., $25.5\ Mm = 25.5(10^6)m = 2.55(10^7)m$.

The use of SI units is becoming more commonplace in the United States, but a complete changeover from English units will probably not take place for many years. During the transition period engineers and scientists will need to be familiar with both systems and will frequently find it necessary to convert quantities from one system to the other. Table A-3 contains selected factors useful in engineering mechanics.

TABLE A-3
CONVERSION FACTORS

	English to SI	*SI to English*
Length	1 in. = 25.40 mm 1 ft = 0.3048 m 1 mile = 1.609 km	1 mm = 0.03937 in. 1 m = 39.37 in. = 3.281 ft 1 km = 0.6214 mile
Mass	1 slug (1 lb-sec^2/ft) = 14.59 kg	1 kg = 0.06853 slugs = 0.06853 lb-sec^2/ft
Area	1 in.^{2} = 6.452 cm^2 1 ft^2 = 0.09290 m^2	1 cm^2 = 0.1550 in.^{2} 1 m^2 = 10.76 ft^2
Volume	1 in.^{3} = 16.39 cm^3 1 ft^3 = 0.02832 m^3	1 cm^3 = 0.06102 in.^{3} 1 m^3 = 35.31 ft^3
Force	1 lb = 4.448 N = 4.448 kg · m/s^2	1 N = 0.2248 lb
Moment of force	1 in-lb = 0.1130 N · m 1 ft-lb = 1.356 N · m	1 N · m = 0.7376 ft-lb = 8.851 in-lb
Work or energy	1 ft-lb = 1.356 N · m = 1.356 J	1 J = 0.7376 ft-lb
Power	1 ft-lb/sec = 1.356 W 1 hp = 745.7 W	1 W = 0.7376 ft-lb/sec 1 kW = 1.341 hp
Velocity	1 in/sec = 25.40 mm/s 1 ft/sec = 0.3048 m/s 1 mile/hr = 0.4470 m/s = 1.609 km/hr	1 m/s = 3.281 ft/sec 1 km/hr = 0.6214 mph = 0.9113 ft/sec
Acceleration	1 in/sec^2 = 25.40 mm/s^2 1 ft/sec^2 = 0.3048 m/s^2	1 m/s^2 = 3.281 ft/sec^2
Density	1 slug/ft^3 = 515.3 kg/m^3 1 lb-sec/ft^4 = 515.3 kg/m^3	1 Mg/m^3 = 1.941 slug/ft^3
Spring modulus	1 lb/in. = 175.1 N/m 1 lb/ft = 14.59 N/m	1 N/m = 0.06852 lb/ft 1 kN/m = 5.710 lb/in.
Second moment of mass	1 slug-ft^2 = 1.355 kg · m^2 1 lb-sec^2-ft = 1.355 kg · m^2	1 kg · m^2 = 0.7378 slug-ft^2
Load intensity	1 lb/in.^{2} = 6.895 kN/m^2 1 lb/ft^2 = 47.88 N/m^2	1 kN/m^2 = 0.1450 lb/in.^{2} 1 kN/m^2 = 20.89 lb/ft^2

Answers to Problems

CHAPTER 1

1-2 150 lb →, 80 lb ↑, both through A

1-4 320 kN (3, 4 slope), 240 kN (4, 3 slope), both through C

1-6 (a) 940 lb →, 342 lb ↓,
(b) 500 lb ←, 866 lb ↓, all through A

1-8 187.9 lb →, 128.6 lb (30°), both through A

1-10 280 lb →, 300 lb (4, 3 slope), both through A

1-12 3606 lb, (2, 3 slope), 3000 lb →, both through B

1-14 420 lb ←, 260 lb (12, 5 slope), both through intersection of AB and the 400 lb force

1-16 2.79 lb (12, 5 slope), through intersection of **F** and CB

1-18 21.3 N (4, 3 slope)

1-20 $\cos\theta_x = -0.4706$, $\cos\theta_y = 0.5294$, $\cos\theta_z = -0.7059$
$\mathbf{F} = (-80\mathbf{i} + 90\mathbf{j} - 120\mathbf{k})$ lb through 0.

1-22 $\cos\theta_x = -0.7778$, $\cos\theta_y = 0.4444$, $\cos\theta_z = -0.4444$
$\mathbf{F} = (-280\mathbf{i} + 160\mathbf{j} - 160\mathbf{k})$ lb, through any point on its line of action

1-24 $\mathbf{F} = (50.9\mathbf{i} - 152.7\mathbf{j} - 118.7\mathbf{k})$ lb through the origin

1-26 $\cos\theta_x = -0.425$, $\cos\theta_y = 0.242$
$\cos\theta_z = 0.872$

1-28 -13, $-13\mathbf{i}+\mathbf{j}+9\mathbf{k}$

1-30 (a) $(-178.9\mathbf{i}+74.6\mathbf{j})$ N
(b) $(-44.8\mathbf{i}-33.6\mathbf{j})$ N

1-32 $(84.1\mathbf{i}+126.1\mathbf{j}+252\mathbf{k})$ lb through origin

1-34 $(23.3\mathbf{i}-34.9\mathbf{j}+69.8\mathbf{k})$ lb through A
$(36.7\mathbf{i}+4.90\mathbf{j}-9.80\mathbf{k})$ lb through A

1-36 $(15\mathbf{i}+30\mathbf{j}+10\mathbf{k})$ lb through A

1-38 $\mathbf{P}=(-30.6\mathbf{i}+61.2\mathbf{j}-20.4\mathbf{k})$ lb through any point 0 on line of action of $\mathbf{F}$
$\mathbf{Q}=(400\mathbf{i}+200\mathbf{j})$ lb through 0
$\mathbf{R}=(-69.4\mathbf{i}+138.8\mathbf{j}+520\mathbf{k})$ lb through 0

1-40 4

1-42 $-7\mathbf{i}+8\mathbf{j}+5\mathbf{k}$

1-44 (a) 40 N · m ↺
(b) 0.2 m

1-46 (a) 390 lb through upper right hand corner
(b) 270 ft-lb clockwise
(c) 0.692 ft

1-48 304,900 ft-lb clockwise

1-50 420 ft-lb 

1-52 270 ft-lb 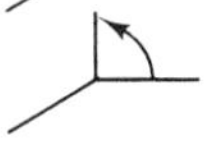

1-54 0, $(-10.8\mathbf{i}+3.6\mathbf{j})$ N · m

1-56 (a) $(-80\mathbf{i}-240\mathbf{j}+120\mathbf{k})$ ft-lb
(b) 2 ft

1-58 (a) $(-9.6\mathbf{i}+12.6\mathbf{j}+24.0\mathbf{k})$ N · m
(b) $4.8\mathbf{k}$ N · m
(c) $(19.58\mathbf{j}+14.69\mathbf{k})$ N · m

1-60 (a) $\mathbf{M}_D = 3.21F(-0.440\mathbf{i}+0.880\mathbf{j}-0.176\mathbf{k})$ in-lb
(b) 3.21 in.

1-62 (a) 123.6 in-lb 

(b) 1231 in-lb magnitude

1-64 (a) $(10\mathbf{i}-15\mathbf{j}+30\mathbf{k})$ lb
(b) $(150\mathbf{i}+210\mathbf{j}+55\mathbf{k})$ ft-lb

1-66 $\mathbf{F}_2 = 250$ lb 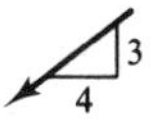

$x = 6.67$ ft to the right of the origin

1-68 $\mathbf{F}=(-87.8\mathbf{i}-58.5\mathbf{j}-175.6\mathbf{k})$ lb through C
$\mathbf{n}=(0.702\mathbf{i}-0.702\mathbf{j}-0.117\mathbf{k})$

1-70 $x = 2$ ft, $z = 5$ ft

1-72 $x = -0.02$m, $y = 0.03$ m

1-74 8000 lb 

and 59,500 in-lb ↺

1-76 $-1000\mathbf{k}$ lb through 0,
$\mathbf{C}=(-3000\mathbf{i}-4000\mathbf{j})$ in-lb

1-78 2 kN 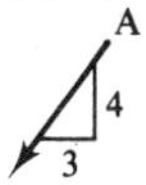

through A and 40 N · m ↺

1-80 $\mathbf{M}_{BC} = 0$
$\mathbf{M}_{AB} = 2000$ in-lb counterclockwise looking from B to A
$\mathbf{M}_{OA} = -5830\mathbf{i}$ in-lb

1-82 (a) $-2000\mathbf{i}$ lb through A and $(10{,}000\mathbf{i}-6000\mathbf{j}+8000\mathbf{k})$ in-lb
(b) $-6000\mathbf{j}$ in-lb, 12,810 in-lb in direction of $5\mathbf{i}+4\mathbf{k}$

1-84 $50\mathbf{k}$ lb through B, and $(283\mathbf{i}-990\mathbf{j})$ in-lb,
$\mathbf{M}_{AB} = -990\mathbf{j}$ in-lb
$\mathbf{M}_{BC} = 500$ in-lb clockwise looking from C to B

1-86 $M' = 300\mathbf{i}$ in-lb
$\mathbf{P}=(-80\mathbf{j}+40\mathbf{k})$ lb through 0
$\mathbf{Q}=(60\mathbf{i}+120\mathbf{j}+80\mathbf{k})$ lb through A

1-88 $c = FT/L$, $a = FT^2/L$

1-90 $P = F$, $I = FT^2L$

1-92 $y = L$, $b = T^{-1}$, a and c are dimensionless

1-94 $m_f = FT^2/L$, $I = T$

1-96 No

1-98 $c = FT/L$, $\omega = T^{-1}$, $k = F/L$, $P = F$

CHAPTER 2

2-2 162.8 lb 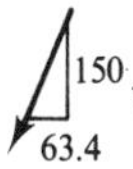

through point of intersection

2-4 800 lb→ through point of intersection

2-6 $\mathbf{P} = 656$ lb, $\mathbf{Q} = 344$ lb

2-8 31.6 kN through origin in direction $30\mathbf{j} + 10\mathbf{k}$

2-10 135.1 lb through origin in direction of $0.527\mathbf{i} + 0.542\mathbf{j} + 0.655\mathbf{k}$

2-12 (a) 141.4 kN through origin in direction of $80\mathbf{i} - 60\mathbf{j} + 100\mathbf{k}$
(b) 4800 N·m clockwise looking from *B* to *A*

2-14 (a) 104.9 lb through 0 in direction of $10\mathbf{i} + 30\mathbf{j} + 100\mathbf{k}$
(b) 304 in-lb clockwise looking from *A* to *B*

2-16 150 lb downward 5.47 ft to the right of A

2-18 750 lb upward 5.33 ft to the right of *A*

2-20 850 lb downward 7.94 ft to the right of A

2-22 258 lb

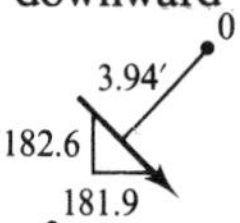

2-24 342 lb

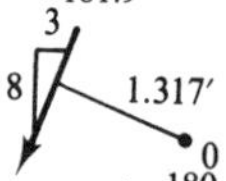

2-26 269 N

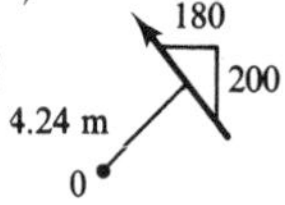

2-28 90 lb↓, through *B*,
11.18 lb through *A*

2-30 581 lb

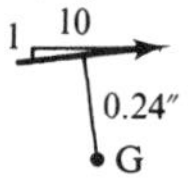

2-32 (a) 2500 lb

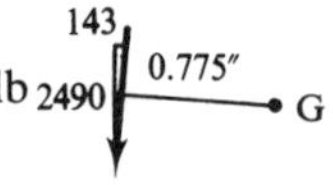

(b) $L = 2490$ lb, $D = 143$ lb as shown
$\mathbf{M} = 3070$ in-lb↺

2-34 100 lb↗ through (1.4, 0.6) ft

2-36 40 lb downward through $x = -1.5$ m, $z = 2.25$ m

2-38 100 lb upward through $x = 5$ ft, $y = -4.9$ ft

2-40 $(-6.85\mathbf{i} - 5.15\mathbf{j} + 2.14\mathbf{k})$ N·m

2-42 $(-30\mathbf{i} + 20\mathbf{j})$ lb through *A* and the couple $(200\mathbf{i} + 90\mathbf{j} + 260\mathbf{k})$ ft-lb

2-44 $(60\mathbf{i} + 50\mathbf{j} + 40\mathbf{k})$ lb through 0 and the couple $(-140\mathbf{i} - 240\mathbf{j})$ ft-lb

2-46 $(140\mathbf{i} - 220\mathbf{j} + 80\mathbf{k})$ lb through 0 and the couple $(640\mathbf{i} - 660\mathbf{j} - 1860\mathbf{k})$ ft-lb

2-48 $(1000\mathbf{i} + 100\mathbf{j} + 200\mathbf{k})$ lb through *A* and the couple $(200\mathbf{i} + 2400\mathbf{j} + 2400\mathbf{k})$ ft-lb

CHAPTER 3

3-2 $x = -0.333$ m
$y = 0.467$ m

3-4 (1.000, 1.071, 0.571) ft

3-6 (3, 5) in.

3-8 (1.210, 1.653) cm

3-10 (0.574, 0.410) ft

3-12 0.417 in.^{3}

3-14 $(\frac{2}{5}b, b)$

3-16 $(\frac{11}{5}a, \frac{8}{25}a)$

3-18 2.40 m

3-20 $x = 3.88$ in.
$y = 3.88$ in.

3-22 $\frac{3}{8}r$

3-24 $\left(0, \frac{3a}{4}, \frac{h}{4}\right)$

3-26 (0.2, 1, 0) ft

3-28 (5, 0.882, 0) in.

3-30 (3.23, 10, 0) cm

3-32 $(0.702b, 0, 0)$

3-34 $0.75L$ from left end on the axis of symmetry

3-36 (−0.670, 3.99) in.

3-38 (1.641, 2.5) in.

3-40 (33.8, 55.4) mm

3-42 2.90 in. above center of *WF* section on axis of symmetry

3-44 −0.466 in.

3-46 $(x_1+x_2+x_3)/3$

3-48 (30.8, 44.1, 30.8) mm

3-50 (0.571, 0.414) ft

3-52 (0.934, 3.44, 0.560) in.

3-54 (2.21, 2.86) cm

3-56 $4\pi^2 ar$

3-58 $44.9r^2$

3-60 (a) 2.4 in.
(b) (1.875, 4, 0) in.

3-62 4830 in.^{2}

3-64 (a) 169.6 in^3
(b) (2.17, 0, 0) in.

3-66 (a) 50 lb-ft$^{-3/2}$
(b) 900 lb downward 5.4 ft to the right of 0.

3-68 18.4 kN downward 2.95 m to the right of *A*.

3-70 3090 lb (2400 left, 1950 down) through a point 17.10 ft to the right of *A*.

3-72 2.15 kN (1.919 right 0.961 upward) through *B*.

3-74 250 lb downward 3 in. to the right of the origin on the *x* axis.

3-76 3000 lb, 4.5 ft below water surface on axis of symmetry.

3-78 67,500 lb through $x = 96.5$ in., $z = 76$ in.

3-80 $1.049a$

3-82 0.550 in. above web of channel on vertical axis of symmetry

3-84 -5.25 cm^3

3-86 -0.275 in.

3-88 928 cm^3

3-90 (4, 2.56, 0) in.

3-92 1.144 ft from left side of plate
7.87 ft below water surface

3-94 (4.00, 5.69, 3.05) in.

3-96 (1.00, 0.286, 0) ft

CHAPTER 4

4-8 $T_A = 851$ lb, $T_B = 1015$ lb

4-10 $\mathbf{P} = 173.2$ lb

4-12 $m_1 \doteq 357$ kg

4-14 $T_{AC} = 3167$ lb

4-16 (a) 6
(b) $3.11W$ lb

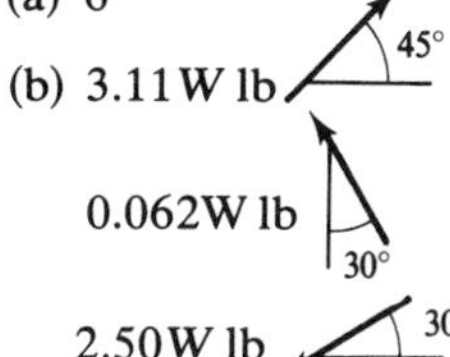

$0.062W$ lb

$2.50W$ lb

W lb ↓

4-18 $T_{AB} = 3727$ lb

4-20 (a) $\mathbf{P} = 420$ lb→, $\mathbf{N} = 160$ lb↑
(b) $\theta = 20.9°$, $\mathbf{P} = 450$ lb

4-22 $T_A = 313$ lb; $T_B = 335$ lb; $T_C = 187.5$ lb

4-24 $\mathbf{P} = 375\mathbf{i}$ lb; $T_A = 781$ lb; $T_C = 469$ lb

4-26 $\mathbf{F}_1 = 982\mathbf{i}$ N; $\mathbf{F}_2 = 736\mathbf{k}$ N,
$\mathbf{F}_3 = 1323(-0.742\mathbf{i} + 0.371\mathbf{j} - 0.557\mathbf{k})$ N

4-28 $\mathbf{A} = 40$ lb→ through *A*
$\mathbf{B} = 64$ lb (4, 5) through *B*

4-30 $\mathbf{T}_{DC} = 0.319W$ lb (1, 2) through *C*
$\mathbf{A} = 0.357W$ lb (3, 4) through *A*

4-32 $\mathbf{F}_M = 27.5$ *kN* ← through *M*
$\mathbf{F}_O = 34.3$ kN (3, 4) through *O*

4-34 $\mathbf{F}_E = F_E$ 60°
$\mathbf{F}_A = F_A$ 15°

4-36 $D = 5.61$ ft
20.9° counterclockwise from vertical

4-38 $\mathbf{A} = 9$ kN↑ at *A*
$\mathbf{M}_A = 38$ kN·m↺

4-40 On *BC*: $\mathbf{C} = 375$ lb↑ at *C*
$\mathbf{B} = 625$ lb↑ at *B*
On *AQ*: $\mathbf{B} = 625$ lb↓ at *B*
$\mathbf{A} = 2625$ lb↑ at *A*
$\mathbf{M}_A = 16{,}500$ ft-lb↺

4-42 (a) $\mathbf{P} = 144$ lb 45° on AB

(b) $\mathbf{E} = 166.5$ lb 52.3° on AB

4-44 $\mathbf{A} = 896$ lb 60°;
$\mathbf{B} = 1303$ lb 9.9°

4-46 $W = 120$ lb

4-48 (a) 40.8 in. behind front wheels
(b) 8040 lb ↑

4-50 $\mathbf{T} = 400$ lb ←; $\mathbf{Q} = 162$ lb ↓;
$d = 1.155$ ft

4-54 $\mathbf{A} = 279$ N 40.5° 40.5°; $\mathbf{N} = 226$ N ↑

4-56 $L = 31{,}700$ lb, $D = 10{,}480$ lb, 1.675 ft below the G axis.

4-58 $\mathbf{F}_A = 181.6$ lb 59.6°

$\mathbf{F}_C = (90.7$ 30° $+ 341$ 60° $)$ lb on AB

4-60 $\mathbf{B}_X = 696$ lb →, $\mathbf{B}_Y = 23{,}800$ lb ↑,
$\mathbf{A} = 13{,}850$ lb ↓

4-62 (a) $W = 16.77$ lb
(b) $F = 2.97$ lb
(c) $\theta = 22.2°$

4-64 $\mathbf{E} = 10.90$ kN 4 3

$T = 9.08$ kN

4-66 $\mathbf{R}_C = 1645$ lb 76.0°

$\mathbf{R}_E = 2240$ lb 79.7°

4-68 $\mathbf{R}_D = 133$ lb 55.7° on A

$\mathbf{R}_D = 61.1$ lb 79.2 on B.

4-70 $\mathbf{R}_A = 900$ lb ← on AB

$\mathbf{R}_A = 361$ lb 3.18° on AE

$\mathbf{R}_D = 722$ lb 41.6° on DC

4-72 $\mathbf{R}_B = 8480$ N 12.22° on AB

$\mathbf{R}_B = 5350$ N 14.30° on BC

4-74 $T_U = 9.51$ lb, $T_L = 19.51$ lb
$\mathbf{A} = (28.1 \rightarrow +22.5 \uparrow)$ lb

4-76 $AD = CD = 4.35$ kN T
$AB = 3.57$ kN T
$BC = 4.93$ kN C
$BD = 4.0$ kN T

4-78 $AB = 756$ lb T
$AE = 1717$ lb T
$CD = 1667$ lb C

4-80 $AB = DE = P\,C$
$AF = FE = 1.333P\,C$
$AC = CE = 1.667P\,C$
$BC = CD = CF = 0$

4-82 $BF = 0$
$BC = 62.5$ k T
$DG = 6.71$ k C

4-84 $CE = 1000$ lb T
$EF = 1667$ lb C

4-86 $AC = 20.8$ kN T
$BC = 6.25$ kN C
$DF = 19.2$ kN C
$EF = 12.0$ kN T

4-88 $BE = 100$ lb C
$DE = 262$ lb C

4-90 $EG = 10{,}000$ lb T
$EH = 37{,}300$ lb C
$FH = 10{,}000$ lb C

4-92 $BC = 13$ kN T
$BI = 2.5$ kN T

4-94 $BD = 0.500$ k C
$BE = 0.417$ k T
$DG = 0$

4-96 $BI = 16.58$ kN C
$BK = 58.4$ kN T

4-98 $T = 8000$ ft-lb
$BE = 0$
$DG = 4250$ lb T
$FG = 0$

4-100 $BC = 3125$ lb C
$DE = 7290$ lb T
$GH = 4170$ lb C

4-102 $AC = 71.4$ kN C
$BC = 25.2$ kN T
$DF = 43.75$ kN C

4-104 $DM = 542$ lb T

4-106 $BC = 800$ lb C

$\mathbf{R}_A = 1767$ lb, 64.9°

4-108 $T_D = 7730$ lb
$T_A = 8490$ lb
$L = 154$ ft

4-110 $f = 12.91$ ft
$l = 217$ ft

4-112 $T_{max} = q_O L\sqrt{\frac{1}{4}+\left(\frac{L}{3f}\right)^2}$

4-114 $f_A = 29.2$ ft
$T_{max} = 38.1$ lb
$T_{min} = 23.5$ lb

4-116 $2a = 277$ ft

4-118 Greatest tendency to tip occurs when load is midway between two feet.

4-120 $T = 6.15$ kN · m
$\mathbf{R}_1 = 12.33$ kN
$\mathbf{R}_2 = 1.333$ kN

4-122 $\mathbf{R}_B = (6.67\mathbf{i} + 31.67\mathbf{j} - 6.67\mathbf{k})$ lb

4-124 $\mathbf{F} = 60\mathbf{j}$ lb
$\mathbf{R}_A = (-120\mathbf{i} - 40\mathbf{j})$ lb

4-126 (a) $\mathbf{AE} = 9.61(0.640\mathbf{i} + 0.480\mathbf{j} - 0.600\mathbf{k})$ ft
(b) $\mathbf{R}_E = -2.54(0.780\mathbf{j} + 0.624\mathbf{k})$ lb

4-128 $\mathbf{R}_F = 374(0.845\mathbf{i} + 0.535\mathbf{j})$ lb

4-130 $\mathbf{R}_A = 306(0.689\mathbf{i} + 0.435\mathbf{j} + 0.580\mathbf{k})$ lb
$\mathbf{M}_A = (1761\mathbf{i} + 956\mathbf{k})$ ft-lb

4-132 $\mathbf{R}_A = 12.42(-0.469\mathbf{i} + 0.535\mathbf{j} - 0.703\mathbf{k})$ kN

4-134 $\mathbf{F} = 228(-0.316\mathbf{i} + 0.822\mathbf{j} - 0.474\mathbf{k})$ lb

4-136 $\mathbf{P} = 34.7$ lb

4-138 $\mathbf{F}_C = -9000\mathbf{k}$ lb
$BF = 13{,}800$ lb T
$BG = 26{,}250$ lb T

4-140 $\mathbf{C} = 136.4(-0.8\mathbf{j} + 0.6\mathbf{k})$ lb
$\mathbf{B} = 655\mathbf{i}$ lb
$\mathbf{A} = (-655\mathbf{i} + 509\mathbf{j} - 81.8\mathbf{k})$ lb

4-142 $\mathbf{R}_A = 4.91(0.8\mathbf{j} + 0.6\mathbf{k})$ N
$\mathbf{M}_A = 0.1717(0.515\mathbf{i} + 0.858\mathbf{j})$ N · m

4-144 f.s. = 6.06

4-146 f.s. = 2.77

4-148 $b = 10.42$ ft

4-150 $H = 3.44$ ft

4-152 $d = 0.6\left(\frac{0.8+h}{0.6+h}\right)$ m

4-154 (a) 8.04 in. right of center
(b) 3.34 in.

4-156 0.1814 ft rise

4-158 0.150 m above top of tank

4-160 $\theta = 15.5°$

4-162 (a) 0.197 ft^3
(b) 142.9 lb

4-164 (a) 25.4°
(b) 6.33 ft

4-166 0.620 m

4-168 $r_1/r_2 = 0.900$

4-170 $\mathbf{A} = 1718$ lb, 20.4°
$\mathbf{B} = 2350$ lb →

4-172 $\mathbf{F} = 1359$ N, 30°, 348 N, 30°

4-174 (a) $\mathbf{R} = 52{,}700$ lb/ft
(b) f.s. = 2.44

49,530
17,970
12.55′
B

4-176 $d = 0.420$ ft

4-178 $\mathbf{F}_B = (752 \leftarrow + 94.0 \uparrow)$ lb

4-180 $\mathbf{T}_{GE} = 818$ N, 3, 4 on A

$\mathbf{F}_R = 1097$ N, 2, 1 on A

4-182 $e = \left(\frac{d}{b}\right)(a+b)$

4-184 $\mathbf{B} = (25.5\rightarrow + 136.9\downarrow)$ lb on DE
$\mathbf{E} = (25.5\leftarrow + 8.9\uparrow)$ lb on DE

4-186 $BE = 100.0$ lb C
$DE = 262$ lb C

4-188 $AK = 0$
$CJ = 60$ k T
$FH = 0$

4-190 $T = 1.594$ kN
$\mathbf{F}_B = 1.962$ kN $\rightarrow$

4-192 $\mathbf{F}_B = 1.569\mathbf{i}$ kN 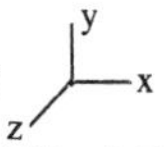

$\mathbf{F}_R = 1.930(0.640\mathbf{j} - 0.769\mathbf{k})$ kN

4-194 (a) $R = 197.7$ ft
(b) $T_A = 2220$ lb
$T_B = 2213$ lb
$T_C = 0$

4-196 $T = 2651$ lb

4-198 $d = 4.83$ ft

4-200 $\mathbf{B} = 173.2\mathbf{j}$ lb
$\mathbf{C} = (197.2\mathbf{i} - 146.4\mathbf{j} - 7.3\mathbf{k})$ lb
$\mathbf{D} = (102.8\mathbf{i} - 192.7\mathbf{k})$ lb

4-202 $\mathbf{P} = 133.3\mathbf{k}$ lb
$\mathbf{F}_E = (200\mathbf{i} - 33.3\mathbf{j} - 267\mathbf{k})$ lb

CHAPTER 5

5-2 (a) 22.6 lb
(b) Yes

5-4 (a) 0.150W
(b) 0.30W

5-6 $\theta = 26.6°$
$\mu = 0.1277$

5-8 0.371

5-10 0.357

5-12 $36.0 \text{ lb} < F_P < 46.7 \text{ lb}$
$13.33 \text{ lb} < F_Q < 24.0 \text{ lb}$

5-14 495 N

5-16 30.0 lb

5-18 245 N$\downarrow$ +61.3 N$\rightarrow$ at top
343 N$\uparrow$ +61.3 N$\leftarrow$ at bottom

5-20 0.150

5-22 0.1786

5-24 54.6 N

5-26 33.5 lb$\rightarrow$, 31.3 lb$\uparrow$

5-28 60.95°

5-30 $T = 0.375\,P\dfrac{\mu(5+7\mu)}{(7-5\mu)(5+2\mu)}$

5-32 16.50 lb

5-34 2.56 lb

5-38 3.33 lb

5-40 $-39.5\mathbf{k}$ N

5-42 32.8 lb

5-44 300 ft-lb↻

5-46 $\beta = \tan^{-1}(\mu \cot \theta)$

5-48 46.4°

5-50 26.6°

5-52 5

5-54 22.7 N·m

5-56 85.6 lb

5-58 114.5 lb

5-60 0.1014

5-62 0.655

5-64 216 in-lb

5-66 1.95 m

5-68 $63.1 \text{ lb} < W_C < 108.7 \text{ lb}$

5-70 1.259 kN

5-72 4.05 lb 

5-74 111.1 lb$\leftarrow$

5-76 8.76 lb

5-78 150.5 N

5-80 27.3 lb

5-82 0.261

5-84 65.9°

5-86 (a) Belt slips on A
(b) 563 N in bottom belt.

5-88 63.7 in-lb

5-90 1020 lb

5-92 0.324

5-94 $\frac{L}{d} = \pi\left(\frac{1-\mu R}{R+\mu}\right)$; $R = \frac{W}{Q}$

5-96 116.1 lb

5-98 $M_O = \mu P(R_2 + R_1)/2$

CHAPTER 6

6-2 68 in.^{4}

6-4 815 cm^4

6-6 0.781 in.^{4}

6-8 (a) 10.00 cm^4
(b) 30.0 cm^4

6-10 0.267 in.^{4}

6-12 $\pi/8$ in.^{4}

6-14 2.00 cm^4

6-16 7.89 in.^{4}

6-18 $4a^3b/33$

6-20 1.752 in.^{4}

6-22 (a) 0.816 ft
(b) 2.49 ft

6-24 0.478 ft

6-26 $I_x = 291$ cm^4, $I_y = 90.7$ cm^4

6-28 90.1 in.^{4}

6-30 2.40 in.

6-32 1336 in.^{4}

6-34 2510 ft^4

6-36 (a) 12.46 in.
(b) $I_x = 1248$ in.^{4}, $I_y = 350$ in.^{4}

6-38 $I_x = 15{,}460$ in.^{4}, $I_y = 6200$ in.^{4}

6-40 1299 in.^{4}

6-42 7.20 cm^4

6-44 44.7 cm^4

6-46 161.2 in.^{4}

6-48 192.2 in.^{4}

6-50 67.5 in.^{4}

6-52 0.333 cm^4

6-54 3.38 cm^4

6-56 −40.5 in.^{4}

6-58 42.0 in.^{4}

6-60 $I_{x'} = 139.9$ in.^{4}, $I_{y'} = 14.44$ in.^{4}

6-62 39.8 cm^4

6-64 $I_u = 130.3$ in.^{4}, $I_v = 9.17$ in.^{4}

6-66 $I_u = 0.393$ in.^{4}, $I_v = 0.0263$ in.^{4}

6-68 $I_u = 509$ cm^4, $I_v = 34.0$ cm^4

6-70 684 slug-ft^2

6-72 $I_{xy} = 150$ slug-ft^2,
$I_{xz} = -100$ slug-ft^2,
$I_{yz} = -60.0$ slug-ft^2

6-74 −0.0721 lb-sec^2-in.

6-76 $h/D = 8.62$ (min)

6-78 $(mL^2 \sin 2\theta)/24$

6-80 $\frac{1}{2}m(r_2^2 + r_1^2)$

6-82 (a) 0.611 kg · m^2
(b) 0.396 kg · m^2

6-84 (a) 0.036 m
(b) 0.017 m

6-86 629 slug-ft^2

6-88 11.47 m

6-90 140.7 slug-ft^2

6-92 5.60 mb^2

6-94 0.490 lb-sec^2-in.

6-96 $I_{xy} = mR^2/2\pi$, $I_{yz} = 2mRL/3\pi$

6-98 (a) $0.289L$
(b) $0.577L$
(c) d

6-100 $0.598r$

6-102 79.5 slug-ft^2

6-104 88.5 slug-ft^2

6-106 2180 slug-ft^2

6-108 (a) 9.01 ft
(b) 680 slug-ft^2

6-110 150.2 slug-ft^2

6-112 1.217 slug-ft^2

6-114 14.60 in.

6-116 $I_y = 0.1775$ kg · m^2, Error = 4.05%

CHAPTER 7

7-2 616 ft-lb

7-4 45.0 lb

7-6 15 lb per ft

7-8 15.77 ft-lb

7-10 120.0 ft-lb

7-12 200 lb →

7-14 24.8 N

7-16 0° or 66.4°

7-18 3.46 lb per in.

7-20 197.4 mm

7-22 202 lb

7-24 66.8° or 246.8°

7-26 50 lb

7-28 0.800 m

7-30 (a) 54.3°
(b) 58.7°

7-32 Unstable

7-34 Stable for $\beta = 76.4°$
Unstable for $\beta = 230.5°$

7-36 Stable for $\theta = 0$
Unstable for $\theta = 66.4°$

7-38 Stable for $\theta = 246.8°$
Unstable for $\theta = 66.8°$

7-40 $W/4L$

7-42 Stable for $\theta = 0$
Unstable for $\theta = 75,5°$

CHAPTER 8

8-2 (a) 11.0 fps ←, (b) 24.0 fps^2 →,
(c) 16.0 ft →

8-4 (a) 0, 4 sec, (b) 40 rad ↻, (c) 68 rad

8-6 (a) 9 ft →, (b) 15.83 ft
(c) $\mathbf{a}_{0\text{ sec}} = 8$ fps^2 ←, $\mathbf{a}_{1.6\text{ sec}} = 8$ fps^2 →

8-8 99.9 cm

8-10 (a) 7.0 rad per sec^2 ↺, (b) 6.0 rad ↺

8-12 4.0 m right of origin

8-14 $\mathbf{x} = 1.746\mathbf{i}$ in., $\mathbf{v} = 4.87\mathbf{i}$ ips, $\mathbf{a} = 5.44\mathbf{i}$ ips^2

8-16 39.0 rad

8-18 (a) 0.244 ft →, (b) 0.1189 fps ←,
(c) 0.1161 fps^2 →

8-20 0.613 sec

8-22 $\mathbf{x} = 5.06\mathbf{i}$ ft, $\mathbf{v} = 6.75\mathbf{i}$ fps,
$\mathbf{a} = 6.75\mathbf{i}$ fps^2

8-24 $x = x_0 e^{v_0^2/2k}$, $\mathbf{a} = -(k/x_0)e^{-v_0^2/2k}\mathbf{i}$

8-26 $\mathbf{x} = 52.1\mathbf{i}$ ft, $\mathbf{v} = 3.70\mathbf{i}$ fps, $\mathbf{a} = -0.686\mathbf{i}$ fps^2

8-28 $x = [(2kt + v_0^2)^{3/2} - v_0^3]/3k,\ k \rightarrow L^2T^{-3}$

8-30 $\boldsymbol{\omega} = 0.1875$ rad/s ↻, $\boldsymbol{\alpha} = 0.0406$ rad/s ↺

8-32 $\boldsymbol{\omega} = 1.732$ rad per sec ↻, $\boldsymbol{\alpha} = 2.89$ rad per sec^2 ↻

8-34 (a) $v = 2620$ fps, $a = 0$,
(b) $v = 0$ $a = 3.49(10^4)$ fps^2

8-38 $v_i^2 + v_f^2 = 2aQ$

8-40 52.0 cm/s^2 ←

8-42 $\Delta t = 2.42$ sec, $\mathbf{v}_f = 44.5$ ips ↑

8-44 $\mathbf{v}_i = 20$ fps←, $\mathbf{v}_f = 30$ fps→, $\mathbf{a} = 10$ fps²→

8-46 (a) $\mathbf{a} = 18.89$ fps²←, $\mathbf{v}_f = 13.33$ fps← (b) $\mathbf{a} = 47.8$ fps²←, $\mathbf{v}_f = 186.7$ fps←

8-48 (a) 327 ips² ↓, (b) 6.69 in.

8-50 (a) 9 fps²→, (b) 72 ft, (c) 12 sec

8-52 12 sec

8-54 15.15 sec

8-56 61.8 sec

8-58 4 sec

8-60 1.677 m 26.60°

8-62 24.0 ips²→

8-64 16.67 mm/s²→

8-66 $v_A s_A + v_B(s_B + \sqrt{s_A^2 + s_B^2})$

8-68 (a) 0, (b) 6.00 ips ↑, (c) 5.00 ips² ↑

8-70 (a) 10.00 ips ↑, (b) 18.82 ips ↑

8-72 $v_A(2s_A + s_B) + v_B(s_A + 2s_B)$

8-74 7.47 ips→

8-76 40 mm/s←

8-78 $\mathbf{v}_A = 7.75$ ips ↓, $\mathbf{v}_B = 7.75$ ips ↑, $\mathbf{v}_C = 11.62$ ips ↑

8-80 62.5 ft

8-82 (a) 4.00 sec, (b) 96.0 ft

8-84 (a) *B* wins by 198 ft, (b) 660 ft, (c) 2.40 miles

8-86 (a) 16.87 fps ↓, (b) 11.21 ft (c) 18.40 ft

8-88 (a) 107.7 ft 21.8°, (b) 4.71 fps 45°

8-90 $v = 91.4$ mph at 5500 ft

8-92 30.8° east of north

8-94 133.3 ft

8-96 (a) $(6\mathbf{i} - 3\mathbf{j})$ m, (b) $(8\mathbf{i} - \mathbf{j})$ m/s (c) $(4\mathbf{i} + 1.5\mathbf{j})$ m/s²

8-98 (a) $(-\mathbf{i} + \mathbf{j} + 0.368\mathbf{k})$ in. (b) $(2\mathbf{i} + 3\mathbf{j} - 0.368\mathbf{k})$ ips (c) $(2\mathbf{i} + 6\mathbf{j} + 0.368\mathbf{k})$ ips²

8-100 $(0.1338\mathbf{i} + 0.233\mathbf{j})$ m/s²

8-102 (a) $\left(\frac{x-b}{b}\right)^2 + \left(\frac{y}{c}\right)^2 = 1$ (b) $\mathbf{r} = (6\mathbf{i} - 3\mathbf{j})$ in., $\mathbf{v} = 9.42\mathbf{i}$ ips

8-104 $3.60\mathbf{i}$ ips²

8-106 $\mathbf{v}_A = b[(\cos\theta - \sin\theta)\mathbf{i} + \cos\theta\,\mathbf{j}]\dot{\theta}$
$\mathbf{a}_A = b\{[\ddot{\theta}(\cos\theta - \sin\theta) - \dot{\theta}^2(\cos\theta + \sin\theta)]\mathbf{i} + (\ddot{\theta}\cos\theta - \dot{\theta}^2\sin\theta)\mathbf{j}\}$

8-108 25.0 fps 7 24

8-110 $-27.7\mathbf{i}$ ips²

8-114 14.14%

8-116 40.2°

8-118 Home run (Cleared fence 1.94 ft)

8-120 34.2° or 69.8°

8-122 170.0 fps 31.2°

8-124 68.2 fps 61.0°

8-126 20.5 fps

8-128 73.1 fps 50.9°

8-130 $\mathbf{v} = 3.00$ fps→, $\mathbf{a} = 2.69$ fps² 42.0°

8-132 35.8 m/s² 26.6°

8-134 12.00 fps ↓

8-136 22.0 m/s² 74.90°

8-138 (a) $(19.08\mathbf{i} + 15.90\mathbf{j})$ ips² (b) $(4.92\mathbf{i} - 5.90\mathbf{j})$ ips², (c) 127.0 in.

8-140 825 ft

8-142 $\mathbf{a}_t = (37.5\mathbf{i} + 36.0\mathbf{j})$ fps^2, $\mathbf{a}_n = (-37.5\mathbf{i} + 39.0\mathbf{j})$ fps^2

8-144 (a) $\mathbf{a}_t = (3.33\mathbf{i} + 5.00\mathbf{j} - 0.612\mathbf{k})$ ips^2, $\mathbf{a}_n = (-1.329\mathbf{i} + 1.006\mathbf{j} + 0.980\mathbf{k})$ ips^2
(b) $\mathbf{r}_c = (-5.67\mathbf{i} + 4.53\mathbf{j} + 3.81\mathbf{k})$ in.

8-146 $\mathbf{a}_R = -4330\mathbf{e}_r$ ips^2, $\mathbf{a}_T = 2000\mathbf{e}_\theta$ ips^2

8-148 166.6 cm/s^2 19.22°

8-150 (a) 147.1 ips^2, (b) 56.0 ips^2

8-152 (a) 15.78 ips 17.66°
(b) 114.8 ips^2 60.0°

8-154 $\theta = 120°$, $\mathbf{v} = -6.00\mathbf{i}$ ips

8-156 $\mathbf{v} = 11.18$ ips 71.6°,
$\mathbf{a} = 103.1$ ips^2 31.0°

8-158 $\mathbf{v} = (-1.25\mathbf{e}_r + 23.6\mathbf{e}_\theta - 2.5\mathbf{k})$ ips, $\mathbf{a} = (-74.0\mathbf{e}_r - 7.85\mathbf{e}_\theta)$ ips^2

8-160 (a) $-c_2\mathbf{e}_r + (bc_1 - c_1c_2t)\mathbf{e}_\theta - \dfrac{2hc_2}{b^2}(b - c_2t)\mathbf{k}$
(b) $\mathbf{v}_0 = (-0.5\mathbf{e}_r + 2\pi\mathbf{e}_\theta - 2\mathbf{k})$ ips
$\mathbf{v}_{10} = (-0.5\mathbf{e}_r + \pi\mathbf{e}_\theta - \mathbf{k})$ ips

8-162 6.24 fps ←

8-164 $v = -r\sin\theta\dot{\theta}$, $a = -r(\sin\theta\ddot{\theta} + \cos\theta\dot{\theta}^2)$

8-166 (a) 131.1 ft, (b) 15.90 ft

8-168 $\mathbf{a} = (2\mathbf{i} - 4.5\mathbf{j})$ cm/s^2

8-170 $s = 3\tan(3t + \pi/4)$

8-172 (a) $\boldsymbol{\omega}_{CE} = 6.0$ rad/s ↻, $\boldsymbol{\omega}_{CD} = 15$ rad/s ↺
(b) 110.4 ips 42.8°

8-174 0.028 rad per sec ↻

8-176 106.4 ips →

8-178 2.89 cm/s 60°

8-180 $\mathbf{v}_B = 14$ ips ←, $\mathbf{v}_C = 18$ ips ←, $\mathbf{v}_D = 6$ ips →, $\mathbf{v}_E = 13.42$ ips 63.4°

8-182 13.33 ips →

8-184 (a) 0.640 rad per sec ↺
(b) 8.00 ips 16.26°
(c) 11.20 ips 36.9°

8-186 16.79 ips 35.9°

8-188 $\boldsymbol{\omega}_{BC} = 0$, $\boldsymbol{\omega}_{CD} = 3.00$ rad/sec ↺

8-190 (a) 23.1 ips 30°
(b) 21.4 ips 21.05°

8-192 (a) 48.8 ips (slope 5/12), (b) 6.25 rad/sec ↻

8-194 $\mathbf{v}_A = 25$ ips (slope 3/4), $\mathbf{v}_B = 40.3$ ips (slope 7/4), $\mathbf{v}_D = 15$ ips →

8-196 $\mathbf{v}_P = 10.17$ ips ←, $\mathbf{v}_Q = 17.07$ ips 35.9°

8-198 $\mathbf{v}_A = 1.00$ m/s ←, $\mathbf{v}_B = 3.81$ m/s 23.2°

8-200 (a) 25 rpm ↻. (b) 18.51 cm/s 45°

8-202 (a) 6.67 rpm ↺, (b) 10.47 cm/s ↓

8-204 $\mathbf{v}_A = 90$ ips (slope 4/3), $\mathbf{v}_B = 54$ ips ↑

8-206 56.0 cm/s ←

8-208 24.0 cm/s ←

8-210 $\boldsymbol{\omega}_{AB} = 0$, $\mathbf{v}_Q = 132.0$ cm/s →

8-212 104.8 cm/s 13.24°

8-214 $\mathbf{v}_B = 33.0$ ips ←, $\mathbf{v}_F = 14.71$ ips →

8-216 $\mathbf{v}_C = (2v_A/9)(3\mathbf{i}-\mathbf{j})$,
$\boldsymbol{\omega} = (v_A/27)(2\mathbf{i}-3\mathbf{j}-3\mathbf{k})$

8-218 $(2.50\mathbf{i}-0.833\mathbf{j}-0.833\mathbf{k})$ rad per sec

8-220 (a) $(-8\mathbf{i}+4\mathbf{j})$ rad per sec
(b) $-88\mathbf{k}$ ips

8-222 $\omega_1(-\mathbf{i}+\mathbf{j}+\mathbf{k})/3$ rad per sec

8-224 $\mathbf{v}_A = 52.9$ ips 40.90°

$\mathbf{a}_A = 381$ ips^2 41.3°

8-226 $\mathbf{v}_0 = r\omega \rightarrow$, $\mathbf{a}_0 = \vec{r\alpha} + r^2\overset{\uparrow}{\omega}{}^2/R$.

8-228 48.3 cm/s^2 24.4°

8-230 170.9 ips^2 35.0°

8-232 30.0 rad per sec^2 ↺

8-234 $\mathbf{v}_C = 0$, $\mathbf{a}_C = 50.0$ ips^2 ↓

8-236 $\boldsymbol{\omega} = 6.00$ rad per sec ↻,
$\boldsymbol{\alpha} = 11.43$ rad per sec^2 ↻

8-238 68.4 ips^2 ↓

8-240 (a) 4.00 sec, (b) 22.1 ips^2 26.6°

(c) 33.0 ips^2 32.8°

8-242 $\boldsymbol{\alpha}_{AB} = 18.33$ rad per sec^2 ↺,
$\boldsymbol{\alpha}_{BC} = 4.67$ rad per sec^2 ↺

8-244 $\mathbf{v}_A = 13.71$ cm/s ←,
$\mathbf{a}_A = 20.4$ cm/s^2 ←

8-246 $(\frac{4}{33})(27\mathbf{i}+80\mathbf{j}-\mathbf{k})$ rad per sec

8-248 (a) $\frac{4}{3}\omega_2^2\mathbf{k}$, (b) $\frac{4}{75}\omega_2^2 r(-24\mathbf{i}-25\mathbf{j})$

8-250 $-\frac{5}{27}r\omega_1^2(\mathbf{j}+2\mathbf{k})$

8-252 (a) $(25.7\mathbf{i}-58\mathbf{j}+20\mathbf{k})$ rad/s^2
(b) $(42\mathbf{i}+67\mathbf{j})$ cm/s^2

8-254 $\boldsymbol{\omega}_{AB} = 3.52$ rad/s ↻, $\boldsymbol{\alpha}_{AB} = 9.50$ rad/s^2 ↺

8-256 $\boldsymbol{\omega} = 0.769$ rad/s ↺, $\boldsymbol{\alpha} = 0.917$ rad/s^2 ↺

8-258 $\boldsymbol{\alpha}_A = 170$ rad/s^2 ↺, $\mathbf{a}_{P/A} = 850$ *ips*^{2} ↓

8-260 $\mathbf{v}_P = (-32\mathbf{i}+15\mathbf{j})$ fps,
$\mathbf{a}_P = (-20\mathbf{i}-134\mathbf{j})$ fps^2

8-262 (a) 0.896 rad/sec^2 ↺

(b) 108.9 ips^2 64.2°

8-264 16.55 rad per sec^2 ↻

8-266 $\mathbf{v} = 24$ cm/s →, $\mathbf{a} = 14$ cm/s^2 →

8-268 (a) 12.15 ips ←, (b) 4.91 rad per sec ↺
(c) 176.5 ips^2 ←

8-270 $\mathbf{v}_C = (7.5\mathbf{i}+18.0\mathbf{j}-25.2\mathbf{k})$ fps
$\mathbf{a}_C = (-57.2\mathbf{i}+11.25\mathbf{j}-18.00\mathbf{k})$ fps^2

8-272 $\mathbf{v}_P = (-22\mathbf{i}+4\mathbf{j}+6\mathbf{k})$ fps,
$\mathbf{a}_P = (-4\mathbf{i}-40\mathbf{j})$ fps^2

8-274 (a) $(-8\mathbf{i}+5\mathbf{j}+3\mathbf{k})$ rad per sec^2
(b) $(20\mathbf{i}+4\mathbf{j}-20\mathbf{k})$ ips
(c) $(-46\mathbf{i}+55\mathbf{j}-25\mathbf{k})$ ips^2

8-276 (a) $(8\mathbf{i}+3\mathbf{j}+6\mathbf{k})$ rad per sec^2
(b) $(-2\mathbf{i}+30\mathbf{j}+9\mathbf{k})$ ips^2

8-278 $\mathbf{v} = b\omega_0(\mathbf{i}-2\mathbf{j})$,
$\mathbf{a} = (b\omega_0^2/2)(3\mathbf{i}+4\mathbf{j})$

8-280 $\mathbf{v}_P = (2b\omega_0/9)(2\mathbf{i}+\mathbf{j}-2\mathbf{k})$,
$\mathbf{a}_P = (4b\omega_0^2/27)(-\mathbf{i}+\mathbf{j}-2\mathbf{k})$

8-282 $\mathbf{v}_C = 50$ ips (slope 3, 4)

$\mathbf{v}_P = 36.1$ ips (slope 3, 2)

8-284 116.5 ips^2 65.8°

8-286 $\mathbf{v}_C = (-12\mathbf{i}-16\mathbf{j}-12\mathbf{k})$ ips
$\mathbf{a}_C = (-60\mathbf{i}-80\mathbf{j}+76\mathbf{k})$ ips^2

8-288 $\mathbf{v}_C = 30.0$ ips ←, $\mathbf{a}_C = 651$ ips^2

8-290 3.75 rad per sec^2 ↺

8-292 (a) 33.8 ips →, (b) 55.6 ips^2 → 7.94°

8-294 $\boldsymbol{\omega}_{AB} = 1.479$ rad/s ↻,
$\boldsymbol{\alpha}_{AB} = 25.0$ rad/s^2 ↻

8-296 $\mathbf{v}_A = 214$ ips 30°,

$\mathbf{a}_A = 38{,}700$ ips^2 49.1°

8-298 $\boldsymbol{\omega}_C = (10\mathbf{i}+5\mathbf{j}+3\mathbf{k})$ rad per sec,
$\boldsymbol{\alpha}_C = (-15\mathbf{i}+30\mathbf{j}-50\mathbf{k})$ rad per sec^2

CHAPTER 9

9-2 (a) $m = 5.28$ slugs, $W = 134.3$ lb
(b) 24,600 fps = 16,760 mph

9-4 (a) 26.8 fps² (slope 3/4, ↙)
(b) 5.37 fps² (slope 3/4, ↗), (c) $a = 0$

9-6 (a) 3.19 fps↗, (b) 0.658 fps↙

9-8 (a) $\tan^{-1}(-\mu)$
(b) $g[\mu + (P/W)(1+\mu^2)^{1/2}]$

9-10 (a) $T_{AB} = 139.8$ N, $T_{BC} = 103.0$ N
(b) 0.490 m/s² ←

9-12 (a) 3.22 fps² ↓, (b) 13.50 lb, (c) 6.00 lb

9-14 24.5 lb

9-16 1.44 lb

9-18 11.59 fps² (slope 3/4, ↙)

9-20 $T = 4.64$ lb, $\boldsymbol{\alpha} = 2.01$ rad/s² ↺

9-22 (a) $T_{AB} = m(9\omega^2 - 24.15)$,
(b) 1.638 rad per sec
(c) 1.465 rad per sec

9-24 0.259

9-26 3.29 rad per sec

9-28 $\Delta t = 0.509$ sec, $\Delta s_B = 0.833$ ft

9-30 $\mathbf{a}_A = 8.00$ fps² →, $\mathbf{a}_B = 4.00$ fps² →,
$\mathbf{a}_C = 6.00$ fps² →

9-32 3.27 fps² (slope 5/12, ↗)

9-34 (a) $\omega = 11.35$ rad/sec, $T = 2.33$ lb
(b) 7.59 rad/sec

9-36 (a) 2.548 lb, (b) 14.77 fps² →
(c) 24.0 fps² ↓

9-38 $\mathbf{a}_B = 3.63$ m/s² ↓, $T = 12.37$ N

9-40 353 N

9-42 0.932 lb

9-44 24.4 lb ←

9-46 4.72 rad per sec

9-48 $\mathbf{a}_A = 4.20$ m/s² ↑, $T_A = 140.1$ N,
$T_B = 224$ N

9-50 Body has translation. $\mathbf{a}_P$ is parallel to line HG

9-52 (a) 27.9 fps² (↘, 60°),
$\mathbf{R}_A = \mathbf{R}_B = 2.50$ lb (↗, 30°)

9-54 (a) $T_B = 30$ lb, $T_C = 60$ lb
(b) 6.44 fps² ↑

9-56 (a) 70.0 lb →, (b) 21.5 fps² →

9-58 $T_A = 3.09$ lb, $T_B = 20.1$ lb,
$T_C = 51.0$ lb, $\mathbf{a} = 30.8$ fps² (↗, 14.30°)

9-60 $\mathbf{a} = 3.22$ fps² →, $\mathbf{R}_L = 420$ lb ↑,
$\mathbf{R}_R = 480$ lb ↑

9-62 $\mathbf{a} = 6.44$ fps² (slope 3/4, ↙), $\theta = 10.30°$

9-64 $\mathbf{B}_x = 3.00$ lb ←, $\mathbf{B}_y = 5.02$ lb ↑

9-66 $a = 1.200$ g. Not resonable since tires would slip unless $\mu > 1.2$

9-68 $A_x = A_z = 0$, $B_x = ma$,
$B_y = m(4g - 3a)/8$, $C_y = m(4g + 3a)/24$,
$D_y = m(4g + 3a)/12$

9-70 112.5 lb ←

9-72 $\mathbf{C}_x = 174.2$ lb ←, $\mathbf{D}_x = 18.94$ lb →
$\mathbf{D}_y = 231$ lb ↑

9-74 0.577

9-76 (a) 4.57 rad per sec ↻
(b) $\mathbf{N} = 53.8$ lb (slope 5/12, ↗)
$\mathbf{F} = 8.46$ lb (slope 12/5, ↖)

9-78 11.47 rad per sec² ↺

9-80 187.5 rad per sec² ↺

9-82 0.677 ft

9-84 $\mathbf{P} = 121.2$ N ←, $\mathbf{C}_x = 82.8$ N ←
$\mathbf{C}_y = 304$ N ↑

9-86 0.0344 in.

9-88 164.4 N

9-90 3.28 rad per sec ↺

9-92 0.496 m/s^2 ↑

9-94 2.52 ft

9-96 (a) 1.502 rad/s^2 ↻, (b) 286 lb 81.8°

9-98 $T_A = 5$ mg/8, $T_B = 3$ mg/8

9-100 21.16 lb 70.9°

9-102 23.9 rad/sec^2

9-104 (a) 59.4 rad/s^2
(b) $\mathbf{C}_x = 3.40$ lb ↓, $\mathbf{C}_y = 2.37$ lb↙,
$\mathbf{D}_x = 1.687$ lb ↓, $\mathbf{D}_y = 0.088$ lb↙

9-106 $T = 18.55$ lb,
$\mathbf{R}_D = 7.45(0.987\mathbf{i} + 0.161\mathbf{k})$ lb

9-108 (a) 4.00 fps^2 ←, (b) 0.373

9-110 (a) 37.2 rad/s^2 ↻, (b) 12.19 fps ↓

9-112 3.01 rad per sec^2 ↺

9-114 $\mathbf{a}_B = 6.71$ fps^2 →, $\mathbf{a}_C = 25.5$ fps^2 ↓

9-116 (a) $\boldsymbol{\alpha} = 2.40$ rad/s^2 ↻,

$\mathbf{a}_G = 0.720$ m/s^2 5 12

(b) $\boldsymbol{\alpha} = 10.72$ rad/s^2 ↻,

$\mathbf{a}_G = 0.684$ m/s^2 5 12

9-118 107.5 kg

9-120 (a) 60.0 lb →, (b) 1360 ft-lb ↻

9-122 (a) 427 N ←, (b) 1.885 m/s^2 →
(c) 2.26 rad/s^2 ↺

9-124 (a) 96.6 rad per sec^2 ↺, (b) 64.4 fps^2 ↑

9-126 $\mathbf{a}_G = 12.08$ fps^2 30°,

$\boldsymbol{\alpha} = 8.05$ rad/s^2 ↻

9-128 (a) 372 ft-lb ↻, (b) 4.32 fps^2 →

9-130 28.3 ft 4 3

9-132 40.0 rad per sec ↻

9-134 $\mathbf{a}_A = 4.32$ fps^2 3 4,

$\boldsymbol{\alpha}_B = 10$ rad/s^2 ↺

9-136 (a) 10.59 rad/s^2 ↻, (b) 1.918 lb

9-138 8.49 rad/s^2 ↻

9-140 0.1697 rad/s^2 ↺

9-142 67.1 lb 72.6°

9-144 0.275 ft

9-146 (a) 10.00 lb →, (b) 15.85 fps →

9-148 164.4 N

9-150 (a) 372 ft-lb ↻, (b) 4.32 fps^2 →

9-152 $\mathbf{P} = 40$ lb ←, $\mathbf{C} = 50.7$ ft-lb ↺

9-154 2.40 m/s^2 ←

9-156 $\mathbf{C}_x = 0.552$ lb →, $\mathbf{C}_y = 34.7$ lb ↑,
$\mathbf{D}_x = 0.868$ lb →, $\mathbf{D}_y = 33.4$ lb ↑

9-158 6.58 fps^2 →

9-160 (a) 2.87 rad/s^2 ↺

(b) 382 lb 40.4°

9-162 12.15 fps^2 3 4

9-164 $\mathbf{A}_x = 3590$ lb →, $\mathbf{A}_y = 1498$ lb ↑

9-166 (a) 4.42 rad/s ↻, (b) 78.2 lb 14.59°

9-168 (a) $\mathbf{a}_G = 0$, $T = 8.00$ lb
(b) $\mathbf{a}_G = 2.58$ fps^2 →, $T = 4.00$ lb

CHAPTER 10

10-2 14.30 fps to the right

10-4 0.738 m to the left of original starting position

10-6 2160 lb/ft

10-8 3 fps

10-10 6.89 m/s down the slot

10-12 (a) $v = \sqrt{36.3x - 4x^2}$
(b) 8.08 ft

10-14 2.68 m/s to the right

10-16 (a) 38.6 in-lb
(b) 0.879 in.
(c) 11.58 lb

10-18 35.1 lb/ft

10-20 204 lb/ft

10-22 88.5 lb/ft

10-24 8.27 fps to the right

10-26 29.5 ips down the slot

10-28 (a) $v = \sqrt{2gR(1 - \cos\theta)}$
(b) 48.2°

10-30 17.94 fps

10-32 (a) 12.40 fps to the left
(b) 8.87 lb

10-34 16.05 fps to the left

10-38 (a) $v = \sqrt{2gR(1 - \cos\theta)}$
(b) 48.2°

10-40 (a) 16 ft-lb
(b) 16 ft-lb
(c) 16 ft-lb

10-42 only (b) $V = -\left(\frac{x^2}{2} + \frac{y^2}{2} + \frac{z^2}{2}\right) + C$

10-44 yes, 5.39 rad/sec clockwise

10-46 89.0 lb/ft

10-48 1.485 ft downward

10-50 66.8 lb upward

10-52 281 N, 4 to the left and 3 downward

10-54 32.5 lb

10-56 126.7 lb, 4 to the left and 3 downward

10-58 5.42 rad/sec clockwise

10-60 174.8 lb/ft

10-62 counterclockwise, 11.31 rad/sec clockwise

10-64 691 N/m

10-66 1.644 slugs

10-68 3.49 rad/sec clockwise

10-70 30.7 lb/ft

10-72 11.23 lb/ft

10-74 110.4 N/m

10-76 4.99 fps to the right

10-78 6.37 m/s downward

10-80 11.94 N · m

10-82 76.2%

10-84 (a) 36.4 hp
(b) 40.9 hp

10-86 (a) 1.894 hp
(b) 78.9%

10-88 (a) 77.8 kg
(b) 2290 N · m/sec

10-90 83.1%

10-92 33.6°

10-94 75.5°

10-96 (a) $T = 28.3$ lb along cord
$N = 47.4$ lb upward
$F = 7.56$ lb to the left
(b) $T = 58.7$ lb upward
$N = 5.67$ lb upward
$F = 0$

10-98 (a) 39.2 N
(b) 40.6 m/s downward

10-100 41.8°

CHAPTER 11

11-2 (a) 140 N · s up the plane
(b) 398 N · s up the plane

11-4 34.4 fps to the right

11-6 (a) 18 lb to the right
(b) $s = t^4/4$
(c) 729 ft-lb
(d) 36 lb to the right

11-8 83.2 N · s, 74 to the right and 38 downward

11-10 2.74 sec

11-12 29.8 fps to the right

11-14 $(22.8\mathbf{i} - 17.6\mathbf{j})$ m/s

11-16 0.545

11-18 (a) 1.605 sec
(b) 7.25 fps to the right

11-20 28.2 fps up the plane

11-22 20.7 fps to the left

11-24 (a) No, $\mathbf{v}_{min} = 17.12$ fps to the right
(b) 30 fps to the right

11-26 $\mathbf{v}_A = 17.92$ fps right
$\mathbf{v}_B = 14.33$ fps right
$\mathbf{v}_C = 10.75$ fps left

11-28 136.1 lb, 60.9 right, 121.8 downward

11-30 79.4 fps

11-32 (a) 909 lb right
(b) 82.7 hp
(c) 131.2 ft-lb/lb

11-34 59.8°

11-36 (a) 6.67 fps right
(b) 99.5 ft-lb

11-38 $v = \dfrac{M+m}{m}[2gL(1-\cos\theta)]^{1/2}$ m/s

11-40 0.1391 ft

11-42 (a) 26 fps left
(b) 27,960 lb/ft

11-44 2.30 ft

11-46 $\left(\dfrac{gr}{21}\right)^{1/2}$

11-48 $(29.7\mathbf{i} - 4.28\mathbf{k})$ m/s

11-50 (a) $\mathbf{v}_A = 4$ m/s right, $\mathbf{v}_B = 22.0$ m/s right
(b) 12800 N right

11-52 0.789

11-54 6.34 ft

11-56 16.39 fps, 6.6 right and 15 upward

11-58 $\mathbf{v}_A = 2$ fps left
$\mathbf{v}_B = 13.42$ fps, 6 right and 12 upward

11-60 $\mathbf{v}_A = 13.58$ fps, 3.2 left and 13.2 upward
$\mathbf{v}_B = 8.53$ fps, 6.6 right and 5.4 upward

11-62 17.55 m/s, 17.28 right and 3.04 upward

11-66 $\mathbf{v}_A = 15.13$ m/s, 7.44 right and 13.17 upward
$\mathbf{v}_B = 17.56$ m/s right

11-68 $\dfrac{m_A m_B (v_A^2 - v_B^2)(1-e^2)}{(m_A + m_B)(m_A v_A^2 + m_B v_B^2)}$

11-70 (a) $\mathbf{H}_O = 13mL^2\dot{\theta}\mathbf{k}$,
(b) $\ddot{\theta} = -5g\sin\theta/13L$

11-72 (a) 6 fps ↓
(b) 2 fps 19.47°
(c) (5.66 m) lb-sec 70.5°

11-74 15 lb/ft

11-76 9 ft

11-78 (a) 0.327 ft
(b) 8.97 ft

11-80 (a) 19.46 Mm, (b) 17.72 Mm/hr

11-82 (a) 15,800 mph, (b) 12,920 miles

11-84 (a) 0.450, (b) 32.2 Mm/hr

11-86 (a) 14.40 Mm, (b) 3.64 hr

11-88 $r = r_0 \sec \dfrac{\sqrt{h^2 - k}}{h}\theta$

11-90 5.32 fps^2

11-92 (a) 3.67 hr, (b) 7070 miles

11-94 6.00 kN

11-96 (a) 3610 fps, (b) 14,580 fps

11-98 813 ft

11-100 14.65 fps ↑

11-102 16.01 fps ↓

11-104 31.6 ft

11-106 $25.4[\ln(5+y)-0.1y-1.708]^{1/2}$

11-108 31.7 fps ↓

11-110 12.92 lb ↓

11-112 37.6 lb/ft^3

11-114 $F = 139.8$ lb, $R = 242$ lb

11-116 (a) 10.00 fps ←, (b) 0.500

11-118 (a) 354**k** ft-lb-sec
(b) 3.22 ft below O

11-120 555**k** $kg \cdot m^2/s$

11-122 (a) 45.0 $kg \cdot m^2/s$ ↻
(b) 3, 4, G, 0.180m, G

11-126 1.408 ft

11-128 0.00659

11-130 7.07 sec

11-132 (a) 27.7 rad/s ↺, (b) 0.97 m/s ↑

11-134 (a) 12.0 fps →
(b) 9.94 lb-sec →
(c) 1.333 rad per sec ↻
(d) 2.48 lb-sec →

11-136 5.00 fps ↑

11-138 (a) 35.2 rad per sec ↺, (b) 8.97 sec

11-140 39.7 kg

11-142 13.92 ft-lb per sec^2

11-144 2.84 lb

11-146 4.44 sec

11-148 1.810 in.

11-150 1.886 sec

11-152 (a) 0.874 s, (b) 857 m/s →
(c) 0.374 m

11-154 0.874 m

11-156 25.7 fps →

11-158 5.00 fps

11-160 $\boldsymbol{\omega} = 20.3$ rad per sec ↺,
$\mathbf{v}_G = 11.95$ fps ←

11-162 $\boldsymbol{\omega}_A = 113.0$ rad per sec,
$\boldsymbol{\omega}_B = 67.0$ rad per sec

11-164 1.246 rad per sec ↺

11-166 (a) 14.75 rad/s, (b) $P_i = 120.0$ N,
$P_f = 544$ N, (c) 295 N · m

11-168 5.43 ft

11-170 $\boldsymbol{\omega} = 23.4$ rad per sec ↺,
$\mathbf{v}_G = 33.7$ fps ↓

11-172 $\boldsymbol{\omega}_{AB} = 0.857$ rad per sec ↻
$\boldsymbol{\omega}_{CD} = 10.43$ rad per sec ↻

11-174 (a) 0.491 rad per sec
(b) 0.1638 rad per sec

11-176 $\mathbf{A}_x = 120$ lb↖, $\mathbf{A}_y = 234$ lb ↓
$\mathbf{B}_x = 120$ lb↖, $\mathbf{B}_y = 395$ lb ↑

11-178 $\mathbf{A}_x = 87.9\mathbf{i}$ lb, $\mathbf{A}_y = 0$,
$\mathbf{B}_x = -71.8\mathbf{i}$ lb, $\mathbf{B}_y = 0$

11-182 0.882**k**′ rad/s

11-184 75.5° from vertical
64.3° from north

11-186 $\mathbf{R}_A = 10.06(0.965\mathbf{i}' + 0.261\mathbf{j}')$ N,
$\mathbf{R}_B = 53.6(-0.500\mathbf{i}' + 0.866\mathbf{j}')$ N

CHAPTER 12

12-4 $A = 0.550\ \mu m$, $v_{max} = 1.658$ mm/s

12-6 27.0 ips^2

12-8 $f = 0.459$ Hz, $x = 0.2 \cos 2.89t$

12-10 (a) 0.500 ft, (b) 32.0 fps^2

12-12 0.857 fps

12-14 Amp = 0.763 in., $f = 3.13$ Hz

12-16 $W = 6.52$ lb, $k = 2.67$ lb/in.

12-18 0.459 m

12-20 $\theta = 0.1333 \sin 4.5t + 0.1000 \cos 4.5t$

12-22 $\theta = \theta_0 \cos 3.04t$, $T = 2.07$ s

12-24 6.00 lb

12-26 (b) 12.04 ips

12-28 $f = 0.633$ Hz, Amp $= 4.23$ in.

12-30 $\theta_{wheel} = -0.219 \cos 5.16t$ rad
$f = 0.822$ Hz

12-32 $T = 0.201$ sec, $\theta_m = 0.1111$ rad $= 6.37°$

12-34 35.8 N

12-36 (a) Yes, (b) 0.466 Hz

12-38 4.02 Hz

12-40 (a) 5.11 Hz, (b) 3.27 Hz

12-42 $(v_A)_{max} = 2.63$ ips, $f = 0.374$ Hz

12-44 (a) 20 N · s/m
(b) $\mathbf{x} = 0$, $\dot{\mathbf{x}} = 92.0$ mm/s $\leftarrow$
(c) $\mathbf{x} = 1.348$ mm $\leftarrow$,
$\dot{\mathbf{x}} = 5.05$ mm/s $\rightarrow$

12-46 (a) Underdamped,
(b) 0.477 Hz, (c) 0.316

12-48 (a) 0.818, (b) 0.1291

12-50 $c_{cr} = 21.6$ lb-sec per ft, $\delta = 4.19$

12-52 13 cycles

12-54 $2[mf^2(2e^2k - fmg)]^{1/2}/d^2$

12-56 (a) 0.0220 lb-sec/in., (b) 63.2 lb/in.
(c) 2.51 lb-sec/in., (d) 0.0549

12-58 (a) $x = 2.48 \cos 2t$, (b) 4.96 ips

12-60 (a) $\mathbf{O}_x = 24.2$ lb $\leftarrow$, $\mathbf{O}_y = 50.0$ lb $\uparrow$
(b) $\mathbf{O}_x = 29.6$ lb $\leftarrow$, $\mathbf{O}_y = 57.7$ lb $\uparrow$

12-62 (a) $0 < k < 234$ lb/in., (b) 81.8 lb

12-64 (a) 265 rpm
(b) 0.918 in. (out of phase)

12-66 Amp $= 35.4$ mm, B lags A 135°

12-68 1.588 lb

12-70 (c) Most improvement,
(b) Least improvement

12-72 (a) 88.9 N · s/m (b) 88.9 N

12-74 $\omega > 56.6$ rad per sec

12-76 $x = 0.0357 \cos(3t - 0.697)$,
$R_{max} = 28.1$ N

12-78 (a) $r\sqrt{2k/I}$, (b) $\omega < \sqrt{2}\,\omega_n$

Index